THE
WRITER'S
HANDBOOK
2010

Barry Turner has worked on both sides of publishing, as a journalist and author, editor and marketing director. He started his career as a teacher before joining *The Observer* and then moving on to radio and television. His first book, a study of British politics in the early twentieth century, was published in 1970. He has written over twenty books including *A Place in the Country*, which inspired a television series, and a best-selling biography of the actor, Richard Burton. For many years he wrote on travel for *The Times* and now reviews and serializes books for the paper. Among his recent books are *Countdown to Victory* on the last months of World War II and *Suez 1956: The First Oil War*. As founding editor of *The Writer's Handbook* he has taken this annual reference title through to its twenty-third edition. For eleven years, Barry has edited the annual *Statesman's Yearbook*. He is a founder and current chairman of the National Academy of Writing at Birmingham City University, where he is also a visiting professor.

The Writer's Handbook
2010 Competition

→ Your chance to get published

PLUS

Up to 50 creative writing prizes to be won!

HOW TO ENTER

→ **Purchase your copy of *The Writer's Handbook 2010*** from your local bookstore or via www.palgrave.com

→ **Register for your free access to a wealth of information on www.thewritershandbook.com** including the exclusive comprehensive directory, resources and advice for writers. The access code can be found on the sticker on the front cover of this book.

→ **Submit your article** of 1,500-2,000 words entitled **'What Makes a Bestseller?'** on the competition webpage at www.thewritershandbook.com
see the competition page for terms and conditions.

All entries will be judged by a panel of experts. First prize: the winning article will be published in *The Writer's Handbook 2011* and the winner will also receive a free creative writing course of their choice from a vast range of courses from The Complete Creative Writing Course at the Groucho Club in Soho, London. Forty-nine randomly selected entrants will receive creative writing prizes. For full terms and conditions of this prize draw competition please visit: www.thewritershandbook.com. Competition deadline: 31 Jan 2010

THE
WRITER'S
HANDBOOK
2010

EDITOR

BARRY TURNER

MACMILLAN

First published 2009 by
MACMILLAN PUBLISHERS LTD
Houndmills, Basingstoke, Hampshire RG21 6XS and
175 Fifth Avenue, New York, N.Y. 10010
Companies and representatives throughout the world

ISBN-13: 978-0230-57324-6
ISBN-10: 0230-57324-X

Inclusion in *The Writer's Handbook* is entirely at the editor's discretion. Macmillan Publishers Ltd makes no
recommendation whatsoever by inclusion or omission of any agency, publisher or organization.

   While every effort has been made to ensure all of the information contained in this publication is
correct and accurate, the publisher cannot accept any responsibility for any omission or errors that may
occur or for any consequences arising therefrom.

This book is printed on paper suitable for recycling and made from fully
managed and sustained forest sources. Logging, pulping and manufacturing
processes are expected to conform to the environmental regulations of the
country of origin.

A catalogue record for this book is available from the British Library.

A catalog record for this book is available from the Library of Congress.

10 9 8 7 6 5 4 3 2 1
18 17 16 15 14 13 12 11 10 09

Credits
Editor *Barry Turner*
Assistant Editor *Jill Fenner*
Editorial Assistant *Kenneth Hadley*
Poetry Editor *Chris Hamilton-Emery*
Contributors   *Nick Hodder*
                Jon Slack
Tax and Finance Adviser *Ian Spring*

Printed in the UK by
CPI William Clowes Beccles NR34 7TL
Data management and typesetting by
HWA Text and Data Management, London

www.thewritershandbook.com

Contents

Introduction:
Looking on the bright side

Barry Turner

The headless chickens are on the run again. It happens at every economic upset (and I've seen a few). While there is plenty of evidence to show that books, along with other entertainments, can buck the trend there are publishers and their tame commentators in the trade press who tell us that the only way to react to a general downturn is to cut, cut and cut again.

A lot of this is window dressing. A recession is a good excuse to discard unwanted baggage including authors who are past their prime and editors who fall short on ideas. But while there are publishers too close to the margin to keep up the rate of investment in new projects, those in funds know that the opportunities are there to catch a public eager for distraction. On top of that, they need to speculate now to have products to sell when the good times come round again.

So, authors, newcomers and established writers alike, have every reason to think positively. What books will appeal at this time of stress? If you are planning a misery memoir, forget it. There is enough depression around without you adding to it. For the next year or two, readers will be looking for a comfort zone.

It is surely no coincidence that Peter Mayle scored with his lyrical evocation of Provence (Shangri-la by another name) when the economy was in freefall. It occurred to me then that the last time Provence had been top of the bestseller list was in the 1930s when Lady Winifred Fortescue, a clergyman's daughter and chorus girl who had married well, beguiled readers with stories of a French idyll where unemployment and poverty were unknown. Where now for an escape from redundancy and negative equity? You tell me. Armchair travellers are waiting to hear from you.

Bad times are good for adventure, romance and popular history in fact and fiction (expect a book about Merrie England any day now). I am not saying that serious literature has no place, merely that readers are disinclined to be lectured on where we, they, have gone wrong. Leave that to the newspaper columnists who, incidentally, are having a hard time keeping up their circulations.

And what about the comic novel? Not the television spinoff with its heard-them-all-before jokes but a real honest to God comic novel. I am told that when the financial pressure rises, the sale of Wodehouse titles also goes up. It is surely time for the master, who died over thirty years ago, to clear some shelf space.

And then there was Google

Barry Turner

In forty years' association with the book trade, I have learned a lot about the blanket power of complacency. Every major change I can think of, from the end of price fixing to the rise of Amazon to the appearance of the e-book, has been greeted with a dead men's chorus from booksellers, authors and publishers:

'It can't happen here.'
'Oh my God, it has happened here.'
(Altogether now:) 'Help.'

With the e-book revolution (forget the other examples, they are ancient history unless you happen to belong to one of those memory lane literary dining clubs) we are somewhere midway in the chorus. Just a few months ago I was told by an arch practitioner of complacency (a writer as it happens), who even contrived to lean back in his chair, fold his arms behind his head and adopt a smug smile just to show how complacent he really is, that American lobbies for the creative arts would never surrender to Google's demand to digitize every written word in sight. All we had to do to protect our interests was, and I quote, 'keep a watching brief'.

The Google age

Well, things have moved on since then as any half-attentive observer of the scene might have predicted. Having secured the lead in transferring out of copyright material to the screen, Google has now come to a $125 million deal with American publishers and authors to extend their service to copyright books that are no longer in print, i.e. the vast majority of books in the public domain. The agreement gives all parties a bite of the cherry. Google will keep 37 per cent of revenue from online book sales plus anything they get for advertisements, leaving the remainder to a specially created Books Registry to hand on to publishers and authors.

But it is the readers who have most to gain. After previewing up to 20 per cent of pages of out-of-print books for free they will then be able to purchase a download or follow the conventional route to acquiring what they read between hard covers. Licences to access complete texts of millions of titles will be available to libraries and other organizations. As for copyright books that are still readily available, permission to include them in the service will have to be sought from their authors and publishers.

What happens in the US, inevitably ...

All this applies only to the US. It hasn't happened here, yet. But it will. Already the siren voices of booksellers are warning of Google creating a retail monopoly which is a bit of a laugh coming from a trade association that for years fought to restrict the number of book outlets.

Anyway, to argue that the US settlement threatens higher prices and will harm 'cultural diversity' is sheer nonsense. Access to the internet has done more for cultural awareness than all the book browsing through the ages, however pleasurable that activity may be to a coterie of literary purists.

As for higher prices, the book is not an isolated or, indeed, an insulated product. In the open market it is in competition with a whole range of other leisure activities. The value of the book, in monetary terms, will continue to be judged accordingly, whether it is sold by Google, by Amazon, directly by a publisher or by the high street bookshop. It might even be argued that Google is helping to break down barriers by opening a new market for out-of-print books, many of which are hard to track down.

Not all reactions have been doom laden. Mark Le Fanu, secretary general of the Society of Authors, has it right when he speaks of the Google deal as 'a major development in the evolutionary process of digital publishing that reconciles the needs of writers, creators and users'. The challenge is to begin now to prepare for the changes that will inevitably occur when digitization really takes off in Britain.

An author's rights

For a start, authors deserve a clearer definition of their rights when their books go out of print. It used to be that after a few years in the archive, the rights to a title reverted to the author. Recently, however, publishers have taken to protecting their interests, and potential revenue, by keeping books in circulation long after their best sell-by-date by setting aside a few copies or by offering print on demand. There is nothing inherently wrong in this as long as there is a mutually acceptable agreement on the revenue split between authors and publishers. Less than 50 per cent to the author would be hard to justify.

Then there is the impact on libraries and the contribution of the Public Lending Right (PLR) to authors' incomes. At least twenty library authorities have launched or are in the process of launching e-book programmes. This is only the start. Before long library users will be equipped with a card to download e-books and audiobooks at home. But these services are not subject to the PLR which is restricted to loans of printed books. To ensure that authors do not lose out, the PLR scheme needs to be updated; either that or an entirely new system of collecting loan payments, possibly via the Authors' Licensing & Collecting Society (ALCS), will have to evolve, though this will be hard to negotiate. If libraries and other major users are already paying Google and, indirectly, publishers and authors for site licences, are they likely to agree to a second layer of fees? This could be a problem for the ALCS which currently takes the bulk of its revenue from photocopying licences, a contracting source. The days of the PLR and the ALCS may be numbered.

Electronic reading? Surely not!

The future is made more complicated and more exciting by the appearance of e-book readers like the Kindle or the Sony Reader. These gadgets are still quite pricey but they are taking off in the States where sales of e-books are increasing by 50 per cent year on year while conventional book sales are static. An indication of what is in store was provided at last year's Frankfurt Book Fair where a survey of book-buying professionals found that 40 per cent believe that digital sales, regardless of format, will surpass ink on paper within a decade.

When this happens (and please don't say 'if') those of us who make any sort of living from writing will have got used to a whole new way of reaching out to readers. Start with the novel. Most fiction comes in king-sized packages. In selling hard print publishers demand a product that looks value for money. The production cost of a novel of 100,000 words is not much reduced by halving its length but it will carry a much higher purchase price. Authors are happy to go along with this because, as we know, verbosity comes naturally. To write succinctly is hard work.

The writer's future

But all will be different when we get into e-books. There will be no obvious advantage in stretching out a novel because size will not be immediately apparent. Indeed, the demand by technically sophisticated readers is more likely to be for snappier, quick reaction narratives that are more in tune with their other entertainments. Expect the short story to make a comeback.

What of non-fiction? One of the greatest risks of books online is the ease of cross referencing. The days of sitting in a library with a stack of volumes to pull together diverse bits of information, which is what non-fiction writers do for much of their time, will soon end. What we need to know, on or within any particular subject from any library at home or abroad, will be easily accessible on-screen. As a result, academics will have to try harder for originality. Filling half a book or more on what has gone before will no longer be acceptable.

Children's books present the greatest challenge. They are among the top borrowings from public libraries and a favoured gift for birthdays and Christmas. But the junior reader of tomorrow will be among the most enthusiastic online consumers with their e-readers rivalling the iPod and the BlackBerry. The demand for compelling stories will be as strong as ever, though how they will adapt to podcasts remains to be seen. Maybe someone is working on the problem but if so I have yet to hear of it.

A part of the future is already with us. The recent decline in non-fiction sales, though blamed on the recession, almost certainly has more to do with online competition. It does not come as any surprise to find that the two categories of books in the forefront of change are reference and travel. Their survival strategy, at least for the medium term, is tied to a combination of online and hard print. Any reference or travel book (armchair travel excepted) without a website is in trouble, maybe not now, but soon. To paraphrase G. B. Shaw, those who can do, those who can't go bust.

Writing for laughs

Nick Hodder

OK, let's get it out of the way. In order to write comedy you do need an element of natural talent. It might not be apparent immediately, in fact the chances are it will take quite a lot of hard work and effort to develop, but you will need to know what is funny, and what isn't, and not everyone does. So, let's assume you've got that natural talent.

If you take nothing else from this article, remember these important things:

- Comedy is entirely subjective, and although some types of comedy can be more accessible than others, originality and accessibility rarely go hand in hand. It's important to find your own comedy, and write what you think is funny.
- The best comedy is about you, even if it isn't delivered in the first person. It will be about people you've met, your opinions of them, things you've seen them do. Coming up with funny situations is good, but if you haven't established characters or personas it probably won't work.
- Comedy is difficult. In my view it's the most difficult genre to write for. Your first attempt will be rubbish. If your first five pages are pure gold, then stop reading this. You are a genius, put this book down, I can't teach you anything.

Finding your 'comedy voice'

In the world of stand-up comedy you will frequently hear the phrase 'finding your comedy voice'. Although this term isn't applied much to stage or screen, it is a really useful concept, and an important element in learning to write comedy. You won't find 'your comedy voice' straight away, and it's likely that it will take a lot of trial and error to find it. The more you write and edit your comedy, the more you will learn to hone your craft, and your work will improve. Eventually, you will find something that is innately funny, which will strike a chord with audiences. When that happens, you know you're on the right track. It may take you ten years to find your own style, but you can only find it if you keep writing.

Motivation

I, like most people, write comedy because I enjoy writing comedy. It is a mystery then, that for some reason I and many of my fellow writers seem to be constantly putting off doing any of it. I'm enjoying writing this piece, but I have to tell you that the first three days that I tried to start this article, things started to crop up. I had to go to the shops because I'd run out of drawing pins; I had to clean my cutlery drawer; there was a film on, that I had seen before, and hated. All that I can tell you here is that once you start you'll enjoy yourself – and that *Showgirls* is an awful movie.

Being original

Don't steal other people's material. It is not good for either friendships or careers. For the benefit of readers, here is a list of the well known comedians that have stolen material:

OK, the editor may have taken that out.

Joke writing and punchlines

I won't go into them in any detail here, but suffice to say that there is a basic 'set-up–punchline' model, which is at work in virtually everything that is funny, from silent films and sit-coms to stand-up and everything in-between. Punchlines are where the laughs are. Take a look at some of your own writing. Can you identify where the punchlines are?

Something to try...

Record a piece of comedy, and listen to it again. It could be radio, film, podcast or TV. It could be stand-up, sitcom, sketch, it doesn't really matter. Review the piece and isolate the punchline. Sometimes it's obvious, sometimes it's a little more difficult. Sometimes a punchline is a line delivered by an actor, but sometimes it's a cut to another scene, a camera movement or a sound cue. If you want a challenge, look for some comedy drama or another more subtle comedy genre. It will be there. If you can work out where the punchlines are in other people's work, you can find them in your own. This will tighten up your script and make it funnier.

The great news is, you can learn to write jokes. If a joke is funny, it's funny for a reason. There is no single methodology behind joke writing. There are lots of 'joke writing systems' in comedy writing publications, and many of them can be very effective. However, I found them much more useful early on. After a while, I found my own way. It's important to find a method of writing that you're comfortable with. If you're comfortable and relaxed, it's much easier to find the funny.

Characters

All great comedy comes from characters. In stand-up that character is normally you, but in most other forms of comedy it's someone else. It really doesn't matter who they are, but the personas of your characters need to be distinct. If you're writing for drama, you can have mysterious characters, but it is very difficult to make them funny. Shows like *Little Britain* and *Catherine Tate* created characters with obvious faults. Familiarity with these characters provided them with a cult following. Also remember, familiarity breeds contempt. So after nine series of the same punchline, try to knock it on the head.

Your script

A commissioning editor won't penalize you too much if your script isn't formatted correctly, but a well formatted script will ensure that you look a little more professional. It's worth downloading a piece of free software from BBC *writersroom* called 'Script Smart', and also looking at other scripts in their archive to get an idea of formatting. As a breed, we are mostly rubbish at spelling and grammar. For example I just spelt proofreading with no 'p' and a couple of extra 'k's (if that made it through the editorial process it's worrying. I think there is an address for complaints at the back of the book). If, for example, you're writing for stand-up, the only necessity is that you understand what you've written. However, for a script submission, make sure you use a spellcheck.

Read it out loud

At some point your script will have to be performed. The only way to know whether your script truly works is to say it out loud. Wait until everyone else has left the house, shut all the doors and windows, and read it out loud. Make sure you stand-up and 'perform' it. You may not be a professional actor, but you'll give yourself the best chance of truly understanding whether it's 'actually funny'. If you've got the courage, try it in front of one or more of your most honest friends, but not in front of your mum. Her opinion will be swayed by love, or possibly hate. Delete as applicable.

Commissioning

Just to give you an idea, a 30-minute script equates to about 7,500 words. If it's a sitcom, it may be worth writing a short summary, describing how the series will develop. Don't worry too much about perfect English. After all, the dialogue needs to be natural. However, do make sure that you spellcheck it, as it is probably best to make sure that it can, at the very least, be understood. Make sure you hand your script to a friend who isn't an illiterate clown for some proofreading prior to submission.

Critics

Vacuous/contemptible.

Constructive criticisms

Unconstructive criticism will normally be as simple as 'I hated it!' (see 'Critics') and therefore should be ignored. Constructive criticism (generally not received from critics) is the kind of stuff that you should really listen to. If someone is talking about 'why' they didn't think the material worked, it will be something that you can take away and think about. However, always remember that comedy is subjective. Try to solicit as many opinions as possible.

Still staring at a blank piece of paper?

Your own ideas will come from your own life. You write about the things that you do or the things that you see. If the thing that you do is 'sit down', and the thing that you see is 'a blank piece of paper', you have writer's block. The cure for writer's block is to go outside and live your life for a bit. Ideas come from you, but it's the you that is 'living life', not the you 'not living life'. Does that make sense?

Key advice

- Comedy is difficult, but keep going, it's the only way to get better.
- Comedy is entirely subjective.
- The best comedy is about you.
- Follow good tips, but try to find your own way of writing.
- It's all about character. Yours and other people's.
- Get out and live life, but make time to write – you'll enjoy it!
- Be original – don't be a hack.
- Read out loud, and get lots of opinions, but not from the critics!
- Relax, that's when the comedy will come.

Something to try...

Get a girlfriend, be an idiot, get dumped. That terrible, heart wrenching pain you feel – write about that. That's funny.

Nick Hodder is a stand-up comedian, writer and producer. Nick has written comedy for radio, film and stage. In 2006 he was a winner of BBC New Talent's Funny Hunt and his short film 'Gloves' won BBC Big Screen – Best Comedy.

The name is the game

The role of branding in the writing industry

Barry Turner

Publishers have been slow to catch on to branding. A few names command instant recognition and the instinctive purchase that goes with it – Penguin Classics and Mills & Boon are obvious examples – and there are smaller publishers like Pen and Sword that rely on the association of name with product. But if consumers know anything about HarperCollins, Macmillan, Random House or Hodder Headline it is as publishers of lots of books of varying content and quality. Breaking up a conglomerate into imprints may help a little but to ask a cross section of book buyers what Picador, Century or Abacus means to them is to invite a look of blank incomprehension.

The publisher brand

Does it matter? There are many publishers who think not. They see their job as selling a wide range of books on their multifarious merits, focusing, albeit briefly, on each product with its own peculiar characteristics. But this is to ignore what one book can do for another if the two are even vaguely related, which is the same thing as saying that if a publisher does well by a book, creating a bestseller, the book can do well by the publisher, assuming that consumers make the link across. Sadly, they rarely do.

The author brand

A case in point is the relationship between J.K. Rowling and Bloomsbury. Whatever one thinks of the Rowling talent – and she may well be a genius for all I know – there is no denying Bloomsbury's magnificent contribution in making her a world megaseller. Yet little of this has rubbed off on the publisher. Writers are in on the secret: just ask how many Harry Potter manuscript lookalikes are buried in the Bloomsbury slush pile. And yet there is scarcely any bookshop recognition of Bloomsbury as the publisher with a sharp eye for the sort of reading matter that grabs young people.

It has been suggested to me that the criticism is unfair since no publisher can expect more than one Harry Potter in a lifetime. But that is to miss the point. Perception is all and here we are talking about the public perception that a publisher smart enough to discover J.K. Rowling may well come up with other goodies to keep young readers happy. Assuming, of course, that they have even heard of Bloomsbury which, given the lost opportunity in marketing, is unlikely.

Legacy

Closer to home (*The Writer's Handbook* is published by Macmillan) I recall how the company cleaned up on the reading habits of Harold Macmillan, the most famous holder of its name and an active participant in the family firm. When, as prime minister, the old rogue confessed to a liking for Anthony Trollope, hundreds of thousands of some of the most tedious books ever written were foisted on a public only too pleased to think that they were in illustrious company. A confidence trick? Well, no more than selling expensive coffee or fizzy drinks on the strength of a name. And our Harold was not short of other ideas. One of his happiest inspirations was to recreate the Edwardian Christmas in a series of seasonal drawings that appealed equally across the generations. All this said, I can't help wondering why Macmillan does not make more use of a famous name. When, to critical acclaim, a dramatized version of Harold Macmillan's life was recently presented at the National Theatre, there was a display of books not one of which, not even Alistair Horne's multivolume biography, carried the Macmillan logo.

Beyond books

While publishers choose to hide their identity, refusing, after all these years, to acknowledge the power of advertising and public relations, best-selling authors, or those lucky enough to inherit their copyright, have taken their own route to branding fame and fortune. Enid Blyton, Agatha Christie and Roald Dahl are among the literary brand images that carry over from books into movies, theatre, comic strips, games and a profusion of associated products. Robert Ludlum died seven years ago but that has not stopped the release of thirteen other books bearing his name, including two bestsellers by thriller writer Eric Van Lustbader who has extended the double life of the CIA assassin Jason Bourne, creating enviable fortunes for himself and the Ludlum estate. Ludlum novels, his own and those of his ghosts, have sold over 200 million copies. Perhaps ghost is the wrong word since there is no attempt by the publisher to disguise the fact that Ludlum has departed. His successors are given full credit on the title page albeit in print a fraction of the size of the Ludlum imprimatur.

Another megabrand is attached to James Patterson who, though still very much with us, has other writers working under his name to promote an image that has broadened out from thrillers and romance to take in children's books and, latterly, graphic novels. James Patterson Entertainment is a company that has its own brand manager who is now focusing on film and television.

Other author brands are in the making. The work of Dennis Wheatley, the now largely forgotten but bestselling occult and thriller author of the 1950s, is set for a revival. If the reprints planned by Chorion take off expect a whole raft of new spin-off titles trading on the Wheatley reputation. With hard times slicing back on the commissioning of new books, publishers might occupy their spare time scouring their backlists for potential born-again authors.

Since publishers take their share of revenues generated by branded authors, it might be argued that they are already alongside James Hammond, a guru on the subject, who says that 'a strong brand is the most resilient asset a company can own'. But if this is so, why do publishers not make more profitable use of their illustrious names? Modest possibilities are opening up. Faber is fronting events and seminars, Penguin sells distinctive mugs, clothes and even deckchairs, Haynes gives the stamp of authenticity to T-shirts, bags and toys, while Wiley, of all companies, is endorsing dart boards and cricket sets.

All are encouraging initiatives in scraping the surface. The next move is to follow in the wake of the newspaper industry which is honing its skills in promoting off-the-page sales of merchandise ranging from gardening tools to package holidays. What about Macmillan as a purveyor of household appliances and travel vouchers? Now there's a thought.

The ins and outs of a publishing house

A tale of two sides

Jon Slack

A good agent will help you coordinate the various stages of publishing your book with your publisher, but there is significant value in understanding what goes on behind the scenes yourself. The publishing house can often seem cloaked in mystery to many new authors – as well as to aspiring publishing professionals, who also make use of *The Writer's Handbook* – so we've prepared a helpful guide to the average publishing house with Jon Slack, the former Chair of the Society of Young Publishers.

The many sectors

Trade (consumer) publishers produce books found in high street retail outlets. The books are separated into fiction and non-fiction.

Children's publishing is a lively and growing sector of trade publishing, and ranges from baby through to young adult.

Educational and academic publishers produce high-level titles aimed at students, academics and researchers for niche bookshops, libraries and educational institutions.

STM (scientific, technical, medical) publishers manage publications in a range of specialized and often fast-moving fields.

Large vs small

Publishing houses offer advantages and drawbacks depending on their size.

Information for aspiring authors

Smaller publishing houses can mean you are given more individual treatment, rather than being one of many. But a larger publishing house probably has more expertise, and a bigger budget – although don't forget that even they don't give every book the star treatment. If you are in the enviable position of having the choice of publishers, decide what your priorities are and what's best for you as a writer.

Information for aspiring publishing professionals

Working at a smaller house is the most effective way to learn the overall business, as you can often be involved in a number of different roles at once. It is also a great way to get a feel for which type of career you may wish to pursue. Alternatively, larger houses tend to use more sophisticated systems and operate on a bigger scale, with larger lists and revenues to work with.

Who does what?

Editorial

Commissioning or acquisition editors are responsible for acquiring or seeking out new titles, researching the market and often managing title budgets. Copy-editors have a keen eye to spot bad grammar, punctuation and inconsistencies.

Information for aspiring authors

Your editor will help you shape your manuscript. They will undoubtedly expect changes to be made, which can be hard to accept, but remember that they have many years' experience and you hear of very few successful authors who look back and wish they hadn't trusted their editor's judgement. Having said that, if you do feel very strongly about a proposed change, do speak up, after all, it's your name that is going to be on the front cover.

Information for aspiring publishing professionals

The position most jobseekers apply for, editors are broadly involved in both acquiring and shaping manuscripts. Most entry positions are for editorial assistant. A specialist role continues to grow for digital editors, converting archives and new titles into e-book format, and there is an increasing trend for them to be involved with e-books from a commissioning level. Other editorial positions can include acquiring picture rights and arranging layout, typesetting and proofreading services.

Marketing and publicity

This area of publishing helps bring the book to the masses. Traditionally defined separately as 'marketing equals advertising' and 'publicity equals word of mouth', these departments are now increasingly integrated online. Smaller houses tend not to have a marketing budget, but publicists persuade journalists to arrange book reviews and features and author interviews, as well as to coordinate public events such as book signings and talks. Marketing departments, alternatively, tend to focus on paid advertising and building web presence, usually via author websites, social networks and other online communities.

Information for aspiring authors

Not all books are advertised on the billboards at the tube and train stations, or in the Sunday newspapers. If your book is one of the lucky few, it will be one of the publisher's key titles and you may find yourself whisked around the country on book-signing tours and invited to attend a variety of events. For the rest, you can expect the marketing and publicity guys to be working creatively with small budgets to give your book the best opportunity to succeed.

Information for aspiring publishing professionals

Typically, entry-level positions are for marketing and/or publicity assistants, before progressing to marketing executive or publicity manager. To succeed in marketing you need to have creative ideas, know your market and have the ability to work to tight budgets. To climb to heady heights in publicity, you also need to be confident and persuasive enough to convince those coffee-fuelled journalists to profile the latest title, as well as being able to develop good relationships with writers and their agents.

Production

Responsible for producing the final, physical book, production staff are continually involved throughout the process. At the commissioning stage they influence book schedules and budget, along with decisions such as the size of an initial print run. They liaise constantly with editors, printers and shipping to help get books printed on time and to a high standard. Today, production teams are under increasing pressure to ensure books are created with as little environmental impact as possible.

Information for aspiring authors
Authors tend not to have much contact with the production department, but if you have any queries your editor should be able to help.

Information for aspiring publishing professionals
To succeed in production you need to be a good project manager, someone who cares about the detail and be able to work under pressure.

Design

This department makes sure the look and feel of your book suits the marketplace, but also grabs the attention of the consumer.

Information for aspiring authors
Most publishers have set page designs, so don't be surprised if you can't have much influence here. If you do have any particular requests or ideas, let your publisher know early on, so they can look to accommodate your requests. One of the most exciting things for many authors is getting the cover design in place. The designer should work with you, as well as your editor and marketing sponsors, to create something that you are all happy with. Always feel free to make your own suggestions.

Information for aspiring publishing professionals
For illustrated titles, a job in design means working with the editor and author in agreeing on a layout template, then arranging the text and images on the page. All books naturally need a jacket designer as well. In smaller companies this is usually freelanced out, but larger companies tend to keep this in-house. This is a very desirable job in publishing, so be prepared to come across a lot of competition if this is the job for you.

Sales and distribution

The two main positions in sales are representatives, or 'reps', and administrators. Sales reps visit bookstores within their allocated regions in the UK or internationally, and 'sell in' the next round of book titles to be released, typically several months in advance.

Information for aspiring authors
Most writers want (a) to get their writing out there in the world and/or (b) ideally to make a nice amount in royalties. Of course, none of this is possible without sales reps persuading bookshops – online and traditional bricks and mortar – to stock your book. With over 100,000 new titles published each year in the UK alone, shelf space is vital, and getting your book positioned face-out or on the enticing tables at the front of the store is the key to success.

Information for aspiring publishing professionals
The role of the rep is, in some way, being reduced as book chains consolidate their buying teams to head office. However, as retailers without a history in bookselling are now diversifying it's an exciting time with lots of opportunities – did you really think ten years ago that today you'd be picking up books with your weekly shop or at the cinema? And, of course, much of the appeal of a role in sales is travel: either within the UK or elsewhere in the world. Sales administrators support reps by processing orders through distribution channels and preparing sales kits, so this can be a good first step into sales.

Rights and contracts

The majority of publishing profit is often made in selling publishing rights into external markets, such as the US or Europe. In larger houses rights staff will also look after serializations for newspapers and magazines, though in smaller houses this task usually falls to the publicity team.

In a small house contracts are often handled by senior editors, but in larger houses a separate department prepares the legal small print for agreements with authors and other publishers.

Information for aspiring authors
Effectively another type of sales, this team of people sell your content to foreign publishers. What works well in the UK doesn't always work well in other countries or in different languages, but if your book does grab the attention of a foreign publishing house it can be extremely lucrative for you and your publisher.

Information for aspiring publishing professionals
Often an overlooked area of publishing by recent graduates, working in rights can be extremely rewarding. Much of your time is spent on activities surrounding the big international book fairs (see Barry Turner's article on page 23 for more information) and brokering deals. The ability to speak multiple languages is extremely helpful, as is the ability to negotiate hard.

Accounting

The financial backbone of any publishing operation, accountants employ similar skills here as they would in most other industries. They are responsible for managing accounts, customer credit, the staff payroll and author royalties.

Information for aspiring authors
Responsible for sending those precious royalty statements to you or your agent once or twice a year, the accounts department can make all the hard work worthwhile. Night on the town anyone?

Information for aspiring publishing professionals
Accounting in a publishing house could provide a great new role for those disheartened by the finance sector and wanting to be part of the creative industry – and who could blame them in the current climate?

Jon Slack is a freelance publisher who has previously worked in sales and publicity at Taschen publishers and Aurum Press, and prior to that spent some time working at Borders, South Australia. He was the 2008 Chair of the Society of Young Publishers and now works on special projects for the group. The SYP is a voluntary organization that helps members break into the industry and progress throughout their publishing career.   See www.thesyp.org.uk for more information.

⚷ Key advice for aspiring publishing professionals

- Whether you are a writer or simply passionate about books, it's an exciting time to start working in the publishing industry. One of the reasons is the change brought about by digital publishing. Mass internet use and cheaper print-on-demand services are having a big impact, and publishers must react rapidly to modernize and meet the needs of their authors and readers. For recruiters, this means finding people that can, along with all the usual skills, bring to the position substantial creativity and initiative. This is where you can come in.

- Publishing is extremely competitive to get into, so ask yourself right from the outset how determined you are to work in it, and then get to know how the industry works: behind the attractive image of turning manuscripts into books there is a serious, demanding business that needs to make money, with long hours and comparatively small pay on offer when you start.

- Your CV will join the equivalent of the recruitment 'slush pile', so you'll need to beef up your credentials wherever you can and find ways to make your application stand out. Work experience is the best way to do this and is often unpaid, but it looks great on paper and you'll pick up an invaluable reference. Publishing training courses and a publishing MA can be effective. Make a genuine effort to understand the structure and issues of the book industry and this will put you in good shape should you reach interview stage. Once you get there, it's important to stress how your love of books is only overshadowed by your will to succeed in the publishing *process*, and this will prove your commitment to the *business* of making books.

- Try to understand your targeted field; read newspapers and magazines which include reviews of the books or journals you want to work with.

- Read books about getting into publishing: they'll have some great advice about preparing your CV and more information about specific roles. Read industry news from sources such as *The Bookseller* or *BookBrunch.co.uk*, and browse the internet for publishing-focused blogs to help better understand the industry.

- Attend relevant information seminars and events and talk to others already working. Joining a network like the Society of Young Publishers is a great way to get your foot in the door and show you are serious about finding work.

- Show initiative and phone publishers directly; ask if there are any vacancies or at the very least work experience positions available, even if only for a week or two. Research any companies as much as possible before interview by obtaining brochures or catalogues and searching online.

- Publishing is difficult to break into but if you are determined and resourceful in your attempts you will succeed. And once you're in you can set your sights on a long and rewarding career working in the UK's largest creative sector.

Twenty things I didn't know before becoming a poetry publisher

Chris Hamilton-Emery

1. Getting published these days is about the triumph of first lines

Editors read defensively and have the attention span of small children in Ikea on a bank holiday. This is a major asset in business. If you don't grab my attention with line one I am already reaching for my blueberry muffin and latte, yawning with indifference. Remember that the slush pile is measured in 'body weight per postal delivery' in most publishing houses. Everything is about first lines, the first verse, the first poem. I *desperately* want to reject you and you have to convince me in that first piece of writing that I shouldn't. Poets should buy special pairs of 'Am I Banal?' glasses and read their work twice over. If your first poem isn't as inductive as the first paragraph of any Dickens novel then you've got it wrong. Lead me in.

2. Good writing is obvious even if bad writing is pervasive

Despite what many editors imply, good writing is easy to spot; it takes 4.2 seconds to discover you want to read something. You can certainly spot good writing (it's what you buy). Anyone can do it. It has no camps, no schools, no special tools or techniques. It's often startlingly perfect, like a clean egg in the mud. Actually, it leaps out like a cerise shirt at a funeral when you're going through the slush pile. However, it is very hard to see the real benefits of your own writing, especially when starting out. This is because poetry is social and you need readers for your writing to become literature. There is actually a lot of good writing. But there's far, far more bad writing. Oh yes. This is because the pain threshold of poetry is very high when it comes to writing. But it's very, very, very low when it comes to reading.

3. Great writing can make for a bad book

Great writing often sells poorly. Great books often fail. I come across lots of terrific writing that no one will buy. If I cannot sell it, sadly I cannot commit to publish it. In this way, poetry, like any other art, is built from the collision of fashion, desire and money in any given age. It lies inside society's tastes and predilections and not outside of it. If you write what people want, they'll buy it. No one can effectively guess what people want when it comes to poems. Time marches on. Memory is short. Fashion is fickle. Clever writers stay in touch with the public taste. Effective writing is often pragmatic.

4. Lots of smart civil servants think poetry can succeed through a desirable and perfectible model of skills-based arts administration and long-term public subsidy: it's the planned-economy model of poetry

Even Stalin didn't make this work. You won't be able to persecute enough people to pull this off. Unless you want your poetry to be about regional metal filing quotas or cities of refuge, it's best to keep the public sector out of your writing. Many will not agree with this, largely because they misunderstand the need for a cosmological split between the State and the Artist. In the free world poetry should be adversarial.

5. The 'Super Modern Transactional Model of Poetry®'

Don't write for the dead, they never pay on time. Poems have value within a million daily transactions. Each day, every single transaction forms the foundation of the poetry economy in which poets live, like purple terrapins in a caged pond. If the transaction doesn't take place – the reader buying a printed poem, a paid performance, an MP3 download, even an e-book – then poetry hasn't happened. Poems live inside this world of transactions. This is why poetry is a social art and not a form of embalming. Writing for the future born is also like writing for the dead, we only have 'now' in which to write and read and, besides, you'll never enjoy the eulogy of the new births of 3009. Stick to the breathing world.

6. Some poets believe they ought to be paid by the State to learn how to write, then be paid to write, and then have people be paid to publish it, to distribute it, to write about it and to read it

It's the gravy train model of poetry: Welfare Verse; a dependency model – and there's quite a lot of it in Britain and elsewhere. Once upon a time, in the 1970s, which were mediaeval and orange-coloured, many people believed you owed them a living. They stood around outside smoking Silk Cut beside steel dustbins, pretty much like office workers now on their breaks near the receptions of glass megaliths. Some poets yearn for a return to this condition, like they want a giant duvet filled with the cream of eternity, preferably in a heated, licensed, seaside campus. Don't knock it though. You know, our arts policies are the direct result of the war against the Nazis. It was all 1940s propaganda. State culture had reached its zenith by the late 1960s and Harold Wilson, his policies and trajectory of arts management, hangs over us like his concrete ziggurats in the fag-strewn wasteland of county council quads. All good poetry is born out of necessity. It is beyond specification.

7. Literary development officers don't

Develop literature that is. Not one jot. It's a common misnomer. Readers, writers, publishers: they create literature, not civil servants. Of course, civil servants believe in reader development like Christians believe in the Second Coming. They see readers as brown field sites and are busy ringing the Department of Works for the JCBs. They are often social engineers. Write for real people and avoid bureaucrats.

8. The biggest force in poetry is enervation, the second entropy

Most published poetry is dull. Even famous poets can be dull. Poetry progresses in relation to how dull we find our best poets. If you are boring, then experimenting with language will only make you experimentally boring. Where poetry is concerned having something interesting to say is actually more important than how you say it. Most poets don't realize this.

9. You will make more money busking for a day with a kazoo in Fitzwilliam, West Yorkshire, than selling poetry around the world for a month

Many writers fail to consider if anyone would want to part with their cash to read their verses. It's a big ask. Most people will choose a non-anaesthetized lobotomy over reading a poem. Sales are hard to come by. Harder than business loans from Barclays. Some administrators think that marketing is the key, misunderstanding that marketing is the art of finding what people actually (and bizarrely) want and giving it to them, not making people want stuff they especially hate, like cinnamon-flavoured toothpaste. Don't waste time taking poetry to people who don't want it. It's a fool's errand.

10. Publishers are not giant lending banks with New Lehmans stencilled on the front in big gold letters

Despite having less useful money than Zimbabwe, poetry publishers are often asked to spend it all on things that don't contribute to making sales. This is the equivalent of all those day-time ads for 'Save this drowning polar bear' or 'Don't let this cute dog die in lonely pain' that get pumped out on children's TV. There's simply not much cash in poetry to save the planet or pay for your tour of this sceptred isle. No poetry reading has ever made me money. And remember, everyone loves a free book, as it beats paying for one. Give away enough books and you'll need to borrow money to file for bankruptcy.

11. If Cheops had been a publisher, he'd have needed a bigger pyramid for all the unsold stocks of his books

For many publishers, books return more frequently than a sprout and baked bean balti. The book trade is inherently wasteful with stocks. Print three to sell two. However, the nature of poetry sales is slow and long. This does mean that stocks can hang around. Our house lost 75 per cent of its available space to book stocks. We had to move out of some rooms and lock them up like some scene from a Frances Hodgson Burnett novel. In fact, our children loved the secret tunnels formed by printer's parcels stored at home. Despite all this, the aim is actually to hold just enough stock to keep your business moving. We aim to hold two months' stock. Any more is robbing your business.

12. The weight–value ratio (WVR) of poetry is 54.579

Poetry has a pretty poor weight to value ratio, and yet people think that it should get cheaper and cheaper to make people buy it. The same model is used for clothing by using child labour in Asia to stitch together fifty shirts an hour. Poetry is an expensive art and it's a mistake to worry about its price more than its value. People value expensive goods over cheap ones. This is the basis of poetry's survival in our lives. All knowledge and spiritual value is expensive.

13. Many people don't have time for poetry, so target those who do

Poetry has a high time to benefit ratio, yet the one thing many people don't have is productive leisure time. This may be because they're holding down three jobs to feed their kids or find their time is vacuum packed with debt. Poetry, as art, requires a high level of literacy and a decent slice of time to consume it in. This is partly why poetry has been regarded as an elite art: it occupies the same cultural and economic space as ballet, theatre and opera. Attempts to broaden the audience have had the same success as the redistribution of wealth. It's also true to say that poetry is often relevant to one's life at specific points of emotional intensity. Poetry's audience is largely an audience which can afford leisure time. One may find that this is the young and the old, for in-between we're all busy having children, building careers and paying the mortgage. Anyone can enjoy poetry, but not everyone can afford the time and effort to do so. Identifying those who can afford it is a key to success.

14. Moore's Law and poetry: 'My name is Michael K. Pankovic and I am the 21,011th person to join Facebook's registry of poets'

Roughly speaking, the developed world's population will all become poets in 2033. Poets are the plankton of the literary world. There are lots of us. But this needs qualifying. Some people call themselves poets because they write stuff. This is like describing yourself as an athlete because you jogged to catch the No. 33 bus this morning. Most 'poets' are frauds and, like rogue traders, will turn up to mend your boiler with a sack of bread knives and a tub of Vaseline. We know who you are. Poets are the products of their paying readership.

15. Pundits always like to know which school of poetry is on the ascendant

It's the read kind.

16. Thirty seconds: the lightning speed of forgetfulness

Many literary editors and critics will celebrate tardiness in writing poems as a measure of one's adherence to absolute quality, as if restraint were a form of social service where writing is concerned. This is especially true where critics think that quality is to do with paucity of endeavour: the less is more brigade. We forget writers with tremendous ease. It's cruel, but our culture is built around discarding. Being a writer does require delivering the goods every few years. Once forgotten, a comeback tour will not secure your place in the literary canon. People will just look for crows feet and the signs of the last facelift. Staying in the public mind is a full-time job.

17. Literature can seem like a private members' club with a no admittance policy

Literary editing for some is a special form of being really cross. Being cross marks out 'proper' editors from the amateur rabble at every Important Party. Being cross also shows how utterly engaged one is with the infinitesimally small amount of serious writing there is in the world (the less one attends to the more serious one is about quality). In fact, when talent abounds, some editors almost implode with derision. This is largely due to the exposure of their miserable largesse and their shockingly dull publishing and largely ignored acquisitions.

Some editors match their derision with even greater adherence to the writing of monumental bores. Thank heavens art is breakable.

18. It's all showbiz really

Rooms have not been built large enough for most editorial egos; egos are often displayed in the form of giant ornate wigs which may be purchased from Ede & Ravenscroft. Okay, this isn't strictly true. When brought together at awards ceremonies, editors may often vie for status by displaying, on the one hand, preposterous levels of aloofness and, on the other, an oily fawning attraction to the fashionably unread Writer of the Moment. Thus, some appear as small galaxies rushing away from us, others as dilettantes in some provincial members' club outing, fresh from their charabanc. Who you know matters.

19. In the private sector, you'll need over £300,000 of pure profit to compete against the State-funded collective of presses

Or, roughly speaking, you would need to sell 150,000 poetry books a year, every year, just to catch up with your State-funded competitors being pumped with public cash.

20. 'Why should I buy this book if I've never heard of her?'

This is probably the most frequent question I receive from booksellers. Remember that you buy what you know. What this reveals is that your book, your years of writing, your passion, your perseverance, your conviction, aspiration, skill, importance, even your good looks won't matter if no one has heard of you. Books don't sell themselves. Profile and publicity are key components in a writing life. Ignore authors that explain in gentle nasal interviews that they regard publicity as a little bit vulgar as they have (a) already made a name for themselves after years of gruelling publicity, which is why they are being interviewed by Horatio in the first place; or (b) have a wincing publicist working for them behind the scenes who arranged the interview for them six weeks ago after twenty-five separate pitches. Think of good publishers as being in the 'choice' business, not the book business. They construct choices for book buyers and you can support them, too. Do this by making a good name for yourself, which will almost certainly mean working beyond the boundaries of poetry itself, so take your writing life beyond the barricades and barbed wire surrounding verse and write upon everything you can wherever you can. Good luck.

Chris Emery's poetry has appeared in The Age, Jacket, Magma, Poetry London, Poetry Review, Poetry Wales, PN Review, Quid *and* The Rialto. *He was anthologized in* New Writing 8 *in 1999. A first full-length poetry collection,* Dr Mephisto, *was published by Arc in 2002 and a collection of poetry,* Radio Nostalgia, *was published by Arc in 2006. Chris Emery is publishing director of Salt Publishing, an independent literary press based in Cambridge. He was awarded an American Book Award in 2006 and Salt Publishing won the Nielsen Innovation of the Year prize in 2008*

Making an unsolicited approach to a literary agent

Any active agent is on the lookout for new talent. It's their seed corn. Fostering and encouraging new talent is an important part of their work. Furthermore, no reputable agent charges a reading fee and authors should never agree to pay one. This system allows an agent to react as briefly or as fully to new work as he likes. So don't expect a detailed report as a matter of course.

Most importantly, before approaching an agent you must have *completed* the work in a draft you are happy with. No agent wants to see a draft you plan to rewrite; and no publisher will commission an unknown author to finish a book.

Agents, however, do have other work to do. So to save their time and yours, follow this simple procedure. Write – don't telephone – to one agent at a time (or state it if you are writing to several agents), sending a brief covering letter addressed to the agent concerned. Too often authors send out photocopies addressed 'Dear Sir/Madam' – so it's obvious all the agents in this book are being approached at the same time. These may go straight into an agent's bin. (Do *you* answer all the circulars that come through the letter box?) This letter should state who you are and any writing or other media experience you have (such as previous books published, along with the publishers, and journalism); and the names of the publishers (if any) who have already declined the work you are submitting. If it is non-fiction it would be helpful to give your view of the market for your work and the complementary work by other authors that has already been published. Don't write a blurb. Your novel may potentially be the best since *Gone with the Wind*, but let it speak for itself.

Secondly, with the covering letter send the first two sample chapters, or at least the first fifty pages, and a synopsis of the whole work, together with a note of its length. It's no good sending only a synopsis as it gives no idea of how you write. All work submitted *must* be typed double-spaced and on one side of A4 paper only.

Finally, send return postage (and a stamped postcard if you want an acknowledgement). Agents can't be expected to subsidize a service they are giving free. They are not the literary extension of the Citizens Advice Bureau. *Always* keep a copy of any material submitted – don't rely solely on registered post or recorded delivery – and allow at least a month for a response. By the way, many agencies don't accept submissions as an email attachment and so usually will not bother to respond, or will just suggest you are welcome to submit by post. You may check an individual agency's submissions procedure on the agency's website or in this handbook.

On the way to the fair

Barry Turner

A word of advice to any author who is thinking of attending a book fair. Don't. Unless, that is, you want to find good reasons to give up writing and retire to a sanctuary.

Book fairs are for salespeople. Publishers, agents, rights directors – whatever hat each happens to be wearing – they are all wheeler-dealing in products which just happen to be books but could just as easily be cars or washing machines. The only authors who have any part to play in this ritual are the celebs, the big, big names whose presence, usually fleeting, may help to promote the big, big titles they have put their names to.

For the rest of us, our appearance is a diversion in a foot-slogging schedule that does not allow for minutes wasted in talk about what happens next. What happens right now is the question. On miles of shelving, glossy new titles are lined up like a fairground duck shoot. The analogy is unfair because they are not there to be shot down but to be adapted for publication in another country, for translation or for adaptation to another medium. Future projects can wait on a return to the office.

I speak as an observer of the scene having represented *The Times* at a dozen Frankfurts and, long before that, having held down a proper job in publishing which required slotting book fairs into the calendar. As gamekeeper turned poacher, I have sympathy for the professionals who spend long days trying to give substance to the bottom line. They may not want authors around them at this critical time but they are working on behalf of authors who are only too happy when their books are sold for other languages or formats.

What to expect

But this raises another question. If, as I readily admit, fairs like Frankfurt, London and Bologna (for children's publications) exercise a fascination for anyone interested in books, is there any virtue in an author appearing, as it were, incognito, as part of this visiting throng?

As a cultural experience, Frankfurt, the granddaddy of international book fairs and by far the world's largest, is hugely impressive. It is now in its sixtieth year. The latest feedback has close on 7,400 exhibitors showing off over 400,000 titles. For the big publishers, the Frankfurt experience is a major expense, and I mean major. Add up the cost of exhibition space, enticing displays, meeting areas, accommodation (Frankfurt hotels charge high rates) plus all the extras and the bill for the likes of Random House, HarperCollins or Macmillan must run well into six figures.

For authors who are more accustomed to the small room and the crowded desk, the sheer spectacle is hard to beat. On a practical level, it may be useful to check out the competition. Are there any trends to be spotted? Frankfurt was well ahead of the game with its displays of comic novels; but now it is ebooks that are heading the promotions. Yet, on departing the exhibition halls, taking a deep breath of fresh air (it gets hot in there) and nipping off in search of a half-decent meal (unless you happen to love hot dogs or hamburgers), a thought inevitably occurs. Is this extravagant jamboree value for money?

All the fun of the fair

Some insiders share these doubts. When times are hard (and we may be in for one or more of those years) publishers are heard to grumble about heavy outgoings, suggesting that, possibly, trade-fair deals could be struck more economically by email and telephone.

Old hands at making business friends argue for personal contact as the only sure way of securing long term relationships but it is at least arguable that the heady atmosphere and frenetic pace of the book fair is not necessarily the best environment for influencing people.

Major international books fairs

- Frankfurt
- Beijing
- London
- BookExpo America
- Bologna

See listing on p 620.

There are successful publishers like Ernest Hecht of Souvenir Press who believe that the expense is out of all proportion to the likely returns. The few who argue the opposing view with any confidence are packagers of highly illustrated books who depend on the international market.

For the rest the economics of trade fairs are increasingly hard to justify in hard financial terms. Yet for most publishers the pull of the trade fair remains as strong as ever, stronger if the rise in the number of these events is anything to go by. There are now around thirty worldwide that may be said to be of some consequence. There were dire predictions for the London Book Fair when it was launched in 1972. It has certainly had a roller coaster ride but if its latest incarnation at Earls Court is anything to go by, it has a legitimate claim to be in the major league.

The explanation must be in business psychology. To be seen at Frankfurt or at the other leading book fairs is to be taken seriously as a prime mover and shaker in the world of books. To stay away? No: the publisher shudders and pays up.

The Writer's Handbook recommends: How Not to Write a Novel

Apart from writing, is there any intellectual activity where a 'How not to' book would be useful? I can't imagine whole chapters on How not to Do Mathematics or How not to Do French. I may not be much good at either but my errors are only too obvious without some smart-arse delivering a treatise on them.

Writing is different, however. This is because people who write badly or not too well rarely learn from their mistakes. As the rejection slips pile up they take comfort in the belief that it is not they who are wrong but those who judge them. And so they write on in the same old way in the hope that one day an enlightened agent, publisher or critic will recognize their genius.

Why? Unlike mathematics or French, writing has few fixed rules. Beyond basic punctuation, grammar and syntax, it is a matter of opinion, informed or otherwise, whether your creative efforts are better than mine. It is no coincidence that favourable comments from family and friends do most to bolster false confidence. Even when, against the odds, an inadequate book finds a publisher only to disappear into the remainder shelves, there is always a let out. The publicity was inadequate, the jacket design all wrong, the reviewers were out to lunch.

The myth of the lonely, misunderstood writer against the world, destined eventually to emerge triumphant, endures partly by virtue of the occasional exceptions that prove the rule. More powerful and insidious is the influence of those teachers of creative writing, themselves failed writers, who continue to promote the errors that they have been unable to detect in their own work. If this seems unlikely just try ploughing through a selection of the self-help books churned out by the writing schools. As examples of turgid, cliché-ridden ineptitude they are hard to beat. Which is why *How not to Write a Novel* deserves a five-star welcome. Author Sandra Newman and author-editor Howard Mittelmark, both experienced and, on the evidence of this book, enlightened teachers, take pot shots at every known pretension of third-rate authors. The results are instructive and often hilarious with examples of 'how not to' based, I would guess, on agonizing experience. In a feast of delights my favourite is the mashing of 'legalese, officialese and psychobabble', said by the authors, and how right they are, to be an increasing blight on amateur fiction:

> Having spent long years writing business letters, filling in application forms and suing neighbours, the fledgling writer sits down to write a love scene, and out comes something that reads like a translation from the Vulcan. The language of the psychiatric assessment is mixed with that of the tax form in a

heady brew of four-syllable words meaning 'nice'. The author's long struggle to master the stilted, deracinated language of officialdom has left him believing that writing must be flat and lifeless to be smart.

Newman and Mittelmark itemize 200 mistakes to avoid if you ever want to get published. Illustrations of bad writing that will switch off any reader who is not a literary masochist range from the ill-conceived plot to the clumsy use of metaphors. Be honest. If you see any part of your writing reflected here, stop, learn and start again. Much time and anguish will be saved. More, you will have absorbed the first rule of composition, found here on the opening line of the book: 'As a writer you have only one job: to make the reader turn the page'.

May Newman and Mittelmark now turn their attention to non-fiction, with special attention to the illiteracy of academia. Down with Jargon might serve as a title.

Barry Turner

How Not to Write a Novel *by Sandra Newman and Howard Mittelmark is published by Penguin at £9.99 (ISBN 9780141038544)*

Settling accounts

Ian Spring takes an expert look at the latest Budget and explains how writers can be tax-wise

When he was Chancellor of the Exchequer, Norman Lamont said

Tax should be

- Simple and certain
- Fair and reasonable
- Easy to collect

Unfortunately, over the years, tax law has become ever more complicated and more and more difficult for the 'ordinary' person to understand it. Until recently it was inconceivable that tax changes would be passed which had retrospective effect, but this is no longer the case and is happening. With the merger of the Inland Revenue and Customs and Excise, the former have taken on some of the powers and practices of the latter and have become more aggressive in enforcing the law, as they see it, in their never ending endeavour to increase the tax take.

A taxpayer cannot be certain that a tax planning action legally taken now will still be legal in years to come. Tax and its ramifications, such as tax credits, are increasingly viewed by the general public as being unfair and unreasonable and, whilst tax collection is regarded by the Revenue as being easier, this is only because taxpayers are now responsible for paying rather than the Revenue being responsible for collecting.

It is against this background that authors have to comply with the law and complete and submit a tax return each year.

Income tax

What is a professional writer for tax purposes?

Writers are professionals while they are writing regularly with the intention of making a profit; or while they are gathering material, researching or otherwise preparing a publication.

A professional freelance writer is taxed under section 5 Income Tax (Trading and Other Income) Act 2005. The taxable income is the amount receivable, either directly or by an agent on his or her behalf, less expenses wholly and exclusively laid out for the purpose of the profession. If expenses exceed income, the loss can either be set against other income of the same or the preceding year or carried forward and set against future income from writing. If tax has been paid on that other income, a repayment can be obtained, or the sum can be offset against other tax liabilities. Special loss relief can apply in the opening years of the profession. Losses made in the first four years can be set against income of up to three earlier years.

Where a writer receives very occasional payments for isolated articles, it may not be possible to establish that these are profits arising from carrying on a continuing profession. In such circumstances these 'isolated transactions' may be assessed under section 687 Income Tax (Trading and Other Income) Act 2005. Again, expenses may be deducted in arriving at the taxable income but, if expenses exceed income, the loss can only be set against the profits from future isolated transactions or other income assessable under section 687.

Authors are taxed under a system called 'Self Assessment', where the onus is on the individual to declare income and expenses correctly. Each writer therefore has to decide whether profits arise from a professional or occasional activity. The consequences of getting it wrong can be expensive by way of interest, penalties and surcharges on additional tax subsequently found to be due. If in any doubt the writer should seek professional advice.

Income

A writer's income includes fees, advances, royalties, commissions, sale of copyrights, reimbursed expenses, etc. from any source anywhere in the world, whether or not brought to the UK (non-UK resident or domiciled writers should seek professional advice).

Agents

It should be borne in mind that the agent stands in the shoes of the principal. It is not always realized that when the agent receives royalties, fees, advances, etc. on behalf of the author those receipts became the property of the author on the date of their receipt by the agent. This applies for income tax and value added tax purposes.

Expenses

A writer can normally claim the following expenses:

(a) Secretarial, typing, proofreading, research. Where payment for these is made to the author's wife or husband they should be recorded and entered in the spouse's tax return as earned income which is subject to the usual personal allowances. If payments reach relevant levels, PAYE should be operated.

(b) Telephone, faxes, internet costs, computer software, postage, stationery, printing, equipment maintenance, insurance, dictation tapes, batteries, any equipment or office requisites used for the profession.

(c) Periodicals, books (including presentation copies and reference books) and other publications necessary for the profession; however, amounts received from the sale of books should be deducted.

(d) Hotels, fares, car running expenses (including repairs, petrol, oil, garaging, parking, cleaning, insurance, road fund tax, depreciation), hire of cars or taxis in connection with:

(i) Business discussions with agents, publishers, co-authors, collaborators, researchers, illustrators, etc.

(ii) Travel at home and abroad to collect background material.

As an alternative to keeping details of full car running costs, a mileage rate can be claimed for business use. This rate depends on the engine size and varies from year to year. This is known as the Fixed Profit Car Scheme and is available to writers whose turnover does not exceed the VAT registration limit.

(e) Publishing and advertising expenses, including costs of proof corrections, indexing, photographs, etc.

(f) Subscriptions to societies and associations, press cutting agencies, libraries, etc. incurred wholly for the purpose of the profession.

(g) Rent, council tax, water rates, etc., the proportion being determined by the ratio of the number of rooms, used exclusively for the profession, to the total number of rooms in the residence. But see note on capital gains tax below.

(h) Lighting, heating, cleaning. A carefully calculated figure of the business use of these costs can be claimed as a proportion of the total.

(i) Agent's commission, accountancy charges and legal charges incurred wholly in the course of the profession including the cost of defending libel actions, damages insofar as they are not covered by insurance, and libel insurance premiums. However, where in a libel case damages are awarded to punish the author for having acted maliciously, the action becomes quasi-criminal and costs and damages may not be allowed.

(j) TV and video rental (which may be apportioned for private use), and cinema or theatre tickets, if wholly for the purpose of the profession.

(k) Capital allowances are a means of giving businesses tax relief for money spent on equipment, furniture, machinery, etc. There is one set of rules for the purchase of cars and another for all other items. From 6 April 2009, capital allowances on cars will depend on the car's emissions level.

(i) (Cars with very low carbon dioxide emissions (up to 110g/km) will qualify for 100 per cent-year allowances, i.e. the cost is written off in the year of purchase.

(ii) Cars with carbon dioxide emissions between 110g/km and 160g/km will be treated like all other equipment and attract a 20 per cent allowance, i.e. 20 per cent of the original cost is written off in the first year and then 20 per cent of the written down value in subsequent years.

(iii) Cars with emission levels over 160g/km will have only a 10 per cent allowance.

(iv) For all other business equipment, e.g. computers, televisions, radios, hi-fi sets, tape and video recorders, dictaphones, office furniture, photographic equipment, etc., all businesses, including authors, are able to claim an Annual Investment Allowance (AIA). The AIA allows relief at 100 per cent on the first £50,000 spent on business equipment each year. This should be more than enough to meet the needs of most authors. If expenditure in a year does exceed £50,000, an allowance of 20 per cent is given against the balance.

All allowances for both cars and equipment will be reduced to exclude personal (non-professional) use where necessary.

(l) Lease rent. The cost of lease rent of equipment is allowable; as it also is on cars, subject to restrictions for private use, and for cars with high emission levels.

(m) Other expenses incurred wholly and exclusively for professional purposes. (Entertaining expenses are not allowable in any circumstances.)

Note: It is essential to keep detailed records. Diary entries of appointments, notes of fares and receipted bills are much more convincing to HM Revenue & Customs who are very reluctant to accept estimates.

The Self Assessment regime makes it a legal requirement for proper accounting records to be kept. These records must be sufficient to support the figures declared in the tax return.

In addition to the above, tax relief is available on:

(a) Premiums to a pension scheme. On 6 April 2006 a new pension-scheme tax regime came into effect which replaced all previous rules for occupational, personal pension and retirement annuity schemes. An individual can make, and tax relief is available on, contributions up to the higher of the full amount of relevant income or £3,600, subject to a maximum annual limit. That limit for 2008/09 was £235,000 and for 2009/10 is £245,000.

(b) Gift Aid payments to charities – for any amount.

Capital gains tax

The exemption from capital gains tax which applies to an individual's main residence does not apply to any part of that residence which is used exclusively for business purposes. The appropriate proportion of any increase in value of the residence, since 31 March 1982, can be taxed when the residence is sold, at the capital gains tax rate of 18 per cent.

Writers who own their houses should bear this in mind before claiming expenses for the use of a room for writing purposes. Arguments in favour of making such claims are that they afford some relief now, while capital gains tax in its present form may not stay for ever. Also, where a new house is bought in place of an old one, the gain made on the sale of the first study may be set off against the cost of the study in the new house, thus postponing the tax payment until the final sale. For this relief to apply, each house must have a study and the author must continue his profession throughout. On death there is an exemption of the total capital gains of the estate.

Alternatively, writers can claim that their use is non-exclusive and restrict their claim to the cost of extra lighting, heating and cleaning to avoid any capital gains tax liability.

Can a writer average out his or her income over a number of years for tax purposes?

Writers are able to average their profits (made wholly or mainly from creative works) over two or more consecutive years. If the profits of the lower year are less than 70 per cent of the profits of the higher year, or the profits of one year (but not both) are nil, then the author will be able to claim to have the profits averaged. Where the profits of the lower year are more than 70 per cent but less than 75 per cent of the profits of the higher year, a pro rata adjustment is made to both years to reduce the difference between them.

It is also possible to average out income within the terms of publishers' contracts, but professional advice should be taken before signature. Where a husband and wife collaborate as writers, advice should be taken as to whether a formal partnership agreement should be made or whether the publishing agreement should be in joint names.

Is a lump sum paid for an outright sale of the copyright, or part of it, exempt from tax?

No. All the money received from the marketing of literary work, by whatever means, is taxable. Some writers, in spite of clear judicial decisions to the contrary, still seem to think that an outright sale of, for instance, the film rights in a book is not subject to tax. The averaging relief described above should be considered.

Remaindering

To avoid remaindering, authors can usually purchase copies of their own books from the publishers. Monies received from sales are subject to income tax but the cost of books sold should be deducted because tax is only payable on the profit made.

Is there any relief where old copyrights are sold?

No. There was relief available until April 2001 but it was then withdrawn. The averaging relief described above should be considered.

Are royalties payable on the publication of a book abroad subject to both foreign tax as well as UK tax?

Where there is a Double Taxation Agreement between the country concerned and the UK, then on the completion of certain formalities no tax is deductible at source by the foreign payer, but such income is taxable in the UK in the ordinary way. Where there is no Double Taxation Agreement, credit will be given against UK tax for overseas tax paid. A complete list of countries with which the UK has conventions for the avoidance of double taxation may be obtained from the Centre for Non-Residents, HM Revenue & Customs, St John's House, Merton Road, Bootle, Merseyside L69 9BB, or a local tax office.

Residence abroad

Writers residing abroad will, of course, be subject to the tax laws ruling in their country of residence, and as a general rule royalty income paid from the UK can be exempted from deduction of UK tax at source, providing the author is carrying on his or her profession abroad. A writer who is intending to go and live abroad should make early application for future royalties to be paid without deduction of tax to the Centre for Non-Residents (address as above). In certain circumstances writers resident in the Irish Republic are exempt from Irish income tax on their authorship earnings.

Are grants or prizes taxable?

The law is uncertain. Some Arts Council grants are now deemed to be taxable, whereas most prizes and awards are not, though it depends on the conditions in each case. When submitting the Self Assessment annual returns, such items should be excluded but reference made to them in the 'Additional Information' box on the self-employment (or partnership) pages.

What is the item 'Class 4 N.I.C.' which appears on my Self Assessment return?

All taxpayers who are self-employed pay an additional national insurance contribution if their earned income exceeds a figure which varies each year. This contribution is described as Class 4 and is calculated when preparing the return. It is additional to the self-employed Class 2 contribution but confers no additional benefits and is a form of levy. It applies to men aged under 65 and women under 60.

Should an author use a limited company?

The tax regime has made the incorporation of businesses attractive and there can be advantages in so doing even for businesses with relatively low levels of profit, although these advantages have been reduced in recent Budgets. However, there are disadvantages in all cases and there are particular considerations for authors. The advice of an accountant, knowledgeable about the affairs of authors, should be sought if incorporation is contemplated.

Value added tax

Value added tax (VAT) is a tax levied on:

(a) The total value of taxable goods and services supplied to consumers
(b) The importation of goods into the UK
(c) Certain services or goods from abroad if a taxable person receives them in the UK for business purposes

The normal rate of tax is 17.5 per cent but for a period from 1 December 2008 to 31 December 2009 this has been reduced to a rate of 15 per cent. The 17.5 per cent is set to apply again from 1 January 2010.

Who is taxable?

A writer resident in the UK whose turnover from writing and any other business, craft or art on a self-employed basis is greater than £67,000 annually, before deducting an agent's commission, must register with HM Revenue & Customs as a taxable person. This figure of £67,000 is the registration limit for 2008/09 and should be increased by inflation for subsequent years. Turnover includes fees, royalties, advances, commissions, sale of copyright, reimbursed expenses, etc.

A business is required to register:

- At the end of any month if the value of taxable supplies in the past 12 months has exceeded the annual threshold; or
- If there are reasonable grounds for believing that the value of taxable supplies in the next 12 months will exceed the annual threshold.

Penalties will be claimed in the case of late registration. A writer whose turnover is below these limits is exempt from the requirements to register for VAT but may apply for voluntary registration, which will be allowed at the discretion of HM Revenue & Customs.

A taxable person collects VAT on outputs (turnover) and deducts VAT paid on inputs (taxable expenses) and, where VAT collected exceeds VAT paid, must remit the difference to HM Revenue & Customs. In the event that input exceeds output, the difference will be refunded by HM Revenue & Customs.

Inputs (expenses)

Taxable at the standard rate if supplier is registered	Taxable at the zero or special rate	Not liable to VAT
Rent of certain commercial premises	Books (zero)	Rent of non-commercial premises
Advertisements in newspapers, magazines, journals and periodicals	Coach, rail and air travel (zero)	Postage
Agent's commission (unless it relates to monies from overseas)	Agent's commission (on monies from overseas)	Services supplied by unregistered persons
Accountant's and solicitor's fees for business matters	Domestic gas and electricity (5%)	Subscriptions to the Society of Authors, PEN, NUJ, etc.
Agency services (typing, copying, etc.)	Insurance	*Outside the scope of VAT*
Word processors, typewriters and stationery		PLR (Public Lending Right)
Artists' materials		Profit shares
Photographic equipment		Investment income
Tape recorders and tapes		
Hotel accommodation		
Taxi fares		
Motorcar expenses		
Telephone		
Theatres and concerts		

Note: This table is not exhaustive.

Outputs (turnover)

A writer's outputs are taxable services supplied to publishers, broadcasting organizations, theatre managements, film companies, educational institutions, etc. A taxable writer must invoice all the persons (either individuals or organizations) in the UK to whom supplies have been made, for fees, royalties or other considerations, plus VAT. An unregistered writer cannot and must not invoice for VAT. A taxable writer is not obliged to collect VAT on royalties or other fees paid by publishers or others overseas. In practice, agents usually collect VAT for the registered author, but these authors must remember to do so if they receive income that does not go through their agent, e.g. payments for public readings.

Remit to Customs

The taxable writer adds up the VAT which has been paid on taxable inputs, deducts it from the VAT received and remits the balance to Customs. Business with HM Customs is conducted through the local VAT offices of HM Revenue & Customs which are listed in local telephone directories, except for VAT returns which are sent direct to the HM Revenue & Customs VAT Central Unit, Alexander House, 21 Victoria Avenue, Southend on Sea, Essex SS99 1AA.

Accounting

A taxable writer is obliged to account to HM Revenue & Customs at quarterly intervals. Returns must be completed and sent to VAT Central Unit by the dates shown on the return. Penalties can be charged if the returns are late.

It is possible to account for the VAT liability under the Cash Accounting Scheme (leaflet 731), whereby the author accounts for the output tax when the invoice is paid or royalties etc. are received. The same applies to the input tax, but as most purchases are probably on a 'cash basis', this will not make a considerable difference to the author's input tax. This scheme is only applicable to those with a taxable turnover of less than £1,350,000 and therefore is available to the majority of authors. The advantage of this scheme is that the author does not have to account for VAT before receiving payments, thereby relieving the author of a cash flow problem.

If turnover is less than or expected to be less than £1,350,000 it is possible to pay VAT by monthly instalments, with a final balance at the end of the year (see leaflet 732). This annual accounting method also means that only one VAT return is submitted.

Flat rate scheme

Small businesses can elect to pay VAT under a flat rate scheme (FRS). This is open to businesses with a taxable turnover of up to £150,000 a year. Under the normal VAT accounting rules, each item of turnover and every claimed expense must be recorded and supported by evidence, e.g. invoices, receipts, etc. Under the FRS, detailed records of sales and purchases do not have to be kept. A record of gross income (including zero rated and excepted income) is maintained and a flat rate percentage is applied to the total. This percentage is then paid over to HM Revenue & Customs. The percentage varies from one profession or business to another, but for authors it is 11 per cent when the general VAT rate is 17.5 per cent and 9.5 per cent when the general VAT rate is 15 per cent. In the first year of registration this percent is reduced by 1.

The aim of the scheme is to reduce the amount of time and money spent in complying with VAT regulations, and this is to be welcomed. However, there are disadvantages:

• The detailed records of income and expenses are still going to be required for taxation purposes
• Invoices on sales are issued in the normal way
• The percentage is applied to all business income. So 11 per cent VAT will effectively be paid on income from abroad, zero rated under the normal basis, and PLR, otherwise exempt.

For many authors, normal VAT accounting has imposed a good, timely discipline for dealing with accounting and taxation matters.

Registration

A writer will be given a VAT registration number which must be quoted on all VAT correspondence. It is the responsibility of those registered to inform those to whom they make supplies of their registration number. The taxable turnover limit, which determines whether a person who is registered for VAT may apply for cancellation of registration, is £65,000 in 2008/09, increased by inflation in later years.

Voluntary registration

A writer whose turnover is below the limits may apply to register. If the writer is paying a relatively large amount of VAT on taxable inputs – agent's commission, accountant's fees, equipment, materials or agency services, etc. – it may make a significant improvement in the net income to be able to offset the VAT on these inputs. A writer who pays relatively little VAT may find it easier, and no more expensive, to remain unregistered.

Fees and royalties

A taxable writer must notify those to whom he or she makes supplies of the VAT registration number at the first opportunity. One method of accounting for and paying VAT on fees and royalties is the use of multiple stationery for 'self-billing', one copy of the royalty statement being used by the author as the VAT invoice. A second method is for the recipient of taxable outputs to pay fees, including authors' royalties, without VAT. The taxable writer then renders a tax invoice for the VAT element, for which a second payment of this element will be made. This scheme is cumbersome but does involve only taxable authors. Fees and royalties from abroad will count as payments for exported services and will accordingly be zero-rated.

Agents and accountants

A writer is responsible to HM Revenue & Customs for making VAT returns and payments. Neither an agent nor an accountant nor a solicitor can remove the responsibility, although they can be helpful in preparing and keeping VAT returns and accounts. Their professional fees or commission will, except in rare cases where the adviser or agent is unregistered, be taxable at the standard rate and will represent some of a writer's taxable inputs.

Income tax

An unregistered writer can claim some of the VAT paid on taxable inputs as a business expense allowable against income tax. However, certain taxable inputs fall into categories which cannot be claimed under the income tax regulations. A taxable writer, who has already claimed VAT on inputs, cannot charge it as a business expense for the purposes of income tax.

Certain services from abroad

A taxable author who resides in the UK and who receives certain services from abroad must account for VAT on those services at the appropriate tax rate on the sum paid for them. Examples of the type of services concerned include the services of lawyers, accountants, consultants and the provision of information and copyright permissions.

Inheritance tax

Inheritance tax was introduced in 1984 to replace capital transfer tax, which had in turn replaced estate duty, the first of the death taxes of recent times. Paradoxically, inheritance tax has reintroduced a number of principles present under the old estate duty.

The general principle now is that all assets owned at death are chargeable to tax (currently 40 per cent) except for the first £312,000 (the nil rate band) of the estate and any assets passed to a surviving spouse, civil partner or a charity. The proportion of any unused nil rate band on a person's death can be transferred to the estate of their surviving spouse or civil partner, if he or she died after 9 October 2007. This means that wills, which contain provisions for the nil rate band to go to a discretionary trust (a common tax planning procedure in recent years), should be reviewed.

Gifts made more than seven years before death are exempt, but those made within this period may be taxed on a sliding scale. No tax is payable at the time of making the gift.

In addition, each individual may currently make gifts of up to £3,000 in any year, and these will be considered to be exempt. A further exemption covers any number of annual gifts not exceeding £250 to any one person.

If the £3,000 is not fully utilized in one year, any unused balance can be carried forward to the following year (but no later). Gifts out of income, which do not reduce one's living standards, are also exempt if they are part of normal expenditure.

At death, all assets are valued. They will include any property, investments, life policies, furniture and personal possessions, bank balances and, in the case of authors, the value of copyrights. All, with the sole exception of copyrights, are capable (as assets) of accurate valuation and, if necessary, can be turned into cash. The valuation of copyright is, of course, complicated and frequently gives rise to difficulty. Except where they are bequeathed to the owner's husband or wife, very real problems can be left behind by the author.

Experience has shown that a figure based on two to three years' past royalties may be proposed by HM Revenue & Customs in their valuation of copyright. However, this may not be reasonable and may require negotiation. If a book is running out of print or if, as in the case of educational books, it may need revision at the next reprint, these factors must be taken into account. In many cases the fact that the author is no longer alive and able to make personal appearances, or provide publicity, or write further works, will result in lower or slower sales. Obviously, this is an area in which help can be given by the publishers, and in particular one needs to know what their future intentions are, what stocks of the books remain, and what likelihood there will be of reprinting.

There is a further relief available to authors who have established that they have been carrying on a business for at least two years prior to death. It has been possible to establish that copyrights are treated as business property and in these circumstances inheritance tax 'business property relief' is available. This relief at present is 100 per cent so that the tax saving can be quite substantial. HM Revenue & Customs may wish to be assured that the business is continuing and consideration should therefore be given to (i) the appointment, in the author's will, of a literary executor, who should be a qualified business person; or (ii), in certain circumstances, the formation of a partnership between the author and spouse, or other relative, to ensure that it is established that the business is continuing after the author's death.

If the author has sufficient income, consideration should be given to building up a fund to cover future inheritance tax liabilities. One of a number of ways to do this would be to take out a whole life assurance policy which is assigned to the children, or other beneficiaries, the premiums for which are within the annual exemption of £3,000. The capital sum payable on the death of the assured is exempt from inheritance tax.

Tax credits

Working authors may also be eligible to claim Child Tax Credit and Working Tax Credit. Child tax credit is the main way that families receive money for their younger children and for 16–19-year-olds in education. A claim is based on your income and you can claim whether or not you are working. All families (with children) with an income up to £58,175 a year (or up to £66,350 a year if your child is under one) can claim. To qualify, you must be aged 16 or over, usually live in the UK and, if single (or separated), claim according to your individual circumstances. If you are married or a man and woman living together (or a same sex couple from December 2005) you must claim together based on joint circumstances and income.

Working Tax Credit is a payment to top up earnings of the lower paid (whether employed or self-employed, including those who do not have children). To qualify, you must be aged 16 or over, you or your partner must be responsible for a child, and you must work at least 16 hours a week. If you are part of a couple with children, greater amounts may be claimed if you jointly work at least 30 hours a week, provided one of you works for at least 16 hours. Childless couples cannot add their hours together to qualify for the 30-hour element. If you do not have children and you do not have a disability, you must be aged 25 or over and work at least 30 hours a week. It is also possible that your working tax credit will be increased to assist with the cost of registered or approved childcare.

* * *

NB: VAT registration limits and tax rates are due to change after *Writer's Handbook* has gone to press. For latest information, please refer to the website – www.thewritershandbook.com

* * *

Anyone wondering how best to order his or her affairs for tax purposes should consult an accountant with specialized knowledge in this field. Experience shows that a good accountant is well worth the fee which, incidentally, so far as it relates to professional matters, is an allowable expense.

The information contained in this section has been prepared by Ian Spring of Moore Stephens, Chartered Accountants, who will be pleased to answer questions on tax problems. Please write to Ian Spring, c/o The Writer's Handbook, 34 Ufton Road, London N1 5BX.

UK Publishers

AA Publishing

The Automobile Association, Fanum House, Basingstoke RG21 4EA

☎ 01256 491524 🗏 01256 491974

AAPublish@TheAA.com

www.theAA.com

Publisher *David Watchus*

Publishes maps, atlases and guidebooks, motoring, travel and leisure; also illustrated reference. About 100 titles a year.

Abacus

▷ Little, Brown Book Group UK

ABC–CLIO

▷ entry under US Publishers

Absolute Press

Scarborough House, 29 James Street West, Bath BA1 2BT

☎ 01225 316013 🗏 01225 445836

office@absolutepress.co.uk

www.absolutepress.co.uk

Managing Director *Jon Croft*

Founded 1979. Publishes food and wine-related subjects. About 15 titles a year. No unsolicited mss. Synopses and ideas for books welcome. £ Royalties twice-yearly.

Abson Books London

5 Sidney Square, London E1 2EY

☎ 020 7790 4737 🗏 020 7790 7346

absonbooks@aol.com

www.absonbooks.co.uk

Publisher *M.J. Ellison*

Manager *Sharon Wright*

Founded 1971 in Bristol. Publishes language and dialect glossaries. 2 titles in 2008. No unsolicited mss; synopses and ideas for books welcome. £ Royalties twice-yearly.

Academic Press

▷ Elsevier Ltd

Acair Ltd

7 James Street, Stornoway, Isle of Lewis HS1 2QN

☎ 01851 703020 🗏 01851 703294

info@acairbooks.com

www.acairbooks.com

Specializing in matters pertaining to the Gaidhealtachd, Acair publishes books in Gaelic and English on Scottish history, culture and the Gaelic language. 75% of their children's books are targeted at primary school usage and are published exclusively in Gaelic. £ Royalties twice-yearly.

Acumen Publishing Limited

Stocksfield Hall, Stocksfield NE43 7TN

☎ 01661 844865 🗏 01661 844865

steven.gerrard@acumenpublishing.co.uk

www.acumenpublishing.co.uk

Managing Director *Steven Gerrard*

Founded in 1998 as an independent publisher for the higher education market. Publishes academic books on philosophy, classics and history of ideas. No unsolicited mss. Synopses and ideas for books welcome; send written proposal as per Acumen guidelines. £ Royalties annually.

Addison Wesley

▷ Pearson

Adlard Coles Nautical

▷ A.&C. Black Publishers Ltd

African Books Collective

PO Box 721, Oxford OX1 9EN

www.africanbookscollective.com

Founded 1990. Collectively owned by its 17 founder publishers. Exclusive distribution in the UK, Europe and Commonwealth countries outside Africa and in North America through Michigan State University Press for 120 African participating publishers. Concentration is on scholarly/academic, literature and children's books. Mainly concerned with the promotion and dissemination of African-published material outside Africa. Supplies African-published books to African libraries and organizations, and publishes resource books on the African publishing industry. TITLES (published by ABC)*The African Writers' Handbook*; *African Publishers Networking Directory*; *The Electronic*

African Bookworm: A Web Navigator; Courage and Consequence: Women Publishing in Africa; African Scholarly Publishing.

Age Concern Books

1268 London Road, London SW16 4ER
☎ 020 8765 7200   🖷 020 8765 7211
books@ace.org.uk
www.ageconcern.org.uk/bookshop
Approx. Annual Turnover £400,000+

Publishing arm of Age Concern England. Publishes non-fiction only (no fiction or biographies), including the *Your Rights* series, computing titles for the over fifties, financial handbooks and a series of *Carers Handbooks*; also a wide range of books and training packs for professionals. About 10 titles a year. Synopses and ideas welcome.

Airlife Publishing
▷ **The Crowood Press Ltd**

Akros Publications

33 Lady Nairn Avenue, Kirkcaldy KY1 2AW
☎ 01592 651522
www.akrospublications.co.uk
Publisher *Duncan Glen*

Founded 1965. Publishes poetry collections, pamphlets and anthologies; literary essays and studies; local histories. Also *Z2O* poetry magazine. No fiction. About 10 titles a year. No unsolicited mss.
🖷 Royalties twice-yearly.

Alcemi
▷ **Y Lolfa Cyf**

Ian Allan Publishing Ltd

Riverdene Business Park, Molesey Road, Hersham KT12 4RG
☎ 01932 266600   🖷 01932 266601
info@ianallanpublishing.co.uk
www.ianallanpublishing.com
Chairman *David Allan*
Managing Director *Tristan Hilderley*

Specialist transport publisher – atlases, maps, railway, aviation, road transport – military, history, maritime, reference, Freemasonry also sport and leisure. Manages distribution and sales for third party publishers. IMPRINTS **Ian Allan Publishing**; **Classic** Aviation titles; **Midland Publishing** (see entry); **OPC** Railway titles; **Lewis**; **Lewis Masonic**. About 100 titles a year. Send sample chapter and synopsis with s.a.e.

J.A. Allen & Co.

An imprint of Robert Hale Ltd, Clerkenwell House, 45–47 Clerkenwell Green, London EC1R 0HT
☎ 020 7251 2661   🖷 020 7490 4958

allen@halebooks.com
www.halebooks.com
Publisher *Lesley Gowers*
Approx. Annual Turnover £1 million

Founded 1926 as part of J.A. Allen & Co. (The Horseman's Bookshop) Ltd. Bought by **Robert Hale Ltd** in 1999. Publishes equine and equestrian non-fiction. TITLES *British Horse and Pony Breeds* Claire Richardson; *Equine Biomechanics for Riders* Karin Blignault. About 20 titles a year. Mostly commissioned but willing to consider unsolicited mss of technical/instructional material related to all aspects of horses and horsemanship.
🖷 Royalties twice-yearly.

Allen Lane
▷ **Penguin Group (UK)**

Allison & Busby

13 Charlotte Mews, London W1T 4EJ
☎ 020 7580 1080   🖷 020 7580 1180
susie@allisonandbusby.com
www.allisonandbusby.com
Publishing Director *Susie Dunlop*
Approx. Annual Turnover £1 million

Founded 1967. Publishes fiction, historical fiction, crime fiction, true crime, pop culture, romance and fantasy. TITLES *Republic* trilogy by Jack Ludlow; *The Railway Detective* series Edward Marston; *Boomsday* Christopher Buckley; *The Morganville Vampire* series Rachel Caine. About 90 titles a year. No unsolicited mss.

✱*Authors' update* 'A testament to the ability of smaller publishers to survive in the retail jungle', according to the late lamented *Publishing News*.

Allyn & Bacon
▷ **Pearson**

Alma Books Ltd

London House, 243–253 Lower Mortlake Road, Richmond TW9 2LL
☎ 020 8948 9550   🖷 020 8948 5599
info@almabooks.com
www.almabooks.com
Chairman *Elisabetta Minervini*
Managing Director *Alessandro Gallenzi*
Approx. Annual Turnover £450,000

Established in 2005 by founders of **Hesperus Press**. Publishes general fiction, biography, history. 20 titles in 2008. Accepts submissions by post.
🖷 Royalties annually.

✱*Authors' update* Alma is Spanish for 'soul', an apt name say the company founders for a publisher that 'regards a book as an aesthetic artefact rather than a mass produced commodity'.

Alpha Press
▷ Sussex Academic Press

Amber Lane Press Ltd
Cheorl House, Church Street, Charlbury OX7 3PR
☎ 01608 810024   ▣ 01608 810024
info@amberlanepress.co.uk
www.amberlanepress.co.uk
Chairman *Brian Clark*
Managing Director/Editorial Head *Judith Scott*

Founded 1979 to publish modern play texts. Publishes plays and books on the theatre. TITLES *Whose Life is it Anyway?* Brian Clark; *The Dresser* Ronald Harwood; *Once a Catholic* Mary O'Malley (play texts); *Strindberg and Love* Eivor Martinus. No longer actively publishing. Submissions no longer accepted.
℗ Royalties twice-yearly.

Amberley Publishing Plc
The Hampton Centre, Cirencester Road, Chalford, Stroud GL6 8PE
☎ 01285 760030
www.amberley-books.com
Chairman *Charles Fairbairn*
Managing Director *Alan Sutton*
Approx. Annual Turnover £2 million

Founded in 2008 by Alan Sutton, original founder of Alan Sutton Publishing and Tempus Publishing Limited. Publishes local interest and specialist history. Archaeology, military, transport, maritime and aviation. No fiction. DIVISIONS **General History** *Jonathan Reeve*; **Archaeology** *Peter Kemmis Betty*; **Transport & Maritime** *Campbell McCutcheon*; **Local & Regional Interest** *Sarah Flight*. TITLE *The Complete Diary of a Cotswold Parson.* 65 titles in 2008. Unsolicited mss, synopses and ideas welcome by post. OVERSEAS ASSOCIATE Ediciones Amberley S.L., Madrid.
℗ Royalties twice-yearly.

Amsco
▷ Omnibus Press

Andersen Press Ltd
20 Vauxhall Bridge Road, London SW1V 2SA
☎ 020 7840 8701   ▣ 020 7233 6263
anderseneditorial@randomhouse.co.uk
andersenrights@randomhouse.co.uk
www.andersenpress.co.uk
Managing Director/Publisher *Klaus Flugge*
Editorial Director *Rona Selby*
Editor, Fiction *Elizabeth Maude*

Founded 1976 by Klaus Flugge and named after Hans Christian Andersen. Publishes children's high-quality picture books and fiction sold in association with Random House Children's Books. Seventy per cent of the books are sold as co-productions abroad. TITLES *Elmer* David McKee; *I Want My Potty* Tony Ross; *little.com* Ralph Steadman; *Cat in the Manger* Michael Foreman; *Preston Pig Books* Colin McNaughton; *Junk* Melvin Burgess; *You* Sandra Glover; *I Love You, Blue Kangaroo* Emma Chichester Clark. Unsolicited mss welcome for picture books; synopsis and first three chapters in the first instance for books for young readers up to age 12. No poetry or short stories.
℗ Royalties twice-yearly.

❋*Authors' update* The secret of Andersen's success is publishing children's books that are loved by adults. That and a wicked sense of fun that has been known to upset the education establishment – and hooray for that.

Chris Andrews Publications Ltd (trading as Oxford Picture Library)
15 Curtis Yard, North Hinksey Lane, Oxford OX2 0LX
☎ 01865 723404   ▣ 01865 725294
chris.andrews1@btclick.com
www.cap-ox.co.uk
Directors *Chris Andrews, Virginia Andrews*

Founded 1982. Publishes coffee-table, scenic books, calendars, diaries, cards and posters of the *central* area of the UK and the Channel Islands. TITLES *Romance of Oxford*; *Romance of the Cotswolds*; *Romance of the Thames & Chilterns*. Also owns the **Oxford Picture Library** (see entry under *Picture Libraries*). Unsolicited synopses and ideas for travel/guide books in the areas listed above will be considered; phone in the first instance.

The Angels' Share
▷ Neil Wilson Publishing Ltd

Anness Publishing Ltd
Hermes House, 88–89 Blackfriars Road, London SE1 8HA
☎ 020 7401 2077   ▣ 020 7633 9499
info@anness.com
www.aquamarinebooks.com
www.lorenzbooks.com
www.southwaterbooks.com
Chairman/Managing Director *Paul Anness*
Publisher/Partner *Joanna Lorenz*
Approx. Annual Turnover £15.7 million

Founded 1989. Publishes highly illustrated co-edition titles: general non-fiction – cookery, crafts, interior design, gardening, photography, decorating, lifestyle, health, reference, military, transport and children's. IMPRINTS **Lorenz Books**; **Aquamarine**; **Hermes House**; **Southwater**. About 200 titles a year.

Anova Books

10 Southcombe Street, London W14 0RA
☎ 020 7605 1400   🖷 020 7605 1401
info@anovabooks.com
firstinitialsurname@anovabooks.com
www.anovabooks.com porticobooks.co.uk
CEO *Robin Wood*
Managing Director *Polly Powell*
Approx. Annual Turnover £12.6 million

Anova Books, formerly Chrysalis Books Group, specializes in illustrated non-fiction, publishing under the IMPRINTS **B.T. Batsford**; **Collins & Brown**; **Conway**; **National Trust Books**; **Paper Tiger**; **Pavilion**; **Portico**; **Robson Books**; **Salamander**.

IMPRINTS **B.T. Batsford** Associate Publisher *Tina Persaud* Founded in 1843 as a bookseller, and began publishing in 1874. 'A world leader in books on chess, arts and craft.' Publishes non-fiction, architecture, heritage, bridge and chess, film and entertainment, fashion, crafts and hobbies. About 100 titles a year.

Collins & Brown Associate Publisher *Katie Cowan* Publishes a range of lifestyle categories especially in the areas of practical art, photography and needlecrafts.

Conway Associate Publisher *John Lee* Publishes naval history, maritime culture, ship modelling and military history.

National Trust Books Associate Publisher *Tina Persaud* Imprint of Anova Books since 2005. Publishes heritage, gardening, architecture, gift and cookery-based titles. Unsolicited mss not welcome.

Paper Tiger Associate Publisher *Katie Cowan* (Part of Collins & Brown) Publishes science fiction and fantasy art.

Pavilion Associate Publisher *Polly Powell* Publishes illustrated books in biography, cookery, gardening, humour, art, interiors, music, sport and travel. Ideas and synopses for non-fiction titles considered.

Pavilion Children's Books Associate Publisher *Ben Cameron* Publishes innovative, informative and fun books for children of all ages.

Portico Associate Publisher *Tom Bromley* Publishes cutting edge non-fiction which is fresh, funny and forthright.

Robson Books Associate Publisher *Tom Bromley* Founded 1973. Publishes general non-fiction including biography, cookery, gardening, sport and travel. About 60 titles a year. Unsolicited synopses and ideas for books welcome (s.a.e. essential for reply).

Salamander Publisher *Polly Powell* Founded 1973. Publishes colour illustrated books on collecting, cookery, interiors, gardening, music, crafts, military, aviation, sport and transport. Also a wide range of books on American interest subjects. No unsolicited mss but synopses and ideas for the above subjects welcome.

Anthem Press

75–76 Blackfriars Road, London SE1 8HA
☎ 020 7401 4200   🖷 020 7401 4201
www.anthempress.com
Managing Director *Tej Sood*

Founded in 1992 as a publisher of school texts, Anthem Press also publishes academic and trade books focusing on world history, politics, economics, international affairs, literature and culture. Welcomes mss, synopses and ideas for books.
Ⓔ Royalties annually.

Antique Collectors' Club

Sandy Lane, Old Martlesham, Woodbridge IP12 4SD
☎ 01394 389950   🖷 01394 389999
sales@antique-acc.com
www.antiquecollectorsclub.com
Managing Director *Diana Steel*

Founded 1966. Publishes specialist books on art, antiques decorative arts, architecture, gardening and fashion. The Price Guide series was introduced in 1968 with the first edition of *Price Guide to Antique Furniture*. TITLES *Paul & John Nash Design* Brian Webb and Peyton Skipwith; *Biba.The Biba Experience* Alwyn Turner; *Understanding Jewellery* David Bennett and Daniela Mascetti; *Edward Bawden & His Circle* Malcolm Yorke; *Follies of Europe* Caroline Holmes. Also publishes subscription-only magazine, *Antique Collecting*, published 10 times per year. 30–40 titles a year. Unsolicited synopses and ideas for books and magazine articles welcome; no mss.
Ⓔ Royalties twice-yearly as a rule, but can vary.

Anvil Press Poetry Ltd

Neptune House, 70 Royal Hill, London SE10 8RF
☎ 020 8469 3033   🖷 020 8469 3363
anvil@anvilpresspoetry.com
www.anvilpresspoetry.com
Managing Director *Peter Jay*

Founded 1968. England's oldest independent publishing house dedicated to poetry. Publishes new poets in English and in translation as well as translated classics. AUTHORS include Paul Celan, Carol Ann Duffy, Martina Evans, Michael Hamburger, Dennis O'Driscoll, A.B. Jackson, Stanley Moss and Greta Stoddart. Classic translations include Apollinaire, Baudelaire, Dante, Hikmet, Neruda, Tagore and Verlaine. Preliminary enquiry required for translations.

Unsolicited book-length collections of poems are welcome from writers whose work has appeared in poetry magazines or literary journals. Please enclose adequate return postage.

✳*Authors' update* **With a little help from the Arts Council, Anvil has become one of the foremost publishers of living poets.**

Apex Publishing Ltd

PO Box 7086, Clacton on Sea CO15 5WN
📞 01255 428500
mail@apexpublishing.co.uk
www.apexpublishing.co.uk
Publishing & Production Manager *Chris Cowlin*

Founded 2002. Independent publishing company for unknown and established authors, specializing in autobiography, sport and true crime. 150 titles in print. Welcomes unsolicited mss, synopses and ideas for books. Approach in writing to the address above with a copy of the full ms.
💷 Royalties twice-yearly.

Apollos
▷ Inter-Varsity Press

Apple Press Ltd

7 Greenland Street, London NW1 0ND
📞 020 7284 7168   📠 020 7485 4902
www.apple-press.com
Publisher *Liane Stark*

Founded 1984. Part of **The Quarto Group**. Publishes highly illustrated lifestyle books with particular emphasis on food and drink, home decoration and craft, beauty and fashion, sports and leisure, and gardening.

Appletree Press Ltd

The Old Potato Station, 14 Howard Street South, Belfast BT7 1AP
📞 028 9024 3074   📠 028 9024 6756
reception@appletree.ie
www.appletree.ie
Managing Director *John Murphy*

Founded 1974. Publishes gift books, plus general non-fiction of Irish and Scottish interest. TITLES *In St Patrick's Footsteps; Ireland's Ancient Stones; Scottish Lighthouses; Northern Ireland – International Football Facts.* No unsolicited mss; send initial letter, e-mail or synopsis.
💷 Royalties twice-yearly in the first year, annually thereafter.

Aquamarine
▷ Anness Publishing Ltd

Arabia Books
▷ Haus Publishing

Arc Publications Ltd

Nanholme Mill, Shaw Wood Road, Todmorden OL14 6DA
📞 01706 812338   📠 01706 818948
arc.publications@btconnect.com
www.arcpublications.co.uk
Editorial Director *Tony Ward*
Associate Editors *John Kinsella (international), John W. Clarke (UK), Jean Boase-Beier (translations), Angela Jarman (music)*

Founded in 1969 to specialize in the publication of contemporary poetry from new and established writers both in the UK and worldwide. AUTHORS Philip Salom & Katherine Gallagher (Australia), Meta Kusar (Slovenia), Regina Derieva & Vladimir Mayakovsky (Russia), Mourid Barghouti (Palestine), Kunwar Narain (India), Mary O'Donnell (Ireland), Brian Henry & Jill Bialosky (USA), Marcelijus Martinaitus (Lithuania), Razmic Davoyan (Armenia), Doris Kareva (Estonia), Ludwig Steinherr (Germany), Valerie Rouzeau (France), Douglas Jarman, Liz Almond, Michael Haslam & Michael Hulse (UK). IMPRINT **Arc Music** specializes in profiles of contemporary composers (particularly where none have existed hitherto) and symposia which take a 'new approach' to well-visited territory. Commissioned work only. About 20 titles a year. Unable to accept unsolicited material for the foreseeable future. 'Please consult our submissions policy on the website before submitting any work.'
💷 Royalties as per contracts.

Arcadia Books

15–16 Nassau Street, London W1W 7AB
📞 020 7436 9898
info@arcadiabooks.co.uk
www.arcadiabooks.co.uk
Managing Director *Gary Pulsifer*
Associate Publisher *Daniela de Groote*
Managing Editor *Angeline Rothermundt*
Approx. Annual Turnover £500,000

Independent publishing house established in 1996. Specializes in quality translated fiction from around the world. IMPRINTS **Arcadia**; **BlackAmber**; **Bliss**; **EuroCrime**. TITLES *L'Oréal Took My Home: The Secrets of a Theft* Monica Waitzfelder; *The Hite Report on Women Loving Women* Shere Hite; *The Autobiography of the Queen* Emma Tennant; *No Trace* Barry Maitland. Does not welcome unsolicited material.
💷 Royalties twice-yearly.

✳*Authors' update* **BlackAmber is dedicated to getting more black writers into the marketplace. Arcadia translations are among the best.**

Arcane
▷ Omnibus Press

Architectural Press
▷ Elsevier Ltd

Ardis
▷ Gerald Duckworth & Co. Ltd

Argentum
▷ Aurum Press Ltd

Aris & Phillips
▷ Oxbow Books

Armchair Traveller
▷ Haus Publishing

Arnefold Publishing

PO Box 22, Maidstone ME14 1AH

☎ 01622 759591   ᐧ 01622 209193

Chairman/Managing Director *I.G. Mann*

Publishes original non-fiction and selected fiction, and non-fiction reprints. No new fiction. IMPRINTS **Arnefold**; **George Mann Books**; **Recollections**. 'Will consider non-fiction titles (which authors find hard to place) on a joint venture/shared profit basis if they are worthy of publication or can be made so. Unsolicited material is welcomed provided it is understood that if not accompanied by return postage it will neither be read nor returned; that no new fiction is wanted and that joint-venture publications may be suggested. We are not vanity publishers.'
Ⓔ Royalties annually.

Arris Publishing Ltd

12 Main Street, Adlestrop, Moreton in Marsh GL56 0YN

☎ 01608 658758   ᐧ 01608 659345

gcs@arrisbooks.com

www.arrisbooks.com

Joint Managing Directors *Geoffrey Smith, Michel Moushabeck*

Publishing Director *Victoria Huxley*

Founded in 2001. Publishes politics, history, biography, travel, art and culture. IMPRINTS **Arris Books** *Geoffrey Smith* TITLES *9/11: The New Pearl Harbour*; *Disfigured*; **Chastleton Travel** TITLES *Traveller's Wildlife Guides*; *A Traveller's History of Cyprus*; *100 Best Paintings in London*; *World Heritage Sites: Great Britain & Ireland*. About 20 titles a year. No unsolicited mss; synopses and ideas for books welcome. Send introductory letter, synopsis and specimen chapter by post; no e-mail submissions or CDs. No poetry or fiction.
Ⓔ Royalties annually.

Arrow
▷ The Random House Group Ltd

Artech House

16 Sussex Street, London SW1V 4RW

☎ 020 7596 8750   ᐧ 020 7630 0166

splumtree@artechhouse.co.uk

www.artechhouse.com

CEO (USA) *William M. Bazzy*

Commissioning Editor *Dr Simon Plumtree*

Founded 1969. European office of Artech House Inc., Boston. Publishes books for practising professionals in biomedical engineering, bioinformatics, built environment, computer security, MEMS, nanotechnology, microwaves, radar, antennas and propagation, electromagnetic analysis, wireless communications, telecommunications, space applications, remote sensing, solid state technology and devices, technology management. 60–70 titles a year. Unsolicited mss and synopses in the specialized areas listed are considered.
Ⓔ Royalties twice-yearly

Ashgate Publishing Ltd

Wey Court East, Union Road, Farnham GU9 7PT

☎ 01252 331551   ᐧ 01252 344405 Ashgate and Gower)/ 020 7440 7530 (Lund Humphries)

info@ashgatepublishing.com

www.ashgate.com

www.gowerpub.com

www.lundhumphries.com

Chairman *Nigel Farrow*

Founded 1967. Publishes business and professional titles under the Gower imprint, humanities, social sciences, law and legal studies under the Ashgate imprint, and art and art history under the Lund Humphries imprint. Acquired Lund Humphries, publisher of art books and exhibition catalogues in 1999. DIVISIONS **Ashgate** *Adrian Shanks* Social sciences; *John Smedley* History/Variorum collected studies; *Heidi May* Music; *Erika Gaffney* Literary studies; *Tom Gray* History; *Sarah Lloyd* Theology and religious studies; *Guy Loft* Aviation studies. **Gower** *Jonathan Norman* Business, management and training; *Alison Kirk* Law and legal studies. **Lund Humphries** *Lucy Myers*. Access the websites for information on submission of material.

Ashgrove Publishing

27 John Street, London WC1N 2BX

☎ 020 7831 5013   ᐧ 020 7831 5011

gm073@dial.pipex.com

www.ashgrovepublishing.com

Chairman/Managing Director *Brad Thompson*

Acquired by Hollydata Publishers in 1999, Ashgrove has been publishing for over 20 years. Publishes mind, body, spirit, health, cookery, sports. 3 titles in 2007. No unsolicited mss; approach with letter and outline in the first instance.
Ⓔ Royalties twice-yearly.

Ashmolean Museum Publications

Beaumont Street, Oxford OX1 2PH
☎ 01865 288070   ✆ 01865 278106
publications@ashmus.ox.ac.uk
www.ashmolean.org
Chairman *Dr Christopher Brown*
Managing Director *Declan McCarthy*
Approx. Annual Turnover £150,000

Owned by the University of Oxford, Ashmolean
Museum Publications was founded in 1972.
Publishes popular and scholarly works with a
direct relationship to the Ashmolean's rich and
diverse collection (art and archaeology, fine art
and exhibition catalogues). 5 titles in 2008. No
unsolicited mss but synopses and ideas for books
welcome; approach by e-mail in the first instance.
No fiction or crime.
Ⓔ Royalties annually.

Ashwell Publishing Limited
▷ Olympia Publishers

Atlantic Books

26–27 Boswell Street, London WC1N 3JZ
☎ 020 7269 1610   ✆ 020 7430 0916
enquiries@groveatlantic.co.uk
www.groveatlantic.co.uk
Chairman & Publisher *Toby Mundy*
Managing Director *Daniel Scott*
Editor-in-Chief *Ravi Mirchandani*

Founded 2000. A subsidiary of **Grove/Atlantic
Inc.**, New York (see entry under *US Publishers*).
Publishes literary and crime fiction, history,
current affairs, biography and memoir, politics,
popular science and reference books. Distributor
for the American imprint McSweeney's and
co-publisher of Tusker Rock Press. 90 titles a
year. Strictly no unsolicited material accepted.
Submissions from literary agents only.
Ⓔ Royalties twice-yearly.

✸*Authors' update* This fine publisher has
gained its just reward with a Man Booker win
for Aravind Adiga's debut novel, *The White
Tiger*. The non-fiction list also deserves an
award.

Atlantic Europe Publishing Co. Ltd

Greys Court Farm, Greys Court, Nr Henley on
Thames RG9 4PG
☎ 01491 628188   ✆ 01491 628189
writers@atlanticeurope.com
www.AtlanticEurope.com
www.curriculumVisions.com
Director *Dr B.J. Knapp*

Publishes full-colour, highly illustrated primary
school level text books. Not interested in any
other material. Main focus is on National
Curriculum titles, especially in the fields of
religion, science, technology, social history and
geography. About 50 titles a year. Unsolicited
synopses and ideas for non-fiction curriculum-
based books welcome by e-mail only. Does not
accept material sent by post.
Ⓔ Fees paid.

Atom
▷ Little, Brown Book Group UK

Aurora Metro

67 Grove Avenue, Twickenham TW1 5HX
☎ 020 3261 0000   ✆ 020 8898 0735
info@aurorametro.com
www.aurorametro.com
Managing Director *Cheryl Robson*
Submissions *Aidan Jenkins*

Founded in 1989 to publish new writing. Over
100 authors published including Germaine
Greer, Meera Syal, Benjamin Zephaniah, Carole
Hayman. International work in translation and
fiction, children's, drama, biography, cookery,
non-fiction. About 15 titles a year. No unsolicited
mss; send synopses and ideas by e-mail first.
Ⓔ Royalties annually.

Aurum Press Ltd

7 Greenland Street, London NW1 0ND
☎ 020 7284 7160   ✆ 020 7485 4902
www.aurumpress.co.uk
Managing Director *Bill McCreadie*
Publisher *Graham Coster*
Approx. Annual Turnover £4.5 million

Founded 1976. Acquired by the The Quarto
Group in August 2005. High-quality, illustrated/
non-illustrated adult non-fiction in the areas of
biography, sport, current affairs, military history,
travel, music and film. IMPRINTS **Argentum**
Practical photography books; **Jacqui Small** High-
quality lifestyle books. About 70 titles a year. No
unsolicited mss or e-mailed submissions.
Ⓔ Royalties twice-yearly.

✸*Authors' update* Non-fiction for the middle
market. If there are still serious readers out
there who appreciate quality, Aurum will find
them.

Austin & Macauley Publishers Limited

CGC-33-01, 25 Canada Square, Canary Wharf,
London E14 5LQ
☎ 020 7038 8212   ✆ 020 7038 8100
editors@austinmacauley.com
www.austinmacauley.com
Chief Editor *Annette Longman*
Executive Editor *David Calvert*

Founded 2005. Publishes fiction and non-fiction,
autobiography, biography, cookery, environmental
issues, gardening, history, humour, memoirs,
post-1945 fiction, true life experiences, children's,
crime, fitness, health, romance, sport, self-help,

women's and marital issues (including divorce). Expanding educational list: reference books, business studies, cultural matters, current affairs, English, mathematics, mythology, politics, human interest, war, music, theatre, social sciences, cinema and TV. About 50 titles a year. Unsolicited synopses and ideas considered. Accepts e-mail enquiries. Submissions must be by post and include return postage.
£ Royalties twice-yearly.

✱*Authors' update* Authors may be asked to cover their own productions costs.

Authentic Media

9 Holdom Avenue, Bletchley, Milton Keynes MK1 1QR
☏ 01908 364200
www.authenticmedia.co.uk
Publishing Director *Mark Finnie*
Approx. Annual Turnover £2 million

A division of IBS-STL UK Ltd. Publishes Christian books on evangelism, discipleship, biography and mission. IMPRINT **Paternoster** Founded 1935. Editorial Director *Robin Parry* Publishes academic, religion and learned/church/life-related books and journals. Over 50 titles a year. No fiction, children's/youth books or poetry. 'All manuscript submissions must be made by e-mail, using the details and form available on the website.'
£ Royalties quarterly.

AuthorHouse UK

500 Avebury Boulevard, Milton Keynes MK9 2BE
☏ 0800 197 4150 🖷 0800 197 4151
www.authorhouse.co.uk
President & CEO *Kevin Weiss*
Vice President *Terry Dwyer*
UK Managing Director *Tim Davies*

Founded 1997. Acquired by Betram Capital in January 2007. Publishes fiction, biography and memoirs, current affairs, history, politics, humour, travel, health, spirituality, sport and academic. Provides a comprehensive range of publishing and promotional services including editorial, design, production, distribution and marketing paid for by the author who retains editorial and creative control.

Authors OnLine

19 The Cinques, Gamlingay, Sandy SG19 3NU
☏ 01767 652005 🖷 01767 652005
theeditor@authorsonline.co.uk
www.authorsonline.co.uk
Submissions: Freephone 0800 107 2423
Managing Director/Editor *Richard Fitt*
Submissions Editor *Mrs Gaynor Johnson*
Approx. Annual Turnover £250,000

Founded 1998. A service for authors wishing to self-publish. Publishes new and reverted rights work in both electronic format via their website and traditional hard-copy mainly using digital print-on-demand technology. 80 titles in 2008. All genres welcome. Submit mss by e-mail or post (disk or CD-ROM) to the Submissions Editor.
£ Payment 60% net, paid quarterly.

Autumn Publishing

Appledram Barns, Birdham Road, Chichester PO20 7EQ
☏ 01243 531660 🖷 01243 538160
autumn@autumnpublishing.co.uk
www.autumnpublishing.co.uk
Managing Director *Michael Herridge*
Editorial Director *Lyn Coutts*

Founded 1976. Division of Bonnier Media Ltd. Publishes baby and toddler books, children's activity, sticker and early learning books. About 200 titles a year. No responsibility accepted for the return of unsolicited mss.

Avon
▷ **HarperCollins Publishers Ltd**

Award Publications Limited

The Old Riding School, The Welbeck Estate, Worksop S80 3LR
☏ 01909 478170 🖷 01909 484632
info@awardpublications.co.uk
www.awardpublications.co.uk

Founded 1958. Publishes children's books, both fiction and reference. IMPRINT **Horus Editions**. No unsolicited mss, synopses or ideas.

Azure
▷ **Society for Promoting Christian Knowledge**

Badger Publishing Ltd

15 Wedgwood Gate, Pin Green Industrial Estate, Stevenage SG1 4SU
☏ 01438 356907 🖷 01438 747015
orders@badger-publishing.co.uk
www.badger-publishing.co.uk

Founded 1989. Educational publisher. 45 titles in 2008. No unsolicited mss; synopses and ideas for books considered. Approach by e-mail.
£ Royalties twice-yearly.

Baillière Tindall
▷ **Elsevier Ltd**

Duncan Baird Publishers

Castle House, 75–76 Wells Street, London W1T 3QH
☏ 020 7323 2229 🖷 020 7580 5692
info@dbairdpub.co.uk
www.dbp.co.uk
Managing Director *Duncan Baird*
Editorial Director *Bob Saxton*

Approx. Annual Turnover £8 million

Founded in 1992 to publish and package co-editions overseas and went on to launch its own publishing operation in 1998. Publishes illustrated cultural reference, world religions, health, mind, body and spirit, lifestyle, graphic design. 75 titles in 2008. No unsolicited mss. Synopses and ideas welcome; approach in writing in the first instance with s.a.e. No fiction or UK-only subjects.
Ⓔ Royalties twice-yearly.

Bantam/Bantam Press
▷ **Transworld Publishers**

Barefoot Books Ltd
124 Walcot Street, Bath BA1 5BG
☎ 01225 322400   ℻ 01225 322401
info@barefootbooks.co.uk
www.barefootbooks.com
Publisher *Tessa Strickland*

Founded 1993. Publishes children's picture books, particularly new and traditional stories from a wide range of cultures. No unsolicited mss. See website for submission guidelines.
Ⓔ Royalties twice-yearly.

✱*Authors' update* **Writers of children's books would do well to keep track of Barefoot which, from small beginnings, is building a quality list that must be the envy of bigger publishers.**

Barrington Stoke
18 Walker Street, Edinburgh EH3 7LP
☎ 0131 225 4113   ℻ 0131 225 4140
barrington@barringtonstoke.co.uk
www.barringtonstoke.co.uk
Chairman *David Croom*
Managing Director *Sonia Raphael*
Editorial Head *Kate Paice*

Founded in 1998 to publish books for 'dyslexic, struggling and reluctant' young readers. DIVISIONS **Fiction** for 8–12-year olds, reading at age 8; **Teenage Fiction** for age 12+, reading at age 8; **4u2read.OK** for children aged 8–13, with a reading age of 7; **gr8reads** for age 13+, with a reading age of 8; **fyi** for ages 10–14, fiction with stacks of facts; **Reality Checks** True-life stories for ages 10–14; **Reloaded Series** Myths retold for ages 10–14, with a reading age of 8; **Most Wanted** Fiction for adults, with a reading age of 8; **Solos** for ages 10–14, with a reading age of 6½; **Gos!** for ages 10–14, with a reading age of 6. Also books for teachers and resources to accompany fiction books. Commissioned via literary agents only. *No unsolicited material.*
Ⓔ Royalties twice-yearly.

B.T. Batsford
▷ **Anova Books**

BBC Active
▷ **Pearson**

BBC Books
▷ **The Random House Group Ltd**

BBC Children's Books
▷ **Penguin Group (UK)**

BBC Worldwide
Media Centre, 201 Wood Lane, London W12 7TQ
☎ 020 8433 2000
www.bbcworldwide.com
CEO *John Smith*
COO *Sarah Cooper*

The main commercial arm and wholly owned subsidiary of the BBC. Consists of seven core businesses: BBC Worldwide Channels, BBC Worldwide Sales & Distribution, BBC Magazines, BBC Worldwide Content & Production, BBC Worldwide Digital Media, BBC Home Entertainment and BBC Worldwide Global Brands. Third largest consumer magazine publisher in the UK with sales of nearly 100 million copies a year. TITLES include *BBC Good Food, Gardeners' World* and *Top Gear.* Acquired a 75% stake in **Lonely Planet Publications** Ltd in October 2007.

Beautiful Books Ltd
36–38 Glasshouse Street, London W1B 5DL
☎ 020 7734 4448   ℻ 020 3070 0764
office@beautiful-books.co.uk
www.beautiful-books.co.uk
Managing Director *Simon Petherick*
Approx. Annual Turnover £500,000

Founded 2004. Publishes adult fiction and non-fiction. IMPRINTS **Bloody Books** Horror fiction; **Burning House** Contemporary fiction; **Beautiful Books** All other categories. 24–30 titles a year. Unsolicited mss welcome. Approach by post in first instance, with covering letter, brief biography, 1–2-page synopsis and 30–50-page sample of the ms. Include s.a.e. for reply (or return of material if required).
Ⓔ Royalties twice-yearly.

Belair
▷ **Folens Limited**

Berg Publishers
1st Floor, Angel Court, 81 St Clements Street, Oxford OX4 1AW
☎ 01865 245104   ℻ 01865 791165
enquiry@bergpublishers.com
www.bergpublishers.com
Managing Director *Kathryn Earle*
Editorial Director *Tristan Palmer*

Acquired by **Bloomsbury Publishing Plc** in 2008. Publishes scholarly books in the fields of

fashion, cultural studies, social sciences, history and humanities. TITLES *Tartan* Jonathan Faiers; *Jews and Shoes* Edna Nahshon; *Making Short Films* Clifford Thurlow; *Bite Me: Food in Popular Culture* Fabio Parasecoli. IMPRINT **Oswald Wolff Books**. About 65 titles a year plus 16 journals: *Fashion Theory, Home Cultures, Textile: The Journal of Cloth and Culture, Cultural Politics, The Senses and Society* and *Material Religion, Food, Culture and Society, Cultural and Social History, Anthrozoos, Time and Mind, The Journal of Modern Craft, Art in Translation, Design and Culture, The Design Journal, Fashion Practice*. No unsolicited mss. Synopses and ideas for books welcome.
Ⓔ Royalties annually.

Berghahn Books

3 Newtec Place, Magdalen Road, Oxford OX4 1RE
☎ 01865 250011   🖷 01865 250056
publisher@berghahnbooks.com
www.berghahnbooks.com
Chairman/Managing Director *Marion Berghahn*
Approx. Annual Turnover £700,000

Founded 1994. Academic publisher of books and journals. TITLES *Imperial Germany, 1871–1918; Children of Palestine; Critical Interventions; A Different Kind of War; Escape From Hell*. About 90 titles a year. No unsolicited mss; will consider synopses and ideas for books. Approach by e-mail in the first instance. No fiction or trade books. OVERSEAS ASSOCIATE Berghahn Books Inc., New York.
Ⓔ Royalties annually.

Berlitz Publishing

58 Borough High Street, London SE1 1XF
☎ 020 7403 0284   🖷 020 7403 0290
berlitz@apaguide.co.uk
www.berlitzpublishing.com
Managing Director *Katharine Leck*

Founded 1970. Acquired by the Langenscheidt Publishing Group in February 2002. Publishes travel and language-learning products only: visual travel guides, phrasebooks and language courses. SERIES *Pocket Guides; Berlitz Complete Guide to Cruising and Cruise Ships; Phrase Books; Pocket Dictionaries; Business Phrase Books; Self-teach: Rush Hour Commuter Cassettes; Think & Talk; Berlitz Kids*. No unsolicited mss.

BFI Publishing

Palgrave Macmillan, Porters North, 4 Crinan Street, London N1 9XW
☎ 020 7833 4000   🖷 020 7843 4650
bfipublishing@palgrave.com
www.palgrave.com/bfi

Founded 1980. The publishing imprint of the **British Film Institute**. Publishes academic, schools and general film/television-related books and resources. TITLE *The Cinema Book* Pam Cook. SERIES *BFI Film Classics; BFI TV Classics; BFI Screen Guide; World Directors*. About 30 titles a year. E-mail submissions preferred.
Ⓔ Royalties annually.

BFP Books

Focus House, 497 Green Lanes, London N13 4BP
☎ 020 8882 3315   🖷 020 8886 3933
mail@thebfp.com
Chief Executive *John Tracy*
Commissioning Editor *Stewart Gibson*

Founded 1982. The publishing arm of the **Bureau of Freelance Photographers**. Publishes illustrated books on photography, mainly aspects of freelancing and marketing pictures. No unsolicited mss but ideas welcome.

Clive Bingley Books
▷ **Facet Publishing**

BIOS Scientific Publishers
▷ **Taylor & Francis Group**

Birlinn Ltd

West Newington House, 10 Newington Road, Edinburgh EH9 1QS
☎ 0131 668 4371   🖷 0131 668 4466
info@birlinn.co.uk
www.birlinn.co.uk
Managing Director *Hugh Andrew*

Founded 1992. Publishes history, military history, adventure, Gaelic, humour, Scottish history, Scottish reference, guidebooks and folklore. IMPRINT **John Donald Publishers** (see entry). Acquired **Polygon** in 2002 (see entry) and Tuckwell Press in 2005 and Mercat Press in 2007. About 170 titles a year across all imprints. No unsolicited mss; synopses and ideas welcome.
Ⓔ Royalties paid.

✴*Authors' update* A thriving regional publisher with titles that hit the wider market.

Bitter Lemon Press

37 Arundel Gardens, London W11 2LW
☎ 020 7727 7927   🖷 020 7460 2164
books@bitterlemonpress.com
www.bitterlemonpress.com
Directors *Laurence Colchester, François von Hurter*

Founded 2003. Specializes in thrillers, *romans noir* and other contemporary fiction from abroad. 8 titles in 2008. No unsolicited material; submissions from agents only.
Ⓔ Royalties annually.

Black & White Publishing Ltd

29 Ocean Drive, Edinburgh EH6 6JL
☎ 0131 625 4500   🖷 0131 625 4501

mail@blackandwhitepublishing.com
www.blackandwhitepublishing.com
Director *Campbell Brown*

Founded 1990. Publishes general non-fiction, biography, sport and humour as well as selected fiction and children's books. IMPRINT **Itchy Coo** Scots language resources for use in schools. Send brief synopsis along with 30 sample pages. 'Do *not* submit full mss. Submission by e-mail preferred.' See the website for full submission guidelines. Royalties twice-yearly.

A.&C. Black Publishers Ltd

36 Soho Square, London W1D 3QY
020 7758 0200   020 7758 0222
enquiries@acblack.com
www.acblack.com
Managing Director *Jill Coleman*
Deputy Managing Director *Jonathan Glasspool*

Publishes children's and educational books, including music, for 0–16 year-olds, visual natural history, ornithology, nautical, reference, business, dictionaries, sport, performing arts and writing books. Acquisitions brought the Herbert Press' art, design and general books, Adlard Coles' and Thomas Reed's sailing lists, plus *Reeds Nautical Almanac*, Christopher Helm, Pica Press and T&AD Poyser's natural history and ornithology lists, children's educational publisher Andrew Brodie and Featherstone Education, and Whitakers Almanack into A.&C. Black's stable. Bought by **Bloomsbury Publishing** in May 2000. Acquired **Methuen**'s drama list in 2006. IMPRINTS **Adlard Coles Nautical; Thomas Reed; Christopher Helm; T&AD Poyser; The Herbert Press; Andrew Brodie; Methuen Drama; Featherstone Education.** TITLES *Who's Who; Writers' & Artists' Yearbook; Children's Writers' & Artists' Yearbook; Whitaker's Almanack; New Mermaids; White Wolves; Methuen Drama Modern Plays.* Around 500 titles in 2009. Initial enquiry appreciated before submission of mss. Royalties vary according to contract.

Black Ace Books

PO Box 7547, Perth PH2 1AU
01821 642822   01821 642101
www.blackacebooks.com
Managing Director *Hunter Steele*

Founded 1991. Publishes new fiction, Scottish and general; some non-fiction including biography, history, philosophy and psychology. IMPRINTS **Black Ace Books; Black Ace Paperbacks** TITLES *Succeeding at Sex and Scotland, Or the Case of Louis Morel* Hunter Steele; *La Tendresse* Ken Strauss MD; *Count Dracula (The Authorized Version)* Hagen Slawkberg; *Caryddwen's Cauldron* Paul Hilton; *The Sinister Cabaret* John Herdman. No children's, poetry, cookery, DIY, religion.

36 titles in print. 'No submissions at all, please, without first checking our website for details of current requirements and submission guidelines.' Royalties twice-yearly.

Black Lace
▷ The Random House Group Ltd

Black Spring Press Ltd

Curtain House, 134–146 Curtain Road, London EC2A 3AR
020 7613 3066   020 7613 0028
general@blackspringpress.co.uk
Director *Robert Hastings*

Founded 1986. Publishes fiction and non-fiction, literary criticism, biography. TITLES *The Lost Weekend* Charles Jackson; *The Big Brass Ring* Orson Welles; *The Gorse Trilogy* Patrick Hamilton; *Julian Maclaren-Ross Collected Memoirs.* About 5 titles a year. No unsolicited mss. Royalties annually.

Black Swan
▷ Transworld Publishers

BlackAmber
▷ Arcadia Books

Blackstaff Press Ltd

4C Heron Wharf, Sydenham Business Park, Belfast BT3 9LE
028 9045 5006   028 9046 6237
info@blackstaffpress.com
www.blackstaffpress.com
Managing Editor *Patsy Horton*

Founded 1971. Publishes mainly, but not exclusively, Irish interest books, fiction, poetry, history, sport, cookery, politics, illustrated editions, natural history and humour. About 20 titles a year. Unsolicited mss considered, but preliminary submission of synopsis plus short sample of writing is essential. Return postage *must* be enclosed. No submissions by e-mail. Royalties twice-yearly.

✱*Authors' update* Regional and proud of it, this Belfast publisher is noted for bringing on young talent and for 'wonderfully well-presented catalogues and promotional material.'

Blackstone Publishers

18 Soho Square, London W1D 3QL
020 7025 8034   020 7025 8100
mail@blackstonepublishers.com
mail@redarrowbooks.com
www.blackstonepublishers.com
www.redarrowbooks.com
Editor *Nina Rogers*
Publisher *John Kurtz*

Founded 2007. Publishes fiction, including romance, women's fiction, mysteries, suspense, science fiction, horror, fantasy, humour, gay, lesbian, erotica; non-fiction, including memoirs, biography, history, war, reference; children's, picture books, young reader and teen fiction. IMPRINT **Red Arrow Books**. 20–40 titles a year. Unsolicited mss, synopses and ideas for books (with s.a.e.) welcome.
Ⓔ Royalties twice-yearly.

Blackwell Publishing
▷ Wiley-Blackwell

John Blake Publishing Ltd
3 Bramber Court, 2 Bramber Road, London W14 9PB
☎ 020 7381 0666   🖷 020 7381 6868
words@blake.co.uk
www.johnblakepublishing.co.uk
Managing Director *John Blake*
Deputy Managing Director *Rosie Ries*

Founded 1991 and expanding rapidly. Bought the assets of Smith Gryphon Ltd in 1997 and acquired Metro Publishing in 2001. Publishes mass-market non-fiction. No fiction, children's, specialist or non-commercial. About 100 titles a year. No unsolicited mss; synopses and ideas welcome. Please enclose s.a.e.
Ⓔ Royalties twice-yearly.

✱*Authors' update* **With a tabloid journalist's talent for knowing what makes the front page and the front of shop display, John Blake is a first call for celebrity memoirs and celebrity brand titles.**

Blandford Press
▷ Octopus Publishing Group

Bliss
▷ Arcadia Books

Bloodaxe Books Ltd
Highgreen, Tarset NE48 1RP
☎ 01434 240500   🖷 01434 240505
editor@bloodaxebooks.com
www.bloodaxebooks.com
Managing/Editorial Director *Neil Astley*

Publishes poetry, literature and criticism, and related titles by British, Irish, European, Commonwealth and American writers. Ninety-five per cent of the list is poetry. TITLES include seven major anthologies: *Staying Alive, Being Alive, Earth Shattering* and *In Person* all edited by Neil Astley; *The Bloodaxe Book of 20th Century Poetry* ed. Edna Longley; *Modern Women Poets* ed. Deryn Rees-Jones; editions of Fleur Adcock, David Constantine, Helen Dunmore, Roy Fisher, Jackie Kay, Brendan Kennelly, Mary Oliver, J.H. Prynne, Peter Reading, Ken Smith,

Anne Stevenson, C.K. Williams and Benjamin Zephaniah. About 30–40 titles a year. Unsolicited poetry mss welcome; send a sample of no more than 10 poems with s.a.e., 'but if you don't read contemporary poetry, don't bother'. No e-mail submissions of any kind.
Ⓔ Royalties annually.

✱*Authors' update* **Assisted by regional Arts Council funding, Bloodaxe is one of the liveliest and most innovative of poetry publishers with a list that takes in some of the best of the younger poets.**

Bloody Books
▷ Beautiful Books Ltd

Bloomsbury Publishing Plc
36 Soho Square, London W1D 3QY
☎ 020 7494 2111   🖷 020 7434 0151
www.bloomsbury.com
Founder and Chief Executive *Nigel Newton*
Executive Director *Richard Charkin*
Publishing Directors *Alexandra Pringle (Editor-in-Chief), Liz Calder, Kathy Rooney, Sarah Odedina, Michael Fishwick*
Approx. Annual Turnover £100 million

Founded 1986. Many of its authors have gone on to win prestigious literary prizes, including J.K. Rowling's *Harry Potter and the Philosopher's Stone, Harry Potter and the Chamber of Secrets* and *Harry Potter and the Prisoner of Azkaban* which won the Nestlé Smarties Book Prize in 1997, 1998 and 1999 respectively. Margaret Atwood's *The Blind Assassin* won the Booker Prize in 2000.

Started Bloomsbury USA in 1998. Acquired **A.&C. Black Publishers Ltd** in May 2000 (see entry), Peter Collin Publishing Ltd in September 2002, Berlin Verlag in 2003, Walker Publishing Inc in New York in 2004, Featherstone Education (A.&C. Black imprint), **Berg Publishers** (see entry) and John Wisden & Co. in 2008. Publishes literary fiction and non-fiction, including general reference; also audiobooks. AUTHORS include Margaret Atwood, T.C. Boyle, Sophie Dahl, Jeffrey Eugenides, Neil Gaiman, Daniel Goleman, David Guterson, Sheila Hancock, John Irving, Jay McInerney, Tim Pears, Celia Rees, J.K. Rowling, Ben Schott, Will Self, Donna Tartt, Rupert Thomson, Barbara Trapido, Benjamin Zephaniah. No unsolicited submissions.
Ⓔ Royalties twice-yearly.

✱*Authors' update* **Cash rich from Harry Potter, Bloomsbury is investing imaginatively in a range of publishing initiatives including an 'on demand' academic imprint which is to publish titles online for free. Turnover has been boosted by the acquisition of Featherstone Education, Berg Publishers and Wisden.**

Announced at the Frankfurt Book Fair was a deal with the Qatar Foundation to launch a Doha-based publisher. All this may seem remote from the jobbing author but think what it is doing for Bloomsbury's marketing reach.

Blue Door
▷ HarperCollins Publishers Ltd

BMM
▷ SportsBooks Limited

Bobcat
▷ Omnibus Press

Bodleian Library
Communications & Publishing Office, Bodleian Library, Broad Street, Oxford OX1 3BG
☎ 01865 277627 📠 01865 277187
publishing@bodley.ox.ac.uk
www.bodleianbookshop.co.uk
Head of Publishing *Samuel Fanous*

Founded 1602. Publishes trade and scholarly titles relating to the Bodleian Library collections of over eight million books and manuscripts. TITLES *The Original Frankenstein*; *Book of Beasts: A Facsimile of MS. Bodley 764*; *Postcards from the Trenches*; *Postcards from the Russian Revolution*; *German Invasion Plans for the British Isles 1940*. 9 titles in 2008. No unsolicited mss. Synopses and ideas for book related to the collections will be considered; approach should be made by post, addressed to the Commissioning Editor.

The Bodley Head/Bodley Head Children's Books
▷ The Random House Group Ltd

Boltneck Publications Limited
Head Office: Westpoint, 78 Queens Road, Clifton, Bristol BS8 1QX
☎ 0117 985 8709
clive.birch@medavia.co.uk
www.boltneck.co.uk
Submissions: King's Cote, Valley Road, Finmere MK18 4AL
Managing Director *David Thomas*
Publishing Director *Clive Birch*

Founded 2004. Linked to Medavia Media Agency. Publishes fiction and popular non-fiction on topical subjects or people of interest. DIVISION **Medavia Publishing** *Clive Birch* TITLES *Married to Albert*; *No Big Deal*; *A Decent Man*; *Kirsty – Angel of Courage*. No unsolicited mss. Initial letter, synopsis, sample chapter, c.v. and s.a.e. 💷 Royalties paid annually.

Bonnier Books (UK)
Appledram Barns, Birdham Road, Chichester PO20 7EQ
☎ 01243 531660 📠 01243 774433

CEO *Des Higgins*
Founded in February 2007. Part of the Swedish media group, Bonnier AB. Publishes illustrated lifestyle and cookery titles. TITLES *Bride and Groom Cookbook*; *Healthy Heart Cookbook*. 32 titles in 2008.

Book Blocks
▷ CRW Publishing Ltd

Book Guild Publishing
Pavilion View, 19 New Road, Brighton BN1 1UF
☎ 01273 720900 📠 01273 723122
info@bookguild.co.uk
www.bookguild.co.uk
Chairman *George M. Nissen, CBE*
Managing Director *Carol Biss*

Founded 1982. Publishes fiction, children's and non-fiction, including human interest and biography/memoirs. Expanding mainstream list. Subject areas: **Human Interest** TITLES *Mud and the City* Jessica Fellowes; *Not for Wimps* John Parry. **Biography** TITLES *Cupboard Love* Laura Lockington; *Home Was a Grand Hotel* Pamela Sydney Wilson; *A PG Wodehouse Treasury* Mark Hichens. **General Non-Fiction** TITLE *The Wives of the Kings of* England Mark Hichens. **Fiction** TITLES *The Tail of Augustus Moon* Melanie Whitehouse; *Ironic* Paul Nagle. **Children's** TITLE *Ma'at's Feather* Juliet Desailly. About 90 titles a year. Unsolicited mss, ideas and synopses welcome.
💷 Royalties twice-yearly.

✳**Authors' update** There are two sides to Book Guild Publishing, not always coexisting in happy harmony. The favoured image is of a conventional publisher with a strong list of general titles. But also on offer is a service where authors cover the costs of publication. Unlike many vanity publishers, Book Guild Publishing clearly sets out its terms of agreement. Still, the outlay can be high relative to sales and first-time authors should not be misled into believing that they have found an easy route to fame and fortune.

Border Lines Biographies
▷ Seren

Boulevard Books & The Babel Guides
71 Lytton Road, Oxford OX4 3NY
☎ 01865 712931
ray.keenoy@gmail.co.uk
www.babelguides.com
Managing Director *Ray Keenoy*

Specializes in contemporary world fiction by young writers in English translation. Existing or forthcoming series of fiction from Brazil, Italy, Latin America, Low Countries, Greece, and elsewhere. The Babel Guides series of popular

guides to fiction in translation started in 1995. SERIES *Babel Guides to Fiction in Translation* Series Editor *Ray Keenoy* TITLES *Eminent Hungarians*; *Babel Guide to Italian Fiction in Translation*; *Babel Guide to the Fiction of Portugal, Brazil & Africa in Translation*; *Babel Guide to French Fiction in English Translation*; *Babel Guide to Jewish Fiction*; *Babel Guide to Brazilian Fiction*; *Babel Guide to Welsh Fiction*. Suggestions and proposals for translations of contemporary fiction welcome. Also seeking contributors to forthcoming Babel Guides (all literatures).
Ⓔ Royalties annually.

Bound Biographies Limited

Heyford Park House, Heyford Park, Bicester OX25 5HD
📞 01869 232911   📠 01869 232698
office@boundbiographies.com
www.boundbiographies.com
Managing Director *Michael Oke*
Editorial Head *Dr A.J. Gray*
Approx. Annual Turnover £250,000

Founded in 1992 to assist in the writing and production of private life stories and corporate histories. Print-on-demand facilities to provide short runs of paper or hardback books to complement the Bound Biographies leather-bound range. Publishes autobiographies predominantly although novels, poetry and special interest books are considered. TITLES *Red Tails in the Sunset* Bryn Williams; *Think of a Card* Don Lewin. 40 titles in 2008. Unsolicited material welcome; approach by post, telephone or e-mail.
Ⓔ Royalties annually.

Bounty
▷ Octopus Publishing Group

Bowker (UK) Ltd

1st Floor, Medway House, Cantelupe Road, East Grinstead RH19 3BJ
📞 01342 310450   📠 01342 310486
sales@bowker.co.uk
www.bowker.co.uk
Managing Director *Doug McMillan*

Part of the Cambridge Information Group (CIG). Publishes bibliographic references used by publishers, libraries and retailers throughout the world to source new book information. TITLES *Books In Print* and *Global Books in Print*. Unsolicited material will not be read.

Boxtree
▷ Macmillan Publishers Ltd

Marion Boyars Publishers Ltd

24 Lacy Road, London SW15 1NL
📞 020 8788 9522   📠 020 8789 8122

rebecca@marionboyars.com
www.marionboyars.co.uk
Editor *Rebecca Gillieron*
Editor, Non-fiction *Kit Maude*

Founded 1975, formerly Calder and Boyars. Publishes biography and autobiography, fiction, literature and criticism, music, philosophy, psychology, sociology and anthropology, theatre and drama, film and cinema, women's studies. AUTHORS include Georges Bataille, Ingmar Bergman, Heinrich Böll, Jean Cocteau, Carlo Gébler, Julian Green, Ivan Illich, Pauline Kael, Ken Kesey, Toby Litt, Kenzaburo Oe, Hubert Selby Jr, Lawrence Potter, Eudora Welty, Judith Williamson, Hong Ying, Elif Shafak and Riverbend. About 30 titles a year. Unsolicited mss not welcome for fiction or poetry; submissions from agents only. Unsolicited synopses and ideas welcome for non-fiction. OVERSEAS ASSOCIATES Marion Boyars Publishers Inc.
Ⓔ Royalties annually.

Bradt Travel Guides

23 High Street, Chalfont St Peter SL9 9QE
📞 01753 893444   📠 01753 892333
info@bradtguides.com
www.bradtguides.com
Managing Director *Donald Greig*
Editorial Director *Adrian Phillips*
Approx. Annual Turnover £1 million

Founded in 1974 by Hilary Bradt. Specializes in travel guides to off-beat places and quirky travel and wildlife-related titles. SERIES Country guides and island guides (Azores, Syria, Namibia); wildlife guides (Madagascar, Galapagos, Arctic, Antarctica, Southern Africa); mini guides to cities (Budapest, Dubrovnik, Bratislava, Tallinn) and the 'Eccentric' series (Britain, London, France, America, Oxford, Cambridge, Edinburgh). Worldwide distribution. 48 titles in 2008. No unsolicited mss; synopses and relevant ideas (not travelogues) welcome.
Ⓔ Royalties twice-yearly.

✱***Authors' update** Judged to be the best of guide publishers by *Wanderlust* magazine, Bradt may well have hit on the antidote to online travel guides. A list of unlikely tourist destinations such as Burkina Faso and Kabul has achieved such success to inspire plans for doubling the number of titles by 2010.*

Nicholas Brealey Publishing

3–5 Spafield Street, London EC1R 4QB
📞 020 7239 0360   📠 020 7239 0370
rights@nicholasbrealey.com
www.nicholasbrealey.com
Managing Director *Nicholas Brealey*

Founded 1992. Independent publishing group focusing on innovative trade/professional books covering business and economics, popular psychology and the increasingly active fields of travel writing and crossing cultures. The group has offices in Boston and includes Intercultural Press. TITLES *Your Writing Coach*; *Get to the Top on Google*; *The Cult of the Amateur*; *The Leader's Way* by the Dalai Lama; *Authentic Happiness*; *50 Psychology Classics*; *The 80/20 Principle*; *It's All Greek to Me!*; *Don't Tell Mum I Work on the Rigs She Thinks I'm a Piano Player in a Whorehouse*. No fiction, poetry or leisure titles. 30 titles a year. No unsolicited mss; synopses and ideas welcome.

✱*Authors' update* **Looks to be succeeding in breaking away from the usual computer-speak business manuals to publish information and literate texts. Lead titles have a distinct trans-Atlantic feel.**

The Breedon Books Publishing Co. Ltd

3 The Parker Centre, Mansfield Road, Derby DE21 4SZ

☎ 01332 384235   ⊞ 01332 292755
steve.caron@breedonpublishing.co.uk
www.breedonbooks.co.uk
Owner/Managing Director *Steve Caron*
Approx. Annual Turnover £1 million

Founded 1983. Publishes football and sport, local history, old photographs, heritage and ghost stories with a local slant. About 60 titles a year. Unsolicited mss, synopses and ideas welcome if accompanied by s.a.e. No poetry, children's stories or fiction.

Martin Breese International

19 Hanover Crescent, Brighton BN2 9SB
☎ 01273 687555
mbreese999@aol.com
www.abracadabra.co.uk
Chairman/Managing Director *Martin R. Breese*

Founded in 1975 to produce specialist conjuring books. No unsolicited submissions.

Brewin Books Ltd

Doric House, 56 Alcester Road, Studley B80 7LG
☎ 01527 854228   ⊞ 01527 852746
admin@brewinbooks.com
www.brewinbooks.com
Chairman/Managing Director *Alan Brewin*
Company Secretary *Julie Brewin*
Director *Alistair Brewin*

Founded 1976. Publishes books on all aspects of Midland life and history including social, hospital, police, military, transport and family histories as well as biographies. TITLES *Witness to War: Correspondence during the Spanish Civil War* Rose Parish; *Carry on Car Making: Life in China after Longbridge* Ian S. Pogson; *Birmingham Back to the Seventies* Alton Douglas. IMPRINTS **Brewin Books**; **Alton Douglas Books**; **History**

Into Print. About 30 titles a year. Not interested in children's, poetry, short stories or novels. Approach by letter, but do not send full mss. Unsolicited synopses and ideas welcome with s.a.e.
⑤ Royalties twice-yearly.

Bristol Classical Press
▷ Gerald Duckworth & Co. Ltd

Bristol Phoenix Press
▷ University of Exeter Press

The British Academy

10 Carlton House Terrace, London SW1Y 5AH
☎ 020 7969 5200   ⊞ 020 7969 5300
pubs@britac.ac.uk
www.britac.ac.uk
Head of Publications *James Rivington*
Book Production Manager *Janet English*
Publications Officer *Amrit Bangard*

Founded 1902. The primary body for promoting scholarship in the humanities and social sciences, the Academy publishes many series stemming from its own long-standing research projects, or series of lectures and conference proceedings. Main subjects include history, philosophy and archaeology. SERIES *Auctores Britannici Medii Aevi*; *Early English Church Music*; *Fontes Historiae Africanae*; *Records of Social and Economic History*. About 20 titles a year. Proposals for these series are welcome and are forwarded to the relevant project committees. The British Academy is a registered charity and does not publish for profit.
⑤ Royalties only when titles have covered their costs.

The British Computer Society

First Floor, Block D, North Star House, North Star Avenue, Swindon SN2 1FA
☎ 0845 300 4417   ⊞ 01793 480270
publishing@hq.bcs.org.uk
www.bcs.org/books
Chief Executive *David Clarke*
Head of Publishing & Information Products *Elaine Boyes*
Book Publisher *Matthew Flynn*
Approx. Annual Turnover £20 million (Society)

Founded 1957. BCS is the leading professional and learned society in the field of computers and information systems. Publishes books which support the professional, academic and practical needs of the IT community. 10 titles in 2008. Unsolicited material welcome; submissions form and guide on website page: www.bcs.org/books/writer
⑤ Royalties annually.

The British Library

96 Euston Road, London NW1 2DB
☎ 020 7412 7535   📠 020 7412 7768
blpublications@bl.uk
www.bl.uk
Publisher *David Way*
Senior Editor *Catherine Britton*

Founded 1979 as the publishing arm of The
British Library to publish works based on
the historic collections and related subjects.
Publishes bibliographical reference, manuscript
studies, illustrated books based on the Library's
collections, and book arts. TITLES *1000 Years
of English Literature*; *The Books of Henry VIII
and His Wives*; *Futurist Typography*. About 50
titles a year. Unsolicited mss, synopses and ideas
welcome if related to the history of the book,
book arts or bibliography. No fiction or general
non-fiction.
© Royalties annually.

The British Museum Press

38 Russell Square, London WC1B 3QQ
☎ 020 7323 1234   📠 020 7436 7315
www.britishmuseum.org
Managing Director *Brian Oldman*
Director of Publishing *Rosemary Bradley*

The book publishing division of The British
Museum Company Ltd. Founded in 1973 as
British Museum Publications Ltd; relaunched 1991
as British Museum Press. Publishes books based
on the British Museum's collection, including
ancient history, archaeology, ethnography, art
history, exhibition catalogues, guides, children's
books and all official publications of the British
Museum. TITLES *The First Emperor: China's
Terracotta Army*; *The Parthenon Sculptures in
the British Museum*; *500 Things to Know About
the Ancient World*; *Indian Art in Detail*. About
40 titles a year. Synopses and ideas for books
welcome.
© Royalties twice-yearly.

Andrew Brodie
▷ A.&C. Black Publishers Ltd

Brown Skin Books

PO Box 57421, London E5 0ZD
☎ 020 8986 1115
info@brownskinbooks.co.uk
www.brownskinbooks.co.uk
Chairman *Dr John Lake*
Managing Director *Vastiana Belfon*

Founded 2002. Publishes 'quality, intelligent
erotic fiction by black women around the world'
and a new series of erotic crime thrillers. TITLES
Body and Soul Jade Williams; *Personal Business*
Isabel Baptiste; *Scandalous* and *A Darker Shade of
Blue* Angela Campion; *Playthings* Faith Graham;
Online Wildfire Crystal Humphries; *Strip Poker*
and *Beg Me* Lisa Lawrence; *The Singer* Aisha
DuQuesne; *Sorcerer* Tamzin Hall. 2 titles in
2008. No unsolicited mss; synopses and sample
chapters welcome. Send by e-mail or post. No
poetry.
© Royalties twice-yearly.

Brown, Son & Ferguson, Ltd

4–10 Darnley Street, Glasgow G41 2SD
☎ 0141 429 1234   📠 0141 420 1694
info@skipper.co.uk
www.skipper.co.uk
Chairman/Joint Managing Director *T. Nigel
Brown*

Founded 1850. Specializes in nautical textbooks,
both technical and non-technical. Also Scottish
one-act/three-act plays. Unsolicited mss,
synopses and ideas for books welcome.
© Royalties annually.

Bryntirion Press

Bryntirion, Bridgend CF31 4DX
☎ 01656 655886   📠 01656 665919
office@emw.org.uk
www.emw.org.uk
Publications Manager *Huw Kinsey*
Welsh Language Publications *Gethin Rhys*
Approx. Annual Turnover £40,000

Owned by the Evangelical Movement of Wales.
Publishes Christian books in English and Welsh.
50 titles in 2007 No unsolicited mss.
© Royalties annually.

Burning House
▷ Beautiful Books Ltd

Burns & Oates
▷ The Continuum International Publishing Group
Limited

Business Education Publishers Ltd

Evolve Business Centre, Cygnet Way, Rainton
Bridge Business Park, Houghton-le-Springs
DH4 5QY
☎ 0191 305 5160
info@bepl.com
www.bepl.com
Managing Director *Mrs A. Murphy*
Approx. Annual Turnover £400,000

Founded 1981. Publishes business education,
economics and law for BTEC and GNVQ reading.
Currently expanding into further and higher
education, computing, IT, business, travel and
tourism, occasional papers for institutions and
local government administration. Unsolicited mss
and synopses welcome.
© Royalties annually.

Business Plus
▷ Hachette UK

Buster Books
▷ Michael O'Mara Books Ltd

Butterworth Heinemann
▷ Elsevier Ltd

Cadogan Guides
▷ New Holland Publishers (UK) Ltd

Calder Publications Ltd

51 The Cut, London SE1 8LF
☎ 020 7633 0599   📠 020 7928 5930
info@calderpublications.com
www.calderpublications.com
Managing Director/Editorial Head *Alessandro Gallenzi*

Acquired in April 2007 by independent publishers OneWorld Classics. A publishing company which has grown around the tastes and contacts of John Calder, the iconoclast of the literary establishment. The list has a reputation for controversial and opinion-forming publications. Publishes autobiography, biography, drama, literary fiction, literary criticism, music, opera, poetry, politics, sociology, ENO opera guides. AUTHORS Antonin Artaud, Marguerite Duras, Martin Esslin, Erich Fried, P.J. Kavanagh, Robert Menasse, Robert Pinget, Luigi Pirandello, Alain Robbe-Grillet, Nathalie Sarraute, L.F. Céline, Eva Figes, Claude Simon, Raymond Queneau . No new material accepted.
🄔 Royalties annually.

✲*Authors' update* After 60 years in publishing and with 25 Nobel laureates to his credit, John Calder has passed on the baton to Alessandro Gallenzi who is rebranding the list for today's market.

California University Press
▷ University Presses of California, Columbia & Princeton Ltd

Cambridge University Press

The Edinburgh Building, Shaftesbury Road, Cambridge CB2 8RU
☎ 01223 312393   📠 01223 315052
www.cambridge.org
Chief Executive *Stephen R.R. Bourne*
Managing Director, Publishing, Academic and Professional (books and journals in H&SS and STM) *A.M.C. Brown*
Managing Director, Publishing, Cambridge Learning (ELT and Education) *J.M. Pieterse*
Managing Director, Publishing, New Directions *J.G. Tuttle*

The oldest printer and publisher in the world, now a global organization with a regional structure operating in the UK, Europe, Middle East & Africa, in the Americas and in Asia-Pacific. Has warehousing centres in Cambridge, New York, Melbourne, Madrid, Cape Town, São Paulo, Singapore, New Delhi and Tokyo, with offices and agents in many other countries. Publishing includes major ELT courses; tertiary textbooks, monographs and journals; scientific and medical reference; professional lists in law, management and engineering; educational coursebooks for the National Curriculum; and e-learning materials for schools. Recent ventures in e-pedagogy include the companies Global Grid for Learning and Cambridge-Hitachi. Publishes at all levels from primary school to postgraduate research and professional. Also Bibles, prayer books and over 230 academic journals. Around 25,000 authors in 116 countries and between 1,500 and 2,000 new titles a year. Synopses and ideas for books are welcome (and preferable to the submission of unsolicited mss). No fiction or poetry.
🄔 Royalties twice-yearly.

✲*Authors' update* A strong year can be credited to English Language Teaching, a series which is making impressive headway in China and Asia.

Camden Press Ltd

43 Camden Passage, London N1 8EA
☎ 020 7226 4673
Chairman *Bob Borzello*

Founded 1985. Publishes social issues; all books are launched in connection with major national conferences. DIVISION **Publishing for Change** *Bob Borzello* TITLE *Living with the Legacy of Abuse*. IMPRINT **Mindfield** TITLES *Hate Thy Neighbour: The Race Issue*; *Therapy on the Couch*. No unsolicited material. Approach by telephone in the first instance.
🄔 Royalties annually.

Campbell Books
▷ Macmillan Publishers Ltd

Candle
▷ Lion Hudson plc

Canongate Books Ltd

14 High Street, Edinburgh EH1 1TE
☎ 0131 557 5111   📠 0131 557 5211
info@canongate.co.uk
www.meetatthegate.com
Also at: Lower Basement, 151 Chesterton Road, London W10 6ET
Publisher/Managing Director *Jamie Byng*
Publishing Director *Anya Serota*
Editorial Director *Nick Davies*
Approx. Annual Turnover £7.95 million

Founded 1973. In 2002, won both 'Publisher of the Year' at the British Book Awards and the Booker Prize with *Life of Pi* by Yann Martel. Publishes a wide range of literary fiction and non-fiction. Historically, there is a strong Scottish slant to the list but its output is increasingly international, especially now with its joint venture with **Grove Atlantic** in the US and co-publishing agreement with Text Publishing in Australia. Canongate also has a growing reputation for originating unusual projects (typified by *The Pocket Canons* and more recently by *The Myths* series, launched in 2005). Key AUTHORS include Michel Faber, Yann Martel, Alasdair Gray, John Fante, James Meek and Louise Welsh.

Ⓔ Royalties twice-yearly.

✹*Authors' update* **With turnover trebling on the Obama autobiography and** *The Mighty Boosh*, **Canongate is using its extra muscle on the e-book market. By the end of the year its entire backlist of 450 titles will be digitized.**

Canterbury Press
▷ SCM – Canterbury Press Ltd

Capall Bann Publishing
Auton Farm, Milverton TA4 1NE
☎ 01823 401528 📠 01823 401529
enquiries@capallbann.co.uk
www.capallbann.co.uk
Chairman *Julia Day*
Editorial Head *Jon Day*

Founded in 1993, Capall Bann now has well over 300 titles in print. Family-owned and run company, 'operated by people with real experience in the topics we publish' which include British traditions, folklore, animals, alternative healing, environmental, Celtic lore, mind, body and spirit. TITLES *The Magical and Mystical World of Faerie*; *The Seer's Guide to Crystal and Gem Divination*; *Nature's Children – Celebrating the Seasons in a Pagan Family*; *Reaching For the Divine – How to Communicate Effectively With Your Spirit Guides & Loved Ones On The Other Side*; *Aphrodisiacs – Aphrodite's Secrets*. About 40 titles a year. Unsolicited proposals for books welcome. No fiction or poetry.

Ⓔ Royalties twice-yearly.

Jonathan Cape/Jonathan Cape Children's Books
▷ The Random House Group Ltd

Capstone Publishing (A Wiley Company)
John Wiley & Sons Ltd, The Atrium, Southern Gate, Chichester PO19 8SQ
☎ 01243 779777
info@wiley-capstone.co.uk
www.wiley.com

Founded 1997. Part of **John Wiley & Sons Ltd**. Publishes business, personal development, lifestyle, health and humour books.

Carcanet Press Ltd
4th Floor, Alliance House, 30 Cross Street, Manchester M2 7AQ
☎ 0161 834 8730 📠 0161 832 0084
info@carcanet.co.uk
www.carcanet.co.uk
Owner *Folio Holdings*
Chairman *Kate Gavron*
Managing Director/Editorial Director *Michael Schmidt, OBE, FRSL*

Having started the company as an undergraduate hobby, Robert Gavron bought the company in 1983 and established strong European, Commonwealth and American links. Winner of the Sunday Times Small Publisher of the Year award in 2000, it took over the Oxford Poets list from **Oxford University Press** in 1999, which it now publishes as a distinct imprint. Primarily a poetry publisher but also publishes academic, literary biography, fiction in translation and translations. AUTHORS John Ashbery, Eavan Boland, Donald Davie, Natalia Ginzburg, Robert Graves, Elizabeth Jennings, Hugh MacDiarmid, Edwin Morgan, Sinead Morrissey, Les Murray, Frank O'Hara, Richard Price, Frederic Raphael, C.H. Sisson, Charles Tomlinson, Jane Yeh. About 40 titles a year, including *PN Review* (six issues yearly). Poetry submissions (hard copy only): 6–10 poems with covering letter and return postage. Prospective writers should familiarize themselves with the Carcanet list before submitting work.

Ⓔ Royalties annually.

✹*Authors' update* **Now one of the UK's leading literary publishers with a diverse list of modern and classic poetry, Carcanet stays close to the source of literary creativity with its own postgraduate Writing School at Manchester Metropolitan University (see entry under** *UK and Irish Writers' Courses*).

Cardiff Academic Press
St Fagans Road, Fairwater, Cardiff CF5 3AE
☎ 029 2056 0333 📠 029 2055 4909
info@drakeed.com
www.drakeed.com
Managing Director *R.G. Drake*

Academic publishers.

Carlton Publishing Group
20 Mortimer Street, London W1T 3JW
☎ 020 7612 0400 📠 020 7612 0401
enquiries@carltonbooks.co.uk
www.carltonbooks.co.uk
Publisher *Jonathan Goodman*
Managing Director *Belinda Rasmussen*
Editorial Director *Piers Murray Hill*
Approx. Annual Turnover £17 million

Founded 1992. Carlton Publishing Group is an independent publishing house. It has three main divisions: **Carlton Books** Illustrated leisure and entertainment books aimed at the mass market. Subjects include history, sport, puzzles, health, popular science, children's non-fiction, popular culture, music, fashion and design; **André Deutsch** Autobiography, biography, history, current affairs, narrative non-fiction and the arts; **Prion Books** Humour, nostalgia, drink and classic literature. 120 titles per year. No unsolicited mss; synopses and ideas welcome. No novels, science fiction, poetry or children's fiction.
Ⓔ Royalties twice-yearly.

✹*Authors' update* **Imaginative illustrated books on popular subjects sell across all language barriers.**

Carroll & Brown Publishers Limited

20 Lonsdale Road, London NW6 6RD
☎ 020 7372 0900   🖷 020 7372 0460
mail@carrollandbrown.co.uk
www.carrollandbrown.co.uk
Editorial Director *Amy Carroll*

Publishes practical parenting, health, fitness, recreations, mind, body and spirit. TITLES *Your Pregnancy Bible*; *Practical Wabi Sabi*; *Beading for Beginners*. Synopses and ideas for illustrated books welcome; approach in writing in the first instance, enclosing s.a.e. for reply. No fiction
Ⓔ Fees or royalties paid.

Frank Cass
▷ Taylor & Francis Group

Cassell
▷ The Orion Publishing Group Limited

Cassell Illustrated
▷ Octopus Publishing Group

Kyle Cathie Ltd

122 Arlington Road, London NW1 7HP
☎ 020 7692 7215   🖷 020 7692 7260
general.enquiries@kyle-cathie.com
www.kylecathie.co.uk
Editorial Assistant *Vicky Murrell*

Founded 1990 to publish and promote 'books we have personal enthusiasm for'. Publishes non-fiction: cookery, food and drink, health and beauty, mind, body and spirit, gardening, homes and interiors, reference and occasional books of classic poetry. TITLES *Clarissa's Comfort Food* Clarissa Dickson Wright; *The Chopsticks Diet* Kimiko Barber; *Gardening for Small Spaces* John Cushnie; *Cycle for Life* Nicole Cooke; *Liz Earle's Skin Secrets* Liz Earle. About 25 titles a year. No unsolicited mss. 'Synopses and ideas are considered in the fields in which we publish.'
Ⓔ Royalties twice-yearly.

Catholic Truth Society (CTS)

40–46 Harleyford Road, London SE11 5AY
☎ 020 7640 0042   🖷 020 7640 0046
f.martin@cts-online.org.uk
www.cts-online.org.uk
Chairman *Most Rev. Paul Hendricks*
General Secretary *Fergal Martin*

Founded originally in 1868 and re-founded in 1884. Publishes religious books – Roman Catholic; a variety of doctrinal, moral, biographical, devotional and liturgical publications, including a large body of Vatican documents and sources. Unsolicited mss, synopses and ideas welcome if appropriate to their list.
Ⓔ Royalties annually.

Catnip Publishing Ltd

14 Greville Street, London EC1N 8SB
☎ 020 7138 3650   🖷 020 7138 3658
andrea.reece@catnippublishing.co.uk
www.catnippublishing.co.uk
Managing Director *Robert Snuggs*
Directors *Andrea Reece, Martin West*
Approx. Annual Turnover £400,000

Independent children's books publisher, founded in 2005. Concentrating on developing fiction for the 8–10 and 9–11 age groups. IMPRINTS **Catnip** New books from Joan Lingard, Dominic Barker, Sarah Matthias; **Happy Cat** TITLE *Scaredy Squirrel at the Beach* Melanie Watt; SERIES *Happy Cat First Readers* (5–7 years); *Talking it Through*. 30 titles a year. Unsolicited submissions for fiction considered; send outline synopsis, including proposed age range and length, by e-mail. No non-fiction and, currently, only limited interest in picture books.
Ⓔ Royalties twice-yearly.

Causeway Press
▷ Pearson

CBA Publishing

St Mary's House, 66 Bootham, York YO30 7BZ
☎ 01904 671417   🖷 01904 671384
catrinaappleby@britarch.ac.uk
www.britarch.ac.uk
Publications Officer *Catrina Appleby*
British Archaeology Editor *Mike Pitts*
Young Archaeologists' Club Communication Officer *Nicky Milsted*
Approx. Annual Turnover £115,000

Publishing arm of the **Council for British Archaeology**. Publishes archaeology reports, practical handbooks, *Internet Archaeology*; *British Archaeology* (bi-monthly magazine), *Young Archaeologist* (Young Archaeologists' Club magazine), monographs, archaeology and education. TITLES *Prehistoric Coastal Communities*; *Londinium and Beyond*; *Human*

Remains in Archaeology – A Handbook. 8–10 titles a year. Please contact by telephone before submitting mss and proposals.

CBD Research Ltd

Chancery House, 15 Wickham Road, Beckenham BR3 5JS

☎ 020 8650 7745   ⊞ 020 8650 0768
cbd@cbdresearch.com
www.cbdresearch.com
Managing Director *S.P.A. Henderson*
Approx. Annual Turnover £500,000

Founded 1961. Publishes directories and other reference guides to sources of information. No fiction. IMPRINT **Chancery House Press** Non-fiction of an esoteric/specialist nature for 'serious researchers and the dedicated hobbyist'. About 6 titles a year.
⒠ Royalties quarterly.

CCV
▷ The Random House Group Ltd

Cengage Learning (EMEA) Ltd

High Holborn House, 50–51 Bedford Row, London WC1R 4LR

☎ 020 7067 2500   ⊞ 020 7067 2600
www.cengage.co.uk
CEO (Worldwide) *Ron Dunn*
CEO/Managing Director (Cengage Learning EMEA) *Jill Jones*
Publishing Director (Cengage Learning EMEA) *John Yates*

Founded 1993. Formerly Thomson Learning, part of Cengage Learning and as such has offices worldwide with the UK office being reported to by Copenhagen (for Europe), Greece (Middle East) and South Africa. Acquired by funds represented by Apax Partners and OMERS Capital Partners in 2007. Publishes higher and vocational education, ELT. TITLES *Management and Cost Accounting* Drury; *Economics* Mankiw and Taylor. Unsolicited material aimed at students is welcome but telephone in the first instance to check out the idea.
⒠ Royalties vary according to contract.

Centaur Press
▷ Open Gate Press

Century
▷ The Random House Group Ltd

CF4K
▷ Christian Focus Publications

Chambers Harrap Publishers Ltd

7 Hopetoun Crescent, Edinburgh EH7 4AY
☎ 0131 556 5929   ⊞ 0131 556 5313
admin@chambersharrap.co.uk
www.chambersharrap.co.uk

Managing Director *Patrick White*
Editorial Director *Vivian Marr*
Office Manager *Esther Fulton*

Specialists in dictionaries and reference publishing in print and online: English dictionaries and thesauruses for everyone from primary school students to professional writers, including an expanding adult learners' list; general reference dictionaries covering everything from biography to the unexplained; bilingual dictionaries, study aids, and phrase books; factfinders; crossword dictionaries; puzzles and games books; books for word lovers; slang dictionaries; and language and writing guides. A recent acquisition is the Brewer's list of reference books. Part of **Hachette-Livre (UK)**. About 75 titles a year. Send synopsis with accompanying letter rather than completed mss.

✳*Authors' update* **Bucking the trend in a declining reference book market, Chambers' distinctive red livery and quirky titles, including a dictionary uniquely noted for a sense of humour, continue to make a strong impact.**

Chancery House Press
▷ CBD Research Ltd

Channel 4 Books
▷ Transworld Publishers

Paul Chapman Publishing Ltd
▷ Sage Publications

Chapman Publishing

4 Broughton Place, Edinburgh EH1 3RX
☎ 0131 557 2207
chapman-pub@blueyonder.co.uk
www.chapman-pub.co.uk
Managing Editor *Dr Joy Hendry*

A venture devoted to publishing works by 'the best of Scottish writers, both up-and-coming and established, published in *Chapman*, Scotland's leading literary magazine'. Only publishing one or two titles per annum, mainly poetry. TITLES *Winter Barley* George Gunn; *Lure* Dilys Rose; *Wild Women Series – Wild Women of a Certain Age* Magi Gibson; *Ye Cannae Win* Janet Paisley. No unsolicited mss.
⒠ Royalties annually.

Charnwood
▷ F.A. Thorpe Publishing

Chartered Institute of Personnel and Development (CIPD)

151 The Broadway, London SW19 1JQ
☎ 020 8612 6200   ⊞ 020 8612 6201
publish@cipd.co.uk
www.cipd.co.uk/bookstore
Publishing Manager *Ruth Lake*
Approx. Annual Turnover £3 million

Part of CIPD Enterprises Limited. Publishes on personnel, training and management. A list of around 160 titles including looseleaf and online subscription products, student textbooks and professional information resources. New product proposals welcome.
Ⓔ Royalties annually or flat fees paid.

Chastleton Travel
▷ Arris Publishing Ltd

Chatto & Windus
▷ The Random House Group Ltd

Cherrytree Books
▷ Evans Brothers Ltd

Chicken House Publishing

2 Palmer Street, Frome BA11 1DS
📞 01373 454488   📠 01373 454499
chickenhouse@doublecluck.com
Chairman/Managing Director *Barry Cunningham*

Children's publishing house founded in 2000. Acquired by Scholastic Inc. in May 2005. Publishes books 'that are aimed at real children' – fiction, original picture books. TITLES Cornelia Funke's *Inkheart* Trilogy. Aiming to publish about 25 titles a year. Due to the large number of submissions received, Chicken House is not accepting any unsolicited picture books or fiction for under 9s; fiction for 9-16 year-olds is only accepted in the form of submissions to the annual The Times/Chicken House Children's Fiction Competition (see entry under *Prizes*).
Ⓔ Royalties twice-yearly.

✱*Authors' update* **Set up as a 'small, creative company', Chicken House is now part of Scholastic. But editorial freedom is preserved 'with no rules about what to publish as long as it's good'.**

Child's Play (International) Ltd

Ashworth Road, Bridgemead, Swindon SN5 7YD
📞 01793 616286   📠 01793 512795
office@childs-play.com
www.childs-play.com
Chief Executive *Neil Burden*

Founded in 1972, Child's Play is an independent publisher specializing in learning through play, whole child development, life-skills and values. Publishes books, games, A-V materials and plush toys. TITLES *Big Hungry Bear*; *There Was an Old Lady*; *Our Cat Cuddles*; *Royston Knapper*; *Sign and Sing-Along*; *Just Like Me*; *The Ding Dong Bag*; *It's Raining, It's Pouring*; *The Flower*; *Max and the Doglins*; *The Ant and the Big Bad Bully Goat*; *The Wim Wom from the Mustard Mill*. Unsolicited mss welcome. Send s.a.e. for return or response. Expect to wait two months for a reply.

Ⓔ Outright or royalty payments are subject to negotiation.

Chimera
▷ Pegasus Elliot Mackenzie Publishers Ltd

Christian Focus Publications

Geanies House, Fearn, Tain IV20 1TW
📞 01862 871000   📠 01862 871699
info@christianfocus.com
www.christianfocus.com
Owner *Balintore Holdings plc*
Chairman *R.W.M. Mackenzie*
Managing Director *William Mackenzie*
Editorial Manager *Willie Mackenzie*
Children's Editor *Catherine Mackenzie*
Approx. Annual Turnover £1.5 million

Founded 1979 to produce children's books for the co-edition market. Now a major producer of Christian books. Publishes adult and children's books, including some fiction for children but not adults. No poetry. Publishes for all English-speaking markets, as well as the UK. IMPRINTS **CF4K** Children's books; **Christian Focus** General books; **Mentor** Study books; **Christian Heritage** Classic reprints. About 90 titles a year. Unsolicited mss, synopses and ideas welcome from Christian writers. See website for submission criteria.
Ⓔ Royalties annually.

Christian Heritage
▷ Christian Focus Publications

Chrome Dreams
▷ entry under Audio Books

Chrysalis Books Group
▷ Anova Books

Churchill Livingstone
▷ Elsevier Ltd

Churchwarden Publications Ltd

PO Box 420, Warminster BA12 9XB
📞 01985 840189   📠 01985 840243
enquiries@churchwardenbooks.co.uk
www.churchwardenbooks.co.uk
Managing Director *John Stidolph*

Founded 1974. Publishes books and stationery for churchwardens and church administrators including *The Churchwarden's Yearbook* (annual).

Cicerone Press

2 Police Square, Milnthorpe LA7 7PY
📞 01539 562069   📠 01539 563417
info@cicerone.co.uk
www.cicerone.co.uk
Managing/Editorial Director *Jonathan Williams*

Founded 1969. Guidebook publisher for outdoor enthusiasts. No fiction or poetry. TITLES

Hillwalker's Guide to Mountaineering; *Tour of Mont Blanc*; *Coast to Coast Trail*. SERIES include *Alpine Walking*; *International Walking*; *British Long-Distance Trails*, *Cycling* and *Winter Activities*. About 30 titles a year. No unsolicited mss; synopses and ideas considered.
Ⓔ Royalties twice-yearly.

Cico Books

20–21 Jockey's Fields, London WC1R 4BW
☎ 020 7025 2280   📠 020 7025 2281
mail@cicobooks.co.uk
Publisher *Cindy Richards*

Founded in 1999 and acquired by **Ryland Peters & Small Limited** in 2006. Publishes highly illustrated lifestyle books covering interiors, crafts, mind, body and spirit and gift. About 30 titles a year. No unsolicited mss but synopses and ideas welcome.
Ⓔ Royalties twice-yearly.

Cisco Press
▷ Pearson

Clairview Books Ltd

Hillside House, The Square, Forest Row RH18 5ES
☎ 0870 486 3526
office@clairviewbooks.com
www.clairviewbooks.com
Managing Director *S. Gulbekian*
Approx. Annual Turnover £100,000

Founded 2000. Publishes non-fiction. General books challenging conventional thinking: mind, body and spirit, current affairs, the arts, science, health and therapy, etc. TITLES *Imperial America* Gore Vidal; *The Biodynamic Food and Cookbook* Wendy Cook; *My Descent Into Death* Howard Storm. 5 titles in 2008. No unsolicited material; send initial letter of enquiry. No poetry or fiction.
Ⓔ Royalties annually.

T&T Clark
▷ The Continuum International Publishing Group Ltd

James Clarke & Co.

PO Box 60, Cambridge CB1 2NT
☎ 01223 350865   📠 01223 366951
publishing@jamesclarke.co.uk
www.jamesclarke.co.uk
Managing Director *Adrian Brink*

James Clarke and Co Ltd was founded in 1859 in Fleet Street, London, mainly as a magazine publisher. It produced the highly influential religious magazine, *Christian World*, which by the outbreak of the First World War was selling over 100,000 copies a week and was the leading nonconformist weekly. It soon began to publish books and by the 1920s it became an exclusively book publishing company, building up a list in religion and in alternative medicine

and spirituality. **The Lutterworth Press** (see entry) became associated with James Clarke & Co. in 1984. James Clarke & Co.'s current list concentrates on academic, scholarly and reference works, specializing in theology, history, literature and related subjects. TITLES *Geoffrey Fisher. Archbishop of Canterbury* David Hein; *The Soul Sleepers. Christian Mortalism from Wycliffe to Priestley* Bryan W. Ball; *Reading Bonhoefer* Geffrey B. Kelly. 20 titles in 2008. Book proposals to be submitted in writing.

✱*Authors' update* A leading theological publisher with high intellectual standards.

Classic
▷ Ian Allan Publishing Ltd

Collector's Library/Collector's Library Editions in Colour
▷ CRW Publishing Ltd

Peter Collin Publishing Ltd
▷ Bloomsbury Publishing Plc

Collins
▷ HarperCollins Publishers Ltd

Collins & Brown
▷ Anova Books

Colourpoint Books

Colourpoint House, Jubilee Business Park, 21 Jubilee Road, Newtownards BT23 4YH
☎ 028 9182 6339   📠 028 9182 1900
sales@colourpoint.co.uk
www.colourpoint.co.uk
Partners *Wesley Johnston, Malcolm Johnston, Norman Johnston (transport editor), Sheila M. Johnston (educational editor)*

Founded 1993. Publishes school textbooks and transport (covering the whole of the British Isles), plus books of Irish, general interest and fiction. About 25 titles a year. Unsolicited material accepted but approach in writing in the first instance; include return postage, please.
Ⓔ Royalties twice-yearly.

Columbia University Press
▷ University Presses of California, Columbia & Princeton Ltd

Communication Ethics
▷ Troubador Publishing Ltd

Compendium Publishing Ltd

43 Frith Street, London W1D 4SA
☎ 020 7287 4570   📠 08451 303173
alan.greene@compendiumpublishing.com
www.compendiumpublishing.com
Managing Director *Alan Greene*
Editorial Director *Simon Forty*
Rights *Christine Delaborde*

Approx. Annual Turnover £4 million

Founded 2000. Publishes and packages for international publishing companies, general illustrated non-fiction: history, reference, hobbies, children's and educational transport and militaria. 70 titles in 2008 No unsolicited mss; synopses and ideas preferred.

Ⓔ Royalties twice-yearly or fees per project.

Condor
▷ Souvenir Press Ltd

Conran Octopus
▷ Octopus Publishing Group

Constable & Robinson Ltd

3 The Lanchesters, 162 Fulham Palace Road, London W6 9ER

📞 020 8741 3663   📠 020 8748 7562

enquiries@constablerobinson.com
www.constablerobinson.com

Chairman *Nick Robinson*
Managing Director *Pete Duncan*
Editorial Director (Crime Fiction) *Krystyna Green*
Manager, Overcoming series *Fritha Saunders*

Archibald Constable, Walter Scott's publisher and the originator of the '3 decker novel', published his first titles in Edinburgh in 1795. Constable & Robinson Ltd are now in the third century of independent publishing under the Constable name. IMPRINTS **Constable** *Becky Hardie, Andreas Campomar* Current affairs, biography, general and military history, psychology, travel, photography and crime fiction mainly in hardback. **Robinson** SERIES *Britannica Guides*; *Brief History* (*Leo Hollis*); Crime fiction (*Krystyna Green*); the 'Mammoth Book Of' (*Duncan Proudfoot*); the Overcoming psychology SERIES (*Fritha Saunders*); Humour (*Hugh Barker*). The very successful Elliot Right Way, publisher of practical author-expert self-help guides, was acquired by Constable & Robinson in 2007. 150 titles. Unsolicited sample chapters, synopses and ideas for books welcome. No mss; *no e-mail submissions*. Enclose return postage.

Ⓔ Royalties twice-yearly.

✱*Authors' update* **A strong list of popular interest titles has proved successful in connecting with the elusive general reader.**

The Continuum International Publishing Group Limited

The Tower Building, 11 York Road, London SE1 7NX

📞 020 7922 0880   📠 020 7922 0881

www.continuumbooks.com

CEO *Oliver Gadsby*
Approx. Annual Turnover £10 million

Founded in 1999 by a buy-out of the academic and religious publishing of Cassell and the acquisition of Continuum New York. Publishes academic, religious and general books.

DIVISIONS **Academic Humanities** Publisher *Anna Fleming* Education, social sciences, literature, film and music, philosophy (including the **Thoemmes** imprint), linguistics, biblical studies and theology printed under the **T&T Clark** imprint. TITLES *Getting the Buggers to Behave*; *Teaching in Further Education*; *What Philosophy Is*; *The Guerrilla Film Makers Handbook*. **General Trade & Continuum Religion** Publishing Director *Robin Baird-Smith* Publishes under **Continuum Burns & Oates** (RC books) TITLES *Seven Basic Plots*; *The Home We Build Together*. IMPRINTS **Continuum**; **T&T Clark Intl**; **Thoemmes Continuum**; **Burns & Oates**; **Network Continuum**. 550 titles in 2008. Unsolicited synopses and ideas within the subject areas listed above are welcome; approach in writing in the first instance. OVERSEAS SUBSIDIARIES The Continuum International Publishing Group Inc., New York and Harrisburg.

Ⓔ Royalties twice-yearly.

✱*Authors' update* **With this publisher what might be thought to be academic books are often written in an accessible style that appeals to the non-specialist.**

Conway
▷ Anova Books

Thomas Cook Publishing

Thomas Cook Business Park, Coningsby Road, Peterborough PE3 8SB

📞 01733 416477   📠 01733 416688

Director of Publishing *Chris Young*

Part of the Thomas Cook Group Plc, publishing commenced in 1873 with the first issue of *Cook's Continental Timetable*. Publishes guidebooks, maps and timetables. About 130 titles a year. No unsolicited mss; synopses and ideas welcome as long as they are travel-related.

Ⓔ Royalties annually.

Leo Cooper
▷ Pen & Sword Books Ltd

Corgi
▷ Transworld Publishers

Corgi Children's Books
▷ The Random House Group Ltd

Corinthian Books
▷ Icon Books Ltd

Cornerstone
▷ The Random House Group Ltd

Country Publications Ltd

The Water Mill, Broughton Hall, Skipton
BD23 3AG
☎ 01756 701381   ⊞ 01756 701326
editorial@dalesman.co.uk
www.dalesman.co.uk
Book Editor *Mark Whitley*

Publishers of *Countryman*, *Dalesman*, *Cumbria*
and *Down Your Way* magazines, and regional
books covering the North of England. Subjects
include walking, guidebooks, history, humour and
folklore. About 10 titles a year. Will consider mss
on subjects listed above, relating to the North.
© Royalties annually.

Countryside Books

2 Highfield Avenue, Newbury RG14 5DS
☎ 01635 43816   ⊞ 01635 551004
info@countrysidebooks.co.uk
www.countrysidebooks.co.uk
Publisher *Nicholas Battle*

Founded 1976. Publishes local interest paperbacks
on regional subjects, generally by English
county. Local history, dialects, genealogy,
walking, photographic, aviation and military,
some transport. Over 400 titles available. About
50 titles a year. Unsolicited mss and synopses
welcome but no fiction, poetry, natural history
or personal memories. Check the website for
suitability before any submission.
© Royalties twice-yearly.

CRC Press
▷ Taylor & Francis Group

Crème de la Crime Ltd

PO Box 523, Chesterfield S40 9AT
☎ 01246 520835   ⊞ 01246 520835
info@cremedelacrime.com
www.cremedelacrime.com
Managing Director *Lynne Patrick*
Approx. Annual Turnover £60,000

Founded in 2003 to discover and publish crime
fiction by new authors. 6 titles scheduled for
2009. Welcomes enquiries and unsolicited
submissions, 'especially from talented but
unpublished authors who are urged to read the
detailed guidelines on the website'.
© Royalties annually.

Cressrelles Publishing Co. Ltd

10 Station Road Industrial Estate, Colwall,
Malvern WR13 6RN
☎ 01684 540154   ⊞ 01684 540154
simonsmith@cressrelles.co.uk
www.cressrelles.co.uk
Managing Director *Leslie Smith*

Publishes a range of local interest books and
drama titles. IMPRINTS **J. Garnet Miller** Plays and
theatre texts; **Kenyon-Deane** Plays and drama
textbooks; **New Playwrights' Network** Plays.
About 6–12 new play titles a year. Submissions
welcome.

Crimson Publishing

Westminster House, Kew Road, Richmond
TW9 2ND
☎ 020 8334 1600   ⊞ 020 8334 1601
info@crimsonpublishing.co.uk
www.crimsonpublishing.co.uk
Chairman *David Lester*

Founded 2006. Part of Crimson Business Ltd.
Non-fiction publisher of books 'to improve the
way you live'. Subjects covered: business, travel,
working abroad, gap year, careers and education,
parenting.
DIVISIONS **Crimson Publishing**; **Trotman** (see
entry). IMPRINTS **Crimson**; **Vacation Work**;
Trotman; **White Ladder**. 30 titles in 2008.
Welcomes carefully considered unsolicited mss,
synopses and ideas for books in the subject areas
listed above. No fiction or books on topics not
listed. Approach by e-mail.
© Royalties twice-yearly.

Crombie Jardine Publishing Limited

Office 2, 3 Edgar Buildings, George Street, Bath
BA1 2FJ
☎ 01225 464445
catriona@crombiejardine.com
www.crombiejardine.com
Publishing Director *David Crombie*
Publishing Director *Catriona Jardine*

Specialists in fun, quirky and topical humour
books. TITLES *The Little Book of Chavs*; *101 Things
to Do at University*; *Shag Yourself Slim*; *Lazy
Sod's Guide to Sex*; *What Shat That?*; *How to be
a Ghastly Grandparent*; *The Arse in Art*. 20–40
titles a year. No postal submissions. Send synopsis
by e-mail in the first instance.
© Flat fee paid; no royalties.

✳*Authors' update* **The ultimate gift books for
those with a quirky sense of humour.**

Crossway
▷ Inter-Varsity Press

Crown House Publishing

Crown Buildings, Bancyfelin, Carmarthen
SA33 5ND
☎ 01267 211345   ⊞ 01267 211882
books@crownhouse.co.uk
www.crownhouse.co.uk
Managing Director *David Bowman*

Founded 1998. Publisher of Neuro Linguistic
Programming (NLP), hypnosis, counselling
and psychotherapy, personal, business and
life coaching, education, accelerated learning,

thinking skills and personalized learning. 25 titles in 2008. Unsolicited submissions, via the website, are welcome.
Ⓔ Royalties twice-yearly.

The Crowood Press Ltd

The Stable Block, Crowood Lane, Ramsbury, Marlborough SN8 2HR
☎ 01672 520320   ⊕ 01672 520280
enquiries@crowood.com
www.crowood.com
Chairman *John Dennis*
Managing Director *Ken Hathaway*

Publishes sport and leisure titles, including animal and land husbandry, climbing and walking, maritime, country sports, equestrian, fishing and shooting; also crafts, dogs, gardening, DIY, theatre, natural history, aviation, military history and motoring. IMPRINT **Airlife Publishing** Specialist aviation titles for pilots, historians and enthusiasts. About 70 titles a year. Preliminary letter preferred in all cases.
Ⓔ Royalties annually.

CRW Publishing Ltd

69 Gloucester Crescent, London NW1 7EG
☎ 020 7485 5764   ⊕ 0870 751 7254
marcus.clapham@crw-publishing.co.uk
www.collectors-library.com
Also at: 6 Turville Barns, Eastleach, Cirencester GL7 3QB
Chairman *Cameron Brown*
Editorial Director *Marcus Clapham*
Approx. Annual Turnover £2 million

Founded 2003. Publishes literary classics, gift, children's, boxed sets. IMPRINTS **Collector's Library**; **Collector's Library Editions in Colour**; **Book Blocks**. 20 titles in 2008. No unsolicited mss.
Ⓔ Royalties annually.

Benjamin Cummings
▷ Pearson

Custom Publishing
▷ The Orion Publishing Group Limited

D&B Publishing

80 Walsingham Road, Hove BN3 4FF
☎ 01273 711443
info@dandbpublishing.com
www.dandbpublishing.com
Joint Managing Directors *Dan Addelman, Byron Jacobs*

Founded 2002. Publishes games books, specializing in gambling, primarily. 8 titles in 2007. Unsolicited mss, synopses and ideas welcome; approach by e-mail in the first instance.
Ⓔ Royalties annually.

Terence Dalton Ltd

Water Street, Lavenham, Sudbury CO10 9RN
☎ 01787 249291   ⊕ 01787 248267
www.terencedalton.com
Director/Editorial Head *Elisabeth Whitehair*

Founded 1967. Part of the Lavenham Group Plc, a family company. Non-fiction only. No unsolicited mss. Ideas welcome; telephone in the first instance.
Ⓔ Royalties annually.

Darton, Longman & Todd Ltd

1 Spencer Court, 140–142 Wandsworth High Street, London SW18 4JJ
☎ 020 8875 0155   ⊕ 020 8875 0133
tradesales@darton-longman-todd.co.uk
www.dltbooks.com
Editorial Director *Brendan Walsh*
Editorial Assistant/Rights Officer *Claudine Nightingale*
Approx. Annual Turnover £1 million

A leading independent publisher of books on spirituality and religion. Unique among UK religious publishers in being jointly owned and managed by all its staff members. While predominantly Christian, DLT publishes books from different backgrounds and traditions. TITLES *Jerusalem Bible*; *New Jerusalem Bible*; *NRSV Bible: Catholic Edition*; *Return of the Prodigal Son*; *God of Surprises*; *The Enduring Melody*; *Hostage in Iraq*. About 45 titles a year. Information on submissions available on the website.
Ⓔ Royalties twice-yearly.

David & Charles Publishers

Brunel House, Forde Close, Newton Abbot TQ12 4PU
☎ 01626 323200   ⊕ 01626 323317
postmaster@davidandcharles.co.uk
www.davidandcharles.co.uk
Managing Director & Publisher *Stephen Bateman*
Director of Editorial and Design *Alison Myer*

Founded 1960, owned by F+W Media. Publishes illustrated non-fiction for international markets, specializing in needlecraft, crafts, art techniques, practical photography, military history, nostalgia, transport and equestrian. No fiction, poetry or memoirs. TITLES *Ghosts Caught on Film*; *Sew Pretty Homestyle*; *John Howe Fantasy Art Workshop*; *Keeping Bees and Making Honey*; *Memories of Steam*. About 100 titles a year. Unsolicited mss will be considered if return postage is included; synopses and ideas welcome for the subjects listed above. 'Please allow at least eight weeks for a response.'
Ⓔ Royalties twice-yearly or flat fees.

Christopher Davies Publishers Ltd

PO Box 403, Swansea SA1 4YF

☎ 01792 648825 📠 01792 648825

chris@cdaviesbookswales.com

Managing Director/Editorial Head *Christopher T. Davies*

Approx. Annual Turnover £20,000

Founded 1949 to promote and expand Welsh-language publications. Publishes biography, cookery, history, sport and literature of Welsh interest. TITLES *An A–Z of Wales and the Welsh*; *Welsh Birthplaces*; *Carwyn: A Personal Memoir*; *Who's Who in Welsh History*. About 2 titles a year. No unsolicited mss. Synopses and ideas for books welcome.

Ⓔ Royalties twice-yearly.

✳*Authors' update* **A favourite for Celtic readers and writers.**

Giles de la Mare Publishers Ltd

PO Box 25351, London NW5 1ZT

☎ 020 7485 2533 📠 020 7485 2534

gilesdelamare@dial.pipex.com

www.gilesdelamare.co.uk

Chairman/Managing Director *Giles de la Mare*

Approx. Annual Turnover £30,000

Founded 1995 and commenced publishing in April 1996. Publishes mainly non-fiction, especially art and architecture, biography, history, music and travel. TITLES *Short Stories, Vols I, II & III* Walter de la Mare; *Handsworth Revolution* David Winkley; *The Life of Henry Moore* Roger Berthoud; *Becoming an Orchestral Musician* Richard Davis; *Blindness and the Visionary* John Coles; *Tricks Journalists Play* Dennis Barker; *Venice: The Anthology Guide* Milton Grundy; *Inherit the Truth 1939–1945* Anita Lasker-Wallfisch; *Calatafimi* Angus Campbell; *Musical Heroes* Robert Ponsonby. Unsolicited mss, synopses and ideas welcome after initial telephone call.

Ⓔ Royalties twice-yearly.

Debrett's Ltd

18–20 Hill Rise, Richmond TW10 6UA

☎ 020 8939 2250 📠 020 8939 2251

people@debretts.co.uk

www.debretts.co.uk

Head of Publishing *Liz Wyse*

Chairman *Conrad Free*

Biographical reference plus etiquette, correct form and other social guides; also diaries. Book TITLES include triennial *Debrett's Peerage & Baronetage* and annual *People of Today* (as book, CD-ROM and online); *Debrett's A-Z of Modern Manners*; *Debrett's Guide for the Modern Gentleman*; *Etiquette for Girls*; *Correct Form*;

Debrett's Wedding Guide. Book and article proposals welcome.

Dedalus Ltd

Langford Lodge, St Judith's Lane, Sawtry PE28 5XE

☎ 01487 832382

info@dedalusbooks.com

www.dedalusbooks.com

Chairman *Juri Gabriel*

Managing Director *Eric Lane*

Approx. Annual Turnover £200,000

Founded 1983. Publishes contemporary European fiction and classics and original literary fiction. TITLES *The Dedalus Book of Absinthe* Phil Baker; *The Arabian Nightmare* Robert Irwin; *Memoirs of a Gnostic Dwarf* David Madsen; *Music in a Foreign Language* Andrew Crumey (winner of the Saltire Best First Book Award in 1994); *The Father of Locks* Andrew Killeen; *Made in Yaroslavl* Jeremy Weingard; *Mr Dick or the Tenth Book* Jean-Pierre Ohl.

DIVISIONS/IMPRINTS **Original Fiction in Paperback**; **Dedalus Europe 1992–2012**; **Literary Concept Books**. Welcomes submissions for original fiction and books suitable for its list but 'most people sending work in have no idea what kind of books Dedalus publishes and merely waste their efforts'. Author guidelines on website. Particularly interested in intellectually clever and unusual fiction. A letter about the author should always accompany any submission. No replies without s.a.e.

Ⓔ Royalties annually.

✳*Authors' update* **Put at risk by the withdrawal of its Arts Council grant, Dedalus was rescued by a cash injection from Informa publishing group. Thus, virtue and originality are rewarded.**

JM Dent
▷ The Orion Publishing Group Ltd

Despatches
▷ Reportage Press

André Deutsch Ltd
▷ Carlton Publishing Group

Digital Press
▷ Elsevier Ltd

Dinas
▷ Y Lolfa Cyf

John Donald Publishers Ltd

West Newington House, 10 Newington Road, Edinburgh EH9 1QS

☎ 0131 668 4371 📠 0131 668 4466

info@birlinn.co.uk

www.birlinn.co.uk

Managing Director *Hugh Andrew*

Bought by **Birlinn Ltd** in 1999. Publishes academic and scholarly, archaeology, architecture, textbooks, guidebooks, local, and social history. New books are published as an imprint of Birlinn Ltd. About 30 titles a year.

Donhead Publishing Ltd

Lower Coombe, Donhead St Mary, Shaftesbury SP7 9LY

☎ 01747 828422   ▨ 01747 828522

jillpearce@donhead.com

www.donhead.com

Contact *Jill Pearce*

Founded 1990 to specialize in publishing how-to books for building practitioners; particularly interested in architectural conservation material. Publishes building, architecture and heritage only. TITLES *Windows*; *Stone Conservation, Plastering*; *Preserving Post-War Heritage*; *Stone Cleaning*; *Architecture 1900*; *Encyclopaedia of Architectural Terms*; *Cleaning Historic Buildings*; *Brickwork*; *Practical Stone Masonry*; *Conservation of Timber Buildings*; *Surveying Historic Buildings*; *Journal of Architectural Conservation* (3 issues yearly). 6 titles a year. Unsolicited mss, synopses and ideas welcome.

Dorling Kindersley Ltd

Part of the Penguin Group, 80 Strand, London WC2R 0RL

☎ 020 7010 3000   ▨ 020 7010 6060

www.dk.com

CEO *Gary June*

Founded 1974. Packager and publisher of illustrated non-fiction: cookery, crafts, gardening, health, travel guides, atlases, natural history and children's information and fiction. Launched a US imprint in 1991 and an Australian imprint in 1997. Acquired Henderson Publishing in 1995 and was purchased by Pearson plc for £311 million in 2000.

DIVISIONS Adult: **Travel/Reference** Publisher *Douglas Amrine*; **Reference** Publisher *Jonathan Metcalf*; **Lifestyle** Publishers *Mary-Clare Jerram, Peggy Vance, Stephanie Jackson*. Children's: **PreSchool** Publisher *Susan Leonard*; **5–8 Years** Publisher *Mary Ling*; **8+ Years** Publisher *Laura Buller*; **Eyewitness** Publisher *Andrew Macintyre*. IMPRINTS **Funfax**; **Eyewitness Guides**; **Eyewitness Travel Guides**. Does not accept unsolicited mss. PUBLISHING PARTNERS include RHS, BMA, RSPB, Fairtrade Foundation.

✱*Authors' update* Most titles are team efforts with writers and illustrators working closely with an in-house editor.

Doubleday
▷ Transworld Publishers

Doubleday Children's Books
▷ The Random House Group Ltd

Alton Douglas Books
▷ Brewin Books Ltd

Drake Educational Associates

St Fagans Road, Fairwater, Cardiff CF5 3AE

☎ 029 2056 0333   ▨ 029 2956 0313

info@drakeed.com

www.drakeed.com

Managing Director *R.G. Drake*

Literacy, phonics and language development games and activities. Ideas and scripts in these fields welcome. Also resources for subject areas in the primary school and modern languages at KS2.

Dref Wen

28 Church Road, Whitchurch, Cardiff CF14 2EA

☎ 029 2061 7860   ▨ 029 2061 0507

gwilym@drefwen.com

Chairman *R. Boore*

Managing Director *G. Boore*

Founded 1970. Publishes Welsh language and bilingual children's books, Welsh and English educational books for Welsh learners.
€ Royalties annually.

Gerald Duckworth & Co. Ltd

First Floor, 90–93 Cowcross Street, London EC1M 6BF

☎ 020 7490 7300   ▨ 020 7490 0080

info@duckworth-publishers.co.uk

www.ducknet.co.uk

Managing Director *Peter Mayer*

Editorial Director (Academic) *Deborah Blake*

Editor (General) *Mary Morris*

Approx. Annual Turnover £2 million

Founded 1898 by Gerald Duckworth. Original publishers of Virginia Woolf. Other early authors include Hilaire Belloc, John Galsworthy, D.H. Lawrence and George Orwell. Duckworth is a general trade publisher whose authors include John Bayley, Mary Warnock, Joan Bakewell and J.J. Connolly. In addition to its trade list, Duckworth has a strong academic division. Acquired by Peter Mayer in 2003.

IMPRINTS/DIVISIONS **Bristol Classical Press** Classical texts and modern languages; **Ardis** Russian literature; **Duckworth Academic**; **Duckworth General**. No unsolicited mss; synopses and sample chapters only. Enclose s.a.e. or return postage for response/return.
€ Royalties twice-yearly.

✱*Authors' update* What this publisher describes as an 'eclectic' list is set for rapid growth.

Dunedin Academic Press Ltd

Hudson House, 8 Albany Street, Edinburgh
EH1 3QB
☎ 0131 473 2397   ℻ 01250 870920
mail@dunedinacademicpress.co.uk
www.dunedinacademicpress.co.uk
Director *Anthony Kinahan*

Founded 2001. Publishes academic and serious
general non-fiction. Considerable experience in
academic and professional publishing. 18 titles in
2008. Not interested in science (other than earth
sciences), poetry, fiction or children's. Check
website for active subject areas. No unsolicited
mss. Synopses and ideas welcome. Approach
first in writing outlining proposal and identifying
market. Proposal guidelines are available via the
website.
ⓔ Royalties annually.

Ebury Publishing/Ebury Press
▷ The Random House Group Ltd

Economist Books
▷ Profile Books

Eden
▷ Transworld Publishers

Edinburgh University Press

22 George Square, Edinburgh EH8 9LF
☎ 0131 650 4218   ℻ 0131 662 0053
www.euppublishing.com
www.eupjournals.com
Chairman *Ivon Asquith*
Chief Executive *Timothy Wright*
Deputy Chief Executive/Head of Book Publishing
Jackie Jones
Head of Journals *Sarah Edwards*
Senior Commissioning Editor *Nicola Carr*
Commissioning Editors *Carol Macdonald,
Esmé Watson*

Publishes academic and scholarly books (and
journals): African studies, ancient history and
classics, film and media studies, Islamic studies,
linguistics, literary criticism, philosophy, politics,
Scottish studies and religious studies. About 120
titles a year. E-mail submissions accepted; consult
website for guidelines.
ⓔ Royalties paid annually.

Egmont Press

3rd Floor, Beaumont House, Avonmore Road,
London W14 8TS
☎ 020 7605 6600   ℻ 020 7605 6601
www.egmont.co.uk
Managing Director *Robert McMenemy*
Director of Press *Cally Poplak*

Founded 1878. Publishes children's books.
AUTHORS include Michael Morpurgo, Jenny
Nimmo, William Nicholson, Helen Oxenbury,
Enid Blyton, Andy Stanton, Julia Golding, Jan
Fearnley and Lydia Monks. 'We accept unsolicited
mss but will not acknowledge or respond to
individual submissions unless successful.'

Egmont Publishing

239 Kensington High Street, London W8 6SA
☎ 020 7761 3500   ℻ 020 7761 3510
initial.surname@euk.egmont.com
www.egmont.co.uk
Managing Director *Robert McMenemy*
Director of Publishing *David Riley*

Publishes children's books: annuals, activity
books, novelty books, film and TV tie-ins.
Licensed character list includes Thomas the Tank
Engine, Barbie, Mr Men, Postman Pat, Ben 10,
Rupert Bear, Winnie the Pooh.

✴*Authors' update* The fourth largest children's
**book publishing group in the UK, Egmont has
bucked the trend with a solid year of growth.**

Eland Publishing Ltd

Third Floor, 61 Exmouth Market, London
EC1R 4QL
☎ 020 7833 0762   ℻ 020 7833 4434
info@travelbooks.co.uk
www.travelbooks.co.uk
Directors *Rose Baring, John Hatt,
Barnaby Rogerson*
Approx. Annual Turnover £250,000

Eland reprints classics of travel literature with
a backlist of 60 titles, including *Naples '44*
Norman Lewis; *Travels with Myself and Another*
Martha Gelhorn; *Portrait of a Turkish Family*
Irfan Orga; *Jigsaw* Sybille Bedford. *92 Acharnon
Street* by John Lucas won the Dolman Best Travel
Book Award in 2008. IMPRINTS **Eland** Classic
travel literature; **Through Writers' Eyes** Travel
literature collections; **Poetry of Place** Booklets;
Sickle Moon History, anthropology, historical
travel journals. About 6–12 titles a year. No
unsolicited mss. Postcards and e-mails welcome.
ⓔ Royalties annually.

Element
▷ HarperCollins Publishers Ltd

11:9
▷ Neil Wilson Publishing Ltd

Edward Elgar Publishing Ltd

The Lypiatts, 15 Lansdown Road, Cheltenham
GL50 2JA
☎ 01242 226934   ℻ 01242 262111
info@e-elgar.co.uk
www.e-elgar.com
Managing Director *Edward Elgar*

Founded 1986. International publisher in
economics, the environment, public policy,
business and management and law. TITLES

Who's Who in Economics (3rd ed.); *Handbook of Environmental and Resource Economics*; *Who's Who in the Management Sciences*. 350 titles in 2008. No unsolicited mss; synopses and ideas in the subject areas listed above welcome. Approach by letter or e-mail; no telephone inquiries.

Elliot Right Way
▷ Constable & Robinson Ltd

Elliott & Thompson

27 John Street, London WC1N 2BX
☎ 020 7831 5013   📠 020 7831 5011
www.elliottthompson.com
Publishers *Brad Thompson, Lorne Forsyth*
Approx. Annual Turnover £100,000

Founded 2001. Publishes literary fiction, biography, belles lettres, reprints of classic male writers. New proposals welcome (send synopsis and sample pages initially) but s.a.e. essential.

Elm Publications/Training (wholly owned subsidiary of Elm Consulting Ltd)

Seaton House, Kings Ripton, Huntingdon
PE28 2NJ
☎ 01487 773254
elm@elm-training.co.uk
www.elm-training.co.uk
Managing Director *Sheila Ritchie*

Founded 1977. Publishes textbooks, teaching aids, educational resources and educational software in the fields of business and management for adult learners. Books and teaching/training resources are generally commissioned to meet specific business, management and other syllabuses. About 30 titles a year. Ideas are welcome; initial approach by e-mail with outline, or by a brief telephone call only.
💷 Royalties annually.

Elsevier Ltd

The Boulevard, Langford Lane, Kidlington, Oxford OX5 1GB
☎ 01865 843000   📠 01865 843010
www.elsevier.com
Managing Director *Anna Moon (Oxford)*
CEO, Science & Technology (books & journals) *Herman van Campenhout*
CEO (Health & Sciences) *Michael Hansen*

Parent company Reed Elsevier, Amsterdam. Incorporates Pergamon Press. Academic and professional reference books, scientific, technical and medical books, journals, CD-ROMs and magazines. IMPRINTS **Academic Press; Architectural Press; Butterworth Heinemann; Digital Press; Elsevier; Elsevier Advanced Technology; Focal Press; Gulf Professional Press; Made Simple Books; Morgan Kauffman; Newnes; North-Holland.**

DIVISION **Elsevier (Health Sciences)** 32 Jamestown Road, London NW1 7BY ☎ 020 7424 4200 📠 020 7483 2293 www.elsevier-health.com Publishes scientific, technical, medical books and journals. IMPRINTS **Baillière Tindall; Churchill Livingstone; Mosby; Pergamon; Saunders.** No unsolicited mss, but synopses and project proposals welcome.
💷 Royalties annually.

Emerging Communications Technologies
▷ Troubador Publishing Ltd

Emissary Publishing

PO Box 33, Bicester OX26 4ZZ
☎ 01869 323447   📠 01869 322552
www.emissary-publishing.com
Editorial Director *Val Miller, BA(Hons)*

Founded 1992. Publishes mainly humorous paperback novels, including the complete set of Peter Pook novels; fiction; non-fiction. No poetry or children's. AUTHORS include Peter Pook, Roscoe Howells, Frederick E. Smith. No unsolicited mss or synopses.
💷 Royalties paid according to contract.

Emma Treehouse Ltd

The Studio, Church Street, Nunney BA11 4LW
☎ 01373 836233   📠 01373 836299
richard.powell4@virgin.net
sales@emmatreehouse.com
www.emmatreehouse.com
Co-Directors *Richard Powell, David Bailey*
Approx. Annual Turnover £22 million

Founded 1992. Publishes children's pre-school novelty books. No mss. Illustrations, synopses and ideas for books welcome; write in the first instance.
💷 Fees paid; no royalties.

Empiricus Books
▷ Janus Publishing Company Ltd

English Heritage (Publishing)

Kemble Drive, Swindon SN2 2GZ
☎ 01793 414453   📠 01793 414769
www.english-heritage.org.uk
Head of Publishing *John Hudson*
Managing Editor *Robin Taylor*

English Heritage's publishing programme reflects the organization's core interests in the historic environment of England, covering archaeology, architecture and general history titles. 20 titles a year. 'All titles relate directly to the work of the organization, so we do not accept unsolicited material.'

Enitharmon Press

26B Caversham Road, London NW5 2DU
☎ 020 7482 5967   📠 020 7284 1787

books@enitharmon.co.uk
www.enitharmon.co.uk
Director *Stephen Stuart-Smith*
Contacts *Isabel Brittain, Jacqueline Gabbitas*

Founded 1967. An independent company with an enterprising editorial policy, Enitharmon has established itself as one of Britain's leading literary presses. 'Patron of the new and the neglected, Enitharmon also prides itself on the success of its collaborations between writers and artists, now published by its associate company', **Enitharmon Editions**. Publishes poetry, literary criticism, fiction, art and photography. TITLES *Selected Poems* Duncan Forbes; *Behold* Nicki Jackowska; *Nothing Like Love* Jenny Joseph; *A Jovial Hullabaloo* Michael Longley; *Plan B* Paul Muldoon. No unsolicited mss.
Ⓔ Royalties according to contract.

Epworth

Methodist Church House, 25 Marylebone Road, London NW1 5JR
☎ 020 7486 5502
www.mph.org.uk
www.scm-canterburypress.co.uk
Chair *The Revd Michael Townsend*

Publishes Christian books only: philosophy, theology, biblical studies, pastoralia, social concern and Methodist studies. No fiction, poetry or children's. About 12 titles a year. Unsolicited mss, synopses and ideas welcome; send sample chapter and contents with covering letter and s.a.e.
Ⓔ Royalties annually.

Erotic Review Books

31 Sinclair Road, London W14 0NS
☎ 020 7371 1532   🖷 020 7603 8378
eros@eroticprints.org
www.eroticprints.org
Managing Director *J. Maclean*

Publishes books of erotic art and photography and erotic literature, and the *Erotic Review* magazine. Unsolicited material accepted but approach in writing in the first instance with a summary or first chapter.

EuroCrime
▷ Arcadia Books

Europa Publications
▷ Taylor & Francis Group

Evans Brothers Ltd

2A Portman Mansions, Chiltern Street, London W1U 6NR
☎ 020 7487 0920   🖷 020 7487 0921
sales@evansbrothers.co.uk
www.evansbooks.co.uk
Managing Director *Stephen Pawley*

International Publishing Director *Brian Jones*
UK Publisher *Su Swallow*
Approx. Annual Turnover £4.5 million

Founded in 1908 by Robert and Edward Evans. Publishes UK children's and educational books, and educational books for Africa, the Caribbean and Latin America. IMPRINTS **Cherrytree Books; Zero to Ten**. OVERSEAS ASSOCIATES in Kenya, Cameroon, Sierra Leone; Evans Bros (Nigeria Publishers) Ltd. About 175 titles a year. 'Submissions welcome but cannot be returned. Unable to respond to individual proposals that are not accepted.'
Ⓔ Royalties annually.

Everyman
▷ The Orion Publishing Group Ltd

Everyman Chess
▷ Gloucester Publishers Plc

Everyman's Library

Northburgh House, 10 Northburgh Street, London EC1V 0AT
☎ 020 7566 6350   🖷 020 7490 3708
books@everyman.uk.com
Publisher *David Campbell*
Approx. Annual Turnover £3.5 million

Founded 1906. Publishes hardback classics of world literature, pocket poetry anthologies, children's books and travel guides. Publishes no new titles apart from poetry anthologies; only classics (no new authors). AUTHORS include Bulgakov, Bellow, Borges, Heller, Marquez, Nabokov, Naipaul, Orwell, Rushdie, Updike, Waugh and Wodehouse. No unsolicited mss.
Ⓔ Royalties annually.

University of Exeter Press

Reed Hall, Streatham Drive, Exeter EX4 4QR
☎ 01392 263066   🖷 01392 263064
uep@exeterpress.co.uk
www.exeterpress.co.uk
Publisher *Simon Baker*

Founded 1956. Publishes academic books: archaeology, classical studies and ancient history, history, maritime studies, English literature (especially medieval), film history, performance studies and books on Exeter and the South West. IMPRINT **Bristol Phoenix Press** Classics and ancient history. About 30 titles a year. Proposals welcomed in the above subject areas.
Ⓔ Royalties annually.

Exley Publications Ltd

16 Chalk Hill, Watford WD19 4BG
☎ 01923 474480   🖷 01923 800440
editorial@helenexley.com
www.helenexleygiftbooks.com
Editorial Director *Helen Exley*

Founded 1976. Independent family company. Publishes giftbooks, quotation anthologies and humour. No submissions, please.

Expert Books
▷ Transworld Publishers

Eyewitness Guides/Eyewitness Travel Guides
▷ Dorling Kindersley Ltd

Faber & Faber Ltd
Bloomsbury House, 74–77 Great Russell Street, London WC1B 3DA
📞 020 7927 3800　📠 020 7927 3801
www.faber.co.uk
Chief Executive *Stephen Page*
Approx. Annual Turnover £14.5 million

Geoffrey Faber founded the company in the 1920s, with T.S. Eliot as an early recruit to the board. The original list was based on contemporary poetry and plays (the distinguished backlist includes Eliot, Auden and MacNeice). Publishes poetry and drama, children's, fiction, film, music, politics, biography. In June 2008 launched **Faber Finds**, a new imprint bringing back classic titles via print on demand. DIVISIONS **Fiction** *Lee Brackstone, Hannah Griffiths, Angus Cargill, Sarah Savitt* AUTHORS P.D. James, Peter Carey, Rachel Cusk, Giles Foden, Michael Frayn, Jane Harris, Tobias Hill, Kazuo Ishiguro, Barbara Kingsolver, Milan Kundera, Hanif Kureishi, John Lanchester, John McGahern, Rohinton Mistry, Lorrie Moore, Andrew O'Hagan, DBC Pierre, Jane Smiley; **Children's** *Julia Wells* AUTHORS Philip Ardagh, Ricky Gervais, Harry Hill, Russell Stannard, G.P. Taylor; **Film** *Walter Donohue* and **Plays** *Dinah Wood* AUTHORS Samuel Beckett, Alan Bennett, David Hare, Brian Friel, Rebecca Lenkiewicz, Patrick Marber, Harold Pinter, Tom Stoppard, Woody Allen, John Boorman, Joel and Ethan Coen, John Hodge, Martin Scorsese; **Music** *Belinda Matthews* AUTHORS Edward Blakeman, John Bridcut, Humphrey Burton, Jonathan Carr, Rupert Christiansen, Nicholas Kenyon, Richard Morrison, Susan Tomes, John Tyrrell, Richard Wigmore, Elizabeth Wilson; **Poetry** *Paul Keegan, Matthew Hollis* AUTHORS Simon Armitage, Douglas Dunn, Seamus Heaney, Ted Hughes, Paul Muldoon, Daljit Nagra, Tom Paulin; **Non-fiction** *Neil Belton, Julian Loose* AUTHORS John Carey, Simon Garfield, John Gray, Jan Morris, Francis Spufford, Frances Stonor Saunders, Jenny Uglow.
🇪 Royalties twice-yearly.

❊*Authors' update* Celebrating its eightieth anniversary, one of the few generally recognizable publishing brands is venturing beyond books to embrace events, seminars and other profitable spinoffs. The conventional output took a bit of a knock last Christmas but the mood is still buoyant with crime leading the expansion. Faber is the prime mover in the successful Independent Alliance of publishers.

Facet Publishing
7 Ridgmount Street, London WC1E 7AE
📞 020 7255 0590/0505 (text phone)
📠 020 7255 0591
info@facetpublishing.co.uk
www.facetpublishing.co.uk
Managing Director *John Woolley*

Publishing arm of **CILIP: The Chartered Institute of Library and Information Professionals** (formerly the Library Association). Publishes library and information science, monographs, reference and bibliography aimed at library and information professionals. IMPRINTS **Library Association Publishing**; **Clive Bingley Books**; **Facet Publishing**. Over 200 titles in print, including *The New Walford* and *AACR2*. 25–30 titles a year. Unsolicited mss, synopses and ideas welcome provided material falls firmly within the company's specialist subject areas.
🇪 Royalties annually.

Featherstone Education
▷ A.&C. Black Publishers Ltd

Fernhurst Books
▷ Wiley Nautical

David Fickling Books
▷ The Random House Group Ltd

Fig Tree
▷ Penguin Group (UK)

Findhorn Press Ltd
305A The Park, Findhorn IV36 3TE
📞 01309 690582　📠 0131 777 2711
info@findhornpress.com
www.findhornpress.com
Director *Thierry Bogliolo*

Founded 1971. Publishes mind, body and spirit, New Age and healing. 24 titles in 2008. Unsolicited synopses and ideas via e-mail only if they come within Findhorn's subject areas. No children's books, fiction or poetry. 'We receive over 1,000 submissions annually so please check our website and look at the type of books we publish prior to contacting us.'
🇪 Royalties twice-yearly.

Firebrand
▷ Infinite Ideas

First & Best in Education Ltd
Hamilton House, Earlstrees Court, Earlstrees Road, Corby NN17 4HH
📞 01536 399005　📠 01536 399012
editorial@firstandbest.co.uk
www.firstandbest.co.uk
www.shop.firstandbest.co.uk

Publisher *Tony Attwood*
Editor *Anne Cockburn*

Publishers of over 1,000 educational books of all types for all ages of children and for parents and teachers. Also small series of books on marketing. All books are published as being suitable for photocopying and/or as electronic books. 'Currently looking for books on school management and administration. Also some secondary subjects. No primary subjects, no fiction.' TITLES *Raising Grades Through Study Skills*; *Business Sponsorship of Secondary Schools*; *Policy Documents for Day Nurseries and Nursery Units*. IMPRINT **School Improvement Reports**. Check the website for submission guidelines.
Ⓔ Royalties twice-yearly.

Fitzgerald Publishing

89 Ermine Road, Ladywell, London SE13 7JJ
☎ 020 8690 0597
fitzgeraldbooks@yahoo.co.uk
Managing Editors *Tim Fitzgerald, Michael Fitzgerald*
General Editor *Andrew Smith*

Founded 1974. Specializes in scientific studies of insects and spiders. TITLES *The Tarantula*; *Keeping Spiders and Insects in Captivity*; *Tarantulas of the USA*; *Scorpions of Medical Importance* (books) and *Earth Tigers – Tarantulas of Borneo*; *Desert Tarantulas* (TV/video documentaries). 1–2 titles a year. Unsolicited mss, synopses and ideas for books welcome. Also considers video scripts for video documentaries.

Fitzjames Press
▷ Motor Racing Publications

Floris Books

15 Harrison Gardens, Edinburgh EH11 1SH
☎ 0131 337 2372   ℻ 0131 347 9919
floris@florisbooks.co.uk
www.florisbooks.co.uk
Managing Director *Christian Maclean*
Editors *Christopher Moore, Sally Martin*
Approx. Annual Turnover £550,000

Founded 1977. Publishes books related to the Steiner movement, including The Christian Community, as well as arts & crafts, children's (including fiction and picture books with a Scottish theme), history, religious, science, social questions and Celtic studies. Synopses and ideas are welcome for all books.
Ⓔ Royalties annually.

Focal Press
▷ Elsevier Ltd

Folens Limited

Waterslade House, Thame Road, Haddenham HP17 8NT
☎ 0870 609 1235   ℻ 0870 609 1236
folens@folens.com
www.folens.com
Chairman *Dirk Folens*
Managing Director *Adrian Cockell*

Founded 1987. A leading educational publisher. IMPRINTS **Folens**; **Belair**. About 150 titles a year. Unsolicited mss, synopses and ideas for educational books welcome.
Ⓔ Royalties annually.

Fort Publishing Ltd

Old Belmont House, 12 Robsland Avenue, Ayr KA7 2RW
☎ 01292 880693   ℻ 01292 270134
fortpublishing@aol.com
www.fortpublishing.co.uk
Chairman *Agnes Jane McCarroll*
Managing Director *James McCarroll*
Approx. Annual Turnover £95,000

Founded 1999. Publishes general non-fiction, history, sport, crime and local interest. TITLES *Glasgow Then and Now*; *Great Hull Stories*; *Evil Scotland*; *Ten Days That Shook Rangers*. 45 titles in 2008. No unsolicited mss; send synopses and ideas for books by post or e-mail.
Ⓔ Royalties twice-yearly.

Foulsham Publishers

The Oriel, Thames Valley Court, 183–187 Bath Road, Slough SL1 4AA
☎ 01753 526769   ℻ 01753 535003
reception@foulsham.com
www.foulsham.com
Chairman/Managing Director *B.A.R. Belasco*
Approx. Annual Turnover £2.5 million

Founded 1800 and now one of the few remaining independent family companies to survive takeover. Publishes non-fiction on most subjects including lifestyle, travel guides, family reference, cookery, diet, health, DIY, business, self improvement, self development, astrology, dreams, MBS. No fiction. IMPRINT **Quantum** Mind, body and spirit titles. TITLES *Classic 1000 Cocktails*; *Brit Guide to Orlando and Walt Disney World*; *Old Moore's Almanack*; *Raphael's Astrological Ephemeris*. Around 60 titles a year. Unsolicited mss, synopses and ideas welcome in hard copy.
Ⓔ Royalties twice-yearly.

Fountain Press

Newpro UK Ltd., Old Sawmills Road, Faringdon SN7 7DS
☎ 01367 242411   ℻ 01367 241124

sales@newprouk.co.uk
www.newprouk.co.uk
Publisher *C.J. Coleman*
Approx. Annual Turnover £800,000

Founded 1923 when it was part of the Rowntree
Trust Social Service. Owned by the British
Electric Traction Group until 1982 when it
was bought out by H.M. Ricketts. Acquired
by Newpro UK Ltd in July 2000. Publishes
mainly photography and natural history. TITLES
Photography and Digital 'Workshop'; *Antique
and Collectable Cameras*; *Camera Manual*
(series). About 2 titles a year. Unsolicited mss and
synopses welcome.
Ⓔ Royalties twice-yearly.

✱*Authors' update* **Highly regarded for
production values, Fountain has the reputation
for involving authors in every stage of the
publishing process.**

Fourth Estate
▷ HarperCollins Publishers Ltd

Free Association Books Ltd
PO Box 37664, London NW7 2XU
☎ 020 8906 0396   📠 020 8906 0006
info@fabooks.com
www.fabooks.com
Managing Director/Publisher *T.E. Brown*

Publishes psychoanalysis and psychotherapy,
psychology, cultural studies, sexuality and
gender, women's studies, applied social sciences.
TITLES *Drug Use and Cultural Context* eds.
Ross Coomber and Nigel South; *Skin Disease*
Ann Maguire; *Caring for the Dying at Home*
Gill Pharoah; *A History of Group Study and
Psychodynamic Organizations* Amy L. Fraher.
Always e-mail in the first instance with an outline.
OVERSEAS ASSOCIATE ISBS, USA.
Ⓔ Royalties twice-yearly.

W. H. Freeman and Company
Palgrave Macmillan, Brunel Road, Houndmills,
Basingstoke RG21 6XS
☎ 01256 329242   📠 01256 330688
www.palgrave.com
President *Elizabeth Widdicombe (W. H. Freeman
and Company, New York)*
Editorial Director *Margaret Hewinson (Palgrave
Macmillan, UK)*

W. H. Freeman and Company publishes academic
educational and textbooks in biochemistry,
biology and zoology, chemistry, economics,
mathematics and statistics, natural history,
neuroscience, palaeontology, physics. The
editorial offices for W. H. Freeman and Company
is in New York (Basingstoke is a sales and
marketing office only) but unsolicited mss can go
through Basingstoke. Those which are obviously

unsuitable will be sifted out; the rest will be
forwarded to New York.
Ⓔ Royalties annually.

Samuel French Ltd
52 Fitzroy Street, London W1T 5JR
☎ 020 7387 9373   📠 020 7387 2161
theatre@samuelfrench-london.co.uk
www.samuelfrench-london.co.uk
Chairman *Leon Embry*
Managing Director *Vivien Goodwin*

Founded 1830 with the object of acquiring acting
rights and publishing plays. Publishes plays only.
About 50 titles a year. Unsolicited mss considered
only after introductory letter addressed to the
Performing Rights Department.
Ⓔ Royalties twice-yearly for books; performing
royalties monthly, subject to a minimum amount.

✱*Authors' update* **Thrives on the amateur
dramatic societies who are forever in need
of play texts. Editorial advisers give serious
attention to new material but a high proportion
of the list is staged before it goes into print. New
writers are advised to try one-act plays, much in
demand by the amateur dramatic societies but
rarely turned out by established playwrights.**

The Friday Project
▷ HarperCollins Publishers Ltd

Frontline Books
5A Accommodation Road, Golders Green,
London NW11 9ED
☎ 020 8455 5559   📠 01226 734438
michael@frontline-books.com
www.frontline-books.com
Managing Director *Charles Hewitt*

Military imprint of **Pen & Sword Books Ltd**.
Publishes a wide range of military history topics
and periods, from Ancient Greece and Rome to
the present day. 20 titles in 2007. Send synopsis
and sample chapter giving an outline of the book's
theme and contents, chapter headings and word
extent, if possible. Indication of sources would
be helpful together with number of and type of
illustrations, if relevant, plus personal details
that underline qualifications for writing the book
rather than a formal c.v. Enclose s.a.e. for return
of material. No fiction.
Ⓔ Royalties twice-yearly.

David Fulton Publishers Ltd
2 Park Square, Milton, Abingdon OX14 4RN
☎ 020 7017 6000   📠 020 8996 3622
info@routledge.co.uk
www.routledge.co.uk
Owner *Taylor & Francis Group*
Senior Publishers *Bruce Roberts (textbooks),
Monika Lee (classroom)*
Approx. Annual Turnover £2.4 million

Founded 1987. Part of Routledge Education. Publishes textbooks and practical books for students, trainee and working teachers at all levels of the curriculum (foundation stage to post compulsory). Special educational needs books also published in collaboration with the National Association for Special Educational Needs (NASEN); some overlap into other areas of health and social care, therapy, educational psychology, and speech and language therapy. A growing list of Early Years books for pre-school and nursery staff. About 100 titles a year. No unsolicited mss; synopses and ideas for books welcome. ⓔ Royalties annually.

Funfax
▷ Dorling Kindersley Ltd

Fusion Press
▷ Vision

Gaia Books
▷ Octopus Publishing Group

Garland Science
▷ Taylor & Francis Group

Garnet Publishing Ltd
8 Southern Court, South Street, Reading RG1 4QS
☎ 0118 959 7847   🖷 0118 959 7356
dan@garnetpublishing.co.uk
www.garnetpublishing.co.uk
Editorial Manager *Dan Nunn*

Founded 1992 and purchased Ithaca Press in the same year. Publishes art, architecture, photography, archive photography, cookery, travel classics, comparative religion, Islamic culture and history, foreign fiction in translation. Core subjects are Middle Eastern but list is rapidly expanding to be more general.

IMPRINTS **Garnet Publishing** TITLES *Simply Lebanese*; *The Art and Architecture of Islamic Cairo*; *What Did We Do To Deserve This?*; **Ithaca Press** Specializes in post-graduate academic works on the Middle East, political science and international relations TITLES *The Making of the Modern Gulf States*; *Reform in the Middle East Oil Monarchies*; *The United States and Persian Gulf Security*. About 20 titles a year. Unsolicited mss not welcome; write with outline and ideas first plus current c.v.
ⓔ Royalties twice-yearly.

Geddes & Grosset
David Dale House, New Lanark ML11 9DJ
☎ 01555 665000   🖷 01555 665694
Publishers *Ron Grosset, R. Michael Miller*
Approx. Annual Turnover £2.5 million

Founded 1989. Publisher of children's and reference books. Unsolicited mss, synopses and ideas welcome. No adult fiction.

The Geological Society Publishing House
Unit 7, Brassmill Enterprise Centre, Brassmill Lane, Bath BA1 3JN
☎ 01225 445046   🖷 01225 442836
sales@geolsoc.org.uk
enquiries@geolsoc.org.uk
www.geolsoc.org.uk
Commissioning Editor *Angharad Hills*

Publishing arm of the Geological Society of London which was founded in 1807. Publishes postgraduate texts in the earth sciences. 30 titles a year. Unsolicited synopses and ideas welcome.

Stanley Gibbons Publications
7 Parkside, Christchurch Road, Ringwood BH24 3SH
☎ 01425 472363   🖷 01425 470247
info@stanleygibbons.com
www.stanleygibbons.com
Also at: 399 Strand, London WC2R 0LX
Chief Executive *M. Hall*
Editorial Head *H. Jefferies*
Publishing Director *Ann-Marie Halligan*
Approx. Annual Turnover £2.9 million

Part of The Stanley Gibbons Group Limited, an AIM listed company. Long-established force in the philatelic and autograph world with 150 years in the business. Publishes philatelic reference catalogues and handbooks. Reference works relating to other areas of collecting may be considered. TITLES include *Stanley Gibbons British Commonwealth Stamp Catalogue*; *Stamps of the World*. Foreign catalogues include Japan and Korea, Portugal and Spain, Germany, Middle East, Balkans, China. Monthly publication: *Gibbons Stamp Monthly* (see entry under *Magazines*). www.collectorcafe.com – Internet site covering full range of collectibles – editorial input always considered. About 20 titles a year. Unsolicited mss, synopses and ideas welcome.
ⓔ Royalties by negotiation.

Gibson Square
47 Lonsdale Square, London N1 1EW
☎ 020 7096 1100   🖷 020 7993 2214
info@gibsonsquare.com
www.gibsonsquare.com
Chairman *Martin Rynja*

Founded 2001. Publishes exclusively non-fiction: biography, current affairs, politics, cultural criticism, psychology, history, travel, art history, philosophy. Books must have high publicity profile. IMPRINT **Gibson Square** TITLES *Londonistan* Melanie Phillips; *House of Bush House of Saud* Craig Unger; *Blowing up Russia* Alexander Litvinenko; *Two Lipsticks & a Lover* Helena Frith Powell; *If I Did It* O.J. Simpson. Welcomes synopses and ideas. Approach by e-mail; reply sent only if interested. No fiction.

Ⓔ Royalties twice-yearly.

✱*Authors' update* Dedicated to publishing books 'on the cutting edge', Gibson Square is one of the few publishers to embrace serious political issues.

Ginn
▷ Pearson (Heinemann)

GL Assessment

The Chiswick Centre, 414 Chiswick High Road, London W4 5TF
☎ 0845 602 1937   🖷 0845 601 5358
info@gl-assessment.co.uk
www.gl-assessment.co.uk
Publishing Director *Andrew Thraves*

'The UK's largest independent provider of educational assessments', GL Assessment has supplied schools with tests and assessment services (which are used by over 85% of all UK primary and secondary schools) for 27 years. 20 titles a year. No submissions as products are usually commissioned by the company.
Ⓔ Royalties twice-yearly.

Global Publishing
▷ Quarto Publishing plc

Gloucester Publishers Plc

10 Northburgh Street, London EC1V 0AT
☎ 020 7253 7887   🖷 020 7490 3708
markbicknell@everyman.uk.com
www.everymanchess.com
Managing Director *Mark Bicknell*
Approx. Annual Turnover £1 million

Publishes exclusively academic and leisure books relating to chess. IMPRINT **Everyman Chess** TITLES *Garry Kasparov: My Great Predecessors, Vols 1–7*; *Play Winning Chess* Yasser Seirawan; *Art of Attack in Chess* Vladimir Vukovich. Over 300 titles. No unsolicited material.
Ⓔ Royalties annually.

GMC Publications Ltd

166 High Street, Lewes BN7 1XU
☎ 01273 477374   🖷 01273 402866
pubs@thegmcgroup.com
www.thegmcgroup.com
Joint Managing Directors *J.A.J. Phillips, J.A.B. Phillips*
Managing Editor *Gerrie Purcell*

Founded 1974. Publisher and distributor of craft and leisure books and magazines, covering topics such as woodworking, DIY, architecture, photography, gardening, cookery, art, puzzles and games, reference, humour, TV and film. About 50 illustrated books and 11 magazines a year. Unsolicited mss, synopses and ideas for books welcome. No fiction.
Ⓔ Royalties twice-yearly.

Godsfield Press
▷ Octopus Publishing Group

Gollancz
▷ The Orion Publishing Group Limited

Gomer Press/Gwasg Gomer

Llandysul Enterprise Park, Llandysul SA44 4JL
☎ 01559 363090   🖷 01559 363758
gwasg@gomer.co.uk
www.gomer.co.uk
www.pontbooks.co.uk
Chairman/Managing Director *J.E. Lewis*
Publishing Director *Mairwen Prys Jones*
Editor (adult books in Welsh) *To be appointed*
Editor (adult books in English) *Ceri Wyn Jones*
Editors (children's books in Welsh) *Sioned Lleinau, Helen Evans, Rhiannon Davies*
Pont Books (children's books in English) Editor *Viv Sayer*

Founded 1892. Publishes adult fiction and non-fiction, children's fiction and educational material in English and in Welsh. IMPRINTS **Gomer**; **Pont Books**. 125 titles a year. Unsolicited mss welcome. Prior enquiry required. Mss must have a Welsh dimension.
Ⓔ Royalties twice-yearly.

Gordon & Breach
▷ Taylor & Francis Group

Gower
▷ Ashgate Publishing Ltd

Graham-Cameron Publishing & Illustration
▷ entry under UK Packagers

Granta Books

Twelve Addison Avenue, London W11 4QR
☎ 020 7605 1360   🖷 020 7605 1361
www.granta.com
Managing Director *David Graham*
Publisher *Philip Gwyn Jones*
Submissions *Amber Dowell (Editorial Assistant)*

Founded 1979. Acquired by Sigrid Rausing in 2006. Publishes largely narrative non-fiction and some literary fiction. About 70 titles a year. No unsolicited mss; synopses and sample chapters welcome.
Ⓔ Royalties twice-yearly.

✱*Authors' update* Largely associated with erudite titles, there are moves to revitalize the Granta list with the commissioning of more mainstream books.

W. Green (Scotland)
▷ Sweet & Maxwell Group

Green Books

Foxhole, Dartington, Totnes TQ9 6EB
☎ 01803 863260   🖷 01803 863843

edit@greenbooks.co.uk
www.greenbooks.co.uk
Chairman *Satish Kumar*
Publisher *John Elford*
Approx. Annual Turnover £500,000

Founded in 1987 with the support of a number of Green organizations. Closely associated with *Resurgence* magazine. Publishes high-quality books on a wide range of Green issues, including economics, politics and the practical application of Green thinking. No fiction or books for children. TITLES *The Transition Handbook* Rob Hopkins; *Timeless Simplicity* John Lane; *Allotment Gardening* Susan Berger. No unsolicited mss. Synopses and ideas welcome but check guidelines on the website in the first instance.

Ⓔ Royalties twice-yearly.

Green Print
▷ The Merlin Press Ltd

Green Profile
▷ Profile Books

Gresham Books Ltd

46 Victoria Road, Summertown, Oxford OX2 7QD
☎ 01865 513582   🖷 01865 512718
info@gresham-books.co.uk
www.gresham-books.co.uk
Managing Director *Paul Lewis*
Approx. Annual Turnover £350,000

A small specialist publishing house. Publishes hymn and service books for schools and churches, school histories, also craft-bound choir and orchestral folders. No unsolicited material but ideas welcome.

Grub Street

4 Rainham Close, London SW11 6SS
☎ 020 7924 3966/7738 1008   🖷 020 7738 1009
post@grubstreet.co.uk
www.grubstreet.co.uk
Managing Director *John Davies*

Founded 1982. Publishes cookery, food and wine, military and aviation history books. About 30 titles a year. Unsolicited mss and synopses welcome in the above categories but please enclose return postage.

Ⓔ Royalties twice-yearly.

Guinness World Records Ltd

3rd Floor, 184–192 Drummond Street, London NW1 3HP
☎ 020 7891 4567   🖷 020 7891 4501
marketing@guinnessworldrecords.com
www.guinnessworldrecords.com
Managing Director *Alistair Richards*

First published in 1955, the annual Guinness World Records book is published in more than 100 countries and 28 languages and is the highest-selling copyright book of all time, with more than three million copies sold annually across the globe. In addition to *Guinness World Records*, the company publishes *Gamer's Edition*. Acquired by Jim Pattison Group (JPG) in February 2008. Contact from prospective researchers, editors and designers welcome.

✷*Authors' update* **It had to happen. With its sale to the Jim Pattison Group, Guinness World Records has joined the creator of Ripley's Believe It or Not, often thought to have been the original inspiration for Guinness. Some claim that World Records has dumbed down in recent years but you have to hand it to an annual that can sell close to 600,000 copies.**

Gulf Professional Press
▷ Elsevier Ltd

Gullane Children's Books

185 Fleet Street, London EC4A 2HS
☎ 020 7400 1092   🖷 020 7400 1037
info@gullanebooks.com
www.gullanebooks.com
Publisher *Simon Rosenheim*
Editor *Sophie Winter-Cole*
Art Director *Paula Burgess*
Rights Director *Katherine Judge*

Owned by D.C. Thomson, Gullane Children's Books was bought by **Meadowside Children's Books** in 2007. Publishes picture books and novelty titles. Unsolicited mss accepted. Picture book mss: 800 words max. (for children up to the age of 7). Response will be sent only if s.a.e. enclosed. Do not send original artwork samples (copies only). Address to *The Submissions Editor*.

Gwasg Carreg Gwalch

12 Iard yr Orsaf, Llanrwst LL26 0EH
☎ 01492 642031   🖷 01492 641502
myrddin@carreg-gwalch.com
www.carreg-gwalch.com
Managing Editor *Myrddin ap Dafydd*

Founded in 1980. Publishes Welsh language; English books of Welsh interest – history, folklore, guides and walks. About 90 titles a year. Unsolicited mss, synopses and ideas welcome if on Welsh history and heritage.

Ⓔ Royalties paid.

Hachette Children's Books

338 Euston Road, London NW1 3BH
☎ 020 7873 6000
www.hachettechildrens.co.uk
www.franklinwatts.co.uk
www.orchardbooks.co.uk
www.hodderchildrens.co.uk
www.waylandbooks.co.uk
Owner *Hachette Book Group Inc.*

Managing Director *Marlene Johnson*
Approx. Annual Turnover £34 million

Formed by the combining of Watts Publishing Group with Hodder Children's Books in April 2005. Publishes children's non-fiction, reference, information, gift, fiction, picture and novelty books and audio books. IMPRINTS **Hodder Children's Books** *Anne McNeil* Fiction, picture books, novelty, general non-fiction and audio; **Orchard Books** Picture book and novelties *Kate Burns*; Fiction *Penny Morris*; **Franklin Watts** *Rachel Cooke* Non-fiction and information books; **Wayland** *Joyce Bentley* Non-fiction and information books. About 1,300 titles a year. Unsolicited material is not considered. Ⓔ Royalties twice-yearly.

Hachette Livre

French publishing conglomerate which embraces **Chambers Harrap Publishers, Hachette UK, The Orion Publishing Group, Octopus Publishing Group, Little, Brown Book Group**. For further information, see imprint entries and entry under *European Publishers*.

Hachette UK

338 Euston Road, London NW1 3BH
🕾 020 7873 6000   🖷 020 7873 6024
www.hachette.co.uk
Group Chief Executive *Tim Hely Hutchinson*

Hodder Headline was formed in June 1993 through the merger of Headline Book Publishing and Hodder & Stoughton. The company was acquired by WHSmith plc in 1999 and then by Hachette Livre S.A. in October 2004. Renamed Hachette Livre UK in August 2007 and subsequently renamed Hachette UK in October 2008. Purchased John Murray (Publishers) Ltd in 2002. About 2,000 titles a year.

DIVISIONS **Headline Publishing Group** CEO *Kate Wilson*, Deputy Managing Director and Chief Operating Officer *Kerr MacRae*, Deputy Managing Director *Jane Morpeth* Publishes commercial and literary fiction (hardback and paperback) and popular non-fiction including autobiography, biography, food and wine, history, humour, popular science, sport and TV tie-ins. IMPRINTS **Business Plus; Headline; Headline Review; Little Black Dress**. AUTHORS Lyn Andrews, Emily Barr, Louise Bagshawe, Martina Cole, Janet Evanovich, Victoria Hislop, Wendy Holden, Jonathan Kellerman, Jill Mansell, Maggie O'Farrell, Sheila O'Flanagan, Andrea Levy, Jennifer Johnson, Simon Scarrow, Jed Rubenfeld, Karen Rose, Pamela Stephenson, Penny Vincenzi.

Hodder & Stoughton General CEO *Jamie Hodder-Williams*, Deputy Managing Director *Lisa Highton*; **Non-fiction** *Rowena Webb*; **Sceptre**

Carole Welch; **Fiction** *Carolyn Mays*; **Audio** (see entry under *Audio Books*). Publishes commercial and literary fiction; biography, autobiography, history, self-help, humour, travel and other general interest non-fiction; audio. IMPRINTS **Hodder & Stoughton; Hodder Paperbacks; Sceptre; Mobius**. AUTHORS Melvyn Bragg, Jeffrey Deaver, Charles Frazier, Elizabeth George, Thomas Keneally, Stephen King, John le Carré, Andrew Miller, David Mitchell, Jodi Picoult, Rosamunde Pilcher, Mary Stewart, Fiona Walker. No unsolicited mss.

Hodder Children's Books (See **Hachette Children's Books**)

Hodder & Stoughton Faith Director *Wendy Grisham* **Bibles** *Ian Metcalfe* Publishes NIV and Today's NIV Bibles, Christian fiction and non-fiction, biographies, autobiographies, Manga. AUTHORS Philip Yancey, Rob Parsons, Joyce Meyer, J. John. IMPRINTS **Hodder & Stoughton; Hodder Paperbacks**.

Hodder Education Group CEO *To be appointed* Publishes in the following areas: **Schools** *Lis Tribe*; **Consumer Education** *Katie Roden*; **Further Education/Higher Education Textbooks** *Alexia Chan*; **Health Sciences** *Joanna Koster*. IMPRINTS **Hodder Education; Teach Yourself; Hodder Arnold; Hodder Gibson** (see entry).

Hachette Books Scotland Publisher *Bob McDevitt* Commercial and literary fiction and non-fiction.

Hachette Books Ireland Managing Director Publishing *Breda Purdue* Commercial and literary fiction and non-fiction (see entry under *Irish Publishers*).

John Murray (Publishers) Ltd (see entry).

Octopus Publishing Group (see entry).

Little, Brown Book Group (see entry).

The Orion Publishing Group Limited (see entry).

✴*Authors' update* The leading British publisher by turnover, Hachette is strong on mainstream bestsellers. The losers are midlist authors. At the same time the search is on for new talent which can benefit from imaginative and energetic marketing.

Halban Publishers

22 Golden Square, London W1F 9JW
🕾 020 7437 9300   🖷 020 7437 9512
books@halbanpublishers.com
www.halbanpublishers.com
Directors *Peter Halban, Martine Halban*

Founded 1986. Independent publisher. Publishes biography, memoirs, history, fiction, books on the Middle East and of Jewish interest. 8–10 titles a

year. No unsolicited material. Approach by letter in the first instance. Unsolicited e-mails will be deleted.
Ⓔ Royalties twice-yearly for first two years, thereafter annually in December.

Robert Hale Ltd

Clerkenwell House, 45–47 Clerkenwell Green, London EC1R 0HT
📞 020 7251 2661   📠 020 7490 4958
enquire@halebooks.com
www.halebooks.com
Chairman/Managing Director *John Hale*

Founded 1936. Family-owned company. Publishes adult fiction (but not interested in category romance or science fiction) and non-fiction. No specialist material (education, law, medical or scientific). Acquired NAG Press Ltd in 1993 with its list of horological, gemmological, jewellery and metalwork titles, and **J.A. Allen & Co.** in 1999 with its extensive list of horse and dog books (see entry). TITLES Non-fiction: *Flies that Catch Trout* Terry Lawton; *RAF Bomber Command at War* Rupert Matthews; Fiction: *A Pity About the Girl and Other Stories* Michael Gilbert; *No Less Than the Journey* E.V. Thompson; *The Queen's Margarine* Wendy Perriam; *Sherlock Holmes in Russia* Alex Auswaks Over 200 titles a year. Unsolicited mss, synopses and ideas for books welcome.
Ⓔ Royalties twice-yearly.

Halsgrove

Halsgrove House, Ryelands Industrial Estate, Bagley Road, Wellington TA21 9PZ
📞 01823 653777   📠 01823 216796
sales@halsgrove.com
www.halsgrove.com
Publishers *Simon Butler, Steven Pugsley*

Founded in 1990 and now a leading publisher and distributor of regional books under the **Halsgrove**, **Halstar**, **PiXZ Books** and **Ryelands** IMPRINTS. Publishes books of regional interest throughout the UK: local history (including the *Community History* series), biography, photography and art, mainly in hardback. No fiction or poetry. 170 titles in 2008. Unsolicited mss, synopses and ideas for books of regional interest welcome.
Ⓔ Royalties annually.

Halstar
▷ Halsgrove

Hamish Hamilton
▷ Penguin Group (UK)

Hamlyn Octopus
▷ Octopus Publishing Group

Hammersmith Press Ltd

496 Fulham Palace Road, London SW6 6JD
📞 020 7736 9132   📠 020 7348 7521
gmb@hammersmithpress.co.uk
www.hammersmithpress.co.uk
Managing Director *Georgina Bentliff*

Founded 2004. Publishes health, nutrition, diet, academic medicine, and 'literary medicine', i.e. literary works that relate to the practice of medicine. TITLES *Trick and Treat* Barry Groves; *Smart Guide to Infertility* Professor Robert Harrison. 6 titles in 2008. No unsolicited mss; synopses and ideas for books welcome; approach by e-mail. No fiction or children's material.
Ⓔ Royalties twice-yearly.

Happy Cat
▷ Catnip Publishing Ltd

Harlequin Mills & Boon Limited

Eton House, 18–24 Paradise Road, Richmond TW9 1SR
📞 020 8288 2800   📠 020 8288 2898
www.millsandboon.co.uk
Managing Director *Guy Hallowes*
Editorial Director *Karin Stoecker*
Approx. Annual Turnover £19.5 million

Founded 1908. Owned by the Canadian-based Torstar Group. Publishes a wide range of women's fiction including romantic novels.

IMPRINTS **Mills & Boon Modern** *Tessa Shapcott* (50–55,000 words) Alpha males and attractive women swept up by intense emotion and passion set against a backdrop of international locations, luxury and wealth. **Mills & Boon Modern Heat** *Bryony Green* (50,000–55,000 words) Delivers a feel-good experience, focusing on the kind of relationship that women aged 18 to 35 aspire to. Young characters in urban settings meet, flirt, share experiences, have great sex and fall in love, finally making a commitment that will bind them forever. Based around emotional issues, other concerns – e.g. jobs and friendships – are also touched upon and resolved in upbeat way. **Mills & Boon Romance** *Kimberley Young* (50–55,000 words) Contemporary upbeat and heroine-focused romance, driven by strongly emotional conflicts which are believable and relevant to today's women. Capturing the depth of emotion and sheer excitement of falling in love in a variety of international settings. **Mills & Boon Medical** *Sheila Hodgson* (50–55,000 words) Modern medical practice provides a unique background to highly emotional contemporary romances. **Mills & Boon Historical Romance** *Linda Fildew* (70–75,000 words) Richly textured, emotionally intense historical novels covering a wide range of time periods from ancient civilizations up to and including the Second World War. About 950 titles a year. Other submissions: also accepting submissions in all general women's fiction genres (contemporary and historical) for longer series and single title imprints. Tip sheets and guidelines are available from the Mills & Boon website (see above), www.eharlequin.com or Harlequin Mills & Boon Editorial Dept. (telephone or write sending s.a.e.). Agented and unagented submissions accepted. Send query letter, synopsis and first three chapters in the first instance.

※*Authors' update* **To celebrate its centenary last year, M&B sent round the country an exhibition on the changing patterns of courtship over the decades. If nothing else, it proved that light romance still provides a good income for writers who can adapt to the clearly defined M&B formula. Just think, M&B has a network of 12,000 authors worldwide and every title scores six-figure sales. But be warned. M&B gets 2,000 unsolicited manuscripts a year. Latest initiatives include a series of rugby romances in partnership with the Rugby Football Union. It seems that one third of rugby supporters are women. There are plans to make up to 2,000 titles available in digital format.**

HarperCollins Publishers Ltd

77–85 Fulham Palace Road, London w6 8jb
🕾 020 8741 7070   🖷 020 8307 4440
www.harpercollins.co.uk
Also at: Westerhill Road, Bishopbriggs, Glasgow G64 2QT 🕾 0141 772 3200 🖷 0141 306 3119
Owner *News Corporation*
CEO/Publisher *Victoria Barnsley*
Managing Director *Amanda Ridout*
Approx. Annual Turnover £173.4 million (UK)

HarperCollins is one of the top book publishers in the UK, with a wider range of books than any other publisher; from cutting-edge contemporary fiction to enduring classics; from school textbooks to celebrity memoirs; and from downloadable dictionaries to prize-winning literature. The wholly-owned division of News Corporation also publishes a wide selection of non-fiction including history, celebrity memoirs, biography, popular science, mind, body and spirit, dictionaries, maps and reference books. HarperCollins is also one of the largest education publishers in the UK. Authors include many award-winning and international bestsellers such as Isabel Allende, Paulo Coelho, Josephine Cox, Michael Crichton, Cathy Kelly, Judith Kerr, Doris Lessing, Frank McCourt, Tony Parsons, Nigel Slater and Diana Wynne Jones. Bestselling licensed properties include Dr Seuss and Noddy. About 1,200 titles a year.

Harper Fiction Publisher *Lynne Drew* IMPRINTS **HarperCollins** Publishing Directors *Susan Watt, Julia Wisdom*; **Voyager** Fantasy/science

fiction Publishing Director *Jane Johnson* General, historical fiction, crime and thrillers and women's fiction.

Harper Non-Fiction Managing Director *Belinda Budge*, Publisher *Carole Tonkinson*, Editorial Directors Kate Latham (cookery & lifestyle), *Natalie Jerome* (popular culture), Publishing Director *Jonathan Taylor* Sporting biographies, guides and histories. IMPRINT **HarperThorsons/Element** *Katy Carrington* (mind, body & spirit), *Sally Annett* (media tie-ins).

Avon Managing Director *Caroline Ridding*, Editorial Director *Maxine Hitchcock* Popular fiction. IMPRINTS **HarperCollins** Publishing Director *David Brawn* (Agatha Christie, J.R.R. Tolkien, C.S. Lewis) Media-related books from film companions to celebrity autobiographies and TV tie-ins.

HarperCollins Children's Books Managing Director *Mario Santos*, Publisher *Ann-Janine Murtagh*, Brand & Properties Developer *Claire Harding* Quality picture books for under-7s; fiction for age 6 up to young adult; TV and film tie-ins and properties Publishing Directors *Gillie Russell* (fiction), *Sue Buswell* (picture books).

PRESS BOOKS DIVISION

Managing Director *John Bond*

IMPRINTS **Fourth Estate** Publishing Director/Publisher *Nick Pearson* Fiction, literary fiction, current affairs, popular science, biography, humour, travel; **HarperPress** General trade non-fiction and fiction Publishing Directors *Arabella Pike* (non-fiction), *Clare Smith* (fiction); **HarperPerennial** Paperback imprint. Publishing Director *Paul Baggaley*; **Blue Door** Publisher *Patrick Janson-Smith* Fiction; **The Friday Project** Publisher *Scott Pack*, Managing Editor *Heather Smith* Fiction, non-fiction and children's.

COLLINS

Managing Director *Katie Fulford* Guides and handbooks, phrase books and manuals on popular reference, art instruction, illustrated, cookery and wine, crafts, DIY, gardening, military, natural history, pet care, pastimes Publishing Directors *Denise Bates* (illustrated reference), Associate Publisher *Myles Archibald* (natural history). IMPRINTS **Collins; Collins Gem; Jane's**.

COLLINS GEO

Managing Director *Sheena Barclay* Times World Atlases, Collins World Atlases, Publishing Director *Helen Gordon* Maps, atlases, street plans and leisure guides. IMPRINTS **Collins; Times Books**.

Collins Language Managing Director *Robert Scriven* Bilingual and English dictionaries, English dictionaries for foreign learners Publishing

Directors *Helen Newstead* (digital development), *Elaine Higgleton* (editorial).

Collins Education Managing Director *Nigel Ward* Books, CD-ROMs and online material for UK primary and secondary schools and colleges.

✱*Authors' update* The first publisher to digitize its content, HarperCollins has plans to publish all major titles in print and digital format simultaneously. A website that encourages would-be authors to upload sample chapters has unearthed some talent. Authors with the Friday Project are to be offered profit-share deals – the first from any UK publisher. The latest publishing initiative is a true life imprint shared with Sky Real Lives and News International papers. Strongest growth is in commercial fiction.

Harrap
⊳ Chambers Harrap Publishers Ltd

Harriman House Ltd

3A Penns Road, Petersfield GU32 2EW
☏ 01730 233870   🖷 01730 233880
contact@harriman-house.com
www.harriman-house.com
Managing Director *Myles Hunt*
Head of Rights *Suzanne Anderson*

Founded 1994; commenced publishing in 2001. Independent publisher of finance, trading, business and economics books, covering a wide range of subjects from personal finance, small business and lifestyle through to stock market investing, trading and professional guides. 20 titles a year. Unsolicited mss, synopses and ideas for books welcome. Initial approach by e-mail. 💷 Royalties twice-yearly.

Harvard University Press

Fitzroy House, 11 Chenies Street, London WC1E 7EY
☏ 020 7306 0603   🖷 020 7306 0604
info@HUP-MITpress.co.uk
www.hup.harvard.edu
Director *William Sisler*
General Manager *Ann Sexsmith*

European office of **Harvard University Press**, USA. Publishes academic and scholarly works in history, politics, philosophy, economics, literary criticism, psychology, sociology, anthropology, women's studies, biological sciences, classics, history of science, art, music, film, reference. All mss go to the American office: 79 Garden Street, Cambridge, MA 02138 (see entry under *US Publishers*).

Harvill Secker
⊳ The Random House Group Ltd

Harwood Academic
⊳ Taylor & Francis Group

Haus Publishing

70 Cadogan Place, London SW1X 9AH
☏ 020 7838 9055   🖷 020 7235 1999
haus@hauspublishing.com
www.hauspublishing.co.uk
Publisher *Barbara Schwepcke*

DIVISIONS/IMPRINTS **Life&Times** Non-academic biographies of well-known and lesser-known personalities TITLES *Tito*; *Simone de Beauvoir*; *Prime Ministers of the 20th Century*; *Patrice Lumumba*; *Mussolini*. **Armchair Traveller** Literary travel series TITLES *Mumbai to Mecca*; *Venice for Lovers*; *The Liquid Continent: A Mediterranean Trilogy*. **HausBooks** Hardback biographies TITLES *Ellen Terry*; *Rommel*; *Diaghilev*. **Arabia Books** Mainly Arab literature in translation. 26 titles in 2008. Unsolicited mss, synopses and ideas within the subject areas above are welcome; approach by e-mail or post only. No children's, fiction or poetry. 💷 Royalties twice-yearly.

Haynes Publishing

Books Division, Sparkford, Near Yeovil BA22 7JJ
☏ 01963 440635   🖷 01963 440823
bookseditorial@haynes.co.uk
www.haynes.co.uk
Chairman *John H. Haynes, OBE*
Approx. Annual Turnover £31.1 million

Founded 1960. Leading publisher of car and motorcycle manuals, with operations in the UK, USA, Australia and Sweden. Alongside the long-established manuals written by in-house technical authors, the UK-based company also publishes a wide range of books by external authors on all areas of transport – cars, motorcycles, motorsport, aviation, military, maritime, railways, cycling and caravans. In addition, manuals in many new subject areas are being developed, with strong ranges being built in computing, family health, home DIY, gardening, musical instruments and sport. Book proposals and submissions are welcome in all the described subject areas, as are any suggestions for manuals on new topics. Contact *Mark Hughes*, Editorial Director, Books Division. 💷 Royalties twice-yearly.

✱*Authors' update* Having sold Sutton Publishing, Haynes is back to its core programme of publishing 'practical and hands-on' books along with tie-in products such as T-shirts and toys.

Headline/Headline Review
⊳ Hachette UK

Heinemann
> Pearson (Heinemann)

William Heinemann
> The Random House Group Ltd

Helicon Publishing

RM plc, New Mill House, 183 Milton Park, Abingdon OX14 4SE
℡ 01235 823816
helicon@rm.com
www.helicon.co.uk
Business Manager *Caroline Dodds*

A division of RM plc. Helicon is a general reference database publisher for the UK, US and Australian markets. Licenses quality reference content for online, print, and CD-ROM publications. TITLES include the flagship single-volume *Hutchinson Encyclopedia.*

Christopher Helm
> A.&C. Black Publishers Ltd

Helm Information

Crowham Manor, Main Road, Westfield, Hastings TN35 4SR
℡ 01424 882422   ℻ 01424 882817
amandahelm@helm-information.co.uk
www.helm-information.co.uk
Proprietor *Amanda Helm*

Founded 1990. Publishes academic books for students and university libraries. SERIES *The Dickens Companions*; and *Icons*, a series which exposes the processes by which a figure, historical or fictional, achieves iconic status, e.g. Faust, Robin Hood. Will consider ideas and proposals provided they are relevant to the series listed above.
Ⓔ Royalties annually.

Helter Skelter Publishing

18A Radbourne Road, London SW12 0DZ
℡ 020 8673 6320
info@helterskelterpublishing.com
www.helterskelterpublishing.com
Director *Graeme Milton*

Founded 1995. Publishes music and film books only. 10 titles a year. Unsolicited mss, synopses and ideas welcome.

Henderson Publishing
> Dorling Kindersley Ltd

Ian Henry Publications Ltd

20 Park Drive, Romford RM1 4LH
℡ 01708 749119   ℻ 01708 736213
info@ian-henry.com
www.ian-henry.com
Managing Director *Ian Wilkes*

Founded 1976. Publishes local history, humour, transport history and Sherlockian pastiches. TITLE *Hornchurch During the Great War.* 1–2 titles a year. No unsolicited mss. Synopses and ideas for books welcome.
Ⓔ Royalties twice-yearly.

The Herbert Press
> A.&C. Black Publishers Ltd

Hermes House
> Anness Publishing Ltd

Nick Hern Books

The Glasshouse, 49a Goldhawk Road, London W12 8QP
℡ 020 8749 4953   ℻ 020 8735 0250
info@nickhernbooks.demon.co.uk
www.nickhernbooks.co.uk
Chairman/Managing Director *Nick Hern*
Submissions Editor *Matt Applewhite*

Founded 1988. Fully independent since 1992. Publishes books on theatre and film: from practical manuals to plays and screenplays. About 60 titles a year. No unsolicited playscripts. Synopses, ideas and proposals for other theatre material welcome. Not interested in material unrelated to the theatre or cinema.

University of Hertfordshire Press

University of Hertfordshire, College Lane, Hatfield AL10 9AB
℡ 01707 284681   ℻ 01707 284666
UHPress@herts.ac.uk
www.herts.ac.uk/UHPress
Contact *Jane Housham*

Founded 1992. Publishes academic books on Gypsies, literature, regional and local history, education. IMPRINTS **University of Hertfordshire Press; Hertfordshire Publications**. TITLES *Here to Stay: The Gypsies and Travellers of Britain* Colin Clark and Margaret Greenfields; *Selling Shakespeare to Hollywood: The marketing of filmed Shakespeare adaptations from 1989 into the new millennium* Emma French; *Teachers' Legal Rights and Responsibilities* Jon Berry. 15 titles a year. New publishing proposals are welcome; proposal form available on the website.

Hesperus Press Limited

4 Rickett Street, London SW6 1RU
℡ 020 7610 3331   ℻ 020 7610 3217
www.hesperuspress.com
Managing Director *Derrick Holman*

Founded 2001. Publishes lesser-known works by classic authors in both original English and English translation by classic authors. TITLES *Hyde Park Gate News* Virginia Woolf; *Sarrasine* Honoré de Balzac; *No Man's Land* Graham Greene; *Hadji Murat* Lev Tolstoy; *Aller Retour*

New York Henry Miller; *The Watsons* Jane Austen. *Modern Voices* series launched in 2005; *Brief Lives* series of newly commissioned short biographies of celebrated literary figures launched in 2008; contemporary European fiction in translation since 2005. 50 titles in 2008. Does not accept unsolicited mss.
🅔 Royalties annually.

✳**Authors' update** Much praised for matching book design with quality content.

History Into Print
▷ Brewin Books Ltd

The History Press
The Mill, Brimscombe, Stroud GL5 2QG
☎ 01453 883300   📠 01453 883233
sales@thehistorypress.co.uk
www.thehistorypress.co.uk
Chairman *Andy Nash*
Managing Director *Stuart Biles*
Editorial Head *Laura Perehinec*
Approx. Annual Turnover £15 million

Founded in 2007 having acquired the assets of NPI Media. The UK's largest local and specialist history publisher. IMPRINTS include **The History Press**; **Jarrold**; **Nonsuch**; **Phillimore**; **Pitkin**; **Spellmount**; **Stadia**; **Sutton**; **Tempus**. 600 titles in 2008. Unsolicited submissions welcome; approach by e-mail in the first instance. No fiction or children's books. OVERSEAS SUBSIDIARIES The History Press Inc; Sutton Verlag; Editions Alan Sutton.
🅔 Royalties twice-yearly.

✳**Authors' update** The History Press is back on course claiming that all royalties are up to date. Three new series will cover local crime, history and the school in the community. There are also plans to launch an e-magazine.

HMSO
▷ TSO

Hodder & Stoughton
▷ Hachette UK

Hodder Arnold
▷ Hachette UK

Hodder Children's Books
▷ Hachette Children's Books

Hodder Gibson
2a Christie Street, Paisley PA1 1NB
☎ 0141 848 1609   📠 0141 889 6315
hoddergibson@hodder.co.uk
www.hoddergibson.co.uk
Managing Director *John Mitchell*

Part of the **Hachette UK** group. Publishes educational textbooks and revision guides specifically for Scotland and mainly for secondary education. About 25 titles a year. Synopses/ideas preferred to unsolicited mss.
🅔 Royalties annually.

Hodder Headline Ltd
▷ Hachette UK

Honeyglen Publishing Ltd
15 Kensington West, Blythe Road, London W14 0JG
☎ 020 7602 2876   📠 020 7602 2876
Directors *N.S. Poderegin, J. Poderegin*

Founded 1983. A small publishing house whose output is 'extremely limited'. Publishes history, philosophy of history, biography and selective fiction. No children's or science fiction. TITLES *The Soul of China*; *The Soul of India*; *Woman and Power in History*; *Lost World – Tibet*; *A Child of the Century* all by Amaury de Riencourt; *With Duncan Grant in South Turkey* Paul Roche; *Vladimir, The Russian Viking* Vladimir Volkoff; *The Dawning* Milka Bajic-Poderegin; *Quicksand* Louise Hide. Unsolicited mss welcome.

Honno Welsh Women's Press
c/o Canolfan Merched y Wawr, Vulcan Street, Aberystwyth SY23 1JH
☎ 01970 623150   📠 01970 623150
post@honno.co.uk
www.honno.co.uk
Editor *Caroline Oakley*

Founded in 1986 by a group of women who wanted to create more opportunities for women in publishing. A co-operative operation which publishes novels, autobiography, biography, popular women's history, general non-fiction, short stories on commission. Must have a Welsh connection.' Contact us or see website for current calls for short story submissions.' 7 titles a year. Welcomes mss and ideas for books from women who are Welsh or have a significant Welsh connection only. Send as hard copy or by e-mail.
🅔 Royalties annually.

Horus Editions
▷ Award Publications Limited

How To Books Ltd
Spring Hill House, Spring Hill Road, Begbroke, Oxford OX5 1RX
☎ 01865 375794   📠 01865 379162
info@howtobooks.co.uk
www.howtobooks.co.uk
Managing Director *Giles Lewis*
Editorial Director *Nikki Read*

An independent publishing house, founded in 1991. Publishes non-fiction, self-help reference books. How To titles are practical, accessible books that enable their readers to achieve their goals in life and work. How To authors must have first-hand experience of the subject about which they are writing. Subjects covered include small business and self employment, study skills, cooking and gardening, management, jobs and careers, living and working abroad, creative writing, weddings and property investment. IMPRINT **Springhill**. 75 new titles and new editions per year. Submit outline followed by sample chapter.

Human Horizons
▷ Souvenir Press Ltd

Hurst Publishers, Ltd

41 Great Russell Street, London WC1B 3PL
🕾 020 7255 2201
hurst@atlas.co.uk
www.hurstpub.co.uk
Chairman/Managing Director/Editorial Head *Michael Dwyer*

Founded 1967. An independent company publishing contemporary history, politics, religion (not theology) and anthropology. About 25 titles a year. No unsolicited mss. Synopses and ideas welcome.
Ⓔ Royalties annually.

Hutchinson/Hutchinson Children's Books
▷ The Random House Group Ltd

Hymns Ancient & Modern Ltd

St Mary's Works, St Mary's Plain, Norwich NR3 3BH
🕾 01603 612914   🖷 01603 624483
admin@hymnsam.co.uk
www.scm-canterburypress.co.uk
Publishing Director, SCM Press and Canterbury Press *Christine Smith*
Publisher, RMEP *Valerie Bingham*
Publisher, SCM *Natalie Watson*
Approx. Annual Turnover £6 million

Publishing subsidiary **SCM-Canterbury Press Ltd** (see entry) controls **SCM Press**, **Canterbury Press** and **RMEP** IMPRINTS which cover hymn books, liturgical material, academic and general religious books, and multi-faith religious and social education resources for pupils and teachers. Media subsidiary **G.J. Palmer & Sons Ltd** includes *Church Times* (see entry under *Magazines*); *The Sign* and *Home Words* (monthly) and *Crucible* (quarterly). About 100 titles a year. Ideas for new titles welcome but no unsolicited mss.
Ⓔ Royalties annually.

Icon Books Ltd

Editorial, Production, Finance: Old Dairy, Brook Road, Thriplow, Cambridge SG8 7RG
🕾 01763 208008   🖷 01763 208080
info@iconbooks.co.uk
www.iconbooks.co.uk
Marketing, Sales, Press, Rights: Omnibus Business Centre, 39–41 North Road, London N7 9DP
🕾 020 7697 9695
Managing Director *Peter Pugh*
Publishing Director *Simon Flynn*
Editorial Director *Duncan Heath*

Founded 1992. SERIES *Introducing* Graphic introductions to key figures and ideas in the history of science, philosophy, psychology, religion and the arts. TITLE *Introducing Quantum Theory*. Publishes 'provocative and intelligent' non-fiction in science, politics and philosophy. TITLES *Atom*; *The Truth that Sticks*. IMPRINTS **Corinthian Books** Principally commercial sport-led non-fiction; **Wizard Books** Children's non-fiction and game books including SERIES *Fighting Fantasy* (adventure gamebooks). Submit synopsis only. OVERSEAS ASSOCIATE Totem Books, USA, distributed by National Book Network.
Ⓔ Royalties twice yearly.

Ilex Press
▷ The Ilex Press Limited under UK Packagers

Imprint Academic

PO Box 200, Exeter EX5 5YX
🕾 01392 851550   🖷 01392 851187
keith@imprint.co.uk
www.imprint-academic.com
Publisher *Keith Sutherland*

Founded 1980. Publishes books and journals in politics, philosophy and psychology for both academic and general readers. Book SERIES include *St. Andrews Studies in Philosophy and Public Affairs* and *Societas: Essays in political and cultural criticism*. 25–30 titles a year. Unsolicited mss, synopses and ideas welcome with return postage only.
Ⓔ Royalties annually.

The In Pinn
▷ Neil Wilson Publishing Ltd

Incomes Data Services
▷ Sweet & Maxwell Group

Independent Music Press

PO Box 69, Church Stretton SY6 6WZ
🕾 01694 720049   🖷 01694 720049
info@impbooks.com
www.impbooks.com
Managing Director *Martin Roach*
Marketing & New Media *David Hanley*

Founded 1992. Publishes music biography and youth culture. No jazz or classical. TITLES include biographies of My Chemical Romance, Green Day, Muse, The Killers, The Cure, Dave Grohl, The Streets, Oasis, Prodigy as well as subculture

classics such as *Scooter Boys* and Ian Hunter's *Diary of a Rock 'n' Roll Star*. 8 titles a year. Approach by e-mail, enclosing biography and synopsis only.
Ⓔ Royalties twice-yearly.

Independent Voices
▷ Souvenir Press Ltd

Infinite Ideas
36 St Giles, Oxford OX1 3LD
Ⓣ 01865 514888 Ⓕ 01865 514777
info@infideas.com
www.infideas.com
Joint Managing Directors *David Grant, Richard Burton*
Approx. Annual Turnover £1.5 million

Publishes self-help SERIES *Brilliant Ideas* covering health and relationships; careers, finance and personal development; sports, hobbies and games; leisure and lifestyle. No fiction, poetry, illustrated or children's books. IMPRINT **Firebrand** Business books. 'We work with companies and consultants, sharing the risk to bring their books to the global market place.' 40 titles a year. No unsolicited mss for *Brilliant Ideas* series; synopses and ideas welcome by e-mail or letter. Mss welcome for business titles.
Ⓔ Flat fees paid for Brilliant Ideas series. Royalties paid annually on Firebrand Business Books.

Inter-Varsity Press
IVP Book Centre, Norton Street, Nottingham NG7 3HR
Ⓣ 0115 978 1054 Ⓕ 0115 942 2694
ivp@ivpbooks.com
www.ivpbooks.com
Chairman *Ralph Evershed*
Chief Executive *Brian Wilson*

Founded mid-1930s as the publishing arm of Universities and Colleges Christian Fellowship, it has expanded to wider Christian markets worldwide. Publishes Christian belief and lifestyle, reference and bible commentaries. No secular material or anything which fails to empathize with orthodox Protestant Christianity. IMPRINTS **IVP**; **Apollos**; **Crossway** TITLES *The Bible Speaks Today* series and New Testament CD-ROM; *Pierced for Our Transgressions* Jeffery, Ovey and Sach; *The Living Church* John Stott. About 50 titles a year. Synopses and ideas welcome.
Ⓔ Royalties twice-yearly.

Isis Publishing
7 Centremead, Osney Mead, Oxford OX2 0ES
Ⓣ 01865 250333 Ⓕ 01865 790358
sales@isis-publishing.co.uk
www.isis-publishing.co.uk

Part of the Ulverscroft Group Ltd. Publishes large-print books – fiction and non-fiction – and unabridged audio books. Together with **Soundings** (see entry under *Audio Books*) produces around 4,000 titles on audio tape and CD. AUTHORS Lee Child, Martina Cole, Terry Pratchett, Susan Sallis. No unsolicited mss as Isis undertakes no original publishing.
Ⓔ Royalties twice-yearly.

Itchy Coo
▷ Black & White Publishing Ltd

Ithaca Press
▷ Garnet Publishing Ltd

IVP
▷ Inter-Varsity Press

J. Garnet Miller
▷ Cressrelles Publishing Co. Ltd

Jacqui Small
▷ Aurum Press Ltd

Jane's Information Group
Sentinel House, 163 Brighton Road, Coulsdon CR5 2YH
Ⓣ 020 8700 3700
info.uk@janes.com
www.janes.com
Managing Director *Scott Key*
Chief Content Officer *Ian Kay*

Founded in 1898 by Fred T. Jane with the publication of *All the World's Fighting Ships*. Acquired by IHS Inc. in 2007. Recent focus has been on growth opportunities in its core business and in enhancing the performance of initiatives like Jane's Information Group available online and on in-depth intelligence centres. Publishes Web journals, magazines and yearbooks on defence, aerospace, security and transport topics, with details of equipment and systems; plus directories. Full consulting arm provides bespoke offerings to clients. Also *Jane's Defence Weekly* (see entry under *Magazines*).

DIVISIONS **Magazines** TITLES *Jane's Defence Weekly*; *Jane's International Defence Review*; *Jane's Airport Review*; *Jane's Navy International*; *Jane's Islamic Affairs Analyst*; *Jane's Missiles and Rockets*. **Publishing for Defence, Aerospace**. TITLES *Defence, Aerospace Yearbooks*. **Transport** TITLE *Transportation Yearbooks*. **Security** *Sean Howe* TITLES *Jane's Intelligence Review*; *Foreign Report*; *Jane's Sentinel* (regional security assessment); *Jane's Police Review*. CD-ROM and electronic development and publication. Over 100 titles a year. Unsolicited mss, synopses and ideas for reference/yearbooks welcome.

Janus Publishing Company Ltd
105–107 Gloucester Place, London W1U 6BY
Ⓣ 020 7486 6633 Ⓕ 020 7486 6090

publisher@januspublishing.co.uk
www.januspublishing.co.uk
Managing Director *Jeannie Leung*

Publishes fiction, human interest, memoirs, philosophy, mind, body and spirit, religion and theology, social questions, popular science, history, spiritualism and the paranormal, poetry and young adults. IMPRINTS **Janus Books** Subsidy publishing; **Empiricus Books** Non-subsidy publishing. TITLES *The Naked Emperor*; *It: The Architecture of Existence*; *From Necessity to Infinity: Interpretation in Language and Translation*; *The Performer's Anthology*; *More Scouse, Vicar?*; *The Galaxy Boys and The Sphere*; *Supergeek*. 'Two of our authors won the European Literary Award in 2004.' About 450 titles in print. Unsolicited mss welcome. Agents in the USA, Europe and Asia.
Ⓔ Royalties twice-yearly.

✱*Authors' update* **Authors may be asked to cover their own productions costs but Janus has moved into conventional publishing with its Empiricus imprint.**

Jarrold
▷ The History Press

Michael Joseph
▷ Penguin Group (UK)

JR Books Ltd
10 Greenland Street, London NW1 0ND
☎ 020 7284 7163   ☒ 020 7485 4902
jeremyr@jrbooks.com
www.jrbooks.com
Managing Director *Jeremy Robson*
Senior Editor *Lesley Wilson*

Founded October 2006. JR Books publishes a wide range of non-fiction subjects, including biography, politics, music, humour, history, sport and self-help. 60 titles, spring 2009. Unsolicited synopses and ideas (with s.a.e.) welcome. Approach by mail. No fiction, children's or academic books.
Ⓔ Royalties twice-yearly.

Kahn & Averill
9 Harrington Road, London SW7 3ES
☎ 020 8743 3278   ☒ 020 8743 3278
kahn@averill23.freeserve.co.uk
Managing Director *Mr M. Kahn*

Founded 1967 to publish children's titles but now specializes in music titles. A small independent publishing house. No unsolicited mss; synopses and ideas for books considered.
Ⓔ Royalties twice-yearly.

Kenilworth Press Ltd (An imprint of Quiller Publishing Ltd)
Wykey House, Wykey, Shrewsbury SY4 1JA
☎ 01939 261616   ☒ 01939 260991
info@quillerbooks.com
www.kenilworthpress.co.uk
Managing Director *Andrew Johnston*

Founded in 1989 and acquired by Quiller Publishing Ltd in 2005. Publishes equestrian books only. 10 titles in 2008. No unsolicited mss; synopses and ideas for books welcome. Send by post or e-mail.
Ⓔ Royalties twice-yearly.

Kenyon-Deane
▷ Cressrelles Publishing Co. Ltd

Laurence King Publishing Ltd
4th Floor, 361–373 City Road, London EC1V 1LR
☎ 020 7841 6900   ☒ 020 7841 6910
enquiries@laurenceking.co.uk
www.laurenceking.co.uk
Chairman *Nick Perren*
Managing Director *Laurence King*
Commissioning Editor, Architecture *Philip Cooper*
Commissioning Editor, Design *Jo Lightfoot*
Approx. Annual Turnover £6 million

Publishes illustrated books in the fields of graphic design, fashion and textiles, architecture, interiors, product design and art. TITLES *A World History of Art*; *The Design Encyclopedia*; *How to Be a Graphic Designer Without Losing Your Soul*; *1000 New Designs and Where to Find Them*; *100 Years of Fashion Illustration* and the *Portfolio* series for students of art and design. About 60 titles a year. Unsolicited material welcome; send synopsis by post or e-mail (commissioning@ laurenceking.co.uk).

Kingfisher
▷ Macmillan Publishers Ltd

Kinglake Publishing
Ashgrove House, Suite G8, Elland, Calderdale HX5 9JB
☎ 0845 508 6032
info@kinglakepublishing.co.uk
www.kinglakepublishing.co.uk
Editorial Director *Harry Taylor*
Approx. Annual Turnover £120,000

Founded 2005. Publishes fiction, non-fiction, life histories, education, history and politics. TITLES *Mussolini and Fascism* Ernest Rohane; *Be a Successful Writer* ed. Harry Taylor; *Poison* Jessica Scott; *Jail Bait* Steve Katz. 32 titles in 2008. No unsolicited mss; synopses and ideas for books welcome. No books for young children,

highly illustrated books or academic maths and physics.
Ⓔ Royalties annually.

Jessica Kingsley Publishers Ltd

116 Pentonville Road, London N1 9JB
Ⓣ 020 7833 2307   Ⓕ 020 7837 2917
post@jkp.com
www.jkp.com
www.singing-dragon.com
Also at: 400 Market Street, Suite 400,
Philadelphia, PA 19106, USA
Managing Director and Publisher *Jessica Kingsley*
Senior Acquisitions Editor *Stephen Jones*

Founded 1987. Independent, international
publisher of books for professionals, academics
and the general reader on autism, disability,
special education, arts therapies, child
psychology, mental health, practical theology
and social work. IMPRINT **Singing Dragon**
Chinese martial arts, Qigong, Traditional
Chinese Medicine and complementary medicine.
150 titles a year. 'We are actively publishing
and commissioning in all these areas. We
welcome suggestions for books and proposals
from prospective authors. Proposals should
consist of an outline of the book, a contents list,
assessment of the market and author's c.v., and
should be addressed to *The Editorial Director*.
Complete manuscript should not be sent, and all
submissions should be addressed to the London
office. E-mail submissions are encouraged. A
proposals form can be downloaded from the JKP
website.' OVERSEAS SUBSIDIARY in Philadelphia,
USA.
Ⓔ Royalties twice-yearly.

Kluwer Law International

250 Waterloo Road, London SE1 8RD
Ⓣ 020 7981 0656   Ⓕ 020 7981 0587
simon.bellamy@kluwerlaw.com
www.kluwerlaw.com
Also at: Zuidpoolsingel 2a, 2408 ZE Alphen aan
den Rijn, The Netherlands
Publisher *Simon Bellamy*

Founded 1995. Parent company: Wolters Kluwer
Group. Publishes international law. About
200 titles a year, including online, CD-ROM,
loose-leaf journals and monographs. Unsolicited
synopses and ideas for books on law at an
international level welcome.
Ⓔ Royalties annually.

Kogan Page Ltd

120 Pentonville Road, London N1 9JN
Ⓣ 020 7278 0433   Ⓕ 020 7837 3768/6348
pkogan@koganpage.com
www.koganpage.co.uk
Chairman *Philip Kogan*
Managing Director *Helen Kogan*

Approx. Annual Turnover £5.4 million

Founded 1967 by Philip Kogan to publish
The Industrial Training Yearbook. Publishes
business and management reference books
and monographs, careers, marketing, personal
finance, personnel, small business, training and
industrial relations, transport. Further expansion
is planned, particularly in the finance and online
development areas, yearbooks and directories.
Has initiated a number of electronic publishing
projects and provision of EP content. About 130
titles a year.
Ⓔ Royalties twice-yearly.

✴*Authors' update* **For a medium sized publisher
it is amazing how much of the business book
market is covered by Kogan Page. Authors have
benefited from strong overseas sales.**

Ladybird
▷ Penguin Group (UK)

Landmark Publishing Ltd

The Oakes, Moor Farm Road West, Ashbourne
DE6 1HD
Ⓣ 01335 347349
landmark@clara.net
Lindsey@landmarkpublishing.co.uk
www.landmarkpublishing.co.uk
Chairman *R. Cork*
Managing Director *C.L.M. Porter*
Approx. Annual Turnover £450,000

Founded in 1996. Publishes itinerary-based travel
guides, regional, industrial countryside and local
history. About 120 titles in 2007. No unsolicited
mss; telephone in the first instance.
Ⓔ Royalties annually.

Lawrence & Wishart Ltd

99A Wallis Road, London E9 5LN
Ⓣ 020 8533 2506   Ⓕ 020 8533 7369
lw@lwbooks.co.uk
www.lwbooks.co.uk
Managing Director/Editor *Sally Davison*

Founded 1936. An independent publisher with
a substantial backlist. Publishes current affairs,
cultural politics, economics, history, politics and
education. TITLES *After Blair*; *Labour Legends
Russian Gold*; *After Iraq*; *Making Sense of New
Labour*. 10 titles a year.
Ⓔ Royalties annually, unless by arrangement.

The Learning Institute

No. 1 Overbrook Business Centre, Blackford,
Wedmore BS28 4PA
Ⓣ 01934 713563   Ⓕ 01934 713492
courses@inst.org
www.inst.org
Managing Director *Kit Sadgrove*

Founded 1994 to publish home-study courses in vocational subjects such as garden design, writing and computing. Publishes subjects that show the reader how to work from home, gain a new skill or enter a new career. Interests include home working and self employment, especially in twenty-first century jobs. TITLES *Diploma in Interior Design*; *Become a Garden Designer*. Author's guidelines sent on receipt of s.a.e. No unsolicited mss; send synopses and ideas only. Ⓔ Royalties quarterly.

Leckie & Leckie

4 Queen Street, Edinburgh EH2 1JE
℡ 0131 220 6831　℻ 0131 225 9987
enquiries@leckieandleckie.co.uk
www.leckieandleckie.co.uk
Publishing Director *Martin Redfern*

Founded 1989. A division of Huveaux Plc. Publishes Scottish educational materials. TITLES *Success Guides*; *Magical Series*; *Course Note Series*. 120 titles in 2008. No unsolicited mss; synopses and ideas for books welcome by post. Ⓔ Royalties annually.

Legend Press Limited

3rd Floor, Unicorn House, 221–222 Shoreditch High Street, London E1 6PJ
℡ 020 7426 0340
info@legend-paperbooks.co.uk
submissions@legend-paperbooks.co.uk
www.legendpress.co.uk
Managing Director *Tom Chalmers*

Publishes contemporary, diverse and cutting-edge fiction. Is 'committed to providing writers with an advantageous platform from which to develop and succeed. We focus strongly on reaching new audiences and on working with other publishers to drive the independent sector and promote new and different work within the mainstream market.' Acquired **Paperbooks Ltd** in April 2008 (see entry). 10 titles in 2008; doubling list year-on-year. Send three chapters and synopsis to the submissions e-mail address above. Ⓔ Royalties twice-yearly.

✱*Authors' update* Credited with the youngest CEO in the business (Tom Chalmers is still under thirty), Legend is driven by enthusiasm for 'fresh new fiction from authors who are fun and inspiring to work with'. Though hit by a cutback by its major investor, Legend now has Arts Council backing for a scheme to help independent publishers sell their titles through independent bookshops.

Dewi Lewis Publishing

8 Broomfield Road, Heaton Moor, Stockport SK4 4ND
℡ 0161 442 9450　℻ 0161 442 9450
mail@dewilewispublishing.com
www.dewilewispublishing.com
Owners *Dewi Lewis, Caroline Warhurst*
Approx. Annual Turnover £280,000

Founded 1994. Publishes photography and visual arts. TITLES *Industry of Souls* Martin Booth (Booker Prize shortlist, 1998); *Wolfy and the Strudelbakers* Zvi Jagendorf (Booker Prize longlist, 2001; Sagittarius Prize, 2002); *Common Sense* Martin Parr; *New York 1954–5* William Klein. IMPRINT **Dewi Lewis Media** Publishes non-fiction, biography, sports and celebrity books. TITLES *David Beckham: Made in Manchester* Eamonn and James Clarke; *500 Flowers* Roger Camp. 17 titles in 2008. Not currently accepting new fiction submissions. For all submissions it is essential to check the website first.
Ⓔ Royalties annually.

Lewis/Lewis Masonic
▷ Ian Allan Publishing Ltd

Library Association Publishing
▷ Facet Publishing

Library of Wales
▷ Parthian

Life&Times
▷ Haus Publishing

Frances Lincoln Ltd

4 Torriano Mews, Torriano Avenue, London NW5 2RZ
℡ 020 7284 4009　℻ 020 7485 0490
firstnameplussurnameinitial@frances-lincoln.com
www.franceslincoln.com
Managing Director *John Nicoll*
Approx. Annual Turnover £7 million

Founded 1977. Publishes highly illustrated non-fiction: gardening, art and interiors, architecture, parenting, walking and climbing, children's picture and information books; also stationery. DIVISIONS **Adult Non-fiction** *Jo Christian* TITLES *Chatsworth* Duchess of Devonshire; *Grow Your Own Vegetables* Joy Larkcom; *Pictorial Guides to the Lakeland Fells* A. Wainwright; **Children's General Fiction and Non-fiction** *Janetta Otter-Barry* TITLE *The Wanderings of Odysseus* Rosemary Sutcliffe, illus. Alan Lee; **Stationery** *Anna Sanderson* TITLES *RHS Diary and Address Book*; *British Library Diary*. About 200 titles a year. Synopses and ideas for books considered. Ⓔ Royalties twice-yearly.

Linden Press
▷ Open Gate Press

Linford Mystery/Linford Romance/Linford Western
> ▷ **F.A. Thorpe Publishing**

Lion Hudson plc

Wilkinson House, Jordan Hill Road, Oxford
OX2 8DR
☎ 01865 302750   🖷 01865 302757
info@lionhudson.com
www.lionhudson.com
Managing Director *Paul Clifford*
Approx. Annual Turnover £8.5 million

Founded 1971. A Christian book publisher, strong
on illustrated books for a popular international
readership, with rights sold in over 150 languages
worldwide. Publishes a diverse list with
Christian viewpoint the common denominator.
All ages, from board books for children to
multi-contributor adult reference, educational,
paperbacks and colour co-editions and gift
books. IMPRINTS **Lion** and **Lion Children's** *Paula
O'Gorman*; **Candle** *Carol Jones*; **Monarch** *Paul
Sweeney* (see **Monarch Books**). About 175 titles
a year. Unsolicited mss accepted provided they
have a positive Christian viewpoint intended for
a Christian or wide general and international
readership.
🖲 Royalties twice-yearly.

Little Black Dress
> ▷ **Hachette UK**

Little Books Ltd

73 Campden Hill Towers, 112 Notting Hill Gate,
London W11 3QW
☎ 020 7792 7929
www.littlebooks.net
Managing Director *Max Hamilton-Little*

Founded 2001. Publishes general trade books,
history, biography, health, natural history.
IMPRINTS **Max Press** (see entry). About 20–25
books a year. No unsolicited material.
🖲 Royalties annually.

Little, Brown Book Group UK

100 Victoria Embankment, London EC4Y 0DY
☎ 020 7911 8000   🖷 020 7911 8100
uk@littlebrown.co.uk
www.littlebrown.co.uk
Owner *Hachette Book Group*
CEO/Publisher *Ursula Mackenzie*
Approx. Annual Turnover £60 million

Founded 1988 as Little, Brown & Co. (UK) and
became Time Warner Book Group UK in 2002.
Purchased by Hachette in March 2006 and
renamed Little, Brown Book Group UK. Began by
importing its US parent company's titles and in
1990 launched its own illustrated non-fiction list.
Two years later the company took over former

Macdonald & Co. In 2007 the company bought
Piatkus Books. Publishes hardback and paperback
fiction, literary fiction, crime, science fiction
and fantasy; and general non-fiction including
true crime, biography and autobiography,
cinema, history, humour, popular science, travel,
reference, sport. IMPRINTS **Little, Brown** *Ursula
Mackenzie, Richard Beswick, Tim Whiting*
Hardback fiction and general non-fiction; **Abacus**
Richard Beswick, Tim Whiting, Jenny Parrott
Literary fiction and non-fiction paperbacks;
Atom *Darren Nash* Young adult/teen paperbacks;
Hachette Digital *Sarah Shrubb* (see entry under
Audio Books); **Orbit** *Darren Nash* Science fiction
and fantasy; **Sphere** *Antonia Hodgson, David
Shelley, Joanne Dickinson, Adam Strange, Nikola
Scott, Rebecca Saunders* Mass-market fiction and
non-fiction hardbacks and paperbacks; **Virago
Press** *Lennie Goodings, Ursula Doyle* (see entry);
Piatkus Books *Gill Bailey, Helen Stanton, Emma
Beswetherick, Donna Condon* (see entry). 300
titles in 2008. No unsolicited mss. Approach in
writing in the first instance.
🖲 Royalties twice-yearly.

✸*Authors' update* **Though now part of the
Hachette group, Little Brown continues to be
run as an independent entity publishing a wide
range of commercial titles, mostly popular by
virtue of quality.**

Little Tiger Press

An imprint of Magi Publications, 1 The Coda
Centre, 189 Munster Road, London SW6 6AW
☎ 020 7385 6333   🖷 020 7385 7333
info@littletiger.co.uk
www.littletigerpress.com
Publisher *Monty Bhatia*
Associate Publisher *Jude Evans*
Editor *Jo Collins*
Submissions Editor *Laura Roberts*

Little Tiger Press imprint publishes children's
picture and novelty books for ages 0–7. No
texts over 1,000 words. About 30 titles a year.
Unsolicited mss, synopses and new ideas
welcome. See website for submission guidelines.
🖲 Royalties annually.

✸*Authors' update* **A thriving small publisher
with imaginative output.**

Liverpool University Press

4 Cambridge Street, Liverpool L69 7ZU
☎ 0151 794 2233   🖷 0151 794 2235
a.cond@liv.ac.uk
www.liverpool-unipress.co.uk
Managing Director/Editorial Head *Anthony Cond*

LUP's primary activity is the publication of
academic and scholarly books and journals but
it also has a limited number of trade titles. Its
principal focus is on the arts and social sciences,

in which it is active in a variety of disciplines. TITLES *Liverpool 800: Culture, Character and History; Irish, Catholic and Scouse: The History of the Liverpool Irish 1800–1939; Writing Liverpool; Liverpool and Transatlantic Slavery; So Spirited a Town; Public Sculptures of South London; sk-interfaces; The Culture of Capital; Cityscape: Ben Johnson's Liverpool; William Roscoe: Commerce and Culture; A Gallery to Play to: The Story of the Mersey Poets; The Cultural Values of Europe.* 30–40 titles a year.
Ⓔ Royalties annually.

Livewire Books for Teenagers
▷ The Women's Press

Lonely Planet Publications Ltd
2nd Floor, 186 City Road, London EC1V 2NT
Ⓣ 020 7106 2100   Ⓕ 020 7106 2101
go@lonelyplanet.co.uk
www.lonelyplanet.com
Owner *Lonely Planet (Australia)*
Editorial Heads *Kat Browning, Imogen Hall*

'For over 30 years, Lonely Planet's on-the-ground research and no-holds-barred opinion by our team of expert travel writers has inspired and guided independent travellers to explore the world around them.' With over 500 titles in print, publications include travel guidebooks, downloadable digital guides, phrasebooks, travel literature, pictorial books and How To guides. No unsolicited mss; synopses and ideas welcome. Author guidelines available at www.lonelyplanet.com/help/guide

Lorenz Books
▷ Anness Publishing Ltd

Luath Press Ltd
543/2 Castlehill, The Royal Mile, Edinburgh EH1 2ND
Ⓣ 0131 225 4326   Ⓕ 0131 225 4324
gavin.macdougall@luath.co.uk
www.luath.co.uk
Director *G.H. MacDougall*

Founded 1981. Publishes mainly books with a Scottish connection. Current list includes fiction, poetry, guidebooks, walking and outdoor, history, folklore, politics and global issues, cartoons, biography, food and drink, environment, music and dance, sport. TITLE *The Bower Bird* Ann Kelley (2007 Costa Children's Book Award). SERIES *On the Trail Of; The Quest For.* About 30–40 titles a year. Over 200 titles in print. Unsolicited mss, synopses and ideas welcome; 'committed to publishing well-written books worth reading'.
Ⓔ Royalties paid.

Lucky Duck Publishing
▷ Sage Publications

Lulu.com
www.lulu.com

Founded 2002. A self publishing, print-on-demand service for individuals and companies. Produces a variety of digital content including books (hardback and paperback), dissertations, movies, brochures, yearbooks, product manuals, portfolios, music, software, calendars, posters and artwork with no set-up fee and no requirement to buy copies. Authors supply a digital file of their book and a print-ready version of the file is created by the company. Lulu handles sales and pays the royalty specified by the client with a small mark-up as commission. Offers two retail distribution services for a fee. Full details available on the website.

✱*Authors' update* Authors who find it hard to get into print now have recourse to an option that is not vanity publishing, at least not in the strict sense of the word. Lulu is a website designed to carry anything that writers can throw at it, and for free. If browsers show an interest in a particular title, the book can be produced to order and shipped direct. Lulu takes a twenty per cent share of any royalties. Publishing services such as editing, proofing and marketing must be contracted separately. Anyone joining the game should not expect too much. John Morrison, a Lulu author, warns 'this is not a solution for those who want to sell their books in commercial quantities into Waterstone's'.

Lund Humphries
▷ Ashgate Publishing Ltd

The Lutterworth Press
PO Box 60, Cambridge CB1 2NT
Ⓣ 01223 350865   Ⓕ 01223 366951
publishing@jamesclarke.co.uk
www.lutterworth.com
Managing Director *Adrian Brink*

The Lutterworth Press was founded as the Religious Tract Society in Georgian London, with its headquarters just off Fleet Street, in order to provide improving literature for young people and adults. Since then it has published many tens of thousands of titles, ranging from small pamphlets for children to erudite academic works. It became well known to generations of British children thanks to the publication of the *Boy's Own Paper* and the *Girl's Own Paper.* Since 1984 it has been an imprint of **James Clarke & Co.** (see entry). The Lutterworth current list concentrates in several areas, and some of the work, although aimed at a general readership, is at quite a high level. TITLES include: *Hanging a Rebel. The Life of C.R.W. Nevison* Michael J.K. Walsh; *Kilvert's Diary and Landscape* John Toman; *The New*

Empire of Malplaquet Andrew Dalton. 25 titles in 2008. Approach in writing with ideas in the first instance.
£ Royalties annually.

Macdonald & Co.
▷ Little, Brown Book Group UK

McGraw-Hill Education
McGraw-Hill House, Shoppenhangers Road, Maidenhead SL6 2QL
☎ 01628 502500   📠 01628 770224
www.mcgraw-hill.co.uk
www.openup.co.uk
General Manager *Shona Mullen*

Owned by US parent company, founded in 1888. Began publishing in Maidenhead in 1965, having had an office in the UK since 1899. Publishes business, economics, finance, accounting, computing science, social sciences and engineering for the academic, student and professional markets. Acquired **Open University Press** in 2002 (see entry). Around 200 titles a year. See website for author guidelines. Unsolicited proposals via website only.

MacLehose Press
▷ Quercus Publishing

Macmillan New Writing
▷ Macmillan Publishers Ltd

Macmillan Publishers Ltd
The Macmillan Building, 4 Crinan Street, London N1 9XW
☎ 020 7833 4000   📠 020 7843 4640
www.macmillan.com
Owner *Verlagsgruppe Georg von Holtzbrinck*
Chief Executive *Annette Thomas*
Approx. Annual Turnover £350 million (Book Publishing Group)

Founded 1843. Macmillan is one of the largest publishing houses in Britain, publishing approximately 1,400 titles a year. In 1995, Verlagsgruppe Georg von Holtzbrinck, a major German publisher, acquired a majority stake in the Macmillan Group and in 1999 purchased the remaining shares. In 1996, Macmillan bought Boxtree, the successful media tie-in publisher; in 1997, it purchased the Heinemann English language teaching list from Reed Elsevier, and it acquired Kingfisher, the popular children's book publisher, in 2007. The educational publishing division was strengthened by the acquisitions of the Mexican list, Ediciones Castillo and the Argentine company Puerto de Palos. No unsolicited material, except for the **Macmillan New Writing** list (see details below).

DIVISIONS

Palgrave Macmillan Brunel Road, Houndmills, Basingstoke, Hampshire RG21 6XS ☎ 01256 329242 📠 01256 328339 Managing Director *Dominic Knight*; **College** *Margaret Hewison*; **Scholarly & Reference** *Sam Burridge*; **Journals** *David Bull* Publishes textbooks, monographs and journals in academic and professional subjects. Publications in both print and electronic format.

Macmillan Education 4 Between Towns Road, Oxford OX4 3PP ☎ 01865 405700 📠 01865 405701 info@macmillan.com www.macmillaneducation.com Chief Executive Officer *Julian Drinkall*, Publishing Directors *Sue Bale* (dictionaries), *Alison Hubert* (Africa, Caribbean, Middle East and East Asia), *Julie Kniveton* (Latin America), *Angela Lilley* (ELT), *Kate Melliss* (Europe) Publishes a wide range of ELT titles and educational materials for the international education market from Oxford and through 30 subsidiaries worldwide.

Pan Macmillan 20 New Wharf Road, London N1 9RR ☎ 020 7014 6000 📠 020 7014 6001 www.panmacmillan.com Managing Director *Anthony Forbes Watson* Publishes under **Macmillan, Pan, Picador, Macmillan New Writing, Sidgwick & Jackson, Boxtree, Macmillan Children's Books, Kingfisher, Macmillan Digital Audio, Campbell Books, Young Picador, Rodale; Tor.** No unsolicited mss. Visit the writers' area on www.panmacmillan.com for useful articles and information on books that may be helpful in getting your book published, or visit the Macmillan New Writing website (see below) for details of the MNW programme, which does accept unsolicited mss of first novels.

IMPRINTS

Macmillan (founded 1843) Fiction: Publishing Director *Maria Rejt*, Editorial Director *Imogen Taylor* Publishes hardback commercial fiction including genre fiction, romantic, crime and thrillers. IMPRINT **Tor** (founded 2003) Senior Commissioning Editor *Julie Crisp* Publishes science fiction, fantasy and thrillers. Non-Fiction: Publisher *Jon Butler*, Editorial Director *Georgina Morley* Publishes serious and general non-fiction: autobiography, biography, economics, history, philosophy, politics and world affairs, psychology, popular science, trade reference titles.

Pan (founded 1947) Paperback imprint for Pan Macmillan.

Picador (founded 1972) Publisher *Paul Baggaley* Publishes literary international fiction, non-fiction and poetry.

Rodale Editorial Director *Liz Gough* Publishes highly acclaimed books on health, fitness, diet, cookery, spirituality.

Sidgwick & Jackson (founded 1908) Editorial Director *Ingrid Connell* Publishes popular non-fiction in hardback and trade paperback with strong personality or marketable identity, from celebrity and show business to music and sport. Also military history list.

Boxtree (founded 1986) Editorial Director *To be appointed* Publishes brand and media tie-in titles, including TV, film, music and Internet, plus entertainment licences, pop culture, humour and event-related books. TITLES *Dilbert*; *James Bond*; *Purple Ronnie*; *Wallace & Gromit*; *The Onion*.

Macmillan Children's Books (New Wharf Road address) Managing Director *Emma Hopkin*, Publishing Director *Rebecca McNally*; Fiction: Editorial Director *Sarah Dudman*; Non-Fiction and Poetry: Editorial Director *Gaby Morgan*; Picture Books and Gift Books: Editorial Director *Suzanne Carnell*. **Kingfisher** Editorial Director *Martina Challis*; **Campbell Books** Editorial Director *Sarah Fabiny*; **Young Picador** Publishes fiction, non-fiction and poetry in paperback and hardback.

Macmillan New Writing (at New Wharf Road address) Publishing Director *Maria Rejt*, Commissioning Editor *Will Atkins* Founded in 2006 as a way of finding talented new writers who might otherwise go undiscovered. MNW publishes full-length novels from authors who have not previously published a novel. All genres considered and all submissions assessed, but mss must be complete. No advance, but the author pays nothing and receives a royalty of 20% on net sales. Submissions must be sent by e-mail only, via the MNW website (www. macmillannewwriting.com).
Royalties annually or twice-yearly depending on contract.

✳*Authors' update* Given Annette Thomas's track record as a digital innovator, before taking over as CEO last year she had made Nature. com one of the biggest science platforms, an acceleration of online activity was expected. Among major initiatives *The New Palgrave Dictionary of Economics* has a double life, print and online, while Pan Macmillan dominates Waterstone's e-book charts. Our own *Writer's Handbook* website (www.thewritershandbook. com) is now up and running. Macmillan New Writing is attracting rising stars. The deal for first-time authors is a fixed contract with no advance but with a twenty per cent royalty. A new title appears each month. Crime and children's lists are thriving.

Made Simple Books
▷ Elsevier Ltd

Mainstream Publishing Co. (Edinburgh) Ltd
7 Albany Street, Edinburgh EH1 3UG
☎ 0131 557 2959   📠 0131 556 8720
bill.campbell@mainstreampublishing.com
www.mainstreampublishing.com
Directors *Bill Campbell, Peter MacKenzie*
Approx. Annual Turnover £3.7 million

Publishes art, autobiography/biography, current affairs, health, sport, history, illustrated and fine editions, photography, politics and world affairs, popular paperbacks. TITLES *Soldier Five* Mike Coburn; *The Real Nureyev* Carolyn Soutar; *Woodward's England* Mick Collins. Over 80 titles a year. Ideas for books considered, but they should be preceded by a letter, synopsis and s.a.e. or return postage.
Royalties twice-yearly.

Management Books 2000 Ltd
Forge House, Limes Road, Kemble, Cirencester GL7 6AD
☎ 01285 771441   📠 01285 771055
info@mb2000.com
www.mb2000.com
Publisher *Nicholas Dale-Harris*
Approx. Annual Turnover £500,000

Founded 1993 to develop a range of books for executives and managers working in the modern world of business. 'Essentially, the books are working books for working managers, practical and effective.' Publishes business, management, self-development and allied topics as well as sponsored titles. Launched the *In Ninety Minutes* series of compact guide books for managers in 2004, offering advice, ideas and practical help across a range of highly relevant business topics in an hour and a half of study. New ideas for this series are welcome. About 24 titles a year. Unsolicited mss, synopses and ideas for books welcome.

Manchester University Press
Oxford Road, Manchester M13 9NR
☎ 0161 275 2310   📠 0161 274 3346
mup@manchester.ac.uk
www.manchesteruniversitypress.co.uk
Publisher/Chief Executive *David Rodgers*
Head of Editorial *Matthew Frost*
Approx. Annual Turnover £2 million

Founded 1904. MUP is Britain's third largest university press, with a list marketed and sold worldwide. Remit consists mainly of A-level and undergraduate textbooks and research monographs. Publishes in the areas of literature, TV, film, theatre and media, history and history of art, design, politics, economics, sociology and international law. DIVISIONS **Humanities** *Matthew Frost*; **History/Art History** *Emma*

Brennan; **Politics and Law** *Tony Mason*. About 140 titles a year. Unsolicited mss welcome. Ⓔ Royalties annually.

George Mann Books
▷ Arnefold Publishing

Manson Publishing Ltd
73 Corringham Road, London NW11 7DL
☏ 020 8905 5150   🖷 020 8201 9233
manson@mansonpublishing.com
www.mansonpublishing.com
Chairman/Managing Director *Michael Manson*

Founded 1992. Publishes highly illustrated books for study and reference. Subject areas covered include medicine, veterinary medicine, earth science, plant science, agriculture and microbiology. About 10 titles a year. No unsolicited mss; synopses and ideas will be considered.
Ⓔ Royalties twice-yearly.

Marshall Cavendish Ltd
5th Floor, 32–38 Saffron Hill, London EC1N 8FH
☏ 020 7421 8120
mcelt@marshallcavendish.co.uk
www.marshallcavendish.co.uk
www.mcelt.com
Managing Director *Chris Jenner*

Founded 1956. Owned by Times Publishing, Singapore, Marshall Cavendish is a major publisher of books. Publishes non-fiction, business and educational books. DIVISIONS **Trade** *Chris Jenner* TITLE *Football Handbook*; **Education** *Catherine Whitaker* TITLE *Just Right*. About 100 titles a year. Unsolicited mss, synopses and ideas for books welcome; initial approach by e-mail.
Ⓔ Royalties annually.

Marshall Editions
▷ Quarto Publishing plc

Martin Books
▷ Simon & Schuster UK Ltd

Kenneth Mason Publications Ltd
The Book Barn, Westbourne, Emsworth PO10 8RS
☏ 01243 377977   🖷 01243 379136
info@kennethmason.co.uk
www.kennethmason.co.uk
Chairman *Kenneth Mason*
Managing Director *Piers Mason*

Founded 1958. Publishes diet, health, fitness, nutrition and nautical. No fiction. IMPRINT **Research Disclosure**. Initial approach by letter with synopsis only.
Ⓔ Royalties twice-yearly in first year, annually thereafter.

Matador
▷ Troubador Publishing Ltd

Max Press Ltd
73 Campden Hill Towers, 112 Notting Hill Gate, London W11 3QW
☏ 020 7792 7929
Managing Director *Max Hamilton-Little*

Founded 2006. Publishes fiction. About 20 titles a year. No unsolicited material.
Ⓔ Royalties annually.

Kevin Mayhew Publishers
Buxhall, Stowmarket IP14 3BW
☏ 01449 737978   🖷 01449 737834
info@kevinmayhewltd.com
sue@kevinmayhewltd.com
www.kevinmayhew.com
Chairman/Commissioning Editor *Kevin Mayhew*

Established in 1976. One of the leading music and Christian book publishers in the UK. Publishes religious titles – liturgy, sacramental, devotional and school resources and both sacred and secular sheet music for all major instruments. Does not publish children's stories, novels or autobiographies. 250 titles a year. See website for submission guidelines.
Ⓔ Royalties annually.

Meadowside Children's Books
185 Fleet Street, London EC4A 2HS
☏ 020 7400 1092   🖷 020 7400 1037
info@meadowsidebooks.com
www.meadowsidebooks.com
Owner *D.C. Thomson*
Publisher *Simon Rosenheim*
Editor *Lucy Cuthew*
Art Director *Sarah Wilson*
Rights Manager *Katherine Judge*

Founded 2003. Acquired **Gullane Children's Books** in 2007 (see entry). Publishes timeless picture books, junior fiction and young adult fiction. Unsolicited mss accepted. Picture book mss: 1,000 words max. Fiction mss: short synopsis and first three chapters; address to *The Submissions Editor*. Only successful submissions answered.
Ⓔ Royalties annually.

Medavia Publishing
▷ Boltneck Publications Limited

Melrose Books
St Thomas Place, Ely CB7 4GG
☏ 01353 646608   🖷 01353 646602
info@melrosebooks.co.uk
www.melrosebooks.com
Chairman *Richard A. Kay*
Managing Director *Nicholas S. Law*
Approx. Annual Turnover £100,000

Independent subsidy publisher established by **Melrose Press** in 2004. To-date has published

general and children's fiction, literature, fantasy, biography, travel, religion and science reference. Welcomes submissions by e-mail or post
Ⓔ Royalties 55%.

Melrose Press Ltd

St Thomas Place, Ely CB7 4GG
Ⓣ 01353 646600   Ⓕ 01353 646601
tradesales@melrosepress.co.uk
www.melrosepress.co.uk
Chairman *Richard A. Kay*
Managing Director *Nicholas S. Law*
Approx. Annual Turnover £2 million

Founded 1960. Took on its present name in 1969. Publishes biographical who's who reference only (not including *Who's Who*, which is published by **A.&C. Black**). About 10 titles a year.

Mentor
▷ Christian Focus Publications

Mercat Press
▷ Birlinn Ltd

The Merlin Press Ltd

6 Crane Street Chambers, Crane Street, Pontypool NP4 6ND
Ⓣ 01495 764100
adrian.howe@merlinpress.co.uk
www.merlinpress.co.uk
Managing Director *Anthony W. Zurbrugg*
Director *Adrian Howe*

Founded 1956. Publishes in the area of history, philosophy and politics. No fiction. IMPRINTS **Merlin Press**; **Green Print**. TITLES *Socialist Register* (annual); the Chartist Studies SERIES; *Positive Education*. About 10 titles a year.
Ⓔ Royalties annually.

Merrell Publishers Ltd

81 Southwark Street, London SE1 0HX
Ⓣ 020 7928 8880   Ⓕ 020 7928 1199
mail@merrellpublishers.com
www.merrellpublishers.com
Managing Director *Hugh Merrell*
Editorial Director *Julian Honer*
Approx. Annual Turnover £2.8 million

Founded 1993. Publishes art, architecture, design, gardening, interiors, photography, cars and motorcycles. 45 titles in 2008. Unsolicited synopses and ideas for books welcome. Send c.v. and synopsis giving details of the book's target markets and funding of illustrations.
Ⓔ Royalties annually.

Methuen Drama
▷ A.&C. Black Publishers Ltd

Methuen Publishing Ltd

8 Artillery Row, London SW1P 1RZ
Ⓣ 020 7802 0018   Ⓕ 020 7828 1244

sales@methuen.co.uk
www.methuen.co.uk
Managing Director *Peter Tummons*

Founded 1889. Methuen was owned by Reed International until it was bought by Random House in 1997. Purchased by a management buy-out team in 1998. Acquired **Politico's Publishing** in 2003 (see entry). Publishes fiction and non-fiction; travel, sport, humour. No unsolicited mss; synopses and ideas welcome. Prefers to be approached via agents or a letter of inquiry. No first novels, cookery books or personal memoirs.
Ⓔ Royalties twice-yearly.

Metro Publishing
▷ John Blake Publishing Ltd

Michelin Maps & Guides

Hannay House, 39 Clarendon Road, Watford WD17 1JA
Ⓣ 01923 205240/205250   Ⓕ 01923 205241
www.Viamichelin.co.uk
Commercial Director *Ian Murray*

Founded 1900 as a travel publisher. Publishes travel guides, maps and atlases.

Midland Publishing

An imprint of Ian Allan Publishing Ltd, 4 Watling Drive, Hinckley LE10 3EY
Ⓣ 01455 25549   Ⓕ 01455 255495
midlandbooks@compuserve.com
www.midlandcountiessuperstore.co
Publisher *Nick Grant*

Imprint of **Ian Allan Publishing Ltd**. Publishes aviation books. No wartime memoirs. No unsolicited mss; synopses and ideas welcome.
Ⓔ Royalties twice-yearly.

Miller's
▷ Octopus Publishing Group

Millivres Prowler Limited

Unit M, Spectrum House, 32–34 Gordon House Road, London NW5 1LP
Ⓣ 020 7424 7400   Ⓕ 020 7424 7401
info@millivres.co.uk
www.millivres.co.uk

Publishes various magazines including *Gay Times*, *Diva Magazine*, *AXM* monthly magazine and *The Pink Paper*, the national free gay newspaper.

✱*Authors' update* No longer interested in books; the focus is entirely on journals.

Mills & Boon
▷ Harlequin Mills & Boon Ltd

Milo Books Limited

The Old Weighbridge, Station Road, Wrea Green, Preston PR4 2PH
Ⓣ 01772 672900   Ⓕ 01772 687727

info@milobooks.com
www.milobooks.com
Managing Director *Peter Walsh*

Founded 1997. Publishes non-fiction: true crime
and sport. 12 titles a year. Unsolicited mss,
synopses and ideas for books. 'Authors should
note our specialist areas. We cannot promise
to return all material submitted.' Approach in
writing in the first instance. 'If we are interested
in pursuing an idea, we will call to talk it through
in more depth.' No fiction or poetry.
Ⓔ Royalties twice-yearly.

Mindfield
▷ Camden Press Ltd

Mitchell Beazley
▷ Octopus Publishing Group

Mobius
▷ Hachette UK

Monarch Books

Lion Hudson plc, Wilkinson House, Jordan Hill
Road, Oxford OX2 8DR
☏ 01865 302750   ⓕ 01865 302757
monarch@lionhudson.com
Editorial Director *Tony Collins*

An imprint of **Lion Hudson plc**. Publishes an
independent list of Christian books across a wide
range of concerns. IMPRINT **Monarch** Upmarket
paperback list with Christian basis and strong
social concern agenda including psychology,
future studies, politics, mission, theology,
leadership, fiction and spirituality. About 35
titles a year. Unsolicited mss, synopses and ideas
welcome.

Morgan Kauffman
▷ Elsevier Ltd

Mosby
▷ Elsevier Ltd

Motor Racing Publications

PO Box 1318, Croydon CR9 5YP
☏ 020 8654 2711   ⓕ 020 8407 0339
john@mrpbooks.co.uk
www.mrpbooks.co.uk
Chairman/Editorial Head *John Blunsden*

Founded soon after the end of World War II
to concentrate on motor-racing titles. Fairly
dormant in the mid 1960s but was reactivated
in 1968 by a new shareholding structure. John
Blunsden later acquired a majority share and
major expansion followed in the 1970s. Publishes
motor-sport history, classic and performance
car collection and restoration, race track driving
and related subjects. IMPRINTS **Fitzjames Press**;
Motor Racing Publications. About 2–4 titles a

year. No unsolicited mss. Send synopses and ideas
in specified subject areas in the first instance.
Ⓔ Royalties twice-yearly.

Murdoch Books UK Ltd

Erico House, 6th Floor North, 93–99 Upper
Richmond Road, London SW15 2TG
☏ 020 8785 5995   ⓕ 020 8785 5985
info@murdochbooks.co.uk
www.murdochbooks.co.uk
Publisher *Kay Scarlett (Australia)*

Owned by Australian publisher Murdoch Books
Pty Ltd. Publishes full-colour non-fiction: homes
and interiors, gardening, cookery, craft, DIY,
history, travel and narrative non-fiction. About
80 titles a year. Contact inquiry@murdochbooks.
com.au in the first instance.

John Murray (Publishers) Ltd

338 Euston Road, London NW1 3BH
☏ 020 7873 6000   ⓕ 020 7873 6446
firstname.lastname@johnmurrays.co.uk
www.hachettelivre.co.uk
Chief Executive *Jamie Hodder-Williams*
Managing Director *Roland Philipps*
Non-fiction *Eleanor Birne*
Fiction *Kate Parkin*

Founded 1768. Part of **Hachette Livre UK**.
Publishes general trade books. 50 titles a year. No
unsolicited material; send preliminary letter.
Ⓔ Royalties twice-yearly.

✱*Authors' update* **Boosted by a Man Booker
shortlisted title, John Murray has had a good
year with fiction doing particularly well.
Though sharing services with Hodder Headline,
Murray has kept its distinctive voice.**

Myriad Editions

59 Lansdowne Place, Brighton BN3 1FL
☏ 01273 720000
info@myriadeditions.com
www.myriadeditions.com
Managing Director *Candida Lacey*
Rights Manager *Sadie Mayne*

Founded in 1993, Myriad Editions is an
independent publisher and packager, 'committed
to mapping the most pressing issues facing the
world today'. Publications include the award-
winning State of the World atlas series. Recently
expanded into publishing original fiction, activist
non-fiction and documentary comic books to
complement their atlases: 'a distinctive and
diverse selection of unconventional viewpoints,
new voices and new ways of seeing'. 5–10 titles
a year. No unsolicited mss. Synopses and ideas
for books in the specified subject areas only.
Approach by e-mail in the first instance.
Ⓔ Royalties annually.

Authors' update A vote of confidence from the Arts Council accompanies a £60,000 award to develop the Myriad fiction list.

Myrmidon Books

Rotterdam House, 116 Quayside, Newcastle upon Tyne NE1 3DY
📞 0191 206 4005 📠 0191 206 4001
submissions@myrmidonbooks.com
www.myrmidonbooks.com
Chairman/Managing Director *Ed Handyside*
Editor *Anne Westgarth*
Approx. Annual Turnover £200,000

Founded 2006. Publishes adult trade fiction. Full mss from literary agents only; first three chapters from unrepresented writers. No short stories, novellas or works under 65,000 words, children's books, graphic novels, non-fiction. Submit hard copy only after checking requirements on the website.
📖 Royalties quarterly.

NAG Press Ltd
▷ Robert Hale Ltd

The National Archives

Kew TW9 4DU
📞 020 8392 5289 📠 020 8487 1974
catherine.bradley@nationalarchives.gov.uk
www.nationalarchives.gov.uk
Publisher *Catherine Bradley*

Publishes non-fiction books relating to the thousand years of historical records held in The National Archives. Specialist areas: general history, family history, military history and intelligence history. TITLES *Princes of Wales: Royal heirs in waiting* David Loades; *Cromwell's Head* Jonathan Fitzgibbons; *British Intelligence: Secrets, spies and sources* Stephen Twigge, Edward Hampshire and Graham Macklin; *Census: The expert guide* Peter Christian and David Annal; *Army Records* William Spencer. SERIES *Crime Archive* TITLES *The Krays*; *The Great Train Robbery*; *Ruth Ellis*; *Jack the Ripper*. Also publishes *Ancestors*, a full-colour family history magazine. 20 titles in 2008.

National Trust Books
▷ Anova Books

Nautical Data Ltd

The Book Barn, Westbourne, Emsworth PO10 8RS
📞 01243 389352 📠 01243 379136
info@nauticaldata.com
www.nauticaldata.com
Managing Director *Piers Mason*

Founded 1999. Publishes pilots and nautical reference. No unsolicited mss; synopses and ideas welcome. No fiction or non-nautical themes.
📖 Royalties twice-yearly.

NCVO Publications

Regent's Wharf, 8 All Saints Street, London N1 9RL
📞 020 7713 6161/HelpDesk: 0800 2798 798
📠 020 7713 6300
ncvo@ncvo-vol.org.uk
www.ncvo-vol.org.uk/publications
Publications and Promotions Coordinator *Daniel Fluskey*
Approx. Annual Turnover £140,000

Founded 1928. Publishing imprint of the National Council for Voluntary Organisations, embracing former Bedford Square Press titles and NCVO's many other publications. The list reflects NCVO's role as the representative body for the voluntary sector. Publishes directories, good practice information on management and trustee development, finance and employment titles and information of primary interest to the voluntary sector. TITLES *The Voluntary Agencies Directory*; *The Good Trustee Guide*; *The Good Campaigns Guide*; *The Good Financial Management Guide*; *The Good Employment Guide*; *The Good Management Guide*; *The Good Membership Guide*; *The UK Voluntary Sector Almanac*. No unsolicited mss as all projects are commissioned in-house.

Thomas Nelson & Sons Ltd
▷ Nelson Thornes Limited

Nelson Thornes Limited

Delta Place, 27 Bath Road, Cheltenham GL53 7TH
📞 01242 267100 📠 01242 221914
info@nelsonthornes.com
www.nelsonthornes.com
Managing Director *Mary O'Connor*

Now part of Infinitas Learning, Nelson Thornes was formed in 2000 by the merger of Thomas Nelson and Stanley Thornes. Educational publisher (AQA exclusively endorsed) of printed and electronic resources, from pre-school to higher education. Unsolicited mss, synopses and ideas for books welcome if appropriate to specialized lists.
📖 Royalties annually.

Network Continuum
▷ The Continuum International Publishing Group Limited

New Beacon Books Ltd

76 Stroud Green Road, London N4 3EN
📞 020 7272 4889 📠 020 7281 4662
newbeaconbooks@btconnect.com
www.newbeaconbooks.co.uk
Managing Director *Sarah White*
Approx. Annual Turnover £120,000

Founded 1966. Publishes fiction, history, politics, poetry and language, all concerning black people. No unsolicited material.

ⓔ Royalties annually.

New Holland Publishers (UK) Ltd

Garfield House, 86–88 Edgware Road, London W2 2EA

ⓣ 020 7724 7773   ⓕ 020 7258 1293 (editorial)
postmaster@nhpub.co.uk
www.newhollandpublishers.com
Managing Director *Steve Connolly*
Publishing Director *Rosemary Wilkinson*
Approx. Annual Turnover £8.5 million

With its headquarters in London, New Holland Publishers (UK) Ltd is the International Publishing Division of Avusa, one of Africa's leading publishing groups, with offices in Australia and New Zealand. Acquired Cadogan Guides in April 2007. Publishes non-fiction, practical and inspirational books across a range of categories, including food and drink, DIY, practical art, crafts, interiors, health and fitness, gardening, pets, humour and gift, reference, history, sports and outdoor pursuits, natural history, general travel books and Globetrotter travel guides. No unsolicited mss; synopses and ideas welcome.

✱*Authors' update* **To meet the challenges of an increasingly competitive market, Cadogan is relaunching its travel guides with the focus more on regions than on countries. The aim is to upstage elementary websites.**

New Playwrights' Network
▷ Cressrelles Publishing Co. Ltd

Newnes
▷ Elsevier Ltd

Nexus
▷ The Random House Group Ltd

Nielsen Book

3rd Floor, Midas House, 62 Goldsworth Road, Woking GU21 6LQ

ⓣ 01483 712200   ⓕ 01483 712201
info.bookdata@nielsen.com
www.nielsenbookdata.co.uk
Senior Manager, Publishing Services *Peter Mathews*

Nielsen Book creates and maintains a unique database of timely, accurate and content rich records for English-language books and other published media, collected from publishers in over 70 countries. The enriched data includes: price, availability, table of contents (where applicable), subject classifications, market rights, publisher and distributor details, cover/jacket images and book descriptions. The data is made available to the book industry via the **BookData**, **BookNet** and **BookScan** services (see entries under *Miscellany*).

Nightingale Books
▷ Pegasus Elliot Mackenzie Publishers Ltd

NMS Enterprises Limited – Publishing

National Museums Scotland, Chambers Street, Edinburgh EH1 1JF

ⓣ 0131 247 4026   ⓕ 0131 247 4012
publishing@nms.ac.uk
www.nms.ac.uk/books
Director *Lesley A. Taylor*
Approx. Annual Turnover £130,000

Publishes non-fiction related to the National Museums of Scotland collections: academic and general; children's; archaeology, history, decorative arts worldwide, history of science, technology, natural history and geology, poetry. TITLES *Silver: Made in Scotland*; *Tartan: The Highland Habit*; *Beyond the Palace Walls: Islamic Art from the State Hermitage Museum*; *Minerals of Northern England*; *Robert Louis Stephenson: The Travelling Mind*; *Bagpipes: A National Collection of a National Instrument*; *Minerals of Scotland*; *Audubon in Edinburgh*; *Commando Country*; *Weights and Measures in Scotland: a European Perspective* (2005 Saltire Society/ National Library of Scotland Research Book of the Year). No unsolicited mss; only interested in synopses and ideas for books which are genuinely related to NMS collections and to Scotland in general.

ⓔ Royalties twice-yearly.

Nonsuch
▷ The History Press

North-Holland
▷ Elsevier Ltd

Northcote House Publishers Ltd

Horndon House, Horndon, Tavistock PL19 9NQ
ⓣ 01822 810066   ⓕ 01822 810034
northcote.house@virgin.net
www.northcotehouse.co.uk
Managing Director *Brian Hulme*

Founded 1985. Publishes a series of literary critical studies, in association with the British Council, *Writers and their Work*; education management, literary criticism, educational dance and drama. 30 titles in 2007. 'Well-thought-out proposals, including contents and sample chapter(s), with strong marketing arguments welcome.'

ⓔ Royalties annually.

W.W. Norton & Company Ltd

Castle House, 75–76 Wells Street, London W1T 3QT
ⓣ 020 7323 1579   ⓕ 020 7436 4553

office@wwnorton.co.uk
www.wwnorton.co.uk
Managing Director *R. Alan Cameron*

Subsidiary of the US parent company founded in 1923 (see entry under *US Publishers*). Publishes non-fiction and academic books.

Nottingham University Press

Manor Farm, Church Lane, Thrumpton NG11 0AX
☎ 0115 983 1011   ℻ 0115 983 1003
orders@nup.com
www.nup.com

Nottingham University Press has gained 'international recognition as a publisher of high quality scientific and technical publications'.
TITLES *Recent Advances in Animal Nutrition 2008*; *Perfecting the Pig Environment*; *Advances in Equine Nutrition IV*; *The Alcohol Textbook – 5th Edition.*
£ Royalties twice-yearly.

NWP
▷ **Neil Wilson Publishing Ltd**

O Books

The Bothy, Deershot Lodge, Park Lane, Ropley SO24 0BE
office1@o-books.net
www.o-books.net
Approx. Annual Turnover £1.5 million

Publishes religion, MBS, philosophy, fiction and other areas. About 100 titles a year. For submissions, go to 'Contact us' section on the website.

Oak
▷ **Omnibus Press**

Oberon Books

521 Caledonian Road, London N7 9RH
☎ 020 7607 3637   ℻ 020 7607 3629
info@oberonbooks.com
www.oberonbooks.com
Publishing Director *James Hogan*
Managing Director *Charles D. Glanville*
Editor *Stephen Watson*

A leading theatre publisher, Oberon publishes play texts (usually in conjunction with a production), and books on theatre and dance. Specializes in contemporary plays and translations of European classics. IMPRINTS **Oberon Modern Plays**; **Oberon Classics**. Publishes for the National Theatre, Royal Opera House, Royal Court, English Touring Theatre, LAMDA, Gate Theatre, Bush Theatre, West Yorkshire Playhouse, Chichester Festival Theatre and many other London and regional theatre companies. Publishes over 300 writers and translators including Rodney Ackland, Tariq Ali, Howard Barker, John Barton, Ranjit Bolt, Pam

Gems, Tanika Gupta, Sir Peter Hall, Bernard Kops, Adrian Mitchell, Sir John Mortimer, Adriano Shaplin, Laura Wade. An extensive classics list includes performance (edited) versions of Shakespeare including *King Lear*; *A Midsummer Night's Dream*; *Romeo and Juliet* and *Twelfth Night*, as well as new translations of plays by Chekhov, Molière, Strindberg and others. About 80 titles a year. Send mss, with information about forthcoming productions to *Daisy Bowie-Sell* (daisy@oberonbooks.com). 'We do not provide readers' reports or feedback about work submitted.'

✳*Authors' update* A first call for theatre books, Oberon has confirmed its lead in a niche market by sponsoring the Sheridan Morley Prize for theatrical biography (see entry under *Prizes*).

Octagon Press Ltd

78 York Street, London W1H 1DP
☎ 020 7193 6456   ℻ 020 7117 3955
admin@octagonpress.com
www.octagonpress.com
Director *Saira Shah*
Approx. Annual Turnover £50,000

Founded 1960. Publishes travel, biography, literature, folklore, psychology, philosophy with the focus on East-West studies. 4–5 titles a year.
£ Royalties annually.

Octopus Publishing Group

2–4 Heron Quays, London E14 4JP
☎ 020 7531 8400   ℻ 020 7531 8650
www.octopus-publishing.co.uk
Chief Executive *Alison Goff*
Approx. Annual Turnover £45 million (Group)

Formed following a management buyout of Reed Consumer Books from Reed Elsevier plc in 1998. Bought by French publishers Hachette Livre in 2001 and acquired Cassell Illustrated and the lists of Ward Lock and Blandford Press; also, Gaia Books acquired in 2004.

Conran Octopus ℻ 020 7531 8627 info-co@conran-octopus.co.uk www.conran-octopus.co.uk Publishing Director *Lorraine Dickey* Quality illustrated lifestyle books, particularly interiors, design, cookery, gardening and crafts.

Hamlyn Octopus ℻ 020 7531 8562 info-ho@hamlyn.co.uk www.hamlyn.co.uk Publisher *Jo Hemmings* Practical books for home and family, cookery, gardening, craft, sport, health, parenting and pet care.

Mitchell Beazley/Miller's ℻ 020 7537 0773 info-mb@mitchell-beazley.co.uk www.mitchell-beazley.co.uk Publisher *David Lamb* Quality illustrated reference books, particularly food and wine, gardening, interior design and architecture, antiques, general reference.

Philip's ⓔ 020 7531 8460 george.philip@philips-maps.co.uk www.philips-maps.co.uk Managing Director *To be appointed* World atlases, globes, astronomy, road atlases, encyclopaedias, thematic reference.

Bounty ⓔ 020 7531 8607 bountybooksinfo-bp@bountybooks.co.uk Publishing and International Sales Director *Polly Manguel* Bargain and promotional books.

Cassell Illustrated ⓔ 020 731 8624 Commissioning Editor *Laura Price*

Godsfield Press ⓔ 020 7531 8562 enquiries@godsfieldpress.com Publisher *Jo Hemmings* Mind, body and spirit.

Gaia Books Publisher *Jo Hemmings* Alternative health and natural/eco titles.

Spruce ⓔ 020 7531 8604 ljiljana.baird@octopus-publishing.co.uk www.octopusbooks.co.uk Publisher *Ljiljana Baird* Gift, cookery, biography, popular culture, photography, mind body spirit, inspiration, parenting.
ⓔ Royalties twice-yearly/annually, according to contract in all divisions.

✱*Authors' update* The emphasis is on **mainstream trade books at the popular end of the market.**

Old Street Publishing Ltd

28–32 Bowling Green Lane, London EC1R 0BJ
ⓣ 020 7837 1600   ⓕ 020 7900 6563
info@oldstreetpublishing.co.uk
www.oldstreetpublishing.co.uk
Managing Director *Ben Yarde-Buller*
Director *David Reynolds*

Founded in 2006 by Ben Yarde-Buller and David Reynolds, co-founder of Bloomsbury Publishing. Publishes eclectic fiction including literary fiction, crime/thriller and translation. General non-fiction and humour 'of the better sort'. No poetry, technical or children's books. 25 titles in 2008. No unsolicited mss. Synopses and ideas welcome; approach by e-mail.
ⓔ Royalties twice-yearly.

Olympia Publishers

60 Cannon Street, London EC4N 6NP
ⓣ 020 7618 6424   ⓕ 020 7002 1100
editors@olympiapublishers.com
www.olympiapublishers.com
Chief Editor *G. Bartlett*
Managing Editor *M.B. Anderson*

Founded 2004. Publishes general fiction, memoirs, human interest, chick lit, military, naval, children's and non-fiction. Unsolicited synopses, poetry and short stories also

considered. IMPRINTS **Olympia Publishers** Subsidy publishers; **Ashwell Publishing Limited** Non-subsidy publishers. 40–50 titles a year. Send three sample chapters and detailed synopsis for consideration. Postal submissions must be accompanied by s.a.e.
ⓔ Royalties twice-yearly.

✱*Authors' update* Note that this is a two-in-one operation. Submissions turned down by **Ashwell for conventional publication are likely to be passed on to Olympia where the author will be asked to contribute to production costs.**

Michael O'Mara Books Ltd

16 Lion Yard, Tremadoc Road, London SW4 7NQ
ⓣ 020 7720 8643   ⓕ 020 7627 8953
firstname.lastname@mombooks.com
www.mombooks.com
www.mombooks.com/busterbooks
Chairman *Michael O'Mara*
Managing Director *Lesley O'Mara*
Editorial Directors *Toby Buchan, Louise Dixon, Lindsay Davies*
Publishing Director (Buster Books) *Philippa Wingate*
Approx. Annual Turnover £7 million

Founded 1985. Independent publisher. Publishes general non-fiction, humour, anthologies, miscellanies, royalty and celebrity autobiographies and biographies. TITLES *My Word Is My Bond* Roger Moore; *Anything Goes* John Barrowman; *I Before E* Judy Parkinson; *The Mums' Book* and *The Dads' Book*; *The Timewaster Letters* Robin Cooper; *The Book of Senior Moments* Shelley Klein. IMPRINT **Buster Books** Children's TITLES *The Boys' Book* and *The Girls' Book*; *The Girls' Book of Glamour*; *Do You Doodle?*; *The Holiday Doodle Book*; board books, picture books, novelty books, humour and quirky non-fiction. 59 titles in 2008. Unsolicited mss, synopses and ideas for books welcome.
ⓔ Royalties twice-yearly.

✱*Authors' update* Expanding his list, Michael O'Mara has hit the jackpot with celeb memoirs. **O'Mara Books is now the 23rd largest publisher by value.**

Omnibus Press

Music Sales Ltd, 14/15 Berners Street, London W1T 3LJ
ⓣ 020 7612 7400   ⓕ 020 7612 7545
chris.charlesworth@musicsales.co.uk
www.omnibuspress.com
Owner *Robert Wise*
Editorial Head *Chris Charlesworth*

Founded 1971. Independent publisher of music books, rock and pop biographies, song sheets, educational tutors, cassettes, videos and software. A Division of the Music Sales Group of

Companies. IMPRINTS **Amsco; Arcane; Bobcat; Oak; Omnibus; Vision On; Wise Publications**. About 25 titles a year. Unsolicited mss, synopses and ideas for music-related titles welcome.

Oneworld Publications

185 Banbury Road, Oxford OX2 7AR
☎ 01865 310597   ☎ 01865 310598
info@oneworld-publications.com
www.oneworld-publications.com
Editorial Director *Juliet Mabey*

Founded 1986. Independent publisher of intelligent adult non-fiction for both the trade and academic markets, across a range of subjects including current affairs, philosophy, history, politics, popular science, psychology, world religions and fiction. SERIES include *Oneworld Beginners' Guides*; *Coping With* (self-help) and *Makers of the Muslim World*. TITLES *Dance with Chance*; *Body Shopping*; *Oilopoly*; *Strange Fruit*; *Can a Robot be Human?*; *Modern Slavery*; *10 Zen Questions*. 60 titles a year. No unsolicited mss; synopses and ideas welcome, preferably via e-mail or through the website, but if by post, an s.a.e. is required for return of material and/or notification of receipt. No autobiographies, poetry or children's. From 2009, fiction submissions welcome.
Ⓔ Royalties annually.

Onlywomen Press Ltd

40 St Lawrence Terrace, London W10 5ST
☎ 020 8354 0796
onlywomenpress@btconnect.com
www.onlywomenpress.com
Editorial Director *Lilian Mohin*

Founded 1974. Publishes feminist lesbian literature: fiction, poetry, literary criticism and children's books. Unsolicited mss and proposals welcome.

OPC
▷ Ian Allan Publishing Ltd

Open Gate Press incorporating Centaur Press

51 Achilles Road, London NW6 1DZ
☎ 020 7431 4391   ☎ 020 7431 5129
books@opengatepress.co.uk
www.opengatepress.co.uk
Managing Directors *Jeannie Cohen,
Elisabeth Petersdorff*

Founded in 1989 to provide a forum for psychoanalytic social and cultural studies. Publishes psychoanalysis, philosophy, social sciences, politics, literature, religion, environment. SERIES *Psychoanalysis and Society*. IMPRINTS **Open Gate Press; Centaur Press; Linden Press**. No unsolicited mss.
Ⓔ Royalties annually.

Open University Press

McGraw-Hill House, Shoppenhangers Road, Maidenhead SL6 2QL
☎ 01628 502500   ☎ 01628 770224
enquiries@openup.co.uk
www.openup.co.uk

Founded 1977 as an imprint independent of the Open University's course materials. Acquired by **McGraw-Hill Education** in 2002. Publishes academic and professional books in the fields of education, sociology, health, psychology, psychotherapy and counselling, higher education and study skills materials. No economics or anthropology. Not interested in anything outside the social sciences. About 100 titles a year. No unsolicited mss; enquiries/proposals only.
Ⓔ Royalties annually.

Orbit
▷ Little, Brown Book Group UK

Orchard Books
▷ Hachette Children's Books

The Orion Publishing Group Limited

Orion House, 5 Upper St Martin's Lane, London WC2H 9EA
☎ 020 7240 3444   ☎ 020 7240 4822
www.orionbooks.co.uk/pub/index.htm
Owner *Hachette Livre*
Chairman *Arnaud Nourry*
Chief Executive *Peter Roche*
Deputy Chief Executive *Malcolm Edwards*
Approx. Annual Turnover £70 million

Founded 1992. Incorporates Weidenfeld & Nicolson, JM Dent, Gollancz and Cassell.

TRADE DIVISION
Trade Managing Director *Lisa Milton* IMPRINTS **Orion Fiction** Publishing Director *Jon Wood* Hardcover fiction; **Orion Non-Fiction** Publishing Director *Alan Samson* Hardcover non-fiction; **Orion Children's** Publisher *Fiona Kennedy* Children's fiction/non-fiction; **Gollancz** Editorial Directors *Simon Spanton, Jo Fletcher* Science fiction and fantasy. **Weidenfeld Non-Fiction** Publisher *Alan Samson* General non-fiction, history and military; **Weidenfeld Fiction** Publisher *Kirsty Dunseath* History; **Weidenfeld Illustrated** Editor-in-Chief *Michael Dover*; **Custom Publishing** Director *Elizabeth Bond*.

PAPERBACK DIVISION Managing Director *Susan Lamb* IMPRINTS **Orion; Phoenix; Everyman**. No unsolicited material.
Ⓔ Royalties twice-yearly.

Osprey Publishing Ltd

Midland House, West Way, Botley, Oxford OX2 0PH
☎ 01865 727022   ☎ 01865 727017/727019

info@ospreypublishing.com
www.ospreypublishing.com
Chairman *Philip Sturrock*
Managing Director *Rebecca Smart*
Editorial Director *Anita Baker*

Publishes illustrated history of war and warfare and military aviation, also general military history. Founded 1969, Osprey became independent from **Reed Elsevier** in February 1998 and launched Osprey Publishing Inc. as a wholly owned subsidiary in North America in 2004. Acquired **Shire Publications Ltd** in 2007 (see entry). SERIES include *Men-at-Arms*; *Fortress*; *Campaign*; *Essential Histories*; *Warrior*; *Elite*; *New Vanguard*; *Osprey Modelling*; *Duel*; *Aircraft of the Aces*; *Aviation Elite Units*; *Combat Aircraft*; *Raid*; *Command*. About 125 titles a year including 20 non-series titles a year. No unsolicited mss; ideas and synopses welcome.
Ⓔ Royalties twice-yearly.

✱*Authors' update* **A new self publishing service is being launched with AuthorHouse. These projects are rarely cheap but specialist writers in search of an outlet may be attracted.**

Peter Owen Publishers

73 Kenway Road, London SW5 0RE
☏ 020 7373 5628/7370 6093   🖷 020 7373 6760
aowen@peterowen.com
www.peterowen.com
Chairman *Peter Owen*
Editorial Director *Antonia Owen*

Founded 1951. Publishes biography, general non-fiction, English-language literary fiction and translations, history, literary criticism, the arts and entertainment. 'No genre or children's fiction; the company is taking on no first novels or short stories at present.' AUTHORS Jane Bowles, Paul Bowles, Yuri Druzhnikov, Shusaku Endo, Philip Freund, Anna Kavan, Jean Giono, Anaïs Nin, Jeremy Reed, Peter Vansittart. About 25 titles a year. Unsolicited synopses welcome for non-fiction material. Please call or e-mail in advance to submit fiction. Mss should be preceded by a descriptive letter and synopsis with s.a.e.
OVERSEAS ASSOCIATES worldwide.
Ⓔ Royalties twice-yearly.

✱*Authors' update* **Peter Owen has been described as 'a publisher of the old and idiosyncratic school'. He has seven Nobel prizewinners on his list to prove it.**

Oxbow Books

10 Hythe Bridge Street, Oxford OX1 2EW
☏ 01865 241249   🖷 01865 794449
editorial@oxbowbooks.com
www.oxbowbooks.com
Editor *Clare Litt*

Founded 1983. Publishes academic archaeology, Egyptology, ancient and medieval history. IMPRINTS **Aris & Phillips** Greek, Latin and Hispanic texts and translations; **Windgather Press** Landscape archaeology and history. About 60 titles a year.

Oxford University Press

Great Clarendon Street, Oxford OX2 6DP
☏ 01865 556767   🖷 01865 556646
enquiry@oup.com
www.oup.com
Managing Director of Academic Divisions and Journals Division *Martin Richardson*
Approx. Annual Turnover £492 million

A department of Oxford University, OUP started as the university's printing business and developed into a major publishing operation in the 19th century. Publishes academic works in all formats (print and online): dictionaries, lexical and non-lexical reference, scholarly journals, student texts, schoolbooks, ELT materials, music, bibles, paperbacks, and children's books. Around 6,000 titles a year.

DIVISIONS **Academic and Journals** *M. Richardson* Academic and higher education titles in major disciplines, dictionaries, non-lexical reference and trade books. OUP welcomes first-class academic material in the form of proposals or accepted theses. Academic and research journals. **Education** *K. Harris* National Curriculum courses and support materials as well as children's literature. **ELT** *P.R.C. Marshall* ELT courses and dictionaries for all levels.

OVERSEAS BRANCHES/SUBSIDIARIES Sister company in USA with branches or subsidiaries in Argentina, Australia, Brazil, Canada, China, East Africa, India, Japan, Malaysia, Mexico, Pakistan, Southern Africa, Spain and Turkey. Offices in France, Greece, Italy, Poland, Taiwan, Thailand.
Ⓔ Royalties twice-yearly.

✱*Authors' update* **With more than eighty per cent of its sales outside the UK, OUP has benefited from market growth in the US. English and bilingual dictionaries are the core activity with OUP capturing over half the English and a third of the bilingual market. Sales of online reference are set to overtake print.**

Palgrave Macmillan
▷ Macmillan Publishers Ltd

G.J. Palmer & Sons Ltd
▷ Hymns Ancient & Modern Ltd

Pan
▷ Macmillan Publishers Ltd

Paper Tiger
▷ Anova Books

Paperbooks Ltd

3rd Floor, Unicorn House, 221–222 Shoreditch High Street, London E1 6PJ

📞 020 7426 0340

info@legend-paperbooks.co.uk
submissions@legend-paperbooks.co.uk
www.paperbooks.co.uk

Managing Director *Tom Chalmers*

Founded in 2006 by Keirsten Clark. Publishes contemporary, diverse and cutting-edge fiction. 'Focuses strongly on reaching new audiences and on working with other publishers to drive the independent sector and promote new and different work within the mainstream market.' Paperbooks was acquired by fellow independent publisher **Legend Press** in April 2008. Send three chapters and synopsis to the submissions e-mail address above.

Ⓔ Royalties twice-yearly.

Parthian

The Old Surgery, Napier Street, Aberteifi (Cardigan) SA43 1ED

📞 01239 612059

parthianbooks@yahoo.co.uk
www.parthianbooks.co.uk
www.parthianbooks.com

Also: **Library of Wales**
Trinity College, Carmarthen SA31 3EP

📞 01267 676633 📠 01239 612059

www.libraryofwales.org

Publisher *Richard Davies*
Fiction/Poetry Editor *Lucy Llewellyn*
Editor, Library of Wales *Professor Dai Smith*

Founded 1993. Publishes contemporary Welsh writing in English including novels, poetry and drama, also translations from Welsh, Spanish, Basque and Catalan. A keen interest in short stories. Its list includes many young writers. TITLES *Oh Dad! A Search for Robert Mitchum* Lloyd Robson; *Fresh Apples* Rachel Trezise (winner of the EDS Dylan Thomas Award); *The Long Dry* Cynan Jones (winner of the Betty Trask Award); *Martha, Jack and Shanco* Caryl Lewis. New non-fiction list and increasing interest in translation with new titles including, in 2009, *To Bury the Dead* Ignacio Martinez de Pison. A developing art and photography list includes *Attraction of Onlookers* Shimon Attie. Also publishes **Library of Wales** classic series including Raymond Williams' *Border Country*, ed. Prof. Dai Smith. 12–15 titles a year. Synopsis with first three sample chapters welcome. 'An idea of what we do is always an advantage.'

Ⓔ Royalties paid annually.

Particular Books
▷ Penguin Group (UK)

Paternoster
▷ Authentic Media

Pavilion/Pavilion Children's Books
▷ Anova Books

Payne–Gallway
▷ Pearson (Heinemann)

Pearson

UK Head office: Edinburgh Gate, Harlow CM20 2JE

📞 01279 623623

www.pearsoned.com

The world's largest educational publisher. Acquired Heinemann (see **Pearson Heinemann**) in May 2007. Publishes across a wide range of curriculum subjects from primary students to professional practitioners. IMPRINTS **Addison Wesley; Allyn & Bacon; BBC Active; Benjamin Cummings; Causeway Press; Cisco Press; Penguin Longman; Prentice Hall; Prentice Hall Business; QUE Publishing; SAM Publishing; Wharton.** OVERSEAS ASSOCIATES worldwide. Unsolicited mss should be addressed to the appropriate department: **ELT** *Alissa Heilbron*; **Schools** *Carol Hayward*; **University & Professional Level** *Louise Smith*.

Ⓔ Royalties twice-yearly.

Pearson (Heinemann)

Halley Court, Jordan Hill, Oxford OX2 8EJ

📞 01865 888000 (customer services)/311366
📠 01865 314641

enquiries@pearson.com
www.heinemann.co.uk

Heinemann, now part of **Pearson**, produces educational products under the following IMPRINTS: **Heinemann, Rigby, Ginn, Payne-Gallway** and **Raintree**. Pearson provides a range of published resources, teachers' support, and pupil and student material in all core subjects for all ages. For submissions, write to 'Market Planning' at the address above.

Ⓔ Royalties twice-yearly/annually, according to contract.

Pegasus Elliot Mackenzie Publishers Ltd

Sheraton House, Castle Park, Cambridge CB3 0AX

📞 01223 370012　📠 01223 370040

editors@pegasuspublishers.com
www.pegasuspublishers.com

Senior Editor *D.W. Stern*
Editor *R. Sabir*

Publishes fiction and non-fiction, general interest, biography, autobiography, children's, history, humour, fantasy, self-help, educational,

science fiction, poetry, travel, war, memoirs, crime and eroticism. IMPRINTS **Vanguard Press; Nightingale Books; Chimera** TITLES *The Prince's Thorn* Mary Mackie; *Syd Barrett – Legend* Bob Cairns; *The Apes That Sing Their Souls* Dr Mairi Rennie; *Wearing Purple* Dr Moyra Evans; *Spirit & Fire* Sue Hampton; *Field of Graves* Alex Mabon. Unsolicited mss, synopses and ideas considered if accompanied by return postage.
Ⓔ Royalties annually.

✱*Authors' update* **Liable to ask authors to contribute to production costs.**

Pen & Sword Books Ltd
47 Church Street, Barnsley S70 2AS
☎ 01226 734222 📠 01226 734438
editorial@pen-and-sword.co.uk
www.pen-and-sword.co.uk
Commissioning Editors *Peter Coles (aviation), Henry Wilson (military & maritime), Rupert Harding (history, local history & family history), Philip Sidnell (ancient history)*

One of the leading military history publishers in the UK. Publishes non-fiction only, specializing in naval and aviation history, WW1, WW2, Napoleonic, autobiography and biography. Also publishes *Battleground* series for battlefield tourists. IMPRINTS **Leo Cooper; Frontline Books** (see entry); **Pen & Sword Aviation; Pen & Sword Family History; Pen & Sword Maritime; Pen & Sword Military; Remember When** (see entry); **Seaforth Publishing** (see entry); **Wharncliffe Books** (see entry). About 250 titles a year. Unsolicited synopses and ideas welcome; no unsolicited mss.
Ⓔ Royalties twice-yearly.

Penguin Group (UK)
A Pearson Company, 80 Strand, London WC2R ORL
☎ 020 7010 3000 📠 020 7010 6060
www.penguin.co.uk
Group Chairman & Chief Executive *John Makinson*
CEO Penguin UK *Peter Field*
Managing Director: Penguin *Helen Fraser*
Approx. Annual Turnover £121 million

Owned by Pearson plc. The world's best known book brand and for more than 70 years a leading publisher whose adult and children's lists include fiction, non-fiction, poetry, drama, classics, reference and special interest areas. Reprints and new work.

DIVISIONS

Penguin General Books Managing Director *Tom Weldon* Adult fiction and non-fiction is published in hardback under **Michael Joseph, Viking, Fig Tree** and **Hamish Hamilton** imprints.

Paperbacks come under the **Penguin** imprint. IMPRINTS **Viking** Publisher *Venetia Butterfield* Publishing Director *Tony Lacey;* **Hamish Hamilton** Publishing Director *Simon Prosser;* **Michael Joseph** Managing Director *Louise Moore* Does not accept unsolicited mss; **Fig Tree** *Juliet Annan;* **Rough Guides**.

Penguin Press Managing Director *Stefan McGrath* Publishing Directors *Stuart Proffitt, Simon Winder* Academic adult non-fiction, reference, specialist and classics. IMPRINTS **Allen Lane; Particular Books; Penguin Reference; Penguin Classics** Publishing Director *Adam Freudenheim* No unsolicited mss.

Dorling Kindersley Ltd (see entry).

Rough Guides Publishing Director *Martin Dunford* Travel and music guides, phrase books and reference.

Frederick Warne Managing Director *Stephanie Barton* Classic children's publishing and merchandizing including *Beatrix Potter™; Flower Fairies; Orlando.* **Ventura** Publisher *Stephanie Barton* Producer and packager of *Spot* titles by Eric Hill.

Ladybird Managing Director *Stephanie Barton*

BBC Children's Books Managing Director *Stephanie Barton* Children's imprint launched in 2004.

Puffin Managing Director *Francesca Dow* (poetry and picture books) Publishing Director *Sarah Hughes* (fiction), Editorial Director *Louise Bolongaro* (Puffin Picture Book Boutique series), Associate Publisher *Mandy Suhr* (picture books) Leading children's paperback list, publishing in virtually all fields including fiction, non-fiction, poetry, picture books, media-related titles. No unsolicited mss.

Penguin Audiobooks (see entry under *Audio Books*).

ePenguin Penguin Digital Publisher *Jeremy Ettinghausen* E-books list launched in 2001. OVERSEAS ASSOCIATES worldwide.
Ⓔ Royalties twice-yearly.

✱*Authors' update* **One of the few publishers to boast a strong brand image, Penguin is in third place behind Random House and Hachette in terms of UK market share. Of the imprints, Michael Joseph had the best year. Dorling Kindersley and Rough Guides both fell back, a reflection of on-screen competition. An ambitious digitization programme is under way.**

Penguin Longman
▷ Pearson

Pennant Books Ltd

PO Box 5675, London W1A 3FB
☎ 020 7287 1841
info@pennantbooks.com
www.pennantbooks.com
Managing Director *Cass Pennant*
Approx. Annual Turnover £250,000

Independent publisher, founded in 2005.
Publishes non-fiction, biography, sport, true
crime and fan culture. No fiction or children's
books. 16 titles in 2008. No unsolicited mss; send
synopses and ideas for books by post.
Ⓔ Royalties twice-yearly.

Pergamon
▷ Elsevier Ltd

Persephone Books

59 Lamb's Conduit Street, London WC1N 3NB
☎ 020 7242 9292   📠 020 7242 9272
sales@persephonebooks.co.uk
www.persephonebooks.co.uk
Managing Director *Nicola Beauman*

Founded 1999. Publishes reprint fiction and non-
fiction, mostly 'by women, for women and about
women'. TITLES *Miss Pettigrew Lives for a Day*
Winifred Watson; *Kitchen Essays* Agnes Jekyll;
The Far Cry Emma Smith; *The Priory* Dorothy
Whipple. 6 titles a year. No unsolicited material.
Ⓔ Royalties twice-yearly.

Phaidon Press Limited

Regent's Wharf, All Saints Street, London N1 9PA
☎ 020 7843 1000   📠 020 7843 1010
initialandname@phaidon.com
www.phaidon.com
Group Chairman & Publisher *Richard Schlagman*
Chairman *Andrew Price*
Editorial Directors *Amanda Renshaw,
Emilia Terragni*

Phaidon Press is one of the world's leading
publishers of books on the visual arts, culture
and creativity. DIVISIONS (with Editorial Heads)
Architecture and Design *Emilia Terragni*;
Contemporary Art *Craig Garrett*; **Photography**
Amanda Renshaw; **Art & Academic** *David
Anfam*; **Cooking/Food** *Emilia Terragni*;
Children's *Amanda Renshaw*. About 100 titles a
year. Unsolicited mss welcome but 'only a small
amount of unsolicited material gets published'.
Ⓔ Royalties twice-yearly.

✴*Authors' update* **Competition is hotting up
in what used to be a specialist market. Sales
have dropped but with French and German
subsidiaries, Phaidon's European scope gives it
added appeal in the US for titles that might not
otherwise enter the English language market.
Commended for tight management and strong
marketing.**

Philip's
▷ Octopus Publishing Group

Phillimore
▷ The History Press

Phoenix
▷ The Orion Publishing Group Limited

Piatkus Books

100 Victoria Embankment, London EC4Y 0DY
☎ 020 7911 8030   📠 020 8911 8100
info@littlebrown.co.uk
www.piatkus.co.uk
www.littlebrown.co.uk
Owner *Hachette Book Group*
CEO/Publisher *Ursula Mackenzie*
COO *David Kent*
Approx. Annual Turnover £10 million

Founded 1979 by Judy Piatkus, the company
was sold to Hachette in autumn 2007. Piatkus
non-fiction specializes in lifestyle, health, self-
help, mind, body and spirit, popular psychology,
biography and business titles. Piatkus fiction
includes new novels from many of the world's
bestselling international authors. DIVISIONS **Non-
fiction** Editorial Director *Gill Bailey* TITLES *The
Holford Low GL Diet Made Easy* Patrick Holford;
Women Who Think Too Much Susan Nolan-
Hoeksema; *Mind Body Bible* Mark Atkinson.
Fiction Senior Editor *Emma Beswetherick* TITLES
High Noon Nora Roberts; *It's in His Kiss* Julia
Quinn; *Creation in Death* J.D. Robb. About 250
titles a year. Piatkus is expanding its range of
books and welcomes synopses together with first
three chapters.
Ⓔ Royalties twice-yearly.

✴*Authors' update* **Being part of a larger group
does not seem to have dented the Piatkus
enthusiasm for bright ideas. Publisher-author
relations are said to be excellent.**

Pica Press
▷ A.&C. Black Publishers Ltd

Picador
▷ Macmillan Publishers Ltd

Piccadilly Press

5 Castle Road, London NW1 8PR
☎ 020 7267 4492   📠 020 7267 4493
books@piccadillypress.co.uk
www.piccadillypress.co.uk
Publisher/Managing Director *Brenda Gardner*
Approx. Annual Turnover £1.4 million

Founded 1983. Independent publisher of children's
and parental books. 39 titles in 2008. Welcomes
approaches from authors 'but we would like them
to know the sort of books we do. It is frustrating
to get inappropriate material. They should check

in their local libraries, bookshops or look at our website. We will send a catalogue (please enclose s.a.e.)'. No adult or cartoon-type material.
Ⓔ Royalties twice-yearly.

Pictorial Presentations
▷ Souvenir Press Ltd

Pimlico
▷ The Random House Group Ltd

Pitkin
▷ The History Press

PiXZ Books
▷ Halsgrove

Pluto Press Ltd
345 Archway Road, London N6 5AA
☎ 020 8348 2724   ℻ 020 8348 9133
pluto@plutobooks.com
www.plutobooks.com
Chair *Roger Van Zwanenberg*
Managing Director/Publisher *Anne Beech*

Founded 1970. Has developed a reputation for innovatory publishing in the field of non-fiction. Publishes academic and scholarly books across a range of subjects including politics and world affairs and more general titles on key aspects of currrent affairs. About 70 titles a year. Prospective authors are encouraged to consult the website for guidelines on submitting proposals.

Pocket Books
▷ Simon & Schuster UK Limited

Pocket Mountains Ltd
6 Church Wynd, Bo'ness EH51 0AN
☎ 01506 500402   ℻ 01506 500405
info@pocketmountains.com
www.pocketmountains.com
Managing Directors *Robbie Porteous,*
April Simmons
Approx. Annual Turnover £150,000

Founded in 2003 to publish a series of guidebooks to the Scottish hills and now publishing other guides covering Europe: mountaineering, walking, cycling and wildlife books. 23 titles in 2008. Unsolicited mss, synopses and ideas welcome; approach by e-mail or letter. No fiction, children's or cookery.

Poetry of Place
▷ Eland Publishing Ltd

The Policy Press
Fourth Floor, Beacon House, Queen's Road, Bristol BS8 1QU
☎ 0117 331 4054   ℻ 0117 331 4093
tpp-info@bristol.ac.uk
www.policypress.org.uk
Director *Alison Shaw*

Specialist social science publisher 'committed to publishing work that will impact on research, learning, policy and practice'.

Politico's Publishing
8 Artillery Row, London SW1P 1RZ
☎ 020 7802 0018   ℻ 020 7828 1244
sales@methuen.co.uk
www.politicospublishing.co.uk
Managing Director *Peter Tummons*
Publishing Consultant *Alan Gordon Walker*

Founded 1998. Acquired by **Methuen Publishing Ltd** in April 2003. Publishes political books. 20 titles in 2008. Unsolicited mss, synopses and ideas welcome; telephone in the first instance.
Ⓔ Royalties twice-yearly.

Polity Press
65 Bridge Street, Cambridge CB2 1UR
☎ 01223 324315   ℻ 01223 461385
www.polity.co.uk

Founded 1984. Publishes anthropology, criminology, economics, feminism, history, human geography, literature, media and cultural studies, medicine and society, philosophy, politics, psychology, religion and theology, social and political theory, sociology. 'Please do not send mss.'
Ⓔ Royalties annually.

Polygon
West Newington House, 10 Newington Road, Edinburgh EH9 1QS
☎ 0131 668 4371   ℻ 0131 668 4466
info@birlinn.co.uk
www.birlinn.co.uk
Publishing Director *Neville Moir*
Managing Editor *Alison Rae*

Bought by **Birlinn Ltd** in 2002. Publishes literary fiction, crime fiction and poetry. About 40 titles a year. 'We do not accept unsolicited work from authors.'
Ⓔ Royalties twice-yearly.

Pont Books
▷ Gomer Press

Pop Universal
▷ Souvenir Press Ltd

Portico
▷ Anova Books

Portland Press Ltd
3rd Floor, Eagle House, 16 Procter Street, London WC1V 6NX
☎ 020 7280 4110   ℻ 020 7280 4169
editorial@portlandpress.com
www.portlandpress.com
Chairman *Professor John Clark*
Managing Director *Rhonda Oliver*

Managing Editor *Pauline Starley*
Approx. Annual Turnover £2.5 million

Founded 1990 to expand the publishing activities of the Biochemical Society (1911). Publishes biochemisty and medicine for graduate, postgraduate and research students. Expanding the list to include schools and general readership. TITLES *Essays in Biochemistry Volume 45: Systems Biology*; *Biochemical Society Symposia Volume 75: Structure and Function in Cell Adhesion*; *Wenner-Gren Volume 84: The University in the Market.* 3 titles in 2008. Unsolicited mss, synopses and ideas welcome. No fiction.
Ⓔ Royalties twice-yearly.

Portobello Books

Twelve Addison Avenue, London W11 4QR
☎ 020 7605 1380   📠 020 7605 1361
mail@portobellobooks.com
www.portobellobooks.com
Owners *Sigrid Rausing, Eric Abraham, Philip Gwyn Jones*
Managing Director *David Graham*
Senior Editor *Laura Barber*
Associate Publisher *Tasja Dorkofikis*
Rights Director *Angela Rose*
Approx. Annual Turnover £500,000

Founded in 2005 by Philip Gwyn Jones, formerly publishing director of HarperCollins' literary imprint Flamingo. Owners acquired **Granta** in 2005. Publishes international literature and activist non-fiction. About 25 original titles a year. No unsolicited mss; synopses and ideas for books welcome. Approach by post to the submissions editor in the first instance. No science fiction, romance or genre fiction of any kind, no gardening, cookery, mind, body and spirit or reference.
Ⓔ Royalties twice-yearly.

T&AD Poyser
▷ A.&C. Black Publishers Ltd

Preface Publishing
▷ The Random House Group Ltd

Prentice Hall/Prentice Hall Business
▷ Pearson

Prestel Publishing Limited

4 Bloomsbury Place, London WC1A 2QA
☎ 020 7323 5004   📠 020 7636 8004
sales@prestel-uk.co.uk
www.prestel.com
UK Editor *Philippa Hurd*
Editor-in-Chief (Germany) *Curt Holtz*

Founded 1924. Publishes art, architecture, photography, design, fashion, children's and general illustrated books. No fiction. Unsolicited mss, synopses and ideas welcome. Approach by post or e-mail.

Princeton University Press
▷ University Presses of California, Columbia & Princeton Ltd

Prion Books
▷ Carlton Publishing Group

Profile Books

3a Exmouth House, Pine Street, Exmouth Market, London EC1R 0JH
☎ 020 7841 6300   📠 020 7833 3969
info@profilebooks.com
www.profilebooks.com
Publisher & Managing Director *Andrew Franklin*
Editorial Director *Stephen Brough*
Publisher, GreenProfile *Mark Ellingham*
Publisher, Serpent's Tail *Pete Ayrton*
Rights Manager *Penny Daniel*
Approx. Annual Turnover £9.09 million

Founded 1996. Publishes general non-fiction: history, biography, current affairs, popular science, politics, business management. Also publishers of *The Economist* books. IMPRINTS **Serpent's Tail** (see entry); **Green Profile** Environmental books. No unsolicited mss; phone or send preliminary letter.

✱*Authors' update* **From small beginnings not much more than a decade ago, Profile has built up an impressive non-fiction list, matched now by Serpent's Tail's innovative fiction.** *New Scientist* **tie-ins are set to expand.**

Psychology Press (An imprint of Taylor & Francis Group, an Informa plc business)

27 Church Road, Hove BN3 2FA
☎ 020 7017 6000   📠 020 7017 6717
www.psypress.co.uk
Managing Director *Mike Forster*
Deputy Managing Director *Rohays Perry*

Publishes psychology textbooks, monographs, professional books, tests and numerous journals which are available in both printed and online formats. Send detailed proposal plus c.v. and sample chapters. Proposal requirements and submission procedures available on the website. No poetry, fiction, travel or astrology.
Ⓔ Royalties according to contract.

Publishing House

Trinity Place, Barnstaple EX32 9HG
☎ 01271 328892   📠 01271 328768
info@vernoncoleman.com
www.vernoncoleman.com
Managing Director *Vernon Coleman*
Publishing Director *Sue Ward*
Approx. Annual Turnover £750,000

Founded 1989. Self-publisher of fiction, health, humour, animals, politics. Over 90 books published. TITLES *Mrs Caldicot's Cabbage War*; *England our England*; *Bodypower*; *Bilbury Chronicles*; *Second Innings*; *It's Never Too Late*; *Alice's Diary*; *Food for Thought*; *Gordon is a Moron*, all by Vernon Coleman. No submissions.

Puffin
▷ Penguin Group (UK)

Pushkin Press Ltd
12 Chester Terrace, London NW1 4ND
📞 020 7730 0750   📠 020 7730 1341
books@pushkinpress.com
www.pushkinpress.com
Chairman *Melissa Ulfane*

Publishes novels and essays in translation drawn from the best of classic and contemporary European literature.

QED Publishing
▷ Quarto Publishing plc

Qu:id Publishing
▷ Quarto Publishing plc

Quadrille Publishing Ltd
Alhambra House, 27–31 Charing Cross Road, London WC2H OLS
📞 020 7839 7117   📠 020 7839 7118
info@quadrille.co.uk
www.quadrille.co.uk
Chairman *Sir David Cooksey*
Managing Director *Alison Cathie*
Editorial Director *Jane O'Shea*
Approx. Annual Turnover £8.09 million (2006)

Founded in 1994 with a view to producing a small list of top-quality illustrated books. Publishes non-fiction, including cookery, gardening, interior design and decoration, craft, health. TITLES *Life's Too F***ing Short* Janet Street Porter; *Maze, The Cookbook* Jason Atherton; *Urban Garden Handbook* Joe Swift; *Men's Knits* Erika Knight. 38 titles in 2008. Synopses and ideas for books welcome. No fiction or children's books. © Royalties twice-yearly.

Quantum
▷ Foulsham Publishers

Quantum Publishing
▷ Quarto Publishing plc

Quartet Books
27 Goodge Street, London W1T 2LD
📞 020 7636 3992   📠 020 7637 1866
david@quartetbooks.com
Chairman *Naim Attallah*

Founded 1972. Independent publisher. Publishes contemporary literary fiction including translations, popular culture, biography, music, history, politics and some photographic books. Unsolicited sample chapters with return postage welcome; no poetry, romance or science fiction. Submissions by disk or e-mail are not accepted. © Royalties twice-yearly.

Quarto Publishing plc
The Old Brewery, 6 Blundell Street, London N7 9BH
📞 020 7700 6700   📠 020 7700 4191
www.quarto.com
Chairman & CEO *Laurence Orbach*
Director of Co-Edition Publishing *Piers Spence*

The UK's most prolific creators of illustrated co-editions, Quarto Publishing plc consists of nine autonomous business units that conceive and produce leisure, lifestyle and reference titles and license them to publishers in 30 languages and in 44 countries worldwide.

DIVISIONS

Quarto Books Publisher *Paul Carslake* paulc@quarto.com High quality, practical non-fiction with an emphasis on crafts and hobbies: from knitting and crochet to beadweaving and calligraphy TITLE *The World's Most Influential Painters*. **Quarto Children's Books** (incorporating Design Eye) Publisher *Sue Grabham* sueg@quarto.com Intricate and ingenious activity and novelty books for children TITLE *Spinning Globe World Atlas*. **Quintet Publishing** Publisher *James Tavendale* jamest@quarto.com Illustrated reference across the board, specializing in food and drink TITLE *500 Pies and Tarts*. **Marshall Editions** (acquired 2002) Publisher *To be appointed* A big name from the past reinvigorated and producing illustrated family reference TITLE *1001 Ideas for Bathrooms*. **QED Publishing** Publisher *Steve Evans* stevee@quarto.com Fun and clever list of first reading and first concept books TITLE *Dinosaur Families*. **Quintessence Editions** Publisher *Tristan De Lancey* tristandl@quarto.com Best known for the '1001 Series' with sales of over three million copies in 27 languages TITLE *1001 Books You Must Read Before You Die*. **Quantum Publishing** Publisher *Anastasia Cavouras* anastasiac@quarto.com Backlist and value-priced books. **Qu:id Publishing** Publisher *Nigel Browning* nigelb@quid.com Quirky, sideways take on reference publishing: serious information presented in a memorable and fun way TITLE *101 Things to do in a Shed*; **Global Publishing** Publisher *Cheryl Campbell* ccampbell@globalpub.com.au Large format, content rich, illustrated reference books TITLE *Antarctica*.

The Quarto Group, Inc.

226 City Road, London EC1V 2TT
☎ 020 7700 6700 📠 020 7700 4191
www.quarto.com
Chairman & CEO *Laurence Orbach*
CFO *Mick Mousley*
Group Creative Director *Bob Morley*
Director of Co-edition Publishing *Piers Spence*
Approx. Annual Turnover £100 million

Founded in 1976, the Quarto Group has an international portfolio of trade and co-edition publishing interests in the UK, USA, Australia and Asia. UK entities include **Aurum Press**, acquired in 2005 (see entry), **RotoVision**, acquired in 1999 (see entry); **Apple Press** (see entry); the magazine, *The World of Fine Wine* (see entry under *Magazines*); and **Quarto Publishing plc** (see entry). Other Quarto-owned concerns include the Quayside Publishing Group in the US; Lifetime Distributors, Australia's largest direct marketing bookseller; and Regent Publishing Services in Hong Kong.

✻*Authors' update* **Restructuring has proved expensive but a leaner Quarto is well placed to grow on exports and international co-editions.**

QUE Publishing
▷ Pearson

Quercus Publishing

21 Bloomsbury Square, London WC1A 2NS
☎ 020 7291 7200 📠 020 7291 7201
firstname.surname@quercusbooks.co.uk
www.quercusbooks.co.uk
Chief Executive *Mark Smith*
Approx. Annual Turnover £10 million

Founded in 2005 as a commercial publisher of quality fiction and non-fiction. DIVISIONS **Trade Publishing** Managing Director *David North*, Editor-in-Chief *Jon Riley* TITLE *The Tenderness of Wolves*; **Children's Publishing** Editorial Head *Roisin Heycock* TITLE *The Magic Thief*; **Contract Publishing** Managing Director *Wayne Davies* TITLE *Speeches That Changed the World*; **MacLehose Press** Publisher *Christopher MacLehose* TITLE *The Girl With the Dragon Tattoo*; **Quercus Audiobooks** *Laura Palmer* TITLE *Genesis*. 140 titles in 2008. No unsolicited material.
Ⓔ Royalties twice-yearly.

✻*Authors' update* **Focusing on new and underdeveloped writers – 'refugees from big groups who weren't high enough up the pecking order to get into the sunshine of the big market spends' – has produced mixed results. Expansion is on hold while the publishing programme is reassessed.**

Quiller Press (An imprint of Quiller Publishing Ltd)

Wykey House, Wykey, Shrewsbury SY4 1JA
☎ 01939 261616 📠 01939 261606
info@quillerbooks.com
www.countrybooksdirect.com
Managing Director *Andrew Johnston*

Specializes in sponsored books and publications sold through non-book trade channels as well as bookshops. Publishes architecture, biography, business and industry, collecting, cookery, DIY, gardening, guidebooks, humour, reference, sports, travel, wine and spirits. IMPRINTS **Kenilworth Press** (see entry); **The Sportsman's Press** (see entry); **Swan Hill Press** (see entry). About 20 titles a year. Unsolicited mss only if the author sees some potential for sponsorship or guaranteed sales.
Ⓔ Royalties twice-yearly.

Quince Books Limited

158 Hermon Hill, South Woodford, London E18 1QH
info@quincebooks.com
www.quincebooks.com
Editor *Sara Hibbard*

Founded 2005. Publishes commercial ficton. About 15 books a year. Welcomes submissions; send synopsis and three chapters by e-mail.
Ⓔ Royalties annually.

Quintessence Editions
▷ Quarto Publishing plc

Quintet Publishing
▷ Quarto Publishing plc

Radcliffe Press
▷ I.B. Tauris & Co. Ltd

Radcliffe Publishing Ltd

18 Marcham Road, Abingdon OX14 1AA
☎ 01235 528820 📠 01235 528830
contact.us@radcliffemed.com
www.radcliffe-oxford.com
Managing Director *Gregory Moxon*
Editorial Director *Gillian Nineham*
Approx. Annual Turnover £2 million

Founded 1987. Medical publishing house which began by specializing in books for general practice and health service management. Publishes clinical, management, health policy books, training materials, journals, CD-ROMs, DVDs and online. Approximately 80 titles in 2008. Unsolicited mss, synopses and ideas welcome. No non-medical books.
Ⓔ Royalties twice-yearly.

Raintree
▷ Pearson (Heinemann)

The Random House Group Ltd

Random House, 20 Vauxhall Bridge Road,
London SW1V 2SA
☎ 020 7840 8400   📠 020 7233 6058
enquiries@randomhouse.co.uk
www.randomhouse.co.uk
Chief Executive/Chairman *Gail Rebuck*
Deputy CEO *Ian Hudson*
Group Managing Director *To be appointed*
Approx. Annual Turnover £279 million (UK)

The Random House Group is the UK's leading trade publisher comprising over 40 diverse imprints in six separate substantially autonomous companies: CCV, Random House Enterprises, Cornerstone, Ebury Publishing, Transworld (see entry) and Random House Children's Books.

Random House Audio Books ☎ 020 7840 8519
Commissioning Editor *Zoe Howes* (see entry under *Audio Books*)

Random House Digital ☎ 020 7840 8400
Director of Digital *Fionnuala Duggan*

CCV
Managing Director *Richard Cable*, Publisher *Dan Franklin*

IMPRINTS **Jonathan Cape** ☎ 0207 840 8836 📠 020 7233 6117 Publisher *Dan Franklin* Biography and memoirs, current affairs, fiction, history, photography, poetry, politics and travel. **Yellow Jersey Press** ☎ 020 7840 8621 📠 020 7223 6117 Editorial Director *Tristan Jones* Narrative sports books. **Harvill Secker** ☎ 020 7840 8893 📠 020 7233 6117 Publishing Director *Liz Foley* Principally literary fiction with some non-fiction. Literature in translation, English literature, quality thrillers. **Chatto & Windus** ☎ 020 7840 8745 📠 020 7233 6117 Publishing Director *Alison Samuel* Memoirs, current affairs, essays, literary fiction, history, poetry, politics, philosophy and translations. **Pimlico** ☎ 020 7840 8608 📠 020 7233 6117 Publishing Director *Rachel Cugnoni* Quality non-fiction paperbacks, specializing in history. **Vintage** ☎ 020 7840 8608 📠 020 7233 6117 Publishing Director *Rachel Cugnoni* Quality paperback fiction and non-fiction. Vintage (founded in 1990) has been described as one of the 'greatest literary success stories in recent British publishing'. **The Bodley Head** ☎ 020 7840 8553 📠 020 7233 6117 Publishing Director *Will Sulkin* History, science, current affairs, politics, biography, popular culture. **Square Peg** ☎ 020 7840 8621 📠 020 7233 6117 Editorial Director *Rosemary Davidson* Eclectic and commercial non-fiction, humour, novelties and gift books. Unsolicited mss with s.a.e.

RANDOM HOUSE ENTERPRISES
IMPRINTS **Preface Publishing** 1 Queen Anne's Gate, London SW1H 9BT ☎ 020 7840 8443

Publisher *Trevor Dolby*, Publishing Director *Rosie de Courcy* Commercial fiction and non-fiction. Unsolicited mss accepted at prefacesubmissions@randomhouse.co.uk

Virgin Books (Vauxhall Bridge Road address)
☎ 020 7840 8400 📠 020 7840 8383 Chairman *Richard Cable*, Managing Director *John Sadler*, Editorial *Carolyn Thorne, Edward Faulkner, Louisa Joyner* General non-fiction: biography and autobiography, memoir, popular reference, health and lifestyle, sport, travel, film and TV, music, humour, true crime. IMPRINTS **Black Lace** Erotic fiction by women for women; **Nexus** *Adam Nevill* Erotic fiction.

CORNERSTONE
Managing Director *Susan Sandon*

IMPRINTS **Century** ☎ 020 7840 8569 📠 020 7233 6127 Publishing Director *Mark Booth* General fiction and non-fiction including commercial fiction, autobiography, biography, history and self-help. **William Heinemann** ☎ 020 7840 8400 📠 020 7233 6127 Publishing Director *Jason Arthur* Fiction and general non-fiction: literary fiction, quality crime and thrillers, current affairs, translations, biography, popular science. No unsolicited mss. **Hutchinson** ☎ 020 7840 8400 📠 020 7233 7870 Publishing Director *Caroline Gascoigne* Fiction: literary and women's fiction, crime, thrillers. Non-fiction: current affairs, biography, memoirs, history, politics, travel, adventure. **Random House Books** (incorporating **Random House Business Books**) ☎ 020 7840 8451 📠 020 7233 6127 Publisher *Nigel Wilcockson* Non-fiction: social and cultural history, current affairs, popular culture, reference, business and economics. **Arrow** ☎ 020 7840 8414 📠 020 7840 6127 Publishing Director *Kate Elton* Mass-market paperback fiction and non-fiction. **Windmill Books** ☎ 020 7840 8414 📠 020 7840 6127 Editorial Director *Stephanie Sweeney* B-format paperback fiction and non-fiction. No unsolicited mss accepted by Cornerstone.

EBURY PUBLISHING
(Vauxhall Bridge Road address) ☎ 020 7840 8400 📠 020 7840 8406

Managing Director *Fiona MacIntyre*, Publisher *Jake Lingwood*

IMPRINTS **Ebury Press** Senior Publishing Director (non-illustrated & fiction) *Hannah MacDonald*, Publishing Director (illustrated, including cookery, family reference, history, culture and gift) *Carey Smith*, Editorial Director (fiction) *Gillian Green*, Publishing Director (popular non-fiction) *Andrew Goodfellow*. **Vermilion** Publishing Director *Clare Hulton* Personal development, health, diet, relationships and parenting. **Rider** Publishing Director *Judith Kendra* Psychology, biography, philosophy, personal development, spirituality, the paranormal. **Time Out Guides** Senior Publishing Brand Manager *Luthfa Begum* Travel and walking guides, film, reference and lifestyle published in a unique partnership with *Time Out* Group. **BBC Books** Senior Publishing Director *Shirley Patton.*

RANDOM HOUSE CHILDREN'S BOOKS DIVISION
61–63 Uxbridge Road, London W5 5SA
☎ 020 8579 2652 ☒ 020 8231 6737

Managing Director *Philippa Dickinson*, Publisher (fiction) *Annie Eaton*, Editorial Directors (fiction) *Charlie Sheppard*, *Kelly Hurst*, Publisher (Colour and Custom Publishing) *Fiona Macmillan*, Editorial Director (picture books) *Helen Mackenzie-Smith*, Senior Commissioning Editor (picture books) *Natascha Biebow* Picture books, fiction, poetry, non-fiction and audio cassettes under IMPRINTS **Bodley Head Children's Books**; **Corgi Children's Books**; **Doubleday Children's Books**; **Hutchinson Children's Books**; **Jonathan Cape Children's Books**; **Red Fox Children's Books**; **Tamarind Books** and **David Fickling Books** (31 Beaumont Street, Oxford OX1 2NP ☎ 01865 339000 ☒ 01865 339009 dfickling@ randomhouse.co.uk www.davidficklingbooks. co.uk) Publisher *David Fickling*, Senior Editor *Bella Pearson* Children's fiction and picture books.
💷 Royalties twice-yearly for the most part.

✱*Authors' update* Chasing hard on Hachette for the biggest share of the UK market, Random has grabbed a large share of the market for celebrity non-fiction. Restructuring has given a boost to Vintage Classics. Bodley Head has relaunched with a quality non-fiction list and, new on the street, Preface has been set up to capture readable books, fiction and non-fiction, that might otherwise escape the notice of the corporate scouts; newcomers should take heart. Gail Rebuck says that Random is keen to invest in new writers: 'There are no successes without risk.'

Ransom Publishing Ltd

51 Southgate Street, Winchester SO23 9EH
☎ 01962 862307 ☒ 05601 148881
rebecca@ransom.co.uk
www.ransom.co.uk
Managing Director *Jenny Ertle*

Ransom's book programme is designed to develop reading skills (and an enthusiasm for reading) in everyone from young children through to adults. Many of Ransom's books are 'high interest age, low reading age'; both fact and fiction. The list includes well known authors such as David Orme, whose *Boffin Boy* series has proved hugely popular with reluctant readers. Another success

is *Dark Man*. Aimed at adults with a very low reading age, the series won the 2006 Educational Resources Award. Approx. 70 titles in 2009. Interested in hearing from authors writing for reluctant and struggling readers. Initial enquiries by e-mail with details of the submission and three sample chapters. 'A response cannot be guaranteed but we try to respond to all submissions within a month.' Materials returned if supplied with s.a.e.

Reader's Digest Association Ltd

11 Westferry Circus, Canary Wharf, London E14 4HE
☎ 020 7715 8000   ☒ 020 7715 8181
gbeditorial@readersdigest.co.uk
www.readersdigest.co.uk
Managing Director *Chris Spratling*
Editorial Director, Books *Julian Browne*

Editorial office in the USA (see entry under *US Publishers*). Publishes health, gardening, natural history, cookery, history, DIY, computer, travel and word books. About 30 titles a year through trade and direct marketing channels.

Reaktion Books

33 Great Sutton Street, London EC1V 0DX
☎ 020 7253 1071   ☒ 020 7253 1208
info@reaktionbooks.co.uk
www.reaktionbooks.co.uk
Managing Director *Michael R. Leaman*

Founded 1985. Publishes art history, architecture, animal studies, Asian studies, cultural studies, design, film, food history, geography, history, natural history, photography and travel literature (*not* travel guides or journals). TITLE *Travels in the History of Architecture* Robert Harbison; SERIES TITLES *Edible* Pie Janet Clarkson; *Hot Dog* Bruce Kraig; *Spices* Fred Czarra. *Critical Lives* TITLES *Simone de Beauvoir* Ursula Tidd. *Animal* TITLES *Eel* Richard Schweid; *Hare* Simon Carnell. *Exposures* TITLE *Photography and Literature* François Brunet. About 45 titles a year. Written submissions only, to address above or e-mail to ian@reaktionbooks.co.uk Please enclose s.a.e. if you would like to have unsuccessful submitted material returned.
💷 Royalties twice-yearly.

Reardon Publishing

PO Box 919, Cheltenham GL50 9AN
☎ 01242 231800
reardon@bigfoot.com
www.reardon.co.uk
www.cotswoldbookshop.com
www.antarcticbookshop.com
www.nicholasreardon.com
Managing Editor *Nicholas Reardon*

Founded in the mid-1970s. Family-run publishing house specializing in local interest and tourism in the Cotswold area and books on Antarctica. Member of the Outdoor Writers & Photographers Guild and the National Union of Journalists. Publishes walking and driving guides. TITLES *The Cotswold Way*; *The Cotswold Way Map*; *Cotswold Walkabout*; *Cotswold Driveabout*; *The Donnington Way*; *The Haunted Cotswolds*. Also distributes for other publishers such as Ordnance Survey. 10 titles a year. Unsolicited mss, synopses and ideas welcome with return postage only. Ⓔ Royalties twice-yearly.

Recollections
▷ Arnefold Publishing

Red Arrow Books
▷ Blackstone Publishers

Red Fox Children's Books
▷ The Random House Group Ltd

Red Squirrel Publishing

Suite 235, 77 Beak Street, London W1F 9DB
☎ 0845 123 3803
info@redsquirrelbooks.com
www.redsquirrelbooks.com
Managing Director *Henry Dillon*
Approx. Annual Turnover £300,000

Red Squirrel Publishing specializes in citizenship test study guides. Its range of Life in the UK Test study guides has helped thousands of people achieve their goal of becoming British since the first edition was published in January 2006. Now in its third edition, the original definitive guide is expressly designed to help people pass the Life in the UK Test. 4 titles in 2008. Welcomes postal submissions. No fiction, poetry or children's books. Ⓔ Royalties annually.

Thomas Reed
▷ A.&C. Black Publishers Ltd

William Reed Directories

Broadfield Park, Crawley RH11 9RT
☎ 01293 613400　📠 01293 610322
directories@william-reed.co.uk
www.therightinfo.co.uk

William Reed Directories, a division of William Reed Business Media, was established in 1990. Its portfolio includes 13 titles covering the food, drink, non-food, catering, hospitality, retail and export industries, including European versions. The titles are produced as directories, market research reports, exhibition catalogues and electronic publishing.

Reed Elsevier Group plc

1–3 Strand, London WC2N 5JR
☎ 020 7930 7077
www.reedelsevier.com
Also at: 125 Park Avenue, 23rd Floor, New York, NY 10017, USA
☎ 001 212 309 5498　📠 001 212 309 5480
And: Raderweg 29, 1043 NX Amsterdam, The Netherlands
☎ 00 31 20 485 2222　📠 00 31 20 618 0325
Chief Executive Officer UK *Ian Smith*
Approx. Annual Turnover £4.58 billion

Reed Elsevier Group plc is a world-leading publisher and information provider. It is owned equally by its two parent companies, Reed Elsevier Plc and Reed Elsevier NV. No unsolicited material.

Regency House Publishing Limited

Niall House, 24–26 Boulton Road, Stevenage SG1 4QX
☎ 01438 314488　📠 01438 311303
regency-house@btconnect.com
www.regencyhousepublishing.com
Managing Director *Nicolette Trodd*

Founded 1991. Publisher and packager of mass-market non-fiction. No fiction. No unsolicited material.

Remember When

47 Church Street, Barnsley S70 2AS
☎ 01226 734222　📠 01226 734438
enquiries@pen-and-sword.co.uk
www.pen-and-sword.co.uk
Commissioning Editor *Fiona Shoop*

An imprint of **Pen & Sword Books Ltd**, Remember When covers the nostalgia and antiques collecting range of titles. No unsolicited mss; synopses and idea welcome. Ⓔ Royalties twice-yearly.

Reportage Press

26 Richmond Way, London W12 8LY
☎ 07971 461935
info@reportagepress.com
www.reportagepress.com
Chair *Rosie Whitehouse*

Founded in May 2007 to publish books on foreign affairs and books set in foreign countries. **Non-Fiction** TITLES *Bikila: Ethiopia's Barefoot Olympian* Tim Judah; *Passport to Enclavia* Vitali Vitaliev; *Memoirs of a Survivor* Sergei Golitsyn; *Franco's International Brigades* Christopher Othen; **Fiction** TITLE *A Girl in the Film* Charlotte Eager; **Despatches** TITLES *Red Zone: Five Bloody Days in Baghdad* Oliver Poole; *A Week in Waterloo* Magdalene De Lancey. 7 titles in 2008.

Unsolicited mss, synopses and ideas welcome; send synopses by e-mail and mss by post.

Research Disclosure
▷ **Kenneth Mason Publications Ltd**

Richmond House Publishing Company Ltd
70–76 Bell Street, Marylebone, London NW1 6SP
📱 020 7224 9666   📠 020 7224 9688
sales@rhpco.co.uk
www.rhpco.co.uk
Managing Directors *Gloria Gordon, Spencer Block*

Publishes directories for the theatre and entertainment industries. TITLES *British Theatre Directory*; *Artistes and Agents*; *London Seating Plan Guide*. Synopses and ideas welcome.

Rider
▷ **The Random House Group Ltd**

Rigby
▷ **Pearson (Heinemann)**

Rising Stars UK Ltd
22 Grafton Street, London W1S 4EX
📱 020 7495 6793   📠 020 7495 6796
info@risingstars-uk.com
www.risingstars-uk.com
Chairman *David Fulton*
Managing Director *Andrea Carr (andreacarr@risingstars-uk.com)*
Publisher *Camilla Erskine (camillaerskine@risingstars-uk.com)*
Approx. Annual Turnover £4 million

Founded 2002. Educational publisher of books, readers and software for schools. Winner of the ERA Supplier of the Year award for 2007. Unsolicited mss, synopses and ideas welcome; approach by e-mail.

✳*Authors' update* A 'lively publishing programme' and a 'clear understanding of its schools market' persuaded the Independent Publishers Guild to name Rising Stars as 2008 Education Publisher of the Year.

RMEP
▷ **SCM – Canterbury Press Ltd**

Robinson
▷ **Constable & Robinson Ltd**

Robson Books
▷ **Anova Books**

Rodale
▷ **Macmillan Publishers Ltd**

RotoVision
Sheridan House, 114 Western Road, Hove
BN3 1DD
📱 01273 727268   📠 01273 727269
sales@rotovision.com
www.rotovision.com
www.myspace.com/rotovision
Managing Director *Piers Spence*
Publisher *April Sankey*

Founded 1971. Acquired by **The Quarto Group, Inc.** in 1999. International publishers of graphic design, photography, digital design, advertising, film, product design and packaging design books. TITLES *World's Top Photographers: Nudes*; *What Is Graphic Design For?*; *Designs of the Times*; *First Steps in Digital Design*; *500 Digital Photography Hints, Tips and Techniques*; *Print and Production Techniques for Brochures and Catalogues*. 35 titles a year. No unsolicited mss; written synopses and ideas welcome. No phone calls, please. No academic or fiction.
💷 Flat fee paid.

Rough Guides
▷ **Penguin Group (UK)**

Round Hall (Ireland)
▷ **Sweet & Maxwell Group**

Roundhouse Group
Maritime House, Basin Road North, Hove
BN41 1WR
📱 01273 704962   📠 01273 704963
sales@roundhousegroup.co.uk
www.roundhousegroup.co.uk
CEO & Publisher *Alan Goodworth*
Director, Roundabout (Children's Book Division) *Bill Gills*
Director, Windsor Books International (Fine Arts Division) *Geoff Cowen*

Founded 1991. Represents and distributes a broad range of non-fiction publishers from the UK, Europe, USA, Canada, Australia, and elsewhere. 1,000+ titles a year. No unsolicited mss.

Routledge (An imprint of Taylor & Francis Group, an Informa plc business)
2 Park Square, Milton Park, Abingdon OX14 4RN
📱 020 7017 6000   📠 020 7017 6699
www.routledge.com
CEO *Roger Horton*
Managing Director *Jeremy North*
Publishing Directors *Claire L'Enfant, Alan Jarvis*

Routledge is an international academic imprint publishing books primarily for university students, researchers, academics and professionals. Most Routledge authors are academics with specialist knowledge of their subject. Publishes everything from core text books to research monographs and journals in the following subject areas: humanities, including anthropology, archaeology, classical studies, communications and media studies, cultural studies, history, language and literature, music, philosophy, religious studies, theatre studies and sports studies; social sciences, including geography, politics, military and strategic studies, sociology and urban studies; business, management, law and economics; education; Asian and Middle Eastern studies. Over 2,000 titles a year. Send a detailed proposal plus c.v. and sample chapters. Proposal requirements and submission procedures can be found on the website. No poetry, fiction, travel or astrology. Ⓔ Royalties annually or twice-yearly, according to contract.

Routledge-Cavendish/RoutledgeCurzon/RoutledgeFalmer
▷ Taylor & Francis Group

RYA (Royal Yachting Association)
RYA House, Ensign Way, Hamble, Southampton SO31 4YA
☎ 023 8060 4100   ⊞ 023 8060 4299
phil.williamsellis@rya.org.uk
www.rya.org.uk
Chief Executive *Rod Carr*
Publishing Editor/Publications Manager *Phil Williams-Ellis*
Approx. Annual Turnover £12million+

The Royal Yachting Association, the UK governing body representing sailing, windsurfing, motorboating, powerboat racing both at sea and on inland waters, was founded in 1875. Publishes sports instruction books relating to specific training courses, technical, legal and boating advice, all written by experts in their field. TITLES *RYA Weather Handbook*; *RYA Navigation Exercises*; *RYA Powerboat Handbook*; *RYA Go Sailing!* 10 titles a year. No unsolicited mss; synopses and ideas for books welcome. Approach by e-mail or post in the first instance. Ⓔ Royalties twice-yearly.

Ryelands
▷ Halsgrove

Ryland Peters & Small Limited
20–21 Jockey's Fields, London WC1R 5BW
☎ 020 7025 2200   ⊞ 020 7025 2201
info@rps.co.uk
www.rylandpeters.com
Managing Director *David Peters*
Publishing Director *Alison Starling*
Art Director *Leslie Harrington*

Founded 1996. Publishes highly illustrated lifestyle books aimed at an international market, covering home and garden, food and drink, babies and kids and gift. No fiction. Acquired **Cico Books** in 2006. No unsolicited mss; synopses and ideas welcome. Ⓔ Royalties twice-yearly.

Sage Publications
1 Oliver's Yard, 55 City Road, London EC1Y 1SP
☎ 020 7324 8500   ⊞ 020 7324 8600
info@sagepub.co.uk
www.sagepub.co.uk
Managing Director *Stephen Barr*
Editorial Director *Ziyad Marar*

Founded 1971. Publishes academic books and journals in humanities and the social sciences, education and science, technology and medicine. Bought academic and professional books publisher Paul Chapman Publishing Ltd in 1998 and Lucky Duck Publishing in 2004. Ⓔ Royalties twice-yearly.

St Pauls Publishing
187 Battersea Bridge Road, London SW11 3AS
☎ 020 7978 4300   ⊞ 020 7978 4370
editions@stpauls.org.uk
www.stpaulspublishing.com
Editor *Fr. Celso Godilano*

Publishing division of the Society of St Paul. Began publishing in 1914 but activities were fairly limited until around 1948. Publishes religious material mainly: theology, scripture, catechetics, prayer books and biography. About 20 titles a year. Unsolicited mss, synopses and ideas welcome with s.a.e.

Salamander
▷ Anova Books

SAM Publishing
▷ Pearson

Sandstone Press Ltd
PO Box 5725, One High Street, Dingwall IV15 9WJ
☎ 01349 862583   ⊞ 01349 862583
info@sandstonepress.com
www.sandstonepress.com
Managing Director *Robert Davidson*

Founded 2002. Publishes education and non-fiction. 15 titles in 2009. No unsolicited mss; synopses and ideas for books welcome. Initial approach by e-mail, please. Ⓔ Royalties twice-yearly.

Saqi Books
26 Westbourne Grove, London W2 5RH
☎ 020 7221 9347   ⊞ 020 7229 7492
info@saqibooks.com
www.saqibooks.com
Chairman/Managing Director *André Gaspard*
Commissioning Editor *Lara Frankena*

Founded 1981, initially as a specialist publisher of books on the Middle East and Arab world but now includes Central Asia, South Asia and Europe as well as some general titles. Publishes non-fiction – academic and illustrated, including art and food and drink. IMPRINT **Telegram** (see entry). About 20 titles a year. Welcomes unsolicited material; approach by post. Ⓔ Royalties annually.

❋*Authors' update* **A blunt warning from the managing editor that unsolicited manuscripts are not returned. Saqi is not alone in this. Authors should retain copies of everything they send out.**

Saunders
▷ Elsevier Ltd

Savitri Books Ltd
▷ entry under UK Packagers

SB Publications
14 Bishopstone Road, Seaford BN25 2UB
Ⓣ 01323 893498　Ⓕ 01323 893860
sbpublications@tiscali.co.uk
www.sbpublications.co.uk
Owner *Mrs Lindsay Woods*

Founded 1987. Specializes in local history, including themes illustrated by old picture postcards and photographs; also travel, guides (town, walking) and railways. TITLES *Natural Kent*; *Walking the Riversides of Sussex*; *Sussex As She Wus Spoke*; *The Neat and Nippy Guide to Brighton*; *Pre-Raphaelite Trail in Sussex*. Also provides marketing and distribution services for local authors. 12 titles a year. Ⓔ Royalties annually.

Sceptre
▷ Hachette UK

Scholastic Ltd
Euston House, 24 Eversholt Street, London NW1 1DB
Ⓣ 020 7756 7756　Ⓕ 020 7756 7795
www.scholastic.co.uk
Chairman *M.R. Robinson*
Acting Group Managing Director *Alan Hurcombe*
Approx. Annual Turnover £40 million

Founded 1964. Owned by US parent company, Scholastic Inc. who acquired **Chicken House Publishing** in May 2005. Publishes children's fiction and non-fiction and education resources for primary schools.

DIVISIONS

Scholastic Children's Books (Euston House address) Managing Director *Elaine McQuade* IMPRINT **Scholastic** TITLES *Horrible Histories* Terry Deary, illus. Martin Brown; *His Dark Materials Trilogy* Philip Pullman; *Mortal Engines* Philip Reeve; *Stickman* Julia Donaldson and Axel Scheffler.

Educational Publishing (Villiers House, Clarendon Avenue, Leamington Spa CV32 5PR

Ⓣ 01926 887799) Managing Director *Denise Cripps* Professional books and classroom materials for primary teachers, plus magazines such as *Child Education Plus*, *Junior Education Plus*, *Junior Focus*, *Nursery Education*, *Literacy Time Plus*.

Scholastic Book Clubs (Euston House address) Head of Book Clubs *Julie Randles* 'SBC is the UK's number one Schools Book Club, providing the best books at the best prices and supports teachers in the process.'

School Book Fairs (Dolomite Avenue, Coventry Business Park, Coventry CV5 6UE) Head of Book Fairs *Steve Thompson* The Book Fair Division sells directly to children, parents and teachers in schools through 25,000 week-long book events held in schools throughout the UK. Ⓔ Royalties twice-yearly.

SCM – Canterbury Press
13–17 Long Lane, London EC1A 9PN
Ⓣ 020 7776 7540　Ⓕ 020 7776 7556
admin@scm-canterburypress.co.uk
www.scm-canterburypress.co.uk
Publishing Director *Christine Smith*
Commissioning Editors *Natalie K. Watson (SCM)*, *Valerie Bingham (RMEP)*
Approx. Annual Turnover £2.6 million

Part of **Hymns Ancient & Modern Ltd** (see entry). Has three publishing imprints: **SCM Press** Text and reference books for the study of theology, philosophy of religion, ethics and religious studies; **Canterbury Press** Religious titles for the general market, liturgy, spirituality and church resources; **RMEP** Classroom resources for religious education in schools. AUTHORS include Rowan Williams, Ronald Blythe, Helen Prejean. 120 titles a year. Refer to website for guidelines on submitting proposals. Ⓔ Royalties annually.

❋*Authors' update* **Leading publisher of religious ideas with well-deserved reputation for fresh thinking. At SCM, 'questioning theology is the norm'.**

Scottish Cultural Press/Scottish Children's Press
Unit 6, Newbattle Abbey Business Park, Newbattle Road, Dalkeith EH22 3LJ
Ⓣ 0131 660 6366　Ⓕ 0870 285 4846
info@scottishbooks.com
www.scottishbooks.com
Directors *Avril Gray, Brian Pugh*

Founded 1992. Publishes Scottish interest titles, including cultural literature, poetry, archaeology, local history. DIVISION **S.C.P. Children's Ltd** (trading as **Scottish Children's Press**) Children's fiction and non-fiction. Unsolicited mss, synopses

and ideas accepted provided return postage is included, 'but *always* telephone before sending material, please. Mss sent without advance telephone call and return postage will be destroyed.'
Ⓔ Royalties paid.

Seafarer Books Ltd

102 Redwald Road, Rendlesham, Woodbridge
IP12 2TE
☎ 01394 420789   🖷 01394 461314
info@seafarerbooks.com
www.seafarerbooks.com
Managing Director *Patricia M. Eve*

Founded 1968. Publishes sailing titles and stories of the sea in a greatly expanded list reflecting a broader understanding of sailing narrative, maritime history and the crafts, arts and letters of the sea. AUTHORS Erskine Childers, Maldwin Drummond, Tony Groom, Tristan Lovering, Ian Tew, Peter Villiers, Ewen Southby-Tailyour. About 6 titles a year. No unsolicited mss; preliminary letter essential before making any type of submission.
Ⓔ Royalties twice-yearly.

Seaforth Publishing (an imprint of Pen & Sword Books)

47 Church Street, Barnsley S70 2AS
☎ 01226 734222   🖷 01226 734438
info@seaforthpublishing.com
www.seaforthpublishing.com
Managing Director *Charles Hewitt*

Founded 2007. Specialist maritime imprint of **Pen & Sword Books Ltd**. Publishes maritime history and general maritime books. TITLES *The Wooden Walls*; *Iron & Steel Navies*; *Sail & Traditional Craft*; *The Merchant Marine*; *Ship Modelling*. Unsolicited submissions welcome; send synopsis and sample chapter in printed format (double line spacing, single side). No fiction.
Ⓔ Royalties twice-yearly.

Search Press Ltd

Wellwood, North Farm Road, Tunbridge Wells
TN2 3DR
☎ 01892 510850   🖷 01892 515903
searchpress@searchpress.com
www.searchpress.com
Managing Director *Martin de la Bédoyère*
Commissioning Editor *Rosalind Dace*

Founded 1970. Publishes full-colour art, craft, needlecrafts – papermaking and papercrafts, painting on silk, art techniques and embroidery. No unsolicited mss; synopsis with sample chapter welcome.
Ⓔ Royalties annually.

Seren

57 Nolton Street, Bridgend CF31 3AE
☎ 01656 663018
general@seren-books.com
www.seren-books.com
Chairman *Richard Houdmont*
Publisher *Mick Felton*
Approx. Annual Turnover £150,000

Founded 1981 as a specialist poetry publisher but has now moved into general literary publishing with an emphasis on Wales. Publishes poetry, fiction, literary criticism, drama, biography, art, history and translations of fiction.
DIVISIONS **Poetry** *Amy Wack* AUTHORS Owen Sheers, Pascale Petit, Sheenagh Pugh, Kathryn Simmonds, Tiffany Atkinson, John Haynes, Carol Rumens. **Fiction** *Penny Thomas* AUTHORS Richard Collins, Lloyd Jones, Emyr Humphreys, Robert Minhinnick, Nia Wyn. **Art, Literary Criticism, History, Translations** *Mick Felton* IMPRINT **Border Lines Biographies** TITLES *Bruce Chatwin*; *Dennis Potter*; *Mary Webb*; *Wilfred Owen*; *Raymond Williams*. About 25 titles a year. Unsolicited mss, synopses and ideas for books accepted. No children's fiction.
Ⓔ Royalties twice-yearly or annually.

Serpent's Tail

3a Exmouth House, Pine Street, London EC1R 0JH
☎ 020 7841 6300   🖷 020 7833 3969
info@serpentstail.com
www.serpentstail.com
Publisher *Pete Ayrton*

Founded 1986. Since 2007, an imprint of **Profile Books**. Publishes fiction and non-fiction; literary and mainstream work, and work in translation. No unsolicited mss.

Severn House Publishers

9–15 High Street, Sutton SM1 1DF
☎ 020 8770 3930   🖷 020 8770 3850
info@severnhouse.com
www.severnhouse.com
Chairman *Edwin Buckhalter*
Editorial *Amanda Stewart*

Founded 1974. A leader in library fiction publishing. Publishes hardback fiction: romance, horror, crime. Also large print and trade paperback editions of their fiction titles. About 140 titles a year. No unsolicited material. Synopses/proposals preferred through *bona fide* literary agents only. OVERSEAS ASSOCIATE Severn House Publishers Inc., New York.
Ⓔ Royalties twice-yearly.

Sheldon Press
▷ Society for Promoting Christian Knowledge

Sheldrake Press

188 Cavendish Road, London SW12 0DA
☎ 020 8675 1767   📠 020 8675 7736
enquiries@sheldrakepress.co.uk
www.sheldrakepress.co.uk
Publisher *Simon Rigge*

Founded in 1979 as a book packager and commenced publishing under its own imprint in 1991. Publishes illustrated non-fiction: history, travel, style, cookery and stationery. TITLES *The Victorian House Book*; *The Shorter Mrs Beeton*; *The Power of Steam*; *The Railway Heritage of Britain*; *Wild Britain*; *Wild France*; *Wild Spain*; *Wild Italy*; *Wild Ireland*; *Amsterdam: Portrait of a City* and Kate Greenaway stationery books. 1 title in 2008. Synopses and ideas for books welcome but not interested in fiction.

Shepheard-Walwyn (Publishers) Ltd

15 Alder Road, London SW14 8ER
☎ 020 8241 5927
books@shepheard-walwyn.co.uk
www.shepheard-walwyn.co.uk
Managing Director *Anthony Werner*
Approx. Annual Turnover £100,000

Founded 1972. 'We regard books as food for heart and mind and want to offer a wholesome diet of original ideas and fresh approaches to old subjects.' Publishes general non-fiction in four main areas: gift books in calligraphy and/or illustrated; biography, ethical economics, perennial philosophy. About 8 titles a year. Synopses and ideas for books welcome.
📖 Royalties twice-yearly.

The Shetland Times Ltd

Gremista, Lerwick ZE1 0PX
☎ 01595 693622   📠 01595 694637
publishing@shetland-times.co.uk
www.shetland-books.co.uk
Managing Director *Brian Johnston*
Publications Manager *Charlotte Black*

Founded 1872 as publishers of the local newspaper. Book publishing followed thereafter plus publication of monthly magazine, *Shetland Life*. Publishes anything with Shetland connections – local and natural history, music, crafts, maritime. Prefers material with a Shetland theme/connection. 8 titles in 2007.
📖 Royalties annually.

Shire Publications Ltd

Midland House, West Way, Botley, Oxford
OX2 0PH
☎ 01865 727022   📠 01865 727017/727019
shire@shirebooks.co.uk
www.shirebooks.co.uk
Managing Director *Rebecca Smart*
Publisher *Nicholas Wright*

Founded 1962. Publishes original non-fiction paperbacks. Acquired by **Osprey Publishing Ltd** in 2007. 32 titles in 2008. No unsolicited material; send introductory letter with detailed outline of idea.
📖 Royalties annually.

✱*Authors' update* **You don't have to live in the country to write books for Shire but it helps. With titles like** *Church Fonts*, *Haunted Inns* **and** *Discovering Preserved Railways* **there is a distinct rural feel to the list. Another way of putting it, to quote owner John Rotheroe, Shire specializes in 'small books on all manner of obscure subjects'.**

Short Books

3A Exmouth House, Pine Street, Exmouth Market, London EC1R 0JH
☎ 020 7833 9429   📠 020 7833 9500
info@shortbooks.co.uk
www.shortbooks.co.uk
Editorial Directors *Rebecca Nicolson, Aurea Carpenter*

Founded 2001. Publishes general adult and children's non-fiction and fiction; biography, history and politics, humour and children's narrative history. No unsolicited mss.

Sickle Moon
▷ **Eland Publishing Ltd**

Sidgwick & Jackson
▷ **Macmillan Publishers Ltd**

Sigma Press

Stobart House, Pontyclerc, Penybanc Road, Ammanford SA18 3HP
☎ 01269 593100   📠 01269 596116
info@sigmapress.co.uk
www.sigmapress.co.uk
Owners *Nigel Evans, Jane Evans*
Chairman/Managing Director *Nigel Evans*

Founded in 1980 as a publisher of technical books, Sigma Press now publishes mainly in the leisure area. Publishes outdoor, local heritage, adventure and biography. DIVISION **Sigma Leisure** TITLES *The Bluebird Years: Donald Campbell and the pursuit of speed* (biography); *In Search of Swallows & Amazons* (biography); *Walks in Ancient Wales*; *All-Terrain Pushchair Walks*; *Discovering Manchester*. About 10 titles a year. No unsolicited mss; synopses and ideas welcome. No poetry or novels required.
📖 Royalties twice-yearly.

Simon & Schuster UK Limited

1st Floor, 222 Gray's Inn Road, London WC1X 8HB
☎ 020 7316 1900   📠 020 7316 0331
www.simonsays.co.uk
CEO/Managing Director *Ian S. Chapman*

Publishing Director *Suzanne Baboneau*
Publishers, Non-fiction *Mike Jones, Angela Herlihy, Kerri Sharp*
Publishers, Fiction *Suzanne Baboneau, Kate Lyall-Grant*
Pocket Paperback Publisher *Julie Wright*
Approx. Annual Turnover £31.2 million

Founded 1986. Sister company of the leading American company. Publishes hardback (**Simon & Schuster**) and paperback (**Pocket Books**) commercial and literary fiction, media tie-ins and non-fiction, including autobiography, biography, sport, business, history, popular science, politics, health, self-help, mind, body and spirit, humour, travel and other general interest non-fiction.

DIVISIONS **Martin Books** Director *Janet Copleston* Specializes in bespoke publishing and branded editions, and the Simon & Schuster UK cookery list. **Simon & Schuster Children's Books** Publishing Director *Ingrid Selberg*, Editor (fiction) *Venetia Gosling*, Editor (picture books) *Emma Blackburn* Picture books, novelty and fiction titles. No unsolicited mss.
Ⓔ Royalties twice-yearly.

Singing Dragon
▷ Jessica Kingsley Publishers Ltd

Smith Gryphon Ltd
▷ John Blake Publishing Ltd

Colin Smythe Ltd
38 Mill Lane, Gerrards Cross SL9 8BA
☎ 01753 886000   📠 01753 886469
cs@colinsmythe.co.uk
www.colinsmythe.co.uk
Managing Director *Colin Smythe*
Approx. Annual Turnover £2.8 million

Founded 1966. Publishes Anglo-Irish literature, drama; criticism and Irish history. Acts as agent for Terry Pratchett. About 3 titles a year. No unsolicited mss.
Ⓔ Royalties annually/twice-yearly.

Snowbooks Ltd
120 Pentonville Road, London N1 9JN
☎ 0790 406 2414
manuscripts@snowbooks.com
www.snowbooks.com
Managing Director *Emma Barnes*

Founded 2003. Independent publisher of mainly fiction and some non-fiction. 'Keen on promoting open, friendly, productive relationships with authors and retailers.' TITLES *Adept* Robert Finn; *Needle in the Blood* Sarah Bower; *The Idle Thoughts of an Idle Fellow* J.K. Jerome; *Boxing Fitness* Ian Oliver; *Monster Trilogy* David Wellington; *City Cycling* Richard Ballantine; *Plotting for Beginners* Sue Hepworth and Jane Linfoot; *The Other Eden* and *Sand Daughter*

Sarah Bryant. 15–30 titles a year. Unsolicited mss 'very welcome', by e-mail only. Read the submission guidelines in the authors' section of the website first: www.snowbooks.com/submissions.html
Ⓔ Royalties are payable on *net* receipts.

✷*Authors' update* There are only two conditions for taking on a book: an editor must like it and believe it will sell. Snowbooks has already chalked up a 'Nibbie' award as Small Publisher of the Year and in 2008 was named Trade Publisher of the Year by the Independent Publishers Guild.

Society Books
17–21 Wyfold Road, London SW6 6SE
☎ 020 7385 6137   📠 020 7381 8444
sbimprints@societybooks.com
www.societybooks.com
Owner *Books & Life Limited*
Managing Director *Hom Paribag*
Approx. Annual Turnover £500,000

Launched in 2007 by the owners of *Society Today Magazine* to publish books with a social message by victims of injustice, the disabled and ethnic minority authors, both new and established. Publishes fiction and non-fiction, arts and culture and academic titles. 1 title in 2008. Will consider 60-word synopsis with introductory letter but material via literary agent preferred.
Ⓔ Royalties twice-yearly.

Society for Promoting Christian Knowledge (SPCK)
36 Causton Street, London SW1P 4ST
☎ 020 7592 3900   📠 020 7592 3939
www.spck.org.uk
www.sheldonpress.co.uk
Chief Executive Officer *Simon Kingston*

Founded 1698, SPCK is the third oldest publisher in the country. IMPRINTS **SPCK** Publishing Director *Joanna Moriarty* Theology, academic, liturgy, prayer, spirituality, biblical studies, educational resources, mission, pastoral care, gospel and culture, worldwide. **Sheldon Press** Editor *Fiona Marshall* Popular medicine, health, self-help, psychology. **Azure** Senior Editor *Alison Barr* General spirituality.
Ⓔ Royalties annually.

✷*Authors' update* Theology with a liberal agenda.

Sophia Books
▷ Rudolf Steiner Press

Southwater
▷ Anness Publishing Ltd

Souvenir Press Ltd

43 Great Russell Street, London WC1B 3PD
☎ 020 7580 9307/7637 5711   ⊞ 020 7580 5064
souvenirpress@ukonline.co.uk
Chairman/Managing Director *Ernest Hecht*

Independent publishing house. Founded 1951.
Publishes academic and scholarly, animal
care and breeding, antiques and collecting,
archaeology, autobiography and biography,
business and industry, children's, cookery,
crafts and hobbies, crime, educational, fiction,
gardening, health and beauty, history and
antiquarian, humour, illustrated and fine editions,
magic and the occult, medical, military, music,
natural history, philosophy, poetry, psychology,
religious, sociology, sports, theatre and women's
studies. Souvenir's Human Horizons series for
the disabled and their carers is one of the most
pre-eminent in its field and recently celebrated 33
years of publishing for the disabled.

IMPRINTS/SERIES **Condor**; **Human Horizons**;
Independent Voices; **Pictorial Presentations**;
Pop Universal; **The Story-Tellers**. TITLES *Out
of the Shadows* Francois Maspero; *The Writing
Workshop Notebook* Alan Ziegler; *The Light
Jewish Cookery Book* Sylvie Jouffa and Annick
Champetier de Ribes; *The Complete Martial Arts
Student* Martina Sprague; *Fairy Spells* Claire
Nahmad; *The Bullying Problem* Alan Train;
Improve Your Eyesight Jonathan Barnes. About 55
titles a year. Unsolicited mss considered but initial
letter of enquiry and outline always required in
the first instance.
ⓔ Royalties twice-yearly.

❋*Authors' update* One of the last of the great
independents, Souvenir covers every subject
which happens to appeal to Ernest Hecht.
But authors should not expect sentimentality.
Commercial potential is the first priority.

SPCK
▷ Society for Promoting Christian Knowledge

Special Interest Model Books Ltd

PO Box 327, Poole BH15 2RG
☎ 01202 649930   ⊞ 01202 649950
chrlloyd@globalnet.co.uk
www.specialinterestmodelbooks.co.uk
Contact *Chris Lloyd*

Publishes aviation, model engineering, leisure and
hobbies, modelling, electronics, wine and beer
making. Send synopses rather than completed
mss.
ⓔ Royalties twice-yearly.

Speechmark Publishing Ltd

70 Alston Drive, Bradwell Abbey, Milton Keynes
MK13 9HG
☎ 01908 326944   ⊞ 01908 326960

info@speechmark.net
www.speechmark.net
Managing Director *Liz Lane*
Publishing Manager *Sue Christelow*

Speechmark publishes practical books, games
and *ColorCards*® for health, education and special
needs practitioners and students. Unsolicited
mss, synopses and ideas welcome. Approach by
letter or e-mail.
ⓔ Royalties twice-yearly.

Spellmount
▷ The History Press

Sphere
▷ Little, Brown Book Group UK

Spon Press
▷ Taylor & Francis Group

SportsBooks Limited

PO Box 422, Cheltenham GL50 2YN
☎ 01242 256755   ⊞ 05603 108126
randall@sportsbooks.ltd.uk
www.sportsbooks.ltd.uk
Chairman & Managing Director *Randall
Northam*
Approx. Annual Turnover £150,000

Founded 1995. Publishes sports books including
biographies, statistical and practical. TITLES
include *Talking of Sport*; *The Nationwide Football
Annual* (formerly *The News of the World Football
Annual*); *Sporting Justice*; *The Lion and The Eagle*;
Modern Football is Rubbish. IMPRINT **BMM**. 15
titles in 2008. No unsolicited mss. Synopses and
ideas welcome. Approach by letter or e-mail.

The Sportsman's Press (An imprint of Quiller Publishing Ltd)

Wykey House, Wykey, Shrewsbury SY4 1JA
☎ 01939 261616   ⊞ 01939 261606
info@quillerbooks.com
www.countrybooksdirect.com
Managing Director *Andrew Johnston*

Specializes in books on all country subjects
and general sports including fishing, fencing,
shooting, equestrian, cookery, gunmaking and
wildlife art. About 5 titles a year.
ⓔ Royalties twice-yearly.

Springer-Verlag London Ltd

Ashbourne House, The Guildway, Old
Portsmouth Road, Guildford GU3 1LP
☎ 01483 734666   ⊞ 01483 734411
www.springer.com
General Manager *Beverley Ford*
Approx. Annual Turnover £5 million

The UK subsidiary of Springer Science & Business
Media. Publishes science, technical and medical
books and journals. Specializes in computing,

engineering, medicine, mathematics, chemistry, biosciences. All UK published books are sold through Springer's German and US companies as well as in the UK. Globally: 4,000 books and 1,250 journals; UK: 150 titles plus 27 journals. Not interested in social sciences, fiction or school books but academic and professional science mss or synopses welcome.
Ⓔ Royalties annually.

Springhill
▷ How To Books Ltd

Spruce
▷ Octopus Publishing Group

Square Peg
▷ The Random House Group Ltd

Stadia
▷ The History Press

Stainer & Bell Ltd
PO Box 110, 23 Gruneisen Road, London N3 1DZ
☎ 020 8343 3303　🖷 020 8343 3024
post@stainer.co.uk
www.stainer.co.uk
Managing Directors *Carol Y. Wakefield, Keith M. Wakefield*
Publishing Director *Nicholas Williams*
Approx. Annual Turnover £774,000

Founded 1907 to publish sheet music. Publishes music and religious subjects related to hymnody. Unsolicited synopses/ideas for books welcome. Send letter enclosing brief précis.
Ⓔ Royalties annually.

The Stationery Office
▷ TSO

Rudolf Steiner Press
Hillside House, The Square, Forest Row RH18 5ES
☎ 01342 824433　🖷 01342 826437
office@rudolfsteinerpress.com
www.rudolfsteinerpress.com
Chairman *P. Martyn*
Manager *S. Gulbekian*
Approx. Annual Turnover £170,000

Founded in 1925 to publish the work of Rudolf Steiner and related materials. Publishes non-fiction: spirituality and philosophy. IMPRINT **Sophia Books** *S. Gulbekian* TITLES *From Stress to Serenity; Homemaking as a Social Art.* 15 titles in 2008. No unsolicited material.
Ⓔ Royalties annually.

The Story-Tellers
▷ Souvenir Press Ltd

Straightline Publishing Ltd
29 Main Street, Bothwell G71 8RD
☎ 01698 853000　🖷 01698 854208

Director *Patrick Bellew*
Editor *James Hunt*

Founded 1989. Publishes magazines and directories – trade and technical – books of local interest. TITLES *Wirein; enterprising.* No unsolicited material.
Ⓔ Royalties annually.

Studymates Limited
PO Box 225, Abergele, Conwy County LL18 9AY
☎ 01745 832863　🖷 01745 826606
info@studymates.co.uk
www.studymates.co.uk
Managing Editor *Dr Graham Lawler, MA*

Authors need to be practising or recently-retired teachers/lecturers, or published authors for the writers' guides. Ideas in writing first; unsolicited mss will not be considered. List of business books under the IMPRINT **Studymates Professional**. New list of education books planned for ABER but again Studymates will not accept unsolicted mss. 'Please approach in writing in the first instance and outline how your idea will complement the current list.'

Summersdale Publishers Ltd
46 West Street, Chichester PO19 1RP
☎ 01243 771107　🖷 01243 786300
submissions@summersdale.com
www.summersdale.com
Owners/Managing Directors *Stewart Ferris, Alastair Williams*
Editorial Director *Jennifer Barclay*
Approx. Annual Turnover £2.5 million

Founded 1990. Publishes travel literature, self-help, humour and gift books, general non fiction. TITLES *Pimp Your Pet; Watch My Back* (movie tie-in edition); *Get Past Driving Test Nerves; Narrowboat Dreams – A Journey North by England's Waterways; Secret Societies – their mysteries revealed.* 85 titles a year. No unsolicited mss; synopses and ideas welcome by e-mail.
Ⓔ Royalties paid.

Sussex Academic Press
PO Box 139, Eastbourne BN24 9BP
☎ 01323 479220　🖷 01323 478185
edit@sussex-academic.co.uk
www.sussex-academic.co.uk
Managing Director *Anthony Grahame*
Approx. Annual Turnover £350,000

Founded 1994. Academic publisher. DIVISION/IMPRINT **Sussex Academic** TITLES *Chinese Religions: Beliefs & Practices; Picture Imperfect: Photography and Eugenics, 1879-1940.* **Alpha Press** TITLE *Watercraft on World Coins, Volume I: Europe 1800-2006.* 45 titles in 2008. No

unsolicited material; send letter of inquiry in the first instance.
Ⓔ Royalties annually.

Sutton
▷ The History Press

Swan Hill Press (An imprint of Quiller Publishing Ltd)
Wykey House, Wykey, Shrewsbury SY4 1JA
☎ 01939 261616   📠 01939 261606
info@quillerbooks.com
www.countrybooksdirect.com
Managing Director *Andrew Johnston*

Specializes in practical books on all country and field sports activities. Publishes books on fishing, shooting, gundog training, falconry, equestrian, deer, cookery, wildlife art. About 30 titles a year. Unsolicited mss must include s.a.e.
Ⓔ Royalties twice-yearly.

Sweet & Maxwell
100 Avenue Road, London NW3 3PF
☎ 020 7393 7000   📠 020 7393 7010
www.sweetandmaxwell.co.uk
Managing Director *Peter Lake*

Founded 1799. Part of Thomson Reuters. Publishes materials in all media; looseleaf works, journals, law reports, CD-ROMs and online. The legal and professional list is varied and contains academic titles as well as treatises and reference works in the legal and related professional fields.

IMPRINTS **Sweet & Maxwell; W. Green (Scotland); Incomes Data Services; Round Hall (Ireland)**. Over 1,100 products, including 180 looseleafs, 80 periodicals, more than 40 digital products and online information services and 200 new titles each year. Ideas welcome. Writers with legal/professional projects in mind are advised to contact the Legal Business Unit at the earliest possible stage in order to lay the groundwork for best design, production and marketing of a project.
Ⓔ Royalties and fees according to contract.

T2
▷ Troubador Publishing Ltd

Tamarind Books
▷ The Random House Group Ltd

Tango Books Ltd
PO Box 32595, London W4 5YD
☎ 020 8996 9970   📠 020 8996 9977
sales@tangobooks.co.uk
www.tangobooks.co.uk
Directors *David Fielder, Sheri Safran*
Approx. Annual Turnover £1.1 million

Founded 1981. Children's books publisher with international co-edition potential: pop-ups, three-dimensional, novelty, picture and board books; 1,000 words maximum. About 20 titles a year. Preliminary letter or e-mail with sample material. Material posted must include s.a.e. for return.
Ⓔ Primarily, fees paid.

Taschen UK
Fifth Floor, 1 Heathcock Court, 415 Strand, London WC2R 0NS
☎ 020 7845 8585   📠 020 7836 3696
www.taschen.com

UK office of the German photographic, art and architecture publisher. Editorial office in Cologne (see entry under *European Publishers*).

I.B.Tauris & Co. Ltd
6 Salem Road, London W2 4BU
☎ 020 7243 1225   📠 020 7243 1226
mail@ibtauris.com
www.ibtauris.com
Chairman/Publisher *Iradj Bagherzade*
Managing Director *Jonathan McDonnell*

Founded 1984. Independent publisher. Publishes general non-fiction and academic in the fields of international relations, religion, current affairs, history, politics, cultural, media and film studies, Middle East studies. Distributes **Philip Wilson Publishers** and The Federal Trust worldwide.
IMPRINTS **Tauris Academic Studies** Academic monographs in the humanities and social sciences; **Tauris Parke Paperbacks** Trade titles, including history, travel and biography; **Radcliffe Press** Colonial history and biography. Unsolicited synopses and book proposals welcome.
Ⓔ Royalties twice-yearly.

✴*Authors' update* **Imprint Radcliffe Press may require authors to contribute towards costs of publication.**

Taylor & Francis Group (an Informa plc business)
2 Park Square, Milton Park, Abingdon OX14 4RN
☎ 020 7017 6000   📠 020 7017 6699
www.tandf.co.uk
CEO *Roger Horton*
Managing Director (UK) *Jeremy North*
President (US) *Emmett Dages*
Vice President (US) *John Lavender*

Publishes academic and reference books and journals. IMPRINTS **Routledge** (see entry); **Routledge-Cavendish; RoutledgeCurzon; RoutledgeFalmer; Taylor & Francis** (see entry); **Europa Publications; Garland Science; Spon Press; Psychology Press** (see entry); **Gordon & Breach; Harwood Academic; CRC Press; Frank Cass; BIOS Scientific Publishers**. About 1,800 titles a year.

Taylor & Francis (An imprint of Taylor & Francis Group, an Informa plc business)

2 Park Square, Milton Park, Abingdon OX14 4RN
☎ 020 7017 6000   📠 020 7017 6699
www.tandf.co.uk
CEO *Roger Horton*
Managing Director (UK) *Jeremy North*
President (US) *Emmett Dages*
Vice President (US) *John Lavender*

Specializes in scientific and technical books for university students and academics in the following subjects: built environment, architecture, planning and civil engineering; biology, including genetics, plant architecture, molecular biology and biotechnology. About 600 titles a year. Send detailed proposal plus c.v. and sample chapters. Proposal requirements and submission procedures can be found on the website. No poetry, fiction, travel or astrology.
Ⓔ Royalties annually.

⊛*Authors' update* **An early leader in digitization, T&F has 16,000 titles online. Roger Horton sees publishing as a filtering process – searching out quality in the information overload. Very much into higher education and reference. Further expansion is predicted, particularly in the US market.**

Teach Yourself
▷ Hachette UK

Telegram

26 Westbourne Grove, London W2 5RH
☎ 020 7229 2911   📠 020 7229 7492
info@telegrambooks.com
www.telegrambooks.com
Chairman/Managing Director *André Gaspard*
Commissioning Editor *Rebecca O'Connor*

Imprint of **Saqi Books**. Independent publisher of literary fiction from around the world. 15 titles a year. Welcomes unsolicited material; approach by post.
Ⓔ Royalties annually.

Telegraph Books

111 Buckingham Palace Road, London SW1W 0DT
☎ 020 7931 2887   📠 020 7931 2929
www.books.telegraph.co.uk
Owner *Telegraph Media Group Ltd*
Publisher *Caroline Buckland*

Concentrates on Telegraph branded books in association/collaboration with other publishers. Publishes general non-fiction: reference, personal finance, health, gardening, cookery, heritage, guides, humour, puzzles and games. Mainly interested in books if a Telegraph link exists. About 50 titles a year. No unsolicited material.
Ⓔ Royalties twice-yearly.

Templar Publishing

The Granary, North Street, Dorking RH4 1DN
☎ 01306 876361   📠 01306 889097
editorial@templarco.co.uk
www.templarco.co.uk
Managing Director/Editorial Head *Amanda Wood*
Approx. Annual Turnover £17 million

Publisher and packager of quality novelty and gift books, picture books, children's illustrated non-fiction and fiction for the 8 to 12 age range. 80 titles a year. Synopses and ideas for books welcome. 'We are particularly interested in picture book mss and ideas for new novelty concepts.'
Ⓔ Royalties by arrangement.

⊛*Authors' update* **Winner of the Independent Publisher of the Year award, Templar has been praised for its 'Ology' series of books and their accompanying website, said to be 'polished and professional'.**

Temple Lodge Publishing Ltd

Hillside House, The Square, Forest Row RH18 5ES
☎ 01342 824000
office@templelodge.com
www.templelodge.com
Chairman *R. Pauli*
Chief Editor *S. Gulbekian*
Approx. Annual Turnover £65,000

Founded in 1990 to develop the work of Rudolf Steiner (see also **Rudolf Steiner Press**). Publishes non-fiction: mind, body and spirit, current affairs, health and therapy. 12 titles in 2008. No unsolicited material; send initial letter of enquiry. No poetry or fiction.
Ⓔ Royalties annually.

Tempus
▷ The History Press

Thames and Hudson Ltd

181A High Holborn, London WC1V 7QX
☎ 020 7845 5000   📠 020 7845 5050
mail@thameshudson.co.uk
www.thamesandhudson.com
Chairman *Thomas Neurath*
Managing Director *Jamie Camplin*
Approx. Annual Turnover £27 million

Publishes art, archaeology, architecture and design, biography, decorative arts, fashion and textiles, garden and landscape design, graphics, history, illustrated reference, mythology, photography, popular culture, style and interior design, travel. SERIES *World of Art*; *Hip Hotels*; *StyleCity*; *New Horizons*; *Family LifeStyle*; *Earth From the Air*; *The Way We Live*; *Photofile*. TITLES *Art Since 1900*; *The Artist's Yearbook*; *Magnum Magnum*; *The Great LIFE Photographers*;

Hockney's Pictures; Sensation; Germaine Greer's The Boy; The Urban Cookbook; The Eco-Travel Handbook; Fashion Illustration by Fashion Designers; Cartier; Factory Records; XS Green; World Architecture: The Masterworks; The Seventy Great Mysteries of the Natural World; Advertising Next; Avant Gardeners; Great Military Leaders and their Campaigns; Reuters: Our World Now. 200 titles a year. Send preliminary letter and outline before mss.
Ⓔ Royalties twice-yearly.

✱*Authors' update* Operating in a tough environment, Thames and Hudson holds its own in the production of high quality illustrated books with titles that reach beyond the specialist art market.

Third Millennium Publishing

Third Millennium Information Ltd, 2–5 Benjamin Street, London EC1M 5QL
☎ 020 7336 0144   🖷 020 7608 1188
info@tmiltd.com
www.tmiltd.com
Managing Director *Dr Joel Burden*
Approx. Annual Turnover £1.1 million

Founded 1999 as a highly illustrated book publisher in the international museum, heritage and art gallery markets. Also some 'coffee-table' titles in association with various universities, Oxbridge colleges, independent schools and military organizations. 20 titles annually. Unsolicited material welcome; approach in writing with c.v., book synopsis, target market and any supporting information. No fiction.
Ⓔ Royalties annually.

Thoemmes Continuum
▷ **The Continuum International Publishing Group Limited**

Thomson Learning
▷ **Cengage Learning (EMEA) Ltd**

Stanley Thornes
▷ **Nelson Thornes Ltd**

F.A. Thorpe Publishing

The Green, Bradgate Road, Anstey LE7 7FU
☎ 0116 236 4325   🖷 0116 234 0205
Group Chief Executive *Robert Thirlby*
Approx. Annual Turnover £6 million

Founded in 1964 to supply large print books to libraries. Part of the Ulverscroft Group Ltd. Publishes fiction and non-fiction large print books. No educational, gardening or books that would not be suitable for large print. IMPRINTS **Charnwood; Ulverscroft; Linford Romance; Linford Mystery; Linford Western**. No unsolicited material.

Thorsons
▷ **HarperCollins Publishers Ltd**

Through Writers' Eyes
▷ **Eland Publishing Ltd**

Time Out Guides
▷ **The Random House Group Ltd**

Times Books
▷ **HarperCollins Publishers Ltd**

Timewell Press

10 Porchester Terrace, London W2 3TL
☎ 0870 760 5250   🖷 0870 760 5250
info@timewellpress.com
www.timewellpress.com
Chairman *Gerard Noel*

Founded 1997. Publishes literary international fiction and non-fiction. AUTHORS include Anthony Blond, Francis Fulford, Leslie Grantham, Ralph Lownie, C.S. Nicholls, M-J. Lancaster, Jonathan Gathorne-Hardy (J.R. Ackerley Prize, 2005), Charles Campion (Gourmand World Cookbook Awards, 2004), Anthony Powell, George Orwell, D.J. Taylor. No unsolicited mss; send synopses and ideas with letter and s.a.e.
Ⓔ Royalties twice-yearly.

Tindal Street Press Ltd

217 The Custard Factory, Gibb Street, Birmingham B9 4AA
☎ 0121 773 8157
alan@tindalstreeet.co.uk
www.tindalstreet.co.uk
Publishing Director *Alan Mahar*
Editor *Luke Brown*

Founded in 1998 to publish contemporary original fiction from the English regions. Publishes literary fiction only – novels and short story collections. No local history, memoirs or poetry. Published Man Booker 2003 shortlisted *Astonishing Splashes of Colour* and 2009 longlisted *Girl in a Blue Dress* and Costa First Novel 2007 *What Was Lost.* 7 titles in 2009. Approach with a letter, synopsis and three chapters, with s.a.e. if return required.
Ⓔ Royalties paid twice-yearly.

✱*Authors' update* One of the most encouraging success stories in recent publishing, Tindal has had three authors longlisted for the Man Booker Prize. Arts Council funding supports literary excellence.

Titan Books

144 Southwark Street, London SE1 0UP
☎ 020 7620 0200   🖷 020 7803 1990
editorial@titanemail.com
www.titanbooks.com
Managing Director *Nick Landau*

Editorial Director *Katy Wild*

Founded 1981. Now a leader in the publication of graphic novels and film and television tie-ins. Publishes comic books/graphic novels, film and television titles. IMPRINT **Titan Books** TITLES *Batman*; *Bones: The Official Companion*; *The Simpsons*; *Sweeney Todd: The Demon Barber of Fleet Street*; *Star Wars*; *Superman*; *The Winston Effect: The Art and History of Stan Winston Studio*; *Transformers*; *24*; *The Official Companion Season 6*; *Wallace & Gromit*. About 200 titles a year. No unsolicited fiction or children's books. Ideas for film and TV titles considered; send synopsis/outline with sample chapter. No e-mail submissions. Author guidelines available.
Ⓔ Royalties twice-yearly.

Tor
> Macmillan Publishers Ltd

Transworld Publishers, A division of the Random House Group Ltd

61–63 Uxbridge Road, London W5 5SA
☎ 020 8579 2652   🖷 020 8579 5479
info@transworld-publishers.co.uk
www.booksattransworld.co.uk
Managing Director *Larry Finlay*
Publisher *Bill Scott-Kerr*
Approx. Annual Turnover £85 million

Founded 1950. A subsidiary of **Random House, Inc.**, New York, which in turn is a wholly-owned subsidiary of **Bertelsmann AG**, Germany. Publishes general fiction and non-fiction, gardening, sports and leisure. IMPRINTS **Bantam**; **Bantam Press** *Sally Gaminara*; **Corgi** & **Black Swan** *Linda Evans*; **Doubleday** *Marianne Velmans*; **Channel 4 Books** *Doug Young*; **Eden** *Susanna Wadeson*; **Expert Books** *Susanna Wadeson*; **Transworld Ireland** *Eoin McHugh*. AUTHORS Monica Ali, Kate Atkinson, Dan Brown, Bill Bryson, Lee Child, Jilly Cooper, Richard Dawkins, Ben Elton, Frederick Forsyth, Tess Gerritsen, Robert Goddard, Joanne Harris, Stephen Hawking, Mo Hayder, D.G. Hessayon, John Irving, Simon Kernick, Sophie Kinsella, Kathy Lette, Paul McKenna, Andy McNab, John O'Farrell, Terry Pratchett, Patricia Scanlan, Danielle Steel, Joanna Trollope, Robert Winston. No unsolicited mss. OVERSEAS ASSOCIATES Random House Australia Pty Ltd; Random House New Zealand; Random House (Pty) Ltd (South Africa).
Ⓔ Royalties twice-yearly.

Travel Publishing Ltd

Airport Business Centre, 10 Thornbury Road, Estover, Plymouth PL6 7PP
☎ 01752 697280   🖷 01752 697299
info@travelpublishing.co.uk
www.travelpublishing.co.uk

Directors *Peter Robinson, Chris Day*

Founded in 1997 by two former directors of Reed Elsevier plc. Publishes travel guides covering places of interest, accommodation, food, drink and specialist shops in Britain and Ireland. SERIES *Hidden Places*; *Hidden Inns*; *Country Pubs*; *Country Living Rural Guides* (in conjunction with *Country Living* magazine); *Off the Motorway*. Over 30 titles in print. Welcomes unsolicited material; send letter in the first instance.
Ⓔ Royalties twice-yearly.

Trentham Books Ltd

Westview House, 734 London Road, Stoke-on-Trent ST4 5NP
☎ 01782 745567/844699   🖷 01782 745553
tb@trentham-books.co.uk
www.trentham-books.co.uk
Directors *Dr Gillian Klein, Barbara Wiggins*
Approx. Annual Turnover £1 million

Publishes education (nursery, school to higher), social sciences, intercultural studies, gender studies and law for professional readers *not* for children and parents. Also academic and professional journals. No fiction, biography or poetry. Over 30 titles a year. Unsolicited mss, synopses and ideas welcome if relevant to their interests. Material only returned if adequate s.a.e. sent.
Ⓔ Royalties annually.

Trident Press Ltd

Empire House, 175 Piccadilly, London W1J 9TB
☎ 020 7491 8770   🖷 020 7491 8664
admin@tridentpress.com
www.tridentpress.com
Managing Director *Peter Vine*
Approx. Annual Turnover £550,000

Founded 1997. Publishes TV tie-ins, natural history, travel, geography, underwater/marine life, history, archaeology and culture. DIVISIONS **Fiction/General Publishing** *Paula Vine*; **Natural History** *Peter Vine* TITLES *Red Sea Sharks*; *The Elysium Testament*; *BBC Wildlife Specials*; *UAE in Focus*. No unsolicited mss; synopses and ideas welcome, particularly TV tie-ins. Approach in writing or *brief* communications by e-mail, fax or telephone.
Ⓔ Royalties annually.

Trotman

Westminster House, Kew Road, Richmond TW9 2ND
☎ 020 8334 1600   🖷 020 8334 1601
www.trotman.co.uk
Commercial Director *Tom Lee*
Approx. Annual Turnover £3 million

Division of **Crimson Publishing**. Publishes general careers books, higher education guides, teaching support material, employment and training resources. TITLES *Degree Course Offers*; *The Student Book*; *Which Degree?*; *Trotman Green Guides*; *Students' Money Matters*; *Careers Uncovered*; *Real Life Guides*; *Real Life Issues*; *Careers 2010*; *You're Hired: CV and Interview Guide*. About 50 titles a year. Unsolicited material welcome.

Royalties twice-yearly.

Troubador Publishing Ltd

9 De Montfort Mews, Leicester LE1 7FW
01116 255 9311   0116 244 9323
books@troubador.co.uk
www.troubador.co.uk
Chairman *Jane Rowland*
Managing Director *Jeremy Thompson*
Approx. Annual Turnover £750,000

Founded 1998. Publishes fiction, children's, academic, non-fiction, translation and poetry. Offers self-publishing agreement for fiction or non-fiction under its Matador imprint. DIVISIONS **Academic** *Jane Rowland*; **Fiction/ Non-Fiction** *Jeremy Thompson*. IMPRINTS **Matador; Troubador Italian Studies; T2; Emerging Communications Technologies; Communication Ethics** TITLES *Energy Beyond Oil*; *Grazia Deledda: A Biography*; *The Ethics of Teaching Practice*. 150 titles in 2008. Unsolicited mss welcome, by e-mail only. No synopses or ideas for books.

Royalties quarterly.

TSO (The Stationery Office)

St Crispins, Duke Street, Norwich NR3 1PD
01603 622211
www.tso.co.uk
Chief Executive *Richard Dell*
Approx. Annual Turnover £250 million

Formerly HMSO, which was founded in 1786. Became part of the private sector in October 1996. Publisher of material sponsored by Parliament, government departments and other official bodies. Also commercial publishing in the following broad categories: business and professional, environment, transport, education and law. 11,000 new titles each year with 50,000 titles in print.

Tuckwell Press
▷ **Birlinn Ltd**

Tusker Rock Press
▷ **Atlantic Books**

Twenty First Century Publishers Ltd

Braunton Barn, Kiln Lane, Isfield TN22 5UE
01892 522802

TFCP@btinternet.com
www.twentyfirstcenturypublishers.com
Chairman *Fred Piechoczek*

Founded 2002. Publishes general fiction, financial thrillers and crime. Welcomes submissions by e-mail (manuscripts@connectfree.co.uk). Send brief synopsis in the body of the e-mail with extracts from the book in a file attachment ('as much as you wish to send'). No non-fiction, young adult or children's. Hard copy submissions will not necessarily be returned.

Royalties twice-yearly.

Two Ravens Press Ltd

Green Willow Croft, Rhiroy, Lochbroom, Ullapool IV23 2SF
01854 655307
info@tworavenspress.com
www.tworavenspress.com
Managing Director *Sharon Blackie*

Founded 2006. Publishes literary fiction, poetry, selected literary non-fiction. No commercial non-literary fiction, genre fiction (unless it's literary), anything connected with celebrities, children's books, local history, gift books. **Fiction** *Sharon Blackie* TITLES *One True Void* Dexter Petley; *The Falconer* Alice Thompson; **Poetry** *David Knowles* TITLES *The Atlantic Forest* George Gunn; *The Zig Zag Woman* Maggie Sawkins; **Non-Fiction** *Sharon Blackie* TITLE *Blazing Paddles, Dances with Waves* Brian Wilson. 24 titles in 2008. Unsolicited material welcome subject to initial e-mail inquiry with synopsis and short biog in body of e-mail. 'We do not accept unsolicited mss by mail and do not have time to respond to telephone enquiries (all information about how to submit is on our website).'

Royalties twice-yearly.

✸*Authors' update* One to watch. The fiction list has attracted plaudits from the trade press. It looks as if Scotland is to be blessed with another strong independent publisher.

Ulverscroft
▷ **F.A. Thorpe Publishing**

University Presses of California, Columbia & Princeton Ltd

1 Oldlands Way, Bognor Regis PO22 9SA
01243 842165   01243 842167
lois@upccp.demon.co.uk

Publishes academic titles only. US-based editorial offices. Enquiries only. Over 200 titles a year.

Usborne Publishing Ltd

83–85 Saffron Hill, London EC1N 8RT
020 7430 2800   020 7430 1562
mail@usborne.co.uk
www.usborne.com

Managing Director *Peter Usborne*
Publishing Director *Jenny Tyler*
Approx. Annual Turnover £28 million

Founded 1973. Publishes non-fiction, fiction, art and activity books, puzzle books and music for children, young adults and pre-school. Some titles for parents. Up to 250 titles a year. Non-fiction books are written in-house to a specific format and therefore unsolicited mss are not normally welcome. Ideas which may be developed in-house are sometimes considered. Fiction for children will be considered. Keen to hear from new illustrators and designers.
Royalties twice-yearly.

✸*Authors' update* From specializing in children's reference books, Usborne has moved on to 'all round' publishing with, for example, the recent appearance of a popular History of Britain. But the main growth is in fiction and baby books.

Vacation Work
▷ Crimson Publishing

Vanguard Press
▷ Pegasus Elliot Mackenzie Publishers Ltd

Ventura
▷ Penguin Group (UK)

Vermilion
▷ The Random House Group Ltd

Verso
6 Meard Street, London W1F 0EG
☎ 020 7437 3546   ☎ 020 7734 0059
enquiries@verso.co.uk
www.versobooks.com
Approx. Annual Turnover £2 million

Formerly New Left Books which grew out of the *New Left Review*. Publishes politics, history, sociology, economics, philosophy, cultural studies, feminism. TITLES *The Occupation* Patrick Cockburn; *Planet of Slums* Mike Davis; *Polemics* Alain Badiou; *The Soviet Century* Moshe Lewin; *NHS plc: The Privatisation of our Health Care* Allyson M. Pollock; *Street-Fighting Years* Tariq Ali; *Planet of Slums* Mike Davis. No unsolicited mss; synopses and ideas for books welcome. OVERSEAS OFFICE in New York.
Royalties annually.

✸*Authors' update* Dubbed by the *Bookseller* as 'one of the most successful small independent publishers'.

Viking
▷ Penguin Group (UK)

Vintage
▷ The Random House Group Ltd

Virago Press
Little, Brown Book Group UK, 100 Victoria Embankment, London EC4Y 0DY
☎ 020 7911 8000   ☎ 020 7911 8100
virago.press@littlebrown.co.uk
www.virago.co.uk
Publisher *Lennie Goodings*
Editor, Virago Modern Classics *Donna Coonan*
Editor, Virago *Elise Dillsworth*

An imprint of **Little, Brown Book Group UK**. Founded in 1973 by Carmen Callil, Virago publishes women's literature, both fiction and non-fiction. IMPRINT **Virago Modern Classics** 19th and 20th-century fiction reprints by writers such as Daphne du Maurier, Angela Carter and Edith Wharton. Virago AUTHORS include Margaret Atwood, Maya Angelou, Jennifer Belle, Waris Dirie, Sarah Dunant, Germaine Greer, Michèle Roberts, Gillian Slovo, Talitha Stevenson, Natasha Walter, Sarah Waters. 50 titles a year. 'We do not currently accept any unsolicited submissions.'
Royalties twice-yearly.

✸*Authors' update* Still the first name in women's publishing. A team of young editors favours writers with something new to say.

Virgin Books
▷ The Random House Group Ltd

Vision
101 Southwark Street, London SE1 0JF
☎ 020 7928 5599   ☎ 020 7928 8822
info@visionpaperbacks.co.uk
www.visionpaperbacks.co.uk
Managing Director *Sheena Dewan*

Founded 1996. Non-fiction publisher. IMPRINTS **Vision; Vision Paperbacks; Fusion Press**. 18 titles in 2007. Unsolicited mss, synopses and ideas for books welcome. Initial approach by e-mail, without attachments. No fiction or poetry.
Royalties annually.

Vision On
▷ Omnibus Press

The Vital Spark
▷ Neil Wilson Publishing Ltd

Voyager
▷ HarperCollins Publishers Ltd

University of Wales Press
10 Columbus Walk, Brigantine Place, Cardiff CF10 4UP
☎ 029 2049 6899   ☎ 029 2049 6108
post@uwp.co.uk
www.uwp.co.uk
Commissioning Editor *Sarah Lewis*
Assistant Commissioning Editor *Ennis Akpinar*

Approx. Annual Turnover £1 million

Founded 1922. Primarily a publisher of academic and scholarly books in English and Welsh. Subject areas include: history, political philosophy, Welsh and Celtic Studies, Literary Studies, European Studies and Medieval Studies. SERIES include *Political Philosophy Now*; *Religion and Culture in the Middle Ages*; *Iberian and Latin American Studies*; *Gender Studies in Wales*; *Gothic Literary Studies*; *Kantian Studies*; *Histories of Europe*; *French and Francophone Studies*; *European Crime Fiction*. 85 titles a year. Unsolicited mss considered but see website for further guidance. Ⓔ Royalties annually.

Walker Books Ltd

87 Vauxhall Walk, London SE11 5HJ
☎ 020 7793 0909   📠 020 7587 1123
editorial@walker.co.uk
www.walker.co.uk
Managing Director *Helen McAleer*
Publisher *Jane Winterbotham*
Editors *Gill Evans (fiction)*, *Deirdre McDermott (picture books)*, *Caroline Royds (early learning, non-fiction & gift books)*, *Denise Johnstone-Burt (picture books, board & novelty)*
Approx. Annual Turnover £47.7 million (Group)

Founded 1979. Publishes illustrated children's books, children's fiction and non-fiction. TITLES *Maisy* Lucy Cousins; *Where's Wally?* Martin Handford; *We're Going on a Bear Hunt* Michael Rosen and Helen Oxenbury; *Can't You Sleep, Little Bear?* Martin Waddell and Barbara Firth; *Guess How Much I Love You* Sam McBratney and Anita Jeram; *Alex Rider* series by Anthony Horowitz. About 300 titles a year. Ⓔ Royalties twice-yearly.

⊞*Authors' update* Dedicated to 'groundbreaking books'. Noted for close working relationships with authors.

Wallflower Press

6 Market Place, London W1W 8AF
☎ 020 7436 9494
info@wallflowerpress.co.uk
www.wallflowerpress.co.uk
Commissioning Editor *Yoram Allon*
Editorial Manager *Jackie Downs*
Approx. Annual Turnover £300,000

Founded 1999. Publishes trade and academic books devoted to cinema and the moving image. SERIES *Short Cuts* Introductory undergraduate books; *Director's Cuts* Studies on significant international filmmakers; *24 Frames* Anthologies focusing on national and regional cinemas. Over 35 titles a year. Unsolicited mss, synopses and proposals welcome. No fiction or academic material not related to the moving image. Ⓔ Royalties annually.

Ward Lock
▷ **Octopus Publishing Group**

Ward Lock Educational Co. Ltd

BIC Ling Kee House, 1 Christopher Road, East Grinstead RH19 3BT
☎ 01342 318980   📠 01342 410980
wle@lingkee.com
www.wardlockeducational.com
Owner *Ling Kee (UK) Ltd*

Founded 1952. Publishes educational books (primary, middle, secondary, teaching manuals) for all subjects, specializing in maths, science, geography, reading and English and currently focusing on Key Stages 1 and 2.

Frederick Warne
▷ **Penguin Group (UK)**

Franklin Watts
▷ **Hachette Children's Books**

Wayland
▷ **Hachette Children's Books**

Weidenfeld & Nicolson
▷ **The Orion Publishing Group Ltd**

Wharncliffe Books

47 Church Street, Barnsley S70 2AS
☎ 01226 734222   📠 01226 734438
enquiries@wharncliffebooks.co.uk
www.wharncliffebooks.co.uk
Commissioning Editor *Rupert Harding*

An imprint of **Pen & Sword Books Ltd**. Wharncliffe is the book and magazine publishing arm of an old-established, independently owned newspaper publishing and printing house. Publishes local history throughout the UK, focusing on nostalgia and old photographs. SERIES *Foul Deeds*; *Local History Companions*; *Pals*; *Aspects*. No unsolicited mss; synopses and ideas welcome. Ⓔ Royalties twice-yearly.

Wharton
▷ **Pearson**

Which? Books

2 Marylebone Road, London NW1 4DF
☎ 020 7770 7000   📠 020 7770 7660
rebecca.leach@which.co.uk
www.which.co.uk
Head of Book Publishing *Angela Newton*

Founded 1957. Publishing arm of Which? Publishes non-fiction: information and reference on personal finance, property, divorce, consumer law. All titles offer direct value to the consumer. IMPRINT **Which? Books** TITLES *The Good Food Guide*; *Wills and Probate*; *Renting & Letting*.

10–12 titles a year. No unsolicited mss; send synopses and ideas only.

🅔 Royalties, if applicable, twice-yearly.

White Ladder
▷ Crimson Publishing

Wild Goose Publications

Iona Community, 4th Floor, The Savoy Centre, 140 Sauchiehall Street, Glasgow G2 3DH

☎ 0141 332 6292   📠 0141 332 1090

admin@ionabooks.com

www.ionabooks.com

Editorial Head *Sandra Kramer*

The publishing house of the Iona Community, established in the Celtic Christian tradition of St Columba, publishes books and CDs on holistic spirituality, social justice, political and peace issues, healing, innovative approaches to worship, song and material for meditation and reflection.

Wiley Europe Ltd

The Atrium, Southern Gate, Chichester PO19 8SQ

☎ 01243 779777   📠 01243 775878

europe@wiley.co.uk

www.wiley.com

Founded 1807. US parent company. Publishes professional, reference trade and text books, scientific, technical and biomedical.

Wiley Nautical

John Wiley and Sons Ltd., The Atrium, Southern Gate, Arundel BN19 8SQ

☎ 01243 779777

nautical@wiley.co.uk

www.wileynautical.com

Executive Editor *David Palmer*

Books for people who love watersports. Publishes over 100 watersports and sailing books, including 'Cruising Companions' and the series of books previously published under the Fernhurst Books imprint. Synopses and ideas welcome.

Wiley-Blackwell

9600 Garsington Road, Oxford OX4 2DQ

☎ 01865 776868   📠 01865 714591

www.blackwellpublishing.com

Senior Vice President, Wiley-Blackwell *Eric Swanson*

Wiley-Blackwell was formed from the acquisition of Blackwell Publishing (Holdings) Ltd by John Wiley & Sons, Inc. Blackwell's publishing programme has merged with Wiley's global scientific, technical and medical business. The merged operation is now the largest of the three of John Wiley & Sons, Inc.; its other businesses are Professional/Trade and Higher Education.

✱*Authors' update* Wiley has joined the top ten publishers reflecting healthy sales for its academic titles.

Neil Wilson Publishing Ltd

19 Netherton Avenue, Glasgow G13 1BQ

☎ 0141 954 8007   📠 0560 150 4806

info@nwp.co.uk

www.nwp.co.uk

Managing Director/Editorial Director *Neil Wilson*

Approx. Annual Turnover £110,000

Founded 1992. Publishes Scottish interest and history, biography, humour and hillwalking, whisky; also Scottish cookery and travel. IMPRINTS **The In Pinn** Outdoor pursuits; **The Angels' Share** Whisky, drink and food-related subjects; **The Vital Spark** Humour; **NWP** History, biography, reference, true crime; **11:9** Scottish fiction. About 6 titles a year. Unsolicited mss, synopses and ideas welcome. No politics, academic or technical.

🅔 Royalties twice-yearly.

Philip Wilson Publishers Ltd

109 Drysdale Street, The Timber Yard, London N1 6ND

☎ 020 7033 9900   📠 020 7033 9922

pwilson@philip-wilson.co.uk

www.philip-wilson.co.uk

Chairman *Philip Wilson*

Founded 1976. Publishes art, art history, antiques and collectibles. About 14 titles a year.

Windgather Press
▷ Oxbow Books

Windmill Books
▷ The Random House Group Ltd

Wingedchariot Press

7 Court Royal, Eridge Road, Tunbridge Wells TN4 8HT

info@wingedchariot.com

www.wingedchariot.com

Editor *Ann Arscott*

Founded in 2005 'to bring the best of children's books in translation to the UK for the first time'. About 4 titles a year. Will consider books in their original language with illustrations. E-mail synopsis and scan of drawings. No English language books.

John Wisden & Co.
▷ Bloomsbury Publishing Plc

Wise Publications
▷ Omnibus Press

WIT Press

Ashurst Lodge, Ashurst, Southampton SO40 7AA

☎ 023 8029 3223   📠 023 8029 2853

marketing@witpress.com
www.witpress.com
Owner *Computational Mechanics International Ltd, Southampton*
Chairman/Managing Director/Editorial Head *Professor C.A. Brebbia*

Founded in 1980 as Computational Mechanics Publications to publish engineering analysis titles. Changed to WIT Press to reflect the increased range of publications. Publishes scientific and technical, mainly at postgraduate level and above, including architecture, environmental engineering, bioengineering. 50 titles in 2007. Unsolicited mss, synopses and ideas welcome; approach by post or e-mail. No non-scientific or technical material or lower level (school and college-level texts). OVERSEAS SUBSIDIARY Computational Mechanics, Inc., Billerica, USA.
ⓔ Royalties annually.

Wizard Books
▷ Icon Books Ltd

Oswald Wolff Books
▷ Berg Publishers

The Women's Press
27 Goodge Street, London W1T 2LD
☎ 020 7636 3992   ℻ 020 7637 1866
www.the-womens-press.com
Chairman *Naim Attallah*

Part of the Namara Group. First title published in 1978. Publishes women only: quality fiction and non-fiction. Fiction usually has a female protagonist and a woman-centred theme. International writers and subject matter encouraged. Non-fiction: books for and about women generally; gender politics, race politics, disability, feminist theory, health and psychology, literary criticism. IMPRINTS **Women's Press Classics; Livewire Books for Teenagers** Fiction and non-fiction series for young adults. No mss without previous letter, synopsis, sample material and return postage. Submissions by disk or e-mail not accepted.
ⓔ Royalties twice-yearly.

Woodhead Publishing Ltd
Abington Hall, Abington, Cambridge CB21 6AH
☎ 01223 891358   ℻ 01223 893694
wp@woodheadpublishing.com
www.woodheadpublishing.com
Chairman *Richard Dawes*
Managing Director *Martin Woodhead*
Approx. Annual Turnover £2.2 million

Founded 1989. Publishes engineering, materials technology, textile technology, finance and investment, food technology, environmental science. DIVISION **Woodhead Publishing** *Martin Woodhead*. About 60 titles a year. Unsolicited material welcome.
ⓔ Royalties annually.

Wordsworth Editions Ltd
8B East Street, Ware SG12 9HJ
☎ 01920 465167   ℻ 01920 462267
enquiries@wordsworth-editions.com
www.wordsworth-editions.com
Directors *E.G. Trayler, Derek Wright*
Approx. Annual Turnover £2.5 million

Founded 1987. Publishes classics of English and world literature, reference books, poetry, mystery and the supernatural. SERIES include *Children's Classics; Tales of Mystery and The Supernatural; Wordsworth Literary Lives; Wordsworth Library Collection; Wordsworth Special Editions.* About 40 titles a year. No unsolicited mss.

Y Lolfa Cyf
Talybont, Ceredigion SY24 5AP
☎ 01970 832304   ℻ 01970 832782
ylolfa@ylolfa.com
www.ylolfa.com/
Managing Director *Garmon Gruffudd*
General Editor *Lefi Gruffudd*
English Editor *Gwen Davies*
Approx. Annual Turnover £1 million

Founded 1967. Small company which publishes mainly in Welsh; has its own four-colour printing and binding facilities. Publishes Welsh language publications; language tutors; English language books for the Welsh and Celtic tourist trade. Expanding slowly. TITLES *Roberto* Roberto Martinez; *The Welsh Learner's Dictionary* Heini Gruffudd; *The Fight for Welsh Freedom* Gwynfor Evans; *Celtic Vision* John Merion Morris. IMPRINTS **Dinas** Part-author-subsidized imprint for non-mainstream books of Welsh interest in English and Welsh. **Alcemi** English language fiction. About 50 titles a year. Write first with synopses or ideas.
ⓔ Royalties twice-yearly.

Yale University Press (London)
47 Bedford Square, London WC1B 3DP
☎ 020 7079 4900   ℻ 020 7079 4901
sales@yaleup.co.uk
www.yalebooks.co.uk

Founded 1961. Owned by US parent company. Publishes art history, architecture, decorative arts, history, religion, music, politics, current affairs, biography and history of science. About 350 titles (worldwide) a year. Unsolicited mss and synopses welcome if within specialized subject areas.
ⓔ Royalties annually.

✱*Authors' update* **A publisher with a talent for turning out serious books which find a general market.**

Yellow Jersey Press
▷ **The Random House Group Ltd**

Young Picador
▷ **Macmillan Publishers Ltd**

Zambezi Publishing Ltd

PO Box 221, Plymouth PL2 2YJ
🕾 01752 367300   🖷 01752 350453
info@zampub.com
www.zampub.com
Chair *Sasha Fenton*
Managing Director *Jan Budkowski*

Founded 1999. Publishes non-fiction: mind, body and spirit, self-help, finance and business, including the 'Simply' series. About 12 titles a year. Send biography, synopsis and sample chapter by mail. Brief e-mail communication acceptable but *no* attachments, please.
💷 Royalties twice-yearly.

Zed Books Ltd

7 Cynthia Street, London N1 9JF
🕾 020 7837 4014   🖷 020 7833 3960
sales@zedbooks.net
www.zedbooks.co.uk
Approx. Annual Turnover £1.4 million

Founded 1976. Publishes international and Third World affairs, development studies, gender studies, environmental studies, human rights and specific area studies. No fiction, children's or poetry. DIVISIONS **Development & Environment** *Tamsine O'Riordan, Ellen McKinlay*; **Gender Studies**. TITLES *Rogue State: A Guide to the World's Only Superpower* William Blum; *Staying Alive* Vandana Shiva; *The Autobiography of Nawal* Nawal El Saadawi. About 60 titles a year. No unsolicited mss; synopses and ideas welcome.
💷 Royalties annually.

Zero to Ten
▷ **Evans Brothers Ltd**

Irish Publishers

🔑 Key advice

International Reply Coupons (IRCs) For return postage, IRCs are required (*not* UK postage stamps). These are available from post offices: letters, 60 pence; mss according to weight. To calculate postage from the Republic of Ireland to the UK go to www.anpost.ie/AnPost and click on 'Calculate the Postage'.

Oblurb

International Reply Coupons (IRCs) For return postage, IRCs are required (*not* UK postage stamps). These are available from post offices: letters, 60 pence; mss according to weight. To calculate postage from the Republic of Ireland to the UK go to www.anpost.ie/AnPost and click on 'Calculate the Postage'.

An Gúm

27 Sr. Fhreidric Thuaidh, Baile Átha Cliath 1
📞 00 353 1 889 2800   📠 00 353 1 873 1140
angum@forasnagaeilge.ie
www.gaeilge.ie
Senior Editor *Seosamh Ó Murchú*
Editor *Máire Nic Mheanman*

Founded 1926. Formerly the Irish language publications branch of the Department of Education and Science. Has now become part of the North/South Language Body, Foras na Gaeilge. Publishes educational, children's, young adult, music, lexicography and general. Little adult fiction or poetry. About 50 titles a year. Unsolicited mss, synopses and ideas for books welcome. Also welcomes reading copies of first and second level school textbooks with a view to translating them into the Irish language.
💷 Royalties annually.

Anvil Books

45 Palmerston Road, Dublin 6
📞 00 353 1 497 3628
Managing Director *Rena Dardis*

Founded 1964 with emphasis on Irish history and biography. IMPRINT **The Children's Press**. Publishes Irish history, biography (particularly 1916–22), folklore and children's fiction. No adult fiction, poetry, fantasy, short stories or full-colour books for children. About 4 titles a year. 'Because of promotional requirements, only books by Irish-based authors considered and only books of Irish interest.' Send synopsis only with IRCs (no UK stamps); unsolicited mss not returned.
💷 Royalties annually.

Ashfield Press

30 Linden Grove, Blackrock, Co. Dublin
📞 00 353 1 288 9808
susanwaine@ashfieldpress.com
www.ashfieldpress.com
Directors *Susan Waine, John Davey, Gerry O'Connor*

Publishes Irish-interest non-fiction books. TITLES *Lady Icarus*; *The Weather is a Good Storyteller*; *Secret Sights II*. Unsolicited mss and synopses welcome. ASSOCIATE COMPANY Ashfield Press Publishing Services.
💷 Royalties twice-yearly.

Atrium
▷ Cork University Press

Attic Press
▷ Cork University Press

Blackhall Publishing Ltd

Lonsdale House, Evoca Avenue, Blackrock, Co. Dublin
📞 00 353 1 278 5090
info@blackhallpublishing.com
www.blackhallpublishing.com
Managing Director *Gerard O'Connor*
Commissioning Editor *Elizabeth Brennan*
Editor *Eileen O'Brien*

Publishes business, management, law, social studies and life issues. Main subject areas include accounting, finance, management, social studies and law books aimed at both students and professionals in the industry. Also provides legal publishing services including Statute Law Revision, Law Reporting, Legislative Drafting and Subscription Services. 20 titles a year. Unsolicited mss and synopses welcome.
💷 Royalties annually.

Bradshaw Books

Tigh Filí, Civic Trust House, 50 Popes Quay, Cork
☎ 00 353 21 421 5175
info@tighfili.com
www.tighfili.com
Managing Director *Maire Bradshaw*
Project Manager *Leslie Ryan*

Founded 1985. Publishes poetry, women's issues, children's books. Organizers of the annual Cork Literary Review poetry manuscript competition. SERIES *Cork Literary Review* and *Eurochild Anthology of Poetry and Art*.
Ⓡ Royalties not generally paid.

Brandon/Mount Eagle Publications

Dingle, Co. Kerry
☎ 00 353 66 915 1463   📠 00 353 66 915 1234
www.brandonbooks.com
Publisher *Stephen MacDonogh*
Approx. Annual Turnover €500,000

Founded in 1997, taking over Brandon Book Publishers (founded 1982). Publishes Irish fiction, biography, memoirs and other non-fiction. About 15 titles a year. Not seeking unsolicited mss.

Brookside

▷ New Island

Edmund Burke Publisher

Cloonagashel, 27 Priory Drive, Blackrock, Co. Dublin
☎ 00 353 1 288 2159   📠 00 353 1 283 4080
deburca@indigo.ie
www.deburcararebooks.com
Managing Director *Eamonn De Búrca*
Publications Managers *Regina De Búrca-McAuley, William De Búrca*
Approx. Annual Turnover €320,000

Small family-run business publishing historical and topographical and fine limited-edition books relating to Ireland. TITLES *Michael Collins and The Making of a New Ireland*, 2 vols; *The Great Book of Irish Genealogies*, 5 vols.; *The Annals of the Four Masters*, 7 vols; *Flowers of Mayo* (illus. Wendy Walsh); *O'Curry's Manners and Customs of the Ancient Irish*; *Joyce's Irish Names of Places*. Unsolicited mss welcome. No synopses or ideas.
Ⓡ Royalties annually.

The Children's Press

▷ Anvil Books

Church of Ireland Publishing

Church of Ireland House, Church Avenue, Rathmines, Dublin 6
☎ 00 353 1 492 3979   📠 00 353 1 492 4770
susan.hood@rcbdub.org
www.cip.ireland.anglican.org
Owner *Representative Church Body of the Church of Ireland*

Chairman *Dr Kenneth Milne*
Publications Officer *Dr Susan Hood*

Founded 2004. Publishes official publications of the Church of Ireland: theological, doctrinal, historical and administrative, and facilitates publication for other church-related bodies. 6 titles in 2008. No unsolicited material.

Cló Iar-Chonnachta

Indreabhán, Connemara, Galway
☎ 00 353 91 593307   📠 00 353 91 593362
cic@iol.ie
www.cic.ie
General Manager *Deirdre Ní Thuathail*
Approx. Annual Turnover €500,000

Founded 1985. Publishes fiction, poetry, plays, teenage fiction and children's, mostly in Irish, including translations. Also publishes cassettes of writers reading from their own works. TITLES *Seosamh Ó hÉanaí: Nár fhágha mé bás choíche* Liam Mac Con Iomaire; *Mise an Fear Ceoil: Séamus Ennis Dialann Taistil 1942-1946* Ríonach Uí Ógáin; *Rí an Fhocail & Máirtín Ó Cadhain sa gCnocán Glas* (DVD); *Lá an Phaoraigh* Seán Óg and Aoife de Paor; *On a Rock in the Middle of the Ocean* Lillis Ó Laoire. 12 titles in 2008.
Ⓡ Royalties annually.

Cois Life

62 Rose Park, Kill Avenue, Dún Laoghaire, Co. Dublin
☎ 00 353 1 280 7951   📠 00 353 1 280 7951
eolas@coislife.ie
www.coislife.ie
Chairman *Dr Seán Ó Cearnaigh*
Managing Director *Dr Caoilfhionn Nic Pháidín*
Approx. Annual Turnover €200,000

Founded 1996. General publishers of literature, academic and children's books, all in the Irish language. 11 titles in 2008. Mss, synopses and ideas for Irish-language books welcome; submissions by e-mail.
Ⓡ Royalties annually.

The Collins Press

West Link Park, Doughcloyne, Wilton, Cork
☎ 00 353 21 434 7717
enquiries@collinspress.ie
www.collinspress.ie
Managing Director *Con Collins*
Approx. Annual Turnover €750,000

Founded 1989. Publishes general non-fiction, Irish interest and children's non-fiction books. 25–30 titles a year. No unsolicited mss; synopses and ideas welcome; approach by e-mail. See website for submission guidelines. No academic, technical, professional, fiction, poetry, literary criticism or short stories.

The Columba Press

55A Spruce Avenue, Stillorgan Industrial Park,
Blackrock, Co. Dublin
℡ 00 353 1 294 2556   ℻ 00 353 1 294 2564
sean@columba.ie (editorial)
info@columba.ie (general)
www.columba.ie
Chairman *Neil Kluepfel*
Managing Director *Seán O'Boyle*
Approx. Annual Turnover €1.02 million

Founded 1985. Small company committed to
growth. Publishes religious and counselling titles.
TITLES *Already Within* Donal O'Leary; *Celtic
Prayers and Reflections* Jenny Child. IMPRINT
Currach Press (see entry). Backlist of 250 titles.
Unsolicited ideas and synopses preferred rather
than full mss.
Ⓔ Royalties twice-yearly.

Cork University Press

Youngline Industrial Estate, Pouladuff Road,
Togher, Cork
℡ 00 353 21 490 2980   ℻ 00 353 21 431 5329
corkuniversitypress@ucc.ie
www.corkuniversitypress.com
Owner *University College Cork*
Publications Director *Mike Collins*
Editor *Sophie Watson*
Production Editor *Maria O'Donovan*

Founded 1925. Relaunched in 1992, the Press
publishes academic and some trade titles. TITLES
Ivor Browne: Music and Madness; *Wexford: A
Town and its Landscape*; *Newgrange*; *Knock:
The Virgin's Apparition in Nineteenth-Century
Ireland*. Also a biannual journal, *The Irish Review*,
an interdisciplinary cultural review. No fiction.
IMPRINTS **Attic Press**; **Atrium**. 15 titles in 2008.
Unsolicited Irish interest related synopses
and ideas welcome for textbooks, academic
monographs, trade non-fiction, illustrated
histories and journals.
Ⓔ Royalties annually.

Currach Press

55A Spruce Avenue, Stillorgan Industrial Park,
Blackrock, Co. Dublin
℡ 00 353 1 294 2556   ℻ 00 353 1 294 2564
jo@currach.ie
www.currach.ie
Publisher *Jo O'Donoghue*

Founded 2002. An imprint of **The Columba
Press**. Publishes non-fiction titles of Irish interest,
including politics, history, biography, sport, music
and criticism. TITLES *A Memoir* Terry de Valera;
The Harcourt Street Line Brian Mac Aongusa;
Short Hands, Long Pockets Eddie Hobbs.
Unsolicited ideas welcome; synopses rather than
full mss preferred.

A.&A. Farmar Ltd

78 Ranelagh Village, Dublin 6
℡ 00 353 1 496 3625
afarmar@iol.ie
www.aafarmar.ie
Managing Director *Anna Farmar*
Production Director *Tony Farmar*

Founded 1992. Publishes Irish social history,
business, food and wine. About 10 titles a year.
No unsolicited mss. Synopses and ideas welcome;
approach by letter or e-mail in the first instance.

Four Courts Press Ltd

7 Malpas Street, Dublin 8
℡ 00 353 1 453 4668   ℻ 00 353 1 453 4672
info@fourcourtspress.ie
www.fourcourtspress.ie
Chairman/Managing Director *Michael Adams*
Director *Martin Healy*
Senior Editor *Martin Fanning*

Founded 1969. Publishes mainly scholarly books
in the humanities. About 70 titles a year. Synopses
and ideas for books welcome.
Ⓔ Royalties annually.

The Gallery Press

Loughcrew, Oldcastle, Co. Meath
℡ 00 353 49 854 1779   ℻ 00 353 49 854 1779
gallery@indigo.ie
www.gallerypress.com
Managing Director *Peter Fallon*

Founded 1970. Publishes Irish poetry and drama.
13 titles in 2008. Currently, only interested in
work from Irish poets and dramatists (plays must
have had professional production). Unsolicited
mss welcome but prefers to send submission
notes to potential writers in the first instance.
Ⓔ Royalties paid.

Gill & Macmillan

10 Hume Avenue, Park West, Dublin 12
℡ 00 353 1 500 9500   ℻ 00 353 1 500 9599
www.gillmacmillan.ie
Chairman *M.H. Gill*
Managing Director *Dermot O'Dwyer*
Approx. Annual Turnover €12 million

Founded 1968 when M.H. Gill & Son Ltd and
Macmillan Ltd formed a jointly owned publishing
company. Publishes biography/autobiography,
history, current affairs, literary criticism (all
mainly of Irish interest), guidebooks and cookery.
Also educational textbooks for secondary and
tertiary levels. Contacts: *Anthony Murray*
(educational), *Marion O'Brien* (college), *Fergal
Tobin* (general). About 80 titles a year. Unsolicited
synopses and ideas welcome.
Ⓔ Royalties subject to contract.

Hachette Books Ireland

8 Castlecourt Centre, Castleknock, Dublin 15
☎ 00 353 1 824 6288   ☏ 00 353 1 824 6289
info@hbgi.ie
www.hbgi.ie
Managing Director, Publishing *Breda Purdue*
Senior Commissioning Editors *Ciara Considine,*
Claire Rourke
Editorial Manager *Ciara Doorley*

Founded 2002. Irish division of **Hachette
UK**. Publishes general fiction and non-fiction:
memoirs, biography, sport, politics, current
affairs, humour, food and drink. 29 titles in 2008.
Consult the website for submission guidelines.
ⓔ Royalties twice-yearly.

Institute of Public Administration

57–61 Lansdowne Road, Dublin 4
☎ 00 353 1 240 3600   ☏ 00 353 1 269 8644
sales@ipa.ie
www.ipa.ie
Director-General *John Cullen*
Publisher *Eileen Kelly*
Approx. Annual Turnover €1.3 million

Founded in 1957 by a group of public servants,
the Institute of Public Administration is the
Irish public sector management development
agency. The publishing arm of the organization
is one of its major activities. Publishes academic
and professional books and periodicals: history,
law, politics, economics and Irish public
administration for students and practitioners.
TITLES *A Review of the Civil Service Grading
and Pay System* Joanna O'Riordan; *Politics in a
Changing Ireland 1960-2007: A Tribute to Seamus
Pattison* Tom O'Connor and Anthony O'Halloran;
Belongings: Shaping Identity in Modern Ireland
Mary P. Corcoran and Perry Share; *Government
in Modern Ireland* Muiris MacCarthaigh; *Social
Policy, Ageing and Voluntary Action* Nicholas
Acheson and Brian Harvey; *Understanding and
Managing Organisational Culture* Orla O'Donnell
and Richard Boyle. About 10 titles a year. No
unsolicited mss; synopses and ideas welcome. No
fiction or children's publishing.
ⓔ Royalties annually.

Irish Academic Press Ltd

44 Northumberland Road, Ballsbridge, Dublin 4
☎ 00 353 1 668 8244   ☏ 00 353 1 660 1610
info@iap.ie
www.iap.ie
Chairman *Stewart Cass (London)*
Dublin Office *Rachel Milotte*

Founded 1974. Publishes academic monographs
and humanities. Unsolicited mss, synopses and
ideas welcome.
ⓔ Royalties annually.

Liberties Press

Guinness Enterprise Centre, Taylor's Lane,
Dublin 8
☎ 00 353 1 415 1224
info@libertiespress.com
www.libertiespress.com
Joint Managing Directors *Sean O'Keeffe,*
Peter O'Connell

Founded 2003. Publishes non-fiction: history,
sport, memoirs, biography, health, business and
finance, politics and current affairs. 15 titles in
2008 Mss, synopses and ideas for books welcome;
submissions by e-mail or post. No fiction, poetry,
true crime or science fiction.
ⓔ Royalties twice-yearly.

The Liffey Press Ltd

Ashbrook House, 10 Main Street, Raheny,
Dublin 5
☎ 00 353 1 851 1458   ☏ 00 353 1 851 1459
info@theliffeypress.com
www.theliffeypress.com
Managing Director/Publisher *David Givens*

Founded 2001. Publishes general interest, Irish-
focused non-fiction. 20 titles in 2008. Synopses,
ideas and proposals with sample chapters
welcome; approach by letter or e-mail in the first
instance. No fiction, children's or books not of
interest to Irish readers.
ⓔ Royalties twice-yearly.

The Lilliput Press

62–63 Sitric Road, Arbour Hill, Dublin 7
☎ 00 353 1 671 1647   ☏ 00 353 1 671 1233
info@lilliputpress.ie
www.lilliputpress.ie
Chair *Kathy Gilfillan*
Managing Director *Antony Farrell*
Approx. Annual Turnover €400,000

Founded 1984. Publishes non-fiction: literature,
history, autobiography and biography, ecology,
essays; criticism; fiction. TITLES *What the Curlew
Said* John Moriaty; *Bad Day in Blackrock* Kevin
Power; *Ireland's Other Poetry: Anonymous
to Zozimus* John Wyse Jackson and Hector
McDonnell; *More Myers* Kevin Myers; *The
Dublin Edition of Ulysses* James Joyce. About 18
titles a year. Unsolicited mss, synopses and ideas
welcome. No children's or sport titles.
ⓔ Royalties annually.

Marino Books
▷ Mercier Press Ltd

Maverick House Publishers

Office 19, Dunboyne Business Park, Dunboyne,
Co. Meath
☎ 00 353 1 825 5717   ☏ 00 353 1 686 5036

info@maverickhouse.com
www.maverickhouse.com
Managing Director *Jean Harrington*

Founded 2001. Publishes biography and autobiography, history, current affairs, non-fiction, politics, sport. TITLES *Welcome to Hell: One Man's Fight for Life Inside the Bangkok Hilton* Colin Martin; *The Commandant* Ian Baxter; *With Bare Hands* Alain Robert; *Hell in Barbados* Terry Donaldson. Unsolicited material welcome; send by post. No fiction, children's books or poetry. Ⓔ Royalties twice-yearly.

Mercier Press Ltd

Unit 3B, Oak House, Bessboro Road, Blackrock, Cork
Ⓣ 00 353 21 461 4700
info@mercierpress.ie
www.mercierpress.ie
Chairman *John Spillaner*
Managing Director *Clodagh Feehan*

Founded 1944. One of Ireland's largest publishers with a list of approx 250 Irish interest titles. IMPRINTS **Mercier Press**; **Marino Books** Editorial Director *Mary Feehan* Children's, politics, history, folklore, biography, mind, body and spirit, current affairs, women's interest. TITLES *The Course of Irish History*; all of John B. Keane's works; *Beyond Prozac*; *It's a Long Way from Penny Apples*; *Ireland's Master Storyteller*. Unsolicited synopses and ideas welcome.
Ⓔ Royalties annually.

Merlin Publishing

Newmarket Hall, Cork Street, Dublin 8
Ⓣ 00 353 1 453 5866 Ⓕ 00 353 1 453 5930
publishing@merlin.ie
www.merlinwolfhound.com
Managing Director *Chenile Keogh*
Editorial Director *Aoife Barrett*

Founded 2000. Member of **Clé**. Publishes true crime, film, music, art, biography, general non-fiction, history and gift books. IMPRINT **Wolfhound Press** Founded 1974. Non-fiction. TITLES *Famine*; *Eyewitness Bloody Sunday*; *Father Browne's Titanic Album*. About 15 titles a year. Two sample chapters of unsolicited mss (with synopses and s.a.e.) welcome by e-mail (aoife@merlin.ie). See Merlin website for submission guidelines and proposal form.

Mount Eagle
▷ Brandon/Mount Eagle Publicaions

National Library of Ireland

Kildare Street, Dublin 2
Ⓣ 00 353 1 603 0200 Ⓕ 00 353 1 676 6690
info@nli.ie
www.nli.ie

Founded 1877. Publishes books and booklets based on the library's collections; folders of historical documents; academic and specialist books; reproduction folders and CD-ROMs. TITLES *Ulysses Unbound: A Reader's Companion to James Joyce's Ulysses* Terence Killeen; *Cooper's Ireland: Drawings and Notes from an Eighteenth Century Gentleman* Peter Harbison; *Into the Light: An Illustrated Guide to the Photographic Collections in the National Library of Ireland* Sarah Rouse; *WB Yeats: Works and Days* James Quin, Eilis Ni Dhuibhne, Ciara McDonnell; *Librarians, Poets and Scholars: A festschrift for Donall O Luanaigh* ed. Felix M. Larkin; *Stranger to Citizens: The Irish in Europe 1600–1800* Mary Ann Lyons and Thomas O'Connor. 1 title in 2008.

New Island

2 Brookside, Dundrum Road, Dundrum, Dublin 14
Ⓣ 00 353 1 298 9937/298 3411
Ⓕ 00 353 1 298 7912
inka.hagen@newisland.ie
www.newisland.ie
Managing Director/Editorial Head *Edwin Higel*
Editorial Manager *Deirdre O'Neill*

Founded 1990. Publishes fiction, Irish non-fiction, poetry and drama. Branching out into popular fiction and memoirs. IMPRINT **Brookside**. About 25 titles a year. Unsolicited mss, synopses and ideas welcome. Send three chapters and synopsis by post.
Ⓔ Royalties twice-yearly

The O'Brien Press Ltd

12 Terenure Road East, Rathgar, Dublin 6
Ⓣ 00 353 1 492 3333 Ⓕ 00 353 1 492 2777
books@obrien.ie
www.obrien.ie
Managing Director *Ivan O'Brien*
Publisher *Michael O'Brien*

Founded 1974. Publishes business, true crime, biography, music, travel, sport, Celtic subjects, food and drink, history, humour, politics, reference. Children's publishing – mainly fiction for every age from tiny tots to teenage. Illustrated fiction SERIES *Panda Cubs* (3 years+); *Pandas* (5 years+); *Flyers* (6 years+); *Red Flag* (8 years+). Novels (10 years+): contemporary, historical, fantasy. Some non-fiction: mainly historical, and art and craft, resource books for teachers. No poetry, adult fiction or academic. Unsolicited mss (sample chapters only), synopses and ideas for books welcome. No e-mail submissions. Submissions will not be returned.
Ⓔ Royalties annually.

Oak Tree Press

19 Rutland Street, Cork
Ⓣ 00 353 21 431 3855 Ⓕ 00 353 21 431 3496

info@oaktreepress.com
www.oaktreepress.com
Owners *Brian O'Kane, Rita O'Kane*
Managing Director *Brian O'Kane*

Founded 1991. Specialist publisher of business and professional books with a focus on small business start-up and development. Unsolicited mss and synopses welcome; send to the managing director, at the address above.
Ⓔ Royalties annually.

On Stream Publications Ltd

Currabaha, Cloghroe, Co. Cork
☎ 00 353 21 438 5798
info@onstream.ie
www.onstream.ie
Chairman/Managing Director *Roz Crowley*

Founded 1992. Formerly Forum Publications. Publishes academic, cookery, wine, general health and fitness, local history, railways, photography and practical guides. TITLES *The Health Squad Guide to Health and Fitness; A Kingdom of Wine: A Celebration of Ireland's Winegeese; At Home in Renvyle; Stop Howling at the Moon; 101 Bedtime Stories for Managers.* About 3 titles a year. Synopses and ideas welcome. No children's books.
Ⓔ Royalties annually.

Penguin Ireland

25 St Stephen's Green, Dublin 2
☎ 00 353 1 661 7695   🅕 00 353 1 661 7696
info@penguin.ie
www.penguin.ie
Managing Director *Michael McLoughlin*

Founded 2002. Part of the **Penguin Group UK**. DIVISIONS **Commercial fiction/non-fiction** Editorial Director *Patricia Deevy* TITLES *In My Sister's Shoes* Sinéad Moriaty; *This Champagne Mojito is the Last Thing I Own* Ross O'Carroll Kelly; *Secret Diary of a Demented Housewife* Niamh Greene; *Under My Skin* Alison Jameson. **Literary fiction/non-fiction** Editor *Brendan Barrington* TITLES *Connemara: Listening to the Wind* Tim Robinson; *Notes from a Turkish Whorehouse* Philip Ó Ceallaigh; *Keys to the Kingdom* Jack O'Connor; *In Your Face* Lia Mills; *A Book of Uncommon Prayer* Theo Dorgan. 18 titles in 2008. Unsolicited mss, synopses and ideas for books welcome; send by post.
Ⓔ Royalties twice-yearly.

Poolbeg Press Ltd

123 Grange Hill, Baldoyle, Dublin 13
☎ 00 353 1 832 1477   🅕 00 353 1 832 1430
poolbeg@poolbeg.com
www.poolbeg.com
Managing Director *Kieran Devlin*
Publisher *Paula Campbell*
Non-Fiction Commissioning Editor *Brian Langan*

Founded 1976 to publish the Irish short story and has since diversified to include all areas of fiction (literary and popular), children's fiction and non-fiction, and adult non-fiction: history, biography and topics of public interest. AUTHORS discovered and first published by Poolbeg include Maeve Binchy, Marian Keyes, Sheila O'Flanagan, Cathy Kelly, Patricia Scanlan and Melissa Hill. 'Our slogan is Poolbeg.com – The *Irish* for Bestsellers!' IMPRINTS **Poolbeg** (paperback and hardback); **Poolbeg For Children**. About 40 titles a year. Unsolicited mss, synopses and ideas welcome (mss preferred). No drama.
Ⓔ Royalties twice-yearly.

Royal Dublin Society

Science Section, Ballsbridge, Dublin 4
☎ 00 353 1 240 7217   🅕 00 353 1 660 4014
science@rds.ie
www.rds.ie/science
Development Executive, Science & Technology *Dr Claire Mulhall*

Founded 1731 for the promotion of agriculture, science and the arts, and throughout its history has published books and journals towards this end. Publishes conference proceedings and books on the history of Irish science. TITLES *Agricultural Development for the 21st Century; The Right Trees in the Right Places; Agriculture & the Environment; Water of Life; Science, Blueprint for a National Irish Science Centre; Science, Technology and Realism; Sheets of Many Colours – The Mapping of Ireland's Rock's 1750–1890; Science in the Service of the Fishing Industry;* occasional papers in *Irish Science & Technology* series. SERIES *Science and Irish Culture* TITLES *Science and Irish Culture: Why the History of Science Matters; Science and Ireland – Value for Society; It's Part of What We Are.*
Ⓔ Royalties not generally paid.

Royal Irish Academy

19 Dawson Street, Dublin 2
☎ 00 353 1 676 2570   🅕 00 353 1 676 2346
publications@ria.ie
www.ria.ie/publications
Executive Secretary *Patrick Buckley*
Managing Editor *Ruth Hegarty*

Founded in 1785, the Academy has been publishing since 1787. Core publications are journals but more books published in last 15 years. Publishes academic, Irish interest and Irish language. About 12 titles a year. Welcomes mss, synopses and ideas of an academic standard, but currently encourages publications which will popularize some aspect of the sciences and humanities. Address correspondence to the managing editor of publications.

Tír Eolas

Newtownlynch, Doorus, Kinvara, Co. Galway
☎ 00 353 91 637452 🖷 00 353 91 637452
info@tireolas.com
www.tireolas.com
Publisher/Managing Director *Anne Korff*

Founded 1987. Publishes books and guides on ecology, archaeology, folklore and culture. TITLES *The Book of the Burren*; *The Shores of Connemara*; *Not a Word of a Lie*; *The Book of Aran*; *Kinvara, A Seaport Town on Galway Bay*; *A Burren Journal*; *Alive, Alive-O, The Shellfish and Shellfisheries of Ireland*; *The Burren Wall*; *Corrib Country, Guide & Map*; *Shannon Valley, Guide & Map*. Unsolicited mss, synopses and ideas for books welcome. No specialist scientific and technical, fiction, plays, school textbooks or philosophy. € Royalties annually.

University College Dublin Press

Newman House, 86 St Stephen's Green, Dublin 2
☎ 00 353 1 477 9812/9813 🖷 00 353 1 477 9821
ucdpress@ucd.ie
www.ucdpress.ie
Executive Editor *Barbara Mennell*
Assistant Editor *Noelle Moran*

Founded 1995. Academic publisher. 17 titles in 2008. Unsolicited mss, synopses and ideas welcome by post. Approach in writing. No non-academic, journals, poetry, novels or books based on academic conferences. € Royalties annually.

Veritas Publications

7–8 Lower Abbey Street, Dublin 1
☎ 00 353 1 878 8177 🖷 00 353 1 878 6507
publications@veritas.ie
www.veritas.ie
Director *Maura Hyland*
Managing Editor *Ruth Kennedy*

Founded 1969 to supply religious textbooks to schools and later introduced a wide-ranging general list. Part of the Catholic Communications Institute. Publishes books on religious, ethical, moral, societal and social issues. 40 titles a year. Unsolicited mss, synopses and ideas for books welcome. € Royalties annually.

Wolfhound Press
▷ Merlin Publishing

The Woodfield Press

17 Jamestown Square, Dublin 8
☎ 00 353 1 454 7991
terrimcdonnell@ireland.com
www.woodfield-press.com
Managing Director *Terri McDonnell*

Founded in 1995 and since then has published 21 books in a mix of biography, women's history, local history and industrial history. Terri McDonnell divides her time between a full-time job in legal publishing and managing Woodfield. 2 titles in 2008. No unsolicited mss; synopses and ideas for books welcome. Approach by e-mail. No fiction. € Royalties annually.

European Publishers

Austria

Springer-Verlag GmbH

Sachsenplatz 4–6, A–1201 Vienna
☏ 00 43 1 3302415   🖷 00 43 1 3302426
springer@springer.at
www.springer.at

Founded 1924. Austria's largest scientific
publisher. Publishes academic and textbooks,
journals and translations: architecture, art,
cell biology, chemistry, computer science, law,
medicine, neurosurgery, neurology, nursing,
psychiatry, psychotherapy.

Verlag Carl Ueberreuter GmbH

Alser Strasse 24, A–1090 Vienna
☏ 00 43 1 404440
www.ueberreuter.de

Founded 1548. Austria's largest privately-owned
publishing house. Publishes children's and young
adult titles; adult non-fiction includes health and
nutrition, history, culture and politics.

Paul Zsolnay Verlag GmbH

Prinz Eugen Strasse 30, A–1040 Vienna
☏ 00 43 1 505 7661-0   🖷 00 43 1 505 7661-10
info@zsolnay.at
www.zsolnay.at

Founded 1923. Publishes biography, fiction,
general non-fiction, history, poetry.

Belgium

Brepols Publishers NV

Begijnhof 67, B–2300 Turnhout
☏ 00 32 14 44 80 20   🖷 00 32 14 42 89 19
info@brepols.net
www.brepols.net

Founded 1796. Independent, academic humanities
publisher of books, monographs, journals, CD-
ROMs and online. Bibliography, encyclopedias,
reference, history, archaeology, language,
literature, art, art history, music, philosophy,
architecture.

Uitgeverij Lannoo NV

Kasteelstraat 97, B–8700 Tielt
☏ 00 32 51 42 42 11   🖷 00 32 51 40 11 52
lannoo@lannoo.be
www.lannoo.be

Founded 1909. Publishes general non-fiction,
art, architecture and interior design, cookery,
biography, children's, gardening, health and
nutrition, history, management, self-help,
religion, travel, young adult fiction, photography,
psychology.

Standaard Uitgeverij

Mechelsesteenweg 203, B–2018 Antwerp
☏ 00 32 3 285 72 00   🖷 00 32 3 285 72 99
info@standaarduitgeverij.be
www.standaarduitgeverij.be

Founded 1919. Publishes education, fiction,
poetry, humour, maps, children's and young adult.

Universitaire Pers Leuven

Minderbroedersstraat 4, B-3000 Leuven
☏ 00 32 16 32 53 45   🖷 00 32 16 32 53 52
info@upers.kuleuven.be
www.lup.be

Founded 1971. Academic publisher of
anthropolgy, archaeology, architecture, art, arts,
economics, geology, history, law, philosophy,
political science, psychology, sociology, science
and technology.

Bulgaria

Kibea Publishing Company

Komplex Lulin–3, Block 329, Entrance 7, App.
130, BG–1336 Sofia
☏ 00 359 2 925 0152   🖷 00 359 2 925 0748
office@kibea.net
www.kibea.net

Founded 1991. Publishes fiction and illustrated
books; encyclopedias, children's, anthologies, art,
history, health and lifestyle guides, alternative
and popular medicine, psychology and
psychoanalysis, philosophy, religion.

Lettera

62 Rhodope Street, POBox 802, BG–4000 Polvdiv
☎ 00 359 3 262 8027   ᵮ 00 359 3 262 8027
office@lettera.bg
www.lettera.bg

Founded 1991. Educational publisher of textbooks and school aids, dictionaries, languages, audio and video; reference and contemporary fiction.

Naouka I Izkoustvo

11 Slaveikov Square, BG–1000 Sofia
☎ 00 359 2 987 4790
nauk_izk@sigma-bg.com

Founded 1948. Publishes dictionaries, grammar books, language textbooks, translations, psychology, philosophy, essays.

Czech Republic

Atlantis Ltd

PS 374 Ceská 15, CZ–602 00 Brno
☎ 00 420 542 213 221   ᵮ 00 420 542 213 221
atlantis-brno@volny.cz
www.volny.cz/atlantis

Publishes fiction, language and linguistics, literary studies, poetry, religion, social science.

Brána a.s.

Jankovcova 18/938, CZ–170 37 Prague 7
☎ 00 420 220 191 313   ᵮ 00 420 220 191 313
info@brana-knihy.cz
www.brana-knihy.cz

Publishes fiction, non-fiction, art, architecture, encyclopedias, language and linguistics, literary studies, medicine, health, social science, sport, tourism.

Dokorán Ltd

Zborovská 40, CZ–150 00 Prague 5
☎ 00 420 257 320 803   ᵮ 00 420 257 320 805
dokoran@dokoran.cz
www.dokoran.cz

Founded 2001. Publishes fiction, non-fiction, translations, natural sciences (physics, mathematics, nature), arts, history, social science.

Olympia Publishing Co. Inc.

Klimentská 1, CZ–110 15 Prague 1
☎ 00 420 224 810 146   ᵮ 00 420 222 312 137
olympia@mbox.vol.cz
www.iolympia.cz

Publishes fiction, non-fiction, children's, entertainment, leisure, popular science, science fiction, sport, textbooks, tourism.

Denmark

Forlaget Apostrof ApS

Postboks 2580, DK–2100 Copenhagen OE
☎ 00 45 3920 8420   ᵮ 00 45 3920 8453
apostrof@apostrof.dk
www.apostrof.dk

Founded 1980. Publishes psychology and psychiatry.

Blackwell Munksgaard

1 Rosenørns Allé, DK–1970 Frederiksberg C
☎ 00 45 7733 3333   ᵮ 00 45 7733 3377
info@msk.blackwellpublishing.com
www.blackwellmunksgaard.com

Founded 1917. Part of Wiley-Blackwell. Publishes academic books and journals for higher education, research and professional markets.

Borgens Forlag A/S

Valbygardsvej 33, DK–2500 Valby
☎ 00 45 3615 3615   ᵮ 00 45 3615 3616
post@borgen.dk
www.borgen.dk

Founded 1948. Publishes fiction and general non-fiction: biography, children's and young adult, history, thrillers, humour, professional, psychology.

Forlaget Forum

Postbox 2252, DK–1019 Copenhagen K
☎ 00 45 3341 1800   ᵮ 00 45 3341 1801
gb@gb-forlagene.dk
www.forlagetforum.dk

Founded 1940. Part of Gyldendalske Boghandel-Nordisk Forlag A/S. Publishes fiction, bibliography, history, humour, mysteries, children's and young adults.

Gyldendal

Klareboderne 3, DK–1001 Copenhagen K
☎ 00 45 3375 5555
gyldendal@gyldendal.dk
www.gyldendal.dk

Founded 1770. Publishes fiction, children's and young adult, directories, art, biography, education, history, how-to, medicine, music, poetry, philosophy, psychiatry, reference, general and social sciences, sociology, textbooks.

Høst & Søn Publishers Ltd

Købmagergade 62, DK–1150 Copenhagen K
☎ 00 45 3341 1800   ᵮ 00 45 3341 1801
gb@gb-forlagene.dk
www.hoest.dk

Founded 1836. Publishes fiction, biography, children's, crafts, environmental studies, history.

Lindhardt og Ringhof Forlag A/S

Vognmagergade 11, DK–1148 Copenhagen K
☎ 00 45 3369 5000   ⊞ 00 45 3369 5001
info@lindhardtogringhof.dk
www.lrforlag.dk

Founded 1971. Merged with Aschehoug Dansk
Forlag A/S in 2008. Publishes fiction and general
non-fiction.

Nyt Nordisk Forlag Arnold Busck A/S

Landemærket 11, 5. sal, DK–1119 Copenhagen K
☎ 00 45 3373 3575   ⊞ 00 45 3314 0115
nnf@nytnordiskforlag.dk
www.nytnordiskforlag.dk

Founded 1896. Publishes fiction, architecture, art,
academic, cookery, design, education, gardening,
health, natural history, philosophy, reference,
religion, leisure, medicine, nursing, psychology,
textbooks, travel.

Det Schønbergske Forlag

Landemærket 11, 5. sal, DK–1119 Copenhagen K
☎ 00 45 3373 3585   ⊞ 00 45 3373 3586
schoenberg@nytnordiskforlag.dk
www.nytnordiskforlag.dk

Founded 1857. Part of **Nyt Nordisk Forlag
Arnold Busck A/S**. Publishes directories, fiction,
non-fiction, reference, textbooks.

Tiderne Skifter Forlag A/S

Læderstræde 5, 1. sal, DK–1201 Copenhagen K
☎ 00 45 3318 6390   ⊞ 00 45 3318 6391
tiderneskifter@tiderneskifter.dk
www.tiderneskifter.dk

Founded 1973. Publishes fiction, literature
and literary criticism, essays, poetry, drama,
translations, history, art, politics, psychoanalysis,
sexual politics, ethnicity, photography.

Finland

Gummerus Oy

PO Box 749, SF–00101 Helsinki
☎ 00 358 9 584301   ⊞ 00 358 9 5843 0200
www.gummerus.fi

Founded 1872. Publishes fiction and general
non-fiction.

Karisto Oy

PO Box 102, SF–13101 Hämeenlinna
☎ 00 358 3 63151   ⊞ 00 358 3 616 1565
kustannusliike@karisto.fi
www.karisto.fi

Founded 1900. Publishes fiction (including
crime, thrillers and fantasy), general non-fiction
(including health, history, family and childcare,
self improvement, pets, fishing), juvenile and
young adult.

Otava Publishing Co. Ltd

Uudenmaankatu 8–12, SF–00120 Helsinki
☎ 00 358 9 19961
otava@otava.fi
www.otava.fi

Founded 1890. Publishes fiction, translations,
general non-fiction, children's and young adult,
multimedia, education, textbooks.

Tammi Publishers

PO Box 410, SF–00101 Helsinki
☎ 00 358 9 6937 621   ⊞ 00 358 9 6937 6266
tammi@tammi.fi
www.tammi.fi

Founded 1943. Finland's third largest publisher.
Publishes fiction, general non-fiction, children's
and young adult, education, translations. Part of
the Bonnier Group.

WSOY (Werner Söderström Osakeyhtiö)

PO Box 222, SF–00121 Helsinki
☎ 00 358 9 616 81   ⊞ 00 358 9 6168 3560
firstname.lastname@wsoy.fi
www.wsoy.fi

Founded 1878. Publishes fiction, translationS,
general non-fiction, education, dictionaries,
encyclopedias, textbooks, children's and young
adult.

France

Éditions Arthaud

Flammarion Groupe, 87 quai Panhard-et-
Levassor, F–75647 Paris Cedex 13
☎ 00 33 1 40 51 31 00
contact@arthaud.fr
www.arthaud.fr

Founded 1890. Imprint of **Éditions Flammarion**.
Publishes illustrated books, photography, travel.

Éditions Belfond

12 avenue d'Italie, F–75627 Paris Cedex 13
☎ 00 33 1 44 16 05 00
www.belfond.fr

Founded 1963. Publishes fiction and general
non-fiction.

Éditions Bordas

89 blvd Auguste-Blanqui, F–75013 Paris
☎ 00 33 1 44 39 54 45
www.editions-bordas.com

Founded 1946. Publishes education and
general non-fiction: dictionaries, directories,
encyclopedias, reference.

Éditions Calmann-Lévy

31 rue de Fleurus, F–75006 Paris
☎ 00 33 1 49 54 36 00   📠 00 33 1 45 44 86 32
www.editions-calmann-levy.com

Founded 1836. Part of **Hachette Livre**. Publishes fiction, science fiction, fantasy, biography, history, humour, economics, philosophy, psychology, psychiatry, social sciences, sociology, sport.

Éditions Denoël

9 rue du Cherche-Midi, F–75278 Paris Cedex 06
☎ 00 33 1 44 39 73 73   📠 00 33 1 44 39 73 90
denoel@denoel.fr
www.denoel.fr

Founded 1932. Part of Gallimard Groupe. Publishes fiction, science fiction, fantasy, thrillers, art, art history, biography, directories, government, history, philosophy, political science, psychology, psychiatry, reference.

Librairie Arthème Fayard

13 due du Montparnasse, F-75006
☎ 00 33 1 45 49 82 00
www.fayard.fr

Founded 1857. Part of **Hachette Livre**. Publishes fiction, biography, directories, maps, atlases, history, dance, music, philosophy, religion, reference, sociology, general science.

Éditions Flammarion

87 quai Panhard-et-Levassor, F–75647 Paris Cedex 13
☎ 00 33 1 40 51 31 00
www.editions.flammarion.com

Imprint of Groupe Flammarion. Publishes general fiction, science fiction, thrillers, essays, adventure, literature, literary criticism; non-fiction: art, architecture, biography, children's, cookery, crafts, design, gardening, history, lifestyle, natural history, plants, popular science, reference, medicine, nursing, wine.

Éditions Gallimard

5 rue Sébastien-Bottin, F–75328 Paris Cedex 07
☎ 00 33 1 49 54 42 00   📠 00 33 1 45 44 94 03
www.gallimard.fr

Founded 1911. Publishes fiction, poetry, art, biography, history, music, philosophy, children's and young adult.

Éditions Grasset & Fasquelle

61 rue des Saints-Pères, F–75006 Paris
☎ 00 33 1 44 39 22 00   📠 00 33 1 42 22 64 18
www.edition-grasset.fr

Founded 1907. Part of **Hachette Livre**. Publishes fiction and general non-fiction, biography, essays, literature, literary criticism, thrillers, translations, philosophy.

Hachette Livre

43 quai de Grenelle, F–75905 Paris Cedex 15
☎ 00 33 1 43 92 30 00   📠 00 33 1 43 92 30 30
www.hachette.com

Founded 1826. With the acquisition of Time Warner Book Group, Hachette, already the owner of Hodder Headline and Orion, is now the biggest publisher in the British market and the third largest publisher in the world behind Pearson and Bertelsmann. Publishes fiction and general non-fiction, bibliographies, bilingual, children's, directories, education, encyclopedias, general engineering, government, language and linguistics, reference, general science, textbooks.

Éditions Robert Laffont

24 avenue Marceau, F–75381 Paris Cedex 08
☎ 00 33 1 53 67 14 00
www.laffont.fr

Founded 1987. Publishes fiction and non-fiction.

Éditions Larousse

21 rue du Montparnasse, F–75283 Paris Cedex 06
☎ 00 33 1 44 39 44 00
www.larousse.fr

Founded 1852. Publishes animals, art, bilingual, children's and young adult, childcare, cinema, cookery, dictionaries, directories, drama, encyclopedias, food, games, gardening, health, home interest, history, medicine, natural history, nursing, music, dance, self-help, psychology, reference, general science, language arts, linguistics, literature, religion, sport, technology, textbooks.

Éditions Jean-Claude Lattès

17 rue Jacob, F–75006 Paris
☎ 00 33 1 44 41 74 00   📠 00 33 1 43 25 30 47
www.editions-jclattes.fr

Founded 1968. Part of **Hachette Livre**. Publishes fiction and general non-fiction.

Éditions Magnard

20 rue Berbier-du-Mets, F–75647 Paris Cedex 13
☎ 00 33 1 44 08 85 85   📠 00 33 1 44 08 49 79
contact@magnard.fr
www.magnard.fr

Founded 1933. Part of Groupe Albin-Michel. Publishes education, foreign language and bilingual books, juvenile and young adults, textbooks.

Michelin Éditions des Voyages

46 ave de Breteuil, F–75007 Paris
☎ 00 33 1 45 66 11 70

Founded 1900. Publishes travel: maps and atlases.

Les Éditions de Minuit SA

7 rue Bernard-Palissy, F–75006 Paris
☎ 00 33 1 44 39 39 20 📠 00 33 1 45 44 82 36
www.leseditionsdeminuit.fr

Founded 1942. Publishes fiction, essays, literature, literary criticism, philosophy, social sciences, sociology.

Les Éditions Nathan

9 rue Méchain, F–75686 Paris
☎ 00 33 1 45 87 50 00 📠 00 33 1 45 87 53 43
www.nathan.fr

Founded 1881. Publishes fiction and non-fiction for children and young adults. Adult reference, leisure and natural history; education, technical and training.

Presses de la Cité

12 ave d'Italie, F–75627 Paris Cedex 13
☎ 00 33 1 44 16 05 00 📠 00 33 1 44 16 05 01
pressesdelacite@placedesediteurs.com
www.pressesdelacite.com

Founded 1947. Imprint of **Editions Belfond**. Publishes fiction and general non-fiction, science fiction, fantasy, history, mysteries, romance, biography.

Presses Universitaires de France (PUF)

6 ave Reille, F–75685 Paris Cedex 14
☎ 00 33 1 58 10 31 00
www.puf.com

Founded 1921. Academic publisher of scientific books, dictionaries, economics, encyclopedias, geography, history, law, linguistics, literature, philosophy, psychology, psychiatry, political science, reference, sociology, textbooks.

Éditions du Seuil

27 rue Jacob, F–75261 Paris Cedex 06
☎ 00 33 1 40 46 50 50
www.seuil.com

Founded 1935. Part of the Groupe Martinière. Publishes fiction and non-fiction, translations, literature, literary criticism, essays, poetry, biography, classics, illustrated, history.

Les Editions de la Table Ronde

14 rue Séguier, F–75006 Paris
☎ 00 33 1 40 46 70 70 📠 00 33 1 40 46 71 01
editionslatableronde@editionslatableronde.fr
www.editionslatableronde.fr

Founded 1944. Part of Gallimard Groupe. Publishes fiction and general non-fiction, biography, history, psychology, psychiatry, religion, sport.

Librairie Vuibert

20 rue Berbier-du-Mets, F–75647 Paris Cedex 13
☎ 00 33 1 44 08 49 00 📠 00 33 1 44 08 49 69

www.vuibert.com

Founded 1877. Part of Groupe Albin-Michel. Publishes textbooks.

Germany

Verlag C.H. Beck (OHG)

Wilhelmstrasse 9, D–80801 Munich
☎ 00 49 89 38 189-0 📠 00 49 89 38 189-398
www.beck.de

Founded 1763. Publishes general non-fiction, anthropology, archaeology, art, CD-ROMs, dance, dictionaries, directories, economics, essays, encyclopedias, history, illustrated, journals, language arts, law, linguistics, literature, literary criticism, music, philosophy, professional, reference, social sciences, sociology, textbooks, theology.

C. Bertelsmann

Neumarkter Strasse 28, D–81673 Munich
☎ 00 49 89 4136-0

Founded 1835. Imprint of the **Random House Group**. Publishes fiction and general non-fiction, art, autobiography, biography, government and politics.

Carlsen Verlag GmbH

Postfach 50 03 80, D–22703 Hamburg
☎ 00 49 40 39804-0 📠 00 49 40 39804-390
info@carlsen.de
www.carlsen.de

Founded 1953. Publishes children's and comic books.

Deutscher Taschenbuch Verlag GmbH & Co. KG

Friedrichstrasse 1a, D–80801 Munich
☎ 00 49 89 38167-0 📠 00 49 89 34 64 28
verlagdtv.de
www.dtv.de

Founded 1961. Publishes fiction and general non-fiction; biography, business, child care and development, children's, dictionaries, directories, economics, education, encyclopedias, fantasy, government, health and nutrition, history, poetry, psychiatry, psychology, philosophy, politics, reference, religion, literature, literary criticism, essays, thrillers, translations, travel, young adult.

S. Fischer Verlag GmbH

Hedderichstrasse 114, D–60596 Frankfurt am Main
☎ 00 49 69 6062-0 📠 00 49 69 6062-214
www.fischerverlage.de

Founded 1886. Part of the **Holtzbrinck Group**. Publishes fiction, general non-fiction, biography, business, children's and young adult, history, literature, natural history, politics, psychology, reference.

Carl Hanser Verlag GmbH & Co. KG

Postfach 86 04 20, D–81631 Munich
℡ 00 49 89 998 30 0   🖷 00 49 89 98 48 09
info@hanser.de
www.hanser.de/verlag

Founded in 1928. Publishes international and German contemporary literature; children's and juveniles; specialist books on engineering and technology, natural science, computers and computer science, economics and management, CD-ROMs, journals, academic, textbooks, translations.

Heyne Verlag

Bayerstrasse 71–73, D–80335 Munich
℡ 00 49 89 4136-0
www.heyne.de

Founded 1934. Part of **Random House Group**. Publishes fiction, mystery, romance, humour, science fiction, fantasy, biography, cookery, film, history, how-to, occult, psychology, psychiatry.

Hoffmann und Campe Verlag GmbH

Harvestehuder Weg 42, D–20149 Hamburg
℡ 00 49 40 44188-0   🖷 00 49 40 44188-202
email@hoca.de
www.hoca.de

Founded 1781. Publishes fiction and general non-fiction; art, audio, biography, dance, history, illustrated, journals, music, poetry, philosophy, psychology, psychiatry, general science, social sciences.

Verlagsgruppe Georg von Holtzbrinck GmbH

Gänsheidestrasse 26, D–70184 Stuttgart
℡ 00 49 711 2150-0   🖷 00 49 711 2150-269
info@holtzbrinck.com
www.holtzbrinck.com

Founded 1948. One of the world's largest publishing groups with 12 book publishing houses and 40 imprints. Also publishes the newspapers, *Handelsblatt* and *Die Zeit*.

Hüthig GmbH & Co. KG

Postfach 10 28 69, D–69018 Heidelberg
℡ 00 49 6221 489-0   🖷 00 49 6221 489-279
www.huethig.de

Founded 1925. Germany's fourth largest professional publisher: CD-ROMs, multimedia, journals, academic, textbooks, translations.

Ernst Klett Verlag GmbH

Rotebühlstrasse 77, D–70178 Stuttgart
℡ 00 49 711 6672 1333
www.klett.de

Founded 1897. Educational publisher: dictionaries, encyclopedias, textbooks.

Verlagsgruppe Lübbe GmbH & Co. KG

Scheidtbachstrasse 23–31, D–51469 Bergisch Gladbach
℡ 00 49 2202 121-0   🖷 00 49 2202 121-928
www.luebbe.de

Founded 1963. Independent publisher of fiction and general non-fiction, audio, art, biography, history, how-to.

Rowohlt Verlag GmbH

Hamburgerstr 17, D–21465 Reinbeck
℡ 00 49 40 72 72-0   🖷 00 49 40 72 72-319
info@rowohlt.de
www.rowohlt.de

Founded 1908. Publishes general non-fiction.

Springer GmbH

Tiergartenstrasse 17, D–69121 Heidelberg
℡ 00 49 6221 487-0
www.springer.de

Founded 1842. Part of Springer Science+Business Media Deutschland GmbH. Leading specialist publisher of science, technology and medical books, journals, CD-ROMs, databases. Subjects include astronomy, engineering, computer science, economics, geography, law, linguistics, mathematics, philosophy, psychology, architecture, construction and transport – with eighty per cent of publications in English.

Suhrkamp Verlag

Postfach 101945, D–60019 Frankfurt am Main
℡ 00 49 69 75601-0   🖷 00 49 69 75601-314
www.suhrkamp.de

Founded 1950. Publishes fiction, poetry, biography, cinema, film, theatre, philosophy, psychology, psychiatry, general science.

Taschen GmbH

Hohenzollernring 53, D–50672 Cologne
℡ 00 49 221 201 80-0   🖷 00 49 221 25 49 19
contact@taschen.com
www.taschen.com

Founded 1980. Publishes photography, art, erotica, architecture and interior design.

Ullstein Buchverlage GmbH

Friedrichstrasse 126, D–10117 Berlin
℡ 00 49 30 23456-300   🖷 00 49 30 23456-303
info@ullstein-buchverlage.de
www.ullsteinbuchverlage.de

Founded 1903. Publishes general fiction and non-fiction; literature, mystery, biography, business, economics, health, gift books, memoirs, politics.

WEKA Holding GmbH & Co. KG

Postfach 13 31, D–86438 Kissing

📞 00 49 8233 23-0   📠 00 49 8233 23-195

info@weka-holding.de

www.weka-holding.de

Founded 1973. Germany's largest professional publisher.

Hungary

Akadémiai Kiadó

Prielle Kornélia u. 19/D, H–1516 Budapest

📞 00 36 1 464 8282   📠 00 36 1 464 8251

info@akkrt.hu

www.akkrt.hu

Hungary's oldest publishing house. Founded 1828. Publishes scholarly reference, economics, education, dictionaries, languages, journals, scientific.

Jelenkor Kiadó Szolgáltató Kft.

Munkácsy Mihály u. 30/A, H–7621 Pécs

📞 00 36 72 314 782   📠 00 36 72 532 047

editors@jelenkor.com

www.jelenkor.com

Founded 1993. Independent publishing house. Publishes fiction, philosophy, poetry.

Kiss József Kónyvkiadó, Kereskedel mi és Reklám Kft.

Hamzsabégi út 31, H–1114 Budapest

📞 00 36 209 9140   📠 00 36 466 0703

konyvhet@alexero.hu

www.konyv7.hu

Publishes fiction, crime, film, history, politics, theatre, translations.

Kráter Association and Publishing House

Búzavirág Street, H–2013 Pomáz

📞 00 36 26 328 491   📠 00 36 26 328 491

info@krater.hu

www.krater.hu

Publishes poetry and literature.

Móra Könyvkiadö Zrt. (Móra Publishing House)

Váci út 19, H–1134 Budapest

📞 00 36 1 320 4740   📠 00 36 1 320 5382

mora@mora.hu

www.mora.hu

Founded 1950. Publishes books for children of all age groups – fiction and school books, popular science, poems, bath books and teenage horror. Recently extended range to include adult fiction and non-fiction.

Italy

Adelphi Edizioni SpA

Via S. Giovanni sul Muro 14, I–20121 Milan

📞 00 39 2 725731   📠 00 39 2 89010337

info@adelphi.it

www.adelphi.it

Founded 1962. Publishes literature, literary criticism, anthropology, art, autobiography, archaeology, architecture, biography, biology, cinema, economics, history, linguistics, mathematics, medicine, music, philosophy, photography, politics, psychiatry, religion, general science, sociology, theatre, translations.

Bompiani

Via Mecenate 91, I–20138 Milan

📞 00 39 2 5095 2876   📠 00 39 2 5065 2788

www.rcslibri.it

Founded 1929. Part of RCS Libri publishing group. Publishes fiction and general non-fiction, autobiography, biography, art, illustrated, memoirs, philosophy.

Bulzoni Editore SRL

Via dei Liburni 14, I–00185 Rome

📞 00 39 6 4455207   📠 00 39 6 4450355

bulzoni@bulzoni.it

www.bulzoni.it

Founded 1969. Publisher of college textbooks.

Cappelli Editore

Via Farini 14, I–40124 Bologna

📞 00 39 51 239060   📠 00 39 51 239286

info@cappellieditore.com

www.cappellieditore.com

Founded 1851. Publishes reference, textbooks, juvenile and young adult.

Garzanti Libri SpA

Via Gasparotto 1, I–20124 Milan

📞 00 39 2 0062 3201

info@garzantilibri.it

www.garzantilibri.it

Founded 1861. Publishes fiction and non-fiction, literature, literary criticism, encyclopedias, essays, biography, cookery, crime, dictionaries, history, memoirs, philosophy, poetry, reference, textbooks.

Giunti Editoriale SpA

Via Bolognese 165, I–50139 Florence

📞 00 39 55 5062 1   📠 00 39 55 5062 398

www.giunti.it

Founded 1841. Publishes fiction and non-fiction, literature, essays, art, archaeology, alternative medicine, children's, cookery, crafts, dictionaries, education, health and beauty, history, illustrated, journals, language and linguistics, leisure, multimedia, music, psychology, reference, science, textbooks, tourism. Italian publishers of National Geographical Society books.

Casa Editrice Longanesi SpA

Via Gherardini 10, I–20145 Milan
☏ 00 39 2 34597620 🖷 00 39 2 34597212
info@longanesi.it
www.longanesi.it

Founded 1946. Mass market paperback publisher of fiction, adventure, archaeology, art, biography, essays, fantasy, history, journalism, maritime, philosophy, popular science, thrillers.

Arnoldo Mondadori Editore SpA

Via Mondadori 1, I–20090 Segrate (Milan)
☏ 00 39 2 75421 🖷 00 39 2 75422302
www.mondadori.it

Founded 1907. Italy's largest publisher. Publishes fiction and non-fiction, mystery, romance, art, biography, children's and young adult, directories, history, how-to, journals, medicine, music, poetry, philosophy, psychology, reference, religion, general science, education, textbooks.

Società editrice il Mulino

Strada Maggiore 37, I–40125 Bologna
☏ 00 39 51 256011 🖷 00 39 51 256034
info@mulino.it
www.mulino.it

Founded 1954. Publishes economics, government, history, law, language arts, linguistics, literature, literary criticism, philosophy, political science, psychology, psychiatry, religion, social sciences, sociology, textbooks, journals.

Gruppo Ugo Mursia Editore SpA

Via Melchiorre Gioia 45, I–20124 Milan
☏ 00 39 2 6737 8500 🖷 00 39 2 6737 8605
info@mursia.com
www.mursia.com

Founded 1941. Publishes fiction, art, biography, directories, education, history, maritime, philosophy, reference, religion, sport, general science, social sciences, textbooks, juvenile and young adult.

Rizzoli Editore

Via Mecenate 91, I–20138 Milan
☏ 00 39 2 50951
www.rcslibri.it

Founded 1945. Part of RCS Libri publishing group. Publishes fiction and non-fiction, juvenile and young adult.

Società Editrice Internazionale – SEI

Corso Regina Margherita 176, I–10152 Turin
☏ 00 39 11 5227 1 🖷 00 39 11 5211 320
www.seieditrice.com

Founded 1908. Publishes children's, dictionaries, education, encyclopedias, textbooks.

Sonzogno

Via Mecenate 91, I–20138 Milan
☏ 00 39 2 50951 🖷 00 39 2 5065361
www.rcslibri.it

Founded 1818. Part of RCS Libri publishing group. Publishes fiction, mysteries, and general non-fiction.

Sperling e Kupfer Editori SpA

Via Marco D'Aviano 2, I–20131 Milan
☏ 00 39 2 28523 1 🖷 00 39 2 28523 277
info@sperling.it
www.sperling.it

Founded 1899. Publishes fiction and general non-fiction; young adult, biography, health, history, science, sport, thrillers.

Sugarco Edizioni Srl

Via don Gnocchi 4, I–20148 Milan
☏ 00 39 2 407 8370 🖷 00 39 2 407 8493
info@sugarcoedizioni.it
www.sugarcoedizioni.it

Founded 1956. Publishes fiction, biography, history, how-to, philosophy.

Todariana Editrice

Via Gardone 29, I–29139 Milan
☏ 00 39 2 5681 2953 🖷 00 39 2 5521 3405
toeura@tin.it
www.todariana-eurapress.it

Founded 1967. Publishes fiction, poetry, science fiction, fantasy, literature, literary criticism, language arts, linguistics, psychology, psychiatry, social sciences, sociology, travel.

The Netherlands

A.W. Bruna Uitgevers BV

Postbus 40203, NL–3504 AA Utrecht
☏ 00 31 30 247 0411 🖷 00 31 30 241 0018
info@awwbruna.nl
www.awbruna.nl

Founded 1868. Publishes fiction and general non-fiction; mysteries, thrillers, computer science, CD-ROMs, history, philosophy, psychology, psychiatry, general science, social sciences, sociology.

Uitgeverij BZZTÔH BV

Laan van Meerdervoort 10, NL–2517 AJ The Hague
☏ 00 31 70 363 2934 🖷 00 31 70 363 1932
info@bzztoh.nl
www.bzztoh.nl

Founded 1970. Publishes fiction, mysteries, romance, literature, literary criticism, foreign language, bilingual, general non-fiction, astrology, biography, cookery, finance, self-help, health and nutrition, lifestyle, occult, sport, thrillers, translations, travel.

Uitgeversmaatschappij J. H. Kok BV

Postbus Box 5018, NL–8260 GA Kampen
① 00 31 38 339 2555
www.kok.nl

Founded 1894. Publishes fiction, history, religion, directories, encyclopedias, reference, psychology, general science, social sciences, sociology, textbooks, juvenile and young adult.

Uitgeverij J.M. Meulenhoff BV

Postbus 100, NL–1000 AC Amsterdam
① 00 31 20 553 3500   ⑤ 00 31 20 625 1135
info@meulenhoff.nl
www.meulenhoff.nl

Founded 1895. Publishes international co-productions, fiction and general non-fiction. Specializes in Dutch and translated literature.

Pearson Education Benelux

Postbus 75598, NL–1070 AN Amsterdam
① 00 31 20 575 5800   ⑤ 00 31 20 664 5334
www.pearsoneducation.nl

Founded 1942. Publishes education, business, computer science, directories, economics, management, reference, textbooks, technology.

Uitgeverij Het Spectrum BV

Postbus 2073, NL–3500 GB Utrecht
① 00 31 30 265 0650   ⑤ 00 31 30 262 0850
het@spetrum.nl
www.spectrum.nl

Founded 1935. Publishes general non-fiction, children's, dictionaries, encyclopedias, health, illustrated, language, management, parenting, reference, science, sport, travel.

Springer Science+Business Media BV

Van Godewijckstraat 30, NL–3311 GX Dordrecht
① 00 31 78 657 6000   ⑤ 00 31 78 657 6555
www.springer.com

Publishes ebooks, CD-ROMs, databases and journals; biomedicine, computer science, economics, engineering, humanities, life sciences, medicine, social sciences, mathematics, physics.

Uitgeverij Unieboek BV

Postbus 97, NL–3990 DB Houten
① 00 31 30 799 8300   ⑤ 00 31 30 799 8398
info@unieboek.nl
www.unieboek.nl

Founded 1891. Publishes fiction: mysteries, romance, thrillers; general non-fiction, business, lifestyle, travel, reference, juvenile and young adults.

Norway

H. Aschehoug & Co (W. Nygaard)

Postboks 363 Sentrum, N–0102 Oslo
① 00 47 22 400 400   ⑤ 00 47 22 206 395
epost@aschehoug.no
www.aschehoug.no

Founded 1872. Publishes fiction and general non-fiction, children's, reference, popular science, hobbies, education and textbooks, translations.

Cappelen Damm AS

Postboks 350 Sentrum, N–0101 Oslo
① 00 47 21 616 500   ⑤ 00 47 21 365 040
www.cappelen.no

Founded 1829. Owned by the Egmont Group. Publishes fiction (including foreign fiction), general non-fiction, maps and atlases, juvenile and young adult, dictionaries, directories, encyclopedias, reference, religion, textbooks.

Gyldendal Norsk Forlag

Postboks 6860, St Olavs Plass, N–0130 Oslo
① 00 47 22 034 100   ⑤ 00 47 22 034 105
gnf@gyldendal.no
www.gyldendal.no

Founded 1925. Publishes fiction, science fiction, children's, dictionaries, directories, encyclopedias, journals, reference, textbooks.

Tiden Norsk Forlag

Postboks 6704, St Olavs Plass, N–0130 Oslo
① 00 47 23 327 660   ⑤ 00 47 23 327 697
tiden@tiden.no
www.tiden.no

Founded 1933. Part of **Gyldendal Norsk Forlag** group. Publishes Norwegian and translated fiction and general non-fiction.

Poland

Ameet Sp. zo.o.

Przybyszewskiego 176/178, PL–93–120 Lodz
① 00 48 42 676 27 78   ⑤ 00 48 42 676 28 19
ameet@ameet.com.pl
www.ameet.pl

Children's publisher.

Nasza Ksiegarnia Publishing House

Sarabandy 24c, PL–02–868 Warsaw
① 00 48 22 643 93 89   ⑤ 00 48 22 643 70 28
www.nk.com.pl

Founded 1921. Publishes books for children and young adults .

Polish Scientific Publishers PWN

ul. Miodowa 10, PL–00–251 Warsaw

📞 00 48 22 69 54 180 📠 00 48 22 69 54 288

www.pwn.pl

Part of the PWN Group. Leading publisher of scientific, educational, professional and reference material.

Publicat SA

ul. Chlebowa 24, PL–61–003 Poznan

📞 00 48 61 652 92 52 📠 00 48 61 652 92 00

office@publicat.pl

www.publicat.pl

Publishes fiction and non-fiction; biography, children's, crime, education, geography, history, religion.

Portugal

Bertrand Editora Lda

Rua Anchieta 15, P–1249–060 Lisbon

📞 00 351 213 476 122

info@bertrand.pt

www.bertrand.pt

Founded 1727. Publishes bilingual, dictionaries, encyclopedias, juvenile and young adult.

Editorial Caminho SARL

Avenida Almirante Gago Coutinho 121, P–17100–029 Lisbon

📞 00 351 21 842 9830 📠 00 351 21 842 9849

www.editorial-caminho.pt

Founded 1975. Publishes fiction and non-fiction, children's and young adult; art, biography, economics, history, linguistics, music, philosophy, photography, psychology, political science, religion.

Civilização Editora

Rua Alberto Aires de Gouveia 27, P–4050–023 Porto

📞 00 351 22 605 0917 📠 00 351 22 605 0999

info@civilizacao.pt

www.civilizacao.pt

Founded 1921. Publishes fiction, art, biography, cookery, dictionaries, economics, encyclopedias, history, political science, religion, sociology, social sciences, theatre, children's and young adult.

Publicações Europa-América Lda

Rua Francisco Lyon de Castro 2, Apartado 8, P–2725–354 Mem Martins

📞 00 351 21 926 7700 📠 00 351 21 926 7771

www.europa-america.pt

Founded 1945. Publishes fiction and non-fiction, children's, directories, reference, textbooks.

Gradiva-Publicações S.A.

Rua Almeida e Sousa, 21 R/C Esq, P–1399–041 Lisbon

📞 00 351 21 397 4067 📠 00 351 21 395 3471

geral@gradiva.mail.pt

www.gradiva.pt

Founded 1981. Publishes academic, fiction, literature, literary criticism, children's, juvenile and young adult, directories, illustrated, reference, translations.

Livros Horizonte Lda

Rua das Chagas 17–1 Dto, P–1200–106 Lisbon

📞 00 351 21 346 6917

www.livroshorizonte.pt

Founded 1953. Publishes children's and young adult, education, textbooks.

Editorial Verbo SA

Av António Augusto de Aguiar 148, P–1069–019 Lisbon

📞 00 351 21 380 1100 📠 00 351 21 386 5397

www.editorialverbo.pt

Founded 1958. Publishes dictionaries, education, encyclopedias, history, general science, juvenile and young adult.

Spain

Alianza Editorial SA

Juan Ignacio Luca de Tena 15, E–28027 Madrid

📞 00 34 91 393 8890 📠 00 34 91 320 4408

www.alianzaeditorial.es

Founded 1966. Part of **Grupo Anaya**. Publishes fiction, poetry, art, history, mathematics, philosophy, government, political science, social sciences, sociology, general science.

Groupo Anaya

Juan Ignacio Luca de Tena 15, E–28027 Madrid

📞 00 349 1 393 8600 📠 00 349 1 742 6631

www.anaya.es

Founded 1959. Publishes education, reference, literature, multimedia, children's and young adult.

EDHASA (Editora y Distribuidora Hispano – Americana SA)

Av Diagonal 519–521, 2 piso, E–08029 Barcelona

📞 00 34 93 494 9720 📠 00 34 93 419 4584

info@edhasa.es

www.edhasa.es

Founded 1946. Publishes fiction, literature, literary criticism, fantasy, essays, biography, history.

Editorial Espasa-Calpe SA

Carreterade Irún Km 12, 200, E–28049 Madrid
℡ 00 34 91 358 9689   📠 00 34 91 358 9364
www.espasa.com

Founded 1926. Part of Groupo Planeta. Publishes fiction, children's, biography, essays, education, literature, CD-ROMs, dictionaries, encyclopedias, reference, maps and atlases.

Grijalbo

Travessera de Gracia 47–49, E–08021 Barcelona
℡ 00 34 93 366 0300   📠 00 34 93 366 0449
www.grijalbo.com

Founded 1962. An imprint of **Random House Mondadori**. Publishes fiction (historical and thrillers) and non-fiction (self-help and inspirational).

Ediciónes Hiperión SL

Calle de Salustiano Olózaga 14, E–28001 Madrid
℡ 00 34 91 557 6015   📠 00 34 91 435 8690
info@hiperion.com
www.hiperion.com

Founded 1976. Publishes literature, essays, poetry, bilingual, foreign language, children's, religion, translations.

Grupo Editorial Luis Vives

Xaudaró 25, E–28034 Madrid
℡ 00 34 91 334 4883   📠 00 34 91 334 4882
dediciones@edelvives.es
www.edelvives.es

Founded 1890. Educational publisher.

Editorial Molino SL

Pérez Galdós 36, E–08012 Barcelona
℡ 00 34 93 226 0625
www.rba.es/holding

Founded 1933. Part of Grupo RBA. Publishes children's and young adult books.

Pearson Educacion SA

Ribera del Loira 28, E–28042 Madrid
℡ 00 34 91 382 8300   📠 00 34 91 382 8329
pearson.educacion@pearson.com
www.pearsoneducacion.com

Founded 1942. Educational publishers.

Editorial Planeta SA

Calle Còrsega 273–279, E–08008 Barcelona
℡ 00 34 93 228 5800
www.planeta.es

Founded 1952. Part of Groupo Planeta SA. Publishes fiction and general non-fiction.

Random House Mondadori

Travessera de Gracia 47–49, E–08021 Barcelona
℡ 00 349 3 366 0300   📠 00 349 3 366 0449
www.randomhousemondadori.es

Publishes fiction and non-fiction; children's and young adult, art, bilingual dictionaries, biography, economics, essays, guides, illustrated, literature, memoirs, poetry, reference, science, self-help, sociology, technology, thrillers,

Editorial Seix Barral SA

Avda Diagonal 662–664, 7º, E–08034 Barcelona
℡ 00 349 3 496 7003   📠 00 349 3 496 7004
editorial@seix-barral.es
www.seix-barral.es

Founded 1911. Part of Grupo Planeta SA. Foreign language publisher of literature.

Tusquets Editores

Cesare Cantù 8, E–08023 Barcelona
℡ 00 349 3 253 0400
www.tusquets-editores.com

Founded 1968. Publishes fiction, art, autobiography, biography, cinema, cookery, food, history, essays, literature, memoirs, philosophy, poetry, social sciences, theatre.

Sweden

Bonnier Books

PO Box 3159, SE–103 63 Stockholm
℡ 00 46 8 696 8000
info@bok.bonnier.se
www.bok.bonnier.se

Founded 2002. Part of the Bonnier AB group. Publishes fiction, non-fiction and children's books.

Albert Bonniers Förlag

PO Box 3159, SE–103 63 Stockholm
℡ 00 46 8 696 8620
info@abforlag.bonnier.se
www.albertbonniersforlag.com

Founded 1837. Part of Bonnier Books. Publishes fiction and general non-fiction.

Bra Böcker AB

PO Box 892, SE–20180 Malmö
℡ 00 46 40 665 4693
www.bbb.se

Founded 1965. Publishes fiction and non-fiction.

Brombergs Bokförlag AB

PO Box 12886, SE–112 98 Stockholm
℡ 00 46 8 562 620 80   📠 00 46 8 562 620 85
info@brombergs.se
www.brombergs.se

Founded 1975. Publishes literary fiction and general non-fiction.

Bokförlaget Forum AB

PO Box 70321, SE-107 23 Stockholm
📞 00 46 8 696 8440 📠 00 46 8 696 8367
www.forum.se

Founded 1944. Publishes fiction (including foreign) and general non-fiction.

Bokförlaget Natur och Kultur

PO Box 27 323, SE-102 54 Stockholm
📞 00 46 8 453 8600 📠 00 46 8 453 8790
info@nok.se
www.nok.se

Founded 1922. Publishes fiction and general non-fiction; education and textbooks.

Norstedts Förlag

Box 2052, SE-103 12 Stockholm
📞 00 46 8 769 88 50 📠 00 46 8 769 88 64
info@norstedts.se
www.norstedts.se

Founded 1823. Sweden's oldest publishing house. Publishes fiction and general non-fiction.

Rabén och Sjögren Bokförlag

PO Box 2052, SE-103 12 Stockholm
📞 00 46 8 769 8800 📠 00 46 8 769 8813
info@raben.se
www.raben.se

Founded 1941. Part of **Norstedts Förlagsgrupp**. Publishes children's and young adult books.

B. Wåhlströms Bokförlag AB

PO Box 6630, SE-113 84 Stockholm
📞 00 46 8 728 2300 📠 00 46 8 618 9761
info@wahlstroms.se
www.wahlstroms.se

Founded 1911. Part of Forma Publishing Group AB. Publishes fiction and general non-fiction, juvenile and young adult.

Switzerland

Arche Verlag AG

Niederdorfstr 90, CH-8001 Zurich
📞 00 41 1 252 24 10 📠 00 41 1 261 11 15
info@arche-verlag.com
www.arche-verlag.com

Founded 1944. Publishes literature and literary criticism, essays, biography, fiction, poetry, music, dance, travel.

Diogenes Verlag AG

Sprecherstr 8, CH-8032 Zurich
📞 00 41 1 254 85 11 📠 00 41 1 252 84 07
info@diogenes.ch
www.diogenes.ch

Founded 1952. Publishes fiction, essays, literature, literary criticism, art, children's.

Neptun-Verlag

Erlenstrasse 2, CH-8280 Kreuzlingen
📞 00 41 71 677 96 55 📠 00 41 72 677 96 50
neptun@bluewin.ch
www.neptunart.ch

Founded 1946. Publishes non-fiction and children's books.

Orell Füssli Verlag

Dietzingerstr 3, CH-8036 Zurich
📞 00 41 1 466 77 11 📠 00 41 1 466 74 12
www.ofv.ch

Founded 1519. Publishes accountancy, architecture, art, art history, business, leisure, photography, politics, psychology.

Payot Libraire

Case Postale 6730, CH-1002 Lausanne
📞 00 41 21 341 32 52 📠 00 41 21 341 32 17
www.payot-libraire.ch

Founded 1875. Publishes fiction and general non-fiction; arts and entertainment, audio, business, children's, education, language, lifestyle, professional, reference, travel.

Sauerländer Verlage AG

Ausserfeldstr. 9, CH-5036 Oberentfelden
📞 00 41 62 836 86 86 📠 00 41 62 836 86 20
verlag@sauerlaender.ch
www.sauerlaender.ch

Founded 1807. Publishes education and general non-fiction.

US Publishers

Anyone corresponding with the big American publishers should know that they are wary of any package or letter that looks in any way suspicious. This, of course, begs the question what is or what is not suspicious which is as difficult to answer as what makes or does not make a good novel. The best policy is an initial enquiry by e-mail.

International Reply Coupons (IRCs)

For return postage, send IRCs, available from post offices. Letters, 60 pence; mss according to weight. For current postal rates to the UK, go to the United States Postal Service website at www.usps.com and click on 'Calculate Postage'.

ABC–CLIO

PO Box 1911, Santa Barbara CA 93116–1911
☎ 001 805 968 1911   📠 001 805 685 9685
www.abc-clio.com
President/CEO *Ronald J. Boehm*

Founded 1953. Publishes academic and reference, focusing on history and social studies; books and multimedia. UK SUBSIDIARY ABC-Clio Ltd, Oxford. No unsolicited mss; synopses and ideas welcome.

Abingdon Press

201 Eighth Avenue South, PO Box 801, Nashville TN 37202–0801
☎ 001 615 749 6290
www.abingdonpress.com

Founded 1789. Publishes non-fiction: religious (Protestant), reference, professional and academic texts. IMPRINTS **Dimensions for Living**; **Kingswod Books**. Over 100 titles a year.

Harry N. Abrams, Inc.

115 West 18th Street, New York NY 10011
☎ 001 212 206 7715   📠 001 212 229 0312
www.hnabooks.com
VP/Editor-in-Chief *Eric Himmel*

Founded 1950. Publishes illustrated books: art, natural history, photography, popular culture. No fiction. About 100 titles a year.

Academy Chicago Publishers

363 West Erie Street, 7E, Chicago IL 60654
☎ 001 312 751 7300   📠 001 312 751 7306
info@academychicago.com
www.academychicago.com
President & Senior Editor *Anita Miller*

Founded 1975. Publishes mainstream fiction; non-fiction: art, history, mysteries. No romance, children's, young adult, religious, sexist or avant-garde. See website for submission guidelines.

Ace Books
▷ Penguin Group (USA) Inc.

University of Alabama Press

Box 870380, Tuscaloosa AL 35487–0380
☎ 001 205 348 5180   📠 001 205 348 9201
www.uapress.ua.edu
Director *Daniel J.J. Ross*
Managing Editor *Suzette Griffith*

Founded 1945. Publishes American history, Latin American history, American religious history; Caribbean archaeology; historical archaeology; ethnohistory; anthropology; Native American studies; Judaic studies, American literature and criticism; history and environmental studies; American social and cultural history. About 70 titles a year. See 'Submissions' page on the website for guidelines and editorial contacts.

Aladdin Paperbacks
▷ Simon & Schuster Children's Publishing

University of Alaska Press

PO Box 756240, University of Alaska, Fairbanks AK 99775
☎ 001 907 474 5831   📠 001 907 474 5502
fypress@uaf.edu
www.uaf.edu/uapress

Publishes scholarly works about Alaska and the North Pacific rim, with a special emphasis on circumpolar regions, history, anthropology, natural history. 12 titles a year. Queries and proposals welcome (see 'For Authors' page on the website for proposal guidelines).

Alpha Books
▷ Penguin Group (USA) Inc.

Amistad
▷ HarperCollins Publishers, Inc.

Anchor Books
▷ The Knopf Doubleday Publishing Group

Julie Andrews Collection
▷ HarperCollins Publishers, Inc.

Anvil
▷ Krieger Publishing Co.

Shaye Areheart Books
▷ Crown Publishing Group

University of Arizona Press
355 S. Euclid Avenue, Suite. 103, Tucson AZ 85719
☎ 001 520 621 1441   🖷 001 520 621 8899
uapress@uapress.arizona.edu
www.uapress.arizona.edu
Director *Christine Szuter*

Founded 1959. Publishes academic non-fiction, particularly with a regional/cultural link, plus Native-American and Hispanic literature. About 55 titles a year.

University of Arkansas Press
McIlroy House, 105 N. McIlroy Avenue, Fayetteville AR 72701
☎ 001 479 575 3246   🖷 001 479 575 6044
www.uapress.com
Director *Lawrence J. Malley*

Founded 1980. Publishes scholarly monographs, poetry and general trade books. Particularly interested in scholarly works in history, politics and civil rights. About 20 titles a year.

Atheneum Books for Young Readers
▷ Simon & Schuster Children's Publishing

Atlantic Monthly Press
▷ Grove/Atlantic Inc.

Atria Books
▷ Simon & Schuster Adult Publishing Group

Avalon Travel
▷ Perseus Books Group

Avery
▷ Penguin Group (USA) Inc.

Avon/Avon A/Avon Inspire/Avon Red
▷ HarperCollins Publishers, Inc.

Back Bay Books
▷ Hachette Book Group USA

Ballantine Books
▷ The Random House Publishing Group

Bantam
▷ The Random House Publishing Group

Barron's Educational Series, Inc.
250 Wireless Boulevard, Hauppauge NY 11788
☎ 001 631 434 3311   🖷 001 631 434 3723
barrons@barronseduc.com
www.barronseduc.com
Chairman/CEO *Manuel H. Barron*
President *Ellen Sibley*

Founded 1941. Publishes adult non-fiction, children's fiction and non-fiction, test preparation materials and language materials, cookbooks, pets, hobbies, sport, photography, health, business, law, computers, art and painting. No adult fiction. Over 300 titles a year. E-mail queries to *Wayne Barr* (waynebarr@barronseduc.com). Ⓔ Royalties twice-yearly.

Basic Books/Basic Civitas
▷ Perseus Books Group

Beacon Press
25 Beacon Street, Boston MA 02108
☎ 001 617 742 2110   🖷 001 617 723 3097
www.beacon.org
Director *Helene Atwan*

Founded 1854. Publishes general non-fiction. About 60 titles a year. See submission guidelines on the website. Currently, not accepting new poetry or fiction.

Berkley
▷ Penguin Group (USA) Inc.

Bison Books
▷ University of Nebraska Press

Black Cat
▷ Grove/Atlantic Inc.

Blackbirch Press
▷ Gale

Boyds Mills Press
815 Church Street, Honesdale PA 18431
☎ 001 570 253 1164
contact@boydsmillspress.com
www.boydsmillspress.com
Publisher *Stephen Roxburgh*

A subsidiary of Highlights for Children, Inc. Founded 1990 as a publisher of children's picture and activity books. Publishes children's and young adult's fiction, non-fiction and poetry under five IMPRINTS: **Boyds Mills Press**; **Calkins Creek**; **Front Street**; **Lemniscaat**; **Wordsong**. About 80 titles a year. Unsolicited material welcome; submission guidelines available on the website.

Brassey's, Inc.
▷ Potomac Books, Inc.

Broadway
▷ Crown Publishing Group

Bulfinch Press
▷ Hachette Book Group USA

Business Plus
▷ Hachette Book Group USA

Caedmon
▷ HarperCollins Publishers, Inc.

University of California Press

2120 Berkeley Way, Berkeley CA 94704
℡ 001 510 642 4247   🖷 001 510 643 7127
askucp@ucpress.edu
www.ucpress.edu
Director *Lynne Withey*

Founded 1893. Publishes scholarly and scientific non-fiction; some fiction in translation. About 200 books and 50 journals annually. Preliminary letter with outline preferred.

Calkins Creek
▷ Boyds Mills Press

Canongate US
▷ Grove/Atlantic Inc.

Carolrhoda Books

241 First Avenue N, Minneapolis MN 55401
℡ 001 612 332 3344   🖷 001 612 332 7615
www.lernerbooks.com

Founded 1969. Children's book publisher. An imprint of the **Lerner Publishing Group, Inc**, Carolrhoda publishes hardcover originals. Averages 8–10 picture books each year for ages 3–8; 6 fiction titles for ages 7–18; 2–3 non-fiction titles for various ages. No longer accepting unsolicited submissions.

Center Street
▷ Hachette Book Group USA

Charlesbridge Publishing

85 Main Street, Watertown MA 02472
℡ 001 617 926 0329   🖷 001 617 926 5720
books@charlesbridge.com
www.charlesbridge.com
President/Publisher *Brent Farmer*

Publishes both picture books and transitional 'bridge books' (books ranging from early readers to middle-grade chapter books); fiction; and non-fiction with a focus on nature, science, social studies and multicultural topic.

The University of Chicago Press

1427 East 60th Street, Chicago IL 60637
℡ 001 773 702 7700   🖷 001 773 702 9756
www.press.uchicago.edu

Founded 1891. Publishes academic non-fiction only.

Children's Press
▷ Scholastic Library Publishing

Christian Large Print
▷ Gale

Chronicle Books LLC

680 Second Street, San Francisco CA 94107
℡ 001 415 537 4200   🖷 001 415 537 4460
frontdesk@chroniclebooks.com
www.chroniclebooks.com
President/Publisher *Jack Jensen*

Founded 1967. Publishes illustrated and non-illustrated adult trade and children's books as well as stationery and gift items. About 175 titles a year. Query or submit outline/synopsis and sample chapters and artwork.

Arthur H. Clarke Company
▷ University of Oklahoma Press

Clarkson Potter
▷ Crown Publishing Group

Collins/Collins Design
▷ HarperCollins Publishers, Inc.

Columbia University Press

61 West 62nd Street, New York NY 10023
℡ 001 212 459 0600   🖷 001 212 459 3678
www.cup.columbia.edu
President & Director *James D. Jordan*
Associate Director *Jennifer Crewe*

Founded 1893. Publishes scholarly and general interest non-fiction, reference, translations of Asian literature. No fiction or poetry. Welcomes unsolicited material if the subject fits their programme (see website). Approach by regular mail only.

Joanna Cotler Books
▷ HarperCollins Publishers, Inc.

Crocodile Books, USA
▷ Interlink Publishing Group, Inc.

Crown Publishing Group

1745 Broadway, New York NY 10019
℡ 001 212 782 9000
www.randomhouse.com/crown
President & Publisher *Jenny Frost*

Founded 1933. Division of the **Random House Publishing Group**. Publishes popular trade fiction and non-fiction.

IMPRINTS **Broadway; Clarkson Potter; Crown Business; Crown; Doubleday Business; Doubleday Religion; Harmony Books; Shaye Areheart Books; Three Rivers Press; WaterBrook Multnomah**. Mss submissions by agents only.

DaCapo
▷ Perseus Books Group

DAW Books, Inc.

375 Hudson Street, 3rd Floor, New York
NY 10014–3658
🕾 001 212 366 2096   📠 001 212 366 2090
daw@penguingroup.com
www.dawbooks.com
Publishers *Elizabeth R. Wollheim,*
Sheila E. Gilbert
Associate Editor *Peter Stampfel*

Founded 1971 by Donald and Elsie Wollheim
as the first mass-market publisher devoted
to science fiction and fantasy. An imprint of
Penguin Group (USA) Inc. Publishes science
fiction/fantasy, and some horror. No short stories,
anthology ideas or non-fiction. About 42 titles
a year. Unsolicited mss, synopses and ideas for
books welcome.
💷 Royalties twice-yearly.

Del Rey
▷ The Random House Publishing Group

Delacorte Press
▷ The Random House Publishing Group

Dell
▷ The Random House Publishing Group

Dial Books for Young Readers

345 Hudson Street, New York NY 10014–3657
🕾 001 212 366 2000
Vice President & Publisher *Laura Hornik*
Queries *Submissions Coordinator*

Founded 1961. A division of Penguin Books
for Young Readers. Publishes children's books,
including picture books, beginning readers,
fiction and non-fiction for middle grade and
young adults. 50 titles a year. Accepts unsolicited
picture book mss and up to ten pages for
longer works with query letter. 'Do not send
self-addressed envelope with submissions. Dial
will respond within four months if interested in a
manuscript.'
💷 Royalties twice-yearly.

Dial Press
▷ The Random House Publishing Group

Dimensions for Living
▷ Abingdon Press

Doubleday
▷ The Knopf Doubleday Publishing Group

Doubleday Business/Doubleday Religion
▷ Crown Publishing Group

Downtown Press
▷ Simon & Schuster Adult Publishing Group

Thomas Dunne Books
▷ St Martin's Press LLC

Dutton/Dutton's Children's Books
▷ Penguin Group (USA) Inc.

East Gate Books
▷ M.E. Sharpe, Inc.

Ecco
▷ HarperCollins Publishers, Inc.

Eerdmans Publishing Company

2140 Oak Industrial Drive, NE, Grand Rapids
MI 49505
🕾 001 616 459 4591   📠 001 616 459 6540
info@eerdmans.com
www.eerdmans.com
President *William B. Eerdmans Jr*

Founded in 1911 as a theological and reference
publisher. Gradually began publishing in other
genres with authors like C.S. Lewis, Dorothy
Sayers and Malcolm Muggeridge on its lists.
Publishes religious: theology, biblical studies,
ethical and social concern, social criticism and
children's, religious history, religion and literature.
About 120 titles a year. Unsolicited mss, synopses
and ideas welcome.
💷 Royalties twice-yearly.

Eos
▷ HarperCollins Publishers, Inc.

Faber & Faber, Inc.
▷ Farrar, Straus & Giroux, Inc.

FaithWords
▷ Hachette Book Group USA

FalconGuides
▷ The Globe Pequot Press

Farrar, Straus & Giroux, Inc.

18 West 18th Street, New York NY 10011
🕾 001 212 741 6900
www.fsgbooks.com

Founded 1946. Publishes general and literary
fiction, non-fiction, poetry and children's books.
IMPRINTS **North Point Press; Hill and Wang;
Faber & Faber, Inc.; FSG Books for Young
Readers**. No unsolicited material.

Firebird
▷ Penguin Group (USA) Inc.

Fireside
▷ Simon & Schuster Adult Publishing Group

First Avenue Editions
▷ Lerner Publishing Group

Five Star
▷ Gale

5-Spot
▷ Hachette Book Group USA

Footprint Guides
▷ The Globe Pequot Press

Forever
▷ Hachette Book Group USA

Free Press
▷ Simon & Schuster Adult Publishing Group

Samuel French, Inc.
45 West 25th Street, New York NY 10010–2751
☎ 001 212 206 8990   📠 001 212 206 1429
info@samuelfrench.com
www.samuelfrench.com
Managing Editor *Roxane Heinze-Bradshaw*

Founded 1830. Publishes plays in paperback:
London, Broadway and off-Broadway hits,
light comedies, mysteries, one-act plays and
plays for young audiences. About 50–60 titles
a year. Unsolicited mss welcome. No synopses.
OVERSEAS ASSOCIATES in London, Toronto,
Sydney and Johannesburg.
📖 Royalties annually (books); twice-yearly
(amateur productions); monthly (professional
productions).

Front Street
▷ Boyds Mills Press

FSG Books for Young Readers
▷ Farrar, Straus & Giroux, Inc.

Gale
27500 Drake Road, Farmington Hills MI 48331
☎ 001 248 699 4253   📠 001 248 699 8070
www.gale.com
President *Patrick C. Sommers*

Part of Cengage Learning, Gale is a world leader
in information and educational publishing.
Addresses all types of information needs,
from homework help to health questions and
business profiles, in a variety of formats: books,
web-based material and microfilm. BRANDS/
IMPRINTS **Blackbirch Press; Five Star; Christian
Large Print; Greenhaven Press; G.K. Hall;
KidHaven Press; Large Print Press; Lucent
Books; Macmillan Reference USA; Primary
Source Media; Schirmer Reference; Scholarly
Resources, Inc.; Charles Scribner's Sons;
Sleeping Bear Press; St James Press; Thorndike
Press; Twayne Publishers; UXL; Wheeler
Publishing.**

Laura Geringer Books
▷ HarperCollins Publishers, Inc.

The Globe Pequot Press
PO Box 480, Guilford CT 06437
☎ 001 203 458 4500   📠 001 203 458 4601
info@gobepequot.com
www.globepequot.com

Founded 1947. Publishes regional and
international travel, lifestyle, current events,
how-to and outdoor recreation. IMPRINTS
include **Footprint Guides; The Lyons Press;
FalconGuides; Insiders' Guides.** Unsolicited mss,
synopses, and ideas welcome; growth categories
include books for women and lifestyle. See
website for submission guidelines.

Gotham Books
▷ Penguin Group (USA) Inc.

Grand Central Publishing
▷ Hachette Book Group USA

Graphic Universe
▷ Lerner Publishing Group

Green Willow Books
▷ HarperCollins Publishers, Inc.

Greenhaven Press
▷ Gale

Greenwood-Heinemann
▷ Houghton Mifflin Harcourt Publishing Company

Griffin
▷ St Martin's Press LLC

Grolier/Grolier Online
▷ Scholastic Library Publishing

Grosset & Dunlap
▷ Penguin Group (USA) Inc.

Grove/Atlantic Inc.
841 Broadway, 4th Floor, New York NY 10003
☎ 001 212 614 7850   📠 001 212 614 7886
www.groveatlantic.com
President/Publisher *Morgan Entrekin*
Managing Editor *Michael Hornburg*

Founded 1952. Publishes general fiction and
non-fiction. IMPRINTS **Atlantic Monthly Press;
Black Cat; Canongate US; Grove Press.** Mss
submissions by agents only.

Hachette Book Group USA
237 Park Avenue, New York NY 10017
☎ 001 212 364 1100
www.hachettebookgroupusa.com
CEO & Chairman *David Young*

Founded in 2006 following the acquisition of the
Time Warner Book Group by Hachette Livre.
Publishes fiction and non-fiction, children's and
young adult, audio books.

DIVISIONS **Grand Central Publishing;
FaithWords; Center Street; Little, Brown and
Company; Little, Brown Books for Young
Readers; Orbit; Yen Press; Hachette Book
Group Digital Media.**

IMPRINTS **Back Bay Books; Bulfinch Press; Business Plus; 5-Spot; Forever; Hachette Audio; LB Kids; Springboard Press; Poppy; Twelve; Vision; Wellness Central**. About 660 titles a year. No unsolicited mss.

G.K. Hall
▷ Gale

Harcourt
▷ Houghton Mifflin Harcourt Publishing Company

Harmony Books
▷ Crown Publishing Group

HarperCollins Publishers, Inc.
10 East 53rd Street, New York NY 10022
☎ 001 212 207 7000
www.harpercollins.com
President/Publisher US General Books *Michael Morrison*

Founded 1817. Subsidiary of News Corporation. Publishes general and literary fiction, general non-fiction, business, children's, reference and religious books. About 1,700 titles a year.

GENERAL BOOKS GROUP IMPRINTS

Amistad; Avon; Avon A; Avon Inspire; Avon Red; Caedmon; Collins; Collins Design; Ecco; Eos; Harper; Harper Paperbacks; Harper Perennial; Harper Perennial Modern Classics; HarperAudio; HarperEntertainment; HarperCollins; HarperCollins e-Books; HarperLuxe; Harper One; Rayo; William Morrow; Morrow Cookbooks.

CHILDREN'S IMPRINTS

Amistad; Eos; Greenwillow Books; HarperCollins Children's Audio; HarperCollins Children's Books; HarperFestival; HarperEntertainment; HarperTeen; HarperTrophy; Joanna Cotler Books; Julie Andrews Collection; Katherine Tegen Books; Laura Geringer Books; Rayo. No unsolicited material.

Harvard University Press
79 Garden Street, Cambridge MA 02138
☎ 001 401 531 2800   ▤ 001 401 531 2801
contact_HUP@harvard.edu
www.hup.harvard.edu
Editor-in-Chief *Michael Fisher*

Founded 1913. Publishes scholarly non-fiction only; humanities, social sciences, sciences.

University of Hawai'i Press
2840 Kolowalu Street, Honolulu HI 96822–1888
☎ 001 808 956 8255   ▤ 001 808 988 6052
uhpbooks@hawaii.edu
www.uhpress.hawaii.edu
Director *William H. Hamilton*
Executive Editor *Patricia Crosby*

Founded 1947. Publishes scholarly books pertaining to East Asia, Southeast Asia, Hawaii and the Pacific. 75 titles a year. Unsolicited synopses and ideas welcome; approach by mail. No poetry, children's books or any topics other than Asia and the Pacific.
▣ Royalties twice-yearly.

Hill and Wang
▷ Farrar, Straus & Giroux, Inc.

Hill Street Press LLC
191 E. Broad Street, Suite 216, Athens GA 30601–2848
☎ 001 706 613 7200   ▤ 001 706 613 7204
info@hillstreetpress.com
hillstreetpress.com
VP/Editor-in-Chief *Tom Payton*

Founded 1998. Publishes non-fiction only, dealing with the American South as well as crossword puzzle books, college trivia and graphic novels. Some children's books. About 15 titles a year. Unsolicited material may be submitted by e-mail only; check the website for guidelines.

Hippocrene Books, Inc.
171 Madison Avenue, New York NY 10016
info@hippocrenebooks.com
www.hippocrenebooks.com
President/Editorial Director *George Blagowidow*

Founded 1970. Publishes general non-fiction and reference books. Particularly strong on foreign language dictionaries, language studies and international cookbooks, illustrated history. No fiction.

Holiday House, Inc.
425 Madison Avenue, New York NY 10017
☎ 001 212 688 0085   ▤ 001 212 421 6134
www.holidayhouse.com
Vice President/Editor-in-Chief *Mary Cash*

Publishes children's general fiction and non-fiction (pre-school to secondary). About 50 titles a year. Send query letters only. IRCs for reply must be included. Submission guidelines available on the website.

Henry Holt & Company, Inc.
175 Fifth Avenue, New York NY 10010
☎ 001 646 307 5095   ▤ 001 646 307 5285
www.henryholt.com
President/Publisher *Dan Farley*
Editor-in-Chief, Adult Trade *Marjorie Braman*

Founded in 1866, Henry Holt is one of the oldest publishers in the United States. Part of Holtzbrinck Publishing Holdings. Publishes fiction, by both American and international authors, biographies, children's, health, science, history, politics, ecology and psychology.

DIVISIONS/IMPRINTS **Adult Trade; Books for Young Readers; Metropolitan Books; Times Books; Holt Paperbacks; Roaring Brook Press**. About 350 titles a year. No unsolicited mss.

Holt McDougal
▷ Houghton Mifflin Harcourt Publishing Company

Houghton Mifflin Harcourt Publishing Company
222 Berkeley Street, Boston MA 02116
☎ 001 617 351 5000   ℻ 001 617 351 1125
www.hmco.com
Contact *Submissions Editor*

Founded 1832. Acquired Harcourt in 2008. Publishes literary fiction and general non-fiction, including autobiography, biography and history. Also school textbooks; children's fiction and non-fiction.

DIVISIONS/SUBSIDIARY COMPANIES **Houghton Mifflin Harcourt School Division; Houghton Mifflin Harcourt Trade & Reference Division; Houghton Mifflin Harcourt Learning Technology; Houghton Mifflin Harcourt International Publishers; Houghton Mifflin Harcourt Supplemental Publishers; Greenwood-Heinemann; Holt McDougal; The Riverside Publishing Company**. About 100 titles a year. Unsolicited adult mss no longer accepted. Send synopses, outline and sample chapters for children's non-fiction; complete mss for children's fiction. Do not include s.a.e. 'You will not hear from us regarding the status of your submission unless we are interested, in which case you can expect to hear back within 16 weeks. We cannot track your manuscript and regret that we cannot respond personally to each submission, but we do consider each and every submission we receive. Mss we are not interested in will be recycled.'

Howard Books
▷ Simon & Schuster Adult Publishing Group

HPBooks
▷ Penguin Group (USA) Inc.

Hudson Street Press
▷ Penguin Group (USA) Inc.

Humanity Books
▷ Prometheus Books

University of Illinois Press
1325 South Oak Street, Champaign IL 61820–6903
☎ 001 217 333 0950   ℻ 001 217 244 8082
uipress@uillinois.edu
www.press.uillinois.edu
Director *Willis G. Regier*
Associate Director and Editor-in-Chief *Joan Catapano*

Founded 1918. Publishes non-fiction, scholarly and general, with special interest in Americana, women's studies, African–American studies, film, religion, American music and regional books. About 140–150 titles a year.

Indiana University Press
601 North Morton Street, Bloomington IN 47404–3797
☎ 001 812 855 8817   ℻ 001 812 855 8507
iupress@indiana.edu
www.indiana.edu/~iupress
Director *Janet Rabinowitch*

Founded 1950. Publishes scholarly non-fiction in the following subject areas: African studies, anthropology, Asian studies, African-American studies, bioethics, environment and ecology, film, folklore, history, Jewish studies, Middle East studies, military history, music, paleontology, philanthropy, philosophy, politics, religion, Russian and East European studies, women's studies. About 140 books and 30 journals a year. Query in writing in first instance.

Insiders' Guides
▷ The Globe Pequot Press

Interlink Publishing Group, Inc.
46 Crosby Street, Northampton MA 01060–1804
☎ 001 413 582 7054   ℻ 001 413 582 7057
info@interlinkbooks.com
www.interlinkbooks.com
Owner/Publisher *Michel Moushabeck*
Editor *Pam Thompson*

Founded 1987. Independent publisher of international fiction, travel, politics, cookbooks. Specializes in Middle East titles and ethnicity. IMPRINTS **Crocodile Books, USA** Editorial Head *Ruth Lane Moushabeck* Children's books; **Olive Branch Press** Editorial Head *Phyllis Bennis* Political books. About 50-60 titles a year. See submission guidelines on the website before making an approach.
£ Royalties annually.

University of Iowa Press
100 Kuhl House, 119 West Park Road, Iowa City IA 52242–1000
☎ 001 319 335 2000   ℻ 001 319 335 2055
uipress@uiowa.edu
www.uiowapress.org
Director *Holly Carver*

Founded 1969 as a small scholarly press publishing about five books a year. Now publishing about 35 annually in a variety of scholarly fields, plus local interest, short stories, creative non-fiction and poetry anthologies. No unsolicited mss; query first. Unsolicited ideas and synopses welcome.
£ Royalties annually.

Jove
▷ Penguin Group (USA) Inc.

University Press of Kansas
2502 Westbrooke Circle, Lawrence
KS 66045–4444
☎ 001 785 864 4154   📠 001 785 864 4586
upress@ku.edu
www.kansaspress.ku.edu
Director *Fred M. Woodward*

Founded 1946. Became the publishing arm for all
six state universities in Kansas in 1976. Publishes
scholarly books in American history, legal studies,
presidential studies, American studies, political
philosophy, political science, military history
and environmental history. About 55 titles a year.
Proposals welcome.
Ⓔ Royalties annually.

Kar-Ben Publishing
▷ Lerner Publishing Group

The Kent State University Press
307 Lowry Hall, PO Box 5190, Kent OH 44242
☎ 001 330 672 7913   📠 001 330 672 3104
ksupress@kent.edu
www.kentstateuniversitypress.com
Director *Will Underwood*
Acquiring Editor *Joyce Harrison*

Founded 1965. Publishes scholarly works in
history, literary studies and general non-fiction
with an emphasis on American history and
literature. 30–35 titles a year. Queries welcome;
no mss.

The University Press of Kentucky
663 South Limestone Street, Lexington
KY 40508–4008
☎ 001 859 257 8434   📠 001 859 323 1873
www.kentuckypress.com
Director *Stephen M. Wrinn*

Founded 1943. Publishes scholarly and general
interest books in the humanities and social
sciences. No unsolicited mss; send synopses and
ideas for books by post or e-mail. No fiction,
drama, poetry, translations, memoirs or children's
books.
Ⓔ Royalties annually.

KidHaven Press
▷ Gale

Kingswood Books
▷ Abingdon Press

The Knopf Doubleday Publishing Group
1745 Broadway, New York NY 10019
☎ 001 212 782 9000
www.randomhouse.com/knopf
President *Tony Chirico*
Senior VP Publishing *Suzanne Herz*

A division of **The Random House Publishing
Group**. Publishes hardcover fiction and
non-fiction. IMPRINTS **Anchor Books; Alfred
A. Knopf; Doubleday; Nan A. Talese Books;
Pantheon Books; Schocken Books**.

Krieger Publishing Co.
PO Box 9542, Melbourne FL 32950
☎ 001 321 724 9542   📠 001 321 951 3671
info@krieger-publishing.com
www.krieger-publishing.com
CEO *Robert E. Krieger*
President *Donald E. Krieger*
Vice-President *Maxine D. Krieger*

Founded 1970. Publishes science, education,
ecology, humanities, history, mathematics,
chemistry, space science, technology and
engineering.

IMPRINTS/SERIES **Anvil; Exploring Community
History; Open Forum; Orbit; Professional
Practices; Public History**. Unsolicited mss
welcome. Not interested in synopses/ideas or
trade type titles.

Large Print Press
▷ Gale

Lark Books
▷ Sterling Publishing Co. Inc.

Latino Voices
▷ Northwestern University Press

LB Kids
▷ Hachette Book Group USA

Lemniscaat
▷ Boyds Mills Press

Lerner Publishing Group
241 First Avenue N., Minneapolis MN 55401
☎ 001 612 332 3344   📠 001 612 332 7615
info@lernerbooks.com
www.lernerbooks.com
Chairman *Harry J. Lerner*
President & Publisher *Adam M. Lerner*

Founded 1959. Publishes children's non-fiction,
fiction and curriculum material for all grade
levels.

DIVISIONS/IMPRINTS **Lerner Publications;
Carolrhoda Books** (see entry); **LernerClassroom;
First Avenue Editions; Millbrook Press; Twenty-
First Century Books; Kar-Ben Publishing;
Lerner U.K.; Graphic Universe**. No puzzle,
song or alphabet books, text books, workbooks,
or plays. No longer accepting unsolicited
submissions.

Little Simon
▷ Simon & Schuster Children's Publishing

Little, Brown and Company/Little, Brown Books for Young Readers
> Hachette Book Group USA

Llewellyn Worldwide

2143 Wooddale Drive, Woodbury MN 55125
☎ 001 651 291 1970   📠 001 612 291 1908
www.llewellyn.com
President & Publisher *Carl L. Weschcke*

Division of Llewellyn Worldwide Ltd. Founded 1901. Publishes self-help and how-to: astrology, alternative health, tantra, Fortean studies, tarot, yoga, Santeria, dream studies, metaphysics, magic, witchcraft, herbalism, shamanism, organic gardening, women's spirituality, graphology, palmistry, parapsychology. Also fiction with an authentic magical or metaphysical theme. About 100 titles a year. Unsolicited mss welcome; proposals preferred. IRCs essential in all cases. Submission guidelines available on the website.

Louisiana State University Press

3990 West Lakeshore Drive, Baton Rouge LA 70808
☎ 001 225 578 6295   📠 001 225 578 6461
lsupress@lsu.edu
www.lsu.edu/lsupress
Director *MaryKatherine Callaway*

Publishes fiction and non-fiction: Southern and American history, Southern and American literary criticism, regional trade books, biography, poetry, Atlantic World studies, media studies, environmental studies, geography, political science and music (jazz). About 85 titles a year. See website for submission guidelines.

Lucent Books
> Gale

The Lyons Press
> The Globe Pequot Press

Margaret K. McElderry Books
> Simon & Schuster Children's Publishing

McFarland & Company, Inc., Publishers

PO Box 611, Jefferson NC 28640
☎ 001 336 246 4460   📠 001 336 246 5018
info@mcfarlandpub.com
www.mcfarlandpub.com
Executive Vice President *Rhonda Herman*
Editorial Director *Steve Wilson*
Acquisitions Editor *Gary Mitchem*

Founded 1979. A library reference and upper-end speciality market press publishing scholarly books in many fields: international studies, performing arts, popular culture, sports, automotive history, women's studies, music and fine arts, chess, history and librarianship. Specializes in general reference. Especially strong in cinema studies.

No fiction, poetry, children's, New Age or inspirational/devotional works. About 320 titles a year. No unsolicited mss; send query letter first. Synopses and ideas welcome; submissions by mail preferred.
💲 Royalties annually.

The McGraw-Hill Companies, Inc.

1221 Avenue of the Americas, New York NY 10020–1095
☎ 001 212 904 2000
www.mcgraw-hill.com
Chairman/President/CEO *Harold W. McGraw*

Founded 1888. Parent of **McGraw-Hill Education** which has offices in the UK (see entry under *UK Publishers*). Publishes a wide range of educational, professional, business, science, engineering and computing books.

Macmillan Reference USA
> Gale

University of Massachusetts Press

PO Box 429, Amherst MA 01004
☎ 001 413 545 2217   📠 001 413 545 1226
info@umpress.umass.edu
www.umass.edu/umpress
Director *Bruce Wilcox*
Senior Editor *Clark Dougan*

Founded 1963. Publishes scholarly, general interest, African-American, ethnic, women's and gender studies, cultural criticism, economics, fiction, literary criticism, poetry, philosophy, biography, history. About 40 titles a year. Unsolicited mss considered but query letter preferred in the first instance. Synopses and ideas welcome.

Metropolitan Books
> Henry Holt & Company, Inc.

The University of Michigan Press

839 Greene Street, Ann Arbor MI 48104–3209
☎ 001 734 764 4388   📠 001 734 615 1540
www.press.umich.edu

Founded 1930. Publishes scholarly and trade non-fiction, textbooks, new media studies, music, political science, English as a second language, classical studies, American history, theatre, literary criticism, gender studies, law studies and fiction.

Milet Publishing LLC

333 North Michigan Avenue, Suite 530, Chicago IL 60601
info@milet.com
www.milet.com
Contact *Editorial Director*

Founded 1995. Publishes children's picture books in English and dual language, and language learning books. Welcomes synopses and ideas for books. Queries may be made by e-mail but all submissions – proposal, outline or synopsis with sample text and/or artwork – by post only. 'Please review submission guidelines and existing titles on our website before submitting. We like bold, original stories and artwork, universal and/or multicultural.'

Millbrook Press
▷ Lerner Publishing Group

University of Minnesota Press
111 Third Avenue South, Suite 290, Minneapolis MN 55401
☎ 001 612 627 1970　🖷 001 612 627 1980
ump@umn.edu
www.upress.umn.edu
Executive Editor *Richard Morrison*

Founded 1925. Publishes academic books for scholars and selected general interest titles: American studies, anthropology, art and aesthetics, cultural theory, film and media studies, gay and lesbian studies, geography, literary theory, political and social theory, race and ethnic studies, sociology and urban studies. No original fiction or poetry. About 110 titles a year. No unsolicited mss. Welcomes synopses and proposals sent by mail (*no e-mails*). See website for submission details.

Minotaur
▷ St Martin's Press LLC

University Press of Mississippi
3825 Ridgewood Road, Jackson MS 39211
☎ 001 601 432 6205　🖷 001 601 432 6217
press@ihl.state.ms.us
www.upress.state.ms.us
Director *Leila W. Salisbury*
Assistant Director/Editor-in-Chief *Craig Gill*

Founded 1970. Non-profit book publisher partially supported by the eight state universities. Publishes scholarly and trade titles in literature, history, American culture, Southern culture, African-American studies, women's studies, popular culture, film and comics studies, folklife, ethnic, performance, art, architecture, photography and other liberal arts. Represented worldwide. UK REPRESENTATIVE: Roundhouse Group. About 65 titles a year. Send letter of enquiry, prospectus, table of contents and sample chapter prior to submission of full mss.
🖹 Royalties annually.

University of Missouri Press
2910 LeMone Boulevard, Columbia MO 65201
☎ 001 573 882 7641　🖷 001 573 884 4498
upress@umsystem.edu
www.system.missouri.edu/upress

Associate Director *Dwight Browne*

Founded 1958. Publishes academic: history, intellectual history, journalism, literary criticism and related humanities disciplines, and occasional volumes of short stories. About 70 titles a year. Best approach is by letter. Send synopses for academic work together with author c.v.

Modern Library
▷ The Random House Publishing Group

William Morrow/Morrow Cookbooks
▷ HarperCollins Publishers, Inc.

NAL
▷ Penguin Group (USA) Inc.

University of Nebraska Press
1111 Lincoln Mall, Lincoln NE 68588–0630
☎ 001 402 472 3581　🖷 001 402 472 6214
pressmail@unl.edu
www.nebraskapress.unl.edu
Interim Director *Joan Giesecke*

Founded 1941. Publishes scholarly, Native American studies, history of the American West, literary and cultural studies, music, Jewish studies, military history, sports history, environmental history. IMPRINT **Bison Books**. No unsolicited mss; welcomes synopses and ideas for books. Send enquiry letter in the first instance with description of project and sample material. No original fiction, children's books or poetry.
🖹 Royalties annually.

University of Nevada Press
Morrill Hall Mail Stop 0166, Reno NV 89557–0166
☎ 001 775 784 6573
www.nvbooks.nevada.edu
Director/Editor-in-Chief *Joanne O'Hare*

Founded 1961. Publishes scholarly and popular books; serious fiction, Native American studies, natural history, Western Americana, Basque studies and regional studies. Unsolicited material welcome if it fits in with areas published, or offers a 'new and exciting' direction.

University Press of New England
One Court Street, Suite 250, Lebanon NH 03766
☎ 001 603 448 1533　🖷 001 603 448 7006
university.press@dartmouth.edu
www.upne.com
Director *Michael Burton*
Editor-in-Chief *Phyllis Deutsch*

Founded 1970. A scholarly books publisher sponsored by six institutions of higher education in the region: Brandeis, Dartmouth, University of Vermont, Tufts, the University of New Hampshire and Northeastern. Publishes general and scholarly non-fiction. OVERSEAS ASSOCIATES Canada: University of British Columbia; UK, Europe,

Middle East: Eurospan University Press Group; Australia, New Zealand, Asia & the Pacific: East-West Export Books. About 80 titles a year. Unsolicited material welcome.
Ⓔ Royalties annually.

University of New Mexico Press

1312 Basehart Road SE, Albuquerque
NM 87106–4363
Ⓣ 001 505 277 2346   Ⓕ 001 505 272 7141
unmpress@unm.edu
www.unmpress.com
Director *Luther Wilson*
Editor-in-Chief *William 'Clark' Whitehorn*

Founded 1929. Publishes scholarly, regional books, biography and fiction. No how-to, humour, self-help or technical. 80 titles a year. Receives around 3,600 submissions a year.
Ⓔ Royalties annually.

New Seeds
▷ **Shambhala Publications, Inc.**

North Point Press
▷ **Farrar, Straus & Giroux, Inc.**

University of North Texas Press

1155 Union Circle, #311336, Denton
TX 76203–5017
Ⓣ 001 940 565 2142   Ⓕ 001 940 565 4590
ronald.chrisman@unt.edu
www.unt.edu/untpress
Director *Ronald Chrisman*
Managing Editor *Karen DeVinney*

Founded 1987. Publishes regional interest, contemporary, social issues, Texas history, military history, music, women's issues, multicultural. Publishes annually the winners of the Vassar Miller Poetry Prize and the Katherine Anne Porter Prize in Short Fiction. About 15 titles a year. Synopses and ideas welcome. Do not send full ms until requested.
Ⓔ Royalties annually.

Northwestern University Press

629 Noyes Street, Evanston IL 60208–4210
Ⓣ 001 847 491 2046   Ⓕ 001 847 491 8150
nupress@northwestern.edu
www.nupress.northwestern.edu
Chairman *Peter Hayes*
Managing Director *Donna Shear*

Founded 1893. Publishes general and academic books: philosophy, drama and performance studies, history, biography, religion, Slavic studies, Latino fiction and contemporary fiction, poetry. No children's, art books. IMPRINTS **Latino Voices, TriQuarterly** Fiction and poetry Editorial Head *Henry Carrigan*. Unsolicited mss, synopses and ideas welcome; approach by mail. No unsolicited mss in fiction or poetry, please.
Ⓔ Royalties annually.

W.W. Norton & Company, Inc.

500 Fifth Avenue, New York NY 10110
Ⓣ 001 212 354 5500   Ⓕ 001 212 869 0856
www.wwnorton.com
VP/Editor-in-Chief *Starling R. Lawrence*

Founded 1923. Publishes fiction and non-fiction, college textbooks and professional books. About 400 titles a year. Submission guidelines available on the website.

Ohio University Press

19 Circle Drive, The Ridges, Athens
OH 45701–2979
Ⓣ 001 740 593 1155   Ⓕ 001 740 593 4536
www.ohioswallow.com
Director *David Sanders*

Founded 1964. Publishes academic, regional and general trade books. IMPRINT **Swallow Press**. Synopses and ideas for books welcome. Send detailed synopsis, sample chapters and c.v. with covering letter. See website for author guidelines. No children's, how-to or genre fiction.
Ⓔ Royalties annually.

University of Oklahoma Press

2800 Venture Drive, Norman OK 73069–8216
Ⓣ 001 405 325 2000   Ⓕ 001 405 325 4000
www.oupress.com
Director *B. Byron Price*
Publisher *John Drayton*

Founded 1928. Publishes general scholarly non-fiction only: American Indian studies, American West, classical studies, anthropology, natural history and political science. IMPRINT **Arthur H. Clarke Company** Publisher *Robert Clark*. About 100 titles a year. See website for submissions policy.

Olive Branch Press
▷ **Interlink Publishing Group, Inc.**

One World
▷ **The Random House Publishing Group**

Open Forum
▷ **Krieger Publishing Co.**

Orbit
▷ **Krieger Publishing Co.**

Orbit
▷ **Hachette Book Group USA**

Palgrave
▷ **St Martin's Press LLC**

Pantheon Books
▷ **The Knopf Doubleday Publishing Group**

Paragon House

1925 Oakcrest Avenue, Suite 7, St Paul
MN 55113–2619

☎ 001 651 644 3087 🖷 001 651 644 0997
paragon@paragonhouse.com
www.paragonhouse.com
Executive Director *Dr Gordon L. Anderson*

Founded 1982. Publishes non-fiction: reference
and academic. Subjects include history, religion,
philosophy, spirituality, society, political science,
psychology.
Ⓔ Royalties twice-yearly.

Paulist Press

997 Macarthur Road, Mahwah NJ 07430–0990
☎ 001 201 825 7300 🖷 001 201 836 3161
pmcmahon@paulistpress.com
www.paulistpress.com
Managing Editor *Paul McMahon*

Founded in 1858 and owned by Paulist Fathers.
Publishes religion, spirituality, theology, pastoral
resources, ecumenical and interfaith works. No
fiction, memoirs or poetry. Planning to increase
children's (spirituality) and catechetical list of
titles. About 85 titles a year. Submissions welcome
for the Paulist Press imprint only. Initial approach
by e-mail but accepts hard copy submissions only.
Ⓔ Royalties twice-yearly.

Pearson Education

One Lake Street, Upper Saddle River NJ 07458
☎ 001 201-236-7000
www.pearson.com

The world's largest educational publisher.
Publishes across a wide range of curriculum
subjects from primary students to professional
practitioners.

Pelican Publishing Company, Inc.

1000 Burmaster Street, Gretna LA 70053
☎ 001 504 368 1175
www.pelicanpub.com
Editor-in-Chief *Nina Kooij*

Publishes general non-fiction: popular history,
cookbooks, travel, art, business, biography,
children's, architecture, collectibles guides
and motivational. About 75 titles a year. Initial
inquiries required for all submissions.
Ⓔ Royalties twice-yearly.

Penguin Group (USA) Inc.

375 Hudson Street, New York NY 10014
☎ 001 212 366 2000 🖷 001 212 366 2666
www.penguingroup.com
CEO *David Shanks*
President *Susan Petersen Kennedy*

The second-largest trade book publisher in
the world. Publishes fiction and non-fiction in
hardback and paperback; adult and children's.

ADULT DIVISION IMPRINTS Hardcover: **Avery;
Dutton; G.P. Putnam's Sons; Gotham Books;
Hudson Street Press; Tarcher; The Penguin
Press; Portfolio; Riverhead Books; Sentinel;
Viking;** Trade Paperback: **Ace Books; Alpha
Books; Berkley; HPBooks; NAL; Perigee;
Penguin Books; Plume; Riverhead.** Mass Market
Paperback: **Ace; Berkley; DAW Books** (see
entry); **Jove.**

YOUNG READERS DIVISION Hardcover: **Dial
Books for Young Readers** (see entry); **Dutton
Children's Books; Frederick Warne; G.P.
Putnam's Sons; Philomel Books; Viking
Children's Books.** Paperback: **Firebird; Puffin
Books; Razorbill; Speak.** Mass Merchandise:
Grosset & Dunlap; Price Stern Sloan. No
unsolicited mss.
Ⓔ Royalties twice-yearly.

University of Pennsylvania Press

3905 Spruce Street, Philadelphia PA 19104–4112
☎ 001 215 898 6261 🖷 001 215 898 0404
www.pennpress.org
Director *Eric Halpern*
Editor-in-Chief *Peter Agree*

Founded 1890. Publishes serious non-fiction:
scholarly, reference, professional, textbooks and
semi-popular trade. No original fiction or poetry.
Over 100 titles a year. No unsolicited mss but
synopses and ideas for books welcome.
Ⓔ Royalties vary according to sales prospects and
other relevant factors.

Perennial
▷ HarperCollins Publishers, Inc.

Perigee Books
▷ Penguin Group (USA) Inc.

Perseus Books Group

387 Park Avenue S., 12th Floor, New York
NY 10016
☎ 001 212 340 8100 🖷 001 212 340 8115
www.perseusbooksgroup.com
President/CEO *David Steinberger*

Founded 1997. IMPRINTS **Avalon Travel** Guide
books; **Basic Books** Psychology, science,
politics, sociology, current affairs, history; **Basic
Civitas** African and African-American studies;
DaCapo Art, biography, film, history, humour,
music, photography, sport; **PublicAffairs**
Current affairs, biography, history, journalism;
Running Press General non-fiction, children's,
illustrated books; **Seal Press** 'Books by women
for women'; **Vanguard Press** Fiction and non-
fiction; **Westview** Undergraduate and graduate
textbooks. No unsolicited mss.

Philomel Books
▷ Penguin Group (USA) Inc.

Picador USA
> St Martin's Press LLC

Players Press
PO Box 1132, Studio City CA 91614–0132
☎ 001 818 789 4980
Vice President, Editorials *Robert W. Gordon*

Founded 1965 as a publisher of plays; now
publishes across the entire range of performing
arts: plays, musicals, theatre, film, cinema,
television, costume, puppetry, plus technical
theatre and cinema material. No unsolicited mss;
synopses/ideas welcome. Send query letter s.a.e.
for response.
Ⓔ Royalties twice-yearly.

Plume
> Penguin Group (USA) Inc.

Pocket Books/Pocket Star
> Simon & Schuster Adult Publishing Group

Poppy
> Hachette Book Group USA

Portfolio
> Penguin Group (USA) Inc.

Potomac Books, Inc.
22841 Quicksilver Drive, Dulles VA 20166
☎ 001 703 661 1548   ℻ 001 703 661 1547
www.potomacbooksinc.com
Publisher *Sam Dorrance*
Senior Editors *Hilary Claggett, Elizabeth Demers*

Formerly Brassey's, Inc., founded 1983 (acquired
in 1999 by Books International of Dulles,
Virginia). Publishes non-fiction titles on topics
of history (especially military and diplomatic
history), world and US affairs, US foreign policy,
defence, intelligence, biography and sport. About
80 titles a year. No unsolicited mss; query letters/
synopses welcome.
Ⓔ Royalties annually.

Presidio Press
> The Random House Publishing Group

Price Stern Sloan
> Penguin Group (USA) Inc.

Primary Source Media
> Gale

Princeton University Press
41 William Street, Princeton NJ 08540–5237
☎ 001 609 258 4900   ℻ 001 609 258 6305
www.pup.princeton.edu
Director *Peter J. Dougherty*

Founded 1905. Publishes academic and trade
books. About 200 titles a year. No unsolicited
mss. Synopses and ideas considered. Send letter
in the first instance. No e-mail submissions.

No fiction. OVERSEAS ASSOCIATE Princeton
University Press, 3 Market Place, Woodstock
OX20 1SY, UK.
Ⓔ Royalties annually.

Professional Practices
> Krieger Publishing Co.

Prometheus Books
59 John Glenn Drive, Amherst NY 14228–2197
☎ 001 716 691 0133   ℻ 001 716 691 0137
marketing@prometheusbooks.com
www.prometheusbooks.com
www.pyrsf.com
Editor-in-Chief *Steven L. Mitchell*

Founded 1969. Publishes books and journals:
educational, scientific, professional, library,
popular science, science fiction and fantasy,
philosophy and young readers. IMPRINTS
Humanity Books; Pyr™. About 100 titles a year.
Unsolicited mss, synopses and ideas for books
welcome.
Ⓔ Royalties twice-yearly.

Public History
> Krieger Publishing Co.

PublicAffairs
> Perseus Books Group

Puffin Books
> Penguin Group (USA) Inc.

G.P. Putnam's Sons
> Penguin Group (USA) Inc.

Pyr™
> Prometheus Books

Quayside Publishing Group
> The Quarto Group, Inc. under UK Publishers

The Random House Publishing Group
1745 Broadway, New York NY 10019
☎ 001 212 782 9000
www.randomhouse.com
President/Publisher *Gina Centrello*

Publishes fiction, non-fiction, science fiction and
graphic novels in hardcover, trade paperback and
mass market.

IMPRINTS

**Ballantine Books; Bantam; Del Rey; Delacorte
Press; Dell; Dial Press; Modern Library; One
World; Presidio Press; Random House; Random
House Trade Paperbacks; Spiegel & Grau;
Triumph Books; Villard**. 800 titles in 2008. No
unsolicited mss.

Rayo
> HarperCollins Publishers, Inc.

Razorbill
> Penguin Group (USA) Inc.

Reader's Digest Association Inc
Reader's Digest Road, Pleasantville
NY 10570−7000
☎ 001 914 238 1000
www.rd.com
President/CEO *Mary Berner*

Publishes home maintenance reference, cookery, DIY, health, gardening, children's books; videos and magazines.

Riverhead
> Penguin Group (USA) Inc.

The Riverside Publishing Company
> Houghton Mifflin Harcourt Publishing Company

Roaring Brook Press
> Henry Holt & Company, Inc.

The Rosen Publishing Group, Inc.
29 East 21st Street, New York NY 10010
☎ 001 212 777 3017   📠 001 212 777 0277
www.rosenpublishing.com
President *Roger Rosen*

Founded 1950. Publishes non-fiction books (supplementary to the curriculum, reference and self-help) for a young adult audience. Reading levels are years 7−12, 4−6 (books for teens with literacy problems), and 5−9. Subjects include conflict resolution, character building, health, safety, drug abuse prevention, self-help and multicultural titles. 100 titles a year. For all imprints, write with outline and sample chapters.

Running Press
> Perseus Books Group

Rutgers University Press
100 Joyce Kilmer Avenue, Piscataway
NJ 08854−8099
☎ 001 732 445 7762   📠 001 732 445 7039
rutgerspress.rutgers.edu
Director *Marlie Wasserman*
Associate Director/Editor-in-Chief *Leslie Mitchner*

Founded 1936. Publishes scholarly books, regional, social sciences and humanities. About 80 titles a year. Unsolicited mss, synopses and ideas for books welcome. No original fiction or poetry.

S&S Libros eñ Espanol
> Simon & Schuster Adult Publishing Group

St James Press
> Gale

St Martin's Press LLC
175 Fifth Avenue, New York NY 10010
☎ 001 646 307 5151   📠 001 212 420 9314
inquiries@stmartins.com
www.stmartins.com
CEO (Macmillan) *John Sargent*
President/Publisher (Trade Division) *Sally Richardson*

Founded 1952. A division of **Macmillan** in the US, owned by German publishing group Holtzbrinck, GmbH, St Martin's Press made its name and fortune by importing raw talent from the UK to the States and has continued to buy heavily in the UK. Publishes general fiction, especially mysteries and crime; and adult non-fiction: history, self-help, political science, travel, biography, scholarly, popular reference, college textbooks.

IMPRINTS

Picador USA; **Griffin** (trade paperbacks); **St Martin's Paperbacks** (mass market); **Thomas Dunne Books**; **Minotaur**; **Palgrave**; **Truman Talley Books**. All submissions via legitimate literary agents only.

Scarecrow Press Inc.
4501 Forbes Boulevard, Suite 200, Lanham
MD 20706
☎ 001 301 459 3366   📠 001 301 429 5748
www.scarecrowpress.com
Publisher/ Editorial Director *Edward Kurdyla*

Founded 1950 as a short-run publisher of library reference books. Acquired by **University Press of America, Inc.** in 1995 which is now part of Rowman and Littlefield Publishing Group. Publishes reference, scholarly and monographs (all levels) for libraries. Reference books in all areas except sciences, specializing in the performing arts, music, cinema and library science. About 160 titles a year. Unsolicited mss welcome but material will not be returned unless requested and accompanied by return postage. Unsolicited synopses and ideas for books welcome.

Schirmer Reference
> Gale

Schocken Books
> The Knopf Doubleday Publishing Group

Scholarly Resources, Inc.
> Gale

Scholastic Library Publishing
90 Sherman Turnpike, Danbury CT 06816
☎ 001 800 621 1115
www.scholastic.com/aboutscholastic/divisions/slp.htm
President *Greg Worrell*

Publishes print and online juvenile non-fiction, encyclopedias, speciality reference sets and picture books.

DIVISIONS/IMPRINTS **Children's Press; Grolier; Grolier Online; Franklin Watts**. About 500 titles a year.

Charles Scribner's Sons
▷ Gale

Scribner/Scribner Classics
▷ Simon & Schuster Adult Publishing Group

Seal Press
▷ Perseus Books Group

Sentinel
▷ Penguin Group (USA) Inc.

Seven Stories Press

140 Watts Street, New York NY 10013
☎ 001 212 226 8760   🖷 001 212 226 1411
www.sevenstories.com
Publisher *Daniel Simon*
Subsidiary Rights *Anna Lui*

Founded 1995. Independent publisher of literature, translations, politics, media studies, popular culture, health. About 50 titles a year. No unsolicited mss; synopses and ideas from agents welcome or send query letter. No e-mail submissions. No children's or business books. OVERSEAS SUBSIDIARIES Turnaround Publishing Services, UK and EU; Palgrave Macmillan, Australia; Macmillian Publishers NZ Ltd., New Zealand; Liberty Books (Pte.) Ltd., Pakistan; Pen International Pte. Ltd., Singapore; Stephan Phillips (Pty.) Ltd., Southern Africa; Publishers Group Canada, Canada.

Shambhala Publications, Inc.

PO Box 308, Boston MA 02117
www.shambhala.com

Publishes contemplative religion, Buddhism, Christianity, philosophy, self-help, health, creativity and fiction. IMPRINTS **Trumpeter; New Seeds; Weatherhill**. 60 titles in 2008. Welcomes submissions – by post only; no e-mails. No original poetry.

M.E. Sharpe, Inc.

80 Business Park Drive, Armonk NY 10504
☎ 001 914 273 1800   🖷 001 914 273 2106
info@mesharpe.com
www.mesharpe.com
President/CEO *M.E. Sharpe*
Snr VP/COO *Vincent Fuentes*

Founded 1958. Publishes books and journals in the social sciences and humanities, both original works and translations in Asian and East European studies. IMPRINTS **East Gate Books;**

Sharpe Reference *Patricia Kolb*. No unsolicited material.
🖹 Royalties annually.

Simon & Schuster Adult Publishing Group (Division of Simon & Schuster, Inc)

1230 Avenue of the Americas, New York NY 10020
☎ 001 212 698 7000   🖷 001 212 698 7007
www.simonsays.com
President & CEO *Carolyn K. Reidy*

Publishes fiction and non-fiction.

DIVISIONS

Atria Books VP & Executive Editorial Director *Emily Bestler*; **Free Press** VP & Editorial Director *Dominick Anfuso*; **Touchstone-Fireside** VP & Editor-in-Chief *Trish Todd*; **Howard Books** EVP & Publisher *John Howard*; **Pocket Books** VP & Editorial Director *Maggie Crawford*; **Scribner** VP & Editor-in-Chief *Nan Graham*; **Simon &Schuster** VP & Editorial Director *Alice Mayhew*; **Simon Spotlight Entertainment** Editorial Director *Tricia Boczkowski*.

IMPRINTS

Atria Books; Downtown Press; Fireside; Free Press; Howard Books; Pocket Books; Pocket Star; Scribner; Scribner Classics; S&S Libros eñ Espanol; Simon & Schuster; Simon Spotlight Entertainment; Strebor Books; Threshold Editions; Touchstone; Washington Square Press. No unsolicited mss.
🖹 Royalties twice-yearly.

Simon & Schuster Children's Publishing

1230 Avenue of the Americas, New York NY 10020
☎ 001 212 698 7200
www.simonsayskids.com
President/Publisher *Rick Richter*

A division of the Simon & Schuster Consumer Group. Publishes pre-school to young adult, picture books, hardcover and paperback fiction, non-fiction, trade, library and mass-market titles.

IMPRINTS

Aladdin Paperbacks Picture books, paperback fiction and non-fiction reprints and originals, and limited series for ages pre-school to young adult; **Atheneum Books for Young Readers** Picture books, hardcover fiction and non-fiction books across all genres for ages three to young adult; **Little Simon** Mass-market novelty books (pop-ups, board books, colouring & activity) and merchandise (book and audio cassette) for ages birth through eight; **Margaret K. McElderry Books** Picture books, hardcover fiction and non-fiction trade books for children ages three to young adult; **Simon & Schuster Books for Young Readers** Picture books, hardcover fiction and non-fiction for children ages three to young adult; **Simon Pulse** Paperbacks for teenagers; **Simon**

Scribbles Colouring and activity books. **Simon Spotlight** Devoted exclusively to children's media tie-ins and licensed properties. About 480 titles a year. No unsolicited material.

Simon Pulse/Simon Scribbles/Simon Spotlight
▷ Simon & Schuster Children's Publishing

Sleeping Bear Press
▷ Gale

Southern Illinois University Press

1915 University Press Drive, SIUC Mail Code 6806, Carbondale IL 62901
☎ 001 618 453 2281   🖷 001 618 453 1221
www.siu.edu/~siupress
Director *Lain Adkins*

Founded 1956. Publishes scholarly and general interest non-fiction books and educational materials. 50 titles a year.

Speak
▷ Penguin Group (USA) Inc.

Spiegel & Grau
▷ The Random House Publishing Group

Springboard Press
▷ Hachette Book Group USA

Stackpole Books

5067 Ritter Road, Mechanicsburg PA 17055
☎ 001 717 796 0411   🖷 001 717 796 0412
www.stackpolebooks.com
President *M. David Detweiler*
Vice President/Publisher *Judith Schnell*

Founded 1933. Publishes outdoor sports, fishing, hunting, nature, crafts, Pennsylvania and regional, military reference, history. About 130 titles a year.
🖻 Royalties twice-yearly.

Stanford University Press

1450 Page Mill Road, Palo Alto CA 94304–1124
☎ 001 650 723 9434   🖷 001 650 725 3457
info@www.sup.org
www.sup.org
Director *Dr Geoffrey R.H. Burn*

Founded 1925. Publishes non-fiction: scholarly works in all areas of the humanities and social sciences, also professional lists in business, economics, security studies and law. About 160 titles a year. No unsolicited mss; query in writing first.

State University of New York Press

194 Washington Avenue, Suite 305, Albany NY 12210–2384
☎ 001 518 472 5000   🖷 001 518 472 5038
info@sunypress.edu
www.sunypress.edu
Associate Director *James Peltz*

Founded 1966. Part of the Research Foundation of SUNY, the Press is one of the largest university presses in the USA. Publishes scholarly and trade books in the humanities, social sciences and books of regional interest on New York State. Strengths in philosophy, religious studies and Asian studies. 194 titles in 2008. Unsolicited submissions welcome. Send material to *Jane Bunker*, Editor-in-Chief.
🖻 Royalties annually.

Sterling Publishing Co. Inc.

387 Park Avenue South, New York NY 10016
☎ 001 212 532 7160
www.sterlingpub.com
VP/Publisher *Andrew Martin*
President/CEO *Charles Nurnberg*

Founded 1949. A subsidiary of Barnes & Noble, Inc. Publishes illustrated non-fiction: reference and how-to books on arts and crafts, home improvement, history, photography, children's, woodworking, pets, hobbies, gardening, games and puzzles, general non-fiction. IMPRINTS **Lark Books** Crafts, children's, photography. Submission guidelines available.

Strebor Books
▷ Simon & Schuster Adult Publishing Group

Swallow Press
▷ Ohio University Press

Syracuse University Press

621 Skytop Road, Suite 110, Syracuse NY 13244–5290
☎ 001 315 443 5534   🖷 001 315 443 5545
www.SyracuseUniversityPress.syr.edu
Director *Alice R. Pfeiffer*

Founded 1943. Publishes scholarly books in the following areas: Middle East studies, Middle East literature in translation, Irish studies, Jewish studies, sports, gender and globalization, Iroquois studies, women and religion, medieval studies, religion and politics, television and popular culture, geography and regional books. Distributes for the Moshe Dayan Center, Jusoor. About 50 titles a year. No unsolicited mss. Send query letter with IRCs.
🖻 Royalties annually.

Nan A. Talese Books
▷ The Knopf Doubleday Publishing Group

Tarcher
▷ Penguin Group (USA) Inc.

Katherine Tegen Books
▷ HarperCollins Publishers, Inc.

Temple University Press

1601 N. Broad Street, 083–42, Philadelphia
PA 19122–6099
☎ 001 215 204 8787　📠 001 215 204 4719
www.temple.edu/tempress
Editor-in-Chief *Janet M. Francendese*

Founded 1969. Publishes scholarly books. Authors
generally academics. About 60 titles a year. Letter
or e-mail of inquiry with brief outline. Include
fax/e-mail address.

University of Texas Press

PO Box 7819, Austin TX 78722
☎ 001 512 471 7233/471 4278 (editorial)
📠 001 512 232 7178
utpress@uts.cc.utexas.edu
www.utexas.edu/utpress/
www.utexaspress.com
Director *Joanna Hitchcock*
Assistant Director/Editor-in-Chief *Theresa J. May*

Publishes scholarly and regional non-fiction:
anthropology, architecture, classics,
environmental studies, humanities, social
sciences, geography, language studies, literary
modernism; Latin American/Latino/Mexican
American/Middle Eastern/Native American
studies, natural history and ornithology, regional
books (Texas and the southwest). About 100
titles a year and 11 journals. Unsolicited material
welcome in above subject areas only. Author
guidelines available on the website.
💷 Royalties annually.

Thorndike Press
▷ Gale

Three Rivers Press
▷ Crown Publishing Group

Threshold Editions
▷ Simon & Schuster Adult Publishing Group

Times Books
▷ Henry Holt & Company, Inc.

Touchstone
▷ Simon & Schuster Adult Publishing Group

Transaction Publishers Ltd

c/o Rutgers – The State University of New Jersey,
35 Berrue Circle, Piscataway NJ 08854–8042
☎ 001 732 445 2280　📠 001 732 445 3138
trans@transactionpub.com
www.transactionpub.com
Chairman/Editorial Director *I.L. Horowitz*
President *Mary E. Curtis*

Founded 1962. Independent publisher of
academic social scientific books, periodicals
and serials. Publisher of record in international
social research. All unsolicited material should
be submitted in hard copy and full draft. E-mail

submissions are not accepted. 'Decision-making
is rapid.'

TriQuarterly
▷ Northwestern University Press

Triumph Books
▷ The Random House Publishing Group

Truman Talley Books
▷ St Martin's Press LLC

Trumpeter
▷ Shambhala Publications, Inc.

Twayne Publishers
▷ Gale

Twelve
▷ Hachette Book Group USA

Twenty-First Century Books
▷ Lerner Publishing Group

Tyndale House Publishers, Inc.

351 Executive Drive, Carol Stream IL 60188
☎ 001 630 668 8303
www.tyndale.com
President *Mark D. Taylor*

Founded 1962. Books cover a wide range of
categories from non-fiction to gift books,
theology, doctrine, Bibles, fiction, children's and
youth. Also produces video material, calendars
and audio books for the same market. No
poetry. Non-denominational religious publisher
of around 250 titles a year for the evangelical
Christian market. No unsolicited mss. Synopses
and ideas considered. Writer's guidelines available
on the website.

University Press of America, Inc.

4501 Forbes Boulevard, Suite 200, Lanham
MD 20706
☎ 001 301 459 3366
www.univpress.com
VP/Director *Judith L. Rothman*

Founded 1975. An imprint of Rowman &
Littlefield Publishing Group. Publishes scholarly
monographs, college and graduate level
textbooks. No children's, elementary or high
school. About 300 titles a year. Submit outline or
request proposal questionnaire.

UXL
▷ Gale

Vanguard Press
▷ Perseus Books Group

Viking/Viking Children's Books
▷ Penguin Group (USA) Inc.

Villard
▷ The Random House Publishing Group

University of Virginia Press

PO Box 400318, Charlottesville VA 22904–4318
☎ 001 434 924 3468
vapress@virginia.edu
www.upress.virginia.edu
Director *Penelope J. Kaiserlian*

Founded 1963. Publishes academic books in humanities and social science with concentrations in American history, African-American studies, architecture, Victorian literature, Caribbean literature and ecocriticism. 50–60 titles a year. No unsolicited mss; will consider synopses and ideas for books if in their specific areas of concentration. Approach by mail for full proposal; e-mail for short inquiry. No fiction, children's, poetry or academic books outside interests specified above.
Ⓔ Royalties annually.

Vision

▷ Hachette Book Group USA

Walch Education

40 Walch Drive, Portland ME 04103
☎ 001 207 772 2846　Ⓕ 001 207 772 3105
www.walch.com
President *Al Noyes*
Editor-in-Chief *Erica Varney*

Founded 1927. Publishes supplementary educational materials for middle and secondary schools across a wide range of subjects, including English/language arts, literacy, special needs, mathematics, social studies, science and school-to-career. Always interested in ideas from secondary school teachers who develop materials in the classroom. Proposal letters, synopses and ideas welcome.

Walker & Co.

175 Fifth Avenue, 8th Floor, New York NY 10010
☎ 001 646 438 6075　Ⓕ 001 212 727 0984
firstname.lastname@bloomsburyusa.com
www.walkerbooks.com (trade non-fiction)
www.walkeryoungreaders.com (BFYR)
Contact *Submissions Editor*

Founded 1959. A division of **Bloomsbury Publishing**. Publishes children's fiction and non-fiction and adult non-fiction. Please contact the following editors in advance before sending any material to be sure of their interest, then follow up as instructed: **Trade Non-fiction** *George Gibson* Permission and documentation must be available with mss. Submit prospectus first, with sample chapters and marketing analysis. **Books for Young Readers** *Emily Easton* Fiction and non-fiction for all ages. Query before sending non-fiction proposals. Especially interested in picture books – fiction and non-fiction, historical and contemporary fiction for middle grades and young adults. BFYR will consider unsolicited submissions. See website for guidelines.

Frederick Warne

▷ Penguin Group (USA) Inc.

Washington Square Press

▷ Simon & Schuster Adult Publishing Group

Washington State University Press

PO Box 645910, Washington State University, Pullman WA 99164–5910
☎ 001 509 335 3518　Ⓕ 001 509 335 8568
wsupress@wsu.edu
wsupress.wsu.edu
Editor-in-Chief *Glen Lindeman*

Founded 1928. Publishes hardcover originals, trade paperbacks and reprints. Publishes mainly on the history, prehistory, culture, politics and natural history of the Northwest United States (Washington, Idaho, Oregon, Montana, Alaska and British Columbia). Subjects include Native American, maritime, railroad, biography, cooking/food history, nature/environment, politics. 4–6 titles a year. Submission guidelines are available on the website.

WaterBrook Multnomah

▷ Crown Publishing Group

Franklin Watts

▷ Scholastic Library Publishing

Weatherhill

▷ Shambhala Publications, Inc.

Wellness Central

▷ Hachette Book Group USA

Westview

▷ Perseus Books Group

Wheeler Publishing

▷ Gale

John Wiley & Sons, Inc.

111 River Street, Hoboken NJ 07030–5774
☎ 001 201 748 6000　Ⓕ 001 201 748 6088
info@wiley.com
www.wiley.com
President/CEO *William J. Pesce*

Founded 1807. Wiley's core publishing programme includes scientific, technical, medical and scholarly journals, encyclopedias, books and online products and services; professional and consumer books, and subscription and information services in all media; educational materials for undergraduate, graduate students and life-long learners. Submission guidelines available on the website under 'Resources for Authors'.

The University of Wisconsin Press

1930 Monroe Street, 3rd Floor, Madison
WI 53711–2059

📞 001 608 263 1110   📠 001 608 263 1120
uwiscpress@uwpress.wisc.edu
www.wisc.edu/wisconsinpress
Director *Sheila Leary*
Senior Acquisitions Editor *Raphael Kadushin*

Founded 1936. Publishes scholarly, general interest non-fiction and regional books about Wisconsin and the mid-west. About 50 titles a year. No unsolicited mss. Detailed author guidelines available on the website.

Wordsong

▷ Boyds Mills Press

Yen Press

▷ Hachette Book Group USA

Zondervan

5300 Patterson Avenue SE, Grand Rapids
MI 49530

📞 001 616 698 6900
www.zondervan.com
President/CEO *Maureen Girkins*

Founded 1931. Subsidiary of **HarperCollins Publishers, Inc.** Publishes Protestant religion, Bibles, books, audio & video, computer software, calendars and speciality items.

Other International Publishers

Australia

ACER Press

19 Prospect Hill Road, Camberwell, Victoria 3124
📞 00 61 3 9277 5555   📠 00 61 3 9277 5500
www.acer.edu.au

Founded 1930. The publishing arm of the Australian Council for Educational Research. Publishes education, human resources, psychology, parent education and mental health, occupational therapy.

Allen & Unwin Pty Ltd

PO Box 8500, St Leonards, Sydney, NSW 1590
📞 00 61 2 8425 0100   📠 00 61 2 9906 2218
info@allenandunwin.com
www.allenandunwin.com

Founded 1976. Independent publisher of fiction, general non-fiction, academic (social sciences, health) and children's. IMPRINTS **Arena** Fiction (adventure, crime, fantasy, thrillers, chick lit) and non-fiction (biography, DIY, health, humour, self-help, travel); **Jacana Books** Natural history; **Inspired Living** Mind, body and spirit, psychology, spirituality. 250 titles in 2008.

Arena
▷ Allen & Unwin Pty Ltd

Cliffs Notes
▷ John Wiley & Sons Australia Ltd

Currency Press Pty Ltd

PO Box 2287, Strawberry Hills, NSW 2012
📞 00 61 2 9319 5877   📠 00 61 2 9319 3649
enquiries@currency.com.au
www.currency.com.au

Founded 1971. Performing arts publisher – handbooks, biography, crictial studies, directories and reference – drama, theatre, music, dance, film and video.

For Dummies
▷ John Wiley & Sons Australia Ltd

Frommer's
▷ John Wiley & Sons Australia Ltd

Hachette Livre Australia

Level 17, 207 Kent Street, Sydney, NSW 2000
📞 00 61 2 8248 0800   📠 00 61 2 8248 0810
www.hha.com.au

Founded 1958. Part of the **Hachette Livre Publishing Group**. Publishes popular fiction, literature, general non-fiction, autobiography, biography, current affairs, health, history, lifestyle, memoirs, self-help, sport, travel.

Margaret Hamilton Books
▷ Scholastic Australia Pty Limited

HarperCollins Publishers (Australia) Pty Ltd

PO Box 321, Pymble, NSW 2073
📞 00 61 2 9952 5000   📠 00 61 2 9952 5555
www.harpercollins.com.au

Founded 1989. Part of the News Corporation. Publishes fiction and non-fiction, children's and illustrated books.

Inspired Living
▷ Allen & Unwin Pty Ltd

Jacana Books
▷ Allen & Unwin Pty Ltd

Jacaranda
▷ John Wiley & Sons Australia Ltd

Kangaroo Press
▷ Simon & Schuster (Australia) Pty Ltd

Macmillan Publishers Australia Pty Ltd

1–3/15–19 Claremont Street, South Yarra, Victoria 3141
📞 00 61 3 9825 1000   📠 00 61 3 9825 1010
www.macmillan.com.au

Founded 1904. Part of **Macmillan Publishers**, UK. Publishes fiction and non-fiction, academic, children's, reference, education, professional, scientific, technical, medical. DIVISIONS **Macmillan Education** Primary, secondary and ELT; **Palgrave Macmillan** Academic and reference; **Pan Macmillan** (see entry); **Macquarie Dictionary Publishers** Reference (print and online);

Macquarie Dictionary Publishers
▷ Macmillan Publishers Australia Pty Ltd

McGraw-Hill Australia & New Zealand Pty Ltd
Locked Bag 2233, Business Centre, North Ryde, NSW 1670
📞 00 61 2 9900 1800
www.mcgraw-hill.com.au

Founded 1964 (Australia), 1974 (NZ). Owned by **The McGraw-Hill Companies, Inc**. Publishes textbooks and educational material for primary, secondary and higher education. Also professional (medical, general and reference).

Melbourne University Publishing Ltd
187 Grattan Street, Carlton, Victoria 3053
📞 00 61 3 9342 0300   📠 00 61 3 9342 0399
mup-info@unimelb.edu.au
www.mup.unimelb.edu.au

Founded 1922. Academic publishers; general non-fiction, archaeology, architecture, art, biography, business, economics, education, finance, history, language, linguistics, law, literature, maritime, military history, natural history, philosophy, politics, psychology, psychiatry, reference, religion, science and technology, sociology, sport, travel, zoology. IMPRINTS **Melbourne University Press**; **Miegunyah Press**. About 70 titles a year.

Miegunyah Press
▷ Melbourne University Publishing Ltd

Omnibus Books
▷ Scholastic Australia Pty Limited

Oxford University Press (Australia, New Zealand & the Pacific)
GPO Box 2784, Melbourne, Victoria 3001
📞 00 61 3 9934 9123   📠 00 61 3 9934 9100
www.oup.com.au

Founded 1908. Owned by **Oxford University Press**, UK. Publishes for the college, school and trade markets.

Palgrave Macmillan
▷ Macmillan Publishers Australia Pty Ltd

Pan Macmillan Australia Pty Ltd
Level 25, 1 Market Street, Sydney, NSW 2000
📞 00 61 2 9285 9100   📠 00 61 2 9285 9190
www.panmacmillan.com.au

Founded 1983. Part of **Macmillan Publishers**, UK. Publishes fiction, crime fiction, thrillers, literature; general non-fiction, biography, children's, history, self-help, travel.

Pearson Education Australia Pty Ltd
Locked Bag 507, Frenchs Forest, NSW 1640
📞 00 61 3 9454 2200   📠 00 61 3 9453 0089
www.pearson.com.au

Part of Pearson Plc. Australia's largest educational publisher, formed in 1998 from a merger of Addison Wesley Longman Australia and Prentice Hall Australia.

Penguin Group (Australia)
PO Box 701, Hawthorn, Victoria 3124
📞 00 61 3 9811 2400   📠 00 61 3 9811 2620
www.penguin.com.au

Founded 1946. Publishes adult and children's general non-fiction and fiction.

University of Queensland Press
PO Box 6042, St Lucia, Queensland 4067
📞 00 61 7 3365 2127
uqp@uqp.uq.edu.au
www.uqp.uq.edu.au

Founded 1948. Publishes academic, textbooks, children's and young adult, general non-fiction and fiction, literature, literary criticism, short stories, essays, poetry, art, architecture, biography, current affairs, history, indigenous studies, memoirs, military history, reference.

Random House Australia
Level 3, 100 Pacific Highway, North Sydney, NSW 2060
📞 00 61 2 9954 9966   📠 00 61 2 9954 4562
random@randomhouse.com.au
www.randomhouse.com.au

Part of the **Random House Group**. Publishes popular and literary fiction, non-fiction, children's and illustrated.

Scholastic Australia Pty Limited
PO Box 579, Gosford, NSW 2250
📞 00 61 2 4328 3555
www.scholastic.com.au

Founded 1968. Part of **Scholastic Inc.** Educational and children's publisher of books and multimedia. IMPRINTS **Scholastic Press; Omnibus Books; Margaret Hamilton Books**. About 100 titles a year.

Science Press
Bag 7023, Marrickville, NSW 1475
📞 00 61 2 9516 1122   📠 00 61 2 9550 1915
www.sciencepress.com.au

Founded 1948. Educational publisher.

Simon & Schuster (Australia) Pty Ltd
PO Box 33, Pymble, NSW 2073
📞 00 61 2 9983 6600   📠 00 61 2 9988 4293

Founded 1987. Part of **Simon & Schuster Inc.**, USA. Publishes fiction and general non-fiction, children's books. IMPRINTS **Simon & Schuster Australia** Biography, health, parenting, relationships; **Kangaroo Press** Crafts, gardening, military history.

UWA Press

(M419) University of Western Australia, 35 Stirling Highway, Crawley, WA 6009
☏ 00 61 8 6488 3670 📠 00 61 8 6488 1027
admin@uwapress.uwa.edu.au
www.uwapress.uwa.edu.au

Founded 1935. Publishes academic, general non-fiction, essays, literature, literary criticism, history, natural history, children's, maritime history, reference, literary studies.

John Wiley & Sons Australia Ltd

PO Box 1226, Milton, Queensland 4064
☏ 00 61 7 3859 9755 📠 00 61 7 3859 9715
brisbane@johnwiley.com.au
www.johnwiley.com.au

Founded 1954. Owned by **John Wiley & Sons Inc.**, USA. Publishes general non-fiction and education books, dictionaries, encyclopedias, maps and atlases, reference, textbooks. IMPRINTS **Wiley; Jacaranda; WrightBooks; For Dummies; Frommer's; Cliffs Notes**.

WrightBooks
▷ John Wiley & Sons Australia Ltd

Canada

Anchor Canada
▷ Random House of Canada Ltd

Bond Street
▷ Random House of Canada Ltd

Canadian Scholars' Press, Inc

180 Bloor Street West, Suite 801, Toronto, Ontario M5S 2V6
☏ 001 416 929 2774
info@cspi.org
www.cspi.org

Founded 1986. Independent publisher of academic and trade books. IMPRINTS **Women's Press; Kellom Books**.

Doubleday Canada
▷ Random House of Canada Ltd

Emblem Editions
▷ McClelland & Stewart Ltd

H.B. Fenn & Company Ltd

34 Nixon Road, Bolton, Ontario L7E 1W2
☏ 001 905 951 6600 📠 001 905 951 6601
www.hbfenn.com

Founded 1977. Publishes adult and young adult sports books.

Firefly Books Ltd

66 Leek Crescent, Richmond Hill, Ontario L4B 1H1
☏ 001 416 499 8412 📠 001 416 499 8313
www.fireflybooks.com

Founded 1977. Publishes children's and young adult non-fiction. No unsolicited submissions.

Fitzhenry & Whiteside Limited

195 Allstate Parkway, Markham, Ontario L3R 4T8
☏ 001 905 477 9700
godwit@fitzhenry.ca
www.fitzhenry.ca

Founded 1966. Publishes reference, biography, history, poetry, photography, sport, children's and young adult books. About 200 titles a year.

Douglas Gibson Books
▷ McClelland & Stewart Ltd

Harlequin Enterprises Ltd

225 Duncan Mill Road, Don Mills, Ontario M3B 3K9
☏ 001 416 445 5860
www.eharlequin.com

Founded 1949. Canada's second largest publishing house. Publishes romantic and women's fiction.

HarperCollins Canada Ltd

2 Bloor Street E, 20th Floor, Toronto, Ontario M4W 1A8
☏ 001 416 975 9334 📠 001 416 975 5223
www.harpercanada.com

Founded 1989. Publishes commerical and literary fiction, non-fiction, children's, reference, cookery and religious books.

Kellom Books
▷ Canadian Scholars' Press, Inc

Knopf Canada
▷ Random House of Canada Ltd

LexisNexis Canada Ltd

Suite 200, 181 University Avenue, Toronto, Ontario M5H 3M7
☏ 001 416 862 7656 📠 001 416 862 8073
www.lexisnexis.ca

Founded 1912. Publishes law books, CD-ROMs, journals, newsletters, law reports.

McClelland & Stewart Ltd

75 Sherbourne Street, Fifth Floor, Toronto, Ontario M5A 2P9
☏ 001 416 598 1114 📠 001 416 598 7764
mail@mcclelland.com
www.mcclelland.com

Founded 1906. Publishes fiction and general non-fiction, poetry. DIVISIONS/IMPRINTS **Douglas**

Gibson Books; Emblem Editions; Tundra Books (see entry). About 100 titles a year.

McGraw-Hill Ryerson Ltd

300 Water Street, Whitby, Ontario L1N 9B6
☎ 001 905 430 5000
www.mcgrawhill.ca

Founded 1944. Subsidiary of **The McGraw-Hill Companies, Inc.** Publishes education and professional.

Nelson Education

1120 Birchmount Road, Scarborough, Ontario M1K 5G4
☎ 001 416 752 9448   ✆ 001 416 752 8101
inquire@nelson.com
www.nelson.com

Part of the Thomson Corporation. Publishes directories, education, professional, reference, textbooks.

New Star Books Ltd

107–3477 Commercial Street, Vancouver, British Columbia V5N 4E8
☎ 001 604 738 9429   ✆ 001 604 738 9332
info@newstarbooks.com
www.newstarbooks.com

Founded 1974. Publishes fiction and non-fiction, literature, poetry, history, sociology, social sciences.

Oxford University Press, Canada

70 Wynford Drive, Don Mills, Ontario M3C 1J9
☎ 001 416 441 2941   ✆ 001 416 444 0427
www.oup.com/ca

Founded 1904. Owned by **Oxford University Press**, UK. Publishes for college, school and trade markets.

Pearson Canada

26 Prince Andrew Place, Toronto, Ontario M3C 2T8
☎ 001 416 447 5101   ✆ 001 416 443 0948
www.pearsoncanada.ca/index.html

Founded 1966. A division of Pearson Plc and Canada's largest publisher. Publishes fiction, non-fiction and educational material in English and French.

Penguin Group (Canada)

90 Eglinton Avenue East, Suite 700, Toronto, Ontario M4P 2Y3
☎ 001 416 925 2249   ✆ 001 416 925 0068
info@penguin.ca
www.penguin.ca

Founded 1974. Owned by Pearson Plc. Publishes fiction and non-fiction books and audio cassettes.

Random House of Canada Ltd

One Toronto Street, Unit 300, Toronto, Ontario M5C 2V6
☎ 001 416 364 4449   ✆ 001 416 364 6863
www.randomhouse.ca

Founded 1944. Publishes fiction and non-fiction and children's. IMPRINTS **Anchor Canada** Fiction and non-fiction; **Bond Street** Fiction and non-fiction; **Doubleday Canada** Fiction (literary and commercial), business, history, memoir, journalism, young adult books; **Knopf Canada** Fiction and non-fiction; **Seal Books** Primarily reprints of fiction and some non-fiction; **Vintage Canada** Paperback fiction and non-fiction.

Scholastic Canada Ltd

175 Hillmount Road, Markham, Ontario L6C 1Z7
☎ 001 905 887 7323   ✆ 001 905 887 3639
www.scholastic.ca

Founded 1957. Publishes (in English and French) children's books and educational material.

Seal Books
▷ Random House of Canada Ltd

University of Toronto Press, Inc.

10 St Mary Street, Suite 700, Toronto, Ontario M4Y 2W8
☎ 001 416 978 2239
info@utpress.utoronto.ca
www.utpress.utoronto.ca

Founded 1901. Publishes scholarly books and journals. About 150 titles a year.

Tundra Books

75 Sherbourne Street, 5th Floor, Toronto, Ontario M5A 2P9
☎ 001 416 598 4786
tundra@mcclelland.com
www.tundrabooks.com

Founded 1967. Division of **McClelland & Stewart Ltd**. Publishes (in English and French) children's and young adult books.

Vintage Canada
▷ Random House of Canada Ltd

John Wiley & Sons Canada Ltd

5353 Dundas Street W., Suite 400, Toronto, Ontario M9B 6H8
www.wiley.ca

Founded 1968. Subsidiary of **John Wiley & Sons Inc.**, USA. Publishes professional, reference and textbooks.

Women's Press
▷ Canadian Scholars' Press, Inc

India

Affiliated East West Press Pvt Ltd

G–1/16 Ansari Road, Darya Ganj, New Delhi
110 002

☎ 00 91 11 2327 9113
affiliate@vsnl.com
www.aewpress.com

Founded 1962. Publishes textbooks, CD-ROMs,
electronic books.

Caring

▷ **Vision Books Pvt. Ltd**

S. Chand & Co Ltd

7361 Ram Nagar, Qutub Road, New Delhi 110 055

☎ 00 91 11 2367 2080   🖷 00 91 11 2367 7446
info@schandgroup.com
www.schandgroup.com

Founded 1917. Publishes computer science,
dictionaries, encyclopedias, engineering,
reference, textbooks and educational CDs.

HarperCollins Publishers India Ltd

A 53, Sector 57, Noida, UP

☎ 00 91 120 404 4800
contact@harpercollins.co.in
www.harpercollins.co.in

A joint venture between the India Today Group
and **HarperCollins Publishers**. Publishes fiction
and non-fiction; autobiography, astrology,
biography, children's, cinema, cookery, crime,
current affairs, dictionaries, fantasy, gardening,
graphic novels, home interest, literature,
memoirs, mind, body and spirit, poetry, politics,
popular science, philosophy, psychology, puzzles,
reference, religion, science fiction, short stories,
sport, thrillers, travel.

Hind Pocket Books Pvt. Ltd

18–19 Dilshad Garden, G.T. Road, Shahadara,
Delhi 110 095

☎ 00 91 11 229 7792   🖷 00 9111 228 2332

Founded 1957. Publishes fiction and general non-
fiction; literature, health and self-help.

Jaico Publishing House

127 M.G. Road, Mumbai 400 023

☎ 00 91 22 267 6702   🖷 00 91 22 265 6412
jaicobom@bom5.vsnl.net.in
www.jaicobooks.com/home.asp

Founded 1946. Publishes computer science,
engineering, finance, health and nutrition,
information technology, law, reference,
literature, self-help, philosophy, religion,
history, government, political science, sociology,
textbooks.

LexisNexis India

C–33 Inner Circle, Connaught Place, New Delhi
110 001

☎ 00 91 11 4355 9999   🖷 00 91 11 4151 3388
help.in@lexisnexis.co.in
www.lexisnexis.co.in

Part of Reed Elsevier. Publishes law, taxation,
government and business books in print form,
online and CD-ROM.

Macmillan India Ltd

315–316 Raheja Chambers, 12 Museum Road,
Bangalore 560 011

☎ 00 91 80 2558 7878
bgledit@macmillan.co.in
www.macmillanindia.com

Founded 1893. Part of **Macmillan Publishers**,
UK. Publishes education, non-fiction, reference,
dictionaries, encyclopedias.

McGraw-Hill Education (India)

B-4, Sector 63, Distt. Gautam Budh Nagar, Noida,
UP 201 301

☎ 00 91 9512 0438 3400
www.tatamcgrawhill.com

Founded 1970. Educational publisher.

Munshiram Manoharlal Publishers Pvt Ltd

PO Box 5715, 54 Rani Jhansi Road, New Delhi
110 055

☎ 00 91 11 2367 1668   🖷 00 91 11 2361 2745
info@mrmlonline.com
www.mrmlbooks.com

Founded 1952. Publishes academic, dictionaries,
anthropology, art and art history, autobiography,
architecture, archaeology, biography,
environment, religion, philosophy, geography,
history, humanities, language arts and linguistics,
law, literature, medicine, numismatics, music,
dance, theatre, Asian studies, sociology, travel.

National Publishing House

2/35 Ansari Road, New Delhi 110 002

☎ 00 91 11 2327 5267
info@nationalpublishinghouse.com
www.nationalpublishinghouse.com

Founded 1945. Publishes non-fiction, children's,
history, literature, science, textbooks.

Orient Paperbacks

5 A/8 Ansari Road, First Floor, Darya Ganj, Delhi
110 002

☎ 00 91 11 2327 8877   🖷 00 91 11 2327 8879
mail@orientpaperbacks.com
www.orientpaperbacks.com

Founded 1975. Division of **Vision Books Pvt.
Ltd**. Publishes (in English and Hindi) fiction and
general non-fiction; humour, health and nutrition,

literature, New Age, puzzles, reference, religion, self-help.

Oxford University Press India

1st Floor, YMCA Library Building, 1 Jai Singh Road, New Delhi 110 001
℡ 00 91 11 4360 0300 ℻ 00 91 11 2336 0897
admin.in@oup.com
www.oup.co.in

Founded 1912. Owned by **Oxford University Press**, UK. Publishes academic, education, dictionaries. Reference subjects include arts, architecture, biography, natural history, memoirs.

Rajpal
▷ Vision Books Pvt. Ltd

Vision Books Pvt. Ltd

24, Feroze Gandhi Road, Lajpat Nagar III, New Delhi 110 024
℡ 00 91 11 2983 6470 ℻ 00 91 11 2983 6490
editor@visionbooksindia.com
www.visionbooksindia.com

Founded 1975. Specializes in books on business and management, finance and taxation, careers, current affairs, religion. DIVISIONS/IMPRINTS **Vision Books**; **Orient Paperbacks** (see entry); **Caring**; **Rajpal**.

New Zealand

Auckland University Press

University of Auckland, Private Bag 92019, Auckland
℡ 00 64 9 373 7528 ℻ 00 64 9 373 7465
aup@auckland.ac.nz
www.auckland.ac.nz/aup

Founded 1966. Publishes academic, biography, government, political science, history, social sciences, sociology, poetry, literature, literary criticism, Maori studies.

Canterbury University Press

University of Canterbury, Private Bag 4800, Christchurch
℡ 00 64 3 364 2914 ℻ 00 64 3 364 2044
mail@cup.canterbury.ac.nz
www.cup.canterbury.ac.nz

Founded 1964. Academic publishers; general non-fiction; history, marine biology, natural history.

The Caxton Press

113 Victoria Street, Christchurch
℡ 00 64 3 353 0734 ℻ 00 64 3 365 7840
www.caxton.co.nz

Founded 1935. Publishes general non-fiction.

Hachette Livre New Zealand Ltd

4 Whetu Place, Mairangi Bay, Auckland
℡ 00 64 9 478 1000 ℻ 00 64 9 478 1010
www.hha.com.au

Founded 1971. Part of the **Hachette Livre Publishing Group**. Publishes fiction and general non-fiction; biography, business, children's, cookery, humour, sport.

HarperCollins Publishers (New Zealand) Ltd

PO Box 1, Auckland
℡ 00 64 9 443 9400 ℻ 00 64 9 443 9403
editors@harpercollins.co.nz
www.harpercollins.co.nz

Founded 1888. Publishes fiction, literature, children's, business, cookery, gardening, reference, religion.

Huia (NZ) Ltd

PO Box 17-335, 39 Pipitea Street, Wellington, Aotearoa
℡ 00 64 4 473 9262 ℻ 00 64 4 473 9265
info@huia.co.nz
www.huia.co.nz

Founded 1991. Independent publisher of Maori cultural history and language, children's books in Maori and English, fiction.

LexisNexis New Zealand

PO Box 472, Wellington
℡ 00 64 4 385 1479 ℻ 00 64 4 385 1598
www.lexisnexis.co.nz

Founded 1914. Part of the LexisNexis Group. Publishes law, professional and textbooks in book and electronic formats.

Macmillan Publishers New Zealand

6 Ride Way, Albany, Auckland
℡ 00 64 9 414 0350 ℻ 00 64 9 414 0351
www.macmillan.co.nz

Education publishers – general, school and academic books; fiction and non-fiction for all ages.

McGraw-Hill New Zealand Pty Ltd
▷ McGraw-Hill Australia & New Zealand Pty Ltd

Oxford University Press New Zealand
▷ Oxford University Press (Australia)

Pearson Education New Zealand Ltd

46 Hillside Road, Auckland 10
℡ 00 64 9 444 4968 ℻ 00 64 9 444 4957
www.pearson.co.nz

Founded in 1998 from the merger of Addison Wesley Longman New Zealand and Prentice Hall New Zealand. Educational publishers.

Penguin Books (NZ) Ltd

Private Bag 102 902, North Shore Mail Centre,
Auckland 0745

📞 00 64 9 442 7400 📠 00 64 9 442 7401
publishing@penguin.co.nz
www.penguin.co.nz

Founded 1978. Owned by **Penguin UK**. Adult and
children's fiction and non-fiction.

Random House New Zealand

Private Bag 102950, North Shore Mail Centre,
Auckland

📞 00 64 9 444 7197 📠 00 64 9 444 7524
editor@randomhouse.co.nz
www.randomhouse.co.nz

Publishes fiction and non-fiction (cooking,
gardening, art, natural history) and children's.

Raupo Publishing (NZ) Ltd

Private Bag 102950, Birkenhead, Auckland

📞 00 64 9 441 2960 📠 00 64 9 480 4999
info@raupopublishing.co.nz
www.raupopublishing.co.nz

Founded 1988. Publishes fiction and general
non-fiction, biography, children's, cookery,
health, history, natural history, regional interests,
reference, self help, travel.

Victoria University Press

PO Box 600, Wellington

📞 00 64 4 463 6580 📠 00 64 4 463 6581
victoria-press@vuw.ac.nz
www.vuw.ac.nz/vup

Founded 1979. Academic publishers; new fiction,
biography, poetry, literature, essays, New Zealand
history, Maori topics. About 25 titles a year.

Bridget Williams Books Ltd

PO Box 5482, Wellington

📞 00 64 4 473 8128
info@bwb.co.nz
www.bwb.co.nz

Founded 1990. Independent publisher of New
Zealand history, Maori experience, contemporary
issues and women's studies.

South Africa

Heinemann International Southern Africa

PO Box 781940, Sandton 2146

📞 00 27 11 322 8600 📠 00 27 11 322 8716
www.heinemann.co.za

Founded 1986. Educational publisher.

University of Kwa-Zulu-Natal Press

Private Bag X01, Scottsville 3209

📞 00 27 33 260 5226 📠 00 27 33 260 5801

books@ukzn.ac.za
www.ukznpress.co.za

Founded 1947. Publishes academic and general
books; children's, African literature, poetry,
economics, military history, natural sciences,
social sciences.

LexisNexis Butterworths (Pty) Ltd South Africa

PO Box 792, Durban 4000

📞 00 27 31 268 3111 📠 00 27 31 268 3109
www.lexisnexis.co.za

Publishes professional, accountancy, finance,
business, law, taxation and economics books in
print form, online and CD-ROM.

Macmillan South Africa

Head office: 2nd Floor, The Pavilion, Wanderers
Office Park, 52 Corlett Drive, Illovo 2196

📞 00 27 11 731 3300 📠 00 27 11 731 3500
info@macmillan.co.za
www.macmillan.co.za

Founded 1972. Part of **Macmillan Publishers Ltd**,
UK. Educational publisher.

Maskew Miller Longman

PO Box 396, Cape Town 8000

📞 00 27 21 532 6000 📠 00 27 21 531 8103
www.mml.co.za

Founded 1893. Owned jointly by **Pearson
Education** and Caxton Publishers and Printers
Ltd. Publishes education, ELT and teacher
support books.

Oshun

▷ **Random House Struik (Pty) Ltd**

Oxford University Press Southern Africa (Pty) Ltd

PO Box 12119, N1 City, Cape Town 7463

📞 00 27 21 596 2300 📠 00 27 21 596 1234
oxford.za@oup.com
www.oup.com/za

Founded 1915. Parent company: **Oxford
University Press**, UK. Publishes academic,
educational and general books.

Random House Struik (Pty) Ltd

PO Box 1144, Cape Town 8000

📞 00 27 21 462 4360 📠 00 27 21 462 4377
info@struik.co.za
www.struik.co.za

Formed by the merger of Random House SA and
Struik Publishing in September 2008. Publishes
general non-fiction, illustrated, art and culture,
lifestyle, natural history, travel. IMPRINTS **Oshun**
Fiction, biography, health, humour, parenting,
reference, self help, South African interest; **Umuzi**

Literary fiction with a South African flavour and non-fiction; **Zebra** Fiction and non-fiction.

Shuter & Shooter (Pty) Ltd

21c Cascades Crescent, Pietermaritzburg, KwaZulu-Natal 3201

📞 00 27 33 347 6100   📠 00 27 33 346 6120

www.shuter.co.za

Founded 1925. Publishes educational and general books.

Umuzi

▷ Random House Struik (Pty) Ltd

Wits University Press

PO Wits, Johannesburg 2050

📞 00 27 11 484 5907   📠 00 27 11 484 5971

witspress@wup.wits.ac.za

witspress.wits.ac.za

Founded 1922. Publishes academic, biography and memoirs, economics, heritage, history, business, politics, popular science, women's writing.

Zebra

▷ Random House Struik (Pty) Ltd

Poetry Presses

Agenda Editions

The Wheelwrights, Fletching Street, Mayfield
TN20 6TL
☎ 01435 873703
editor@agendapoetry.co.uk
www.agendapoetry.co.uk
Editor *Patricia McCarthy*

Separate collections of poetry. Very few published in a year and not many unsolicited. Seek publication in **Agenda Poetry Magazine** first – see entry under *Poetry Magazines*.

Akros Publications

33 Lady Nairn Avenue, Kirkcaldy KY1 2AW
☎ 01592 651522
www.akrospublications.co.uk
Contact *Duncan Glen*

Publisher of Scottish poetry and literary criticism. See also *ZED20* magazine and entry under *UK Publishers*.

Anvil Press Poetry Ltd

Neptune House, 70 Royal Hill, London SE10 8RF
☎ 020 8469 3033   ☒ 020 8469 3363
anvil@anvilpresspoetry.com
www.anvilpresspoetry.com
Contact *Peter Jay*

Contemporary British poetry and poetry in translation. See entry under *UK Publishers*.

Arc Publications Ltd.

Nanholme Mill, Shaw Wood Road, Todmorden
OL14 6DA
☎ 01706 812338   ☒ 01706 818948
arc.publications@btconnect.com
www.arcpublications.co.uk
Editorial Director *Tony Ward*

Contemporary poetry from new and established writers using English as their first language from the UK and abroad. See entry under *UK Publishers*.

Arrowhead Press

70 Clifton Road, Darlington DL1 5DX
editor@arrowheadpress.co.uk
roger.collett@ntlworld.com
www.arrowheadpress.co.uk

Managing Editor *Roger Collett*

Quality books and pamphlets of contemporary poetry.

Atlantean Publishing

38 Pierrot Steps, 71 Kursaal Way, Southend on Sea
SS1 2UY
atlanteanpublishing@hotmail.com
www.geocities.com/dj_tyrer/atlantean_pub.html
Contact *David John Tyrer*

Single author broadsheets and poetry/prose booklets. See also **Awen**, **Monomyth**, **Bard**, **Garbaj** and **The Supplement** magazines.

Barque Press

26 Allerton Road, London N16 5UJ
www.barquepress.com
Contact *Andrea Brady*

Poetry. 'Barque does not accept or return unsolicited mss.'

BB Books

Spring Bank, Longsight Road, Copster Green, Blackburn BB1 9EU
☎ 01254 249128
Contact *Dave Cunliffe*

Post-Beat poetics and counterculture theoretic. Iconoclastic rants and anarchic psycho-cultural tracts. See also **Global Tapestry Journal**.

Between the Lines

See entry under *Small Presses*.

BeWrite Books

See entry under *Small Presses*.

Beyond the Cloister Publications

14 Lewes Crescent, Brighton BN2 1FH
☎ 01273 687053
www.beyondthecloister.com
Contact *Hugh Hellicar*

Anthologies of poetry and slim volumes to launch poets.

Bloodaxe Books Ltd

Highgreen, Tarset NE48 1RP
☎ 01434 240500   ☒ 01434 240505

editor@bloodaxebooks.com
www.bloodaxebooks.com
Contact *Neil Astley*

Britain's leading publisher of new poetry. No submissions by e-mail attachments. See entry under *UK Publishers*.

Blue Butterfly Publishers

13 Irvine Way, Inverurie AB51 4ZR
☎ 01467 625986
blue7butterfly@which.net
madill.prodigynet.co.uk/bbp
Contact *Betty Madill*

Christian poetry. 'Poetry to be submitted through our poetry competitions only.' Entry forms from address above or from the website.

Bluechrome Publishing

PO Box 109, Portishead, Bristol BS20 7ZJ
www.bluechrome.co.uk
Contact *Anthony Delgrado*

Founded 2002. Poetry, literary and experimental fiction. Submission details available on the website.

Bradshaw Books

See entry under *Irish Publishers*.

Bridge Pamphlets

PO Box 309, Aylsham, Norwich NR11 6LN
mail@therialto.co.uk
www.therialto.co.uk
Contact *Michael Mackmin*

An imprint of **Rialto Publications**.

The Brodie Press

See entry under *Small Presses*.

Calder Wood Press

1 Beachmont Court, Dunbar EH42 1YF
☎ 01368 864953
colin.will@zen.co.uk
www.calderwoodpress.co.uk
Contact *Colin Will*

Founded 1997. Publishes pamphlets and short-run publications. No unsolicited mss.

Carcanet Press Ltd

Major poetry publisher. See entry under *UK Publishers*; also publishes **PN Review** magazine.

Chapman Publishing

See entry under *UK Publishers*.

Cinnamon Press

Ty Meirion, Glan yr afon, Tanygrisiau, Blaenau Ffestiniog LL41 3SU
☎ 01766 832112
jan@cinnamonpress.com
www.cinnamonpress.com

Editor *Dr Jan Fortune-Wood*

Poetry, fiction, some non-fiction. Growing list from Wales, UK & international. Regular literary competitions. See also **Envoi** magazine.

The Collective Press

c/o Penlanlas Farm, Llantilio Pertholey, Y-fenni NP7 7HN
☎ 01873 856350   🖷 01873 858357
jj@jojowales.co.uk
www.welshwriters.com
Contact *John Jones & Frank Olding*

Non-profit promoters and publishers of contemporary poetry.

Day Dream Press

39 Exmouth Street, Swindon SN1 3PU
☎ 01793 523927
Contact *Kevin Bailey*

Small poetry collections. Now associated with **Bluechrome Press**. See also **HQ Poetry Magazine (Haiku Quarterly)**.

The Dedalus Press

13 Moyclare Road, Baldoyle, Dublin 13, Republic of Ireland
☎ 00 353 1 839 2034   🖷 0870 127 2089
office@dedaluspress.com
www.dedaluspress.com
Publisher *Pat Boran*

Contemporary Irish poetry and poetry from around the world in English translation.

Diehard Publishers

91–93 Main Street, Callander FK17 8BQ

Poetry and drama, mainly Scottish. See also **Poetry Scotland** magazine.

Dionysia Press

127 Milton Road West, 7 Duddingston House Courtyard, Edinburgh EH15 1JG
Contact *Denise Smith*

Collections of poetry, short stories, novels and translations. See also **Understanding** magazine.

DogHouse

PO Box 312, Tralee, Co. Kerry, Republic of Ireland
☎ 00 353 667 137547   🖷 00 353 667 137547
info@doghousebooks.ie
www.doghousebooks.ie
Contact *Noel King*

Poetry and short story collections by individuals (Irish born only). About 4–6 titles a year.

Dream Catcher

4 Church Street, Market Rasen, Lincoln LN8 3ET
☎ 01673 844325
www.dreamcatchermagazine.co.uk
Contact *Paul Sutherland*

Short fiction, poetry, reviews, artwork, biographies. See also **Dream Catcher** magazine.

Driftwood Publications

Apartment 47, 8 William Jessop Way, Liverpool L3 1EJ
☎ 0151 236 7550   📠 0151 524 0216
janet.speedy@tesco.net
Contact *Brian Wake*

New work by new and established poets more suited to the page than the stage.

Egg Box Publishing Ltd

c/o The New Writing Partnership, 14 Princes Street, Norwich NR3 1AE
www.eggboxpublishing.com
Contact *Nathan Hamilton & Gordon Smith*

New poetry and fiction.

Enitharmon Press

26B Caversham Road, London NW5 2DU
☎ 020 7482 5967   📠 020 7284 1787
books@enitharmon.co.uk
www.enitharmon.co.uk
Contacts *Stephen Stuart-Smith, Isabel Brittain, Jacqueline Gabbitas*

Poetry and literary criticism. See entry under *UK Publishers*.

Essence Press

8 Craiglea Drive, Edinburgh EH10 5PA
jaj@essencepress.co.uk
www.essencepress.co.uk
Contact *Julie Johnstone*

Handbound editions of poetry, poetry postcards. Interested in concrete poetry and nature. See also **Island** magazine.

Fal Publications

PO Box 74, Truro TR1 1XS
☎ 07887 560018
info@falpublications.co.uk
www.falpublications.co.uk
Contact *Victoria Field*

Poetry and literature from Cornwall. Please e-mail for submissions policy.

Feather Books

PO Box 438, Shrewsbury SY3 0WN
☎ 01743 872177   📠 01743 872177
john@waddysweb.freeuk.com
www.waddysweb.freeuk.com
Contact *Rev. J. Waddington-Feather*
General Manager *Paul Evans*

Christian poetry, drama, hymns, novels. See entry under *Small Presses*.

Five Leaves Publications

PO Box 8786, Nottingham NG1 9AW
☎ 0115 969 3597
info@fiveleaves.co.uk
www.fiveleaves.co.uk
See entry under *Small Presses*.

Flambard Press

Stable Cottage, East Fourstones, Hexham NE47 5DX
☎ 01434 674360   📠 01434 674178
flambardpress@btinternet.com
www.flambardpress.co.uk
Contact *Peter Lewis & Will Mackie*

Concentrates on poetry and literary fiction. Consult the website for submission information and for forthcoming changes of address and phone number.

Flarestack Poets Pamphlets

PO Box 14479, Birmingham B13 3GU
meria@btinternet.com
jacquirowe@hotmail.co.uk
www.flarestackpoets.co.uk
Contact *Meredith Andrea & Jacqui Rowe*

See also **Obsessed With Pipework** magazine.

Forward Press

Remus House, Woodston, Peterborough PE2 9JX
☎ 01733 890099   📠 01733 313524
info@forwardpress.co.uk
www.forwardpress.co.uk
Contact *The Editorial Team*

General poetry and short fiction anthologies.

The Frogmore Press

42 Morehall Avenue, Folkestone CT19 4EF
☎ 07751 251689
J.N.Page@sussex.ac.uk
www.frogmorepress.co.uk
Managing Editor *Jeremy Page*

A forum for contemporary poetry, prose and artwork. See also **The Frogmore Papers** magazine.

The Gallery Press

Loughcrew, Oldcastle, Co. Meath, Republic of Ireland
☎ 00 353 49 854 1779   📠 00 353 49 854 1779
gallery@indigo.ie
www.gallerypress.com
Editor & Publisher *Peter Fallon*

Poems and plays by contemporary Irish writers. Write for submission notes.

Gomer Press/Gwasg Gomer

Llandysul Enterprise Park, Llandysul SA44 4JL
☎ 01559 363090   📠 01559 363758
www.gomer.co.uk
English Books for Adults *Ceri Wyn Jones (ceri@gomer.co.uk)*

English Books for Children *Viv Sayer (viv@gomer. co.uk)*
Publishing Director *Mairwen Prys Jones (mairwen@gomer.co.uk)*

Welsh interest. See entry under *UK Publishers.*

Green Arrow Publishing

6 Green Bank, Stacksteads, Bacup OL13 8LQ
☎ 01706 870357/07967 315270
mail@johndench.demon.co.uk
Contact *John Dench*

Collaborative publishing scheme and other services for writers. See also **Scriptor** magazine.

HappenStance Press

21 Hatton Green, Glenrothes KY7 4SD
nell@happenstancepress.com
www.happenstancepress.com
Editor *Helena Nelson*

Poetry publishing; mainly chapbooks and usually first collections. Submission guidelines on website. 'Reading "windows" December and July'. See also **Sphinx** magazine.

Headland Publications

Ty Coch, Galltegfa, Ruthin LL15 2AR
☎ 0151 625 9128
headlandpublications@hotmail.co.uk
www.headlandpublications.co.uk
Contact *Gladys Mary Coles*

Fine editions of poetry; anthologies.

Hearing Eye

c/o 99 Torriano Avenue, London NW5 2RX
hearing_eye@torriano.org
www.torriano.org/hearing_eye/
Contact *John Rety*

Poetry/literature publishers based in Kentish Town.

Hilltop Press

4 Nowell Place, Almondbury, Huddersfield HD5 8PD
booksmusicfilmstv.com/HilltopPress.htm
Contact *Steve Sneyd*

Specialist publisher of science fiction and dark fantasy poetry, and background material.

Hippopotamus Press

22 Whitewell Road, Frome BA11 4EL
☎ 01373 466653   ℻ 01373 466653
rjhippopress@aol.com
Contact *Roland John & Mansell Pargitter*

First collections of verse from those with a track record in the magazines.

Honno Welsh Women's Press
▷ entry under UK Publishers

1*D Books

Connah's Quay Library, Wepre Drive, Connah's Quay, Deeside
☎ 01244 830485

Poetry, short fiction, local history – mainly self-publishing by associated writers' group.

Iron Press

5 Marden Terrace, Cullercoats, North Shields NE30 4PD
☎ 0191 253 1901
ironpress@blueyonder.co.uk
www.ironpress.co.uk
Contact *Peter Mortimer*

Poetry and fiction.

Katabasis

10 St Martin's Close, London NW1 0HR
☎ 020 7485 3830   ℻ 020 7485 3830
katabasis@katabasis.co.uk
www.katabasis.co.uk
Contact *Dinah Livingstone*

Down-to-earth and Utopian poetry and prose from home an abroad – English and Latin American. No unsolicited mss.

Kates Hill Press

126 Watsons Green Road, Kates Hill, Dudley DY2 7LG
kateshillpress@blueyonder.co.uk
www.kateshillpress.co.uk

Fiction and social history from the West Midlands or with a West Midlands theme. Some poetry.

Kettillonia

24 South Street, Newtyle PH12 8UQ
☎ 01828 650615
james@kettillonia.co.uk
www.kettillonia.co.uk
Editor *James Robertson*

Occasional publisher of pamphlets. No unsolicited manuscripts.

The King's England Press

Cambertown House, Commercial Road, Goldthorpe, Rotherham S63 9BL
☎ 01484 663790   ℻ 01484 663790
sales@kingsengland.com
www.kingsengland.com
www.pottypoets.com
Contact *Steve Rudd*

Founded 1989. History, folklore, adult and children's poetry.

KT Publications

16 Fane Close, Stamford PE9 1HG
☎ 01780 754193
Editor *Kevin Troop*

Always looking for new material of the highest standard. See also **The Third Half** magazine.

Leaf Books

GTi Suite, Valleys Innovation Centre, Navigation Park, Abercynon CF45 4SN
☎ 029 20810726
contact@leafbooks.co.uk
www.leafbooks.co.uk

Poetry and Micro-Fiction competition anthologies: getting new writers into print.

Leafe Press

4 Cohen Close, Chilwell, Nottingham NG9 6RW
www.leafepress.com
Contact *Alan Baker*

Pamphlets of contemporary poetry with occasional full-length collections. Currently, not accepting unsolicited submissions.

Mare's Nest

41 Addison Gardens, London W14 0DP
☎ 020 7603 3969
www.maresnest.co.uk
Contact *Pamela Clunies-Ross*

Icelandic literature.

Mariscat Press

10 Bell Place, Edinburgh EH3 5HT
☎ 0131 343 1070
hamish.whyte@btinternet.com
Contact *Hamish Whyte & Diana Hendry*

Currently publishing poetry pamphlets only.

Masque Publishing

PO Box 3257, Littlehampton BN16 9AF
masque_pub@btinternet.com
http://myweb.tiscali.co.uk/masquepublishing
Contact *Lisa Stewart*

Self publishing service. See also **Decanto** poetry magazine.

Maypole Editions

See entry under *Small Presses*.

Mews Press

English Department, Owen 11, Sheffield Hallam University, Howard Street, Sheffield S1 1WB
☎ 0114 225 4163
s.l.earnshaw@shu.ac.uk
extra.shu.ac.uk/mews-press
Contact *Professor Steven Earnshaw*

Work by writers associated with Sheffield Hallam University's MA Writing, including *Matter* and the *Ictus*.

Mudfog Press

Arts Development, The Stables, Stewart Park, The Grove, Marton, Middlesborough TS7 8AR
www.mudfog.co.uk

Community press publishing poetry and short fiction only from writers in the Tees Valley and adjoining rural areas.

New Hope International

20 Werneth Avenue, Gee Cross, Hyde SK14 5NL
www.geraldengland.co.uk
Contact *Gerald England*

Poetry booklet publisher. No longer considering unsolicited mss. No new publications planned. Most back titles still available at discount price.

New Island

2 Brookside, Dundrum Road, Dublin 14, Republic of Ireland
www.newisland.ie

See entry under *Irish Publishers*.

nthposition

38 Allcroft Road, London NW5 4NE
☎ 020 7485 5002
val@nthposition.com
www.nthposition.com
Editor *Val Stevenson*
Poetry Editor *Todd Swift (toddswift@clara.co.uk)*

Eclectic and award-winning mix of politics and opinion, travel writing, fiction and poetry, reviews and interviews and some high weirdness.

Object Permanence

First Floor, 16 Ruskin Terrace, Glasgow G12 8DY
robinpurves@yahoo.co.uk
www.objectpermanence.co.uk
Contact *Robin Purves*

Originally a poetry magazine, now poetry publishers.

The Old Stile Press

Catchmays Court, Llandogo, Nr Monmouth NP25 4TN
☎ 01291 689226
oldstile@dircon.co.uk
www.oldstilepress.com
Contact *Frances & Nicolas McDowall*

Fine, hand-printed books with text and images. No unsolicited mss.

The One Time Press

Model Farm, Linstead Magna, Halesworth IP19 0DT
☎ 01986 785422
www.onetimepress.com
Contact *Peter Wells*

Poetry of the forties and occasionally some contemporary work in limited editions. Illustrated.

Original Plus

17 High Street, Maryport CA15 6BQ
☎ 01900 812194

smithsssj@aol.com
members.aol.com/smithsssj/index.html
Contact *Sam Smith*

See also **The Journal** magazine.

Parthian

The Old Surgery, Napier Street, Aberteifi
(Cardigan) SA43 1ED
☎ 01239 612059
www.parthianbooks.co.uk

New Welsh writing. See entry under *UK
Publishers*.

Partners

289 Elmwood Avenue, Feltham TW13 7QB
partners_writing_group@hotmail.com
Contact *Ian Deal*

Competitions and poetry pamphlets. See also
A Bard Hair Day, **Imagenation** and **Poet Tree**
magazines.

Peepal Tree Press Ltd

17 King's Avenue, Leeds LS6 1QS
☎ 0113 245 1703
contact@peepaltreepress.com
www.peepaltreepress.com
Contact *Jeremy Poynting*

Peepal Tree Press is the home of challenging and
inspiring literature from the Caribbean and Black
Britain. Online bookstore.

Penniless Press

100 Waterloo Road, Ashton, Preston PR2 1EP
☎ 01772 736421
www.pennilesspress.co.uk
Contact *Alan Dent*

Novels, short fiction, poetry, plays. See also **The
Penniless Press** magazine.

Peterloo Poets

The Old Chapel, Sand Lane, Calstock PL18 9QX
☎ 01822 833473
publisher@peterloopoets.com
www.peterloopoets.com
Publishing Director *Harry Chambers*

Pigasus Press

13 Hazely Combe, Arreton, Isle of Wight PO30 3AJ
mail@pigasuspress.co.uk
www.pigasuspress.co.uk
Contact *Tony Lee*

Publisher of science fiction magazine and genre
poetry.

Pighog Press

PO Box 145, Brighton BN1 6YU
info@pighog.co.uk
www.pighog.co.uk
Contact *The Editor*

'Small press. Big impact.' Award-winning small
press based in Brighton. No unsolicited mss.

Pikestaff Press

Ellon House, Harpford, Sidmouth EX10 0NH
☎ 01395 568941
Contact *Robert Roberts*

Contemporary poetry belonging to the English
tradition.

Pipers' Ash Ltd

Pipers Ash, Church Road, Christian Malford,
Chippenham SN15 4BW
☎ 01249 720563   ☏ 0870 056 8916
pipersash@supamasu.com
www.supamasu.com
Contact *Mr A. Tyson*

Collections of 60 poems arranged in their best
order, grouped on common themes, with an
appropriate title. See entry under *Small Presses*.

Poems in the Waiting Room

PO Box 488, Richmond TW9 4SW
pitwr@blueyonder.co.uk
www.poemsinthewaitingroom.org
Joint Editors *Michael Lee & Isobel Montgomery
Campbell*

Arts in Health charity providing poetry cards
for NHS patients. 'It is now the most widely read
national poetry publication with a circulation
reaching 30,000 each quarter.'

Poetry Now

Remus House, Coltsfoot Drive, Woodston,
Peterborough PE2 8JX
☎ 01733 898101   ☏ 01733 313524
poetrynow@forwardpress.co.uk
www.forwardpress.co.uk
Editorial Manager *Michelle Afford*

Lively, personal and contemporary. Imprint of
Forward Press.

Poetry Powerhouse Press
▷ Performance Poetry Society under Organizations of
Interest to Poets

Poetry Press Ltd

26 Park Grove, Edgware HA8 7SJ
☎ 020 8958 6499
poetrypress@yahoo.co.uk
www.judyk.co.uk
Contact *Judy Karbritz*

Anthologies – rhyming poetry welcome.

Poetry Salzburg

Dept. of English, University of Salzburg,
Akademiestr. 24, A–5020 Salzburg, Austria
☎ 00 43 662 8044 4424   ☏ 00 43 662 8044 167
editor@poetrysalzburg.com
www.poetrysalzburg.com

Co-editors *Wolfgang Görtschacher & Andreas Schachermayr*

Publishes books of poetry, poetry in translation, poetry anthologies and literary criticism. See also **Poetry Salzburg Review** magazine.

Poetry Wednesbury

25 Griffiths Road, West Bromwich B71 2EH
☎ 07950 591455
ppatch66@hotmail.com
www.poetrywednesbury.co.uk
Contact *Geoff Stevens*

Occasional CDs, DVDs, leaflets and booklets. See also **Purple Patch** magazine. Public meetings for readings, workshops, etc. monthly at St Paul's Church, Wood Green, Wednesbury.

Poets Anonymous

70 Aveling Close, Purley CR8 4DW
poets@poetsanon.org.uk
www.poetsanon.org.uk
Contact *Peter L. Evans*

Anthologies and collections of predominantly south London poets. See also **Poetic Licence** magazine.

The Potty Poets

www.pottypoets.com

Children's poetry. See **The King's England Press**.

Precious Pearl Press

41 Grantham Road, Manor Park, London E12 5LZ
Contact *P.G.P. Thompson*

Romantic, lyrical, spiritual, inspirational and mystical poetry; a traditionalist poetry press. See also **Rubies in the Darkness** magazine.

PS Avalon

Box 1865, Glastonbury BA6 8YR
will@willparfitt.com
www.psavalon.com
Contact *Will Parfitt*

Quality books of psychospiritual poetry and for personal and spiritual development.

Puppet State Press

40/1 Woodhall Road, Edinburgh EH13 0DU
☎ 0131 441 9693
richard@puppetstate.com
www.puppetstate.com
Contact *Richard Medrington*

QQ Press

York House, 15 Argyle Terrace, Rothesay, Isle of Bute PA20 0BD
Contact *Alan Carter*

Collections of poetry plus poetry anthologies. Enquiries should be marked 'Collections' and sent with an s.a.e. or two IRCs if from abroad. See also **Quantum Leap** magazine.

Rack Press

The Rack, Kinnerton, Presteigne LD8 2PF
☎ 01547 560411
rackpress@nicholasmurray.co.uk
www.nicholasmurray.co.uk/RackPress.html
Blog: www.rackpress.blogspot.com
Contact *Nicholas Murray*

Welsh poetry pamphlet imprint with an international vision.

Ragged Raven Press

1 Lodge Farm, Snitterfield, Stratford-upon-Avon CV37 0LR
☎ 01789 730320
raggedravenpress@aol.com
www.raggedraven.co.uk
Contact *Bob Mee & Janet Murch*

Poetry. See also **Iota** magazine.

Raunchland Publications

26 Alder Grove, Dunfermline KY11 8RP
raunchland@hotmail.com
www.raunchland.co.uk
Contact *John Mingay*

Limited edition poetry/graphics booklets and online publications.

Reality Street Editions

63 All Saints Street, Hastings TN34 3BN
info@realitystreet.co.uk
www.realitystreet.co.uk
Contact *Ken Edwards*

New writing from Britain, Europe and America. No unsolicited mss.

Red Candle Press

77 Homegrove House, Grove Road North, Southsea PO5 1HW
www.members.tripod.com/redcandlepress
Contact *M.L. McCarthy*

Founded 1970. Traditionalist poetry press. See also **Candelabrum Poetry Magazine**.

Rialto Publications

PO Box 309, Aylsham, Norwich NR11 6LN
mail@therialto.co.uk
www.therialto.co.uk
Contact *Michael Mackmin*

'We want to publish first collections by poets of promise.' See also **The Rialto** magazine and **Bridge Pamphlets** press.

Route Publishing

PO Box 167, Pontefract WF8 4WW
☎ 0845 158 1565
info@route-online.com
www.route-online.com

Contact *Ian Daley & Isabel Galan*

Contemporary fiction and performance poetry. Online magazine publishes downloadable books.

Salt Publishing

PO Box 937, Great Wilbraham, Cambridge CB1 5JX

☎ 01223 882220   📠 01223 882260

sales@saltpublishing.com

www.saltpublishing.com

Contact *Chris Hamilton-Emery*

International poetry and poetics. 2008 winner of the Nielsen Innovation of the Year prize.

Second Light Publications

9 Greendale Close, London SE22 8TG

dilyswood@tiscali.co.uk

www.secondlightlive.co.uk

www.poetrypf.co.uk/secondlight.html

www.esch.dircon.co.uk/second/second.htm

Contact *Dilys Wood*

Administrator *Anne Stewart (editor@poetrypf. co.uk)*

Publishes high-calibre women's poetry, mainly anthologies, occasional collections (by invitation). See **Second Light** under *Organizations of Interest to Poets.*

Seren

57 Nolton Street, Bridgend CF31 3AE

☎ 01656 663018

seren@seren-books.com

www.seren-books.com

Contact *Mick Felton*

Poetry, fiction, lit crit, biography, essays. See entry under *UK Publishers*; also **Poetry Wales** magazine.

Shearsman Books

58 Velwell Road, Exeter EX4 4LD

☎ 01392 434511

editor@shearsman.com

www.shearsman.com

Contact *Tony Frazer*

Publishes poetry almost exclusively. Contact editor before submitting or consult the submissions page at the Shearsman website for the most up-to-date instructions. See also **Shearsman** magazine.

Shoestring Press

19 Devonshire Avenue, Beeston, Nottingham NG9 1BS

☎ 0115 925 1827   📠 0115 925 1827

info@shoestringpress.co.uk

www.shoestring-press.com

Contact *John Lucas*

Poetry. No unsolicited mss.

Sixties Press

89 Connaught Road, Sutton SM1 3PJ

☎ 020 8286 0419

Contact *Barry Tebb*

Small press concentrating on poetry and novellas. See also **Leeds Poetry Quarterly** and **Literature and Psychoanalysis** magazines.

Smith/Doorstop

▷ The Poetry Business under Organizations of Interest to Poets

Smokestack Books

PO Box 408, Middlesbrough TS5 6WA

☎ 01642 813997

info@smokestack-books.co.uk

www.smokestack-books.co.uk

Contact *Andy Croft*

Champions poets who are unconventional, unfashionable, radical or left-field.

Spectacular Diseases

83b London Road, Peterborough PE2 9BS

Contact *Paul Green*

Innovative poetry and some prose. Translations of both.

The Stinging Fly Press

PO Box 6016, Dublin 8, Republic of Ireland

stingingfly@gmail.com

www.stingingfly.org

Contact *Declan Meade*

Founded 2005. New Irish and international writing. No e-mail submissions. See also **The Stinging Fly** magazine.

Talking Pen

12 Derby Crescent, Moorside, Consett DH8 8DZ

☎ 01207 505724

Contact *Steve Urwin*

See also **Moodswing** magazine.

Templar Poetry

PO Box 7082, Bakewell DE45 9AF

☎ 01629 582500

info@templarpoetry.co.uk

www.templarpoetry.co.uk

Contact *Alexander McMillen*

Publishes new poets and poetry via an annual open pamphlet and collection competition, and launches new titles at the Derwent Poetry Festival in Derbyshire in October. Poets are also occasionally commissioned to submit work for publication. 'Templar Poetry is committed to the professional publication and development of excellent contemporary poetry.'

Vane Women Press

9 Alice Row, Stockton TS18 1JU

☎ 01642 606421

anne.hine@ntlworld.com
www.vanewomen.co.uk
Contact *Anne Hine*

Poetry and short stories by women of the North East.

Waterloo Press

126 Furze Croft, Furze Hill, Hove BN3 1PF
🕾 01273 202876
www.waterloopresshove.co.uk
Contact *Simon Jenner*

Poetry and periodical publisher; please phone for submissions. See also **Eratica** magazine.

Waywiser Press

The Cottage, 14 Lyncroft Gardens, Ewell KT17 1UR
🕾 020 8393 7055   🖷 020 8393 7055
waywiserpress@aol.com
www.waywiser-press.com

See entry under *Small Presses.*

Wendy Webb Books

9 Walnut Close, Taverham, Norwich NR8 6YN
tipsforwriters@yahoo.co.uk
Contact *Wendy Webb*

New forms, traditional forms, books, anthologies, Margaret Munro Gibson Memorial Poetry Competition (annual). See also **Norfolk Poets and Writers – Tips Newsletter** magazine.

West House Books

40 Crescent Road, Nether Edge, Sheffield S7 1HN
🕾 0114 258 6035
alan@nethedge.demon.co.uk
www.westhousebooks.co.uk
Contact *Alan Halsey*

Contemporary poetry, poets' prose and related work.

Worple Press

See entry under *Small Presses.*

Poetry Magazines

Acumen

6 The Mount, Higher Furzeham, Brixham
TQ5 8QY
☎ 01803 851098
pwoxley@aol.com
www.acumen-poetry.co.uk
Contact *Patricia Oxley*

Good poetry, intelligent articles and wide-ranging reviews.

Aesthetica

PO Box 371, York YO23 1WL
☎ 01904 527560 ☑ 01904 527 451
info@aestheticamagazine.com
www.aestheticamagazine.com
Managing Editor *Cherie Federico*
Development Officer *Shona Fairweather*

Contemporary glossy bi-monthly: visual arts, literature, music, film, theatre, news, reviews, interviews and features. Distribution throughout the UK in WH Smith and Borders.

Agenda Poetry Magazine

The Wheelwrights, Fletching Street, Mayfield
TN20 6TL
☎ 01435 873703
editor@agendapoetry.co.uk
www.agendapoetry.co.uk
Editor *Patricia McCarthy*

Poems, essays, reviews. Visit the website for further information and supplements to Agenda. Submissions: up to five poems, with s.a.e. and e-mail address. Young poets (and young artists) and essayists (age 15 to late thirties) also encouraged in the journal and in the online Broadsheets. Subscriptions: (one vol. = 4 issues = one subscription) £28 individual (£22 OAPs/students); £35 libraries/institutions. See **Agenda Editions** press.

Aireings

submissions@aireings.co.uk
www.aireings.co.uk/
Contact *Lesley Quayle & Linda Marshall*

Online poetry mag: poetry, prose, reviews, b&w artwork.

Aquarius

Flat 4, 116 Sutherland Avenue, London W9 2QP
☎ 020 7289 4338
www.geocities.com/eddielinden
Contact *Eddie S. Linden*

Literary magazine; prose and poetry. No emailed submissions.

Areopagus

48 Cornwood Road, Plympton, Plymouth PL7 1AL
www.areopagus.org.uk
Contact *Julian Barritt*

A Christian-based arena for creative writers.

ARTEMISpoetry

⊳ **Second Light under Organizations of Interest to Poets**

Awen

38 Pierrot Steps, 71 Kursaal Way, Southend on Sea
SS1 2UY
atlanteanpublishing@hotmail.com
www.geocities.com/dj_tyrer/awen.html
Contact *David John Tyrer*

Poetry and vignette-length fiction of any style/genre. See also **Bard**, **Monomyth**, **Garbaj** and **The Supplement** magazines and **Atlantean Publishing** press.

Bard

38 Pierrot Steps, 71 Kursaal Way, Southend on Sea
SS1 2UY
atlanteanpublishing@hotmail.com
www.geocities.com/dj_tyrer/bard.html
Contact *David John Tyrer*

Short poetry. See also **Awen**, **Garbaj**, **Monomyth** and **The Supplement** magazines and **Atlantean Publishing** press.

A Bard Hair Day

289 Elmwood Avenue, Feltham TW13 7QB
Contact *Ian Deal*

Contemporary poetry magazine. See also **Partners** press, **Imagenation** and **Poet Tree** magazines.

Blithe Spirit

12 Eliot Vale, Blackheath, London SE3 0UW
☎ 020 8318 4677
www.britishhaikusociety.org
Contact *Graham High*

Journal of the British Haiku Society. Haiku and related forms.

Brittle Star

PO Box 56108, London E17 0AY
☎ 020 8802 1507
magazine@brittlestar.org.uk
www.brittlestar.org.uk
Contact *Louise Hooper*

Poetry, short stories and articles on contemporary poetry.

Candelabrum Poetry Magazine

77 Homegrove House, Grove Road North, Southsea PO5 1HW
www.members.tripod.com/redcandlepress
Contact *M.L. McCarthy*

Formalist poetry magazine for people who like poetry rhythmic and shapely. See also **Red Candle Press**.

Cannon's Mouth

22 Margaret Grove, Harborne, Birmingham B17 9JH
☎ 0121 426 6413
info@cannonpoets.co.uk
www.cannonpoets.co.uk
Contact *Greg Cox*

New poetry, criticism, articles, information.

Carillon

19 Godric Drive, Brinsworth, Rotherham S60 5NA
☎ 01709 372875
editor@carillonmag.org.uk
grippon1@tiscali.co.uk
www.carillonmag.org.uk
Editor *Graham Rippon*

Eclectic poetry and prose. Thrice-yearly. UK subscription: £9.50.

Chanticleer Magazine

6/1 Jamaica Mews, Edinburgh EH3 6HN
Contact *Richad Livermore*

Poetry and ideas – not necessarily in that order.

Chapman

4 Broughton Place, Edinburgh EH1 3RX
☎ 0131 557 2207
chapman-pub@blueyonder.co.uk
www.chapman-pub.co.uk

The best in Scottish and international writing, well-established writers and the up-and-coming. See entry under *Magazines*.

Current Accounts

Apartment 2D, Hardcastle Gardens, Bradshaw Hall, Bolton BL2 4NZ
☎ 01204 598913
bswscribe@aol.com
hometown.ao.co.uk/bswscribe/myhomepage/writing.html
Co-Editor *James Hartnell*

Magazine of the Bank Street Writers' Group. Poetry, short fiction, articles. Submissions also accepted from non-members (by e-mail or by post with s.a.e.).

Dandelion Arts Magazine

24 Frosty Hollow, East Hunsbury NN4 0SY
☎ 01604 701730   ✆ 01604 701730
Editor/Publisher *Jacqueline Gonzalez-Marina*

Biannual, international publication. Poetry, stories, interviews, reviews and illustrations welcome. No religious or political material. Essential to become a subscriber when seeking publication: UK, £17 p.a.; Europe, £30; RoW, US$90.

Decanto

Masque Publishing, 84 Dinsdale Gardens, Littlehampton BN16 9AF
masque_pub@btinternet.com
http://myweb.tiscali.co.uk/masquepublishing
Contact *Lisa Stewart*

Non-conformist poetry magazine; any style considered, not just contemporary. See also **Masque Publishing** (self-publishing service).

Dream Catcher

4 Church Street, Market Rasen, Lincoln LN8 3ET
☎ 01673 844325
www.dreamcatchermagazine.co.uk
www.inpressbooks.co.uk
Editor *Paul Sutherland*

Short fiction, poetry, reviews, artwork, biographies. See also **Dream Catcher** press.

Earth Love

PO Box 11219, Paisley PA1 2WH
www.earthlovepoetrymagazine.co.uk
Contact *Tracy Patrick*

Poetry magazine for the environment; proceeds to conservation charities.

Eastern Rainbow

17 Farrow Road, Whaplode Drove, Spalding PE12 0TS
p_rance@yahoo.co.uk
uk.geocities.com/p_rance/pandf.htm
Contact *Paul Rance*

Focuses on 20th century culture via poetry, prose and art. See also **Peace and Freedom** magazine.

The Engine

Flat 57, 5–7 Hornsey Street, London N7 8GB
☎ 07972 593189
clitophon@yahoo.com
www.theengine.com
Contact *Paul Murphy*

Journal of art, poetry, articles and reviews. E-mail submissions only.

Envoi

Ty Meirion, Glan yr afon, Tanygrisiau, Blaenau Ffestiniog LL41 3SU
☎ 01766 832112
Contact *Jan Fortune-Wood*

Poetry, sequences, features, reviews, competitions. Now in it's 51st year. See also **Cinnamon Press**.

Equinox

134b Joy Lane, Whitstable CT5 4ES
☎ 01227 282718
dordi.barbara@neuf.fr
www.poetrymagazines.org.uk/equinox
Contact *Barbara Dordi*

Contemporary poetry invited. No e-mail submissions.

Eratica

126 Furze Croft, Furze Hill, Hove BN3 1PF
☎ 01273 202876
drjenner@ntlworld.com
Contact *Simon Jenner*

Biannual journal with colour plates – focus on poetry, strong on music and art. See also **Waterloo Press**.

Feather's Miscellany

▷ Feather Books under Small Presses

The Firing Squad

50 Lichfield Road, Walsall WS4 2DJ
☎ 07736 411089
editor@the-firing-squad.com
www.the-firing-squad.com
Editor *Alex Barzdo*

Protest poetry website – 'or the grumpy old website'. E-mail submissions only.

First Offense

Snails Field, Mare Hill Road, Pulborough RH20 2DS
tim@firstoffense.co.uk
www.firstoffense.co.uk
Contact *Tim Fletcher*

Magazine for contemporary poetry; not traditional but is received by most groundbreaking poets.

The Frogmore Papers

21 Mildmay Road, Lewes BN7 1PJ
☎ 07751 251689
J.N.Page@sussex.ac.uk
www.frogmorepress.co.uk
Editor *Jeremy Page*

Founded 1983. Biannual. Poetry, prose and artwork. See also **The Frogmore Press**.

Gabriel: A Christian Poetry Magazine

27 Headingley Court, North Grange Road, Leeds LS6 2QU
Editor *Thelma Laycock*

Annual Christian poetry magazine. Unsolicited submissions by post welcome.

Garbaj

38 Pierrot Steps, 71 Kursaal Way, Southend on Sea SS1 2UY
atlanteanpublishing@hotmail.com
www.geocities.com/dj_tyrer/garbaj.html
Contact *D.J. Tyrer*

Humourous/non-pc poetry, vignette-length fiction, fake news, etc. See also **Awen**, **Bard**, **Monomyth** and **The Supplement** magazines and **Atlantean Publishing** press.

Global Tapestry Journal

Spring Bank, Longsight Road, Copster Green, Blackburn BB1 9EU
☎ 01254 249128
Contact *Dave Cunliffe*

Global Bohemia, post-Beat and counterculture orientation. See also **BB Books** press.

Green Queen

BM Box 5700, London WC1N 3XX
eandk2@btinternet.com
Editor *Elsa Wallace*

Fiction, poetry, reviews, news, articles, opinions – serious and humorous – dealing with issues of humane living, ecology, veganism from a LG-BT viewpoint. Unsolicited contributions welcome; approach by mail.

Haiku Scotland

2 Elizabeth Gardens, Stoneyburn EH47 8PB
haiku.scotland@btinternet.com
Contact *Frazer Henderson*

Haiku, senryu, aphorism, epigram, reviews and articles.

Handshake

5 Cross Farm, Station Road North, Fearnhead, Warrington WA2 0QG
Contact *John Francis Haines*

Newsletter of **The Eight Hand Gang**, an association of UK sci-fi poets. See entry under *Organizations of Interest to Poets*.

Harlequin

PO Box 23392, Edinburgh EH8 7YZ
columbine@harlequinmagazine.com
www.harlequinmagazine.com
Contact *Jim Sinclair*

Poetry and artwork of mysticism and wisdom.

HQ Poetry Magazine (Haiku Quarterly)

39 Exmouth Street, Swindon SN1 3PU
☎ 01793 523927
Editor *Kevin Bailey*

General poetry mag with slight bias towards imagistic/haikuesque work

Imagenation

289 Elmwood Avenue, Feltham TW13 7QB
Contact *Ian Deal*

Poetry and artwork magazine. See also **Partners** press, **A Bard Hair Day** and **Poet Tree** magazines.

Inclement

White Rose House, 8 Newmarket Road, Fordham, Ely CB7 5LL
inclement_poetry_magazine@hotmail.com
www.inclementpoetrymagazine.webs.com
Contact *Michelle Foster*

All forms and styles of poetry. Responses within a month.

The Interpreter's House

19 The Paddox, Oxford OX2 7PN
www.interpretershouse.org.uk
Contact *Merryn Williams*

Poems and stories up to 2,500 words; new and established writers. Check website for subscription details.

Iota

1 Lodge Farm, Snitterfield, Stratford-upon-Avon CV37 0LR
☎ 01789 730320
iotapoetry@aol.com
www.iotapoetry.co.uk
Contact *Bob Mee & Janet Murch*

Poetry, reviews; long and short poems welcome. No epics. See also **Ragged Raven Press**.

Irish Pages

The Linen Hall Library, 17 Donegall Square North, Belfast BT1 5GB
☎ 028 9043 4800
irishpages@yahoo.co.uk
www.irishpages.org

'Ireland's premier literary journal.' See entry under *Magazines*.

Island

8 Craiglea Drive, Edinburgh EH10 5PA
jaj@essencepress.co.uk
www.essencepress.co.uk
Contact *Julie Johnstone*

A distinctive space for writing inspired by nature and exploring our place within the natural world. See also **Essence Press**.

The David Jones Journal

The David Jones Society, 22 Gower Road, Sketty, Swansea SA2 9BY
☎ 01792 206144
contact@davidjonessociety.org,
www.davidjonessociety.org/default.aspx
Contact *Anne Price-Owen*

Articles, poetry, information, reviews and inspired works.

The Journal

17 High Street, Maryport CA15 6BQ
☎ 01900 812194
smithsssj@aol.com
members.aol.com/smithsssj/index.html
Contact *Sam Smith*

Poems in translation alongside poetry written in English. See also **Original Plus** press.

Krax

63 Dixon Lane, Wortley, Leeds LS12 4RR
Contact *Andy Robson*

Light-hearted, contemporary poetry, short fiction and graphics. 'No devotional writing or smutty limericks.'

Leeds Poetry Quarterly

89 Connaught Road, Sutton SM1 3PJ
☎ 020 8286 0419
Contact *Barry Tebb*

New poems, reviews, articles on literary and psychoanalytic matters. See also **Sixties Press** and **Literature and Psychoanalysis** magazine.

Literature and Psychoanalysis

89 Connaught Road, Sutton SM1 3PJ
☎ 020 8286 0419
Contact *Barry Tebb*

New poems, reviews, articles on literary and psychoanalytical matters. See also **Sixties Press** and **Leeds Poetry Quarterly** magazine.

The London Magazine
▸ entry under Magazines

Magma

43 Keslake Road, London NW6 6DH
magmapoetry@ntlworld.com
www.magmapoetry.com
Contact *David Boll*

New poetry plus poetry reviews and interviews.
Submissions to: contributions@magmapoetry.
com or address above.

Markings

▷ entry under Magazines

Modern Poetry in Translation

The Queen's College, Oxford OX1 4AW
www.mptmagazine.com
Contact *David & Helen Constantine*

Publishes and promotes poetry in English
translation. E-mail submissions not accepted.

Monkey Kettle

monkeykettle@hotmail.com
www.monkeykettle.co.uk
www.myspace.com/monkeykettle
Editor *Matthew M. Taylor*

Biannual poetry, prose, photos, articles in Milton
Keynes and further afield.

Monomyth

38 Pierrot Steps, 71 Kursaal Way, Southend on Sea
SS1 2UY
atlanteanpublishing@hotmail.com
www.geocities.com/dj_tyrer/monomyth.html
Contact *David John Tyrer*

Poetry, prose and articles; all genres, styles and
lengths considered. New writers welcome. See
also **Awen**, **Bard**, **Garbaj** and **The Supplement**
magazines and **Atlantean Publishing** press.

Moodswing

12 Derby Crescent, Moorside, Consett DH8 8DZ
☎ 01207 505724
Contact *Steve Urwin*

Pocket broadsheet – light/dark – psychologically
charged. See also **Talking Pen** press

Mslexia

PO Box 656, Newcastle upon Tyne NE99 1PZ
☎ 0191 261 6656   📠 0191 261 6636
postbag@mslexia.co.uk
www.mslexia.co.uk

National quarterly magazine for women who
write. See entry under *Magazines*.

Neon Highway Poetry/Art Magazine

37 Grinshill Close, Liverpool L8 8LD
poetshideout@yahoo.com
www.geocities.com/poetshideout/Neon.html
Editors *Alice Lenkiewicz (at postal/e-mail
address above)*, *Dee McMahon (14 Tower Hill,*

*Ormskirk L39 2EG; dee_mcbuc@yahoo.
co.uk)*, *Matt Fallaize (54 Chapel Street,
Ormskirk L39 4QF; matt.fallaize@gmail.com)*

BIANNUAL. Features work by poets and artists.
Interviews, art and information on readings
listed. Submissions of innovative poetry should
be sent to the editors; 'always enclose s.a.e'. Single
issue, £3 (two issues, £5.50); cheque to be made
out to Alice Lenkiewicz.

Never Bury Poetry

Bracken Clock, Troutbeck Close, Hawkshaw,
Bury BL8 4LJ
☎ 01204 884080
n.b.poetry@zen.co.uk
Contact *Jean Tarry*

Founded 1989. Biannual. International reputation.
Each issue has a different theme. Submissions by
post with s.a.e.

New Welsh Review

PO Box 170, Aberystwyth SY23 1WZ
☎ 01970 628410
editor@newwelshreview.com
www.newwelshreview.com
Editor *Kathryn Gary*
Development Manager *Sue Fisher*

Vibrant literary quarterly magazine which
showcases the best new writing from Wales. Also
includes features and reviews. See entry under
Magazines.

Norfolk Poets and Writers – Tips Newsletter

9 Walnut Close, Taverham, Norwich NR8 6YN
tipsforwriters@yahoo.co.uk
Editor *Wendy Webb*

Tips for Writers; six magazines per year;
competitions, themes and forms. See also **Wendy
Webb Books** press.

The North

The Poetry Business, Bank Street Arts, 32–40
Bank Street, Sheffield S1 2DS
office@poetrybusiness.co.uk
www.poetrybusiness.co.uk
Contact *Peter Sansom & Ann Sansom*

Contemporary poetry and articles, extensive
reviews.

Obsessed With Pipework

PO Box 14479, Birmingham B13 3GU
meria@btinternet.com
jacquirowe@hotmail.co.uk
www.flarestackpoets.co.uk
Contacts *Meredith Andrea & Jacqui Rowe*

Quarterly magazine of new poetry 'to surprise
and delight'. See also **Flarestack Publishing** press.

Open Wide Magazine

40 Wingfield Road, Lakenheath IP27 9HR
☎ 07790 962317
contact@openwidemagazine.co.uk
www.openwidemagazine.co.uk
Contact *James Quinton*

'The UK's best arts magazine', publishing short fiction, punchy poetry, reviews and interviews plus OWM Extra, a new online supplement.

Other Poetry

Fourlawshill Top, Bellingham, Hexham NE48 2EY
editors@otherpoetry.com
www.otherpoetry.com
Contact *The Editors*

Founded 1979. Thrice-yearly, 60–70 poems per issue. Process of selection involves all the editors. Token payment. Submit up to four poems (with name on each sheet) plus s.a.e.

Parameter

PO Box 443, Sale M33 0AH
editor@parametermagazine.org
www.parametermagazine.org

Partners Annual Poetry Competition

289 Elmwood Avenue, Feltham TW13 7QB
partners_writing_group@hotmail.com
Contact *Ian Deal*

Showcases poems entered into the Partners annual open poetry competitions.

Peace and Freedom

17 Farrow Road, Whaplode Drove, Spalding
PE12 0TS
☎ 01406 330242
p_rance@yahoo.co.uk
uk.geocities.com/p_rance/pandf.htm
Contact *Paul Rance*

Poetry, prose, art mag; humanitarian, environmental, animal welfare. See also **Eastern Rainbow** magazine.

Peer Poetry International

26 (wr.Hndbk) Arlington House, Bath Street, Bath
BA1 1QN
☎ 01225 445298
peerpoetry@yahoo.co.uk
Editor *Paul Amphlett*

Peerpoetry EMagazine, and printed haibun etc. 'Moderator, Newcomers/Beginners World Haiku Club at above e-mail address if you are interested in free tuition and material regarding Haiku/Senryu.' PeerPoetry EMagazine: verse in all styles, via e-mail. Published biannually or more frequently. Will help with publishing collections. Also competition; winners in Ebook compilation.

The Penniless Press

100 Waterloo Road, Ashton, Preston PR2 1EP
☎ 01772 736421
www.pennilesspress.co.uk
Contact *Alan Dent*

Quarterly for the poor pocket and the rich mind. Poetry, fiction, essays, reviews. See also **Penniless Press**.

Pennine Ink Magazine

The Gallery, Mid-Pennine Arts, Yorke Street, Burnley BB11 1HD
☎ 01282 432992
sheridansdandl@yahoo.co.uk
Contact *Laura Sheridan*

Quality poetry reflecting traditional and modern trends.

Pennine Platform

Frizingley Hall, Frizinghall Road, Bradford
BD9 4LD
☎ 01274 541015
www.pennineplatform.co.uk
Contact *Nicholas Bielby*

Biannual poetry magazine; eclectic and serious-minded. Considers hard copy submissions only.

Planet

PO Box 44, Aberystwyth SY23 3ZZ
☎ 01970 611255   ✉ 01970 611197
planet.enquiries@planetmagazine.org.uk
www.planetmagazine.org.uk
Editor *Helle Michelsen*

The Welsh Internationalist: current affairs, arts, environment.

PN Review

▷ **Carcanet Press Ltd under UK Publishers**

Poet Tree

289 Elmwood Avenue, Feltham TW13 7QB
Contact *Ian Deal*

Contemporary poetry magazine. See also **Partners** press, **A Bard Hair Day** and **Imagenation** magazines.

Poetic Hours

Admin: 43 Willow Road, Carlton, Nottingham
NG4 3BH
erranpublishing@hotmail.com (all communications)
www.poetichours.homestead.com
Editor *Nick Clark*

Non-profit supporter of Third World charities.

Poetic Licence

70 Aveling Close, Purley CR8 4DW
poets@poetsanon.org.uk
www.poetsanon.org.uk

Contact *Peter L. Evans*

Original unpublished poems and drawings. See also **Poets Anonymous** press.

Poetry Combination Module

P.E. F Productions, 196 High Road, London
N22 8HH
page84direct@yahoo.co.uk
www.geocities.com/andyfloydplease
Contact *Mr P.E. F*

Available late spring, early autumn and mid-winter. Printable from the website. Poetry, anti-poetry, aphorism and artwork magazine by P.E. F and guest artists. Submissions welcome, preferably by e-mail.

Poetry Cornwall/Bardhonyeth Kernow

11a Penryn Street, Redruth TR15 2SP
☎ 01209 218209
poetrycornwall@yahoo.com
www.poetrycornwall.freeservers.com
Editor *Les Merton*

Publishes poets from around the world, poetry in its original language with English translation and poetry in Kernewek and Cornish dialect. Three issues a year. Submission guidelines on the website and back cover of every issue.

Poetry Express

Studio 11, Bickerton House, 25–27 Bickerton Road, London N19 5JT
☎ 020 7281 4654
simon@survivorspoetry.org.uk
www.survivorspoetry.com
Contact *Dr Simon Jenner & Phil Ruthen*

Biannual newsletter from **Survivors' Poetry** (see entry under *Organizations of Interest to Poets*).

Poetry Ireland News

2 Proud's Lane, Dublin 2, Republic of Ireland
☎ 00 353 1 478 9974   📠 00 353 1 478 0205
publications@poetryireland.ie
www.poetryireland.ie
Publications Officer *Paul Lenehan*

Bi-monthly newsletter. See also **Poetry Ireland Review** magazine.

Poetry Ireland Review/Eigse Eireann

2 Proud's Lane, Dublin 2, Republic of Ireland
☎ 00 353 1 478 9974   📠 00 353 1 478 0205
www.poetryireland.ie
Editor *Caitríona O'Reilly*

Quarterly journal of poetry and reviews. E-mail submissions not accepted. See also **Poetry Ireland News** magazine.

Poetry London

81 Lambeth Walk, London SE11 6DX
☎ 020 7735 8880   📠 020 7735 8880
editors@poetrylondon.co.uk (enquiries only)
www.poetrylondon.co.uk
Poetry Editor *Maurice Riordan (6 Daniels Road, London SE15 3LR)*
Listings *Gyonghi Vegh (listings@poetrylondon. co.uk)*
Reviews *Scott Verner*

Published three times a year, *Poetry London* includes poetry by new and established writers, reviews of recent collections and anthologies, articles on issues relating to poetry, and an encyclopædic listings section of virtually everything to do with poetry in the capital and elsewhere in the UK. E-mail submissions not accepted.

Poetry Nottingham

11 Orkney Close, Stenson Fields, Derby DE24 3LW
adrian.buckner@btopenworld.com
Editor *Adrian Buckner*

Poetry, articles, reviews, published thrice-yearly. Subscription: £12 p.a. (£4 individual issues).

Poetry Reader

Scottish Poetry Library, 5 Crichton's Close, Canongate, Edinburgh EH8 8DT
☎ 0131 557 2876   📠 0131 557 8393
inquiries@spl.org.uk
www.spl.org.uk
www.readingroom.spl.org.uk
Marketing Officer *Jane Alexander*

Newsletter of the **Scottish Poetry Library**.

Poetry Review

Poetry Society, 22 Betterton Street, London WC2H 9BU
☎ 020 7420 9883   📠 020 7240 4818
poetryreview@poetrysociety.org.uk
www.poetrysoc.org.uk
Editor *Fiona Sampson*

The senior review of contemporary poetry. Quarterly.

Poetry Salzburg Review

Dept. of English, University of Salzburg, Akademiestr. 24, A–5020 Salzburg, Austria
☎ 00 43 662 8044 4424   📠 00 43 662 8044 167
editor@poetrysalzburg.com
www.poetrysalzburg.com
Co-editors *Wolfgang Görtschacher & Andreas Schachermayr*
Editorial board: *David Miller (99 Mitre Road, London SE1 8PT, katermurr@btinternet. com); William Bedford (The Croft, 3 Manor Farm Court, Purton Stoke SN5 4LA, w.bedford@ btopenworld.com); Robert Dassanowsky (Dept. of Languages & Cultures, University of Colorado, 1420 Austin Bluffs, Colorado CO 80933, USA, rvondass@uccs.edu)*

Poetry magazine, formerly *The Poet's Voice*. Published at the University of Salzburg. Publishes new poetry, translations, interviews, review-essays, essays on poetics and poetry, artwork.

Poetry Scotland

91–93 Main Street, Callander FK17 8BQ
www.poetryscotland.co.uk
Contact *Sally King*

All-poetry broadsheet with Scottish emphasis. See also **Diehard Publishers** press.

Poetry Wales

57 Nolton Street, Bridgend CF31 3AE
☎ 01656 663018
www.poetrywales.co.uk
Editor *Zöe Skoulding (School of English, Bangor University, Bangor LL57 2DG)*

An international magazine with a reputation for fine writing and criticism. E-mail submissions not accepted. See also **Seren Books** under *UK Publishers*.

Premonitions

13 Hazely Combe, Arreton, Isle of Wight PO30 3AJ
☎ 01983 865668
mail@pigasus.press.co.uk
www.pigasuspress.co.uk
Contact *Tony Lee*

Magazine of science fiction and horror stories, with genre poetry and artwork. See also **Pigasus Press**.

Presence

12 Grovehall Avenue, Leeds LS11 7EX
martin.lucas2@btinternet.com
http://haiku-presence.50webs.com
Contact *Martin Lucas*

Haiku, senryu, tanka, renku and related poetry in English.

Pulsar Poetry Magazine

34 Lineacre Close, Grange Park, Swindon SN5 6DA
☎ 01793 875941
pulsar.ed@btopenworld.com
www.pulsarpoetry.com
Editor *David Pike*

Hard-hitting/inspirational poetry with a message and meaning. Biannual.

Purple Patch

25 Griffiths Road, West Bromwich B71 2EH
ppatch66@hotmail.com
www.purplepatchpoetry.co.uk
www.poetrywednesburyinternational.org.uk
Contact *Geoff Stevens*

Poetry mag founded 1976. Includes reviews and gossip column. See also **Poetry Wednesbury** press.

Quantum Leap

York House, 15 Argyle Terrace, Rothesay, Isle of Bute PA20 0BD
Contact *Alan Carter*

User-friendly magazine – encourages new writers; all types of poetry – pays. Mark envelope 'Guidelines' and enclose s.a.e. or two IRCs if from abroad. See also **QQ Press**.

Rainbow Poetry News

74 Marina, St Leonards-on-Sea TN38 0BJ
☎ 01424 444072
beyondcloister@hotmail.co.uk
Contact *Hugh C. Hellicar*

Quarterly with poems, articles and news of poetry recitals in London and the South East but with overseas readership.

Read the Music

20 Wharfedale Street, Wednesbury WS10 9AG
☎ 07970 441110
mooncrow@tiscali.co.uk
mooncrow@blueyonder.co.uk
www.poetrywednesbury.co.uk
Contact *Brendan Hawthorne*

Original poetry based on music and its influences.

The Reater

Wrecking Ball Press, 24 Cavendish Square, Hull HU3 1SS
editor@wreckingballpress.com
www.wreckingballpress.com
Contact *Shane Rhodes*

No flowers, just blunt, chiselled poetry.

The Recusant

www.therecusant.moonfruit.com
Editor *Alan Morrison*

Non-conformist poetry, prose, polemic, reviews and articles. Seeks to provide a home for contemporary writing that goes against the grain, in subject more than style. Socially-inclined work with a left-field slant particularly welcome. Visit the website for submission guidelines.

Red Poets – Y Beirdd Coch

26 Andrew's Close, Heolgerrig, Merthyr Tydfil CF48 1SS
☎ 01685 376726
marcjones@phonecoop.coop
Contact *Mike Jenkins & Marc Jones*

A magazine of left-wing poetry from Wales and beyond: socialist, republican and anarchist.

The Rialto

PO Box 309, Aylsham, Norwich NR11 6LN
mail@therialto.co.uk
www.therialto.co.uk
Contact *Michael Mackmin*

Excellent poetry in a clear environment. Buy online at www.impressbooks.co.uk See also **Rialto Publications** press.

Roundyhouse

3 Crown Street, Port Talbot SA13 1BG
srjones@alunbooks.co.uk
Contact *Sally R. Jones*

Poems and articles on poetry. Reviews. Eclectic approach.

Route Magazine

See **Route Publishing** press.

Rubies In the Darkness

41 Grantham Road, Manor Park, London E12 5LZ
Contact *P.G.P. Thompson*

Romantic, lyrical, spiritual, inspirational, traditional and mystical poetry. See also **Precious Pearl Press**.

Sable

PO Box 33504, London E9 7YE
info@sablelitmag.org
www.sablelitmag.org
Also at: PO Box 2803, Upper Darby, PA 19082, USA
Contact *Kadija Sesay*

Literary magazine for writers of African, Caribbean and Asian descent. Poetry, prose, memoirs, travel, etc.

Scar Tissue

Pigasus Press, 13 Hazley Combe, Arreton, Isle of Wight PO30 3AJ
Contact *Tony Lee*

SF genre poetry, short fiction, cartoons and artwork.

Scribbler!

Remus House, Coltsfoot Drive, Woodston, Peterborough PE2 9JX
☎ 01733 890066　📠 01733 313524
youngwriters@forwardpress.co.uk
www.youngwriters.co.uk
Editor *Donna Samworth*

Thrice-yearly magazine by 7–11 year-olds: poetry, stories, guest authors, workshops, features.

Scriptor

6 Green Bank, Stacksteads, Bacup OL13 8LQ
☎ 01706 870357/07967 315270 (mobile)
mail@johndench.demon.co.uk
Contact *John Dench*

Poetry, short stories, essays from the UK; published biennally (next issue 2010); guidelines available from July 2009. See also **Green Arrow Publishing** press.

The Seventh Quarry – Swansea Poetry Magazine

Dan-Y-Bryn, 74 Cwm Level Road, Brynhyfryd, Swansea SA5 9DY
www.peterthabitjones.com
Contact *Peter Thabit Jones*
Consulting Editor, America *Vince Clemente*

Quality poetry from Swansea and beyond. Submission details available on the website.

Shearsman

58 Velwell Road, Exeter EX4 4LD
☎ 01392 434511
editor@shearsman.com
www.shearsman.com
Contact *Tony Frazer*

Mainly poetry, some prose, some reviews. Poetry in the modernist tradition. No fiction. See also **Shearsman Books** press.

The ShOp: A Magazine Of Poetry

Skeagh, Schull, Co. Cork, Republic of Ireland
theshop@theshop-poetry-magazine.ie (not for submissions)
www.theshop-poetry-magazine.ie
Contact *John Wakeman & Hilary Wakeman*

International but with emphasis on Irish poetry.

Smiths Knoll

Goldings, Goldings Lane, Leiston IP16 4EB
Contact *Joanna Cutts & Michael Laskey*

Poems worth re-reading; two-week turnaround for submissions.

Smoke

The Windows Project, Liver House, 96 Bold Street, Liverpool L1 4HY
☎ 0151 709 3688
www.windowsproject.co.uk
Contact *Dave Ward*

Poetry, graphics, short prose. Biannual; 24pp.

South

PO Box 3744, Cookham, Maidenhead SL6 9UY
www.southpoetry.org
Contact *Chrissie Williams*

Poetry magazine from the southern counties of England that welcomes poets from across the world. Poems judged anonymously.

Southword

Frank O'Connor House, 84 Douglas Street, Cork, Republic of Ireland
www.munsterlit.ie
Contact *Rosemary Canavan*

Biannual, 120 pages of poetry, fiction, visual art and book reviews. E-mail submissions not accepted.

Sphinx

21 Hatton Green, Glenrothes KY7 4SD
nell@happenstancepress.com
www.happenstancepress.com
Editor *Helena Nelson*

A feature and review-based magazine focusing on poetry publishing: who, how and why. No unsolicited submissions. See also **Happenstance Press**.

Springboard

Corrimbla, Ballina, Co. Mayo, Republic of Ireland
bobgroom@eircom.net
Contact *Robert Groom*

Short fiction, poetry and articles.

The Stinging Fly

PO Box 6016, Dublin 8, Republic of Ireland
stingingfly@gmail.com
www.stingingfly.org
Contact *Declan Meade*

New Irish and international writing. Poetry, short fiction, author interviews, essays and book reviews. See also **The Stinging Fly Press**. Submissions are considered from January to March each year. No e-mail submissions.

The Supplement

38 Pierrot Steps, 71 Kursaal Way, Southend on Sea SS1 2UY
atlanteanpublishing@hotmail.com
www.geocities.com/dj_tyrer/mms.html
Contact *D.J. Tyrer*

Non-fiction: articles, news, reviews, letters, etc. See also **Awen**, **Bard**, **Garbaj** and **Monomyth** magazines and **Atlantean Publishing** press.

Taliesin

Academi, Mount Stuart House, Mount Stuart Square, Cardiff CF10 5FQ
029 2047 2266   029 2049 2930
taliesin@academi.org
www.academi.org
Joint Editors *Manon Rhys & Christine James*

Wales' leading Welsh language literary journal, published three times a year.

Tears in the Fence

38 Hod View, Stourpaine, Nr Blandford Forum DT11 8TN
01258 456803   01258 454026
david@davidcaddy.wanadoo.co.uk
www.myspace.com/tearsinthefence
Contact *David Caddy*

'An international magazine that appreciates social and poetic awareness and writing that prompts close and divergent readings.'

10th Muse

33 Hartington Road, Southampton SO14 0EW
a.jordan@surfree.co.uk
Editor *Andrew Jordan*

Publishes poetry and reviews as well as short prose (usually no more than 2,000 words) and graphics. E-mail for information only; no submissions by e-mail.

The Third Half

16 Fane Close, Stamford PE9 1HG
01780 754193
Editor *Kevin Troop*

Searches for good poetry and fiction. See also **KT Publications** press.

Time Haiku

Basho-an, 105 King's Head Hill, London E4 7JG
020 8529 6478
facey@aol.com
Contact *Erica Facey*
Editor *Doreen King*

Founded 1994. Promotes haiku, tanka and haiku-related forms. Aims to create accessibility to all people interested in this form of poetry as well as established poets. A biannual journal and newsletter are available to subscribers.

Understanding

127 Milton Road West, 7 Duddingston House Courtyard, Edinburgh EH15 1JG
Editor *Denise Smith*

Original poetry, short stories, parts of plays, reviews and articles. See also **Dionysia Press**.

Wasafiri

The Open University in London, 1–11 Hawley Crescent, London NW1 8NP
020 7556 6110   020 7556 6187
wasafiri@open.ac.uk
www.wasafiri.org
Editor *Susheila Nasta*
Deputy Editor *Sharmilla Beezmohun*
Assistant Editor *Nisha Jones*
Editorial Manager *Teresa Palmiero*
Reviews Editor *Mark Stein*

The magazine of international contemporary writing.

Zed20

33 Lady Nairn Avenue, Kirkcaldy KY1 2AW
01592 651522
Contact *Duncan Glen*

See also **Akros Publications** press.

Organizations of Interest to Poets

Academi (Yr Academi Gymreig)

3rd Floor, Mount Stuart House, Mount Stuart Square, Cardiff Bay CF10 5FQ

📱 029 2047 2266 📠 029 2049 2930

post@academi.org

www.academi.org

Glyn Jones Centre: Wales Millennium Centre, Bute Place Cardiff CF10 5FQ

North Wales Office: Tŷ Newydd, Llanystumdwy, Cricieth, Gwynedd LL52 0LW

West Wales Office: Dylan Thomas Centre, Somerset Place, Swansea SA1 1RR

Chief Executive *Peter Finch*

The writers' organization of Wales with special responsibility for literary activity, writers' residencies, writers on tour, festivals, writers' groups, readings, tours, exchanges and other development work. Academi awards financial bursaries annually, runs a criticism service and organizes the **Wales Book of the Year Award**. **Yr Academi Gymreig/The Welsh Academy** operates the Arts Council of Wales franchise for Wales-wide literature development. The Academi sponsors a range of annual contests including the John Tripp Award For Spoken Poetry and the prestigious **Academi Cardiff International Poetry Competition**. Publishes *A470*, a quarterly literary information magazine. For further information, see entry under *Professional Associations and Societies*.

Apples & Snakes

The Albany, Deptford, London SE8 4AG

📱 0845 521 3460 📠 0845 521 3461

info@applesandsnakes.org

www.myspace.com/applesandsnakespoetry

www.applesandsnakes.org

Director *Geraldine Collinge*

Works nationwide to promote popular, high-quality and cross-cultural poetry; programmes live events for new and established poets including open mic events; operates a Poets in Education scheme where poets run workshops and perform in schools, prisons and community settings; coordinates the professional development of poets, with opportunities for training mentoring, national touring and residencies.

The Arvon Foundation

▷ entry under UK and Irish Writers' Courses

The British Haiku Society

📱 01772 251827

www.britishhaikusociety.org

General Secretary *Doreen King*

Formed in 1990. Promotes the appreciation and writing of haiku and related forms; welcomes overseas members. Provides tutorials, workshops, readings, critical comment and information. Specialist advisers are available. Runs a haiku library and administers a haiku contest. The journal, *Blithe Spirit*, is produced quarterly. The Society has active local groups and issues a regular newsletter. Current membership details can be obtained from the website.

Camelford Poetry Workshops

The Indian King, Garmoe Cottage, 2 Trefrew Road, Camelford PL32 9TP

📱 01840 212161

indianking@btconnect.com

Director *Helen Wood*

A weekly half-day workshop where regular participants write together, bring work in progress for criticism and share experience of publication. They also organize a programme of visiting poets who run a writing workshop one Saturday every other month followed by a public reading. This is also the home of he Poetry Society's Camelford Poetry Stanza. The alternate months see a Saturday Novel Writing Surgery led by Karen Hayes.

Contemporary Poetics Research Centre

School of English and Humanities, Birkbeck
College, Malet Street, London WC1E 7HX
w.rowe@bbk.ac.uk
c.watts@bbk.ac.uk
www.bbk.ac.uk/cprc
Contacts *Will Rowe & Carol Watts*

Dedicated to fostering research, performance
and practice in all modes and formats of
contemporary poetry. It runs a Creative Reading
Workshop, holds readings, performances, talks
and debates and organizes conferences, the latest
of which was Text and Music 2009. Collaborates
closely with Royal Holloway University of
London and the Centre for Cultural Poetics
at the University of Southampton, e.g. on the
British Electronic Poetry Centre at www.soton.
ac.uk/~bepc Poets associated with the Centre
include Ulli Freer, Caroline Bergvall and Sean
Bonney. The Centre publishes the web journals,
Pores (www.pores.bbk.ac.uk) and *Readings* (www.
bbk.ac.uk/readings/).

The Eight Hand Gang

5 Cross Farm, Station Road North, Fearnhead,
Warrington WA2 0QG
Secretary *John F. Haines*

An association of SF poets. Publishes *Handshake*,
a single-sheet newsletter of SF poetry and
information available free in exchange for an s.a.e.

57 Productions

57 Effingham Road, Lee Green, London SE12 8NT
☎ 020 8463 0866   ▣ 020 8463 0866
info@57productions.com
www.57productions.com
Contact *Paul Beasley*

Specializes in the promotion of poetry and
its production through an agency service, a
programme of events and a series of audio
publications. Services are available to event
promoters, festivals, education institutions and
the media. Poets represented include Jean 'Binta'
Breeze, Adrian Mitchell, Patience Agbabi and
many others. 57 Productions' series of audio
cassettes, CDs and Poetry in Performance
compilations offer access to some of the most
exciting poets working in Britain today – check
their Poetry Jukebox on their excellent website.

The Football Poets

4 The Retreat, Butterow, Stroud GL5 2LS
☎ 01453 757376
editors@footballpoets.org
ctm@crispinthomas.orangehome.co.uk
www.footballpoets.org
Editor & Performance Poet *Crispin Thomas*
Co-Editors *Simon Williams, Peter Goulding*

The Football Poets exist to promote and
encourage the writing, reading and performing
of football poetry. Formed collectively in 1995,
they perform extensively and also provide
comprehensive football poetry workshops in
schools, football clubs, prisons and communities.
Their website, which was launched in 2000, is a
fast, free and entertaining mix of literature and
soccer poetry from around the world. Anyone
may contribute. The site has been archived by the
British Library and currently hosts over 11,000
football poems.

Performance Poetry Society

PO Box 11178, Birmingham B11 4WP
☎ 0121 242 6644
performancepoetry@yahoo.com
Branch contact & rehearsal room: The Old
Meeting House, behind 20/22 Wolverhampton
Street, Dudley
Contact *Sandra Dennis*
Dudley Branch Contact *Jim MacCool (01384
258191)*

PPS was founded in 1999 by members of the
Birmingham Midland Institute to serve the
interests of performance poets and poetry in
performance throughout the UK and Ireland.
Initiated October as National Poetry Month in
2000. Organizes an extensive national tour each
autumn, visiting small halls, colleges, schools and
prisons. The Society has an in-house publishing
company, Poetry Powerhouse Press.

The Poetry Archive

PO Box 286, Stroud GL6 1AL
☎ 01453 832090   ▣ 01453 836450
info@poetryarchive.org
www.poetryarchive.org
Directors *Richard Carrington, Andrew Motion*

A permanent online archive of audio recordings
of work by poets who write in English. New
60-minute recordings are made for the Archive
by a wide-ranging list of contemporary poets;
extracts from those recordings are available
on the website launched in 2005 at www.
poetryarchive.org and the full-length recordings
are available on CD for purchase by mail order.
The site also contains historic recordings by now-
dead poets, a separate archive of recordings for
young children, and a wealth of educational and
intepretative information. A registered charity,
the Archive intends to ensure that all significant
poets are recorded for posterity and that their
recordings are widely valued and enjoyed.

The Poetry Book Society

Fourth Floor, 2 Tavistock Place, London
WC1H 9RA
☎ 020 7833 9247   ▣ 020 7833 5990

info@poetrybooks.co.uk
www.poetrybooks.co.uk
www.poetrybookshoponline.com
www.childrenspoetrybookshelf.co.uk
Director *Chris Holifield*
Editorial/Marketing Officer *Pinda Bryars*
Sales/Membership Officer *David McDonagh*

For readers, writers, students and teachers
of poetry. Founded in 1953 by T.S. Eliot and
funded by the Arts Council, the PBS is a
unique membership organization and book
club providing up-to-date and comprehensive
information about poetry from publishers
in the UK and Ireland. Members receive the
quarterly *Selectors' Choice* and the quarterly
PBS Bulletin packed with articles by poets,
poems, news, listings and access to discounts
of at least 25% off featured titles. Subscriptions
start at £12. The PBS website (www.poetrybooks.
co.uk) has a recruitment area and a closed
section for members. The PBS also offers a
range of over 30,000 poetry titles at www.
poetrybookshoponline.com, sells the Poetry
Archive CDs and has a new section called
SoundBlast for performance poets' CDs. During
2005 the PBS relaunched the Children's Poetry
Bookshelf (www.childrenspoetrybookshelf.co.uk)
for 7–11-year-olds with new membership schemes
for parents, grandparents and libraries, adding an
annual competition for child poets, now called
Old Possum's Children's Poetry Competition, a
year later. Also runs the annual **T.S. Eliot Prize**
for the best collection of new poetry, which has a
Shadowing Scheme for schools. In 2004 the PBS
ran the Next Generation Poets promotion.

The Poetry Business

Bank Street Arts, 32–40 Bank Street, Sheffield
S1 2DS
office@poetrybusiness.co.uk
www.poetrybusiness.co.uk
Directors *Peter Sansom, Ann Sansom*

Founded in 1986, the Business publishes *The
North* magazine and books and pamphlets under
the Smith/Doorstop imprint. It runs an annual
competition and organizes writing days and a
Writing School. Send an s.a.e. for full details.

Poetry Can

12 Great George Street, Bristol BS1 5RH
☎ 0117 933 0900
admin@poetrycan.co.uk
www.poetrycan.co.uk
Director *Colin Brown*
Administrator *Peter Hunter*

Based in Bristol and founded in 1995, Poetry
Can is a poetry development agency working
across Bristol and the South West Region. It
organizes events such as the annual Bristol Poetry

Festival, an education programme and provides
information and advice concerning all aspects of
poetry to individuals, agencies and organizations.
It also hosts the Literature South West website.

Poetry Ireland/Eigse Eireann

2 Proud's Lane, Dublin 2, Republic of Ireland
☎ 00 353 1 478 9974   ☏ 00 353 1 478 0205
poetry@iol.ie
www.poetryireland.ie
Education: 00 353 1 475 8605
Writers in Schools scheme: 00 353 1 475 8601
Director *Joseph Woods*

Poetry Ireland is the national organization for
poetry in Ireland, with its four core activities
being readings, publications, education and an
information and resource service. Organizes
readings by Irish and international poets
countrywide. Through its website, telephone,
post and public enquiries, Poetry Ireland operates
as a clearing house for everything pertaining
to poetry in Ireland. Operates the Writers in
Schools scheme. Publishes *Poetry Ireland News*,
a bi-monthly newsletter containing information
on events, competitions and opportunities.
Poetry Ireland Review appears quarterly and is the
journal of record for poetry in Ireland; current
editor: *Caitríona O'Reilly*.

The Poetry Library

Royal Festival Hall, Level 5, London SE1 8XX
☎ 020 7921 0943/0664   ☏ 020 7921 0607
info@poetrylibrary.org.uk
www.poetrylibrary.org.uk
www.poetrymagazines.org.uk
Librarians *Chris McCabe, Miriam Valencia*

Founded by the Arts Council in 1953. A collection
of 100,000 books, pamphlets, CDs, magazines
and videos of modern poetry since 1912, from
Georgian to Rap, representing all English-
speaking countries and including translations
into English by contemporary poets. Two copies
of each title are held, one for loan and one for
reference. A wide range of poetry magazines and
ephemera from all over the world are kept along
with cassettes, CDs and videos for consultation,
with many available for loan. There is a children's
poetry section with teacher's resource collection.

An information service compiles lists of poetry
magazines, competitions, publishers, groups and
workshops which are available from the Library
on receipt of a large s.a.e., or direct from the
website. The main website also has a noticeboard
for lost quotations through which it tries to
identify lines or fragments of poetry which have
been sent in by other readers.

Poetry pf

20 Clovelly Way, Orpington BR6 0WD
☎ 01689 811394

editor@poetrypf.co.uk
www.poetrypf.co.uk
Contact *Anne Stewart*

Poet showcase site. Creates and manages searchable and contactable internet presence for poets. Publishes Poem Cards for public sale. Poetry-related print and design services and project work.

The Poetry School

81 Lambeth Walk, London SE11 6DX
📞 020 7582 1679   📠 020 7625 5391
programme@poetryschool.com
www.poetryschool.com
Contact *The Administrator*

Founded in 1997 and funded by Arts Council England, London, The Poetry School offers a wide range of courses and workshops on writing and reading poetry and is open to anyone regardless of experience or formal qualifications. Tutors include Mimi Khalvati, Linda Chase, Graham Fawcett, Pascale Petit, Peter Sansom and Myra Schneider. The School provides a forum for practitioners to share experiences, develop skills and extend appreciation of the traditional and innovative aspects of their art. There are Special Events with visiting poets such as Marilyn Hacker, Jorie Graham, Molly Peacock, Galway Kinnell, Paul Muldoon, Alfred Corn and Sharon Olds. The Poetry School is a Registered Charity No. 1069314.

The Poetry Society

22 Betterton Street, London WC2H 9BX
📞 020 7420 9880   📠 020 7240 4818
info@poetrysociety.org.uk
www.poetrysociety.org.uk
Director *Judith Palmer*

Founded in 1909 which ought to make it venerable, the Society exists to help poets and poetry thrive in Britain. Reaching out from its Covent Garden base, it promotes the national health of poetry in a range of imaginative ways. Membership costs £40 for individuals (concessions available). *Poetry News* membership is from £18. Current activities include:

* Quarterly magazine of new verse, views and criticism, *Poetry Review*.

* Quarterly newsletter, *Poetry News*, with lively relevant articles for members and competitions.

* Promotions, events and cooperation with Britain's many literature festivals, poetry venues and poetry publishers.

* Competitions and awards, including the annual **National Poetry Competition** with a £5,000 first prize.

* Improve your poetry with *Poetry Prescription*. Reduced rates for members.

* Members' web pages including photos and profiles.

* Stanzas – poetry groups run by Poetry Society members in your area.

* Monthly e-bulletins – latest news, competitions and events – sent to your inbox.

* Provides information and advice, publishes books, posters and resources for schools and libraries. Education membership costs £50 (secondary) or £30 (primary) which includes free poetry anthologies, lesson plans and a subscription to Poems on the Underground. Publications include *The Poetry Book for Primary Schools* and *Jumpstart Poetry for the Secondary School*, colourful poetry posters for Keystages 1, 2, 3 and 4. Many of Britain's most popular poets – including Michael Rosen, Roger McGough and Jackie Kay – contribute, offering advice and inspiration.

*The Society's online poetry classroom is at www.poetryclass.net

* The Poetry Café serving snacks and drinks to members, friends and guests, part of The Poetry Place, a venue for many poetry activities – readings, workshops and poetry launches. This space is available for bookings (020 7420 9887).

Poetry Landmarks of Britain, a free resource on the Society's website, packed with regional poetry places, publishers, venues and regular events.

The Poetry Trust

The Cut, 9 New Cut, Halesworth IP9 8BY
📞 01986 835950   📠 01986 874524
info@thepoetrytrust.org
www.thepoetrytrust.org
Director *Naomi Jaffa*

One of the UK's flagship poetry organizations, delivering a year-round programme of live events, creative education opportunities, courses, prizes and publications. Major events include the international **Aldeburgh Poetry Festival** held annually in early November (6th–8th in 2009) and the **Aldeburgh First Collection Prize**. Registered Charity No. 1102893.

Point Editions

Ithaca, Apdo. 125, E–03590 Altea, Spain
📞 00 34 96 584 2350   📠 00 34 96 584 2350
elpoeta@point-editions.com
www.point-editions.com
Also at: Rekkemsestraat 167, B–8510 Marke, Belgium
Director *Germain Droogenbroodt*

Founded as POetry INTernational in 1984, Point is based in Spain and Belgium. A multilingual publisher of contemporary verse from *established* poets, the organization has brought out more than 80 titles in at least eight languages, including

English. Point Editions runs the original work alongside a verse translation into Dutch made in cooperation with the poet. The organization's website features the world's best known and unknown poets in English, Spanish and Dutch. Point also co-organizes several international poetry festivals.

The Railway Prism

☎ 01202 624334
myabbott@tiscali.co.uk
myweb.tiscali.co.uk/railwayprism
Contact *Horace Jay*

The Railway Prism website is a focal point for lovers of British railway poetry. Preferably unpublished submissions welcome about railways old and new, reflecting industrial and social history, childhood nostalgia and modern rail travel both above and below ground. 'We're after atmosphere rather than dry facts.' New feature: lost railway poems help page.

Regional Arts Councils

See **Arts Councils and Regional Offices**, page 552.

Scottish Pamphlet Poetry

25 Market Square, Masham, Ripon HG4 4EG
hazel@scottish-pamphlet-poetry.biz
www.scottish-pamphlet-poetry.com
Administrator *Hazel Cameron*

A publisher collective of more than 50 mainly Scottish-based presses promoting poetry published as pamphlets. A pamphlet is defined as not less than six pages but no more than 30 and with a first edition of less than 300 copies. The organization has an excellent website, publishes a video, runs book fairs and events, and administers the annual Callum Macdonald Memorial Award created to recognize publishing skill and effort in the pamphlet form.

Scottish Poetry Library

5 Crichton's Close, Canongate, Edinburgh EH8 8DT
☎ 0131 557 2876   ℻ 0131 557 8393
inquiries@spl.org.uk
www.spl.org.uk
Director *Robyn Marsack*
Librarian *Julie Johnstone*

A comprehensive reference and lending collection of work by Scottish poets in Gaelic, Scots and English, plus the work of British, European and international poets. Stock includes books, audio, videos, news cuttings and magazines. Borrowing is free to all. Services include a postal lending scheme, for which there is a small fee; enquiries; schools workshops throughout Scotland on application; exhibitions; bibliographies;

publications; general information in the field of poetry. Also available is an online catalogue and index to Scottish poetry periodicals. There is a Friends' scheme costing £25 annually. Friends receive a newsletter and other benefits and support the library.

Second Light

9 Greendale Close, London SE22 8TG
dilyswood@tiscali.co.uk
www.secondlightlive.co.uk
www.poetrypf.co.uk/secondlight.html
www.esch.dircon.co.uk/second/second.ht
Director *Dilys Wood*
Administrator *Anne Stewart (editor@poetrypf. co.uk)*

A network of over 350 women poets, major names and less well-known. With its publishing arm, Second Light Publications, the network aims to develop and promote women's poetry. Issues a twice-yearly poetry journal, *ARTEMISpoetry*, open to submissions for non-members; runs an annual poetry competition; holds residential workshops, readings; has published three anthologies of women's poetry.

Spiel Unlimited

20 Coxwell Street, Cirencester GL7 2BH
☎ 01285 640470/07814 830031
spiel@arbury.freeserve.co.uk
info@spiel.wanadoo.co.uk
Directors *Marcus Moore, Sara-Jane Arbury*

'Spoken word, written word, anywhere, everywhere' with two writers who put a positive charge in live literature by organizing quirky and original events such as Slam!Fests, Spontaneity Days and Living Room Poetry performances, hosting UK poetry slams and running workshops for schools and adults. Also produce *SPIEL*, a monthly e-mail newsletter. Specialists in breathing new life into literature.

Survivors' Poetry

Studio 11, Bickerton House, 25–27 Bickerton Road, London N19 5JT
☎ 020 7281 4654   ℻ 020 7281 7894
info@survivorspoetry.org.uk
www.survivorspoetry.com
National Outreach: royb@survivorspoetry.org.uk
National Mentoring Scheme & *Poetry Express*: simon@survivorspoetry.org.uk
Director *Dr Simon Jenner*
Outreach Coordinator *Roy Birch*
Administrator *Blanche Donnery*

A unique national literature organization promoting poetry by survivors of mental distress through workshops, readings and performances to audiences all over the UK. It was founded in 1991 by four poets with first-hand experience of the mental health system. Survivors' community outreach work provides training and performance workshops and publishing projects. For the organization's quarterly magazine, see *Poetry Express*. Survivors work with those who have survived mental distress and those who empathize with their experience.

Tŷ Newydd Writers' Centre

Llanystumdwy, Cricieth LL52 0LW
01766 522811   01766 523095
post@tynewydd.org
www.tynewydd.org
Executive Director *Sally Baker*

Run by the Taliesin Trust, an independent, Arvon-style residential writers centre established in the one-time home of Lloyd George in North Wales. The programme (in both Welsh and English) has a regular poetry content. (See also *UK and Irish Writers' Courses*.) Fees start at £215 for weekends and £420 for week-long courses. Bursaries available. Among the many tutors to-date have been: Gillian Clarke, Owen Sheers, Robert Minhinnick, Liz Lochhead, Michael Longley, Ian Macmillan and Jo Shapcott. Send for the centre's descriptive leaflets and programme of courses.

The Western Writers' Centre/Ionad Scriobhneoirí Chaitlin Maud

27 Nuns Island, Galway, Republic of Ireland
00 353 91 533594
westernwriters@eircom.net
www.twwc.ie
Director *Fred Johnston*
Administration *Marvelle Maguire*

Founded 2001.The only writers' centre in the West Shannon region, named after the late Caitlin Maud, prominent Irish-language activist, poet and singer. Courses, organized readings and workshops are held in the Galway City and County area and the West of Ireland generally. Small library of literary periodicals. News and reviews of books in the 'Kiosque!' section of the website. Produces a monthly literary newsletter, *The Word Tree*. Accepts original work for online publication in Gaelic, English and French. No payment for publication.

Small Presses

Aard Press

c/o Aardverx, 31 Mountearl Gardens, London
SW16 2NL
Managing Editors *Dawn Redwood, D. Jarvis*

Founded 1971. Publishes artists' bookworks, experimental/visual poetry and art theory, topographics, ephemera, international mail-art and performance art documentation. Very small editions. No unsolicited material or proposals.
Ⓔ Royalties not paid. No sale-or-return deals.

Accent Press

The Old School, Upper High Street, Bedlinog
CF46 6RY
☎ 01443 710930   🖷 0870 130 6934
info@accentpress.co.uk
www.accentpress.co.uk
Managing Editor *Hazel Cushion*

Founded 2002. Publishes general fiction, erotic fiction, health, humour and biography. 24 titles a year. Visit the website for submission guidelines.
Ⓔ Royalties twice-yearly.

Allardyce, Barnett, Publishers

14 Mount Street, Lewes BN7 1HL
☎ 01273 479393
www.abar.net
Publisher *Fiona Allardyce*
Managing Editor *Anthony Barnett*

Founded 1981. Publishes art, literature and music. IMPRINT **Allardyce Book**. About 3 titles a year. Unsolicited mss and synopses cannot be considered.

Alloway Publishing Limited
▷ Stenlake Publishing Limited

Anglo-Saxon Books

25 Brocks Road, Swaffham PE37 7XG
☎ 0845 530 4200/0845 430 4200
www.asbooks.co.uk
Managing Editor *Tony Linsell*

Founded 1990 to promote a greater awareness of and interest in early English history, language and culture. Publishes Anglo-Saxon history, culture, language. About 5 titles a year. Please phone before sending mss.
Ⓔ Royalties at standard rate.

Antitype Press

595A Commercial Road, London E1 0HJ
☎ 020 7001 1366
admin@antitype-press.co.uk
www.antitype-press.co.uk
Managing Editor *Miss A. White*

Founded 2009. Publishes fiction, lesbian fiction, crime. 1 title in 2009. No unsolicited mss; synopses and ideas for books welcome. Send query letter and first five pages in a Word attachment.
Ⓔ Royalties annually.

M.&M. Baldwin

24 High Street, Cleobury Mortimer, Kidderminster DY14 8BY
☎ 01299 270110
mb@mbaldwin.free-online.co.uk
Managing Editor *Dr Mark Baldwin*

Founded 1978. Publishes local interest/history, WW2 codebreaking and inland waterways books. Up to 5 titles a year. Unsolicited mss, synopses and ideas for books welcome (not general fiction).
Ⓔ Royalties paid.

Barny Books

The Cottage, Hough on the Hill, near Grantham
NG32 2BB
☎ 01400 250246   🖷 01400 251737
www.barnybooks.biz
Managing Director/Editorial Head *Molly Burkett*
Business Manager *Jayne Thompson*

Founded with the aim of encouraging new writers and illustrators. Publishes a wide variety of books including adult fiction and non-fiction. Offers schools' projects where students help to produce books. About 12–15 titles a year. Too small a concern to have the staff/resources to deal with unsolicited mss. Writers with strong ideas should approach Molly Burkett by letter in the first instance. Also runs a readership and advisory service for new writers (£20 fee for short stories or illustrations; £50 for full-length stories).
Ⓔ Division of profits 50/50.

BB Books
▷ entry under Poetry Presses

Between the Lines

The Cottage, 14 Lyncroft Gardens, Ewell KT17 1UR
📞 020 8393 7055 📠 020 8393 7055
btluk@aol.com
waywiser-press.com/imprints/betweenthelines.
html
Editorial Board *Peter Dale, Philip Hoy,
J.D. McClatchy*
Editorial Assistant *Ryan Roberts*

Founded in 1998 and since 2002 an imprint of
the **Waywiser Press**. Publishes in book form
extended interviews with leading contemporary
poets, amongst them Peter Dale, Dick Davis,
Rachel Hadas, Timothy Steele, John Ashbery,
Charles Simic, Ian Hamilton, Donald Justice,
Seamus Heaney, Richard Wilbur, Thom Gunn,
Donald Hall, Anthony Hecht, Anthony Thwaite,
Michael Hamburger and W.D. Snodgrass. Most
volumes include a career sketch, a comprehensive
bibliography and a representative selection of
quotations from the poets' critics and reviewers;
some also include photographs as well as
previously uncollected poems.
📧 Royalties not paid.

BeWrite Books

32 Bryn Road South, Ashton-in-Makerfield,
Wigan WN4 8QR
contact@bewrite.net
www.bewrite.net
Managing Editor *Neil Marr*

Founded 2002. Publishes fiction and poetry in
paperback and eBook form. 12 titles in 2008.
Unsolicited mss and synopses welcome using the
online form on the website (www.bewrite.net/
bookshop/submission_requirements.htm).
📧 Royalties quarterly.

Bluemoose Books Limited

25 Sackville Street, Hebden Bridge HX7 7DJ
📞 01422 842731
dufk@aol.com
www.bluemoosebooks.com
Managing Editor *Hetha Duffy*
Publisher *Kevin Duffy*

Founded 2006. Publishes fiction and non-fiction.
2 titles in 2008. No unsolicited mss; send first
three chapters and synopsis either by post
(include s.a.e.) or e-mail.
📧 Royalties annually.

Book Castle Publishing

2a Sycamore Business Park, Dishforth Road, Copt
Hewick HG4 5DF
📞 07837 217893
bc@book-castle.co.uk
www.book-castle.co.uk
Managing Editor *Paul Bowes*
Assistant Editor *Sally Siddons*

Founded 1986. Publishes non-fiction of
local interest (Bedfordshire, Hertfordshire,
Buckinghamshire, Oxfordshire, the Chilterns).
Over 120 titles in print. 12 titles a year.
Unsolicited mss, synopses and ideas for books
welcome.
📧 Royalties paid.

Brilliant Publications

Unit 10, Sparrow Hall Farm, Edlesborough,
Dunstable LU6 2ES
📞 01525 222292 📠 01525 222720
info@brilliantpublications.co.uk
www.brilliantpublications.co.uk
Publisher *Priscilla Hannaford*

Founded 1993. Publishes practical resource
books for teachers, parents and others working
with 0–13-year-olds. About 10–15 titles a year.
Submit synopsis and sample pages in the first
instance. Most of the books have black and white
insides and many have pupil sheets that may
be photocopied. Potential authors are strongly
advised to look at the format of existing books
before submitting synopses. 'We do not publish
children's picture books.'
📧 Royalties twice-yearly.

The Brodie Press

thebrodiepress@hotmail.com
www.brodiepress.co.uk
Project Editor *Penny Price*

Founded 2002. Publishes poetry and anthologies.
SERIES *The Brodie Poets*, launched in 2003,
including Julie-ann Rowell, Poetry Book Society
Pamphlet Choice, Winter 2003. Send synopses
and proposals by e-mail in the first instance.

Charlewood Press

7 Weavers Place, Chandlers Ford SO53 1TU
📞 023 8026 1192
gponting@clara.net
www.home.clara.net/gponting/index-page11.html
Managing Editors *Gerald Ponting, Anthony Light*

Founded 1987. Publishes local history books
and walks booklets on Hampshire and adjacent
counties, especially around the Fordingbridge
area. Mss in field of Hampshire local history
(only) considered.

CNP Publications

▷ Lyfrow Trelyspen

Contact Publishing Ltd

Unit 346, 176 Finchley Road, London NW3 6BT
📞 020 7193 1782
info@contact-publishing.co.uk
www.contact-publishing.co.uk
Managing Director *Anne Kontoyannis*

Founded 2003. Publishes general fiction, non-
fiction, self help and New Age, poetry, thrillers,

action and detective novels. No children's, sports or westerns. 'Interested in seeing mss by new authors with original angles and new ideas.' See website for author guidelines. Send synopsis, three sample chapters and outline in the first instance.
Ⓔ Royalties twice-yearly.

The Cosmic Elk

68 Elsham Crescent, Lincoln LN6 3YS
☏ 01522 820922
post@cosmicelk.co.uk
www.cosmicelk.co.uk
www.cosmicelk.com
www.cosmicelk.org.uk
Contact *Heather Hobden*

Founded 1988. Publishes easily updated print-on-demand books on science, history and the history of science. Websites designed and maintained, with or without associated books, printed or pdf format. 'Please check the CosmicElk website first where you will find the information you need and contact us by e-mail through the website.'

Crescent Moon Publishing and Joe's Press

PO Box 393, Maidstone ME14 5XU
☏ 01622 729593
cresmopub@yahoo.co.uk
www.crescentmoon.org.uk
Managing Editor *Jeremy Robinson*

Founded 1988 to publish critical studies of figures such as D.H. Lawrence, Thomas Hardy, André Gide, Walt Disney, Rilke, Leonardo da Vinci, Mark Rothko, C.P. Cavafy and Hélène Cixous. Publishes literature, criticism, media, art, feminism, painting, poetry, travel, guidebooks, cinema and some fiction. Literary magazine, *Passion*, launched February 1994 (quarterly). *Pagan America*, twice-yearly anthology of American poetry. About 15–20 titles per year. Unsolicited synopses and ideas welcome but approach in writing first and send an s.a.e. Do not send whole mss.
Ⓔ Royalties negotiable.

Crime Express
▷ **Five Leaves Publications**

Day Books

Orchard Piece, Crawborough, Charlbury OX7 3TX
☏ 01608 811196
diaries@day-books.com
www.day-books.com
Managing Editor *James Sanderson*

Founded in 1997 to publish a series of great diaries from around the world. Unsolicited mss, synopses and ideas about diaries only, please. Include postage if return of material is required.

The Dragonby Press

30 King Street, Winterton DN15 9TP
☏ 01724 737254
rich@rahwilliams.orangehome.co.uk
Managing Editor *Richard Williams*

Founded 1987 to publish affordable bibliography for reader, collector and dealer. About 3 titles a year. Unsolicited mss, synopses and ideas welcome for bibliographical projects only.
Ⓔ Royalties paid.

Dramatic Lines

PO Box 201, Twickenham TW2 5RQ
☏ 020 8296 9502   ☏ 020 8296 9503
mail@dramaticlinespublishers.co.uk
www.dramaticlines.co.uk
Managing Editor *John Nicholas*

Founded to promote drama for all. Publications with a wide variety of theatrical applications including classroom use and school assemblies, drama examinations, auditions, festivals, theatre group performance, musicals and drama resource handbooks. Unsolicited drama-related mss, proposals and synopses welcome; enclose s.a.e.
Ⓔ Royalties paid.

Edgewell Publishing

5A Front Street, Prudhoe NE42 5HJ
☏ 01661 835330   ☏ 01661 835330
keithminton@btconnect.com
www.edgewell-publishing.co.uk
Editor *Keith Minton*

Quarterly short story and poetry magazine (*New Lively Tales*) to encourage writing from new and established writers, also children. Writers from the North East especially welcome. Please send by e-mail, where possible.
Ⓔ 'No royalties; writers maintain copyright.'

Educational Heretics Press

113 Arundel Drive, Bramcote, Nottingham NG9 3FQ
☏ 0115 925 7261   ☏ 0115 925 7261
www.edheretics.gn.apc.org
Directors *Janet & Roland Meighan*

Non-profit venture which aims to question the dogmas of schooling in particular and education in general, and establish the logistics of the next personalized learning system. No unsolicited material. Enquiries only.
Ⓔ Royalties paid on recent contracts.

Eilish Press

4 Collegiate Crescent, Broomhall Park, Sheffield S10 2BA
☏ 07973 353964
eilishpress@hotmail.co.uk
eilishpress.tripod.com

Head of Marketing *Suzi Kapadia*

Specializes in academic work and popular non-fiction in the areas of women's studies, human rights and anti-racism. Also produces children's literature with humanitarian themes. No adult fiction. Accepting no new projects in 2009.
Ⓔ Royalties annually.

Fand Music Press

Glenelg, 10 Avon Close, Petersfield GU31 4LG
☎ 01730 267341   📠 01730 267341
paul@fandmusic.com
www.fandmusic.com
Managing Editor *Peter Thompson*

Founded in 1989 as a sheet music publisher, Fand Music Press has expanded its range to include CD recordings and books on music. Also publishes poetry and short stories. 10 titles a year. No unsolicited mss. Write with ideas in the first instance.

Feather Books

PO Box 438, Shrewsbury SY3 0WN
☎ 01743 872177   📠 01743 872177
john@waddysweb.freeuk.com
www.waddysweb.freeuk.com
Managing Director *Rev. John Waddington-Feather*
Music Director *David Grundy*
Drama Director/Recordings Manager *Tony Reavil*

Founded 1980 to publish writers' group work. All material has a strong Christian ethos. Publishes poetry (mainly, but not exclusively, religious); Christian mystery novels (the Revd. D.I. Blake Hartley series); Christian children's *Quill Hedgehog* novels; seasonal poetry collections and *Feather's Miscellany*. Produces poetry, drama and hymn CD/cassettes; also the anthology 'Pilgrimages' ed. Professor Walter Nash ('a milestone in Christian poetry publishing'). About 20 titles a year. All correspondence to include s.a.e., please.

Fidra Books

219 Bruntsfield Place, Edinburgh EH10 4DH
☎ 0131 447 1917
vanessa@fidrabooks.com
www.fidrabooks.com
Managing Editor *Vanessa Robertson*

Founded 2005. Publishes children's fiction, specializing in reprints of sought-after 20th century children's fiction. No unsolicited material; initial approach by letter or e-mail.
Ⓔ Royalties quarterly.

Five Leaves Publications

PO Box 8786, Nottingham NG1 9AW
☎ 0115 969 3597
info@fiveleaves.co.uk
www.fiveleaves.co.uk

Contact *Ross Bradshaw*

Founded 1995 (taking over the publishing programme of Mushroom Bookshop). Publishes fiction, poetry, social history, young adult fiction and Jewish interest. Also publishes for the Anglo-Catalan Society. IMPRINT **Crime Express** Crime fiction novellas. Titles normally commissioned. About 15 titles a year.
Ⓔ Royalties paid.

Frontier Publishing

Windetts, Kirstead NR15 1EG
☎ 01508 558174
frontier.pub@macunlimited.net
www.frontierpublishing.co.uk
Managing Editor *John Black*

Founded 1983. Publishes non-fiction only: travel, photography, sculptural history and literature. 2–3 titles a year. No unsolicited mss; synopses and ideas welcome.
Ⓔ Royalties paid.

Galore Park Publishing Ltd

19/21 Sayer's Lane, Tenterden TN30 6BW
☎ 01580 764242   📠 01580 764242
info@galorepark.co.uk
www.galorepark.co.uk
Managing Director *Nicholas Oulton*

Founded 1999. Publishes school textbooks and children's books. 19 titles in 2008. No unsolicited mss. Synopses and ideas welcome; send by e-mail.
Ⓔ Royalties twice-yearly.

Glosa Education Organisation

PO Box 18, Richmond TW9 2GE
www.glosa.org
Managing Editor *Wendy Ashby*

Founded 1981. Publishes textbooks, dictionaries and translations for the teaching, speaking and promotion of Glosa (an international, auxiliary language); also a newsletter, *Plu Glosa Nota* and journal, *PGN*. Unsolicited mss and ideas for Glosa books welcome.

Godstow Press

60 Godstow Road, Wolvercote, Oxford OX2 8NY
☎ 01865 556215
info@godstowpress.co.uk
www.godstowpress.co.uk
Managing Editors *Linda Smith, David Smith*

Founded in 2003 for the publication of creative work with a philosophical/spiritual content. 2–3 titles a year. Considers works self-published by the author for inclusion in the catalogue. The press works on direct sales rather than through usual trade outlets. Synopses and ideas welcome but no unsolicited mss.

Graffeg

2 Radnor Court, 256 Cowbridge Road East,
Cardiff CF5 1GZ
☎ 029 2037 7312   ⊞ 029 2039 8101
info@graffeg.com
www.graffeg.com
Managing Editor *Peter Gill*

Founded 2002. Publishes travel, food and drink,
Wales, landscapes, photography. 8 titles in 2008.
Unsolicited material welcome by post.
Ⓔ Royalties twice-yearly.

Grant Books

The Coach House, New Road, Cutnall Green,
Droitwich WR9 0PQ
☎ 01299 851588   ⊞ 01299 851446
golf@grantbooks.co.uk
www.grantbooks.co.uk
Managing Editor *H.R.J. Grant*

Founded 1978. Publishes golf-related titles only:
course architecture, history, biography, etc., but
not instructional, humour or fiction material.
About 3 titles a year. Mss, synopses and ideas
welcome.
Ⓔ Royalties paid.

Great Northern Publishing

PO Box 202, Scarborough YO11 3GE
☎ 01723 581329   ⊞ 01723 581329
books@greatnorthernpublishing.co.uk
www.greatnorthernpublishing.co.uk
Production Manager/Senior Editor *Mark Marsay*

Small, independent, award-winning, family-
owned company founded in 1999. Publishers of
books, magazines and journals. Also provides
full book, magazine and print production
services to individuals, businesses, charities,
museums and other small publishers. Publishes
non-fiction (mainly military), adult erotic
material and, occasionally, fiction. No romance,
religious, political, feminist, New Age, medical
or children's books. Publishers of the bi-monthly
magazines, *The Great War* and *Jade* (see entries
under *Magazines*). Mail order and Internet-based
bookshop stocking selected titles alongside its
own. Website provides submission guidelines and
current publishing requirements. No unsolicited
phone calls or mss; send letter or text-only e-mail
in the first instance.
Ⓔ Royalties twice-yearly.

Griffin Publishing Associates

61 Grove Avenue, London N10 2AL
☎ 020 8444 0815/0778 842 5262
griffin.publishing.associates@googlemail.com
Managing Editor *Ali Ismail*

Founded 2007. Publishes all kinds of fiction
(plays, poetry and prose) with an emphasis on
science fiction, fantasy and horror. Unsolicited

mss, synopses and ideas for books welcome.
Approach by post (return postage must be
included) or telephone. (For Australian and New
Zealand writers and literary agents, postal contact
address: Seeham Moheed, GPA local agent, 58
Challenger Street, St Helier, Auckland, New
Zealand.
Ⓔ Royalties annually.

GSSE

11 Malford Grove, Gilwern, Abergavenny NP7 0RN
☎ 01873 830872
GSSE@zoo.co.uk
www.gsse.org.uk
Owner/Manager *David P. Bosworth*

Publishes books and booklets describing
classroom practice (at all levels of education
and training). Ideas welcome – particularly
from practising teachers, lecturers and trainers
describing how they use technology in their
teaching.
Ⓔ Royalties by arrangement.

Haunted Library

Flat 1, 36 Hamilton Street, Hoole, Chester
CH2 3JQ
☎ 01244 313685
pardos@globalnet.co.uk
www.users.globalnet.co.uk/~pardos/GS.html
Managing Editor *Rosemary Pardoe*

Founded 1979. Publishes the *Ghosts and Scholars
M.R. James Newsletter* twice a year, featuring
articles, news and reviews (no fiction).
Ⓔ Royalties not paid.

Headpress

Suite 306, The Colourworks, 2a Abbot Street,
London E8 3DP
☎ 0845 330 1844
office@headpress.com
www.headpress.com
Managing Editor *David Kerekes*

Founded 1991. Publishes *Headpress* journal and
books devoted to film, music, popular culture,
the strange and esoteric. No fiction or poetry.
Unsolicited material welcome; send letter and
outline in the first instance.
Ⓔ Royalties/flat fees vary.

Heart of Albion Press

62 Wartnaby Street, Market Harborough LE16 9BE
☎ 01858 431717
albion@indigogroup.co.uk
www.hoap.co.uk
Managing Editor *R.N. Trubshaw*

Founded 1990 to publish local history. Publishes
folklore, local history and mythology. Synopses
relating to these subjects welcome.
Ⓔ Royalties negotiable.

Ignotus Press

BCM—Writer, London WC1N 3XX
☎ 00 353 625 6582
ignotuspress@eircom.net
www.ignotuspress.com
Commissioning Editor *Suzanne Ruthven*

Founded 1994. Publishes metaphysical non-fiction
and fiction. 5 titles in 2008. No unsolicited mss;
send outline by e-mail or post after reading
submission guidelines on the website.
🄯 Royalties twice-yearly.

Immanion Press

8 Rowley Grove, Stafford ST17 9BJ
☎ 01785 613299
editorial@immanion-press.com
www.immanion-press.com
Managing Editor *Storm Constantine*

Founded 2003. Publishes genre fiction (science
fiction, fantasy, horror), slipstream fiction;
esoteric non-fiction. IMPRINT **Megalithica Books**
Commissioning Editor *Taylor Ellwood* (info@
immanion-press.com) Non-fiction, specifically
cutting edge esoteric work. Submissions by
e-mail or post (enclose s.a.e.). Website gives full
submission details.
🄯 Royalties paid.

Indepenpress

▷ **Pen Press & Indepenpress Publishing Ltd**

Infinity Junction

PO Box 64, Neston DO CH64 0WB
infin-info@infinityjunction.com
www.infinityjunction.com
Managing Editor *Neil Gee*

Established originally as a self-publishing,
self-help organization in 1999, Infinity Junction
became a publisher in 2001. Currently offers a
variety of services to authors, including advice,
free web display space and full commercial
publication. Authors are strongly advised to read
the detailed information on the website before
making contact. E-mail communication preferred.
'We cannot undertake to read unsolicited whole
mss.'

Inner Sanctum Publications

75 Greenleaf Gardens, Polegate BN26 6PQ
☎ 01323 484058
mary@maryhession.com
www.maryhession.com/
innersanctumpublications.htm
Managing Editor *Mary Hession*

Founded in 1999 to publish spiritual books.

Iolo

38 Chaucer Road, Bedford MK40 2AJ
☎ 01234 301718/07909 934866   🖷 01234 301718

dedwydd1@ntlworld.com
www.dedwyddjones.co.uk
Managing Director *Dedwydd Jones*

Publishes Welsh theatre-related material and
campaigns for a Welsh National Theatre. Ideas on
Welsh themes welcome; approach in writing.

Ivy Publications

72 Hyperion House, Somers Road, London
SW2 1HZ
☎ 020 8671 6872
Proprietor *Ian Bruton-Simmonds*

Founded 1989. Publishes educational, science,
fiction, philosophy, children's, travel, literary
criticism, history, film scripts. No unsolicited
mss; send two pages, one from the beginning and
one from the body of the book, together with
synopsis (one paragraph) and s.a.e. No cookery,
gardening or science fiction.
🄯 Royalties annually.

The Jupiter Press

The Coach House, Gatacre, Claverley, Nr
Wolverhampton WV5 7AW
☎ 01746 710694   🖷 01746 710158
gordonthomasdrury@btinternet.com
Managing Editor *Gordon Thomas Drury*

Founded 1995. Looking for niche market and
information publications, also quiz and game
subjects. Interested in holistic, clairvoyance and
esoteric subjects. Synopses and ideas for books
welcome. Send idea and sample chapter (include
contact telephone number and return postage).
Unsolicited mss welcome 'but may request a
reader's fee'. Currently seeking books and ideas
for SERIES entitled *How To Make Money At ....*
🄯 Royalties paid.

Kittiwake

3 Glantwymyn Village Workshops, Glantwymyn,
Nr Machynlleth SY20 8LY
☎ 01650 511314   🖷 01650 511314
perrographics@btconnect.com
www.kittiwake-books.com
Managing Editor *David Perrott*

Founded 1986. Publishes guidebooks only, with an
emphasis on careful design/production. Specialist
research, writing, cartographic and electronic
publishing services available. Unsolicited mss,
synopses and ideas for guidebooks welcome.
🄯 Royalties paid.

The Lindsey Press

Unitarian Headquarters, 1–6 Essex Street, Strand,
London WC2R 3HY
☎ 020 7240 2384   🖷 020 7240 3089
info@unitarian.org.uk
Convenor *Kate Taylor*

Established at the end of the 18th century as a vehicle for disseminating liberal religion. Adopted the name of The Lindsey Press at the beginning of the 20th century (after Theophilus Lindsey, the great Unitarian Theologian). Publishes books reflecting liberal religious thought or Unitarian denominational history. Also worship material – hymn books, collections of prayers, etc. No unsolicited mss; synopses and ideas welcome. Ⓔ Royalties not paid.

The Linen Press

75c (13) South Oswald Road, Edinburgh EH9 2HH
lynn@linenpressbooks.co.uk
www.linenpressbooks.co.uk
Managing Director *Lynn Michell*

Founded 2006. Small, independent publishing house based in Edinburgh run by women for women. 'We are keen to champion new literary fiction, international writers, innovative prose and women writers from minority groups. We will consider fiction and non-fiction but not short stories, children's books, science fiction, romance or poetry'. 4 titles a year. Happy to receive submissions from emerging as well as established writers as long as they fulfil the criteria and ethos of the Linen Press. Please read the submission guidelines before sending in a proposal. Synopsis only in the first instance.

Logaston Press

Little Logaston, Logaston, Woonton, Almeley HR3 6QH
Ⓣ 01544 327344
logastonp@aol.com
www.logastonpress.co.uk
Managing Editors *Andy Johnson, Karen Stout*

Founded 1985. Publishes history, social history, archaeology, art, architecture, guides to rural West Midlands, Central and South Wales. 20 titles in 2008. Unsolicited material welcome but prefers initial phone call or e-mail to ensure subject is one they might consider.
Ⓔ Royalties paid on publication.

Lyfrow Trelyspen

The Roseland Institute, Gorran, St Austell PL26 6NT
Ⓣ 01726 843501   Ⓕ 01726 843501
trelispen@care4free.net
www.theroselandinstitute.co.uk
Managing Editor *Dr James Whetter*

Founded 1975. Publishes works on Cornish history, biography, essays, etc. Also **CNP Publications** which publishes the quarterly journal *The Cornish Banner/An Baner Kernewek*. 1–2 titles a year. Unsolicited mss, synopses and ideas welcome.
Ⓔ Royalties not paid.

The Maia Press

82 Forest Road, London E8 3BH
Ⓣ 020 7683 8141   Ⓕ 020 7683 8141
maggie@maiapress.com
www.maiapress.com
Contacts *Maggie Hamand, Jane Havell*

Founded 2002. Publishes fiction by known and new writers. 6 titles a year. No unsolicted mss, write with c.v. and synopsis only. S.a.e. essential.
Ⓔ Royalties twice-yearly.

Maypole Editions

18 The Lindens, Shifnal TF11 8AB
Managing Editor *Barry Taylor*

Founded 1987. Publishes fiction, drama, filmscripts and poetry. No submissions until 2010 at which time an s.a.e. will be required. 'No reply means no. Sounds harsh but I'm drowning in poets'.

Megalithica Books
▷ **Immanion Press**

Mercia Cinema Society

29 Blackbrook Court, Durham Road, Loughborough LE11 5UA
Ⓣ 01509 218393
Mervyn.Gould@virgin.net
www.merciacinema.org
Managing Editor *Kate Taylor*
Series Editor *Mervyn Gould*

Founded 1980 to foster research into the history of picture houses. Publishes books and booklets on the subject, including cinema circuits and chains. Books are often tied in with specific geographical areas. Quarterly journal: *Mercia Bioscope*, editor *Paul Smith* (Paul@Smith2004. fs.net.co.uk). 63 titles to date. Unsolicited mss (preferably on disk); synopses and ideas.
Ⓔ Royalties paid. Six free copies to the author.

Meridian Books

40 Hadzor Road, Oldbury B68 9LA
Ⓣ 0121 429 4397
meridian.books@btopenworld.com
Managing Editor *Peter Groves*

Founded 1985 as a small home-based enterprise following the acquisition of titles from Tetradon Publications Ltd. Publishes walking and regional guides. 4–5 titles a year. Unsolicited mss, synopses and ideas welcome if relevant. Send s.a.e. if mss to be returned.
Ⓔ Royalties paid.

Neil Miller Publications

c/o Ormonde House, 49 Ormonde Road, Hythe CT21 6DW
contact@neilmillersbooks.co.uk
www.neilmillersbooks.co.uk

Also at: 4 Rue D'Equirre Bergueneuse, Pas de Calais, France 62134
Managing Editor *Neil Miller*

Founded 1994. Publishes novels and collections of short stories on any subject, fact or fiction: memoirs, autobiography, mystery, horror, war, comedy, science fiction. Interested in working with new and published writers. Evaluation and critique service available. 'Send your novel or selection of short stories for consideration plus our reader's fee of £9.75. For this you receive a brief critique plus two of our latest authors' books and have the opportunity of being published by us. European authors receive one novel.' Enclose s.a.e. for return of work. No unrequested tales.

Mirage Publishing

PO Box 161, Gateshead NE8 1WY
☎ 0870 720 9498
sales@miragepublishing.com
www.miragepublishing.com
Artistic Director *Sharon Anderson*

Founded 1998. Publishes mind, body and spirit, autobiography, biography. 9 titles in 2008. No unsolicited mss; send synopses and ideas by mail or e-mail.
⊞ Royalties twice-yearly.

Natzler Enterprises

1 Wakeford Cottages, Selden Lane, Worthing BN11 2LQ
☎ 01903 211785
info@natzler.com
www.natzler.com
Managing Editor *Paul Gordon*

Founded 1993. Publishes magic and conjuring books. 9 titles in 2008. No unsolicited material; letter of enquiry only.

Need2Know

Remus House, Coltsfoot Drive, Woodston PE2 9JX
☎ 01733 898103   ☏ 01733 313524
sales@n2kbooks.com
www.n2kbooks.com

Founded 1995 'to fill a gap in the market for self-help books'. Need2Know is an imprint of **Forward Press** (see entry under *Poetry Presses*). Publishes contemporary health and lifestyle issues. No unsolicited mss. Call in the first instance.
⊞ Advance plus 15% royalties.

Jane Nissen Books

Swan House, Chiswick Mall, London W4 2PS
☎ 020 8994 8203   ☏ 020 8742 8198
jane@nissen.demon.co.uk
www.janenissenbooks.co.uk
Publisher *Jane Nissen*

Founded 2000. Publishes reprints of children's fiction only. Winner of the 2007 Eleanor Farjeon Award. 3 titles in 2008; list of 30 titles. *Does not consider submitted material* as all titles published are reprints of classics or forgotten children's books.

North Staffordshire Press Ltd

Staffordshire University, Business Village, 72 Leek Road, Stoke on Trent ST4 2AR
☎ 01782 442831
malcolm.henson@editorial.staffs.com
www.northstaffordshiriepress.com
Managing Editor *Malcolm Henson*
E-learning Consultant *Maria De Marks*

Founded 2006. Publishes sociology, history, sport, poetry, fiction and academic books. 2 titles in 2008. Unsolicited mss, synopses and ideas welcome; approach by letter, e-mail or phone call.
⊞ Royalties annually.

The Nostalgia Collection

Silver Link Publishing Ltd, The Trundle, Ringstead Road, Great Addington, Kettering NN14 4BW
☎ 01536 330588   ☏ 01536 330588
sales@nostalgiacollection.com
www.nostalgiacollection.com
Managing Editor *Will Adams*

Founded 1985. Small independent company specializing in illustrated post-war nostalgia titles including railways, trams, ships and other transport subjects, towns and cities, villages and rural life, rivers and inland waterways, and industrial heritage under the **Silver Link** and **Past and Present** imprints.
⊞ Fees/royalties paid.

Nyala Publishing

4 Christian Fields, London SW16 3JZ
☎ 020 8764 6292/0115 981 9418
☏ 020 8764 6292/0115 981 9418
nyala.publishing@geo-group.co.uk
www.geo-group.co.uk
Editorial Director *J.F.J. Douglas*

Founded 1996. Publishing arm of Geo Group. Publishes a limited range of biography, travel and general non-fiction. Offers a wide range of printing and publishing services. Reading, editing and proofreading. No unsolicited mss; synopses and ideas considered.
⊞ Royalties annually.

Orpheus Publishing House

4 Dunsborough Park, Ripley Green, Ripley, Guildford GU23 6AL
☎ 01483 225777   ☏ 01483 225776
orpheuspubl.ho@btinternet.com
Managing Editor *J.S. Gordon*

Founded 1996. Publishes 'well-researched and properly argued' books in the fields of occult

science, esotericism and comparative philosophy/ religion. 'Keen to encourage good (but sensible) new authors.' In the first instance, send maximum three-page synopsis with s.a.e.
Ⓔ Royalties by agreement.

Packard Publishing Limited

Forum House, Stirling Road, Chichester PO19 7DN
Ⓣ 01243 537977 Ⓕ 01243 537977
info@packardpublishing.co.uk
www.packardpublishing.com
Chairman/Managing Director *Michael Packard*

Founded in 1977 to distribute overseas publishers' lists in biology, biochemistry and ecology. Publishes academic & professional: school/ university interface, postgraduate – mainly in land management and applied ecology; agriculture, forestry, nature conservation, rural studies, landscape architecture and garden design; some languages (French, Arabic). SERIES Instructional books on garden and landscape design, and monographs or critical biographies in garden and landscape design and history. 32 titles to date. No unsolicited mss. Synopses and ideas in relevant areas welcome. Telephone first.
Ⓔ Royalties twice-yearly in first year, then annually.

Panacea Press Limited

86 North Gate, Prince Albert Road, London NW8 7EJ
Ⓣ 020 7722 8464 Ⓕ 020 7586 8187
ebrecher@panaceapress.net
www.panaceapress.net
Managing Editor *Erwin Brecher, PhD*

Founded as a self-publisher but now open for non-fiction from other authors. Material of academic value considered provided it commands a wide general market. No unsolicited mss; synopses and ideas welcome. No telephone calls. Approach by e-mail, fax or letter.
Ⓔ Royalties annually.

Parapress

The Basement, 9 Frant Road, Tunbridge Wells TN2 5SD
Ⓣ 01892 512118 Ⓕ 01892 512118
office@parapress.myzen.co.uk
www.parapress.co.uk
Managing Editor *Elizabeth Imlay*

Founded 1993. Publishes animals, autobiography, biography, history, literary criticism, military and naval, music, self-help. Some self-publishing. About 2 titles a year.

Past and Present
▷ The Nostalgia Collection

Paupers' Press

37 Quayside Close, Turney's Quay, Trent Bridge, Nottingham NG2 3BP
Ⓣ 0115 986 3334 Ⓕ 0115 986 3334
books@pauperspress.com
www.pauperspress.com
Managing Editor *Colin Stanley*

Founded 1983. Publishes extended essays in booklet form (about 15,000 words) on literary criticism and philosophy. 'Sometimes we stray from these criteria and produce full-length books, but only to accommodate an exceptional manuscript.' Limited hardback editions of bestselling titles. Centre for Colin Wilson studies. Critical essays on his work encouraged. About 6 titles a year. No unsolicited mss but synopses and ideas for books welcome.
Ⓔ Royalties paid annually.

Peepal Tree Press Ltd
▷ entry under Poetry Presses

Pen Press & Indepenpress Publishing Ltd

25 Eastern Place, Brighton BN2 1GJ
Ⓣ 0845 108 0530 Ⓕ 01273 261434
info@penpress.co.uk
www.penpress.co.uk
Managing Director *Lynn Ashman*

Founded 1996. Publishing across a wide range of categories, Pen Press helps new authors to self-publish. Distribution, promotion and marketing included in self-publishing deal. Publisher, not author, pays for reprints and shares in sales revenue. Now offering traditional publishing under the **Indepenpress** and **Pulp Press** IMPRINTS. Ove 1,000 titles to date. Write, phone or e-mail for submission form and full details. Return form with full ms (digital copy preferred).
Ⓔ Royalties 45%, payable twice-yearly.

Perfect Publishers Ltd

23 Maitland Avenue, Cambridge CB4 1TA
Ⓣ 01223 424422 Ⓕ 01223 424414
editor@perfectpublishers.co.uk
www.perfectpublishers.co.uk
Managing Editor *S.N. Rahman*

A print on demand book publishing company, founded 2005. Offers self-publishing services. Publishes books of all genres. Submissions by e-mail or letter.
Ⓔ 'We pay 100% royalties twice-yearly.'

Pipers' Ash Ltd

'Pipers' Ash', Church Road, Christian Malford, Chippenham SN15 4BW
Ⓣ 01249 720563 Ⓕ 0870 0568916
pipersash@supamasu.com
www.supamasu.com
Managing Editor *Mr A. Tyson*

Founded 1976. The company's publishing activities include individual collections of contemporary short stories, science fiction short stories, poetry, plays, short novels, local histories, children's fiction, philosophy, biographies, translations and general non-fiction. 12 titles a year. Synopses and ideas welcome; 'new authors with potential will be actively encouraged'. Offices in New Zealand and Australia.
Ⓔ Royalties annually.

Playwrights Publishing Co.

70 Nottingham Road, Burton Joyce, Nottingham NG14 5AL
℡ 0115 931 3356
playwrightspublishingco@yahoo.com
geocities.com/playwrightspublishingco
Managing Editors *Liz Breeze, Tony Breeze*

Founded 1990. Publishes one-act and full-length plays online. Unsolicited scripts welcome. No synopses or ideas. Reading fees: £15 one act; £30 full length (waived if evidence of professional performance or if writer is unwaged). S.a.e. required.
Ⓔ Royalties paid.

Pomegranate Press

Dolphin House, 51 St Nicholas Lane, Lewes BN7 2JZ
℡ 01273 470100   📠 01273 470100
pomegranatepress@aol.com
www.pomegranate-press.co.uk
Managing Editor *David Arscott*

Founded in 1992 by writer/broadcaster David Arscott, who also administers the **Sussex Book Club**. Specializes in books about Sussex and self publishing. IMPRINT **Pomegranate Practicals** How-to books.
Ⓔ Royalties twice-yearly.

David Porteous Editions

PO Box 5, Chudleigh, Newton Abbot TQ13 0YZ
℡ 01626 853310
editorial@davidporteous.com
www.davidporteous.com
Publisher *David Porteous*

Founded 1992 to produce colour illustrated books on hobbies and leisure for the UK and international markets. Publishes crafts, hobbies, art techniques and needlecrafts. No poetry or fiction. 3–4 titles a year. Unsolicited mss, synopses and ideas welcome if return postage included.
Ⓔ Royalties twice-yearly.

Praxis Books

Crossways Cottage, Walterstone HR2 0DX
℡ 01873 890695
rebeccatope@btinternet.com
www.rebeccatope.com

Proprietor *Rebecca Smith*

Founded 1992. Publishes reissues of the works of Sabine Baring-Gould, memoirs, diaries and local interest. 22 titles to date. Unsolicited mss accepted with s.a.e. No fiction. Editing and advisory service available. Funding negotiable. 'I am most likely to accept work with a clearly identifiable market.'

Pulp Press
▷ Pen Press & Indepenpress Publishing Ltd

QueenSpark Books

Room 211, 10–11 Pavilion Parade, Brighton BN2 1RA
℡ 01273 571710
info@queensparkbooks.org.uk
www.queensparkbooks.org.uk

A community writing and publishing group. Since the early 1970s, QueenSpark Books has published 90 titles, mainly featuring the lives of local people. Writing workshops and groups held on a regular basis. No unsolicited mss, please.

Radikal Phase Publishing House Ltd

Willow Court, Cordy Lane, Underwood NG16 5FD
℡ 01773 764288   📠 01773 764282
sales@phaseprint.com
www.phaseprint.com
Managing Director *Kevin Marks*

Founded 2001. Publishes radical revelation and technical electrical books. IMPRINTS **William Ernest** *Kevin Marks*; **Radikal Phase** *Philip Gardiner*. Welcomes unsolicited material; approach in writing in the first instance.
Ⓔ Royalties twice-yearly.

Robinswood Press

30 South Avenue, Stourbridge DY8 3XY
℡ 01384 397475   📠 01384 440443
info@robinswoodpress.com
www.robinswoodpress.com
Managing Editor *Christopher J. Marshall*

Founded 1985. Publishes children's, educational, teaching resources, SEN. Also collaborative publishing, e.g., with Camphill Foundation. About 30–50 titles a year. Unsolicited mss, synopses and ideas welcome. See website.
Ⓔ Royalties paid.

Rockpool Children's Books Ltd

15 North Street, Marton CV23 9RJ
℡ 01926 633114
stuart@rockpoolchildrensbooks.com
www.rockpoolchildrensbooks.com
Managing Director *Stuart Trotter*

Established 2006. Publishes children's picture books, novelty, board books. 17 titles published to date. Unsolicited mss, synopses and ideas for books welcome; send by e-mail but ensure all illustration samples are low res files (PDF or JPEG). No short stories, novels or non-fiction.

Serif

47 Strahan Road, London E3 5DA
☏ 020 8981 3990 📠 020 8981 3990
stephen@serif.books.co.uk
www.serifbooks.co.uk
Managing Editor *Stephen Hayward*

Founded 1993. Publishes cookery, Irish and African studies, travel writing and modern history; no fiction.
ⓔ Royalties paid.

Silver Link

▷ **The Nostalgia Collection**

Spacelink Books

115 Hollybush Lane, Hampton TW12 2QY
☏ 020 8979 3148
www.spacelink.fsworld.co.uk
Managing Director *Lionel Beer*

Founded 1967. Named after a UFO magazine published in the 1960/70s. Publishes non-fiction titles connected with UFOs, Fortean phenomena and paranormal events. Publishers of *TEMS News* for the Travel and Earth Mysteries Society. Distributors of a wide range of related titles and magazines. No unsolicited mss; send synopses and ideas – non-fiction.
ⓔ Royalties/fees according to contract.

Stenlake Publishing Limited

54–58 Mill Square, Catrine KA5 6RD
☏ 01290 552233
enquiries@stenlake.co.uk
www.stenlake.co.uk

Publishes illustrated local history, railways, shipping, aviation and industrial. IMPRINT **Alloway Publishing Limited** acquired in January 2008. 22 titles in 2008. Unsolicited mss, synopses and ideas welcome if accompanied by s.a.e. or sent by e-mail. Freelance writers with experience in above fields also sought for specific commissions.
ⓔ Royalties or fixed fee paid.

Stone Flower Limited

PO Box 1513, Ilford IG1 3QU
renalomen@hotmail.com
Managing Editor *L.G. Norman*

Founded 1989. Publishes humour and general fiction. No submissions. All material commissioned or generated in-house.

Superscript

404 Robin Square, Newtown SY16 1HP
☏ 01588 650452
editor@superscript.org.uk
www.superscript.org.uk
www.dubsolution.org
Editor *Max Fuller*
Chair *Anand Somesh*
Secretary *Ray Pahl*

Founded 2002. Publishes literary, philosophical and political fiction, humanities and social sciences. 15 titles in 2008. Initial enquiries by letter, telephone or e-mail to the editor.
ⓔ Royalties usually 10% RRP but vary with each contract.

Sylph Editions

5 St Swithun's Terrace, Lewes BN7 1UJ
☏ 01273 471706 📠 01273 808247
or@sylpheditions.com
www.sylpheditions.com
Publisher *Ornan Rotem*
Associate Publisher *Num Stibbe*

Founded 2006. Publishes limited-edition art and photography books, monographs, theoretical essays and different forms of experimental writing. 9 titles in 2008. Ideas welcome; approach by e-mail.
ⓔ Royalties twice-yearly.

Tarquin Publications

99 Hatfield Road, St Albans AL1 4JL
☏ 01727 833866 📠 0845 456 6385
info@tarquinbooks.com
www.tarquinbooks.com
Managing Editor *Andrew Griffin*

Founded 1970 as a hobby which gradually grew and now publishes mathematical, cut-out models, teaching and pop-up books. Other topics covered if they involve some kind of paper cutting or pop-up scenes. About 5 titles a year. No unsolicited mss; letter with 1–2 page synopsis welcome.
ⓔ Royalties paid.

Tartarus Press

Coverley House, Carlton-in-Coverdale, Leyburn DL8 4AY
☏ 01969 640399 📠 01969 640399
tartarus@pavilion.co.uk
www.tartaruspress.com
Proprietor *Raymond Russell*
Editor *Rosalie Parker*

Founded 1987. Publishes fiction, short stories, reprinted classic supernatural fiction and reference books. About 10 titles a year. 'Please do not send submissions. We cater to a small, collectable market and commission the fiction we publish.'

Tharpa Publications

Conishead Priory, Ulverston LA12 9QQ
☏ 01229 588599 📠 01229 483919

info.uk@tharpa.com
www.tharpa.com
Managing Editor *Neil Elliott*

Founded 1980. Publishes non-fiction, religion, Buddhism and meditation. No unsolicited material.

Tlön Books Publishing Ltd

64 Arthurdon Road, London SE4 1JU
☎ 020 8690 0642　📠 020 8690 0642
mark.reid@tlon.co.uk
www.tlon.co.uk
Managing Editor *Jason Shelley*

Founded 2002. Publishes fiction, art, photography, poetry and children's books. Unsolicited mss, synopses and ideas welcome; approach by e-mail.
💷 Royalties twice-yearly.

Wakefield Historical Publications

19 Pinder's Grove, Wakefield WF1 4AH
☎ 01924 372748
kate@airtime.co.uk
Managing Editor *Kate Taylor*

Founded 1977 by the Wakefield Historical Society to publish well-researched, scholarly works of regional (namely West Riding) historical significance. 1–2 titles a year. Unsolicited mss, synopses and ideas for books welcome.
💷 Royalties not paid.

Watling Street Publishing Ltd

33 Hatherop, Nr Cirencester GL7 3NA
☎ 01285 750212
chris.mclaren@saltwaypublishing.co.uk
Managing Director *Chris McLaren*
Editorial Head *Christine Kidney*

Founded 2001. Publishes non-fiction titles on London and its history; children's and adult. No unsolicited mss. Ideas and synopses welcome. Approach in writing or by e-mail with a one-page proposal. 'Not interested in anything not related to London.'

Waywiser Press

The Cottage, 14 Lyncroft Gardens, Ewell KT17 1UR
☎ 020 8393 7055　📠 020 8393 7055
waywiser-press@aol.com
waywiser-press.com/
Managing Editor *Philip Hoy*
Editorial Advisers *Joseph Harrison, Clive Watkins, Greg Williamson*

Founded 2002. An independent press, publishing poetry, fiction, criticism and other kinds of literary work by new as well as established writers, as well as a series of in-depth interviews with senior contemporary poets (under the press's IMPRINT **Between The Lines**). The press also runs an annual literary award, the **Anthony Hecht**

Poetry Prize (see entry under *Prizes*). Submission details available on the website.
💷 Royalties paid.

Whitchurch Books Ltd

67 Merthyr Road, Whitchurch, Cardiff CF14 1DD
☎ 029 2052 1956
whitchurchbooks@btconnect.com
Managing Director *Gale Canvin*

Founded 1994. Publishes local interest, particularly local history. Welcomes unsolicited mss, synopses and ideas on relevant subjects. Initial approach by phone or in writing.
💷 Royalties twice-yearly.

Whittles Publishing

Dunbeath Mains Cottages, Dunbeath KW6 6EY
☎ 01593 731333　📠 01593 731400
info@whittlespublishing.com
www.whittlespublishing.com
Publisher *Dr Keith Whittles*

Publisher in geomatics, civil and structural engineering, geotechnics, manufacturing and materials technology and fuel and energy science. Also publishes non-technical books within the following areas: Asian connections, maritime, pharology, military history, landscape and nature writing. Unsolicited mss, synopses and ideas welcome on appropriate themes.
💷 Royalties annually.

William Ernest
▷ **Radikal Phase Publishing House Ltd**

Willow Bank Publishers Ltd

E–Space North, 181 Wisbech Road, Littleport, Ely CB6 1RA
☎ 01353 687997
editorial@willowbankpublishers.co.uk
www.willowbankpublishers.co.uk
Managing Editor *Christopher Sims*

Founded 2005. Publishes all genres, fiction and non-fiction. Unsolicited mss, synopses and ideas for books welcome; send via post in the first instance.
💷 Royalties annually.

Witan Books

Cherry Tree House, 8 Nelson Crescent, Cotes Heath, via Stafford ST21 6ST
☎ 01782 791673
witan@mail.com
Director & Managing Editor *Jeff Kent*

Founded in 1980 for self-publishing and commenced publishing other writers in 1991. Publishes general books, including biography, education, environment, geography, history, politics, popular music and sport. 1 or 2 titles

a year. Unsolicited mss, synopses and ideas welcome (include s.a.e.).
€ Royalties paid.

Worple Press

PO Box 328, Tonbridge TN9 1WR
☎ 01732 368958
theworpleco@aol.com
www.theworplepress.co.uk
Managing Editors *Peter Carpenter, Amanda Knight*

Founded 1997. Independent publisher specializing in poetry, art and alternative titles. Write or phone for catalogue and flyers. 4 titles a year. No unsolicited mss.

Zymurgy Publishing

Hoults Estate, Walker Road, Newcastle upon Tyne NE6 2HL
☎ 0191 276 2425   ⊞ 0191 276 2425
martin.ellis@ablibris.com
zymurgypublishing.com
Chairman *Martin Ellis*

Founded 2000. Publishes adult non-fiction, ranging from full colour illustrated hardbacks to mass market paperbacks. Synopses and ideas for books welcome; initial contact by telephone or e-mail. No unsolicited mss.
€ Royalties twice-yearly.

Audio Books

BBC Audiobooks Ltd

St James House, The Square, Lower Bristol Road, Bath BA2 3BH
☎ 01225 878000   ℻ 01225 310771
bbcaudiobooks@bbc.co.uk
www.bbcworldwide.com
Managing Director *Paul Dempsey*
Publishing Director *Jan Paterson*

BBC Audiobooks Ltd was established in 2003 with the integration of BBC Radio Collection, Cover to Cover and Chivers Audio Books. Since then it has become a leading trade and library publisher in the UK, publishing a wide range of entertainment on a variety of formats including CD, cassette, MP3-CD and downloads, podcasts, etc.

DIVISIONS

Trade Titles range from original books and full-cast radio dramas through to abridged readings and full, unabridged recordings. TITLES *War and Peace*; *The Odyssey*; *Captain Corelli's Mandolin*; the complete Sherlock Holmes canon. IMPRINTS **BBC Audio – Children's** From pre-school nursery rhymes to modern classics. TITLES *The Chronicles of Narnia*; *His Dark Materials* trilogy. **BBC Audio – Radio Collection** Comedy classics from TV and radio. TITLES *The Goons*; *Hancock's Half Hour*; *Fawlty Towers*; *The News Quiz*; *Just a Minute*. **BBC Audio** Contemporary comedy. TITLES *The Mighty Boosh*; *Have I Got News For You*; *Little Britain*. **BBC Audio – Lifestyle** A range of titles to help the listener improve or learn. TITLES include the *Glenn Harrold's Ultimate Guide to ...* series. **BBC Audio – Radio 4** TITLES *This Sceptred Isle*; *Letter from America*; *Churchill Remembered*. **Science Fiction & Fantasy** TITLES *The Hitchhiker's Guide to the Galaxy*; *Doctor Who*; *Journey into Space*; *The Lord of the Rings*.

Library IMPRINTS **Chivers Audio Books**; **Chivers Children's Audio Books** Founded 1980. Publishes over 500 titles annually. Complete and unabridged books for adults and children are sold to libraries and to the public by direct mail through **The Audiobook Collection**. CDs and cassettes covering bestselling fiction and popular non-fiction.

Bloomsbury Publishing
▷ entry under UK Publishers

Chivers Audio Books/Chivers Children's Audio Books
▷ BBC Audiobooks Ltd

Chrome Dreams

12 Seaforth Avenue, New Malden KT3 6JP
☎ 020 8715 9781   ℻ 020 8241 1426
mail@chromedreams.co.uk
www.chromedreams.co.uk
Managing Director *Rob Johnstone*

A small record company and publisher founded 1998 to produce audio-biographies of current rock and pop artists and legendary performers on CD and, more recently, books on the same subjects. Ideas for biographies welcome.

Corgi Audio

Transworld Publishers, 61–63 Uxbridge Road, London W5 5SA
☎ 020 8579 2652
www.booksattransworldpublishers.co.uk
Managing Director *Larry Finlay*
Publisher *Bill Scott-Kerr*

A division of **The Random House Group Ltd**. Publishes fiction, autobiography, humour and travel writing on audio. TITLES include Terry Pratchett's *Discworld* series; Bill Bryson's *The Life and Times of the Thunderbolt Kid* and *A Short History of Nearly Everything*; Sophie Kinsella's *Shopaholic* series and *The Lollipop Shoes* by Joanne Harris.

Creative Content

8 The Cornfields, Hatch Warren, Basingstoke RG22 4QB
☎ 01256 466641
ali@creativecontentdigital.com
www.creativecontentdigital.com
Managing Director *Alison Muirden*

Established in 2008, Creative Content is a joint partnership company owned by Ali Muirden, former Audio Publisher at Macmillan and Lorelei

King, a multi award winning voice artist and actor. Publishes business, self improvement/self help digital download audio publications. TITLES *Business Etiquette – China* Florian Loloum; *Business Etiquette – Russia* Charles McCall and Slava Katamidze; *Improve Your Speech – British English* David Gwillam and Deirdra Morris; *Improve Your Speech – US English* Mark Caven; *Top Tips for Wannabe CEOs* Richard Charkin and Richard Pettinger; *Business Etiquette India* Mike Barnard. 4 titles in 2008. Ideas welcome from agents and authors.

CSA Word

6a Archway Mews, 241a Putney Bridge Road, London SW15 2PE
☎ 020 8871 0220   📠 020 8877 0712
info@csaword.co.uk
www.csaword.co.uk
Managing Director *Clive Stanhope*
Audio Director *Victoria Williams*
Editorial and Production Assistant *Vanessa Brown*

Founded 1989. Publishes fiction, non-fiction, children's, short stories, poetry, travel, biographies, classics. Over 170 titles to date on CD and available as downloads. Tends to favour quality/classic/nostalgic/timeless literature. TITLES include *Carry on Jeeves* P.G. Wodehouse; *Room with a View* E.M. Forster; *Brideshead Revisited* Evelyn Waugh; *William's Happy Days* Richmal Crompton; *Lady Chatterley's Lover* D.H. Lawrence; *Bunter Does His Best* Frank Richards; *The Prince* Nicolo Machiavelli; *Seven Pillars of Wisdom* T.E. Lawrence. Also produces programmes for the BBC – ideas welcome – see entry under *Film, TV and Radio Producers*.

CYP

The Fairway, Bush Fair, Harlow CM18 6LY
☎ 01279 444707   📠 01279 445570
enquiries@cyp.co.uk
www.kidsmusic.co.uk
Operations Director *Mike Kitson*

Founded 1978. Publishes music DVDs and audiobooks for young children; educational, entertainment. TV music and soundtrack production.

57 Productions
▶ entry under Organizations of Interest to Poets

Hachette Digital

100 Victoria Embankment, London EC4Y 0DY
☎ 020 7911 8044   📠 020 7911 8100
sarah.shrubb@littlebrown.co.uk
www.littlebrown.co.uk
Publisher *Ursula Mackenzie*
Editor *Sarah Shrubb*

Launched in 2003 with titles from bestselling authors Alexander McCall Smith and Mitch Albom. Publishes fiction, humour, poetry and non-fiction. TITLES include new fiction from authors such as Patricia Cornwell and Sarah Waters, non-fiction such as Sharon Osbourne's bestselling autobiography *Extreme*, and unabridged classics such as Patricia Highsmith's *The Talented Mr Ripley* and Joseph Heller's *Catch 22*. A list of over 200 titles.

HarperCollins AudioBooks

77–85 Fulham Palace Road, London W6 8JB
☎ 020 8741 7070
www.harpercollins.co.uk
Head of Audio *Jo Forshaw*

The HarperCollins audio list was launched in 1990. Publishes a wide range including popular and classic fiction, non-fiction, children's, Shakespeare, poetry and self help on cassette, CD and digital download. AUTHORS Agatha Christie, Ian McEwan, C.S. Lewis, Roald Dahl, Dr Seuss, Lemony Snicket, J.R.R. Tolkien, Bernard Cornwell, Enid Blyton, Nick Butterworth, Judith Kerr.

Hodder & Stoughton Audiobooks

338 Euston Road, London NW1 3BH
☎ 020 7873 6000   📠 020 7873 6194
rupert.lancaster@hodder.co.uk
www.hodder.co.uk
Publisher *Rupert Lancaster*

Launched in 1994 with the aim of publishing outstanding authors from within the Hodder group and commissioning independent audio titles. Publishes fiction and non-fiction. AUTHORS include Stephen King, John LeCarré, Alan Titchmarsh, Elizabeth George, Joanne Harris, Haruki Murakami, Jodi Picoult, Pam Ayres, David Mitchell, Linda Smith, John Humphries, Al Murray, Dickie Bird, Peter Robinson and Charles Frazier.

Macmillan Digital Audio

20 New Wharf Road, London N1 9RR
☎ 020 7014 6000   📠 020 7014 6141
www.panmacmillan.co.uk
Owner *Macmillan Publishers Ltd*
Digital Director *Sara Lloyd*

Founded 1995. Publishes adult fiction, non-fiction and autobiography, focusing mainly on lead book titles and releasing audio simultaneously with hard or paperback publication. Also publishes children's audio titles. Won Audio Publisher of the Year at the 2003 Spoken Word Awards and has won many other Spoken Word Awards for individual audio titles in previous years About 40–60 titles a year. Submissions not accepted.

Naxos AudioBooks

40a High Street, Welwyn AL6 9EQ
☎ 01438 717808   📠 01438 717809
naxos_audiobooks@compuserve.com
www.naxosaudiobooks.com
Owner *HNH International, Hong Kong/Nicolas Soames*
Managing Director *Nicolas Soames*

Founded 1994. Part of Naxos, the classical budget CD company. Publisher of the Year in the 2001 Spoken Word Awards. Publishes classic and modern fiction, non-fiction, young adult and junior classics, drama and poetry. TITLES *Ulysses* Joyce; *King Lear* Shakespeare; *History of the Musical* Fawkes; *Just So Stories* Kipling. A list of 400 titles.

Oakhill Publishing Ltd

PO Box 3855, Bath BA1 3WW
☎ 01225 874355   📠 01225 874442
info@oakhillpublishing.com
Managing Director *Julian Batson*

Founded in 2005 by Julian Batson, formerly with Chivers Press (now trading as **BBC Audiobooks**) to offer a wider selection of titles for the library market. Produces complete and unabridged audiobooks of popular fiction and non-fiction for adults and children. TITLES *The Gathering* Anne Enright (Man Booker Prize 2007); *The Visible World* Mark Slouka; *The Sixth Wife* Suzannah Dunn; *Dirty Bertie* David Roberts and Alan Macdonald; *Free Lance* Paul Stewart and Chris Riddell; *My Swordhand is Singing* Marcus Sedgwick; *The White Tiger* Aravind Adiga (Man Booker Prize 2008). 99 titles in 2008. Suggestions for titles from literary agents only.

Orion Audiobooks (Division of the Orion Publishing Group Ltd)

Orion House, 5 Upper St Martin's Lane, London WC2H 9EA
☎ 020 7520 4425   📠 020 7240 4822
pandora.white@orionbooks.co.uk
www.orionbooks.co.uk
Publisher *Pandora White*

Orion Audio has released over 500 titles since it was founded in 1998 and over the years has won many audio awards. It publishes fiction, non-fiction, humour, autobiographies, children's, science fiction, crime and thrillers. AUTHORS include Maeve Binchy, Ian Rankin, Robert Crais, Michael Connelly, Francesca Simon (*Horrid Henry* series), Harlan Coben, Kate Mosse, Dan Brown, Sally Gardner, Michelle Paver, Meg Cabot, Linwood Barclay and Richard Hammond. 60 titles a year. In 2006, Orion expanded into the digital download market via www.audible.co.uk

Penguin Audiobooks

80 Strand, London WC2R 0RL
☎ 020 7010 3000   📠 020 7010 6060
audio@penguin.co.uk
www.penguin.co.uk
Head of Audio Publishing *Jeremy Ettinghausen*

Launched in November 1993 and has rapidly expanded since then to reflect the diversity of Penguin Books' list. Publishes mostly fiction, both classical and contemporary, non-fiction, autobiography and an increasing range of digital audiobooks as well as children's titles under the **Puffin Audiobooks** imprint. Contemporary AUTHORS include: Zadie Smith, Eoin Colfer, John Mortimer, Roald Dahl, Nicci French, Gervase Phinn, Charlie Higson, Marina Lewycka, Niall Fergusson, Claire Tomalin. About 30 titles a year.

Puffin Audiobooks
▷ **Penguin Audiobooks**

Quercus Audiobooks
▷ **Quercus Publishing under UK Publishers**

Random House Audio Books

20 Vauxhall Bridge Road, London SW1V 2SA
☎ 020 7840 8529
Owner *The Random House Group Ltd.*
Commissioning Editor *Zoe Howes*

The audiobooks division of Random House started in 1991. Acquired the Reed Audio list in 1997. Publishes fiction and non-fiction. AUTHORS include Monica Ali, Louis de Bernières, Sebastian Faulks, John Grisham, Mark Haddon, Robert Harris, Thomas Harris, Ian McEwan, Andy McNab, Ruth Rendell, Kathy Reichs, Chris Ryan.

Rickshaw Productions Limited

Suite 125, 99 Warwick Street, Leamington Spa CV32 4RB
☎ 0780 3553214   📠 01926 402490
info@thelisteningzone.com
www.thelisteningzone.com
Commissioning Editor *Ms L.J. Fairgrieve*

Founded in 1998 as audiobook publishers, Rickshaw Productions Limited are now concentrating solely on their audiobook/music downloading website (www.thelisteningzone.com). Apart from their mainstream suppliers, the website also includes work by established and non-established writers and musicians. Uniquely, their audiobook and music titles are acquired from publishers and individuals from around the world. Available for download are full-length audiobooks, short stories, poetry, non-fiction, self-help, lectures, interviews, new music titles, read/performed in Dutch, English, French, German, Italian, Corsican and Spanish, with more languages, including Chinese, being added at a later date. Enquiries by e-mail or telephone. Submissions by snail-mail *only*; salacious material will not be accepted. Two sample chapters, one-page synopsis and s.a.e. to cover return by Recorded Delivery.

Shakespeare Appreciated
▷ SmartPass Ltd

Simon & Schuster Audio

1st Floor, 222 Gray's Inn Road, London WC1X 8HB
☎ 020 7316 1900   ℻ 020 7316 0332
info@simonandschuster.co.uk
www.simonsays.co.uk
Audio Manager *Rhedd Lewis*

Simon & Schuster Audio began by distributing their American parent company's audio products. Moved on to repackaging products specifically for the UK market and in 1994 became more firmly established in this market with a huge rise in turnover. Publishes adult fiction, self help and business titles. TITLES *Above Suspicion* Lynda la Plante; *The 7 Habits of Highly Effective People* Stephen R. Covey; *Chronicles* Bob Dylan; *Angels and Demons* and *Deception Point* Dan Brown; *The Kite Runner* and *A Thousand Splendid Suns* Khaled Hosseini; *The Secret* Rhonda Byrne.

SmartPass Ltd

15 Park Road, Rottingdean, Brighton BN2 7HL
☎ 01273 300742
info@smartpass.co.uk
www.smartpass.co.uk
www.shakespeareappreciated.com
www.spaudiobooks.com
Managing Director *Phil Viner*
Creative Director *Jools Viner*

Founded 1999. SmartPass produces audio education study resources, full-cast drama with commentary analysis and study strategies for English Literature set texts – novels, plays and poetry. Available in student and teacher formats, including audio linked study materials. TITLES TITLES include *A Kestrel for a Knave*; *Animal Farm*; *Pride and Prejudice*; *Great Expectations*; *The Mayor of Casterbridge*; *Shakespeare: the works*; *War Poetry*; *An Inspector Calls*; *Twelfth Night*; *Romeo and Juliet*. IMPRINTS **Shakespeare Appreciated** Full-cast unabridged dramas with commentary explaining who's who and what's going on plus historical insights and background information; created for adults rather than students, for enjoyment rather than education. TITLES *Twelfth Night*; *Othello*; *King Lear*. **SPAudiobooks** Full-cast unabridged dramas of classic and cult texts TITLES *The Antipope* Robert Rankin; *Macbeth*; *Henry V*; *Romeo and Juliet*. Welcomes ideas from authors who are teachers and from agents who represent the authors of studied texts.

Soundings Audio Books

Isis House, Kings Drive, Whitley Bay NE26 2JT
☎ 0191 253 4155   ℻ 0191 251 0662
www.isis-publishing.co.uk

Founded in 1982. Part of the Ulverscroft Group Ltd. Together with **Isis Publishing** publishes fiction and non-fiction; crime, romance. AUTHORS include Lyn Andrews, Rita Bradshaw, Lee Child, Catherine Cookson, Alexander Fullerton, Joanne Harris, Anna Jacobs, Robert Ludlum, Patrick O'Brian, Pamela Oldfield, Susan Sallis, Judith Saxton, Mary Jane Staples, Sally Worboyes. About 190 titles a year. Many also available on CD.

SPAudiobooks
▷ SmartPass Ltd

UK Packagers

Aladdin Books Ltd

PO Box 53987, London SW15 2SF
☎ 020 3174 3090   ℻ 020 8780 3939
sales@aladdinbooks.co.uk
www.aladdinbooks.co.uk
Managing Director *Charles Nicholas*

Founded in 1979 as a packaging company but with joint publishing ventures in the UK and USA. Commissions children's fully illustrated, non-fiction reference books. IMPRINTS **Aladdin Books** *Bibby Whittaker* Children's reference; **Nicholas Enterprises** *Charles Nicholas* Adult non-fiction; **The Learning Factory** *Charles Nicholas* Early learning concepts 0–4 years. TITLES *World Issues*; *Science Readers*; *The Atlas of Animals*. About 40 titles a year. Will consider synopses and ideas for children's non-fiction with international sales potential only. No fiction.
🖎 Fees usually paid instead of royalties.

The Albion Press Ltd

Spring Hill, Idbury OX7 6RU
☎ 01993 831094
Chairman/Managing Director *Emma Bradford*

Founded 1984. Commissions illustrated trade titles, particularly children's. TITLES *The Great Circle: A History of the First Nations* Neil Philip; *Fairy Tales of Hans Christian Andersen* Isabelle Brent. About 2 titles a year. Unsolicited synopses and ideas for books not welcome.
🖎 Royalties paid; fees paid for introductions and partial contributions.

Amber Books Ltd

Bradleys Close, 74–77 White Lion Street, London N1 9PF
☎ 020 7520 7600   ℻ 020 7520 7606/7
enquiries@amberbooks.co.uk
www.amberbooks.co.uk
Managing Director *Stasz Gnych*
Rights Director *Sara Ballard*
Publishing Manager *Charles Catton*

Founded 1989. Commissions history, encyclopedias, military, aviation, transport, sport, combat, survival and fitness, cookery, lifestyle, naval history, crime, childrens, schools and library sets and general reference. No fiction, poetry or biography. IMPRINT **Tiptoe Books** Picture books, sticker books, calendars, reference and educational for children of all ages. 80 titles in 2008. No unsolicited material outside subject areas listed above.
🖎 Fees paid.

BCS Publishing Ltd

2nd Floor, Temple Court, 109 Oxford Road, Cowley OX4 2ER
☎ 01865 770099   ℻ 01865 770050
bcs-publishing@dsl.pipex.com
Managing Director *Steve McCurdy*
Approx. Annual Turnover £150,000

Commissions general interest non-fiction for the international co-edition market.

Bender Richardson White

PO Box 266, Uxbridge UB9 5BD
☎ 01895 832444   ℻ 01895 835213
brw@brw.co.uk
Partners *Lionel Bender, Kim Richardson, Ben White*

Founded 1990 to produce illustrated non-fiction for children, adults and family reference for publishers in the UK and USA. 60–70 titles a year. Unsolicited material not welcome. Also offers professional advice to authors, editors and illustrators.
🖎 Fees paid.

Book House
▷ **Salariya Book Company Ltd**

Breslich & Foss Ltd

Unit 2A, Union Court, 20–22 Union Road, Clapham, London SW4 6JP
☎ 020 7819 3990   ℻ 020 7819 3998
sales@breslichfoss.com
Directors *Paula Breslich, K.B. Dunning*
Approx. Annual Turnover £650,000

Packagers of adult non-fiction titles, including interior design, crafts, gardening, health and children's non-fiction and picturebooks. Unsolicited mss welcome but synopses preferred. Include s.a.e. with all submissions.
🖎 Royalties paid twice-yearly.

Brown Wells and Jacobs Ltd

Forresters Hall, 25–27 Westow Street, London
SE19 3RY
☎ 020 8771 5115   ⊡ 020 8771 9994
graham@bwj-ltd.com
www.bwj.org
Managing Director *Graham Brown*
Production Manager & PA *Jenny Broom*

Founded 1979. Commissions non-fiction, novelty,
pre-school and first readers. About 40 titles a
year. Unsolicited synopses and ideas for books
welcome.
Ⓔ Fees paid.

Cameron & Hollis

PO Box 1, Moffat DG10 9SU
☎ 01683 220808   ⊡ 01683 220012
cameronhollis@cameronbooks.co.uk
www.cameronbooks.co.uk
Directors *Ian A. Cameron, Jill Hollis*
Approx. Annual Turnover £400,000

Commissions contemporary art, design,
collectors' reference, decorative arts, architecture.
About 3 titles a year.
Ⓔ Payment varies with each contract.

Compendium Publishing Ltd
▷ entry under UK Publishers

Diagram Visual Information Ltd

34 Elaine Grove, London NW5 4QH
☎ 020 7485 5941   ⊡ 020 7485 5941
brucerobertson@diagramgroup.com
Managing Director *Bruce Robertson*

Founded 1967. Producer of library, school,
academic and trade reference books. About 10
titles a year. 'We no longer seek submissions.'
Ⓔ Fees paid; no payment for sample material.

Eddison Sadd Editions

St Chad's House, 148 King's Cross Road, London
WC1X 9DH
☎ 020 7837 1968   ⊡ 020 7837 2025
info@eddisonsadd.com
www.eddisonsadd.com
Managing Director *Nick Eddison*
Editorial Director *Ian Jackson*
Approx. Annual Turnover £2.5 million

Founded 1982. Produces a wide range of popular
illustrated non-fiction with books and interactive
kits published in 30 countries. Many titles suit gift
markets. Incorporates Bookinabox Ltd. Ideas and
synopses, rather than mss, are welcome but titles
must have international appeal.
Ⓔ Royalties paid twice-yearly; fees paid when
appropriate.

Elwin Street Productions

144 Liverpool Road, London N1 1LA
☎ 020 7700 6785   ⊡ 020 7700 6031
info@elwinstreet.com
www.elwinstreet.com
Managing Director *Silvia Langford*
Commissioning Editor *Olivia Skinner*
Approx. Annual Turnover £1 million

Founded 2003. Produces illustrated non-fiction
in quirky humour, gift, reference and practical
life including self-help and parenting. 22 titles
in 2007. Synopses and ideas for books welcome;
approach by post only with s.a.e. No cookery or
gardening. All books must have an international
angle and not of UK interest only.
Ⓔ Fees paid.

Erskine Press

The White House, Sandfield Lane, Eccles,
Norwich NR16 2PB
☎ 01953 887277   ⊡ 01953 888361
erskpres@aol.com
www.erskine-press.com
Chief Executive *Crispin de Boos*
Approx. Annual Turnover £65,000

Specialist publisher of books on Antarctic
exploration – facsimiles, diaries, previously
unpublished works and first English translations
of expeditions of the late 19th and early
20th centuries. Publisher of general interest
biographies with special reference to Norfolk.
Recent publications include a number of Second
World War related subjects. Also produces
scholarly reprints and limited edition publications
for academic/business organizations ranging from
period print reproductions to facsimiles of rare
and important books no longer available. 7 titles
in 2008. No unsolicited mss. Ideas welcome.
Ⓔ Royalties paid twice-yearly.

Expert Publications Ltd

Sloe House, Halstead CO9 1PA
☎ 01787 474744   ⊡ 01787 474700
expert@lineone.net
Chairman *Dr. D.G. Hessayon*

Founded 1993. Produces the *Expert* series of
books by Dr. D.G. Hessayon. Currently 24 titles
in the series, including *The Flower Expert*; *The
Evergreen Expert*; *The Vegetable & Herb Expert*;
The Flowering Shrub Expert; *The Container
Expert*; *The Orchid Expert*. No unsolicited
material.

Graham-Cameron Publishing & Illustration

The Studio, 23 Holt Road, Sheringham NR26 8NB
☎ 01263 821333   ⊡ 01263 821334
enquiry@graham-cameron-illustration.com
www.graham-cameron-illustration.com

Also at: 59 Hertford Road, Brighton BN1 7GG
☏ 01273 385890
Editorial Director *Mike Graham-Cameron*
Art Director *Helen Graham-Cameron*

Founded 1984. Packages illustrated information and educational books. TITLES *Up From the Country*; *In All Directions*; *The Holywell Story*; *Flashbacks*; *Let's Look at Dairying*. Has 37 contracted book illustrators concentrating on information, educational and children's books. Absolutely *no* unsolicited mss.
Ⓕ Royalties annually.

Haldane Mason Ltd

PO Box 34196, London NW10 3YB
☏ 020 8459 2131   ☏ 020 8728 1216
info@haldanemason.com
www.haldanemason.com
Editorial Director *Sydney Francis*
Art Director *Ron Samuel*

Founded 1995. Commissions mainly children's illustrated non-fiction. Children's books are published under the **Red Kite Books** imprint. Adult list consists mainly of mind, body and spirit plus alternative health books under the Neal's Yard Remedies banner; children's age range 0–15. Unsolicited synopses and ideas welcome but approach by phone or e-mail first. No fiction, please.
Ⓕ Fees paid.

The Ilex Press Limited

The Old Candlemakers, West Street, Lewes BN7 2NZ
☏ 01273 403124   ☏ 01273 487441
[surname]@ilex-press.com
www.ilex-press.com
Managing Director *Stephen Paul*
Rights Director *Roly Allen*

Founded 1999. Sister company of **The Ivy Press Limited**. IMPRINTS **Ilex Photo** Commissions titles on photography and graphics software; **Ilex Press** Publishes reference works on popular culture, comics, some crafts and practical arts. No fiction. Unsolicited synopses and ideas welcome; send a *brief* idea outline (3 or 4 pages) and a letter.
Ⓕ Fees paid.

The Ivy Press Limited

210 High Street, Lewes BN7 2NS
☏ 01273 487440   ☏ 01273 487441
[surname]@ivy-group.co.uk
www.ivy-group.co.uk
Managing Director *Stephen Paul*
Publisher *Jason Hook*

Founded 1996. Sister company of **The Ilex Press Limited**. Commissions illustrated non-fiction books covering subjects such as art, lifestyle, popular culture, health, self-help and humour. No

fiction. Unsolicited synopses and ideas welcome; send a *brief* idea outline (3 or 4 pages) and a letter.
Ⓕ Fees paid.

The Learning Factory
▷ **Aladdin Books Ltd**

Lennard Associates Ltd

Windmill Cottage, Mackerye End, Harpenden AL5 5DR
☏ 01582 715866   ☏ 01582 715866
stephenson@lennardqap.co.uk
Chairman/Managing Director *Adrian Stephenson*

Founded 1979. Now operating as a packager/production company only with no trade distribution. 1 title in 2008. No unsolicited mss.
Ⓕ Both fees and royalties by arrangement.

Lexus Ltd

60 Brook Street, Glasgow G40 2AB
☏ 0141 556 0440   ☏ 0141 556 2202
peterterrell@lexusforlanguages.co.uk
www.lexusforlanguages.co.uk
Managing/Editorial Director *P.M. Terrell*

Founded 1980. Compiles bilingual reference, language and phrase books. TITLES Rough Guide phrasebooks; Langenscheidt dictionaries; *HarperCollins English-Chinese*; *Oxford Italian Pocket Dictionary*. Own series of *Travelmates* published in 2004; launched a new beginner's course for learning Chinese, *The Chinese Classroom*, in 2007. About 5 titles a year. Unsolicited material considered although books are mostly commissioned. Freelance contributors employed for a wide range of languages.
Ⓕ Generally flat fee.

Lionheart Books

10 Chelmsford Square, London NW10 3AR
☏ 020 8459 0453   ☏ 020 8451 3681
lionheart.brw@btinternet.com
Senior Partner *Lionel Bender*
Partner *Madeleine Samuel*

A design/editorial packaging team. Titles are primarily commissioned from publishers. Highly illustrated non-fiction for children aged 8+, mostly natural history, history and general science. See also **Bender Richardson White**. About 20 titles a year.
Ⓕ Generally flat fee.

M&M Publishing Services

33 Warner Road, Ware SG12 9JL
☏ 01920 466003
mikemoran@moran01.wanadoo,co.uk
www.photography-london.co.uk
Proprietors *Mike Moran, Maggie Copeland*

Project managers, production and editorial, packagers and publishers. TITLES *MM Publisher*

Database; *MM Printer Database* (available in UK, European and international editions).

Market House Books Ltd

Suite B, Elsinore House, 43 Buckingham Street, Aylesbury HP20 2NQ
☎ 01296 484911
books@mhbref.com
www.markethousebooks.com
Directors *John Daintith, Peter Sapsed, Anne Kerr*

Founded 1970. Compiles dictionaries, encyclopedias and other reference books. TITLES *APA Dictionary of Psychology*; *Collins English Dictionary*; *Collins Encyclopedia* (10 vols.); Facts on File Dictionaries of: *Astronomy, Chemistry*, etc.; *Larousse Thematica* (6 vols); *The Macmillan Dictionary of Philosophy*, etc.; Oxford Dictionaries of: *Business, Science, Medicine*, etc.; *Penguin Dictionary of Electronics*; *Penguin Rhyming Dictionary*, etc.; *Taylor & Francis Biographical Encyclopedia of Scientists*. About 15 titles a year. Unsolicited material not welcome as most books are compiled in-house.
⬡ Fees paid.

Monkey Puzzle Media Ltd

Little Manor Farm, Brundish IP13 8BL
☎ 01379 388990
info@monkeypuzzlemedia.com
Chairman/Managing Director *Roger Goddard-Coote*
Commissioning and List Development *Paul Mason*

Founded 1998. Packager of adult and children's non-fiction for trade, school, library and mass markets. No fiction or textbooks. About 80 titles a year. Synopses and ideas welcome. Include s.a.e. for return.
⬡ Fees paid; no royalties.

Nicholas Enterprises
▷ Aladdin Books Ltd

Orpheus Books Limited

6 Church Green, Witney OX28 4AW
☎ 01993 774949　📠 01993 700330
info@orpheusbooks.com
www.orpheusbooks.com
Directors *Nicholas Harris, Sarah Hartley*

Founded 1993. Commissions children's non-fiction. About 20 titles a year. No unsolicited material.
⬡ Fees paid.

Mathew Price Ltd

Albury Court, Albury, Thame OX9 2LP
mathewp@mathewprice.com
www.mathewprice.com
Chairman/Managing Director *Mathew Price*

Main office: 5013 Golden Circle, Denton, TX 76208, USA
☎ 001 940 565 9594　📠 001 940 565 9313

Founded 1983. Commissions full-colour novelty and picture books and fiction for young children. 'We accept submissions by e-mail, only. We are looking for stories for the very young and intelligent novelties, whether fiction or non-fiction. We do not accept stories in rhyme. Please understand that we cannot reply at length to all submissions, so if we can't publish it, our response will probably be brief and impersonal.'

Quarto Publishing plc
▷ entry under UK Publishers

Red Kite Books
▷ Haldane Mason Ltd

Regency House Publishing Limited
▷ entry under UK Publishers

Salariya Book Company Ltd

25 Marlborough Place, Brighton BN1 1UB
☎ 01273 603306　📠 01273 693857
salariya@salariya.com
www.salariya.com
www.book-house.co.uk
www.scribblersbook.com
Managing Director *David Salariya*
Editor *Stephen Haynes*
Art Director *Carolyn Franklin*

Founded 1989. Books for children: history, art, music, science, architecture, education, nature, environment, fantasy, folktales, multicultural, fiction, young readers and picture books. IMPRINTS **Book House**; **Scribblers** Highly illustrated books in all subjects for babies and children, from pre-school board books to teenage graphic novels. For fiction, submit complete c.v., ms for picture books; outline/synopsis and one sample story or chapter for collections, novels or graphic novels. Response in four months if s.a.e. included. Illustrations: no original artwork. Send c.v., promotion sheet to be kept on file. Samples returned if s.a.e. included.
⬡ Payment by arrangement.

Savitri Books Ltd

25 Lisle Lane, Ely CB7 4AS
☎ 01353 654327　📠 01353 654327
munni@savitribooks.demon.co.uk
Managing Director *Mrinalini S. Srivastava*
Approx. Annual Turnover £200,000

Founded 1983 and since 1998, Savitri Books has also become a publisher in its own right (textile crafts). Keen to work 'very closely with authors/ illustrators and try to establish long-term relationships with them, doing more books with the same team of people'. Commissions illustrated non-fiction: biography, history, travel. About 7 titles a year. Unsolicited synopses and ideas for books 'very welcome'.

Scribblers
▷ Salariya Book Company Ltd

Stonecastle Graphics Ltd

Highlands Lodge, Chartway Street, Sutton Valence ME17 3HZ
📞 01622 844414
paul@stonecastle-graphics.com
sue@stonecastle-graphics.com
www.stonecastle-graphics.com
Director *Paul Turner*
Editorial Head *Sue Pressley*
Approx. Annual Turnover £300,000

Founded 1976. Formed additional design/packaging partnership, **Touchstone**, in 1983 (see entry). Commissions illustrated non-fiction general books: motoring, health, sport, leisure, lifestyle and children's non-fiction. TITLES *Skyscrapers*; *Snakes*; *Baby Animals*; *Learn to Play the Piano*; *Learn to Dance*; *Spirit of China*; *Spirit of India*; *Spirit of the Elephant*; *Spirit of Ireland*; *Spirit of Latin America*; *Learn to Play the Drums*; *Funny Animals.* 14 titles in 2008. Unsolicited synopses and ideas for books welcome
💷 Fees paid.

Templar Publishing
▷ entry under UK Publishers

Tiptoe Books
▷ Amber Books Ltd

Toucan Books Ltd

Third Floor, 89 Charterhouse Street, London EC1M 6HR
📞 020 7250 3388   📠 020 7250 3123
Managing Director *Ellen Dupont*

Founded 1985. Specializes in international co-editions and fee-based editorial, design and production services. Commissions illustrated non-fiction for children and adults. No fiction or non-illustrated titles. About 15 titles a year.
💷 Royalties paid twice-yearly; fees paid in addition to or instead of royalties.

Touchstone Books Ltd

Highlands Lodge, Chartway Street, Sutton Valence ME17 3HZ
📞 01622 844414   📠 01622 844414

paul@touchstone-books.net
sue@touchstone-books.net
www.touchstone-books.net
Directors *Paul Turner, Sue Pressley, Nick Wigley, Trevor Legate*

Founded in 2006, Touchstone Books Ltd is an independent publisher of lavishly-illustrated books on motoring, aviation, art, photography and general-interest subjects. TITLES *Cobra: The First 40 Years*; *100 Years of Grand Prix*; *The Very Best of Volkswagen*; *In the Company of Dali*; *100 Years of Brooklands: The Birthplace of British Motorsport and Aviation*; *Fighting Force: The 90th Anniversary of the Royal Air Force* (officially endorsed by the RAF); *The Red Arrows.* 6 titles in 2008. Unsolicited synopses and ideas for books welcome.
💷 Fees paid.

David West Children's Books

7 Princeton Court, 55 Felsham Road, London SW15 1AZ
📞 020 8780 3836   📠 020 8780 9313
dww@btinternet.com
www.davidwestchildrensbooks.com

Founded 1992. Commissions children's illustrated reference books. No fiction or adult books. Unsolicited ideas and synopses welcome; approach in writing in the first instance
💷 Fees/royalties paid annually.

Working Partners Ltd

Stanley House, St Chad's Place, London WC1X 9HH
📞 020 7841 3939   📠 020 7841 3940
enquiries@workingpartnersltd.co.uk
www.workingpartnersltd.co.uk
Chairman *Ben Baglio*
Managing Director *Chris Snowdon*
Contact *James Noble*

Specializes in quality mass-market fiction for leading children's publishers including Bloomsbury, HarperCollins, Hodder, Macmillan, Orchard, OUP, Random House and Scholastic. First chapter books to young adult. No picture books. **Working Partners Two** handles adult fiction across all popular genres on a similar basis. Welcomes approaches from interested writers.
💷 Both fees and royalties by arrangement.

Electronic Publishing and Other Services

www.ABCtales.com

8 Clifton Road, Brighton BN1 3HP
☎ 07899 901023
tcook@abctales.com
www.abctales.com
Owner *Burgeon Creative Ideas Ltd*
Editor *Tony Cook*

Founded by A. John Bird, MBE, co-founder of the *Big Issue* magazine, Tony Cook and Gordon Roddick. ABCtales is a free website dedicated to publishing and developing new writing. Content is predominantly short stories, autobiography and poetry. Anyone can upload creative writing to the website for free. Authors such as Joe Dunthorne and Drew Gummerson started out on ABCtales and still write regularly on the site, testing out new work. There are frequent ABCtales evenings around the country, competitions and excellent feedback on work from fellow writers.

Authors OnLine
▷ entry under UK Publishers

BeWrite Books
▷ See entry under Small Presses

Books 4 Publishing

Rural Enterprise Centre, Eco Park Road, Ludlow SY8 1ES
☎ 0870 777 1113
editor@books4publishing.com
www.books4publishing.com
Owner *Corvedale Media Ltd*
Managing Director *Mark Oliver*

Founded 2000. Specializes in showcasing unpublished books and helping new authors gain recognition for their work by displaying synopses and up to 5,000 words on the Books 4 Publishing website and promoting titles to publishers and literary agents. Mss, synopses and ideas welcome; initial enquiries by e-mail, post or submission form on website.

e-book-downloads-uploads.com

63 Lulworth Avenue, Preston PR2 2BE
☎ 07722 209009
editor@e-book-downloads-uploads.com
info@e-book-downloads-uploads.com
www.e-book-downloads-uploads.com
Contact *H. Pritchard*

e-book publisher, established in 2006, 'willing to give opportunities to writers who by no fault of their own cannot get their work to market'. Will consider all submissions and all categories. Approach by e-mail with short synopsis attached. ⓔ Royalties twice-yearly.

Fern HousE-Publishing

19 High Street, Haddenham, Ely CB6 3XA
☎ 01353 740222
epub@fernhouse.com
www.fernhouse.com
Managing Editor *Rodney Dale*

Founded 1995. Has published mainly non-fiction (13 titles in print) but now entering the field of e-publishing. Open to submissions – 'but only of the highest quality'.

50connect – www.50connect.co.uk

Morley House, Badminton Court, Church Street, Old Amersham HP7 0DD
☎ 01494 736130   ▤ 0800 634 2250
admin@50connect.com
www.50connect.co.uk
Editor *Emily Bird (ebird@50connect.com)*

Launched in 2000, and awarded 'esuperbrand status' in 2006, www.50connect.co.uk is a leading editorial-led lifestyle website for today's over-45s. Covers finance, health, travel, overseas retirement, living abroad, retirement, food, wine, theatre, music, gardening, pets and much more. Also strong community element including forums and a chat room. Submissions welcome but please contact via e-mail in the first instance with some previous examples of work published.

Fledgling Press

7 Lennox Street, Edinburgh EH4 1QB
☎ 0131 343 2367
info@fledglingpress.co.uk
www.fledglingpress.co.uk
Director *Zander Wedderburn*

Founded 2000. Internet publisher, aiming to be a launching pad for new authors. Special interest in authentic writing about the human condition, including autobiography, diaries, poetry and fictionalized variations on these. Monthly online competition (www.canyouwrite.com) with small prizes for short pieces. Also free books and reports in the areas of shiftwork and working time. Send mss and other details by e-mail from the website or by post. Links into short-run book production. 24 titles to date.

Justis Publishing Ltd

Grand Union House, 20 Kentish Town Road, London NW1 9NR

📞 020 7267 8989   📠 020 7267 1133
communications@justis.com
www.justis.com

Founded 1986. Electronic publisher of UK and European legal and offical information on CD-ROM, online and the Internet.

Lulu

▷ entry under UK Publishers

New Authors Showcase

▷ Barrie James Literary Agency under UK Literary Agents

Online Originals

Priory Cottage, Wordsworth Place, London NW5 4HG

📞 020 7267 4244
editor@onlineoriginals.com
www.onlineoriginals.com
Managing Director *David Gettman*
Technical Director *Neill Sanders*

Publishes book-length works as e-books on the web and occasionally as print on demand or short-run library editions. Acquires global electronic rights in literary fiction, intellectual non-fiction, drama, and youth fiction (ages 8–16). No poetry, fantasy, how-to, self-help, picture books, cookery, hobbies, crafts or local interest. TITLES *Being and Becoming* Christopher Macann (4 vols.); *The Jealous God* Martin Boyland; *Lard and Speck* Andrew Cogan; *Quintet* Frederick Forsyth; *The Angels of Russia* Patricia Leroy. 5 titles in 2008. 'Unsolicited mss, synopses and proposals for books welcome via our unique, peer-review, automated, online submissions system, accessed via the website address above. No submissions on paper or disk please.' 💷 Royalties paid annually (50% royalties on e-book price of £6).

WritersServices

▷ entry under Useful Websites

Wuggles Publishing

185 Kingrosia Park, Clydach, Swansea SA6 5PF

📞 01792 844445
wuggles@ic24.net
www.wugglespublishing.co.uk
Contact *Chris Thomas*

Established in 2001 to assist unpublished writers 'ignored by the big boys to get off the ground'. Only interested in non-PC work of any genre except children's, religious, gay, academic or porn. No mss; submit by phone or e-mail in first instance. See website for further details.

UK Literary Agents

* = Members of the **Association of Authors' Agents**

Sheila Ableman Literary Agency*

48–56 Bayham Place, London NW1 0EU
☎ 020 7388 7222
sheila@sheilaableman.co.uk
Contact *Sheila Ableman*

Founded 1999. Handles non-fiction including history, science, biography and autobiography. Specializes in celebrity ghostwriting and TV tie-ins. No poetry, children's, gardening or sport. COMMISSION Home 15%; US & Translation 20%. Unsolicited mss welcome. Approach in writing with publishing history, c.v., synopsis, three chapters and s.a.e. for return. No reading fee.

The Agency (London) Ltd*

24 Pottery Lane, London W11 4LZ
☎ 020 7727 1346   ℻ 020 7727 9037
info@theagency.co.uk
Contacts *Stephen Durbridge, Leah Schmidt, Norman North, Julia Kreitman, Bethan Evans, Hilary Delamere (children's books), Katie Haines, Faye Webber, Nick Quinn, Fay Davies, Ian Benson, Jago Irwin*

Founded 1995. Deals with writers and rights for TV, film, theatre, radio scripts and children's books: picture books and fiction, 0–teen. Also handles directors. Only existing clients for adult fiction or non-fiction. COMMISSION Home 10%; US by arrangement; Children's books: Home 15%; Overseas 20%. Send letter with s.a.e. No unsolicited mss. No reading fee.

Aitken Alexander Associates Ltd*

18–21 Cavaye Place, London SW10 9PT
☎ 020 7373 8672   ℻ 020 7373 6002
reception@aitkenalexander.co.uk
www.aitkenalexander.co.uk
Contacts *Gillon Aitken, Clare Alexander, Lesley Thorne (also film/TV)*
Associated Agents *Anthony Sheil, Mary Pachnos*

Founded 1977. Handles fiction and non-fiction. No plays, scripts or children's fiction unless by existing clients. CLIENTS include Clare Allan, Pat Barker, Sarah Bradford, Gordon Burn, Jung Chang, John Cornwell, Sarah Dunant, Susan Elderkin, Diana Evans, Roopa Farooki, Sebastian Faulks, Helen Fielding, Germaine Greer, Mark Haddon, Mohammed Hanif, Nicky Haslam, Susan Howatch, Liz Jensen, V.S Naipaul, Charles Moore, Jonathan Raban, Piers Paul Read, Louise Rennison, Michèle Roberts, Jennie Rooney, Nicholas Shakespeare, Gillian Slovo, Rory Stewart, Nick Stone, Colin Thubron, Sally Vickers, Penny Vincenzi, Willy Vlautin, A.N. Wilson, Robert Wilson. COMMISSION Home 10%; US & Translation 20%; Film/TV 10%. Send preliminary letter, with half-page synopsis and first 30pp of sample material, and adequate return postage, in the first instance. No reading fee.

The Ampersand Agency Ltd*

Ryman's Cottages, Little Tew OX7 4JJ
☎ 01608 683677   ℻ 01608 683449
info@theampersandagency.co.uk
www.theampersandagency.co.uk
Blog: www.theampersandagency.co.uk/blog
Contacts *Peter Buckman, Anne-Marie Doulton*
Consultants *Peter Janson-Smith, Patrick Neale*
Blog Manager *Amy Wigelsworth*

Founded 2003. Handles literary and commercial fiction and non-fiction; contemporary and historical novels, crime, thrillers, biography, women's fiction, history, memoirs. No scripts unless by existing clients. No poetry, science fiction or fantasy. CLIENTS include Kerry Bailey, Helen Black, S.J. Bolton, Georgette Heyer estate, Druin Burch, Martin Conway, Vanessa Curtis, Will Davis, Tracy Gilpin, Cora Harrison, Michael Hutchinson, Beryl Kingston, Miriam Morrison, Philip Purser, Niamh Shaw, P. Robert Smith, Ivo Stourton, Vikas Swarup, Nick van Bloss, Michael Walters. COMMISSION Home 12½–15%; US 15–20%; Translation 20%. Translation rights handled by **The Buckman Agency**. Unsolicited mss, synopses and ideas welcome, as are e-mail enquiries, but send sample chapters and synopsis by post (s.a.e. required if material is to be returned). No reading fee.

Darley Anderson Children's Literary Agency Ltd*

Estelle House, 11 Eustace Road, London SW6 1JB
☎ 020 7386 2674   ℻ 020 7386 5571
enquiries@darleyanderson.com
www.darleyandersonchildrens.com

Proprietor *Darley Anderson*
Director *Peter Colegrove*
Children's Agent *Becky Stradwick*
Head of Rights *Madeleine Buston*

Handles children's fiction and non-fiction. No poetry or academic books. CLIENTS Cathy Cassidy, Lisa Clark, Helen Grant, Michelle Harrison, Adrienne Kress, Carmen Reid, Rob Stevens, Kate Wild, Ahmet Zappa. COMMISSION Home 15%; US & Translation 20%. OVERSEAS ASSOCIATES APA Talent and Literary Agency (LA/Hollywood); Liza Dawson Literary Agency (New York); and leading foreign agents worldwide. Postal queries preferred. Send preliminary letter, synopsis and first three chapters. Return postage/s.a.e. essential for reply. No reading fee.

Darley Anderson Literary, TV & Film Agency*

Estelle House, 11 Eustace Road, London SW6 1JB
☎ 020 7385 6652 📠 020 7386 5571
enquiries@darleyanderson.com
www.darleyanderson.com
Contacts *Darley Anderson (thrillers), Zoe King (non-fiction), Camilla Bolton (crime and mystery), Ella Andrews (general and women's fiction), Madeleine Buston (Head of Rights), Kasia Benke (Rights Assistant), Rosanna Bellingham (Finance)*

Founded 1988. Run by an ex-publisher with a knack for spotting talent and making great deals – many for six and seven figure advances. Handles commercial fiction and non-fiction, children's fiction and non-fiction; also selected scripts for film and TV. No academic books or poetry. Special fiction interests: all types of thrillers and crime (American/hard boiled/cosy/historical); women's fiction (sagas, chick-lit, contemporary, love stories, 'tear jerkers', women in jeopardy) and all types of American and Irish novels. Non-fiction interests: celebrity autobiographies, biographies, sports books, 'true life' women in jeopardy, revelatory history and science, popular psychology, self improvement, diet, beauty, health, finance, fashion, animals, humour/cartoon, gardening, cookery, inspirational and religious. CLIENTS include Anne Baker, Alex Barclay, Constance Briscoe, Paul Carson, Lee Child, Martina Cole, John Connolly, Margaret Dickinson, Clare Dowling, Tana French, Helen Grant, Milly Johnson, Danny King, Patrick Lennon, Lesley Pearse, Lynda Page, Adrian Plass, Rebecca Shaw, Graeme Sims.

COMMISSION Home 15%; US & Translation 20%; Film/TV/Radio 20%. Established the **Darley Anderson Children's Literary Agency Ltd** in 2008 (see entry). OVERSEAS ASSOCIATES APA Talent and Literary Agency (LA/Hollywood); Liza Dawson Literary Agency (New York); and leading foreign agents throughout the world. Send letter, synopsis and first three chapters; plus s.a.e. for return. No reading fee.

Anubis Literary Agency

7 Birdhaven Close, Lighthorne, Warwick CV35 0BE
☎ 01926 642588 📠 01926 642588
Contact *Steve Calcutt*

Founded 1994. Handles genre fiction in the following categories: science fiction, fantasy and horror. No other material considered. COMMISSION Home 15%; US & Translation 20%. Works with **The Marsh Agency** on translation rights. In the first instance send 50 pages with a one-page synopsis (s.a.e. essential). No telephone calls. No reading fee.

Artellus Limited

30 Dorset House, Gloucester Place, London NW1 5AD
☎ 020 7935 6972 📠 020 8609 0347
www.artellusltd.co.uk
Chairman *Gabriele Pantucci*
Director *Leslie Gardner*
Associates *Darryl Samaraweera, Eleanor Day*

Founded 1986. Full-length and short mss. Handles crime, science fiction, historical, contemporary and literary fiction; non-fiction: art history, current affairs, biography, general history, science. Works directly in the USA and with agencies internationally. COMMISSION Home 10%; Overseas 20%. Sample chapters in the first instance. No reading fee. Return postage essential.

Author Literary Agents

53 Talbot Road, London N6 4QX
☎ 020 8341 0442/07767 022659
📠 020 8341 0442
agile@authors.co.uk
Contact *John Havergal*

Founded 1997. 'We put to leading publishers and producers strong new novels and graphic media ideas.' COMMISSION (VAT extra) Book, Screen & Internet production rights: Home 15%; Overseas & Translation 25%. Send s.a.e. with first chapter, scene or section writing sample, plus half-to-one page outline. Please include graphics samples, if applicable. No reading fee.

Diane Banks Associates Ltd

submissions@dianebanks.co.uk
www.dianebanks.co.uk
Contact *Diane Banks*

Founded 2006. Handles commercial fiction and non-fiction. Fiction: women's, crime, thrillers, literary fiction with a strong storyline; non-fiction: memoir, real-life stories, celebrity, autobiography, biography, popular history, popular science, self-help, popular psychology, fashion, health and beauty. No poetry, children's, academic books, plays, scripts or short stories. CLIENTS include Miss S, Andy Taylor (Duran Duran), Alex Higgins, Elise Lindsay, Elizabeth Burton-Phillips, Hannah Sandling, Narinder Kaur, Emily Winterburn, Sasha Wagstaff and Polly Courtney. COMMISSION Home 15%; Overseas 20%. Send brief c.v., synopsis and first three chapters as Word attachments to the e-mail address above. Aims to give initial response within two weeks. No reading fee.

The Bell Lomax Moreton Agency

James House, 1 Babmaes Street, London SW1Y 6HF

☎ 020 7930 4447   📠 020 7925 0118
agency@bell-lomax.co.uk
Executives *Eddie Bell, Pat Lomax, Paul Moreton, June Bell*

Established 2002. Handles quality fiction and non-fiction, biography, children's, business and sport. No unsolicited mss without preliminary letter. No scripts. No reading fee.

Lorella Belli Literary Agency (LBLA)*

54 Hartford House, 35 Tavistock Crescent, Notting Hill, London W11 1AY

☎ 020 7727 8547   📠 0870 787 4194
info@lorellabelliagency.com
www.lorellabelliagency.com
Contact *Lorella Belli*

Founded 2002. Handles full-length fiction (from literary to genre) and general non-fiction. Particularly interested in first novelists, journalists, multi-cultural and international writing, books on or about Italy. No children's, fantasy, science fiction, short stories, poetry, plays, screenplays or academic books. CLIENTS include Michael Bess, Sean Bidder, Zöe Brân, Diablo Cody, Annalisa Coppolaro-Nowell, Emily Giffin, William Little, Nisha Minhas, Alanna Mitchell, Rick Mofina, Angela Murrills, Jennifer Ouellette, Robert Ray, Sheldon Rusch, Grace Saunders, Dave Singleton, Rupert Steiner, Justine Trueman, Diana Winston, Carol Wright. COMMISSION Home 15%; US, Translation & Dramatic Rights 20%. Works in conjunction with leading associate agencies in most countries and a film/TV agency in London. Also represents American, Canadian, Australian and European agencies in the UK. Welcomes approaches from new authors. Send full proposal plus two chapters for non-fiction, and short synopsis plus first three chapters for fiction. S.a.e. essential. No reading fee. Revision suggested where appropriate.

Blake Friedmann Literary Agency Ltd*

122 Arlington Road, London NW1 7HP
☎ 020 7284 0408   📠 020 7284 0442

firstname@blakefriedmann.co.uk
www.blakefriedmann.co.uk
Contacts *Carole Blake (books), Julian Friedmann (film/TV), Conrad Williams (original scripts/radio), Isobel Dixon (books), Oli Munson (books)*

Founded 1977. Handles all kinds of fiction from genre to literary; a varied range of specialist and general non-fiction, plus scripts for TV, radio and film. No poetry, science fiction or short stories (unless from existing clients). Special interests: commercial women's fiction, intelligent thrillers, literary fiction, upmarket non-fiction, teenage fiction. CLIENTS include Gilbert Adair, Jane Asher, Andy Briggs, Edward Carey, Elizabeth Chadwick, Anne de Courcy, Anna Davis, Barbara Erskine, David Gilman, Ann Granger, Billy Hopkins, Peter James, Deon Meyer, Lawrence Norfolk, Gregory Norminton, Joseph O'Connor, Sheila O'Flanagan, Michael Ridpath, Craig Russell, Tess Stimson, Michael White. COMMISSION Books: Home 15%; US & Translation 20%. Radio/TV/Film: 15%. OVERSEAS ASSOCIATES 24 worldwide. Unsolicited mss welcome but initial letter with synopsis and first two chapters preferred. Letters should contain as much information as possible on previous writing experience, aims for the future, etc. No reading fee.

Luigi Bonomi Associates Limited (LBA)*

91 Great Russell Street, London WC1B 3PS
☎ 020 7637 1234   📠 020 7637 2111
info@bonomiassociates.co.uk
www.bonomiassociates.co.uk
Contacts *Luigi Bonomi, Amanda Preston, Molly Stirling*

Handles commercial and literary fiction, thrillers, crime and women's fiction; non-fiction: history, science, parenting, lifestyle, diet, health; teen and young adult fiction. CLIENTS include James Barrington, Chris Beardshaw, Linda Blair, Jo Carnegie, Gennaro Contaldo, Josephine Cox, Nick Foulkes, David Gibbins, Richard Hammond, Bruno Hare, Jane Hill, Matt Hilton, John Humphrys, Graham Joyce, Simon Kernick, James May, Gavin Menzies, Niki Monaghan, Mike Morley, Gillian McKeith, Richard Madeley & Judy Finnigan, Sue Palmer, William Petre, Melanie Phillips, Jem Poster, Mitch Symons, Alan Titchmarsh, Jon Trigell, Sarah Tucker, Charlotte Ward, Sally Worboyes, Sir Terry Wogan. COMMISSION Home 15%; US & Translation 20-25%. 'We are very keen to find new authors and help them develop their careers.' No scripts, poetry, science fiction/fantasy, children's storybooks. *Send synopsis and first chapter only.* 'We do *not* reply by e-mail so return postage/s.a.e. essential for reply.' No reading fee.

BookBlast Ltd

PO Box 20184, London W10 5AU

☎ 020 8968 3089

www.bookblast.com

Contact *Address material to the Company*

Handles fiction and non-fiction. Memoirs, travel, popular culture, multicultural writing only. Reading very selectively. No new authors taken on except by recommendation. Editorial advice given to own authors. Initiates in-house projects. Also offers translation consultancy. Film, TV and radio rights sold in works by existing clients. COMMISSION Home 12%; US & Translation 20%; Film, TV & Radio 20%. No reading fee.

Bookseeker Agency

PO Box 7535, Perth PH2 1AF

☎ 01738 620688

bookseeker@blueyonder.co.uk

Contact *Paul Thompson*

Founded 2005. Handles poetry and any creative writing except non-fiction. COMMISSION negotiable 'but not exceeding 15%'. Approach in writing, giving a personal introduction, a summary of work so far and any current projects. No mss or sample chapters; synopsis or half-a-dozen poems acceptable. No reading fee.

Alan Brodie Representation Ltd

6th Floor, Fairgate House, 78 New Oxford Street, London WC1A 1HB

☎ 020 7079 7990   🖷 020 7079 7999

info@alanbrodie.com

www.alanbrodie.com

Contacts *Alan Brodie, Sarah McNair, Lisa Foster, Harriet Pennington Legh*

Founded 1989. Handles theatre, film and TV scripts. No books. COMMISSION Home 10%; Overseas 15%. Preliminary letter plus professional recommendation and c.v. essential. No reading fee but s.a.e. required.

Jenny Brown Associates*

33 Argyle Place, Edinburgh EH9 1JT

☎ 0131 229 5334

info@jennybrownassociates.com

www.jennybrownassociates.com

Contacts *Jenny Brown, Mark Stanton*

Children's *Lucy Juckes*

Crime *Allan Guthrie*

Founded 2002. Handles literary fiction, crime writing, writing for children; non-fiction: memoir, sport, music, nature, popular culture. No poetry, short stories, science fiction, fantasy or academic. CLIENTS include Lin Anderson, Alex Gray, Keith Gray, Laura Hird, Roger Hutchinson, Sara Maitland, Richard Moore, Suhayl Saadi, Paul Torday, Esther Woolfson. COMMISSION Home 12½%; US & Translation 20%. See website for full submission guidelines.

Felicity Bryan*

2A North Parade, Banbury Road, Oxford OX2 6LX

☎ 01865 513816   🖷 01865 310055

agency@felicitybryan.com

www.felicitybryan.com

Agents *Felicity Bryan, Catherine Clarke, Caroline Wood*

Associate Agent *Sally Holloway*

Founded 1988. Handles fiction of various types and non-fiction with emphasis on history, biography, science and current affairs. No scripts for TV, radio or theatre. No crafts, picture books, how-to, science fiction or light romance. CLIENTS include Carlos Acosta, David Almond, Karen Armstrong, Simon Blackburn, Artemis Cooper, Isla Dewar, John Dickie, Sally Gardner, A.C. Grayling, Tim Harford, Sadie Jones, Diarmaid MacCulloch, John Man, Martin Meredith, James Naughtie, Linda Newbery, John Julius Norwich, Iain Pears, Robin Pilcher, Rosamunde Pilcher, Matt Ridley, Meg Rosoff, Miriam Stoppard, Roy Strong, Adrian Tinniswood, Colin Tudge, Eleanor Updale, Lucy Worsley and the estates of Robertson Davies and Humphrey Carpenter. COMMISSION Home 15%; US & Translation 20%. OVERSEAS ASSOCIATES **Andrew Nurnberg**, Europe; several agencies in US. See website for submission guidelines. No e-mail submissions. No reading fee.

The Buckman Agency

Ryman's Cottages, Little Tew OX7 4JJ

☎ 01608 683677/020 8544 2674

🖷 01608 683449/020 8543 9653

r.buckman@talk21.com

j.buckman@talk21.com

Partners *Rosie Buckman, Jessica Buckman*

Founded in the early 1970s, the agency specializes in foreign rights and represents leading authors and agents from the UK and US. COMMISSON 20% (including sub-agent's commission). No unsolicited mss.

Juliet Burton Literary Agency*

2 Clifton Avenue, London W12 9DR

☎ 020 8762 0148   🖷 020 8743 8765

juliet.burton@btinternet.com

Contact *Juliet Burton*

Founded 1999. Handles fiction and non-fiction. Special interests crime and women's fiction. No plays, film scripts, articles, poetry or academic material. COMMISSION Home 15%; US & Translation 20%. Approach in writing in the first instance; send synopsis and two sample chapters with s.a.e. No e-mail submissions. No unsolicited mss. No reading fee.

Campbell Thomson & McLaughlin Ltd*

50 Albemarle Street, London w1s 4bd
📞 020 7493 4361   📠 020 7495 8961
submissions@ctmcl.co.uk
Contact *Charlotte Bruton*

Founded 1931. Handles fiction and general non-fiction, excluding children's. No plays, film/TV scripts, articles, short stories or poetry. Translation rights handled by **The Marsh Agency**. No unsolicited mss or synopses. Preliminary enquiry essential, by letter or e-mail. No reading fee.

Capel & Land Ltd*

29 Wardour Street, London w1d 6ps
📞 020 7734 2414   📠 020 7734 8101
rosie@capelland.co.uk
www.capelland.com
Contact *Georgina Capel*

Handles fiction and non-fiction. Also film and TV writers and broadcasters; not scripts. CLIENTS Kunal Basu, John Gimlette, Andrew Greig, Dr Tristram Hunt, Andrew Roberts, Simon Sebag Montefiore, Stella Rimington, Diana Souhami, Louis Theroux, Fay Weldon. COMMISSION Home, US & Translation 15%. Send sample chapters and synopsis with covering letter and e-mail address in the first instance. 'We are unable to return material so be sure to keep a hard copy. We will contact you within six weeks if we are interested in taking things further.' No reading fee.

Casarotto Ramsay and Associates Ltd

Waverley House, 7–12 Noel Street, London
w1f 8gq
📞 020 7287 4450   📠 020 7287 9128
agents@casarotto.co.uk
www.casarotto.co.uk
Film/TV/Radio *Jenne Casarotto, Charlotte Kelly, Jodi Shields, Elinor Burns, Miriam James, Rachel Holroyd, Mark Casarotto, Abby Singer, Lucinda Prain*
Stage *Tom Erhardt, Mel Kenyon*

Handles scripts for TV, theatre, film and radio. CLIENTS include: **Theatre** Alan Ayckbourn, Caryl Churchill, Christopher Hampton, David Hare, Sarah Kane estate, Mark Ravenhill. **Film** Laura Jones, Neil Jordan, Nick Hornby, Shane Meadows, Purvis & Wade, Lynne Ramsay. **TV** Howard Brenton, Amy Jenkins, Susan Nickson, Jessica Hynes. COMMISSION Home 10%. OVERSEAS ASSOCIATES worldwide. No unsolicited material without preliminary letter.

Celia Catchpole

56 Gilpin Avenue, London sw14 8qy
📞 020 8255 4835
www.celiacatchpole.co.uk
Contact *Celia Catchpole*

Founded 1996. Handles children's books – artists and writers. No TV, film, radio or theatre scripts. No poetry. COMMISSION Home 10% (writers) 15% (artists); US & Translation 20%. Works with associate agents abroad. No unsolicited mss.

Chapman & Vincent*

7 Dilke Street, London sw3 4je
📞 020 7352 5582   📠 020 7352 5582
chapmanvincent@hotmail.co.uk
Directors *Jennifer Chapman, Gilly Vincent*

A small agency handling only non-fiction work – often highly illustrated – and whose clients come mainly from personal recommendation. The agency is not actively seeking clients but is happy to consider really original work. CLIENTS include George Carter, Leslie Geddes-Brown, Lucinda Lambton, Rowley Leigh and Eve Pollard. COMMISSION Home 15%; US & Europe 20%. Works with The Elaine Markson Agency. Please do not telephone or submit by fax. Submissions: two sample chapters and s.a.e. Will consider e-mail submissions without attachments. No reading fee.

Mic Cheetham Associates

50 Albemarle Street, London w1s 4bd
📞 020 7495 2002   📠 020 7495 8961
info@miccheetham.com
www.miccheetham.com
Contacts *Mic Cheetham, Simon Kavanagh*

Established 1994. Handles general and literary fiction, crime and science fiction, and some specific non-fiction. No film/TV scripts apart from existing clients. No children's, illustrated books or poetry. CLIENTS include Iain Banks, Simon Beckett, Carol Birch, Alan Campbell, Paul Cornell, Laurie Graham, M. John Harrison, Toby Litt, Ken MacLeod, China Miéville, Antony Sher. COMMISSION Home 15%; US & Translation 20%. No unsolicited mss. Approach in writing with publishing history, first two chapters and return postage. No reading fee.

Judith Chilcote Agency*

8 Wentworth Mansions, Keats Grove, London
nw3 2rl
📞 020 7794 3717
judybks@aol.com
Contact *Judith Chilcote*

Founded 1990. Handles commercial fiction, TV tie-ins, health and nutrition, popular psychology, biography and celebrity autobiography and current affairs. COMMISSION Home 15%; Overseas 20–25%. *No* academic, science fiction, short stories, film scripts or poetry. *No approaches by e-mail.* Send letter with c.v., synopsis, three chapters and s.a.e. for return. No reading fee.

Teresa Chris Literary Agency Ltd*

43 Musard Road, London W6 8NR
☎ 020 7386 0633
Contacts *Teresa Chris, Charles Brudenell-Bruce*

Founded 1989. Handles crime, general, women's, commercial and literary fiction, and non-fiction: history, biography, cookery, lifestyle, gardening, etc. Specializes in crime fiction and commercial women's fiction. No scripts. Film and TV rights handled by co-agent. No poetry, short stories, fantasy, science fiction or horror. CLIENTS include Stephen Booth, Martin Davies, Tamara McKinley, Marguerite Patten, Eileen Ramsay, Debby Holt. COMMISSION Home 10%; US & Translation 20%. OVERSEAS ASSOCIATES Patty Moosbrugger Literary Agency, USA; representatives in all other countries. Unsolicited mss welcome. Send query letter with first two chapters plus two-page synopsis (s.a.e. *essential*) in first instance. No reading fee.

Mary Clemmey Literary Agency*

6 Dunollie Road, London NW5 2XP
☎ 020 7267 1290   🖷 020 7813 9757
mcwords@googlemail.com
Contact *Mary Clemmey*

Founded 1992. Handles fiction and non-fiction – high-quality work with an international market. No science fiction, fantasy or children's books. TV, film, radio and theatre scripts from existing clients only. US clients: Frederick Hill Bonnie Nadell Inc., Lynn C. Franklin Associates Ltd, The Miller Agency, Roslyn Targ, Weingel-Fidel Agency Inc, Betsy Amster Literary Enterprises. COMMISSION Home 15%; US & Translation 20%. OVERSEAS ASSOCIATE Elaine Markson Literary Agency, New York. No unsolicited mss and no e-mail submissions. Approach by letter only in the first instance giving a description of the work (include s.a.e.). No reading fee.

Jonathan Clowes Ltd*

10 Iron Bridge House, Bridge Approach, London NW1 8BD
☎ 020 7722 7674   🖷 020 7722 7677
Contacts *Ann Evans, Lisa Thompson*

Founded 1960. Pronounced 'clewes'. Now one of the biggest fish in the pond and not really for the untried unless they are true high-flyers. Fiction and non-fiction, plus scripts. No textbooks or children's. Special interests: situation comedy, film and television rights. CLIENTS include David Bellamy, Len Deighton, Elizabeth Jane Howard, Doris Lessing, David Nobbs, Gillian White and the estate of Kingsley Amis. COMMISSION Home & US 15%; Translation 19%. OVERSEAS ASSOCIATES **Andrew Nurnberg Associates**; Sane Töregard Agency. No unsolicited mss; authors come by recommendation or by successful follow-ups to preliminary letters.

Elspeth Cochrane Personal Management

16 Trinity Close, The Pavement, London SW4 0JD
☎ 020 7622 3566
elspethcochrane@talktalk.net
Contact *Elspeth Cochrane*

Founded 1960. Handles fiction, non-fiction, biographies, screenplays and plays. No children's fiction. CLIENTS include Royce Ryton, Robert Tanitch. COMMISSION 12½%. No unsolicited mss. Phone before submitting work.

Rosica Colin Ltd

1 Clareville Grove Mews, London SW7 5AH
☎ 020 7370 1080   🖷 020 7244 6441
Contact *Joanna Marston*

Founded 1949. Handles all full-length mss, plus theatre, film, television and sound broadcasting but few new writers being accepted. COMMISSION Home 10%; US & Translation 20%. Preliminary letter with return postage essential; writers should outline their writing credits and whether their mss have previously been submitted elsewhere. May take 3–4 months to consider full mss; synopsis preferred in the first instance. No reading fee.

Conville & Walsh Limited*

2 Ganton Street, London W1F 7QL
☎ 020 7287 3030   🖷 020 7287 4545
info@convilleandwalsh.com
www.convilleandwalsh.com
Directors *Clare Conville, Patrick Walsh*
Book Rights *Ben Mason*
Children's *Jo Unwin*
Finance *Alan Oliver*
Foreign Rights *Jake Smith-Bosanquet*
Book Development *Susan Armstrong, Robert Dinsdale*

Established in 2000 by Clare Conville (ex-A.P. Watt) and Patrick Walsh (ex-Christopher Little Literary Agency). Handles all genres of fiction, non-fiction and children's worldwide but no poetry, screenplays or short stories. 'Agency taste is generally upmarket to mass-market: literary and commercial fiction.' Fiction CLIENTS range from the Booker winner DBC Pierre to John Llewellyn Rhys prize-winner Sarah Hall and the Orwell Prize winner Delia Jarrett-Macauley. Commercial novelists include Isabel Wolff, Nick Harkaway, Matt Dunn and Michael Cordy. Non-fiction CLIENTS Tom Holland, Helen Castor, Simon Singh, Arthur Potts Dawson, Kate Spicer, Richard Wiseman, Ian Stewart, Ruth Padel, Misha Glenny, Michael Bywater. Also handles the estate of artist Francis Bacon. Prestigious children's and young adult list ranges from John Burningham to Steve Voake and the estate of Astrid Lindgren. COMMISSION Home 15%; US & Translation 20%. Send first three chapters, one-page synopsis, covering letter and s.a.e. No reading fee.

Jane Conway-Gordon Ltd*

1 Old Compton Street, London W1D 5JA
☎ 020 7494 0148   🖷 020 7287 9264
jconway_gordon@dsl.pipex.com
Contact *Jane Conway-Gordon*

Founded 1982. Handles fiction and general non-fiction. No poetry, science fiction, children's or short stories. COMMISSION Home 15%; US & Translation 20%. OVERSEAS ASSOCIATES Lyons Literary LLC, New York; plus agencies throughout Europe and Japan. Unsolicited mss welcome; preliminary letter and return postage essential. No reading fee.

Rupert Crew Ltd*

1A King's Mews, London WC1N 2JA
☎ 020 7242 8586   🖷 020 7831 7914
info@rupertcrew.co.uk (correspondence only)
www.rupertcrew.co.uk
Contacts *Doreen Montgomery, Caroline Montgomery*

Founded 1927. International representation, handling volume and subsidiary rights in fiction and non-fiction properties. No screenplays, plays or poetry, journalism or short stories, science fiction or fantasy. COMMISSION Home 15%; Elsewhere 20%. Preliminary letter and return postage essential. No reading fee.

Curtis Brown Group Ltd*

Haymarket House, 28/29 Haymarket, London SW1Y 4SP
☎ 020 7393 4400   🖷 020 7393 4401
cb@curtisbrown.co.uk
www.curtisbrown.co.uk
CEO *Jonathan Lloyd*
Director of Operations *Ben Hall*
Directors *Jacquie Drewe, Jonny Geller, Nick Marston, Sarah Spear*
Books *Jonny Geller (MD, Book Division), Jonathan Lloyd, Camilla Hornby, Vivienne Schuster, Elizabeth Sheinkman, Janice Swanson, Gordon Wise, Karolina Sutton*
Joint Head of Foreign Rights *Betsy Robbins, Kate Cooper*
Foreign Rights Agents *Carol Jackson, Katie McGowan, Elizabeth Iveson (on behalf of ICM books), Daisy Meyrick (on behalf of IC books)*
Film/TV/Theatre *Nick Marston (MD, Media Division), Tally Garner, Ben Hall, Joe Phillips, Amanda Davis*
Actors *Grace Clissold, Maxine Hoffman, Sarah MacCormick, Sarah Spear, Kate Staddon, Claire Stannard, Lucy Johnson, Mary Fitzgerald*
Presenters *Jacquie Drewe*

Founded 1899. Agents for the negotiation in all markets of novels, general non-fiction, children's books and associated rights (including multimedia) as well as plays, film, theatre, TV and radio scripts. OVERSEAS ASSOCIATES in Australia and the US. Send outline for non-fiction and short synopsis for fiction with two/three sample chapters and autobiographical note. No reading fee. Return postage essential. No submissions by e-mail. See further submission guidelines on website. Also represents playwrights, film and TV writers and directors, theatre directors and designers, TV and radio presenters and actors.

Judy Daish Associates Ltd

2 St Charles Place, London W10 6EG
☎ 020 8964 8811   🖷 020 8964 8966
Contacts *Judy Daish, Tracey Elliston, Howard Gooding*

Founded 1978. Theatrical literary agent. Handles scripts for film, TV, theatre and radio. No books. Preliminary letter essential. No unsolicited mss.

Caroline Davidson Literary Agency*

5 Queen Anne's Gardens, London W4 1TU
☎ 020 8995 5768   🖷 020 8994 2770
cdla@ukgateway.net
www.cdla.co.uk
Contact *Caroline Davidson*

Founded 1988. Handles a wide range of original fiction and non-fiction. Non-fiction includes archaeology, architecture, art, biography, climate change, current affairs, design, gardening, health, history, medicine, natural history, reference, science. Finished, polished first novels positively welcomed. No thrillers, crime, fantasy, science fiction. Short stories, children's, plays, poetry not considered. CLIENTS Peter Barham, Andrew Beatty, Andrew Dalby, Emma Donoghue, Chris Greenhalgh, Ed Husain, Tom Jaine, Simon Unwin. COMMISSION Home & Commonwealth, US, Translation 12½%; 20% if sub-agents are involved in foreign sales or films. Send an initial letter giving details of the project, plus c.v. and s.a.e. Non-fiction writers should include a detailed proposal with chapter synopsis. Fiction writers should submit the first 50 and last ten pages of their novel. No response to submissions sent by fax or e-mail and those without return postage/s.a.e. Absolutely no telephone enquiries.

Merric Davidson Literary Agency
▷ MBA Literary Agents Ltd

Felix de Wolfe

Kingsway House, 103 Kingsway, London WC2B 6QX
☎ 020 7242 5066   🖷 020 7242 8119
info@felixdewolfe.com
Contact *Felix de Wolfe*

Founded 1938. Handles quality fiction only, and scripts. No non-fiction or children's. CLIENTS include Jan Butlin, Jeff Dowson, Brian Glover, Sheila Goff, Aileen Gonsalves, John Kershaw, Ray Kilby, Bill MacIlwraith, Angus Mackay, Gerard McLarnon, Malcolm Taylor, David Thompson, Paul Todd, Dolores Walshe. COMMISSION Home 12½%; US 20%. No unsolicited mss. No reading fee.

The Dench Arnold Agency

10 Newburgh Street, London W1F 7RN
📧 020 7437 4551   📠 020 7439 1355
contact@dencharnold.co.uk
www.dencharnold.co.uk
Contacts *Elizabeth Dench, Michelle Arnold*

Founded 1972. Handles scripts for TV and film. CLIENTS include Peter Chelsom. COMMISSION Home 10–15%. OVERSEAS ASSOCIATES include William Morris/Sanford Gross and C.A.A., Los Angeles. Unsolicited mss will be read, but a letter with sample of work and c.v. (plus s.a.e.) is required.

Dorian Literary Agency (DLA)*

Upper Thornehill, 27 Church Road, St Marychurch, Torquay TQ1 4QY
📧 01803 312095
Contact *Dorothy Lumley*

Founded 1986. Handles popular genre fiction for adults: romance, historicals, sagas; crime and thrillers; science fiction, fantasy, horror. No short stories, poetry, autobiography, film/TV scripts or plays. CLIENTS include Gillian Bradshaw, Gary Gibson, Kate Hardy, Stephen Jones, Brian Lumley, Amy Myers, Rosemary Rowe, Lyndon Stacey. COMMISSION Home 12½% for new clients; US 15%; Translation 20–25%. Works with agents in most countries for translation. Introductory letter with outline and 1–3 chapters (with return postage/s.a.e.) only, please. No enquiries or submissions by telephone, fax or e-mail. No reading fee.

Toby Eady Associates Ltd

9 Orme Court, London W2 4RL
📧 020 7792 0092   📠 020 7792 0879
toby@tobyeady.demon.co.uk
www.tobyeadyassociates.co.uk
Contacts *Toby Eady, Laetitia Rutherford*

Handles fiction, and non-fiction. Special interests: China, Middle East, Africa, India. CLIENTS include Mark Burnell, Michaela Clarke, Bernard Cornwell, John Carey, Yasmin Crowther, Rana Dasgupta, Gavin Esler, Fadia Faqir, Sophie Gee, Samson Kambalu, Richard Lloyd Parry, Julia Lovell, Francesca Marciano, Lawrence Potter, Deborah Scroggins, Rachel Seiffert, Samia Serageldin, John Stubbs, Diane Wei Liang, Robert Winder, Fan Wu, Xinran Xue. COMMISSION Home 15%; Elsewhere 20%. OVERSEAS ASSOCIATES USA: ICM; France: La Nouvelle Agence; Holland: Jan Michael; deal direct: Italy, Spain, Germany, Portugal, Scandinavia; China, Taiwan: The Bardon Chinese Media Agency; Japan: The English Agency; Korea: The Eric Yang Agency; Eastern Europe: Prava I Prevodi; Russia: Permissions & Rights Ltd; Turkey: Akcali Copyright; Czech Republic: Kristin Olson; Hungary: Katai & Bolza; Greece: JLM. No film/TV scripts or poetry.

Eddison Pearson Ltd*

West Hill House, 6 Swains Lane, London N6 6QS
📧 020 7700 7763   📠 020 7700 7866
mail@eddisonpearson.com
Contact *Clare Pearson*

Founded 1996. Small, personally-run agency. Handles children's books, fiction and non-fiction, poetry. CLIENTS include Sue Heap, Robert Muchamore, Valerie Bloom. COMMISSION Home 10%; US & Translation 15–20%. Enquiries and submissions by e-mail only. Please e-mail for up-to-date submission guidelines by return. No reading fee.

Edwards Fuglewicz*

49 Great Ormond Street, London WC1N 3HZ
📧 020 7405 6725   📠 020 7405 6726
Contacts *Ros Edwards, Helenka Fuglewicz, Julia Forrest*

Founded 1996. Handles literary and commercial fiction (not children's, science fiction, horror or fantasy); non-fiction: biography, history, popular culture. COMMISSION Home 15%; US & Translation 20%. No unsolicited mss or e-mail submissions.

Faith Evans Associates*

27 Park Avenue North, London N8 7RU
📧 020 8340 9920   📠 020 8340 9410
faith@faith-evans.co.uk

Founded 1987. Small agency. CLIENTS include Melissa Benn, the estate of Madeleine Bourdouxhe, Eleanor Bron, Caroline Conran, Alicia Foster, Midge Gillies, Ed Glinert, Vesna Goldsworthy, Cate Haste, Jim Kelly, Helena Kennedy, Seumas Milne, Tom Paulin, Sheila Rowbotham, the estate of Lorna Sage, Rebecca Stott, Harriet Walter, Elizabeth Wilson. COMMISSION Home 15%; US & Translation 20%. OVERSEAS ASSOCIATES worldwide. List full; no submissions, please.

The Feldstein Agency

123–125 Main Street, 2nd Floor, Bangor BT20 4AE
📧 028 9147 2823
paul@thefeldsteinagency.co.uk
www.thefeldsteinagency.co.uk

Contacts *Paul Feldstein, Susan Feldstein*

Founded 2007. Handles commercial fiction, literary fiction and criticism, romantic fiction, science fiction, thrillers, short stories, essays, adventure, autobiography/biography, business, cookery, crime, current affairs, football, golf, guidebooks, heritage, history, humanities, humour, Celtic and Irish interest, journalism, leisure, lifestyle, local history, media, memoirs, military, music, mysteries, philosophy, politics, sociology, travel, war, wine, women's interest and world affairs. CLIENTS Mary O'Sullivan, Ray Strobel, Adrian White. COMMISSION Home 15%; US & Translation 20%. Unsolicited mss, sample chapters and synopses welcome; approach by e-mail. No children's books. No reading fee.

Laurence Fitch Ltd

Mezzanine, Quadrant House, 80–82 Regent Street, London W1B 5AU
☎ 020 7734 9911
information@laurencefitch.com
www.laurencefitch.com
Contact *Brendan Davis*

Founded 1952, incorporating the London Play Company (1922) and in association with Film Rights Ltd (1932). Handles children's and horror books, scripts for theatre, film, TV and radio only. CLIENTS include Carlo Ardito, Hindi Brooks, John Chapman & Ray Cooney, Jeremy Lloyd, Dave Freeman, John Graham, Robin Hawdon, Glyn Robbins, Lawrence Roman, Gene Stone, the estate of Dodie Smith, Edward Taylor. COMMISSION UK 10%; Overseas 15%. OVERSEAS ASSOCIATES worldwide. No unsolicited mss. Send synopsis with sample scene(s)/first chapters in the first instance. No reading fee.

Jill Foster Ltd

9 Barb Mews, Brook Green, London W6 7PA
☎ 020 7602 1263   ☎ 020 7602 9336
www.jflagency.com
Contacts *Alison Finch, Simon Williamson, Dominic Lord, Gary Wild, Jill Foster*

Founded 1976. Handles scripts for TV, radio, film and theatre. No fiction, short stories or poetry. CLIENTS include Ian Brown, Grant Cathro, Phil Ford, Nev Fountain and Tom Jamieson, Rob Gittins, Wayne Jackman, David Lane, Tony Millan and Mike Walling, Jim Pullin and Fraser Steele, Pete Sinclair, Paul Smith, Peter Tilbury, Roger Williams, Susan Wilkins. COMMISSION Home 12½%; US & Translation 15%. No unsolicited mss; approach by letter in the first instance. No approaches by e-mail. No reading fee.

Fox & Howard Literary Agency*

4 Bramerton Street, London SW3 5JX
☎ 020 7352 8691   ☎ 020 7352 8691

Contacts *Chelsey Fox, Charlotte Howard*

Founded 1992. A small agency, specializing in non-fiction, that prides itself on 'working closely with its authors'. Handles biography, history and popular culture, reference, business, mind, body and spirit, health and personal development, popular psychology. COMMISSION Home 15%; US & Translation 20%. No unsolicited mss; send letter and synopsis with s.a.e. for response. No reading fee.

Fraser Ross Associates

6 Wellington Place, Edinburgh EH6 7EQ
☎ 0131 657 4412/0131 553 2759
kjross@tiscali.co.uk
lindsey.fraser@tiscali.co.uk
www.fraserross.co.uk
Contacts *Lindsey Fraser, Kathryn Ross*

Founded 2002. Handles children's books, adult literary and mainstream fiction. No poetry, short stories, adult fantasy and science fiction, academic, scripts. CLIENTS Thomas Bloor, Sally J. Collins, Vivian French, Janey Louise Jones, Tanya Landman, Joan Lennon, Iain MacIntosh, Jack McLean, Michaela Morgan, Dugald Steer, Linda Strachan. Send preliminary letter, synopsis, first three chapters or equivalent, c.v. and s.a.e. No reading fee.

Futerman, Rose & Associates*

91 St Leonards Road, London SW14 7BL
☎ 020 8255 7755   ☎ 020 8286 4860
enquiries@futermanrose.co.uk
www.futermanrose.co.uk
Contact *Guy Rose*

Founded 1984. Handles fiction, biography, show business, current affairs, teen fiction and scripts for TV and film. No children's, science fiction or fantasy. CLIENTS include Larry Barker, Christian Piers Betley, David Bret, Tom Conti, Iain Duncan Smith, Sir Martin Ewans, Susan George, Paul Hendy, Russell Warren Howe, Keith R. Lindsay, Eric MacInnes, Paul Marx, Max Morgan-Witts, John Repsch, Liz Rettig, Peter Sallis, Paul Stinchcombe, Gordon Thomas, Dr Mark White, Toyah Willcox, Simon Woodham, Allen Zeleski. No unsolicited mss. Send preliminary letter with brief resumé, synopsis, first 20pp and s.a.e.

Jüri Gabriel

35 Camberwell Grove, London SE5 8JA
☎ 020 7703 6186
Contact *Jüri Gabriel*

Handles quality fiction and non-fiction and (almost exclusively for existing clients) film, TV and radio rights. Jüri Gabriel worked in television, wrote books for 20 years and is chairman of Dedalus publishers. No short stories, articles, verse or books for children. CLIENTS include Maurice Caldera, Diana Constance, Miriam Dunne, Matt Fox, Paul Genney, Pat Gray, Mikka Haugaard, Robert Irwin, Andrew Killeen, John Lucas, David Madsen, Richard Mankiewicz, David Miller, Andy Oakes, John Outram, Philip Roberts, Dr Stefan Szymanski, Frances Treanor, Jeremy Weingard, Dr Terence White, Dr Robert Youngson. COMMISSION Home 10%; US & Translation 20%. Unsolicited mss ('two-page synopsis and three sample chapters in first instance, please') welcome if accompanied by return postage and letter giving sufficient information about author's writing experience, aims, etc.

Eric Glass Ltd

25 Ladbroke Crescent, London W11 1PS
☎ 020 7229 9500   📠 020 7229 6220
eglassltd@aol.com
Contact *Janet Glass*

Founded 1934. Handles fiction, non-fiction and scripts for publication or production in all media. No poetry, short stories or children's works. CLIENTS include Pierre Chesnot, Charles Dyer, Henry Fleet, Tudor Gates, Philip Goulding, Pauline Macaulay, Brendan Murray and the estates of Rodney Ackland, Marc Camoletti, Jean Cocteau, Warwick Deeping, William Douglas Home, Philip King, Robin Maugham, Alan Melville, Beverley Nichols, Jean-Paul Sartre. COMMISSION Home 15%; US & Translation 20%. OVERSEAS ASSOCIATES in the US, Australia, Czech Republic, France, Germany, Greece, Holland, Italy, Japan, Poland, Scandinavia, Slovakia, South Africa, Spain. No unsolicited mss. Return postage required. No reading fee.

David Godwin Associates

55 Monmouth Street, London WC2H 9DG
☎ 020 7240 9992   📠 020 7395 6110
assistant@davidgodwinassociates.co.uk
www.davidgodwinassociates.co.uk
Contact *David Godwin*

Founded 1996. Handles literary and general fiction, non-fiction, biography. No scripts, science fiction or children's. No reading fee. COMMISSION Home 15%; Overseas 15%. 'Please contact the office before making a submission.'

Graham Maw Christie Literary Agency

19 Thornhill Crescent, London N1 1BJ
☎ 020 7737 4766
enquiries@grahammawchristie.com
www.grahammawchristie.com
Contacts *Jane Graham Maw, Jennifer Christie*

Founded 2005. Handles general non-fiction. No children's or poetry.

Annette Green Authors' Agency*

1 East Cliff Road, Tunbridge Wells TN4 9AD
☎ 01892 514275   📠 01892 558262
david@annettegreenagency.co.uk
www.annettegreenagency.co.uk
Contact *Address material to the Agency*

Founded 1998. Handles literary and general fiction and non-fiction, popular culture and current affairs, science, music, film, history, biography, older children's and teenage fiction. No dramatic scripts or poetry. CLIENTS include Andrew Baker, Bill Broady, Meg Cabot, Simon Conway, Justin Hill, Mary Hogan, Anvar Khan, Maria McCann, Adam Macqueen, Ian Marchant, Professor Charles Pasternak, Kirsty Scott, Bernadette Strachan, Elizabeth Woodcraft. COMMISSION Home 15%; US & Translation 20%. Letter, synopsis, sample chapters and s.a.e. essential. No reading fee.

Christine Green Authors' Agent*

6 Whitehorse Mews, Westminster Bridge Road, London SE1 7QD
☎ 020 7401 8844   📠 020 7401 8860
info@christinegreen.co.uk
www.christinegreen.co.uk
Contact *Christine Green*

Founded 1984. Handles fiction (general and literary) and general non-fiction. No scripts, poetry or children's. COMMISSION Home 10%; US & Translation 20%. Initial letter, synopsis and first three chapters preferred. No reading fee but return postage essential.

Louise Greenberg Books Ltd*

The End House, Church Crescent, London N3 1BG
☎ 020 8349 1179   📠 020 8343 4559
louisegreenberg@msn.com
Contact *Louise Greenberg*

Founded 1997. Handles full-length literary fiction and serious non-fiction only. COMMISSION Home 15%; US & Translation 20%. CHILDREN'S ASSOCIATE **Sarah Manson Literary Agent**. *No telephone approaches.* No reading fee. S.a.e. essential.

Greene & Heaton Ltd*

37 Goldhawk Road, London W12 8QQ
☎ 020 8749 0315   📠 020 8749 0318
info@greeneheaton.co.uk
www.greeneheaton.co.uk
Contacts *Carol Heaton, Judith Murray, Antony Topping, Linda Davis*

A medium-sized agency with a broad range of clients. Handles all types of fiction and non-fiction. No original scripts for theatre, film or TV. CLIENTS include Poppy Adams, Mark Barrowcliffe, Bill Bryson, Jan Dalley, Marcus du Sautoy, Suzannah Dunn, Hugh Fearnley-

Whittingstall, Jane Fearnley-Whittingstall, Michael Frayn, P.D. James, William Leith, James McGee, Shona MacLean, Thomasina Miers, The Royal Society, C.J. Sansom, William Shawcross, Sarah Waters. COMMISSION Home 15%; US & Translation 20%. OVERSEAS ASSOCIATES worldwide. No reply to unsolicited submissions without e-mail address and/or return postage.

The Greenhouse Literary Agency Ltd

Stanley House, St Chad's Place, London
WC1X 9HH
📞 020 7841 3959   📠 020 7841 3940
submissions@greenhouseliterary.com
www.greenhouseliterary.com
Contacts *Julia Churchill, Sarah Davies*

Founded January 2008. Handles children's and teen fiction: high-concept young series; 8–12 and young adult fiction. No picture books, pre-school texts or illustrators, stand-alone very young novels of less than 10,000 words, non-fiction, inspirational/religious work, short stories or writing for adults. COMMISSION UK & US 15%; Translation 25%. OVERSEAS ASSOCIATE The Greenhouse Literary Agency Ltd USA (see entry under *US Literary Agents*). See website for submission details. Welcomes query e-mails (no snail mail) with up to five pages (first five) pasted into the body of the e-mail. No reading fee.

Gregory & Company Authors' Agents*
(formerly Gregory & Radice)

3 Barb Mews, London W6 7PA
📞 020 7610 4676   📠 020 7610 4686
info@gregoryandcompany.co.uk (general enquiries)
www.gregoryandcompany.co.uk
Contact *Jane Gregory*
Editorial Manager *Stephanie Glencross*
Rights Manager *Claire Morris*
Rights Executive *Jemma McDonagh*

Founded 1987. Handles all kinds of fiction and general non-fiction. Special interest fiction – literary, commercial, women's fiction, crime, suspense and thrillers. 'We are particularly interested in books which will also sell to publishers abroad.' No original plays, film or TV scripts (only published books are sold to film and TV). No science fiction, fantasy, poetry, academic or children's books. No reading fee. Editorial advice given to own authors. COMMISSION Home 15%; US, Translation, Radio/TV/Film 20%. Is well represented throughout Europe, Asia and US. No unsolicited mss; send a preliminary letter with c.v., synopsis, first three chapters and future writing plans (plus return postage). Short submissions by fax or e-mail (maryjones@gregoryandcompany.co.uk) in the first instance; maximum five pages.

David Grossman Literary Agency Ltd

118b Holland Park Avenue, London W11 4UA
📞 020 7221 2770   📠 020 7221 1445
Contact *Submissions Dept.*

Founded 1976. Handles full-length fiction and general non-fiction – good writing of all kinds and anything healthily controversial. No verse or technical books for students. No original screenplays or teleplays (only works existing in volume form are sold for performance rights). Generally works with published writers of fiction only but 'truly original, well-written novels from beginners' will be considered. COMMISSION Rates vary for different markets. OVERSEAS ASSOCIATES throughout Europe, Asia, Brazil and the US. Best approach by preliminary letter giving full description of the work and, in the case of fiction, with the first 50 pages. 'Please ensure material is printed in at least 1½ spacing on one side of paper only; that all pages are numbered consecutively throughout and in minimum 11pt typeface.' All material must be accompanied by return postage. No approaches or submissions by fax or e-mail. No unsolicited mss. No reading fee.

Gunnmedia Ltd

50 Albemarle Street, London W1S 4BD
📞 020 7529 3745
ali@gunnmedia.co.uk
sarah@gunnmedia.co.uk
www.gunnmedia.co.uk
Contacts *Ali Gunn, Sarah McFadden, Georgina Bean, Jonny McCune*
Head of Foreign Rights *Diana McKay*

Founded 2005. Handles commercial fiction and non-fiction; TV and film scripts. CLIENTS include Jenny Colgan, Sarah Webb, Brian Freeman, Mil Millington, Michael Bilton, Carole Malone, Carol Thatcher, Martina Navratilova, Lowri Turner, Fiona McIntosh, Nabeel Yasin. COMMISSION Home 15%; US & Translation 20%. OVERSEAS ASSOCIATES Gelfman Schneider; Fletcher Parry. Unsolicited mss, sample chapters and synopses welcome; approach in writing. No reading fee.

The Rod Hall Agency Limited

6th Floor, Fairgate House, 78 New Oxford Street, London W1A 1HB
📞 020 7079 7987   📠 0845 638 4094
office@rodhallagency.com
www.rodhallagency.com
Contact *Charlotte Knight*

Founded 1997. Handles drama for film, TV and theatre and writers-directors. Does not represent writers of episodes for TV series where the format is provided but represents originators of series. CLIENTS include Simon Beaufoy (*The Full Monty*), Jeremy Brock (*Mrs Brown*), Liz Lochhead (*Perfect Days*), Martin McDonagh (*The*

Pillowman), Simon Nye (*Men Behaving Badly*). COMMISSION Home 10%; US & Translation 15%. No reading fee.

The Hanbury Agency*

28 Moreton Street, London SW1V 2PE
📞 020 7630 6768
maggie@hanburyagency.com
www.hanburyagency.com
Contact *Margaret Hanbury*

Personally run agency representing quality fiction and non-fiction. No plays, scripts, poetry, children's books, fantasy, horror. CLIENTS include George Alagiah, J.G. Ballard, Simon Callow, Jane Glover, Bernard Hare, Judith Lennox, Katie Price. COMMISSION Home 15%; Overseas 20%. No unsolicited approaches via e-mail.

Roger Hancock Ltd

4 Water Lane, London NW1 8NZ
📞 020 7267 4418   📠 020 7267 0705
Contact *Material should be addressed to the Submissions Department*

Founded 1960. Special interests: comedy drama and light entertainment. COMMISSION Home 10%; Overseas 15%. Unsolicited mss not welcome. Initial phone call required.

Antony Harwood Limited*

103 Walton Street, Oxford OX2 6EB
📞 01865 559615   📠 01865 310660
mail@antonyharwood.com
www.antonyharwood.com
Contacts *Antony Harwood, James Macdonald Lockhart*

Founded 2000. Handles fiction and non-fiction. CLIENTS Amanda Craig, Peter F. Hamilton, Alan Hollinghurst, A.L. Kennedy, Douglas Kennedy, Dorothy Koomson, Chris Manby, George Monbiot, Garth Nix, Tim Parks. COMMISSION Home 15%; US & Translation 20%. Send letter and synopsis with return postage in the first instance. No reading fee.

A.M. Heath & Co. Ltd*

6 Warwick Court, London WC1R 5DJ
📞 020 7242 2811   📠 020 7242 2711
www.amheath.com
Contacts *Bill Hamilton, Sarah Molloy, Victoria Hobbs, Euan Thorneycroft*

Founded 1919. Handles fiction, general non-fiction and children's. No dramatic scripts, poetry, picture books or short stories. CLIENTS include Christopher Andrew, Nadeem Aslam, Bella Bathurst, Anita Brookner, Geoff Dyer, R.J. Ellory, Katie Fforde, Graham Hancock, Conn Iggulden, Marina Lewycka, Lucy Magan, Hilary Mantel, Joanna Nadin, Maggie O'Farrell, Susan Price, John Sutherland, Barbara Trapido. COMMISSION Home

10–15%; US & Translation 20%; Film & TV 20%. OVERSEAS ASSOCIATES in the US, Europe, South America, Japan and the Far East. Preliminary letter and synopsis essential. No reading fee.

Rupert Heath Literary Agency

50 Albemarle Street, London W1S 4BD
📞 020 7788 7807   📠 020 7691 9331
emailagency@rupertheath.com
Contact *Rupert Heath*

Founded 2000. Handles literary and general fiction and non-fiction, including history, biography and autobiography, current affairs, popular science, the arts and popular culture. No scripts or poetry. COMMISSION Home 15%; US & Translation 20%. OVERSEAS ASSOCIATES worldwide. Approach with e-mail initially ('telling us a bit about yourself and the book you wish to submit'). No reading fee.

Henser Literary Agency

174 Pennant Road, Llanelli SA14 8HN
📞 01554 753520   📠 01554 753520
Contact *Steve Henser*

Founded 2002. Handles fiction: mystery and general; science fiction and fantasy; translation between English and Japanese; non-fiction book on Japan. Also TV, film, radio and theatre scripts. No horror or sadism. No reading fee. COMMISSION Home 15%; US 20%. No unsolicited material. Not looking for additional clients at present. No reading fee.

hhb agency ltd*

122 Arlington Road, London N1 7HP
📞 020 7485 0044
heather@hhbagency.com
elly@hhbagency.com
www.hhbagency.com
Contacts *Heather Holden-Brown, Elly James*

Founded in 2005 by Heather Holden-Brown, a publishing editor for 20 years with Waterstone's, Harrap, BBC Books and Headline. Handles non-fiction: history and politics, contemporary autobiography/biography, popular culture, entertainment and TV, business, family memoir, food and cookery a speciality. No scripts. COMMISSION Home 15%. No unsolicited mss; please contact by e-mail or telephone before sending any material. No reading fee.

David Higham Associates Ltd*

5–8 Lower John Street, Golden Square, London W1F 9HA
📞 020 7434 5900   📠 020 7437 1072
dha@davidhigham.co.uk
www.davidhigham.co.uk
Books *Anthony Goff, Bruce Hunter, Veronique Baxter, Lizzy Kremer, Caroline Walsh (children's), Andrew Gordon, Georgia Glover, Alice Williams*

Scripts *Nicky Lund, Georgina Ruffhead*

Founded 1935. Handles fiction, general non-fiction (biography, history, current affairs, etc.) and children's books. Also scripts. CLIENTS include J.M. Coetzee, Stephen Fry, Jane Green, James Herbert, Alexander McCall Smith, Lynne Truss, Jacqueline Wilson. COMMISSION Home 15%; US & Translation 20%; Scripts 10%. Preliminary letter with synopsis essential in first instance. No e-mail submissions. No reading fee. See website for further information.

Vanessa Holt Ltd*

59 Crescent Road, Leigh-on-Sea SS9 2PF

☎ 01702 473787   📠 01702 473787

Contact *Vanessa Holt*

Founded 1989. Handles general fiction, non-fiction and non-illustrated children's books. No scripts, poetry, academic or technical. Specializes in crime fiction, commercial and literary fiction, and particularly interested in books with potential for sales abroad and/or to TV. COMMISSION Home 15%; US & Translation 20%; Radio/TV/Film 15%. Represented in all foreign markets. Approach by letter or telephone only; no unsolicited mss.

Kate Hordern Literary Agency

18 Mortimer Road, Clifton, Bristol BS8 4EY

☎ 0117 923 9368

katehordern@blueyonder.co.uk

Contact *Kate Hordern*

Founded 1999. Handles quality fiction and general non-fiction. CLIENTS Richard Bassett, Jeff Dawson, Leigh Eduardo, Kylie Fitzpatrick, Duncan Hewitt, J.T. Lees, Will Randall. COMMISSION Home 15%; US & Translation 20%. OVERSEAS ASSOCIATES VVV Agency, France; Carmen Balcells Agency, Spain; Synopsis Agency, Russia and various agencies in Asia. Approach in writing (or by e-mail) in the first instance with details of project. New clients taken on very selectively. Synopsis required for fiction; proposal/chapter breakdown for non-fiction. Sample chapters on request only. S.a.e. essential. No reading fee.

Valerie Hoskins Associates Limited

20 Charlotte Street, London W1T 2NA

☎ 020 7637 4490   📠 020 7637 4493

vha@vhassociates.co.uk

Contacts *Valerie Hoskins, Rebecca Watson*

Founded 1983. Handles scripts for film, TV and radio. Special interests feature films, animation and TV. COMMISSION Home 12½%; US 20% (maximum). No unsolicited scripts; preliminary letter of introduction essential. No reading fee.

Independent Talent Group Limited*

Oxford House, 76 Oxford Street, London W1D 1BS

☎ 020 7636 6565   📠 020 7323 0101

Contacts *Sue Rodgers, Jessica Sykes, Cathy King, Greg Hunt, Hugo Young, Michael McCoy, Duncan Heath, Paul Lyon-Maris*

Founded 1973. Specializes in scripts for film, theatre, TV and radio. No reading fee.

The Inspira Group

5 Bradley Road, Enfield EN3 6ES

☎ 020 8292 5163   📠 0870 139 3057

darin@theinspiragroup.com

www.theinspiragroup.com

Managing Director *Darin Jewell*

Founded 2001. Handles children's books, crime fiction, lifestyle/relationships, science fiction/fantasy, humour, non-fiction. No scripts. CLIENTS Stephanie Baudet, Fergus O'Connell, Michael G.R. Tolkien, Simon Brown, Simon Hall. COMMISSION Home & US 15%. Unsolicited mss and synopses welcome; approach by e-mail or telephone. No reading fee, but may charge admin fees.

Intercontinental Literary Agency*

Centric House, 390–391 Strand, London WC2R 0LT

☎ 020 7379 6611   📠 020 7240 4724

ila@ila-agency.co.uk

www.ila-agency.co.uk

Contacts *Nicki Kennedy, Sam Edenborough, Tessa Girvan, Katherine West, Mary Esdaile, Jenny Robson*

Founded 1965. Represents translation rights only on behalf of Lucas Alexander Whitley; Luigi Bonomi Associates; Mulcahy Conway Associates; The Viney Agency; Mary Cunnane Agency (NSW); Wade & Doherty; John Beaton Writers' Agent; Jane Conway-Gordon; Short Books; Free Agents; Bell Lomax Moreton; The Agency (Hilary Delamere clients only); Stimola Literary Studio (New York); Elyse Cheney Literary Associates (New York); Barer Literary (New York); Park Literary Group (New York); Harold Matson Company (New York); The Steinberg Agency (New York); The Turnbull Agency (John Irving); and selected authors on behalf of various other agencies. No unsolicited submissions accepted.

Barrie James Literary Agency

Rivendell, Kingsgate Close, Torquay TQ2 8QA

☎ 01803 326617

mail@newauthors.org.uk

www.newauthors.org.uk

Contact *Barrie E. James*

Founded 1997. Includes New Authors Showcase an internet site for new writers and poets to display their work to publishers and others. *No unsolicited mss.* First contact: s.a.e. or e-mail.

Janklow & Nesbit (UK) Ltd*

33 Drayson Mews, London W8 4LY
☎ 020 7376 2733   📠 020 7376 2915
queries@janklow.co.uk
www.janklowandnesbit.co.uk
Agents *Tif Loehnis, Claire Paterson, Will Francis*
Head of Rights *Rebecca Folland*

Founded 2000. Handles fiction and non-fiction; commercial and literary. Send full outline (non-fiction), synopsis and three sample chapters (fiction) plus informative covering letter and return postage. No reading fee. No poetry, plays or film/TV scripts. US rights handled by **Janklow & Nesbit Associates** in New York (see entry under *US Literary Agents*).

Johnson & Alcock*

Clerkenwell House, 45–47 Clerkenwell Green, London EC1R 0HT
☎ 020 7251 0125   📠 020 7251 2172
info@johnsonandalcock.co.uk
www.johnsonandalcock.co.uk
Contacts *Andrew Hewson, Michael Alcock, Anna Power, Ed Wilson*

Founded 1956. Handles literary and commercial fiction, children's 9+ fiction; general non-fiction including current affairs, biography, memoirs, history, culture, design, lifestyle and health, film and music, graphic novels. No poetry, screenplays, technical or academic material. COMMISSION Home 15%; US & Translation 20%. For fiction, send first three chapters, full synopsis and covering letter. For non-fiction, send covering letter with details of writing and other relevant experience, plus full synopsis. No response to e-mail submissions. No reading fee but return postage essential. Will give editorial advice on work with exceptional promise.

Jane Judd Literary Agency*

18 Belitha Villas, London N1 1PD
☎ 020 7607 0273   📠 020 7607 0623
www.janejudd.com
Contact *Jane Judd*

Founded 1986. Handles general fiction and non-fiction: women's fiction, crime, thrillers, literary fiction, humour, biography, investigative journalism, health, women's interests and travel. 'Looking for good contemporary women's fiction but not Mills & Boon-type; also historical fiction and non-fiction.' No scripts, academic, gardening, short stories or DIY. CLIENTS include Andy Dougan, Cliff Goodwin, Jill Mansell, Jonathon Porritt, Rosie Rushton, Manda Scott, David Winner. COMMISSION Home 10%; US & Translation 20%. Approach with letter, including

synopsis, first chapter and s.a.e. Initial telephone call helpful in the case of non-fiction.

Michelle Kass Associates*

85 Charing Cross Road, London WC2H 0AA
☎ 020 7439 1624   📠 020 7734 3394
office@michellekass.co.uk
Contacts *Michelle Kass, Andrew Mills*

Founded 1991. Handles literary fiction, film and television primarily. COMMISSION Home 10%; US & Translation 10–20%. No mss accepted without preliminary phone call. No e-mail submissions. No reading fee.

Frances Kelly*

111 Clifton Road, Kingston upon Thames KT2 6PL
☎ 020 8549 7830   📠 020 8547 0051
Contact *Frances Kelly*

Founded 1978. Handles non-fiction, including illustrated: biography, history, art, self-help, food & wine, complementary medicine and therapies, finance and business books; and academic non-fiction in all disciplines. No scripts except for existing clients. COMMISSION Home 10%; US & Translation 20%. No unsolicited mss. Approach by letter with brief description of work or synopsis, together with c.v. and return postage.

Knight Features

20 Crescent Grove, London SW4 7AH
☎ 020 7622 1467   📠 020 7622 1522
peter@knightfeatures.co.uk
Contacts *Peter Knight, Samantha Ferris, Gaby Martin, Andrew Knight*

Founded 1985. Handles motor sports, cartoon books, puzzles, business, history, factual and biographical material. No poetry, science fiction or cookery. CLIENTS include Frank Dickens, Gray Jolliffe, Angus McGill, Chris Maslanka, Barbara Minto. COMMISSION dependent upon authors and territories. OVERSEAS ASSOCIATES United Media, US; Auspac Media, Australia; Puzzle Co., New Zealand. No unsolicited mss and no e-mail submissions. Send letter accompanied by c.v. and s.a.e. with synopsis of proposed work.

LAW*

14 Vernon Street, London W14 0RJ
☎ 020 7471 7900   📠 020 7471 7910
Contacts *Mark Lucas, Julian Alexander, Araminta Whitley, Alice Saunders, Peta Nightingale, Philippa Milnes-Smith (children's), Aisha Mobin (children's)*

Founded 1996. Handles full-length commercial and literary fiction, non-fiction and children's books. No fantasy (except children's), plays, poetry or textbooks. Film and TV scripts handled for established clients only. COMMISSION Home 15%; US & Translation 20%. OVERSEAS

ASSOCIATES worldwide. Unsolicited mss considered; send brief covering letter, short synopsis and two sample chapters. S.a.e. essential. No e-mailed or disk submissions.

LBA
▷ Luigi Bonomi Associates Limited

Barbara Levy Literary Agency*
64 Greenhill, Hampstead High Street, London NW3 5TZ
☎ 020 7435 9046   📠 020 7431 2063
Contacts *Barbara Levy, John Selby*

Founded 1986. Handles general fiction, non-fiction, and film and TV rights. COMMISSION Home 15%; US 20%; Translation by arrangement, in conjunction with **The Buckman Agency**. US ASSOCIATE Marshall Rights. No unsolicited mss. No reading fee.

Limelight Management*
33 Newman Street, London W1T 1PY
☎ 020 7637 2529   📠 020 7637 2538
limelight.management@virgin.net
www.limelightmanagement.com
Contacts *Fiona Lindsay, Mary Bekhait*

Founded 1991. Specialists in general non-fiction in the areas of cookery, gardening, antiques, interior design, wine, art and crafts, fashion, beauty and health. No academic text, no poetry, short stories, plays, musicals or short films. COMMISSION Home 15%; US & Translation 20%. Unsolicited mss welcome; send preliminary letter (s.a.e. essential). No reading fee.

The Lindsay Literary Agency
East Worldham House, East Worldham, Alton GU34 3AT
☎ 01420 83143
info@lindsayliteraryagency.co.uk
Contact *Becky Bagnell*

Founded 2005. Handles literary fiction, non-fiction and children's fiction. No scripts. AUTHORS include Sue Lloyd Roberts, Stella Wiseman. COMMISSION Home 12%. Send letter with first three chapters and s.a.e. No reading fee.

The Christopher Little Literary Agency*
Eel Brook Studios, 125 Moore Park Road, London SW6 4PS
☎ 020 7736 4455   📠 020 7736 4490
info@christopherlittle.net
[firstname]@christopherlittle.net
www.christopherlittle.net
Contacts *Christopher Little, Romily Must*

Founded 1979. Handles commercial and literary full-length fiction and non-fiction. No poetry, plays, science fiction, fantasy, textbooks, illustrated children's or short stories. Film scripts for established clients only. AUTHORS include

Paul Bajoria, Janet Gleeson, Carol Hughes, Alastair MacNeill, Robert Mawson, Haydn Middleton, A.J. Quinnell, Christopher Matthew, Robert Radcliffe, J.K. Rowling, Darren Shan, D.B. Shan, Wladyslaw Szpilman, John Watson, Pip Vaughan-Hughes, Christopher Hale, Peter Howells, Gen. Sir Mike Jackson, Shiromi Pinto, Anne Zouroudi. COMMISSION UK 15%; US, Canada, Translation, Audio, Motion Picture, TV & Electronic 20%.

London Independent Books
26 Chalcot Crescent, London NW1 8YD
☎ 020 7706 0486   📠 020 7724 3122
Proprietor *Carolyn Whitaker*

Founded 1971. A self-styled 'small and idiosyncratic' agency. Handles fiction and non-fiction reflecting the tastes of the proprietor. All subjects considered (except computer books and young children's), providing the treatment is strong and saleable. Scripts handled only if by existing clients. Special interests: boats, travelogues, commercial fiction, science fiction and fantasy, teenage fiction. COMMISSION Home 15%; US & Translation 20%. No unsolicited mss; letter, synopsis and first two chapters with return postage the best approach. No reading fee.

The Andrew Lownie Literary Agency*
36 Great Smith Street, London SW1P 3BU
☎ 020 7222 7574   📠 020 7222 7576
lownie@globalnet.co.uk
www.andrewlownie.co.uk
Contact *Andrew Lownie*

Founded 1988. Specializes in non-fiction, especially history, biography, current affairs, military history, UFOs, reference and packaging celebrities and journalists for the book market. No poetry, short stories or science fiction. Formerly a journalist, publisher and himself the author of 12 non-fiction books, Andrew Lownie's CLIENTS include Richard Aldrich, Juliet Barker, the Joyce Cary estate, Tom Devine, Duncan Falconer, Jonathan Fryer, Laurence Gardner, Cathy Glass, Timothy Good, David Hasselhoff, Lawrence James, Damien Lewis, Julian Maclaren-Ross estate, Patrick MacNee, Norma Major, Sir John Mills, Nick Pope, Martin Pugh, Desmond Seward, David Stafford. COMMISSION Worldwide 15%. Preferred approach by e-mail in format suggested on website.

Lucas Alexander Whitley
▷ LAW

Lucy Luck Associates (In association with Aitken Alexander Associates Ltd)
18–21 Cavaye Place, London SW10 9PT
☎ 020 7373 8672

lucy@lucyluck.com
www.lucyluck.com
Contact *Lucy Luck*

Founded 2006. Handles quality fiction and non-fiction. No TV, film, radio or theatre scripts; no illustrated or children's books. CLIENTS include Jon Hotten, Doug Johnstone, Ewan Morrison, Philip Ó Ceallaigh, Catherine O'Flynn, Rebbecca Ray, Adam Thorpe. COMMISSION Home 15%; US & Translation 20%. Send sample chapters by post. No reading fee.

Lutyens and Rubinstein*

231 Westbourne Park Road, London W11 1EB
☎ 020 7792 4855   📠 020 7792 4833
submissions@lutyensrubinstein.co.uk
www.lutyensrubinstein.co.uk
Partners *Sarah Lutyens, Felicity Rubinstein*

Founded 1993. Handles adult fiction and non-fiction books. No TV, film, radio or theatre scripts. COMMISSION Home 15%; US & Translation 20%. Unsolicited mss accepted by e-mail; send introductory letter, c.v. and two chapters. No reading fee.

Luxton Harris Ltd*

2 Deanery Street, London W1K 1AU
☎ 020 7629 8325
rebecca.winfield@btopenworld.com
Contacts *Rebecca Winfield, Jonathan Harris, David Luxton*

Founded 2003. Specializes in sport and general non-fiction; biography, popular culture, history. No children's books, science fiction, fantasy, horror or poetry. COMMISSION Home 15%; US & Translation 20%. No unsolicited mss. Please send a letter of enquiry or e-mail in the first instance.

Duncan McAra

28 Beresford Gardens, Edinburgh EH5 3ES
☎ 0131 552 1558
duncanmcara@mac.com
Contact *Duncan McAra*

Founded 1988. Handles fiction (literary fiction) and non-fiction, including art, architecture, archaeology, biography, history, military, travel and books of Scottish interest. COMMISSION Home 10%; Overseas 20%. Preliminary letter, synopsis and sample chapter (including return postage) essential. No reading fee.

McKernan Literary Agency & Consultancy

5 Gayfield Square, Edinburgh EH1 3NW
☎ 0131 557 1771
info@mckernanagency.co.uk
www.mckernanagency.co.uk
Agents *Maggie McKernan, Edwin Hawkes*

Founded 2005. Maggie McKernan handles literary and commercial fiction and non-fiction: Scottish,

biography, history, current affairs, memoirs; Edwin Hawkes handles science fiction, fantasy and non-fiction (popular history, politics, travel, and science.) Sells to the UK market, and works with **Capel & Land Ltd** for translation, US, film and TV sales. CLIENTS include Michael Collins, Alan Taylor, Belinda Seaward, Michael Fry, Michael Schmidt and Carlos Alba. No reading fee. See website for submission guidelines. No reading fee.

Bill McLean Personal Management

23B Deodar Road, London SW15 2NP
☎ 020 8789 8191
Contact *Bill McLean*

Founded 1972. Handles scripts for all media. No books. CLIENTS include Dwynwen Berry, Graham Carlisle, Phil Clark, Pat Cumper, Jane Galletly, Patrick Jones, Tony Jordan, Bill Lyons, John Maynard, Michael McStay, Les Miller, Ian Rowlands, Jeffrey Segal, Richard Shannon, Ronnie Smith, Barry Thomas, Garry Tyler, Frank Vickery, Laura Watson, Mark Wheatley. COMMISSION Home 10%. No unsolicited mss. Phone call or introductory letter essential. No reading fee.

Eunice McMullen Ltd

Low Ibbotsholme Cottage, Off Bridge Lane, Troutbeck Bridge, Windermere LA23 1HU
☎ 01539 448551
eunicemcmullen@totalise.co.uk
www.eunicemcmullen.co.uk
Contact *Eunice McMullen*

Founded 1992. Handles all types of children's fiction. Especially interested in 9plus and has 'an excellent' list of picture book authors and illustrators. CLIENTS include Wayne Anderson, Jon Berkeley, Sam Childs, Caroline Jayne Church, Jason Cockcroft, Ross Collins, Charles Fuge, Angela McAllister, David Melling, Angie Sage, Gillian Shields, David Wood. COMMISSION Home 10%; US 15%; Translation 20%. *No unsolicited scripts.* Telephone enquiries only.

Andrew Mann Ltd*

1 Old Compton Street, London W1D 5JA
☎ 020 7734 4751   📠 020 7287 9264
info@andrewmann.co.uk
www.andrewmann.co.uk
Director/Agent *Anne Dewe*
Director/Agent *Tina Betts*
Agent *Louise Burns*

Founded 1975. Handles fiction, general non-fiction, children's and film, TV, theatre, radio scripts. COMMISSION Home 15%; US & Translation 20%. OVERSEAS ASSOCIATES various. Unsolicited mss accepted but only first three chapters together with preliminary letter, and synopsis (s.a.e. essential). E-mail submissions for synopses only; 'we do not open attachments'. No reading fee.

Sarah Manson Literary Agent

6 Totnes Walk, London N2 0AD
☎ 020 8442 0396
info@sarahmanson.com
www.sarahmanson.com
Contact *Sarah Manson*

Founded 2002. Handles quality fiction for children and young adults. COMMISSION Home 10%; Overseas & Translation 20%. Please consult website for submission guidelines. No reading fee.

Marjacq Scripts Ltd

34 Devonshire Place, London W1G 6JW
☎ 020 7935 9499   📠 020 7935 9115
subs@marjacq.com
www.marjacq.com
Books *Philip Patterson*
Film/TV *Luke Speed*

Handles fiction and non-fiction, literary and commercial, graphic novels and comic books as well as film, TV, screenwriters and directors. No poetry. COMMISSION Home 10%; Overseas 20%. New work welcome; send brief letter, synopsis and approx. first 50 pages plus s.a.e. No reading fee.

The Marsh Agency Ltd*

50 Albemarle Street, London W1S 4BD
☎ 020 7493 4361   📠 020 7495 8961
enquiries@marsh-agency.co.uk
www.marsh-agency.co.uk
Managing Director *Paul Marsh*
Contacts *Geraldine Cooke, Jessica Woollard, Caroline Hardman, Hannah Ferguson*
Foreign Rights *Camilla Ferrier, Annina Meyerhans*

Founded in 1994 as international rights specialist for British, American and Canadian agencies. Expanded to act as agents handling fiction and non-fiction. Specializes in authors with international potential. See website for further information on individual agents' areas of interest. COMMISSION Home 15%; Elsewhere 20%; Film & TV 15%. See also **Paterson Marsh Ltd**. No plays, scripts or poetry. Unsolicited mss considered. Electronic submissions to go through the website.

MBA Literary Agents Ltd*

62 Grafton Way, London W1T 5DW
☎ 020 7387 2076   📠 020 7387 2042
[firstname]@mbalit.co.uk
www.mbalit.co.uk
Contacts *Diana Tyler, Meg Davis, Laura Longrigg, David Riding, Sophie Gorell Barnes, Susan Smith, Jean Kitson*

Founded 1971. Handles fiction and non-fiction, TV, film, radio and theatre scripts. Works in conjunction with agents in most countries. Also UK representative for Donald Maass Agency, Frances Collin Literary Agency, Martha Millard Literary Agency. In November 2006, the Merric Davidson Literary Agency became part of MBA. COMMISSION Home 15%; Overseas 20%; Theatre/TV/Radio 10%; Film 10–20%. See website for submission details.

Miles Stott Children's Literary Agency*

East Hook Farm, Lower Quay Road, Hook, Haverfordwest SA62 4LR
☎ 01437 890570
nancy@milesstottagency.co.uk
Contact *Nancy Miles*

Established in 2003. Handles children's novelty books, fiction for all ages including teenage fiction and series fiction. No non-fiction or poetry. CLIENTS Ronda Armitage, Dominic Barker, Sebastien Braun, Chris Fisher, Stacy Gregg, Frances Hardinge, Hiawyn Oram, Justin Richards. Publisher clients include Roaring Brook Press (USA) and Allen & Unwin (Australia). COMMISSION Home 15%; USA & Translation 20%. Welcomes unsolicited submissions by post or e-mail; send covering letter, brief synopsis and first three chapters (s.a.e. essential). No reading fee. OVERSEAS ASSOCIATES Barry Goldblatt Literary LLC, USA.

Christy Moore Ltd
▷ Sheil Land Associates Ltd

William Morris Agency (UK) Ltd*

Centre Point, 103 New Oxford Street, London WC1A 1DD
☎ 020 7534 6800   📠 020 7534 6900
www.wma.com
Books *Eugenie Furniss (Head of Literary UK), Rowan Lawton, Cathryn Summerhayes, Raffaella De Angelis (translation rights)*
TV *Holly Pye (Head of TV UK), Sophie Laurimore, Isabella Zoltowski*
Film *Caroline Taylor*

Worldwide talent and literary agency with offices in New York, Beverly Hills, Nashville, Miami and Shanghai. UK office handles fiction and general non-fiction and TV formats. COMMISSION Film & TV 10%; UK Books 15%; USA Books & Translation 20%. Accepts *postal* submissions only. Send query letter, synopsis and three chapters (no longer than 50 pages). No poetry, plays, screenplays or picture books. No reading fee.

Michael Motley Ltd

The Old Vicarage, Tredington, Tewkesbury
GL20 7BP
☎ 01684 276390　📠 01684 297355
Contact *Michael Motley*

Founded 1973. Handles only full-length mss
(adult: 60,000+; children's – all ages – and
humour: length variable). No short stories or
journalism. No science fiction, horror, poetry or
original dramatic material. COMMISSION Home
10–15%; US 15%; Translation 20%. New clients by
referral only: no unsolicited material considered.
No reading fee.

Judith Murdoch Literary Agency*

19 Chalcot Square, London NW1 8YA
☎ 020 7722 4197
Contact *Judith Murdoch*

Founded 1993. Handles full-length fiction only,
especially accessible literary and popular women's
fiction. No science fiction/fantasy, children's,
poetry or short stories. CLIENTS include Anne
Bennett, Anne Berry, Tony Black, Alison Bond,
Leah Fleming, Meg Hutchinson, Lola Jaye, Pamela
Jooste, Jessie Keane, Catherine King, Eve Makis,
Kitty Neale. Translation rights handled by The
Marsh Agency Ltd. COMMISSION Home 15%; US
& Translation 20%. No unsolicited mss; approach
in writing only enclosing first two chapters and
brief synopsis. Submissions by e-mail cannot
be considered. Return postage/s.a.e. essential.
Editorial advice given. No reading fee.

The Narrow Road Company

182 Brighton Road, Coulsdon CR5 2NF
☎ 020 8763 9895　📠 020 8763 2558
richardireson@narrowroad.co.uk
www.narrowroad.co.uk
Also at: 3rd Floor, 76 Neal Street, London
WC2H 9PL
Agent/Director *Richard Ireson*
Agent *Amy Ireson (020 7379 9598)*

Founded 1986. Part of the Narrow Road Group.
Theatrical literary agency. Handles scripts for
TV, theatre, film and radio. No novels or poetry.
CLIENTS include Ron Aldridge, Geoff Aymer,
Gil Brailey, Steve Gooch, Joe Graham, Renny
Krupinski, Simon McAllum, Tim Sanders. No
unsolicited mss or e-mail attachments; approach
by letter with c.v. only. Interested in writers with
some experience in television, theatre and radio

The Maggie Noach Literary Agency*

7 Peacock Yard, Iliffe Street, London SE17 3LH
☎ 020 7708 3073
info@mnla.co.uk
www.mnla.co.uk
Contact *Address material to the Company.*

Pronounced 'no-ack'. Handles fiction, general
non-fiction and children's. COMMISSION Home
15%; US & Translation 20%. No unsolicited mss;
submissions by arrangement only.

Andrew Nurnberg Associates Ltd*

Clerkenwell House, 45–47 Clerkenwell Green,
London EC1R 0QX
☎ 020 7417 8800　📠 020 7417 8812
www.andrewnurnberg.com
Directors *Andrew Nurnberg, Sarah Nundy,
D. Roger Seaton, Vicky Mark*

Agent for UK and international authors, including
children's writers. Sells rights internationally
on behalf of UK/US agencies and publishers.
Branches in Moscow, Budapest, Prague, Sofia,
Warsaw, Riga, Beijing and Taipei. COMMISSION
Home 15%; US & Translation 20%. See website for
full information and submission guidelines.

Alexandra Nye

'Craigower', 6 Kinnoull Avenue, Dunblane
FK15 9JG
☎ 01786 825114
Contact *Alexandra Nye*

Founded 1991. Handles fiction and topical
non-fiction. Special interests: literary fiction
and history. CLIENTS include Dr Tom Gallagher,
Harry Mehta, Robin Jenkins. COMMISSION
Home 10%; US 20%; Translation 15%. Unsolicited
mss welcome (s.a.e. essential for return). No
phone calls. Preliminary approach by letter, with
synopsis, preferred. Reading fee for supply of
detailed report.

Deborah Owen Ltd*

78 Narrow Street, Limehouse, London E14 8BP
☎ 020 7987 5119/5441　📠 020 7538 4004
do@deborahowen.co.uk
Contact *Deborah Owen*

Founded 1971. Small agency specializing in only
representing three authors direct around the
world. *No new authors.* CLIENTS Amos Oz, David
Owen and Delia Smith. COMMISSION Home 15%;
US & Translation 15%.

Paterson Marsh Ltd*

50 Albemarle Street, London W1S 4BD
☎ 020 7297 4312　📠 020 7495 8961
www.patersonmarsh.co.uk
Contact *Stephanie Ebdon*

World rights representatives of authors
and publishers primarily in the UK and
US, specializing in books for mental health
professionals. CLIENTS include the estates of
Sigmund and Anna Freud, Donald Winnicott,
Michael Balint and Wilfred Bion. Also interested
in serious and popular non-fiction. Visit website
for further information and submissions.

COMMISSION 20% (including sub-agent's commission). See also **The Marsh Agency Ltd**. No fiction, scripts, poetry, children's, articles or short stories. Unsolicited mss considered. Send outline and sample chapter.

John Pawsey

8 Snowshill Court, Giffard Park, Milton Keynes MK14 5QG
☎ 01908 217179
Contact *John Pawsey*

Founded 1981. Handles non-fiction only: biography, current affairs, popular culture, sport. Special interests: sport and biography. No fiction, children's, drama scripts, poetry, short stories, journalism, academic or submissions from outside the UK. CLIENTS include David Rayvern Allen, William Fotheringham, Don Hale, Patricia Hall, Roy Hudd OBE, Gary Imlach, Anne Mustoe, Gavin Newsham, Matt Rendell. COMMISSION Home 12½%; US & Translation 19–25%. OVERSEAS ASSOCIATES in the US, Japan, South America and throughout Europe. Preliminary letter with s.a.e. essential (no e-mail submissions). No reading fee.

Maggie Pearlstine Associates*

31 Ashley Gardens, Ambrosden Avenue, London SW1P 1QE
☎ 020 7828 4212   ⊞ 020 7834 5546
maggie@pearlstine.co.uk
Contact *Maggie Pearlstine*

Founded 1989. Small agency representing a select few authors. CLIENTS include Matthew Baylis, Roy Hattersley, Charles Kennedy, Quentin Letts, Claire Macdonald, Prof. Raj Persaud, Prof. Lesley Regan, Winifred Robinson, Henrietta Spencer-Churchill, Prof. Kathy Sykes, Prof. Robert Winston. No new authors. Translation rights handled by **Aitken Alexander Associates Ltd**.

Peters Fraser & Dunlop Group Ltd
▷ PFD

PFD*

Drury House, 34–43 Russell Street, London WC2B 5HA
☎ 020 7344 1000   ⊞ 020 7836 9539/7836 9541
postmaster@pfd.co.uk
www.pfd.co.uk
Chief Executive Officer *Caroline Michel*
Managing Director *Lesley Davey*
Books *Caroline Michel, Michael Sissons, Annabel Merullo, Marcella Edwards*
Foreign Rights *Louisa Pritchard*
Film, TV, Theatre, Presenters, Public Speakers *Gemma Hirst, Jessica Cooper, Alexandra Henderson*
Children's Books *Suzy Jenvey*

PFD represents authors of fiction and non-fiction, children's writers, screenwriters, playwrights, documentary makers, technicians, presenters and public speakers throughout the world. PFD has 85 years of international experience in all media. Outline sample chapters and author biographies should be addressed to the books department. Material should be submitted on an exclusive basis; or in any event disclose if material is being submitted to other agencies or publishers. Return postage essential. No reading fee. Response to e-mail submissions cannot be guaranteed. See website for submission guidelines.

Pollinger Limited*

9 Staple Inn, London WC1V 7QH
☎ 020 7404 0342   ⊞ 020 7242 5737
info@pollingerltd.com
www.pollingerltd.com
Managing Director *Lesley Pollinger*
Agents *Lesley Pollinger, Joanna Devereux, Tim Bates, Ruth Needham*
Permissions *Electronic form available on website*
Rights Manager *rightsmanager@pollingerltd.com*
Consultants *Leigh Pollinger, Joan Deitch*

CLIENTS include Michael Coleman, Catherine Fisher, Philip Gross, Kelly McKain, Jeremy Poolman, Nicholas Rhea and Michal Snunit. Also the estates of H.E. Bates, Erskine Caldwell, Rachel Carson, D.H. Lawrence, John Masters, Carson McCullers, Alan Moorehead, W. Heath Robinson, Eric Frank Russell, Clifford D. Simak and other notables. COMMISSION Home 15%; Translation 20%. Overseas, theatrical and media associates. Submissions policy outlined on the website.

Shelley Power Literary Agency Ltd*

13 rue du Pré Saint Gervais, 75019 Paris, France
☎ 00 33 1 42 38 36 49   ⊞ 00 33 1 40 40 70 08
shelley.power@wanadoo.fr
Contact *Shelley Power*

Founded 1976. Shelley Power works between London and Paris. This is an English agency with London-based administration/accounts office and the editorial office in Paris. Handles general commercial fiction, quality fiction, business books, self-help, true crime, investigative exposés, film and entertainment. No scripts, short stories, children's or poetry. COMMISSION Home 12½%; US & Translation 20%. Preliminary letter with brief outline of project (plus return postage as from UK or France) essential. 'We do not consider submissions by e-mail.' No reading fee.

PVA Management Limited

Hallow Park, Worcester WR2 6PG
☎ 01905 640663   ⊞ 01905 641842
maggie@pva.co.uk
www.pva.co.uk
Managing Director *Paul Vaughan*

Founded 1978. Handles non-fiction only. COMMISSION 15%. Send synopsis and sample chapters together with return postage.

Redhammer Management Ltd*

186 Bickenhall Mansions, Bickenhall Street, London W1U 6BX

📠 020 7486 3465　📠 020 7000 1249

info@redhammer.info

www.redhammer.info

Vice President *Peter Cox*

'Provides in-depth management for a small number of highly talented authors.' Willing to take on unpublished authors who are professional in their approach and who have major international potential, ideally, for book, film and/or TV. CLIENTS Donna Ball, Martin Bell OBE, Garry Bushell, Brian Clegg, Audrey Eyton, Maria Harris, Lucy Johnson, Senator Orrin Hatch, Amanda Lees, Hon. Nicholas Monson, Michelle Paver, Kellie Santin, Carolyn Soutar, Carole Stone, Donald Trelford, David Yelland. Submissions considered only if the guidelines given on the website have been followed. Do not send unsolicited mss by post. No radio or theatre scripts. No reading fee.

Robinson Literary Agency Ltd*

Block A511, The Jam Factory, 27 Green Walk, London SE1 4TT

📠 020 7096 1460

info@rlabooks.co.uk

www.rlabooks.co.uk

Managing Director *Peter Robinson*

Agent *Sam Copeland*

Founded 2005. Handles fiction, general non-fiction, popular culture. Also, TV documentaries and presenters. CLIENTS Joanne Harris, Steve Jones, Ian Rankin, David Starkey. COMMISSION Home 15%; US & Translation 20%. Unsolicited material welcome. No reading fee.

Rogers, Coleridge & White Ltd*

20 Powis Mews, London W11 1JN

📠 020 7221 3717　📠 020 7229 9084

www.rcwlitagency.com

Chairman *Deborah Rogers*

Managing Director *Peter Straus*

Directors *Gill Coleridge, Pat White, David Miller, Zoe Waldie, Laurence Laluyaux, Stephen Edwards* Other Agents *Catherine Pellegrino, Hannah Westland*

Founded 1967. Handles full-length mss fiction, non-fiction and children's books. COMMISSION Home 15%; US & Translation 20%. No submissions by fax or e-mail.

Uli Rushby-Smith Literary Agency

72 Plimsoll Road, London N4 2EE

📠 020 7354 2718　📠 020 7354 2718

Contact *Uli Rushby-Smith*

Founded 1993. Handles fiction and non-fiction, commercial and literary, both adult and children's. Film and TV rights handled in conjunction with a sub-agent. No plays or poetry. COMMISSION Home 15%; US & Translation 20%. Represents UK rights for Columbia University Press. Approach with an outline, two or three sample chapters and explanatory letter in the first instance (s.a.e. essential). No disks. No reading fee.

Rosemary Sandberg Ltd

6 Bayley Street, London WC1B 3HB

📠 020 7304 4110

rosemary@sandberg.demon.co.uk

Contact *Rosemary Sandberg*

Founded 1991. In association with **Ed Victor Ltd**. Specializes in children's writers and illustrators. COMMISSION 10%. No unsolicited mss as client list is currently full.

The Sayle Literary Agency*

1 Petersfield, Cambridge CB1 1BB

📠 01223 303035　📠 01223 301638

www.sayleliteraryagency.com

Contact *Rachel Calder*

Handles literary, crime and general fiction, current affairs, social issues, travel, biography, history, general non-fiction. No plays, poetry, textbooks, technical, legal or medical books. COMMISSION Home 15%; US & Translation 20%. OVERSEAS ASSOCIATES Dunow & Carlson & Lerner Literary Agency; Darhansoff, Verrill Feldman; Anne Edelstein Literary Agency; New England Publishing Associates, USA; translation rights handled by **The Marsh Agency Ltd**; film rights by **Sayle Screen Ltd**. No unsolicited mss. Preliminary letter essential, including a brief biographical note and a synopsis plus two or three sample chapters. 'Can only respond to successful submissions.' No reading fee.

Sayle Screen Ltd

11 Jubilee Place, London SW3 3TD

📠 020 7823 3883　📠 020 7823 3363

info@saylescreen.com

www.saylescreen.com

Agents *Jane Villiers, Matthew Bates, Toby Moorcroft*

Specializes in writers and directors for film and television. Also deals with theatre and radio. Works in association with **The Sayle Literary Agency, Greene & Heaton Ltd, Robinson Literary Agency Ltd** representing film and TV rights in novels and non-fiction. CLIENTS include Andrea Arnold, Shelagh Delaney, Marc Evans, Margaret Forster, John Forte, Rob Green, Mark Haddon, Christopher Monger, Paul Morrison, Gitta Sereny, Sue Townsend. No unsolicited material without preliminary letter. No e-mail submissions.

The Science Factory Limited

2 Twyford Place, Tiverton EX16 6AP
☎ 020 7193 7296
info@sciencefactory.co.uk
www.sciencefactory.co.uk
Contact *Peter Tallack*

Established in 2008 by Peter Tallack, ex-**Conville & Walsh** director and **Weidenfeld & Nicolson** publishing director. Handles all areas of non-fiction, including history, biography, memoirs, politics, current affairs, travel and science. Specializes in science, technology, medicine and natural history. No poetry, science fiction and fantasy; no TV, film, theatre or radio scripts. COMMISSION Home 15%; US & Translation 20%. Unsolicited mss, sample chapters or synopses welcome; approach by e-mail with covering letter and proposal, including c.v., chapter summaries and sample chapter. No reading fee.

The Sharland Organisation Ltd

The Manor House, Manor Street, Raunds
NN9 6JW
☎ 01933 626600
tso@btconnect.com
www.sharlandorganisation.co.uk
Contacts *Mike Sharland, Alice Sharland*

Founded 1988. Specializes in national and international film and TV negotiations. Also negotiates multimedia, interactive TV deals and computer game contracts. Handles scripts for film, TV, radio and theatre; also non-fiction. Markets books for film and handles stage, radio, film and TV rights for authors. No scientific, technical or poetry. COMMISSION Home 15%; US & Translation 20%. OVERSEAS ASSOCIATES various. No unsolicited mss. Preliminary enquiry by letter or phone essential.

Sheil Land Associates Ltd* (incorporating Richard Scott Simon Ltd 1971 and Christy Moore Ltd 1912)

52 Doughty Street, London WC1N 2LS
☎ 020 7405 9351   📠 020 7831 2127
info@sheilland.co.uk
Agents, UK & US *Sonia Land, Vivien Green, Ian Drury, Piers Blofeld*
Film/Theatrical/TV *Sophie Janson, Lucy Fawcett*
Foreign *Gaia Banks, Emily Dyson*

Founded 1962. Handles full-length, literary and commercial fiction and non-fiction, including: biography, politics, history, military history, gardening, science, thrillers, crime, romance, drama, science fiction, fantasy, business, travel, cookery, memoirs, humour, UK and foreign estates. Also theatre, film, radio and TV scripts. CLIENTS include Peter Ackroyd, Benedict Allen, Charles Allen, Roy Apps, Pam Ayres, Melvyn Bragg, Steven Carroll, Mark Chadbourn, David Cohen, Anna del Conte, Judy Corbalis, Elizabeth Corley, Seamus Deane, Greg Dyke, Rosie Goodwin, Chris Ewan, Alan Gilby, Jean Goodhind, Robert Green, Susan Hill, Richard Holmes, HRH The Prince of Wales, Iain Johnstone, Irene Karafilly, Richard Mabey, Mary Morris, Rachel Murrell, Graham Rice, Robert Rigby, Steve Rider, Martin Riley, Diane Setterfield, Tom Sharpe, Martin Stephen, Jeffrey Tayler, Andrew Taylor, Rose Tremain, Barry Unsworth, Kevin Wells, Prof. Stanley Wells, Neil White, John Wilsher, Paul Wilson and the estates of Catherine Cookson, Patrick O'Brian, Penelope Mortimer, Jean Rhys and F.A. Wesley. COMMISSION Home 15%; US & Translation 20%. OVERSEAS ASSOCIATES Georges Borchardt, Inc. (Richard Scott Simon). US film and TV representation: CAA, APA, and others. Welcomes approaches from new clients either to start or to develop their careers. Preliminary letter with s.a.e. essential. No reading fee.

Caroline Sheldon Literary Agency Ltd*

71 Hillgate Place, London W8 7SS
☎ 020 7727 9102
carolinesheldon@carolinesheldon.co.uk
pennyholroyde@carolinesheldon.co.uk
www.carolinesheldon.co.uk
Contacts *Caroline Sheldon, Penny Holroyde*

Founded 1985. Handles fiction and children's books. Special interests: all fiction for women, sagas, suspense, contemporary, chick-lit, historical fiction, fantasy and comic novels. Non-fiction: true life stories, memoirs, humour, quirky, animal interest. Children's books: fiction for all age groups, contemporary, comic, fantasy and all major genres including fiction series. Picture book stories. Non-fiction. Children's illustration: all quality illustrations. COMMISSION Home 15%; US & Translation 20%; Film & TV 15%. Send submissions by post or e-mail. 'Include introductory information about yourself and ambitions and the first three chapters of your book.' If sending by e-mail, type 'Submission' in subject line. Send large s.a.e. for postal submissions. No reading fee. 'We do not represent TV or film scripts except for current clients. Editorial advice given on work of exceptional promise.'

Dorie Simmonds Agency*

Riverbank House, 1 Putney Bridge Approach, London SW6 3JD
☎ 020 7736 0002   📠 020 7736 0010
dhsimmonds@aol.com
Contact *Dorie Simmonds*

Handles a wide range of subjects including commercial fiction and non-fiction, children's books and associated rights. Specialities include contemporary personalities and historical biographies, commercial women's fiction and children's books. COMMISSION Home & US 15%; Translation 20%. Outline required for non-fiction; a short synopsis for fiction with 2–3 sample chapters, and a c.v. with writing experience/publishing history. No reading fee. Return postage essential.

Jeffrey Simmons

15 Penn House, Mallory Street, London NW8 8SX
☎ 020 7224 8917
jasimmons@unicombox.co.uk
Contact *Jeffrey Simmons*

Founded 1978. Handles biography and autobiography, cinema and theatre, fiction (both quality and commercial), history, law and crime, politics and world affairs, parapsychology and sport (but not exclusively). No science fiction/fantasy, children's books, cookery, crafts, hobbies or gardening. Film scripts handled only if by book-writing clients. Special interest in personality books of all sorts and fiction from young writers (i.e. under 40) with a future. COMMISSION Home 10–15%; US & Foreign 15%. Writers become clients by personal introduction or by letter, enclosing a synopsis if possible, a brief biography, a note of any previously published books, plus a list of any publishers and agents who have already seen the mss.

Richard Scott Simon Ltd
▷ Sheil Land Associates Ltd

Sinclair-Stevenson

3 South Terrace, London SW7 2TB
☎ 020 7581 2550   ☎ 020 7581 2550
Contact *Christopher Sinclair-Stevenson*

Founded 1995. Handles biography, current affairs, travel, history, fiction, the arts. No scripts, children's, academic, science fiction/fantasy. CLIENTS include Jennifer Johnston, J.D.F. Jones, Ross King, Christopher Lee, Andrew Sinclair and the estates of Alec Guinness, John Cowper Powys and John Galsworthy. COMMISSION Home 10%; US 15%; Translation 20%. OVERSEAS ASSOCIATE T.C. Wallace Ltd, New York. Translation rights handled by **David Higham Associates Ltd**. Send synopsis with s.a.e. in the first instance. No reading fee.

Robert Smith Literary Agency Ltd*

12 Bridge Wharf, 156 Caledonian Road, London N1 9UU
☎ 020 7278 2444   ☎ 020 7833 5680
robertsmith.literaryagency@virgin.net
Contact *Robert Smith*

Founded 1997. Handles non-fiction; biography, memoirs, current affairs, history, entertainment, sport, true crime, cookery and lifestyle. No scripts, fiction, poetry, academic or children's books. CLIENTS Kate Adie (serializations), Martin Allen, Amanda Barrie (serializations), Peta Bee, Jamie Cameron Stewart, Martyn and Michelle Compton, Judy Cook, Stewart Evans, Neil and Christine Hamilton, Lenny Harper, Bob Harris, James Haspiel, Muriel Jakubait, Lois Jenkins, Siobhan Kennedy-McGuinness, Roberta Kray, Mark Lawrenson, Jean MacColl, Ann Ming, Michelle Morgan, Theo Paphitis, Frances Reilly, Keith Skinner, Jayne Sterne. COMMISSION Home 15%; US & Translation 20%. No unsolicited mss. Send a letter and synopsis in the first instance. No reading fee.

The Standen Literary Agency

53 Hardwicke Road, London N13 4SL
☎ 020 8889 1167   ☎ 020 8889 1167
yasmin@standenliteraryagency.com
www.standenliteraryagency.com
Contact *Mrs Yasmin Standen*

Founded 2004. Handles fiction, both adult and children's. 'We are interested in first time writers.' No TV/radio scripts. CLIENTS include Jonathan Yeatman Biggs, Zara Kane, Zoe Marriott, Andrew Murray. COMMISSION Home 15%; Overseas 20%. Follow the submissions procedure on the website. 'With regard to non-fiction, please contact us in the first instance.' No reading fee.

Elaine Steel

110 Gloucester Avenue, London NW1 8HX
☎ 020 7483 2681
ecmsteel@aol.com
Contact *Elaine Steel*

Founded 1986. Handles scripts, screenplays and books. No technical or academic. CLIENTS include Les Blair, Anna Campion, Michael Eaton, Pearse Elliott, Gwyneth Hughes, Brian Keenan, Troy Kennedy Martin, James Lovelock, Vince O'Connell, Albie Sachs, Ben Steiner. COMMISSION Home 10%; US & Translation 20%. Initial phone call preferred.

Abner Stein*

10 Roland Gardens, London SW7 3PH
☎ 020 7373 0456   ☎ 020 7370 6316
Contact *Arabella Stein*

Founded 1971. Mainly represents US agents and authors but handles some full-length fiction and general non-fiction. No scientific, technical, etc. No scripts. COMMISSION Home 10%; US & Translation 20%.

Micheline Steinberg Associates

104 Great Portland Street, London W1W 6PE
☎ 020 7631 1310

info@steinplays.com
www.steinplays.com
Contacts *Micheline Steinberg, Helen MacAuley*

Founded 1988. Specializes in drama for stage, TV, film, animation and radio. Represents writers for film and TV rights in fiction and non-fiction on behalf of book agents. COMMISSION Home 12½%; Overseas 15–20%. Works in association with agents in the USA and overseas. Return postage essential.

Shirley Stewart Literary Agency*

3rd Floor, 4A Nelson Road, London SE10 9JB
☎ 020 8293 3000
Director *Shirley Stewart*

Founded 1993. Handles fiction and non-fiction. No scripts, children's, science fiction, fantasy or poetry. COMMISSION Home 10–15%; US & Translation 20%. OVERSEAS ASSOCIATES Curtis Brown Ltd, New York. Will consider unsolicited material; send letter with two or three sample chapters in the first instance. S.a.e. essential. Submissions on disk not accepted. No reading fee.

Sarah Such Literary Agency

81 Arabella Drive, London SW15 5LL
☎ 020 8876 4228   ✉ 020 8878 8705
sarah@sarahsuch.com
www.sarahsuch.com
Director *Sarah Such*

Founded 2006. Former publisher with 18 years' experience. Handles high-quality literary and commercial non-fiction and fiction including children's books. 'Always looking for original work and new talented writers.' COMMISSION Home 15%; US & Translation 20%. OVERSEAS ASSOCIATES worldwide. Film and TV representation: **Aitken Alexander Associates Ltd**. Works by recommendation but welcomes unsolicited synopses and sample chapter submissions. TV/film scripts for established clients only. No radio or theatre scripts, poetry, fantasy, self-help or short stories. Contact by e-mail with brief synopsis, author biog and first two chapters (as Word attachment). S.a.e. essential if postal submission. No unsolicited mss and no telephone submissions. No reading fee. Will suggest revision.

The Susijn Agency Ltd

3rd Floor, 64 Great Titchfield Street, London W1W 7QH
☎ 020 7580 6341   ✉ 020 7580 8626
info@thesusijnagency.com
www.thesusijnagency.com
Agents *Laura Susijn, Nicola Barr*

Founded April 1998. Specializes in selling rights worldwide in literary fiction and non-fiction. Also represents non-English language authors and publishers for UK, US and translation rights worldwide. COMMISSION Home 15%; US & Translation 15–20%. Send preliminary letter, synopsis and first two chapters by post (plus s.a.e. if material is to be returned). No reading fee.

The Tennyson Agency

10 Cleveland Avenue, Wimbledon Chase, London SW20 9EW
☎ 020 8543 5939
submissions@tenagy.co.uk
www.tenagy.co.uk
Contacts *Christopher Oxford, Adam Sheldon*

Founded 2001. Specializes in theatre, radio, television and film scripts. Related material considered on an ad hoc basis. No short stories, children's, poetry, travel, military/historical, academic, fantasy, science fiction or sport. CLIENTS include Vivienne Allen, Tony Bagley, Kristina Bedford, Alastair Cording, Caroline Coxon, Iain Grant, Jonathan Holloway, Philip Hurd-Wood, Joanna Leigh, Steve MacGregor, Antony Mann, Ken Ross, John Ryan, Walter Saunders, Graeme Scarf, Diane Speakman, Diana Ward and the estate of Julian Howell. COMMISSION Home 12½–15%; Overseas 20%. No unsolicited material; send introductory letter with author's résumé and proposal/outline of work. No reading fee.

J.M. Thurley Management

Archery House, 33 Archery Square, Walmer CT14 7JA
☎ 01304 371721   ✉ 01304 371416
jmthurley@aol.com
www.thecuttingedge.biz
Contacts *Jon Thurley, Patricia Preece*

Founded 1976. Handles fiction and non-fiction, including academic, adventure, anthropology, archaeology, art and art history, autobiography, biotechnology, China, cinema, crime, current affairs, defence, documentary, economics, film, history, journalism, law, medical, memoirs, mysteries, mythology, occult, politics, psychology, romantic and science fiction, television, theatre, thrillers, world affairs. Also offers a paid manuscript consultancy service for assessment and creative input (see website for details).

Tibor Jones & Associates

30 Percy Street, London W1T 2DB
☎ 020 7631 0032
www.tiborjones.com
Contact *Kevin Conroy Scott*

Founded 2008. Handles fiction, literary biography, black writing/issues, children's, Commonwealth, EU, Far East and Latin American literature, autobiography, commercial fiction. Unsolicited material welcome; send covering letter, three chapters and a synopsis.

Lavinia Trevor Literary Agency*

29 Addison Place, London W11 4RJ
☎ 020 7603 5254 📠 0870 129 0838
www.laviniatrevor.co.uk
Contact *Lavinia Trevor*

Founded 1993. Handles general fiction (literary and commercial) and non-fiction, including popular science. No fantasy, science-fiction, poetry, academic, technical or children's books. No TV, film, radio, theatre scripts. COMMISSION rate by agreement with author. 'Sorry – no unsolicited submissions.'

Jane Turnbull*

Mailing address: Barn Cottage, Veryan, Truro TR2 5QA
☎ 01872 501317
jane.turnbull@btinternet.com
www.janeturnbull.co.uk
London office: 58 Elgin Crescent, London W11 2JJ
☎ 020 7727 9409
Contact *Jane Turnbull*

Founded 1986. Handles fiction and non-fiction but specializes in biography, history, current affairs, self-help and humour. Translation rights handled by **Aitken Alexander Associates Ltd**. COMMISSION Home 15%; US & Foreign 20%. No unsolicited mss. Approach with letter in the first instance. No reading fee.

TVmyworld

14 Dean Street, London W1D 3RS
☎ 020 7437 4188
business@tvmyworld.com
www.TVmyworld.com
Contacts *Mark Maco, Malcolm Rasala*

Specializes in books, television, movies, internet tv, games, second life. Has a production arm making motion pictures, television, commercials and internet television. Send first 30 pages only initially.

United Agents Limited*

12–26 Lexington Street, London W1F 0LE
www.unitedagents.co.uk
Contacts *Simon Trewin, Caroline Dawnay, Jim Gill, Sarah Ballard, Rosemary Scoular, Rosemary Canter, Anna Webber, Jane Willis, Jessica Craig*

Founded 2007. Handles fiction, non-fiction, biography, children's. Also TV, film, radio and theatre scripts. Unsolicited material welcome; see submissions policy on the website. No reading fee.

Ed Victor Ltd*

6 Bayley Street, Bedford Square, London WC1B 3HE
☎ 020 7304 4100 📠 020 7304 4111
mary@edvictor.com
Contacts *Ed Victor, Maggie Phillips, Sophie Hicks, Charlie Campbell*
Foreign Rights Manager *Morag O'Brien*

Founded 1976. Handles a broad range of material but leans towards the more commercial ends of the fiction and non-fiction spectrums. Excellent children's book list. No poetry, scripts or academic. Takes on very few new writers and does not accept unsolicited submissions. After trying his hand at book publishing and literary magazines, Ed Victor, an ebullient American, found his true vocation. Strong opinions, very pushy and works hard for those whose intelligence he respects. Loves nothing more than a good title auction. CLIENTS include John Banville, Herbie Brennan, Eoin Colfer, Frederick Forsyth, A.A. Gill, Josephine Hart, Jack Higgins, Nigella Lawson, Kathy Lette, Allan Mallinson, Andrew Marr, Danny Scheinmann, Janet Street-Porter and the estates of Douglas Adams, Raymond Chandler, Dame Iris Murdoch, Sir Stephen Spender and Irving Wallace. COMMISSION Home & US 15%; Translation 20%. No unsolicited mss.

Wade & Doherty Literary Agency Ltd

33 Cormorant Lodge, Thomas More Street, London E1W 1AU
☎ 020 7488 4171 📠 020 7488 4172
rw@rwla.com
bd@rwla.com
www.rwla.com
Contacts *Robin Wade, Broo Doherty*

Founded 2001. Handles general fiction and non-fiction including children's books. Specializes in military history and crime books. No scripts, poetry, plays or short stories. COMMISSION Home 10%; Overseas & Translation 20%. 'Fees negotiable if a contract has already been offered.' Send detailed synopsis and first 10,000 words by e-mail with a brief biography. No reading fee.

Cecily Ware Literary Agents

19C John Spencer Square, London N1 2LZ
☎ 020 7359 3787 📠 020 7226 9828
info@cecilyware.com
Contacts *Cecily Ware, Gilly Schuster, Warren Sherman*

Founded 1972. Primarily a film and TV script agency representing work in all areas: drama, children's, series/serials, adaptations, comedies, etc. No books or theatre. COMMISSION Home 10%; US 10–20% by arrangement. No unsolicited mss or phone calls. Approach in writing only. No reading fee.

Watson, Little Ltd*

48–56 Bayham Place, London NW1 0EU
📞 020 7388 7529   📠 020 7388 8501
office@watsonlittle.com
Contacts *Mandy Little, James Wills, Sally Anne Sweeney*

Handles fiction, commercial women's fiction, crime and literary fiction. Non-fiction special interests include history, science, popular psychology, self-help and general leisure books. Also children's fiction and non-fiction. No short stories, poetry, TV, play or film scripts. Not interested in purely academic writers. COMMISSION Home 15%; US & Translation 20%. OVERSEAS ASSOCIATE The Marsh Agency Ltd; FILM & TV ASSOCIATES The Sharland Organisation Ltd; MBA Literary Agents Ltd; USA Howard Morhaim (adult); The Chudney Agency (children). Informative preliminary letter and synopsis with return postage essential.

A.P. Watt Ltd*

20 John Street, London WC1N 2DR
📞 020 7405 6774   📠 020 7831 2154
apw@apwatt.co.uk
www.apwatt.co.uk
Directors *Caradoc King, Linda Shaughnessy, Derek Johns, Georgia Garrett, Natasha Fairweather*

Founded 1875. The oldest-established literary agency in the world. Handles full-length typescripts, including children's books, screenplays for film and TV. No poetry, academic or specialist works. CLIENTS include Monica Ali, Trezza Azzopardi, David Baddiel, Sebastian Barry, Quentin Blake, Melvin Burgess, Marika Cobbold, Michael Cox, Helen Dunmore, Nicholas Evans, Giles Foden, Esther Freud, Janice Galloway, Martin Gilbert, Nadine Gordimer, Linda Grant, Philip Hensher, Reginald Hill, Michael Holroyd, Michael Ignatieff, Mick Jackson, Philip Kerr, Dick King-Smith, India Knight, John Lanchester, Alison Lurie, Jan Morris, Andrew O'Hagan, Tony Parsons, Caryl Phillips, Philip Pullman, James Robertson, Jancis Robinson, Jon Ronson, Elaine Showalter, Zadie Smith, Graham Swift, Fiona Walker and the estates of Graves and Maugham. COMMISSION Home 15%; US & Translation 20%. No unsolicited mss accepted.

Josef Weinberger Plays

12–14 Mortimer Street, London W1T 3JJ
📞 020 7580 2827   📠 020 7436 9616
general.info@jwmail.co.uk
www.josef-weinberger.com
Contact *Michael Callahan*

Josef Weinberger is both agent and publisher of scripts for the theatre. CLIENTS include Ray Cooney, John Godber, Peter Gordon, Debbie Isitt, Arthur Miller, Sam Shepard, John Steinbeck. OVERSEAS REPRESENTATIVES in the USA, Canada, Australia, New Zealand and South Africa. No unsolicited mss; introductory letter essential. No reading fee.

John Welch, Literary Consultant & Agent

Mill Cottage, Calf Lane, Chipping Camden GL55 6JQ
📞 01386 840237   📠 01386 840568
johnwelch@cyphus.co.uk
Contact *John Welch*

Founded 1992. Handles military aviation and naval history, and history in general. No fiction, poetry, children's books or scripts for radio, TV, film or theatre. Already has a full hand of authors so no new authors being considered. CLIENTS include Alexander Baron, Michael Calvert, Patrick Delaforce, David Holbrook, Timothy Jenkins, Sybil Marshall, Ewart Oakeshott, Peter Reid, Norman Scarfe, Anthony Trew, Peter Trew, David Wragg. COMMISSION Home 10%.

Whispering Buffalo Literary Agency Ltd

97 Chesson Road, London W14 9QS
📞 020 7565 4737
mariam@whisperingbuffalo.com
www.whisperingbuffalo.com
Director *Mariam Keen*

Founded 2007. Handles commercial and literary fiction and non-fiction including children's and teenage fiction. Subjects include adventure, anthropology, art, arts and entertainment, astronomy, autobiography, cartoons, classical studies, etiquette, fashion, film, graphic novels, health and beauty, humour, interior design, lifestyle, magic, memoirs, music, mythology, natural history, occult, politics, romantic and science fiction, self help, thrillers. Special interest: book to film adaptations. No TV, film, radio or theatre scripts and no poetry or academic material. COMMISSION Home 12½%; Film, US & Translation 20%. Welcomes unsolicited material; send preliminary letter, c.v., synopsis and three sample chapters together with s.a.e. for return. No reading fee.

Eve White Literary Agent*

1a High Street, Kintbury RG17 9TJ
📞 01488 657656
eve@evewhite.co.uk
www.evewhite.co.uk
Contact *Eve White*

Founded 2003. Handles full-length adult and children's fiction and non-fiction. No poetry, short stories or textbooks. CLIENTS Susannah

Corbett, Carolyn Ching, Jimmy Docherty, Rae Earl, Shanta Everington, David Flavell, Marguerite Hann Syme, Abie Longstaff, Vijay Medtia, Gillian Rogerson, Andy Stanton, Tabitha Suzuma. COMMISSION Home 15%; US & Translation 20%. 'Authors must see the website for up-to-date submission requirements.' No initial approach by e-mail or telephone necessary. No reading fee.

Dinah Wiener Ltd*

12 Cornwall Grove, Chiswick, London W4 2LB
020 8994 6011 020 8994 6044
Contact *Dinah Wiener*

Founded 1985. Handles fiction and general non-fiction: auto/biography, popular science, cookery. No scripts, children's or poetry. CLIENTS include Valerie-Anne Baglietto, Malcolm Billings, Guy Burt, David Deutsch, Wendy K. Harris, Jenny Hobbs, Mark Jeffery, Michael Lockwood, Daniel Snowman, Rachel Trethewey, Marcia Willett. COMMISSION Home 15%; US & Translation 20%. Approach with preliminary letter in first instance, giving full but brief c.v. of past work and future plans. Mss submitted must include s.a.e. and be typed in double-spacing.

The Wylie Agency (UK) Ltd

17 Bedford Square, London WC1B 3JA
020 7908 5900 020 7908 5901
mail@wylieagency.co.uk

Handles fiction and non-fiction. No scripts or children's books. COMMISSION Home 10%; US 15%; Translation 20%. The Wylie Agency does not accept unsolicited submissions. Enquire by letter or e-mail before submitting. Any submission must be accompanied by return postage/s.a.e.

Susan Yearwood Literary Agency

2 Knebworth House, Londesborough Road, London N16 8RL
020 7503 0954
susan@susanyearwood.com
www.susanyearwood.com
Contact *Susan Yearwood*

Founded May 2007. Handles adult commercial and literary fiction and non-fiction, children's fiction (not picture books); also TV and film scripts. COMMISSION Home 15%. Unsolicited mss, sample chapters or synopses welcome; approach by e-mail in the first instance. No reading fee.

Agency Consultants

Agent Research & Evaluation, Inc. (AR&E)

425 North 20th Street, Philadelphia, PA 19130
001 215 563 1867 001 215 563 6797
info@agentresearch.com
www.agentresearch.uk.net
Contact *Bill Martin*

US consultancy founded in 1996. Works with authors seeking a new or first literary agent in the USA, UK, or Canada. 'A hands-on, customized service with hard data – who sells what to whom for how much, going back over twenty-five years – presented from a writer's point of view. An individually designed approach to the right agents for what you're doing.' See the website for 'Who We Are', details of pricing on 'Our Services' page and of the authors worked with on the 'Links' page.

UK Literary Scouts

Lucy Abrahams Literary Scouting

📱 0781 000 8243
lucy@literaryscout.co.uk
www.literaryscout.co.uk
Contact *Lucy Abrahams*

Scouts for Books in the Attic, Israel; Carrera, The Flying Dutchman and Lebowski/ Dutch Media, The Netherlands.

Louise Allen-Jones Literary Scouts

5c Old Town, London SW4 0JT
📱 020 7720 2453   📠 020 7627 3510
Contacts *Louise Allen-Jones (louise@louiseallenjones.com), Sally Page (sally@louiseallenjones.com), Naomi Tongue (naomi@louiseallenjones.com)*

Scouts for Ullstein Buchverlage (Ullstein hc/ppbk, List, Claassen, Marion von Schroeder, Econ, Propyläen), Germany; AW Bruna (Bruna, VIP, Signature) and Meulenhoff Boekerij (De Boekerij, Forum, Mynx, Arena, J.M. Meulenhoff), The Netherlands; Editora Record, Brazil; Sperling & Kupfer, Italy; Keter Books, Israel; Oceanida, Greece; The English Agency, Japan; Jentas, Denmark; Alma Littera, Lithuania.

Badcock & Rozycki Literary Scouts

1 Old Compton Street, London W1D 5JA
📱 020 7734 7997   📠 020 7734 6886
Contacts *June Badcock (june@litscouts.co.uk), Barbara Rozycki (barbara@litscouts.co.uk), Claire Holt (claire@litscouts.co.uk), Hollie Paterson (hollie@litscouts.co.uk)*

Scouts for Wilhelm Heyne Verlag and Diana Verlag, Germany; Unieboek BV and Het Spectrum, The Netherlands; RCS Libri Group, Italy; Editions Jean-Claude Lattès, France; Ediciones Salamandra, Spain; Ellinika Grammata, Greece; Werner Söderström Osakeyhtiö, Finland; Forum, Sweden; Cappelen Damm, Norway; Lindhardt & Ringhof Forlag, Denmark; Verold division of Bjartur-Verold, Iceland; Rupa & Co Publishers, India; Blueprint Pictures, UK.

Natasha Farrant

77 Starfield Road, London W12 9SN
📱 020 8746 1857
natasha.farrant@btinternet.com
Contact *Natasha Farrant*

Scouts for children's and young adult fiction on behalf of Hachette Jeunesse, France; Edizione Piemme, Italy; Carlsen Verlag, Germany; Jane Southern Literary Scout; Laika Entertainment (animation).

Anne Louise Fisher

29 D'Arblay Street, London W1F 8EP
📱 020 7494 4609   📠 020 7494 4611
annelouise@alfisher.co.uk
Contacts *Anne Louise Fisher, Catherine Eccles*

Scouts for Doubleday/Broadway Books/Nan Talese/Spiegel & Grau, US; Doubleday/Bond Street Books, Canada; Librarie Plon, Univers Poche/Pocket Jeunesse, France; C. Bertelsmann, Knaus, Blanvalet, DVA/Siedler/Pantheon, C. Bertelsmann Jugenbuch, Germany; Arnoldo Mondadori Editore, Oscar and Mondadori Ragazzi, Italy; Albert Bonniers Forlag, Sweden; Otava, Finland; Gyldendal Norsk Forlag, Norway; Gyldendal, Denmark; Mouria, Holland; Random House Mondadori and Montena, Spain; Editora Nova Fronteira, Brazil; Patakis Publications, Greece; Heyday Films, UK.

Folly Marland, Literary Scout

📱 020 8986 0111   📠 020 8986 0111
fmarland@pobox.com
Contact *Folly Marland*

Scout for S. Fischer, Scherz Verlag, Krüger Verlag, Germany; Livani, Greece; FMG: Truth & Dare, Trademark, Mistral, Pimento, The Netherlands; Village Books, Japan; Kowalski, Italy.

Rosalind Ramsay Limited

Third Floor, 23 Monmouth Street, London
WC2H 9DD
📞 020 7836 0054
ros@rosalindramsay.com
www.rosalindramsay.com
Contacts *Rosalind Ramsay, Françoise Higson*
Assistants *Ben Fowler, Carla Burnett*
Office Manager *Lindsey Clarke*

Scouts for Ambo Anthos and Uitgeverij Artemis,
The Netherlands; Kadokawa Shoten, Japan;
Kiepenheuer and Witsch, Germany; Droemer
Knaur, Germany; Kinneret Zmora Bitan Dvir,
Israel; Norstedts Forlagsgrupp, Sweden; Il
Saggiatore, Edizioni EL, Italy; Santillana Ediciones
General, Spain; Editora Objetiva, Brazil; Channel
4 and FilmFour, Framestore CFC, UK.

Heather Schiller

heather.schiller@virgin.net
Contact *Heather Schiller*

Scouts for Prometheus, The Netherlands;
Metaichmio, Greece; Piper, Germany; Bompiani,
Italy; Urano, Spain.

Petra Sluka

3 Dry Bank Road, Tonbridge TN10 3BS
📞 01732 353694
petrasluka@petrasluka.plus.com
Contact *Petra Sluka*

Scouts for Verlagsgruppe Lübbe, Gemany; De
Fontein/De Kern, Holland; Ediciones B, Spain;
Bra Böcker, Sweden; Japan Uni Agency, Inc.,
Japan; Harlenic Hallas, Greece.

Jane Southern

11 Russell Avenue, Bedford MK40 3TE
📞 01234 400147
jane@janesouthern.com
Contact *Jane Southern*

UK Scout for Der Club, Germany; The House
of Books, The Netherlands; Belfond, Presses
de la Cité and France Loisirs, France; Newton
Compton, Italy; Damm Forlag, Sweden; Planeta
Group, Spain; Greyhawk Agency, Taiwan;
Minoas, Greece; Owls Agency, Japan.

Mira Trenchard, Literary Scout

45 Mitford Road, London N19 4HJ
📞 020 7686 4452
Contact *Mira Trenchard*

Scout for Karl Blessing Verlag, Germany; Edizioni
Piemme, Italy; Achuzat Bayit, Israel.

Sylvie Zannier-Betts

114 Springfield Road, Brighton BN1 6DE
📞 01273 557370   📠 01273 557370
szannier@lineone.net
Contact *Sylvie Zannier-Betts*

Scouts for Uitgeverij De Geus, The Netherlands;
S. Fischer Verlag, Germany; Gummerus
Publishers, Finland; Alfabeta Bokfoerlag, Sweden;
Borgen Forlag, Denmark; Garzanti Libri, Italy;
RBA Libros, Spain. Exclusive agent in the UK for
S. Fischer Verlag, Germany.

PR Consultants

Claire Bowles Publicity

Ashley Court, Ashley, Market Harborough
LE16 8HF

☎ 01858 565800　📠 01858 565811
www.clairebowlespr.co.uk
Contact *Claire Bowles*

Represents non-fiction: The Good Pub Guide; Oz Clarke. CLIENTS include Workman (US); Artisan (US); Ebury Press; BBC Books; David & Charles; Readers Digest; Anova Books

Maria Boyle Communications Limited

36 Shalstone Road, Mortlake, London SW14 7HR
☎ 020 8876 8444
maria@mbcomms.co.uk
www.mbcomms.co.uk
Director *Maria Boyle*

Handles PR projects for some of the UK's leading publishers. Named as one of the top three independent PR consultants in the UK by *PR Week*. Works for a range of consumer and corporate clients. 'One area of specialism includes developing and delivering media campaigns that jump off the books pages and feature in the main body of newspapers and magazines, on national high profile TV and radio programmes and online.' CLIENTS include Penguin Classics; Pearson Booktime; DK Rough Guides; HarperCollins; The Book People and a number of high-profile bestselling authors.

Cameron Publicity and Marketing

35 Edward Road, Farnham GU9 8NP
☎ 07903 951957
ben@cameronpm.co.uk
www.cameronpm.co.uk
Contact *Ben Cameron*

Represents publishers and authors for both adult and children's books. CLIENTS include Anova Books; Hachette Children's Books (Hodder and Orchard); The National Maritime Museum; Pavilion Books; Smith TCI Publicity.

Colbert Macalister PR

7 Killieser Avenue, London SW2 5NU
☎ 020 8671 6615

ailsa@colmacpr.co.uk
www.colbertmacalister.co.uk
Contacts *Ailsa Macalister, Diana Colbert*

Represents publishers' lists and authors. CLIENTS include John Blake Publishing; Random House; Faber & Faber; Hay House; Bonnie; Robert Hale; Absolute Press.

Colman Getty

28 Windmill Street, London W1T 2JJ
☎ 020 7631 2666　📠 020 7631 2699
info@colmangetty.co.uk
www.colmangetty.co.uk
Chief Executive *Dotti Irving*
Managing Director *Liz Sich*

Handles prizes, book campaigns, authors, publishers' lists, professional associations. CLIENTS include The Man Booker Prize for Fiction; World Book Day; National Poetry Day; Frankfurt Book Fair; Pan Macmillan; Penguin; Quick Reads.

Day Five Ltd

1 Archibald Road, London N7 0AN
☎ 020 7619 0098
eobr@blueyonder.co.uk
Contact *Emma O'Bryen*

Represents publishers' lists and individual book projects. CLIENTS include Frances Lincoln Publishers; Macmillan; Natural History Museum.

FMcM Associates

3rd Floor, Colonial Buildings, 59–61 Hatton Garden, London EC1N 8LS
☎ 020 7405 7422　📠 020 7405 7424
publicity@fmcm.co.uk
www.fmcm.co.uk
Director *Fiona McMorrough*

FMcM is an award-winning PR and marketing consultancy specializing in publishing, literature, the arts and events. The agency works closely with new and established authors and artists and across the breadth of UK publishing. CLIENTS include: Penguin; Hachette; Random House; HarperCollins; The National Archives; British Council; English Pen; Guardian Books; Althorp Literary Festival and London Review of Books.

Author campaigns include Marian Keyes; Alastair Campbell; Margaret Atwood; Bernard Cornwell; Amanda Foreman and Lionel Shriver

Cathy Frazer PR

Old Hayle Barn, Hayle Farm, Marle Place Road, Horsmonden TN12 8DZ

☎ 01892 724156   📠 01892 724155

cathyfrazer@hotmail.com

Contact *Cathy Frazer*

Handles illustrated non-reference, consumer education, children's books; subjects include wine, food, antiques, sport, languages, health, parenting. Individual authors also represented. Over 20 years of publishing experience. CLIENTS include Hodder Education; Hodder Children's Books; Virgin; Mitchell Beazley; Millers; DBP; Kingfisher; Carol Vorderman; Uri Geller; Viscount Linley; Sir Patrick Moore; Carol Smillie; Lorraine Kelly; Peter Allis; Jonty Hearnden; Eric Knowles; Judith Miller; Marguerite Patten; Sarah Kennedy.

Giant Rooster PR

9 Hereford Road, London W2 4AB

☎ 020 7792 4701

gina@giantroosterpr.co.uk

Contact *Gina Rozner*

Established 1997. Handles authors, publishers' lists and prizes. CLIENTS include Weidenfeld & Nicolson; Simon & Schuster; Poetry Book Society (T.S. Eliot Prize); Constable & Robinson; Short Books; Gallic Books.

Katie Hambly PR

6 Lockington Avenue, Hartley, Plymouth PL3 5QP

☎ 01752 778211   📠 01752 778211

katie.hambly1@btinternet.com

Contact *Katie Hambly*

Freelance publicist, formerly Publicity Director at Weidenfeld & Nicolson, part of the Orion Publishing Group.

Idea Generation

11 Chance Street, London E2 7JB

☎ 020 7749 6850

hector@ideageneration.co.uk

www.ideageneration.co.uk

Managing Director *Hector Proud*

Arts and entertainment PR consultancy, working across the arts, photography, publishing, museums, film, TV, media and lifestyle sectors. Specializes in PR for illustrated, arts and photographic books. CLIENTS include Gloria Books; Genesis Publications; Vision On Publishing; Dazed Publishing; Waddell Publishing.

Andrea Marks Public Relations

27 London House, Canon's Corner, Edgware HA8 8AX

☎ 020 8958 5128   📠 020 8958 5128

info@andreamarks.co.uk

www.andreamarks.co.uk

Contact *Andrea Marks*

Corporate PR for publishers, booksellers, publishers' lists (not individual titles), prizes and awards, events, projects and initiatives, professional associations.

MGA (Macdougall Gabriel Associates)

190 Shaftesbury Avenue, London WC2H 8JL

☎ 020 7836 4774   📠 020 7836 4775

mga@mga-pr.com

www.mga-pr.com

Managing Director *Beth Macdougall*

Director *Bethan Jones*

Publishing consultancy which specializes in promoting publishers' lists and authors. CLIENTS include (publishers' lists) Arris Books; English Heritage; Good Life Press; Psychology News Press; Sibling Press; Souvenir Press. Selected books for Alastair Sawday; Constable & Robinson; Harper Collins; Michael Joseph; Malavan Books; Oberon Books; Penguin Press; Weidenfeld & Nicolson. Authors include Brian Aldiss, L P Berger, Bowvayne and Dinah Lampitt/Deryn Lake.

Louise Page PR

☎ 020 8741 5663   📠 020 8741 5663

louise@lpagepr.co.uk

Contact *Louise Page*

PR for fiction and non-fiction authors. CLIENTS include Maeve Binchy; Martina Cole; Wendy Holden.

Platypus PR

2 Tidy Street, Brighton BN1 4EL

☎ 01273 692215

info@platypuspr.com

www.platypuspr.com

Contact *Jeff Scott*

Represents publishers, authors, including self-published authors. CLIENTS include Cambridge University Press; Simon & Schuster; Dorling Kindersley; Pearson Education; Continuum; Clairview Books; Gibson Square; John Wiley & Sons; The Disinformation Company Ltd; various independent publishers. Books handled have won the Sunday Times Political Book of the Year 2005, Economist Best Book 2006: Politics & Current Affairs, Economist Biography Book of the Year 2008, Economist Economics and Business Book of the Year 2008 and Financial Times Best Business Book of 2007.

Nicky Potter

181 Alexandra Park Road, London N22 7UL
☎ 020 8889 9735
nicpot@dircon.co.uk
Contact *Nicky Potter*

Represents publishers' lists, prizes and professional organizations. CLIENTS include Booktrust (for the Children's Laureate); School Library Association (SLA School Librarian of the Year Award); Marsh Award for Children's Literature in Translation; Frances Lincoln; Barrington Stoke.

Publishing Services

9 Curwen Road, London W12 9AF
☎ 020 8222 6800   🖷 020 8222 6799
susanne@publishing-services.co.uk
www.publishing-services.co.uk

Handles publishers and authors. Third-party associates include book distributor Central Books; sales agency Signature Book Services. Publishers: mainly niche publishers such as educational/reference publisher Lucas Publications and history specialist Peter Lang. Authors: the emphasis is on media-friendly authors capable of taking part in the marketing process. Areas of interest include politics/current affairs; English language/reference; history, biography/autobiography; quirky idiosyncratic books; country/green. No children's books, poetry, plays or screenplays. In the spring and autumn of 2008, held a series of Society of

Authors seminars (How to Work With Your Publisher to Sell More Books).

Sonia Pugh PR

☎ 01375 891063
sonia.pugh@virgin.net
Contact *Sonia Pugh*

Independent PR consultancy which handles illustrated non-fiction books in all areas including launches, etc.

Sally Randall PR

Forty Acre Oast, Woodchurch, Ashford TN26 3PW
☎ 01233 860670
sallyrandall@reynoldsm.f2s.com
Contact *Sally Randall*

Handles general fiction and non-fiction, independent publishers' lists, individual authors/ titles, arts. CLIENTS include Meet the Author; Galore Park; Simon & Schuster; Michael O'Mara.

Claire Sawford PR

Studio One, 7 Chalcot Road, London NW1 8LH
☎ 020 7722 4114
claire@clairsawford.net
Contact *Claire Sawford*

Handles authors, publishers, prizes and professional associations. Specializes in non-fiction: art, architecture, fashion, design, biography. CLIENTS include V&A Publications; RIBA Bookshops; Royal Academy of Arts; National Portrait Gallery; Gill Hicks.

Irish Literary Agents

The Book Bureau Literary Agency

7 Duncairn Avenue, Bray, Co. Wicklow
☎ 00 353 1 276 4996   📠 00 353 1 276 4834
thebookbureau@oceanfree.net
Contact *Ger Nichol*

Handles general and literary fiction. Special interest in women's fiction, crime and thrillers and some non-fiction. COMMISSION Home 10%; Overseas 20%. Works with foreign associates. Will suggest revision. Send preliminary letter, synopsis and first three chapters (single line spacing preferred); return postage essential (IRCs only from UK and abroad). No reading fee.

Font International Literary Agency

Hollyville House, Hollybrook Road, Clontarf, Dublin 3
☎ 00 353 1 853 2356
info@fontlitagency.com
Contact *Ita O'Driscoll*

Founded 2003. Handles book-length adult fiction and non-fiction from previously published writers (no children's, drama, sci-fi, erotic, technical or poetry). COMMISSION Home 15–20%; Translation 20–25%. ASSOCIATE **The Marsh Agency Ltd**, London for translation rights. No unsolicited mss. 'As an initial contact, please query our interest in your property either by e-mail or post only.' Include details of writing and other media experience. S.a.e. required. 'We regret that we cannot discuss queries over the phone.' No reading fee.

Marianne Gunn O'Connor Literary Agency

Morrison Chambers, Suite 17, 32 Nassau Street, Dublin 2
☎ 00 353 1 677 9100   📠 00 353 1 677 9101
mgoclitagency@eircom.net
Contact *Marianne Gunn O'Connor*

Founded 1996. Handles commercial and literary fiction, non-fiction: biography, health and children's fiction. CLIENTS include Patrick McCabe, Cecelia Ahern, Claudia Carroll, Morag Prunty, Gisele Scanlon (Goddess Guide), Chris Binchy, Anita Notaro, Noelle Harrison, Julie Dam, Mike McCormack, Paddy McMahon, Thrity Engineer, John Lynch, Helen Falconer, Brinda Charry, Peter Murphy, Louise Douglas, Carla Gylnn, Caroline McFarlane-Watts. COMMISSION UK 15%; Overseas 20%; Film & TV 20%. Translation rights handled by Vicki Satlow Literary Agency, Milan. No unsolicited mss; send preliminary enquiry letter plus half-page synopsis per e-mail.

The Lisa Richards Agency

108 Upper Leeson Street, Dublin 4
☎ 00 353 1 637 5000   📠 00 353 1 667 1256
faith@lisarichards.ie
www.lisarichards.ie
Contact *Faith O'Grady*

Founded 1998. Handles fiction and general non-fiction. CLIENTS include, Helena Close, Susan Connolly, June Considine, Matt Cooper, Denise Deegan, Christine Dwyer Hickey, Robert Fannin, Karen Gillece, Tara Heavey, Paul Howard (aka Ross O'Carroll-Kelly), Arlene Hunt, Roisin Ingle, Alison Jameson, Declan Lynch, Roisin Meaney, Pauline McLynn, Aisling McDermot, Anna McPartlin, David O'Doherty, Damien Owens, Trisha Rainsford, Kevin Rafter, Jarlath Regan, Eirin Thompson. COMMISSION Home 10%; UK 15%; US & Translation 20%; Film & TV 15%. OVERSEAS ASSOCIATE **The Marsh Agency Ltd** for translation rights. Approach with proposal and sample chapter for non-fiction, and 3–4 chapters and synopsis for fiction (s.a.e. essential). No reading fee.

Walsh Communications

75 Holywell Drive, Holywell, Swords, Co. Dublin
☎ 00 353 86 606 9313
info@walshcommunications.ie
www.walshcommunications.ie
Contact *Emma Walsh*

Founded 2006. Handles fiction, non-fiction, children's literature. Also TV, film, theatre and radio scripts. No poetry or science fiction. COMMISSION Home 15%; US 20%. Unsolicited mss, synopses and sample chapters welcome; approach by e-mail in the first instance. No reading fee 'for clients that I initially take on in a representative capacity'.

Jonathan Williams Literary Agency

Rosney Mews, Upper Glenageary Road, Glenageary, Co. Dublin

☎ 00 353 1 280 3482 📠 00 353 1 280 3482

Contact *Jonathan Williams*

Founded 1980. Handles general trade books: fiction, auto/biography, travel, politics, history, music, literature and criticism, gardening, cookery, sport and leisure, humour, reference, social questions, photography. Some poetry. No plays, science fiction, children's books, mind, body and spirit, computer books, theology, multimedia, motoring, aviation. COMMISSION Home 10%; US & Translation 20%. OVERSEAS ASSOCIATES Piergiorgio Nicolazzini Literary Agency, Italy; Linda Kohn, International Literature Bureau, Holland and Germany; Antonia Kerrigan Literary Agency, Spain; Tuttle-Mori Agency Inc., Japan. No reading fee 'unless the author wants a very fast opinion'. Initial approach by phone or letter. UK submissions should be accompanied by Irish postage stamps or International Reply Coupons.

US Literary Agents

* = Members of the **Association of Authors' Representatives, Inc.**

Dominick Abel Literary Agency, Inc.*
146 West 82nd Street, Suite 1B, New York
NY 10024
☎ 001 212 877 0710
Contact *Dominick Abel*

Established 1975. No scripts, children's books or
poetry. COMMISSION Home 15%; Translation 20%.
No unsolicited material. Send query letter with
s.a.e. in the first instance. No reading fee.

Miriam Altshuler Literary Agency*
53 Old Post Road North, Red Hook NY 12571
☎ 001 845 758 9408
www.miriamaltshulerliteraryagency.com/

Founded 1994. Handles literary and commercial
fiction; general non-fiction, narrative non-fiction,
memoirs, psychology, biography, young adult
fiction. No romance, science fiction, mysteries,
poetry, westerns, fantasy, how-to, techno thrillers
or self-help. COMMISSION Home 15%; Foreign &
Translation 20%. Send query letter and synopsis
in the first instance with an e-mail address.
'Should we wish to contact you to see material
we can no longer respond without an s.a.e.' (no
e-mail or fax queries). No reading fee.

The Axelrod Agency, Inc.*
55 Main Street, PO Box 357, Chatham NY 12037
☎ 001 518 392 2100
steve@axelrodagency.com
Contact *Steven Axelrod*

Founded 1983. Handles commercial women's
fiction, romantic fiction and mysteries. No
non-fiction or scripts. COMMISSION Home 15%;
Translation 20%. Welcomes submissions by post
(include s.a.e.) or e-mail. No reading fee.

Malaga Baldi Literary Agency
233 West 99th Street, Suite 19C, New York
NY 10025
☎ 001 212 222 3213
baldibooks@gmail.com
Contact *Malaga Baldi*

Founded 1986. Handles quality fiction and non-
fiction. No scripts. No westerns, men's adventure,
science fiction/fantasy, romance, how-to, young
adult or children's. COMMISSION 15%. OVERSEAS
ASSOCIATES **Abner Stein**, UK; Owl Agency;
Eliane Benisti, France; Marsh Agency. Writers of
fiction should send query letter describing the
novel plus IRCs. For non-fiction, approach in
writing with a proposal, table of contents and two
sample chapters. No reading fee.

The Balkin Agency, Inc.*
PO Box 222, Amherst MA 01004
☎ 001 413 548 9835   ⊜ 001 413 548 9835
rick62838@crocker.com
Contact *Richard Balkin*

Founded 1973. Handles adult non-fiction only.
COMMISSION Home 15%; Foreign 20%. No reading
fee for outlines and synopses.

Loretta Barrett Books, Inc.*
101 Fifth Avenue, New York NY 10003
☎ 001 212 242 3420
www.lorettabarrettbooks.com
Contact *Loretta Barrett*

Founded 1990. Handles all non-fiction and fiction
genres except children's, poetry, science fiction/
fantasy and historical romance. Specializes in
women's fiction, history, spirituality. No scripts.
COMMISSION Home 15%; Foreign 20%. No e-mail
submissions. Send query letter with biography
and return postage only in the first instance.

Meredith Bernstein Literary Agency, Inc.*
2095 Broadway, Suite 505, New York NY 10023
☎ 001 212 799 1007   ⊜ 001 212 799 1145
Contacts *Meredith Bernstein*

Founded 1981. Handles commercial and literary
fiction, mysteries and non-fiction (women's
issues, biography, memoirs, health, current
affairs, crafts). COMMISSION Home & Dramatic

15%; Translation 20%. OVERSEAS ASSOCIATES
Abner Stein, UK; Lennart Sane, Holland,
Scandinavia and Spanish language; Thomas
Schluck, Germany; Bardon Chinese Media
Agency; William Miller, Japan; Frederique
Porretta, France; Agenzia Letteraria, Italy.

Bleecker Street Associates, Inc.*

532 LaGuardia Place, #617, New York NY 10012
📞 001 212 677 4492   📠 001 212 388 0001

Founded 1984. Handles fiction: women's, mystery,
suspense, literary; non-fiction: history, women's
interests, parenting, health, relationships,
psychology, sports, sociology, current events,
biography, spirituality, New Age, business. No
poetry, children's, westerns, science fiction,
professional, academic. COMMISSION Home
15%; Foreign 25%. Send query letter in the first
instance. Will only respond if envelope and
return postage enclosed. No phone calls, faxes or
e-mails. No reading fee.

Georges Borchardt, Inc.*

136 East 57th Street, New York NY 10022
📞 001 212 753 5785   📠 001 212 838 6518

Founded 1967. Works mostly with established/
published authors. Specializes in fiction,
biography, and general non-fiction of unusual
interest. COMMISSION Home, UK & Dramatic
15%; Translation 20%. UK ASSOCIATE **Sheil Land
Associates Ltd (Richard Scott Simon)**, London.
Unsolicited mss not read.

Barbara Braun Associates, Inc.*

151 West 19th Street, 4th Floor, New York
NY 10011
📞 001 212 604 9023   📠 001 212 604 9041
barbara@barbarabraunagency.com
www/barbarabraunagency.com
Contacts *Barbara Braun, John Baker*

Founded 1995. Handles literary and mainstream,
women's and historical fiction, also serious
non-fiction, including psychology and biography.
Also young adult and mystery books. Specializes
in art history, archaeology, architecture and
cultural history. No scripts, poetry, science
fiction. COMMISSION Home 15%; Foreign 20%.
No unsolicited mss; send query letter in the first
instance. See website for submission guidelines.
No reading fee.

Browne & Miller Literary Associates*

410 S. Michigan Avenue, Suite 460, Chicago
IL 60605
mail@browneandmiller.com
www.browneandmiller.com
President *Danielle Egan-Miller*

Founded 1971. Formerly known as Multimedia
Product Development. Handles commercial and
literary fiction, and practical non-fiction with
wide appeal. No scripts, juvenile, science fiction
or poetry. COMMISSION Home 15%; Foreign
& Translation 20%. OVERSEAS ASSOCIATES in
Europe, Latin America, Japan and Asia. In the
first instance, send query e-mail or letter with
IRCs. No unsolicited material. No reading fee.

Pema Browne Ltd, Literary Agents

11 Tena Place, Valley Cottage NY 10989
ppbltd@optonline.net
www.pemabrowneltd.com
Contact *Pema Browne*

Founded 1966. ('Pema rhymes with Emma.')
Handles mass-market mainstream and hardcover
fiction: romance, business, children's picture
books and young adult; non-fiction: how-to and
reference. COMMISSION Home 20%; Translation
20%; Overseas authors 20%. No unsolicited mss;
send query letter with IRCs. No fax or e-mail
queries. No longer accepting screenplays. 'We are
accepting very few new clients at this time.'

Sheree Bykofsky Associates, Inc.*

PO Box 706, Brigantine NJ 08203
www.shereebee.com
Contact *submitbee@aol.com*

Founded 1991. Handles adult fiction and non-
fiction. No scripts. No children's, young adult,
horror, science fiction, romance, westerns, occult
or supernatural. COMMISSION Home 15%; UK
(including sub-agent's fee) 20%. No unsolicited
mss. Send query letter first with brief synopsis or
outline and writing sample (1–3 pp) for fiction.
IRCs essential for reply or return of material.
'Please do not send material via methods that
require signature, such as FedEx, etc.' No phone
calls. See website for submission guidelines. No
reading fee.

Maria Carvainis Agency, Inc.*

Rockefeller Center, 1270 Avenue of the Americas,
Suite 2320, New York NY 10020
📞 001 212 245 6365   📠 001 212 245 7196
mca@mariacarvainisagency.com
President *Maria Carvainis*

Founded 1977. Handles fiction: literary and
mainstream, contemporary women's, mystery,
suspense, historical, young adult novels;
non-fiction: business, women's issues, memoirs,
health, biography, medicine. No film scripts
unless from writers with established credits.
No science fiction. COMMISSION Domestic &
Dramatic 15%; Translation 20%. No faxed or
e-mailed queries. No unsolicited mss; they will
be returned unread. Queries only, with IRCs for
response. No reading fee.

Castiglia Literary Agency*

1155 Camino del mar, Suite 510, Del Mar CA 92014
☎ 001 858 755 8761   ☏ 001 858 755 7063
www.castigliaagency.com
Contacts *Julie Castiglia, Winifred Golden, Sally Van Haitsma, Deborah Ritchken*

Founded 1993. Handles fiction: literary, mainstream, ethnic; non-fiction: narrative, biography, business, science, health, parenting, memoirs, psychology, women's and contemporary issues. Specializes in science, biography and literary fiction. No scripts, poetry, horror or fantasy. COMMISSION Home 15%; Foreign & Translation 25%. No unsolicited material; send query letter only in the first instance. No reading fee.

Jane Chelius Literary Agency, Inc.*

548 Second Street, Brooklyn, New York NY 11215
☎ 001 718 499 0236   ☏ 001 718 832 7335
queries@janechelius.com
www.janechelius.com
Contact *Jane Chelius*

Founded 1995. Handles popular and literary fiction; narrative and how-to non-fiction. No children's, young adult, screenplays, scripts or poetry. COMMISSION Home 15% Foreign 20%. Submission guidelines available on the website.

Linda Chester & Associates*

Rockefeller Center, 630 Fifth Avenue, Suite 2662, New York NY 10111
www.lindachester.com
Contact *Linda Chester*

Founded 1978. Handles literary and commercial fiction and non-fiction in all subjects. No scripts, children's or textbooks. COMMISSION Home & Dramatic 15%; Translation 25%. No unsolicited mss or queries. No reading fee for solicited material.

William Clark Associates*

154 Christopher Street, Suite 3C, New York NY 10014
☎ 001 212 675 2784
www.wmclark.com
Contact *William Clark*

Founded 1997. Handles non-fiction, mainstream literary fiction and some young adult books. No scripts, horror, science fiction, fantasy, diet or mystery. COMMISSION Home 15%; Foreign & Translation 20%. OVERSEAS ASSOCIATES **Ed Victor Ltd**; **Andrew Nurnberg Associates Ltd** (translation rights). 'William Clark began accepting representation queries via e-mail in 1993. Since that time the query mailbox address has propagated across the net and receives an overwhelming amount of spam e-mail daily. The agency longer accept queries via any other means

than the form on our Query Guidelines page at www.wmclark.com/queryguidelines.html and queries sent any other way will be deleted unread. The agency responds to all queries sent according to its guidelines, whether or not it is interested in the work or idea.'

Frances Collin Literary Agent*

PO Box 33, Wayne PA 19087–0033
☎ 001 610 254 0555   ☏ 001 610 254 5029
www.francescollin.com
Contacts *Frances Collin, Sarah Yake*

Founded 1948. Successor to Marie Rodell. Handles general fiction and non-fiction. No scripts. OVERSEAS ASSOCIATES worldwide. No unsolicited mss. E-mail queries preferred (queries@francescollin.com) but with no attachments. Or send query letter only, with IRCs for reply, for the attention of *Sarah Yake*. No fax or telephone queries, please. No reading fee. Rarely accepts non-professional writers or writers not represented in the UK.

Don Congdon Associates, Inc.*

156 Fifth Avenue, Suite 625, New York NY 10010–7002
☎ 001 212 645 1229   ☏ 001 212 727 2688
Contacts *Michael Congdon, Susan Ramer, Cristina Concepcion, Maura Kye-Casella, Katie Kotchman*

Founded 1983. Handles fiction and non-fiction. No academic, technical, romantic fiction or scripts. COMMISSION Home 15%; UK & Translation 19%. OVERSEAS ASSOCIATES worldwide. No unsolicited mss. Queries via e-mail or snailmail letter with return postage. No reading fee.

Curtis Brown Ltd*

10 Astor Place, New York NY 10003
☎ 001 212 473 5400
Book Rights *Laura Blake Peterson, Katherine Fausset, Peter L. Ginsberg, Emilie Jacobson, Ginger Knowlton, Maureen Walters, Mitchell Waters, Elizabeth Harding, Ginger Clark*
Film & TV Rights *Timothy Knowlton, Holly Frederick*
Translation *Dave Barbor*

Founded 1914. Handles general fiction and non-fiction. Also some scripts for film, TV and theatre. OVERSEAS ASSOCIATES Representatives in all major foreign countries. No unsolicited mss; queries only, with IRCs for reply. No reading fee.

Sandra Dijkstra Literary Agency*

1155 Camino Del Mar, PMB 515, Del Mar CA 92014
☎ 001 858 755 3115   ☏ 001 858 794 2822
www.dijkstraagency.com

Founded 1981. Handles quality and commercial non-fiction and fiction, including some genre fiction. No scripts. No westerns, science fiction or poetry. Specializes in quality fiction including women's and multicultural fiction, mystery/thrillers, romance, suspense, children's literature, narrative non-fiction, psychology, self-help, science, health, business, memoirs, biography, current affairs and history. 'Dedicated to promoting new and original voices and ideas.' COMMISSION Home 15%; Translation 20%. OVERSEAS ASSOCIATES **Abner Stein**, UK; Agence Hoffman, Germany; Licht & Burr, Scandinavia; Luigi Bernabo, Italy; Sandra Bruna, Spain/Portugal; Caroline Van Gelderen, Netherlands; La Nouvelle Agence, France; The English Agency, Japan; Bardon-Chinese Media Agency, China/Taiwan; Prava I Prevodi, Eastern Europe; Tuttle Mori, Thailand; Synopsis, Russia/Baltic States; Maxima Creative Agency, Indonesia; Graal, Poland. For fiction send brief synopsis (one page) and first 50 pages; for non-fiction send proposal with overview, chapter outline, author biog, 1–2 sample chapters and profile of competition. All submissions should be accompanied by IRCs. No reading fee.

Dunham Literary Inc.*

156 Fifth Avenue, Suite 625, New York NY 10010
📞 001 212 929 0994 📠 001 212 929 0904
www.dunhamlit.com
Contacts *Jennie Dunham, Blair Hewes*

Founded 2000. Handles literary fiction, non-fiction and children's (from picture books through young adult). No scripts, romance, westerns, horror, science-fiction/fantasy or poetry. COMMISSION Home 15%; Foreign & Translation 20%. OVERSEAS ASSOCIATE UK & Europe: **A.M. Heath & Co. Ltd**. The Rhoda Weyr Agency is now a division of Dunham Literary Inc. No unsolicited material. Send query letter (with return postage) giving information on the author and ms. No e-mail or faxed queries; see the website for further contact information. No reading fee.

Dystel & Goderich Literary Management*

One Union Square West, Suite 904, New York NY 10003
📞 001 212 627 9100 📠 001 212 627 9313
www.dystel.com
Contacts *Jane Dystel, Miriam Goderich, Stacey Glick, Michael Bourret, Jim McCarthy, Lauren E. Abramo, Adina Kahn, Chasya Milgrom*

Founded 1994. Handles non-fiction and fiction. Specializes in politics, history, biography, cookbooks, current affairs, celebrities, commercial and literary fiction. No reading fee.

Educational Design Services, LLC

5750 Bou Avenue, Ste. 1508, N. Bethesda MD 20852
📞 001 301 881 8611
blinder@educationaldesignservices.com
www.educationaldesignservices.com
President *Bertram Linder*

Founded 1979. Specializes in texts and professional development materials for the education and school market. COMMISSION Home 15%; Foreign 25%. 'E-submissions' welcome. IRCs must accompany mail submissions.

Ethan Ellenberg Literary Agency*

548 Broadway, Suite 5E, New York NY 10012
📞 001 212 431 4554 📠 001 212 941 4652
agent@ethanellenberg.com
www.ethanellenberg.com
Contact *Ethan Ellenberg*

Founded 1984. Handles fiction: commercial, genre, literary and children's; non-fiction: history, biography, science, health, cooking, current affairs. Specializes in commercial fiction, thrillers, suspense and romance. No scripts, poetry or short stories. COMMISSION Home 15%; Translation 20%. Prefers submissions via e-mail. Will accept submissions by post but does not return submission material from outside the USA. For fiction send synopsis and first 50 pages; for non-fiction send proposal and sample chapters, if available. No reading fee. For e-mail submissions send query letter only; no attachments.

Jeanne Fredericks Literary Agency, Inc.*

221 Benedict Hill Road, New Canaan CT 06840
📞 001 203 972 3011 📠 001 203 972 3011
jeanne.fredericks@gmail.com
www.jeannefredericks.com
Contact *Jeanne Fredericks*

Founded 1997. Handles quality adult non-fiction, usually of a practical and popular nature by authorities in their fields. Specializes in health, gardening, science/nature/environment, business, self-help, reference. No fiction, juvenile, poetry, essays, politics, academic or textbooks. COMMISSION Home 15%; Foreign 25% (with co-agent) or 20% (direct). No unsolicited material; send query by e-mail (no attachments) or letter with return postage in the first instance. No reading fee.

Robert A. Freedman Dramatic Agency, Inc.*

Suite 2310, 1501 Broadway, New York NY 10036
📞 001 212 840 5760
President *Robert A. Freedman*
Senior Vice-President *Selma Luttinger*
Vice-President *Marta Praeger*

Founded 1928 as Brandt & Brandt Dramatic Department, Inc. Took its present name in 1984.

Works mostly with established authors. Send letter of enquiry first with s.a.e. Specializes in plays, film and TV scripts. COMMISSION Dramatic 10%. Unsolicited mss not read.

Gelfman Schneider Literary Agents, Inc.*

250 West 57th Street, Suite 2122, New York NY 10107

℡ 001 212 245 1993   ℻ 001 212 245 8678
Contacts *Deborah Schneider, Jane Gelfman*

Founded 1919 (London), 1980 (New York). Formerly John Farquharson Ltd. Works mostly with established/published authors. Specializes in general trade fiction and non-fiction. No poetry, short stories or screenplays. COMMISSION Home 15%; Dramatic 15%; Foreign 20%. OVERSEAS ASSOCIATE **Curtis Brown Group Ltd**, UK. No reading fee for outlines. Submissions must be accompanied by IRCs. No e-mail queries please.

Mollie Glick – Foundry Literary + Media*

33 West 17th Street, PH, New York City NY 10011
mglick@foundrymedia.com
www.foundrymedia.com
Literary Agent *Mollie Glick*

Represents literary fiction, narrative non-fiction and some practical non- fiction. Particularly interested in fiction that bridges the literary/commercial divide, combining strong writing with a great plot, and non-fiction dealing with popular science, medicine, psychology, cultural history, memoir and current events. Send e-mail query. No reading fee.

Sanford J. Greenburger Associates, Inc.*

55 Fifth Avenue, 15th Floor, New York NY 10003
℡ 001 212 206 5600   ℻ 001 212 463 8718
www.greenburger.com
Contacts *Heide Lange, Faith Hamlin, Daniel Mandel, Matt Bialer, Jeremy Katz, Tricia Davey, Teri Tobias*

Handles fiction and non-fiction. No unsolicited mss. First approach with query letter, sample chapter and synopsis. No reading fee.

The Greenhouse Literary Agency Ltd

11308 Lapham Drive, Oakton VA 22124
℡ 001 703 865 4990
submissions@greenhouseliterary.com
www.greenhouseliterary.com
Vice President *Sarah Davies*

Founded January 2008. Handles children's and teen fiction: high-concept chapter book series; middle grade and YA fiction. No picture books, pre-school texts or illustrators, stand-alone very short chapter books of less than 10,000 words, non-fiction, inspirational/religious work, short stories or writing for adults. COMMISSION US & UK 15%; Translation 25%. OVERSEAS ASSOCIATE

The Greenhouse Literary Agency Ltd UK (see entry under *UK Literary Agents*) See website for submission details. Welcomes query e-mails (no snailmail) with up to five pages (first five) pasted into the body of the e-mail. No reading fee.

The Charlotte Gusay Literary Agency

10532 Blythe Avenue, Los Angeles CA 90064
℡ 001 310 559 0831   ℻ 001 310 559 2639
gusay1@ca.rr.com (queries only)
www.gusay.com
Owner/President *Charlotte Gusay*

Founded 1989. Handles fiction, non-fiction, young adult/teen and screenplays (feature scripts/plays to film). No short stories, poetry. Specializes in novels, narrative non-fiction, gardening and travel. COMMISSION Home 15%; Foreign & Translation 25%. No unsolicited mss; send one-page query letter and s.a.e. in the first instance. No reading fee.

Joy Harris Literary Agency, Inc.*

156 Fifth Avenue, Suite 617, New York NY 10010
℡ 001 212 924 6269   ℻ 001 212 924 6609
Contact *Joy Harris*

Handles adult non-fiction and fiction. COMMISSION Home 15%; Foreign 20%. Query letter in the first instance (s.a.e. required). No reading fee.

John Hawkins & Associates, Inc.*

71 West 23rd Street, Suite 1600, New York NY 10010
℡ 001 212 807 7040   ℻ 001 212 807 9555
www.jhalit.com
Contacts *John Hawkins, William Reiss, Anne Hawkins, Warren Frazier, Moses Cardona*

Founded 1893. Handles film and TV rights. COMMISSION Apply for rates. No unsolicited mss; send queries with 1–3-page outline and one-page c.v. IRCs necessary for response. No reading fee.

The Jeff Herman Agency, LLC

PO Box 1522, Stockbridge MA 01262
℡ 001 413 298 0077   ℻ 001 413 298 8188
jeff@jeffherman.com
www.jeffherman.com
Contact *Jeffrey H. Herman*

Handles all areas of non-fiction, textbooks and reference, business, spiritual and psychology. No scripts. COMMISSION Home 15%; Translation 10%. No unsolicited mss. Query by letter with IRCs in the first instance or by e-mail. No reading fee. Jeff Herman publishes a useful reference guide to the book trade called *Jeff Herman's Guide to Book Editors, Publishers & Literary Agents* (Three Dog Press).

The Barbara Hogenson Agency, Inc.*

165 West End Avenue, Suite 19–C, New York
NY 10023
☎ 001 212 874 8084   📠 001 212 362 3011
Owner *Barbara Hogenson*
Contracts Manager *Nicole Verity*

Founded 1994. Handles non-fiction, literary fiction and plays. No science fiction, romance or children's books. Specializes in theatre-related books and literary fiction. COMMISSION Home 15%; Foreign & Translation 20%. OVERSEAS ASSOCIATES Lora Fountain, France; Andrew Nurnberg, Eastern Europe; Japan Uni, Japan. No unsolicited mss; query letter only. Recommendation from current clients preferred or reference to a client's work. No reading fee.

Janklow & Nesbit Associates

445 Park Avenue, 13th Floor, New York
NY 10022–2606
☎ 001 212 421 1700   📠 001 212 980 3671
info@janklow.com
foreignrights@janklow.com
Partners *Morton L. Janklow, Lynn Nesbit*
Senior Vice-President *Anne Sibbald*
Directors *Tina Bennett, Eric Simonoff*
Agents *Priscilla Gilman, Rebecca Gradinger, Luke Janklow, Richard Morris*
Foreign Rights Director *Cullen Stanley*

Founded 1989. Handles fiction and non-fiction; commercial and literary. See also **Janklow & Nesbit (UK) Ltd** under *UK Literary Agents*. No unsolicited mss.

JCA Literary Agency, Inc.

174 Sullivan Street, New York NY 10012
www.jcalit.com/
Contact *Tom Cushman*

Founded 1978. Handles general fiction and non-fiction. No scripts, poetry, science fiction/fantasy or children's books. COMMISSION Home 15%; Foreign 20%. OVERSEAS ASSOCIATE **Vanessa Holt Ltd**, UK. No unsolicited mss. No reading fee.

Kirchoff/Wohlberg, Inc.*

866 United Nations Plaza, Suite 525, New York
NY 10017
☎ 001 212 644 2020   📠 001 212 223 4387
www.kirchoffwohlberg.com
Authors' Representative *Liza Pulitzer-Voges*

Founded 1930. Handles books for children and young adults. No adult material. No scripts for TV, radio, film or theatre. Send letter of enquiry with synopsis or outline and writing sample plus IRCs for reply or return. No reading fee.

Harvey Klinger, Inc.*

300 West 55th Street, Suite 11V, New York
NY 10019
queries@harveyklinger.com
www.harveyklinger.com
Contact *Harvey Klinger*

Founded 1977. Handles mainstream fiction and non-fiction. Specializes in commercial and literary fiction, psychology, health, science and children's books. No scripts, poetry or computer books. COMMISSION Home 15%; Foreign 25%. OVERSEAS ASSOCIATES in all principal countries. Welcomes unsolicited material; send by e-mail or post (no faxes). No reading fee.

Linda Konner Literary Agency*

10 West 15 Street, Suite 1918, New York NY 10011
ldkonner@cs.com
Contact *Linda Konner*

Founded 1996. Handles non-fiction only, specializing in health, self-help, diet/fitness, pop psychology, relationships, parenting, personal finance. Also some pop culture/celebrities. COMMISSION Home 15%; Foreign 25%. Books must be written by or with established experts in their field. No scripts, fiction, poetry or children's books. No unsolicited material; send one-page query with return postage or via e-mail. No reading fee.

Peter Lampack Agency, Inc.

551 Fifth Avenue, Suite 1613, New York NY 10176
☎ 001 212 687 9106   📠 001 212 687 9109
alampack@verizon.net
Contact *Andrew Lampack*
International Rights *Ripsime Dilanyan*

Founded in 1977. Handles commercial fiction: male action and adventure, contemporary relationships, mysteries and suspense, literary fiction; also non-fiction from recognized experts in a given field, plus biographies. Handles theatrical, motion picture and TV rights from book properties. No original scripts or screenplays, series or episodic material. COMMISSION Home & Dramatic 15%; Translation & UK 20%. Best approach by letter in first instance. No reply without s.a.e. 'We will respond within four weeks and invite the submission of manuscripts which we would like to examine.' No reading fee. No unsolicited mss.

Michael Larsen/Elizabeth Pomada Literary Agency*

1029 Jones Street, San Francisco CA 94109
☎ 001 415 673 0939
larsenpoma@aol.com
laurie@agentsavant.com
www.larsenpomada.com
www.agentsavant.com

Non-Fiction *Michael Larsen*
Fiction *Elizabeth Pomada*
Genre Fiction/Young Adult *Laurie McLean*

Founded 1972. Handles adult fiction and non-fiction – literary and commercial. No scripts, poetry, children's, science fiction. COMMISSION Home 15%; Foreign 20–30%. ASSOCIATE Laurie McLean (represents romance, science fiction and fantasy and young adult books). OVERSEAS ASSOCIATES **David Grossman Literary Agency Ltd**, British rights; Chandler Crawford (foreign rights). Fiction: send first ten pages with two-page synopsis and s.a.e.; non-fiction: e-mail first ten pages with two-page synopsis in the body of the e-mail (no attachments) to larsenpoma@aol.com; genre fiction and YA: e-mail first ten pages and two-page synopsis in the body of the e-mail to laurie@agentsavant.com (consult website for guidelines).

The Ned Leavitt Agency*

70 Wooster Street, Suite 4F, New York NY 10012
www.nedleavittagency.com
President *Ned Leavitt*
Agent *Britta Steiner Alexander*

Ned Leavitt specializes in creativity, spirituality, health and literary fiction; Britta Alexander specializes in smart non-fiction for 20 and 30-somethings (dating, relationships, business) and commercial fiction. No screenplays or genre fiction. COMMISSION Home 15%. Submissions by recommendation only; see guidelines on the website. No reading fee.

Lescher & Lescher Ltd*

346 East 84th Street, New York NY 10028
☎ 001 212 396 1999   ☎ 001 212 396 1991
Contacts *Robert Lescher, Susan Lescher*

Founded 1964. Handles a broad range of serious non-fiction including current affairs, history, biography, memoirs, politics, law, contemporary issues, popular culture, food and wine, literary and commercial fiction including mysteries and thrillers; some children's books. Specializes in wine books and cookbooks. No poetry, science fiction, New Age, spiritual, romance. COMMISSION Home 15%; Foreign 20%. No unsolicited mss – please query first. No reading fee.

Sterling Lord Literistic, Inc.

65 Bleecker Street, New York NY 10012
☎ 001 212 780 6050   ☎ 001 212 780 6095
info@sll.com
www.sll.com
Contacts *Philippa Brophy, Chris Calhoun, Laurie Liss*

Founded 1979. Handles all genres, fiction and non-fiction. COMMISSION Home 15%; UK &

Translation 20%. No unsolicited mss. Prefers letter outlining all non-fiction. No reading fee.

Lowenstein-Yost Associates Inc.*

121 West 27th Street, Suite 601, New York NY 10001
☎ 001 212 206 1630   ☎ 001 212 727 0280
www.lowensteinyost.com
Agents *Barbara Lowenstein, Nancy Yost, Zoe Fishman, Natanya Wheeler*

Founded 1976. Handles fiction: upmarket, commercial and multicultural, thrillers, mysteries, women's fiction of all kinds including historicals, paranormals and romantic suspense, fantasy and young adult. Non-fiction: narrative, business, health, spirituality, psychology, relationships, politics, history, memoirs, natural science, women's issues, social issues, pop culture, parenting, personal finance. COMMISSION Home 15%; Foreign & Translation 20%. OVERSEAS ASSOCIATES in all major countries. Fiction: send query letter with short synopsis, first chapter and s.a.e. Include any prior literary credits (previously published titles and reviews, writing courses, awards/grants); non-fiction: send query letter, project overview, list of credentials, media appearances, previous titles, reviews and s.a.e. No reading fee.

The Margret McBride Literary Agency*

7744 Fay Avenue, Suite 200, La Jolla CA 92037
☎ 001 858 454 1550   ☎ 001 858 454 2156
staff@mcbridelit.com
www.mcbrideliterary.com
Submissions Manager *Michael Daley*
Assistant to Margret McBride *Faye Atchison*
Vice President/Associate Agent *Donna DeGutis*

Founded 1981. Handles non-fiction mostly, plus business and some fiction. No scripts, science fiction, fantasy, romance or children's books. Specializes in business and management. COMMISSION Home 15%; Overseas 15–25%; Translation 25%. No unsolicited mss. Send query letter with brief synopsis and s.a.e. only. Check the submission guidelines on the website before making contact.

Carol Mann Agency*

55 Fifth Avenue, New York NY 10003
☎ 001 212 206 5635   ☎ 001 212 675 4809
www.carolmannagency.com
Contacts *Carol Mann*

Founded 1977. Handles literary and commercial fiction and non-fiction. No scripts or genre fiction. COMMISSION Home 15%; Foreign 20%. No unsolicited material; send query letter with return postage. No reading fee.

Manus & Associates Literary Agency, Inc.*

425 Sherman Avenue, Suite 200, Palo Alto
CA 94306
ManusLit@ManusLit.com
www.ManusLit.com
Contacts *Jillian Manus, Jandy Nelson,
Stephanie Lee, Penny Nelson, Dena Fischer*

Handles commercial and literary fiction, also
young adult and middle grade; non-fiction,
including true crime, self-help, memoirs, history,
pop culture and popular science. *No* scripts,
science fiction/fantasy, westerns, romance,
horror, poetry or children's books. COMMISSION
Home 15%; Foreign 25%. No unsolicited mss.
Fiction: send query letter and first 30 pages; non-
fiction: query letter and proposal. Include return
postage. No reading fee.

The Evan Marshall Agency*

Six Tristam Place, Pine Brook NJ 07058–9445
📱 001 973 882 1122   📠 001 973 822 3099
evanmarshall@thenovelisit.com
www.theevanmarshallagency.com
President *Evan Marshall*

Founded 1987. Handles fiction (adult and young
adult). No scripts, children's books or non-fiction.
COMMISSION Home 15%; Foreign & Translation
20%. No unsolicited material. 'We are currently
taking clients by professional referral only. Do not
query us at this time.'

Mews Books Ltd

c/o Sidney B. Kramer, 20 Bluewater Hill,
Westport CT 06880
📱 001 203 227 1836   📠 001 203 227 1144
mewsbooks@aol.com (initial contact only;
submission by regular mail)
Contacts *Sidney B. Kramer, Fran Pollak*

Founded 1970. Handles adult fiction and
non-fiction, children's, pre-school and young
adult. No scripts, short stories or novellas
(unless by established authors). Specializes
in cookery, medical, health and nutrition,
scientific non-fiction, children's and young adult.
COMMISSION Home 15%; Film & Translation
20%. Unsolicited material welcome. Presentation
must be professional and should include brief
summary of plot/characters ('avoid a reviewer's
point of view when describing plot'), one or
two sample chapters, personal credentials and
targeted market, all suitable for forwarding to a
publisher. Send by regular mail, not e-mail. No
reading fee. Requests exclusivity while reading
and information if material has been circulated.
Charges for photocopying, postage expenses,
telephone calls and other direct costs. Principal
agent is an attorney and former publisher (a
founder of Bantam Books, Corgi Books, London).
Offers consultation service through which writers
can get advice on a contract or on publishing
problems.

Doris S. Michaels Literary Agency Inc.*

1841 Broadway, Suite 903, New York NY 10023
📱 001 212 265 9474   📠 001 212 265 9480
query@dsmagency.com
www.dsmagency.com
Contact *Doris Michaels*

Handles literary fiction that has a commercial
appeal and strong screen potential and women's
fiction; non-fiction: business, current affairs,
biography and memoirs, self-help, humour,
history, health, classical music, sports, women's
issues, social sciences and pop culture. No action/
adventure, suspense, science fiction, romance,
New Age, religion/spirituality, gift books, art
books, fantasy, thrillers, mysteries, westerns,
occult and supernatural, horror, historical fiction,
poetry, children's literature, humour or travel
books. COMMISSION Home 15% Send query letter
via e-mail with a one-page synopsis and include a
short paragraph detailing credentials. No reading
fee. No unsolicited calls.

Howard Morhaim Literary Agency*

30 Pierrepont Street, Brooklyn NY 11201
📱 001 718 222 8400   📠 001 718 222 5056
www.morhaimliterary.com
Contact *Howard Morhaim*

Founded 1979. Handles general adult and
young adult fiction and non-fiction. No scripts
poetry or religious. COMMISSION Home 15%;
UK & Translation 20%. OVERSEAS ASSOCIATES
worldwide. Send query letter with synopsis and
sample chapters for fiction; query letter with
outline or proposal for non-fiction. Include return
postage. No unsolicited mss. No reading fee.

B.K. Nelson Literary Agency

84 Woodland Road, Pleasantville NY 10570
📱 001 914 741 1322   📠 001 914 741 1324
bknelson4@cs.com
www.bknelson.com
Also at: 1565 Paseo Vida, Palm Springs, CA 92264
📱 001 760 778 8800   📠 001 914 778 0034
President *Bonita K. Nelson*
Vice President *Leonard 'Chip' Ashbach*
Editorial Director *John W. Benson*

Founded 1979. Specializes in novels, business,
self-help, how-to, political, autobiography,
celebrity biography. Major motion picture and TV
documentary success. COMMISSION 20%. Lecture
Bureau for Authors founded 1994; Foreign
Rights Catalogue established 1995; BK Nelson
Infomercial Marketing Co. 1996, primarily for
authors and endorsements, and BKNelson, Inc.
for motion picture production in 1998. Signatory
to Writers Guild of America, West (WGAW). No

unsolicited mss. Letter of inquiry. Reading fee charged.

Richard Parks Agency*

PO Box 693, Salem NY 12865
☎ 001 518 854 9466   📠 001 518 854 9466
rp@richardparksagency.com
Contact *Richard Parks*

Founded 1989. Handles general trade fiction and non-fiction: literary novels, mysteries and thrillers, commercial fiction, science fiction, biography, pop culture, psychology, self-help, parenting, medical, cooking, gardening, history, etc. No scripts. No technical or academic. COMMISSION Home 15%; UK & Translation 20%. OVERSEAS ASSOCIATES **The Marsh Agency Ltd; Barbara Levy Literary Agency**. No unsolicited mss. Fiction read by referral only. Non-fiction query with s.a.e. Faxed or e-mail queries will *not* be considered. No reading fee.

James Peter Associates

PO Box 358, New Canaan CT 06840
☎ 001 203 972 1070
gene_brissie@msn.com
Contact *Gene Brissie*

Founded 1971. Handles adult non-fiction. No scripts, fiction or children's books. COMMISSION Home 15%; Foreign & Translation 20%. Send query letter and proposal in first instance. No reading fee.

Alison J. Picard Literary Agent

PO Box 2000, Cotuit MA 02635
☎ 001 508 477 7192
📠 001 508 477 7192 (notify before faxing)
ajpicard@aol.com
Contact *Alison Picard*

Founded 1985. Handles mainstream and literary fiction, contemporary and historical romance, children's and young adult, mysteries and thrillers; plus non-fiction. No short stories or poetry. Rarely any science fiction and fantasy. Particularly interested in expanding non-fiction titles. COMMISSION 15%. OVERSEAS ASSOCIATE **John Pawsey**, UK. Approach with written query. No reading fee.

Pinder Lane & Garon-Brooke Associates Ltd*

159 West 53rd Street, Suite 14–C, New York NY 10019
☎ 001 212 489 0880
Owner Agents *Dick Duane, Robert Thixton*

Founded 1951. Fiction and non-fiction. No category romance, westerns or mysteries. COMMISSION Home 15%; Dramatic 10–15%; Foreign 30%. No unsolicited mss. First approach by query letter. No reading fee.

PMA Literary & Film Management, Inc.

PO Box 1817, Old Chelsea Station, New York NY 10113
☎ 001 212 929 1222   📠 001 212 206 0238
queries@pmalitfilm.com
www.pmalitfilm.com
President *Peter Miller*

Founded 1976. Commercial fiction and non-fiction. Specializes in books with motion picture and television potential and in true crime. No poetry, pornography, non-commercial or academic. COMMISSION Home 15%; Dramatic 10–15%; Foreign 20–25%. No unsolicited mss. Approach by query letter (with s.a.e.) or, preferred, e-mail (no attachments).

The Aaron M. Priest Literary Agency*

708 Third Avenue, 23rd Floor, New York NY 10017–4201
☎ 001 212 818 0344
www.aaronpriest.com

Founded 1974. Handles literary and commercial fiction. No scripts, children's fantasy or sci-fi. Submissions by e-mail only; refer to the website for full details. No reading fee.

Susan Ann Protter Literary Agent*

320 Central Park West, Suite #12E, New York NY 10025
☎ 001 212 362 7920   📠 001 212 362 4562
sapla@aol.com
www.susanannprotter.com
Contact *Susan Ann Protter*

Founded 1971. Handles general fiction, mysteries, thrillers, science fiction and fantasy; non-fiction: history, general reference, biography, science, health, current affairs and parenting. No romance, poetry, westerns, religious, children's or sport manuals. No scripts. COMMISSION Home & Dramatic 15%. OVERSEAS ASSOCIATES **Abner Stein**, UK; agents in all major markets. First approach with letter, including IRCs. No reading fee.

Quicksilver Books, Literary Agents

508 Central Park Avenue, Suite 5101, Scarsdale NY 10583
☎ 001 914 722 4664   📠 001 914 722 4664
quickbooks@optonline.net
President *Bob Silverstein*

Founded 1973. Handles literary fiction and mainstream commercial fiction: blockbuster, suspense, thriller, contemporary, mystery and historical; and general non-fiction, including self-help, psychology, holistic healing, ecology, environmental, biography, fact crime, New Age, health, nutrition, cookery, enlightened wisdom and spirituality. No scripts, science fiction and fantasy, pornography, children's or romance.

COMMISSION Home & Dramatic 15%; Translation 20%. UK material being submitted must have universal appeal for the US market. Unsolicited material welcome but must be accompanied by IRCs for response, together with biographical details, covering letter, etc. No reading fee.

Raines & Raines*

103 Kenyon Road, Medusa NY 12120
📞 001 518 239 8311   📠 001 518 239 6029
Contacts *Theron Raines, Joan Raines, Keith Korman*

Founded 1961. Handles general non-fiction. No scripts. COMMISSION Home 15%; Foreign & Translation 20%. No unsolicited material; send one-page letter in the first instance. No reading fee.

Reece Halsey North Literary Agency*

98 Main Street, #704, Tiburon CA 94920
📞 001 415 789 9191   📠 001 415 789 9177
info@reecehalseynorth.com (AmyBurkhardt)
www.reecehalseynorth.com
Contacts *Kimberley Cameron, Elizabeth Evans, April Eberhardt*

The Reece Halsey Agency was founded 1957. Aldous Huxley, Upton Sinclair and William Faulkner have been among their clients. Represents literary and mainstream fiction, non-fiction. No scripts, poetry or children's books. COMMISSION Home 15%; Foreign 20%. Send query letter with first 10–50 pages by e-mail. No reading fee.

Helen Rees Literary Agency*

376 North Street, Boston MA 02113–2103
📞 001 617 227 9014   📠 001 617 227 8762
reesagency@reesagency.com
Contact *Helen Rees*
Associates *Ann Collette, Lorin Rees*

Founded 1982. Specializes in books on health and business; also handles biography, autobiography and history; quality fiction. No scholarly or technical books. No scripts, science fiction, children's, poetry, photography, short stories, cookery. COMMISSION Home 15%; Foreign 20%. No e-mail queries or attachments. Send query letter with IRCs. No reading fee.

The Angela Rinaldi Literary Agency*

PO Box 7877, Beverly Hills CA 90212–7877
📞 001 310 842 7665   📠 001 310 837 8143
amr@rinaldiliterary.com
www.rinaldiliterary.com
Contact *Angela Rinaldi*

Founded 1995. Handles fiction: commercial, literary, upmarket contemporary women's, suspense, literary historical thrillers, gothic suspense, children's and young adult fiction.

Non-fiction: memoirs, women's issues/studies, current issues, biography, psychology, popular reference, self-help, health books on specific issues, business, career, personal finance. No scripts, humour, techno/espionage/drug thrillers, category romance, science fiction, fantasy, horror, westerns, cookbooks, poetry, film scripts, religion or occult. COMMISSION Home 15%; Foreign 20%. E-mail enquiry (no attachments). Novels: send brief synopsis and first three chapters only; non-fiction queries: send detailed covering letter or proposal. 'Please tell me if I have your work exclusively or if it is a multiple submission. Allow 4–6 weeks for response.' No reading fee.

Ann Rittenberg Literary Agency, Inc.*

30 Bond Street, New York NY 10012
📞 001 212 684 6936   📠 001 212 684 6929
www.rittlit.com
Contacts *Ann Rittenberg, Penn Whaling*

Founded 1991. Handles literary fiction, serious narrative non-fiction, biography/memoirs, cultural history, upmarket women's fiction and thrillers. No scripts, romance, science fiction, self-help, inspirational and nothing at genre level. COMMISSION Home 15%; Foreign 20%. Send query letter, sample chapters and synopses by post; no e-mails. Enclose s.a.e. No reading fee.

B.J. Robbins Literary Agency*

5130 Bellaire Avenue, North Hollywood CA 91607
📞 001 818 760 6602
robbinsliterary@aol.com
Contact *B.J. Robbins*

Founded 1992. Handles literary fiction, narrative and general non-fiction. No scripts, genre fiction, romance, horror, science fiction or children's books. COMMISSION Home 15%; Foreign & Translation 20%. OVERSEAS ASSOCIATES **Abner Stein** and **The Marsh Agency Ltd**, UK. Send covering letter with first three chapters or e-mail query in the first instance. No reading fee.

Linda Roghaar Literary Agency, LLC*

133 High Point Drive, Amherst MA 01002
📞 001 413 256 1921
contact@LindaRoghaar.com
www.lindaroghaar.com
Contact *Linda L. Roghaar*

Founded 1997. Handles fiction and specializes in non-fiction titles. No romance, science fiction or horror. COMMISSION Home 15%; Foreign & Translation rate varies. Send query by e-mail or letter with return postage (for fiction, include the first five pages). No reading fee.

The Rosenberg Group*

23 Lincoln Avenue, Marblehead MA 01945
📞 001 781 990 1341   📠 001 781 990 1344
www.rosenberggroup.com

Contact *Barbara Collins Rosenberg*

Founded 1998. Handles non-fiction (please see website for areas of interest), fiction, specializing in romance (single title and category) and women's; trade non-fiction with a special interest in sports, also handles college textbooks for the first and second year courses. No scripts, science fiction, true crime, inspirational fiction, children's and young adult. COMMISSION Home 15%; Foreign 25%. No unsolicited material; send query letter by post only. No reading fee.

Russell & Volkening, Inc.*

50 West 29th Street, #7E, New York NY 10001
☎ 001 212 684 6050   ℻ 001 212 889 3026
Contacts *Tim Seldes, Jesseca Salky (adult), Carrie Hannigan (children's), Joy Ami (humour & fiction)*

Founded 1940. Handles literary fiction and non-fiction; science, history, current affairs, politics, mystery, thriller, true crime, children's. Specializes in politics and current affairs. No scripts, science fiction/fantasy or very prescriptive non-fiction. COMMISSION Home 15%; Foreign 20%. No unsolicited mss; sample chapters or synopsis with query and covering letter in the first instance. No reading fee.

The Sagalyn Literary Agency*

4922 Fairmont Avenue, Suite 200, Bethesda MD 20814
☎ 001 301 718 6440
query@sagalyn.com
www.sagalyn.com

Founded 1980. Handles mostly upmarket nonfiction with some fiction. No screenplays, romance, science fiction/fantasy, children's literature. No unsolicited material. See website for submissions procedure. No reading fee.

Victoria Sanders & Associates LLC*

241 Avenue of the Americas, Suite 11H, New York NY 10014
☎ 001 212 633 8811   ℻ 001 212 633 0525
queriesvsa@hotmail.com
www.victoriasanders.com
Contacts *Victoria Sanders, Diane Dickensheid*

Founded 1992. Handles general trade fiction and non-fiction, plus ancillary film and television rights. COMMISSION Home & Dramatic 15%; Translation 20%. Please send all queries via e-mail.

Jack Scagnetti Talent & Literary Agency

5118 Vineland Avenue, Suite 106, North Hollywood CA 91601
☎ 001 818 762 3871/761 0580
www.jackscagnetti.com
Contact *Jack Scagnetti*

Founded 1974. Works mostly with established/published authors. Handles non-fiction, fiction, film and TV scripts. No reading fee. COMMISSION Home & Dramatic 10% (scripts), 15% (books); Foreign 15%.

Schiavone Literary Agency, Inc.

Corporate Offices: 236 Trails End, West Palm Beach FL 33413–2135
☎ 001 561 966 9294   ℻ 001 561 966 9294
profschia@aol.com
www.publishersmarketplace.com/members/profschia
New York branch office: 3671 Hudson Manor Terrace, Suite 11H, Bronx, NY 10463
☎ ℻ 001 718 548 5332
CEO *James Schiavone (Florida office)*
President *Jennifer Duvall (New York; jendu77@aol.com)*
Executive VP *Kevin McAdams (New York; kvn.mcadams@yahoo.com)*

Founded 1996. Handles fiction and non-fiction (all genres). Specializes in biography, autobiography, celebrity memoirs. No poetry. COMMISSION Home 15%; Foreign & Translation 20%. OVERSEAS ASSOCIATES in Europe and Asia. No unsolicited mss; send query letter only with s.a.e. and IRCs. No response without s.a.e. For fastest response, e-mail queries of one page (no attachments) are acceptable and encouraged. No reading fee.

Susan Schulman, A Literary Agency*

454 West 44th Street, New York NY 10036
☎ 001 212 713 1633   ℻ 001 212 581 8830
schulman@aol.com
Submissions Editor (books) *Emily Uhry*
Submissions Editor (plays) *Linda Kiss*
Rights & Permissions Editor *Eleanora Tevís*

Founded 1980. Specializes in non-fiction of all types but particularly in health and psychology-based self-help for men, women and families. Other interests include business, memoirs, the social sciences, biography, language and international law. Fiction interests include contemporary fiction, including women's, mysteries, historical and thrillers with a cutting edge. 'Always looking for something original and fresh.' Represents properties for film and theatre, and works with agents in appropriate territories for translation rights. COMMISSION Home & Dramatic 15%; Translation 20%. No unsolicited mss. Query first, including outline and three sample chapters with IRCs. No reading fee.

Scovil Galen Ghosh Literary Agency, Inc.*

276 Fifth Avenue, Suite 708, New York NY 10001
☎ 001 212 679 8686   ℻ 001 212 679 6710
info@scglit.com
www.scglit.com

Contacts *Russell Galen, Anna M. Ghosh, Jack Scovil*

Founded 1993. Handles all categories of books. No scripts. COMMISSION Home 15%; Foreign 20%. No unsolicited material; send query by e-mail or letter. No reading fee.

Rosalie Siegel, International Literary Agency, Inc.*

1 Abey Drive, Pennington NJ 08534
☎ 001 609 737 1007   📠 001 609 737 3708
Contact *Rosalie Siegel*

Founded 1978. A one-woman, highly selective agency that takes on only a limited number of new projects. Handles fiction and non-fiction (especially narrative). Specializes in French history and literature; Europe in general; American history, social history. No science fiction, photography, illustrated art books or children's. COMMISSION Home 15%; Foreign 20%. No unsolicited material.

Michael Snell Literary Agency

PO Box 1206, Truro MA 02666–1206
☎ 001 508 349 3718
snellliteraryagency@yahoo.com
President *Michael Snell*
Vice President *Patricia Smith*

Founded 1978. Adult non-fiction, especially psychology, health, parenting, science, business and women's issues. Specializes in business and pet and animal books (professional and reference to popular trade how-to); general how-to and self-help on all topics, from diet and exercise to parenting, relationships, health, sex, psychology, personal finance, and dogs, cats and horses, plus literary and suspense fiction. COMMISSION Home 15%. No unsolicited mss. Send outline and sample chapter with return postage for reply. No reading fee for outlines. Brochure available on how to write a book proposal. Model proposal also for sale directly to prospective clients. Author of *From Book Idea to Bestseller*, published by Prima. Rewriting, developmental editing, collaborating and ghostwriting services available on a fee basis. Send IRCs.

Spectrum Literary Agency*

320 Central Park West, Suite 1-D, New York NY 10025
☎ 001 212 362 4323   📠 001 212 362 4562
www.spectrumliteraryagency.com
President and Agent *Eleanor Wood*
Agent *Justin Bell*

Founded 1976. Handles science fiction, fantasy, mystery, suspense and romance. No scripts. Not interested in self-help, New Age, religious fiction/non-fiction, children's books, poetry, short stories, gift books or memoirs. COMMISSION

Home 15%; Translation 20%. Send query letter in first instance with s.a.e. Faxed or electronic submissions are not accepted. No reading fees.

Philip G. Spitzer Literary Agency, Inc.*

50 Talmage Farm Lane, East Hampton NY 11937
☎ 001 631 329 3650
Contact *Philip Spitzer*

Founded 1969. Works mostly with established/published authors. Specializes in general non-fiction and fiction – thrillers. COMMISSION Home & Dramatic 15%; Foreign 20%. No reading fee for outlines.

Gloria Stern Literary

12535 Chandler Boulevard #3, Valley Village CA 91607
☎ 001 818 508 6296   📠 001 818 508 6296
Contact *Gloria Stern*

Founded 1984. Handles fiction (both mainstream and literary; TV, film, radio and theatre scripts. Specializes in chick lit. No material containing gratuitous violence. COMMISSION Home 10%; Foreign 15%. No unsolicited mss. Send sample material with covering letter. Reading fee charged.

Stimola Literary Studio, LLC*

306 Chase Court, Edgewater NJ 07020
☎ 001 201 945 9353   📠 001 201 945 9353
info@stimolaliterarystudio.com
www.stimolaliterarystudio.com
Contact *Rosemary B. Stimola*

Founded 1997. Handles children's books – pre-school through young adult – fiction and non fiction. Specializes in picture books, middle/young adult novels. No adult fiction. COMMISSION Home 15%; Foreign 20%. No unsolicited material; send query e-mail (no attachments). No reading fee.

Barbara W. Stuhlmann Author's Representative

PO Box 276, Becket MA 01223–0276
☎ 001 413 623 5170
Contact *Barbara Ward Stuhlmann*

Founded 1954. COMMISSION Home 10%; Foreign 15%; Translation 20%. Query first with IRCs, including sample chapters and synopsis of project. No reading fee. 'No new clients at this time.'

Patricia Teal Literary Agency*

2036 Vista del Rosa, Fullerton CA 92831
☎ 001 714 738 8333   📠 001 714 738 8333
Contact *Patricial Teal*

Founded 1978. Handles women's fiction: series and single-title works; commercial and popular non-fiction. Specializes in romantic fiction. No scripts, science fiction, fantasy, horror, academic texts. COMMISSION Home 15%; Foreign 20%. No unsolicited material. Send query letter with s.a.e. in the first instance. No reading fee. 'Accepting queries in fiction from published novelists, only.'

S©ott Treimel NY*

434 Lafayette Street, New York NY 10003
001 212 505 8353
jmc.st.ny@verizon.net

Founded 1995. Handles children's books only, from concept/board books to teen fiction. No picture book texts or filmscripts. COMMISSION Home 15%; Foreign 20%. No unsolicited material. Query first. 'Prefer writers and illustrators recommended by professional authors or editors.' No reading fee.

2M Communications Ltd*

121 West 27th Street, Suite 601, New York
NY 10001
001 212 741 1509   001 212 691 4460
morel@bookhaven.com
www.2mcommunications.com
Contact *Madeleine Morel*

Founded 1982. Specializes in representing ghostwriters. Represents ghostwriters, only. No unsolicited mss.

Wales Literary Agency, Inc.*

PO Box 9428, Seattle WA 98109–0428
001 206 284 7114
waleslit@waleslit.com
www.waleslit.com
Contacts *Elizabeth Wales, Neal Swain*

Founded 1988. Handles quality fiction and non-fiction. No westerns, romance, self help, science fiction or horror. Special interest in 'Pacific Rim', West Coast and Pacific Northwest stories. COMMISSION Home 15%; Dramatic & Translation/Foreign 20%. No unsolicited mss; send query letter. No e-mail queries longer than one page and no attachments. No reading fee.

John A. Ware Literary Agency

392 Central Park West, New York NY 10025
001 212 866 4733   001 212 866 4734
Contact *John Ware*

Founded 1978. Specializes in non-fiction: biography, history, current affairs, investigative journalism, science, nature, inside looks at phenomena, medicine and psychology (academic credentials required). No personal memoirs. Also handles literary fiction, mysteries/thrillers, sport, oral history, Americana and folklore. COMMISSION Home & Dramatic 15%; Foreign 20%. Unsolicited mss not read. Send query letter first with IRCs to cover return postage. No reading fee.

The Wendy Weil Agency, Inc.*

232 Madison Avenue, Suite 1300, New York
NY 10016
001 212 685 0030   001 212 685 0765
wweil@wendyweil.com
www.wendyweil.com
Agents *Wendy Weil, Emily Forland*
Rights *Emma Patterson*

Founded 1987. Handles fiction (commercial and literary), non-fiction (journalism, memoirs, etc.). No scripts, science fiction, romance, visual books, self-help, cookery. OVERSEAS ASSOCIATES **David Higham Associates Ltd**; Paul & Peter Fritz Agency. Send query letter with synopsis. No reading fee.

Cherry Weiner Literary Agency

28 Kipling Way, Manalapan NJ 07726
001 732 446 2096   001 732 792 0506
Cherry8486@aol.com
Contact *Cherry Weiner*

Founded 1977. Handles all types of fiction: science fiction and fantasy, mainstream, romance, mystery, westerns and some non-fiction. COMMISSION 15%. No submissions except through referral. No reading fee.

Rhoda Weyr Agency
▷ Dunham Literary Inc.

Writers House, LLC.*

21 West 26th Street, New York NY 10010
001 212 685 2400   001 212 685 1781
www.writershouse.com
Contacts *Albert Zuckerman, Amy Berkower, Merrilee Heifetz, Susan Cohen, Susan Ginsburg, Robin Rue, Simon Lipskar, Steven Malk, Jodi Reamer, Dan Lazar, Rebecca Sherman, Ken Wright, Dan Conaway, Michele Rubin*

Founded 1973. Handles all types of fiction, including children's and young adult, plus narrative non-fiction: history, biography, popular science, pop and rock culture as well as how-to, business and finance, and New Age. Specializes in popular and literary fiction, women's novels, thrillers and children's. No scripts. No professional or scholarly. COMMISSION Home & Dramatic 15%; Foreign 20%. Albert Zuckerman is author of *Writing the Blockbuster Novel*, published by Little, Brown & Co. and Warner Paperbacks. For consideration of unsolicited mss, send a one-page letter of enquiry, 'explaining why your book is wonderful, briefly what it's about and outlining your writing background'. No reading fee.

Susan Zeckendorf Associates, Inc.*

171 West 57th Street, Suite 11B, New York
NY 10019
📱 001 212 245 2928
Contact *Susan Zeckendorf*

Founded 1979. Handles non-fiction of all kinds:
self help, social history, biography; fiction:
commercial historical, mystery and literary. No
scripts, romance, science fiction or children's
books. COMMISSION Home 15%; Foreign &
Translation 20%. ASSOCIATES Rosemarie
Buckman (translation); **Abner Stein**, UK. No
unsolicited material; send query letter with s.a.e.
in the first instance. No e-mails. No reading fee.

Agency Consultants

Agent Research & Evaluation, Inc. (AR&E)
▷ entry under Agency Consultants (p.247)

Book Clubs

David Arscott's Sussex Book Club

Dolphin House, 51 St Nicholas Lane, Lewes
BN7 2JZ
☎ 01273 470100   📠 01273 470100
sussexbooks@aol.com

Founded January 1998. Specializes in books
about the county of Sussex. Represents all the
major publishers of Sussex books and offers a
wide range of titles. Free membership without
obligation to buy.

Artists' Choice

PO Box 3, Huntingdon PE28 0QX
☎ 01832 710201   📠 01832 710488
www.artists-choice.co.uk

Specializes in books for the amateur artist at all
levels of ability.

Baker Books

Manfield Park, Cranleigh GU6 8NU
☎ 01483 267888   📠 01483 267409
bakerbooks@dial.pipex.com
www.bakerbooks.co.uk

Book clubs for schools: Funfare for ages 3–8,
Bookzone, ages 9–12 and My Word, ages 13–16.
Four issues per year operated in the UK and
overseas.

BCA (Book Club Associates)

Hargreaves Road, Groundwell, Swindon SN25 5BG
☎ 01793 723547
www.bca.co.uk

BCA, a wholly-owned subsidiary of Bertelsmann
AG, is Britain's largest book club organization.
Clubs include: Books For Children (BFC),
BooksDirect, Fantasy & SF, World of Mystery
& Thriller, Railway Book Club, The Softback
Preview (TSP), WorldBooks.

Bibliophile Books

Unit 5, Datapoint, 6 South Crescent, London
E16 4TL
☎ 020 7474 2474   📠 020 7474 8589
orders@bibliophilebooks.com
www.bibliophilebooks.com

New books covering a wide range of subjects at
discount prices. Write, phone, fax or e-mail for
free catalogue of 3,000 titles issued 10 times a
year.

Book Club Associates
▷ BCA

Cygnus Books

PO Box 15, Llandeilo SA19 6YX
☎ 01558 825500   📠 01550 825517
enquiries@cygnus-books.co.uk
www.cygnus-books.co.uk

'Bookseller offering books for your next step in
spirituality and complementary health care.' Over
1,000 hand-picked titles. Publishes *The Cygnus
Review* magazine (free of charge) which features
30–40 reviews on new mind, body and spirit titles
each month.

The Folio Society

44 Eagle Street, London WC1R 4FS
☎ 020 7400 4222   📠 020 7400 4242
enquiries@foliosoc.co.uk
www.foliosociety.com

Fine editions of classic fiction, history and
memoirs; also some children's classics.

Letterbox Library

71–73 Allen Road, London N16 8RY
☎ 020 7503 4801   📠 020 7503 4800
info@letterboxlibrary.com
www.letterboxlibrary.com

Children's book cooperative. Hard and softcover
specialist in multicultural and inclusive books for
children from one to teenage.

Poetry Book Society

See entry under *Organizations of Interest to Poets.*

Readers' Union Ltd

Brunel House, Forde Close, Newton Abbot
TQ12 4PU
☎ 01626 323200   📠 01626 323318
www.readersunion.co.uk

Book clubs with specific interests: Anglers' Book
Club, The Craft Club, Craftsman's Book Club,
Country Sports Book Club, Equestrian Society,
The Gardeners Society, Needlecrafts with Cross
Stitch, Painting for Pleasure, Photographers' Book
Club, Puzzles Plus.

Scholastic Book Clubs

See **Scholastic Ltd** under *UK Publishers.*

Magazines

Acclaim
> The New Writer

Accountancy
145 London Road, Kingston Upon Thames
KT2 6SR
☎ 020 8247 1379 📠 020 8247 1424
accountancynews@cch.co.uk
www.accountancymagazine.com
Owner *Wolters Kluwer UK*
Editor *Sally Percy*
Circulation 149,236

Founded 1889. MONTHLY. Written ideas welcome.
FEATURES *Sally Percy* Accounting, tax, business-
related articles of high technical content aimed at
professional/managerial readers. Maximum 2,000
words. 💷 Payment by arrangement.

Accountancy Age
32–34 Broadwick Street, London W1A 2HG
☎ 020 7316 9000 📠 020 7316 9250
accountancy.age@incisivemedia.com
www.accountancyage.com
Owner *Incisive Media*
Editor-in-Chief *Damian Wild*
Editor *Gavin Hinks* (020 7316 9608)
Circulation 56,794

Founded 1969. WEEKLY. Unsolicited mss welcome.
Ideas may be suggested in writing provided they
are clearly thought out.
FEATURES *Kevin Reed* Topics right across the
accountancy, business and financial world. Max.
1,400 words. 💷 Payment negotiable.

ACE Tennis Magazine
Advantage Publishing (UK) Ltd, First Floor, Barry
House, 20–22 Worple Road, London SW19 4DH
☎ 020 8947 0100 📠 020 8947 0117
paul.morgan@acemag.co.uk
Owner *LTA*
Editor *Paul Morgan*
Circulation 47,766

Founded 1996. MONTHLY specialist tennis
magazine. News and features. No unsolicited
mss; send feature synopses by e-mail in the first
instance. No tournament reports.

Acoustic
Oyster House Media Ltd, Oyster House, Hunter's
Lodge, Kentisbeare EX15 2DY
☎ 01884 266100 📠 01884 266101
editor@acousticmagazine.com
www.acousticmagazine.com
Owner *Oyster House Media Ltd*
Assistant Editor *Russell Welton*
Consulting Editor *Mark Tucker*
Circulation 25,000

Launched 2004. BI-MONTHLY. Reviews, news,
features and interviews related to acoustic guitars,
acoustic instruments and acoustic instrument
manufacturing. Approach by e-mail.

Acumen
> under Poetry Magazines

Aeroplane
IPC Media Ltd., Blue Fin Building, 110 Southwark
Street, London SE1 0SU
☎ 020 3148 4100
aeroplane_monthly@ipcmedia.com
www.aeroplanemonthly.com
Owner *IPC Country & Leisure Media Ltd*
Editor *Michael Oakey*
Circulation 35,704

Founded 1973. MONTHLY. Historic aviation and
aircraft preservation from its beginnings to the
1960s. No poetry. Will consider news items and
features written with authoritative knowledge
of the subject, illustrated with good quality
photographs.
NEWS *Tony Harmsworth* Max. 500 words.
FEATURES *Michael Oakey* Max. 3,000 words.
Approach by letter or e-mail in the first instance.
💷 £60 per 1,000 words; £10–40 per picture used.

AIR International
PO Box 100, Stamford PE9 1XQ
☎ 01780 755131 📠 01780 757261
malcolm.english@keypublishing.com
www.airinternational.com
Owner *Key Publishing Ltd*
Editor *Malcolm English*

Circulation 25,000

Founded 1971. MONTHLY. Civil and military aircraft magazine. Unsolicited mss welcome but initial approach by phone or in writing preferred.

AirForces Monthly

PO Box 100, Stamford PE9 1XQ
☎ 01780 755131　📠 01780 757261
edafm@keypublishing.com
Owner *Key Publishing Ltd*
Editor *Alan Warnes*
Circulation 20,439

Founded 1988. MONTHLY. Modern military aviation magazine. Unsolicited contributions welcome but initial approach by phone or in writing preferred.

All About Soap

Hachette Filipacchi (UK) Ltd, 64 North Row, London W1K 7LL
☎ 020 7150 7000　📠 020 7150 7681
allaboutsoap@hf-uk.com
www.allaboutsoap.co.uk
Editor *Johnathon Hughes*
Circulation 105,198

Founded 1999. FORTNIGHTLY. Soap opera storylines, television gossip and celebrity features. No unsolicited contributions; make initial contact by phone or e-mail.

Amateur Gardening

Westover House, West Quay Road, Poole BH15 1JG
☎ 01202 440840　📠 01202 440860
Owner *IPC Media (A Time Warner Company)*
Editor *Tim Rumball*
Circulation 46,128

Founded 1884. WEEKLY. Topical and practical gardening articles. News items are compiled and edited in-house, generally. Approach with news stories by e-mail in the first instance.
💷 Payment negotiable.

Amateur Photographer

IPC Media Ltd., Blue Fin Building, 110 Southwark Street, London SE1 0SU
☎ 020 3148 4138
amateurphotographer@ipcmedia.com
Owner *IPC Media (A Time Warner Company)*
Editor *Damien Demolder*
Circulation 24,567

Founded 1884. WEEKLY. For the competent amateur with a technical interest. Freelancers are used but writers should be aware that there is ordinarily no use for words without pictures.

Ancestors
▷ The National Archives under UK Publishers

Angler's Mail

IPC Media Ltd., Blue Fin Building, 110 Southwark Street, London SE1 0SU
☎ 020 3148 5000
anglersmail@ipcmedia.com
www.anglersmail.com
Owner *IPC Media (A Time Warner Company)*
Editor *Tim Knight*
Circulation 32,400

Founded 1965. WEEKLY. Angling news and matches. Interested in pictures, stories, tip-offs and features. Approach the news desk by telephone. 💷 £10–40 per 200 words; pictures, £20–40.

Animal Action

Wilberforce Way, Southwater, Horsham RH13 9RS
☎ 0300 123 0392　📠 0303 123 0392
publications@rspca.org.uk
www.rspca.org.uk
Owner *RSPCA*
Editor *Sarah Evans*
Circulation 30,655

BI-MONTHLY (SEVEN ISSUES YEARLY). RSPCA youth membership magazine. Articles (pet care, etc.) are written in-house. Good-quality animal photographs welcome.

Animals and You

D.C. Thomson & Co. Ltd, 2 Albert Square, Dundee DD1 9QJ
☎ 01382 223131　📠 01382 322214
animalsandyou@dcthomson.co.uk
Owner *D.C. Thomson & Co. Ltd*
Editor *Margaret Monaghan*
Circulation 38,503

Founded 1998. THREE-WEEKLY magazine aimed at girls, aged 7–11 years who love animals. Contains cute 'n' cuddly pin-ups, pet tips and advice, animal charity information, photo stories, puzzles, readers' pictures and drawings. Most work done in-house but will consider short features, photographic or with good illustrations. Approach by e-mail.

Antique Collecting
▷ Antique Collectors' Club under UK Publishers

Antiquesnews

PO BOX 3369, Chippenham SN15 9DU
☎ 01225 742240
mail@antiquesnews.co.uk
www.antiquesnews.co.uk
Owner/Editor *Gail McLeod*

Founded 1997. WEEKLY. Up-to-date information online for the British antiques and art trade. News, photographs, gossip and controversial views on all aspects of the fine art and antiques world welcome. Articles on antiques and fine arts themselves are also featured. Approach in writing with ideas.

Apex Advertiser

PO Box 7086, Clacton-on-Sea CO15 5WN
☎ 01255 428500
mail@apexpublishing.co.uk
www.apexadvertiser.co.uk
Owner *Apex Publishing Ltd*
Editor *Chris Cowlin*
Circulation 12,000

Founded 2002. MONTHLY advertising magazine including guides and information, features, listings and reviews. Approach by post, only.

Apollo

22 Old Queen Street, London SW1H 9HP
☎ 020 7961 0107
editorial@apollomag.com
www.apollo-magazine.com
Owner *The Spectator (1828) Ltd*
Chief Executive *Andrew Neil*
Editor *Michael Hall*

Founded 1925. MONTHLY. Specialist articles on art and antiques, collectors and collecting, exhibition and book reviews, information on dealers and auction houses. Unsolicited mss welcome but please make initial contact by e-mail. Interested in new research on fine and decorative arts of all eras, and architecture.

The Architects' Journal

Greater London House, Hampstead Road, London NW1 7EJ
☎ 020 7728 4574   ▤ 020 7728 4666
[firstname].[surname]@emap.com
www.architectsjournal.co.uk
Owner *Emap Construct*
Editor *Kieran Long*
Circulation 11,694

WEEKLY trade magazine dealing with all aspects of the industry. No unsolicited mss. Approach in writing with ideas.

Architectural Design

John Wiley & Sons, 3rd Floor, International House, Ealing Broadway Centre, London W5 5DB
☎ 020 8326 3800   ▤ 020 8326 3801
Owner *John Wiley & Sons Ltd*
Editor *Helen Castle*
Circulation 5,000

Founded 1930. BI-MONTHLY. Sold as a book and journal as well as online, *AD* charts theoretical and topical developments in architecture. Format consists of a minimum of 128pp, the first part dedicated to a theme compiled by a specially commissioned guest-editor; the back section (AD+) carries series and more current one-off articles. Unsolicited mss not welcome generally, though journalistic contributions will be considered for the back section.

The Architectural Review

Greater London House, Hampstead Road, London NW1 7EJ
☎ 020 7728 4589
www.arplus.com
Editor *Paul Finch*
Circulation 20,733

MONTHLY international professional magazine dealing with architecture and all aspects of design. No unsolicited mss. Approach in writing with ideas.

The Armourer Magazine

PO Box 161, Congleton CW12 3WJ
☎ 01260 278044   ▤ 01260 278044
editor@armourer.co.uk
www.armourer.co.uk
Owner *Beaumont Publishing Ltd*
Editor *Irene Moore*
Circulation 15,000

Founded 1994. BI-MONTHLY. Militaria, military antiques, military history, events and auctions, with an emphasis on the First and Second World Wars. Unsolicited military-related contributions are welcome but 'we do not have an editorial budget'. Approach by e-mail in the first instance.

Art Monthly

4th Floor, 28 Charing Cross Road, London WC2H 0DB
☎ 020 7240 0389   ▤ 020 7497 0726
info@artmonthly.co.uk
www.artmonthly.co.uk
Owner *Britannia Art Publications Ltd*
Editor *Patricia Bickers*
Circulation 6,000

Founded 1976. TEN ISSUES YEARLY. News and features of relevance to those interested in modern and contemporary visual art. Unsolicited mss welcome. Contributions should be addressed to the deputy editor, accompanied by s.a.e.
FEATURES Always commissioned. Interviews and articles of up to 2,000 words on art theory, individual artists, contemporary art history and issues affecting the arts (e.g. funding and arts education). Exhibition and book reviews: 750–1,000 words.
NEWS Brief reports (250–300 words) on art issues.
€ Payment negotiable.

The Art Newspaper

70 South Lambeth Road, London SW8 1RL
☎ 020 7735 3331   ▤ 020 7735 3332
contact@theartnewspaper.com
www.theartnewspaper.com
www.theartnewspaper.tv

Owner *Umberto Allemandi & Co. Publishing*
Editor *Jane Morris*
Circulation 22,600

Founded 1990. MONTHLY. Tabloid format with hard news on the international art market, news, museums, exhibitions, archaeology, conservation, books and current debate topics. Length 250–2,000 words. No unsolicited mss. Approach with ideas in writing. Commissions only.

Art Review

1 Sekforde Street, London EC1R 0BE
📞 020 7107 2760 📠 020 7107 2761
office@artreview.com
www.artreview.com
Owner *Dennis Hotz*
Editor *Mark Rappolt*
Executive Editor *David Terrien*
Circulation 35,000

Founded 1949. MONTHLY magazine of international contemporary art. Features, profiles, previews and reviews of artists and exhibitions worldwide. 'We have a very international focus so it's important that the artist is exhibiting internationally and has good gallery representation.'
NEWS All items written in-house.
FEATURES *Mark Rappolt* Max. 1,500 words. E-mail pitch and c.v. at least three months in advance of related show/publication. 💷 35p. per word.

Art World Magazine

502 Clerkenwell Workshops, 27/31 Clerkenwell Close, London EC1R 0AU
📞 020 7014 3440 📠 020 7014 3435
info@artworldmagazine.com
www.artworldmagazine.com
Owner *Steve Bush*
Editor *Victoria Macdonald*
Circulation 20,000

Launched 2007. BI-MONTHLY. Contemporary art-based publication that aims to present art in the voice of the artists and up-to-date news. Welcomes suggestion of artists to cover, also feature and news ideas. 'We are strictly contemporary art; no design, craft, etc.' Approach by e-mail.

The Artist

Caxton House, 63–65 High Street, Tenterden TN30 6BD
📞 0158076 3673 📠 0158076 5411
www.painters-online.co.uk
Owner/Editor *Sally Bulgin*
Circulation 23,000

Founded 1931. MONTHLY. Art journalists, artists, art tutors and writers with a good knowledge of art materials are invited to write to the editor with ideas for practical and informative features about art, techniques and artists.

Athletics Weekly

83 Park Road, Peterborough PE1 2TN
📞 01733 808550 📠 01733 808535
results@athletics-weekly.com
www.athleticsweekly.com
Owner *Descartes Publishing*
Editor *Jason Henderson*
Results Editor *Steve Smythe*
Circulation 14,000

Founded 1945. WEEKLY. Covers track and field, road, fell, cross-country, race walking. Results, fixtures, interviews, technical coaching articles and news surrounding build up to 2012 Olympics.
NEWS *Paul Halford* Max. 400 words.
FEATURES *Jason Henderson* Max. 2,000 words. Approach in writing.
💷 Payment negotiable.

Attitude Magazine

211 Old Street, London EC1V 9NR
attitude@attitude.co.uk
www.attitude.co.uk
Owner *Trojan Publishing*
Editor *Matthew Todd*
Circulation 90,000

Launched 1994. MONTHLY. Top selling gay magazine with global cover interview exclusives including Madonna, David Beckham, Heath Ledger. Welcomes feature suggestions and opinion pieces. 'No sex and the city style columns.' Approach by e-mail.

The Author

84 Drayton Gardens, London SW10 9SB
📞 020 7373 6642
theauthor@societyofauthors.org
Owner *The Society of Authors*
Editor *Andrew Rosenheim*
Manager *Kate Pool*
Circulation 9,000

Founded 1890. QUARTERLY journal of the **Society of Authors**. Most articles are commissioned.

Auto Express

30 Cleveland Street, London W1T 4JD
📞 020 7907 6000 📠 020 7907 6234
editorial@autoexpress.co.uk
www.autoexpress.co.uk
Owner *Dennis Publishing*
Editor *David Johns*
Deputy Editor *Graham Hope*
Acting News and Features Editor *Julie Sinclair*
Circulation 77,587

Founded 1988. WEEKLY consumer motoring title with news, drives, tests, investigations, etc.
NEWS News stories and tip-offs welcome. Fillers

100 words max.; leads 300 words.
FEATURES Ideas welcome. No fully-written articles – features will be commissioned if appropriate. Max. 2,000 words.

Autocar

Haymarket Consumer Media, Teddington Studios, Broom Road, Teddington TW11 9BE
☎ 020 8267 5630   🖷 020 8267 5759
autocar@haymarket.com
www.autocar.co.uk
Owner *Haymarket Consumer Media,*
Editor *Chas Hallett*
News Editor *Dan Stevens*
Circulation 58,091

Founded 1895. WEEKLY. All news stories, features, interviews, scoops, ideas, tip-offs and photographs welcome. 💷 Payment negotiable.

Aviation News

HPC, Drury Lane, St Leonards on Sea TN38 9BJ
☎ 01424 205537   🖷 01424 443693
editor@aviation-news.co.uk
www.aviation-news.co.uk
Owner *Hastings Printing Company*
Editor *David Baker*
Circulation 21,000

Founded 1939. MONTHLY review of aviation for those interested in military, commercial and business aircraft and equipment – old and new. Will consider articles on military and civil aircraft, airports, air forces, current and historical subjects. No fiction.
FEATURES 'New writers always welcome and if the copy is not good enough it is returned with guidance attached.' Max. 5,000 words.
NEWS Items are always considered from new sources. Max. 300 words.
💷 Payment negotiable.

Balance

Diabetes UK, 10 Parkway, London NW1 7AA
☎ 020 7424 1000   🖷 020 7424 1001
balance@diabetes.org.uk
Owner *Diabetes UK*
Editor *Martin Cullen*
Circulation 250,000

Founded 1935. BI-MONTHLY. Unsolicited mss are not accepted. Writers may submit a brief proposal in writing. Only topics relevant to diabetes will be considered.
NEWS Short pieces about activities relating to diabetes and the lifestyle of people with diabetes. Max. 150 words.
FEATURES Medical, diet and lifestyle features written by people with diabetes or with an interest and expert knowledge in the field.

General features are mostly based on experience or personal observation. Max. 1,500 words.
💷 NUJ rates.

The Banker

Number One Southwark Bridge, London SE1 9HL
☎ 020 7873 3000
stephen.timewell@ft.com
www.thebanker.com
Owner *FT Business*
Editor-in-Chief *Stephen Timewell*
Circulation 28,771

Founded 1926. MONTHLY. News and features on banking, finance and capital markets worldwide and technology.

BBC Gardeners' World Magazine

Media Centre, 201 Wood Lane, London W12 7TQ
☎ 020 8433 2000   🖷 020 8433 3986
adam.pasco@bbc.com
www.gardenersworld.com
Owner *BBC Worldwide Publishing Ltd*
Editor *Adam Pasco*
Deputy Editor *Lucy Hall*
Commissioning Editor *Kevin Smith*
Circulation 238,673

Founded 1991. MONTHLY. Gardening advice, ideas and inspiration. No unsolicited mss. Approach by phone or in writing with ideas – interested in features about exceptional small gardens. Also interested in any exciting new gardens showing good design and planting ideas as well as new plants and the stories behind them. Including sections on Grow & Eat, problem solving, wildlife gardening, shopping and buyers' guides, plus practical projects.

BBC Good Food

Media Centre, 201 Wood Lane, London W12 7TQ
☎ 020 8433 2000   🖷 020 8433 3931
goodfood@bbc.com
www.bbcgoodfood.com
Owner *BBC Worldwide Publishing Ltd*
Editor *Gillian Carter*
Circulation 340,057

Founded 1989. MONTHLY recipe magazine featuring celebrity chefs. No unsolicited mss.

BBC Good Homes Magazine

Media Centre, 201 Wood Lane, London W12 7TQ
☎ 020 8433 2391   🖷 020 8433 2691
deargoodhomes@bbc.co.uk
www.bbcgoodhomes.com
Owner *BBC Worldwide Publishing Ltd*
Editor *Bernie Herlihy*
Circulation 127,202

Founded 1998. MONTHLY. Decorating, interiors, readers' homes, shopping, makeovers, property and consumer information. No fiction.

PROPERTY & READERS' HOMES *Rachel Watson*
Property features from specialists. Approach by
letter with cuttings.

BBC History Magazine

Tower House, Fairfax Street, Bristol BS1 3BN
☎ 0117 927 9009
historymagazine@bbcmagazinesbristol.com
www.bbchistorymagazine.com
Owner *Bristol Magazines Ltd*
Editor *Dave Musgrove*
Circulation 58,530

Founded 2000. MONTHLY. General news and
features on British and international history,
with books and CD reviews, listings of history
events, TV and radio history programmes and
regular features for those interested in history
and current affairs. Winner, British Society of
Magazine Editors 'Editor of the Year Award', 2002,
for general interest and current affairs titles. Will
consider submissions from academic or otherwise
expert historians/archaeologists as well as,
occasionally, from historically literate journalists
who include expert analysis and historiography
with a well-told narrative. Ideas for regular
features are welcome. Also publishes cartoons,
quizzes and crosswords. 'We cannot guarantee to
acknowledge all unsolicited mss.'
FEATURES should be pegged to anniversaries or
forthcoming books/TV programmes, current
affairs topics, etc. 750–3,000 words.
NEWS 400–500 words. Send short letter or e-mail
with synopsis, giving appropriate sources, pegs
for publication dates, etc.
Ⓔ Payment negotiable.

BBC Music Magazine

9th Floor, Tower House, Fairfax Street, Bristol
BS1 3BN
☎ 0117 927 9009   ℻ 0117 934 9008
music@bbcmagazinesbristol.com
www.bbcmusicmagazine.com
Owner *BBC Worldwide Publishing Ltd*
Editor *Oliver Condy*
Circulation 47,104

Founded 1992. MONTHLY. All areas of classical
music. Not interested in unsolicited material.
Approach with ideas by e-mail or fax.

BBC Top Gear Magazine

Second Floor A, Energy Centre, Media Village,
201 Wood Lane, London W12 7TQ
☎ 020 8433 2000
www.topgear.com
Owner *BBC Magazines*
Editorial Director *Michael Harvey*

Circulation 200,471

Founded 1993. MONTHLY general interest
motoring magazine. No unsolicited material as
most features are commissioned.

BBC Wildlife Magazine

14th Floor, Tower House, Fairfax Street, Bristol
BS1 3BN
☎ 0117 927 9009   ℻ 0117 927 9008
wildlifemagazine@bbcmagazinesbristol.com
www.bbcwildlifemagazine.com
Owner *Bristol Magazines Ltd*
Editor *Sophie Stafford*
Features Editor *Fergus Collins*
News & Travel Editor *James Fair*
Circulation 47,114

Founded 1963. Formerly *Wildlife*; became *BBC
Wildlife Magazine* in 1983. THIRTEEN ISSUES A
YEAR. Welcomes idea pitches but not unsolicited
mss. Read guidelines on website to improve
chances of success.
FEATURES Most features commissioned from
excellent writers with expert knowledge of
wildlife or conservation subjects. Max. 2,000
words.
NEWS Most news stories commissioned from
known freelancers. Max. 400 words.
Ⓔ Features, £200–450; news, £80–120.

Bee World
▷ Journal of Apicultural Research

Bella

H. Bauer Publishing, Academic House, 24–28
Oval Road, London NW1 7DT
☎ 020 7241 8000   ℻ 020 7241 8056
Owner *H. Bauer Publishing*
Editor *Julia Davis*
Circulation 221,661

Founded 1987. WEEKLY. Women's magazine
specializing in real-life, human interest stories
plus celebrity, fashion, beauty, home and
health. No fiction. Does not accept unsolicited
submissions.

Best

33 Broadwick Street, London W1F 0DQ
☎ 020 7339 4500   ℻ 020 7339 4580
best@natmags.co.uk
Owner *National Magazine Company/ACP*
Editor *Michelle Hather*
Circulation 301,437

Founded 1987. WEEKLY women's magazine.
Multiple features, news, short stories on all topics
of interest to women. Important for would-be
contributors to study the magazine's style which
differs from many other women's weeklies.
Approach in writing with s.a.e.
FEATURES *Helen Garston, Charlotte Seligman*

Real life stories only. Max. 1,500 words. No unsolicited mss.
FICTION *Pat Richardson* Short story slot: 900–1,200 words. Send s.a.e. for guidelines.
Ⓔ Payment negotiable.

Best of British

Church Lane Publishing Ltd, Bank Chambers, 6 Market Place, Market Deeping PE6 8DL
☎ 01778 342814
mail@british.fsbusiness.co.uk
www.bestofbritishmag.co.uk
Owner *Church Lane Publishing Ltd*
Editor-in-Chief *Ian Beacham*
Assistant Editor *Linne Matthews*

Founded 1994. MONTHLY magazine celebrating all things British, both past and present. Emphasis on nostalgia – memories from the 1940s, 1950s and 1960s. Study of the magazine is advised in the first instance. All preliminary approaches should be made in writing or by e-mail.

The Big Issue and The Big Issue South West

1–5 Wandsworth Road, London SW8 2LN
☎ 020 7526 3200   Ⓕ 020 7526 3201
helena.drakakis@bigissue.com
www.bigissue.com
Editor-in-Chief *A. John Bird*
Editor *Charles Howgego*
Associate Editor *Helena Drakakis*
Circulation 67,641; 12,582 (South West); 134,456 (group)

Founded 1991. WEEKLY. An award-winning campaigning and street-wise general interest magazine sold in London, the Midlands, the North East and South of England. Separate regional editions sold in Manchester, Scotland, Wales and the South West.
NEWS *Helena Drakakis* Hard-hitting exclusive stories with emphasis on social injustice aimed at national leaders.
ARTS *Charles Howgego* Interested in interviews and analysis ideas. Reviews written in-house. Send synopses to arts editor.
FEATURES Interviews, campaigns, comment, opinion and social issues reflecting a varied and informed audience. Balance includes social issues but mixed with arts and cultural features. Freelance writers used each week commissioned from a variety of contributors. Best approach in the first instance is to e-mail or post synopses to Editor *Charles Howgego* with examples of work. Max. 1,500 words.

The Big Issue Cymru

55 Charles Street, Cardiff CF10 2GD
☎ 029 2025 5670   Ⓕ 029 2025 5672
alex.donohue@bigissuecymru.co.uk
www.bigissuecymru.co.uk
Editor *Paul McNamee*

Circulation 8,451

Founded 1993. WEEKLY. Hard-hitting social and news features plus arts and entertainment.
FEATURES/NEWS *Paul McNamee* Welcomes contributions but must be relevant to a Welsh readership. Approach by e-mail. Lead in of four to six weeks. Max. 1,000 words (features); 350 (news). No poetry or fiction.

The Big Issue in Scotland

2nd Floor, 43 Bath Street, Glasgow G2 1HW
☎ 0141 352 7260   Ⓕ 0141 333 9049
editorial@bigissuescotland.com
www.bigissuescotland.com
Editor *Paul McNamee*
Circulation 31,401

WEEKLY. Coverage of social issues, arts and culture, sport, interviews; news and hard-hitting journalism.

The Big Issue in the North

10 Swan Street, Manchester M4 5JN
☎ 0161 831 5550   Ⓕ 0161 831 5577
kevin.gopal@bigissuenorth.co.uk
www.thebiglifecompany.com
Owner *The Big Life Company*
Editor *Kevin Gopal*
Circulation 30,192

Launched 1992. WEEKLY. Entertainment, celebrity interviews, political and environmental issues and issues from the social sector as well as homelessness. Welcomes news stories not covered by the national press; approach by e-mail.

The Big Issue South West

Founded 1994. WEEKLY. Hard-hitting, socially aware news and features, local arts and entertainment. Welcomes news features with a regional relevance. No poetry or fiction. Editorial dealt with by London office – see entry for *The Big Issue*.

Bike

Bauer Automotive Ltd, Media House, Lynch Wood, Peterborough PE2 6EA
☎ 01733 468099   Ⓕ 01733 468196
bike@bauermedia.co.uk
www.bikemagazine.co.uk
Owner *Bauer Automotive Ltd*
Editor *Steve Rose*
Circulation 81,579

Founded 1971. MONTHLY. Broad-based motorcycle magazine. Approach by e-mail.

Bird Life Magazine

The RSPB, UK Headquarters, The Lodge, Sandy SG19 2DL
☎ 01767 680551   Ⓕ 01767 683262

derek.niemann@rspb.org.uk
Owner *Royal Society for the Protection of Birds*
Editor *Derek Niemann*
Circulation 90,000

Founded 1965. BI-MONTHLY. Bird, wildlife and nature conservation for 8–12-year-olds (RSPB Wildlife Explorers members). No unsolicited mss. No 'captive/animal welfare' articles.
FEATURES Unsolicited material rarely used.
NEWS News releases welcome. Approach in writing in the first instance.

Bird Watching

Media House, Lynchwood, Peterborough PE2 6EA
☎ 01733 468000   📠 01733 468387
birdwatching@bauermedia.co.uk
www.birdwatching.co.uk
Editor *Sheena Harvey*
Circulation 19,763

Founded 1986. MONTHLY magazine for birdwatchers at all levels of experience. Offers practical advice of where and when to see birds, how to identify them, product reviews (particularly binoculars, telescopes and digiscoping equipment), extensive bird sightings, 'Go Birding' walks (10 each issue), news and general articles.
FEATURES Interested in authoratative articles on bird behaviour and bird watching/sites in the UK. Please send synopsis, not the whole piece. Max. 1,000 words.
NEWS UK news pre-eminent, preferably with pictures. Max. 350 words.
💷 Fees negotiable.

Birds

The RSPB, UK Headquarters, The Lodge, Sandy SG19 2DL
☎ 01767 680551   📠 01767 683262
rob.hume@rspb.org.uk
www.rspb.org.uk
Owner *Royal Society for the Protection of Birds*
Editor *R.A. Hume*
Circulation 616, 197

QUARTERLY magazine which covers not only wild birds but also other wildlife and related conservation topics. No interest in features on pet birds or 'rescued' sick/injured/orphaned ones. Content refers mostly to RSPB work so opportunities for freelance work on other subjects are limited but some freelance submissions are used in most issues. Phone or e-mail to discuss.

Birdwatch

B403A The Chocolate Factory, 5 Clarendon Road, London N22 6XJ
☎ 020 8881 0550   📠 020 8881 0990
editorial@birdwatch.co.uk
www.birdwatch.co.uk

Owner *Solo Publishing*
Editor *Dominic Mitchell*
Circulation 14,000

Founded 1992. MONTHLY magazine featuring illustrated articles on all aspects of birds and birdwatching, especially in Britain. No unsolicited mss. Approach in writing with synopsis of 100 words max. Annual **Birdwatch Bird Book of the Year** award (see entry under *Prizes*).
FEATURES *Dominic Mitchell* Unusual angles/personal accounts, if well-written. Articles of an educative or practical nature suited to the readership. Max. 2,000 words.
FICTION *Dominic Mitchell* Very little opportunity although occasional short story published. Max. 1,500 words.
NEWS *Dominic Mitchell* Very rarely use external material.
💷 £50 per 1,000 words.

Bizarre

30 Cleveland Street, London W1T 4JD
☎ 020 7907 6000   📠 020 7907 6516
bizarre@dennis.co.uk
www.bizarremag.com
Owner *Dennis Publishing*
Editor *David McComb*
Deputy Editor *Kate Hodges*
Circulation 50,132

Founded 1997. THIRTEEN ISSUES YEARLY. Amazing stories and images from around the world. No fiction, poetry, illustrations, short snippets.
FEATURES *Kate Hodges* No unsolicited mss. Send synopsis by post or e-mail.
💷 Payment negotiable.

Black Beauty & Hair

2nd Floor, Culvert House, Culvert Road, Battersea, London SW11 5DH
☎ 020 7720 2108   📠 020 7498 3023
info@blackbeautyandhair.com
www.blackbeautyandhair.com
Owner *Hawker Consumer Publications Ltd*
Publisher *Pat Petker (ext. 207)*
Editor *Irene Shelley (ext. 215)*
Circulation 19,697

BI-MONTHLY with one annual special: *The Hairstyle Book* in October; and a *Bridal Supplement* in the April/May issue. Black hair and beauty magazine with emphasis on authoritative articles relating to hair, beauty, fashion, health and lifestyle. Unsolicited contributions welcome.
FEATURES Beauty and fashion pieces welcome from writers with a sound knowledge of the Afro-Caribbean beauty scene plus bridal features. Minimum 1,000 words.
💷 £100 per 1,000 words.

Black Static

TTA Press, 5 Martins Lane, Witcham, Ely CB6 2LB
info@ttapress.demon.co.uk
www.ttapress.com
Owner *TTA Press*
Editor *Andy Cox*

Formerly *The Third Alternative*, founded 1993.
BIMONTHLY colour horror magazine: short
fiction, news and gossip, film and television,
manga and anime books. Publishes talented
newcomers alongside famous authors.
Unsolicited mss welcome if accompanied by
s.a.e. or e-mail address for overseas submissions
(max. 8,000 words). Fiction submissions as hard
copy only; feature suggestions can be e-mailed.
Potential contributors are advised to study the
magazine. Contracts are exchanged and payment
made upon acceptance. Winner of British Fantasy
Awards, International Horror Guild Award.

Bliss Magazine

Panini UK Ltd, Brockbourne House, 77 Mount
Ephraim, Tunbridge Wells TN4 8BS
☎ 01892 500106   🖷 01892 545666
bliss@panini.co.uk
www.mybliss.co.uk
Owner *Panini UK Ltd*
Editor *Leslie Sinoway*
Circulation 151,729

Founded 1995. MONTHLY teenage lifestyle
magazine for girls. No unsolicited mss; e-mail
the Features Editor, *Angeli Milburn* (amilburn@
panini.co.uk) with an idea and then follow up
with a call.
NEWS Worldwide teenage news. Max. 200 words.
FEATURES Real life teenage stories with subjects
willing to be photographed. Reports on teenage
issues. Max. 1,500 words.

Blueprint

Progressive Media Publishing, 1st Floor,
Boundary House, 91–93 Charterhouse Street,
London EC1M 6HR
☎ 020 7336 5200   🖷 020 7366 5201
vrichardson@blueprintmagazine.co.uk
www.blueprintmagazine.co.uk
Owner *Wilmington Media*
Editor *Vicky Richardson*
Deputy Editor *Tim Abrahams*
Assistant Editor *Peter Kelly*
Circulation 6,453

Founded 1983. MONTHLY. 'Crossing the
boundaries of design and architecture (from
graphics and product design to interiors and
urban development), Blueprint examines their
impact on society in general.' Features, news and
reviews. Welcomes ideas and pitches for articles
on topics relevant to the magazine; no completed
features. Approach by e-mail.

The Book Collector

PO Box 12426, London W11 3GW
☎ 020 7792 3492   🖷 020 7792 3492
editor@thebookcollector.co.uk (editorial)
www.thebookcollector.co.uk
Owner *The Collector Ltd*
Editor *Nicolas J. Barker*

Founded 1950. QUARTERLY magazine on
bibliography and the history of books, book-
collecting, libraries and the book trade.

Book World Magazine

2 Caversham Street, London SW3 4AH
☎ 020 7351 4995   🖷 020 7351 4995
leonard.holdsworth@btinternet.com
Owner *Christchurch Publishers Ltd*
Editor *James Hughes*
Circulation 5,500

Founded 1980. MONTHLY news and reviews for
serious book collectors, librarians, antiquarian
and other booksellers. No unsolicited mss.
Interested in material relevant to literature,
art and book collecting. Send letter in the first
instance.

The Bookseller

5th Floor, Endeavour House, 189 Shaftesbury
Avenue, London WC2H 8TJ
☎ 020 7420 6006   🖷 020 7420 6103
www.theBookseller.com
Owner *Nielsen Business Media*
Editor-in-Chief *Neill Denny*

Trade journal of the book trade, publishing, retail
and libraries – the essential guide to the book
business since 1858. Trade news and features,
including special features, company news,
publishing trends, bestseller data, etc. Unsolicited
mss rarely used as most writing is either done
in-house or commissioned from experts within
the trade. Approach in writing first.

Boxing Monthly

40 Morpeth Road, London E9 7LD
☎ 020 8986 4141   🖷 020 8986 4145
mail@boxing-monthly.demon.co.uk
www.boxing-monthly.co.uk
Owner *Topwave Ltd*
Editor *Glyn Leach*
Circulation 25,000

Founded 1989. MONTHLY. International coverage
of professional boxing; previews, reports and
interviews. Unsolicited material welcome.
Interested in small hall shows and grass-roots
knowledge. No big fight reports. Approach in
writing in the first instance.

Boyz

18 Brewer Street, London W1F 0SH
☎ 020 7025 6100   🖷 020 7025 6109

davidb@boyz.co.uk
Managing Editor *David Bridle*
Circulation 55,000

Founded 1991. WEEKLY entertainment and
features magazine aimed at a gay readership
covering clubs, fashion, TV, films, music, theatre,
celebrities and the London gay scene in general.
Unsolicited mss welcome.

Brand Literary Magazine

Creative Writing Programme, EPS, King William
Court, University of Greenwich, Maritime
Campus, Park Row, London SE10 9LS
📞 020 8331 8952
info@brandliterarymagazine.co.uk
www.brandliterarymagazine.co.uk
Editor *Nina Rapi*
Circulation 1,000

Biannual literary magazine, launched in 2007,
focusing on the short form; left of field work
and international writing. Welcomes stories,
plays, poems and creative non-fiction; no genre
or historical material. Approach by post; see
submission guidelines on the website.
FICTION *Nina Rapi* 'Looking for writing that
takes risks and has a strong voice'. Max. 2,500
words.
POETRY *Cherry Smyth* Max. 40 lines.
PLAYS *Nina Rapi, Harry Derbyshire* (Advisor)
Bold, short plays and performance text. Max. ten
minutes' performance time.
💰 £30.

Brides

Vogue House, Hanover Square, London W1S 1JU
📞 020 7499 9080   📠 020 7152 3369
www.bridesmagazine.co.uk
Owner *Condé Nast Publications Ltd*
Editor *Deborah Joseph*
Circulation 68,259

Founded 1955. BI-MONTHLY. Much of the
magazine is produced in-house, but good,
relevant emotional and wedding planning features
always welcome. Max. 1,000 words. Prospective
contributors should make initial contact by
e-mail.

British Birds

4 Harlequin Gardens, St Leonards-on-Sea
TN37 7PF
📞 01424 755155   📠 01424 755155
editor@britishbirds.co.uk
Editor *Dr R. Riddington*
Circulation 6,000

Founded 1907. MONTHLY ornithological
journal. Features main papers on topics such as
behaviour, distribution, ecology, identification
and taxonomy; annual *Report on Rare Birds in
Great Britain* and *Rare Breeding Birds in the*

UK; sponsored competition for Bird Photograph
of the Year. Unsolicited mss welcome from
ornithologists.
FEATURES Well-researched, original material
relating to Western Palearctic birds welcome.
NEWS *Adrian Pitches* (adrian.pitches@
blueyonder.co.uk) Items ranging from
conservation to humour. Max. 200 words.
💰 Payment for photographs, drawings, paintings
and main papers.

British Chess Magazine

The Chess Shop, 44 Baker Street, London
W1U 7RT
📞 020 7486 8222   📠 020 7486 3355
bcmchess@compuserve.com
www.bcmchess.co.uk
Director/Editor *John Saunders*

Founded 1881. MONTHLY. Emphasis on
tournaments, the history of chess and chess-
related literature. Approach in writing with
ideas. Unsolicited mss not welcome unless from
qualified chess experts and players.

British Journalism Review

Sage Publications Ltd, 1 Oliver's Yard, 55 City
Road, London EC1Y 1SP
📞 020 7324 8500   📠 020 7324 8756
bill.hagerty@btinternet.com
www.bjr.org.uk
www.bjr.sagepub.com
Owner *BJR Publishing Ltd*
Editor *Bill Hagerty*

Founded 1988. QUARTERLY. Media magazine
providing a forum for the news media. Freelance
contributions welcome: non-academic articles on
all aspects of the news media only. No academic
papers. Approach by e-mail. Max. 2,000 words.

British Medical Journal

BMA House, Tavistock Square, London WC1H 9JR
📞 020 7387 4499   📠 020 7383 6418
editor@bmj.com
www.bmj.com
Owner *British Medical Association*
Editor *Dr Fiona Godlee*
Circulation 122,982

One of the world's leading general medical
journals.

British Philatelic Bulletin

35–50 Rathbone Place, London W1T 1HQ
📞 020 7441 4744
www.royalmail.com/stamps
Owner *Royal Mail*
Editor *John Holman*
Circulation 15,000

Founded 1963. MONTHLY bulletin giving details
of forthcoming British stamps, features on older

stamps and postal history, and book reviews. Welcomes photographs of interesting, unusual or historic letter boxes.

FEATURES Articles on all aspects of British philately. Max. 1,500 words.

NEWS Reports on exhibitions and philatelic events. Max. 500 words. Approach in writing in the first instance.

Ⓔ £75 per 1,000 words.

British Railway Modelling

The Maltings, West Street, Bourne PE10 9PH
Ⓣ 01778 391027   Ⓕ 01778 425437
johne@warnersgroup.co.uk
www.brmodelling.co.uk
Owner *Warners Group Publications Plc*
Managing Editor *David Brown*
Editor *John Emerson*
Assistant Editor *Tony Wright*
Circulation 26,000

Founded 1993. MONTHLY. A general magazine for the practising modeller. Ideas are welcome. Interested in features on quality models, from individual items to complete layouts. Approach in writing first.

FEATURES Articles on practical elements of the hobby, e.g. locomotive construction, kit conversions, etc. Layout features and articles on individual items which represent high standards of the railway modelling art. Max. length 6,000 words (single feature).

NEWS News and reviews containing the model railway trade, new products, etc. Max. length 1,000 words.

Broadcast

Greater London House, Hampstead Road, London NW1 7EJ
Ⓣ 020 7728 5542
www.broadcastnow.co.uk
Owner *Emap Communications*
Acting Editor *Emily Booth*
Circulation 9,702

Founded 1959. WEEKLY. Opportunities for freelance contributions. Write to the relevant editor in the first instance.

FEATURES *Poppy Brech* Any broadcasting issue. Max. 1,500 words.

NEWS *Chris Curtis* Broadcasting news. Max. 350 words.

Ⓔ £270 per 1,000 words.

Build It

1 Canada Square, Canary Wharf, London E14 5AP
Ⓣ 020 7772 8452   Ⓕ 020 7772 8599
buildit@oceanmedia.co.uk
www.buildit-online.co.uk
Owner *Ocean Media*
Editor *Sarah Herbert*

Circulation 16,613

Founded 1990. MONTHLY magazine covering self-build, conversion and renovation. Unsolicited material welcome on self-build case studies as well as articles on technical construction, architecture and design, home improvements and dealing with builders. Max. length 2,500 words. Approach by post or e-mail.

Building Magazine

3rd Floor, Ludgate House, 245 Blackfriars Road, London SE1 9UY
Ⓣ 020 7560 4141   Ⓕ 020 7560 4080
denise.chevin@ubm.com
www.building.co.uk
Owner *UBM Builder Group*
Editor *Denise Chevin*
Circulation 24,241

WEEKLY magazine for the whole construction industry, from architects and engineers to contractors and house builders. News and features. No unsolicited contributions as all material is commissioned, although legal items welcome.

The Burlington Magazine

14–16 Duke's Road, London WC1H 9SZ
Ⓣ 020 7388 1228   Ⓕ 020 7388 1230
editorial@burlington.org.uk
www.burlington.org.uk
Owner *The Burlington Magazine Publications Ltd*
Managing Director *Kate Trevelyan*
Editor *Richard Shone*
Deputy Editor *Bart Cornelis*
Associate Editor *Jane Martineau*
Contributing Editor *John-Paul Stonard*

Founded 1903. MONTHLY. Unsolicited contributions welcome on the subject of art history provided they are previously unpublished. All preliminary approaches should be made in writing.

EXHIBITION REVIEWS Usually commissioned, but occasionally unsolicited reviews are published if appropriate. Max. 1,000 words.

ARTICLES Max. 4,500 words. Ⓔ Max. £150.

SHORTER NOTICES Max. 2,000 words. Ⓔ Max. £80.

Buses

PO Box 14644, Leven KY9 1WX
Ⓣ 01333 340637   Ⓕ 01333 340608
buseseditor@btconnect.com
www.busesmag.com
Owner *Ian Allan Publishing*
Editor *Alan Millar*
Circulation 13,000

Founded 1949. MONTHLY magazine of the bus and coach industry; vehicles, companies and people. Specialist articles welcome; no general articles with no specific bus content. E-mail enquiry welcome.

Business Brief

PO Box 582, Five Oaks, St Saviour JE4 8XQ
📞 01534 611600   📠 01534 611610
mspeditorial@msppublishing.com
Owner *MSP Publishing*
Editor *Peter Body*
Circulation 6,500 (Jersey & Guernsey)

Founded 1989. MONTHLY magazine covering business developments in the Channel Islands and how they affect the local market. Styles itself as the magazine for business people rather than just a magazine about business. Interested in business-orientated articles only – 800 words max. Approach the editor by e-mail in the first instance with telephone follow-up.
💷 Payment negotiable.

Business Traveller

2nd Floor, Cardinal House, Albemarle Street, London W1S 4TE
📞 020 7647 6330   📠 020 7647 6331
editorial@businesstraveller.com
www.businesstraveller.com
Owner *Panacea Publications*
Editorial Director *Tom Otley*
Circulation 500,000 (worldwide)

MONTHLY. Consumer publication. Opportunities exist for freelance writers. Would-be contributors are strongly advised to study the magazine or the website first. Approach in writing with ideas.
💷 Payment varies.

Buzz Extra

IBRA, 16 North Road, Cardiff CF10 3DY
📞 029 2037 2409   📠 05601 135640
mail@ibra.org.uk
www.ibra.org.uk
Owner *International Bee Research Association*
Editor *Richard Jones*

Informative, full-colour newsletter of the International Bee Research Association. QUARTERLY (March, June, September and December).

Cage & Aviary Birds

IPC Media Ltd., Blue Fin Building, 110 Southwark Street, London SE1 0SU
📞 020 3148 4171   📠 020 3148 8129
birds@ipcmedia.com
Owner *IPC Media (A Time Warner Company)*
Editor *Kim Forrester*
Circulation 17,000

Launched 1902. WEEKLY. 'Everything you need to know to keep, breed and show captive birds, including news and opinion.' No wild birds, bird watching, holidays/travel trips about birds.

FEATURES Happy to consider features based on first-hand experience of keeping birds or profiles of successful birdkeepers. Approach by e-mail. Max. 1,200 words. 💷 Payment varies but from £40 per 1,000 words.

The Cambridgeshire Journal Magazine

Cambridge Newspapers Ltd, Winship Road, Milton CB24 6PP
📞 01223 434409   📠 01223 434415
Owner *Cambridge Newspapers Ltd*
Editor *Debbie Tweedie*
Circulation 11,000

Founded 1990. MONTHLY magazines covering items of local interest – history, people, conservation, business, places, food and wine, fashion, homes and gardens.
FEATURES 750–1,500 words max., plus pictures. Approach the editor with ideas in the first instance.

Campaign

174 Hammersmith Road, London W6 7JP
📞 020 8267 4683   📠 020 8267 4927
campaign@haymarket.com
www.brandrepublic.com/campaign
Owner *Haymarket Publishing Ltd*
Editor *Claire Beale*
Circulation 10,112

Founded 1968. WEEKLY. Lively magazine serving the advertising and related industries. Freelance contributors are best advised to write in the first instance.
FEATURES Ideas welcome.
NEWS No submissions as all news handled in-house.
💷 Payment negotiable.

Camping and Caravanning

Greenfields House, Westwood Way, Coventry CV4 8JH
📞 0845 130 7631   📠 024 7647 5413
www.campingandcaravanningclub.co.uk
Owner *The Camping and Caravanning Club*
Features Editor *Sue Taylor*
Circulation 209,042

Founded 1901. MONTHLY. Interested in journalists with camping and caravanning knowledge. Write with ideas for features in the first instance.
FEATURES Outdoor pieces in general, plus items on specific regions of Britain. Max. 1,200 words. Illustrations to support text essential.

Canals and Rivers

PO Box 618, Norwich NR7 0QT
📞 01603 708930
chris@themag.fsnet.co.uk
www.canalsandrivers.co.uk
Owner *A.E. Morgan Publications Ltd*
Editor *Chris Cattrall*

Circulation 14,000

Covers all aspects of waterways, narrow boats and cruisers. Contributions welcome. Make initial approach by post or e-mail.
FEATURES Waterways, narrow boats and motor cruisers, cruising reports, practical advice, etc. Unusual ideas and personal comments are particularly welcome. Max. 2,000 words. Articles should be supplied in PC Windows format disk or e-mailed with JPEG images.
NEWS Items of up to 200 words welcome on the Inland Waterways System, plus photographs if possible.
⊕ Features, around £60 per page; news, £20.

Candis

Newhall Lane, Hoylake CH47 4BQ
☎ 0844 545 8102   ⊞ 0844 545 8103
helen@candis.co.uk
www.candis.co.uk
Owner *Newhall Publications*
Editor *Debbie Atwell*
Circulation 296,103

MONTHLY. General features on health and family lifestyle including makeover and celebrity – no fashion or first-person accounts of illness. No unsolicited contributions as most features are individually commissioned but fiction will be considered (max. 2,000 words). Send by e-mail (fiction@candis.co.uk).

Car Mechanics

Kelsey Publishing Group, Cudham Tithe Barn, Berry's Hill, Cudham TN16 3AG
☎ 01959 541444
cm.ed@kelsey.co.uk
www.carmechanicsmag.co.uk
Editor *Martyn Knowles*
Circulation 32,000

MONTHLY. Practical guide to maintenance and repair of post-1988 cars for DIY and the motor trade. Initial approach by e-mail, letter or phone strongly recommended, 'but please study a recent copy for style and content first'. No non-technical material, 'travelogues', cartoons or fiction.
FEATURES. Good, entertaining and well-researched technical material welcome, especially anything presenting complex matters clearly and simply.
PAYMENT negotiable 'but generous for the right material. We normally require words and pictures and can rarely use words alone.'

Caravan Magazine

IPC Media Ltd, Leon House, 233 High Street, Croydon CR9 1HZ
☎ 020 8726 8000   ⊞ 020 8726 8299
caravan@ipcmedia.com
www.caravanmagazine.co.uk

Owner *IPC Media (A Time Warner Company)*
Editor *Victoria Bentley*
Circulation 15,774

Founded 1933. MONTHLY. Unsolicited mss welcome. Approach in writing with ideas. All correspondence should go direct to the editor.
FEATURES Touring with strong caravan bias, technical/DIY features and how-to section. Max. 1,500 words.
⊕ Payment by arrangement.

Cat World

Ancient Lights, 19 River Road, Arundel BN18 9EY
☎ 01903 884988   ⊞ 01903 885514
laura@ashdown.co.uk
www.catworld.co.uk
Owner *Ashdown Publishing Ltd*
Editor *Laura Quiggen*

Founded 1981. MONTHLY. Unsolicited mss welcome but initial approach in writing preferred. No poems or fiction.
FEATURES Lively, first-hand experience features on every aspect of the cat. Breeding features and veterinary articles by acknowledged experts only. Preferred length 750 or 1,700 words. Accompanying pictures should be good quality and sharp. Submissions by e-mail (MS Word attachment) if possible, or by disk with accompanying hard copy and s.a.e. for return.

Caterer and Hotelkeeper

Quadrant House, The Quadrant, Sutton SM2 5AS
☎ 020 8652 3500   ⊞ 020 8652 8213
caterernews@rbi.co.uk
www.caterersearch.com
Owner *Reed Business Information*
Editor *Mark Lewis*
Circulation 17,641

Founded 1878. WEEKLY journal of the catering industry: news, case studies, products. Contributions welcome; approach by e-mail.

The Catholic Herald

Lamb's Passage, Bunhill Row, London EC1Y 8TQ
☎ 020 7448 3603   ⊞ 020 7256 9728
editorial@catholicherald.co.uk
www.catholicherald.co.uk
Editor *Luke Coppen*
Features Editor *Ed West*
Literary Editor *Stav Sherez*
Circulation 22,000

WEEKLY. Interested mainly in straight Catholic issues but also in general humanitarian matters, social policies, the Third World, the arts and books. ⊕ Payment by arrangement.

Celebs on Sunday
▷ Sunday Mirror under National Newspapers

Chapman

4 Broughton Place, Edinburgh EH1 3RX
📞 0131 557 2207
chapman-pub@blueyonder.co.uk
www.chapman-pub.co.uk
Owner/Editor *Dr Joy Hendry*
Circulation 2,000

Founded 1970. Scotland's quality literary magazine. Features poetry, short works of fiction, criticism, reviews and articles on theatre, politics, language and the arts. Gives priority to Scottish writers but also has work from UK and international authors. Unsolicited material welcome if accompanied by s.a.e. Approach in writing unless discussion is needed. Priority is given to full-time writers.

FEATURES Topics of literary interest, especially Scottish literature, theatre, culture or politics. Max. 5,000 words.

FICTION Short stories, occasional novel extracts if self-contained. Shortage of space means that longer work is less likely to be used.

SPECIAL PAGES Poetry, both UK and non-UK in translation (mainly, but not necessarily, European). Featured artists in every issue.
💷 Payment by negotiation.

Chat

IPC Media Ltd., Blue Fin Building, 110 Southwark Street, London SE1 0SU
📞 020 3148 5000
www.ipcmedia.com/pubs/chat.htm
Owner *IPC Media (A Time Warner Company)*
Editor *Gilly Sinclair*
Circulation 519,413

Founded 1985. WEEKLY general interest women's magazine. Unsolicited mss considered; approach in writing with ideas. Not interested in contributors 'who have never bothered to read *Chat* and therefore don't know what type of magazine it is. *Chat* does not publish fiction.'

FEATURES *Devinder Bains* Human interest and humour. Max. 1,000 words.
💷 Up to £500 maximum.

Chemist & Druggist

Riverbank House, Angel Lane, Tonbridge TN9 1SE
📞 01732 377487   📠 01732 367065
chemdrug@cmpmedica.com
www.chemistanddruggist.co.uk
Owner *CMP Medica*
Editor *Gary Paragpuri*
Circulation 13,675

Founded 1859. WEEKLY news magazine for the UK community pharmacy.

FEATURES Practice, politics, professional matters, clinical articles and business advice. Contact by phone or e-mail to discuss ideas.

NEWS Contact news editor with articles relating to local pharmacy matters, local pharmaceutical industry events and pharmacists in the news. Max. 200 words.
💷 Payment standard rates.

Choice

First Floor, 2 King Street, Peterborough PE1 1LT
📞 01733 555123   📠 01733 427500
editorial@choicemag.co.uk
www.choicemag.co.uk
Owner *Choice Publishing Ltd*
Editor *Norman Wright*
Circulation 85,000

Monthly full-colour, lively and informative magazine for people aged 50 plus which helps them get the most out of their lives, time and money after full-time work.

FEATURES Real-life stories, hobbies, interesting (older) people, British heritage and countryside, involving activities for active bodies and minds, health, relationships, book/entertainment reviews. Unsolicited mss read (s.a.e. for return of material); write with ideas and copies of cuttings if new contributor. No phone calls, please.

RIGHTS/MONEY All items affecting the magazine's readership are written by experts. Areas of interest include pensions, state benefits, health, finance, property, legal.
💷 Payment by arrangement.

Church Music Quarterly

19 The Close, Salisbury SP1 2EB
📞 01722 424848   📠 01722 424849
cmq@rscm.com
www.rscm.com
Owner *Royal School of Church Music*
Guest Editor *Julian Elloway*
Circulation 14,000

QUARTERLY. Contributions welcome. Phone in the first instance.

FEATURES Articles on church music or related subjects considered. Max. 2,100 words. 💷 £60 per page.

Church of England Newspaper

14 Great College Street, London SW1P 3RX
📞 020 7878 1001   📠 020 7878 1031
cen@churchnewspaper.com
www.churchnewspaper.com
Owner *Religious Intelligence*
Editor *Colin Blakely*
Circulation 9,500

Founded 1828. WEEKLY. Almost all material is commissioned but unsolicited mss are considered.

FEATURES Preliminary enquiry essential. Max. 1,200 words.

NEWS Items must be sent promptly to the e-mail address above and should have a church/

Christian relevance. Max. 200–400 words.
ⓔ Payment negotiable.

Church Times

13–17 Long Lane, Smithfield, London EC1A 9PN
☎ 020 7776 1060   ☏ 020 7776 1086
news@churchtimes.co.uk
features@churchtimes.co.uk
www.churchtimes.co.uk
Owner *Hymns Ancient & Modern*
Editor *Paul Handley*
Circulation 27,000

Founded 1863. WEEKLY. Unsolicited mss
considered (no poetry).
FEATURES *Christine Miles* Articles and pictures
(any format) on religious topics. Max. 1,500
words.
NEWS *Helen Saxbee* Occasional reports
(commissions only) and up-to-date photographs.
ⓔ Features, £100 per 1,000 words; news, by
arrangement.

Classic & Sports Car

Teddington Studios, Broom Road, Teddington
TW11 9BE
☎ 020 8267 5399   ☏ 020 8267 5318
letters.c&sc@haymarket.com
www.classicandsportscar.com
Owner *Haymarket Consumer Media*
Editor *James Elliott*
Circulation 76,508

Founded 1982. MONTHLY. Best-selling classic car
magazine covering all aspects of buying, selling,
owning and maintaining historic vehicles. Will
consider news stories and features involving
classic cars but 'be patient because hundreds of
unsolicited material received monthly'. Initial
contact by e-mail (james.elliott@haymarket.com).

Classic Bike

Bauer Automotive, Media House, Lynchwood,
Peterborough Business Park, Peterborough
PE2 6EA
☎ 01733 468000   ☏ 01733 468092
classic.bike@bauermedia.co.uk
www.classicbike.co.uk
Owner *Bauer Automotive*
Editor *Hugo Wilson*
Deputy Editor *Jim Moore*
Circulation 41,285

Founded 1978. MONTHLY. Mainly pre-1990 classic
motorcycles. Freelancers must contact the editor
before making submissions. Approach in writing
or by e-mail.
NEWS Genuine news with good illustrations, if
possible, suitable for a global audience. Max. 400
words.
FEATURES British motorcycle industry inside
stories, technical features 'that can be understood

by all', German, Spanish and French machine
features, people and classic Japanese bikes from
the 1950s to the 1990s. Restoration features, riding
features; 'good reads'. Max. 2,000 words.
SPECIAL PAGES How-to features, oddball
machines, stunning pictures, features with a fresh
slant.
ⓔ Payment negotiable.

Classic Boat

IPC Media Ltd, Leon House, 233 High Street,
Croydon CR9 1HZ
☎ 020 8726 8000   ☏ 020 8726 8195
cb@ipcmedia.com
www.classicboat.co.uk
Owner *IPC Media (A Time Warner Company)*
Editor *Dan Houston*
Circulation 11,000

Founded 1987. MONTHLY. Traditional boats and
classic yachts, old and new; maritime history.
Unsolicited mss, particularly if supported by
good photos, are welcome. Sail and power boat
pieces considered. Approach in writing with
ideas. Interested in well-researched stories on
all nautical matters. News reports welcome.
Contributor's notes available online.
FEATURES Boatbuilding, boat history and design,
events, yachts and working boats. Material must
be well-informed and supported where possible
by good-quality or historic photos. Max. 3,000
words. Classic Boat is defined by excellence of
design and construction – the boat need not be
old and wooden!
NEWS New boats, restorations, events,
boatbuilders, etc. Max. 500 words.
ⓔ Features, £75–100 per published page; news,
according to merit.

Classic Cars

Media House, Peterborough Business Park,
Lynchwood, Peterborough PE2 6EA
☎ 01733 468582   ☏ 01733 468888
classic.cars@bauermedia.co.uk
www.classiccarsmagazine.co.uk
Owner *Bauer*
Editor *Phil Bell*
Circulation 34,589

Founded 1973. TWELVE ISSUES YEARLY.
International classic car magazine containing
entertaining and informative articles about
classic cars, events and associated personalities.
Contributions welcome. ⓔ Payment negotiable.

Classic Land Rover World
▷ Land Rover World

Classic Rock

Future Publishing Ltd, 2 Balcombe Street, London
NW1 6NW
☎ 020 7042 4000

classicrock@futurenet.co.uk
www.classicrockmagazine.co.uk
Owner *Future Publishing Ltd*
Editor *Sian Llewellyn*
Circulation 67,000

Launched 1998. MONTHLY rock music magazine, both old and new.

FEATURES Welcomes star interviews, well-written and relevant features. E-mail the editor. 'Read the magazine first.' Payment varies.

Classic Stitches

D.C. Thomson & Co. Ltd, 80 Kingsway East, Dundee DD4 8SL
01382 223131   01382 452491
editorial@classicstitches.com
www.classicstitches.com
Owner *D.C. Thomson & Co. Ltd*
Editor *Mrs Bea Neilson*
Circulation 9,000

Founded 1994. BI-MONTHLY magazine with exclusive designs to stitch, textile-based features, designer profiles, book reviews and 'what's new shopping' information. Will consider features covering textiles – historical and modern, designer profiles and original designs in context with the style of the magazine (no items on knitting or crochet). Approach by e-mail or letter with short synopsis to avoid clashes. For designs, send sketches, swatches and brief description of the idea.

FEATURES *Bea Neilson* Three features per issue with accompanying photography. Max. 2,000 words.
NEWS *Liz O'Rourke* Max. 300 words.
SPECIAL PAGES *Liz O'Rourke* Details only.
WEBSITE *Allison Hay* Sub-editor. Short, craft-orientated features. Max 500 words.

Classical Guitar

1 & 2 Vance Court, Trans Britannia Enterprise Park, Blaydon on Tyne NE21 5NH
0191 414 9000   0191 414 9001
classicalguitar@ashleymark.co.uk
www.classicalguitarmagazine.com
Owner *Ashley Mark Publishing Co.*
Editor *Maurice Summerfield*

Founded 1982. MONTHLY.
FEATURES *Colin Cooper* Usually written by staff writers. Max. 1,500 words.
NEWS *Thérèse Wassily Saba* Small paragraphs and festival concert reports welcome.
REVIEWS *Tim Panting* Concert reviews of up to 250 words are usually written by staff reviewers. Features, by arrangement; no payment for news.

Classical Music

241 Shaftesbury Avenue, London WC2H 8TF
020 7333 1742   020 7333 1769
classical.music@rhinegold.co.uk
www.rhinegold.co.uk
Owner *Rhinegold Publishing Ltd*
Editor *Keith Clarke*

Founded 1976. FORTNIGHTLY. A specialist magazine using precisely targeted news and feature articles aimed at the music business. Most material is commissioned but professionally written unsolicited mss are occasionally published. Freelance contributors may approach in writing with an idea but should familiarize themselves beforehand with the style and market of the magazine. Payment negotiable.

Climber

Warners Group Publications plc, West Street, Bourne PE10 9PH
01778 391000
climberscomments@warnersgroup.co.uk
www.climber.co.uk
Owner *Warners Group Publications plc*
Editor *Kate Burke*

Founded 1962. MONTHLY. Features articles and news on climbing and mountaineering in the UK and abroad. No unsolicited mss.

Closer

Endeavour House, 189 Shaftesbury Avenue, London WC2H 8JG
020 7437 9011   020 7859 8600
closer@bauermedia.co.uk
www.closeronline.co.uk
Owner *Bauer Consumer Media*
Editor *Lisa Burrow*
Circulation 527,325

Launched 2002. WEEKLY celebrity and real life magazine. No unsolicited material; approach by e-mail with idea in the first instance.

Club International

1–2 Archer Street, London W1D 7AW
020 7292 8000   020 7734 5030
mattb@paulraymond.com
Editor *Matt Berry*
Circulation 80,000

Founded 1972. MONTHLY. Features and short humorous items aimed at young male readership aged 18–30.
FEATURES Max. 1,000 words.
SHORTS 200–750 words. Payment negotiable.

Coach and Bus Week

Rouncy Media Ltd, 3 The Office Village, Forder Way, Cygnet Park, Hampton, Peterborough PE7 8GX

☎ 01733 293240   🖷 0845 280 2927
jacqui.grobler@rouncymedia.co.uk
www.CBWonline.com
Owner *Rouncy Media Ltd*
Editorial Director *Andrew Sutcliffe*
Circulation 4,000

WEEKLY magazine, aimed at coach and bus operators. Interested in coach and bus industry-related items only: business, financial and legal. Contact by letter, e-mail, telephone or via website.

Coin News

Token Publishing Ltd, Orchard House, Duchy Road, Heathpark, Honiton EX14 1YD

☎ 01404 46972   🖷 01404 44788
info@tokenpublishing.com
www.tokenpublishing.com
Owners *J.W. Mussell, Carol Hartman*
Editor *J.W. Mussell*
Circulation 10,000

Founded 1964. MONTHLY. Contributions welcome. Approach by phone in the first instance.
FEATURES Opportunity exists for well-informed authors 'who know the subject and do their homework'. Max. 2,000 words.
ⓔ Payment by arrangement.

Company

National Magazine House, 72 Broadwick Street, London W1F 9EP

☎ 020 7312 3775   🖷 020 7312 3797
company.mail@natmags.co.uk
www.getlippy.com
Owner *National Magazine Co. Ltd*
Editor *Victoria White*
Circulation 250,075

MONTHLY. Glossy women's magazine appealing to the independent and intelligent young woman. A good market for freelancers: 'We look for great newsy features relevant to young British women'. Keen to encourage bright, new, young talent, but uncommissioned material is rarely accepted. Feature outlines are the only sensible approach in the first instance. Max. 1,500–2,000 words. Send to *Emma Justice*, Features Editor. ⓔ £250 per 1,000 words.

Computer Arts

30 Monmouth Street, Bath BA1 2BW

☎ 01225 442244   🖷 01225 732295
ca.mail@futurenet.co.uk
www.computerarts.co.uk
Owner *Future Publishing*
Publisher *Liz Taylor*
Editor *Rob Carney*

Circulation 21,452

Founded 1995. MONTHLY. The world of computer arts and digital creative design. 3D, web design, Photoshop, digital video. No unsolicited mss. Interested in tutorials, profiles, tips, software and hardware reviews. Approach by post or e-mail.

Computer Weekly

Quadrant House, The Quadrant, Sutton SM2 5AS

☎ 020 8652 3122   🖷 020 8652 8979
computer.weekly@rbi.co.uk
www.computerweekly.com
Owner *Reed Business Information*
Editor *Brian McKenna*
Circulation 143,000

Founded 1966. Freelance contributions welcome.
FEATURES Always looking for good new writers with specialized industry knowledge. Max. 1,800 words.
NEWS *Bill Goodwin* Some openings for regional or foreign news items. Max. 300 words.
ⓔ Payment negotiable.

Computeractive

Incisive Media Ltd, 32–34 Broadwick Street, London W1A 2HG

☎ 020 7316 9000   🖷 020 7316 9520
letters@computeractive.co.uk
www.computeractive.co.uk
Owner *Incisive Media Ltd*
Editor *Paul Allen*
Circulation 200,307

Launched 1998. FORTNIGHTLY. Consumer and computer technology magazine. No unsolicited material.

Computing

Incisive Media Ltd, 32–34 Broadwick Street, London W1A 2HG

☎ 020 7316 9000   🖷 020 7316 9160
feedback@computing.co.uk
www.computing.co.uk
Owner *Incisive Media Ltd*
Editor *Bryan Glick*
Circulation 100,000

Founded 1973. WEEKLY newspaper for IT professionals. Unsolicited articles welcome. Please enclose s.a.e. for return.
NEWS *Madeline Bennett*
ⓔ Up to £300 per 1,000 words.

Condé Nast Traveller

Vogue House, Hanover Square, London W1S 1JU

☎ 020 7499 9080   🖷 020 7493 3758
editorcntraveller@condenast.co.uk
www.cntraveller.com
Owner *Condé Nast Publications*
Editor *Sarah Miller*

Circulation 85,262

Founded 1997. MONTHLY travel magazine. Proposals rather than completed mss preferred. Approach in writing in the first instance. No unsolicited photographs. 'The magazine has a no freebie policy and no writing can be accepted on the basis of a press or paid-for trip.'

Construction News

Greater London House, Hampstead Road, London NW1 7EJ
℡ 020 7728 4632   🖷 020 7728 4400
cneditorial@emap.com
www.cnplus.co.uk
Owner *Emap*
Editor *Nick Edwards*
Associate Editor (News) *Alex Hawkes*
Circulation 19,776

WEEKLY. Since 1871, *Construction News* has provided news to the industry. Readership ranges from construction site to the boardroom, from building and specialist construction through to civil engineering. Initial approach by e-mail.

Contemporary Review

PO Box 1242, Oxford OX1 4FJ
℡ 01865 201529   🖷 01865 201529
editorial@contemporaryreview.co.uk
www.contemporaryreview.co.uk
Owner *Contemporary Review Co. Ltd*
Editor *Dr Richard Mullen*

Founded 1866. QUARTERLY. Independent review dealing with questions of the day, chiefly politics, international affairs, religion, literature and the arts.
FEATURES Limited scope for freelancers. Authoritative articles of 2,000–3,000 words. Send by e-mail. 💷 Payment £5 per page.

Coop Traveller

River Publishing, Victory House, 14 Leicester Place, London WC2H 7BZ
℡ 020 7306 0304   🖷 020 7306 0314
travellereditorial@riverltd.co.uk
www.riverltd.co.uk
Owner *River Publishing*
Editor *Heather Farmbrough*
Circulation 220,000

Founded 2001. THREE ISSUES YEARLY. Magazine for Co-op Travel customers. Travel articles and travel-related news for the independent traveller. Approach by telephone, fax, letter or e-mail in the first instance.
FEATURES Ideas for exciting travel features, including destinations, accommodation, activities etc. Max. 1,000 words.
NEWS Stories that reveal fascinating facts for holidaymakers, including interesting products, new destinations and developments in the travel

market. Max. 300 words.
💷 Payment negotiable.

Cosmopolitan

National Magazine House, 72 Broadwick Street, London W1F 9EP
℡ 020 7439 5000   🖷 020 7439 5016
cosmo.mail@natmags.co.uk
www.cosmopolitan.co.uk
Owner *National Magazine Co. Ltd*
Editor *Louise Court*
Deputy Editor *Claire Askew*
Circulation 470,735

MONTHLY. Designed to appeal to the mid-twenties, modern-minded female. Popular mix of articles, with emphasis on relationships and careers, and hard news. No fiction. Will rarely use unsolicited mss but always on the look-out for 'new writers with original and relevant ideas and a strong voice'. All would-be writers should be familiar with the magazine. E-mail for submission guidelines.

Counselling at Work

Association for Counselling at Work, BACP, BACP House, 15 St John's Business Park, Lutterworth LE17 4HB
℡ 01455 883300   🖷 01455 550243
acw@bacp.co.uk
www.counsellingatwork.org.uk
Owner *British Association for Counselling and Psychotherapy*
Editor *Rick Hughes*
Circulation 1,600

Founded 1993. QUARTERLY official journal of the Association for Counselling at Work, a Division of BACP. Looking for well-researched articles (500–2,400 words) about *any* aspect of workplace counselling and employee support. Mss from those employed as counsellors or in welfare posts are particularly welcome. Photographs accepted. No fiction or poetry. Contact the editor for submission details 💷 No payment.

Country Homes and Interiors

IPC Media Ltd., Blue Fin Building, 110 Southwark Street, London SE1 0SU
℡ 020 3148 5000
countryhomes@ipcmedia.com
Owner *IPC Media (A Time Warner Company)*
Editor *Rhoda Parry*
Circulation 87,056

Founded 1986. MONTHLY. The best approach for prospective contributors is with an idea in writing as unsolicited mss are not welcome.
FEATURES *Jean Carr* Monthly personality interviews of interest to an intelligent, affluent readership (women and men), aged 25–44. Max. 1,200 words.

HOUSES *Vivienne Ayers* Country-style homes with excellent design ideas. Length 1,000 words.
Ⓔ Payment negotiable.

Country Life

IPC Media Ltd., Blue Fin Building, 110 Southwark Street, London SE1 0SU
☎ 020 3148 4444
www.countrylife.co.uk
Owner *IPC Media (A Time Warner Company)*
Editor-in-Chief *Mark Hedges*
Circulation 40,821

Established 1897. WEEKLY. *Country Life* features articles which relate to architecture, countryside, wildlife, rural events, sports, arts, exhibitions, current events, property and news articles of interest to town and country dwellers. Strong informed material rather than amateur enthusiasm. 'Uncommissioned material is rarely accepted. We regret we cannot be liable for the safe custody or return of any solicited or unsolicited materials.'

Country Living

National Magazine House, 72 Broadwick Street, London W1F 9EP
☎ 020 7439 5000   📠 020 7439 5093/5077
www.allaboutyou.com
Owner *National Magazine Co. Ltd*
Editor *Susy Smith*
Circulation 195,159

Magazine aimed at both country dwellers and town dwellers who love the countryside. Covers people, conservation, wildlife, houses (gardens and interiors) and rural businesses. No unsolicited mss

Country Smallholding

Archant Publishing, Fair Oak Close, Exeter Airport Business Park, Clysts Honiton, Nr Exeter EX5 2UL
☎ 01392 888481   📠 01392 888499
editorial.csh@archant.co.uk
www.countrysmallholding.com
Owner *Archant Publishing*
Editor *Simon McEwan*
Circulation 18,372

Founded 1975. MONTHLY magazine for small farmers, smallholders, practical landowners and for anyone interested in self sufficiency at all levels both in town and in the country. Articles welcome on organic growing, keeping poultry, livestock and other animals, crafts, cookery, herbs, building and energy. Articles should be detailed and practical, based on first-hand knowledge and experience. Approach in writing or by e-mail.

Country Walking

Media House, Lynchwood, Peterborough PE2 6EA
☎ 01733 468000
country.walking@bauermedia.co.uk
www.livefortheoutdoors.com
Owner *Bauer Media*
Editor *Jonathan Manning*
Circulation 44,549

Founded 1987. MONTHLY magazine focused on walking in the finest landscapes of the UK and Europe. News, gear tests and profiles of celebrity walkers also feature. Special monthly pull-out section of detailed walking routes. Original high quality photography and ideas for features considered, although very few unsolicited mss accepted. Approach by letter or e-mail.
SPECIAL PAGES Routes section of magazine, including outline Ordnance Survey map and step-by-step directions. Accurately and recently researched walk and fact file including photographs and points of interest en route. Please contact *Kerry Leahy* for guidelines (unsolicited submissions generally not accepted for this section).
Ⓔ Payment not negotiable.

The Countryman

The Water Mill, Broughton Hall, Skipton BD23 3AG
☎ 01756 701381   📠 01756 701326
editorial@thecountryman.co.uk
www.thecountryman.co.uk
Owner *Country Publications Ltd*
Editor *Paul Jackson*
Circulation 19,817

Founded 1927. Monthly. Countryside feature ideas welcome by e-mail, phone or in writing. Articles supplied with top quality illustrations (digital, colour transparencies, archive b&w prints and line drawings) are far more likely to be used. Max. article length 1,200 words.

The Countryman's Weekly (incorporating Gamekeeper and Sporting Dog)

Unit 2, Lynher House, 3 Bush Park, Estover, Plymouth PL6 7RG
☎ 01752 762990   📠 01752 771715
editorial@countrymanswweekly.com
www.countrymansweekly.com
Owner *Diamond Publishing Ltd*
Editor *David Venner*

Founded 1895. WEEKLY. Unsolicited material welcome.
FEATURES on any country sports topic. Max. 1,000 words.
Ⓔ Payment rates available on request.

The Cricketer International
> The Wisden Cricketer

Crimewave

TTA Press, 5 Martins Lane, Witcham, Ely CB6 2LB
info@ttapress.demon.co.uk
www.ttapress.com
Owner *TTA Press*
Editor *Andy Cox*
Circulation 8,000

Founded 1998. BIANNUAL large paperback format
magazine of cutting edge crime and mystery
fiction. 'The UK's only magazine specializing in
crime short stories, publishing the very best from
across the spectrum.' Every issue contains stories
by authors who are household names in the crime
fiction world but room is found for lesser known
and unknown writers. Submissions welcome
(not via e-mail) with appropriate return postage.
Potential contributors are advised to study the
magazine. Contracts exchanged upon acceptance.
Ⓔ Payment on publication.

Crucible
> Hymns Ancient & Modern Ltd under UK Publishers

Cumbria

The Water Mill, Broughton Hall, Skipton
BD23 3AG
Ⓣ 01535 637983
editorial@cumbriamagazine.co.uk
www.dalesman.co.uk
Owner *Country Publications Ltd*
Editor *Terry Fletcher*
Circulation 12,505

Founded 1951. MONTHLY. County magazine of
strong regional and countryside interest, focusing
on the Lake District. Unsolicited mss welcome.
Max. 1,500 words. Approach in writing, by
phone or e-mail with feature ideas. Ⓔ Payment
negotiable.

Cycle Sport

IPC Media Ltd, Leon House, 233 High Street,
Croydon CR9 1HZ
Ⓣ 020 8726 8460   Ⓕ 020 8726 8499
cyclesport@ipcmedia.com
cycling@ipcmedia.com
www.cyclesport.co.uk
Owner *IPC Media (A Time Warner Company)*
Managing Editor *Robert Garbutt*
Deputy Editor *Edward Pickering*
Circulation 18,996

Founded 1993. MONTHLY magazine dedicated to
professional cycle racing. Unsolicited ideas for
features welcome.

Cycling Weekly

IPC Media Ltd, Leon House, 233 High Street,
Croydon CR9 1HZ
Ⓣ 020 8726 8462   Ⓕ 020 8726 8499
cycling@ipcmedia.com
www.cyclingweekly.co.uk
Owner *IPC Media (A Time Warner Company)*
Publishing Director *Keith Foster*
Editor *Robert Garbutt*
Circulation 27,609

Founded 1891. WEEKLY. All aspects of cycle sport
covered. Unsolicited mss and ideas for features
welcome. Approach in writing with ideas. Fiction
rarely used. 'It is important that you are familiar
with the format so read the magazine first.'
FEATURES Cycle racing, coaching, technical
material and related areas. Max. 2,000 words.
Most work commissioned but interested in seeing
new work.
NEWS Short news pieces, local news, etc. Max.
300 words.
Ⓔ Features, around £60–120 per 1,000 words
(quality permitting); news, £15 per story.

Dalesman

The Water Mill, Broughton Hall, Skipton
BD23 3AG
Ⓣ 01756 701381   Ⓕ 01756 701326
editorial@dalesman.co.uk
www.dalesman.co.uk
Owner *Country Publications Ltd*
Editor *Paul Jackson*
Circulation 37,687

Founded 1939. Now the biggest-selling regional
publication of its kind in the country. MONTHLY
magazine with articles of specific Yorkshire
interest. Unsolicited mss welcome; receives
approximately ten per day. Initial approach in
writing, by phone or e-mail. Max. 1,000 words. Ⓔ
Payment negotiable.

Dance Today

Dancing Times Ltd, Clerkenwell House, 45–47
Clerkenwell Green, London EC1R 0EB
Ⓣ 020 7250 3006   Ⓕ 020 7253 6679
dancetoday@dancing-times.co.uk
www.dance-today.co.uk
Owner *The Dancing Times Ltd*
Editor *Katie Gregory*
Editorial Assistants *Alison Kirkman, Jon Gray*
Circulation 3,500

Founded in 1956 as the *Ballroom Dancing
Times* and relaunched as *Dance Today* in 2001.
MONTHLY magazine for anyone interested in
social and competitive dance. Styles include
ballroom and Latin American, salsa, tango,
flamenco, tea dancing and sequence. Related
topics covered include health and fitness and

dance fashion. Contributions welcome; send c.v. with sample work by post or e-mail to the editor.

The Dancing Times

The Dancing Times Ltd, 45–47 Clerkenwell Green, London EC1R OEB

📞 020 7250 3006   📠 020 7253 6679

editorial@dancing-times.co.uk

www.dancing-times.co.uk

Owner *The Dancing Times Ltd*

Editor *Jonathan Gray*

Founded 1910. MONTHLY. Freelance suggestions welcome from specialist dance writers and photographers only. Approach in writing.

The Dandy

D.C. Thomson & Co. Ltd, Albert Square, Dundee DD1 8QJ

📞 01382 223131   📠 01382 322214

editor@dandy.co.uk

www.dandy.com

Owner *D.C. Thomson & Co. Ltd*

Editor *Craig Graham*

Founded 1937. WEEKLY comic. Original, humorous comic strips for 7–12-year-olds, featuring established characters such as Desperate Dan as well as new ones like Jak. Generally, unsolicited contributions are welcome but 'please read the comic before sending any material. Try a one-off script for an established, current *Dandy* character. Send by e-mail or post but never send your only copy.' Include s.a.e. for reply. No non-script material such as puzzles, poetry, etc.

Darts World

81 Selwood Road, Croydon CR0 7JW

📞 020 8650 6580   📠 020 8654 4343

dartsworld@blueyonder.co.uk

www.dartsworld.com

Owner *World Magazines Ltd*

Editor *A.J. Wood*

Circulation 12,263

FEATURES Single articles or series on technique and instruction. Max. 1,200 words.

FICTION Short stories with darts theme. Max. 1,000 words.

NEWS Tournament reports and general or personality news required. Max. 800 words.

💷 Payment negotiable.

Day by Day

Woolacombe House, 141 Woolacombe Road, London SE3 8QP

📞 020 8856 6249

Owner *Loverseed Press*

Editor *Patrick Richards*

Circulation 25,000

Founded 1963. MONTHLY. News commentary and digest of national and international affairs, with reviews of the arts (books, plays, art exhibitions, films, opera, musicals) and county cricket and Test reports among regular slots. Unsolicited mss welcome (s.a.e. essential). Approach in writing with ideas. Contributors are advised to study the magazine in the first instance. (Specimen copy £1.50) UK subscription £15; Europe £19; RoW £23.

NEWS *Ronald Mallone* Interested in themes connected with non-violence and social justice only. Max. 600 words.

FEATURES No scope for freelance contributions here.

POEMS *Michael Gibson* Short poems in line with editorial principles considered. Max. 20 lines.

💷 Payment negotiable.

Dazed & Confused

112–116 Old Street, London EC1V 9BG

📞 020 7336 0766   📠 020 7336 0966

rod@dazedgroup.com

www.dazeddigital.com

Owner *Waddell Ltd*

Editor *Rod Stanley*

Circulation 93,000

Founded 1992. MONTHLY. Cutting edge fashion, music, film, art, interviews and features. No unsolicited material. Approach by e-mail with ideas in the first instance.

Decanter

Blue Fin Building, 110 Southwark Street, London SE1 0SU

📞 020 3148 4488   📠 020 3148 8524

editor@decanter.com

www.decanter.com

Editor *Guy Woodward*

Circulation 40,000

FOUNDED 1975. Glossy wines magazine. Feature ideas welcome; send to the editor by post, fax or e-mail. No fiction.

NEWS/FEATURES All items and articles should concern wines, food and related subjects. 💷 £250 per 1,000 words.

Delicious

Seven Publishing Group, Sea Container's House, 20 Upper Ground, London SE1 9PD

📞 020 7775 7757   📠 020 7775 7705

info@7publishing.co.uk

www.deliciousmagazine.co.uk

Owner *Seven Publishing Group*

Editor *Matthew Drennan*

Circulation 103,041

Launched 2003. MONTHLY publication for people who love food and cooking. Recipes – from the simple to the sophisticated – plus features on food and where it comes from. Welcomes freelance contributions; approach by e-mail.

Derbyshire Life and Countryside

61 Friar Gate, Derby DE1 1DJ
☎ 01332 227850   🖷 01332 227860
www.derbyshirelife.co.uk
Owner *Archant Life*
Editor *Joy Hales*
Circulation 15,793

Founded 1931. MONTHLY county magazine for Derbyshire. Unsolicited mss and photographs of Derbyshire welcome, but written approach with ideas preferred.

Descent

Wild Places Publishing, PO Box 100, Abergavenny NP7 9WY
☎ 01873 737707
descent@wildplaces.co.uk
www.caving.uk.com
Owner *Wild Places Publishing*
Editor *Chris Howes*
Assistant Editor *Judith Calford*

Founded 1969. BI-MONTHLY magazine for cavers and mine enthusiasts. Submissions welcome from freelance contributors who can write accurately and knowledgeably and in the style of existing content on any aspect of caves, mines or underground structures.
FEATURES General interest articles of under 1,000 words welcome, as well as short foreign news reports, especially if supported by photographs/illustrations. Suitable topics include exploration (particularly British, both historical and modern), expeditions, equipment, techniques and regional British news. Max. 2,000 words.
🖸 Payment on publication according to page area filled.

Desire

211 Old Street, London EC1V 9NR
☎ 020 7608 6383   🖷 020 7608 6380
editorial@trojanpublishing.co.uk
www.desire.co.uk
Owner *Trojan Publishing*
Editor *Ian Jackson*

Founded 1994. SIX ISSUES YEARLY. Britain's only erotic magazine for both women and men, celebrating sex and sensuality with a mix of features, reviews, interviews, photography and erotic fantasy. For a sample copy of the magazine plus contributors' guidelines and rates, please enclose four loose first class stamps.

Destination France

151 Station Street, Burton on Trent DE14 1BG
☎ 01283 742950   🖷 01283 742957
carmen.konopka@wwonline.co.uk
Owner *Waterways World*
Editor *Carmen Konopka*

Circulation 15,000

A stylish, contemporary, QUARTERLY magazine aimed at Francophiles. Covers food, wine, art, architecture, books and lifestyle as well as travel within France. Features include sports, general tourism, property and regional information. Photographs to accompany articles. Approach in writing, preferably by e-mail, in the first instance.
🖸 Payment negotiable.

Devon Life

Archant House, Babbage Road, Totnes TQ9 5JA
☎ 01803 860910   🖷 01803 860926
devonlife@archant.co.uk
www.devonlife.co.uk
Owner *Archant*
Editor *Jane Fitzgerald*
Circulation 16,217

Originally launched in 1963; relaunched 1996. MONTHLY. Articles on all aspects of Devon – places of interest, personalities, events, arts, history and food. Contributions welcome; send by e-mail to the editor.
FEATURES Words/picture package preferred on topics listed above. Max. 1,000–2,000 words.
FICTION Occasional short stories with a distinctly Devon flavour. Max 1,600 words.
🖸 Payment by arrangement.

Director

116 Pall Mall, London SW1Y 5ED
☎ 020 7766 8950   🖷 020 7766 8840
director-ed@iod.com
Group Editor *Richard Cree*
Circulation 58,404

1991 Business Magazine of the Year. Published by Director Publications Ltd for members of the Institute of Directors and subscribers. Wide range of features from political and business profiles and management thinking to employment and financial issues. Also book reviews. Regular contributors used. Send letter with synopsis/published samples rather than unsolicited mss.
🖸 £350 per 1,000 words.

Disability Now

6 Market Road, London N7 9PW
☎ 020 7619 7323   🖷 020 7619 7331
editor@disabilitynow.org.uk
www.disabilitynow.org.uk
Owner *SCOPE*
Editor *Ian Macrae*
Circulation 20,735

Founded 1984. Leading MONTHLY magazine for disabled people in the UK – people with all disabilities, as well as their families, carers and relevant professionals. Freelance contributions welcome. No fiction. Approach in writing or by e-mail.

FEATURES Covering new initiatives and services, personal experiences and general issues of interest to a wide national readership. Max. 900 words. Disabled contributors welcome.
NEWS Max. 300 words.
SPECIAL PAGES Possible openings for cartoonists.
£ Payment by arrangement.

Diva, lesbian life and style

Spectrum House, 32–34 Gordon House Road, London NW5 1LP
☎ 020 7424 7400   ℻ 020 7424 7401
edit@divamag.co.uk
www.divamag.co.uk
Owner *Millivres-Prowler Group*
Editor *Jane Czyzselska*

Founded 1994. MONTHLY glossy magazine featuring lesbian news and culture including fashion, lifestyle and satire. Welcomes news, features and photographs. No poetry. Contact the news editor at the e-mail address above with news items and feature ideas and photo/fashion photographs.

Dog World

Somerfield House, Wotton Road, Ashford TN23 6LW
☎ 01233 621877   ℻ 01233 645669
editor@dogworld.co.uk
www.dogworld.co.uk
Owner *DW Media Holdings Ltd*
Editor *Stuart Baillie*
Circulation 17,828

Founded 1902. WEEKLY newspaper for people who are seriously interested in pedigree dogs. Unsolicited mss occasionally considered but initial approach in writing preferred.
FEATURES Well-researched historical items or items of unusual interest concerning dogs. Max. 2,000 words. Photographs of unusual 'doggy' situations occasionally of interest.
NEWS Freelance reports welcome on court cases and local government issues involving dogs.
£ Features, up to £75; photos, £15.

Dogs Monthly

ABM Publishing Ltd, 61 Great Whyte, Ramsey, Huntingdon PE26 1HJ
☎ 0845 094 8958   ℻ 08707 662273
info@dogsmonthly.co.uk
www.dogsmonthly.co.uk
Owner *ABM Publishing Ltd*
Editor *Caroline Davis*
Assistant Editor *Hannah Roche*
Sub-editor/Writer *Jenny Chafer*
Circulation 20,000

MONTHLY magazine containing expert, practical advice on caring for and training pet dogs by

'the best of the canine world's writers, vets and trainers'.

Dorset Life – The Dorset Magazine

7 The Leanne, Sandford Lane, Wareham BH20 4DY
☎ 01929 551264   ℻ 01929 552099
editor@dorsetlife.co.uk
www.dorsetlife.co.uk
Owner/Editor *John Newth*
Circulation 10,000

Founded 1968. MONTHLY magazine about Dorset, both present and past. Unsolicited contributions welcome if specifically about Dorset.

Early Music Today

Rhinegold Publishing, 241 Shaftesbury Avenue, London WC2H 8TF
☎ 020 7333 1744   ℻ 020 1733 1769
emt@rhinegold.co.uk
www.rhinegold.co.uk
Owner *Rhinegold Publishing*
Editor *Jonathan Wikeley*
Circulation 10,000

Launched 1993. BI-MONTHLY. Articles on early music, concerts, CD and book reviews, comprehensive concert listings, early music news.
FEATURES All contributors are experts in the early music field (pre-1800). Relevant articles and news welcome. Approach by e-mail or post.

Eastern Eye

Unit 2, 65 Whitechapel Road, London E1 1DU
☎ 020 7650 2000   ℻ 020 7650 2001
editor@easterneyeuk.co.uk
www.easterneyeonline.co.uk
Owner *Ethnic Media Group*
Editor *Hamant Verma*
Circulation 20,844

Founded 1989. WEEKLY community paper for the Asian community in Britain. Interested in relevant general, local and international issues. Approach in writing with ideas for submission.

Easy Living

68 Old Bond Street, London W1S 4PH
☎ 020 7499 9080   ℻ 020 7399 2625
easylivingeditorial@condenast.co.uk
www.easylivingmagazine.com
Owner *Condé Nast*
Editor *Susie Forbes*
Features Director *Andriena Cordani*
Health and Beauty Director *Catherine Turner*
Fashion Director *Liz Thody*
Decorating Editor *Sophie Martell*
Food Editor *David Herbert*
Circulation 196,150

Launched 2005. MONTHLY magazine aimed at women aged 30–50-plus. Focuses on aspects of fashion, home life, beauty and health, food, relationships; stylish and glossy with real-life, practical and useful items.

The Ecologist

Unit 102, Lana House Studios, 116-118 Commercial Street, London E1 6NF
☎ 020 7422 8100   020 7422 8101
editorial@theecologist.org
www.theecologist.org
Owner *Ecosystems Ltd*
Managing Editor *Pat Thomas*
Director *Zac Goldsmith*
Circulation 30,000

Founded 1970. TEN ISSUES YEARLY. Unsolicited mss welcome but best approach is a brief proposal to the editor by e-mail (address above), outlining experience and background and summarizing suggested article.
FEATURES Radical approach to political, economic, social and environmental issues, with an emphasis on rethinking the basic assumptions that underpin modern society. Articles of between 500 and 3,000 words.
💷 Payment negotiable.

The Economist

25 St James's Street, London SW1A 1HG
☎ 020 7830 7000   020 7839 2968
www.economist.com
Owner *Pearson/individual shareholders*
Editor *John Micklethwait*
Circulation 182,539 (UK)

Founded 1843. WEEKLY. Worldwide circulation. Approaches should be made in writing to the editor. No unsolicited mss.

The Edge

65 Guinness Buildings, Hammersmith, London W6 8BD
☎ 0845 456 9337
www.theedge.abelgratis.co.uk
Editor *David Clark*

QUARTERLY. Reviews and features: film (indie, arts), books, popular culture.
FICTION 2,000+ words, almost any 'type' of story but see site or magazine to e-mail and for examples/tone. Sample issue £5 post free (cheques payable to 'The Edge'). Guidelines on website or for s.a.e.
💷 Payment 'negotiable after we express interest'.

Edinburgh Review

22A Buccleugh Place, Edinburgh EH8 9LN
☎ 0131 651 1415
edinburgh.review@ed.ac.uk
www.edinburghreview.org.uk
Owner *English Literature, University of Edinburgh*
Editor *Brian McCabe*

Founded 1802. THREE ISSUES YEARLY. Cultural and literary magazine. Articles on Scottish and international culture, history, art, literature and politics. Each issue is broadly geographically themed; recent issues include the Carribean, Poland and Northern Ireland. Unsolicited contributions welcome, particularly articles, fiction and poetry; send by post or e-mail. 'Writers offering unsolicited essays should contact us by e-mail in the first instance.'

Electrical Times

Purple Media Solutions Ltd., 2nd Floor, 207–215 High Street, Orpington BR6 0PL
☎ 01689 899170   01689 899171
ben.cronin@purplems.com
Owner *Purple Media Solutions Ltd*
Editor *Ben Cronin*
Circulation 10,613

Founded 1891. MONTHLY. Aimed at electrical contractors, designers and installers. Unsolicited mss welcome but initial approach preferred.

Elle

64 North Row, London W1K 7LL
☎ 020 7150 7000   020 7150 7670
Owner *Hachette Filipacchi*
Editor *Lorraine Candy*
Features Director *Kerry Potter*
Circulation 203,471

Founded 1985. MONTHLY fashion glossy. Prospective contributors should approach the relevant editor in writing in the first instance, including cuttings.
FEATURES Max. 2,000 words.
PREVIEW Short articles on hot trends, events, fashion and beauty. Max. 500 words.
💷 Payment negotiable.

Elle Decoration

64 North Row, London W1K 7LL
☎ 020 7150 7000   020 7150 7671
Owner *Hachette Filipacchi*
Editor-in-Chief *Michelle Ogundehin*
Circulation 64,030

Launched 1989. MONTHLY style magazine for the home – 'beautiful homes, great designs, inspiration and ideas'. No unsolicited contributions; approach by e-mail to the appropriate department.

Embroidery

The Embroiderers' Guild, Apt. 41 Hampton Court Palace, East Molesey KT8 9AU
☎ 01260 273891
jo.editor@btinternet.com
www.embroiderersguild.com
Owner *The Embroiderers' Guild*
Editor *Joanne Hall*

Circulation 6,000

Founded 1933. BI-MONTHLY. Features articles on contemporary practice within art textiles and embroidery. Insights into historical and world embroidery. Also includes news, exhibition and book reviews. Unsolicited material welcome. Max. 1,000 words. Authors must source or identify photography/images.

Ⓔ Payment negotiable.

Empire

4th Floor, Mappin House, 4 Winsley Street, London W1W 8HF

☎ 020 7182 8781   📠 020 7182 8703

empire@bauermedia.com
www.empireonline.com
Owner *Bauer Media*
Editor *Mark Dinning*
Circulation 187,202

Founded 1989. Launched at the Cannes Film Festival. MONTHLY guide to the movies which aims to cover the world of films in a 'comprehensive, adult, intelligent and witty package'. Although most of *Empire* is devoted to films and the people behind them, it also looks at the developments and technology behind television and DVDs plus music, multimedia and books. Wide selection of in-depth features and stories on all the main releases of the month, and reviews of over 100 films and videos. Contributions welcome but approach in writing first.

FEATURES Behind-the-scenes features on films, humorous and factual features.

Ⓔ Payment by agreement.

The Engineer

50 Poland Street, London W1F 7AX

☎ 020 7970 4154   📠 0870 600 8399

tepr@centaur.co.uk
www.theengineer.co.uk
Owner *Centaur Media*
Editor *Andrew Lee*
Circulation 31,469

FOUNDED 1856. FORTNIGHTLY news magazine for technology and innovation.

FEATURES Most outside contributions are commissioned but good ideas are always welcome. Max. 2,000 words.

NEWS Scope for specialist regional freelancers, and for tip-offs. Max. 500 words.

TECHNOLOGY Technology news from specialists. Max. 500 words.

Ⓔ Payment by arrangement.

The English Garden

Archant House, Oriel Road, Cheltenham
GL50 1BB

☎ 01242 211080   📠 01242 211081

theenglishgarden@archant.co.uk
www.theenglishgarden.co.uk
Owner *Archant Specialist*
Editor *Tamsin Westhorpe*
Editor-at-Large *Jackie Bennett (01366 328606)*
Circulation 34,814 (UK), 64,892 (worldwide)

Founded 1996. MONTHLY. Features on beautiful gardens with practical ideas on design and planting. No unsolicited mss.

FEATURES *Jackie Bennett* Max. 800 words. Approach in writing in the first instance; send synopsis of 150 words with strong design and planting ideas, feature proposals, or sets of photographs of interesting gardens.

ES

▷ Evening Standard under Regional Newspapers

Esquire

National Magazine House, 72 Broadwick Street, London W1F 9EP

☎ 020 7439 5000   📠 020 7439 5675

www.esquire.com
Owner *National Magazine Co. Ltd*
Editor *Jeremy Langmead*
Circulation 58,136

Founded 1991. MONTHLY. Quality men's general interest magazine. No unsolicited mss or short stories.

Essentials

IPC Media Ltd., Blue Fin Building, 110 Southwark Street, London SE1 0SU

☎ 020 3148 7211

Owner *IPC Media (A Time Warner Company)*
Editor *Jules Barton-Breck*
Circulation 100,214

Founded 1988. MONTHLY women's lifestyle magazine. Initial approach by e-mail or telephone preferred. Prospective contributors should study the magazine thoroughly before submitting anything. No fiction.

FEATURES Max. 2,000 words (A4/double-spaced).

Ⓔ Payment negotiable.

Eventing

▷ Horse and Hound

Evergreen

PO Box 52, Cheltenham GL50 1YQ

☎ 01242 537900   📠 01242 537901

Editor *Stephen Garnett*
Circulation 48,000

Founded 1985. QUARTERLY magazine featuring articles and poems about Britain. Unsolicited contributions welcome.

FEATURES Britain's natural beauty, towns and villages, nostalgia, wildlife, traditions, odd customs, legends, folklore, crafts, etc. Length 250–2,000 words.

💷 £15 per 1,000 words; poems £4.

Eye: The international review of graphic design

Eye Magazine Ltd., Studio 6, The Lux Building, 2–4 Hoxton Square, London N1 6NU

📞 020 7684 6530   📠 020 7684 6531

john.walters@eyemagazine.com

www.eyemagazine.com

www.blog.eyemagazine.com

Owner *Eye Magazine Ltd*

Editor *John L. Walters*

Circulation 9,000

Founded 1990. QUARTERLY. 'The world's most beautiful and authoritative magazine about graphic design and visual culture.' Unsolicited contributions not generally welcomed.

FEATURES Specialist articles about graphic design and designers. Design history by experts. Max. 3,000 words.

REVIEWS Max. 200–1,000 words.

💷 Payment negotiable.

Fabulous

▷ **News of the World under National Newspapers**

Fairgame Magazine

The Baltic Business Centre, Gateshead NE8 3DA

📞 0191 442 4003   📠 0191 442 4002

jen@fgmag.com

www.fgmag.com

Editor *Jennifer O'Neill*

Circulation 15,000

Founded 2003. BI-MONTHLY magazine for women football players. Contributions welcome.

FEATURES *Jennifer O'Neill* International reports, player and team profiles, interviews, diet, health and fitness, tactics, training advice, play improvement, fund-raising. Max. 1,500 words.

NEWS *Wilf Frith* (wilf@fgmag.com) Match reports, team news, transfers, injuries, results and fixtures. Max. 600 words.

Families First

Mary Sumner House, 24 Tufton Street, London SW1P 3RB

📞 020 7222 5533   📠 020 7227 9731

Owner *MS Enterprises Ltd*

Editor *Catherine Butcher*

Circulation 46,798

Founded 1976. Formerly *Home & Family*. BI-MONTHLY. Unsolicited mss very rarely considered. No fiction or poetry. Features on family life, social issues, marriage, Christian faith, etc. Max. 1,000 words. 💷 Payment 'modest'.

Family Tree Magazine

61 Great Whyte, Ramsey, Huntingdon PE26 1HJ

📞 01487 814050   📠 01487 711361

sue.f@family-tree.co.uk

www.family-tree.co.uk

Owner *ABM Publishing Ltd*

Editor *Helen Tovey*

Circulation 40,000

Founded 1984. THIRTEEN ISSUES YEARLY. News and features on matters of genealogy. Articles with a social or military history angle, or with useful research steps for fellow readers favoured. Approach in writing with ideas. All material should be addressed to *Helen Tovey*.

FEATURES Any genealogically related subject. Max. 2,000 words.

💷 Features payment negotiable.

Fancy Fowl

The Publishing House, Station Road, Framlingham IP13 9EE

📞 01728 622030   📠 01728 622031

fancyfowl@todaymagazines.co.uk

www.fancyfowl.com

Owner *Today Magazines Ltd*

Editor *Kevin Davis*

Circulation 3,000

Founded 1979. MONTHLY. Devoted entirely to poultry, waterfowl, turkeys, geese, pea fowl, etc. – management, breeding, rearing and exhibition. Outside contributions of knowledgeable, technical poultry-related articles and news welcome. Max. 1,000 words. Approach by letter or e-mail.

💷 Payment negotiable.

Farmers Weekly

Quadrant House, Sutton SM2 5AS

📞 020 8652 4911   📠 020 8652 4005

farmers.weekly@rbi.co.uk

www.fwi.co.uk

Owner *Reed Business Information*

Editor *Jane King*

Circulation 70,315

WEEKLY. PPA Awards winner – 2006 Business Magazine of the Year. For practising farmers and those in the ancillary industries. Unsolicited mss considered.

FEATURES A wide range of material relating to running a farm business: specific sections on arable and livestock farming, farm life, practical and general interest, machinery and business.

NEWS General farming news, technical stories and case studies.

💷 Payment negotiable.

Fast Car

Future Publishing Ltd, 30 Monmouth Street, Bath
BA1 2BW
☎ 01225 442244
www.fastcar.co.uk
Owner *Future Publishing Ltd*
Editor *Steve Chalmers*
Circulation 51,414

Founded 1987. THIRTEEN ISSUES YEARLY.
Lad's magazine about performance tuning and
modifying cars. Covers all aspects of this youth
culture including the latest street styles and
music. Features cars and their owners, product
tests and in-car entertainment. Also includes a
free reader ads section.
NEWS Any item in line with the above.
FEATURES Innovative ideas in line with the above
and in the *Fast Car* writing style. Generally two
to four pages in length. No Kit-car features, race
reports or road test reports of standard cars.
Copy should be as concise as possible.
Ⓔ Payment negotiable.

FHM

Mappin House, 4 Winsley Street, London
W1W 8HF
☎ 020 7182 8000   Ⓕ 020 7182 8021
general@fhm.com
www.fhm.com
Owner *Bauer Media*
Editor-in-Chief *Anthony Noguera*
Circulation 280,392

Founded in 1986 as a free fashion magazine,
FHM evolved to become more male oriented.
Best-selling men's magazine covering all areas of
men's lifestyle. Published MONTHLY with editions
worldwide. No unsolicited mss. Synopses and
ideas welcome by e-mail. Ⓔ Payment negotiable.

The Field

IPC Media Ltd., Blue Fin Building, 110 Southwark
Street, London SE1 0SU
☎ 020 3148 4772   Ⓕ 020 3148 8179
field_secretary@ipcmedia.com
www.thefield.co.uk
Owner *IPC Media (A Time Warner Company)*
Editor-in-Chief *Jonathan Young*
Circulation 32,366

Founded 1853. MONTHLY magazine for those who
are serious about the British countryside and its
pleasures. Unsolicited mss welcome but initial
approach should be made in writing or by phone.
FEATURES Exceptional work on any subject
concerning the countryside. Most work tends to
be commissioned.
Ⓔ Payment varies.

50 Connect – www.50connect.co.uk
▷ entry under Electronic Publishing and Other Services

Film Review

Visual Imagination Ltd, 9 Blades Court, Deodar
Road, London SW15 2NU
☎ 020 8875 1520   Ⓕ 020 8875 1588
filmreview@visimag.com
www.visimag.com
Owner *Visual Imagination Ltd*
Editor *Nikki Baughan*
Circulation 50,000

MONTHLY. Reviews, profiles, interviews and
special reports on films. Unsolicited material
considered. Ⓔ Payment negotiable.

Fishing News

IntraFish, 6th Floor, Eldon House, 2 Eldon Street,
London EC2M 7UA
☎ 020 7650 1026
tim.oliver@fishingnews.co.uk
www.fishingnews.co.uk
Owner *IntraFish Media*
Editor *Tim Oliver*
Circulation 8,320

Founded 1913. WEEKLY. All aspects of the
commercial fishing industry in the UK and
Ireland. No unsolicited mss; e-mail enquiry in the
first instance. Max. 600 words for news and 1,500
words for features. Ⓔ £100 per 1,000 words.

The Fix

info@ttapress.demon.co.uk
www.thefix-online.com
Owner *TTA Press*
Editor *Eugie Foster*

Founded 1994. Free online review of short fiction
(magazines, anthologies, collections, webzines,
podcasts etc), plus related interviews and regular
columns. Interested reviewers and feature writers
should visit the website and follow the contact
information.

Flash: The International Short-Short Story Magazine

Department of English, University of Chester,
Parkgate Road, Chester CH1 4BJ
☎ 01244 392715
flash.magazine@chester.ac.uk
www.chester.ac.uk/flash.magazine
Editors *Dr Peter Blair, Dr Ashley Chantler*

Launched 2008. BIANNUAL magazine of short
stories of up to 360 words (title included).
Articles and reviews considered. Full details on
the website.

Flight International

Quadrant House, The Quadrant, Sutton SM2 5AS
☎ 020 8652 3842   Ⓕ 020 8652 3840
flight.international@flightglobal.com
www.flightglobal.com
Owner *Reed Business Information*

Editor *Murdo Morrison*
Circulation 41,000

Founded 1909. WEEKLY. International trade magazine for the aerospace industry, including civil, military and space. Unsolicited mss considered. Commissions preferred – phone with ideas and follow up with letter. E-mail submissions encouraged.

FEATURES *Murdo Morrison* Technically informed articles and pieces on specific geographical areas with international appeal. Analytical, in-depth coverage required, preferably supported by interviews. Max. 1,800 words.

NEWS *Dan Thisdell* Opportunities exist for news pieces from particular geographical areas on specific technical developments. Max. 350 words.
Ⓔ NUJ rates.

Flora International

The Fishing Lodge Studio, 77 Bulbridge Road, Wilton, Salisbury SP2 0LE
Ⓣ 01722 743207   Ⓕ 01722 743207
floramag@aol.com
Publisher/Editor *Maureen Foster*
Circulation 16,000

Founded 1974. BI-MONTHLY magazine for flower arrangers and florists. Unsolicited mss welcome. Approach in writing with ideas. Not interested in general gardening articles.

FEATURES Fully illustrated: colour photographs, transparencies or CD. Flower arranging, flower arrangers' gardens and flower crafts. Floristry items written with practical knowledge and well illustrated are particularly welcome. Average 1,000 words.
Ⓔ £60 per 1,000 words plus additional payment for suitable photographs, by arrangement.

FlyPast

PO Box 100, Stamford PE9 1XQ
Ⓣ 01780 755131   Ⓕ 01780 757261
flypast@keypublishing.com
www.flypast.com
Owner *Key Publishing Ltd*
Editor *Ken Ellis*
Circulation 39,620

Founded 1981. MONTHLY. Historic aviation and aviation heritage, mainly military, First and Second World War period 1914–1970. Unsolicited mss welcome.

Focus

Bristol Magazines Ltd, Tower House, Fairfax Street, Bristol BS1 3BN
Ⓣ 0117 927 9009 (switchboard)   Ⓕ 0117 934 9008
focus@bbcmagazinesbristol.com
www.bbcfocusmagazine.com
Owner *Bristol Magazines Ltd, a subsidiary of BBC Worldwide*

Editor *Jheni Osman*
Deputy Editor *IanTaylor*
Circulation 70,122

Founded 1992. FOUR-WEEKLY. Popular science, technology, the future and discovery. Welcomes *relevant* summaries of original feature ideas by e-mail.

Fortean Times: The Journal of Strange Phenomena

Dennis Publishing, 30 Cleveland Street, London W1T 4JD
Ⓣ 020 7907 6235   Ⓕ 020 7907 6139
david_sutton@dennis.co.uk
www.forteantimes.com
Editor *David Sutton*
Circulation 21,529

Founded 1973. THIRTEEN ISSUES YEARLY. Accounts of strange phenomena and experiences, curiosities, mysteries, prodigies and portents. Unsolicited mss welcome. Approach by e-mail with ideas. No fiction, poetry, rehashes or politics.

FEATURES Well-researched and referenced material on current or historical mysteries, or first-hand accounts of oddities. Max. 4,000 words, preferably with good relevant photos/ illustrations.

NEWS Concise copy with full source references essential.
Ⓔ Payment negotiable.

Foundation: the international review of science fiction

Journal of the **Science Fiction Foundation** (see entry under *Professional Associations and Societies*).

FQ Magazine

Seymour House, South Street, Bromley BR1 1RH
Ⓣ 020 8460 6060   Ⓕ 020 8460 6050
andy@touchline.com
www.fqmagazine.co.uk
Owner *3D Media Ltd*
Editor *Andy Tongue*

Launched 2003. BI-MONTHLY. Lifestyle magazine for young fathers. Interviews, reviews, finance, kids' toys, child issues, parenting tips and general fatherhood advice. Ideas welcome; contact the editor by e-mail.

France Magazine

Archant House, Oriel Road, Cheltenham GL50 1BB
Ⓣ 01242 216050   Ⓕ 01242 216094
editorial@francemag.com
www.francemag.com
Owner *Archant Life France*
Editor *Carolyn Boyd*

Circulation 37,082 (group, including US edition); 19,351 (UK edition)

FOUNDED 1989. MONTHLY magazine containing all things of interest to Francophiles – in English. Approach by e-mail in the first instance.

Freelance Market News

Sevendale House, 7 Dale Street, Manchester M1 1JB

☎ 0161 228 2362 ⊠ 0161 228 3533
fmn@writersbureau.com
www.freelancemarketnews.com
Editor *Angela Cox*

MONTHLY. News and information on the freelance writers' market, both inland and overseas. Includes market information, competitions, seminars, courses, overseas openings, etc. Short articles (700–1,500 words). Unsolicited contributions welcome.
⑤ Payment by negotiation.

The Freelance

▷ National Union of Journalists under Professional Associations and Societies

Front Magazine

Flip Media, 19 Waterside, 44–48 Wharf Road, London N1 7UX

☎ 020 7689 3916
front@frontmag.co.uk
Owner *Flip Media*
Editor *Joe Barnes*
Circulation 85,000

Founded 1998. MONTHLY men's interest magazine. Includes features on sport (including extreme sport), music, crime, war, humour, sex, alternative culture, real life stories and irreverent celebrity interviews. Welcomes contributions. Approach by e-mail.

Garden Answers

Bushfield House, Orton Centre, Peterborough PE2 5UW

☎ 01733 237111
Owner *Bauer Consumer Media*
Editor *Geoff Stebbings*
Circulation 28,018

Founded 1982. MONTHLY. 'It is unlikely that unsolicited manuscripts will be used, as articles are usually commissioned and must be in the magazine style.' Prospective contributors should approach the editor in writing. Interested in hearing from gardening writers on any subject, whether flowers, fruit, vegetables, houseplants or greenhouse gardening.

Garden News

Bushfield House, Orton Centre, Peterborough PE2 5UW

☎ 01733 237111 ⊠ 01733 465779

Owner *Bauer Consumer Media*
Editor *Neil Pope*
Circulation 36,248

Founded 1958. Britain's biggest-selling, full-colour gardening WEEKLY. News and advice on growing flowers, fruit and vegetables, plus colourful features on all aspects of gardening especially for the committed gardener. News and features welcome, especially if accompanied by top-quality photos or illustrations. Contact the editor before submitting any material.

The Garden, Journal of the Royal Horticultural Society

RHS Media, 4th Floor, Churchgate New Road, Peterborough PE1 1TT

☎ 0845 260 0909 ⊠ 01733 341633
thegarden@rhs.org.uk
www.rhs.org.uk
Owner *The Royal Horticultural Society*
Editor *Ian Hodgson*
Circulation 350,817

Founded 1866. MONTHLY journal of the Royal Horticultural Society. Covers all aspects of the art, science and practice of horticulture and garden making. 'Articles must have depth and substance.' Approach by letter or e-mail with a synopsis in the first instance.

Gardens Illustrated

Bristol Magazines Ltd, Tower House, Fairfax Street, Bristol BS1 3BN

☎ 0117 314 8774 ⊠ 0117 933 8032
gardens@bbcmagazinesbristol.com
www.gardensillustrated.com
Owner *Bristol Magazines Ltd*
Editor *Juliet Roberts*
Circulation 30,753

Founded 1993. MONTHLY. Dubbed the 'World of Exteriors'. 'Internationally regarded as the most inspiring, authoritative and thought-provoking guide to gardens, plants and design.' Unsolicited mss are rarely used and it is best that prospective contributors approach the editor with ideas in writing, supported by photographs.

Gay Times

▷ GT

Gibbons Stamp Monthly

Stanley Gibbons, 7 Parkside, Christchurch Road, Ringwood BH24 3SH

☎ 01425 472363 ⊠ 01425 470247
hjefferies@stanleygibbons.co.uk
www.gibbonsstampmonthly.com
Owner *Stanley Gibbons Ltd*
Editor *Hugh Jefferies*

Circulation 20,000

Founded 1890. MONTHLY. News and features. Unsolicited mss welcome. Make initial approach in writing by telephone or e-mail to avoid disappointment.

FEATURES *Hugh Jefferies* Unsolicited material of specialized nature and general stamp features welcome. Max. 3,000 words but longer pieces can be serialized.

NEWS *John Moody* Any philatelic news item. Max. 500 words.

Ⓔ Features, £40–50 per 1,000 words; news, no payment.

Girl Talk/Girl Talk Extra

Media Centre, 201 Wood Lane, London W12 7TQ
☎ 020 8433 1010
girltalk.magazine@bbc.co.uk
www.bbcgirltalk.com
Owner *BBC Worldwide*
Editor *Sam Robinson*
Circulation 81,912 (Girl Talk) 37,211 (Girl Talk Extra)

Founded 1995. FORTNIGHTLY/MONTHLY. Lifestyle magazine for girls aged 8–12. Friendship and belonging, celebrity. No unsolicited submissions; send c.v. by e-mail.

Glamour

6–8 Old Bond Street, London W1S 4PH
☎ 020 7499 9080   📠 020 7491 2551
letters@glamourmagazine.co.uk
www.glamour.com
Owner *Condé Nast*
Editor *Jo Elvin*
Circulation 551,351

Founded 2001. Handbag-size glossy women's magazine – fashion, beauty and celebrities. No unsolicited mss; send ideas for features in synopsis form to the Features Editor, *Corrie Jackson*.

Go Caravan

Warners Group Publications plc, The Maltings, West Street, Bourne PE10 9PH
☎ 01778 391111   📠 01778 425437
sallyp@warnersgroup.co.uk
www.outandaboutlive.co.uk
Editor *Sally Pepper*

Launched in August 2008. Caravanning lifestyle magazine. News, reviews and features.

Gold Dust Magazine

55 Elmdale Road, London E17 6PN
☎ 020 8765 4665
davidgardiner@worldonline.co.uk
www.golddustmagazine.co.uk
Owner *Omma Velada*
Editors *David Gardiner, Claire Tyne*

Circulation 500

Launched 2004. Biannual magazine of short stories, flash fiction, short plays, film scripts and novel extracts. Articles about writing, book and film reviews. Unsolicited submissions welcome; approach by e-mail.

FICTION *David Gardiner* 'We are always looking for quality work from new or experienced writers. Please read all our submission guidelines before submitting work.' Max. 3,000 words. Ⓔ £25 best prose.

POETRY *Claire Tyne* Max. 50 lines. Ⓔ £25 best poem.

FEATURES *David Gardiner* Book reviews. Max. 2,000 words.

Golf Monthly

IPC Media Ltd., Blue Fin Building, 110 Southwark Street, London SE1 0SU
☎ 020 3148 4530   📠 020 3148 8130
golfmonthly@ipcmedia.com
www.golf-monthly.co.uk
Owner *IPC Media (A Time Warner Company)*
Editor *Michael Harris*
Assistant Editor *Alex Narey*
Deputy Editor *Neil Tappin*
Senior Staff Writer *Luke Norman*
Circulation 82,264

Founded 1911. MONTHLY. Player profiles, golf instruction, general golf features and columns. Not interested in instruction material from outside contributors. Approach in writing with ideas.

FEATURES Max. 1,500–2,000 words.
Ⓔ Payment by arrangement.

Golf World

Media House, Lynchwood, Peterborough Business Park, Peterborough PE2 6EA
☎ 01733 468000   📠 01733 468671
chris.jones@bauermedia.co.uk
www. todaysgolfer.co.uk/golfworld
Owner *Bauer Media*
Editor *Chris Jones*
Editorial Assistant *Linda Manigan*
Circulation 30,034

Founded 1962. MONTHLY. No unsolicited mss. Approach in writing with ideas.

Good Holiday Magazine

Parman House, 30–36 Fife Road, Kingston-upon-Thames KT1 1SY
☎ 020 8547 9822   📠 020 8546 0984
edit@goodholidayideas.com
www.goodskiguide.com
Editor *John Hill*
Circulation 100,000

Founded 1985. QUARTERLY aimed at affluent holiday-makers (income of £120,000+) rather

than travellers/backpackers. Worldwide destinations including Europe and domestic. No unsolicited material but approach in writing or by e-mail with ideas and/or synopsis.
Ⓔ Payment negotiable.

Good Housekeeping

National Magazine House, 72 Broadwick Street, London W1F 9EP
☎ 020 7439 5000   📠 020 7439 5616
firstname.lastname@natmags.co.uk
www.allaboutyou.com/goodhousekeeping
Owner *National Magazine Co. Ltd*
Editor *Lindsay Nicholson*
Circulation 430,930

Founded 1922. MONTHLY glossy. No unsolicited mss. Write with ideas in the first instance to the appropriate editor.
FEATURES *Lucy Moore* Most work is commissioned but original ideas are always welcome. No ideas are discussed on the telephone. Send short synopsis, plus relevant cuttings, showing previous examples of work published. No unsolicited mss.
HEALTH *Julie Powell*. Submission guidelines as for features; no unsolicited mss.

Good Motoring

Station Road, Forest Row RH18 5EN
☎ 01342 825676   📠 01342 824847
editor@motoringassist.com
www.roadsafety.org.uk
Owner *Gem Motoring Assist*
Editor *James Luckhurst*
Circulation 62,000

Founded 1932. QUARTERLY motoring, road safety, travel and general features magazine. 1,500 words max. Prospective contributors should approach in writing only.

Good News

9 Spinney Close, West Bridgford, Nottingham NG2 6HH
☎ 0115 923 3424
goodnewseditor@ntlworld.com
www.goodnews-paper.org.uk
Owner *Good News Fellowship*
Editor *Andrew Halloway*
Circulation 60,000

Founded 2001. MONTHLY evangelistic newspaper which welcomes contributions. No fiction. Send for sample copy of writers' guidelines in the first instance. No poetry or children's stories.
NEWS Items of up to 500 words (preferably with pictures) showing God at work and human interest photo stories. 'Churchy' items not wanted. Testimonies of how people have come to personal faith in Jesus Christ and the difference it has made are always welcome.

WOMEN'S PAGE Relevant items of interest welcome.
Ⓔ Payment negotiable.

The Good Ski Guide

Parman House, 30–36 Fife Road, Kingston-upon-Thames KT1 1SY
☎ 0208 547 9822   📠 0208 546 0984
edit@goodholidayideas.com
www.goodskiguide.com
Owner *Good Holiday Group*
Editors *John Hill, Nick Dalton*
Circulation 250,000

Founded 1976. SIX ISSUES (October–Easter). Unsolicited mss welcome from writers with a knowledge of skiing and ski resorts. Prospective contributors are best advised to make initial contact in writing as ideas and work need to be seen before any discussion can take place.
Ⓔ Payment negotiable. Good Ski Guide is distributed worldwide by British Airways, so features must have international flavour, e.g. not British interest only.

GQ

Vogue House, Hanover Square, London W1S 1JU
☎ 020 7499 9080   📠 020 7495 1679
www.gq.com
Owner *Condé Nast Publications Ltd*
Editor *Dylan Jones*
Circulation 130,008

Founded 1988. MONTHLY. Men's style magazine. No unsolicited material. Write or fax with an idea in the first instance.

Granta

12 Addison Avenue, London W11 4QR
☎ 020 7605 1360   📠 020 7605 1361
editorial@granta.com
www.granta.com
Editor *Alex Clark*

QUARTERLY magazine of new writing, including fiction, memoirs, reportage and photography published in paperback book form. Highbrow, diverse and contemporary, often with a thematic approach. Unsolicited mss (including fiction) considered. A lot of material is commissioned. Vital to read the magazine first to appreciate its very particular fusion of cultural and political interests. No reviews or news articles. Access the website for submission guidelines.
Ⓔ Payment negotiable.

Grazia

Bauer Media, Endeavour House, 189 Shaftesbury Avenue, London WC2H 8JG
☎ 020 7437 9011   📠 020 7520 6589
www.graziamagazine.co.uk
Owner *Bauer Consumer Media*
Editor *Jane Bruton*

Circulation 227,083

Launched February 2005. Britain's first WEEKLY glossy aimed at women aged 25 to 45. Features articles on fashion, celebrity, news, beauty and lifestyle.

The Great Outdoors
▷ TGO

The Great War (1914–1918)
PO Box 202, Scarborough YO11 3GE
☎ 01723 581329   📠 01723 581329
books@greatnorthernpublishing.co.uk
www.greatnorthernpublishing.co.uk
Owner *Great Northern Publishing*
Editor *Mark Marsay*

Founded 2001. BI-MONTHLY subscription only, non-academic magazine published in A5 format. 'The little magazine dedicated to the Great War (1914–18) and to those who perished and those who returned.' Articles, personal stories and accounts of those who served (men and women of all nationalities) and their families: diaries, anecdotes, letters, postcards, poetry, unit histories, events and memorials, etc. Absolutely no fiction or long academic works expounding historian's personal views. New material welcome but contact editor prior to sending. See website for submission guidelines and editorial content. 'Open door policy: all welcome regardless of ability to write to high standard as all work is carefully edited. No subject or topic excluded.' Sample copy £5.

Grow Your Own
25 Phoenix Court, Hawkins Road, Colchester CO2 8JY
☎ 01206 505979   📠 01206 505945
craig.drever@aceville.co.uk
www.growfruitandveg.co.uk
Owner *Helen Tudor*
Editor *Craig Drever*

Launched 2005. MONTHLY magazine aimed at aspiring self-sufficients giving information on the best ways to grow (and cook) seasonal produce. Contributions (features and news) are welcome on anything relevant to fruit, vegetables and herbs. Approach the editor by e-mail.

GT (Gay Times)
Unit M, Spectrum House, 32–34 Gordon House Road, London NW5 1LP
☎ 020 7424 7400   📠 020 7424 7401
edit@gaytimes.co.uk
www.gaytimes.co.uk
Owner *Millivres-Prowler Group*
Editor *Joseph Galliano*

Circulation 68,000

Covers all aspects of gay life, plus general interest likely to appeal to the gay community, art reviews, fashion, style and news. Regular freelance writers used. 💷 Payment negotiable.

Guardian Weekend
▷ The Guardian under National Newspapers

Guiding magazine
17–19 Buckingham Palace Road, London SW1W 0PT
☎ 020 7834 6242   📠 020 7828 8317
guiding@girlguiding.org.uk
www.girlguiding.org.uk
Owner *Girlguiding UK*
Editor *To be appointed*
Circulation 79,000

Founded 1914. MONTHLY. Unsolicited mss welcome provided topics relate to the Movement and/or women's role in society. Ideas in writing appreciated in first instance. No nostalgic, 'when I was a Guide', pieces, please.
ACTIVITY IDEAS Interesting, contemporary ideas and instructions for activities for girls aged 5 to 18+ to do during unit meetings – crafts, games (indoor/outdoor), etc.
FEATURES Topics relevant to today's women. 500 words.
NEWS Items likely to be of interest to members. Max. 100–150 words.
💷 Payment negotiable.

H&E Naturist
Burlington Court, Carlisle Street, Goole DN14 5EG
☎ 01405 760298/766769   📠 01405 763815
editor@henaturist.net
www.henaturist.net
Owner *New Freedom Publications Ltd*
Editor *Sam Hawcroft*
Circulation 20,000

Founded 1898. MONTHLY naturist magazine.
FEATURES Will consider short features on social nudism, longer features on nudist holidays and nudist philosophy. 90% of every issue is by freelance contributors. 800–1,500 words.
NEWS 'We are always on the lookout for national and international nudist news stories.' 250–500 words. No soft porn, 'sexy' stories or sleazy photographs. Approach by post, e-mail or telephone.

Hair
IPC Media Ltd., Blue Fin Building, 110 Southwark Street, London SE1 0SU
☎ 020 3148 7274
Owner *IPC Media (A Time Warner Company)*
Editor *Louise White*

Circulation 80,319

Founded 1977. TWELVE ISSUES YEARLY. Hair and beauty magazine. No unsolicited mss, but always interested in good photographs. Approach with ideas in writing.

FEATURES Fashion pieces on hair trends and styling advice.

£ Payment negotiable.

Hair Ideas

Origin Publishing, 9th Floor, Tower House, Fairfax Street, Bristol BS1 3BN

☎ 0117 314 8805

michelletiernan@originpublishing.co.uk

www.loveyourhair.com

Owner *Origin Publishing*

Editor *Michelle Tiernan*

Circulation 76,426

Launched 2003. MONTHLY guide for women looking for a new hairstyle – hundreds of new ideas, celebrity features and product tests. No unsolicited material.

Hairflair & Beauty

Haversham Publications Ltd, Freebournes House, Freebournes Road, Witham CM8 3US

☎ 01376 534540   🖷 01376 534546

Owner *Haversham Publications Ltd*

Editor *Ruth Page*

Circulation 27,628

Founded 1982. BI-MONTHLY. Original and interesting hair and beauty-related features written in a young, lively style to appeal to a readership aged 16–35 years. Freelancers are not used.

Harper's Bazaar

National Magazine House, 72 Broadwick Street, London W1F 9EP

☎ 020 7439 5000   🖷 020 7439 5506

www.harpersbazaar.co.uk

Owner *National Magazine Co. Ltd*

Editor *Lucy Yeomans*

Circulation 109,033

MONTHLY. Up-market glossy combining the stylish and the streetwise. Approach in writing (not by phone) with ideas.

FEATURES *Francesca Martin* Ideas only in the first instance.

NEWS Snippets welcome if very original.

£ Payment negotiable.

Health & Fitness Magazine

Swan House, 37–39 High Holborn, London WC1V 6AA

☎ 020 7042 4000

mary.comber@burdamagazines.co.uk

www.healthandfitnessonline.co.uk

Owner *Hubert Burda Media*

Editor *Mary Comber*

Circulation 31,346

Founded 1983. MONTHLY. Target reader: active, health-conscious women aged 25–40.

Features news and articles on nutrition, exercise, healthy eating, holistic health and well-being. Will consider ideas; approach in writing in the first instance.

Healthy

River Publishing, Victory House, 14 Leicester Place, London WC2H 7BZ

☎ 020 7413 9447   🖷 020 7306 0314

healthy@riverltd.co.uk

www.healthy-magazine.co.uk

Owner *River Publishing*

Editor *Heather Beresford*

Circulation 188,171

EIGHT ISSUES YEARLY. Customer health magazine for Holland & Barrett health food shops. Only mentions products sold in store and will *not* mention any products by brand name, even those sold in store. Authoritative and accessible information and advice. Will consider holistic health and lifestyle features; approach by e-mail or post.

Heat

Endeavour House, 189 Shaftesbury Avenue, London WC2H 8JG

☎ 020 7859 8657   🖷 020 7859 8670

heat@bauerconsumer.co.uk

www.heatworld.co.uk

Owner *Bauer Consumer Media*

Acting Editor *Julian Linley*

Circulation 533,034

Founded 1999. WEEKLY entertainment magazine dealing with TV, film and radio information, fashion and features, with an emphasis on celebrity interviews and news. Targets 18 to 40-year-old readership, male and female. Articles written both in-house and by trusted freelancers. No unsolicited mss.

Hello!

Wellington House, 69–71 Upper Ground, London SE1 9PQ

☎ 020 7667 8700   🖷 020 7667 8716

www.hellomagazine.com

Owner *Hola! (Spain)*

Deputy Editor *Ruth Sullivan*

Assistant Editor, Features *Rosie Nixon*

News Editor *Thomas Whitaker*

Circulation 427,054

WEEKLY. Owned by a Madrid-based publishing family, *Hello!* has grown faster than any other British magazine since its launch here in 1988. The magazine has editorial offices both in Madrid and London. Major colour features plus regular

news pages. Although much of the material is provided by regulars, good proposals do stand a chance. Approach the commissioning editor with ideas in the first instance. No unsolicited mss.
FEATURES Interested in celebrity-based features, with a newsy angle, and exclusive interviews from generally unapproachable personalities.
💷 Payment by arrangement.

Hi-Fi News

IPC Media Ltd, Leon House, 233 High Street, Croydon CR9 1HZ
📞 020 8726 8311 📠 020 8726 8397
hi-finews@ipcmedia.com
www.hifinews.co.uk
Owner *IPC Media (A Time Warner Company)*
Editor *Paul Miller*
Circulation 11,353

Founded 1956. MONTHLY. Write in the first instance with suggestions based on knowledge of the magazine's style and subject. All articles must be written from an informed technical or enthusiast viewpoint. 💷 Payment negotiable, according to technical content.

High Life

85 Strand, London WC2R 0DW
📞 020 7550 8000 📠 020 7550 8250
high.life@cedarcom.co.uk
www.cedarcom.co.uk
Owner *Cedar Communications*
Editor *Kerry Smith*
Circulation 200,298

Founded 1973. MONTHLY glossy. British Airways in-flight magazine. Almost all the content is commissioned. No unsolicited mss. Few opportunities for freelancers.

History Today

20 Old Compton Street, London W1D 4TW
📞 020 7534 8000
p.lay@historytoday.com
www.historytoday.com
Owner *History Today Trust for the Advancement of Education*
Editor *Peter Lay*
Circulation 25,274

Founded 1951. MONTHLY. General history and archaeology worldwide, history behind the headlines. Serious submissions only. Approach by e-mail.

Home & Family
▷ Families First

Homes & Antiques

9th Floor, Tower House, Fairfax Street, Bristol BS1 3BN
📞 0117 927 9009

natashagoodfellow@bbcmagazinesbristol.co
www.homesandantiques.com
Owner *BBC Magazines Bristol*
Editor *Angela Linforth*
Circulation 85,121

Founded 1993. MONTHLY traditional home interest magazine with a strong bias towards antiques and collectibles. 'There are plenty of opportunities for freelancers and *Homes & Antiques* is always looking for interesting people and stories to feature in the fields of heritage, collecting, crafts and home design.' No fiction or health and beauty. Approach with ideas by phone or in writing.
FEATURES *Natasha Goodfellow* Pieces commissioned on recce shots and cuttings. Guidelines available on request. Send cuttings of relevant work published. Max. 1,500 words.
💷 Payment negotiable.

Homes & Gardens

IPC Media Ltd., Blue Fin Building, 110 Southwark Street, London SE1 0SU
📞 020 3148 5000 📠 020 3148 8165
www.homesandgardens.com
Owner *IPC Media (A Time Warner Company)*
Editor *Deborah Barker*
Circulation 143,900

Founded 1919. MONTHLY. Almost all published articles are specially commissioned. No fiction or poetry. Best to approach in writing with an idea, enclosing snapshots, if appropriate.

Horse

IPC Media Ltd., Blue Fin Building, 110 Southwark Street, London SE1 0SU
📞 020 3148 4609
joanna_browne@ipcmedia.com
www.horsemagazine.com
Owner *IPC Media (A Time Warner Company)*
Editor *Joanna Browne*
Circulation 20,062

Founded 1997. MONTHLY magazine aimed at serious leisure riders and keen competitors. Each month the magazine includes easy-to-follow training tips from some of the world's top riders, alongside horsecare features and the latest news and gossip from the horseworld. Send feature ideas with a short synopsis to the editor.

Horse and Hound

IPC Media Ltd., 9th Floor, Blue Fin Building, 110 Southwark Street, London SE1 0SU
📞 020 3148 4562
jenny_sims@ipcmedia.com
www.horseandhound.co.uk
Owner *IPC Media (A Time Warner Company)*
Editor *Lucy Higginson*

Circulation 65,631

Founded 1884. WEEKLY. The oldest equestrian magazine on the market. Contains regular veterinary advice and instructional articles, as well as authoritative news and comment on fox hunting, international and national showjumping, horse trials, dressage, driving and endurance riding. Also weekly racing and point-to-points, breeding reports and articles. Regular books and art reviews, and humorous articles and cartoons are frequently published. Plenty of opportunities for freelancers. Unsolicited contributions welcome.

Also publishes a sister monthly publication, *Eventing*, which covers the sport of horse trials comprehensively; Editor *Julie Harding*.
Ⓔ NUJ rates.

Horse and Rider

Headley House, Headley Road, Grayshott
GU26 6TU
☏ 01428 601020   🖷 01428 601030
djm@djmurphy.co.uk
www.horseandrideruk.com
Owner *D.J. Murphy (Publishers) Ltd*
Editor *Nicky Moffatt*
Deputy Editor *Jane Gazzard*
Circulation 45,327

Founded 1949. MONTHLY. Adult readership, largely horse-owning. News and instructional features, which make up the bulk of the magazine, are almost all written in-house or commissioned. New contributors and unsolicited articles are occasionally used. Approach the editor in writing with ideas.

Hotline

Ink Publishing, 141–143, Shoreditch High Street, London E1 6JE
☏ 020 7613 8777
www.ink-publishing.com
www.ink-live.com/hotline/
Editor *Madeleine Ensor*
Deputy Editor *Sophy Grimshaw*

Founded 1997. MONTHLY events-led magazine in the style of *Time Out* with a readership of 1.7 million. Film reviews, home, dining out, celebrity interviews, and stories featuring the 'best of the best' across the Virgin Trains network.

House & Garden

Vogue House, Hanover Square, London W1S 1JU
☏ 020 7499 9080   🖷 020 7629 2907
www.houseandgarden.co.uk
Owner *The Condé Nast Publications Ltd*
Editor *Susan Crewe*
Circulation 143,089

Founded 1947. MONTHLY. Most feature material is produced in-house but occasional specialist features are commissioned from qualified freelancers, mainly for the interiors, wine and food sections and travel.
FEATURES *Hatta Byng* Suggestions for features, preferably in the form of brief outlines of proposed subjects, will be considered.

House Beautiful

National Magazine House, 72 Broadwick Street, London W1F 9EP
☏ 020 7439 5000
Owner *National Magazine Co. Ltd*
Editor *Julia Goodwin*
Circulation 173,129

Founded 1989 and relaunched in November 2003. MONTHLY. Lively magazine offering sound, practical information and plenty of inspiration for those who want to make the most of where they live. Over 100 pages of easy-reading editorial. Regular features about decoration, DIY, food, gardening and occasionally property, plus topical features with a newsy slant to fit the 'Hot Topics' slot. Approach in writing with synopses in the first instance including a brief outline of areas of specialism and examples of previously published articles in a similar vein.

How to Spend It
▷ *Financial Times* under National Newspapers

i-D Magazine

124 Tabernacle Street, London EC2A 4SA
☏ 020 7490 9710   🖷 020 7490 9737
editor@i-dmagazine.co.uk
www.i-dmagazine.co.uk
Owner *Levelprint*
Editor *Ben Reardon*
Circulation 85,000

Founded 1980. MONTHLY lifestyle magazine for both sexes with a fashion bias. International. Very hip. Does not accept unsolicited contributions but welcomes new ideas from the fields of fashion, music, clubs, art, film, technology, books, sport, etc. No fiction or poetry. 'We are always looking for freelance non-fiction writers with new or unusual ideas.' A different theme each issue – past themes have included Green politics, taste, films, sex, love and loud dance music – means it is advisable to discuss feature ideas in the first instance.

Ideal Home

IPC Media Ltd., Blue Fin Building, 110 Southwark Street, London SE1 0SU
☏ 020 3148 7357
Owner *IPC Media (A Time Warner Company)*
Editorial Director *Isobel McKenzie-Price*
Circulation 214,483

Founded 1920. MONTHLY glossy. Unsolicited feature ideas are welcome only if appropriate to the magazine. Prospective contributors wishing to submit ideas should do so in writing to the editor. No fiction.

FEATURES Furnishing and decoration of houses, kitchens or bathrooms; interior design, soft furnishings, furniture and home improvements, lifestyle, consumer, gardening, property, food and readers' homes. Length to be discussed with editor.

Ⓔ Payment negotiable.

Image Magazine

Upper Mounts, Northampton NN1 3HR
📞 01604 467000   📠 01604 467190
image@northantsnews.co.uk
www.northamptontoday.co.uk
Owner *Northamptonshire Newspapers Ltd*
Editor *Ruth Supple*
Circulation 14,000

Founded 1905. MONTHLY general interest regional magazine. No unsolicited mss.

FEATURES Local issues, personalities, businesses, etc., of Northamptonshire, Bedfordshire, Buckinghamshire interest. Max. 500 words.

NEWS No hard news as such, just monthly diary column.

OTHER Regulars on motoring, fashion, beauty, lifestyle and travel. Max. 500 words.

Ⓔ Payment for features negotiable.

In Britain

The Chelsea Magazine Co. Ltd., 30 Old Church Street, London SW3 5BY
📞 020 7349 3150   📠 020 7349 3160
inbritain@chelseamagazines.com
Editor *Andrea Spain*
Circulation 27,882

Founded in the 1930s. BI-MONTHLY. Travel magazine of 'VisitBritain'. Not much opportunity for unsolicited work – approach with ideas and samples. Also interested in article ideas with good quality digital images or transparencies.

Independent Magazine
▷ The Independent under National Newspapers

Inspire

CPO, Garcia Estate, Canterbury Road, Worthing BN13 1BW
📞 01903 264556   📠 01903 821081
russbravo@cpo.org.uk
www.inspiremagazine.org.uk
Owner *Christian Publishing & Outreach Ltd*
Editor *Russ Bravo*
Circulation 65,000

MONTHLY. Upbeat, good news Christian magazine featuring human interest stories, growing churches and community transformation. Limited freelance articles used. Contributor's guidelines available.

InStyle

IPC Media Ltd., Blue Fin Building, 110 Southwark Street, London SE1 0SU
📞 020 3148 5000   📠 020 3148 8166
firstname_lastname@instyleuk.com
Owner *IPC Media (A Time Warner Company)*
Senior Editor *Kate O'Donnell*
Editor *Trish Halpin*
Deputy Editor *Charlotte Moore*
Fashion Director *Fiona Rubie*
Circulation 180,879

Launched March 2001. MONTHLY. UK edition of US fashion, beauty, celebrity and lifestyle magazine. No unsolicited material.

Interzone: Science Fiction & Fantasy

TTA Press, 5 Martins Lane, Witcham, Ely CB6 2LB
info@ttapress.demon.co.uk
www.ttapress.com
Owner *TTA Press*
Editor *Andy Cox*
Circulation 15,000

Founded 1982. BI-MONTHLY magazine of science fiction and fantasy. Unsolicited mss are welcome 'from writers who have a knowledge of the magazine and its contents'. S.a.e. essential for return.

FICTION 2,000–6,000 words.

FEATURES Book/film reviews, interviews with writers and occasional short articles. Length by arrangement.

Ⓔ Fiction, £30 per 1,000 words; features, negotiable.

Investors Chronicle

Number One, Southwark Bridge, London SE1 9HL
📞 020 7873 3000   📠 020 7382 8105
www.investorschronicle.co.uk
Owner *Pearson*
Editor *Oliver Ralph*
Deputy Editor *Rosie Carr*
Circulation 33,514

Founded 1860. WEEKLY. Opportunities for freelance contributors in the survey section only. All approaches should be made in writing. Over forty surveys are published each year on a wide variety of subjects, generally with a financial, business or investment emphasis. Copies of survey list and synopses of individual surveys are obtainable from the surveys editor.

Ⓔ Payment negotiable.

Irish Pages

The Linen Hall Library, 17 Donegall Square North, Belfast BT1 5GB
📞 028 9043 4800
irishpages@yahoo.co.uk
www.irishpages.org
Editor *Chris Agee*

Circulation 2,800

Founded 2002. BIANNUAL non-partisan, non-sectarian journal publishing writing from Ireland and abroad. 'The most important cultural journal in Ireland at the present moment' (Jonathan Allison, Director of the Yeats Summer School). FEATURES Poetry, short fiction, essays, non-fiction, memoirs, nature-writing, translated work, literary journalism and other autobiographical, historical and scientific writing of literary distinction. Irish language and Ulster Scots writing are published in the original, with English translations. Equal editorial attention is given to established, emergent and new writers. Send submissions to the editor by post.

IRRV Valuer

IRRV, 41 Doughty Street, London WC1N 2LF
☎ 01843 290919 📠 01843 290919
jcroberts54@hotmail.com
www.irrv.org.uk
Owner *Institute of Revenues, Rating and Valuation*
Managing Editor *John Roberts*
Circulation 2,000

Founded 2001. FOUR issues yearly for property valuation professionals. Well-qualified commentary, analysis and news covering issues affecting valuers, including reforms, legislation and technology. Small amount of appropriate lifestyle coverage. No unsolicited material; approach by telephone, fax, letter or e-mail in the first instance.
FEATURES Ideas for stories must be well sourced and informed.
💷 Payment negotiable.

Jade – The International Erotic Art and Literature Magazine

PO Box 202, Scarborough YO11 3GE
☎ 01723 581329 📠 01723 581329
books@greatnorthernpublishing.co.uk
www.greatnorthernpublishing.co.uk
Owner *Great Northern Publishing*
Editor *Mark Marsay*

Founded 2002. MONTHLY adult subscription-only magazine. Uncensored, high-res PDF download format since April 2008. Official magazine of The Guild of Erotic Artists. Features new and established international photographers, artists, sculptors and writers in all erotic genres from around the world. Some editorial articles and features within the genre. No advice columns or non-erotica related submissions. Contributors (photographers, artists, sculptors and writers – articles and fiction) should consult the website for submission guidelines, style and current requirements; contact the editor prior to sending material. Sample download copy £5 (contains strong adult content).

Jane's Defence Weekly

Sentinel House, 163 Brighton Road, Coulsdon CR5 2YH
☎ 020 8700 3700 📠 020 8763 1007
jdw@janes.com
www.janes.com
Owner *Jane's Information Group*
Editor *Peter Felstead*
Circulation 28,100

Founded 1984. WEEKLY. No unsolicited mss. Approach in writing with ideas in the first instance.
FEATURES Current military and defence industry topics of worldwide interest. No history pieces. Max. 4,000 words.

Jazz Journal International

3 & 3A Forest Road, Loughton IG10 1DR
☎ 020 8532 0456 📠 020 8532 0440
Owner *Jazz Journal Ltd*
Editor *Janet Cook*
Circulation 8,000+

Founded 1948. MONTHLY. A specialized jazz magazine, for record collectors, principally using expert contributors whose work is known to the editor. Unsolicited mss not welcome, with the exception of news material (for which no payment is made). It is not a gig guide, nor a free reference source for students.

Jersey Now

PO Box 582, Five Oaks, St Saviour JE4 8XQ
☎ 01534 611743 📠 01534 611610
eperchard@msppublishing.com
Owner *MSP Publishing*
Managing Editor *Peter Body*
Editor *Elisabeth Perchard*
Circulation 26,000

Founded 1987. QUARTERLY lifestyle magazine for Jersey covering homes, gardens, the arts, Jersey heritage, motoring, boating, fashion, beauty, food and drink, health, travel and technology. Upmarket glossy aimed at an informed and discerning readership. Interested in Jersey-orientated articles only; 1,200 words max. Approach the editor initially. 💷 Payment negotiable.

Jewish Chronicle

25 Furnival Street, London EC4A 1JT
☎ 020 7415 1500 📠 020 7405 9040
editorial@thejc.com
www.thejc.com
Owner *Kessler Foundation*
Editor *Stephen Pollard*

Circulation 32,875

WEEKLY. Unsolicited mss welcome if 'the specific interests of our readership are borne in mind by writers'. Approach in writing, except for urgent current news items. No fiction. Max. 1,500 words for all material.

FEATURES *Alan Montague*
NEWS EDITOR *Jenni Frazer*
FOREIGN NEWS *Daniella Peled*
SUPPLEMENTS *Angela Kiverstein*
Ⓔ Payment negotiable.

Jewish Quarterly

Haskell House, 152 West End Lane, London
NW6 1SD
Ⓣ 020 7443 5155 (editorial)
editor@jewishquarterly.org
admin@jewishquarterly.org
www.jewishquarterly.org
Publisher *Jewish Literary Trust*
Editor *Rachel Lasserson*

Founded 1953. QUARTERLY illustrated magazine featuring Jewish literature and fiction, politics, art, music, film, poetry, history, dance, community, autobiography, Hebrew, Yiddish, Israel and the Middle East, Judaism, interviews, Zionism, philosophy and holocaust studies. Features a major books and arts section. Unsolicited mss welcome but letter or e-mail preferred in the first instance.

Jewish Telegraph Group of Newspapers

Telegraph House, 11 Park Hill, Bury Old Road, Prestwich, Manchester M25 0HH
Ⓣ 0161 740 9321　Ⓕ 0161 740 9325
manchester@jewishtelegraph.com
www.jewishtelegraph.com
Owner *Jewish Telegraph Ltd*
Editor *Paul Harris (pharris@jewishtelegraph.com
0161 741 2633 Fax 0161 740 5555)*
Circulation 16,000

Founded 1950. WEEKLY publication with local, national and international news and features. (Separate editions published for Manchester, Leeds, Liverpool and Glasgow.) Unsolicited features on Jewish humour and history welcome. NEWS Newsdesk@jewishtelegraph.com or 0161 741 2631.

Journal of Apicultural Research (incorporating Bee World)

IBRA, 16 North Road, Cardiff CF10 3DY
Ⓣ 029 2037 2409　Ⓕ 05601 135640
mail@ibra.org.uk
www.ibra.org.uk
Owner *International Bee Research Association*
Editor *Norman Carreck*

Circulation 1,700

Bee World, founded 1919; Journal of Apicultural Research, 1962. QUARTERLY. High-quality factual journal, including peer-reviewed articles, with international readership. Features on apicultural science and technology. Unsolicited mss welcome but authors should write to the editor for guidelines before submitting material.

Junior Education PLUS

Scholastic Ltd, Villiers House, Clarendon Avenue, Leamington Spa CV32 5PR
Ⓣ 01926 887799　Ⓕ 01926 883331
vpaley@scholastic.co.uk
www.scholastic.co.uk/magazines
Owner *Scholastic Ltd*
Editor *Michelle Guy*
Assistant Editor *Kirstin McCreadie*
Circulation 9,654

'UK's leading magazine, in print and online', aimed at teachers of children aged 7–11 years. Practical articles from teachers about education for this age group are welcome. Max. 900 words. Approach in writing with synopsis.

Kerrang!

Mappin House, 4 Winsley Street, London
W1W 8HF
Ⓣ 020 7182 8000　Ⓕ 020 7182 8910
feedback@kerrang.com
www.kerrang.com
Owner *Bauer Consumer Media*
Editor *Paul Brannigan*
Circulation 76,937

Founded 1981. WEEKLY rock, punk and metal magazine. 'Written by fans for fans'. Will consider ideas for features but not actively seeking new contributors unless expert in specialist fields such as black metal and hardcore, emo, etc.

Koi Magazine

Origin Publishing, Tower House, Fairfax Street, Bristol BS1 3BN
Ⓣ 0117 927 9009　Ⓕ 0117 934 9008
koi@originpublishing.co.uk
www.koimag.co.uk
Owner *Origin Publishing*
Editor *Beckie Rodgers*
Circulation 15,000

Launched 1999. FOUR WEEKLY. Practical guide to koi-keeping featuring everything you need to know about the hobby, including koi appreciation and pond construction. Contributions welcome; send press releases, articles and photographs by e-mail.

FEATURES *Beckie Rodgers* Max. 4,000 words.
Ⓔ £150–£400.
NEWS *Amy Schroeter* Max. 200 words.

The Lady

39–40 Bedford Street, London WC2E 9ER
☎ 020 7379 4717   📠 020 7836 4620
www.lady.co.uk
Editor *Arline Usden*
Circulation 28,911

Founded 1885. WEEKLY. Unsolicited mss are accepted provided they are not on the subject of politics or religion, or on topics covered by staff writers or special correspondents, i.e. fashion and beauty, health, cookery, household, gardening, finance and shopping.
FEATURES Well-researched pieces on British and foreign travel, historical subjects or events; interviews and profiles and other general interest topics. Max. 1,200 words for illustrated two-page articles; 900 words for one-page features; 420 words for first-person 'Viewpoint' pieces. All material should be addressed to the editor with s.a.e. enclosed. Photographs supporting features may be supplied as colour transparencies, b&w prints or on disk. Telephone enquiries about features are not encouraged.

Lake District Life

3 Tustin Court, Port Way, Preston PR2 2YQ
☎ 01772 722022   📠 01772 736496
paul.mackenzie@archant.co.uk
www.lakedistrict-life.co.uk
Owner *Archant Life*
Editor *Paul Mackenzie*
Circulation 23,000

Founded 1947. MONTHLY. A celebration of life in the Lake District, both past and present. Unsolicited contributions welcome; approach the editor by e-mail.

Lancashire Life

3 Tustin Court, Port Way, Preston PR2 2YQ
☎ 01772 722022   📠 01772 736496
roger.borrell@lancashirelife.co.uk
www.lancashirelife.co.uk
Owner *Archant Life*
Editor *Roger Borrell*
Circulation 23,455

Founded 1947. MONTHLY county magazine. Features and pictures about Lancashire. Unsolicited contributions welcome; approach the editor by e-mail.

Land Rover World

Kelsey Publishing Ltd, Cudham Tithe Barn, Berry's Hill, Cudham TN16 3AG
☎ 01959 541444   📠 01959 541400
www.landroverworld.co.uk
Owner *IPC Media (A Time Warner Company)*
Editor *John Carroll*

Circulation 30,000

Founded 1994. MONTHLY. Incorporates *Practical Land Rover World* and *Classic Land Rover World*. Unsolicited material welcome, especially if supported by high-quality illustrations.
FEATURES All articles with a Land Rover theme of interest. Potential contributors are strongly advised to examine previous issues before starting work.
£ Payment negotiable.

Lawyer 2B

50 Poland Street, London W1F 7AX
☎ 0207 970 4000
www.lawyer2b.com
Owner *Centaur Media plc*
Editor *Husnara Begum*
Circulation 30,000

Launched in 2001, *Lawyer 2B* is a sister publication of *The Lawyer*, circulated to graduate recruitment professionals, law students and trainees. FOUR ISSUES PER ACADEMIC YEAR. Regular opportunities for profiles/features relating to legal education and law students.

The Lawyer

50 Poland Street, London W1F 7AX
☎ 020 7970 4000
editorial@thelawyer.com
www.thelawyer.com
Owner *Centaur Media plc*
Editor *Catrin Griffiths*
News Editor *Margaret Taylor*
Web Editor *Jon Parker*
Special Reports Editor *Tom Phillips*
Associate Editor (New York) *Matt Byrne*
Circulation 31,571 (print); 134,040 (online)

Launched 1987. WEEKLY plus daily news e-mail alerts and commentaries. Leading news magazine for City and commercial lawyers. Strong international coverage and leading-edge research.
FEATURES *Steve Hoare* Technical legal content, written in an accessible style. No unsolicited contributions; initial approach by e-mail with specific idea.

Lincolnshire Life

County House, 9 Checkpoint Court, Sadler Road, Lincoln LN6 3PW
☎ 01522 527127   📠 01522 842000
editorial@lincolnshirelife.co.uk
www.lincolnshirelife.co.uk
Publisher *C. Bingham*
Executive Editor *Josie Thurston*
Circulation 10,000

Founded 1961. MONTHLY county magazine featuring geographically relevant articles on local culture, history, personalities, etc. Max. 1,500 words. Contributions supported by three or four good-quality photographs welcome. Approach in writing. ⊞ Payment varies.

The Lincolnshire Poacher

County House, 9 Checkpoint Court, Sadler Road, Lincoln LN6 3PW
☎ 01522 527127　⊞ 01522 842000
editorial@lincolnshirelife.co.uk
www.lincolnshirelife.co.uk
Publisher *C. Bingham*
Executive Editor *Josie Thurston*
Circulation 5,000

QUARTERLY county magazine featuring geographically relevant but nostalgic articles on history, culture and personalities of Lincolnshire. Max. 2,000 words. Contributions supported by three or four good-quality photographs/illustrations appreciated. Approach in writing. ⊞ Payment varies.

The List

14 High Street, Edinburgh EH1 1TE
☎ 0131 550 3050　⊞ 0131 557 8500
claire.prentice@list.co.uk
www.list.co.uk
Owner *The List Ltd*
Publisher *Robin Hodge*
Editor *Claire Prentice*
Circulation 9,686

Founded 1985. FORTNIGHTLY. Events guide covering Glasgow and Edinburgh. Interviews and profiles of people working in film, theatre, music and the arts. Max. 1,200 words. No unsolicited mss. News material tends to be handled in-house.

Literary Review

44 Lexington Street, London W1F 0LW
☎ 020 7437 9392　⊞ 020 7734 1844
editorial@literaryreview.co.uk
www.literaryreview.co.uk
Editor *Nancy Sladek*
Circulation 15,000

Founded 1979. MONTHLY. Publishes book reviews (commissioned), features and articles on literary subjects. Prospective contributors are best advised to contact the editor in writing. Unsolicited mss not welcome. Runs a monthly competition for 'poems which rhyme, scan and make sense'. ⊞ Payment 'minuscule'.

Litro Magazine

1 Hippodrome Place, London W11 4NG
☎ 020 7792 0750
editor@litro.co.uk
www.litro.co.uk
Owner *Ocean Media Books Ltd*
Editor *Tom Chivers*

Circulation 100,000

A free MONTHLY magazine distributed online, in libraries, galleries, bars, cafes and other venues in London and beyond. Publishes new, original short fiction. Contributors have included Irvine Welsh, Yiyun Li, Glyn Maxwell, Benjamin Zephaniah and Andrew Crumey. Unsolicited contributions are welcome; send to the editor by e-mail.
FICTION Used for Litro Magazine; max. 2,000 words.
FEATURES Used for Litro Online; max. 500 words.
NEWS Used for Litro Online; max. 200 words.
⊞ No payment.

Living France

Archant House, Oriel Road, Cheltenham GL50 1BB
☎ 01242 216086　⊞ 01242 211081
editorial@livingfrance.com
www.livingfrance.com
Owner *Archant Life*
Editor *Eleanor O'Kane*

Founded 1989. FOUR-WEEKLY. A magazine for people who are considering or actively purchasing a home in France. Editorial consists of travel features as well as practical, friendly information and advice on living in France or becoming an owner of French property. Will consider articles describing different regions of France, interviews with owners of property or new-build and aspects of living in France. Also interested in submissions on interiors in France and gardens and gardening in France. Welcomes contact from professional writers. No unsolicited mss; approach in writing or by e-mail with an idea.

Loaded

IPC Media Ltd., Blue Fin Building, 110 Southwark Street, London SE1 0SU
☎ 020 3148 5000　⊞ 020 3148 8107
andy_sherwood@ipcmedia.com
Owner *IPC Media (A Time Warner Company)*
Editor *Martin Daubney*
Features Editor *Andy Sherwood*
Circulation 95,371

Founded 1994. MONTHLY men's lifestyle magazine featuring music, sport, sex, humour, travel, fashion, hard news and popular culture. Will consider material which comes into these categories; approach the features editor in writing or by e-mail in the first instance. No fiction or poetry.

Logos

5 Beechwood Drive, Marlow SL7 2DH
☎ 01628 483371　⊞ 01628 477577
logos-marlow@dial.pipex.com
www.logos-journal.org

Owner *LOGOS International Educational Publishing Foundation*
Editor Emeritus *Gordon Graham*
Business Manager *Betty Graham*

Founded 1990. QUARTERLY. Aims to 'deal in depth with issues which unite, divide, excite and concern the world of books', with an international perspective. Each issue contains six to eight articles of between 3,500 and 7,000 words. 'Logos is a non-profit making professional forum, not a scholarly journal.' Suggestions and ideas for contributions are welcome, and should be addressed to the editor. 'Guidelines for Contributors' available. Contributors write from their experience as authors, publishers, booksellers, librarians, etc.
£ Contributors receive off-prints, a copy of the issue in which their article appears and a 50% concession on the subscription rate.

The London Magazine

Editorial office: 32 Addison Grove, London W4 1ER
☎ 020 8400 5882 (admin. office)
🖷 020 8994 1713
admin@thelondonmagazine.net
www.thelondonmagazine.net
Publisher *Christopher Arkell*
Editor *Sara-Mae Tuson*
Circulation 3,000

Founded originally in 1732. BI-MONTHLY review of literature and the arts. Publishes poems, short stories, features, memoirs, and book and performance reviews. Not interested in overtly political material. Postal submissions should include an s.a.e. (plus IRCs if overseas) to the editorial office.
FEATURES Max. 3,000 words.
FICTION Max. 3,000 words.
POEMS 'Any length within reason.'
£ No payment.

London Review of Books

28 Little Russell Street, London WC1A 2HN
☎ 020 7209 1101   🖷 020 7209 1102
edit@lrb.co.uk
www.lrb.co.uk
Owner *LRB Ltd*
Editor *Mary-Kay Wilmers*
Circulation 45,905

Founded 1979. FORTNIGHTLY. Reviews, essays and articles on political, literary, cultural and scientific subjects. Also poetry. Unsolicited contributions welcome (approximately 50 received each week). No pieces under 2,000 words. Contact the editor in writing. Please include s.a.e.
£ £200 per 1,000 words; poems, £125.

Lothian Life

Larick House, Whitehouse, Tarbert PA29 6XR
☎ 01880 730360/07734 699607
info@lothianlife.co.uk
www.lothianlife.co.uk
Owner *Pages Editorial & Publishing Services*
Editor *Suse Coon*

Online county magazine for people who live, work or have an interest in the Lothians. Features on successful people, businesses or initiatives. Regular articles by experts in the Lifestyle section including Homes & Gardens, Tastebuds, Out & About, The Arts and Health & Fitness. Phone first to discuss content and timing.
£ Payment: see website.

Machine Knitting Monthly

PO Box 1479, Maidenhead SL6 8YX
☎ 01628 783080   🖷 01628 633250
mail@machineknittingmonthly.net
www.machineknittingmonthly.net
Owner *RPA Publishing Ltd*
Editor *Anne Smith*

Founded 1986. MONTHLY. Unsolicited mss considered 'as long as they are applicable to this specialist publication. We have our own regular contributors each month but we're always willing to look at new ideas from other writers.' Approach in writing in the first instance.

MacUser

30 Cleveland Street, London W1T 4JD
☎ 020 7907 6000
mailbox@macuser.co.uk
www.macuser.co.uk
Owner *Dennis Publishing*
Editor *Nik Rawlinson*
Circulation 14,502

Launched 1985. FORTNIGHTLY computer magazine.

Management Today

174 Hammersmith Road, London W6 7JP
☎ 020 8267 4610
editorial@managementtoday.co.uk
Owner *Haymarket Media*
Managing Director *Nick Simpson*
Editor *Matthew Gwyther*
Circulation 100,003

Launched in 1966. General business topics and features. Ideas welcome. Send brief synopsis to the features editor.

marie claire

IPC Media Ltd., Blue Fin Building, 110 Southwark Street, London SE1 0SU
☎ 020 3148 7513   🖷 020 3148 8120

marieclaire@ipcmedia.com
www.marieclaire.co.uk
Owner *European Magazines Ltd*
Editor *Trish Halpin*
Circulation 330,182

Founded 1988. MONTHLY. An intelligent glossy magazine for women, with strong international features and fashion. No unsolicited mss. Approach with ideas in writing or by e-mail (marieclaireideas@ipcmedia.com). No fiction. FEATURES *Miranda McMinn* Detailed proposals for feature ideas should be accompanied by samples of previous work.

Marketing Week

St Giles House, 50 Poland Street, London W1F 7AX
📞 020 7970 4000   📠 020 7970 6722
mw.editorial@centaur.co.uk
www.marketingweek.co.uk
Owner *Centaur Communications*
Editor *Stuart Smith*
Circulation 39,093

WEEKLY trade magazine of the marketing industry. Features on all aspects of the business, written in a newsy and up-to-the-minute style. Approach with ideas in the first instance. FEATURES *Daney Parker*
💷 Payment negotiable.

Markings

The Bakehouse, 44 High Street, Gatehouse of Fleet DG7 2HP
📞 01557 814175
info@markings.org.uk
www.markings.org.uk
Editors *John Hudson, Chrys Salt*
Circulation 500+

Launched 1995. BIANNUAL Scottish literary magazine with an international readership. Poetry, contemporary art, short stories, criticism and reviews. Occasionally publishes special features on established or emergent literary figures. Unsolicited submissions are invited in any language but must be accompanied by English translation. 'Markings wishes to be challenging, controversial and loyal to honest and humane writing.' Black and white artwork appears in every issue; artists and illustrators are encouraged to submit work. Electronic submissions preferred but if sending by post enclose an s.a.e.
FEATURES Any subject relating to the arts. Max. 2,500 words.
FICTION Short stories on any subject. Max. 3,000 words.
POETRY Max. six submissions.

Match!

Media House, Peterborough Business Park, Lynchwood, Peterborough PE2 6EA
📞 01733 468000   📠 01733 468724
match.magazine@bauermedia.co.uk
www.matchmag.co.uk
Owner *Bauer Media*
Editor *James Bandy*
Circulation 113,049

Founded 1979. WEEKLY. The UK's biggest-selling football magazine aimed at 8–14-year-olds. All material is generated in-house by a strong news and features team. Work experience placements often given to trainee journalists and students; several staff have been recruited through this route. Approach in writing or by phone.

Maxim

30 Cleveland Street, London W1T 4JD
📞 020 7907 6000   📠 020 7907 6439
editorial.maxim@dennis.co.uk
www.maxim.co.uk
Owner *Dennis Publishing*
Editor *Ben Raworth*

Established in 1995, the print edition ceased publication in April 2009. Available as an online men's lifestyle magazine featuring sex, grooming, girls, fun-stuff, cars and fashion. No fiction or poetry. Send outlines of ideas only together with examples of published work.

Mayfair

2 Archer Street, Piccadilly Circus, London W1D 7AW
📞 020 7292 8000   📠 020 7734 5030
mayfair@paulraymond.com
www.paulraymond.com
Owner *Paul Raymond Publications*
Editor *David Spenser*
Circulation 85,000

Founded 1966. THIRTEEN ISSUES YEARLY. Unsolicited material accepted if pertinent to the magazine and if accompanied by suitable illustrative material. 'We will *only* publish work if we can illustrate it.' Interested in features and humour aimed at men aged 18 to 80; 800–1,000 words. Punchy, bite-sized humour, top ten features, etc. 'Must make the editor laugh.' Also considers erotica.

Mayfair Times

Blandel Bridge House, 56 Sloane Square, London SW1W 8AX
📞 020 7259 1050   📠 020 7901 9042
mayfair.times@pubbiz.com
www.pubbiz.com
Owner *Publishing Business Ltd*
Editor *Selma Day*

Circulation 17,000

Founded 1985. MONTHLY. Features on Mayfair of interest to residents, local workers, visitors and shoppers.

Medal News

Token Publishing, Orchard House, Duchy Road, Heathpark, Honiton EX14 1YD
📞 01404 46972   📠 01404 44788
info@tokenpublishing.com
www.tokenpublishing.com
Owners *J.W. Mussell, Carol Hartman*
Editor *J.W. Mussell*
Circulation 6,000

Founded 1989. TEN ISSUES YEARLY. Unsolicited material welcome but initial approach by phone or in writing preferred.
FEATURES 'Opportunities exist for well-informed authors who know the subject and do their homework. It would help if a digital copy of the contribution is provided in the form of an e-mail or CD. Relevant illustrations welcome.' Max. 2,500 words.
💷 £30 per 1,000 words.

Media Week

Griffin House, 161 Hammersmith Road, London W6 8BS
📞 020 8267 5000   📠 020 8267 8020
firstname.surname@haymarket.com
www.mediaweek.co.uk
Owner *Haymarket Publishing Ltd*
Editor *Steve Barrett*
Circulation 11,494

Founded 1986. WEEKLY trade magazine. UK and international coverage on all aspects of commercial media. Approach in writing with ideas. See website for e-mail addresses.

Melody Maker
▷ **New Musical Express**

Men's Health

National Magazine House, 33 Broadwick Street, London W1F 0DQ
📞 020 7339 4400   📠 020 7339 4444
www.menshealth.co.uk
Owner *Natmag-Rodale Publishing*
Editor *Morgan Rees*
Circulation 245,212

Founded 1994. MONTHLY men's healthy lifestyle magazine covering health, fitness, nutrition, stress and sex issues. No unsolicited mss; will consider ideas and synopses tailored to men's health. No fiction or extreme sports. Approach in writing in the first instance.

MiniWorld Magazine

IPC Media Ltd, Leon House, 233 High Street, Croydon CR9 1HZ
📞 020 8726 8000   📠 020 8726 8399
miniworld@ipcmedia.com
www.miniworld.co.uk
Owner *IPC Media (A Time Warner Company)*
Editor *Monty Watkins*
Circulation 37,000

Founded 1991. THIRTEEN ISSUES YEARLY. Car magazine devoted to the Mini. Unsolicited material welcome but prospective contributors are advised to contact the editor.
FEATURES Maintenance, tuning, restoration, technical advice, classified, sport, readers' cars and social history of this cult car.
💷 Payment negotiable.

Mixmag

Development Hell Ltd, 90–92 Pentonville Road, London N1 9HS
📞 020 7078 8400
www.mixmag.net
Owner *Development Hell Ltd*
Editor *Nick DeCosemo*
Circulation 34,073

Originally launched in 1982 and relaunched by new owners, Development Hell Ltd in 2006. Bestselling MONTHLY dance music magazine and 'leading authority on club culture'. News, features, night-life reviews, fashion, club listings and free mix CDs.

Mizz

Panini UK Ltd, Broxbourne House, 77 Mount Ephraim, Tunbridge Wells TN4 8BS
📞 01892 500100   📠 01892 545666
mizz@panini.co.uk
www.mizz.com
Owner *Panini UK Ltd*
Editor *Karen O'Brien*
Circulation 71,092

Founded 1985. FORTNIGHTLY magazine for the 10–14-year-old girl.
FEATURES 'We have a full features team and thus do not accept freelance features.'

Mobilise

Mobilise Organisation, Ashwellthorpe, Norwich NR16 1EX
📞 01508 489449   📠 01508 488173
editor@mobilise.info
www.mobilise.info
Owner *Mobilise Organisation*
Editor *Lisa Crosby*
Circulation 20,000

MONTHLY publication of the Mobilise Organisation, which aims to promote and protect the interests and welfare of disabled people and help and encourage them in gaining increased mobility. Various discounts available for members; membership costs £14 p.a. (single), £18 (joint). The magazine includes information for members plus members' letters. Approach in writing with ideas. Unsolicited mss welcome.

Model Collector

IPC Focus Network, Leon House, 233 High Street, Croydon CR9 1HZ
℡ 020 8726 8000   📠 020 8726 8299
modelcollector@ipcmedia.com
www.modelcollector.com
Owner *AOL Time Warner/IPC Media*
Editor *Lindsey Amrani*
Circulation 10,254

Founded 1987. THIRTEEN ISSUES YEARLY. Britain's best selling die cast magazine. From the latest models to classic 1930's Dinkys. Interested in historical articles about particular models or ranges and reviews of the latest products. Photographs welcome. Not interested in radio controlled models or model railways.
FEATURES Freelancers should contact the editor. Unsolicited material may be considered.
NEWS Trade news welcome. Call *Lindsey Armrani* (020 8726 8238) to discuss details.

Mojo

Mappin House, 4 Winsley Street, London W1W 8HF
℡ 020 7182 8616   📠 020 7182 8596
mojo@bauermedia.co.uk
www.mojo4music.com
Owner *Bauer Consumer Media*
Editor-in-Chief *Phil Alexander*
Circulation 106,367

Founded 1993. MONTHLY magazine containing features, reviews and news stories about all types of music and its influences. Receives about five mss per day. No poetry, think-pieces on dead rock stars or similar fan worship.
FEATURES Amateur writers discouraged except as providers of source material, contacts, etc.
NEWS All verifiable, relevant stories considered.
REVIEWS Write to Reviews Editor, *Jenny Bulley* with relevant specimen material.
💷 Payment negotiable.

Moneywise

1st Floor, Standon House, Mansell Street, London E1 8AA
℡ 020 7680 3600   📠 020 7702 0710
rachel.lacey@moneywise.co.uk
www.moneywise.co.uk
Owner *Moneywise Publishing Ltd*
Editor *Rachel Lacey*

Circulation 25,399

Founded 1990. MONTHLY. No unsolicited mss; ideas welcome. Make initial approach in writing to the editorial department (c.v. preferred).

More

Endeavour House, 189 Shaftesbury Avenue, London WC2H 8JG
℡ 020 7208 3165   📠 020 7208 3595
www.moremagazine.co.uk
Owner *Bauer Publishing Consumer Media*
Editor *Lisa Smosarski*
Circulation 162,544

Founded 1988. WEEKLY women's magazine aimed at the working woman aged 18–25. Useful features on men, sex and relationships plus celebrity relationships gossip, and highstreet fashion, shocking real lives and practical beauty. Most items are commissioned; approach features editor and news editor with ideas. Prospective contributors are strongly advised to study the magazine's style before submitting anything.

Mother and Baby

Endeavour House, 189 Shaftesbury Avenue, London WC2H 8JG
℡ 020 7295 5560
mother&baby@bauermedia.co.uk
Owner *Bauer Media*
Editor *Miranda Levy*
Editorial Assistant *Lucy Quick*
Circulation 65,856

Founded 1956. MONTHLY. Welcomes suggestions for feature ideas from national magazine and newspaper writers about pregnancy, newborn basics and baby development. Approaches may be made by e-mail or in writing to the Senior Writer, *Louise Evans*.

Motor Boat & Yachting

IPC Media Ltd., Blue Fin Building, 110 Southwark Street, London SE1 0SU
℡ 020 3148 4651
mby@ipcmedia.com
www.mby.com
Owner *Time Warner*
Editor *Hugo Andreae*
Technical Editor *David Marsh*
Circulation 17,504

Founded 1904. MONTHLY for those interested in motor boats and motor cruising.
FEATURES *Hugo Andreae* Cruising features and practical features especially welcome. Illustrations/photographs (colour) are just as important as text. Max. 3,000 words.
NEWS *Rob Peake* Factual pieces. Max. 200 words.
💷 Features: from £125 per 1,000 words or by arrangement; news: up to £50 per item.

Motor Boats Monthly

IPC Media Ltd., 9th Floor, Blue Fin Building, 110 Southwark Street, London SE1 0SU
☎ 020 3148 4664 📠 020 3148 8128
carl_richardson@ipcmedia.com
Owner *IPC Media (A Time Warner Company)*
Editor *Carl Richardson*
Circulation 20,000

Founded 1987. MONTHLY magazine devoted to motor boating in the UK. Unsolicited contributions welcome: cruising articles, DIY and human interest. No powerboat racing or any sailing-related material. Approach by e-mail.

Motor Caravan Magazine

IPC Media Ltd, Leon House, 233 High Street, Croydon CR9 1HZ
☎ 020 8726 8000
motorcaravan@ipcmedia.com
www.motorcaravanmagazine.co.uk
Owner *IPC Media (A Time Warner Company)*
Editor *Victoria Bentley*
Assistant Editor *Naomi Leach*
Circulation 10,000

Founded 1986. MONTHLY magazine with ideas about where to go in your motor caravan, expert advice to keep trips fun and practical advice on vans and kit. Interested in features on touring, unusual motorhomes, and activities people pursue while out and about in their van. E-mail the editor with ideas. 💷 £60 a page.

Motor Cycle News

Media House, Lynchwood, Peterborough Business Park, Peterborough PE2 6EA
☎ 01733 468000 📠 01733 468028
MCN@motorcylenews.com
www.motorcylenews.com
Owner *Bauer Consumer Media*
Editor *Marc Potter*
Circulation 128,801

Founded 1955. WEEKLY. Interested in short news stories and features on motorcyles and motorcycle racing. Contact relevant desks direct.

Motorcaravan Motorhome Monthly (MMM)

PO Box 88, Tiverton EX16 7ZN
mmmeditor@warnersgroup.co.uk
www.outandaboutlive.co.uk
Owner *Warners Group Publications Plc*
Editor *Mike Jago*
Circulation 40,408

Founded 1966. MONTHLY. 'There's no money in motorcaravan journalism but for those wishing to cut their first teeth ...' Unsolicited mss welcome if relevant, but ideas in writing preferred in first instance.
FEATURES Caravan site reports. Max. 500 words.
TRAVEL Motorcaravanning trips (home and overseas). Max. 2,000 words.
NEWS Short news items for miscellaneous pages. Max. 200 words.
SPECIAL PAGES DIY – modifications to motorcaravans. Max. 1,500 words.
OWNER REPORTS Contributions welcome from motorcaravan owners. Contact the editor for requirements. Max. 2,000 words.
💷 Payment varies but starts at £70 per published page (incl.).

Mountain Magic

Parman House, 30–36 Fife Road, Kingston-upon-Thames KT1 1ST
☎ 020 8547 9822 📠 020 8546 0984
info@goodholidayideas.com
Editor *John Hill*
Circulation 150,000

QUARTERLY magazine for those seeking healthy holidays away from beaches and cities, such as hill walking, trekking, fishing, windsurfing, etc. No backpacking or serious climbing. No unsolicited material but approach in writing or by e-mail with ideas and/or synopsis. 💷 Payment negotiable.

Mslexia (For Women Who Write)

PO Box 656, Newcastle upon Tyne NE99 1PZ
☎ 0191 261 6656
postbag@mslexia.co.uk
www.mslexia.co.uk
Owner *Mslexia Publications Limited*
Editor *Daneet Steffens*
Circulation 10,700

Founded 1997. QUARTERLY. Articles, advice, reviews, interviews, events for women writers plus new poetry and prose. Will consider fiction, poetry, features and letters but contributors *must* send for guidelines first or see website for details.

Music Week

Ludgate House, 245 Blackfriars Road, London SE1 9UY
☎ 020 7921 5000
ben@musicweek.com
www.musicweek.com
Owner *CMP Information*
Publisher *Joe Hosken*
Editor *Paul Williams*
Circulation 7,960

Britain's only WEEKLY music business magazine. Also produces a news and data website. Annual music industry contacts book – the *Music Week Directory*. Free daily news e-mail. No unsolicited mss. Approach in writing with ideas.
FEATURES Analysis of specific music business events and trends.
NEWS Music industry news only.

Musical Opinion

50 Collinswood Drive, St Leonards on Sea
TN38 0NX
☎ 01424 715167
musicalopinion2@aol.com
www.musicalopinion.com
Owner *Musical Opinion Ltd*
Editor *Denby Richards*
Circulation 8,500

Founded 1877. Glossy, full-colour BI-MONTHLY
magazine. International classical music content,
with topical features on music, musicians,
festivals, etc., and reviews (concerts, festivals,
opera, ballet, jazz, CDs, DVDs, videos, books
and printed music). International readership. No
unsolicited mss; commissions only. Ideas always
welcome though; approach by phone or e-mail,
giving telephone number. Visit the website for full
information.
€ Payment negotiable.

My Weekly

80 Kingsway East, Dundee DD4 8SL
☎ 01382 223131   01382 452491
myweekly@dcthomson.co.uk
www.dcthomson.co.uk
Owner *D.C. Thomson & Co. Ltd*
Editor *Sally Hampton*
Deputy Editor *Fiona Brown (fbrown@dcthomson.
co.uk)*
Fiction Editor *Liz Smith (lsmith@dcthomson.
co.uk)*
Circulation 162,874

WEEKLY title aimed at women of 50-plus. Mix
of fiction, true-life stories, health, fashion, food,
finance, gardening, travel. Ideas welcome. Pitch
ideas in writing or via e-mail.
FEATURES Accepted on commission only basis.
Pitch ideas in writing or via e-mail.
FICTION E-mail for current guidelines.
€ Payment negotiable.

My Weekly Pocket Novels

D.C. Thomson & Co. Ltd, 80 Kingway East,
Dundee DD4 8SL
☎ 01382 575540   01382 452491
tsteele@dcthomson.co.uk
Owner *D.C. Thomson & Co. Ltd*
Editor *Tracey Steele*

WEEKLY. Publishes 25,000–30,000-word
romantic stories aimed at the adult market. Send
first three chapters and synopsis by post.

The National Trust Magazine

Heelis, Kemble Drive, Swindon SN2 2NA
☎ 01793 817716   01793 817401
magazine@nationaltrust.org.uk
Owner *The National Trust*
Editor *Sue Herdman*

Circulation 1.82 million

Founded 1968. THREE ISSUES YEARLY.
Conservation of historic houses, coast and
countryside in England, Northern Ireland and
Wales. No unsolicited mss. Approach in writing
with ideas.

The Naturalist

c/o University of Bradford, Bradford BD7 1DP
☎ 01274 234212   01274 234231
m.r.d.seaward@bradford.ac.uk
Owner *Yorkshire Naturalists' Union*
Editor *Prof. M.R.D. Seaward*
Circulation 5,000

Founded 1875. QUARTERLY. Natural history,
biological and environmental sciences for a
professional and amateur readership. Unsolicited
mss and b&w illustrations welcome. Particularly
interested in material – scientific papers –
relating to the north of England. € No payment.

Nature

The Macmillan Building, 4–6 Crinan Street,
London N1 9XW
☎ 020 7833 4000   020 7843 4596
nature@nature.com
www.nature.com/nature
Owner *Nature Publishing Group*
Editor-in-Chief *Philip Campbell*
Circulation 65,000

Covers all fields of science, with articles and news
on science and science policy only. Scope only
for freelance writers with specialist knowledge in
these areas.

NB

105 Judd Street, London WC1H 9NE
☎ 020 7388 1266
www.rnib.org.uk/nbmagazine
Owner *Royal National Institute of Blind People*
Editor *Ann Lee*
Circulation 1,500

Founded 1917. MONTHLY. Published in
print, braille and on audio CD and e-mail.
Unsolicited mss accepted. Authoritative items by
professionals or volunteers working in the field
of eye health and sight loss welcome. Max. 1,500
words.
€ Payment negotiable.

New!

The Northern & Shell Building, 10 Lower Thames
Street, London EC3R 6EN
☎ 0871 520 7016
karmel.doughty@express.co.uk
Owner *Richard Desmond*
Editor *Kirsty Mouatt*

Circulation 434,218

Founded 2003. Celebrity WEEKLY with true life content plus fashion and beauty. Celebrity interviews and true life submissions welcome. Approach by e-mail.

New Humanist

1 Gower Street, London WC1E 6HD
☎ 020 7436 1151 📠 020 7079 3588
editor@newhumanist.org.uk
www.newhumanist.org.uk
Owner *Rationalist Association*
Editor *Caspar Melville*
Circulation 5,000

Founded 1885. BI-MONTHLY. Unsolicited mss welcome. No fiction.
FEATURES Articles with a humanist perspective welcome in the following fields: religion (critical), humanism, human rights, philosophy, current events, literature, history and science. 2,000 words.
BOOK REVIEWS 750–1,000 words, by arrangement with the editor.
£ Payment for features is nominal, but negotiable.

New Internationalist

55 Rectory Road, Oxford OX4 1BW
☎ 01865 811400 📠 01865 793152
ni@newint.org
www.newint.org/
Owner *New Internationalist Trust*
Co-Editors *Vanessa Baird, David Ransom, Adam Ma'anit, Jess Worth, Chris Brazier*
Circulation 75,000

Radical and broadly leftist in approach, but unaligned. Concerned with world poverty and global issues of peace and politics, feminism and environmentalism, with emphasis on the Third World. Difficult to use unsolicited material as they work to a theme each month and features are commissioned by the editor on that basis. The way in is to send examples of published or unpublished work; writers of interest are taken up.

New Musical Express

IPC Media Ltd., Blue Fin Building, 110 Southwark Street, London SE1 0SU
☎ 020 3148 5000 📠 020 3148 8107
www.nme.com
Owner *IPC Media (A Time Warner Company)*
Editor *Conor McNicholas*
Circulation 56,284

Britain's best-selling musical WEEKLY. Now incorporates *Melody Maker*. Freelancers used, but always for reviews in the first instance. Specialization in areas of music is a help.
REVIEWS: ALBUMS *Julian Marshall* Send in

examples of work, either published or specially written samples.

New Nation

Unit 2, 65 Whitechapel Road, London E1 1DU
☎ 020 7650 2000 📠 020 7650 2001
theeditor@newnation.co.uk
www.newnation.co.uk
Owner *Ethnic Media Group*
Editor *Lester Holloway*
Circulation 73,000

Founded 1996. WEEKLY community paper for the black community in Britain. Interested in relevant general, local and international issues. Approach in writing with ideas for submission.
GOSPEL *Antonia Hines*
BEAUTY *Juliana Olapido*
FILM *Elsa O'Toole*
ARTS *Andrea Enisouh*
LISTINGS *Sophia A. Jackson*
FITNESS *Mark Jenkins*
BUSINESS *Sonia Brown*

New Scientist

8th Floor, Lacon House, 84 Theobalds Road, London WC1X 8NS
☎ 020 8652 3500 📠 020 7611 1250 (news)
www.newscientist.com
Owner *Reed Business Information Ltd*
Editor *Jeremy Webb*
Circulation 173,820 (worldwide)

Founded 1956. WEEKLY. No unsolicited mss. Approach with ideas (one A4-page synopsis) by fax or e-mail.
FEATURES Commissions only, but good ideas welcome. Max. 3,500 words.
NEWS *Shaoni Bhattacharya* Mostly commissions, but ideas for specialist news welcome. Max. 1,000 words.
REVIEWS are commissioned.
OPINION Unsolicited material welcome if of general/humorous interest and related to science. Max. 1,000 words.
£ Payment negotiable.

The New Shetlander

Market House, 14 Market Street, Lerwick ZE1 0JP
☎ 01595 743902 📠 01595 696787
scss@shetland.org
www.shetlandcss.co.uk
Owner *Shetland Council of Social Service*
Editors *Brian Smith, Laureen Johnson*
Circulation 1,400

Founded 1947. QUARTERLY literary magazine containing short stories, essays, poetry, historical articles, literary criticism, political comment, arts and books. The magazine has two editors and an editorial committee who all look at submitted material. Interested in considering

short stories, poetry, historical articles with a northern Scottish or Scandinavian flavour, literary pieces and articles on Shetland. As a rough guide, items should be between 1,000 and 2,000 words although longer mss are considered. Initial approach in writing, please.
Ⓔ Complimentary copy.

New Statesman

3rd Floor, 52 Grosvenor Gardens, London SW1W 0AU
☎ 020 7730 3444   🖷 020 7259 0181
info@newstatesman.co.uk
www.newstatesman.com
Publisher *Spencer Neal*
Editor *Jason Cowley*
Circulation 26,208

WEEKLY magazine, the result of a merger (1988) of *New Statesman* and *New Society*. Coverage of news, book reviews, arts, current affairs, politics and social reportage. Unsolicited contributions with s.a.e. will be considered. No short stories.
Arts Editor *Alice O'Keefe*
Books Editor *Ian Irvine*

New Theatre Quarterly

Oldstairs, Kingsdown, Deal CT14 8ES
☎ 01304 373448
simontrussler@btinternet.com
www.uk.cambridge.org
Publisher *Cambridge University Press*
Editors *Simon Trussler, Maria Shevtsova*

Founded 1985 (originally launched in 1971 as *Theatre Quarterly*). Articles, interviews, documentation and reference material covering all aspects of live theatre. Recommend preliminary e-mail enquiry before sending contributions. No theatre reviews or anecdotal material. Online submission preferred.

New Welsh Review

PO Box 170, Aberystwyth SY23 1WZ
☎ 01970 628410   🖷 01970 628410
editor@newwelshreview.com
www.newwelshreview.com
Owner *New Welsh Review Ltd*
Editor *Kathryn Gray*
Circulation 3,500

Founded 1988. QUARTERLY Welsh literary magazine in the English language. Welcomes material of literary and cultural relevance to Welsh readers and those with an interest in Wales. Approach in writing in the first instance.
FEATURES Max. 3,000 words.
FICTION Max. 3,000 words.
REVIEWS Max. 800 words.
Ⓔ Average of £165 (features); £80 (fiction); £45 (reviews); £27 per poem.

The New Writer

PO Box 60, Cranbrook TN17 2ZR
☎ 01580 212626
editor@thenewwriter.com
www.thenewwriter.com
Publisher *Merric Davidson*
Editor *Suzanne Ruthven*

Founded 1996. Published BI-MONTHLY following the merger between *Acclaim* and *Quartos* magazines. Available by subscription only, TNW continues to offer practical 'nuts and bolts' advice on creative writing but with the emphasis on *forward-looking* articles and features on all aspects of the written word that demonstrate the writer's grasp of contemporary writing and current editorial/publishing policies. Plenty of news, views, competitions, reviews; writers' guidelines available with s.a.e. Monthly e-mail news bulletin is included free of charge in the subscription package.
FEATURES Unsolicited mss welcome. Interested in lively, original articles on writing in its broadest sense. Approach with ideas in writing in the first instance. No material is returned unless accompanied by s.a.e.
FICTION Publishes short-listed entries from competitions and subscriber-only submissions.
POETRY Unsolicited poetry welcome. Both short and long unpublished poems, providing they are original and interesting.
Ⓔ Features, £20 per 1,000 words; fiction, £10 per story; poetry, £3 per poem.
Full guidelines and details of annual Prose & Poetry Prizes at the website.

New Writing Scotland

Association for Scottish Literary Studies, c/o Department of Scottish Literature, 7 University Gardens, University of Glasgow G12 8QH
☎ 0141 330 5309   🖷 0141 330 5309
nws@asls.org.uk
www.asls.org.uk
Contact *Duncan Jones*

ANNUAL anthology of contemporary poetry and prose in English, Gaelic and Scots, produced by the **Association for Scottish Literary Studies** (see entry under *Professional Associations and Societies*). Will consider poetry, drama, short fiction or other creative prose but not full-length plays or novels, though self-contained extracts are acceptable. Contributors should be Scottish by birth or upbringing, or resident in Scotland. Max. length of 3,500 words is suggested. Send no more than one short story and no more than four poems. Submissions should be accompanied by two s.a.e.s (one for receipt, the other for return of mss). Mss, which must be sent by 30 September, should be typed, double-spaced, on one side of the paper only with the sheets secured at top left-hand corner. Provide covering letter with full contact details but do not put name or address on individual work(s). Prose pieces should carry an approximate word count. Ⓔ £20 per printed page.

newbooks

4 Froxfield Close, Winchester SO22 6JW
guy@newbooksmag.com
www.newbooksmag.com
Owner/Editor *Guy Pringle*
Circulation 80,000

Founded 2000. BI-MONTHLY magazine for readers and reading groups with extracts from the best new fiction and free copies to be claimed. No unsolicited contributions. Also publishes *tBkmag* for 8–12-year-olds.

The North
▷ under Poetry Magazines

Notes from the Underground

23 Sutherland Square, London SE17 3EQ
Ⓣ 020 7701 2777
info@notesfromtheunderground.co.uk
www.notesfromtheunderground.co.uk
Joint Editors *Christopher Vernon, Tristan Summerscale*

Free creative writing paper, aimed at bridging the gap between niche literary journals and those free newspapers that have a low quality of editorial content. Publishes short stories, cartoons and original illustrations as well as arts-based non-fiction. A print run of 100,000 copies. All the content from the print edition is available online along with web exclusive stories, videos and podcasts. Contributor information available on the website.

Now

IPC Media Ltd., Blue Fin Building, 110 Southwark Street, London SE1 0SU
Ⓣ 020 3148 6373
Owner *IPC Media (A Time Warner Company)*
Editor *Abigail Blackburn*
Circulation 470,290

Founded 1996. WEEKLY magazine of celebrity gossip, news and topical features aimed at the working woman. Unlikely to use freelance contributions due to specialist content – e.g. exclusive showbiz interviews – but ideas will be considered. Approach in writing; no faxes.

Nursery Education

Scholastic Ltd, Villiers House, Clarendon Avenue, Leamington Spa CV32 5PR
Ⓣ 01926 887799   Ⓕ 01926 883331
earlyyears@scholastic.co.uk
www.scholastic.co.uk
Owner *Scholastic Ltd*
Publishing Director *Helen Freeman*

Circulation 14,506

Founded 1997. MONTHLY print and online magazine for early years professionals. Welcomes contributions from freelance journalists specializing in education (print and online).
FEATURES Early years issues explored. Max. 1,500 words.
NEWS Relevant and timely news for the early years sector. Max. 400 words.
PRACTICAL IDEAS Themed activities that support the Foundation Stage curriculum and Birth-to-Three activites. Max. 800 words. Approach by letter or e-mail.

Nursing Times

Greater London House, Hampstead Road, London NW1 7EJ
Ⓣ 020 7728 3702   Ⓕ 020 7728 3700
nt@emap.com
www.nursingtimes.net
Owner *Emap Public Sector*
Editor *Rachel Downey*
Circulation 48,388

Some of *Nursing Times*' feature content is from unsolicited contributions sent on spec. Articles on all aspects of nursing and health care, including clinical information, written in a contemporary style, are welcome.

Nuts

IPC Media Ltd., Blue Fin Building, 110 Southwark Street, London SE1 0SU
Ⓣ 020 3148 5000
nutsmagazine@ipcmedia.com
www.nuts.co.uk
Owner *IPC Media (A Time Warner Company)*
Editor *Dominic Smith*
Circulation 250,061

Launched 2004. WEEKLY magazine for men aged 16–45 'who like chat, women, gossip, cars and football'. Welcomes true life stories and items on gadgets. Approach by e-mail in the first instance.

OK! Magazine

The Northern & Shell Building, 10 Lower Thames Street, London EC3R 6EN
Ⓣ 0871 434 1010   Ⓕ 0871 434 7305
firstname.lastname@express.co.uk
Owner *Northern & Shell Media/Richard Desmond*
Editor *Lisa Byrne*
Circulation 683,451

Founded 1996. WEEKLY celebrity-based magazine. Welcomes interviews and pictures on well known personalities, and ideas for general features. Approach the editor by phone or fax in the first instance.

Old Glory

Mortons Heritage Media, Mortons Way,
Horncastle LN9 6JR
☎ 01507 529306   ✆ 01507 529495
editor@oldglory.co.uk
www.oldglory.co.uk
Owner *Mortons Media Ltd*
Editor *Colin Tyson*

Founded 1988. MONTHLY magazine covering all
aspects of transport and industrial heritage both
in the UK and overseas. Specializes in vintage
vehicles and steam preservation. No unsolicited
articles. Approach with idea by e-mail to the
editor.
FEATURES Restoration projects from steam rollers
to windmills. Max. 2,000 words.
NEWS Steam and transport rally reports. News
concerning industrial and heritage sites and
buildings.

The Oldie

65 Newman Street, London W1T 3EG
☎ 020 7436 8801   ✆ 020 7436 8804
theoldie@theoldie.co.uk
www.theoldie.co.uk
Owner *Oldie Publications Ltd*
Editor *Richard Ingrams*
Circulation 30,000

Founded 1992. MONTHLY general interest
magazine with a strong humorous slant for the
older person. Submissions welcome; enclose s.a.e.
No poetry.

Olive

Media Centre, 201 Wood Lane, London W12 7TQ
☎ 020 8433 1402
firstname.surname@bbc.co.uk
www.olivemagazine.co.uk
Owner *BBC Worldwide Publishing Ltd*
Editorial Coordinator *Danielle Theunissen*
Deputy Editor *Lulu Grimes*
Features Editor *Jessica Gunn (jess.gunn@bbc.
co.uk)*
Food Editor *Janine Ratcliffe*
Travel Editor *Alison Bowles*
Circulation 85,196

Launched 2003. MONTHLY food magazine.
Recipes, restaurants and food lovers' travel.
Unsolicited pitches are welcome 'but please
note that we will reply *only* to those we plan to
commission'.

Opera

36 Black Lion Lane, London W6 9BE
☎ 020 8563 8893   ✆ 020 8563 8635
editor@opera.co.uk
www.opera.co.uk
Owner *Opera Magazine Ltd*
Editor *John Allison*

Circulation 11,500

Founded 1950. MONTHLY review of the current
opera scene. Almost all articles are commissioned
and unsolicited mss are not welcome. All
approaches should be made in writing.

Opera Now

241 Shaftesbury Avenue, London WC2H 8TF
☎ 020 7333 1740   ✆ 020 7333 1769
opera.now@rhinegold.co.uk
www.rhinegold.co.uk
Publisher *Rhinegold Publishing Ltd*
Editor-in-Chief *Ashutosh Khandekar*
Deputy Editor *Antonia Couling*
Assistant Editor *Hannah Beynon*

Founded 1989. BI-MONTHLY. News, features and
reviews aimed at those involved as well as those
interested in opera. No unsolicited mss. All work
is commissioned. Approach with ideas in writing.

Organic Gardening

Mortons Media Group, Media Centre, Morton
Way, Horncastle LN9 6JR
slawson@mortons.co.uk
Editor *Sarah Lawson*
Circulation 20,000

Founded 1988. MONTHLY. Articles and features
on all aspects of gardening based on organic
methods. Unsolicited material welcome;
800–2,000 words for features and 100–300 for
news items. Poetry not published. Prefers 'hands-
on' accounts of projects, problems, challenges and
how they are dealt with. Approach in writing.
💷 Payment by arrangement.

OS (Office Secretary) Magazine

15 Grangers Place, Witney OX28 4BS
☎ 01993 775545   ✆ 01993 778884
paul.ormond@peeblesmedia.com
www.peeblesmedia.com
Owner *Peebles Media Group Ltd*
Editor *Paul Ormond*
Circulation 24,977

Founded 1986. BI-MONTHLY. Features articles
of interest to secretaries and personal assistants
aged 25–60. No unsolicited mss.
FEATURES Informative pieces on technology and
practices, office and employment-related topics.
Length 1,000 words. Approach with ideas by
telephone or in writing.
💷 Payment by negotiation.

Out & About

Rouncy Media Ltd, 3 The Office Village, Forder
Way, Cygnet Park, Hampton, Peterborough
PE7 8GX
☎ 01733 293240   ✆ 0845 280 2927
bernard.horton@rouncymedia.co.uk
Owner *Rouncy Media Ltd*

Editorial Director *Andrew Sutcliffe*
Editor *Benard Horton*
Circulation 4,000

WEEKLY magazine, aimed at coach and bus operators. Interested in coach and bus industry-related items only: business, financial and legal. Contact by letter, e-mail, telephone or via website.

Park Home & Holiday Caravan

IPC Media Ltd, Leon House, 233 High Street, Croydon CR9 1HZ
☎ 020 8726 8252   🖷 020 8726 8299
alex_melvin@ipcmedia.com
www.phhc.co.uk
Owner *IPC Inspire*
Editor *Emma Bartlett*
Circulation 15,000

Founded 1960. THIRTEEN ISSUES YEARLY. News of parks, models, legislation, accessories, lifestyle for those living in residential park homes or owning holiday caravans. Welcomes material specifically related to park home and holiday caravan sites; no general touring features. Approach by e-mail.

PC Format

Future Publishing, 30 Monmouth Street, Bath BA1 2BW
☎ 01225 442244   🖷 01225 732295
pcfmail@futurenet.co.uk
www.pcformat.co.uk
Owner *Future Publishing*
Publisher *Stuart Anderton*
Editor *Alan Dexter*
Circulation 23,764

Founded 1991. FOUR-WEEKLY magazine covering performance hardware and gaming. Welcomes feature and interview ideas in the first instance; approach by e-mail only.

Pensions World

2 Addiscombe Road, Croydon CR9 5AF
☎ 020 8212 1946   🖷 020 8212 1920
stephanie.hawthorne@lexisnexis.co.uk
www.pensionsworld.co.uk
Owner *Reed Elsevier*
Editor *Stephanie Hawthorne*
Circulation 7,685

Launched 1972. MONTHLY magazine for pensions professionals. Pensions, law, investment and retirement issues. Specialist contributors only. No consumer pension stories. Approach by e-mail.
FEATURES Max. 1,500 words.
NEWS Max. 400 words.
💷 Payment negotiable.

People Management

Personnel Publications Ltd., 17–18 Britton Street, London EC1M 5TP
☎ 020 7324 2729   🖷 020 7324 2791
editorial@peoplemanagement.co.uk
www.peoplemanagement.co.uk
Editor *Rima Evans*
Circulation 128,904

FORTNIGHTLY magazine on human resources, industrial relations, employment issues, etc. Welcomes submissions but apply for 'Guidelines for Contributors' in the first instance.
FEATURES *Jane Pickard/Claire Warren*
NEWS *Anna Scott*
LAW AT WORK *Jill Evans*

People's Friend Pocket Novels

D.C. Thomson & Co. Ltd, 80 Kingsway East, Dundee DD4 8SL
☎ 01382 575539   🖷 01382 452491
smonks@dcthomson.co.uk
Owner *D.C. Thomson & Co. Ltd*
Editor *Sandra Monks*

Founded 1938. FORTNIGHTLY. Family and romantic stories of 50,000–55,000 words, aimed at the 30+ age group. Unsolicited contributions welcome. Phone, write or e-mail for guidelines; synopsis and first three chapters to be sent initially.

The People's Friend

80 Kingsway East, Dundee DD4 8SL
☎ 01382 223131   🖷 01382 452491
peoplesfriend@dcthomson.co.uk
Owner *D.C. Thomson & Co. Ltd*
Editor *Angela Gilchrist*
Circulation 330,093

The *Friend* is a fiction magazine, with two serials and several short stories each week. Founded in 1869, it has always prided itself on providing 'a good read for all the family'. Stories should be about ordinary, identifiable characters with the kind of problems the average reader can understand and sympathize with. 'We look for the romantic and emotional developments of characters, rather than an over-complicated or contrived plot. We regularly use period serials and, occasionally, mystery/adventure.' Guidelines on request with s.a.e.
SHORT STORIES Can vary in length from 1,000 words or less to 4,000.
SERIALS Serials of 8–12 instalments.
ARTICLES Short fillers welcome.
💷 Payment on acceptance.

Period Living

St Giles House, 50 Poland Street, London W1F 7AX
☎ 020 7970 4433   🖷 020 7970 4438
period.living@centaur.co.uk
Owner *Centaur*
Editor-in-Chief *Michael Holmes*
Editor *Sarah Whelan*

Circulation 52,834

Founded 1992. Formed from the merger of *Period Living* and *Traditional Homes*. Covers interior decoration in a period style, period house profiles, traditional crafts, renovation of period properties. Ⓔ Payment varies according to length/type of article.

The Philosopher

6 Craghall Dene Avenue, Newcastle upon Tyne NE3 1QR
m.c.bavidge@ncl.ac.uk
www.the-philosopher.co.uk
Owner *The Philosophical Society of England*
Editor *Martin Cohen*

Founded 1913. BIANNUAL journal of the Philosophical Society of Great Britain with an international readership made up of members, libraries and specialist booksellers. Wide range of philosophical interests, but leaning towards articles that present philosophical investigation which is relevant to the individual and to society in our modern era. Accessible to the non-specialist. Will consider articles and book reviews. 'Notes for Contributors' available; see website. As well as short philosophical papers, will accept:
NEWS about lectures, conventions, philosophy groups. Ethical issues in the news. Max. 1,000 words.
REVIEWS of philosophy books (max. 600 words); discussion articles of individual philosophers and their published works (max. 2,000 words).
MISCELLANEOUS items, including graphics, of philosophical interest and/or merit.
Ⓔ Free copies.

Piano

241 Shaftesbury Avenue, London WC2H 8TF
☎ 020 7333 1744   📠 020 7333 1769
piano@rhinegold.co.uk
www.rhinegold.co.uk
Owner *Rhinegold Publishing*
Editor *Jeremy Siepmann*
Deputy Editor *Jonathan Wikeley*
Circulation 11,000

Founded 1993. BI-MONTHLY magazine containing features, profiles, technical information, news, reviews of interest to those with a serious amateur or professional concern with pianos or their playing. No unsolicited material. Approach with ideas in writing only.

Picture Postcard Monthly

15 Debdale Lane, Keyworth, Nottingham NG12 5HT
☎ 0115 937 4079   📠 0115 937 6197
reflections@postcardcollecting.co.uk
www.postcardcollecting.co.uk

Owners *Brian & Mary Lund*
Editor *Brian Lund*
Circulation 4,000

Founded 1978. MONTHLY. News, views, clubs, diary of fairs, sales, auctions, and well-researched postcard-related articles. Might be interested in general articles supported by postcards. Unsolicited mss welcome. Approach by phone, e-mail or in writing with ideas.

Pilot

Archant Specialist, The Mill, Bearwalden Business Park, Wendens Ambo, Saffron Walden CB11 4GB
☎ 01799 544200   📠 01799 544201
nick.bloom@pilotweb.co.uk
www.pilotweb.co.uk
Publisher *Archant Specialist*
Editor *Nick Bloom*
Circulation 19,181

Founded 1966. MONTHLY magazine for private plane pilots. Much of the magazine is written by invitation from regular contributors. Unsolicited mss welcome but ideas in writing preferred. Non-pilot authors unlikley to be considered. Perusal of any issue of the magazine will reveal the type of material bought. See the website for advice to would-be contributors.
NEWS Contributions, preferably with photographs, need to be as short as possible. See *Pilot Notes* and *Old-Timers* in the magazine.
FEATURES Many articles are unsolicited personal experiences/travel accounts from pilots of private planes; good photo coverage is very important. Max. 5,000 words.

Pink Paper

Unit M, Spectrum House, 32–34 Gordon House Road, London NW5 1LP
☎ 020 7424 7400   📠 020 7424 7401
news@pinkpaper.com
www.pinkpaper.com
Owner *Millivres Prowler Group*
Editor *Tris Reid-Smith*
Circulation 55,000

Founded 1987. FORTNIGHTLY. 'Britain's biggest publication' for lesbians and gay men covering politics, social issues, health, the arts, celebrity interviews and all areas of concern to lesbian/gay people. Unsolicited mss welcome. Initial approach by post with an idea preferred. Interested in profiles, reviews, in-depth features and short news pieces.
Ⓔ Payment by arrangement.

Poetry Ireland Review
▷ under Poetry Magazines

Poetry Review
▷ under Poetry Magazines

Poetry Scotland
▷ under Poetry Magazines

Poetry Wales
▷ under Poetry Magazines

PONY

D.J. Murphy (Publishers) Ltd, Headley House,
Headley Road, Grayshott GU26 6TU
☎ 01428 601020   📠 01428 601030
djm@djmurphy.co.uk (text only)
www.ponymag.com
Owner *D.J. Murphy (Publishers) Ltd*
Editor *Janet Rising*
Assistant Editor *Penny Rendall*
Circulation 42,219

Founded 1948. Lively MONTHLY aimed at
10–16-year-olds. News, instruction on riding,
stable management, veterinary care, interviews.
Approach in writing with an idea.
FEATURES welcome. Max. 900 words.
NEWS Written in-house. Photographs and
illustrations (serious/cartoon) welcome.
💷 £65 per 1,000 words.

post plus
▷ Sunday Post under National Newspapers

PR Week

174 Hammersmith Road, London W6 7JP
☎ 020 8267 4429   📠 020 8267 4509
prweek@haymarket.com
www.prweek.com/uk
Owner *Haymarket Business Publications Ltd*
Editor *Daniel Rogers*
Circulation 16,078

Founded 1984. WEEKLY. Contributions accepted
from experienced journalists. Approach in
writing with an idea.
💷 Payment negotiable.

Practical Boat Owner

IPC Media Ltd, Westover House, West Quay
Road, Poole BH15 1JG
☎ 01202 440820   📠 01202 440860
pbo@ipcmedia.com
www.pbo.co.uk
Owner *IPC Media (A Time Warner Company)*
Editor *Sarah Norbury*
Circulation 48,202

Britain's biggest selling boating magazine.
MONTHLY magazine of practical information for
cruising boat owners. Receives about 1,500 mss
per year.
FEATURES Technical articles about maintenance,
restoration, modifications to cruising boats,
power and sail up to 45ft. European and British
regional pilotage articles and cruising guides.

Approach in writing with synopsis in the first
instance.
💷 Payment negotiable.

Practical Caravan

Teddington Studios, Broom Road, Teddington
TW11 9BE
☎ 020 8267 5629   📠 020 8267 5725
practical.caravan@haymarket.com
www.practicalcaravan.com
Owner *Haymarket Consumer Media*
Editor *Nigel Donnelly*
Circulation 45,158

Founded 1967. MONTHLY. Contains caravan
reviews, travel features, investigations, products,
park reviews. Unsolicited mss welcome on travel
relevant only to caravanning/touring vans. No
motorcaravan or static van stories. Approach with
ideas by phone, letter or e-mail.
FEATURES Must refer to caravanning. Written in
friendly, chatty manner. Pictures essential. Max.
length 2,000 words.
💷 Payment negotiable.

Practical Fishkeeping

Bauer Media, Bushfield House, Orton Centre,
Peterborough PE2 5UW
☎ 01733 237111   📠 01733 465820
www.practicalfishkeeping.co.uk
Owner *Bauer Media*
Editor *Karen Youngs*
Circulation 15,446

THIRTEEN ISSUES YEARLY. Practical articles
on all aspects of fishkeeping. Unsolicited mss
welcome; approach in writing with ideas. Quality
photographs of fish always welcome. No fiction
or verse.

Practical Land Rover World
▷ Land Rover World

Practical Parenting

Magicalia Ltd, Berwick House, 8–10 Knoll Rise,
Orpington BR6 0EL
☎ 01689 899284
Owner *Magicalia Ltd*
Editor *Susie Boone*
Circulation 35,616

Founded 1987. MONTHLY. Practical advice
on pregnancy, birth, babycare and childcare,
0–4 years. Submit ideas/synopses by e-mail.
Interested in feature articles of up to 1,500 words
in length, and in readers' experiences/personal
viewpoint pieces of between 750–1,000 words.
All material must be written for the magazine's
specifically targeted audience and in-house style.
Submissions to: *Daniella Delaney* (daniella.
delaney@magicalia.com)
💷 Payment negotiable.

Practical Photography

Bauer Media, Media House, Lynchwood, Peterborough PE2 6EA

☎ 01733 468000   📠 01733 468387

practical.photography@bauermedia.co.uk

www.photoanswers.co.uk

Owner *Bauer Media*

Editor-in-Chief *Andrew James*

Circulation 62,303

MONTHLY All types of photography, particularly technique-orientated pictures. No unsolicited mss. Preliminary approach may be made by e-mail. Always interested in new, especially unusual, ideas.

FEATURES Anything relevant to the world of photography, but 'not the sort of feature produced by staff writers. Features on technology, digital imaging techniques and humour are three areas worth exploring. Bear in mind that there is a three-month lead-in time.' Max. 2,000 words.

💷 Payment varies.

Practical Wireless

Arrowsmith Court, Station Approach, Broadstone BH18 8PW

☎ 0845 803 1979   📠 01202 659950

rob@pwpublishing.ltd.uk

www.pwpublishing.ltd.uk

Owner *PW Publishing Ltd*

Editor *Rob Mannion (G3XFD/EI5IW)*

Circulation 20,000

Founded 1932. MONTHLY. News and features relating to Amateur Radio, radio construction and radio communications. Author's guidelines available on request and are essential reading (send s.a.e.). Approach by phone or e-mail with ideas in the first instance. Only interested in hearing from people with an interest in and knowledge of Amateur Radio. Copy (on disk) should be supported where possible by artwork, either illustrations, diagrams or photographs.

💷 £54–70 per page.

Practical Woodworking

Myhobbystore Media, Berwick House, 8–10 Knoll Rise, Orpington BR6 0EL

☎ 08444 122262   📠 01689 899266

neilmead@freeuk.com

www.getwoodworking.com

Owner *Myhobbystore Media*

Editor *Neil Mead*

Circulation 7,000

Founded 1965. SIX ISSUES YEARLY. Contains articles relating to woodworking: projects, techniques, tips, etc. Unsolicited mss welcome. Approach with ideas in writing, by phone or e-mail.

FEATURES Projects, techniques, etc.

💷 Payment by arrangement.

The Practitioner

245 Blackfriars Road, London SE1 9UY

☎ 020 7921 8113

cshort@cmpmedica.com

www.pulseportfolio.com

Owner *CMP Mediac*

Editor *Corinne Short*

Circulation 35,656

Founded 1868. MONTHLY publication that keeps General Practitioners up to date on clinical issues. All articles are independently commissioned.

Prediction

IPC Focus Network, Leon House, 233 High Street, Croydon CR9 1HZ

☎ 020 8726 8000   📠 020 8726 8296

prediction@ipcmedia.com

www.predictionmagazine.co.uk

Owner *IPC Media (A Time Warner Company)*

Editor *Marion Williamson*

Circulation 13,718

Founded 1936. MONTHLY. Covering astrology and mind, body, spirit topics. Unsolicited material in these areas welcome (about 200–300 mss received every year). Writers' guidelines available on request.

ASTROLOGY Pieces should be practical and of general interest.

FEATURES Articles on divination, shamanism, alternative healing, psychics and other supernatural phenonema considered. Please read a recent copy of the magazine before sending unsolicited material.

Pregnancy & Birth

Endeavour House, 189 Shaftebury Avenue, London WC2H 8JG

☎ 020 7295 5563

p.andb@bauermedia.co.uk

Owner *Bauer London Lifestyle*

Editor *Ellie Hughes*

Circulation 44,152

MONTHLY magazine covering all aspects of pregnancy from health to fashion. Regularly commissions features from health journalists. Freelancers should approach by e-mail in the first instance.

💷 Payment varies.

Pregnancy, Baby & You

Magicalia Publishing Ltd, Berwick House, 8–10 Knoll Rise, Orpington BR6 0EL

☎ 01689 899284

emma.hartfield@magicalia.com

Owner *Magicalia Publishing Ltd*

Editor *Emma Hartfield*

Deputy Editor *Susannah Parker*

Circulation 18,951

THIRTEEN ISSUES YEARLY focusing on issues affecting health and emotional issues during pregnancy. No unsolicited mss; e-mail the editor with ideas only.

Press Gazette

Paulton House, 8 Shepherdess Walk, London N1 7LB

☎ 020 7324 2385   🖷 020 7566 5769
pged@pressgazette.co.uk
www.pressgazette.co.uk
Owner *Wilmington Business Information*
Editor *Dominic Ponsford*
Circulation 5,010

MONTHLY multi-media magazine and news website for all journalists in regional and national newspapers, magazines, broadcasting and online, containing news, features and analysis of all areas of journalism, print and broadcasting. Unsolicited mss welcome; interested in profiles of magazines, broadcasting companies and news agencies, personality profiles, technical and current affairs relating to the world of journalism. Also welcomes gossip for diary page. Approach with ideas by phone, e-mail, fax or post.

Pride

Pride House, 55 Battersea Bridge Road, London SW11 3AX

☎ 020 7228 3110   🖷 020 7801 6717
info@pridemagazine.com
Owner/Editor *Carl Cushnie Junior*
Circulation 40,000

Founded 1991. MONTHLY lifestyle magazine for black women with features, beauty, arts and fashion. Approach in writing or by e-mail with ideas.
FEATURES Issues pertaining to the black community. 'Ideas and solicited mss are welcomed from new freelancers.' Max. 2,000 words.
FASHION & BEAUTY *Shevelle Rhule* Freelancers used for short features. Max. 1,000 words.

Prima

National Magazine House, 72 Broadwick Street, London W1F 9EP

☎ 020 7439 5000
prima@natmags.co.uk
www.natmags.co.uk
Owner *National Magazine Company*
Editor *Maire Fahey*
Circulation 282,544

Founded 1986. MONTHLY women's magazine.
HEALTH & FEATURES Coordinator *Karen Swayn* Mostly practical and real life, written

by specialists or commissioned from known freelancers. Unsolicited mss not welcome.

Private Eye

6 Carlisle Street, London W1D 3BN

☎ 020 7437 4017   🖷 020 7437 0705
strobes@private-eye.co.uk
www.private-eye.co.uk
Owner *Pressdram*
Editor *Ian Hislop*
Circulation 205,231

Founded 1961. FORTNIGHTLY satirical and investigative magazine. Prospective contributors are best advised to approach the editor in writing. News stories and feature ideas are always welcome, as are cartoons.

Prospect

2 Bloomsbury Place, London WC1A 2QA

☎ 020 7255 1344 (editorial)/1281 (publishing)
🖷 020 7255 1279
editorial@prospect-magazine.co.uk
publishing@prospect-magazine.co.uk
www.prospect-magazine.co.uk
Owner *Prospect Publishing Limited*
Editor *David Goodhart*
Circulation 26,767

Founded 1995. MONTHLY. Essays, reviews, short fiction and research on current/international affairs and cultural issues. No news features. Unsolicited contributions welcome, although more useful to approach in writing with ideas in the first instance.

Psychic News

The Coach House, Stansted Hall, Stansted CM24 8UD

☎ 01279 817050   🖷 01279 817051
pnadverts@btconnect.com
www.psychicnewsbookshop.co.uk
Owner *Psychic Press 1995 Ltd*
Editor *Susan Farrow*
Circulation 40,000

Founded 1932. *Psychic News* is the world's only WEEKLY spiritualist newspaper. It covers subjects such as psychic research, hauntings, ghosts, poltergeists, spiritual healing, survival after death and paranormal gifts. Unsolicited material considered. 🖷 No payment.

Psychologies

64 North Row, London W1K 7LL

☎ 020 7150 7000
psychologies@hf-uk.com
www.psychologies.co.uk
Owner *Hachette Filipacchi UK*
Editor *Maureen Rice*
Features Editors *Rebecca Alexander (rebecca. alexander@hf-uk.com), Catherine Jones*
Deputy Editor *Clare Longrigg*

Circulation 140,162

Founded 2005. MONTHLY. Women's lifestyle magazine covering relationships, family and parenting, health, beauty, social trends, travel, personality and behaviour, spirituality and sex. No fashion items. Consult the submissions guidelines on the website before submitting ideas for features.

Q

Mappin House, 4 Winsley Street, London W1W 8HF

☎ 020 7182 8482   ℻ 020 7182 8547
[firstname].[surname]@qthemusic.com
www.qthemusic.com
Owner *Bauer Media*
Editor-in-Chief *Paul Rees*
Senior Editor *Matt Mason*
Circulation 113,174

Founded 1986. MONTHLY. Glossy aimed at educated popular music enthusiasts of all ages. Few opportunities for freelance writers. Prospective contributors should approach in writing only to the relevant section editor.

Q-News, The Muslim Magazine

PO Box 4295, London W1A 7YH
☎ 0870 224 5908
info@q-news.com
www.q-news.com
Owner *Fuad Nahdi*
Editor *Fareena Alam*
Circulation 40,000

Founded 1992. MONTHLY British Muslim community magazine concerned with shaping the debate about Muslims in Britain. Covers recruitment, news, opinion and reviews.
FEATURES *Abdul-Rehman Malik* 'Writers wishing to focus on areas of interest to our readership will gain access and credibility among the relevant people.' Max. 3,000 words.
OTHER *Fareena Alam* Analysis on news and current affairs – alternative rather than mainstream viewpoint preferred. Max. 2,000 words.
🖭 No payment.

Quality Fiction (QF)

c/o AllWriters' Workplace & Workshop, 234 Brook Street, Unit 2, Waukesha WI 53188, USA
qfsubmissions@yahoo.com
www.allwriters.org (click on QWF)

Quality Fiction magazine, formerly QWF (Quality Women's Fiction), is a semi-annual literary magazine published in January and July published through AllWriters' Workplace & Workshop in the US. 'Evoking emotion is the most important characteristic of a QF story. Any genre, written by any gender, about any situation, is welcome

as long as it is quality writing.' Submissions by e-mail only. Submit only during reading periods: September 1 to November 31 and March 1 to May 31. Submissions received outside of the reading period will be returned unread. For further information and detailed guidelines, please access the website or contact the editor at the e-mail address above.

Quarterly Review

26 Meadow Lane, Sudbury CO10 2TD
☎ 01507 339056
editor@quarterly-review.org
www.quarterly-review.org
Owner *Quarterly Review*
Editor *Derek Turner*

Founded 2007. A revival of the classic journal founded by Sir Walter Scott, Robert Southey and George Canning in 1809. QUARTERLY journal of politics, ideas and culture, with an emphasis on regionalism, ecology and socio-biology. Approach in writing in the first instance. Articles from 1,000 – 8,000 words.
🖭 No payment.

Quartos
▷ The New Writer

QWF Magazine
▷ Quality Fiction

RA Magazine (Royal Academy of Arts Magazine)

Royal Academy of Arts, Burlington House, Piccadilly, London W1J 0BD
☎ 020 7300 5820   ℻ 020 7300 8032
ramagazine@royalacademy.org.uk
www.ramagazine.org.uk
Owner *Royal Academy of Arts*
Editor *Sarah Greenberg*
Circulation 100,000

Founded 1983. QUARTERLY visual arts, architecture and culture magazine distributed to Friends of the RA, promoting exhibitions and publishing articles on art shows, books and news worldwide. 'There are opportunities for freelancers who can write about art in an accessible way. Most articles need to be tied in to forthcoming exhibitions, books and projects.' No unsolicited contributions. Approach by telephone or e-mail.

Racecar Engineering

IPC Media Ltd, Leon House, 233 High Street, Croydon CR9 1HZ
☎ 020 8726 8364   ℻ 0208 726 8398
racecar@ipcmedia.com
www.racecar-engineering.com
Owner *IPC Media (A Time Warner Company)*
Editor *Graham Jones*

Circulation 15,000

Founded 1990. MONTHLY. In-depth features on motorsport technology plus news and products. Interested in receiving news and features from freelancers.
FEATURES Informed insight into current motorsport technology. 3,000 words max.
NEWS New cars, products or business news relevant to motorsport. No items on road cars or racing drivers. 500 words max. Call or e-mail to discuss proposal.

Racing Post

1 Canada Square, Canary Wharf, London E14 5AP
☎ 020 7293 3000   🖷 020 7293 3758
editor@racingpost.com
www.racingpost.com
Owner *Centricomm Ltd*
Editor *Bruce Millington*
Circulation 61,893

Founded 1986. DAILY horse racing paper with some general sport. In 1998, following an agreement between the owners of *The Sporting Life* and the *Racing Post*, the two papers merged.

Radio Times

Media Centre, 201 Wood Lane, London W12 7TQ
☎ 020 8433 3400   🖷 020 8433 3160
radio.times@bbc.com
www.radiotimes.com
Owner *BBC Worldwide Limited*
Editor *Gill Hudson*
Circulation 1.05 million

WEEKLY. UK's leading broadcast listings magazine. The majority of material is provided by freelance and retained writers, but the topicality of the pieces means close consultation with editors is essential. Very unlikely to use unsolicited material. Detailed BBC, ITV, Channel 4, Channel 5 and satellite television and radio listings are accompanied by feature material relevant to the week's output.
🖸 Payment by arrangement.

Radio User

Arrowsmith Court, Station Approach, Broadstone BH18 8PW
☎ 0845 803 1979   🖷 01202 659950
roger@pwpublishing.ltd.uk
www.pwpublishing.ltd.uk
Owner *PW Publishing Ltd*
Editor *Roger Hall*
Circulation 19,000

Founded 1937. In 2006, *Shortwave Magazine* merged with *Radio Active* magazine to become *Radio User*. MONTHLY. Specialist electronics and radio enthusiasts' magazine covering all aspects of listening to radio signals. Features, news, regular columns and projects. Will consider contributions

but recommends that contact be made prior to submission to check publishing plans.

Rail

Brookfield House, Orton Centre, Peterborough PE2 5UW
☎ 01733 237111   🖷 01733 288163
www.rail-magazine.co.uk
Owner *Emap Active Ltd*
Managing Editor *Nigel Harris*
Circulation 21,369

Founded 1981. FORTNIGHTLY magazine dedicated to modern railway. News and features, and topical newsworthy events. Unsolicited mss welcome. Approach by phone with ideas. Not interested in personal journey reminiscences. No fiction.
FEATURES By arrangement with the editor. All modern railway British subjects considered. Max. 2,000 words.
NEWS Any news item welcome. Max. 500 words.
🖸 Features, varies/negotiable; news, up to £100 per 1,000 words.

Rail Express

Foursight Publications Ltd, 20 Park Street, King's Cliffe, Peterborough PE8 6XN
☎ 01780 470086   🖷 01780 470060
editor@railexpress.co.uk
www.railexpress.co.uk
Owner/Editor *P.A. Sutton*
Circulation c. 16,500

Launched 1996. MONTHLY glossy dealing with contemporary railway operations in the UK. Well researched and illustrated news (max. 500 words) and features (max. 4,000 words) that appeal to the core enthusiast readership are considered. Contact with brief in the first instance.

Railway Gazette International

NINE Sutton Court Road, Sutton SM1 4SZ
☎ 020 8652 5200   🖷 020 8652 5210
editor@railwaygazette.com
www.railwaygazette.com
Owner *DVV Media UK*
Editor *Chris Jackson*
News Editor *Andrew Grantham*
Industry Editor *Nick Kingsley*
Circulation 9,945

Founded 1835. MONTHLY magazine for senior railway, metro and tramway managers, engineers and suppliers worldwide. 'No material for railway enthusiast publications.' E-mail to discuss ideas in the first instance.

The Railway Magazine

IPC Media Ltd., Blue Fin Building, 110 Southwark Street, London SE1 0SU
☎ 020 3148 4683   🖷 020 3148 8521
railway@ipcmedia.com
www.ipcmedia.com

Owner *IPC Media (A Time Warner Company)*
Editor *Nick Pigott*
Circulation 34,661

Founded 1897. MONTHLY. Articles, photos and short news stories of a topical nature, covering modern railways, steam preservation and railway history, welcome. Max. 2,000 words, with sketch maps of routes, etc., where appropriate. Unsolicited mss welcome. No poetry.
Ⓔ Payment negotiable.

Random Acts of Writing

11 Alexander Place, Inverness IV3 5BX
randomactsofwriting@fsmail.net
www.randomactsofwriting.co.uk
Owners *Jennifer Thomson, Vikki Trelfer*
Editor *Jennifer Thomson*
Sub-editor *Vikki Trelfer*
Circulation 200

Founded 2005. THREE ISSUES YEARLY. Promotes the short story. 'Unsolicited short fiction (max. 3,000 words) welcomed from all, particularly new writers or those based in or writing about the north of Scotland. Will give advice and feedback.' Approach by e-mail. Ⓔ Complimentary copy of the magazine plus discounts on further copies.

Reader's Digest

11 Westferry Circus, Canary Wharf, London E14 4HE
Ⓣ 020 7715 8000   Ⓕ 020 7715 8716
www.readersdigest.co.uk
Owner *Reader's Digest Association Ltd*
Editor *To be appointed*
Circulation 638,693

Although in theory, a good market for general interest features of around 2,500 words very few are ever accepted. However, 'a tiny proportion' comes from freelance writers, all of which are specially commissioned. Opportunities exist for short humorous contributions.
Ⓔ Up to £100.

The Reader

Reader Office, 19 Abercromby Square, Liverpool L69 7ZG
Ⓣ 0151 794 2830
readers@thereader.org.uk
www.thereader.org.uk
Editor *Philip Davis*
Circulation 1,200

Founded 1997. QUARTERLY. Poetry, short fiction, literary articles and essays, thought, reviews, recommendations for individual readers and reading groups. Contributions from internationally lauded and new voices. Welcomes articles/essays about reading and about issues in modern life, max. 2,000 words. Recommendations for good reading, max. 1,000

words. Short stories, max. 2,500 words. Aims to reach serious readers in non-academic but thoughtful style. Approach in writing.
Ⓔ Payment negotiable.

Readers' Review Magazine
▷ The Self Publishing Magazine

Real People

33 Broadwick Street, London W1F 0DQ
Ⓣ 020 7439 5000
Owner *National Magazine Co.*
Editor *Samm Taylor*
Circulation 230,000

WEEKLY magazine of real-life stories in a popular women's weekly, first-person format. FEATURES Real-life stories for specific sections and tailored to the magazine. No fiction. Max. 1,800 words. Ⓔ Payment varies; £300+ for spread.

Record Collector

Room 101, The Perfume Factory, 140 Wales Farm Road, Acton, London W3 6UG
Ⓣ 020 8752 8080   Ⓕ 0870 732 6060
alan.lewis@metropolis.co.uk
www.recordcollectormag.com
Owner *Metropolis*
Editor *Alan Lewis*

Founded 1979. MONTHLY. Detailed, well-researched articles welcome on any aspect of record collecting or any collectable artist in the field of popular music (1950s to present day), with complete discographies where appropriate. Unsolicited mss welcome. Approach the editor with ideas by phone or e-mail.
Ⓔ Payment negotiable.

Red

64 North Row, London W1K 7LL
Ⓣ 020 7150 7000   Ⓕ 020 7150 7684
may.frost@redmagazine.co.uk
www.redmagazine.co.uk
Owner *Hachette Filipacchi UK Ltd*
Editor *Sam Baker*
Features Director *Lindsay Frankel*
Features & Lifesyle Assistant *May Frost*
Web Editor *Kate McGinley (kate.mcginley@ redmagazine.co.uk)*
Circulation 224,208

Founded 1998. MONTHLY magazine aimed at the 30-something woman. Will consider ideas sent in 'on spec' but tends to rely on regular contributors.

Report

ATL, 7 Northumberland Street, London WC2N 5RD
Ⓣ 020 7930 6441   Ⓕ 020 7782 1618
www.atl.org.uk
Owner *Association of Teachers and Lecturers*
Editor *Alex Tomlin*

Managing Editor *Victoria Poskitt*
Circulation 160,000

Founded 1978. NINE ISSUES YEARLY during academic terms. Contributions welcome. All submissions should go directly to the editor. Articles should be no more than 800 words and must be of practical interest to the classroom teacher and F.E. lecturers.

Retail Week

Emap, Greater London House, Hampstead Road, London NW1 7EJ
☎ 020 7728 5000
editorial@retail-week.com
www.retail-week.com
Owner *Emap*
Editor *Tim Danaher*
Circulation 10,197

Founded 1988. WEEKLY. Leading news publication for the multiple retail sector. No unsolicited contributions. Limited opportunities for both features and news. Approach by e-mail.
FEATURES *Charlotte Hardie* Max. 1,600 words.
NEWS *Jennifer Creevy*

Reveal

National Magazine Co., National Magazine House, 33 Broadwick Street, London W1F 0DQ
☎ 020 7339 4500   ℻ 020 7339 4529
michael@natmags.co.uk
www.natmags.co.uk
Owner *National Magazine Co.*
Editor *Michael Butcher*
Circulation 277,002

Launched 2004. WEEKLY glossy lifestyle magazine with celebrities, real-life stories, cookery, beauty and seven-day TV guide. No unsolicited contributions.

Rugby World

IPC Media Ltd., Blue Fin Building, 110 Southwark Street, London SE1 0SU
☎ 020 3148 4708
paul_morgan@ipcmedia.com
www.rugbyworld.com
Owner *IPC Media (A Time Warner Company)*
Editor *Paul Morgan*
Circulation 48,381

Founded 1960. MONTHLY. Features of special rugby interest only. Unsolicited contributions welcome but s.a.e. essential for return of material. Prior approach by phone or in writing preferred.

Runner's World

33 Broadwick Street, London W1F 0DQ
☎ 020 7339 4400   ℻ 020 7339 4220
rwedit@natmag-rodale.co.uk
www.runnersworld.co.uk
Owner *Natmag – Rodale Ltd*

Editor *Andy Dixon*
Circulation 88,567

Founded 1979. MONTHLY magazine giving practical advice on all areas of distance running including products and training, travel features, news and cross-training advice. Innovative ideas for training and nutrition features, as well as longer 'read' pieces, are welcome. Approach with ideas in writing in the first instance.

Running Fitness

14 Priestgate, Peterborough PE1 1JA
☎ 01733 557235
rf.ed@kelsey.co.uk
www.runningfitnessmag.com
Owner *Kelsey Publishing*
Editor *David Castle*
Circulation 25,000

Founded 1985. MONTHLY. Instructional articles on running, fitness, and lifestyle, plus running-related activities and health.
FEATURES Specialist knowledge an advantage. Opportunities are wide, but approach with ideas in first instance.
NEWS Opportunities for people stories, especially if backed up by photographs.

Saga Magazine

Saga Publishing Ltd, The Saga Building, Enbrook Park, Sandgate, Folkstone CT20 3SE
☎ 01303 771523   ℻ 01303 776699
editor@saga.co.uk
www.saga.co.uk
Owner *Saga Publishing Ltd*
Editor *Kate Bravery*
Circulation 653,657

Founded 1984. MONTHLY magazine that sets out to celebrate the role of older people in society, reflecting their achievements, promoting their skills, protecting their interests, and campaigning on their behalf. A warm personal approach, addressing the readership in an up-beat and positive manner. It has a hard core of celebrated commentators/writers (e.g. Keith Waterhouse, Katharine Whitehorn, Sally Brampton) as regular contributors. Articles mostly commissioned or written in-house but exclusive celebrity interviews welcome if appropriate/relevant. Length 1,000–1,200 words (max. 1,600). No short stories, travel, fiction or poems, please.

Sailing Today

Swanwick Marina, Lower Swanwick, Southampton SO31 1ZL
☎ 01489 585200   ℻ 01489 565054
duncan.kent@sailingtoday.co.uk
www.sailingtoday.co.uk
Owner *Edisey Ltd*
Editor *Duncan Kent*

Founded 1997. MONTHLY practical magazine for cruising sailors. *Sailing Today* covers owning and buying a boat, equipment and products for sailing and is about improving readers' skills, boat maintenance and product tests. Most articles are commissioned but will consider practical features and cruising stories with photos. Approach by telephone or in writing in the first instance.

Sainsbury's Magazine

20 Upper Ground, London SE1 9PD
☎ 020 7775 7775   🖷 020 7775 7705
edit@sainsburysmagazine.co.uk
Owner *Seven Publishing Ltd*
Editor *Sue Robinson*
Consultant Director *Delia Smith*
Circulation 308,175

Founded 1993. MONTHLY featuring a main core of food and cookery with features, health, beauty, travel, home and gardening. No unsolicited mss. Approach in writing with ideas only in the first instance.

The Salisbury Review

33 Canonbury Park South, London N1 2JW
☎ 020 7226 7791   🖷 020 7354 0383
salisburyreview@tiscali.co.uk
www.salisburyreview.co.uk
Managing Editor *Merrie Cave*
Consulting Editors *Roger Scruton, Myles Harris, Lord Charles Cecil, Mark Baillie, Christie Davies*
Literary Editor *Ian Crowther*
Circulation 1,700

Founded 1982. QUARTERLY magazine of conservative thought. Editorials and features from a right-wing viewpoint. Unsolicited material welcome. No fiction or poetry.
FEATURES Max. 2,000 words.
REVIEWS Max. 1,000 words.
£ Small payment.

Scotland in Trust

91 East London Street, Edinburgh EH7 4BQ
☎ 0131 556 2220   🖷 0131 556 3300
editor@scotlandintrust.co.uk
www.scotlandintrust.co.uk
Owner *National Trust for Scotland*
Editor *Iain Gale*
Contact *Neil Braidwood*
Circulation 179,082

Founded 1983. THREE ISSUES YEARLY. Membership magazine of the National Trust for Scotland. Contains heritage/conservation features relating to the Trust, their properties and work. No unsolicited mss.

Scotland Means Business
▷ **Daily Record under National Newspapers**

The Scots Magazine

D.C. Thomson & Co., 2 Albert Square, Dundee DD1 9QJ
☎ 01382 223131
mail@scotsmagazine.com
www.scotsmagazine.com
Owner *D.C. Thomson & Co. Ltd*
Editor *John Methven*
Circulation 37,632

Founded 1739. MONTHLY. Covers a wide field of Scottish interests ranging from personalities to wildlife, climbing, reminiscence, history and folklore. Outside contributions welcome; 'staff delighted to discuss in advance by letter or e-mail'.
EVENTS LISTING *Christina Dolan*
PHOTOGRAPHY/MUSIC *Ian Neilson*
BOOKS *Alison Cook*

The Scottish Farmer

Newsquest Magazines, 200 Renfield Street, Glasgow G2 3QB
☎ 0141 302 7727   🖷 0141 302 7799
ken.fletcher@thescottishfarmer.co.uk
www.thescottishfarmer.co.uk
Owner *Newsquest Magazines*
Editor *Alasdair Fletcher*
Circulation 19,223

Founded 1893. WEEKLY. Farmer's magazine covering most aspects of Scottish agriculture. Unsolicited mss welcome. Approach with ideas in writing, by fax or e-mail.
FEATURES *Ken Fletcher* Technical articles on agriculture or farming units. 1,000–2,000 words.
NEWS *Ken Fletcher* Factual news about farming developments, political, personal and technological. Max. 800 words.
WEEKEND FAMILY PAGES Rural and craft topics.

Scottish Field

Special Publications, Craigcrook Castle, Craigcrook Road, Edinburgh EH4 3PE
☎ 0131 312 4550   🖷 0131 312 4551
editor@scottishfield.co.uk
www.scottishfield.co.uk
Owner *Wyvex Media Ltd*
Editor *Archie Mackenzie*
Circulation 14,106

Founded 1903. MONTHLY. Scotland's quality lifestyle magazine. Unsolicited mss welcome but writers should study the magazine first.
FEATURES Articles of general interest on Scotland and Scots abroad with good photographs or, preferably, e-mailed images at 300 DPI. Approx. 1,000 words.
£ Payment negotiable.

Scottish Home & Country

42 Heriot Row, Edinburgh EH3 6ES
☎ 0131 225 1724   🖷 0131 225 8129

magazine@swri.demon.co.uk
www.swri.org.uk
Owner *Scottish Women's Rural Institutes*
Editor *Liz Ferguson*
Circulation 11,000

Founded 1924. MONTHLY. Scottish or rural-related issues, health, travel, women's issues and general interest. Unsolicited mss welcome. Commissions are rare and tend to go to established contributors only.

Scouting Magazine

Gilwell House, Gilwell Park, Chingford E4 7QW
📞 020 8433 7100   📠 020 8433 7103
scouting.magazine@scout.org.uk
www.scouts.org.uk
Owner *The Scout Association*
Editors *Chris James, Hilary Galloway, Ellis Matthews*
Circulation 80,000

BI-MONTHLY magazine for adults connected to or interested in the Scout Movement. Interested in Scouting-related submissions only. Sent to Members of the Association; not available on subscription
💷 Payment by negotiation.

Screen

Gilmorehill Centre for Theatre, Film and Television, University of Glasgow, Glasgow G12 8QQ
📞 0141 330 5035   📠 0141 330 3515
screen@arts.gla.ac.uk
www.screen.arts.gla.ac.uk
screen.oxfordjournals.org
Publisher *Oxford University Press*
Editorial Office *Caroline Beven, Heather Middleton*
Editors *Annette Kuhn, John Caughie, Simon Frith, Karen Lury, Jackie Stacey, Sarah Street*

QUARTERLY academic journal of film and television studies for a readership ranging from undergraduates to screen studies academics and media professionals. There are no specific qualifications for acceptance of articles. Straightforward film reviews are not normally published. Check the magazine's style and market in the first instance.

Screen International

Greater London House, Hampstead Road, London NW1 7EJ
📞 020 7728 5605   📠 020 7728 5555
michael.gubbins@emap.com
www.screendaily.com
Owner *Emap*
Editor *Michael Gubbins*

Circulation 6,828

'The international voice of the international film industry.' Expert freelance writers used in all areas. No unsolicited mss. Approach by e-mail.
NEWS *Eleanor Kenny*
FEATURES *Louise Tutt, Sarah Cooper*
REVIEWS Editor *Finn Halligan*
BOX OFFICE *Jack Warner*
💷 Negotiable on NUJ basis.

Screentrade Magazine

Screentrade Media Ltd, PO Box 144, Orpington BR6 6LZ
📞 01689 833117   📠 01689 833117
philip@screentrademagazine.com
philip45other@yahoo.co.uk
www.screentrademagazine.com
Owner *Screentrade Media Ltd*
Editor *Philip Turner*
US Representative, Screentrade Media Ltd
Pamala Stanton (001 616 847 0144/616 405 2643; pam@screentrademagazine.com)
Circulation 3,000–5,000

Founded 2000. QUARTERLY journal for British, European and North American exhibitors and film distributors. Now also serving Russia, China, India and Pacific Rim territories.
FEATURES Dedicated sections on the following: cinema management, digital cinema & alternative content, 3D cinema, cinema architecture, seating, point-of-sale/ticketing, concessions, screen technology, health & safety, nostalgia, showmanship, book reviews, box office figures and film columns, career tips and other managerial matters. Some film reviews but no 'film star' interviews. Most contributions are from within the exhibition industry. Items on the state of cinema exhibition, film distribution, cinema architecture, interviews with key industry personnel frequently undertaken. 1,500–3,000 words.
NEWS Topical items (if substantiated) welcome. Events coverage (e.g. festivals, premières) from an exhibitor's viewpoint preferred.

ScriptWriter Magazine

2 Elliott Square, London NW3 3SU
📞 020 7586 4853   📠 020 7586 4853
jonquil@twelvepoint.com
www.twelvepoint.com
Owner *Scriptease Ltd*
Managing Editor *Jonquil Florentin*
Editor *Julian Friedmann*
Circulation 1,500

Launched 2001. SIX ISSUES PER YEAR. Magazine for professional scriptwriters covering all aspects of the business and craft of writing for the small and large screen. Interested in serious, in-depth analysis; max. 1,500–3,500 words. E-mail with synopsis, sample material and c.v.

Sea Breezes

Mannin Media Group Ltd, Media House, Cronkbourne IM4 4SB
☎ 01624 696573 📠 01624 661655
seabreezes@manninmedia.co.im
www.seabreezes.co.im
Owner *Print Centres*
Editor *Captain A.C. Douglas*
Circulation 17,000

Founded 1919. MONTHLY. Covers virtually everything relating to ships and seamen. Unsolicited mss welcome; they should be thoroughly researched and accompanied by relevant photographs. No fiction, poetry, or anything which 'smacks of the romance of the sea'. FEATURES Factual tales of ships, seamen and the sea, Royal or Merchant Navy, sail or power, nautical history, shipping company histories, epic voyages, etc. Length 1,000–4,000 words. 'The most readily acceptable work will be that which shows it is clearly the result of first-hand experience or the product of extensive and accurate research.' 💷 £14 per page (about 800 words).

The Self Publishing Magazine (incorporating Readers' Review Magazine)

9 De Montfort Mews, Leicester LE1 7FW
☎ 0116 255 9311 📠 0116 255 9323
selfpublishing@troubador.co.uk
www.selfpublishingmagazine.co.uk
Owner *Troubador Publishing Ltd*
Editor *Jane Rowland*
Circulation 2,000

Launched 2006. THREE ISSUES YEARLY. Articles on self publishing. Submissions welcome on aspects of authors publishing their own books (design, production, editing, print, marketing). Approach by e-mail. Max. 2,000 words. BOOK REVIEWS Self-published books, only. Max. 300 words.

She Magazine

National Magazine House, 72 Broadwick Street, London W1F 9EP
☎ 020 7439 5000 📠 020 7312 3940
www.allaboutyou.com
Owner *National Magazine Company Ltd*
Editor *Sian Rees*
Circulation 173,377

Glossy MONTHLY for the 35-plus woman, addressing her needs as a modern individual, a parent and a homemaker. Talks to its readers in an intelligent, humorous and informative way. FEATURES Approach with feature ideas in writing or e-mail to Deputy Editor, *Carmen Bruegmann*

(carmen.bruegmann@natmags.co.uk). No unsolicited material.
💷 Payment negotiable.

Ships Monthly

IPC Inspire, 222 Branston Road, Burton-upon-Trent DE14 3BT
☎ 01283 542721 📠 01283 546436
shipsmonthly@ipcmedia.com
www.shipsmonthly.com
Owner *IPC Inspire*
Editor *Iain Wakefield*
Deputy Editor *Nicholas Leach*
Circulation 22,000

Founded 1966. MONTHLY A4 format magazine for ship enthusiasts and maritime professionals. News, photographs and illustrated articles on all kinds of ships – mercantile and naval, past and present. No yachting. Welcomes contributions on company histories, classic and modern ships, ports and harbours. Most articles are commissioned; prospective contributors should telephone or e-mail in the first instance.

Shooting and Conservation

BASC, Marford Mill, Rossett, Wrexham LL12 0HL
☎ 01244 573000 📠 01244 573001
jeffrey.olstead@basc.org.uk
www.basc.org.uk
Owner *The British Association for Shooting and Conservation (BASC)*
Editor *Jeffrey Olstead*
Circulation 128,000

SIX ISSUES PER YEAR. Good articles and stories on shooting, conservation and related areas may be considered although most material is produced in-house. Max. 1,500 words.
💷 Payment negotiable.

The Shooting Gazette

PO Box 225, Stamford PE9 2HS
☎ 01780 485350 📠 01780 485390
will_hetherington@ipcmedia.com
Owner *IPC Media (A Time Warner Company)*
Editor *Will Hetherington*
Circulation 17,200

Launched 1989. MONTHLY aimed at those who enjoy driven game shooting in the UK and around the world.
FEATURES *Will Hetherington* Welcomes articles directly related to interesting or unusual game shooting stories; max. 2,000 words. Approach by e-mail.
NEWS *Martin Puddifer* Max. 500 words.

Shooting Times & Country Magazine

IPC Media Ltd., Blue Fin Building, 110 Southwark Street, London SE1 0SU
☎ 020 3148 4741

steditorial@ipcmedia.com
www.shootingtimes.co.uk
Owner *IPC Media (A Time Warner Company)*
Editor *Camilla Clark*
Circulation 26,614

Founded 1882. WEEKLY. Covers shooting, fishing and related countryside topics. Unsolicited contributions considered.
Ⓔ Payment negotiable.

Shout Magazine

D.C. Thomson & Co., Albert Square, Dundee DD1 9QJ
☎ 01382 223131   ℻ 01382 200880
shout@dcthomson.co.uk
www.shoutmag.co.uk
Owner *D.C. Thomson Publishers*
Editor *Maria T. Welch*
Circulation 84,937

Founded 1993. FORTNIGHTLY. Pop music, quizzes, emotional, beauty, fashion, soap features.

Shout!

PO Box YR46, Leeds LS9 6XG
☎ 0113 248 5700
shout.magazine@ntlworld.com
www.shoutweb.co.uk
Owner/Editor *Mark Michalowski*
Circulation 7,000

Founded 1995. MONTHLY lesbian/gay and bisexual news, views, arts and scene for Yorkshire; lgb health and politics. Interested in reviews of Yorkshire lgb events, happenings, news, analysis – 300 to 1,000 words max. No fiction, fashion or items with no reasonable relevance to Yorkshire and the north.
Ⓔ £40 per 1,000 words.

Showing World

Robin Aldwood Publications Ltd, PO Box 793, Needham IP6 8WN
☎ 01449 722505
sandy@showingworldonline.co.uk
www.showingworldonline.co.uk
Owner *Robin Aldwood Publications Ltd*
Editor *Sandy Wooderson*
Circulation 10,000

Founded 1991. SIX ISSUES YEARLY Features every aspect of showing horses and ponies plus natives, miniatures and donkeys. Knowledgeable and how-to-do-it articles welcome. Photos essential.
Ⓔ Payment negotiable.

Shropshire Magazine

Waterloo Road, Telford TF1 5HU
☎ 01952 241455   ℻ 01952 254605
Owner *Midland News Association*
Editor *Neil Thomas*

Founded 1950. MONTHLY.

FEATURES Personalities, topical items, historical (e.g. family) of Shropshire; also general interest: homes, weddings, antiques, etc. Max. 1,000 words. Unsolicited mss welcome but ideas in writing preferred.
Ⓔ Payment negotiable 'but modest'.

Sight & Sound

BFI, 21 Stephen Street, London W1T 1LN
☎ 020 7255 1444   ℻ 020 7436 2327
s&s@bfi.org.uk
www.bfi.org.uk/sightandsound
Owner *BFI*
Editor *Nick James*
Circulation 20,283

Founded 1932. MONTHLY. Topical and critical articles on international cinema, with regular columns from the USA and Europe. Length 1,000–5,000 words. Relevant photographs appreciated. Also book, film and DVD release reviews. Approach in writing with ideas.
Ⓔ Payment by arrangement.

The Skier and Snowboarder Magazine

Mountain Marketing Ltd., PO Box 386, Sevenoaks TN13 1AQ
☎ 0845 310 8303
frank.baldwin@skierandsnowboarder.co.uk
www.skierandsnowboarder.co.uk
Publisher *Mountain Marketing Ltd*
Editor *Frank Baldwin*
Circulation 30,000

SEASONAL (from July to May). FIVE ISSUES YEARLY. Outside contributions welcome.
FEATURES Various topics covered, including race and resort reports, fashion, equipment update, artificial slope, club news, new products, health and safety. Crisp, tight, informative copy of 800 words or less preferred.
NEWS All aspects of skiing news covered.
Ⓔ Payment negotiable.

Skymag

News Magazines Ltd, 2 Chelsea Manor Gardens, London SW3 5PN
☎ 020 7198 3000   ℻ 020 7198 3232
sky@newsmagazines.co.uk
www.sky.com/skymag
Owner *News Magazines*
Editor *Lysanne Currie*
Circulation 7.5 million

MONTHLY lifestyle and entertainment magazine. Contributions of celebrity interviews and real-life stories welcome. Send by e-mail to the deputy editor, *Nick Chalmers* (nick.chalmers@ newsmagazines.co.uk).

Slightly Foxed: The Real Reader's Quarterly

67 Dickinson Court, 15 Brewhouse Yard, London
EC1V 4JX
☎ 020 7549 2121/2111   🖷 0870 199 1245
all@foxedquarterly.com
www.foxedquarterly.com
Owner *Slightly Foxed Ltd*
Editors *Gail Pirkis, Hazel Wood*
Circulation 5,500

Founded 2004. 'Reviews of books (fiction and
non-fiction) that have stood the test of time or
books that have been published recently and are
of real quality but which have been overlooked
by reviewers and bookshops.' Unsolicited
contributions of 'lively, personal, idiosyncratic
writing of real quality' are welcome but it is
recommended that would-be contributors read
the magazine first to gauge its approach. Send
e-mail with sample work in the first instance. Not
interested in anything that is not actually a review
of a book or author.

Slim at Home

25 Phoenix Court, Hawkins Road, Colchester
CO2 8JY
☎ 01206 505987
laura@aceville.co.uk
www.slimathome.co.uk
Owner *Aceville Publications (2001) Ltd*
Editor *Naomi Abeykoon*
Circulation 110,000

Launched 2007. MONTHLY. National women's
magazine for dieters who want to lose weight in
their own way.
FEATURES *Naomi Abeykoon* Targeted weight-loss
articles must be highly researched and of use to
people who want to lose weight. No unrelated
health features, quick tips or reviews of products
and services. Approach by e-mail.
CELEBS *Naomi Abeykoon* Current celebrity
weight-loss stories commissioned. Pictures or
shoot must be available.

Smallholder

Editorial: Hook House, Wimblington March
PE15 0QL
☎ 01354 741538   🖷 01354 741182
liz.wright1@btconnect.com
www.smallholder.co.uk
Owner *Newsquest Plc*
Editor *Liz Wright*
Circulation 20,000

Founded 1982. MONTHLY. Outside contributions
welcome. Send for sample magazine and editorial
schedule before submitting anything. Follow up
with samples of work to the editor so that style
can be assessed for suitability. No poetry or
humorous, unfocused personal tales; no puzzles.
FEATURES New writers always welcome, but

must have high level of technical expertise – 'not
textbook stuff'. 'How to do it' articles with photos
welcome. Illustrations and photos paid for. Length
750–1,500 words.
NEWS All agricultural and rural news welcome.
Length 200–500 words.
Ⓒ Payment negotiable.

SmartLife International

25 Phoenix Court, Hawkins Road, Colchester
CO2 8JY
☎ 01206 505924   🖷 01206 505929
stuart@smartlifeint.com
www.smartlifeint.com
Owner *Aceville Publications Ltd*
Editor *Stuart Pritchard*
Circulation 18,933

Launched 2000. TEN ISSUES YEARLY. High-end
lifestyle magazine: consumer technology, interior
design, luxury cars, travel, fashion and grooming.
Welcomes proposals by e-mail.
FEATURES *Jake Stow* 'No good feature ideas that
fit the remit are ruled out.' Max. 2,000 words.
NEWS *Callum Fauser* Generally compiled in-
house.

Snooker Scene

Hayley Green Court, 130 Hagley Road, Hayley
Green, Halesowen, Birmingham B63 1DY
☎ 0121 585 9188   🖷 0121 585 7117
clive.everton@talk21.com
www.snookersceneonline.com
Owner *Everton's News Agency*
Editor *Clive Everton*
Circulation 8,000

Founded 1971. MONTHLY. No unsolicited mss.
Approach in writing with an idea.

Somerset Life

Archant House, Babbage Road, Totnes TQ9 5JA
☎ 01803 860910   🖷 01803 860926
nicki.lukehurst@archant.co.uk
www.somerset-life.co.uk
Owner *Archant Life*
Editor *Nicki Lukehurst*
Circulation 14,000

Launched 1994. MONTHLY magazine that
celebrates Somerset with coverage of towns,
villages, the county's history, property, antiques,
fashion, interior design, gardens and motoring.
FEATURES *Nicki Lukehurst* Prefers picture/word
packages, particularly on food and drink, celebrity
interviews, countryside, heritage, health and
beauty, people, business and the arts. 'Always on
the lookout for new voices (particularly if also a
competent photographer) on a regular or one-off
freelance basis.' Max. 1,500 words. Approach by
e-mail. Ⓒ £75 for a 1,000-word/picture package.

Spear's Wealth Management Survey

Luxury Publishing Ltd., 5 Jubilee Place, London
SW3 3TD
☎ 020 7591 2900   📠 020 7591 2929
info@luxurypublishing.com
www.luxurypublishing.com
Editor *William Cash*
CEO *Alan O'Sullivan*
Circulation 25,000

Launched 2005. QUARTERLY specialist guide for
'high and ultra-high net worth' individuals to all
aspects of wealth management and the super-rich
lifestyle. Subscription is by invitation, only. No
unsolicited contributions.

Speciality Food

25 Phoenix Court, Hawkins Road, Colchester
CO2 8JY
☎ 01206 505981
nicola.mallett@aceville.co.uk
www.specialityfoodmagazine.co.uk
Owner *Aceville Publications Ltd*
Editor *Nicola Mallett*
Circulation 8,699

Founded 2002. NINE ISSUES PER YEAR. Trade
magazine focusing on premium food and drink
relevant for the fine food retail sector. Unsolicited
contributions not generally welcome.
FEATURES Subjects such as cheese, dairy and
chocolate; also organic food. Approach by e-mail.
Max. 1,800 words. 💰 Payment negotiable.

The Spectator

22 Old Queen Street, London SW1H 9HP
☎ 020 7961 0200   📠 020 7961 0058
editor@spectator.co.uk
www.spectator.co.uk
Owner *The Spectator (1828) Ltd*
Editor *Matthew d'Ancona*
Books *Mark Amory*
Style Editor *Sarah Standing*
Business Editor *Martin Vander Weyer*
Art Editor *Liz Anderson*
Circulation 76,952

Founded 1828. WEEKLY political and literary
magazine. Prospective contributors should
write in the first instance to the relevant editor.
Unsolicited mss welcome, but no 'follow up'
phone calls, please.
💰 Payment nominal.

Spectrum

▷ Scotland on Sunday under National Newspapers

Staffordshire Life

The Publishing Centre, Derby Street, Stafford
ST16 2DT
☎ 01785 257700   📠 01785 253287
editor@staffordshirelife.co.uk
www.staffordshirelife.co.uk
Owner *Staffordshire Newspapers Ltd*
Editor *Louise Elliott*
Circulation 15,500

Founded 1982. MONTHLY. Full-colour county
magazine devoted to Staffordshire, its
surroundings and people. Contributions welcome.
Approach in writing with ideas.
FEATURES Max. 1,200 words.
💰 NUJ rates.

Stage

Stage House, 47 Bermondsey Street, London
SE1 3XT
☎ 020 7403 1818   📠 020 7939 8478
editor@thestage.co.uk
www.thestage.co.uk
Owner *The Stage Newspaper Ltd*
Editor *Brian Attwood*
Circulation 21,362

Founded 1880. WEEKLY. No unsolicited mss.
Prospective contributors should write with ideas
in the first instance.
FEATURES Preference for middle-market, tabloid-
style articles. 'Puff pieces', PR plugs and extended
production notes will not be considered. Max.
800 words. Profiles: 1,200 words.
NEWS News stories from outside London are
always welcome. Max. 300 words.
💰 £100 per 1,000 words.

Stamp Lover

National Philatelic Society, c/o The British Postal
Museum & Archive, Freeling House, Phoenix
Place, London WC1X 0DL
☎ 020 7239 2571
nps@ukphilately.org.uk
www.ukphilately.org.uk/nps
Owner *National Philatelic Society*
Editor *Michael L. Goodman*
Circulation 800

Founded 1908. SIX ISSUES YEARLY. Magazine of
the National Philatelic Society. Welcomes articles
about the hobby; stamps, old and new. Approach
by e-mail.
NEWS Information about worldwide philately.
Max. 1,000 words. 💰 No payment.

Stamp Magazine

IPC Media Ltd, Leon House, 233 High Street,
Croydon CR9 1HZ
☎ 020 8726 8243   📠 020 8726 8299
guy_thomas@ipcmedia.com
Owner *IPC Media (A Time Warner Company)*
Editor *Guy Thomas*
Circulation 12,000

Founded 1934. MONTHLY news and features on
the world of stamp collecting from the past to

the present day. Interested in articles by experts on particular countries or themes such as subject matter illustrated on stamps. Approach in writing.
NEWS *Julia Lee* News of latest stamp issues or industry news. Max. 500 words.
FEATURES *Guy Thomas* Any features welcome on famous stamps, rarities, postmarks, postal history, postcards, personal collections. Must be illustrated with colour images ('we can arrange for photography of original stamps').
Payment negotiable.

Steam Railway Magazine

Bushfield House, Peterborough PE2 5UW
01733 288026
steam.railway@bauermedia.co.uk
Owner *Bauer Media*
Editor *Danny Hopkins*
News Editor *Gary Boydhope*
Circulation 32,889

Founded 1979. FOUR-WEEKLY magazine targeted at all steam enthusiasts interested in the modern preservation movement. Unsolicited material welcome. News reports, photographs, steam-age reminiscences. Approach in writing or by e-mail.

Stella Magazine

> The Sunday Telegraph under National Newspapers

The Strad

Orpheus Publications, Newsquest Specialist Media Ltd, 2nd Floor, 30 Cannon Street, London EC4M 6YJ
020 7618 3095 020 7618 3483
thestrad@orpheuspublications.com
www.thestrad.com
Owner *Newsquest Specialist Media Group*
Editor *Ariane Todes*
Circulation 17,500

Founded 1890. MONTHLY for classical string musicians, makers and enthusiasts. Unsolicited mss accepted occasionally 'though acknowledgement/return not guaranteed'.
FEATURES Profiles of string players, teachers, luthiers and musical instruments, also relevant research. Max. 2,000 words.
REVIEWS *Matthew Rye*
£150 per 1,000 words.

Stuff

Teddington Studios, Broom Road, Teddington TW11 9BE
020 8267 5036
stuff@haymarket.com
www.stuff.tv
Owner *Haymarket Publishing Ltd*
Editor *Fraser Macdonald*

Circulation 96,866

Launched 1998. MONTHLY guide to new technology with information on all the latest gadgets and gear. Includes a 'Buyers' Guide' with information about the top tried and tested items. No unsolicited contributions.

Suffolk and Norfolk Life

The Publishing House, Framlingham IP13 9EE
01728 622030 01728 622031
Owner *Today Magazines Ltd*
Editor *Kevin Davis*
Circulation 17,000

Founded 1989. MONTHLY. General interest, local stories, historical, personalities, wine, travel, food. Unsolicited mss welcome. Approach by phone or in writing with ideas. Not interested in anything which does not relate specifically to East Anglia.
FEATURES *Kevin Davis* Max. 1,500 words, with photos.
NEWS *Kevin Davis* Max. 1,000 words, with photos.
SPECIAL PAGES *William Locks* Study the magazine for guidelines. Max. 1,500 words.
£45–£70.

Sugar Magazine

64 North Row, London W1K 7LL
020 7150 7087
www.sugarmagazine.co.uk
Owner *Hachette Filipacchi (UK)*
Editor *Annabel Brog*
Circulation 157,261

Founded 1994. MONTHLY. Everything that might interest the teenage girl. No unsolicited mss. Will consider ideas or contacts for real-life features. No fiction. Approach in writing in the first instance.

SuperBike Magazine

IPC Media Ltd, Leon House, 233 High Street, Croydon CR9 1HZ
020 8726 8455 020 8726 8499
kenny_pryde@ipcmedia.com
www.superbike.co.uk
Owner *IPC Media (A Time Warner Company)*
Publishing Director *Keith Foster*
Editor *Kenny Pryde*
Circulation 47,511

Founded 1977. MONTHLY. Dedicated to all that is best and most exciting in the world of motorcycling. Unsolicited mss, synopses and ideas by e-mail are welcome.

Surrey Life

Holmesdale House, 46 Croydon Road, Reigate RH2 0NH
01737 247188 01737 246596

editor@surreylife.co.uk
www.surreylife.co.uk
Owner *Archant Life*
Editor *Caroline Harrap*
Circulation 15,000

Launched 2002. MONTHLY county magazine
featuring the best property, food and drinks and
celebrity interviews. Unsolicited submissions of
Surrey-based feature ideas are welcome; send by
e-mail.

Sussex Life

Baskerville Place, 28 Teville Road, Worthing
BN11 1UG
☎ 01903 703730   📠 01903 703770
jonathan.keeble@archant.co.uk
www.susssexlife.co.uk
Owner *Archant Life*
Editor *Jonathan Keeble*
Circulation 11,000 (paid for)

Founded 1965. MONTHLY. Sussex and general
interest magazine. Regular supplements on
education, fashion, homes and gardens. Interested
in investigative, journalistic pieces relevant to
the area and celebrity profiles. Unsolicited mss,
synopses and ideas welcome; approach by e-mail.

Swimming Times Magazine

41 Granby Street, Loughborough LE11 3DU
☎ 01509 632230   📠 01509 632233
swimmingtimes@swimming.org
www.britishswimming.org
Owner *Amateur Swimming Association*
Editor *P. Hassall*
Circulation 20,000

Founded 1923. MONTHLY about competitive
swimming and associated subjects. Unsolicited
mss welcome.
FEATURES Technical articles on swimming, water
polo, diving or synchronized swimming. Length
and payment negotiable.

The Tablet

1 King Street Cloisters, Clifton Walk, London
W6 0GY
☎ 020 8748 8484   📠 020 8748 1550
thetablet@tablet.co.uk
www.thetablet.co.uk
Owner *The Tablet Publishing Co Ltd*
Publisher *Ignatius Kusiak*
Circulation 23,048

Founded 1840. WEEKLY. International Catholic
journal with an ecumenical outlook. Authoritative
articles on religion, politics, literature and the arts
by writers of international standing.

Take a Break

Academic House, 24–28 Oval Road, London
NW1 7DT
☎ 020 7241 8000
tab.features@bauer.co.uk
Owner *H. Bauer Publishing Ltd*
Editor *John Dale*
Circulation 981,504

Founded 1990. WEEKLY. True-life feature
magazine. Approach with ideas in writing.
NEWS/FEATURES Always on the look-out for
good, true-life stories. Max. 1,200 words.
FICTION Sharp, succinct stories which are
well told and often with a twist at the end.
All categories, provided it is relevant to the
magazine's style and market. Max. 1,000 words.
💷 Payment negotiable.

Take It Easy
▷ The People under National Newspapers

Tate Etc

Millbank, London SW1P 4RG
☎ 020 7887 8724   📠 020 7887 3940
tateetc@tate.org.uk
www.tate.org.uk
Owner *Tate*
Editorial Director *Bice Curiger*
Editor *Simon Grant*
Circulation 100,000

Relaunched May 2004. THREE ISSUES YEARLY.
Visual arts magazine aimed at a broad readership
with articles blending the historic and the
contemporary. Please send material by post.

Tatler

Vogue House, Hanover Square, London W1S 1JU
☎ 020 7499 9080   📠 020 7493 1641
www.tatler.co.uk
Owner *Condé Nast Publications Ltd*
Editor *Catherine Ostler*
Editor-at-Large *Vassi Chamberlain*
Features Editor *Ticky Hedley-Dent*
Editor's Assistant *Fiona Kent*
Features Associate *Richard Dennen*
Circulation 90,128

Up-market glossy from the Condé Nast stable.
New writers should send in copies of either
published work or unpublished material. The
magazine works largely on a commission basis:
they are unlikely to publish unsolicited features.

The Teacher

Hamilton House, Mabledon Place, London
WC1H 9BD
☎ 020 7380 4708   📠 020 7387 8458
teacher@nut.org.uk
www.teachers.org.uk
Owner *National Union of Teachers*

Editor *Elyssa Campbell-Barr*
Circulation 320,000

Journal of the National Union of Teachers, published EIGHT TIMES YEARLY. News, advice, information and special features on educational matters relating to classroom teaching. 'Please send a brief outline of any feature ideas for consideration.' Approach by letter or e-mail.

Tesco Magazine

85 Strand, London WC2R 0DW
☎ 020 7550 8900
tesco@cedarcom.co.uk
www.tesco.com/magazine
Owner *Cedar Communications*
Editor *Dawn Alford*
Circulation 2 million

Launched 2004. BI-MONTHLY glossy women's magazine aimed at 25–44-year-olds. Includes beauty, fashion, health, travel, real-life, entertainment features, practical ideas and tips. No unsolicited material.

TGO (The Great Outdoors)

Newsquest (Herald & Times) Magazines Ltd., 200 Renfield Street, Glasgow G2 3QB
☎ 0141 302 7700   ⊞ 0141 302 7799
cameron.mcneish@tgomagazine.co.uk
www.tgomagazine.co.uk
Owner *Newsquest*
Editor *Cameron McNeish*
Circulation 10,705

Founded 1978. MONTHLY. Deals with walking, backpacking and wild country topics. Unsolicited mss are welcome.
FEATURES Well-written and illustrated items on relevant topics. Max. 2,500 words. Quality high resolution digital colour images only, please.
NEWS Short topical items (or photographs). Max. 300 words.
£ Payment negotiable.

That's Life!

3rd Floor, Academic House, 24–28 Oval Road, London NW1 7DT
☎ 020 7241 8000   ⊞ 020 7241 8008
firstname.lastname@bauer.co.uk
www.bauer.co.uk
Owner *H. Bauer Publishing Ltd*
Editor *Jo Checkley*
Circulation 433,921

Founded 1995. WEEKLY. True-life stories, puzzles, health, homes, parenting, cookery and fun.
FEATURES *Tiffany Sherlock* Max. 1,600 words.

The Third Alternative
▷ Black Static

This England

PO Box 52, Cheltenham GL50 1YQ
☎ 01242 537900   ⊞ 01242 537901
Owner *This England Ltd*
Editor *Roy Faiers*
Circulation 140,000

Founded 1968. QUARTERLY, with a strong overseas readership. Celebration of England and all things English: famous people, natural beauty, towns and villages, history, traditions, customs and legends, crafts, etc. Generally a rural basis, with the 'Forgetmenots' section publishing readers' recollections and nostalgia. Up to a hundred unsolicited pieces received each week. Unsolicited mss/ideas welcome. Length 250–2,000 words.
£ £25 per 1,000 words.

Time

Blue Fin Building, 110 Southwark Street, London SE1 0SU
☎ 020 3148 3000
edit_office@timemagazine.com
www.time.com
Owner *Time Warner*
International Editor *Michael Elliott*
Circulation 136,164 (UK)

Founded 1923. WEEKLY current affairs and news magazine. There are few opportunities for freelancers on *Time* as almost all the magazine's content is written by staff members from various bureaux around the world. No unsolicited mss.

Time Out

Universal House, 251 Tottenham Court Road, London W1T 7AB
☎ 020 7813 3000   ⊞ 020 7813 6001
www.timeout.com
Publisher & Managing Director *Mark Elliott*
Editor *To be appointed*
Deputy Editor *Rachel Halliburton*
Managing Editor *Claire Hojem*
Circulation 76,356

Founded 1968. WEEKLY magazine of news, arts, entertainment and lifestyle in London plus listings.
FEATURES *Nina Caplan* 'Usually written by staff writers or commissioned, but it's always worth submitting an idea if particularly apt to the magazine.' Word length varies; up to 2,000 max.
CONSUMER SECTION Food and drink, shopping, services, travel, design, property, health and fitness.
£ Payment negotiable.

The Times Educational Supplement

26 Red Lion Square, Holborn, London WC1R 4HQ
☎ 020 3194 3000
⊞ 020 3194 3194 (editorial)/3194 3277 (newsdesk)

www.tes.co.uk
Owner *Charterhouse*
Editor *Karen Dempsey*
Circulation 57,714

Founded 1910. WEEKLY. New contributors are
welcome and should fax or e-mail ideas on one
sheet of A4 for news (newsdesk@tes.co.uk),
features (features@tes.co.uk) or reviews. The
main newspaper accepts contributions in the
following sections:
COMMENT Weekly slot for a well-informed and
cogently argued viewpoint. Max. 750 words.
(comment@tes.co.uk).
LEADERSHIP Practical issues for school governors
and managers. Max. 500 words.
FE FOCUS Weekly pull-out section covering
post-16 education and training in colleges, work
and the wider community. Aimed at everyone
from teachers/lecturers to leaders and opinion
formers in lifelong learning. News, features,
comment and opinion on all aspects of college life
welcome. Length from 350 words (news) to 1,000
max. (features). Contact *Alan Thomson* (alan.
thomson@tes.co.uk).
TES MAGAZINE Weekly magazine for teachers
focusing on looking at their lives, inside and
outside the classroom, investigating the key issues
of the day and highlighting good practice. Max.
800 words (features@tes.co.uk).

The Times Educational Supplement Scotland

Thistle House, 21–23 Thistle Street, Edinburgh
EH2 1DF
☎ 0131 624 8332   📠 0131 467 8019
scoted@tes.co.uk
www.tes.co.uk/scotland
Owner *TSL Education Ltd*
Editor *Neil Munro*
Circulation 7,062

Founded 1965. WEEKLY. Unsolicited mss welcome.
FEATURES Articles on education in Scotland.
Max. 1,000 words.
NEWS Items on education in Scotland. Max. 600
words.

The Times Higher Education

26 Red Lion Square, London WC1R 4HQ
☎ 020 3194 3000   📠 020 3194 3300
editor@tsleducation.com
www.tsleducation.co.uk
Owner *Charterhouse Private Equity*
Editor *Ann Mroz*
Circulation 21,140

Founded 1971. WEEKLY. Unsolicited items are
welcome but most articles and almost all book
reviews are commissioned. 'In most cases it is
better to write or e-mail, but in the case of news
stories it is all right to phone.'
FEATURES *Tariq Tahir* Most articles are

commissioned from academics in higher
education.
NEWS *Phil Baty* Freelance opportunities very
occasionally.
SCIENCE *Zoe Corbyn*
BOOKS *Karen Shook*
FOREIGN *John Gill*
💷 Payment by negotiation.

The Times Literary Supplement

Times House, 1 Pennington Street, London
E98 1BS
☎ 020 7782 5000   📠 020 7782 4966
www.the-tls.co.uk
Owner *The Times Literary Supplement Ltd*
Editor *Peter Stothard*
Circulation 33,768

Founded 1902. WEEKLY review of literature,
history, philosophy, science and the arts.
Contributors should approach in writing and be
familiar with the general level of writing in the
TLS.
LITERARY DISCOVERIES *Alan Jenkins*
POEMS *Mick Imlah*
NEWS Reviews and general articles concerned
with literature, publishing and new intellectual
developments anywhere in the world. Length by
arrangement.
💷 Payment by arrangement.

Today's Golfer

Media House, Lynchwood, Peterborough
Business Park, Peterborough PE2 6EA
☎ 01733 468000   📠 01733 468671
editorial@todaysgolfer.co.uk
www.todaysgolfer.co.uk
Owner *Bauer Media*
Editor-in-Chief *Andy Calton*
Associate Editor *Simon Penson*
Circulation 104,000

Founded 1988. MONTHLY. Golf instruction,
features, player profiles and news. Most features
written in-house but unsolicited mss will be
considered. Approach in writing with ideas. Not
interested in instruction material from outside
contributors.
FEATURES/NEWS *Kit Alexander* Opinion, player
profiles and general golf-related features.

Top of the Pops Magazine

Media Centre, 201 Wood Lane, London W12 7TQ
☎ 020 8433 3847
www.totpmag.com
Owner *BBC Worldwide Publishing Ltd*
Editor *Peter Hart*
Circulation 130,174

Founded 1995. FOUR-WEEKLY teenage celebrity
magazine with a lighthearted and humorous
approach. No unsolicited material. Freelance

contributions only after commissioning by the Editor.

Total Coarse Fishing

DHP Ltd, 2 Stephenson Close, Drayton Fields, Daventry NN11 8RF

☎ 01327 311999   📠 01327 311190

kevin.wilmot@dhpub.co.uk

www.tcfmagazine.com

Owner *David Hall Publishing Ltd*

Editor *Kevin Wilmot*

Circulation 35,000

Founded 2006. MONTHLY magazine for anglers who like to keep their fishing varied and target quality coarse fish from exceptional venues. No unsolicited contributions.

FEATURES Overseas freshwater angling features with 'exceptional' photographs. Max. 2,200 words. Approach by e-mail. 💷 £200.

Total Film

2 Balcombe Street, London NW1 6NW

☎ 020 7042 4000   📠 020 7042 4839

totalfilm@futurenet.com

Owner *Future Publishing*

Editor *Aubrey Day (aubrey.day@futurenet.com)*

Reviews *Matthew Leyland (matthew.leyland@futurenet.com)*

Features *Jamie Graham (jamie.graham@futurenet.com), Andy Lowe (andy.lowe@futurenet.com)*

News *Jonathan Dean (jonathan.dean@futurenet.com)*

Circulation 84,520

Founded 1997. MONTHLY reviews-based movie magazine. Interested in ideas for features, not necessarily tied in to specific releases, and humour items. No reviews or interviews with celebrities/directors. Approach by post or e-mail.

Total Flyfisher

DHP Ltd, 2 Stephenson Close, Drayton Fields, Daventry NN11 8RF

☎ 01327 311999   📠 01327 311190

steve.cullen@dhpub.co.uk

www.totalflyfisher.com

Owner *David Hall Publishing Ltd*

Editor *Steve Cullen*

Circulation 17,000

Founded 2003. MONTHLY instructional fly fishing magazine. Welcomes contributions including fishery reports and fly tying articles. No foreign features or advanced skills. Approach by e-mail.

FEATURES Stillwater trout tactics, salmon tactics, river trout/grayling tactics, UK saltwater tactics. Max. 2,500 words. 💷 Payment (including pictures) negotiable.

NEWS Stories relevant to UK fly fishers.

FICTION Funny fishing stories. Max. 400 words. 💷 Payment negotiable.

Total TV Guide

H. Bauer Publishing, Academic House, 24–28 Oval Road, London NW1 7DT

☎ 020 7241 8000   📠 020 7241 8042

barbara.miller@bauer.co.uk

www.bauer.co.uk

Owner *H. Bauer Publishing*

Editor *Jon Peake*

Circulation 92,004

Founded 2003. WEEKLY TV listings for those with access to multichannel television, covering everything from movies to sport, drama to children's programmes.

FEATURES *Ben Lawrence* Commissions features relating to current TV programmes; chiefly celebrity interviews and set visits. No unsolicited contributions.

Total Vauxhall

Future Publishing, 30 Monmouth Street, Bath BA1 2BW

☎ 01225 442244   📠 01225 446019

info@totalvauxhall.co.uk

www.totalvauxhall.co.uk

Owner *Future Publishing*

Editor *Barton Brisland*

Circulation 20,000

Founded 2001. MONTHLY independent newsstand magazine aimed at the Vauxhall enthusiast. Covers new, modified, historical/classic, race and rally Vauxhalls of all kinds. Substantial amount of technical content. Also covers the more interesting parts of the GM family, particularly Opel and Holden. Uses a lot of freelance contributors; 'those who hit deadlines and fulfil the brief get regular work and lots of it.' Feature ideas, news items, historical pieces and potential feature cars welcome but must have a Vauxhall/GM tilt. Call the editor directly for an informal discussion on style, approach and angle.
💷 Payment 'surprisingly generous'.

Trail

Media House, Lynchwood, Peterborough PE2 6EA

☎ 01733 468205

trail@bauer.co.uk

www.trailroutes.com

Owner *Bauer Consumer Media*

Editor *Matthew Swaine*

Circulation 39,349

Founded 1990. MONTHLY. Gear reports, where to walk and practical advice for the hill walker and long distance walker. Inspirational reads on people and outdoor/walking issues. Health, fitness and injury prevention for high level walkers and outdoor lovers. 'We always welcome

ideas from writers who understand our readers' needs. No travel pieces. We are interested in pieces that allow our readers to get out more, to enjoy their time outdoors and improve their skills. So don't tell us what *you've* done. Tell our readers what they could do, why they should do it and how best to do it. Inspiration and practical advice in equal measure.' Approach by phone or in writing in the first instance.
FEATURES *Simon Ingram* Very limited requirement for overseas articles, 'written to our style'. Ask for guidelines. Max. 2,000 words. Limited requirement for guided walks articles. Specialist writers only. Ask for guidelines. 750– 2,000 words (depending on subject).
ⓔ Payment varies.

Traveller

45–49 Brompton Road, London SW3 1DE
☎ 020 7589 0500 ℻ 020 7581 8476
traveller@wexas.com
www.traveller.org.uk
Owner *Wexas International Ltd*
Editor *Amy Sohanpaul*
Circulation 23,817

Founded 1970. QUARTERLY travel magazine.
FEATURES High quality, personal narratives of remarkable journeys. Articles should be off-beat, adventurous, authentic. For guidelines, see website. Articles may be accompanied by professional quality, original slides or high-res digital photographs. Freelance articles considered. Max. 1,000 words. Initial contact by e-mail.
ⓔ £200 per 1,000 words.

Tribune

9 Arkwright Road, London NW3 6AN
☎ 020 7433 6410 ℻ 020 7433 6410
mail@tribunemagazine.co.uk
www.tribunemagazine.co.uk
Owner *Tribune Publications*
Editor *Chris McLaughlin*

Founded 1937. WEEKLY. Independent Labour publication covering parliament, politics, trade unions, public policy and social issues, international affairs and the arts. Welcomes freelance contributions; approach by telephone or e-mail.
FEATURES *George Osgerby* Anything in line with the topics above. Max. 1,000 words.
NEWS *Keith Richmond* Breaking or forthcoming news events/stories. Max. 480 words.
OTHER PAGES Interviews, reviews, cartoons.

Trout Fisherman

Bushfield House, Orton Centre, Peterborough
PE2 5UW
☎ 01733 237111 ℻ 01733 465820
russell.hill@bauermedia.co.uk
www.gofishing.co.uk

Owner *Bauer Media*
Editor *Russell Hill*
Circulation 25,673

Founded 1977. MONTHLY instructive magazine on trout fishing. Most of the articles are commissioned, but unsolicited mss and quality colour images welcome.
FEATURES Max. 2,500 words.
ⓔ Payment varies.

TVTimes

IPC Media Ltd., Blue Fin Building, 110 Southwark Street, London SE1 0SU
☎ 020 3148 5615 ℻ 020 3148 8115
Owner *IPC Media (A Time Warner Company)*
Editor *Ian Abbott*
Circulation 336,929

Founded 1955. WEEKLY magazine of listings and features serving the viewers of independent television, BBC, satellite and radio. Freelance contributions by commission only. No unsolicited contributions.

20x20 magazine

36 Osborne Road, London N4 3SD
info@20x20magazine.com
20x20magazine.com
Owner/Editors *Giovanna Paternò, Francesca Ricci*
Circulation 500

Launched 2008. BIANNUAL. 'A square platform for writings, visuals and cross-bred projects.' Includes three sections – Words: in the shape of fiction, essays, poetry; Visions: drawings and photography; The Blender: where words and visions cross path. Contributions welcome 'although they must fit in our "metawords" spirit'. Max. 1,500 words. Approach by e-mail.

Ulster Business

5b Edgewater Business Park, Belfast Harbour Estate, Belfast BT3 9JQ
☎ 028 9078 3200 ℻ 028 9078 3210
davidelliott@greerpublications.com
www.ulsterbusiness.com
Owner *James Greer*
Editor *David Elliott*
Circulation 7,000

Founded 1987. MONTHLY. General business content with coverage of all sectors: ICT, retail, agriculture, commercial property, etc. 'Opportunities exist for well-written local orientated features, particularly those dealing with local firms or the issues they face; local business personality interviews also useful.' Contact the editor by e-mail. Max. 650 words.

Ulster Tatler

39 Boucher Road, Belfast BT12 6UT
☎ 028 9066 3311 ℻ 028 9038 1915

edit@ulstertatler.com
www.ulstertatler.com
Owner/Editor *Chris Sherry*
Circulation 10,870

Founded 1965. MONTHLY. Articles of local interest and social functions appealing to Northern Ireland's ABC1 population. Welcomes unsolicited material; approach by phone or in writing in the first instance.
FEATURES *James Sherry* Max. 1,500 words.
FICTION *Richard Sherry* Max. 3,000 words

Ultimate Advertiser

PO Box 7086, Clacton on Sea CO15 5WN
☎ 01255 428500
mail@apexpublishing.co.uk
www.ultimateadvertiser.co.uk
Owner *Apex Publishing Ltd*
Editor *Chris Cowlin*
Circulation 5,000

Founded 2005. MONTHLY advertising magazine including guides and information, features, listings and reviews. Approach by post, only.

Uncut

IPC Media Ltd., Blue Fin Building, 110 Southwark Street, London SE1 0SU
☎ 020 3148 6985
www.uncut.co.uk
Owner *IPC Media (A Time Warner Company)*
Editor *Allan Jones*
Associate Editor *Michael Bonner*
Music Editor *John Mulvey*
Art Editor *Marc Jones*
Online News Editor *Farah Ishaq*
Circulation 86,925

Launched 1997. MONTHLY music and film magazine with free CD each issue. Welcomes music and film reviews; approach by e-mail.

The Universe

Gabriel Communications Ltd., 4th Floor, Landmark House, Station Road, Cheadle Hulme SK8 7JH
☎ 0161 488 1700   ⊟ 0161 488 1701
joseph.kelly@totalcatholic.com
www.totalcatholic.com
Owner *Gabriel Communications Ltd*
Editor *Joseph Kelly*
Circulation 60,000

Occasional use of new writers, but a substantial network of regular contributors already exists. Interested in a very wide range of material: all subjects which might bear on Catholic life. Fiction not normally accepted.
💷 Payment negotiable.

The Vegan

Donald Watson House, 21 Hylton Street, Birmingham B18 6HJ
☎ 0121 523 1730
editor@vegansociety.com
www.vegansociety.com
Owner *Vegan Society*
Editor *Rosamund Raha*
Circulation 7,000

Founded 1944. QUARTERLY. Deals with the ecological, ethical and health aspects of veganism. Unsolicited mss welcome. Max. 2,000 words.
💷 Payment negotiable.

Vogue

Vogue House, Hanover Square, London W1S 1JU
☎ 020 7499 9080   ⊟ 020 7408 0559
www.vogue.com
Owner *Condé Nast Publications Ltd*
Editor *Alexandra Shulman*
Circulation 221,090

Launched 1916. Condé Nast Magazines tend to use known writers and commission what's needed, rather than using unsolicited mss. Contacts are useful.
FEATURES *Harriet Quick* Upmarket general interest rather than 'women's'. Good proportion of highbrow art and literary articles, as well as travel, gardens, food, home interest and reviews.

The Voice

GV Media Group Ltd., Northern & Shell Tower, 6th Floor, 4 Selsdon Way, London E14 9GL
☎ 020 7510 0340   ⊟ 020 7510 0341
yourviews@gvmedia.co.uk
www.voice-online.co.uk
Acting Managing Director *George Ruddock*
Editor *Steve Pope*
Circulation 40,000

Founded 1982. Leading WEEKLY newspaper for black Britons and other minority communities. Includes news, features, arts, sport and a comprehensive jobs section. Illustrations: colour and b&w photos. Open to ideas for news and features on sport, business, community events and the arts.

Volkswagen Golf+

Kelsey Publishing Ltd., Cudham Tithe Barn, Berry's Hill, Cudham TN16 3AG
☎ 01959 541444   ⊟ 01959 541400
golf.ed@kelsey.co.uk
www.kelsey.co.uk
Owner *IPC Media (A Time Warner Company)*
Editor *Fabian Cotter*
Circulation 25,000

Founded 1995. MONTHLY. Features Volkswagen Golfs and other watercooled VAG cars. Articles on tuning, styling, performance, technical and potential feature cars. 'Not interested in anything to do with the game of golf.' Submissions should be made directly to the editor. Articles must show excellent specialist knowledge of the subject matter. 🖪 Payment by negotiation.

W.I. Life

104 New Kings Road, London SW6 4LY
🕾 020 7731 5777   🖷 020 7736 4061
Owner *National Federation of Women's Institutes*
Editor *Neal Maidment*
Circulation 201,800

Women's Institute membership magazine. First issue February 2007. EIGHT ISSUES YEARLY. Contains articles on a wide range of subjects of interest to women. Strong environmental country slant with crafts and cookery plus gardening. Contributions, photos and illustrations from WI members welcome. 🖪 Payment by arrangement.

Walk Magazine

2nd Floor, Camelford House, 87–90 Albert Embankment, London SE1 7TW
🕾 020 7339 8500   🖷 020 7339 8501
walkmag@ramblers.org.uk
www.ramblers.org.uk
www.walkmag.co.uk
Owner *Ramblers' Association*
Editor *Dominic Bates*
Assistant Editor *Denise Noble*
Circulation 105,000

QUARTERLY. Official magazine of the Ramblers' Association, available to members and on sale to the public through selected outlets. Unsolicited mss welcome. S.a.e. required for return.
FEATURES Freelance features are invited on any aspect of walking in Britain. Length up to 1,500 words, preferably with good photographs. No general travel articles.

Wallpaper

Blue Fin Building, 110 Southwark Street, London SE1 0SU
🕾 020 3148 5000   🖷 020 3148 8119
contact@wallpaper.com
www.wallpaper.com
Owner *IPC Media (A Time Warner Company)*
Editor *Tony Chambers*
Circulation 112,847

Launched 1996. MONTHLY magazine with the best of international design, architecture, fashion, travel and lifestyle.
FEATURES *Nick Compton* Ideas welcome but contributors must be familiar with the voice and subject matter of the magazine. No celebrity items. Approach by e-mail.

The War Cry

101 Newington Causeway, London SE1 6BN
🕾 020 7367 4900   🖷 020 7367 4710

warcry@salvationarmy.org.uk
www.salvationarmy.org.uk/warcry
Owner *The Salvation Army*
Editor *Major Nigel Bovey*
Circulation 58,000

Founded 1879. WEEKLY full-colour magazine containing Christian comment on current issues. Unsolicited mss welcome if appropriate to contents. No fiction or poetry. Approach by phone with ideas.
NEWS relating to Christian Church or social issues. Max. length 500 words.
FEATURES Human interest articles aimed at the 'person-in-the-street'. Max. 500 words.
🖪 Payment negotiable.

Waterways World

151 Station Street, Burton on Trent DE14 1BG
🕾 01283 742950   🖷 01283 742957
editorial@waterwaysworld.com
www.waterwaysworld.com
Owner *Waterways World Ltd*
Editor *Richard Fairhurst*
Circulation 16,602

Founded 1972. MONTHLY magazine for inland waterway enthusiasts. Unsolicited mss welcome, provided the writer has a good knowledge of the subject. No fiction.
FEATURES *Keith Goss* Articles (preferably illustrated) are published on all aspects of inland waterways in Britain and abroad but predominantly recreational boating on rivers and canals.
NEWS *Chris Daniels* Max. 500 words.
🖪 £70 per published page.

Web User

IPC Media Ltd., Blue Fin Building, 110 Southwark Street, London SE1 0SU
🕾 020 3148 5000   🖷 020 3148 8122
webuser@ipcmedia.com
www.webuser.co.uk
Owner *IPC Media (A Time Warner Company)*
Editor *Claire Woffenden*
Circulation 35,279

Founded 2001. FORTNIGHTLY bestseller Internet magazine for all users of the Internet. No unsolicited material; send e-mail or letter of enquiry in the first instance.

Wedding

IPC Media Ltd., Blue Fin Building, 110 Southwark Street, London SE1 0SU
🕾 020 3148 7800
wedding@ipcmedia.com
www.weddingmagazine.co.uk
Owner *IPC Media (A Time Warner Company)*
Editor *Catherine Westwood*

Circulation 40,000

Founded 1985. BI-MONTHLY offering ideas and inspiration for women planning their wedding. Most features are written in-house or commissioned from known freelancers. Unsolicited mss are not welcome, but approaches may be made in writing.

Weekly News

D.C. Thomson & Co. Ltd., Albert Square, Dundee DD1 9QJ

☏ 01382 223131   🖷 01382 201390
weeklynews@dcthomson.co.uk
Owner *D.C. Thomson & Co. Ltd*
Editor *David Burness*
Circulation 64,358

Founded 1855. WEEKLY. Newsy, family-orientated magazine designed to appeal to the busy housewife. Regulars include true-life stories told in the first person, showbiz, royals and television. Two or three general interest fiction stories each week. Usually commissions, but writers of promise will be taken up. 💷 Payment negotiable.

Welsh Country Magazine

Aberbanc, Llandysul SA44 5NP
☏ 01559 372010   🖷 01559 371995
editor@welshcountry.co.uk
www.welshcountry.co.uk
Owner *Equine Marketing Limited*
Editor *Kath Rhodes*
Circulation 19,671

Founded 2004. BI-MONTHLY. Outside contributions welcome. Contact the editor with samples of work before submitting full article. The magazine is written in English.
FEATURES Pesonal experiences or viewpoints, historical or modern, but there needs to be a strong Welsh connection in some form.
NEWS All Welsh news welcome; anything at all commercial is not paid for.
💷 Payment negotiable.

What Car?

Teddington Studios, Teddington Lock, Broom Road, Teddington TW11 9BE
☏ 020 8267 5688
editorial@whatcar.com
www.whatcar.com
Owner *Haymarket Motoring Publications Ltd*
Editor *Steve Fowler*
Circulation 108,028

MONTHLY. The car buyer's bible, *What Car?* concentrates on road test comparisons of new cars, news and buying advice on used cars, as well as a strong consumer section. Some scope for freelancers. No unsolicited mss.
💷 Payment negotiable.

What Hi-Fi? Sound & Vision

Teddington Studios, Teddington Lock, Broom Road, Teddington TW11 9BE
☏ 020 8943 5000   🖷 020 8267 5019
whathifi@haymarket.com
www.whathifi.com
Owner *Haymarket Consumer Media Ltd*
Editor *Richard Melville*
Circulation 59,147

Founded 1976. THIRTEEN ISSUES YEARLY. Features on hi-fi and home cinema. No unsolicited contributions. Prior consultation with the editor essential.
FEATURES General or more specific items on hi-fi and home cinema pertinent to the consumer electronics market.
REVIEWS Specific product reviews. All material is now generated by in-house staff. Freelance writing no longer accepted.

What Investment

Vitesse Media, Octavia House, 50 Banner Street, London EC1Y 8ST
☏ 020 7250 7026   🖷 020 7250 7011
keiron.root@vitessemedia.co.uk
www.whatinvestment.co.uk
Owner *Vitesse Media plc*
Editor *Keiron Root*
Deputy Editor *Jenny Lowe*
Circulation 21,446

Founded 1982. MONTHLY. Features articles on a variety of savings and investment matters. E-mail ideas to the editor.
FEATURES Max. 2,000 words.
💷 £200 per 1,000 words.

What Satellite and Digital TV

2 Balcombe Street, London NW1 6NW
☏ 020 7042 4000
wotsat@futurenet.co.uk
www.wotsat.com
Owner *Future Publishing Ltd*
Editor *Alex Lane*
Circulation 40,000

Founded 1986. MONTHLY including news, technical information, reviews, equipment tests, programme background, listings. Contributions welcome – phone first.
FEATURES In-depth guides to popular/cult shows. Technical tutorials.
NEWS Industry and programming. Max. 250 words.

Which Caravan

Warners Group Publications plc, The Maltings, West Street, Bourne PE10 9PH
☏ 01778 391111   🖷 01778 425437
sallyp@warnersgroup.co.uk
www.outandaboutlive.co.uk

Editor *Sally Pepper*

Founded 1987. Buyers' guide for first-time and experienced caravanners and enthusiasts providing product information, advice and testing. Opportunities for caravanning, relevant touring and travel material, also monthly reviews of the best tow cars, all with good-quality colour photographs.

Wild Times

The RSPB, UK Headquarters, The Lodge, Sandy SG19 2DL

☎ 01767 680551   🖷 01767 683262

derek.niemann@rspb.org.uk

Owner *Royal Society for the Protection of Birds*
Editor *Derek Niemann*

Founded 1965. BI-MONTHLY. Bird, wildlife and nature conservation for under-8-year-olds (RSPB Wildlife Explorers members). No unsolicited mss. No 'captive/animal welfare' articles.

FEATURES Unsolicited material rarely used.
NEWS News releases welcome. Approach in writing in the first instance.

Wine & Spirit

William Reed Publishing, Broadfield Park, Crawley RH11 9RT

☎ 01293 613400   🖷 01293 610317

david.williams@william-reed.co.uk

www.wine-spirit.com

Owner *William Reed Business Media*
Editor *David Williams*
Circulation 11,275

MONTHLY. New owner, William Reed, has merged *Wine Magazine* with its sister title *Wine & Spirit International* to create *Wine & Spirit*. Still a consumer/specialist magazine, covering wine, spirit and beer, plus food and food/wine-related travel. No unsolicited mss. Prospective contributors should approach in writing or by e-mail.

Wingbeat

The RSPB, UK Headquarters, The Lodge, Sandy SG19 2DL

☎ 01767 680551   🖷 01767 683262

derek.niemann@rspb.org.uk

Owner *Royal Society for the Protection of Birds*
Editor *Derek Niemann*

Founded 1965. QUARTERLY. Bird, wildlife and nature conservation for 13–18-year-olds (RSPB Wildlife Explorers teenage members). No unsolicited mss. No 'captive/animal welfare' articles.

FEATURES Unsolicited material rarely used.
NEWS News releases welcome. Approach in writing in the first instance.

The Wisden Cricketer

123 Buckingham Palace Road, London SW1W 9SL

☎ 020 7705 4911   🖷 020 7921 9151

twc@wisdencricketer.com

www.thewisdencricketer.com

Owner *BSkyB Publications*
Editor *John Stern*
Deputy Editor *Edward Craig*
Assistant Editor *Daniel Brigham*
Editorial Assistant *Beni Moorhead*
Circulation 34,559

Founded 2003. MONTHLY. Result of a merger between *The Cricketer International* (1921) and *Wisden Cricket Monthly* (1979). Very few uncommissioned articles are used, but would-be contributors are not discouraged. Approach in writing. 🖂 Payment varies.

Woman

IPC Media Ltd., Blue Fin Building, 110 Southwark Street, London SE1 0SU

☎ 020 3148 6491

woman@ipcmedia.com

www.ipcmedia.com

Owner *IPC Media (A Time Warner Company)*
Editor *Jackie Hatton*
Deputy Editor *Jenny Vereker*
Circulation 370,289

Founded 1937. WEEKLY. Long-running, popular women's magazine which boasts a readership of over 2.5 million. No unsolicited mss. Most work commissioned. Approach with ideas in writing.

FEATURES *Anna Kingsley* Max. 1,250 words.
BOOKS *Claire Snewin*

Woman and Home

IPC Media Ltd., Blue Fin Building, 110 Southwark Street, London SE1 0SU

☎ 020 3148 7836

www.womanandhome.com

Owner *IPC Media (A Time Warner Company)*
Editor *Sue James*
Circulation 336,888

Founded 1926. MONTHLY. No unsolicited mss. Prospective contributors are advised to write with ideas, including photocopies of other published work or details of magazines to which they have contributed. S.a.e. essential for return of material. All freelance work is specially commissioned.

Woman's Own

IPC Media Ltd., Blue Fin Building, 110 Southwark Street, London SE1 0SU

☎ 020 3148 5000   🖷 020 3148 8112

Owner *IPC Media (A Time Warner Company)*
Editor *Karen Livermore*
Features Editor *Anna Wharton*

Circulation 339,138

Founded 1932. WEEKLY. Prospective contributors should contact the features editor *in writing* in the first instance before making a submission.

Woman's Weekly

IPC Media Ltd., Blue Fin Building, 110 Southwark Street, London SE1 0SU
📞 020 3148 5000
womansweeklypostbag@ipcmedia.com
Owner *IPC Media (A Time Warner Company)*
Editor *Diane Kenwood*
Deputy Editor *Geoffrey Palmer*
Features Editor *Sue Pilkington*
Fiction Editor *Gaynor Davies*
Circulation 36,073

Founded 1911. Mass-market WEEKLY for the mature woman.
FEATURES General features covering any subject of interest to women of forty-plus. Can be newsy and/or emotional but should be informative. Could be campaigning, issue-based, first or third person. Strong real-life stories always of interest. Words: ranging from 800 to 2,000. Only experienced journalists. Synopses and ideas should be submitted by e-mail to susan_archer@ipcmedia.com
FICTION Short stories, 1,000–2,500 words; serials, 12,000–30,000 words. Guidelines: should be contemporary and engaging. Humour and wit also welcome. Short stories up to 5,000 words considered for *Woman's Weekly Fiction Special* (see entry). E-mails to maureen_street@ipcmedia.com

Woman's Weekly Fiction Special

IPC Media Ltd., Blue Fin Building, 110 Southwark Street, London SE1 0SU
📞 020 3148 6600
womansweeklypostbag@ipcmedia.com
www.ipcmedia.com
Owner *IPC Media (A Time Warner Company)*
Editor *Gaynor Davies*

Launched 1998. EIGHT ISSUES A YEAR. Welcomes short stories of between 1,000 and 5,000 words. Guidelines are available from the Fiction Department or via the postbag (above). See also *Woman's Weekly*.

The Woodworker

MyHobbyStore Publishing Ltd., Berwick House, 8–10 Knoll Rise, Orpington BR6 0EL
📞 01689 899027
jon.bentman@magicalia.com
www.getwoodworking.com
Owner *MyHobbyStore Publishing Ltd.*
Editor-in-Chief *Jonathan Bentman*

Founded 1901. MONTHLY. Contributions welcome; approach with ideas in writing.

FEATURES Articles on woodworking with good photo support appreciated. Max. 2,000 words.
💷 Payment negotiable.

World Fishing

Mercator Media Ltd., The Old Mill, Lower Quay, Fareham PO16 0RA
📞 01329 825335　　📠 01329 825330
cwills@worldfishing.net
www.worldfishing.net
Owner *Mercator Media Ltd*
Editor *Carly Wills*
Circulation 3,429

Founded 1952. MONTHLY. Unsolicited mss welcome; approach by phone or in writing with an idea.
NEWS/FEATURES of a technical or commercial nature relating to the commercial fishing and fish processing industries worldwide (the magazine is read in over 100 different countries). Max. 1,500 words.
💷 Payment by arrangement.

The World of Fine Wine

226 City Road, London EC1V 2TT
📞 020 7700 8500
www.finewinemag.com
Editor *Dr Neil Beckett*
Editorial Adviser *Hugh Johnston*
Publisher *Sara Basra*
Managing Director *Piers Spence*

QUARTERLY book-length journal covering fine and noteworthy wine, established in 2004 and owned by the Quarto Group. Prospective contributors should contact the editor in the first instance.

The World of Interiors

Vogue House, Hanover Square, London W1S 1JU
📞 020 7499 9080　　📠 020 7493 4013
interiors@condenast.co.uk
www.worldofinteriors.co.uk
Owner *Condé Nast Publications Ltd*
Editor *Rupert Thomas*
Circulation 67,575

Founded 1981. MONTHLY. Best approach by fax or letter with an idea, preferably with reference snaps or guidebooks.
FEATURES *Rupert Thomas* Most feature material is commissioned. 'Subjects tend to be found by us, but we are delighted to receive suggestions of interiors, archives, little-known museums, collections, etc. unpublished elsewhere, and are keen to find new writers.'

World Ski Guide

Mountain Leisure Limited, Parman House, 30–36 Fife Road, Kingston-upon-Thames KT1 1SY
📞 020 8547 9822　　📠 020 8546 0984
edit@goodholidayideas.com
www.goodskiguide.com

Owner *Mountain Leisure Ltd*
Editors *John Hill, Christian Berger*
Circulation 40,000

Launched 2002. ANNUAL. Translation of *Good Ski Guide* into German. See website for editorial guidelines.
FEATURES *John Hill* Destination features on ski resorts worldwide. Max. 900 words. Ⓔ Fee £200.

World Soccer

IPC Media Ltd., Blue Fin Building, 110 Southwark Street, London SE1 0SU
☎ 020 3148 4817   ℻ 020 3148 8130
world_soccer@ipcmedia.com
www.worldsoccer.com
Owner *IPC Media (A Time Warner Company)*
Editor *Gavin Hamilton*
Circulation 44,020

Founded 1960. MONTHLY. Unsolicited material welcome but initial approach by e-mail or in writing. News and features on world soccer.

Writers' Forum incorporating World Wide Writers

PO Box 6337, Bournemouth BH1 9EH
☎ 01202 586848
editorial@writers-forum.com
www.writers-forum.com
Owner *Select Publisher Services*
Editor *Carl Styants*
Publisher *Tim Harris*

Founded 1993. MONTHLY. Magazine covers all aspects of the craft of writing. Well written articles welcome. Write to the editor in the first instance.
Writers Forum Short Story Competition Prizes: £300 (1st), £150 (2nd), £100 (3rd). Entrance fee, non-subscribers: £10; subscribers: £7. Winners published in every issue.
Writers' Forum Poetry Competition First prize £100; runners up receive a dictionary. Entrance fee: £5 for one poem or £7 for two. Winners published in every issue. Annual subscription: £33 UK; £46 Worldwide. Send s.a.e. for free back issue.

Writers' News/Writing Magazine

Warners Group Publications, Fifth Floor, 31–32 Park Row, Leeds LS1 5JD
☎ 0113 200 2929   ℻ 0113 200 2928
www.writersnews.co.uk
Owner *Warners Group Publications plc*
Publisher *Janet Davison*
Editor, Writers' News *Jonathan Telfer*
Editor, Writing Magazine *Hilary Bowman*

Founded 1989. MONTHLY magazines containing news and advice for writers. *Writers' News* is available exclusively by subscription; *Writing Magazine*, a full-colour glossy publication, is available by subscription or on newsstands. No poetry or general items on 'how to become a writer'. Receive 1,000 mss each year. Approach in writing or by e-mail.
NEWS Exclusive news stories of interest to writers. Max. 350 words.
FEATURES How-to articles of interest to professional writers. Max. 1,000 words.

Yachting Monthly

IPC Media Ltd., Blue Fin Building, 110 Southwark Street, London SE1 0SU
☎ 020 3148 4872   ℻ 020 3148 8128
paul_gelder@ipcmedia.com
www.yachtingmonthly.com
Owner *IPC Media (A Time Warner Company)*
Editor *Paul Gelder*
Circulation 34,402

Founded 1906. MONTHLY magazine for yachting and cruising enthusiasts – not racing. Unsolicited mss welcome, but many are received and not used. Prospective contributors should make initial contact in writing.
FEATURES A wide range of features concerned with maritime subjects and cruising under sail; well-researched and innovative material always welcome, especially if accompanied by high resolution digital images. Max. 1,800 words. Ⓔ £80–£110 per 1,000 words.

Yachting World

IPC Media Ltd., Blue Fin Building, 110 Southwark Street, London SE1 0SU
☎ 020 3148 4846   ℻ 020 3148 8127
yachting_world@ipcmedia.com
www.yachtingworld.com
Owner *IPC Media (A Time Warner Company)*
Editor *Andrew Bray*
Circulation 26,400

Founded 1894. MONTHLY with international coverage of yacht racing, cruising and yachting events. Will consider well researched and written sailing stories. Preliminary approaches should be by phone for news stories and in writing for features.
Ⓔ Payment by arrangement.

The Yellow Room Magazine

1 Blake Close, Bilton, Rugby CV22 7LJ
☎ 01788 334302
jo.derrick@ntlworld.com
www.theyellowroom-magazine.co.uk
Owner/Editor *Jo Derrick*

Launched 2008. BI-ANNUAL magazine aiming to publish quality short stories, only. 'The Yellow Room is a place where female writers can find support, encouragement and friendship.' Welcomes well written short stories which appeal to a predominantly female readership. These are stories which you wouldn't find in a weekly women's magazine, however. Submit one story with either s.a.e. for reply or e-mail address. Max. 4,000 words. 🄯 Payment £10 per story. See website and guidelines for further details. No poetry, articles or novel extracts.

Yorkshire Women's Life Magazine

PO Box 113, Leeds LS8 2WX
ywlmagenquiries@btinternet.com
www.yorkshirewomenslife.co.uk
Editor/Owner *Dawn Maria France*
Fashion *Sky Taylor*
Magazine PA *Anna Jenkins*
Diary *Giles Smith*
Circulation 15,000

Founded 2001. THREE ISSUES YEARLY. Features of interest to women along with regional, national, international news and lifestyle articles. Past issues covered health, stress management, living with domestic abuse, pampering breaks for city women, coping with a difficult boss, Windrush awards. 'It is important to study the style of the magazine before submitting material. Send A4 s.a.e. with 46p stamp for copy of submission guidelines. Unsolicited mss and new writers actively encouraged; approach in writing in the first instance with s.a.e.' Magazine available on subscription at the website address.

You & Your Wedding

National Magazine Company, 72 Broadwick Street, London W1F 9EP
📞 020 7439 5000 (editorial)   📠 020 7439 2985
cathy.howes@natmags.co.uk
www.youandyourwedding.co.uk
Owner *The National Magazine Company Ltd*
Editor *Colette Harris*
Circulation 60,024

Founded 1985. BI-MONTHLY. Anything relating to weddings, setting up home, and honeymoons. No unsolicited mss. Submit ideas by e-mail only. No phone calls.

You – The Mail on Sunday Magazine
▷ The Mail on Sunday under National Newspapers

Young Writer

Fifth Floor, 31–32 Park Row, Leeds LS1 5JD
📞 0113 200 2929   📠 0113 200 2928
ftelfer@writersnews.co.uk
www.youngwriter.org
Editor *Jonathan Telfer*

Describing itself as 'The Magazine for Children with Something to Say', *Young Writer* is issued three times a year, at the back-to-school times of September, January and April. A forum for young people's writing – fiction and non-fiction, prose and poetry – the magazine is an introduction to independent writing for young writers aged five

to 18. 🄯 Payment from £20 to £100 for freelance commissioned articles (these can be from adult writers).

Your Cat Magazine

Roebuck House, 33 Broad Street, Stamford PE9 1RB
📞 01780 766199   📠 01780 766416
yourcat@bournepublishinggroup.co.uk
www.yourcat.co.uk
Owner *BPG (Stamford) Ltd*
Editor *Sue Parslow*

Founded 1994. MONTHLY magazine giving practical information on the care of cats and kittens, pedigree and non-pedigree, plus a wide range of general interest items on cats. Will consider 'true life' cat stories (max. 900 words) and quality fiction by published novelists. Send synopsis in the first instance. 'No articles written as though by a cat and no poetry.'

Your Dog Magazine

Roebuck House, 33 Broad Street, Stamford PE9 1RB
📞 01780 766199   📠 01780 754774
s.wright@bournepublishinggroup.co.uk
www.yourdog.co.uk
Owner *BPG (Stamford) Ltd*
Editor *Sarah Wright*
Circulation 31,299

Founded 1995. MONTHLY. Practical advice for pet dog owners. Will consider practical features and some personal experiences (no highly emotive pieces or fiction). Telephone in the first instance. NEWS Max. 300–400 words. FEATURES Max. 2,500 words; limited opportunities.
🄯 Payment negotiable.

Your Hair

Origin Publishing, 9th Floor, Tower House, Fairfax Street, Bristol BS1 3BN
📞 0117 927 9009   📠 0117 314 8310
michelletiernan@originpublishing.co.uk
www.yourhair.co.uk
Editor *Michelle Tiernan*
Circulation 38,206

Launched 2001. MONTHLY magazine for women aged 21-plus who need some style inspiration, a whole new look or to keep a finger on the pulse of catwalk trends. 'We pride ourselves on wearable, achievable solutions.' No unsolicited contributions but welcome product information from PR agencies.

Your Horse

Bushfield House, Orton Centre, Peterborough PE2 5UW
📞 01733 465644   📠 01733 288163
www.yourhorse.co.uk

Owner *Bauer Media*
Acting Editor *Julie Brown*
Digital Editor *Justine Thompson*
Circulation 40,006

'The magazine that aims to make owning and riding horses more rewarding.' Most writing produced in-house but well-targeted articles will always be considered.

Yours Magazine

Media House, Peterborough Business Park, Peterborough PE2 6EA

☎ 01733 468000
yours@bauermedia.co.uk
Owner *Bauer Media*
Editor *Valery McConnell*
Circulation 327,072

Founded 1973. FORTNIGHTLY. Aimed at a readership aged 50 and over. Submission guidelines on request.
FEATURES Unsolicited mss welcome but must enclose s.a.e. Max. 1,400 words.
FICTION One or two short stories used in each issue. Max. 1,500 words.
£ Payment negotiable.

Zest

National Magazine House, 72 Broadwick Street, London W1F 9EP

☎ 020 7439 5000   ℻ 020 7312 3750
zest.mail@natmags.co.uk
www.zest.co.uk
Owner *National Magazine Company*
Editor *Mandie Gower*
Deputy Editors *Rebecca Frank, Charlotte Bradshaw*
Circulation 96,611

Founded 1994. MONTHLY. Health, beauty, fitness, nutrition and general well-being. No unsolicited mss. Prefers ideas in synopsis form; approach in writing.

ZOO

Bauer Consumer Media, Mappin House, 4 Winsley Street, London W1W 8HF

☎ 020 7182 8355   ℻ 020 7182 8300
info@zooweekly.co.uk
www.zootoday.com
Owner *Bauer Publishing*
Editor *Ben Todd*
Circulation 161,331

Launched 2004. WEEKLY men's lifestyle magazine featuring showbiz, sport, humour, sex, fashion and news.

National Newspapers

Departmental e-mail addresses are too numerous to include in this listing. They can be obtained from the newspaper's main switchboard or the department in question.

Daily Express

The Northern & Shell Building, No. 10 Lower Thames Street, London EC3R 6EN
☎ 0871 434 1010
www.express.co.uk
Owner *Northern & Shell Media/Richard Desmond*
Editor *Peter Hill*
Circulation 728,296

Under owner Richard Desmond, publisher of *OK!* magazine, the paper features a large amount of celebrity coverage. The general rule of thumb is to approach in writing with an idea; all departments are prepared to look at an outline without commitment. Ideas welcome but already receives many which are 'too numerous to count'.
News Editor *Geoff Maynard*
Diary (Hickey Column) *Kathryn Spencer*
Features Editor *Fergus Kelly*
City Editor *Stephen Kahn*
Political Editor *Macer Hall*
Sports Editor *Bill Bradshaw*
Planning Editor (News Desk) should be circulated with copies of official reports, press releases, etc., to ensure news desk cover at all times.

Saturday magazine. Editor *Graham Bailey*
£ Payment negotiable.

Daily Mail

Northcliffe House, 2 Derry Street, London W8 5TT
☎ 020 7938 6000
www.dailymail.co.uk
Owner *Associated Newspapers/Lord Rothermere*
Editor *Paul Dacre*
Circulation 2.14 million

In-house feature writers and regular columnists provide much of the material. Photo-stories and crusading features often appear; it's essential to hit the right note to be a successful Mail writer.

Close scrutiny of the paper is strongly advised. Not a good bet for the unseasoned. Accepts news on savings, building societies, insurance, unit trusts, legal rights and tax.
News Editor *Keith Poole*
City Editor *Alex Brummer*
'Money Mail' Editor *Tony Hazell*
Political Editor *Benedict Brogan*
Education Correspondent *Laura Clark*
Diary Editor *Richard Kay*
Features Editor *Leaf Kalfayan*
Literary Editor *Sandra Parsons*
Head of Sport *Tim Jotischky*

Femail Assistant Editor *Andrew Morrod*
Weekend Saturday supplement. Editor *Lisa Collins*

Daily Mirror

1 Canada Square, Canary Wharf, London E14 5AP
☎ 020 7293 3000   📠 020 7293 3409
www.mirror.co.uk
Owner *Trinity Mirror plc*
Editor *Richard Wallace*
Circulation 1.35 million

No freelance opportunities for the inexperienced, but strong writers who understand what the tabloid market demands are always needed.
Deputy Editor *Conor Hanna*
Head of News *Anthony Harwood*
News Editor *Barry Rabbetts*
Features Editor *Carole Watson*
Political Editor *Bob Roberts*
Business Editor *Clinton Manning*
Showbusiness Diary Editor *Clemmie Moody*
Sports Editor *Dean Morse*

Daily Record

One Central Quay, Glasgow G3 8DA
☎ 0141 309 3000   📠 0141 309 3340
reporters@dailyrecord.co.uk
www.dailyrecord.co.uk
Owner *Trinity Mirror plc*
Editor *Bruce Waddell*
Circulation 349,235

Mass-market Scottish tabloid. Freelance material is generally welcome.
News Editor *Andy Lines*

Features Editor *Melanie Harvey*
Political Editor *Magnus Gardham*
Assistant Editor (Sport) *James Traynor*
Magazine Editor *Jayne Savva*

Scotland Means Business Quarterly business magazine, launched 2002. Editor *Magnus Gardham*

Daily Sport

19 Great Ancoats Street, Manchester M60 4BT
☎ 0161 236 4466   📠 0161 236 4535
www.dailysport.net
Owner *Sport Media Group*
Editor-in-Chief *Murray Morse*
Editor *Pam McVitie*
Circulation 78,000

Tabloid catering for young male readership. Unsolicited material welcome; send to News Editor, *Jane Field-Harris*
Sports Editor *Marc Smith*

Lads Mag Monthly glossy magazine. Editor *Mark Harris.*

Daily Star

The Northern & Shell Building, No. 10 Lower Thames Street, London EC3R 6EN
☎ 0871 434 1010
Owner *Northern & Shell Media/Richard Desmond*
Editor *Dawn Neesom*
Circulation 725,671

In competition with *The Sun* for off-the-wall news and features. Freelance opportunities available.
Deputy Editor *Keiron Saunders*
Sports Editor *Howard Wheatcroft*

Daily Star Sunday

The Northern & Shell Building, No. 10 Lower Thames Street, London EC3R 6EN
☎ 0871 520 7424   📠 0871 434 2941
michael.booker@dailystar.co.uk
www.dailystarsunday.co.uk
Owner *Northern & Shell Media/Richard Desmond*
Editor *Gareth Morgan*
Circulation 342,019

Launched in September 2002 in direct competition with *News of the World* and *The People.*

Supplement: *Take5* Lifestyle/showbiz magazine.

The Daily Telegraph

111 Buckingham Palace Road, London SW1W 0DT
☎ 020 7931 2000
www.telegraph.co.uk
Owner *Telegraph Media Group*
Editor *William Lewis*
Circulation 824,244

Unsolicited mss not generally welcome – 'all are carefully read and considered, but very few published'. Contenders should approach the paper in writing, making clear their authority for writing on that subject. No fiction.
Arts Editor *Sarah Crompton*
Head of Business *Damian Reece*
Online City Editor *Richard Blackden*
Political Editor *Andrew Porter*
Diary Editor *Tim Walker* Always interested in diary pieces.
Education *Graeme Paton*
Environment *Charles Clover*
Features Editor *Liz Hunt* Most material supplied by commission from established contributors. New writers are tried out by arrangement with the features editor. Approach in writing. Maximum 1,500 words.
Assistant Editor (Books) *Brian MacArthur*
Group Managing Editor, Sport *Keith Perry* Occasional opportunities for specialized items.
Style Editor *Georgina Cover*
💷 Payment by arrangement.

Daily Telegraph Weekend Saturday supplement. Editor *Jon Stock*
Telegraph Magazine. Editor *Michele Lavery*

Financial Times

1 Southwark Bridge, London SE1 9HL
☎ 020 7873 3000   📠 020 7873 3076
www.ft.com
Owner *Pearson*
Editor *Lionel Barber*
Circulation 435,319

Founded 1888. UK and international coverage of business, finance, politics, technology, management, marketing and the arts. All feature ideas must be discussed with the department's editor in advance. Not snowed under with unsolicited contributions – they get less than any other national newspaper. Approach by e-mail with ideas in the first instance.
Comment and Analysis Editor *Brian Groom*
Arts Editor *Jan Dalley*
City Editor *Andrew Hill*
UK News Editor *Sarah Neville*
FT.com News Editor *Andrew Slade*
Books and Arts Editor, *FT Magazine Rosie Blau*
Literary Editor *Jan Dalley*
Education *David Turner*
Environment Correspondent *Fiona Harvey*
Political Editor *George Parker*
Leisure Industries *Roger Blitz*

Weekend FT Editor *Andy Davis*
How to Spend It Monthly magazine. Editor *Gillian de Bono*

The Guardian

Kings Place, 90 York Way, London N1 9AG
☎ 020 3353 2000

firstname.secondname@guardian.co.uk
www.guardian.co.uk
Owner *The Scott Trust*
Editor *Alan Rusbridger*
Circulation 343,010

Of all the nationals *The Guardian* probably offers the greatest opportunities for freelance writers, if only because it has the greatest number of specialized pages which use freelance work. Read specific sections before submitting mss.
Deputy Editor *Paul Johnson* No opportunities except in those regions where there is presently no local contact for news stories.
Literary Editor *Claire Armitstead*
City Editor *Julia Finch*
G2 Features Editors *Clare Margitson, Emily Wilson*
'Technology' Editor *Charles Arthur* Thursday supplement.
Diary Editor *Jon Henley*
Education Editor *Claire Phipps*
Environment *John Vidal*
Features Editor *Katharine Viner*
'Guardian Society' *Patrick Butler* Focuses on all public service activity.
Media Editor *Jane Martinson*
Political Editor *Patrick Wintour*
Sports Editor *Ben Clissitt*
Women's Page *Kira Cochrane* Runs every Friday.

The Guardian Weekend Glossy Saturday issue.
Editor *Merope Mills*
The Guide Editor *Malik Meer*

The Herald (Glasgow)

200 Renfield Street, Glasgow G2 3QB
📞 0141 302 7000   📠 0141 302 7007
www.theherald.co.uk
Owner *Gannett UK Ltd*
Editor *Donald Martin*
Circulation 60,976

One of the oldest national newspapers in the English-speaking world, *The Herald*, which dropped its 'Glasgow' prefix in February 1992, was bought by Scottish Television in 1996 and by Newsquest in 2003. Lively, quality, national Scottish daily broadsheet. Approach with ideas in writing or by phone in first instance.
News Editor *Calum Macdonald*
Arts Editor *Keith Bruce*
Business Editor *Ian McConnell*
Diary *Ken Smith*
Features Editor *Mark Smith*
Education *Andrew Denholm*
Sports Editor *Donald Cowey*

Herald Magazine Editor *Kathleen Morgan*

The Independent

Independent House, 191 Marsh Wall, London
E14 9RS (due to move to: Northcliffe House,
2 Derry Street, London W8 5TT)
📞 020 7005 2000
www.independent.co.uk
Owner *Independent Newspapers*
Editor *Roger Alton*
Circulation 200,242

Founded 1986. Particularly strong on its arts/media coverage, with a high proportion of feature material. Theoretically, opportunities for freelancers are good. However, unsolicited mss are not welcome; most pieces originate in-house or from known and trusted outsiders. Ideas should be submitted in writing.
Executive News Editor *Dan Gledhill*
Features *Adam Leigh*
Arts Editor *David Lister*
Business Editor *Jeremy Warner*
Education *Richard Garner*
Environment *Michael McCarthy*
Literary Editor *Boyd Tonkin*
Political Editor *Andrew Grice*
Sports Editor *Matt Tench*
Travel Editor *Simon Calder*

The Independent Magazine Saturday supplement.
Editor *Laurence Earle*
The Independent Traveller 32-page Saturday magazine.
The Information Editor *Jo Ellison*

Independent on Sunday

Independent House, 191 Marsh Wall, London
E14 9RS (due to move to: Northcliffe House,
2 Derry Street, London W8 5TT)
📞 020 7005 2000
www.independent.co.uk
Owner *Independent Newspapers*
Editor *John Mullin*
Circulation 163,545

Founded 1986. Regular columnists contribute most material but feature opportunites exist. Approach with ideas in first instance.
Executive Editor (News) *Peter Victor*
Arts Editor *Mike Higgins*
Comment Editor *James Hanning*
Environment *Geoffrey Lean*
Political Editor *Jane Merrick*
Sports Editor *Marc Padgett*
Travel Editor *Kate Simon*

The New Review supplement. Editor *Bill Tuckey*

International Herald Tribune

6 bis, rue des Graviers, 92521 Neuilly Cedex
📞 0033 1 4143 9322
📠 0033 1 4143 9429 (editorial)

iht@iht.com
www.iht.com
Executive Editor *Martin Gottlieb*
Managing Editor *Alison Smale*
Features Editor *Katherine Knorr*
Circulation 241,000

Published in France, Monday to Saturday, and circulated in Europe, the Middle East, North Africa, the Far East and the USA. General news, business and financial, arts and leisure. Uses regular freelance contributors. Contributor policy can be found on the website at: www.iht.com/contributor.htm

The Mail on Sunday

Northcliffe House, 2 Derry Street, London w8 5TS
☏ 020 7938 6000   🖷 020 7937 3829
Owner *Associated Newspapers/Lord Rothermere*
Editor *Peter Wright*
Circulation 2.06 million

Sunday paper with a high proportion of newsy features and articles. Experience and judgement required to break into its band of regular feature writers.
News Editor *David Dillon*
Financial Editor *Lisa Buckingham*
Diary Editor *Katie Nicholl*
Features Editor/Women's Page *Sian James*
Books *Marilyn Warnick*
Political Editor *Simon Walters*
Sports Editor *Malcolm Vallerius*
'Live Night & Day' Editor *Gerard Greaves*
'Review' Editor *George Thwaites*

You – The Mail on Sunday Magazine Colour supplement. Many feature articles, supplied entirely by freelance writers. Editor *Sue Peart*, Features Editor *Rosalind Lowe*

Morning Star

William Rust House, 52 Beachy Road, London
E3 2NS
☏ 020 8510 0815   🖷 020 8986 5694
newsed@peoples-press.com
www.morningstaronline.co.uk
Owner *Peoples Press Printing Society*
Editor *John Haylett*
Circulation 13,000

Not to be confused with the *Daily Star*, the *Morning Star* is the farthest left national daily. Those with a penchant for a Marxist reading of events and ideas can try their luck, though feature space is as competitive here as in the other nationals.
News Editor *Dan Coysh*
Features/Political Editor *Richard Bagley*
Foreign Editor *Tom Mellen*
Arts Editor *Katie Lambert*
Sports Editor *Greg Leedham*

News of the World

1 Virginia Street, London E98 1NW
☏ 020 7782 1000   🖷 020 7583 9504
www.newsoftheworld.co.uk
Owner *News International plc/Rupert Murdoch*
Editor *Colin Myler*
Circulation 2.99 million

Highest circulation Sunday paper. Freelance contributions welcome. News and features editors welcome tips and ideas.
Deputy Editor *Jane Johnson*
Assistant Editor (News) *Ian Edmondson*
Features Editor *Jules Stenson*
Political Editor *Ian Kirby*
Sports Editor *Paul McArthy*

Fabulous Colour supplement. Editor *Sally Eyden* Glossy magazine, launched in February 2008. Celebrity, fashion and real-life features. Unsolicited mss and ideas welcome.

The Observer

Kings Place, 90 York Way, London N1 9AG
☏ 020 3353 2000
firstname.surname@observer.co.uk
www.observer.co.uk
Owner *Guardian Newspapers Ltd*
Editor *John Mulholland*
Circulation 420,323

Founded 1791. Acquired by Guardian Newspapers from Lonrho in May 1993. Occupies the middle ground of Sunday newspaper politics. Unsolicited material is not generally welcome, 'except from distinguished, established writers'. Receives far too many unsolicited offerings already. No news, fiction or special page opportunities. The newspaper runs annual competitions which change from year to year. Details are advertised in the newspaper.
News Editor *Chris Boffey*
Home News Editor *Lucy Rock*
Foreign News Editor *Julian Coman*
Crime Correspondent *Mark Townsend*
Foreign Affairs Editor *Peter Beaumont*
Home Affairs Editor *Jamie Doward*
Health Correspondent *Denis Campbell*
Social Affairs *Amelia Hill*
Arts and Media Correspondent *Vanessa Thorpe*
Political Editor *Gaby Hinsliff*
Whitehall Correspondent *Toby Helm*
Arts Editor *Sarah Donaldson*
'Review' Editor *Jane Ferguson*
Comment Editor *Ruaridh Nicoll*
Deputy Business Editor/City Editor *Richard Wachman*
Business Editor *Ruth Sunderland*
Personal Finance Editor *Jill Insley*
Science Editor *Robin McKie*
Education Correspondent *Anushka Asthana*

Environment Editor *Juliette Jowit*
Escape Editor *Joanne O'Connor*
Literary Editor *William Skidelsky*
Travel Editor *Joanne O'Connor*
Sports Editor *Brian Oliver*

The Observer Magazine Glossy arts and lifestyle supplement. Editor *Allan Jenkins*
The Observer Sport Monthly Magazine supplement launched in 2000. Editor *Tim Lewis*
The Observer Food Monthly Launched in 2001. Editor *Nicola Jeal*
The Observer Music Monthly Launched in 2003. Editor *Caspar Llewellyn Smith*
Observer Woman Launched in 2006. Editor *Nicola Jeal*
The Observer Film Quarterly Launched in 2008. Editor *Akin Ojumu*

The People

1 Canada Square, Canary Wharf, London E14 5AP
☎ 020 7293 3000   ℻ 020 7293 3517
www.people.co.uk
Owner *Trinity Mirror plc*
Editor *Lloyd Embley*
Circulation 592,306

Slightly up-market version of *The News of the World*. Keen on exposés and big-name gossip. Interested in ideas for investigative articles. Phone in the first instance.
News Editor *Lee Harpin*
Features Editor *Chris Bucktin*
Political Editor *Nigel Nelson*
Sports Editor *James Brown*

Take It Easy Magazine supplement. Editor *Maria Coole* Approach by phone with ideas in first instance.

Scotland on Sunday

Barclay House, 108 Holyrood Road, Edinburgh EH8 8AS
☎ 0131 620 8620   ℻ 0131 620 8491
www.scotlandonsunday.com
Owner *Scotsman Publications Ltd*
Editor *Les Snowdon*
Circulation 65,502

Scotland's top-selling quality broadsheet. Welcomes ideas rather than finished articles.
Assistant Editor *Kenny Farquharson*
News Editor *Peter Laing*
Arts Editor *Fiona Leith*

Spectrum Colour supplement. Features on personalities, etc. Editor *Clare Trodden*

The Scotsman

Barclay House, 108 Holyrood Road, Edinburgh EH8 8AS
☎ 0131 620 8620   ℻ 0131 620 8617
enquiries@scotsman.com
www.scotsman.com

Owner *Johnston Press*
Editor *Mike Gilson*
Circulation 50,558

Scotland's national newspaper. Many unsolicited mss come in, and stand a good chance of being read, although a small army of regulars supply much of the feature material not written in-house. See website for contact details.
News Editor *Frank O'Donnell*
Business Editor *Peter MacMahon*
Education *Fiona MacLeod*
Features Editor *Jackie Hunter*
Arts Editor *Andrew Eaton*
Book Reviews *David Robinson*
Sports Editor *Donald Walker*

The Sun

1 Virginia Street, London E98 1SN
☎ 020 7782 4000   ℻ 020 7782 4108
firstname.lastname@the-sun.co.uk
www.the-sun.co.uk
Owner *News International Ltd/Rupert Murdoch*
Editor *Rebekah Wade*
Deputy Editor *Dominic Mohan*
Circulation 2.9 million

Highest circulation daily. Populist outlook; very keen on gossip, pop stars, TV soap, scandals and exposés of all kinds. No room for non-professional feature writers; 'investigative journalism' of a certain hue is always in demand, however.
Head of News *Chris Pharo*
Head of Features *Victoria Newton*
Head of Sport *Mike Dunn*
Health Editor *Jane Symons*
Fashion Editor *Erica Davies*

Sunday Express

The Northern & Shell Building, Number 10 Lower Thames Street, London EC3R 6EN
☎ 0871 434 1010   ℻ 0871 434 7300
Owner *Northern & Shell Media/Richard Desmond*
Editor *Martin Townsend*
Circulation 638,556

The general rule of thumb is to approach in writing with an idea; all departments are prepared to look at an outline without commitment.
News Editor *Stephen Rigley*
Features Editor *Giulia Rhodes*
Business Editor *Lawrie Holmes*
Political Editor *Jason Groves*
Sports Editor *Scott Wilson*

S Fashion and lifestyle magazine for women. Editor *Louise Robinson* No unsolicited mss. All contributions are commissioned. Ideas in writing only.
💷 Payment negotiable.

Sunday Herald

200 Renfield Street, Glasgow G2 3QB
☎ 0141 302 7800   🖷 0141 302 7963
editor@sundayherald.com
www.sundayherald.com
Owner *Newsquest*
Editor *Richard Walker*
Circulation 45,406

Launched February 1999. Scottish seven-section compact.
Deputy Editor *Stephen Penman*
Head of News *Bill Mackintosh*
Scottish Political Editor *Tom Gordon*
Sports Editor *Jonathan Jobson*
Magazine Editor *Susan Flockhart*

Sunday Mail

One Central Quay, Glasgow G3 8DA
☎ 0141 309 3000   🖷 0141 309 3587
www.sundaymail.co.uk
Owner *Trinity Mirror plc*
Editor *Allan Rennie*
Circulation 440,088

Popular Scottish Sunday tabloid.
News Editor *Brendan McGinty*
Sports Editor *George Cheyne*

7Days Weekly supplement. Editor *Liz Cowan*

Sunday Mirror

1 Canada Square, Canary Wharf, London E14 5AP
☎ 020 7293 3000   🖷 020 7293 3939 (news desk)
www.sundaymirror.co.uk
Owner *Trinity Mirror*
Editor *Tina Weaver*
Circulation 1.2 million

In general terms contributions are welcome, though the paper patiently points out it has more time for those who have taken the trouble to study the market. Initial contact in writing preferred, except for live news situations. No fiction.
News Editor *James Saville* The news desk is very much in the market for tip-offs and inside information. Contributors would be expected to work with staff writers on news stories. Approach by telephone or fax in the first instance.
Finance *Melanie Wright*
Features Editor *Jill Main* 'Anyone who has obviously studied the market will be dealt with constructively and courteously.' Cherishes its record as a breeding ground for new talent.
Sports Editor *David Walker*

Celebs on Sunday Colour supplement. Editor *Mel Brodie*

Sunday Post

2 Albert Square, Dundee DD1 9QJ
☎ 01382 223131   🖷 01382 201064

mail@sundaypost.com
www.sundaypost.com
Owner *D.C. Thomson & Co. Ltd*
Editor *David Pollington*
Circulation 366,275

Contributions should be addressed to the editor.

post plus Monthly colour supplement. Editor *Jan Gooderham*

Sunday Sport

19 Great Ancoats Street, Manchester M60 4BT
☎ 0161 236 4466   🖷 0161 236 4535
www.sundaysport.com
Owner *Sport Media Group*
Editor *Nick Appleyard*
Circulation 70,796

Founded 1986. Sunday tabloid catering for a particular sector of the male 15–35 readership. As concerned with 'glamour' (for which, read: 'page 3') as with human interest, news, features and sport. Unsolicited mss are welcome; receives about 90 a week. Approach should be made by phone in the case of news and sports items, by letter for features. All material should be addressed to the news editor.
News Editor *Neil Goodwin* Off-beat news, human interest, preferably with photographs.
Showbiz Editor *Clare Chapman* Regular items.
Sports Editor *Marc Smith* Hard-hitting sports stories on major soccer clubs and their personalities, plus leading clubs/people in other sports. Strong quotations to back up the news angle essential.
💷 Payment negotiable and on publication.

The Sunday Telegraph

111 Buckingham Palace Road, London SW1W 0DT
☎ 020 7931 2000   🖷 020 7931 2936
stnews@telegraph.co.uk
www.telegraph.co.uk
Owner *Press Holdings Limited*
Editor *Ian MacGregor*
Circulation 597,934

Right-of-centre quality Sunday paper which, although traditionally formal, has pepped up its image to attract a younger readership. Unsolicited material from untried writers is rarely ever used. Contact with idea and details of track record.
Deputy Editor *Dan Roberts*
News Editor *Ben Leapman*
'Seven' Editor *Ross Jones*
City Editor *Damian Reece*
Political Editor *Patrick Hennessy*
Education Editor *Julie Henry*
Arts Editor *Chris Hastings*
Environment Editor *David Harrison*
Literary Editor *Michael Prodger*
Diary Editor *Tim Walker*

Stella Magazine Colour supplement. Editor *Anna Murphy*

The Sunday Times

1 Pennington Street, London E98 1ST
☏ 020 7782 5000   ☐ 020 7782 5658
www.sunday-times.co.uk
Owner *News International plc/Rupert Murdoch*
Editor *John Witherow*
Circulation 1.16 million

Founded 1820. Tendency to be anti-establishment, with a strong crusading investigative tradition. Approach the relevant editor with an idea in writing. Close scrutiny of the style of each section of the paper is strongly advised before sending mss. No fiction. All fees by negotiation.
News Editor *Nicholas Hellen* Opportunities are very rare.
News Review Editor *Susannah Herbert*
Submissions are always welcome, but the paper commissions its own, uses staff writers or works with literary agents, by and large. The features sections where most opportunities exist are *Style* and *The Culture*.
'The Culture' Editor *Helen Hawkins*
Business Editor *John Waples*
City Editor *Jenny Davey*
Science and Environment Editor *Jonathan Leake*
Literary Editor *Andew Holgate*
Sports Editor *Alex Butler*

'Style' Editor *Tiffanie Darke*
Sunday Times Magazine Colour supplement. Editor *To be appointed* No unsolicited material. Write with ideas in first instance.

The Times

1 Pennington Street, London E98 1TT
☏ 020 7782 5000   ☐ 020 7488 3242
www.thetimes.co.uk
Owner *News International plc/Rupert Murdoch*
Editor *James Harding*
Circulation 600,962

Generally right (though features can range in tone from diehard to libertarian). *The Times* receives a great many unsolicited offerings. Writers with feature ideas should approach by letter in the first instance. No fiction.
Deputy Editor *Keith Blackmore*
Deputy Head of News *John Wellman*
Business & City Editor *David Wighton*
Diary Editor *Adam Sherwin*
Arts Editor *Tim Teeman*
Literary Editor *Erica Wagner*
Political Editor *Phil Webster*
Sports Editor *Tim Hallissey*

The Times Magazine Saturday supplement. Editor *Gill Morgan*
Times 2 Editor *Emma Tucker*, Features Editor *Shaun Phillips*

Regional Newspapers

England

Berkshire

Reading Evening Post

8 Tessa Road, Reading RG1 8NS
☏ 0118 918 3000   🖷 0118 959 9363
editorial@reading-epost.co.uk
Owner *Guardian Media Group*
Editor *Andy Murrill*
Circulation 12,879

Unsolicited mss welcome; one or two received every day. Fiction rarely used. Interested in local news features, human interest, well-researched investigations. Special sections include holidays & travel; children's page (Tues); style page (Wed); business (Mon–Fri); motorcycling; food page; gardening; reviews; rock music (Thurs); motoring (Fri). Also magazines, *Food Monthly* and *Business Monthly*.

Cambridgeshire

Cambridge Evening News

Winship Road, Milton, Cambridge CB24 6PP
☏ 01223 434434   🖷 01223 434415
Owner *Cambridge Newspapers Ltd*
Editor *Paul Brackley*
Circulation 25,195

News Editor *Paul Holland*
Business Editor *Jenny Chapman*
Sports Editor *Paul Stimpson*

Evening Telegraph

57 Priestgate, Peterborough PE1 1JW
☏ 01733 555111   🖷 01733 313147 (editorial)
www.peterboroughtoday.co.uk
Owner *East Midlands Newspapers Ltd*
Editor *Mark Edwards*
Circulation 16,428

News Editor *Rose Taylor*
Assistant Editor (Content) *Alex Gordon*
Business Editor *Paul Grinnell*

Cheshire

Chester Chronicle

Chronicle House, Commonhall Street, Chester CH1 2AA
☏ 01244 340151   🖷 01244 606498
eric.langton@cheshirenews.co.uk
www.chesterchronicle.co.uk
Owner *Trinity Mirror Plc*
Editor-in-Chief *Eric Langton*
Circulation 22,070

All unsolicited feature material will be considered. No payment.

Cleveland

Hartlepool Mail

New Clarence House, Wesley Square, Hartlepool TS24 8BX
☏ 01429 239333   🖷 01429 869024
mail.news@northeast-press.co.uk
www.hartlepoolmail.co.uk
www.peterleemail.co.uk
Owner *Johnston Press Plc*
Editor *Joy Yates*
Circulation 17,044

Deputy Editor *Gavin Foster*
News Editor *Ian Willis*
Sports Editor *Roy Kelly*

Cumbria

News & Star

Newspaper House, Dalston Road, Carlisle CA2 5UA
☏ 01228 612600   🖷 01228 612601
Owner *CN Group Ltd*
Editor *Neil Hodgkinson*
Circulation 21,776

Assistant Editor *Andy Nixon*
Associate Editor/Woman's Page *Anne Pickles*
Sports Editor *Phil Rostron*

North West Evening Mail

Abbey Road, Barrow in Furness LA14 5QS

☎ 01229 821835　📠 01229 840164

news@nwemail.co.uk

www.nwemail.co.uk

Owner *Robin Burgess*

Editor *Jonathan Lee*

Circulation 17,654

All editorial material should be addressed to the editor.

Deputy Editor (Content) *Jon Townend*

Assistant Editor (Features) *Peter Leach*

News Editor *Rob Johnson*

Sports Editor *Frank Cassidy*

Derbyshire

Derby Evening Telegraph

Northcliffe House, Meadow Road, Derby DE1 2BH

☎ 01332 291111

newsdesk@derbytelegraph.co.uk

Owner *Northcliffe Media Group Ltd*

Editor *Steve Hall*

Circulation 39,152

Devon

Express & Echo

Heron Road, Sowton, Exeter EX2 7NF

☎ 01392 442211　📠 01392 442294 (editorial)

echonews@expressandecho.co.uk

www.thisisexeter.co.uk

Owner *South West Media Group*

Editor *Marc Astley*

Circulation 19,353

Weekly supplements: *Business Week*; *Property Echo*; *Motoring*; *Weekend Echo*.

Head of Content *Lynne Turner*

Features Editor *Becky Moran*

Sports Editor *Richard Davies* (📠 01392 442416)

The Herald

17 Brest Road, Derriford Business Park, Derriford, Plymouth PL6 5AA

☎ 01752 765500　📠 01752 765527

news@theplymouthherald.co.uk

www.thisisplymouth.co.uk

Owner *South West Media Group Ltd*

Editor *Bill Martin*

Circulation 34,342

All editorial material to be addressed to the editor or the News Editor, *James Garnett*.

Herald Express

Harmsworth House, Barton Hill Road, Torquay TQ2 8JN

☎ 01803 676000　📠 01803 676228 (editorial)

newsdesk@heraldexpress.co.uk

www.thisissouthdevon.co.uk

Owner *Northcliffe Media Group Ltd*

Editor *Andy Phelan*

Circulation 22,980

Drive scene, property guide, *On The Town* (leisure and entertainment guide), Monday sports, special pages, rail trail, Saturday surgery, nature and conservation column, dance scene, shop scene, The Business. Supplements: *Gardening* (weekly) and *Visitors Guide*. Unsolicited mss generally not welcome. All editorial material should be addressed to the editor in writing.

Western Morning News

17 Brest Road, Derriford Business Park, Derriford, Plymouth PL6 5AA

☎ 01752 765500　📠 01752 765535

wmnnewsdesk@westernmorningnews.co.uk

www.thisisdevon.co.uk

www.thisiscornwall.co.uk

www.thisiswesternmorningnews.co.uk

Owner *South West Media Group Ltd*

Editor *Alan Qualtrough*

Circulation 37,819

Unsolicited mss welcome, but must be of topical and local interest and addressed to the Deputy Editor, *Philip Bowern*.

News Editor *Steve Grant*

Features Editor *Su Carroll*

Sports Editor *Mark Stevens*

Dorset

Daily Echo

Richmond Hill, Bournemouth BH2 6HH

☎ 01202 554601　📠 01202 297543

neal.butterworth@bournemouthecho.co.uk

www.bournemouthecho.co.uk

Owner *Newsquest Media Group Ltd (a Gannett company)*

Editor *Neal Butterworth*

Circulation 30,754

Founded 1900. Has a strong news and features content and invites specialist articles, particularly on unusual subjects, either contemporary or historical, but only with a local angle or flavour. Special review sections each day, including sport, property, entertainment and culture, heritage, motoring, gardening, the environment and the coastline. Also *Saturday Magazine* and monthly *Society* glossy magazine. All editorial material should be addressed to the Editor, *Neal Butterworth*. 💷 Payment on publication.

Dorset Echo

Fleet House, Hampshire Road, Granby Industrial Estate, Weymouth DT4 9XD

☎ 01305 830930　📠 01305 830956

www.dorsetecho.co.uk
Owner *Newsquest Media Group Ltd (a Gannett company)*
Editor *Toby Granville*
Circulation 18,026

By-gone days, films, arts, showbiz, brides, motoring, property, weekend leisure and entertainment including computers and gardening and weekend magazine.
News Editor *Paul Thomas*
Sports Editor *Nigel Dean*

Durham

The Northern Echo

Priestgate, Darlington DL1 1NF
☎ 01325 381313   🖷 01325 380539
newsdesk@nne.co.uk
www.thenorthernecho.co.uk
Owner *Newsquest (Yorkshire and North East) Ltd (a Gannett company)*
Editor *Peter Barron*
Circulation 48,783

Founded 1870. Freelance pieces welcome but telephone first to discuss submission. Assistant Editor (News) *Nigel Burton* Interested in reports involving the North East or North Yorkshire. Preferably phoned in.
Features Editor *Lindsay Jennings* Background pieces to topical news stories relevant to the area. Must be arranged with the features editor before submission of any material.
Business Editor *Owen McAteer*
Sports Editor *Nick Loughlin*
💷 Payment and length by arrangement.

Essex

Colchester Gazette

Oriel House, 43–44 North Hill, Colchester CO1 1TZ
☎ 01206 506000   🖷 01206 508274
gazette.newsdesk@nqe.com
www.evening-gazette.co.uk
Owner *Newsquest (Essex) Ltd*
Editor *Irene Kettle*
Circulation 20,523

Monday to Friday daily newspaper servicing north and mid-Essex including Colchester, Harwich, Clacton, Braintree, Witham, Maldon and Chelmsford. Unsolicited mss not generally used. Relies heavily on regular contributors.
Features Editor *Iris Clapp*

Echo Newspapers

Newspaper House, Chester Hall Lane, Basildon SS14 3BL
☎ 01268 522792   🖷 01268 469281

Owner *Newsquest Media Group (a Gannett company)*
Editor *Martin McNeill*
Circulation 33,957

Relies almost entirely on staff and regular outside contributors, but will very occasionally consider material sent on spec. Approach the editor in writing with ideas. Although the paper is Basildon-based, its largest circulation is in the Southend area.

Gloucestershire

The Citizen

6–8 The Oxebode, Gloucester GL1 1RZ
☎ 01452 420621   🖷 01452 420664 (editorial)
citizen.news@glosmedia.co.uk
www.thisisgloucestershire.co.uk
Owner *Northcliffe Newspapers Group Ltd*
Editor *Ian Mean*
Circulation 23,474

News or features material should be addressed to Managing Editor (News & Pictures) *Jenny Eastwood*, Deputy News Editor *Sally Munro* or Assistant Deputy News Editor *Jenni Silver*.

Gloucestershire Echo

1 Clarence Parade, Cheltenham GL50 3NY
☎ 01242 271900   🖷 01242 271848
echo.editor@glosmedia.co.uk
www.thisisgloucestershire.co.uk
Owner *Northcliffe Media Group Ltd*
Editor *Kevan Blackadder*
Circulation 18,850

All material, other than news, should be addressed to the editor.

Hampshire

The News

The News Centre, Hilsea, Portsmouth PO2 9SX
☎ 023 9266 4488   🖷 023 9267 3363
newsdesk@thenews.co.uk
www.portsmouth.co.uk
Owner *Portsmouth Printing & Publishing Ltd*
Editor *Mark Waldron*
Circulation 52,531

Unsolicited mss not generally accepted. Approach by letter.
Assistant Editor *Graeme Patfield* General subjects of S.E. Hants interest. Maximum 600 words. No fiction.
Sports Editor *Howard Frost* Sports background features. Maximum 600 words.

The Southern Daily Echo

Newspaper House, Test Lane, Redbridge,
Southampton SO16 9JX
☎ 023 8042 4777   🖷 023 8042 4545
newsdesk@dailyecho.co.uk
www.dailyecho.co.u
Owner *Newsquest Plc*
Editor *Ian Murray*
Circulation 36,906

Unsolicited mss 'tolerated'. Approach the editor in
writing with strong ideas; staff supply almost all
the material.

Kent

Kent Messenger

6 & 7 Middle Row, Maidstone ME14 1TG
☎ 01622 695666   🖷 01622 664988
messengernews@thekmgroup.co.uk
www.kentonline.co.uk
www.kentmessenger.co.uk
Owner *Kent Messenger Group*
Editor *Bob Bounds*
Circulation 46,733

Very little freelance work is commissioned.

Medway Messenger

Medway House, Ginsbury Close, Sir Thomas
Longley Road, Medway City Estate, Strood,
Rochester ME2 4DU
☎ 01634 227800   🖷 01634 715256
medwaymessenger@thekmgroup.co.uk
www.kentonline.co.uk
www.medwaymessenger.co.uk
Owner *Kent Messenger Group*
Editor *Bob Bounds*
Circulation 11,875 (Mon)/20,000 (Fri)

Published Mondays and Fridays with the free
Medway Extra on Wednesday.
Business Editor *Trevor Sturgess*
Community Editor *David Jones*
Sports Editor *Mike Rees*

Lancashire

The Bolton News

Newspaper House, Churchgate, Bolton BL1 1DE
☎ 01204 522345   🖷 01204 365068
newsdesk@theboltonnews.co.uk
www.theboltonnews.co.uk
Owner *Newsquest Media Group Ltd (a Gannett
company)*
Editor *Ian Savage*
Circulation 27,540

Business, children's page, travel, local services,
motoring, fashion and cookery.

News Editor *James Higgins*
Features Editor/Women's Page *Andrew Mosley*

The Gazette (Blackpool)

Avroe House, Avroe Crescent, Blackpool FY4 2DP
☎ 01253 400888   🖷 01253 361870
jon.rhodes@blackpoolgazette.co.uk
www.blackpoolgazette.co.uk
Owner *Johnston Press plc*
Managing Director *Darren Russell*
Editor *David Helliwell*
Circulation 26,968

Unsolicited mss welcome in theory. Approach in
writing with an idea. Supplements: *Eve* (women,
Tues); *All Stars* (football, youth sport, Wed);
Homes (Thurs); *The Weekend* (entertainment,
Fri); *Wheels* (motoring, Fri); *Life!* magazine
(entertainment & leisure, Sat); *Jobs Plus* (jobs and
careers, Sat).

Lancashire Evening Post

Olivers Place, Eastway, Fulwood, Preston PR2 9ZA
☎ 01772 254841   🖷 01772 880173
www.lep.co.uk
Owner *Johnston Press plc*
Editor *Simon Reynolds*
Circulation 28,275

Unsolicited mss are not generally welcome; many
are received and not used. All ideas in writing to
the editor.

Lancashire Telegraph

Newspaper House, High Street, Blackburn
BB1 1HT
☎ 01254 678678   🖷 01254 680429
lt_editorial@lancashire.newsquest.co.uk
www.lancashiretelegraph.co.uk
Owner *Newsquest Media Group Ltd (a Gannett
company)*
Editor *Kevin Young*
Circulation 28,569

News stories and feature material with an East
Lancashire flavour (a local angle, or written by
local people) welcome. Approach in writing with
an idea in the first instance. No fiction.
News Editor *Ian Singleton*
Features Editor *John Anson*
Sports Editor *Paul Plunkett*
Pictures Editor *Neil Johnson*

Oldham Evening Chronicle

PO Box 47, Union Street, Oldham OL1 1EQ
☎ 0161 633 2121   🖷 0161 652 2111
news@oldham-chronicle.co.uk
www.oldham-chronicle.co.uk
Owner *Hirst Kidd & Rennie Ltd*
Editor *Jim Williams*

Circulation 19,280

Motoring, food and wine, lifestyle supplement, business page.

News Editor *Mike Attenborough*

Leicestershire

Leicester Mercury
St George Street, Leicester LE1 9FQ

📞 0116 251 2512   📠 0116 253 0645

newsdesk@leicestermercury.co.uk

www.thisisleicestershire.co.uk

Owner *Northcliffe Media Group Ltd*

Editor *Keith Perch*

Circulation 64,919

News Editor *Mark Charlton*

Features Editor *Alex Dawson*

Lincolnshire

Grimsby Telegraph
80 Cleethorpe Road, Grimsby DN31 3EH

📞 01472 360360   📠 01472 372257

newsdesk@grimsbytelegraph.co.uk

Owner *Northcliffe Media Group Ltd*

Editor *Michelle Lalor*

Circulation 31,538

Sister paper of the *Scunthorpe Telegraph*. Unsolicited mss generally welcome. Approach in writing. No fiction. Weekly supplements: *Business Telegraph*; *Property Telegraph*; *Motoring*. All material to be addressed to the News Editor, *Lucy Wood*. Particularly welcomes hard news stories; approach in haste by telephone.

Lincolnshire Echo
Brayford Wharf East, Lincoln LN5 7AT

📞 01522 820000   📠 01522 804493

news@lincolnshireecho.co.uk

Owner *Northcliffe Media Group Ltd*

Editor *Jon Grubb*

Circulation 20,181

News and sport, best buys, holidays, motoring, dial-a-service, restaurants, leisure, home improvement, record reviews, gardening corner, stars.

Scunthorpe Telegraph
4–5 Park Square, Scunthorpe DN15 6JH

📞 01724 273273   📠 01724 273101

Owner *Northcliffe Media Group Ltd*

Editor *Mel Cook*

Circulation 18,823

All correspondence should go to the Deputy Editor, *Dave Atkin*.

London

Evening Standard
Northcliffe House, 2 Derry Street, London W8 5EE

📞 020 7938 6000   📠 020 7937 2648

www.standard.co.uk

Owner *Alexander Lebedev*

Editor *Geordie Greig*

Circulation 292,976

Long-established evening paper, serving Londoners with both news and feature material. Genuine opportunities for London-based features. Produces a weekly colour supplement, *ES Magazine* (Fri) and *Homes & Property* (Wed).

Deputy Editor *Andrew Bordiss* (from Aug 09: *Sarah Sands*)

Executive Editor *Anne McElvoy*

Managing Editor *Doug Wills*

Features Editor *Charlotte Ross*

News Editor *Hugh Dougherty*

Arts Editor *Fiona Hughes*

Sports Editor *Steve McKenlay*

ES Editor *To be appointed*

Homes & Property Editor *Janice Morley*

Manchester

Manchester Evening News
Number 1 Scott Place, Manchester M3 3RN

📞 0161 832 7200   📠 0161 211 2030

www.manchestereveningnews.co.uk

Owner *Manchester Evening News Ltd*

Editor *Paul Horrocks*

Circulation 153,724

One of the country's major regional dailies. Initial approach in writing preferred. No fiction. *Personal Finance* (Mon); *Health*; *Holidays* (Tues); *Homes & Property* (Wed); *Small Business* (Thurs); *Lifestyle* (Fri/Sat).

News Editor *Sarah Lester*

Features Editor *Deanna Delamotta* Regional news features, personality pieces and showbiz profiles considered. Maximum 1,200 words.

Sports Editor *Peter Spencer*

Women's Page *Helen Tither*

💷 Payment based on house agreement rates.

Merseyside

Liverpool Daily Post
PO Box 48, Old Hall Street, Liverpool L69 3EB

📞 0151 227 2000   📠 0151 472 2506

www.liverpooldailypost.co.uk

Owner *Trinity Mirror Plc*

Editor *Mark Thomas*

Circulation 12,910

Unsolicited mss welcome. Receives about six a day. Approach in writing with an idea. No

fiction. Local, national/international news,
current affairs, profiles (with pictures). Maximum
800–1,000 words.
Features Editor *Emma Johnson*
News Editor *Andy Kelly*
Business Editor *Bill Gleeson*
Sports Editor *Richard Williamson*

Liverpool Echo

PO Box 48, Old Hall Street, Liverpool L69 3EB
☎ 0151 227 2000   🖷 0151 236 4682
letters@liverpoolecho.co.uk
www.liverpoolecho.co.uk
Owner *Trinity Mirror Merseyside*
Editor *Alastair Machray*
Circulation 97,779

One of the country's major regional dailies.
Unsolicited mss welcome; initial approach with
ideas in writing preferred.
News Editor *Maria Breslin*
Features Editor *Jane Haase*
Sports Editor *John Thompson*
Women's Editor *Susan Lee*

Norfolk

Eastern Daily Press

Prospect House, Rouen Road, Norwich NR1 1RE
☎ 01603 628311   🖷 01603 623872
EDP@archant.co.uk
www.EDP24.co.uk
Owner *Archant Regional*
Editor *Peter Franzen, OBE*
Circulation 61,143

Most pieces by commission only. Supplements:
EDP2 (daily); rental, motoring, business, property
pages, agriculture, employment (all weekly); *Event*
full-colour magazine (Fri); *Saturday* full colour
magazine; *Sunday* 24-page supplement.
Deputy Editor *Peter Waters*
Assistant Editor (News) *Paul Durrant*
Head of Feature Content *Sarah Hardy*
Head of Sport *Richard Willner*
Assistant Editor (Saturday) *Steve Snelling*

Evening News

Prospect House, Rouen Road, Norwich NR1 1RE
☎ 01603 628311   🖷 01603 219060
tim.williams@archant.co.uk
www.eveningnews24.co.uk
Owner *Archant Ltd*
Editor *James Foster*
Circulation 21,030

Includes special pages on local property,
motoring, pop, fashion, arts, entertainments
and TV, gardening, local music scene, home and
family.
Deputy Editor *Tim Williams (tim.williams@
archant.co.uk)*

Features Editor *Derek James (derek.james@
archant.co.uk)*

Northamptonshire

Chronicle and Echo

Upper Mounts, Northampton NN1 3HR
☎ 01604 467000   🖷 01604 467190
Owner *Northamptonshire Newspapers*
Editor *David Summers*
Circulation 19,016

Unsolicited mss are 'not necessarily unwelcome
but opportunities to use them are rare'. Approach
in writing with an idea. No fiction. Supplements:
Sport (Mon); *Property Today* (Wed); *Jobs Today*
(Thurs); *Motors Today* and *The Guide* (Friday);
Weekend Life (Sat).
News Editor *Richard Edmondson*
Features Editor/Women's Page *Lily Canter*
Sports Editor *Jeremy Casey*

Evening Telegraph

Newspaper House, Ise Park, Rothwell Road,
Kettering NN16 8GA
☎ 01536 506100   🖷 01536 506195 (editorial)
et.newsdesk@northantsnews.co.uk
www.northantset.co.uk
Owner *Johnston Press Plc*
Editor *Jeremy Clifford*
Circulation 21,523

Northamptonshire Business Guide (weekly);
Entertainment Guide and *Jobs* supplements
(Thurs); films and eating out (Fri); and *Home &
Garden* – monthly lifestyle supplement including
gardening, etc.
Deputy Editor *Neil Pickford*
Assistant Editor *Nick Tite*
Acting News Editor *David Brennan*
Sports Editor *Jim Lyon*

Nottinghamshire

Evening Post Nottingham

Castle Wharf House, Nottingham NG1 7EU
☎ 0115 948 2000   🖷 0115 964 4032
Owner *Northcliffe Media Group Ltd*
Editor *Malcolm Pheby*
Circulation 51,526

Unsolicited mss occasionally used. Good local
interest only. Maximum 800 words. No fiction.
Send ideas in writing.
News Editor *Steven Fletcher*
Deputy Editor *Martin Done*
Sports Editor *Dave Parkinson*

Oxfordshire

Oxford Mail
Osney Mead, Oxford OX2 OEJ
☎ 01865 425262 📠 01865 425554
news@oxfordmail.co.uk
www.oxfordmail.co.uk
Owner *Newsquest (Oxfordshire) Ltd*
Editor *Simon O'Neill*
Circulation 23,402

Unsolicited mss are considered but a great many unsuitable offerings are received. Approach in writing with an idea, rather than by phone. No fiction.
£ All fees negotiable.

Shropshire

Shropshire Star
Ketley, Telford TF1 5HU
☎ 01952 242424 📠 01952 254605
Owner *Shropshire Newspapers Ltd*
Editor *Sarah-Jane Smith*
Circulation 67,726

No unsolicited mss; approach the editor with ideas in writing in the first instance. No news or fiction.
Head of Features *Carl Jones* Limited opportunities; uses mostly in-house or syndicated material. Maximum 1,200 words.
Sports Editor *Dave Ballinger*

Somerset

The Bath Chronicle
Westpoint, James Street West, Bath BA1 2DA
☎ 01225 322322 📠 01225 322291
www.thisisbath.com
Owner *Northcliffe Media Group Ltd*
Editor *Sam Holliday (s.holliday@bathchron.co.uk)*
Circulation 19,719

Local news and features especially welcomed (news@bathchron.co.uk).
Deputy Editor *Paul Wiltshire*
Features Editor *Jackie Chappell*
Sports Editor *Julie Riegal*

Evening Post
Temple Way, Bristol BS99 7HD
☎ 0117 934 3000 📠 0117 934 3575
epnews@bepp.co.uk
www.thisisbristol.co.uk
Owner *Bristol News & Media Ltd (part of Northcliffe Media Group Ltd)*
Editor *Mike Norton*
Circulation 46,522

News Editor *Robin Perkins*

Features Editor *David Webb*
Sports Editor *Steve Mellen*

Western Daily Press
Temple Way, Bristol BS99 7HD
☎ 0117 934 3000 📠 0117 934 3574
WDEditor orWDNews or features@westpress.co.uk
www.westerndailypress.co.uk
Owner *Bristol News and Media Ltd*
Editor *Andy Wright*
Circulation 38,211

Sports Editor *Steve Mellen*
Women's Page *Susie Weldon*

Staffordshire

Burton Mail
65–68 High Street, Burton upon Trent DE14 1LE
☎ 01283 512345 📠 01283 515351
editorial@burtonmail.co.uk
Owner *Staffordshire Newspapers Ltd*
Editor *Paul Hazeldine*
Circulation 13,734

Fashion, health, wildlife, environment, nostalgia, financial/money; consumer; women's world, rock; property; motoring, farming; what's on, leisure *Weekend Supplement* (Sat).
Sports Editor *Rex Page*

The Sentinel
Sentinel House, Etruria, Stoke on Trent ST1 5SS
☎ 01782 602525 📠 01782 201167
www.thisisthesentinel.co.uk
Owner *Staffordshire Sentinel News & Media Ltd*
Editor-in-Chief *Mike Sassi*
Circulation 58,049

Weekly sports final supplement. All material should be sent to the Assistant Editor, *Martin Tideswell.*

Suffolk

East Anglian Daily Times
Press House, 30 Lower Brook Street, Ipswich IP4 1AN
☎ 01473 230023 📠 01473 324776
news@eadt.co.uk
www.eadt.co.uk
Owner *Archant Ltd*
Editor *Terry Hunt*
Circulation 32,208

Founded 1874. Unsolicited mss generally not welcome; three or four received a week and almost none are used. Approach in writing in the first instance. No fiction. Supplements: sport (Mon); business (Tues); jobs (Wed); property (Thurs); motoring (Fri); magazine (Sat).

News Editor *Brad Jones* Hard news stories involving East Anglia (Suffolk, Essex particularly) or individuals resident in the area are always of interest.

Features *Julian Ford* Mostly in-house, but will occasionally buy in when the subject is of strong Suffolk/East Anglian interest. Photo features preferred (extra payment). Special advertisement features are regularly run. Some opportunities here. Maximum 1,000 words.

Deputy Editor, Sport *Colin Adwent*
Women's Page *Victoria Hawkins*

Evening Star

Press House, 30 Lower Brook Street, Ipswich IP4 1AN
☎ 01473 230023   🖷 01473 324850
Owner *Archant Ltd*
Editor *Nigel Pickover*
Circulation 17,548

News Editor *Jess Nicholls*

Sussex

The Argus

Argus House, Crowhurst Road, Hollingbury, Brighton BN1 8AR
☎ 01273 544544   🖷 01273 505703
editor@theargus.co.uk
www.theargus.co.uk
Owner *Newsquest (Sussex) Ltd*
Editor *Michael Beard*
Circulation 30,070

News Editor *Lee Gibbs*
Sports Editor *Chris Giles*

Teesside

Evening Gazette

Borough Road, Middlesbrough TS1 3AZ
☎ 01642 234242
www.gazettelive.co.uk
Owner *Trinity Mirror plc*
Editor *Darren Thwaites*
Circulation 46,692

Special pages: health, education, family, business, property, entertainment, leisure, motoring, recruitment.
News Editor *Jim Horsley*
Business Features/Commercial Features *Sue Scott*
Sports Editor *Philip Tallentire*
Councils *Sandy McKenzie*
Health *Marie Levy*
Crime *Simon Walton*

Tyne & Wear

Evening Chronicle

Groat Market, Newcastle upon Tyne NE1 1ED
☎ 0191 232 7500   🖷 0191 232 2256
www.chroniclelive.co.uk
Owner *Trinity Mirror Plc*
Editor *Paul Robertson*
Circulation 67,103

Receives a lot of unsolicited material, much of which is not used. Family issues, gardening, pop, fashion, cooking, consumer, films and entertainment guide, home improvements, motoring, property, angling, sport and holidays. Approach in writing with ideas. Limited opportunities for features due to full-time feature staff.

News Editor *James Marley*
Sports Editor *Paul New*
Women's Interests *Liz Lamb*

The Journal

Groat Market, Newcastle upon Tyne NE1 1ED
☎ 0191 232 7500/201 6344 (Newsdesk)
🖷 0191 232 2256/201 6044
jnl.newsdesk@ncjmedia.co.uk
www.journallive.co.uk
Owner *Trinity Mirror Plc*
Editor *Brian Aitken*
Circulation 32,274

Daily platforms include farming and business.
Deputy Editor *Peter Montellier*
Sports Editor *Kevin Dinsdale*
Arts & Entertainment Editor *David Whetstone*
Environment Editor *Tony Henderson*
Business Editor *Iain Laing*

Shields Gazette

Chapter Row, South Shields NE33 1BL
☎ 0191 427 4800   🖷 0191 456 8270
www.southtynesidetoday.co.uk
Owner *Northeast Press Ltd*
Editor *John Szymanski*
Circulation 17,522

News & Features Editor *Helen Charlton*

Sunday Sun

ncjMedia, Groat Market, Newcastle upon Tyne NE1 1ED
☎ 0191 201 6299   🖷 0191 201 6180
colin.patterson@ncjmedia.co.uk
www.sundaysun.co.uk
flnm.co.uk (football website)
Owner *Trinity Mirror Plc*
Editor *Colin Patterson*

Circulation 61,869

All material should be addressed to the appropriate editor (phone to check), or to the editor.

Sports Editor *Neil Farrington*

Sunderland Echo

Echo House, Pennywell, Sunderland SR4 9ER
☎ 0191 501 5800   🖷 0191 534 5975
rob.lawson@northeast-press.co.uk
Owner *Johnston Press Plc*
Editor *Rob Lawson*
Circulation 39,159

All editorial material to be addressed to the News Editor, *Peter Jeffrey* (echo.news@northeast-press.co.uk).

Warwickshire

Coventry Telegraph

Corporation Street, Coventry CV1 1FP
☎ 024 7663 3633   🖷 024 7655 0869
news@coventry-telegraph.co.uk
www.IcCoventry.co.uk
Owner *Trinity Mirror Plc*
Editor *David Brookes*
Circulation 43,594

Unsolicited mss are read, but few are published. Approach in writing with an idea. No fiction. All unsolicited material should be addressed to the editor. Maximum 600 words for features.

News Managers *Steve Chilton, Steve Williams*
Features Manager/Women's Page *Tara Cain*
Sports Editor *Rob Madill*
🖷 Payment negotiable.

Leamington Spa Courier

32 Hamilton Terrace, Leamington Spa CV32 4LY
☎ 01926 457737   🖷 01926 339960
news@leamingtoncourier.co.uk
www.leamingtoncourier.co.uk
Owner *Johnston Press Publishing*
Editor *Martin Lawson*
Circulation 12,296

One of the Leamington Spa Courier series which also includes the *Warwick Courier* and *Kenilworth Weekly News*. Unsolicited feature articles considered, particularly matter with a local angle. Telephone with idea first.

News Editor *Peter Ormerod*

West Midlands

Birmingham Mail

Bpm Media (Midlands) Ltd, Floor 6, Fort Dunlop, Fort Parkway, Birmingham B24 9FF
☎ 0121 236 3366   🖷 0121 233 0271 (editorial)

steve.dyson@birminghammail.net
www.birminghammail.net
Owner *Trinity Mirror Plc*
Editor *Steve Dyson*
Circulation 61,526

Freelance contributions are welcome, particularly topics of interest to the West Midlands and feature pieces offering original and lively comment on family issues, health or education. Also news tips and community picture contributions.

Content Editor *Andy Richards*
Regional Sports Editor *Ken Montgomery*
Regional Images Editor *Steve Murphy*
Regional Business Editor *Alun Thorne*

Birmingham Post

Bpm Media (Midlands) Ltd, Floor 6, Fort Dunlop, Fort Parkway, Birmingham B24 9FF
☎ 0121 234 5301
Owner *Trinity Mirror Plc*
Editor *Marc Reeves*
Circulation 12,791

One of the country's leading regional newspapers. Freelance contributions are welcome. Topics of interest to the West Midlands and pieces offering lively, original comment are particularly welcome.

Content Editor *Sarah Probert*

Express & Star

Queen Street, Wolverhampton WV1 1ES
☎ 01902 313131   🖷 01902 319721
Owner *Midlands News Association*
Editor *Adrian Faber*
Circulation 130,216

Deputy Editor *Keith Harrison*
Assistant Editor *Marc Drew*
Features Editor *Emma Farmer*
Business Editor *Jim Walsh*
Sports Editor *Tony Reynolds*
Internet News Editor *Tim Walters*
Women's Editor *Maria Cusine*

Sunday Mercury (Birmingham)

Bpm Media (Midlands) Ltd, Floor 6, Fort Dunlop, Fort Parkway, Birmingham B24 9FF
☎ 0121 234 5567   🖷 0121 234 5877
sunday.mercury@sundaymercury.net
www.sundaymercury.net
Owner *Trinity Mirror Plc*
Editor *Steve Dyson*
Circulation 54,375

Content Editor *Tony Larner*

Wiltshire

Swindon Advertiser

100 Victoria Road, Swindon SN1 3BE
☎ 01793 501806   🖷 01793 501888

newsdesk@swindonadvertiser.co.uk
www.swindonadvertiser.co.uk
Owner *Newsquest (Wiltshire) Ltd*
Editor *Dave King*
Circulation 20,845

Copy and ideas invited. 'All material must be strongly related or relevant to the town of Swindon or the county of Wiltshire.' Little scope for freelance work. Fees vary depending on material.
Deputy Editor *Pauline Leighton*
News Editor *Kevin Burchall*
Sports Editor *Steve Butt*

Worcestershire

Worcester News

Berrow's House, Hylton Road, Worcester WR2 5JX
☎ 01905 748200   📠 01905 742277
Owner *Newsquest (Midlands South) Ltd*
Editor *Kevin Ward*
Circulation 16,507

Community (Tues); jobs (Wed); property (Thurs); entertainment and motoring (Fri).
News Editor *Stephanie Preece*
Sports Editor *Paul Ricketts*

Yorkshire

The Doncaster Star

Sunny Bar, Doncaster DN1 1NB
☎ 01302 348500   📠 01302 348528
doncaster@sheffieldnewspapers.co.uk
www.thestar.co.uk/doncaster
Owner *Sheffield Newspapers Ltd*
Editor *Alan Powell*
Circulation 3,384

All editorial material to be addressed to the News Editor, *David Kessen*.
Sports Writer *Steve Hossack*

Huddersfield Daily Examiner

Queen Street South, Huddersfield HD1 3DU
☎ 01484 430000   📠 01484 437789
editorial@examiner.co.uk
www.examiner.co.uk
Owner *Trinity Mirror Plc*
Editor *Roy Wright*
Circulation 23,290

Home improvement, home heating, weddings, dining out, motoring, fashion, services to trade and industry.
Deputy Editor *Michael O'Connell*
News Editor *Neil Atkinson*
Features Editor *Andrew Flynn*
Sports Editor *Mel Booth*
Women's Page *Hilarie Stelfox*

Hull Daily Mail

Blundell's Corner, Beverley Road, Hull HU3 1XS
☎ 01482 315350   📠 01482 315353
news@mailnewsmedia.co.uk
www.thisishullandeastriding.co.uk
Owner *Northcliffe Media Group Ltd*
Editor *John Meehan*
Circulation 51,886

News Editor *Rick Lyon*
Features Editor *Brian Marshall*

The Press

PO Box 29, 76–86 Walmgate, York YO1 9YN
☎ 01904 653051   📠 01904 612853
newsdesk@thepress.co.uk
www.yorkpress.co.uk
Owner *Newsquest Media Group (a Gannett company)*
Editor *Kevin Booth*
Circulation 31,569

Unsolicited mss not generally welcome, unless submitted by journalists of proven ability. *Business Press Pages* (daily); *Property Press* (Thurs); *Twenty4Seven* – entertainment (Fri).
Assistant Editor *Fran Clee*
Head of Content *Scott Armstrong*
Picture Editor *Martin Oates*
Sports Editor *Stuart Martel*
💷 Payment negotiable.

Scarborough Evening News

17–23 Aberdeen Walk, Scarborough YO11 1BB
☎ 01723 363636   📠 01723 379033
editorial@scarborougheveningnews.co.uk
www.scarboroughtoday.co.uk
Owner *Yorkshire Regional Newspapers Ltd*
Editor *Ed Asquith*
Circulation 12,676

Special pages include property (Mon); jobs (Thurs); motors (Fri).
Deputy Editor *Sue Wilkinson*
News Editor *Steven Hartley*
Sports Editor *Charles Place*
All other material should be addressed to the editor.

The Star

York Street, Sheffield S1 1PU
☎ 0114 276 7676   📠 0114 272 5978
starnews@sheffieldnewspapers.co.uk
www.thestar.co.uk
Owner *Johnston Press*
Editor *Alan Powell*
Circulation 47,216

Unsolicited mss not welcome, unless topical and local.
News Editor *Charles Smith* Contributions only accepted from freelance news reporters if they relate to the area.

Features Editor *Martin Smith* Very rarely requires outside features, unless on specialized local subjects.
Sports Editor *Bob Westerdale*
Women's Page *Jo Davison*
Ⓔ Payment negotiable.

Telegraph & Argus (Bradford)

Hall Ings, Bradford BD1 1JR
☎ 01274 729511   ℻ 01274 723634
www.thetelegraphandargus.co.uk
Owner *Newsquest Media Group Ltd (a Gannett company)*
Editor *Perry Austin-Clarke*
Circulation 34,042

No unsolicited mss – approach in writing with samples of work. No fiction.
Head of News *Martin Heminway*
Features Editor *Emma Clayton* Local features and general interest. Showbiz pieces. 600–1,000 words (maximum 1,500).

Yorkshire Evening Post

Wellington Street, Leeds LS1 1RF
☎ 0113 243 2701   ℻ 0113 238 8536
eped@ypn.co.uk
www.yorkshireeveningpost.co.uk
Owner *Johnston Press*
Editor *Paul Napier*
Circulation 49,064

Evening sister of the *Yorkshire Post*.
News Editor *Gillian Haworth*
Features Editor *Jayne Dawson*
Sports Editor *Mark Absolom*
Women's Editor *Jayne Dawson*

Yorkshire Post

Wellington Street, Leeds LS1 1RF
☎ 0113 243 2701   ℻ 0113 238 8537
yp.editor@ypn.co.uk
www.yorkshireposttoday.co.uk
Owner *Johnston Press*
Editor *Peter Charlton*
Circulation 45,718

A serious-minded, quality regional daily with a generally conservative outlook. Three or four unsolicited mss arrive each day; all will be considered but initial approach in writing preferred. All submissions should be addressed to the editor. No fiction, poetry or family histories.
Deputy Editor *Duncan Hamilton*
Features Editor *Sarah Freeman* Open to suggestions in all fields (though ordinarily commissioned from specialist writers).
Sports Editor *Matt Reeder*

Northern Ireland

Belfast News Letter

Johnston Press, Northern Ireland Division, 2 Esky Drive, Portadown BT63 5YY
☎ 028 3839 3939   ℻ 028 3839 3940
Owner *Johnston Press*
Editor *Darwin Templeton*

Weekly supplements: *Farming Life*; *Business News Letter*; *Lifestyle*; *PM*; *Sports Ulster*.
News Editors *Karen Grunason, Ben Lowry*
Features Editor *Geoff Hill*
Sports Editor *Brian Millar*
Business Editor *Adrienne McGill*
Agricultural Editor *David McCoy*
Political Editor *Stephen Dempster*

Belfast Telegraph

Royal Avenue, Belfast BT1 1EB
☎ 028 9026 4000   ℻ 028 9055 4506
newseditor@belfasttelegraph.co.uk
www.belfasttelegraph.co.uk
Owner *Independent News & Media (UK)*
Editor *Martin Lindsay*
Circulation 69,457

Weekly business, property and recruitment supplements.
Deputy Editor *Paul Connolly*
News Editor *Ronan Henry*
Features Editor *Gail Walker*
Sports Editor *Steven Beacom*

The Irish News

113/117 Donegall Street, Belfast BT1 2GE
☎ 028 9032 2226   ℻ 028 9033 7505
newsdesk@irishnews.com
www.irishnews.com
Owner *Irish News Ltd*
Editor *Noel Doran*
Circulation 47,819

All material to appropriate editor (phone to check), or to the news desk.
Assistant Editor *Fiona McGarry*
Features Editor *Will Scholes*
Sports Editor *Thomas Hawkins*

Sunday Life

124–144 Royal Avenue, Belfast BT1 1EB
☎ 028 9026 4300   ℻ 028 9055 4507
m.hill@belfasttelegraph.co.uk
www.sundaylife.co.uk
Owner *Independent News & Media (UK)*
Editor *Jim Flanagan*
Circulation 63,528

Deputy Editor *Martin Hill*
Sports Editor *Jim Gracey*

Scotland

The Courier and Advertiser

80 Kingsway East, Dundee DD4 8SL
☎ 01382 223131   📠 01382 454590
courier@dcthomson.co.uk
www.thecourier.co.uk
Owner *D.C. Thomson & Co. Ltd*
Editor *Bill Hutcheon*
Circulation 69,414

Circulates in East Central Scotland. Features occasionally accepted on a wide range of subjects, particularly local/Scottish interest, including finance, insurance, agriculture, motoring, modern homes, lifestyle and fitness. Maximum length: 500 words.
News Editor *Michael Alexander*
Features Editor/Women's Page *Catriona McInnes*
Sports Editor *Graeme Dey*

Daily Record (Glasgow)
▷ National Newspapers

Evening Express (Aberdeen)

PO Box 43, Lang Stracht, Mastrick, Aberdeen
AB15 6DF
☎ 01224 690222   📠 01224 699575
ee.news@ajl.co.uk
Owner *D.C. Thomson & Co. Ltd*
Editor *Damian Bates*
Circulation 52,029

Circulates in Aberdeen and the Grampian region. Local, national and international news and pictures, sport. Family platforms include *What's On*; *Counter* (consumer news); *Eating Out Guide*; *Family Days Out*. Unsolicited mss welcome 'if on a controlled basis'.
Deputy Editor *Richard Prest* Freelance news contributors welcome.
💷 Payment negotiable.

Evening News

Barclay House, 108 Holyrood Road, Edinburgh
EH8 8AS
☎ 0131 620 8620   📠 0131 620 8696
www.edinburghnews.com
Owner *Johnston Press*
Editor *John McLellan*
Circulation 47,129

Founded 1873. Circulates in Edinburgh, Fife, Central and Lothian. Coverage includes: entertainment, gardening, motoring, shopping, fashion, health and lifestyle, showbusiness. Occasional platform pieces, features of topical and/or local interest. Unsolicited feature material welcome. Approach the appropriate editor in writing.
News Editor *Alan Young*
Sports Editor *Graham Lindsay*
💷 NUJ/house rates.

Evening Telegraph

80 Kingsway East, Dundee DD4 8SL
☎ 01382 223131   📠 01382 454590
general@eveningtelegraph.co.uk
www.eveningtelegraph.co.uk
Owner *D.C. Thomson & Co. Ltd*
Editor *Gordon Wishart*
Circulation 23,048

Circulates in Tayside, Dundee and Fife. All material should be addressed to the editor.

Evening Times

200 Renfield Street, Glasgow G2 3QB
☎ 0141 302 7000   📠 0141 302 6677
news@eveningtimes.co.uk
www.eveningtimes.co.uk
Owner *Newsquest*
Editor *Donald Martin*
Circulation 83,010 (Mon.–Fri.)/44,000 (Sat.)

Circulates in Glasgow and the west of Scotland. Supplements: *Job Search*; *City Living*; *Drive Times*; *Extra Times*; *Times Out*; *Life & Times*.
Head of News *Yvonne Flynn*
Features Editor *Garry Scott*
Sports Editor *Graham Shields*

Greenock Telegraph

2 Crawfurd Street, Greenock PA15 1LH
☎ 01475 726511   📠 01475 783734
Owner *Clyde & Forth Press Ltd*
Editor *Wendy Metcalfe*
Circulation 16,349

Circulates in Greenock, Port Glasgow, Gourock, Kilmacolm, Langbank, Inverkip, Wemyss Bay, Skelmorlie, Largs. Unsolicited mss considered 'if they relate to the newspaper's general interests'. No fiction. All material to be addressed to the editor.

The Herald/Sunday Herald (Glasgow)
▷ National Newspapers

Paisley Daily Express

14 New Street, Paisley PA1 1YA
☎ 0141 887 7911   📠 0141 887 6254
pde@s-un.co.uk
www.paisleydailyexpress.co.uk
Owner *Scottish & Universal Newspapers Ltd*
Editor *Anne Dalrymple*
Circulation 8,831

Circulates in Paisley, Linwood, Renfrew, Johnstone, Elderslie, Neilston and Barrhead. Unsolicited mss welcome only if of genuine local (Renfrewshire) interest. The paper does not commission work and will consider submitted material. Maximum 1,000–1,500 words. All submissions to the news editor.
News Editor *Gavin Penny*
Sports Reporter *Paul Behan*

The Press and Journal
PO Box 43, Lang Stracht, Mastrick, Aberdeen
AB15 6DF
☎ 01224 690222   ⊕ 01224 663575
Owner *D.C. Thomson & Co. Ltd*
Editor *Derek Tucker*
Circulation 78,121

A well-established daily that circulates in
Aberdeen, Grampians, Highlands, Tayside,
Orkney, Shetland and the Western Isles. Most
material is commissioned but will consider ideas.
News Editor *Andrew Kellock* Wide variety of hard
or off-beat news and features relating especially,
but not exclusively, to the north of Scotland.
Sports Editor *Alex Martin*
Your Life *Sonja Cox* Saturday lifestyle tabloid
pullout. Features food, fashion, travel, books, arts
and lifestyle.
ⓔ Payment by arrangement.

Scotland on Sunday (Edinburgh)
▷ National Newspapers

The Scotsman (Edinburgh)
▷ National Newspapers

Sunday Mail (Glasgow)
▷ National Newspapers

Sunday Post (Dundee)
▷ National Newspapers

Wales

Daily Post
PO Box 202, Vale Road, Llandudno Junction,
Conwy LL31 9ZD
☎ 01492 574455   ⊕ 01492 574433
welshnews@dailypost.co.uk
www.dailypost.co.uk
Owner *Trinity Mirror Plc*
Editor *Rob Irvine*
Circulation 34,601

Very rarely commissions freelance writers.
Senior Assistant Editor *Mark Brittain*

Evening Leader
Mold Business Park, Wrexham Road, Mold
CH7 1XY
☎ 01352 707707   ⊕ 01352 752180
news@eveningleader.co.uk
Owner *NWN Media*

Editor-in-Chief *Barrie Jones*
Circulation 20,073

Circulates in Wrexham, Flintshire, Deeside
and Chester. Special pages/features: motoring,
travel, arts, women's, children's, local housing,
information and news for the disabled, music and
entertainment.

Deputy Editor *Martin Wright*
Assistant Editor *Jonathon Barnett*
Sports Editor *Nick Harrison*

South Wales Argus
Cardiff Road, Maesglas, Newport NP20 3QN
☎ 01633 810000   ⊕ 01633 777202
www.southwalesargus.co.uk
Owner *Newsquest*
Editor *Gerry Keighley*
Circulation 27,101

Circulates in Newport, Gwent and surrounding
areas.
News Editor *Maria Williams*
Sports Editor *Phil Webb*

South Wales Echo
Media Wales Ltd, Six Park Street, Cardiff
CF10 1XR
☎ 029 2022 3333   ⊕ 029 2058 3624
echo.newsdesk@mediawales.co.uk
www.walesonline.co.uk
Owner *Media Wales (Trinity Mirror Plc)*
Editor *Mike Hill*
Circulation 41,550

Circulates in South and Mid Glamorgan and
Gwent.
Head of News *Nick Machin*
Head of Features *Alison Stokes*
Head of Sport *Delme Parfitt*

South Wales Evening Post
Adelaide Street, Swansea SA1 1QT
☎ 01792 510000   ⊕ 01792 514697
postbox@swwmedia.co.uk
www.thisisswansea.co.uk
Owner *Northcliffe Media Group Ltd*
Editor *Spencer Feeney*
Circulation 47,875

Circulates throughout south west Wales.
News Editor *Chris Davies*
Features Editor *Peter Slee*
Sports Editor *David Evans*

Wales on Sunday
Media Wales Ltd, Six Park Street, Cardiff
CF10 1XR
☎ 029 2022 3333   ⊕ 029 2058 3725
www.walesonline.co.uk
Owner *Media Wales (Trinity Mirror Plc)*
Editor *Tim Gordon*
Circulation 35,955

Launched 1989. Tabloid with sports supplement.
Does not welcome unsolicited mss.
Head of Content *Nick Rippington*
Sports Editor *Paul Banadanato*

Western Mail

Media Wales Ltd, Six Park Street, Cardiff
CF10 1XR
☏ 029 2022 3333   📠 029 2024 3640 (editorial)
www.walesonline.co.uk
Owner *Media Wales (Trinity Mirror Plc)*
Editor *Alan Edmunds*
Circulation 33,693

Circulates in Cardiff, Merthyr Tydfil, Newport,
Swansea and towns and villages throughout
Wales. Mss welcome if of a topical nature,
of Welsh interest. No short stories or travel.
Approach in writing to the features editor. 'Usual
subjects already well covered, e.g. motoring,
travel, books, gardening. We look for the unusual.'
Maximum 1,000 words. Opportunities also on
women's page. Supplements: *Saturday Magazine;*
Education; *WM*; *Box Office*; *Welsh Homes*;
Business; *Sport*; *Motoring*; *Country & Farming*.
Head of Content *Nick Machin*
Head of Features *Margaret O'Reilly*
Head of Sport *Philip Blanche*

Channel Islands

Guernsey Press & Star

Braye Road, Vale, Guernsey GY1 3BW
☏ 01481 240240   📠 01481 240235
newsroom@guernsey-press.com
www.guernsey-press.com
Owner *Guiton Group*
Editor *Richard Digard*
Circulation 15,780

Special pages include children's and women's
interest, gardening and fashion.
News Editor *Gemma Hockey*
Sports Editor *Rob Batiste*
Features Editor *Diane Digard*

Jersey Evening Post

PO Box 582, Jersey JE4 8XQ
☏ 01534 611611   📠 01534 611622
editorial@jerseyeveningpost.com
www.thisisjersey.com
Owner *Claverley Company*
Editor *Chris Bright*
Circulation 20,070

Special pages: gardening, motoring, property,
boating, technology, food and drink, personal
finance, health, business.
News Editor *Carl Walker*
Features Editor *Anna Plunkett-Cole*
Sports Editor *Ron Felton*

News Agencies

Associated Press Limited

The Interchange, 32 Oval Road, London NW1 7DZ

☎ 020 7482 7400   🖷 020 7413 0107

smarshall@ap.org

www.ap.org

Material is either generated in-house or by regulars. Hires the occasional stringer. No unsolicited mss.

Dow Jones Newswires

10 Fleet Place, London EC4M 7QN

☎ 020 7842 9900   🖷 020 7842 9361 (admin)

A real-time financial and business newswire operated by Dow Jones & Co., publishers in the USA of *The Wall Street Journal*. No unsolicited material.

Famous Pictures & Features Agency

See entry under *Picture Libraries*.

National News Press and Photo Agency

4–5 Academy Buildings, Fanshaw Street, London N1 6LQ

☎ 020 7684 3000   🖷 020 7684 3030

news@nationalnews.co.uk

All press releases are welcome. Most work is ordered or commissioned. Coverage includes courts, tribunals, conferences, general news, etc. – words and pictures – as well as PR.

The Press Association Ltd

292 Vauxhall Bridge Road, London SW1V 1AE

☎ 020 7963 7000   🖷 020 7963 7192 (news desk)

copy@pa.press.net

www.pressassociation.co.uk

No unsolicited material. Most items are produced in-house though occasional outsiders may be used. A phone call to discuss specific material may lead somewhere 'but this is rare'.

Reuters

30 South Colonnade, Canary Wharf, London E14 5EP

☎ 020 7250 1122

www.reuters.com

No unsolicited material.

Solo Syndication Ltd

17–18 Hayward's Place, London EC1R 0EQ

☎ 020 7566 0360   🖷 020 7566 0388

wgardiner@solosyndication.com

www.solosyndication.com

Founded 1978. Specializes in worldwide newspaper and magazine syndication of photos, features and cartoons. Professional contributors only. Agents for Associated Newspapers Ltd (UK).

South Yorkshire Sport

6 Sharman Walk, Apperknowle, Sheffield S18 4BJ

☎ 01246 414767/07970 284848   🖷 01246 414767

nickjohnson@uwclub.net

Contact *Nick Johnson*

Provides written/broadcast coverage of sport in the South Yorkshire area.

Space Press News and Pictures

Bridge House, Blackden Lane, Goostrey CW4 8PZ

☎ 01477 533403

Scoop2001@aol.com

Editor *John Williams (07970 213528)*

Pictures *Emma Williams (evwphoto@aol.com 07976 795494)*

Founded 1972. Press and picture agency covering Cheshire and the North West, North Midlands, including Knutsford, Macclesfield, Congleton, Crewe and Nantwich, Wilmslow, Alderley Edge, serving national, regional and local press, TV, radio, and digital picture transmission. Copy and pictures produced for in-house publications and PR. Property, countryside and travel writing. A member of the National Association of Press Agencies (NAPA).

US Media Contacts in the UK

ABC News Intercontinental Inc.
3 Queen Caroline Street, London W6 9PE
☎ 020 8222 5000   📠 020 8222 5020
Director of News Coverage, Europe, Middle East
& Africa *Marcus Wilford*

Associated Press
The Interchange, 32 Oval Road, London NW1 7DZ
☎ 020 7482 7400   📠 020 7410 4999 (news)
Vice President – Global Business for Europe,
Africa & Middle East *Barry Renfrew*

The Baltimore Sun
11 Kensington Court Place, London W8 5BJ
☎ 020 7460 2200
Bureau Chief *Todd Richissin*

Bloomberg News
City Gate House, 39–45 Finsbury Square, London
EC2A 1PQ
☎ 020 7330 7500   📠 020 7330 7797
London Bureau Chief (Print) *Chris Collins*

Business Week
20 Canada Square, Canary Wharf, London
E14 5LH
☎ 020 7176 6060   📠 020 7176 6070
Bureau Chief *Stanley Reed*

CBS News
68 Knightsbridge, London SW1X 7LL
☎ 020 7887 3000   📠 020 7887 3092
Bureau Chief *Jennifer Siebens*

Chicago Tribune Press Service
11 Culmstock Road, London SW11 6LZ
☎ 020 7223 4901
Chief European Correspondent *Laurie Goering*

CNBC
10 Fleet Place, Limeburner Lane, London
EC4M 7QS
☎ 020 7653 9451   📠 020 7653 9393
Assignment Editor *Jennifer Callegher*

CNN International
▷ entry under Television and Radio

Dow Jones Newswires
10 Fleet Place, Limeburner Lane, London
EC4M 7QN
☎ 020 7842 9300
Editor (Europe, Middle East, Africa) *Jan Boucek*

Fairchild Publications of New York
20 Shorts Gardens, London WC2H 9AU
☎ 020 7240 0420
Bureau Chief *Samantha Conti*

Forbes Magazine
Malta House, 36/38 Picadilly, London W1J 0DP
☎ 020 7534 3900
Managing Director & Publisher (Europe, Middle
East & Africa) *Bob Crozier*

Fortune
Blue Fin Building, 110 Southwark Street, London
SE1 0SU
Europe Editor *Nelson Schwartz*

Fox News Channel
6 Centaurs Business Park, Grant Way, Isleworth
TW7 5QD
☎ 020 7805 7143   📠 020 7805 1111
Bureau Chief *Scott Norvell*

The Globe and Mail
16 St Georges Avenue, London N7 0HD
☎ 020 7697 9820
European Bureau Chief *Doug Saunders*

International Herald Tribune
40 Marsh Wall, London E14 9TP
☎ 020 7510 5700
London Correspondent *Eric Pfanner*
See entry under *National Newspapers*.

Los Angeles Times
5 Lancaster Drive, London NW3 4EY
Bureau Chief *Henry Chu*

Market News International
50 Cannon Street, London EC4N 6JJ
☎ 020 7634 1655   📠 020 7236 7122
ukeditorial@marketnews.com
London Bureau Chief *David Thomas*

National Public Radio

Room G-10 East Wing, Bush House, Strand,
London WC2B 4PH
🕾 020 7557 1087   🖷 020 7379 6486
Bureau Chief *Robert Gifford*

NBC News

4th Floor, 3 Shortlands, Hammersmith, London
W6 8HX
🕾 020 8600 6600   🖷 020 8600 6601
Bureau Chief *Chris Hampson*

The New York Times

66 Buckingham Gate, London SW1E 6AU
🕾 020 7799 5050   🖷 020 7799 2962
Bureau Chief *John Burns*

Newsweek

Academy House, 36 Poland Street, London
W1F 7LU
🕾 020 7851 9750   🖷 020 7851 9762
Bureau Chief *Stryker McGuire*

People Magazine

Blue Fin Building, 110 Southwark Street, London
SE1 0SU
🕾 020 3148 3000
Bureau Chief *Simon Perry*

Reader's Digest Association Ltd

11 Westferry Circus, Canary Wharf, London
E14 4HE
🕾 020 7715 8000
See entries under *UK Publishers* and *Magazines.*

Time

Blue Fin Building, 110 Southwark Street, London
SE1 0SU
Bureau Chief *Jef McAllister*
See entry under *Magazines*

Voice of America

London News Centre, 167 Fleet Street, London
EC4A 2EA
🕾 020 7410 0960
Bureau Chief *Michael Drudge*

Wall Street Journal

10 Fleet Place, Limeburner Lane, London
EC4M 7QN
🕾 020 7842 9200
Deputy Bureau Chief *Emily Nelson*

Washington Post

Flat 3, 55–56 Hampstead High Street, London
NW3 1QH
🕾 020 7433 8094
Bureau Chiefs *Mary Jordan, Kevin Sullivan*

Television and Radio

Key advice

For the latest information on broadcast rates for freelancers access the following websites: National Union of Journalists' 'The Freelance Fees Guide' database link (www.gn.apc. org/media); Producers Alliance for Cinema and Television (www.pact.co.uk); Society of Authors (www.societyofauthors.org); and Writers' Guild (www.writersguild.org.uk).

BBC TV and Radio

BBC Vision, BBC Audio & Music

www.bbc.co.uk
Director-General *Mark Thompson*
Deputy Director-General *Mark Byford*

The operational structure of the BBC consists of: BBC Vision Group which runs three main areas: production, commissioning and services and includes the terrestrial and digital television channels; Drama, Entertainment and Children's; Factual and Learning. The Audio & Music Division – responsible for TV Music Entertainment; In-house Factual, Specialist Factual and Drama Audio production. The Journalism Group covers News; Sport; Global News, Nations and Regions.

BBC Television Centre, Wood Lane, London W12 7RJ
☎ 020 8743 8000
Broadcasting House, Portland Place, London W1A 1AA
☎ 020 7580 4468

TV & RADIO COMMISSIONING:
www.bbc.co.uk/commissioning
Director, BBC Vision *Jana Bennett*
Director of Audio & Music Division *Tim Davie*
Controller, BBC One *Jay Hunt*
Controller, BBC Two *Janice Hadlow*
Controller, Daytime *Liam Keelan*
Controller, BBC Three *Danny Cohen*
Controller, BBC Four *Richard Klein*
Controller, BBC Radio 1, 1Xtra, Popular Music, Asian Network and Switch *Andy Parfitt*

Radio 1 documentary submission enquiries: *Joe Harland*, Executive Producer (joe.harland@bbc. co.uk; 020 7765 3552)
1Xtra commissioning enquiries: *Charmaine Cozier*, Commissioning Editor, Documentaries (charmaine.cozier@bbc.co.uk)
1Xtra commissions weekly documentaries from independent producers and an in-house production team. Welcomes proposals that may be suitable as a co-commission with another network such as Radio 1 or Asian Network.
Controller, BBC Radio 2 & 6 Music *Bob Shennan*
Radio 2 commissioning enquiries: *Robert Gallacher*, Editor, Planning & Station Sound (robert.gallacher@bbc.co.uk; 020 7765 4373)
Controller, BBC Radio 3 *Roger Wright*
Radio 3 commissioning enquiries: *David Ireland*, Commissions & Schedules Manager (david. ireland@bbc.co.uk; 020 7765 4943)
Controller, BBC Radio 4 & BBC 7 *Mark Damazer*
Radio 4 proposals must be submitted through an in-house department or a registered independent production company (lists of departments and companies are available at www.bbc.co.uk/ commissioning/radio/network/radio4.shtml).
Controller, BBC Five Live and Five Live Sports Extra *Adrian van Klaveren*
Radio Five Live commissioning enquiries: *Jonathan Wall*, Commissioning Editor (jonathan. wall01@bbc.co.uk; 020 8624 8948)
Head of Programmes, BBC Radio 6 Music *Lewis Carnie*
Radio 6 Music commissioning enquiries: *Paul Rodgers*, Editor (paul.rodgers@bbc.co.uk; 020 7765 4763)

BBC Drama, Comedy and Children's

BBC *writersroom* champions new writers across all BBC platforms for drama, comedy and children's programmes, running targeted schemes and workshops linked directly to production. It accepts and assesses unsolicited scripts for all departments: film, TV drama, radio drama, TV narrative comedy and radio narrative comedy. The *writersroom* website (www. bbc.co.uk/writersroom) offers a diary of events/ opportunities, competitions, interviews with established writers, submission guidelines and free formatting software. BBC *writersroom* also has a Manchester base which focuses on new writing in the north of England. Writers should address hard copies of original, completed scripts to: BBC *writersroom*, Grafton House, 379–381 Euston Road, London NW1 3AU. Before sending in scripts, please go to the website for the latest guidelines on submitting unsolicited work. See also www.bbc.co.uk/commissioning/structure/public.shtml

WRITERSROOM
Creative Director, New Writing *Kate Rowland*
Development Manager *Paul Ashton*
FILMS
Commissioning Editor and Executive Producer *Christine Langan*
Executive Producers *Jamie Laurenson, Joe Oppenheimer*
TV DRAMA
Controller, Drama Commissioning *Ben Stephenson*
Controller, Drama Production & New Talent *John Yorke*
Controller, Series and Serials *Kate Harwood*
Creative Director, Drama Development *Manda Levin*
Executive Producer, EastEnders *Diederick Santer*
RADIO DRAMA
Head of Radio Drama *Alison Hindell*
Executive Producer (Birmingham) & Editor, The Archers *Vanessa Whitburn*
Executive Producer (World Service Drama) *Marion Nancarrow*
Executive Producer (Manchester Radio Drama) *Sue Roberts*
Editor, Silver Street *James Peries*
COMEDY
Head of Comedy *Mark Freeland*
Head of Radio Entertainment *Paul Schlesinger*
Creative Director, Mainstream Comedy *Micheal Jacob*
CHILDREN'S
Controller, BBC Children's *Richard Deverell*
Creative Director, CBBC *Anne Gilchrist*
Creative Director, CBeebies *Michael Carrington*
Head of Children's Drama, Animation & Acquisitions *Steven Andrew*
Head of CBeebies Production, Animation & Acquisitions *Kay Benbow*

BBC News

www.bbc.co.uk
news news.bbc.co.uk

BBC News is the world's largest news-gathering organization with over 2,500 journalists, 45 bureaux worldwide and 15 networks and services across TV, radio and new media.
Director, News *Helen Boaden*

Head of Programmes, News *Stephen Mitchell*
Head of Newsgathering *Francesca Unsworth*
Head of News, BBC World *Richard Porter*
Controller, News 24 *Kevin Bakhurst*
TELEVISION
Editor, 6 o'clock & 10 o'clock News *Craig Oliver*
Editor, Newsnight *Peter Rippon*
Editor, Breakfast & Overnights *Alison Ford*
RADIO
Editor, Today *Ceri Thomas*
Editor, The World at One/World This Weekend/PM/Broadcasting House *Peter Rippon*
Head of News, Five Live *Matt Morris*
Editor, The World Tonight *Alistair Burnett*
Editor, World Service News & Current Affairs *Liliane Landor*
CEEFAX
Room 7540, BBC Television Centre, Wood Lane, London W12 7RJ
☎ 020 8576 1801

SUBTITLING
Red Bee Media, Broadcast Centre, 201 Wood Lane, London W12 7TP www.redbeemedia.com
Available via Ceefax page 888.

BBC Sport

www.bbc.co.uk/sport

Director, Sport *Barbara Slater*
Head of Programmes & Planning *Philip Bernie*
Head of Football *Niall Sloane*

Sports news and commentaries across television, online and Radios 1, 4 and 5 Live.

BBC Religion

New Broadcasting House, Oxford Road, Manchester M60 1SJ
☎ 0161 200 2020   ℻ 0161 244 3183
Head of Religion and Ethics *Michael Wakelin*

Regular programmes for television include *Songs of Praise*. Radio output includes *Good Morning Sunday*; *Sunday Half Hour*; *Thought for the Day*; *The Daily Service*.

BBC New Talent

www.bbc.co.uk/newtalent

The BBC's search for new talent covers a constantly changing range of outlets that has included radio producers, presenters, young storytellers, filmmakers and comedy writers. Access the website for latest information.

BBC World Service

PO Box 76, Bush House, Strand, London WC2B 4PH
☎ 020 7240 3456   ℻ 020 7557 1258
www.bbc.co.uk/worldservice
Director, World Service *Peter Horrocks*
Director, English Networks & News, BBCWS *Gwyneth Williams*

The World Service broadcasts in English and 31 other languages. The English service is round-the-clock, with news and current affairs as the main component. With over 180 million listeners, excluding countries where research is not possible, it reaches a bigger audience than its five closest international broadcasting competitors combined. BBC World Service is increasingly available throughout the world on local FM stations, via satellite and online as well as through short-wave frequencies. Coverage includes world business, politics, people/events/opinions, development issues, the international scene, developments in science and technology, sport, religion, music and the arts. BBC World Service broadcasting is financed by a grant-in-aid voted by Parliament amounting to £265 million for 2008/2009.

BBC writersroom
▷ BBC Drama, Comedy and Children's

BBC Nations

BBC Northern Ireland
Broadcasting House, Ormeau Avenue, Belfast BT2 8HQ
☎ 028 9033 8000
bbc.co.uk/northernireland
Controller *Peter Johnston*
Head of Programmes *Ailsa Orr*
Chief Operating Officer *Mark Taylor*
Head of News & Current Affairs *Andrew Colman*
Head of Drama *Patrick Spence*
Head of Entertainment & Events *Mike Edgar*
Head of Radio *Susan Lovell*
Head of Factual *Paul McGuigan*
Head of Multi Media Commissioning *Fergus Keeling*
Managing Editor Learning, Languages and Social Action *Jane Cassidy*
Senior Producer Irish Language *Antaine O'Donnaile*
Editor, TV News *Angelina Fusco*
Editor, Radio News *Kathleen Carragher*
Editor, News Gathering *Michael Cairns*
Editor TV Current Affairs *Jeremy Adams*
Editor News Online *Seamus Boyd*
Editor – Foyle *Paul McCauley*
Editor, Sport *Shane Glynn*

Regular television programmes include *BBC Newsline 6.30* and a wide range of documentary, popular factual and entertainment programmes. Radio stations: BBC Radio Foyle and BBC Radio Ulster (see entries). For further details on television and film scripts contact: *Susan Carson*, Programme Development Executive Television, BBC Northern Ireland Drama, Room 3.07, Blackstaff House, Great Victoria Street,

Belfast, BT2 7BB (☎ 028 9033 8498; tvdrama.ni@bbc.co.uk). Radio scripts contact: *Anne Simpson*, Manager, Radio Drama at the same address.

BBC Scotland
40 Pacific Quay, Glasgow G51 1DA
☎ 0141 422 6000
www.bbc.co.uk/scotland
Controller *Ken MacQuarrie*
Chief Operating Officer *Bruce Malcolm*
Head of Programme and Services *Donalda MacKinnon*
Head of Drama, Television *Anne Mensah*
Head of Drama, Radio *Patrick Rayner*
Head of Factual Programmes *Andrea Miller*
Head of Service BBC ALBA *Margaret Mary Murray*
Managing Editor, Gaelic *Marion MacKinnon*
Head of Radio *Jeff Zycinski*
Head of News and Current Affairs *Atholl Duncan*
Head of Comedy and Entertainment *Alan Tyler*
Commissioning Editor, Television & Head of Sport *Ewan Angus*
Head of Children's *Simon Parsons*

BBC Scotland provides television and radio programmes for Scotland and the UK networks, online and interactive content and the BBC Scottish Symphony Orchestra. Based in the new digital headquarters building since 2007, there are also centres throughout Scotland, details below:

Aberdeen
Broadcasting House, Beechgrove Terrace, Aberdeen AB15 5ZT
☎ 01224 625233
Dumbarton
Strathleven Bottling Plant, Dumbarton G82 2AP
☎ 01389 736666
Dumfries
Elmbank, Lover's Walk, Dumfries DG1 1NZ
☎ 01387 268008
Dundee
4th Floor, Nethergate Centre, 4th Floor, 66 Nethergate, Dundee DD1 4ER
☎ 01382 202481
Edinburgh
The Tun, 4 Jackson's Entry, 111 Holyrood Road, Edinburgh EH8 8PJ
☎ 0131 557 5888
Glasgow City Halls (BBC Scottish Symphony Orchestra)
87 Albion Street, Glasgow G1 1NQ
☎ 0141 552 0909
Inverness
7 Culduthel Road, Inverness IV2 4AD
☎ 01463 720720
Orkney
Castle Street, Kirkwall, Orkney KW15 1DF
☎ 01856 873939

Portree
Clydesdale Bank Buildings, Somerled Square, Portree, Isle of Skye IV51 9BT
☎ 01478 612005
Selkirk
Unit 1, Ettrick Riverside, Dunsdale Road, Selkirk TD7 5EB
☎ 01750 724567
Shetland
Pitt Lane, Lerwick, Shetland ZE1 0DW
☎ 01595 694747
Stornaway
Radio Nan Gaidheal, Rosebank, 52 Church Street, Stornoway, Isle of Lewis HS1 2LS
☎ 01851 705000

BBC Wales

Broadcasting House, Llandaff CF5 2YQ
☎ 029 2032 2000   🖷 029 2055 2973
www.bbc.co.uk/wales
Controller *Menna Richards*
Head of Programmes (Welsh Language) *Keith Jones*
Head of Programmes (English Language) *Clare Hudson*
Head of News & Current Affairs *Mark O'Callaghan*
Head of Drama *Piers Wenger*
Producer, Pobol y Cwm *Bethan Jones*
Editor, New Media *Iain Tweedale*

Headquarters of BBC Wales, with regional centres in Bangor, Aberystwyth, Carmarthen, Wrexham and Swansea. BBC Wales television produces up to 12 hours of English language programmes a week, 12 hours in Welsh for transmission on **S4C** and an increasing number of hours on network services, including *Doctor Who, Amazon, Tribe* and *Torchwood*. BBC2W, the digital channel, was launched in 2001. Regular programmes include *Wales Today; Newyddion* (Welsh-language daily news); *Week In Week Out, X-Ray* and *Pobol y Cwm* (Welsh-language soap) on television and *Good Morning Wales; Good Evening Wales; Post Cyntaf* and *Post Prynhawn* on Radio Wales and Radio Cymru. Ideas for programmes should be submitted either by post to *Martyn Ingram*, Commissioning Editor, Room 3021 at the address above or by e-mail to commissioning@bbc.co.uk

Bangor
Broadcasting House, Meirion Road, Bangor LL57 2BY
☎ 01248 370880 🖷 01248 351443
Head of Centre *Marian Wyn Jones*

BBC Regions

BBC Asian Network

The Mailbox, Birmingham B1 1RF
☎ 0121 567 6000/0116 201 6772 (newsdesk)
www.bbc.co.uk/asiannetwork

Managing Editor *Vijay Sharma*
Network Manager *Mike Curtis*

Commenced broadcasting in November 1996. Broadcasts nationwide. Programmes in English, Bengali, Gujerati, Hindi, Punjabi and Urdu.

BBC Birmingham

The Mailbox, Birmingham B1 1RF
☎ 0121 567 6767   🖷 0121 567 6875
birmingham@bbc.co.uk
www.bbc.co.uk/birmingham
Head of Regional and Local Programmes *Cath Hearne*
Output Editor *Dave Hart*

BBC Birmingham's Documentaries & Contemporary Factual Department produces a broad range of programming for BBC ONE and BBC TWO, including *Gardener's World*; *Coast; Desi DNA; To Buy or Not to Buy*. Network radio includes: Radio 2, Radio 4 and the Asian Network. Regional & Local Programmes: TV: *Midlands Today; Inside Out; The Politics Show*; Radio: BBC WM (see entry). BBC Birmingham is the headquarters for BBC English Regions. There is a television Drama Village at the University of Birmingham.

BBC Birmingham Drama
Archibald House, 1059 Bristol Road, Selly Oak, Birmingham B29 6LT
☎ 0121 567 7350
Programmes produced include *Dalziel and Pascoe; Doctors; The Afternoon Play* and *Brief Encounters* (which champions new writers).

BBC East Midlands (Nottingham)
East Midlands Broadcasting Centre, London Road, Nottingham NG2 4UU
☎ 0115 955 0500 🖷 0115 902 1983
www.bbc.co.uk/nottingham
Head of Regional Programming *Aziz Rashid*
Output Editor *Sally Bowman*

BBC East (Norwich)
The Forum, Millennium Plain, Norwich NR2 1AW
☎ 01603 284700 🖷 01603 284399
www.bbc.co.uk/norfolk
Head of Regional Programming *Tim Bishop*
Output Editor *Dave Betts*

BBC Bristol

Broadcasting House, Whiteladies Road, Bristol BS8 2LR
☎ 0117 973 2211
www.bbc.co.uk/bristol
Head of Regional and Local Programmes *Lucio Mesquita*
Managing Editor, BBC Bristol *Tim Pemberton*
Controller, Factual Production *Tom Archer*
Head of Natural History Unit *Neil Nightingale*

BBC Bristol is the home of the BBC's Natural History Unit, producing programmes such as *Blue Planet*; *Life of Mammals*; *British Isles, A Natural History*; *Abyss* and *Planet Earth* for BBC ONE and BBC TWO. It also produces natural history programmes for Radio 4 and Radio 5 Live. The Factual department produces a wide range of television programmes, including *DIY SOS*; *Antiques Roadshow*; *Soul Deep* and *Ray Mears* in addition to radio programmes specializing in history, travel, literature and human interest features for Radio 4.

BBC London

35 Marylebone High Street, London W1U 4QA
📞 020 7224 2424
www.bbc.co.uk/london
Executive Editor *Mike MacFarlane* (Responsible for BBC tri-media: BB London News, BBC London 94.9FM and the BBC London website)

BBC South East (Tunbridge Wells)

The Great Hall, Mount Pleasant Road, Tunbridge Wells TN1 1QQ
📞 01892 670000
www.bbc.co.uk/kent
Head of Regional and Local Programmes *Mick Rawsthorne* (Responsible for BBC South East [TV], BBC Radio Kent and BBC Southern Counties Radio)
Editor, South East Today *Quentin Smith*
Managing Editor, Southern Counties Radio *Nicci Holiday*
Managing Editor, Radio Kent *Paul Leaper*
Editor, Newsgathering *Michael Gravesande*
Producer, Inside Out *Linda Bell*

BBC West/BBC South/BBC South West

The three regional television stations, BBC West, BBC South and BBC South West produce breakfast, lunchtime and evening news programmes, as well as *Inside Out*, regular 30-minute local current affairs programmes and regional parliamentary programmes, *The Politics Show*. Each of the regions operates a comprehensive local radio and online service as well as a range of correspondents specializing in subjects such as health, education, business, home affairs and the environment.

BBC West (Bristol)
Broadcasting House, Whiteladies Road, Bristol BS8 2LR
📞 0117 973 2211
Head of Regional and Local Programmes *Lucio Mesquita* (Responsible for the BBC West region. TV: *BBC Points West* (regional news programme), *Inside Out West* (regional current affairs series) and *The Politics Show West*. Local Radio: BBC Radio Bristol, BBC Somerset, BBC Radio Gloucestershire, BBC Radio Wiltshire and the BBC local websites)
Daily TV Output Editor *Stephanie Marshall*
TV Features Editor *Roger Farrant*
Political Editor *Paul Barltrop*

BBC South (Southampton)
Broadcasting House, Havelock Road, Southampton SO14 7PU
📞 023 8022 6201
Head of BBC South *Mike Hapgood* (Responsible for BBC South TV, BBC Radio Berkshire, BBC Radio Oxford, BBC Radio Solent and online content for the region)
TV News Editor, BBC South Today *Lee Desty*
Series Producer, Inside Out *Jane French*
Political Editor *Peter Henley*

BBC South West (Plymouth)
Broadcasting House, Seymour Road, Mannamead, Plymouth PL3 5BD
📞 01752 229201
Head of BBC South West *John Lilley* (Responsible for the BBC South West region. TV: *BBC Spotlight* (regional news programme), *Spotlight Channel Islands* (regional news programme), *Inside Out* (regional current affairs series) and *The Politics Show South West*. Local Radio: BBC Radio Devon, BBC Radio Cornwall, BBC Radio Guernsey, BBC Radio Jersey and the local BBC websites)
Output Editor *Simon Read*
Series Producer, Inside Out *Simon Willis*
Political Editor *Martyn Oates*

BBC Yorkshire/BBC North West/BBC North East & Cumbria

The regional centres at Leeds, Manchester and Newcastle make their own programmes on a bi-media approach, each centre having its own head of regional and local programmes.

BBC Yorkshire
Broadcasting Centre, 2 St Peter's Square, Leeds LS9 8AH
📞 0113 244 1188
Head of Regional and Local Programmes *Helen Thomas*
Editor, News *Tim Smith*
Assistant News Editor, Look North *Denise Wallace*
Political Editor, North of Westminster *Len Tingle*
Weeklies Editor, Close Up North *Ian Cundall*
Producers, Close Up North *Richard Taylor, Paul Greenan*

BBC North West (Manchester)
New Broadcasting House, Oxford Road, Manchester M60 1SJ
📞 0161 200 2020
Head of Regional and Local Programmes *Aziz Rashid*
Editor, Newsgathering *Jim Clarke*

Output Editor *Cerys Griffiths*
Producer, Inside Out *Deborah van Bishop*
Producer, The Politics Show *Michelle Mayman*

BBC North East & Cumbria (Newcastle upon Tyne)
Broadcasting Centre, Barrack Road, Newcastle Upon Tyne NE99 2NE
℡ 0191 232 1313
Head of Regional and Local Programmes *Phil Roberts*
News Editor *Andy Cooper*

BBC Local Radio

BBC Local Radio
bbc.co.uk/england

BBC local radio stations have their own newsroom which supplies local bulletins and national news service. Many have specialist producers. A comprehensive list of programmes for each is unavailable and would soon be out of date. For general information on programming visit the BBC website and select the local station you want or contact the relevant station direct.

BBC Radio Berkshire
PO Box 104.4, Reading RG4 8FH
℡ 0118 946 4200 (news)   📠 0118 946 4555
www.bbc.co.uk/berkshire
Editor *Marianne Bell*

BBC Radio Bristol
PO Box 194, Bristol BS99 7QT
℡ 0117 974 1111   📠 0117 923 8323
radio.bristol@bbc.co.uk
www.bbc.co.uk/bristol/local_radio
Managing Editor *Tim Pemberton*

Wide range of feature material used.

BBC Radio Cambridgeshire
104 Hills Road, Cambridge CB2 1LQ
℡ 01223 259696
cambs@bbc.co.uk
www.bbc.co.uk/cambridgeshire/local_radio
Managing Editor *Jason Horton*

Commenced broadcasting in May 1982.

BBC Radio Cornwall
Phoenix Wharf, Truro TR1 1UA
℡ 01872 275421   📠 01872 240679
radio.cornwall@bbc.co.uk
www.bbc.co.uk/cornwall/local_radio
Managing Editor *Pauline Causey*

On air from 1983 serving Cornwall and the Isles of Scilly. Broadcasts 110 hours of local programmes weekly including news, phone-ins and debate.

BBC Coventry and Warwickshire
Priory Place, Coventry CV1 5SQ
℡ 024 7655 1000   📠 024 7655 2000
coventry@bbc.co.uk
warwickshire@bbc.co.uk
www.bbc.co.uk/coventrywarwickshire
Station Manager *David Clargo*

Commenced broadcasting in January 1990 as CWR. News, current affairs, public service information and community involvement, relevant to its broadcast area. Details of any special events are available on the website.

BBC Radio Cumbria
Annetwell Street, Carlisle CA3 8BB
℡ 01228 592444
radio.cumbria@bbc.co.uk
bbc.co.uk/cumbria
Editor *Nigel Dyson*

Occasional opportunities for plays and short stories are advertised on-air.

BBC Radio Cymru
Broadcasting House, Llandaf, Caerdydd CF5 2YQ
℡ 029 20 322018   📠 029 20 322473
radio.cymru@bbc.co.uk
www.bbc.co.uk/radiocymru
Editor *Siân Gwynedd*

Welsh language station serving Welsh-speaking communities. Rich mix of news, music, chat, sport, features, current affairs, drama, comedy and religion. Writing opportunities: DRAMA During 2009–10, up to 18 new radio plays/monologues; COMEDY Occasional opportunities during the year (for details of independent commissions see www.bbc.co.uk/wales/commissioning/).

BBC Radio Derby
PO Box 104.5, Derby DE1 3HL
℡ 01332 361111   📠 01332 290794
radio.derby@bbc.co.uk
www.bbc.co.uk/derby
Editor *Simon Cornes*

News, sport, information and entertainment.

BBC Radio Devon
Broadcasting House, Seymour Road, Mannamead, Plymouth PL3 5BD
℡ 01752 260323
radio.devon@bbc.co.uk
www.bbc.co.uk/devon
Also at: Walnut Gardens, St David's Hill, Exeter EX4 4DH
Managing Editor *Simon Read*

On air since 1983.

BBC Essex

PO Box 765, Chelmsford CM2 9XB
☎ 01245 616000   🖷 01245 492983
essex@bbc.co.uk
www.bbc.co.uk/essex
Managing Editor *Gerald Main*

Programmes are a mix of news, interviews, expert contributors, phone-ins, sport and special interest topics.

BBC Radio Foyle

8 Northland Road, Londonderry BT48 7GD
☎ 028 7137 8600   🖷 028 7137 8666
radio.foyle@bbc.co.uk
www.bbc.co.uk/northernireland
Managing Editor *Paul McCauley*
News Producer *Chris Capper*

Radio Foyle broadcasts about seven hours of original material a day, seven days a week to the north west of Northern Ireland. Other programmes are transmitted simultaneously with Radio Ulster. The output ranges from news, sport, and current affairs to live music recordings and arts reviews.

BBC Radio Gloucestershire

London Road, Gloucester GL1 1SW
☎ 01452 308585   🖷 01452 309491
radio.gloucestershire@bbc.co.uk
www.bbc.co.uk/radiogloucestershire
Managing Editor *Mark Hurrell*
Assistant Editor *Mark Jones*

News and information covering the large variety of interests and concerns in Gloucestershire. Leisure, sport and music, plus African Caribbean and Asian interests. Regular book reviews and interviews with local authors in the *Steve Kitchen Show*.

BBC Guernsey

Broadcasting House, Bulwer Avenue, St Sampson's, Guernsey GY2 4LA
☎ 01481 200600   🖷 01481 200361
radio.guernsey@bbc.co.uk
www.bbc.co.uk/guernsey
Managing Editor *Robert Wallace*

Opened with its sister station, BBC Radio Jersey, in March 1982. Broadcasts 80 hours of local programming a week.

BBC Hereford & Worcester

Hylton Road, Worcester WR2 5WW
☎ 01905 748485   🖷 01905 748006
bbchw@bbc.co.uk
worcester@bbc.co.uk
www.bbc.co.uk/herefordworcester
Also at: 43 Broad Street, Hereford HR4 9HH
Managing Editor *James Coghill*

Has an interest in writers/writing with local connections.

BBC Radio Humberside

Queen's Court, Queen's Gardens, Hull HU1 3RH
☎ 01482 323232
radio.humberside@bbc.co.uk
www.bbc.co.uk/humber
Managing Editor *Simon Pattern*

On air since 1971. BBC Radio Humberside has broadcast short pieces by local writers.

BBC Radio Jersey

18–21 Parade Road, St Helier, Jersey JE2 3PL
☎ 01534 837200   🖷 01534 732569
radiojersey@bbc.co.uk
www.bbc.co.uk/jersey
Managing Editor *Denzil Dudley*
Assistant Editor *Matthew Price*

Local news, current affairs and community items.

BBC Radio Kent

The Great Hall, Mount Pleasant Road, Tunbridge Wells TN1 1QQ
☎ 01892 670000   🖷 01892 518279
radio.kent@bbc.co.uk
www.bbc.co.uk/kent/local_radio
Managing Editor *Paul Leaper*

Occasional commissions are made for local interest documentaries and other one-off programmes.

BBC Radio Lancashire

20–26 Darwen Street, Blackburn BB2 2EA
☎ 01254 262411
lancashire@bbc.co.uk
www.bbc.co.uk/lancashire/local_radio
Managing Editor *John Clayton*

Journalism-based radio station, interested in interviews with local writers. Contact *Alison Brown*, Daily Programmes Producer, Monday to Friday (alison.brown.01@bbc.co.uk).

BBC Radio Leeds

2 St Peter's Square, Leeds LS9 8AH
☎ 0113 244 2131   🖷 0113 224 7316
radio.leeds@bbc.co.uk
www.bbc.co.uk/leeds/local_radio
Managing Editor *Mike Bettison*

BBC Radio Leeds is the station for West Yorkshire, serving 1.3 million potential listeners since 1968. Broadcasts a mix of news, sport, talk, music and entertainment, 24 hours a day.

BBC Radio Leicester

9 St Nicholas Place, Leicester LE1 5LB
☎ 0116 251 6688   🖷 0116 251 1463
radio.leicesternews@bbc.co.uk
www.bbc.co.uk/leicester
Managing Editor *Kate Squire*

The first local station in Britain. Occasional interviews with local authors.

BBC Radio Lincolnshire

Newport, Lincoln LN1 3XY
☎ 01522 511411   📠 01522 511058
radio.lincolnshire@bbc.co.uk
www.bbc.co.uk/lincolnshire
Managing Editor *Charlie Partridge*

Unsolicited material considered only if locally relevant. Maximum 1,000 words: straight narrative preferred, ideally with a topical content.

BBC London

PO Box 94.9, Marylebone High Street, London W1A 6FL
☎ 020 7224 2424   📠 020 7208 9661 (news)
ldn-planning@bbc.co.uk
www.bbc.co.uk/london
Managing Editor *David Robey*

Formerly Greater London Radio (GLR), launched in 1988, BBC London 94.9 broadcasts news, information, travel bulletins, sport and music to Greater London and the Home Counties.

BBC Radio Manchester

PO Box 951, Oxford Road, Manchester M60 1SJ
☎ 0161 200 2000   📠 0161 236 5804
radiomanchester@bbc.co.uk
www.bbc.co.uk/manchester/local_radio
Managing Editor *John Ryan*
News Editor *Mark Elliot*

One of the largest of the BBC local radio stations, broadcasting news, current affairs, phone-ins, help, advice and sport.

BBC Radio Merseyside

PO Box 95.8, Liverpool L69 1ZJ
☎ 0151 708 5500   📠 0151 794 0988
radio.merseyside@bbc.co.uk
www.bbc.co.uk/liverpool
Managing Editor *Mick Ord*

Launched in 1967. BBC Radio Merseyside's editorial area covers the whole of Merseyside (Liverpool, Bootle, Birkenhead, Southport, St Helens, etc.) much of North Cheshire (Chester, Warrington, Runcorn, Widnes, Ellesmere Port) and West Lancashire (Skelmersdale, Orrell, Burscough, Ormskirk).

BBC Radio Newcastle

Broadcasting Centre, Barrack Road, Newcastle upon Tyne NE99 1RN
☎ 0191 232 4141
radionewcastle.news@bbc.co.uk
www.bbc.co.uk/tyne/local_radio
www.bbc.co.uk/wear/local_radio
Managing Editor *Andrew Robson*
Senior Producer (Programmes) *Sarah Miller*

Commenced broadcasting in January 1971. One of the BBC's big city radio stations in England, Radio Newcastle reaches an audience of 233,000.

BBC Radio Norfolk

The Forum, Millennium Plain, Norwich NR2 1BH
☎ 01603 619331   📠 01603 284488
radionorfolk@bbc.co.uk
www.bbc.co.uk/radionorfolk
Managing Editor *David Clayton*

Good local ideas and material welcome for features and documentaries *if* directly related to Norfolk.

BBC Radio Northampton

Broadcasting House, Abington Street, Northampton NN1 2BH
☎ 01604 239100   📠 01604 230709
northampton@bbc.co.uk
www.bbc.co.uk/northamptonshire
Managing Editor *Louise Dow*

Books of local interest are regularly featured and authors and poets are interviewed on merit.

BBC Radio Nottingham

London Road, Nottingham NG2 4UU
☎ 0115 955 0500   📠 0115 912 1984
radio.nottingham@bbc.co.uk
www.bbc.co.uk/nottingham/local_radio
Editor *Sophie Stewart*

Rarely broadcasts scripted pieces of any kind but interviews with authors form a regular part of the station's output.

BBC Radio Orkney

Castle Street, Kirkwall KW15 1DF
☎ 01856 873939   📠 01856 872908
radio.orkney@bbc.co.uk
Senior Producer *Dave Gray*

Regular programmes include *Around Orkney* (news magazine programme), *Bruck* (swapshop and general features) and *Orky-Ology* (archaeology magazine). As a BBC Community station, all Radio Orkney's programme material is generated from within the Orkney Islands.

BBC Oxford

269 Banbury Road, Oxford OX2 7DW
☎ 08459 311444   📠 08459 311555
oxford@bbc.co.uk
www.bbc.co.uk/oxford/local_radio
Executive Editor *Steve Taschini*

Restored to its original name in 2000 having been merged with BBC Radio Berkshire in 1995 to create BBC Thames Valley. The station occasionally carries interviews with local authors in Jo Thoenes' afternoon show.

BBC Radio Scotland (Dumfries)

Elmbank, Lover's Walk, Dumfries DG1 1NZ
☎ 01387 268008
dumfries@bbc.co.uk
Senior Producer *Willie Johnston*

Previously Radio Solway. The station mainly
outputs news bulletins (four daily) although it
has become more of a production centre with
programmes being made for Radio Scotland as
well as BBC Radio 2 and 5 Live. Freelancers of
a high standard, familiar with Radio Scotland,
should contact the producer.

BBC Radio Scotland (Selkirk)

Unit 1, Ettrick Riverside, Dunsdale Road, Selkirk
TD7 5EB
☎ 01750 724567   🖷 01750 724555
selkirk.news@bbc.co.uk
Senior Broadcaster *Cameron Buttle*

Formerly BBC Radio Tweed. Local news bulletins
only.

BBC Radio Sheffield

54 Shoreham Street, Sheffield S1 4RS
☎ 0114 273 1177   🖷 0114 267 5454
radio.sheffield@bbc.co.uk
www.bbc.co.uk/southyorkshire
Managing Editor *Gary Keown*
Programme Editor *Mike Woodcock*

Writer interviews, writing-related topics and
readings on the Rony Robinson show at 11.00 am.

BBC Radio Shetland

Pitt Lane, Lerwick ZE1 0DW
☎ 01595 694747   🖷 01595 694307
radio.shetland@bbc.co.uk
Senior Producer *John Johnson*

Regular programmes include *Good Evening
Shetland*. An occasional books programme
highlights the activities of local writers and
writers' groups.

BBC Radio Shropshire

2–4 Boscobel Drive, Shrewsbury SY1 3TT
☎ 01743 248484   🖷 01743 237018
radio.shropshire@bbc.co.uk
www.bbc.co.uk/shropshire
Managing Editor *Tim Beech*

On air since 1985. Unsolicited literary material
rarely used, and then only if locally relevant.

BBC Radio Solent

Broadcasting House, Havelock Road,
Southampton SO14 7PU
☎ 023 8063 1311
radio.solent.news@bbc.co.uk
www.bbc.co.uk/radiosolent

Managing Editor *Mia Costello*
Broadcasting since 1970.

BBC Somerset

Broadcasting House, Park Street, Taunton
TA1 4DA
☎ 01823 323956   🖷 01823 332539
somerset@bbc.co.uk
www.bbc.co.uk/somerset
Managing Editor *Jess Rudkin*

Informal, speech-based programming with strong
news and current affairs output and regular local
interest features, including local writing. Poetry
and short stories on the breakfast-time *Adam
Thomas Programme* and Emma Britton's mid-
morning programme; also arts features on Elise
Rayner's *Somerset's Drive* at 5.00 pm.

BBC Southern Counties Radio

Broadcasting House, 40–42 Queens Road,
Brighton BN1 3XB
☎ 01273 320400
southerncounties.radio@bbc.co.uk
www.bbc.co.uk/southerncounties
Also at: Broadcasting Centre, Guildford GU2 7AP
☎ as above
Managing Editor *Nicci Holliday*

BBC Radio Stoke

Cheapside, Hanley, Stoke on Trent ST1 1JJ
☎ 01782 208080   🖷 01782 289115
radio.stoke@bbc.co.uk
www.bbc.co.uk/stoke
Managing Editor *Sue Owen*

On air since 1968, one of the first eight
'experimental' BBC stations. Emphasis on news,
current affairs and local topics. Music represents
one fifth of total output. Unsolicited material of
local interest is welcome – send to News Editor,
James O'Hara.

BBC Radio Suffolk

Broadcasting House, St Matthew's Street, Ipswich
IP1 3EP
☎ 01473 250000   🖷 01473 210887
radiosuffolk@bbc.co.uk
www.bbc.co.uk/suffolk
Managing Editor *Peter Cook*

Strongly locally speech-based, dealing with
news, current affairs, community issues, the arts,
agriculture, commerce, travel, sport and leisure.
Programmes sometimes carry interviews with
writers.

BBC Tees

Broadcasting House, Newport Road,
Middlesbrough TS1 5DG
☎ 01642 225211   🖷 01642 211356
teesnews@bbc.co.uk
www.bbc.co.uk/tees/local_radio

Managing Editor *Matthew Barraclough*
Assistant Editor *Ben Thomas*

Material used is local to Teesside, Co. Durham and North Yorkshire. News, current affairs and features about life in the north east.

BBC Three Counties Radio

1 Hastings Street, Luton LU1 5XL
📞 01582 637400   📠 01582 401467
3CR@bbc.co.uk
www.bbc.co.uk/threecounties/local_radio
Managing Editor *Mark Norman*

Covers Bedfordshire, Hertfordshire and Buckinghamshire. Encourages freelance contributions from the community across a wide range of radio output, including interview and feature material. Interested in local history topics (five minutes maximum).

BBC Radio Ulster

Broadcasting House, Ormeau Avenue, Belfast BT2 8HQ
📞 028 9033 8000
Head of Radio Ulster *Susan Lovell*

Programmes broadcast from 6.30 am to 1.00 am weekdays and from 6.55 am to 1.00 am at weekends. Programmes include: *Good Morning Ulster*; *The Stephen Nolan Show*; *Gerry Anderson*; *Talk Back*; *Evening Extra*; *On Your Behalf*; *Your Place and Mine*; *Sunday Sequence* and *Saturday Magazine*. Comedy, documentary, specialist music and community programming are also included.

BBC Radio Wales

Broadcasting House, Llandaff, Cardiff CF5 2YQ
📞 029 2032 2000   📠 029 2032 2674
radio.wales@bbc.co.uk
www.bbc.co.uk/radiowales
Editor *Sally Collins*
Editor, Radio Wales News *Matthew Evans*

Broadcasts news on the hour and half hour throughout weekday daytime programmes; hourly bulletins in the evenings and at weekends. Programmes include *Good Morning Wales* and *The Richard Evans Programme*.

BBC Radio Wiltshire

BBC Broadcasting House, Prospect Place, Swindon SN1 3RW
📞 01793 513626
radio.wiltshire@bbc.co.uk
www.bbc.co.uk/wiltshire
Editor *Mark Hurrell*

Sister company of BBC Radio Swindon. Regular programmes include Lee Stone's mid-morning show (reviews and author interviews).

BBC WM

The Mailbox, Birmingham B1 1RF
📞 0845 300 9956   📠 0121 567 6025
bbcwm@bbc.co.uk
www.bbc.co.uk/radiowm
Managing Editor *Keith Beech*

Commenced broadcasting in November 1970 as BBC Birmingham and has won four gold Sony awards in recent years. Speech-based station broadcasting to the West Midlands, South Staffordshire and North Worcestershire.

BBC Radio York

20 Bootham Row, York YO30 7BR
📞 01904 641351   📠 01904 610937
northyorkshire.news@bbc.co.uk
www.bbc.co.uk/northyorkshire
Editor *Sarah Drummond*

Books of local interest are sometimes previewed.

Independent Television

Channel 4

124 Horseferry Road, London SW1P 2TX
📞 020 7396 4444
www.channel4.com
Chief Executive *Andy Duncan*
Director of Television and Content *Kevin Lygo*
Head of Channel 4 Commissioning
Julian Bellamy
Head of More4 & Head of Documentaries
Hamish Mykura
Head of E4 *Angela Jain*
Director of Radio *Bob Shennan*
Controller of Film and Drama *Tessa Ross*

Channel 4 launched in 1982 to broadcast programmes that catered to audiences not served by the BBC or ITV. Over the years, its purpose has evolved and grown and today Channel 4 provides online content as well as TV programming while continuing to expand into new areas including radio. There is also a range of 4 channels reaching a variety of audiences including E4, More4, Film4, 4Music, and Channel 4 +1. See channel4.com/nexton4 for more details.

Channel Television

Television Centre, La Pouquelaye, St Helier, Jersey JE1 3ZD
📞 01534 816816   📠 01534 816817
www.channelonline.tv
Also at: Television House, Bulwer Avenue, St Sampsons, Guernsey GY2 4LA 📞 01481 241888
📠 01481 241878
Managing Director, Broadcast *Karen Rankine*
Managing Director, Commercial *Mike Elsey*
Director of Resources & Transmission
Kevin Banner

Channel Television is the independent television broadcaster to the Channel Islands, serving 150,000 residents, most of whom live on the main islands, Jersey, Guernsey, Alderney and Sark. The station has a weekly reach of more than 60% with local programmes (in the region of five and a half hours each week) at the heart of the ITV service to the islands.

Five

22 Long Acre, London WC2E 9LY
☎ 020 7550 5555
www.five.tv
CEO *Dawn Airey*
Director of Programmes *Richard Woolfe*
Controller of Children's Programmes *Nick Wilson*
Senior Programme Controller (News & Current Affairs) *Chris Shaw*

Channel 5 Broadcasting Ltd won the franchise for Britain's third commercial terrestrial television station in 1995 and came on air at the end of March 1997. Regular programmes include *The Wright Stuff* (weekday morning chat show), *The Trisha Goddard Show* (talk show), *CSI: NY* (drama) and *House* (drama).

GMTV

Head Office: The London Television Centre, Upper Ground, London SE1 9TT
☎ 020 7827 7000   ☎ 020 7827 7001
www.gm.tv
Chairman *Clive Jones*
Chief Operating Officer *Clive Crouch*
Director of Programmes *Peter McHugh*
Editor *Martin Frizell*
Deputy Editor *Malcolm Douglas*
Operations Director *Di Holmes*
Head of Entertainment *Corinne Bishop*
Deputy Head of Entertainment *Amy Vosburgh*

Winner of the national breakfast television franchise. Jointly owned by ITV plc and Disney. GMTV took over from TV-am on 1 January 1993, with live programming from 6.00 am to 9.25 am. Regular news bulletins, current affairs, topical features, showbiz and lifestyle. Also competitions, travel and weather reports. The week's news and political issues are reviewed on Sundays, followed by children's programming. Launched its digital service, GMTV2, in January 1999, with daily broadcasts from 6.00 am to 9.25 am. GMTV2 moved from ITV2 to ITV4 in March 2008 and is now known as GMTV Digital. Children's programming with some simulcast with GMTV1.

ITN (Independent Television News Ltd)

200 Gray's Inn Road, London WC1X 8XZ
☎ 020 7833 3000   ☎ 020 7430 4868
contact@itn.co.uk
www.itn.co.uk
Editor-in-Chief, ITV News *David Mannion*

Editor, Channel 4 News *Jim Gray*

Provider of the main national and international news for ITV and Channel 4 and radio news for IRN. Programmes on ITV: *Lunchtime News*; *London Today*; *Evening News*; *London Tonight*; *News at Ten*, plus regular news summaries and three programmes a day at weekends. Programmes on Channel 4 include the in-depth news analysis programmes *Channel 4 News at Noon*, *Channel 4 News* and *More 4 News*.

ITV plc

200 Gray's Inn Road, London WC1X 8HF
☎ 020 7843 8000
www.itvplc.com
www.itv.com
Executive Chairman *Michael Grade*
Director of Television *Peter Fincham*

See individual listings below for ITV Anglia, ITV Border, ITV Central, ITV Granada, ITV London, ITV Meridian, ITV Tyne Tees, ITV Wales, ITV West, ITV Westcountry and ITV Yorkshire.

ITV Network

ITV1 comprises 15 regional networks with ITV plc holding the English and Welsh regional ITV licences as well as the digital channels, ITV2, ITV3, ITV4, CiTV and ITVPlay. The remaining ITV licences belong to: SMG plc (Grampian and Scottish), Ulster Television plc (UTV) and Channel Television. Granada is the production arm of ITV plc.
For commissioning information go to: www.itv.com/AboutITV/Commissioning-Production/

ITV Anglia Television

Anglia House, Norwich NR1 3JG
☎ 01603 615151
www.itvlocal.com/anglia
Managing Director & Regional Director ITV Anglia/ITV Yorkshire *Neil Thompson*

Broadcasting to the east of England, Anglia Television is a major producer of programmes for the ITV network. Its network factual department currently boasts the largest portfolio of North American documentary commissions for a UK-based production company.

ITV Border Television

Television Centre, Durranhill, Carlisle CA1 3NT
☎ 0844 881 5850   ☎ 01228 541384
www.itvlocal.com/border
Regional Director, ITV Border/ITV Tyne Tees *Graeme Thompson*

Border's region covers three different cultures – English, Scottish and Manx. Programming concentrates on documentaries rather than drama. Most scripts are supplied in-house but occasionally there are commissions. Apart from notes, writers should not submit written work until their ideas have been fully discussed.

ITV Central

Gas Street, Birmingham B1 2JT
☎ 0844 881 4000/0808 100 7888 (newsdesk)
www.itvlocal.com/central
Managing Director *Mike Blair*

Formed in 2004 by the merger of Carlton and Granada, ITV Central is the largest of the UK commercial television networks outside of London.

ITV Granada

Quay Street, Manchester M60 9EA
☎ 0161 832 7211   📠 0161 827 2180
granada@itvlocal.com
www.itvlocal.com/central
Director of Production *Claire Poyser*
Controller of Drama *Keiron Roberts*
Regional Director *Mike Blair*

The longest continuous ITC licence holder, broadcasting for over 50 years. Programmes include *Coronation Street* and *Disappearing World*. See also **ITV Productions** under *Film, TV and Radio Producers*.

ITV London

London Television Centre, Upper Ground, London SE1 9LT   ☎ 020 7261 8163
newsdesk@itvlondon.com
www.itvlocal.com/london
Managing Director *Christy Swords*

ITV London, formed in February 2004 by the merger of Carlton and LWT, broadcasts to the Greater London region, extending into the counties on its border.

ITV Meridian

Solent Business Park, Whiteley PO15 7PA
☎ 0844 881 2000
www.www.itvlocal.com/meridian
Managing Director *Mark Southgate*
Head of News *Robin Britton*

Serves viewers across the South and South-East region.

ITV Thames Valley

Solent Business Park, Whiteley PO15 7PA
☎ 0844 881 2000
Managing Director *Mark Southgate*
Head of News *Robin Britton*

Formed in December 2004 from the two former ITV sub-regions, Central South and Meridian West, ITV Thames Valley is ITV's newest regional news service. It covers an area stretching from Banbury in the north to Winchester in the south, from Swindon in the west to Bracknell in the east.

ITV Tyne Tees

Television House, The Watermark, Gateshead NE11 9SZ
☎ 0844 881 5000   📠 0844 881 5010
tynetees@itvlocal.com
www.itvlocal.com/tynetees
Regional Director, ITV Tyne Tees/ITV Border *Graeme Thompson*
Head of News *Catherine Houlihan*
Community Affairs Manager *Brenda Mitchell*
Managing Director, ITV Signpost *Malcolm Wright*

Programming covers politics, news and current affairs, regional documentaries and sport. Regular programmes include *North East Tonight with Jonathan Morrell and Philippa Tomson* and *Around the House* (politics).

ITV Wales

The Television Centre, Culverhouse Cross, Cardiff CF5 6XJ
☎ 0844 881 0100   📠 029 2059 7183
wales@itvlocal.com
www.itvlocal.com/wales
Managing Director *Elis Owen*

Fomerly HTV Wales. Programmes include the consumer affairs magazine, *The Ferret* and current affairs series, *Wales this Week*. The company also makes Welsh language programming for S4C, including the series voted the most popular with viewers on the Welsh channel – *Cefn Gwlad* and the current affairs series, *Y Byd ar Bedwar*.

ITV West

470 Bath Road, Bristol BS4 3HG
☎ 0844 881 2345/2308 (newsdesk)
📠 0844 881 2346
itvwestnews@itv.com (newsdesk)
www.itvlocal.com/west
Regional Director *Jane McCloskey*

Formerly HTV. Broadcasts to over two million viewers in the West of England.

ITV West Country

Langage Science Park, Western Wood Way, Plymouth PL7 5BQ
☎ 0844 881 4900/4800 (newsdesk)
📠 0844 881 4901
news@westcountry.co.uk (newsdesk)
www.itvlocal.com/westcountry
Regional Director *Jane McCloskey*

Broadcasts to Cornwall, Devon, Somerset and west Dorset. Produces regional news, current affairs and features programmes.

ITV Yorkshire

The Television Centre, Leeds LS3 1JS
☎ 0113 222 7000

www.itvlocal.com/yorkshire
Controller of Drama, Leeds *Keith Richardson*
Controller of Comedy Drama and Drama
Features *David Reynolds*

Drama series, comedy drama, single drama,
adaptations and long-running series like
Emmerdale and *Heartbeat*. Always looking for
strong writing in these areas, but prefers to find
it through an agent. Documentary/current affairs
material tends to be supplied by producers;
opportunities in these areas are rare but
adaptations of published work as a documentary
subject are considered. In theory, opportunity
exists within series, episode material but the best
approach is through a good agent.

S4C

Parc Ty Glas, Llanishen, Cardiff CF14 5DU
📞 029 2074 7444   📠 029 2075 4444
s4c@s4c.co.uk
www.s4c.co.uk
Chief Executive *Iona Jones*
Director of Commissioning *Rhian Gibson*

The Welsh 4th Channel, established by the
Broadcasting Act 1980, is responsible for a
schedule of Welsh and English programmes on
the Fourth Channel in Wales. Known as S4C, the
analogue service is made up of about 34 hours
per week of Welsh language programmes and
more than 85 hours of English language output
from Channel 4. S4C digidol broadcasts in Welsh
exclusively for 80 hours per week. Ten hours a
week of the Welsh programmes are provided by
the BBC; the remainder are commissioned from
ITV1 Wales and independent producers. Drama,
comedy, sport, music and documentary are all
part of S4C's programming. Commissioning
guidelines can be viewed on www.s4c.co.uk/
production

stv

Pacific Quay, Glasgow G51 1PQ
📞 0141 300 3000   📠 0141 300 3030
www.stv.tv
Also at: Television Centre, Craigshaw Business
Park, West Tullos, Aberdeen AB12 3QH
📞 01224 848848 📠 01224 848800
Managing Director, Broadcasting, stv *Bobby Hain*
Head of News & Current Affairs (Scotland)
Gordon Macmillan

stv is the ITV (Channel 3) licence holder for
the north and central regions of Scotland.
The company produces a range of television
programmes, covering current affairs, sport,
entertainment, documentary and drama, as well
as flagship news programmes *Scotland Today* and
North Tonight.

Teletext Ltd

10 Chiswick Park, 566 Chiswick High Road,
London W4 5TS
📞 020 8323 3000   📠 020 8323 5204
www.teletext.co.uk
Managing Director *Dr Mike Stewart*
Editor-in-Chief *John Sage*

Teletext operates the digital TV text service on
Freeview (behind channels ITV1, 2, 3 and 4,
Channel 4, E4 and More4), on Sky (behind all
the ITV channels and Channel 4) and on the
analogue TV service behind ITV1, Channel 4
and Five. In addition, Teletext's editorial content
and holiday offers can be accessed online or via
mobile services.

UTV

Ormeau Road, Belfast BT7 1EB
📞 028 9032 8122   📠 028 9024 6695
info@u.tv
www.u.tv
Managing Director, Television *Michael Wilson*
Head of News, Current Affairs and Sport
Rob Morrison
Head of Communications *Orla McKibbin*

Regular programmes on news and current affairs,
sport and entertainment.

Cable and Satellite Television

British Sky Broadcasting Ltd (BSkyB)

Grant Way, Isleworth TW7 5QD
📞 020 7705 3000   📠 020 7705 3030
www.sky.com
Chief Executive *Jeremy Darroch*
Managing Director, Entertainment
Sophie Turner Laing
Chief Operating Officer *Mike Darcey*
Head of Sky News *John Ryley*
Managing Director, Sky Sports and Sky News
Vic Wakeling

Launched 1989. British Sky Broadcasting
gives around 23 million viewers (in 9.2
million households) access to movies, news,
entertainment and sports channels, and
interactive services on Sky digital. In addition, all
customers can receive more than 330 free-to-air
television and radio channels and services.

SKY CHANNELS
Sky News
Launched in 1989 as Europe's first rolling
24-hours news service. Accessible on TV, through
the web, on mobile phones, on big screens, at
railway stations and on radio. Named as 'News
Channel of the Year' by the Royal Television
Society in 2008 (the sixth time in seven years).
Sky 1/Sky 1 HD/Sky 2/Sky 3
The best in entertainment from blockbuster

acquisitions such as *24* and *Lost* to breakthrough commissions such as *Ross Kemp on Gangs* and *Gladiators*.

Sky Sports 1/Sky Sports 2/Sky Sports 3/Sky Sports News/Sky Sports Extra/Sky Sports HD1, HD2 & HD3

Around 40,000 hours of sport are broadcast every year across five Sky Sports channels and three HD channels. Sky Sports 1, 2 and 3 are devoted to live events, support programmes and in-depth sports coverage seven days a week. Sky Sports News provides sports news and the latest results and information 24 hours a day. Sky Sports Extra carries additional sports programming including the award-winning live interactive coverage.

Sky Movies

A choice of movies from the latest Hollywood blockbusters to cult classics. Through agreements with major studios through to leading independents, Sky Movies offers over 850 different titles each month and five premières per week.

Sky Arts 1/Sky Arts 2/Sky Arts 1 HD/Sky Arts 2 HD

The only channels in the UK and Ireland dedicated to the arts.

Sky Real Lives

Launched in 2007. Extraordinary stories of ordinary people, 24 hours a day.

JOINT VENTURES

National Geographic; Nickelodeon; Nick Jr; The History Channel; Paramount Comedy Channel; MUTV; Adventure One; The Biography Channel; Attheraces; Chelsea Digital Media.

CNBC Europe

10 Fleet Place, London EC4M 7QS

☎ 020 7653 9300   ⊞ 020 7653 5956

www.cnbc.com

CEO/President *Mick Buckley*

A service of NBC Universal. 24-hour business and financial news service. Programmes include *Capital Connections*; *Worldwide Exchange*; *Power Lunch Europe*; *European Closing Bell*; *Squawk Box Europe*.

CNN International

Turner House, 16 Great Marlborough Street, London W1P 1DF

☎ 020 7693 1000   ⊞ 020 7693 08921552

edition.cnn.com

Managing Editor, CNN International Europe, Middle East & Africa *Nick Wrenn*

CNN, the leading global 24-hour news network, is available to one billion people worldwide via the 26 CNN branded TV, Internet, radio and mobile services produced by CNN News Group, a Time Warner company. CNN has major production centres in Atlanta, New York, Los Angeles, London, Hong Kong and Mexico City. The London bureau, the largest outside the USA, is CNN's European headquarters and produces over 50 hours of programming per week. Live business and news programmes, including *Business International*; *World News* and *World Business Today*.

MTV Networks Europe

180 Oxford Street, London W1D 1DS

☎ 020 7478 6000

www.mtvne.com

Established 1987. Europe's 24-hour music and youth entertainment channel, available on cable, via satellite and digitally. Transmitted from London in English across Europe.

Travel Channel

64 Newman Street, London W1T 3EF

☎ 020 7636 5401   ⊞ 020 7636 6424

www.travelchannel.co.uk

Launched in February 1994. Broadcasts in 14 languages to over 117 countries across Europe, Africa and the Middle East. Programmes and information on the world of travel; destinations reports, lifestyle programmes plus food and drink, sport and leisure pursuits.

National Commercial Radio

Absolute Radio

1 Golden Square, London W1F 9DJ

☎ 020 7434 1215   ⊞ 020 7434 1197

www.absoluteradio.co.uk

Programme Director *Clive Dickens*

Previously known as Virgin Radio. A music-based station launched in 1973, bought by Chris Evans' Ginger Media Group in December 1997 and acquired by the Scottish Media Group in March 2000.

Classic FM

30 Leicester Square, London WC2H 7LA

☎ 020 7343 9000   ⊞ 020 7766 6100

www.classicfm.com

Managing Director *Darren Henley*

Classic FM, Britain's largest national commercial radio station, started broadcasting in September 1992. Plays accessible classical music 24 hours a day and broadcasts news, weather, travel, business information, political/celebrity/general interest talks, features and interviews. Winner of the 'Station of the Year' Sony Award in 1993, 2000 and 2007.

Digital One

30 Leicester Square, London WC2H 7LA

☎ 020 7288 4600   ⊞ 020 7288 4601

info@digitalone.co.uk
www.ukdigitalradio.com

The UK's national commercial digital radio multiplex operator. Backed by GCap Media and Arqiva, Digital One began broadcasting on 15 November 1999.

talkSPORT

PO Box 1089, London SE1 8WQ
☎ 020 7959 7800
kay.townsend@talksport.co.uk
www.talksport.net
Programme Director *Moz Dee*

Commenced broadcasting in February 1995 as Talk Radio UK. Re-launched January 2000 as TalkSport, the UK's first sports radio station. Acquired by Ulster Television in 2005. Broadcasts 24 hours a day. News items can be e-mailed via the website.

Virgin Radio
▷ Absolute Radio

Independent Local Radio

Capital Radio London

30 Leicester Square, London WC2H 7LA
☎ 020 7766 6000   ℻ 020 7766 6012
www.capitalradio.com
Programme Director *Paul Jackson*

Commenced broadcasting in October 1973 as the country's second commercial radio station (the first being LBC, launched a week earlier). Europe's largest commercial radio station.

Central FM Ltd

201–203 High Street, Falkirk FK1 1DU
☎ 01324 611164   ℻ 01324 611168
email@centralfm.co.uk
www.centralfm.co.uk
Programme Controller *Gary Muircroft*

Broadcasts music, sport and local news to central Scotland, 24 hours a day.

Clyde 1/Clyde 2

Clydebank Business Park, Clydebank G81 2RX
☎ 0141 565 2200   ℻ 0141 565 2265
www.clyde1.com or
www.clyde2.com
Owner *Bauer Radio*
Station Director *Tracey McNellan*

Programmes usually originate in-house or by commission. All documentary material is made in-house. Good local news items always considered.

Forth One/Forth 2

Forth House, Forth Street, Edinburgh EH1 3LE
☎ 0131 556 9255   ℻ 0131 557 4645 (news)

moira.millar@radioforth.com
www.forth.com
Programme Director *Duncan Campbell*
Programme Producer, Forth 2 *Moira Millar*
News Editor *Paul Robertson*
Book Reviewer *Lesley Fraser-Taylor*

News stories welcome from freelancers. Music-based programming.

Isle of Wight Radio

Dodnor Park, Newport PO30 5XE
☎ 01983 822557   ℻ 01983 822109
www.iwradio.co.uk
Station Manager *Dan Bates*
Programme Controller *Paul Topping*

Part of The Local Radio Company, Isle of Wight Radio is the island's only radio station broadcasting local news, music and general entertainment including phone-ins and interview based shows. Music, television, film, popular culture are the main areas of interest. *John Hannam Meets* – long-running interview programme – on Sundays.

kmfm

Head Office: Express House, 34–36 North Street, Ashford TN24 8JR
☎ 01233 895825
www.kmfm.co.uk
KM Group Head of Radio *Steve Fountain*

A wide range of music programming plus news, views and local interest. Part of the KM Group.

LBC Radio Ltd

The Chrysalis Building, Bramley Road, London W10 6SP
☎ 020 7314 7300
www.lbc.co.uk
Editorial Director *Jonathan Richards*

LBC 97.3 FM is a talk-based station broadcasting 24 hours a day, providing entertainment, interviews, celebrity guests, music, chat shows, local interest, news and sport. LBC News 1152 AM, the sister station of LBC 97.3 FM, provides rolling news (7.00 am to 7.15 pm Mon. to Fri.; 7.00 am to 5.00 pm Sat. and Sun.).

NorthSound Radio

Abbotswell Road, Aberdeen AB12 3AJ
☎ 01224 337000
www.northsound1.com
www.northsound2.com
Managing Director *Luke McCullough*
Programme Director *Chris Thomson*

Features and music programmes 24 hours a day including, mid-morning (9.00 am – midday), *Northsound 2* feature programme.

Premier Radio

22 Chapter Street, London SW1P 4NP
📞 020 7316 1300　📠 020 7233 6706
premier@premier.org.uk
www.premier.org.uk
Chief Executive Officer *Peter Kerridge*

Broadcasts programmes that reflect the beliefs
and values of the Christian faith, 24 hours a day
on 1305, 1413, 1332 MW, Sky digital 0123, Virgin
Media 968.

Radio XL 1296 AM

KMS House, Bradford Street, Birmingham
B12 0JD
📞 0121 753 5353　📠 0121 753 3111
Managing Director *Arun Bajaj*

Asian broadcasting for the West Midlands, 24
hours a day. Broadcasts *Love Express* featuring
love stories and poems. Writers should send
material for the attention of *Satinder Dhaiwal*.

Sabras Radio

Radio House, 63 Melton Road, Leicester LE4 6PN
📞 0116 261 0666　📠 0116 266 7776
www.sabrasradio.com
Programme Controller *Don Kotak*

Programmes for the Asian community,
broadcasting 24 hours a day.

Spectrum Radio

4 Ingate Place, London SW8 3NS
📞 020 7627 4433　📠 020 7627 3409
enquiries@spectrumradio.net
www.spectrumradio.net
Managing Director *Toby Aldrich*
Station Director *John Ogden*

Programmes for a broad spectrum of ethnic
groups in London.

Sunrise Radio (Yorkshire)

55 Leeds Road, Bradford BD1 5AF
📞 01274 735043　📠 01274 728534
www.sunriseradio.fm
Programme Controller, Chief Executive &
Chairman *Usha Parmar*

Programmes for the Asian community in West
Yorkshire.

Swansea Sound 1170 MW/96.4 FM The Wave

Victoria Road, Gowerton, Swansea SA4 3AB
📞 01792 511170 (MW)/ 511964 (FM)
📠 01792 511171 (MW)/511965 (FM)
www.swanseasound.co.uk
Station Director *Carrie Mosley*
Programme Director *Steve Barnes*
News Editor *Emma Grant*

Music-based programming on FM while Swansea
Sound is interested in a wide variety of material,
though news items must be of local relevance.
An explanatory letter, in the first instance, is
advisable.

96.4 FM The Wave
▷ Swansea Sound 1170 MW

European Television Companies

Austria

ORF(Österreichisher Rundfunk)1/ORF2
Würzburggasse 30, A–1136 Vienna
☎ 00 43 1 87 878-0
www.orf.at

Belgium

RTBF (Radio-Télévision Belge de la Communauté Française) (French language TV & radio)
Boulevard Auguste Reyers 52, B–1044 Brussels
☎ 00 32 2 737 2111
www.rtbf.be

Vlaamse Televisie Maatschappij (VTM) (cable)
Medialaan 1, B–1800 Vilvoorde
☎ 00 32 2 255 3211
www.vtm.be

VRT (Vlaamse Radio- en Televisieomroep) (Dutch language TV)
Auguste Reyerslaan 52, B–1043 Brussels
☎ 00 32 2 741 3111
www.vrt.be

Bosnia and Herzegovina

PBS BiH (Public Broadcasting Service of Bosnia and Herzegovina)
Bulevar Mese Selimovica 12, 71000 Sarajevo
☎ 00 387 33 45 123/461 101
www.pbsbih.ba

Bulgaria

BNT
29 San Stefano str., BG–1504 Sofia
☎ 00 359 2 944 4999
www.bnt.bg

Croatia

HRT (Hrvatska Televizija)
Prisavlje 3, CR–10000 Zagreb
☎ 00 385 1 634 3683
www.hrt.hr

Czech Republic

CT (Ceska Televize)
Kavci Hory, CZ–140 70 Prague 4
☎ 00 420 2 6113 7000
www.czech-tv.cz

Denmark

DR (Danmarks Radio)
Emil Holms Kanal 20, DK–0999 Copenhagen C
☎ 00 45 35 20 3040
www.dr.dk

SBS TV A/S (cable & satellite)
Mileparken 20A, DK–2740 Skovlunde
☎ 00 45 70 10 1010
www.sbstv.dk

TV-2 Danmark
Rugaardsvej 25, DK–5100 Odense C
☎ 00 45 65 91 9191
www.tv2.dk

TV3 (Viasat) (cable & satellite)
Wildersgåde 8, DK–1408 Copenhagen K
☎ 00 45 77 30 5500
www.viasat.dk

Estonia

ETV (Eesti Televisioon)
Faehlmanni 12, EE–15029 Tallinn
☎ 00 372 628 4100
www.etv.ee

Finland

MTV3 Finland
Ilmalantori 2, SF–00033 Helsinki
☎ 00 358 10 300300
www.mtv3.fi

YLE FST (Swedish language TV)
Yleisradio Oy PO Box 62, SF–00024 Helsinki
☎ 00 358 9 14801
www.yle.fi/fst

YLE TV1
Yleisradio Oy PO Box 90, SF–00024 Helsinki
☎ 00 358 9 14801
www.yle.fi/tv1

YLE TV2
Yleisradio PO Box 196, SF–33101 Tampere
☎ 00 358 3 3456 111
www.yle.fi/tv2

France

Arte France (cable & satellite)
8 rue Marceau, F–92785 Issy-les-Moulineaux
Cedex 09
☎ 00 33 1 55 00 77 77
www.artefrance.fr

Canal+ (pay TV)
25 rue Leblanc, F–75906 Paris Cedex 15
☎ 00 33 1 44 25 10 10
www.canalplus.fr

France 2/France 3
7 Esplanade Henri de France, F–75907 Paris
Cedex 15
☎ 00 33 1 56 22 42 42 (F2)/30 30 (F3)
www.france2.fr
www.france3.fr

France 5
10–16 rue Horace Vernet, F–92136 Issy-les-
Moulineaux
☎ 00 33 1 56 22 91 91
www.france5.fr

Planète (privately-owned documentary channel)
48 quai du Pont du Jour, F–92659 Boulogne
☎ 00 33 1 71 10 10 20
www.planete.com

Voyage (cable & satellite)
241 Boulevard Péreire, F–75838 Paris
☎ 00 33 1 058 05 58 05
www.voyage.fr

Germany

ARD – Das Erste
Arnulfstrasse 42, D–80335 Munich
☎ 00 49 89 5900 3344
www.daserste.de

Arte Germany (cable & satellite)
Postfach 10 02 13, D–76483 Baden-Baden
☎ 00 49 7221 93690
www.arte.de

DW-TV (Deutsche Welle) (international TV service)
Voltastrasse 6, D–13355 Berlin
☎ 00 49 30 4646 0
www.dw-world.de

ZDF (Zweites Deutsches Fernsehen)
ZDF–Strasse 1, D–55127 Mainz
☎ 00 49 61 31/701
www.zdf.de

Greece

ERT S.A. (Elliniki Radiophonia Tileorassi)
Mesogeion 432, PO Box 19, GR–15342 Athens
☎ 00 30 2 10 606 6000
www.ert.gr

Hungary

DUNA TV (Duna Televízió)
Mészáros utca 48, H–1016 Budapest
☎ 00 36 1 489 1200
www.dunatv.hu

MTV 1/MTV 2 (Magyar Televízió)
Szabadság tér 17, H–1810 Budapest
☎ 00 36 1 353 3200
www.mtv.hu

Republic of Ireland

Radio Telefís Éireann (RTÉ One – RTÉ Two)
Donnybrook, Dublin 4
☎ 00 353 1 208 3111
www.rte.ie

TG4 (Teilefís na Gaelige) (Irish language TV)
Baile na hAbhann, Co. na Gaillimhe
☎ 00 353 91 505050
www.tg4.ie

Italy

RAI (RadioTelevisione Italiana)
Viale Mazzini 14, I–00195 Rome
☎ 00 39 06 3878-1
www.rai.it

RTI SpA (Reti Televisive Italiane)
Viale Europa 44, I–20093 Cologno Monzese
Milano
☎ 00 39 02 2514-1
www.rete4.com

Latvia

LTV1 & LTV7
Zakusalas Embankment 3, LV–1509 Riga
☎ 00 371 7 200 314
www.ltv.lv

Lithuania

LRT (Lietuvos radijas ir televizija)
S. Konarskio 49, LT–2600 Vilnius
☎ 00 370 5 236 3209
www.lrt.lt

The Netherlands

AVRO (Algemene Omroep Vereniging)
Postbus 2, NL–1200 JA Hilversum
☎ 00 31 35 671 79 11
www.central.avro.nl/index.asp

EO (Evangelische Omroep) (Evangelical TV)
Postbus 21000, NL–1202 BB Hilversum
☎ 00 31 35 647 4747
www.eo.nl

IKON (Interkerkelijke Omroep Nederland) (church-based TV)
Postbus 10009, NL–1201 DA Hilversum
☎ 00 31 35 672 72 72
www.omroep.nl/ikon

KRO (Katholieke Radio Omroep) (Catholic TV)
PO Box 23000, NL–1202 EA Hilversum
☎ 00 31 35 671 3911
www.kro.nl

NCRV (Nederlandse Christelijke Radio Vereniging)
Postbus 25000, NL–1202 HB Hilversum
☎ 00 31 35 671 99 11
info.omroep.nl/ncrv

NOS (Nederlandse Omroep Stichting) – Nederland 1, 2 & 3
Postbus 26600, NL–1202 JT Hilversum
☎ 00 31 35 677 92 22
www.nos.nl

NPS (Nederlandse Programma Stichting)
Postbus 29000, NL–1202 MA Hilversum
☎ 00 31 35 677 9333
www.omroep.nl/nps

VPRO
Postbus 11, NL–1200 JC Hilversum
☎ 00 31 35 671 2639
www.vpro.nl

Norway

NRK (Norsk Rikskringkasting)
Bjørnstjerne Bjørnsonsplass 1, N–0340 Oslo
☎ 00 47 23 04 7000
www.nrk.no

TV 2 Norge
Postboks 7222, Nøstegatan 72, N–5020 Bergen
☎ 00 47 55 90 8070
www.tv2.no

Poland

TVP (Telewizja Polska S.A.)
ul. J.P. Woronicza 17, PL–00 999 Warsaw
☎ 00 48 22 547 8000
www.tvp.com.pl

Portugal

RTP 1 & RTP 2 (Radiotelevisão Portuguesa)
Avenida Marechal, Gomes da Costa 37,
P–1849–030 Lisbon
☎ 00 351 21 794 7000
www.rtp.pt

SIC (Sociedade Independente de Comunicacão) (cable TV)
Estrada da Outurela 119, P–2799-526 Carnaxide
☎ 00 351 21 417 9406
www.six.pt

TVI (Televisão Independente)
Rua Mário Castelhano 40, Queluz de Baixo,
P–2734–502 Barcarena
☎ 00 351 21 434 7500
www.tvi.pt

Romania

TVR (The Romanian Television Corporation)
Calea Dorobantilor 191, RO–71281 Bucharest 1
☎ 00 40 21 230 6290
www.tvr.ro

Serbia & Montenegro

RTS
Takovska 10, 11000 Beograda
☎ 00 381 11 321 2000
www.rts.co.yu

Slovenia

TV Slovenija
Kolodvorska ulica 2, SL–1550 Ljubljana
☎ 00 386 1 475 211
www.rtvslo.si

Spain

Canal Sur Televisión (Andalucian TV)
Ctra. San Juan de Aznalfarache-Tomares,
E–41920 San Juan de Aznalfarache
☎ 00 34 95 505 4600
www.canalsur.es

Canal+ Espana (pay TV)
Avenida de los Artesanos 6, E–28760 Tres Cantos
Madrid
☎ 00 34 91 736 7940
www.plus.es

ETB-1/ETB-2 (Euskal Telebista S.A.) (Basque TV)
E–48215 Iurreta/Bizkaia
☎ 00 34 94 603 1000
www.eitb.com

TVC (Televisió de Catalunya S.A.) (Catalan TV)
Carrer de la TV 3, E–08970 Saint Joan Despi
☎ 00 34 93 499 9333
www.tvcatalunya.com

TVE-1/ TVE-2 (Television Española)
Prado del Rey, E–28023 Madrid
☎ 00 349 1 346 4000
www.rtve.es

TVV (TVV-Televisio Valenciana) (Valencian TV)
Poligon Acc. Ademuz s/n, E–46100 Burjassot
☎ 00 34 96 318 3000
www.rtvv.es

Sweden

SVT (Sveriges Television)
Oxenstierngatan 26–34, S–105 10 Stockholm
☎ 00 46 8 784 0000
www.svt.se

TV4
Tegeluddsvägen 3–5, S–115 79 Stockholm
☎ 00 46 8 459 4000
www.tv4.se

Switzerland

RTR (Radio e Televisiun Rumantscha) (Romansch language TV)
Via da Masans 2, Plazza dal teater, CH–7002
Cuira
☎ 00 41 81 255 7575
www.rtr.ch

RTSI (Radiotelevisione svizzera di lingua Italiana) (Italian language TV)
Casella postale, CH–6903 Lugano
☎ 00 41 91 803 51 11
www.rtsi.ch

SF DRS (Schweizer Fernsehen) (German language TV)
Fernsehstrasse 1–4, CH–8052 Zurich
☎ 00 41 1 305 66 11
www.sfdrs.tv

SRG SSR idée suisse (Swiss Broadcasting Corp.)
Belpstrasse 48, CH-3000 Berne14
☎ 00 41 31 350 91 11
www.srg-ssr.ch

TSR (Télévision Suisse Romande) (French language TV)
Quai Ernest Ansermet 20, CH–1211 Geneva 8
☎ 00 41 22 708 20 20
www.tsr.ch

Film, TV and Radio Producers

Aardman

Gas Ferry Road, Bristol BS1 6UN

☎ 0117 984 8485   🖷 0117 984 8486

www.aardman.com

Founded 1976. Award-winning animation studio producing feature films and television programmes. Won Oscars for *Creature Comforts* and *Wallace and Gromit*. Other OUTPUT includes *Angry Kid*; *Chicken Run*; *Flushed Away*; *Shaun The Sheep*; *Chop Socky Chooks*; *Purple and Brown*. No unsolicited submissions.

Above The Title

Level 2, 10/11 St Georges Mews, London NW1 8XE

☎ 020 7916 1984   🖷 020 7722 5706

mail@abovethetitle.com

www.abovethetitle.com

CEO *Simon Clegg*

Contact *Rebecca Pinfield*

Producer of radio drama, comedy and factual programmes. OUTPUT includes *Clive Anderson's Chat Room* and *Till the End of the Day – The Kinks Story* (Radio 2); *Unreliable Evidence*; *The Glittering Prizes*; *The Hitchhiker's Guide to the Galaxy*; *Fame & Fortune* and *Dirk Gently's Holistic Dectective Agency* (Radio 4); *Features Like Mine* and *Iain Burnside* (Radio 3). 'We encourage and support new writing in every way we can.' Ideas welcome. Send short synopsis only by e-mail in the first instance (see website for details).

Absolutely Productions Ltd

Unit 19, 77 Beak Street, London W1F 9DB

☎ 020 7644 5575

www.absolutely.biz

Founded in 1988 by a group of writer/performer/producers including Morwenna Banks, Jack Docherty, Moray Hunter, Pete Baikie, John Sparkes and Gordon Kennedy. Long-established producer for TV, has recently moved into feature films. OUTPUT *Absolutely*; *Welcome to Strathmuir* (a new comedy pilot for BBC Scotland) and *Baggage* (Radio 4).

Abstract Images

117 Willoughby House, Barbican, London EC2Y 8BL

☎ 020 7638 5123

productions@abstract-images.co.uk

Contact *Howard Ross*

Television documentary and drama programming. Also theatre productions. OUTPUT includes *Balm in Gilead*; *Road* and *Bent* (all dramas); *God: For & Against* (documentary); *This Is a Man* (drama/doc). New writers should send synopsis in the first instance.

Acacia Productions Ltd

80 Weston Park, London N8 9TB

☎ 020 8341 9392

acacia@dial.pipex.com

www.acaciaproductions.co.uk

Contact *J. Edward Milner*

Producer of award-winning television and video documentaries; also news reports, corporates and programmes for educational charities. Undertakes basic video training and has a video manual in preparation. OUTPUT includes documentary series entitled *Last Plant Standing*; *A Farm in Uganda*; *Montserrat: Under the Volcano*; *Spirit of Trees* (8 programmes); *Vietnam: After the Fire*; *Macroeconomics – the Decision-makers*; *Greening of Thailand*; *A Future for Forests*. No unsolicited mss.

Acrobat Television

107 Wellington Road North, Stockport SK4 2LP

☎ 0161 477 9090   🖷 0161 477 9191

info@acrobat-tv.co.uk

www.acrobat-tv.co.uk

Contact *David Hill*

All script genres for broadcast and coporate television, including training and promotional scripts, comedy and drama-based material. OUTPUT includes *Make a Stand* (Jack Dee, Gina Bellman and John Thompson for Video Arts); *The Customer View* (Roy Barraclough for Air Products); *Serious About Waves* series (Peter Hart for the Royal Yachting Association); *Fat Face Night* series (Extreme). No unsolicited mss.

Actaeon Films Ltd

50 Gracefield Gardens, London SW16 2ST
☎ 020 8769 3339 📠 0870 134 7980
info@actaeonfilms.com
www.actaeonfilms.com
Producer *Daniel Cormack*
Head of Development *Becky Connell*

Founded 2004. Specializes in short and feature
length theatrical motion pictures. OUTPUT
includes *Amelia and Michael* (short drama
starring Anthony Head). In development: *Golden
Apples* (feature-length drama); *The Dead Letters*
and *My Brother's Keeper* (both feature-length
psychological thrillers). Actaeon Films is a
member of the **New Producers Alliance**,
The Script Factory and the London Filming
Partnership. 'We actively encourage new writers
and new writing. However, we can only give
feedback on scripts with strong potential for
developments (all mss are read by a professional
reader within four weeks of submission). Scripts
returned only if accompanied by s.a.e.'

All Out Productions

50 Copperas Street, Manchester M4 1HS
☎ 0161 834 9955 📠 0161 834 6978
mail@allout.co.uk
www.allout.co.uk
Contact *Richard McIlroy*

Producer of documentaries, features and current
affairs programmes for radio. OUTPUT includes
Five Live Report (BBC Five Live – weekly news
documentary); *Lamacq Live* (BBC Radio One –
music and social affairs). Ideas welcome but not
mss. Approach by e-mail.

Alomo Productions
▶ **FremantleMedia Group Ltd**

Anglo/Fortunato Films Ltd

170 Popes Lane, London W5 4NJ
☎ 020 8932 7676 📠 020 8932 7491
anglofortunato@aol.com
Contact *Luciano Celentino*

Film, television and video producer/director of
action comedy and psych-thriller drama. No
unsolicited mss.

Arlington Productions Limited

Cippenham Court, Cippenham Lane, Cippenham,
Nr Slough SL1 5AU
☎ 01753 516767 📠 01753 691785

Television producer. Specializes in popular
international drama, with *occasional* forays into
other areas. 'We have an enviable reputation
for encouraging new writers but only accept
unsolicited submissions via agents.'

Art & Training Films Ltd

PO Box 3459, Stratford-upon-Avon CV37 6ZJ
☎ 01789 294910
andrew.haynes@atf.org.uk
www.atf.org.uk
Contact *Andrew Haynes*

Producer of documentaries, drama, commecials
and corporate films and video. Submissions via
agents considered.

The Ashford Entertainment Corporation Ltd

20 The Chase, Coulsdon CR5 2EG
☎ 020 8668 8188 📠 087 0116 4142
info@ashford-entertainment.co.uk
www.ashford-entertainment.co.uk
Managing Director *Frazer Ashford*

Founded in 1996 by award-winning producer
Frazer Ashford to produce character-led
documentaries for TV. Examples of work
include *Serial Killers*, *Great Little Trains*
(winner of RTS Best Regional Documentary)
and *Streetlife* (winner of bronze award at the
Flagstaff Film Festival). Ashford Entertainment
also incorporates The Reel Thing Ltd, suppliers
of corporate event management, DVD and video
production; also production facilities. Happy to
look at any submissions falling within the remit
of character-led documentaries. Send hard copy
with an s.a.e.

Avalon Television

4a Exmoor Street, London W10 6BD
☎ 020 7598 7280
www.avalonuk.com/tv

Founded in 1993, Avalon Television is a
BAFTA-winning production company supplying
TV programming to all the British terrestrial
channels and most leading satellite channels. Best
known for light entertainment programmes such
as documentary *Frank Skinner on Frank Skinner*
(ITV1) or comedy *Fantasy Football* (BBC2). Now
broadening output to include drama, factual and
formatted shows. Script, treatment or ideas are
welcomed via e-mail.

Baby Cow Productions

33 Foley Street, London W1W 7TL
☎ 020 7612 3370 📠 020 7612 3352
script@babycow.co.uk
www.babycow.co.uk

Founded in 1989 by Steve Coogan and Henry
Normal. Specializes in cutting-edge comedy for
television. Has divisions for animation, radio
and films. OUTPUT *Marion and Geoff*; *Human
Remains*; *The Sketch Show*; *24 Hour Party
People*; *I'm Alan Patridge*; *Nighty Night*; *The
Mighty Boosh*; *Dating Ray Fenwick*. Dedicated
development department always on the lookout
for new talent. In the first instance, send a

synopsis or treatment and a ten-page extract, plus a DVD or video if the idea is filmed, and a covering letter. If the idea is of interest, Baby Cow will contact the writer. Include all contact details, including e-mail.

Beckmann Visual Publishing

Milntown Lodge, Lezayre Road, Ramsey, Isle of Man IM8 2TG
℡ 01624 816585 ℻ 01624 816589
sales@beckmanngroup.co.uk
www.beckmanngroup.co.uk
Contacts *Jo White, Stuart Semark*

Isle of Man-based company. Video and television documentary distributor. OUTPUT *Practical Guide to Europe* (travel series); *Music Maestro* (12-part series on classical composers); *Christian the Lion* (wildlife collection); *Aerial Britain* (British heritage/travel).

Big Heart Media

Flat 4, 6 Pear Tree Court, London EC1R 0DW
℡ 020 7608 0352 ℻ 020 7250 1138
info@bigheartmedia.com
www.bigheartmedia.com
Contact *Colin Izod*

Producer of drama and documetaries for television and video. Ideas/outlines welcome by e-mail, but not unsolicited mss.

Bona Broadcasting Limited

2nd Floor, 9 Gayfield Square, Edinburgh EH1 3NT
℡ 0131 558 1696 ℻ 0131 558 1694
enquiries@bonabroadcasting.com
www.bonabroadcasting.com
Contact *Turan Ali*

Producer of award winning-drama and documentary programmes for BBC radio, TV and film projects. No unsolicited mss but send a one-paragraph summary by e-mail in the first instance. Runs radio and TV drama training courses in the UK and internationally.

Brighter Pictures
▷ Endemol UK Productions

Buccaneer Films

5 Rainbow Court, Oxhey WD19 4RP
℡ 01923 254000
Contact *Michael Gosling*

Corporate video production and still photography specialists in education and sport. No unsolicited mss.

Cactus TV

373 Kennington Road, London SE11 4PS
℡ 020 7091 4900 ℻ 020 7091 4901
cactustv.co.uk

Founded 1994. Specializes in broad-based entertainment, features and chat shows. Produces around 300 hours of broadcast material per year. OUTPUT includes *Richard & Judy*; *Sir Cliff Richard – The Hits I Missed*; *Songs of Bond*; *The British Soap Awards*.

Calon TV

3 Mount Stuart Square, Butetown, Cardiff CF10 5EE
℡ 029 2048 8400 ℻ 029 2048 5962
enquiries@calon.tv
www.calon.tv
Head of Development *Andrew Offiler*

Founded 2005. Animated series, mainly for children. OUTPUT includes *Meeow*; *Hilltop Hospital*; *Hana's Helpline*; *The Hurricanes*; *Tales of the Toothfairies*; *Billy the Cat*; *The Blobs*, as well as the feature films, *Under Milk Wood* and *The Princess and the Goblin*. Write with ideas and sample script in the first instance.

Carnival Film and Television Limited

47 Marylebone Lane, London W1U 2NT
℡ 020 7317 1370 ℻ 020 7317 1380
info@carnivalfilms.co.uk
www.carnivalfilms.co.uk
Managing Director *Gareth Neame*

Film, TV and theatre producer. OUTPUT, FILM: *The Mill on the Floss* (BBC); *Firelight* (Hollywood Pictures/Wind Dancer Productions); *Up on the Roof* (Rank/Granada); *Shadowlands* (Savoy/Spelling); *The Infiltrator* (Home Box Office); *Under Suspicion* (Columbia/Rank/LWT). TELEVISION: *Rosemary & Thyme* (ITV/Granada); *The Grid* (BBC/TNT); *As If* (Ch4/Columbia); *Lucy Sullivan is Getting Married* (ITV); *The Tenth Kingdom* (Sky/NBC); *Agatha Christie's Poirot* (ITV/LWT/A&E); *Every Woman Knows a Secret* and *Oktober* (both for ITV Network Centre); *Hotel Babylon*; *Blott on the Landscape*; *Crime Traveller* and *Bugs 1–4* (all for BBC); *Anna Lee* and *All Or Nothing At All* (LWT); *Head Over Heels* (Carlton); *Jeeves & Wooster* I–IV (Granada); *The Fragile Heart*; *Traffik* and *Porterhouse Blue* (all for Ch4). THEATRE: *How Was it for You?*; *What a Performance*; *Juno & the Paycock*; *Murder is Easy*; *Misery*; *The Ghost Train*; *Map of the Heart*; *Shadowlands*; *Up on the Roof*. No unsolicited mss

Cartwn Cymru

12 Queen's Road, Mumbles, Swansea SA3 4AN
℡ 07771 640400
production@cartwn-cymru.com
Producer *Naomi Jones*

Animation production company. OUTPUT *Toucan 'Tecs* (YTV/S4C); *Funnybones* and *Turandot: Operavox* (both for S4C/BBC); *Testament: The Bible in Animation* (BBC2/S4C); *The Miracle Maker* (S4C/BBC/British Screen/

Icon Entertainment International); *Faeries* (HIT Entertainment plc for CITV); *Otherworld* (animated feature film for S4C Films, British Screen, Arts Council of Wales).

Celador Films

39 Long Acre, London WC2E 9LG
📞 020 7845 6998
www.celador.co.uk
Head of Development *Ivana MacKinnon*

Producer of feature films. OUTPUT includes *Dirty Pretty Things*; *Separate Lies*; *Slumdog Millionaire* and Neil Marshall's second feature, *The Descent*. Submissions from established literary and film agents only or talent already known to the company, or from third party producers seeking possible co-production.

Celador Productions

39 Long Acre, London WC2E 9JT
📞 020 7240 8101
info@celadorproductions.com
www.celadorproductions.com
Director of Production *Heather Hampson*
Head of Radio *Liz Anstee*

Producer of a range of TV and radio programmes, from situation comedy to quizzes, game shows, popular factual entertainment and international co-productions. OUTPUT *You Are What You Eat*; *Perfect Strangers*; *Popcorn*; *It's Been a Bad Week*; *Commercial Breakdown*; *How to Dump Your Mates* and *Three Off the Tee*. 'We are interested in radio scripts but do not accept unsolicited proposals for comedy or entertainment formats. As a relatively small company our script-reading resources are limited.'

Celtic Films Entertainment Ltd

Lodge House, 69 Beaufort Street, London SW3 5AH
📞 020 7351 0909   📠 020 7351 4139
info@celticfilms.co.uk
www.celticfilms.co.uk
Contact *Steven Russell*

Founded in 1986 producing high-end TV drama and mainstream feature films. OUTPUT includes sixteen *Sharpe* television films, *Girl from Rio* and *Stefan Kiszko: A Life for a Life*. No unsolicited submissions.

Chameleon Television Ltd

Church House, 14 Town Street, Horsforth, Leeds LS18 4RJ
📞 0113 205 0040   📠 0113 281 9454
allen@chameleontv.com
Contacts *Allen Jewhurst, Julia Kirby-Smith*

Film and television drama and documentary producer. OUTPUT includes *Edge of the City*; 'Dispatches' – *Channel 4 News*; *The Family Who Vanished*; *Killing for Honour*; *College Girls*; *Ken Dodd in the Dock*; *Unholy War* (all for Ch4); *Diary of a Mother on the Edge*; *Divorces From Hell*; *Shipman*; *Love to Shop*; *The Marchioness* (all for ITV); *Ted & Sylvia – Love, Loss*; *Hamas Bombers* (BBC); *Liverpool Poets* (Ch5). Scripts not welcome unless via agents but new writing is encouraged.

Channel Television Ltd

Television Centre, St Helier, Jersey JE1 3ZD
📞 01534 816873   📠 01534 816889
david@channeltv.co.uk
www.channeltvco.uk
Senior Producer *David Evans*

Producer of TV commercials and corporate material: information, promotional, sales, training and events coverage. CD and DVD production; promotional videos for all types of businesses throughout Europe. No unsolicited mss; new writing/scripts commissioned as required. Interested in hearing from local writers resident in the Channel Islands.

Cheetah Television
▷ **Endemol UK Productions**

The Children's Film & Television Foundation Ltd

c/o Simon George, Head of Finance, Ealing Studios, Ealing Green, London W5 5EP
📞 07887 573479
info@cftf.org
www.cftf.org.uk
Chief Executive *Anna Home*

Involved in the development and co-production of films for children and the family, both for the theatric market and for TV.

Cleveland Productions

5 Rainbow Court, Oxhey, Near Watford WD19 4RP
📞 01923 254000
Contact *Michael Gosling*

Communications in sound and vision A/V production and still photography specialists in education and sport. No unsolicited mss.

COI

Hercules House, Hercules Road, London SE1 7DU
📞 020 7928 2345
www.coi.gov.uk

Government advertising and marketing communications, and public information films.

Collingwood O'Hare Productions Ltd

10–14 Crown Street, London W3 8SB
📞 020 8993 3666   📠 020 8993 9595
info@crownstreet.co.uk
www.collingwoodohare.com

Director *Tony Collingwood*
Head of Development *Helen Stroud*

Film and TV; specializes in children's animation.
OUTPUT *The Secret Show* (BBC/BBC Worldwide);
RARG (award-winning animated film); *Yoko!
Jakamoko! Toto!*; *Daisy-Head Mayzie* (Dr Seuss
animated series for Turner Network and Hanna-
Barbera); *Gordon the Garden Gnome* (CBBC);
Eddy and the Bear and *The King's Beard* (both for
CITV). Unsolicited mss not welcome 'as a general
rule as we do not have the capacity to process
the sheer weight of submissions this creates. We
therefore tend to review material from individuals
recommended to us through personal contact
with agents or other industry professionals. We
like to encourage new writing and have worked
with new writers but our ability to do so is limited
by our capacity for development. We can usually
only consider taking on one project each year, as
development/finance takes several years to put in
place.'

The Comedy Unit

6th Floor 53 Bothwell Street, Glasgow G2 6TS
☎ 0141 220 6400
scripts@comedyunit.co.uk
www.comedyunit.co.uk
Managing Director *April Chamberlain*
Creative Director *Colin Gilbert*

Producers of scripted comedy and entertainment
shows for TV and radio. OUTPUT includes *Rab
C. Nesbitt*; *Fags, Mags & Bags*; *The Incredible Will
& Greg*; *Still Game*; *The Karen Dunbar Show*;
Chewin' the Fat; *Only an Excuse?*; *Watson's Wind
Up*. Unsolicited sketches welcome. Sitcoms
should be submitted as a one-page outline with a
couple of sample scenes.

Company Pictures

Suffolk House, 1–8 Whitfield Place, London
W1T 5JU
☎ 020 7380 3900   📠 020 7380 1166
enquiries@companypictures.co.uk
www.companypictures.co.uk
Managing Directors *George Faber,
Charles Pattinson*

One of the UK's largest independent film
and television drama production companies.
Established in 1998. Won Best Independent
Production Company at the 2008 and 2005
Broadcast Awards and the European Producers
of the Year Award at the 2004 Monte Carlo
Awards. OUTPUT includes *Shameless* (Ch4); *Life
and Death of Peter Sellers* (HBO, two Golden
Globe Awards, 15 Emmies); *Elizabeth I*, starring
Helen Mirren and Jeremy Irons (Ch4/HBO, three
Golden Globe Awards, 9 Emmies). Proposals
accepted via agents only.

Cosgrove Hall Films

ITV Granada, Quay Street, Manchester M60 9EA
☎ 0161 952 1382   📠 0161 952 1381
animation@cosgrovehall.com
www.chf.co.uk

Producer of animated children's television
programmes, founded in 1976. OUTPUT includes
Andy Pandy; *Bill and Ben*; *Count Duckula*;
DangerMouse; *Fifi and the Flowertots*; *Roald
Dahl's BFG*; *The Jungle Book* and *The Wind in the
Willows*.

The Creative Partnership

13 Bateman Street, London W1D 3AF
☎ 020 7439 7762   📠 020 7437 1467
sarah.fforde@thecreativepartnership.co.uk
www.thecreativepartnership.co.uk
Contact *Sarah Fforde*

'Europe's largest "one-stop shop" for advertising
and marketing campaigns for the film and
television industries.' Clients include most major
and independent film companies. No scripts. 'We
train new writers in-house, and find them from
submitted c.v.s. All applicants must have previous
commercial writing experience.'

Cricket Ltd

Medius House, 63–69 New Oxford Street,
London WC1A 1EA
☎ 020 7845 0300   📠 020 7845 0303
team@cricket-ltd.com
www.cricket-ltd.com
Head of Recruitment *Mary McDonnell*

Film and video, live events and conferences,
print and design. 'Communications solutions for
business clients wishing to influence targeted
external and internal audiences.'

CSA Word

6A Archway Mews, 241A Putney Bridge Road,
London SW15 2PE
☎ 020 8871 0220   📠 020 8877 0712
victoria@csaword.co.uk
www.csaword.co.uk
Contacts *Victoria Williams, Clive Stanhope*

Producer of drama, documentaries and readings
for radio. OUTPUT *Alfie Elkins & His Little Life*
(drama for BBC World Service); *The Hungry Years*
(reading for BBC R4); John O'Farrell's *An Utterly
Impartial History of Britain*; *A Stable Relationship*
(features for BBC R4); *Bob Dylan's Chronicles 1*;
Eric Clapton's Autobiography (readings for BBC
R2); *It's a Girl*; *Chat Snaps & Videotape – Two*
(documentaries for BBC World Service); *The
Glenn Miller Story, Berlin...Soundz Decadent* and
Original Soundtrack Recordings (documentaries
for BBC R2), Unsolicited drama ideas welcome by
post or e-mail. 'We encourage new drama ideas.'

Cutting Edge Productions Ltd

27 Erpingham Road, London SW15 1BE
☎ 020 8780 1476 📠 020 8780 0102
juliannorridge@btconnect.com
Contact *Julian Norridge*

Corporate and documentary video and television. OUTPUT includes US series on evangelicalism, 'Dispatches' on US tobacco and government videos. No unsolicited mss; 'we commission all our writing to order, but are open to ideas.'

Diverse Production Limited

6–12 Gorleston Street, London W14 8XS
☎ 020 7603 4567 📠 020 7603 2148
www.diverse.tv

Independent production company specializing in popular prime-time formats, strong documentaries, specialist factual, historical, cultural, religious, arts, music and factual entertainment. Recent OUTPUT includes *Ballet Changed My Life: Ballet Hoo!*; *Mission Africa*; *Codex*; *Bear Grylls' Man vs. Wild*; *Tribal Wife*; *Musicality*; *Operatunity*; *Who Wrote the Bible*; *Who You Callin' Nigger?*; *In Search of Tony Blair*; *Britain AD*; *Shock Treatment*; *Beyond Boundaries*; *Escape to the Legion*.

DoubleBand Films

3 Crescent Gardens, Belfast BT7 1NS
☎ 028 9024 3331 📠 028 9023 6980
www.doublebandfilms.com
Contacts *Michael Hewitt, Dermot Lavery*

Established 1988. Specializes in documentaries and some drama. Recent productions include *Seven Days that Shook the World* and *War in Mind* (both for Ch4); *Christine's Children* (BBC Northern Ireland; nominated for both the RTS and Celtic Film Festival); *D-Day: Triumph and Tragedy* (BBC NI). No unsolicited scripts.

Drake A-V Video Ltd

89 St Fagans Road, Fairwater, Cardiff CF5 3AE
☎ 029 2056 0333
info@drakeav.com
www.drakeav.com

Specialists in corporate and educational films. No unsolicited mss.

Charles Dunstan Communications Ltd

42 Wolseley Gardens, London W4 3LS
☎ 020 8994 2328 📠 020 8994 2328
Contact *Charles Dunstan*

Producer of film, video and TV for documentary and corporate material. OUTPUT *Renewable Energy* for broadcast worldwide in 'Inside Britain' series; *The Far Reaches* travel series; *The Electric Environment*. No unsolicited scripts.

Endemol UK Productions

Shepherds Building Central, Charecroft Way, London W14 0EE
☎ 0870 333 1700 📠 0870 333 1800
info@endemoluk.com
www.endemoluk.com

Major independent producer of TV and digital media content, responsible for over 5,000 hours of programming per year. Incorporates several production brands including Brighter Pictures (factual entertainment shows, reality programming, live events and popular documentaries), Cheetah Television (factual), Initial (prime time entertainment, children's and teens, events and arts shows) and Zeppotron (comedy). OUTPUT includes *Big Brother*; *Deal or No Deal*; *Comic Relief Does Fame Academy*; *Golden Balls*; *Restoration Village*.

Farnham Film Company Ltd

34 Burnt Hill Road, Lower Bourne, Farnham GU10 3LZ
☎ 01252 710313 📠 01252 725855
info@farnfilm.com
www.farnfilm.com
Contact *Ian Lewis*

Television and film: intelligent full-length film. No unsolicited mss; check website for current requirements.

Festival Film and Television Ltd

Festival House, Tranquil Passage, Blackheath, London SE3 0BJ
☎ 020 8297 9999 📠 020 8297 1155
info@festivalfilm.com
www.festivalfilm.com
Managing Director *Ray Marshall*
Producer *Matt Marshall*

Television drama and feature films. Best known for its Catherine Cookson Dramas, which became one of ITV's long-running brands. Now developing both TV drama and features. Only accepts material submitted through an agent. Will not consider unsolicited material.

Fiction Factory

14 Greenwich Church Street, London SE10 9BJ
☎ 020 8853 5100 📠 020 8293 3001
radio@fictionfactory.co.uk
www.fictionfactory.co.uk
Creative Director *John Taylor*

Production company specializing in intelligent entertainments. Recent OUTPUT for BBC Radio includes *A Family Affair* Michael Dobbs; *Chronicles of Art* Michael Batt; *A Dance to the Music of Time* Anthony Powell and *Crime and Trial*. Ideas considered if sent by e-mail. Mss only from agents or writers with a professional track record in the broadcast media.

Film and General Productions Ltd

4 Bradbrook House, Studio Place, London
SW1X 8EL
☏ 020 7235 4495
cparsons@filmgen.co.uk
Contacts *Clive Parsons, Davina Belling*

Film and television drama. Feature films include
True Blue; *Tea with Mussolini* and *I Am David*.
Also *Seesaw* (ITV drama), *The Greatest Store
in the World* (family drama, BBC) and *The
Queen's Nose* (children's series, BBC). Interested
in considering new writing but subject to prior
telephone conversation.

Firehouse Productions

42 Glasshouse Street, London W1B 5DW
☏ 020 7439 2220
postie@firehouse.biz
www.firehouse.biz
Contacts *Peter Granger, Mike MacLeod*

Corporate films and live production events.
OUTPUT includes work for Shell; Vodafone;
government and various agencies.

First Writes Theatre Company

Lime Kiln Cottage, High Starlings, Banham,
Norwich NR16 2BS
☏ 01953 888525   ☏ 01953 888974
ellen@first-writes.co.uk
www.first-writes.co.uk
Director *Ellen Dryden*

Producer of numerous afternoon, Friday and
Saturday plays, classic serials and comedy
narrative series for BBC Radio 3 and Radio 4.

Flannel

21 Berwick Street, London W1F 0PZ
☏ 020 7287 9277   ☏ 020 7287 7785
mail@flannel.net
Contact *Kate Haldane*

Producer of drama, documentaries and comedy
for television and radio. OUTPUT includes
Woman's Hour (BBC R4); *The Hendersons'
Christmas Party* (five-part Christmas drama, BBC
R4). No unsolicited mss. 'Keen to encourage new
writing, but must come via an agent. Particularly
interested in 45–50 minute dramas for radio. Not
in a position to produce plays for stage, but very
happy to consider adaptations. Welcomes comedy
with some track record.'

Flashback Television Limited

58 Farringdon Road, London EC1R 3BP
☏ 020 7253 8768   ☏ 020 7253 8765
mailbox@flashbacktv.co.uk
www.flashbacktelevision.com
Managing Director *Taylor Downing*
Creative Director *David Edgar*

Founded in 1982, award-winning producer of
drama, factual entertainment and documentaries
for television. Recent credits include *Beau
Brummell*. No unsolicited submissions.

Focus Films Ltd

The Rotunda Studios, Rear of 116–118 Finchley
Road, London NW3 5HT
☏ 020 7435 9004   ☏ 020 7431 3562
focus@focusfilms.co.uk
www.focusfilms.co.uk
Managing Director *David Pupkewitz*
Contact *Raimund Berens*

Founded 1982. Independent feature film
development and production company. OUTPUT
The Book of Eve (Canadian drama); *The Bone
Snatcher* (Horror, UK/Can/SA); *Julia's Ghost*
(German co-production); *The 51st State* (feature
film); *Secret Society* (comedy drama feature
film); *Crimetime* (feature thriller); *Diary of a
Sane Man*; *Othello*; *Chemical Wedding*. Projects
in development include *Heaven and Earth*;
Surviving Evil. No unsolicited scripts.

FremantleMedia Group Ltd

1 Stephen Street, London W1T 1AL
☏ 020 7691 6000   ☏ 020 7691 6100
www.fremantlemedia.com
CEO *Tony Cohen*
CEO, talkbackTHAMES *Lorraine Heggessey*

FremantleMedia, formerly known as Pearson
Television, is the production arm of the RTL
Group, Europe's largest TV and radio company.
The company acquired Thames Television in 1993
(producer of *The Bill*) and Grundy Worldwide
(*Neighbours*) in 1995. Further acquisitions were
Witzend Productions (*Lovejoy*) and Alomo
Productions in 1996 and TalkBack Productions
in 2000. FremantleMedia is a global leader in
content production with 60 global formats and
more than 320 titles in production at any one
time. *No unsolicited submissions, please.*

Gaia Communications

20 Pevensey Road, Eastbourne BN21 3HP
☏ 01323 734809/727183
production@gaiacommunications.co.uk
www.gaiacommunications.co.uk
Producer *Robert Armstrong*

Established 1987. Video and TV, corporate and
documentary. OUTPUT *Discovering* (south east
regional tourist and local knowledge series);
Holistic (therapies and general information); local
interest audiobooks.

Noel Gay Television

Shepperton Studios, Studios Road, Shepperton
TW17 0QD
☏ 01932 592569   ☏ 01932 592172

charles.armitage@virgin.net
CEO *Charles Armitage*

OUTPUT *The Fear* (BBC Choice); *Second Chance* and *I-Camcorder* (both for Ch4); *Hububb* Series 1–5 (BBC); *Frank Stubbs Promotes* and *10%ers* Series 2 (both for Carlton/ITV); *Call Up the Stars* (BBC1); *Smeg Outs* (BBC video); *Red Dwarf*; *Dave Allen* (ITV); *Windrush* (BBC2). Joint ventures and companies include a partnership with Odyssey, a leading Indian commercials, film and TV producer, and the Noel Gay Motion Picture Company, whose credits include *Virtual Sexuality*; *Trainspotting* (with Ch4 and Figment Films); *Killer Tongue*; *Dog Soldiers*; *Fast Sofa* and *Pasty Faces*. Associate NGTV companies are Grant Naylor Productions and Pepper Productions. NGTV is willing to accept unsolicited material from writers but 1–2-page treatments only. No scripts, please.

Ginger Productions
▶ SMG Productions & Ginger Productions

Goldcrest Films International Ltd
65–66 Dean Street, London W1D 4PL
☎ 020 7437 8696   ☎ 020 7437 4448
info@goldcrestfilms.com
www.goldcrestfilms.com
Chairman *John Quested*
Contact *Hannah Smith*

Since it was established in 1977, Goldcrest Films has become a leading independent film production company winning many prizes at international festivals including 19 Academy Awards and 28 Baftas. Finances, produces and distributes films and television programmes. OUTPUT includes *Chariots of Fire*; *Gandhi*; *The Killing Fields*; *A Room With a View*; *Local Hero*; *The Mission*; *To End All Wars*. 'We are currently seeking film projects to invest equity in through our Finishing Fund in return for International Sales Rights.' Scripts via agents only.

The Good Film Company Ltd
The Studio, 5–6 Eton Garages, Lambolle Place, London NW3 4PE
☎ 020 7794 6222   ☎ 020 7794 4651
yanina@goodfilms.co.uk
www.goodfilms.co.uk
Contact *Yanina Barry*

Commercials and pop videos. Clients include Suzuki, Burberry, Totally London, Hugo Boss, Cadbury's. *No* unsolicited mss.

Granite Productions
Easter Davoch, Tarland, Aboyne AB34 4US
☎ 01339 880175
Contact *Simon Welfare*

Producer of television documentary programmes such as *Nicholas & Alexandra*; *Victoria & Albert* and *Arthur C. Clarke's Mysterious Universe*.

Grant Naylor Productions
▶ Noel Gay Television

Green Umbrella Ltd
59 Cotham Hill, Cotham, Bristol BS6 6JR
☎ 0117 970 6584   ☎ 0117 923 7003
info@green-umbrella.co.uk
www.green-umbrella.co.uk

Film producers specializing in science and natural history documentaries. OUTPUT includes episodes for *The Natural World*, *Wildlife on One* and original series such as *Living Europe* and *Triumph of Life*. Unsolicited treatments relating to natural history and science subjects are welcome.

Greenwich Village Productions
14 Greenwich Church Street, London SE10 9BJ
☎ 020 8853 5100   ☎ 020 8293 3001
tv@fictionfactory.co.uk
www.fictionfactory.co.uk
Contact *John Taylor*

Features, arts, corporate and educational projects, web-movies and new-media productions. Recent OUTPUT includes *Hot Chocolate* and *The Denis Stuff*. Mss only via agents or from writers with a professional track record in the chosen medium.

H2 Business Communication
Shepperton Studios, Shepperton TW17 0QD
☎ 01932 593717   ☎ 01932 593718
mail@h2bc.co.uk
www.h2bc.co.uk

Conferences, videos, awards presentations and speaker training.

Hammerwood Film Productions
info@filmangels.co.uk
www.filmangel.co.uk

Film, video and TV drama. OUTPUT *Iceni* (film; co-production with Pan-European Film Productions and Boudicca Film Productions Ltd); *Boudicca – A Celtic Tragedy* (TV series). In pre-production: *The Black Egg* (witchcraft in 17th century England); *The Ghosthunter*; *Iceni* (documentary of the rebellion of AD61); *No Case to Answer* (legal series). 'Authors are recommended to access www.filmangel.co.uk' (see **FilmAngels** under *Useful Websites*).

Hartswood Films Ltd
Twickenham Studios, The Barons, St Margaret's TW1 2AW
☎ 020 8607 8736   ☎ 020 8607 8744
films.tv@hartswoodfilms.co.uk
Producer *Elaine Cameron*

Film and TV production for drama, comedy and documentary. OUTPUT *The Cup*; *Jekyll*; *Coupling*; *Men Behaving Badly*; *Border Cafe*; *Carrie & Barry*; *Fear, Stress and Anger* (all for BBC); *After Thomas* (ITV drama).

Hat Trick Productions Ltd
33 Oval Road, London NW1 7EA
☎ 020 7184 7777 📠 020 7184 7778
info@hattrick.com
www.hattrick.com
Managing Director *Jimmy Mulville*

Television programmes.

Healthcare Productions Limited
The Great Barn, Godmersham Park, Canterbury CT4 7DT
☎ 01227 738279
contact@healthcareproductions.co.uk
www.healthcareproductions.co.uk
Contacts *Penny Wilson, Andrew Wilson*

Established in 1989, producing health-based dramas and documentaries for TV and the DVD market. Has won numerous awards from the BMA. Happy to look at new material, which should be submitted by e-mail.

Heritage Theatre Ltd
Unit 1, 8 Clanricarde Gardens, London W2 4NA
☎ 020 7243 2750 📠 020 7792 8584
rm@heritagetheatre.com
www.heritagetheatre.com
Contact *Robert Marshall*

Video recordings of successful stage plays for distribution on DVD and broadcast. 'We cannot deal with scripts of unproduced plays.'

Holmes Associates
The Studio, 37 Redington Road, London NW3 7QY
☎ 020 7813 4333
holmesassociates@blueyonder.co.uk
Contact *Andrew Holmes*

Prolific originator, producer and packager of documentary, drama and music television and films. OUTPUT has included *Ashes and Sand* (Film 4); *Chunky Monkey* (J&V Films); *Prometheus* (Ch4 'Film 4'); *The Shadow of Hiroshima* (Ch4 'Witness'); *The House of Bernarda Alba* (Ch4/WNET/Amaya); *Piece of Cake* (LWT); *The Cormorant* (BBC/Screen 2); *John Gielgud Looks Back*; *Rock Steady*; *Well Being*; *Signals*; *Ideal Home?* (all Ch4); *Timeline* (with MPT, TVE Spain & TRT Turkey); *Seven Canticles of St Francis* (BBC2). Submissions only accepted by e-mail in synopsis form.

Hourglass Productions
27 Prince's Road, London SW19 8RA
☎ 020 8540 8786

productions@hourglass.co.uk
www.hourglass.co.uk
Partners *Martin Chilcott, Jacqueline Chilcott*

Film and video; documentary and drama. OUTPUT BAFTA nominated scientific television documentaries and educational programming. Also current affairs, health and social issues.

Icon Films
1–2 Fitzroy Terrace, Bristol BS6 6TF
info@iconfilms.co.uk
www.iconfilms.co.uk
Contact *Harry Marshall*

Film and TV documentaries. OUTPUT *Nick Baker's Weird Creatures* (five/Animal Planet/Granada International); *Tom Harrisson – The Barefoot Anthropologist* (BBC). Specializes in factual documentaries. Open-minded to new documentary proposals.

Imari Entertainment Ltd
PO Box 158, Beaconsfield HP9 1AY
☎ 01494 677147 📠 01494 677147
info@imarientertainment.com
Contact *Jonathan Fowke*

TV and video producer, covering all areas of drama, documentary and corporate productions.

Isis Productions
387b King Street, Hammersmith, London W6 9NJ
☎ 020 8748 3042 📠 020 8748 7634
hello@isis-productions.com
www.isis-productions.com
Directors *Nick de Grunwald, Jamie Rugge-Price*

Formed in 1991, Isis Productions specializes in the production of music and documentary programmes and are developing entertainment and drama formats. Isis' main production credit is *Classic Albums*, an ongoing documentary series for TV and DVD. Five series to date have included films on albums by Elvis Presley, Stevie Wonder, Elton John, Paul Simon, The Who, Jimi Hendrix, U2, Bob Marley & The Wailers, Lou Reed, Sex Pistols, Steely Dan, Pink Floyd, Nirvana, Queen, Cream, Frank Zappa, Duran Duran, John Lennon, The Doors and Jay-Z. Other major profile films include Elizabeth Taylor, John Simpson and Carla Bruni. Isis produced the first ever televised UK Urban Asian Music Awards, a major opera film, *Dido & Aeneas* and the classical music series, *The Score*.

Isolde Films
28 Twyford Avenue, London W3 9QB
☎ 020 8896 2860
isolde@btinternet.com
www.tonypalmer.org

Film and TV documentaries. OUTPUT *Wagner*; *Margot*; *Menuhin*; *Maria Callas*; *Testimony*; *In*

From the Cold; *Pushkin*; *England, My England* (by John Osborne). Unsolicited material is read, but please send a written outline first.

ITV Productions

The London Television Centre, Upper Ground, South Bank, London SE1 9LT

☎ 020 7620 1620

www.itv.com

Controller of ITV Productions (Drama), London *Kate Bartlett*

Director *John Whiston*

ITV Productions is the largest commercial TV production company in the UK. Produces original programmes, co-productions and TV movies for ITV channels and other broadcasters, both in the UK and abroad. OUTPUT includes *Hornblower*; *Poirot*; *Miss Marple*; *Touching Evil*; *Where the Heart Is*; *The Last Detective*; *Jericho*.

JAM Pictures and Jane Walmsley Productions

8 Hanover Street, London W1S 1YE

☎ 020 7290 2676   ☐ 020 7256 6818

producers@jampix.com

Contacts *Jane Walmsley, Michael Braham*

JAM Pictures was founded in 1996 to produce drama for film, TV and stage. Projects include: *Rain Man* (West End production 2008); *Hillary's Choice* (TV film, A&E Network); *Son of Pocahontas* (TV film, ABC); *Rudy: the Rudy Giuliani Story* (TV film, USA Network); *One More Kiss* (feature, directed by Vadim Jean); *Bad Blood* (UK theatre tour). Jane Walmsley Productions, formed in 1985 by TV producer, writer and broadcaster, Jane Walmsley, has completed award-winning documentaries and features such as *Hot House People* (Ch4). No unsolicited mss. 'Letters can be sent to us, asking if we wish to see mss; we are very interested in quality material, from published or produced writers only, please'.

Justice Entertainment Ltd

PO Box 4377, London W1A 7SX

☎ 020 7287 2355   ☐ 020 7287 2354

info@timwestwood.com

www.timwestwood.com

TV and radio production company. No unsolicited material.

Keo Films.com Ltd

101 St John Street, London EC1M 4AS

☎ 020 7490 3580   ☐ 020 7490 8419

keo@keofilms.com

www.keofilms.com

Head of Production *Maddy Allen*

Television documentaries and factual entertainment. OUTPUT includes programmes for all the main broadcasters including *River Cottage*

Gone Fishing; *River Cottage*; *Meet the Natives* (all for Ch4); *Return to Lullingstone Castle*; *Blizzard: Race to the Pole* (BBC2). No unsolicited mss.

Kingfisher Television Productions Ltd

Martindale House, The Green, Ruddington, Nottingham NG11 6HH

☎ 0115 945 6581   ☐ 0115 921 7750

Contact *Tony Francis*

Broadcast television production.

Kudos Film and Televison Ltd

12–14 Amwell Street, London EC1R 1UQ

☎ 020 7812 3270   ☐ 020 7812 3271

info@kudosfilmandtv.com

www.kudosfilmandtv.com

Joint Managing Directors *Jane Featherstone, Stephen Garrett*

Commercial Director *Daniel Isaacs*

Director of Drama *Simon Crawford Collins*

Television dramas include *Life on Mars*; *Spooks*; *Hustle*; *Tsunami – The Aftermath*; *The Amazing Mrs Pritchard*; *Wide Sargasso Sea*; *Scars*; *MI High*; *Child of Mine*; *Comfortably Numb*; *Pleasureland*; *The Magician's House*; *Psychos*; *Ashes to Ashes*; *The Fixer*; *Moving Wallpaper*; *Echo Beach*; *Plus One*; *Occupation*; *Nearly Famous*; *Spooks: Code 9* and *Law & Order (UK)*. Feature films such as *Among Giants*; *Pure*; *Miss Pettigrew Lives for a Day*; *The Crimson Wing* and *Eastern Promises*. No unsolicited mss.

Ladbroke Productions (Radio) Ltd

17 Leicester Road, East Croydon CR0 6EB

ladbrokeradio@googlemail.com

www.electricairwaves.com

Contact *Neil Gardner*

Producer of radio drama, documentary, corporate, music, music documentaries and readings. OUTPUT includes *The Woman in White*; *The Darling Buds of May*; *The True History of British Pop*; *The World on a String*; *The Colour of Music* (all for BBC Radio 2); *Don Carlos*; *In the Company of Men* (BBC Radio 3 drama); *Sitting in Limbo* (BBC World Service drama); *Your Vote Counts* (Electoral Commision audio CDs). Unsolicited mss and ideas welcome; send letter in the first instance. 'We are willing to help develop for possible submission to the BBC'.

Landseer Productions Ltd

140 Royal College Street, London NW1 0TA

☎ 020 7485 7333

ken@landseerproductions.com

www.landseerfilms.com

Directors *Derek Bailey, Ken Howard*

Film and video production: documentary, drama, music and arts. OUTPUT *Swinger* (BBC2/Arts Council); *Auld Lang Syne* and *Retying the*

Knot – The Incredible String Band (both for BBC Scotland); *Benjamin Zander* ('The Works', BBC2); *Zeffirelli, Johnnie Ray, Petula Clark, Bing Crosby, Maxim Vengerov* (all for 'South Bank Show', LWT); *Death of a Legend – Frank Sinatra* ('South Bank Show' special); *Routes of Rock* (Carlton); *See You in Court* (BBC); *Nureyev Unzipped, Gounod's Faust, The Judas Tree, Ballet Boyz, 4Dance* and *Bourne to Dance* (all for Ch4), *Proms in the Park* (Belfast).

Lilyville Screen Entertainment Ltd

7 Lilyville Road, London SW6 5DP
☎ 020 7471 8989
tony.cash@btclick.com
Contact *Tony Cash*

Drama and documentaries for TV. OUTPUT *Poetry in Motion* (series for Ch4); 'South Bank Show': *Ben Elton* and *Vanessa Redgrave*; *Musique Enquête* (drama-based French language series, Ch4); *Sex and Religion* (ITV); *Landscape and Memory* (arts documentary series for the BBC); Jonathan Miller's production of the *St Matthew Passion* for the BBC; major documentary on the BeeGees for the 'South Bank Show'. Scripts with an obvious application to TV may be considered. Interested in new writing for documentary programmes.

Lime Pictures Limited

Campus Manor, Childwall Abbey Road, Childwall, Liverpool L16 0JP
☎ 0151 722 9122   📠 0151 722 6839
www.limepictures.com
Creative Director *Tony Wood*

Originally known as Mersey Television until All3Media's acquisition in 2005 and a change of name to Lime Pictures in 2006. Creators of television dramas: *Brookside; Hollyoaks; Runners* (Ch4); *Grange Hill; Apparitions; The Well* (BBC); *Living On the Edge* (MTV).

Loftus Audio Ltd

2a Aldine Street, London W12 8AN
☎ 020 8740 4666
ask@loftusaudio.co.uk
www.loftusaudio.co.uk
Contact *Matt Thompson (radio drama*
☎ *01620 893876)*
Head of Radio Drama *Jo Coombs*

Producer of factual radio programmes, documentaries, readings (mostly non-fiction) and drama, as well as audio books and audio guides for museums and galleries. OUTPUT includes titles for Radio 4's 'Book of the Week'. 'We are happy to look at brief emailed synopses of radio plays suitable for broadcast in the available drama slots on BBC Radio 3 or 4, or brief ideas for radio features and documentaries.'

Lucida Productions

5 Alleyn Crescent, London SE21 8BN
☎ 020 8761 6766
pj.lucida@tiscali.co.uk
Contact *Paul Joyce*

Television and cinema: arts, adventure, current affairs, documentary, drama and music. OUTPUT has included *Motion and Emotion: The Films of Wim Wenders; Dirk Bogarde – By Myself; Sam Peckinpah – Man of Iron; Kris Kristofferson – Pilgrim; Wild One: Marlon Brando; Stanley Kubrick: 'The Invisible Man'; 2001: the Making of a Myth* (Ch4); *Mantrap – Straw Dogs, the final cut* (with Dustin Hoffman). Restoration of the Director's Cut of *The Devils* plus the documentary *Hell on Earth* with Ken Russell and Vanessa Redgrave. Currently in development for documentary projects.

Marchmont Films Ltd

24 Three Cups Yard, Sandland Street, London WC1R 4PZ
office@marchmontfilms.com
www.marchmontfilms.com
Development Executives *Beverley Hills, Daniel Hayes, Andrew Cussens*

Producer of short films and feature projects. OUTPUT includes *Out In the Cold; The Green Wave; Punch; Shag & Go*. Welcomes new writers. See website for current submission criteria.

Jane Marshall Productions

The Coach-House, Westhill Road, Blackdown, Leamington Spa CV32 6RA
☎ 01926 831680
jane@janemarshallproductions.co.uk
Contact *Jane Marshall*

Producer of readings of published work both fiction and non-fiction for BBC Radio. Published work only.

Maverick Television

Progress Works, Heath Mill Lane, Birmingham B9 4AL
☎ 0121 771 1812   📠 0121 771 1550
mail@mavericktv.co.uk
www.mavericktv.co.uk
Also at: 40 Church Way, London NW1 1LW
☎ 020 7383 2727
Head of Production *Allan Tott*

Established in 1994, Maverick has a strong reputation for popular factual programming. It is now one of network television's most prolific independent suppliers. OUTPUT includes *How to Look Good Naked; 10 Years Younger; Who'll Age Worst?; Bollywood Star; Fat Chance; Born Too Soon; VeeTV; Trade Secrets; Embarrassing Illnesses; 10 Things You Didn't Know About ...; How To Live Longer; The Property Chain; Male,*

33, Seeks Puberty; Extreme Engineering; Picture This: Accidental Hero; Up Your Street; Motherless Daughters; Highland Bollywood: Black Bag; Health Alert: My Teenage Menopause; Long Haul; Learning to Love the Grey.

Maya Vision International Ltd

6 Kinghorn Street, London EC1A 7HW
☎ 020 7796 4842   🅵 020 7796 4580
www.mayavisionint.com
Contact *Tamsin Ranger*

Film and TV: drama and documentary. OUTPUT *Saddam's Killing Fields* (for 'Viewpoint', Central TV); *3 Steps to Heaven* and *A Bit of Scarlet* (feature films for BFI/Ch4); *A Place in the Sun* and *North of Vortex* (dramas for Ch4/Arts Council); *The Real History Show* (Ch4); *In Search of Myths and Heroes; In Search of Shakespeare; In the Footsteps of Alexander the Great; Conquistadors* (BBC documentaries); *Hitler's Search for the Holy Grail; Once Upon a Time in Iran* (Ch4 documentaries); *The Story of India* (BBC/PBS). Absolutely no unsolicited material; commissions only.

MBP TV

Saucelands Barn, Coolham, Horsham RH13 8QG
☎ 01403 741620   🅵 01403 741647
info@mbptv.com
www.mbptv.com
Contact *Phil Jennings*

Maker of film and video specializing in programmes covering equestrianism and the countryside. No unsolicited scripts, but always looking for new writers who are fully acquainted with the subject.

Melendez Films

Julia House, 44 Newman Street, London W1T 1QD
☎ 020 7323 5273   🅵 020 7323 5373
Contact *Steven Melendez*

Independent production company specializing in 2D animation. Also involved in production and film design for clients in England, Spain, Sweden, India and the US, plus website design and 3D animation on the Web. Clients include book publishers, TV companies and advertisers. Winner of international awards for films, particularly of classic books, stories and comic characters. 'We will look at unsolicited projects in outline or synopsis form only. Enclose s.a.e.'

Mendoza Films

1 Marlborough Court, London W1F 7EE
☎ 020 7209 9617
office@mendozafilms.com
www.mendozafilms.com

Commercials, title sequences (e.g. Alan Bleasdale's *G.B.H.*); party political broadcasts.

Currently in pre-production on a feature-length comedy film. Involved with the Screenwriters' Workshop. Unsolicited mss welcome but 'comedies only, please'. Material will not be returned without s.a.e.

Mersey Television

▷ Lime Pictures Limited

Moonstone Films Ltd

☎ 020 8144 9940
info@moonstonefilms.co.uk
www.moonstonefilms.co.uk
Contact *Tony Stark*

Television: current affairs, science and history documentaries. OUTPUT *Arafat's Authority* and *Arafat Investigated*, both for BBC 'Correspondent'. Plus various Ch4 News commissions. Unsolicited mss welcome.

Neon

Studio Two, 19 Marine Crescent, Kinning Park, Glasgow G51 1HD
☎ 0141 429 6366   🅵 0141 429 6377
stephy@go2neon.com
www.go2neon.com
Contact *Stephanie Pordage*

Television and radio: drama and documentary producers. OUTPUT includes *Brand New Country; Asian Overground; Peeking Past the Gates of Skibo*. Supports and encourages new writing 'at every opportunity'. Welcomes unsolicited material but telephone in the first instance.

Number 9 Films

Linton House, 24 Wells Street, London W1T 3PH
☎ 020 7323 4060   🅵 020 7323 0456
info@number9films.co.uk
Contacts *Stephen Woolley, Elizabeth Karlsen*

Leading feature film producer. OUTPUT includes *Breakfast on Pluto; Stoned; Mrs Harris*. Forthcoming productions: *And When Did You Last See Your Father?; How to Lose Friends and Alienate People; Edith and the Lonely Doll*. No unsolicited material.

Odyssey Productions Ltd

72 Tay Street, Newport-on-Tay DD6 8AP
☎ 01382 542070   🅵 01382 542070
billykay@sol.co.uk
www.billykay.co.uk
Contact *Billy Kay*

Producer of radio documentaries. OUTPUT includes *Scotland's Black History; Gentle Shepherds* (oral history); *Street Kids* (Scottish missionaries working with street kids in Brazil). Ideas for radio documentaries welcome; send letter with one page outlining the idea and programme content.

Omnivision

Pinewood Studios, Iver Heath SL0 0NH
☎ 01753 656329   📠 01753 631145
info@omnivision.co.uk
www.omnivision.co.uk
Contacts *Christopher Morris, Nick Long*

TV and video producers of documentary, corporate, news and sport programming. Also equipment and facilities hire. Interested in ideas; approach by letter or e-mail.

Orlando Television Productions Ltd

Up-the-Steps, Little Tew, Chipping Norton
OX7 4JB
☎ 01608 683218   📠 01608 683364
info@orlandomedia.co.uk
www.orlandomedia.co.uk
Contact *Mike Tomlinson*

Producer of TV documentaries and digital multimedia content, with science, health and information technology subjects as a specialization. Approaches by established writers/journalists to discuss proposals for collaboration are welcome.

Ovation

Upstairs at the Gatehouse, Highgate, London
N6 4BD
☎ 020 8340 4256   📠 020 8340 3466
www.ovation.eu
Contact *John Plews*

Corporate video and conference scripts. Unsolicited mss not welcome. 'We talk to new writers from time to time.' Ovation also runs the fringe theatre, 'Upstairs at the Gatehouse'.

Paper Moon Productions

Wychwood House, Burchetts Green Lane, Littlewick Green, Nr. Maidenhead SL6 3QW
☎ 01628 829819   📠 01628 829819
insight@paper-moon.co.uk
Producer/Director *David Haggas*

Broadcast documentaries and corporate communications. OUTPUT includes *Bilbo & Beyond*, an affectionate glimpse into the life and work of the dedicated philologist and fantasy writer J.R.R. Tolkien.

Parallax East Ltd

Victoria Chambers, St Runwald Street, Colchester
CO1 1HF
☎ 01206 574909   📠 01206 575311
assistant@parallaxindependent.co.uk
Contact *Sally Hibbin*

Feature films/television drama. OUTPUT *A Very British Coup; Riff-Raff; Bad Behaviour; Raining Stones; Ladybird, Ladybird; i.d.; Land and Freedom; The Englishman Who Went up a Hill But Came Down a Mountain; Bliss; Jump the Gun; Carla's Song; The Governess; My Name Is Joe; Stand and Deliver; Dockers; Hold Back the Night; Bread and Roses; Princesa; The Navigators; Sweet Sixteen; Innocence; The Intended; Blind Flight; Yasmin; Almost Adult.*

Passion Pictures

3rd Floor, 33–34 Rathbone Place, London W1T 1JN
☎ 020 7323 9933   📠 020 7323 9030
info@passion-pictures.com
www.passion-pictures.com
Managing Director *Andrew Ruhemann*

Documentary and drama includes: *One Day in September* (Academy Award-winner for Best Feature Documentary, 2000); also commercials and music videos. Unsolicited mss welcome.

Pathé Pictures

14–17 Kenthouse, Market Place, London
W1W 8AR
☎ 020 7323 5151   📠 020 7631 3568
Head of Creative Affairs *Colleen Woodcock*

Produces 4–6 theatrical feature films each year. 'We are pleased to consider all material that has representation from an agent or production company.'

Pearson Television

▷ FremantleMedia Group Ltd

Pelicula Films

59 Holland Street, Glasgow G2 4NJ
☎ 0141 287 9522
Contact *Mike Alexander*

Television producer. Maker of drama documentaries and music programmes for the BBC and Ch4. OUTPUT *As an Eilean (From the Island); The Trans-Atlantic Sessions 1 & 2; Nanci Griffith, Other Voices 2; Follow the Moonstone.*

Pepper Productions

▷ Noel Gay Television

Photoplay Productions Ltd

21 Princess Road, London NW1 8JR
☎ 020 7722 2500   📠 020 7722 6662
info@photoplay.co.uk
www.photoplay.co.uk
Contact *Patrick Stanbury*

Documentaries for film, television and video plus restoration of silent films and their theatrical presentation. OUTPUT includes *The Cat and the Canary; Orphans of the Storm; Cecil B. DeMille: American Epic* and the 'Channel 4 Silents' series of silent film restoration, including *The Wedding March* and *The Iron Mask*. Recently completed *Garbo* and *I'm King Kong!* No unsolicited mss; 'we tend to create and write all our own programmes'.

Picardy Media & Communication

Dalintober Hall, 40 Dalintober Street, Glasgow
G5 8NW
☎ 0141 420 0909   🖷 0141 429 3723
jr@picardy.co.uk
www.picardy.co.uk
Head of Production *John Rocchiccioli*

Produces screen-based content for education, training, sales and marketing, HR and induction. Operates in a variety of areas including health, education, social policy, etc. 'Aims to inform through entertainment.' Unsolicited mss welcome; 'keen to encourage new writing.'

Picture Palace Films Ltd

13 Egbert Street, London NW1 8LJ
☎ 020 7586 8763   🖷 020 7586 9048
info@picturepalace.com
www.picturepalace.com
Producer & Chief Executive *Malcolm Craddock*

Leading independent producer of film and TV drama. OUTPUT *Rebel Heart* (BBC1); *Sharpe's Peril*; *Sharpe's Challenge*; *Extremely Dangerous*; *A Life for A Life* and *Frances Tuesday* (all for ITV); *Sharpe's Rifles* (14 films for Carlton TV); *Little Napoleons* (comedy drama, Ch4); *The Orchid House* (drama serial, Ch4); *Tandoori Nights*; *4 Minutes*; *When Love Dies* (all for Ch4); *Ping Pong* (feature film); *Acid House* (Picture Palace North). Material will only be considered if submitted through an agent.

Planet24 Pictures Ltd

Octagon House, Fir Road, Stockport SK7 2NP
☎ 0870 765 8780
planet24picture@aol.com
www.planet24pictures.co.uk
Contact *Mercedes de Dunewíc*

Producer of television and video documentaries plus community/infomercials. OUTPUT includes *Jack the Ripper: The Conspiracies*; *Truth to Tell*; *Pitmen and Politics: The County Built on Coal*. Mss, plot outlines and screenplays considered but e-mail for information first. 'Welcome the input of new writers with radical points of view.'

Plantagenet Films Limited

Ard-Daraich Studio B, Ardgour, Nr Fort William
PH33 7AB
☎ 01855 841384   🖷 01855 841384
norriemaclaren@btinternet.com
Contact *Norrie Maclaren*

Film and television: documentary and drama programming such as *Dig* (gardening series for Ch4); various 'Dispatches' for Ch4 and 'Omnibus' for BBC. Keen to encourage and promote new writing.

Portobello Pictures

12 Addison Avenue, London W11 4QR
☎ 020 7605 1396   🖷 020 7605 1391
mail@portobellopictures.com
www.portobellopictures.com
Producer *Eric Abraham*
Development Assistant *Kristin Irving*

Oscar-winning film, television and theatre production company. Not accepting unsolicited submissions.

Pozzitive Television

Paramount House, 162–170 Wardour Street, London W1F 8AB
☎ 020 7734 3258   🖷 020 7437 3130
david@pozzitive.co.uk
Contact *David Tyler*

Producer of comedy and entertainment for television and radio. OUTPUT *Dinner Ladies*; *Coogan's Run*; *The 99p Challenge*; *The Comic Side of 7 Days*; *Armando Iannucci's Charm Offensive*. Unsolicited mss of TV and radio comedy welcome. 'No screenplays or stage plays or novels, please. Send hard copy of full sample script. We read everything submitted this way. Sorry, we don't return scripts unless you send an s.a.e.'

Promenade Enterprises Limited

6 Russell Grove, London SW9 6HS
☎ 020 7582 9354
info@promenadeproductions.com
www.promenadeproductions.com
Contact *Nicholas Newton*

Producer of drama for radio and theatre predominantly. Supports new writing but only accepts unsolicited mss via agents or producers.

Redweather

Easton Business Centre, Felix Road, Bristol
BS5 0HE
☎ 0117 941 5854   🖷 0117 941 5851
production@redweather.co.uk
www.redweather.co.uk

Broadcast documentaries on arts and disability, corporate video and CD-ROM, Water Aid, British Oxygen, etc.

The Reel Thing Ltd
▷ The Ashford Entertainment Corporation Ltd

Richmond Films & Television Ltd

PO Box 33154, London NW3 4AZ
☎ 020 7722 6464
mail@richmondfilms.com
Contact *Development Executive*

Film and TV: drama and comedy. OUTPUT *Press Gang*; *The Lodge*; *The Office*; *Wavelength*; *Privates*; *in2minds*. *No* unsolicited scripts.

RS Productions

191 Trewhitt Road, Newcastle upon Tyne NE6 5DY
☎ 07710 064632
info@rsproductions.co.uk
www.rsproductions.co.uk
Contact *Mark Lavender*

Feature films and television: drama series/serials and singles. TV documentaries and series. Working with established and new talent.

Sands Films

119 Rotherhithe Street, London SE16 4NF
☎ 020 7231 2209   🖷 020 7231 2119
info@sandsfilms.co.uk
www.sandsfilms.co.uk
Contacts *Christine Edzard, Olivier Stockman*

Film and TV drama. OUTPUT *Little Dorrit*; *The Fool*; *As You Like It*; *A Dangerous Man*; *The Long Day Closes*; *A Passage to India*; *Topsy Turvy*; *Nicholas Nickleby*; *The Gangs of New York*; *The Children's Midsummer Night's Dream*. No unsolicited scripts.

Scala Productions Ltd

2nd Floor, 37 Foley Street, London W1W 7TN
☎ 020 7637 5720   🖷 020 7637 5734
scalaprods@aol.com
Contacts *Ian Prior, Nik Powell*

Production company set up by ex-Palace Productions Nik Powell and Stephen Woolley, who have an impressive list of credits including *Company of Wolves*; *Absolute Beginners*; *Mona Lisa*; *Scandal*; *The Crying Game*; *Backbeat*; *Neon Bible*; *24:7*; *Little Voice*; *Divorcing Jack*; *The Last September*; *Wild About Harry*; *Last Orders*; *A Christmas Carol – The Movie*; *Black and White*; *Leo*; *The Night We Called It A Day*; *Ladies in Lavender*; *Stoned*. In development: *Kickbacks*; *Celluloid Dream*; *Johnny Bollywood*; *The Rough*; *Black Cockatoo*; *Nightcab*; *White Turks*.

Scope Productions Ltd

180 West Regent Street, Glasgow G2 4RW
☎ 0141 221 4312
laurakingwell@scopeproductions.co.uk
www.scopeproductions.co.uk
Corporate *Laura Kingwell*

Established 1984. Corporate video and multimedia communications for clients across all sectors.

Screen First Ltd

The Studios, Funnells Farm, Down Street, Nutley TN22 3LG
☎ 01825 712034
Producer/Director *Paul Madden*

Founded 1985. Focus on animation, arts, children's programmes, documentary and drama. OUTPUT includes *Ivor the Invisible*. No unsolicited scripts.

Screen Ventures Ltd

31 Tor Court, Tor Gardens, London W8 4HX
☎ 020 7937 5553
info@screenventures.com
www.screenventures.com
Contacts *Christopher Mould, David Chambers*

Film and TV sales and production: documentary, music videos and drama. OUTPUT *Life and Limb* (documentary, Discovery Health Channel); *Pavement Aristocrats* (SABC); *Woodstock Diary*; *Vanessa Redgrave* and *Genet* (both for LWT 'South Bank Show'); *Mojo Working*; *Burma: Dying for Democracy* (Ch4); *Dani Dares* (Ch4 series on strong women); *Pagad* (Ch4 news report).

Screenhouse Productions Ltd

Chapel Allerton House, 114 Harrogate Road, Leeds LS7 4NY
☎ 0113 266 8881   🖷 0113 266 8882
paul.bader@screenhouse.co.uk
www.screenhouse.co.uk
Contact *Paul Bader*

Specializes in science TV, documentary, stunts and events, including outside broadcasts. OUTPUT includes *Cosmos: A Beginner's Guide* (BBC2 astronomy series); *Timewatch: The Hidden Children* (BBC2); *Britain's Tallest Teenagers* (ITV1); TV skills training courses, e.g. Location Directing, Factual TV Research, and Media Skills for Scientists. 'More likely to consider written up proposals.'

September Films Ltd

Glen House, 22 Glenthorne Road, London W6 0NG
☎ 020 8563 9393   🖷 020 8741 7214
september@septemberfilms.com
www.septemberfilms.com
Director of Production *Elaine Day*

Factual entertainment and documentary specialists. Feature film OUTPUT includes *Breathtaking*; *House of America*; *Solomon & Gaenor*. No unsolicited submissions, please.

Shell Like

81 Whitfield Street, London W1T 4HG
☎ 020 7255 5224   🖷 020 7255 5255
enquiries@shelllike.com
www.shelllike.com
Contact *Mike Blunt*

Produces radio commercials.

Sianco Cyf

Pen-y-graig, Llanfaglan, Caernarfon LL54 5RF
☎ 01286 676100/07831 726111
sian@sianco.tv
Contact *Siân Teifi*

Children's, youth and education programmes, children's drama, people-based documentaries for adults. 'Please note, *we do not accept any unsolicited scripts.*'

Silent Sound Films Ltd
Cambridge Court, Cambridge Road, Frinton on Sea CO13 9HN
☎ 01255 676381   📠 01255 676381
thj@silentsoundfilms.co.uk
www.silentsoundfilms.co.uk
www.londonfoodfilmfiesta.co.uk
Contact *Timothy Foster*

Active in European film co-production with mainstream connections in the USA. Special interest in developing stage and film musicals, art house fiction and documentaries on the arts. Initial enquiry via e-mail.

Skyline Productions
10 Scotland Street, Edinburgh EH3 6PS
☎ 0131 556 2026   📠 0131 556 4377
leslie@skyline.uk.com
www.skyline.uk.com
Producer/Writer *Leslie Hills*

Produces film and television drama and documentary.

SMG Productions & Ginger Productions
SMG: Pacific Quay, Glasgow G51 1PQ
☎ 0141 300 3000
www.smgproductions.tv
Also: Ginger Productions, 3 Waterhouse Square, 138–142 Holborn London EC1N 2NY
☎ 020 7882 1020 www.ginger.tv
Managing Director (SMG Productions & Ginger Productions) *Elizabeth Partyka*
Head of Drama *Eric Coulter*

SMG Productions, which incorporates Ginger Productions, makes programmes for the national television networks, including ITV, Ch4 and Sky. Specializes in drama, factual entertainment and children's programming. OUTPUT includes *Taggart*; *Rebus*; *Our Daughter Holly*; *Club Reps*; *Jack Osborne – Adrenaline Junkie*.

Somethin Else
20-26 Brunswick Place, London N1 6DZ
☎ 020 7250 5500   📠 020 7250 0937
info@somethinelse.com
www.somethinelse.com
Contact *Jez Nelson*

Producer of television, video and radio documentaries, DVD and interactive content. Ideas for TV shows welcome; send letter in the first instance.

Specific Films
25 Rathbone Street, London W1T 1NQ
☎ 020 7580 7476   📠 020 7636 6866
info@specificfilms.com
Contact *Michael Hamlyn*

Founded 1991. OUTPUT includes *Mr Reliable* (feature film co-produced by PolyGram and the AFFC); *The Adventures of Priscilla, Queen of the Desert*, co-produced with Latent Image (Australia) and financed by PolyGram and AFFC; *U2 Rattle and Hum*, full-length feature – part concert film/part cinema verité documentary; *Paws* (executive producer); *The Last Seduction 2* (Polygram); and numerous pop promos for major international artists.

Spice Factory (UK) Ltd
14 Regent Hill, Brighton BN1 3ED
☎ 01273 739182   📠 01273 749122
info@spicefactory.co.uk
www.spicefactory.co.uk

Founded 1995. Film producers. OUTPUT *Plots With a View* (Christopher Walken, Brenda Blethyn, Alfred Molina, Lee Evans); *Bollywood Queen* (Preeya Kallidas, James McAvoy, Ian McShane); *The Bridge of San Luis Rey* (Robert De Niro, Kathy Bates, Harvey Keitel); *A Different Loyalty* (Sharon Stone, Rupert Everett); *Head in the Clouds* (Charlize Theron, Penelope Cruz); *The Merchant of Venice* (Al Pacino, Jeremy Irons, Joseph Fiennes, Lynn Collins). No unsolicited material accepted.

'Spoken' Image Ltd
8 Hewitt Street, Manchester M15 4GB
☎ 0161 236 7522   📠 0161 236 0020
info@spoken-image.com
www.spoken-image.com
Contact *Geoff Allman*

Film, video and TV production for documentary and corporate material. Specializes in high-quality brochures and reports, CD-ROMs, exhibitions, conferences, film and video production for broadcast, industry and commerce.

Square Dog Radio LLP
Kilmagadwood Cottage, Scotlandwell, Kinross KY13 9HY
☎ 05600 472247
mike@squaredogradio.co.uk
www.squaredogradio.co.uk
Producers *Mike Hally, Mark Whitaker*

Producer of radio documentaries, features and short story readings. OUTPUT includes *Northern Creative Writing Groups* (returning series, for Radio 4). No unsolicited material.

Tony Staveacre Productions
Channel View, Blagdon BS40 7TP
☎ 01761 462161
newstaving@btinternet.com
Contact *Tony Staveacre*

Producer of dramas and documentaries as well as music, arts and comedy progammes.

Recent OUTPUT *The Wodehouse Notebooks*; *The Liberation of Daphne, Speaking from the Belly*; *Standing Up for Liverpool*; *Kington's Last Tapes*; *Wodehouse in Hollywood* (Radio 4); *The Very Thought of You* and *Jigsy* (theatre); *Tango Maestro* (BBC4); *The Old Boys Band* (BBC1); *Mendip Voices* (CD). No unsolicited mss.

Stirling Film & TV Productions Limited

137 University Street, Belfast BT7 1HP
℡ 028 9033 3848   🖷 028 9043 8644
anne@stirlingtelevision.co.uk
www.stirlingtelevision.co.uk
Contact *Anne Stirling*

Producer of broadcast and corporate programming – documentary, sport, entertainment and lifestyle programmes.

Straight Forward Film & Television Productions Ltd

Building 2, Lesley Office Park, 393 Hollywood Road, Belfast BT4 2LS
℡ 028 9065 1010   🖷 028 9065 1012
enquiries@straightforwardltd.co.uk
www.straightforwardltd.co.uk
Contacts *John Nicholson, Ian Kennedy*

Northern Ireland-based production company specializing in documentary, feature and lifestyle series for both regional and network transmission. OUTPUT includes *We Shall Overcome* (winner of Best Documentary at 1999 Celtic Television Festival for BBC); *Conquering the Normans* (Ch4 Learning – history of Normans in Ireland); *Gift of the Gab* (Ch4 Learning – contemporary Irish writing); *Sportsweek* (BBC Radio Ulster); *On Eagle's Wing* (full stage musical/TV material; story of the Scots/Irish in America); *Fire School*; *Mission Employable*; *A Queer Try*; *Sweet Child of Mine*; *To George with Love*; *Bright Sparks*, science-based schools competition; *School Challenge*, 3rd series; *World Indoor Bowls* (all for BBC NI); *Awash With Colour* (series, BBC Daytime).

Sunset + Vine Productions Ltd

Elsinore House, 77 Fulham Palace Road, London W6 8JA
℡ 020 7478 7300   🖷 020 7478 7403
www.sunsetvine.co.uk

Sports, children's and music programmes for television. No unsolicited mss. 'We hire freelancers only upon receipt of a commission.'

Table Top Productions

1 The Orchard, Chiswick, London W4 1JZ
℡ 020 8742 0507   🖷 020 8742 0507
alvin@tabletopproductions.com
Contact *Alvin Rakoff*

TV and film. OUTPUT *Paradise Postponed*; *The Adventures of Don Quixote*; *A Voyage Round My Father*; *The First Olympics 1896*; *Dirty Tricks*; *A Dance to the Music of Time*; *Too Marvellous for Words*. Also Dancetime Ltd. No unsolicited mss.

talkbackTHAMES

20–21 Newman Street, London W1T 1PG
℡ 020 7861 8000   🖷 020 7861 8001
www.talkbackthames.tv
Chief Executive Officer *Lorraine Heggessey*
Chief Operating Officer *Sara Geater*

talkackTHAMES Productions is a **FremantleMedia** company. OUTPUT includes: comedy, comedy drama, drama, entertainment and factual programmes. OUTPUT *The Apprentice*; *Green Wing*; *The Bill*; *Britain's Got Talent*; *The X Factor*; *Never Mind the Buzzcocks*; *Grand Designs*; *Property Ladder*.

Talkingheads Production Ltd

Argyll House, All Saint's Passage, London SW18 1EP
℡ 020 7292 7575   🖷 020 8875 8301
johnsachs@talkingheadsproductions.com
www.talkingheadsproductions.com
Contact *John Sachs*

Feature films. OUTPUT includes *The Merchant of Venice* starring Al Pacino. Will consider scripts; contact by e-mail. 'Somewhere out there is the new Tom Stoppard.'

Tandem TV & Film Ltd

Suite 206, Charleston House, Hemel Hempstead HP1 3AA
℡ 01442 261576   🖷 01442 219250
info@tandem.tv
www.tandem.tv
Production Director *Barbara Page*
Creative Director *Terry Page*
Production Controller *Jevan Green*

Internal and external communications, documentaries, drama-documentaries, public relations, sales, marketing and training programmes for, amongst others, the construction, civil engineering, transport, local government and charitable sectors. Welcomes unsolicited mss.

Telemagination Ltd

Royalty House, 72–74 Dean Street, London W1D 3SG
℡ 020 7434 1551   🖷 020 7434 3344
mail@tmation.co.uk
www.telemagination.co.uk
Contact *Beth Parker*

Animation production company. CREDITS include *The Animals of Farthing Wood*; *Noah's Island*; *The Last Polar Bears*; *Little Ghosts*; *Pongwiffy*; *Heidi*; *The Cramp Twins ll*; *Pettson and Findus*; *Rudi and Trudi*.

Tern Television Productions Ltd

73 Crown Street, Aberdeen AB11 6EX
☎ 01224 211123 🖷 01224 211199
aberdeen@terntv.com
www.terntv.com
Also at: 4th Floor, 114 Union Street, Glasgow
G1 3QQ ☎ 0141 204 1717 glasgow@terntv.com

1st Floor, Cotton Court, 38–42 Waring Street,
Belfast BT1 2ED ☎ 028 9032 6061 belfast@terntv.
com

3rd Floor, 12–14 Denman Street, London
W1D 7HJ ☎ 020 7287 8000
Contacts *David Strachan, Gwyneth Hardy
(Aberdeen), Harry Bell (Glasgow)*

Broadcast, television and corporate video
productions. Specializes in factual entertainment.
Currently developing drama.

Testimony Films

12 Great George Street, Bristol BS1 5RS
☎ 0117 925 8589 🖷 0117 925 7608
steve.humphries@testimonyfilms.com
Contact *Steve Humphries*

TV documentary producer. Specializes in social
history exploring Britain's past using living
memory. OUTPUT includes *Hooked: History of
Addictions*; *Married Love* (both Ch4 series); *A
Secret World of Sex* (BBC series); *The 50s & 60s
in Living Colour*; *Some Liked It Hot* (both ITV
series). Welcomes ideas from those working on
life stories and oral history.

Tiger Aspect Productions

5 Soho Square, London W1D 3QA
☎ 020 7434 6700 🖷 020 7434 1798
general@tigeraspect.co.uk
www.tigeraspect.co.uk
Head of Drama *Greg Brennan*
Head of Entertainment *Clive Tulloh*
Head of Comedy *Sophie Clarke-Jervoise*

Part of IMG Media. Television producer for
comedy, drama, documentary and entertainment.
OUTPUT *Ross Kemp on Gangs*; *Robin Hood*;
Secret Diary of a Call Girl; *Catherine Tate*;
Horne and Gordon; *Harry and Paul*; *The Vicar of
Dibley*. Sister companies: Tiger Aspect Pictures
(feature films; OUTPUT includes *Billy Elliot*; *The
League of Gentlemen's Apocalypse*) and **Tigress
Productions Ltd** (see entry). Only considers
material submitted via an agent or from writers
with a known track record.

Tigress Productions Ltd

2 St Paul's Road, Bristol BS8 1LT
☎ 0117 933 5600 🖷 0117 933 5666
general@tigressproductions.co.uk
www.tigressproductions.co.uk
Managing Director *Andrew Jackson*

Sister company of **Tiger Aspect Productions**.
Specializes in wildlife, science adventure and
documentary projects. Output includes *Jaguar
Adventure with Nigel Marven*; *Ben's Zoo*; *Everest:
Beyond the Limit*.

Touch Productions Ltd

18 Queen Square, Bath BA1 2HN
☎ 01225 484666 🖷 01225 483620
erica@touchproductions.co.uk
www.touchproductions.co.uk
Contacts *Erica Wolfe-Murray,
Malcolm Brinkworth*

Over the last 20 years, Touch has made a wide
range of programmes including award-winning
investigations, popular documentaries, medical
and science films, revelatory history productions
as well as observational, social, religious and arts
programmes. Current commissions include *The
Human Footprint*, a Ch4 documentary special
and various series and documentaries for the
BBC, National Geographic, TLC and Animal
Planet. Other projects include *Thanks for the
Memories*; *The Boy Who Couldn't Stop Running*;
Parish in the Sun; *Revival* and *Angela's Dying
Wish*.

Transatlantic Films Production and Distribution Company

Cabalva Studios, Whitney-on-Wye HR3 6EX
☎ 01497 831428 🖷 01497 831677
revel@transatlanticfilms.com
www.transatlanticfilms.com
Executive Producer *Revel Guest*

Producer of TV documentaries. OUTPUT *Belzoni*
(Ch4 Schools); *Science of Sleep and Dreams*;
Science of Love and *Extreme Body Parts* (all for
Discovery Health); *Legends of the Living Dead*
(Discovery Travel/S4C International); *2025*
(Discovery Digital); *How Animals Tell the Time*
(Discovery); *Trailblazers* (Travel Channel). No
unsolicited scripts. Interested in new writers to
write 'the book of the series', e.g. for *Greek Fire*
and *History's Turning Points*, but not usually
drama script writers.

TV Choice Ltd

PO Box 597, Bromley BR2 0YB
☎ 020 8464 7402 🖷 020 8464 7845
tvchoiceuk@aol.com
www.tvchoice.uk.com
Contact *Norman Thomas*

Produces a range of educational videos for
schools and colleges on subjects such as history,
geography, business studies and economics. No
unsolicited mss; send proposals only.

Twentieth Century Fox Film Limited

Twentieth Century House, 31–32 Soho Square,
London W1D 3AP
☎ 020 7437 7766
www.fox.co.uk

London office of the American giant. Does not
accept unsolicited material.

Twenty Twenty Television

20 Kentish Town Road, London NW1 9NX
☎ 020 7284 2020   ℻ 020 7284 1810
pch@twentytwenty.tv
www.twentytwenty.tv
Managing Director *Peter Casely-Hayford*
Co-Founder and Consultant Executive Producer
Claudia Milne
Creative Director/Executive Producer
Jamie Isaacs
Head of Factual Programming/Executive
Producer *Tim Carter*

Founded 1982. Large independent television
production company. Concentrates on
documentaries, lifestyle programmes, popular
drama and living history series. OUTPUT includes
The Choir (BBC2); *Bad Lads Army* (ITV1); *How
to Divorce Without Screwing Up Your Children*
(Ch4). Both *Brat Camp* and *That'll Teach Em*
have now aired a third season. Drama enquiries
and scripts should be mailed to *Jamie Isaacs*
(jamieisaacs@twentytwenty.tv).

Twofour Broadcast

5th Floor, 6–7 St Cross Street, London EC1N 8UA
☎ 020 7438 1800   ℻ 020 7438 1850
enq@twofour.co.uk
www.twofour.co.uk
Managing Director *Melanie Leach*
Creative Director *Stuart Murphy*
Director of Programmes *Joe Houlihan*

Television production company founded in 1988.
Specializes in factual entertainment, drama and
comedy. Drama material only accepted when
submitted through an agent.

Tyburn Film Productions Limited

Cippenham Court, Cippenham Lane, Cippenham,
Nr Slough SL1 5AU
☎ 01753 516767   ℻ 01753 691785

Feature films. Subsidiary of **Arlington
Productions Limited**. No unsolicited
submissions.

The UK Film and TV Production Company Plc

3 Colville Place, London W1T 2BH
☎ 020 7255 1650
Head of Development *Henrietta Fudakowski*

In 2006, *Tsotsi*, UKFTV's first wholly-produced
film won the Oscar for best foreign language film.
Other titles include *Bugs! in 3D*. Telephone ahead

of sending a submission to see if script might be
suitable for the company.

Vera Productions

66–68 Margaret Street, London W1W 8SR
☎ 020 7436 6116   ℻ 020 7436 6117
Contact *Phoebe Wallace*

Produces television comedy such as *Bremner, Bird
and Fortune*.

Video Enterprises

12 Barbers Wood Road, High Wycombe HP12 4EP
☎ 01494 534144/07831 875216   ℻ 01494 534144
videoenterprises@ntlworld.com
www.videoenterprises.co.uk
Contact *Maurice R. Fleisher*

Video and TV, mainly corporate: business and
industrial training, promotional material and
conferences. No unsolicited material 'but always
ready to try out good new writers'.

VIP Broadcasting

8 Bunbury Way, Epsom KT17 4JP
☎ 01372 721196   ℻ 01372 726697
mail@vipbroadcasting.co.uk
Contact *Chris Vezey*

Produces a wide range of radio programmes,
particularly interviews, documentaries, music
programmes and live concerts. Awards at Sony's
and New York Festivals. Approach with idea by
e-mail in the first instance; no unsolicited mss.

Wall to Wall

8–9 Spring Place, London NW5 3ER
☎ 020 7485 7424   ℻ 020 7267 5292
mail@walltowall.co.uk
www.walltowall.co.uk
Chief Executive *Alex Graham*

Factual and drama programming. OUTPUT
includes *Who Do You Think You Are?*; *New Tricks*;
A Rather English Marriage; *Glasgow Kiss*; *Sex,
Chips & Rock 'n' Roll*; *The 1940s House*; *Body
Story*; *Neanderthal*; *The Mafia*; *Not Forgotten*;
H.G. Wells; *Dispatches*.

Jane Walmsley Productions
▷ JAM Pictures

Walsh Bros. Limited

4 Trafalgar Grove, London SE10 9TB
☎ 020 8858 6870
john@walshbros.co.uk
www.walshbros.co.uk
Producer/Director *John Walsh*
Producer/Head of Finance *David Walsh, ACA*
Producer/Head of Development *Maura Walsh*

BAFTA-nominated producers of television, film
drama and documentaries. OUTPUT *Sofa Surfers*
(BBC); *Prime Minister's Global Fellowship* (Ch4);
Monarch (feature film); BAFTA-winning *Don't*

Make Me Angry (Ch4); Grierson-nominated *Headhunting the Homeless*; *Trex* (factual series on teenagers at work in China, Mexico, Vancouver and Alaska); *Trex2* (follow-up series covering Romania, India, Iceland and Louisiana); *Boyz & Girlz*, (Derbyshire dairy farm documentary series); *Cowboyz & Cowgirlz* (US sequel to hit series of Brit teens working on a ranch in Montana). Also arts documentaries: *The Comedy Store*; *Ray Harryhausen* (the work of Hollywood special effects legend). Drama: *The Sleeper*; *The Sceptic and the Psychic*; *A State of Mind*. Initial enquiries to: development@walshbros.co.uk

Paul Weiland Film Company

14 Newburgh Street, London W1F 7RT
📞 020 7287 6900   📠 020 7434 0146
info@weilands.co.uk
www.weilands.co.uk

Television commercials and pop promos.

Whistledown Productions Ltd

8A Ayres Street, London SE1 1ES
📞 020 7407 8001
davidprest@whistledown.net
www.whistledown.net
Managing Director *David Prest*

Producer of Sony Award-winning Landmark Series for BBC Radio 4. Features and documentaries on a wide range of social and historical subjects, contemporary issues and popular culture. OUTPUT includes *Questions Questions* and *The Reunion* (Radio 4); music-based documentaries for Radio 2. 'We welcome contributions and ideas, but phone or e-mail first.'

Wise Buddah Creative Ltd

74 Great Titchfield Street, London W1W 7QP
📞 020 7307 1600   📠 020 7307 1601
info@wisebuddah.com
www.wisebuddah.com
Contact *Chris North*

Radio production company: documentaries and commercials. Also studio facilities, sound-to-picture/sound design, talent management. No unsolicited material.

Witzend Productions
▶ FremantleMedia Group Ltd

Working Title Films Ltd

Oxford House, 76 Oxford Street, London W1D 1BS
📞 020 7307 3000   📠 020 7307 3001/2/3
Co-Chairmen *Tim Bevan, Eric Fellner*
Head of Development *Natascha Wharton*
Head of Film *Debra Hayward*
Development Executives *Amelia Granger, Rachael Prior*

Feature films OUTPUT (since 2004) *Shaun of the Dead*; *The Calcium Kid*; *Thunderbirds*; *Wimbledon*; *Bridget Jones: The Edge of Reason*; *Inside I'm Dancing*; *The Interpreter*; *Mickybo & Me*; *Nanny McPhee*; *Pride and Prejudice*; *Catch a Fire*; *United 93*; *Sixty Six*; *Smokin' Aces*; *Gone*; *Mr Bean's Holiday*; *Atonement*; *The Golden Age*; *Definitely, Maybe*; *Hot Fuzz*; *Wild Child*; *Frost/Nixon*; *The Soloist*; *Burn After Reading*; *State of Play*; *Hippie Hippie Shake*; *The Boat that Rocked*; *Green Zone*; *A Serious Man*. No unsolicited mss at present.

Wortman Productions UK

48 Chiswick Staithe, London W4 3TP
📞 020 8994 8886/07976 805976
neville@speakwell.co.uk
Producer *Neville Wortman*

Co-producers with Polestar Pictures Ltd. Feature film and TV production for drama, documentary, entertainment and corporate. OUTPUT 'Lost Ships' series: *White Gold of the Dragon Sea* (Discovery Channel, US); *Neffertiti*; *Resurrected*; *Who Killed Julius Caesar* (Discovery Channel, US/Ch 5, UK); 'Days That Shook the World' series: *Conspiracy to Kill*; *Hitler Bomb Plot*; *Assassinate De Gaulle* (BBC); 'Surviving Disaster' series: *Munich Air Crash*; *Eruption at Mount St Helen*; *Murder in Paradise* (Lion TV/Ch4) *Pevkovsky* ('Nuclear Spies' BBC2 series); *Jimmy's Farming Heroes* (BBC2). Open to new writing, preferably through agents; single page outline, some pages of dialogue; s.a.e. for reply.

Zeppotron
▶ Endemol UK Productions

Theatre Producers

For latest minimum rates for writers of plays for subsidized repertory theatres (not Scotland) access the following websites: Theatrical Management Association at www.tmauk.org; and the Writers' Guild at www.writersguild.org.uk

Actors Touring Company
▷ ATC

Almeida Theatre Company

Almeida Theatre, Almeida Street, London N1 1TA
☎ 020 7288 4900   📠 020 7288 4901
www.almeida.co.uk
Artistic Director *Michael Attenborough*

Founded 1980. The Almeida is a full-time producing theatre, presenting a year-round theatre programme in which new light is shed on an eclectic mix of new plays, classic revivals, adaptations and new versions of international work. Previous productions: *Festen*; *The Goat, or Who is Sylvia?*; *The Lady From the Sea*; *The Late Henry Moss*; *Enemies*. No unsolicited mss.

ATC

The Tab Centre, 3 Godfrey Place, London E2 7NT
☎ 020 7033 7360   📠 020 7739 8298
atc@atctheatre.com
www.atc-online.com
Artistic Director *Bijan Sheibani*

Collaborates with writers on new work, adaptations and/or translations.

Birmingham Repertory Theatre

Centenary Square, Broad Street, Birmingham B1 2EP
☎ 0121 245 2000   📠 0121 245 2100
writing@birmingham-rep.co.uk
www.birmingham-rep.co.uk
Contact *Literary Department*

The Birmingham Repertory Theatre aims to provide a platform for the best work from new writers from both within and beyond the West Midlands region. The Rep is committed to the production of new work which reflects both the diversity of contemporary experience and of approaches to writing for the stage. The commissioning of new plays takes place across the full range of the theatre's activities: in the Main House, The Door (which is a dedicated new writing space) and on tour to community venues in the region. The theatre runs a programme of writers' attachments every year in addition to its commissioning policy and maintains close links with *Script* (the regional writers' training agency) and the MPhil in Playwriting Studies at the University of Birmingham. For more information contact the Literary Department.

Black Theatre Co-op
▷ Nitro

Bootleg Theatre Company

23 Burgess Green, Bishopdown, Salisbury SP1 3EL
☎ 01722 421476
www.bootlettheatre.co.uk
Contact *Colin Burden*

Founded 1984. Tries to encompass as wide an audience as possible and has a tendency towards plays with socially relevant themes. A good bet for new writing since unsolicited mss are very welcome. 'Our policy is to produce new and/or rarely seen plays and anything received is given the most serious consideration.' Actively seeks to obtain grants to commission new writers for the company. New productions to include: *Asking for It*; *King Squealer*; *Prince Charles of Wandsworth Road*; *A Rainy Night in Soho*.

Borderline Theatre Company

Darlington New Church, North Harbour Street, Ayr KA8 8AA
☎ 01292 281010   📠 01292 618685
enquiries@borderlinetheatre.co.uk
www.borderlinetheatre.co.uk
Producer *Eddie Jackson*

Founded 1974. Borderline is one of Scotland's leading touring companies. Tours to main-house theatres and small venues throughout Scotland. Productions include the world premières of Liz Lochhead's work and *The Wall* by D.C. Jackson. Previous writers have included Dario Fo and John Byrne. Borderline is also committed to commissioning and touring new plays for young people. Please contact the company before submitting synopsis or script.

Bristol Old Vic Theatre Company

King Street, Bristol BS1 4ED

📞 0117 949 3993   📠 0117 949 3996

admin@bristol-old-vic.co.uk

www.bristololdvic.org.uk

'Bristol Old Vic places writing for performance at the heart of the organization.' Contact the Literary Department before sending any unsolicited material. Visit the website for more details.

The Bush Theatre

Shepherds Bush Green, London W12 8QD

📞 020 7602 3703   📠 020 7602 7614

info@bushtheatre.co.uk

www.bushtheatre.co.uk

Artistic Director *Josie Rourke*

Founded 1972. The Bush Theatre produces up to nine new plays a year, including up to three visiting companies also producing new work: 'we are a writer's theatre'. Previous productions include: *2000 Feet Away* Anthony Weigh; *Tinderbox* Lucy Kirkwood; *50 Ways to Leave Your Lover* Leah Chillery, Ben Ellis, Stacey Gregg, Lucy Kirkwood and Ben Schiffer; *When You Cure Me* Jack Thorne; *Pumpgirl* Abbie Spallen; *Blackbird* Adam Rapp; *The Wexford Trilogy* Billy Roche; *Love and Understanding* Joe Penhall; *This Limetree Bower* Conor McPherson; *Discopigs* Enda Walsh; *The Pitchfork Disney* Philip Ridley; *Caravan* Helen Blakeman; *Beautiful Thing* Jonathan Harvey; *Killer Joe* Tracy Letts; *Shang-a-Lang* Catherine Johnson; *Howie the Rookie* Mark O'Rowe; *The Glee Club* Richard Cameron; *Adrenalin ... Heart* Georgia Fitch. Scripts are read by a team of Creative Associates then discussed with the core artistic team, a process which takes about two months. The theatre offers a small number of commissions. Writers should send scripts (full-length plays only) with small s.a.e. for acknowledgement and large s.a.e. for return of script.

Carnival Film and Television Limited

▷ entry under Film, TV and Radio Producers

Citizens Theatre

Gorbals, Glasgow G5 9DS

📞 0141 429 5561   📠 0141 429 7374

info@citz.co.uk

www.citz.co.uk

Artistic Directors *Jeremy Raison, Guy Hollands*

Administrative Director *Anna Stapleton*

No formal new play policy. The theatre has a play reader but opportunities to do new work are limited.

Clwyd Theatr Cymru

Mold, Flintshire CH7 1YA

📞 01352 756331   📠 01352 701558

drama@celtic.co.uk

www.clwyd-theatr-cymru.co.uk

Literary Manager *William James (william.james@ clwyd-theatr-cymru.co.uk)*

Clwyd Theatr Cymru produces plays performed by a core ensemble in Mold and tours them throughout Wales (in English and Welsh). Productions are a mix of classics, revivals and contemporary drama. Recent new writing includes: *Drowned Out/Porth y Byddar* Manon Eames; *Two Princes* Meredydd Barker; *Memory* Jonathan Lichtenstein; *Stone City Blue* Ed Thomas; *The Rabbit* Meredydd Barker; *The Journey of Mary Kelly* Siân Evans; *And Now What?* Tim Baker and Sarah Argent; *The Way It Was, Flights of Fancy; Pocketful of Memories; The Ballad of Megan Morgan; Flora's War/ Rhyfel Flora; Word for Word/Gair am Air* and *The Secret/Y Gyfrinach:* all by Tim Baker. Plays by Welsh writers or on Welsh themes will be considered.

Michael Codron Plays Ltd

Aldwych Theatre Offices, Aldwych, London WC2B 4DF

📞 020 7240 8291   📠 020 7240 8467

Michael Codron Plays Ltd manages the Aldwych Theatre in London's West End. The plays it produces do not necessarily go into the Aldwych but always tend to be big-time West End fare. Previous productions: *Bedroom Farce; Blue Orange; Copenhagen; The Invention of Love; Hapgood; Uncle Vanya; Rise and Fall of Little Voice; Arcadia; Dead Funny; My Brilliant Divorce; Dinner; Democracy; Glorious!*. No particular rule of thumb on subject matter or treatment. The acid test is whether 'something appeals to Michael'. Straight plays rather than musicals.

Colchester Mercury Theatre Limited

Balkerne Gate, Colchester CO1 1PT

📞 01206 577006   📠 01206 769607

info@mercurytheatre.co.uk

www.mercurytheatre.co.uk

Chief Executive & Artistic Director *Dee Evans*

Producing theatre with a wide-ranging audience. New writing encouraged. The theatre has a free playwright's group for adults with a serious commitment to writing plays.

The Coliseum, Oldham

Fairbottom Street, Oldham OL1 3SW

📞 0161 624 1731   📠 0161 624 5318

mail@coliseum.org.uk

www.coliseum.org.uk

Chief Executive/Artistic Director *Kevin Shaw*

The artistic policy of the theatre is to present a high quality and diverse theatre programme with the ambition to commission a new play each year. The Coliseum employs a reader to read all submitted scripts; a written report will be sent to the writer within approximately four months. An s.a.e. should be enclosed for return of the script along with an s.a.e. for acknowledgement of receipt, if required. All submitted scripts must be clearly typed, using one side of A4 for each page, and must contain a page of casting requirements.

Concordance

Finborough Theatre, 118 Finborough Road, London SW10 9ED
℡ 020 7244 7439   📠 020 7835 1853
admin@concordance.org.uk
www.concordance.org.uk
Artistic Director *Neil McPherson*

Founded in 1981, Concordance is the resident company based at the **Finborough Theatre** (see entry). Presents world premières of new writing and revivals of neglected work with a special commitment to music theatre as well as integrating music into its work and productions featuring the writing of non-theatrical artists – poets, artists, novelists, etc. – presenting their work in a theatrical setting. We accept unsolicited scripts through the Finborough Theatre's submission department. Details at www.finboroughtheatre.co.uk

Contact Theatre Company

Oxford Road, Manchester M15 6JA
℡ 0161 274 0600   📠 0161 274 0640
search@contact-theatre.org
www.contact-theatre.org
Programme & Projects Co-ordinator *Kate Pearson*

Contact's Search Party is interested in artists and companies with exciting new ideas for theatre through spoken word, audio visuals, music, media, movement, live art, poetry, rap, installations and more. 'Send your ideas to our monthly Search Party and you could end up as part of our Artist In Residence (AIR) programme; as a writer for Verbally Challenged, our new regular writing slam; or get time and space to develop your work here.' Send a sample of your work: ten pages of writing, five minutes of a DVD or CD, and information about you and your idea to the e-mail address above or to Search Party, Contact, at the postal address. For further information, e-mail *Kate Pearson* or call 0161 274 0600.

Crucible Theatre

55 Norfolk Street, Sheffield S1 1DA
℡ 0114 249 5999   📠 0114 249 6003
www.sheffieldtheatres.co.uk

Executive Producer *Mark Feakins*
Literary Associate *Matthew Byam Shaw*

'Although we are interested in all new writing, most of the new work we present will be the result of commissions or prolonged script development with writers in whom we have expressed an interest.'

Druid Theatre Company

Flood Street, Galway, Republic of Ireland
℡ 00 353 91 568660   📠 00 353 91 563109
info@druid.ie
www.druid.ie
Artistic Director *Garry Hynes*
New Writing Manager *Thomas Conway*
General Manager *Tim Smith*

Founded in 1975 and based in Galway, 'Druid has always worked to reinvigorate perceptions of classic dramatic works and to engage with new dramatic works of a challenging, innovative and daring kind'. Druid consistently commissions, develops, and produces new plays by a wide range of emerging and established writers both from Ireland and abroad.

The Dukes

Moor Lane, Lancaster LA1 1QE
℡ 01524 598505   📠 01524 598519
info@dukes-lancaster.org
www.dukes-lancaster.org
Chief Executive/Artistic Director *Joe Sumsion*

Founded 1971. The only producing house in Lancashire. Wide target market for cinema and theatre. Plays in a 313-seater end-on auditorium and in a 198-seater in-the-round studio (The Duke's newly refurbished space). Host for **Litfest** – Lancaster's annual festival of literature. In the summer months open-air promenade performances are held in Williamson Park. DT3 – The Education Centre is the Dukes' newly refurbished space dedicated to work by and for young people.

Dundee Repertory Theatre

Tay Square, Dundee DD1 1PB
℡ 01382 227684   📠 01382 228609
hwatson@dundeereptheatre.co.uk
www.dundeereptheatre.co.uk
Artistic Director *James Brining*

Founded 1939. Plays to a varied audience. Translations and adaptations of classics, and new local plays. Most new work is commissioned. Interested in contemporary plays in translation and in new Scottish writing. No scripts except by prior arrangement.

Eastern Angles Theatre Company

Sir John Mills Theatre, Gatacre Road, Ipswich
IP1 2LQ
☎ 01473 218202   ℻ 01473 384999
admin@easternangles.co.uk
www.easternangles.co.uk
Artistic Director *Ivan Cutting*
General Manager *Jill Streatfeild*

Founded 1982. Plays to a rural audience for the
most part. New work only: some commissioned,
some devised by the company, some researched
documentaries. Unsolicited mss welcome but
scripts or writers need to have some connection
with East Anglia. 'We are always keen to develop
and produce new writing, especially that which is
germane to a rural area.'

Edinburgh Royal Lyceum Theatre
▷ Royal Lyceum Theatre Company

English Touring Theatre

25 Short Street, London SE1 8LJ
☎ 020 7450 1990   ℻ 020 7450 1991
admin@ett.org.uk
www.ett.org.uk
Director *Rachel Tackley*

Founded 1993. National touring company visiting
middle-scale receiving houses and arts centres
throughout the UK. Mostly mainstream. Largely
classical programme, but with increasing interest
to tour one modern English play per year. Strong
commitment to education work. *Unable to accept
or process unsolicited scripts.*

Exeter Northcott

Stocker Road, Exeter EX4 4QB
☎ 01392 223999   ℻ 01392 223996
www.exeternorthcott.co.uk

Founded 1967. The Exeter Northcott is the Southwest's principal subsidized producing theatre.
Situated on the University of Exeter campus, the
Theatre is continually looking to broaden the base
of its audience profile. The company is currently
reviewing its artistic policy and intends to initiate
a new programme to develop new writing for
theatre. Check the website for details.

Finborough Theatre

118 Finborough Road, London SW10 9ED
☎ 020 7244 7439   ℻ 020 7835 1853
admin@finboroughtheatre.co.uk
www.finboroughtheatre.co.uk
Artistic Director *Neil McPherson*

Founded 1980. 'One of London's leading new
writing venues' (*Time Out*). Presents new British
writing, revivals of neglected work from the
19th and 20th centuries, music theatre and
UK premières of foreign work, particularly
from Ireland, the United States and Canada.

The theatre is available for hire and the fee is
sometimes negotiable to encourage interesting
work. Premièred work by Chris Lee, Anthony
Neilson, Naomi Wallace, Tony Marchant, Mark
Ravenhill, Laura Wade, James Graham. Three
times winner of the Pearson Award. Unsolicited
scripts are accepted, but please read carefully
the details on submissions policy on the website
before sending any scripts.

Gate Theatre Company Ltd

11 Pembridge Road, London W11 3HQ
☎ 020 7229 5387   ℻ 020 7221 6055
gate@gatetheatre.co.uk
www.gatetheatre.co.uk
Artistic Directors *Natalie Abrahami,
Carrie Cracknell*

Founded 1979. Plays to a mixed, London-wide
audience, depending on production. Aims
to produce British premières of plays which
originate from abroad and translations of
neglected classics. Recent productions include
The Sexual Neuroses of Our Parents by Lukas
Bärfuss, transl. by Neil Blackadder; *The Car
Cemetery* by Fernando Arrabal, transl. by Barbara
Wright; *I am Falling*, devised in collaboration
with choreographer Anna Williams and dramturg
Jenny Worton; *Press*, co-production created
and performed by Pierre Rigal and his company,
dernière minute; *The Internationalist*, by US
writer Anne Washburn. Positively encourages
writers from abroad to send in scripts or
translations. Most unsolicited scripts are read but
plays by writers from the UK will not be accepted.
Please address submissions to the Artistic
Directors. Always enclose s.a.e. if play needs
returning.

Graeae Theatre Company

LVS Resource Centre, 356 Holloway Road,
London N7 6PA
☎ 020 7700 2455   ℻ 020 7609 7324
info@graeae.org
www.graeae.org
Minicom 020 7700 8184
CEO/Artistic Director *Jenny Sealey*

Europe's premier theatre company of disabled
people, the company tours nationally and
internationally with innovative theatre
productions highlighting both historical and
contemporary disabled experience. Graeae
also runs educational programmes available to
schools, youth clubs, students and disabled adults
nationally and provides vocational training in
theatre arts (including playwriting). Unsolicited
scripts, from disabled writers, welcome. New
work is commissioned.

Hampstead Theatre

Eton Avenue, Swiss Cottage, London NW3 3EU
☎ 020 7449 4200   🖷 020 7449 4201
literary@hampsteadtheatre.com
www.hampsteadtheatre.com
Contact *Katy Silverton*

Hampstead Theatre celebrates its 50th birthday in 2009. Based in a state-of-the art building, the theatre is an intimate space with a flexible stage and an auditorium capable of expanding to seat 325. The artistic policy continues to be the production of British and international new plays and the development of important young writers. 'We are looking for writers who recognize the power of theatre and who have a story to tell. All plays which meet our submission requirements are read and discussed. We give feedback to all writers with potential.' Writers produced in the last five years at Hampstead include Gregory Burke, Dennis Kelly, Nell Leyshon, Sharman Macdonald, Tamsin Oglesby, Debbie Tucker Green and Roy Williams. In earlier years, breakthrough plays by Harold Pinter, David Hare, Mike Leigh, Pam Gems and Stephen Jeffreys were produced.

Harrogate Theatre

Oxford Street, Harrogate HG1 1QF
☎ 01423 502710   🖷 01423 563205
www.harrogatetheatre.co.uk

'Due to recent funding cuts, Harrogate Theatre cannot accept any new writing at present. It is hoped this situation will change in the future.'

Headlong Theatre

Chertsey Chambers, 12 Mercer Street, London WC2H 9QD
☎ 020 7438 9940   🖷 020 7438 9941
info@headlongtheatre.co.uk
www.headlongtheatre.co.uk
Artistic Director *Rupert Goold*

A middle-scale touring company producing established and new plays. On average, one new play or new adaptation a year. Due to forthcoming projects the company is not considering unsolicited scripts at present.

Heritage Theatre Ltd
▷ entry under Film, TV and Radio Producers

The Hiss & Boo Co. Ltd

Nyes Hill, Wineham Lane, Bolney RH17 5SD
☎ 01444 881707
email@hissboo.co.uk
www.hissboo.co.uk
Artistic Director *Ian Liston*

Founded 1976. Produces mainly traditional pantomime, revue, variety and occasional plays for a family audience. No unsolicited scripts; send letter and synopsis in the first instance. No run-of-the mill comedies, historical drama or anything without humour; interested in revue-type shows and unusual thrillers.

Hull Truck Theatre Company

Spring Street, Hull HU2 8RW
☎ 01482 224800   🖷 01482 581182
admin@hulltruck.co.uk
www.hulltruck.co.uk
Interim General Manager *Paul Marshall*
Artistic Directors *John Godber, Gareth Tudor-Price*
Associate Director *Nick Lane*

John Godber, of *Teechers, Bouncers* and *Up 'n' Under* fame (the artistic director of this high-profile Northern company since 1984), has very much dominated the scene in the past with his own successful plays. The emphasis is still on new writing but Godber's works continue to be toured extensively. Recent premières/commissions include *On a Shout; Blue Cross Xmas; Sold; Wilde Boyz; Upon on Roof; Christmas Crackers; Crown Prince; My Favourite Summer; Ladies Down Under; Sully*. Most new plays are commissioned. 'New scripts should be addressed to the Literary Development Manager who will attempt as quick a response as possible. Bear in mind the artistic policy of Hull Truck, which is accessibility and popularity.' In general, not interested in musicals or in plays with casts of more than seven.

Stephen Joseph Theatre

Westborough, Scarborough YO11 1JW
☎ 01723 370540   🖷 01723 360506
enquiries@sjt.uk.com
www.sjt.uk.com
Artistic Director *Chris Monks*

A two-auditoria complex housing a 165-seat end stage theatre/cinema (the McCarthy) and a 400-seat theatre-in-the-round (the Round). Positive policy on new work. Alan Ayckbourn's work features quite strongly but a new writing programme ensures plays from other sources are actively encouraged. Also runs a lunchtime season of one-act plays each summer. Writers are advised however that the SJT is very unlikely to produce an unsolicited script – synopses are preferred. Recent commissions and past productions include: *Touch Wood* Nick Warburton; *The Swing of Things* Torben Betts; *Absolutely Frank* Tim Firth; *Life and Beth* Alan Ayckbourn. 'Writers are welcome to send details of rehearsed readings and productions as an alternative means of introducing the theatre to their work.' Submit to the Literary Department enclosing an s.a.e. for return of mss.

Bill Kenwright Ltd

BKL House, 1 Venice Walk, London W2 1RR
☎ 020 7446 6200　📠 020 7446 6246
Contact *Chloe Saxby*

Presents both revivals and new shows for West End and touring theatres. Although new work tends to be by established playwrights, this does not preclude or prejudice new plays from new writers. The company has an in-house dramaturg. Scripts should be addressed to Bill Kenwright with a covering letter and s.a.e. 'We have enormous amounts of scripts sent to us although we very rarely produce unsolicited work. Scripts are read systematically. Please do not phone; the return of your script or contact with you will take place in time.'

Komedia

44–47 Gardner Street, North Laine, Brighton
BN1 1UN
☎ 01273 647101　📠 01273 647102
admin@komedia.co.uk
www.komedia.co.uk
Contact *Marina Cobler*

Founded in 1994, Komedia promotes, produces and presents new work. Mss of new plays welcome.

Library Theatre Company

St Peter's Square, Manchester M2 5PD
☎ 0161 234 1913　📠 0161 228 6481
ltcadmin@manchester.gov.uk
www.librarytheatre.com
Artistic Director *Chris Honer*

Produces new and contemporary work, as well as the classics. No unsolicited mss. Send outline of the nature of the script first. Encourages new writing through the commissioning of new plays and through a programme of rehearsed readings to help writers' development.

Live Theatre Company

Broad Chare, Quayside, Newcastle upon Tyne
NE1 3DQ
☎ 0191 261 2694　📠 0191 232 2224
info@live.org.uk
www.live.org.uk
Artistic Director *Max Roberts*
Chief Executive *Jim Beirne*

Following a £5.5 million redevelopment, Live Theatre is home to a 200-seat cabaret style theatre, a new studio theatre, renovated rehearsal rooms and a series of dedicated writer's rooms. 'The new space combined with the theatre's commitment to new writing has positioned Live Theatre at the forefront of Newcastle's emergence as one of the country's most exciting artistic and cultural regional capitals.' Productions include *The Pitmen Painters*; *Cooking With Elvis* and *NE1*

by Lee Hall; *A Nightingale Sang* by C.P. Taylor; *Laughter When We're Dead* by Sean O'Brien; *Falling Together* by Tom Hadaway; *Bones* by Peter Straughan

London Bubble Theatre Company

5 Elephant Lane, London SE16 4JD
☎ 020 7237 4434　📠 020 7231 2366
admin@londonbubble.org.uk
www.londonbubble.org.uk
Creative Director *Jonathan Petherbridge*

London Bubble has existed within its community of southeast London for 36 years offering everyone access to enjoying theatre whether as an audience member or theatre-maker. Works with individuals and communities that may otherwise not have a chance to experience theatre, and provides access into the arts for young professionals. London Bubble has an extensive participatory programme, offering the chance to work collaboratively with theatre professionals to create and stage a piece of theatre which is then toured to parks and unusual open spaces around S.E. London. Runs innovative arts education projects in schools with a focus on Greenwich, Lewisham and Southwark.

Lyric Hammersmith

Lyric Square, King Street, London W6 0QL
☎ 0871 221 1722　📠 020 8741 5965
enquiries@lyric.co.uk
www.lyric.co.uk
Executive Director *Jessica Hepburn*

The main theatre stages an eclectic programme of new and revived classics. Interested in developing projects with writers, translators and adaptors. The Lyric does not accept unsolicited scripts. Its 110-seat studio focuses on work for children, young people and families.

mac

Cannon Hill Park, Birmingham B12 9QH
www.macarts.co.uk

Closed in 2008 for a £13.6 million rebuild and refurbishment. Due to re-open in the autumn of 2009. See website for details.

Manchester Library Theatre
▷ Library Theatre Company

New Vic Theatre

Etruria Road, Newcastle under Lyme ST5 0JG
☎ 01782 717954　📠 01782 712885
admin@newvictheatre.org.uk
www.newvictheatre.org.uk
Artistic Director *Theresa Heskins*

The New Vic is a purpose-built theatre-in-the-round. Produces ten in-house plays each year and is active within the education sector and community. New plays produced are the result of specific commissions. Send synopses and the first ten pages of dialogue, *not* unsolicited scripts. 'We cannot guarantee that unsolicited scripts will be read; they will be returned on receipt of an s.a.e.'

Newpalm Productions

26 Cavendish Avenue, London N3 3QN
☎ 020 8349 0802 🖷 020 8346 8257
newpalm@btopenworld.com
Contact *Lionel Chilcott*

Very rarely produces new plays (*As Is* by William M. Hoffman, which came from Broadway to the Half Moon Theatre, was an exception to this). National tours and West End productions such as *Peter Pan (The Musical)*; *Noises Off*; *Seven Brides for Seven Brothers* and *Rebecca*, at regional repertory theatres, are more typical examples of Newpalm's work. Both plays and musicals are, however, welcome; synopses are preferable to scripts.

Nimax Theatres Limited

1 Lumley Court, off 402 Strand, London WC2R 0NB
☎ 0845 434 9290
general@nimaxtheatres.com
www.nimaxtheatres.com
CEO *Nica Burns*

West End theatre managers of the Apollo, Duchess, Garrick, Lyric and Vaudeville theatres. Unsolicited scripts are not considered.

Nitro

6 Brewery Road, London N7 9NH
☎ 020 7609 1331 🖷 020 7609 1221
info@nitro.co.uk
www.nitro.co.uk
Artistic Director *Felix Cross*

Founded 1979. Formerly Black Theatre Co-op. A music theatre company, 'producing dynamic new work that explores stories by black people in Britain and elsewhere. Nitro productions present popular and epic themes, embracing the vibrant spectrum of black music and dance, appealing to diverse audiences of all ages.'

Northumberland Theatre Company

The Playhouse, Bondgate Without, Alnwick NE66 1PQ
☎ 01665 602586 🖷 01665 605837
admin@ntc-touringtheatre.co.uk
www.ntc-touringtheatre.co.uk
Artistic Director *Gillian Hambleton*
General Manager *Anna Flood*

Founded 1978. An Arts Council England revenue funded organization. Predominantly rural, small-scale touring company, playing to village halls and community centres throughout the northern region, the Scottish Borders and countrywide. Productions range from established classics to new work but must be appropriate to their audience. Unsolicited scripts welcome but are unlikely to be produced. All scripts are read and returned with constructive criticism within six months. Writers whose style is of interest may then be commissioned. The company encourages new writing and commissions when possible. Financial constraints restrict casting to a *maximum* of six and minimal set.

Norwich Puppet Theatre

St James, Whitefriars, Norwich NR3 1TN
☎ 01603 615564 🖷 01603 617578
info@puppettheatre.co.uk
www.puppettheatre.co.uk
General Manager *Ian Woods*

Young/family-centred audience (aged 3–12) but developing shows for adult audiences interested in puppetry. All year round programme plus tours to schools and arts venues. Most productions are based on traditional stories but unsolicited mss welcome if relevant.

Nottingham Playhouse

Nottingham Playhouse Trust, Wellington Circus, Nottingham NG1 5AF
☎ 0115 947 4361 🖷 0115 947 5759
www.nottinghamplayhouse.co.uk
Artistic Director *Giles Croft*

Aims to make innovation popular, and present the best of world theatre, working closely with the communities of Nottingham and Nottinghamshire. Unsolicited mss will be read. It normally takes about six months, however, and 'we have never yet produced an unsolicited script. All our plays have to achieve a minimum of 60 per cent audiences in a 732-seat theatre. We have no studio.'

Nottingham Playhouse Roundabout Theatre in Education

Wellington Circus, Nottingham NG1 5AF
☎ 0115 947 4361
andrewb@nottinghamplayhouse.co.uk
www.nottinghamplayhouse.co.uk
Contact *Andrew Breakwell*

Founded 1973. Theatre-in-Education company of the **Nottingham Playhouse**. Plays to a young audience aged 5–18 years of age. 'We are committed to the encouragement of new writing and commission at least two new plays for young people each year. With other major producers in the East Midlands we share the resources of the *Theatre Writing Partnership* which is based at the Playhouse.' See website for philosophy and play details. Please make contact before submitting scripts.

NTC Touring Theatre Company
▷ Northumberland Theatre Company

Nuffield Theatre
University Road, Southampton SO17 1TR
☎ 023 8031 5500   📠 023 8031 5511
alison.thurley@nuffieldtheatre.co.uk
www.nuffieldtheatre.co.uk
Artistic Director *Patrick Sandford*
Script Executive *John Burgess*

Well known as a good bet for new playwrights,
the Nuffield gets an awful lot of scripts. Produces
two new main stage plays every season. Previous
productions: *Exchange* by Yuri Trifonov
(transl. Michael Frayn) which transferred to
the Vaudeville Theatre; *The Floating Light Bulb*
Woody Allen (British première); *Nelson*, a new
play by Pam Gems; *The Dramatic Attitudes of
Miss Fanny Kemble* Claire Luckham; *The Winter
Wife* Claire Tomalin and *Tchaikovsky and the
Queen of Spades* John Clifford. Open-minded
about subject and style, producing musicals as
well as straight plays. Also opportunities for some
small-scale fringe work. Scripts preferred to
synopses in the case of writers new to theatre. All
will, eventually, be read 'but please be patient. We
do not have a large team of paid readers. We read
everything ourselves.'

Octagon Theatre Trust Ltd
Howell Croft South, Bolton BL1 1SB
☎ 01204 529407   📠 01204 556502
info@octagonbolton.co.uk
www.octagonbolton.co.uk
Executive Director *John Blackmore*

Founded in 1967, the award-winning Octagon
Theatre stages at least eight main auditorium
home-produced shows a year, and hosts
UK touring companies such as the Nitro,
Live Theatre, Peshkar Productions and Alan
Ayckbourn's Stephen Joseph Theatre. Also
hosts the work of partner companies as part
of its commitment to creative partnerships.
The Theatre boasts a thriving and constantly
developing participatory department, activ8,
which operates a highly successful youth theatre
as well as initiating and facilitating exciting
education and outreach programmes. The theatre
is keen to encourage new writing but, due to the
lack of a dedicated literary department, is unable
to accept and process new and unsolicited scripts;
the theatre uses **North West Playwrights** as a
reading service instead.

Orange Tree Theatre
1 Clarence Street, Richmond TW9 2SA
☎ 020 8940 0141   📠 020 8332 0369
admin@orangetreetheatre.co.uk
www.orangetreetheatre.co.uk

Artistic Director *Sam Walters*
Executive Director *Gillian Thorpe*

The Orange Tree, a theatre-in-the-round venue,
presents a broad cross section of work. Past
productions have included plays by Rodney
Ackland, John Galsworthy and a new translation
of Lorca as well as new plays by Oliver Ford
Davies, David Lewis, Ben Brown and Kenneth
Jupp. The theatre no longer considers unsolicited
mss; writers who may wish to approach the
theatre are asked to write first.

Out of Joint
7 Thane Works, Thane Villas, London N7 7NU
☎ 020 7609 0207   📠 020 7609 0203
ojo@outofjoint.co.uk
www.outofjoint.co.uk
Director *Max Stafford-Clark*
Producer *Graham Cowley*
Literary Associate *Alex Yates*

Founded 1993. Award-winning theatre company
with new writing central to its policy. Produces
new plays which reflect society and its concerns,
placing an emphasis on education activity to
attract young audiences. Welcomes unsolicited
mss. Productions include: *Blue Heart* Caryl
Churchill; *Our Lady of Sligo*, *The Steward of
Christendom* and *Hinterland* Sebastian Barry;
Shopping and Fucking and *Some Explicit
Polaroids* Mark Ravenhill; *A State Affair* and
Talking to Terrorists Robin Soans; *Sliding with
Suzanne* Judy Upton; *The Positive Hour* and *A
Laughing Matter* April De Angelis; *Duck* and *O go
my Man* Stella Feehily; *The Permanent Way* David
Hare; *The Overwhelming* J.T. Rogers; *Feelgood* and
King of Hearts Alistair Beaton; *Testing the Echo*
David Edgar.

Paines Plough
4th Floor, 43 Aldwych, London WC2B 4DN
☎ 020 7240 4533   📠 020 7240 4534
office@painesplough.com
www.painesplough.com
Artistic Director *Roxana Silbert*
Director *Tessa Walker*

Founded 1975. Award-winning company
commissioning and producing new plays by
British and international playwrights. Tours 2–4
plays a year nationally and worldwide for small
and middle-scale venues. Also runs a range of
projects focused on identifying and launching
emerging playwrights (see website for details).
'We welcome unsolicited scripts and will respond
to all submissions. We are looking for original
plays that engage with the contemporary world
and are written in a distinctive voice.' Recent
productions: *Dallas Sweetman* Sebastian Barry;
Shoot/Get Treasure/Repeat and *Product* Mark
Ravenhill; *Long Time Dead* Rona Munro; *After*

the End Dennis Kelly; *If Destroyed True* Douglas Maxwell; *Pyrenees* David Greig; *Mercury Fur* Philip Ridley; *The Small Things* Enda Walsh.

Perth Theatre

Horsecross Arts Ltd, 185 High Street, Perth PH1 5UW

☏ 01738 472700   🖷 01738 624576

info@horsecross.co.uk

www.horsecross.co.uk

Creative Director – Theatre *Ian Grieve*

Founded 1935. Combination of one to four-weekly repertoire of plays, incoming tours and studio productions. Unsolicited mss are read when time permits, but the timetable for return of scripts is lengthy. New plays staged by the company are usually commissioned under the SAC scheme.

Plymouth Theatre Royal
▷ Theatre Royal

Polka Theatre

240 The Broadway, Wimbledon SW19 1SB

☏ 020 8543 4888 (box office)   🖷 020 8545 8365

admin@polkatheatre.com

www.polkatheatre.com

Artistic Director *Jonathan Lloyd*

Executive Director *Stephen Midlane*

Founded in 1968 and moved into its Wimbledon base in 1979. Leading children's theatre committed to commissioning and producing new plays. Programmes are planned two years ahead and at least three new plays are commissioned each year. 'Many of our scripts are commissioned from established writers. We are, however, keen to develop work from writers new to children's and young people's theatre. We run a new writing programme which includes master classes and workshops. Potential new writers' work is read and discussed on a regular basis; thus we constantly add to our pool of interesting and interested writers.

Queen's Theatre, Hornchurch

Billet Lane, Hornchurch RM11 1QT

☏ 01708 462362   🖷 01708 462363

info@queens-theatre.co.uk

www.queens-theatre.co.uk

Artistic Director *Bob Carlton*

The Queen's Theatre is a 503-seat theatre producing eight main house productions per year, including pantomime. Aims for distinctive and accessible performances in an identifiable house style focused upon actor/musician shows but, in addition, embraces straight plays, classics and comedies. Also runs an occasional writers' group for adults (contact the Education Manager for information). These groups have close links with the Queen's Community Company (amateur actors) as well as the main house company which

workshops and showcases the groups' work. The theatre also runs a biannual New Writing Award scheme which culminates in a festival called Writenow!.

The Really Useful Group Ltd

22 Tower Street, London WC2H 9TW

☏ 020 7240 0880   🖷 020 7240 1204

www.reallyuseful.com

Commercial/West End, national and international theatre producer/co-producer/licensor whose output has included *Joseph and the Amazing Technicolor Dreamcoat*; *Jesus Christ Superstar*; *Cats*; *Evita*; *Song & Dance*; *Daisy Pulls It Off*; *Lend Me a Tenor*; *Starlight Express*; *The Phantom of the Opera*; *Aspects of Love*; *Sunset Boulevard*; *By Jeeves*; *Whistle Down the Wind*; *The Beautiful Game*; *Tell Me On a Sunday*; *Bombay Dreams*; *The Woman in White*; *The Sound of Music*.

Red Ladder Theatre Company

3 St Peter's Buildings, York Street, Leeds LS9 8AJ

☏ 0113 245 5311   🖷 0113 245 5351

rod@redladder.co.uk

www.redladder.co.uk

Artistic Director *Rod Dixon*

Founded 1968. National company touring 1–2 shows a year with a strong commitment to new work and new writers. Aimed at an audience of young people. Performances held in community and theatre venues. Recent productions: *Kaahini* Maya Chowdhry; *Door: This Life Was Given to Me* Madani Younis; *Where's Vietnam?* Alice Nutter; *Forgotten Things* Emma Adams. The company is particularly keen to enter into a dialogue with writers with regard to creating new work for young people. E-mail the artistic director for more information at the address above.

Red Shift Theatre Company

jane@redshifttheatreco.co.uk

www.redshifttheatreco.co.uk

Artistic Director *Jonathan Holloway*

Founded in 1982, formerly a small-scale touring company, Red Shift has recently stepped up to middle scale 'commercial' production. Interested in new plays on subjects and themes with broad audience appeal. Willing to read treatments but full scripts should be sent by invitation only.

Ridiculusmus

c/o BAC, Lavender Hill, London SW11 5TN

enquiries@ridiculusmus.com

www.ridiculusmus.com

Artistic Directors *Jon Haynes, David Woods*

Founded 1992. Touring company which plays to a wide range of audiences. Productions have included adaptations of *The Importance of Being Earnest*; *Three Men In a Boat*; *The Third*

Policeman; *At Swim Two Birds* and original work: *The Exhibitionists*; *Yes, Yes, Yes*; *Say Nothing* and *Ideas Men*. Most recent work: *Tough time, nice time*. Unsolicited scripts not welcome.

Royal Court Theatre

Sloane Square, London sw1w 8as
☎ 020 7565 5050
📠 020 7565 5002 (literary office)
www.royalcourttheatre.com
Literary Manager *Ruth Little*

The Royal Court is a leading international theatre producing up to 17 new plays each year in its 400-seat proscenium theatre and 80-seat studio. In 1956 its first director George Devine set out to find 'hard-hitting, uncompromising writers whose plays are stimulating, provocative and exciting'. This artistic policy helped transform post-war British theatre, with new plays by writers such as John Osborne, Arnold Wesker, John Arden, Samuel Beckett, Edward Bond and David Storey, through to Caryl Churchill, Jim Cartwright, Kevin Elyot and Timberlake Wertenbaker. Since 1994 it has produced a new generation of playwrights such as Joe Penhall, Rebecca Prichard, Sarah Kane, Jez Butterworth, Martin McDonagh, Mark Ravenhill, Ayub Khan-Din, Conor McPherson, Roy Williams and many other first-time writers. The Royal Court has programmes for young writers and international writers, and it is always searching for new plays and new playwrights.

Royal Exchange Theatre Company

St Ann's Square, Manchester m2 7dh
☎ 0161 615 6709   📠 0161 832 0881
jo.combes@royalexchange.co.uk
www.royalexchange.co.uk
Artistic Directors *Braham Murray, Greg Hersov, Sarah Frankcom*
Associate Director (Literary) *Jo Combes*

Founded 1976. The Royal Exchange has developed a new writing policy which it finds is attracting a younger audience to the theatre. Over the past 32 years, Manchester's Royal Exchange has produced over 100 new pieces of work by playwrights ranging from Ronald Harwood, Iain Heggie and Brad Fraser to Debbie Horsfield, Chloe Moss, Simon Stephens and Bridget O'Connor (recently appointed as the RET's writer-in-residence). 'We accept unsolicited work from writers based in the UK, and as we will often include a full script report, we limit our reading to one script per writer per year. If you would like your script returned at the end of the process, please include the appropriate postage. Please note that we do not accept musicals or electronic copies of scripts and our current turnaround time is four months. Send scripts to *Jo Combes*.'

Royal Lyceum Theatre Company

Grindlay Street, Edinburgh eh3 9ax
☎ 0131 248 4800   📠 0131 228 3955
www.lyceum.org.uk
Artistic Director *Mark Thomson*
Administration Manager *Ruth Butterworth*

Founded 1965. Repertory theatre which plays to a mixed urban Scottish audience. Produces classic, contemporary and new plays. Would like to stage more new plays, especially Scottish. No full-time literary staff to provide reports on submitted scripts.

Royal National Theatre

South Bank, London se1 9px
☎ 020 7452 3333   📠 020 7452 3344
www.nationaltheatre.org.uk
Associate Director (Literary) *Sebastian Born*
Deputy Literary Manager *Chris Campbell*
Literary Administrator *Sarah Clarke*

The majority of the National's new plays come about as a result of direct commission or from existing contacts with playwrights. There is no quota for new work, though many of the plays presented have been the work of living playwrights especially in the Cottesloe Theatre. Writers new to the theatre would need to be of exceptional talent to be successful with a script here, however the Royal National Theatre Studio helps a limited number of playwrights, through readings, workshops and discussions. Unsolicited scripts considered from the UK but not by e-mail (send s.a.e).

Royal Shakespeare Company

1 Earlham Street, London wc2h 9ll
☎ 020 7845 0515
Daniel.Usztan@rsc.org.uk
www.rsc.org.uk
Company Dramaturg *Jeanie O'Hare*
Literary Manager *Pippa Ellis*

The RSC is a classical theatre company based in Stratford-upon-Avon. The Company produces a core repertoire of Shakespeare alongside modern classics, new plays and the work of Shakespeare's contemporaries. The RSC works proactively with contemporary writers, encouraging them to write epic plays, seeking out the writers they wish to commission. They are unable to read unsolicited work but do monitor the work of emerging playwrights in production nationally and internationally. The Company is currently undergoing a four-year rebuilding programme transforming the Royal Shakespeare Theatre and the entire Waterside complex of studios, rehearsal rooms, actors' cottages and workshops. The theatre is being re-modelled to create a thrust stage within a 'one room' theatre, while Waterside will house state-of-the-art facilities for artists.

7:84 Theatre Company Scotland

Ceased trading 31 December 2008.

Shared Experience

13 Riverside House, 27/29 Vauxhall Grove,
London SW8 1SY
☎ 020 7587 1596   📠 020 7735 0374
admin@sharedexperience.org.uk
www.sharedexperience.org.uk
Joint Artistic Directors *Nancy Meckler, Polly Teale*

Founded 1975. Varied audience depending on
venue, since this is a touring company. Recent
productions have included: *Anna Karenina,
Mill on the Floss,* and *War and Peace* (1996; all
adapt. by Helen Edmundson); *Jane Eyre* (adapt.
Polly Teale); Euripides' *Orestes* and *War and
Peace* (2008; both adapt. by Helen Edmundson);
Kindertransport Diana Samuels; *A Doll's House*
(transl. Michael Meyer); *The Magic Toyshop*
(adapt. by Bryony Lavery); *The Clearing* and *Gone
to Earth* Helen Edmundson; *A Passage to India*
(adapt. by Martin Sherman); *After Mrs Rochester
and Brontë* Polly Teale. No unsolicited mss.
Primarily not a new writing company. New plays
always commissioned.

Sherman Cymru

Senghennydd Road, Cardiff CF24 4YE
☎ 029 2064 6901   📠 029 2064 6902
www.shermancymru.co.uk
Director *Chris Ricketts*

Originally founded in 1973, April 2007 saw a
merger between The Sherman Theatre Company
and Sgript Cymru to form a new company,
Sherman Cymru, with the aim of becoming a
powerhouse for drama, contemporary theatre and
new writing in Wales and beyond. With a strong
commitment to producing new work by emerging
and experienced writers, Sherman Cymru now
has a Literary Department which provides a
reading service, advice and support, mentorship
and script development for writers at every level.
The company offers bursaries, full commissions
and various one-off events and schemes to
promote and nurture excellence in new writing
for Welsh and Wales-based writers.

Show of Strength

74 Chessel Street, Bedminster, Bristol BS3 3DN
☎ 0117 902 0235
info@showofstrength.org.uk
www.showofstrength.org.uk
Creative Producer *Sheila Hannon*

Founded 1986. Plays to an informal, younger
than average audience. Aims to stage at least one
new play each season with a preference for work
from Bristol and the South West. Unable to read
unsolicted scripts. Working increasingly in non-
theatre venues as in 2008 performing *Trade It? –*
a commissioned work by ten writers – performed
on the streets of Bristol. Output: *The Wills Girls*
Amanda Whittington; *Lags* Ron Hutchinson; *So
Long Life* and *Nicholodeon* Peter Nichols. Also,
rehearsed readings of new work.

Soho Theatre Company

21 Dean Street, London W1D 3NE
☎ 020 7287 5060   📠 020 7287 5061
writers@sohotheatre.com
www.sohotheatre.com
Artistic Director *Lisa Goldman*

The company has an extensive development
programme for writers from all levels of
experience and various disciplines consisting
of a free script-reading service, 'Open Access
Workshops', Masterclasses, talks and events. Soho
Connect holds 'Taster Workshops' and longer
courses for promising playwrights aged 15–26.
The company produces and co-produces year
round. Recent productions include: *The Diver* by
Hideki Noda and Colin Teevan (co-production
with the Setagaya Public Theatre, Tokyo), *Piranha
Heights* by Philip Ridley, *A Couple of Poor-Polish
Speaking Romanians* by Dorota Maslowska,
Joe Guy by Roy Williams (co-production with
Tiata Fahodzi), *Pure Gold* Michael Bhim (co-
production with Talawa); *This Isn't Romance* by
In-Sook Chappell (2007 Verity Bargate Award
winner). Runs the **Verity Bargate Award**, a
biennial competition (see entry under *Prizes*).

Sphinx Theatre Company

25 Short Street, London SE1 8LJ
☎ 020 7401 9993
info@sphinxtheatre.co.uk
www.sphinxtheatre.co.uk
Artistic Director *Sue Parrish*

'Sphinx is a feminist theatre company that places
women at the centre of its artistic endeavour.'
Since 1973 the company has toured groundbreaking productions to small and mid-scale
venues. Synopses and ideas are welcome by e-mail
or post with s.a.e.

Talawa Theatre Company Ltd

53–55 East Road, London N1 6AH
☎ 020 7251 6644   📠 020 7251 5969
hq@talawa.com
www.talawa.com

Founded 1985. 'Aims to provide high quality
productions that reflect the significant creative
role that black theatre plays within the national
and international arena and also to enlarge
theatre audiences from the black community.'
Previous productions include *Pure Gold*; *High
Heel Parrotfish*; *Blues for Mister Charlie*; *The Key
Game.* Seeks to provide a platform for new work
from up and coming black British writers. Send a

copy of the script by post or e-mail. Runs a black script development project and black writers' group. Talawa is funded by Arts Council England, London.

Theatre Absolute

Institute for Creative Enterprise, Technology Park, Puma Way, Coventry CV1 2TT
☎ 024 7615 8340
info@theatreabsolute.co.uk
www.theatreabsolute.co.uk
Artistic Director/Writer *Chris O'Connell*
Producer *Julia Negus*

Founded 1992. An independent theatre company which commissions, produces and tours new plays based on a strong narrative text and aimed at audiences aged 15 and upwards. Productions include: *Car; Raw, Kid* (Street Trilogy); *cloud:burst; Hang Lenny Pope; Zero*.

Theatre Royal Plymouth & Drum Theatre Plymouth

Royal Parade, Plymouth PL1 2TR
☎ 01752 230347 ☏ 01752 230499
david.prescott@theatreroyal.com
www.theatreroyal.com
Artistic Director *Simon Stokes*
Artistic Associate *David Prescott*

Stages small, middle and large-scale drama and music theatre. Commissions and produces new plays. The theatre no longer accepts unsolicited playscripts but will consider plays through known channels, i.e. theatre practitioners, regional and national scriptwriters' groups and agents.

Theatre Royal Stratford East

Gerry Raffles Square, London E15 1BN
☎ 020 8279 1134 ☏ 020 8534 8381
theatreroyal@stratfordeast.com
www.stratfordeast.com
Contact *New Writing Manager*

Situated in the heart of the East End, the theatre caters for a very mixed audience, in terms of both culture and age range. Produces plays and musicals, youth theatre and local community plays/events, all of which is new work. Special interest in plays and musicals reflecting the culturally diverse communities of London and the UK. New initiatives in developing contemporary British musicals. No longer accepts unsolicited scripts but instead asks for a) synopsis of script, b) sample ten pages of writing, c) brief writer's biography. 'From this information we will decide whether or not to ask for full-length script. We only respond to projects of interest. If no reply within three months, assume that the proposal is not right for TRSE.'

Theatre Royal Windsor

Windsor SL4 1PS
☎ 01753 863444 ☏ 01753 831673
info@theatreroyalwindsor.co.uk
www.theatreroyalwindsor.co.uk
Executive Producer *Bill Kenwright*
Theatre Director *Angela Edwards*

Plays to a middle-class, West End-type audience. Produces thirteen plays a year and 'would be disappointed to do fewer than two new plays in a year; always hope to do half a dozen'. Modern classics, thrillers, comedy and farce. Only interested in scripts along these lines.

Theatre Workshop Edinburgh

34 Hamilton Place, Edinburgh EH3 5AX
☎ 0131 225 7942 ☏ 0131 220 0112
www.theatre-workshop.com
Artistic Director *Robert Rae*

First ever professional producing theatre to fully include disabled actors in all its productions. Plays to a young, broad-based audience with many pieces targeted towards particular groups or communities. Output has included *D.A.R.E.*; *Threepenny Opera; Black Sun Over Genoa* and Beckett's *Endgame*. Particularly interested in issues-based work for young people and minority groups. Frequently engages writers for collaborative work and devised projects. Commissions a significant amount of new writing for a wide range of contexts, from large-scale community plays to small-scale professional productions. Favours writers based in Scotland, producing material relevant to a contemporary Scottish audience.

The Torch Theatre

St Peter's Road, Milford Haven SA73 2BU
☎ 01646 694192 ☏ 01646 698919
info@torchtheatre.co.uk
www.torchtheatre.co.uk
Artistic Director *Peter Doran*

Founded 1977. Stages a mixed programme of in-house and middle-scale touring work. Unsolicited scripts will be read and guidance offered but production unlikely due to restricted funding. Please include s.a.e. for return of script and notes; mark clearly, 'FAO Peter Doran'. Torch Theatre Company productions include: *Dancing at Lughnasa; Neville's Island; The Woman in Black; Abigail's Party; Taking Steps; Blue Remembered Hills; A Prayer for Wings; Little Shop of Horrors; The Caretaker; One Flew Over the Cuckoo's Nest; One for the Road; Noises Off; Dead Funny* plus annual Christmas musicals.

Traverse Theatre

Cambridge Street, Edinburgh EH1 2ED
☎ 0131 228 3223 ☏ 0131 229 8443

louise.stephens@traverse.co.uk
www.traverse.co.uk
Artistic Director *Dominic Hill*
Literary Manager *Katherine Mendelsohn*
Literary Assistant *Louise Stephens*

The Traverse is Scotland's only new writing theatre, with a particular commitment to producing new Scottish plays. However, it also has a strong international programme of work in translation and visiting companies. Previous productions include *The People Next Door* Henry Adam; *Dark Earth* David Harrower; *15 Seconds* François Archambarlt, version by Isabel Wright; *Iron* Rona Munro; *Damascus* David Greig; *Gagarin Way* Gregory Burke; *Perfect Days* Liz Lochhead. Please address hard copies of unsolicited scripts to *Louise Stephens*, Literary Assistant.

Trestle Theatre Company

Trestle Arts Base, Russet Drive, St Albans AL4 0JQ
☎ 01727 850950   ℻ 01727 855558
admin@trestle.org.uk
www.trestle.org.uk
Artistic Director *Emily Gray*

Founded 1981. Some devised work, but increasing collaboration with writers to create new writing for touring visual/physical performance work nationally and internationally and also local community projects. No unsolicited scripts.

Tricycle Theatre

269 Kilburn High Road, London NW6 7JR
☎ 020 7372 6611   ℻ 020 7328 0795
www.tricycle.co.uk
Artistic Director *Nicolas Kent*

Founded 1980. Presenting an eclectic, culturally diverse programme that reflects the local area of Kilburn – particularly Black, Irish, Jewish, Asian and South African work. Known for Tribunal Plays, including *Guantanamo*, *Bloody Sunday*, *The Stephen Lawrence Inquiry* and *The Colour of Justice*, adapted from the enquiry transcripts by Richard Norton-Taylor and seen by more than 25 million people worldwide. Previous productions include: *Radio Gulf*; *Joe Turner's Come and Gone*, *The Piano Lesson* and *Two Trains Runnin'* all by August Wilson; and *Let There Be Love* by Kwame Kwei-Armah. New writing welcome. Looks for a strong narrative drive with popular appeal. Fee: £15 per script. Supplies written reader's report. Can only return scripts if s.a.e. is enclosed with original submission.

Unicorn Theatre

147 Tooley Street, London SE1 2HZ
☎ 020 7645 0500   ℻ 020 7645 0550
admin@unicorntheatre.com
www.unicorntheatre.com

Artistic Director *Tony Graham*
Associate Artistic Director *Rosamunde Hutt*
Associate Director (Literary) *Carl Miller*
Executive Director *Chris Moxon*

Founded 1947, resident at the Arts Theatre from 1967 to 1999 and now based at the new Unicorn Theatre with two performance spaces. Produces full-length professionally performed plays for children and adults. Recent work includes *DUCK!* by Philip Osment, *Twin Stars* by Mike Kenny, *The Flying Machine* by Phil Porter, *Red Fortress* by Carl Miller and a new version of *Sleeping Beauty* by Rosy Fordham. Does not produce unsolicited scripts but works with commissioned writers. Writers interested in working with the company should send an e-mail to artistic@unicorntheatre.com requesting further information.

Upstairs at the Gatehouse
▷ **Ovation Productions** under Film, TV and Radio Producers

Warehouse Theatre

Dingwall Road (adjacent to East Croydon Station), Croydon CR0 2NF
☎ 020 8681 1257   ℻ 020 8688 6699
info@warehousetheatre.co.uk
www.warehousetheatre.co.uk
Artistic Director *Ted Craig*

South London's new writing theatre seats 90–100 and produces up to six new plays a year. Also co-produces with, and hosts, selected touring companies who share the theatre's commitment to new work. The theatre continues to build upon a tradition of discovering and nurturing new writers through the **International Playwriting Festival** (see entry under *Festivals*). Also runs a vigorous writers' workshop and hosts youth theatre workshops and Saturday morning children's theatre. Scripts will get a guaranteed response if submitted through the International Playwriting Festival.

Watford Palace Theatre

Clarendon Road, Watford WD17 1JZ
☎ 01923 235455   ℻ 01923 819664
enquiries@watfordpalacetheatre.co.uk
www.watfordpalacetheatre.co.uk
Contact *Assistant Administrator*

An important part of artistic policy is developing new work suitable for this 600-seat proscenium arch theatre, which generally means the writer has some professional production experience. 'We run an open writers' workshop for local playwrights. We regret we are unable to offer a script reading and reporting service for unsolicited scripts and are unable to return unsolicited mss.'

West Yorkshire Playhouse

Playhouse Square, Leeds LS2 7UP
℡ 0113 213 7800   🖷 0113 213 7250
alex.chisholm@wyp.org.uk
www.wyp.org.uk
Associate Director *Alex Chisholm*

Committed to working with new writing originating from the Yorkshire and Humberside region. New writing from outside the region is programmed usually where writer or company is already known to the theatre. For more information, call 0113 213 7286 or e-mail to address above. Scripts should be submitted with s.a.e. for return. 'Not all submitted scripts will be read. You are strongly advised to check guidelines on the website *before* submitting.'

Whirligig Theatre

14 Belvedere Drive, Wimbledon, London SW19 7BY
℡ 020 8947 1732   🖷 020 8879 7648
whirligig-theatre@virgin.net
Contact *David Wood*

Occasional productions and tours to major theatre venues, usually a musical for primary school audiences and weekend family groups. Interested in scripts which exploit the theatrical nature of children's tastes. Previous productions: *The See-Saw Tree*; *The Selfish Shellfish*; *The Gingerbread Man*; *The Old Man of Lochnagar*; *The Ideal Gnome Expedition*; *Save the Human*; *Dreams of Anne Frank*; *Babe, the Sheep-Pig*.

White Bear Theatre Club

138 Kennington Park Road, London SE11 4DJ
www.whitebeartheatre.co.uk

Administration: 3 Dante Road, Kennington, London SE11 4RB ℡ 020 7793 9193
Artistic Director *Michael Kingsbury*

Founded 1988. Output primarily new work for an audience aged 20–35. Unsolicited scripts welcome, particularly new work with a keen eye on contemporary issues, though not agitprop. *Absolution* by Robert Sherwood was nominated by the Writers' Guild for 'Best Fringe Play' and *Spin* by the same author was the *Time Out* Critics' Choice in 2000. The theatre received the *Time Out* award for Best Fringe Venue in 2001 and a Peter Brook award for best up-and-coming venue. In 2004 *Round the Horne ... Revisited* transferred to The Venue, Leicester Square and ran for 14 months.

Windsor Theatre Royal
▷ Theatre Royal Windsor

The Young Vic

66 The Cut, London SE1 8LZ
℡ 020 7922 2800   🖷 020 7922 2801
info@youngvic.org
www.youngvic.org
Artistic Director *David Lan*
Executive Director *Kevin Fitzmaurice*
Dramaturg *Ruth Little*

Founded 1970. Re-opened in 2006 having been rebuilt at a cost of £13 million, The Young Vic is a theatre for everyone. Produces revivals of classics – old and new – as well as new plays and annual events that embrace both young people and adults.

UK and Irish Writers' Courses

England

Berkshire

University of Reading
School of Continuing Education, London Road, Reading RG1 5AQ
☎ 0118 378 8347
Cont-Ed@reading.ac.uk
www.reading.ac.uk/ContEd

An expanding and popular programme of creative writing courses in all genres, including *Life-Writing*; *Making a Start in Poetry*; *Getting Started in Creative Writing*; *Writing Fiction*; *Planning Your Novel*; *Poetry Workshop*; *Publishing Your Writing*; *Scriptwriting*; *Playwriting*; *Article Writing*; *Comedy Writing* and *Writing for Radio*. Additional one-off event includes readings by students of their work and various Saturday workshops. Tutors include novelist Leslie Wilson and poets Jane Draycott, Susan Utting, David Grubb and Paul Bavister. Fees vary depending on the length of course. Concessions available.

The Write Coach
2 Rowan Close, Wokingham RG41 4BH
☎ 0118 978 4904
enquiries@thewritecoach.co.uk
www.thewritecoach.co.uk
Contact *Bekki Hill*

Workshops and one-to-one coaching to assist both professional and aspiring writers to become more successful, break through their blocks, build confidence, increase motivation and expand creativity, define direction and find time and space to write. Free initial consultation available. See website for details.

Buckinghamshire

Missenden Abbey
Evreham Adult Learning Centre, Swallow Street, Iver SL0 0HS
☎ 01296 383582
dcevreham@buckscc.gov.uk
www.arca.uk.net/missendenabbey

Residential and non-residential weekend and summer school workshops include: *Writing for Children*; *Short Story Writing*; *Life Writing*; *Writing to Publish*.

National Film & Television School
Beaconsfield Studios, Station Road, Beaconsfield HP9 1LG
☎ 01494 731425   🖷 01494 674042
info@nfts.co.uk
www.nfts.co.uk

Two-year programme covering all aspects of screenwriting from the development of ideas through to final production. Studying alongside other filmmaking students allows writers to have their work tested in workshops and productions. Unlike a Screenwriting MA set in an academic institution, this course is set in a working studio and emphasizes practical filmmaking and contact with industry personnel through seminars and pitching sessions. The course requires and encourages a high level of dedication and a prolific output. Year One deals with the basic principles of storytelling, the craft of screenwriting for film and television and the collaborative nature of production, and includes several short writing assignments. Year Two focuses on longer writing assignments, with students writing a feature script and TV project as well as their MA dissertation. Opportunities may be available to write short fiction or animation scripts for production. In partnership with **The Script Factory**, The NFTS also offers an 18-month, part-time Diploma in *Script Development*.

Cambridgeshire

National Extension College
Michael Young Centre, Purbeck Road, Cambridge CB2 8HN
☎ 0800 389 2839   🖷 01223 400321
info@nec.ac.uk
www.nec.ac.uk
Contact *Customer Relations Adviser*

The National Extension College (NEC) was set up as a not for profit organization to help people of all ages fit learning into their lives. Supports more than 10,000 learners each year on over 100 distance learning courses. 'If you want to discover or develop writing skills our creative courses help to develop your practical skills or begin to make money for you.' Home study courses include *Essential Editing*; *Creative Writing* and *Writing Short Stories*; *Writing for Money*; *Writing Humour* (new) and *Songwriting* (new). 'You can enrol at any time, study at your own pace, in your own time with the support of a specialist tutor for up to two years.' The NEC is fully accredited by the Open Distance Learning Quality Council. Contact the NEC for a free copy of the *Guide to Courses* or visit the website.

University of Cambridge Institute of Continuing Education

Madingley Hall, Madingley, Cambridge CB23 8AQ
☎ 01954 280399 ℻ 01954 280200
registration@cont-ed.cam.ac.uk
www.cont-ed.cam.ac.uk

A wide range of weekend and five-day creative writing courses for adults is offered by the University at the Institute of Continuing Education's residential headquarters at Madingley Hall. Evening courses are also available in Cambridgeshire and surrounding areas. Details of all courses can be found on the website or phone for a brochure.

Cheshire

Burton Manor

The Village, Burton, Neston CH64 5SJ
☎ 0151 336 5172 ℻ 0151 336 6586
enquiry@burtonmanor.com
www.burtonmanor.com

Wide variety of short courses, residential and non-residential, on writing and literature, including *Creative Writing Workshop* and *Writing for Pleasure*. Full details in brochure.

University of Chester

Parkgate Road, Chester CH1 4BJ
☎ 01244 392715 ℻ 01244 511330
a.chantler@chester.ac.uk
www.chester.ac.uk
Contact *Dr Ashley Chantler*

The Department of English offers two programmes: *Creative Writing* (full-time Combined Honours degree over three years) and *Creative Writing* (full-time MA over one year). See the website for details.

Cornwall

University College Falmouth

Woodlane, Falmouth TR11 4RH
☎ 01326 211077
www.falmouth.ac.uk
www.bloc-online.com (magazine)
Contact *Admissions*

MA *Professional Writing*. An intensive vocational writing programme developing skills in fiction, magazine journalism/features, screenwriting. Students work on an extended writing project and form links with other postgraduate programmes such as television production and broadcast journalism.

Cumbria

Higham Hall College

Bassenthwaite Lake, Cockermouth CA13 9SH
☎ 01768 776276 ℻ 01768 776013
admin@highamhall.com
www.highamhall.com

Winter and summer residential courses. Programme includes *Creative Writing*; *Writing for Radio and TV*. Brochure available.

Derbyshire

Real Writers

PO Box 170, Chesterfield S40 1FE
☎ 01246 520834 ℻ 01246 520834
info@real-writers.com
www.real-writers.com

Appraisal service with personal tuition by post or e-mail from working writers. Send s.a.e. or see website for details.

Swanwick – The Writers' Summer School

The Hayes Conference Centre, Nr Alfreton
☎ 01292 442786
fionamcfadzean@hotmail.com
www.wss.org.uk
Secretary *Fiona McFadzean*

Offers six days of speakers, courses, workshops, informal talks, discussion groups, and evening entertainment. Competitions for free places. All levels welcome. Held mid-August from Saturday to Friday morning. Cost (2009) £336–£478 per person, all inclusive. Full details on website or s.a.e. to the Secretary at 40 Pemberton Valley, Ayr KA7 4UB (telephone and e-mail contact as above).

University of Derby

Admissions Office, Kedleston Road, Derby
DE22 1GB

☎ 01332 622236   ℻ 01332 622754
J.Bains@derby.ac.uk
www.derby.ac.uk

Subject Leader *Professor Carl Tighe*

With upwards of 300 students, this is one of the oldest *Creative Writing* operations at degree level in the UK. The subject offers a BA Hons *Creative Writing* with particular strengths in Short Stories, Writing and New Technologies and Storytelling. All teaching is done through practical workshops led by established and experienced writers. In the final year students produce an independent research project on a subject of their choice. The subject also contributes to the BA Hons Combined Studies.

Devon

Arvon Foundation (Devon)

See entry under Greater London.

Exeter Phoenix

Bradninch Place, Gandy Street, Exeter EX4 3LS
☎ 01392 667080   ℻ 01392 667599
www.exeterphoenix.org.uk

Regular literature events, focusing on performances by living poets and writers, often linked to wider programmes. Tutors in a wide range of writing skills run classes and workshops. 'Uncut Poets' group holds monthly meetings.

Fire in the Head Courses & Writers' Services

PO Box 17, Yelverton PL20 6YF
www.fire-in-the-head.co.uk
Contact *Roselle Angwin*

Comprehensive year-round writing programme in poetry, fiction (short and novel-writing), reflective writing and creative development. Also available, online and snailmail correspondence courses and mentoring.

University College Falmouth – Dartington Campus

Dartington Hall, Totnes TQ9 6EJ
☎ 01803 861600   ℻ 01803 861666
registry@dartington.ac.uk
www.falmouth.ac.uk

Writing Admissions Tutor *Jerome Fletcher* (*j.fletcher@dartington.ac.uk*)

BA (Hons) *Writing* or *Writing (Contemporary Practices)* or *Writing (Scripted Media)*; *Textual Practices* minor award; MA *Performance Writing*: exploratory approaches to writing as it relates to performance, visual arts, sound arts, poetics, narrative and contemporary culture.

Encourages the interdisciplinary, with minor awards and electives at BA level in choreography, music, theatre, digital arts, visual performance, community arts. Contact the Writing Admissions Tutor. (As of 2010 the College will relocate to Falmouth.)

University of Exeter

Department of English, Queen's Building, Queen's Drive, Exeter EX4 4QH
☎ 01392 269305   ℻ 01392 264361
soe.pgoffice@ex.ac.uk
www.sall.ex.ac.uk/english

Offers BA (Hons) in *English* with 2nd and 3rd-year options in creative writing, poetry, short fiction, screenwriting and creative non-fiction. MA *Creative Writing*: poetry, novella, screenwriting, life writing and novel. Also, PhD *Creative Writing*.

Dorset

Bournemouth University

The Media School, Weymouth House, Talbot Campus, Fern Barrow, Poole BH12 5BB
☎ 01202 965553   ℻ 01202 965099

Programme Administrator *Janice Jeffrey*

Three-year, full-time BA(Hons) course in *Scriptwriting for Film and Television*.

Essex

National Council for the Training of Journalists

The New Granary, Station Road, Newport, Saffron Walden CB11 3PL
☎ 01799 544014   ℻ 01799 544015
info@nctj.com
www.nctj.com

Journalism courses, both full-time and via distance learning. For full details, please visit the website.

Gloucestershire

Chrysalis – The Poet In You

5 Oxford Terrace, Uplands, Stroud GL5 1TW
☎ 01453 759436/020 7794 8880
jay@ramsay3892.fsnet.co.uk
www.lotusfoundation.org.uk
Contact *Jay Ramsay, BAHons (Oxon), PGDip, UKCP member*

Two-part correspondence course combining poetry and personal development. Offers postal courses, day and weekend workshops (including `The Sacred Space of the Word'), an on-going poetry group based in London and

Gloucestershire and individual therapy related to the participant's creative process. The Chrysalis course itself consists of Part 1, `for those who feel drawn to reading more poetry as well as wanting to start to write their own', and Part 2, `a more in-depth course designed for those who are already writing and who want to go more deeply into its process and technique, its background and cultural history'. Both courses combine course notes with in-depth individual correspondence and feedback. Editing and information about publication also provided. Brochure and workshop dates available from the address above.

Wye Valley Arts Centre

Hephzibah Gallery, Llandogo NP25 4TW

☎ 01594 530214/01291 689463

info@wyearts.co.uk

www.wyearts.co.uk

Courses ('held at our beautiful gallery in Llandogo in the Wye Valley') include *Writing and Poetry* and *Writing Workshop*. All styles and abilities.

Hampshire

Highbury College, Portsmouth

Dovercourt Road, Cosham, Portsmouth PO6 2SA

☎ 023 9238 3131   ☒ 023 9237 8382

media.journalism@highbury.ac.uk

www.highbury.ac.uk

Course Administrator *Christine Stallard (023 9231 3287)*

The 20-week fast-track postgraduate courses include: *Pre-entry Magazine Journalism*; *Pre-entry Newspaper Journalism*; and Diploma in *Broadcasting Journalism*, run under the auspices of the Periodicals Training Council, the National Council for Training of Journalists and the Broadcast Journalism Training Council respectively.

University of Portsmouth

School of Creative Arts, Film and Media, St George's Building, 141 High Street, Portsmouth PO1 2HY

☎ 023 9284 5138   ☒ 023 9284 5152

creative@port.ac.uk

www.port.ac.uk/departments/academic/scafm

www.hackwriters.com (International Writers Magazine)

BA Course Leader *Alison Habens*

MA Course Leader *Sam North*

PhD Queries *Steven O'Brien*

Offers creative writing and combined creative writing degrees across a wide range of undergraduate programmes and in all genres. BA, MA and PhD study is available. Staff include a number of national and/or international

award-winning writers, short story writers/ novelists, playwrights, screenwriters, new-media writers and poets. The MA annually publishes *Borderlines* – an anthology of best writing from the Creative Writing Department.

University of Winchester

Winchester SO22 4NR

☎ 01962 841515   ☒ 01962 842280

course.enquiries@winchester.ac.uk

www.winchester.ac.uk

Contact *Enquiries (01962 827234)*

The University of Winchester offers a range of courses for budding writers or English enthusiasts both at undergraduate and postgraduate level. Undergraduate Programmes: BA *American Literature*; BA *Creative Writing*; BA *English*; BA *English Literature and Language*; BA *English with American Literature*; BA *Journalism*. Postgraduate: MA *Creative and Critical Writing*; BA *English: Contemporary Literature*; MA *Journalism*; MA *Writing for Children*.

Winchester Writers' Conference, Festival, Bookfair

University of Winchester, Winchester SO22 4NR

☎ 01962 827238

barbara.large@winchester.ac.uk

www.writersconference.co.uk

The 29th Winchester Writers' Conference, Festival and Bookfair will be held in 2009 over the weekend of 3rd–5th July, preceded by the Week-long Writing Workshops (29th June–3rd July). Children's author Michael Morpurgo will give the Plenary Address. This international conference offers all writers the opportunity to harness their creative ideas and to develop technical skills under the guidance and instruction of 65 professional writers, literary agents, commissioning editors and book industry specialists during mini courses, workshops, seminars, lectures and 500 one-to-one appointments. 'Enter the fifteen writing competitions attached to the conference, even if you can't attend.' Book online or download the application form and post. For further information, contact *Barbara Large, MBE*, Conference Director, Winchester Writers' Conference, Faculty of Research and Knowledge Transfer at the University of Winchester address above.

Hertfordshire

Liberato Breakaway Writing Courses

16 Middle King, Braintree CM7 3XY

☎ 01376 551379

liberato@talktalk.net

www.liberato.co.uk

Contact *Maureen Blundell*

Specializes in beginner fiction writers with weekend and week-long courses. Emphasis on individual writing with written ms critiques and one-to-one sessions. Weekends held at Polstead in Suffolk throughout the year. Greek weeks in June and/or September on the Saronic island of Agistri, near Aegina – a relaxing, informal holiday with one-to-one sessions and editing service. B&B accommodation. Also offers postal and e-mail ms critiques on all fiction/autobiography.

Isle of Wight

Annual Writers' Writing Courses & Workshops

F * F Productions, 39 Ranelagh Road, Sandown PO36 8NT
℡ 01983 407772   🖷 01983 407772
felicity@writeplot.co.uk
www.learnwriting.co.uk
Contact *Felicity Fair Thompson*

Regular residential weeks and weekends on creative writing for beginners and experienced writers – individual advice and workshops. Also offers postal ms critiques and one-to-one advice on fiction and film scripts

Kent

North West Kent College

Oakfield Lane, Dartford DA1 2JT
℡ 01322 629436/0800 074 1447 (Freephone helpline)
🖷 01322 629468
www.nwkcollege.ac.uk
Contact *Neil Nixon, Head of School, Media & Communications*

Two-year, full-time course that explores writing from a number of angles, teaching essential skills, market and academic aspects of the subject. Successful students progress to work or the University of Greenwich, the latter option allowing them to gain a BA(Hons) in *Humanities*, BA (Hons) in *Creative Industries* or BA (Hons) in *Media Writing* from a further year of study. Staff include scriptwriters, novelists and a book publisher. Students compile a portfolio in the final year.

University of Kent

The Registry, Canterbury CT2 7NZ
℡ 01227 827272
information@kent.ac.uk
www.kent.ac.uk/studying
www.kent.ac.uk/english
Contact *Dr Emma Bainbridge (01227 823402)*

Diploma and degree level course in English and American Literature with Creative Writing in a vibrant department. Undergraduate certificate courses in *Creative Writing, Practical* and *Imaginative Writing*. Also Combined Studies courses in *English and Creative Writing*.

Lancashire

Alston Hall College

Alston Lane, Longridge, Preston PR3 3BP
℡ 01772 784661   🖷 01772 785835
alston.hall@ed.lancscc.gov.uk
www.alstonhall.com

Holds regular day and residential *Literature* and *Creative Writing* workshops. Full colour brochure available.

Edge Hill University

St Helen's Road, Ormskirk L39 4QP
℡ 01695 575171
shepparr@edgehill.ac.uk
www.edgehill.ac.uk
Contact *Professor Robert Sheppard*

Offers a two-year, part-time MA in *Writing Studies*. Combines advanced-level writers' workshops with closely related courses in the poetics of writing and contemporary writing in English. There is also provision for MPhil and PhD-level research in writing and poetics. A full range of creative writing courses is available at undergraduate level, in poetry, fiction and scriptwriting which may be taken as part of a modular BA (contact: *Daniele Pantano*, daniele. pantano@edgehill.ac.uk).

Lancaster University

English & Creative Writing, County College, Bailrigg, Lancaster LA1 4YD
℡ 01524 594169   🖷 01524 594247
l.kellett@lancaster.ac.uk
www.lancs.ac.uk/fass/english/postgrad/index.htm
Contact *Lyn Kellett*

Offers practical graduate and undergraduate courses in writing fiction, poetry and scripts. All based on group workshops – students' work-in-progress is circulated and discussed. Distance learning MA now available. Graduates include Andrew Miller, Justin Hill, Monique Roffey, Alison MacLeod, Jacob Polley.

Leicestershire

Writing School Leicester

c/o Leicester Adult Education College, 2 Wellington Street, Leicester LE1 6HL
℡ 0116 233 4343   🖷 0116 233 4344

val.moore@writingschoolleicester.co.uk
www.writingschoolleicester.co.uk
Contact *Valerie Moore*

Offers a wide range of quality part-time creative writing and journalism courses throughout the year in Leicester and elsewhere. The writing school is staffed by professional writers and offers a mix of critical workshops, one-day and term-length courses and short craft modules. Supports writers through to publication and has strong links with local media.

Lincolnshire

Creuse Writers' Workshop & Retreat

45 Browning Drive, Lincoln LN2 4HF
☎ 01522 880565   ⊟ 01522 880565
info@cresuewritersworkshopandretreat.com
www.cresuewritersworkshopandretreat.com

A selection of home study is available with personal tuition from professional published writers. One-to-one tutoring. Visit the website for further information.

Greater London

Arvon Foundation

National Administration: 2nd Floor, 42A Buckingham Palace Road, London SW1W 0RE
☎ 020 7931 7611
www.arvonfoundation.org
Devon: Totleigh Barton, Sheepwash, Beaworthy EX21 5NS ☎ 01409 231338 ⊟ 01409 231144
totleighbarton@arvonfoundation.org
Yorkshire: Lumb Bank, The Ted Hughes Arvon Centre, Heptonstall, Hebden Bridge HX7 6DF
☎ 01422 843714 ⊟ 01422 843714
lumbbank@arvonfoundation.org
Inverness-shire: Moniack Mhor, Teavarran, Kiltarlity, Beauly IV4 7HT
☎ 01463 741675
moniackmhor@arvonfoundation.org
Shropshire: The John Osborne Centre, The Hurst, Clunton, Craven Arms SY7 0JA ☎ 01588 640658
⊟ 01588 640509 thehurst@arvonfoundation.org
Joint Presidents *Terry Hands, Sir Robin Chichester-Clark*
Chairman *Nigel Pantling*
National Director *Ariane Koek*

Founded 1968. Offers people of any age (over 16) and any background the opportunity to live and work with professional writers. Four-and-a-half-day residential courses are held throughout the year at Arvon's four centres, covering poetry, fiction, drama, writing for children, songwriting and the performing arts. Bursaries towards the cost of course fees are available for those on low incomes, the unemployed, students and pensioners. Runs a biennial international poetry competition (see entry under *Prizes*).

Birkbeck College, University of London

School of English and Humanities, Malet Street, London WC1E 7HX
☎ 020 7631 6000
a.taylor@english.bbk.ac.uk
www.bbk.ac.uk/eh/eng
Contact *Anne Marie Taylor*

Taught by published writers, Birkbeck offers an MA course in *Fiction Writing*. The course will extend and cultivate existing writing skills, help develop writing to a professional level and is supported by masterclasses and readings from visiting professionals. All classes are held in the evening. Study part-time over two years or full-time over one year. Applications must be supported by a portfolio of creative writing. Download an application form from the website.

The Central School of Speech and Drama

Embassy Theatre, Eton Avenue, London NW3 3HY
☎ 020 7722 8183   ⊟ 020 7722 4132
enquiries@cssd.ac.uk
Contact *Academic Registry*

MA in *Advanced Theatre Practice*. One-year, full-time course aimed at providing a grounding in principal areas of professional theatre practice – *Writing for Performance, Dramaturgy, Directing, Performing, Puppetry* and *Object Theatre*, with an emphasis on collaboration between the various strands. *Writing for Performance* students have the opportunity of working with companies to create new and innovative work for the theatre. Also MA in *Writing for Stage & Broadcast Media*. Focuses on texts for the theatre, television, cinema and radio. Prospectus available.

City Lit

Keeley Street, Covent Garden, London WC2B 4BA
☎ 020 7492 2652   ⊟ 020 7492 2735 (Humanities)
humanities@citylit.ac.uk
www.citylit.ac.uk

The writing school offers a wide range of courses, with workshops catering for every level of student – from beginners to previously published authors – in a variety of disciplines. Classes include *Ways into Creative Writing; Writing for Children; Playwriting (stages 1 & 2); Writing Short Stories; Screenwriting (Stages 1 & 2); Comedy Writing; Writing Novels; Advanced Critical Workshop.* Various lengths of course are available, and the department offers information and advice all year round, with the exception of August.

City University

Northampton Square, London EC1V 0HB
☎ 020 7040 8268   ⊟ 020 7040 8256

ell@city.ac.uk
www.city.ac.uk/cae/cfa/writing_journalism
www.city.ac.uk/journalism

The Courses for Adults evening programme includes: *Certificate in Novel Writing* (three-term course); *Certificate in Short Story Writing* (three-term course); *Writing for Children*; *Writing Poetry*; *Writing Comedy*; *Writing Situation Comedy*; *Food and Drink Journalism*; *Feature Writing*; *Narrative Non-Fiction*; *News Writing*; *Writing about Travel*; *Writing for the Theatre*; *Writing Freelance Articles.* Contact the Education and Lifelong Learning Department at the address above or e-mail ell@city.ac.uk (☏ 020 7040 8268 ☏ 020 7040 8256).

The University's Journalism Department offers an MA in *Creative Writing (Non-fiction)* which focuses on writing and research skills including how to identify a subject, how to use archives, exploring existing genres and structuring material to complete a book. The course offers a taught component, one-to-one tutorials and opportunities to hear guest speakers. The course advisory board, made up of leading non-fiction writers, provides students with expert advice and direct contact with the publishing industry. Contact Programme Director *Julie Wheelwright*, Postgraduate Admissions, Journalism at the address above or e-mail julie.wheelwright.1@city.ac.uk (☏ 020 7040 8221).

The Complete Creative Writing Course at the Groucho Club

☏ 020 7249 3711   ☏ 020 7683 8141
maggie@writingcourses.org.uk
www.writingcourses.org.uk
Contact *Maggie Hamand*

Courses of six three-hour sessions held at the Groucho Club in London's Soho and the New Cavendish Club at Marble Arch, starting in January, April and September, on Mondays, Tuesdays or Saturdays. Beginners, intermediate, advanced and screenwriting courses available. Courses include stimulating exercises, discussion and weekly homework. Fee: £250–£285 for whole course.

Fiction Writing Workshops and Tutorials

5 Queen Elizabeth Close, London N16 0HL
☏ 020 8809 4725
henrietta@writtenwords.net
www.WrittenWords.net
Contact *Henrietta Soames*

Workshops for fiction. Writers led by author Henrietta Soames. 'Lively discussions, stimulating exercises, valuable feedback.' Fiction tutorials/mentoring for writers who require concentrated attention and ongoing, committed support. Short stories/novels, full-length mss welcomed. 'Very reasonable fees.'

London College of Communication

Elephant & Castle, London SE1 6SB
☏ 020 7514 2105
shortcourses@lcc.arts.ac.uk
www.lcc.arts.ac.uk/training

Intensive courses in writing, editing and journalism. Courses include: *News Writing*; *Feature Writing*; *Brush Up Your Grammar*; *Proofreading*; *Introduction to Copy-editing*; *Effective Copywriting*; *Speechwriting*; *Writing Direct Mail*; *Sub-editing.* Prospectus available; telephone or access the website.

London School of Journalism

126 Shirland Road, London W9 2BT
☏ 020 7289 7777   ☏ 020 7432 8141
info@lsjournalism.com
www.home-study.com
www.lsj.org
Contact *Student Administration Office*

Distance learning and attendance courses with an individual and personal approach. Distance learning options include *Novel Writing*; *Short Story Writing*; *Writing for Children*; *Poetry*; *Freelance Journalism* (with *Sport, Travel or Music* as add-ons); *Media Law*; *Improve Your English*; *Cartooning*; *Writing a Personal History*; *English Literature and English History.* Fees range from £220 to £395, dependent upon course. Postgraduate Courses are taught in London (three months, full-time; six months, part-time; nine months, evening classes) as are short courses and evening classes (*Creative Writing*; *News and Features*; *Web Journalism*; *Womens' Magazines*; *Media Law*; *Freelance Writing* and *Travel Writing*). An Online Postgraduate Course (four modules, 24–48 months' duration) is also available.

Middlesex University

Trent Park, Bramley Road, London N14 4YZ
☏ 020 8411 5000   ☏ 020 8411 6652
admissions@mdx.ac.uk
www.mdx.ac.uk

The UK's longest established writing degree offers a Single or Combined Honours programme in *Creative and Media Writing* (full or part-time). This modular programme gives an opportunity to explore creative non-fiction, poetry, prose fiction and dramatic writing for a wide range of genres and audiences. Option for work experience in the media and publishing industries. Contact Admissions or *James Martin Charlton* (j.charlton@mdx.ac.uk).

MA *Writing* (full-time, part-time) concentrates on fiction writing. Includes writing workshops; critical seminars; lectures and workshops from established writers; introduction to agents and publishers. Contact *David Rain* (d.rain@mdx.ac.uk).

Also offers research degrees M.Phil/PhD in *Creative Writing*. Contact: Maggie Butt (m.butt@mdx.ac.uk). The University has a thriving Writing Centre running an annual literary festival, weekly talks, community projects and writers in residence. 'All of our writing courses take place in the setting of a beautiful country park, within easy reach of central London.'

PMA Media Training

The PMA Centre for Media Excellence, 7a Bayham Street, London NW1 0EY
📞 01480 300653/020 7278 0606
📠 01480 496022
training@pma-group.com
www.pma-group.com
www.becomeajournalist.co.uk
Contacts *Vicky Chandler, Melanie Gilbert*

Fast-track nine-week Postgraduate Diploma and more than 150 one and two-day editorial, PR, design and publishing courses held in central London. Intensive workshops run by top journalists, print and online, designers and PR professionals. All short courses are designed to lead to the PMA Gold Standard qualification. Apple, Adobe, Quark and the Periodical Publishers' Association accreditation. NCTJ, NUJ and Communicators in Business recommended. Edexcel Exam Centre. Special discount for freelancers. See website for dates and fees.

The Poetry School
⊡ entry under **Organizations of Interest to Poets**

Roehampton University

Roehampton Lane, London SW15 5PU
📞 020 8392 3232   📠 020 8392 3470
enquiries@roehampton.ac.uk
www.roehampton.ac.uk/postgraduate-courses/creative-and-professional-writing/index.html
Senior Enquiries Officer *Dan Jenkins*

The MA/MRes *Creative and Professional Writing* includes modules in: Creative reading and viewing: research and theory for writers; Research and theory for writers; Fiction: how to grow stories; Writing for a child audience; Poetry: form and innovation; Creative non-fiction and journalism; Knowing and subverting the rules: screenwriting for independent film; Screenwriting for independent film.

Royal Court Theatre – Young Writers Programme and Critical Mass

Royal Court Theatre, Sloane Square, London SW1W 8AS
📞 020 7565 5050

ywp@royalcourttheatre.com
www.royalcourttheatre.com
Associate Director *Ola Animashawun*

Young Writers Programme: an 11-week *Introduction to Playwriting* course led by the Royal Court Theatre's resident playwriting tutor, Leo Butler. The group is open to anyone aged 18–25 and costs £100 or £50 if you are a concession. To apply: submit a play to and the most promising writers will be selected to join the group. There are three intakes a year, with the following scripts deadlines for applicants: 1st December for the January/March group, 1st March for the April/June group and 1st September for the October/December group.

Critical Mass: 'The Royal Court is committed to generating a critical mass of black, Asian and minority ethnic playwrights.' Each autumn they run an 11-week *Introduction to Playwriting* course for ethnic minority writers aged 18+ (no upper age limit). The next Critical Mass playwriting group takes place in autumn 2009.

The Script Factory

The Square, 61 Frith Street, London W1D 3JL
📞 020 7323 1414
general@scriptfactory.co.uk
www.scriptfactory.co.uk

Established in 1996, The Script Factory is a screenwriter and script development organization set up to bridge the gap between writers and the industry, and to promote excellence in screenwriting. Specializing in training, screenings, masterclasses and various services, The Script Factory operates throughout the UK and internationally. For the best source of information on upcoming activities, courses and latest news, and to join the free mailing list, check out the website

Soho Theatre Company
⊡ entry under **Theatre Producers**

Travellers' Tales

92 Hillfield Road, London NW6 1QA
info@travellerstales.org
www.travellerstales.org

'Travellers' Tales is Britain's foremost provider of travel writing and travel photography training for non-professionals.' Offers practical courses with top professionals and editorial consultancy for writers. Tutors include award-winning authors William Dalrymple and Colin Thubron, travel editors Simon Calder (*The Independent*), Lyn Hughes (*Wanderlust*) and Jonathan Lorie (*Traveller*). Courses include beginners' weekends, UK masterclasses and creative holidays overseas.

University of Westminster

School of Media, Arts and Design, Harrow Campus, Watford Road, Harrow HA1 3TP
☎ 020 7911 5903   ⊠ 020 7911 5955
harrow-admissions@wmin.ac.uk
www.wmin.ac.uk

Courses include part-time MAs available in *Journalism Studies*; *Film and Television Studies*; *Screenwriting and Producing*; *Public Communication and Public Relations*; *Communication and Communication Policy*.

Greater Manchester

Manchester Metropolitan University – The Writing School

Department of English, Geoffrey Manton Building, Rosamond Street West, off Oxford Road, Manchester M15 6LL
☎ 0161 247 1732/1   ⊠ 0161 247 6345
a.biswell@mmu.ac.uk (campus course)
h.beck@mmu.ac.uk (online course)
www.hlss.mmu.ac.uk/english/writingschool/
Course Convenor *James Draper*

The Writing School offers three 'routes' for students to follow: *Novel*, *Poetry* and *Children's Writing*. A key feature of the programme is regular readings, lectures, workshops and masterclasses by writers, publishers, producers, booksellers, librarians and agents. Tutors include Simon Armitage, Heather Beck, Andrew Biswell, Carol Ann Duffy, Paul Magrs, Michael Symmons Roberts, Jackie Roy and Nick Royle. The Novel and Poetry routes are available online as well as through the campus).

University of Manchester

Centre for New Writing, School of Arts, Histories and Cultures, Mansfield Cooper Building, Oxford Road, Manchester M13 9PL
☎ 0161 306 1259   ⊠ 0161 275 5987
englishpg@manchester.ac.uk
www.arts.manchester.ac.uk/newwriting
Course Directors *John McAuliffe, Ian McGuire*

Offers a one-year MA in *Creative Writing* (the novel, short story and poetry) and a PhD.

University of Salford

Postgraduate Admissions, School of Media, Music & Performance, Adelphi Building, Peru Street, Salford M3 6EQ
☎ 0161 295 6026   ⊠ 0161 295 6023
r.humphrey@salford.ac.uk
www.smmp.salford.ac.uk

MA in *Television and Radio Scriptwriting*. Two-year, part-time course taught by professional writers and producers. Also offers masterclasses

with leading figures in the radio and television industry.

The Writers Bureau

Sevendale House, 7 Dale Street, Manchester M1 1JB
☎ 0161 228 2362   ⊠ 0161 236 9440
studentservices@writersbureau.com
www.writersbureau.com

Comprehensive home-study writing course with personal tuition service from professional writers. Fiction, non-fiction, articles, short stories, novels, TV, radio and drama all covered in detail. Trial period, guarantee and no time limits. Writing for children, using the Internet to sell your writing and biographies, memoirs and family history courses also available. ODLQC accredited. Quote Ref. EH09. Free enquiry line: 0800 389 7360.

The Writers Bureau College of Journalism

Address etc. as The Writers Bureau above

Home-study course covering all aspects of journalism. Real-life assignments assessed by qualified tutors with the emphasis on getting into print. Comprises 28 modules and three supplements. Ref: EHJ09. Free enquiry line: 0800 389 7360.

The Writers College

Address etc. as The Writers Bureau above

The Art of Writing Poetry Course from The Writers Bureau sister college. A home-study course with a more `recreational' emphasis. The 60,000-word course has 17 modules and lets you complete twelve written assignments for tutorial evaluation. Quote Ref. EHP09. Free enquiry line: 0800 389 7360.

Merseyside

University of Liverpool

Continuing Education, 126 Mount Pleasant, Liverpool L69 3GR
☎ 0151 794 6900/6952 (24 hours)
⊠ 0151 794 2544
conted@liverpool.ac.uk
www.liv.ac.uk/conted
Course Organizer *Dr John Redmond*

Courses include: *Exploring Writing*; *Scriptwriting* (for radio and TV, and for film and TV); *Comedy Writing*; *Writing for Women by Women*; *Writing for the Stage*; *Writing Science Fiction*; *Writing for Children*; *Writing Poetry*. Most courses take place once weekly (evening or daytime) and there is also a series of Saturday courses on aspects of writing. CE courses offer University credits which can be accumulated for personal development purposes, or towards an award such as the Certificate in Higher Education (*Creative Writing*)

– 120 credits. For most courses no previous knowledge is required. Fee concessions available if you receive certain benefits or are retired. Full programme at www.liv.ac.uk/conted (or free printed prospectus on request).

Norfolk

University of East Anglia
School of Literature and Creative Writing, Norwich NR4 7TJ
☎ 01603 592154   📠 01603 250599
pgt.hum.admiss@uea.ac.uk
www.uea.ac.uk/creativewriting
Contact *Jack Camplin, Postgraduate Admissions*

UEA has a history of concern with contemporary literary culture. Among its programmes is the MA in *Creative Writing* (founded by Angus Wilson and Malcolm Bradbury in 1970/71). The course has three parallel entry points: prose fiction, poetry and scriptwriting. A series of weekly workshops provide intensive examination of students' own work, and also draw on aspects of teaching in nineteenth and twentieth century literature, literary theory and film and cultural studies.

Northamptonshire

Knuston Hall Residential College for Adult Education
Irchester, Wellingborough NN29 7EU
☎ 01933 312104   📠 01933 357596
enquiries@knustonhall.org.uk
www.knustonhall.org.uk

Writing courses have included: *Articles for Magazines*; *My Life in Poems and Stories*; *Writer's Workshops* and *Creative Writing*.

Northumberland

Mavis Maureen Raper – Community Creative Writing
'Sea Winds', 2 St Helens Terrace, Berwick-upon-Tweed TD15 1RJ
☎ 01289 305213
mavismaureen@btinternet.com
Author/Tutor/Moderator *Maureen Raper, MBE*

The courses, which are held at the Community Centre College, Berwick-upon-Tweed as well as by distance learning for those unable to travel into class, include: *Creative Writing for Beginners*; *Creative Writing for Intermediates*; *Writing for Radio and Television*. Classes on *Writing Crime and Mystery* and *Writing Romantic Fiction* are held at Berwick Library.

Nottinghamshire

The Nottingham Trent University
College of Arts, Humanities and Education, Clifton Campus, Nottingham NG11 8NS
☎ 0115 848 8977
acc.postgrad@ntu.ac.uk
www.ntu.ac.uk/postgrad/

MA in *Creative Writing*. Hands-on and workshop-based, the course concentrates primarily on the practice and production of writing. A choice of options from *Fiction*, *Poetry*, *Creative Non-Fiction*, *New Media Writing*, *Children's Writing* and *Scriptwriting*. Assignments and a dissertation, but no formal exams. Visiting professors: David Almond and Michael Eaton, MBE. There is a full programme of visiting speakers. Study either full-time or part-time.

Oxfordshire

University of Oxford Department for Continuing Education
Rewley House, 1 Wellington Square, Oxford OX1 2JA
☎ 01865 280973   📠 01865 270309
pp@conted.ox.ac.uk
www.conted.ox.ac.uk
MSt Administrator *Claire Kelly*

Creative writing classes held during the autumn and spring terms. There are also one-week summer school courses, a two-year part-time Diploma and a two-year part-time MSt in *Creative Writing*. Early booking is advised as courses are very popular.

Shropshire

Arvon Foundation (Shropshire)
See entry under Greater London.

Somerset

Ammerdown Conference and Retreat Centre
Ammerdown Park, Radstock, Bath BA3 5SW
☎ 01761 433709   📠 01761 433094
centre@ammerdown.org
www.ammerdown.org
www.ammerdown-conference.co.uk

Courses have included *Creative Journaling* with Ann Beazer, *Writing for Well Being* with Sue Ashby and *Writing the Spirit* with Judy Clinton. En suite residential facilities. Brochure available or full details on the website.

Bath Spa University

Newton Park, Newton St. Loe, Bath BA2 9BN
☎ 01225 875573　📠 01225 875503
r.kerridge@bathspa.ac.uk
www.bathspa.ac.uk
Contact *Elizabeth Rack (01225 875603)*

MA in *Creative Writing*. A course designed to help writers bring their work to a more publishable, performable and broadcastable level held in weekend workshops. Teaching is by published writers in the novel, poetry, short stories and scriptwriting and includes visits by literary agents and publishers. In recent years, several students from this course have won the Manchester Book Prize and been longlisted for the Man Booker, Orange and Costa book prizes, and have had work produced on BBC Radio. MA in *Writing for Young People*. A course for writers for children of all ages, from the picture-book age through to adolescent and 'crossover' writing, which aims at markets among adults as well as young people. An experienced teaching team helps and encourages students to create a significant body of writing, with practical plans for its place in the real world of publishing.

Dillington House

Ilminster TA19 9DT
☎ 01460 52427　📠 01460 52433
dillington@somerset.gov.uk
www.dillington.com

Dillington House is one of the finest historic houses in Somerset. Provides short residential courses across a wide range of subjects, including writing and literary appreciation. Full details of the programme are available in the free brochure and on the website.

Institute of Copywriting

Overbrook Business Centre, Poolbridge Road, Blackford, Wedmore BS28 4PA
☎ 0800 781 1715　📠 01934 713492
copy@inst.org
www.inst.org/copy

Comprehensive copywriting home-study course, including advice on becoming a self-employed copywriter. Each student has a personal tutor who is an experienced copywriter and who provides detailed feedback on the student's assignments. Other courses include: Diploma in *Creative Writing* and Diploma in *Screenwriting*.

University of Bristol

Department of English, 3/5 Woodland Road, Bristol BS8 1TB
☎ 0117 954 6969
www.bris.ac.uk/english
Contact *Lifelong Learning Organizer*

A variety of short courses and day schools designed for both experienced writers and for absolute beginners, including courses exploring writing for therapeutic purposes. Also offers *Diploma in Creative Writing*. Detailed brochure available.

Staffordshire

Keele University

The Centre for Continuing and Professional Education, Keele University, Newcastle under Lyme ST5 5BG
☎ 01782 733436

Daytime, evening and weekend courses on literature and creative writing. The 2008/09 programme included courses on novel writing, writers workshops and sessions on poetry writing.

Surrey

The Guildford Institute

Guildford Institute, Ward Street, Guildford GU1 4LH
☎ 01483 562142　📠 01483 451034
info@guildford-institute.org.uk
www.guildford-institute.org.uk

The Guildford Institute is the venue for various creative writing courses. It also hosts Guildford Writers, a writers' circle meeting on alternate Tuesday evenings from 7.30 pm to 9.30 pm. Members of the group are writing short stories, poetry or novels, and bring their work to read aloud to the group for other members' advice, constructive criticism and general comments. New members are always welcome.

Royal Holloway University of London

Department of Drama and Theatre, Egham Hill, Egham TW20 0EX
☎ 01784 443922　📠 01784 431018
drama@rhul.ac.uk
www.rhul.ac.uk/Drama
www.rhul.ac.uk/english
Contact *Dan Rebellato*

Three-year BA courses in *Drama and Creative Writing* or *English and Creative Writing*, during which students progressively specialize in playwriting, poetry or fiction. Playwriting can be studied as part of the BA *Drama and Creative Writing* degree and poetic practice is an option in the *English* programme. Playwriting can be studied at postgraduate level in the MA in *Theatre* and there are MAs in *Poetic Practice* and *Screenwriting*. 'We can also supervise PhDs that are partly creative writing.'

Sussex

Earnley Concourse

Earnley, Chichester PO20 7JN
☎ 01243 670392   ▣ 01243 670832
info@earnley.co.uk
www.earnley.co.uk

Offers a range of residential and non-residential courses throughout the year. Previous programme has included *Creative Writing: Getting Started*. Brochure available.

The University of Chichester

Bishop Otter Campus, College Lane, Chichester PO19 6PE
☎ 01243 816000   ▣ 01243 816080
S.Norgate@chi.ac.uk
www.chi.ac.uk
MA Programme Coordinator (Creative Writing) *Stephanie Norgate (01243 816296)*

MA in *Creative Writing*, both full and part-time. Also, MPhil/PhD in *Creative Writing, Creative/Critical*. Contact: *Dr Bill Gray* (b.gray@chi.ac.uk).

University of Sussex

Centre for Continuing Education, The Sussex Institute, Essex House, Brighton BN1 9QQ
☎ 01273 606755
cce@sussex.ac.uk
www.sussex.ac.uk/cce
Lecturer in Creative Studies *Dr Sue Roe*

MA in *Creative Writing and Authorship*: a unique opportunity for graduate writers to develop writing practice in the context of the study of cultural and aesthetic issues of authorship, past and present, in workshops, masterclasses and seminars. One year, full-time, two years, part-time. Convenor: *Sue Roe*. Certificate in *Creative Writing*: short fiction, novel and poetry for imaginative writers. Two-years, part-time. Convenor: *Mark Slater*. Both courses include CCE Agents' and Publishers' Day.

Tyne & Wear

University of Sunderland

Centre for Lifelong Learning, 2nd Floor, Bedson Building, Kings Road, Newcastle upon Tyne NE1 7RU
☎ 0191 515 2800   ▣ 0191 515 2890
lifelong.learning@sunderland.ac.uk
cll.sunderland.ac.uk

Courses, held in Newcastle and across the North East, include: *Creative Writing: Feature Writing*; *Screenwriting*; *Writing and Illustrating for Children* and *Writing From the Inside Out*, a workshop for women. Contact the Centre for Lifelong Learning.

Warwickshire

University of Warwick

Open Studies, Centre for Lifelong Learning, Coventry CV4 7AL
☎ 024 7657 3739
k.rainsley@warwick.ac.uk
openstudies@warwick.ac.uk
www.warwick.ac.uk.cll/OpenStudies

Creative writing courses held at the University or in regional centres. Subjects include: *Creative Writing*; *Publishing and Editing*; *Creative Writing for All*; *Journalism*; *Script Writing*; *Life Writing* and *Writing for Pleasure*. One-year certificates in *Creative Writing* and *Journalism* are available; individual modules can be taken.

West Midlands

National Academy of Writing (based at Birmingham City University)

Birmingham City University, School of English, City North Campus, Perry Barr, Birmingham B42 2SU
☎ 0121 331 5471   ▣ 0121 331 6692
samantha.malkin@bcu.ac.uk
www.thenationalacademyofwriting.org.uk

Diploma in *Writing*. This course has a strong professional focus and aims to produce working writers. Modules are available in a number of disciplines, including fiction, life writing and screenwriting. Teaching includes taught modules and supervision by Birmingham City University staff, all of whom are established writers. There are also regular masterclasses with Academy patrons. Applications are welcome from all, regardless of previous academic experience. Admission is based on talent and commitment to a career in writing.

Starz! – Film and Theatre Performing Arts

Christmas House, Chester Road, Castle Bromwich, Solihull B36 0ET
☎ 0121 749 7147
www.gavinprime.mac.com
Creative Director *Gavin Prime*

Teaching disabled, disadvantaged and mainstream children. The only totally free theatre and film school of this type in England. Students range in age from eight to 23 and work in all ways connected to performing arts. This includes writing and showing them the structure of the business and how to try and get work accepted.

University of Birmingham
Department of Drama and Theatre Arts, The Old Library, 998 Bristol Road, Selly Oak, Birmingham B29 6LQ
☎ 0121 414 5998
drama@contacts.bham.ac.uk
www.drama.bham.ac.uk
Course Convenor *Steve Waters (s.waters@bham.ac.uk)*

The MPhil in *Playwriting Studies*, established by playwright David Edgar in 1989, was the UK's first postgraduate course in playwriting. An intensive course which encourages students to think critically about dramatic writing, assisting them to put these insights into practice in their own plays.

Wiltshire

Marlborough College Summer School
Marlborough SN8 1PA
☎ 01672 892388   🖷 01672 892476
admin@mcsummerschool.org.uk
www.mcsummerschool.org.uk

Summer School with literature and creative writing included in its programme. Caters for residential and day students. Brochure available giving full details and prices.

Urchfont Manor College
Urchfont, Devizes SN10 4RG
☎ 01380 840495   🖷 01380 840005
urchfontmanor@wiltshire.gov.uk
www.urchfontmanor.co.uk

Short courses – days, residential weeks and weekends – offered in a varied programme which includes literature and creative writing. 'Beautiful location; historic environment; delicious home cooking.' Send for a brochure.

Yorkshire

Arvon Foundation (Yorkshire)
See entry under Greater London.

Leeds Metropolitan University
School of Film, Television & Performing Arts, Electric Press Building, Leeds LS2 3AD
☎ 0113 812 3860
screenwriting@leedsmet.ac.uk
www.leedsmet.ac.uk
Administrator *Chris Pugh*

Offers a Diploma/MA in *Screenwriting (Fiction).*

Open College of the Arts
Unit 1B, Redbrook Business Park, Wilthorpe Road, Barnsley S75 1JN
☎ 0800 731 2116   🖷 01226 730838

open.arts@ukonline.co.uk
www.oca-uk.com

The OCA correspondence course, *Starting to Write*, offers help and stimulus from experienced writers/tutors. Emphasis is on personal development rather than commercial genre. Subsequent levels available include specialist poetry, fiction, autobiographical and children's writing courses. OCA writing courses are accredited by the Buckinghamshire New University. Prospectus and Guide to Courses available on request.

Sheffield Hallam University
Faculty of Development & Society, City Campus, Sheffield Hallam University, Sheffield S1 1WB
☎ 0114 225 5555   🖷 0114 225 2167
dsmarketing@shu.ac.uk
www.shu.ac.uk

Offers MA in *Writing* (one year to complete seminar courses, two years to full MA, full-time; also part-time) and MA *English Studies* (one year full-time, also part-time).

University of Hull
School of Arts and New Media, Scarborough Campus, Filey Road, Scarborough YO11 3AZ
☎ 01723 362392
a.head@hull.ac.uk
www.hull.ac.uk

BA Single Honours in *Theatre and Performance Studies* incorporates opportunities in writing and other media across each level of the programme. The campus hosts the annual National Student Drama Festival which includes the International Student Playscript Competition (details from The National Information Centre for Student Drama; nsdf@hull.ac.uk).

University of Leeds – Lifelong Learning Centre
Leeds LS2 9JT
☎ 0113 343 3212
part-time@leeds.ac.uk
www.leeds.ac.uk/lifelonglearningcentre
Contacts *Rebecca O'Rourke, Pat Owens*

The Lifelong Learning Centre provides accredited part-time courses in creative writing on the Leeds campus during the day and evening. Introductory workshop courses are offered each year with follow-on and advanced courses providing the opportunity to focus on particular aspects of creative writing. These more specialized courses change from year to year and can include poetry, fiction, script and lifewriting or themed work. These courses form part of the Lifelong Learning Centre's Open Studies programme, which is designed to provide adults with opportunities for continuing education and to create routes into higher education. The Centre also offers a dedicated Part-time Joint Honours Degree where students can study Creative Writing in combination with a choice of other subjects including Arabic and Middle Eastern Studies, Business Studies and Theology and Religious Studies. Tutors include: Ray French, Rebecca O'Rourke, Rommi Smith, Beccy Stirrup, Adam Strickson.

University of Leeds

School of Performance and Cultural Industries, University of Leeds, Leeds LS2 9JT
☎ 0113 343 8735   📠 0113 343 8711
enquiries-pci@leeds.ac.uk
www.leeds.ac.uk/paci
Contact *Garry Lyons or Jane Richardson (Admissions Secretary)*

MA in *Writing for Performance and Publication*. This postgraduate programme (launched in September 2006) is particularly relevant to aspiring writers with professional ambitions in the areas of theatre, film, television and radio drama, while also allowing opportunities to work in prose fiction and other culturally significant genres. Studies are offered over one-year full-time and two-years part-time and the tutors are established authors in their chosen fields. The course director is award winning playwright and screenwriter Garry Lyons (*The Worst Witch, Leah's Trials, Britain's First Suicide Bombers*). The MA is one of a new portfolio of post-graduate degrees to be offered by Leeds University's School of Performance and Cultural Industries and is housed in a brand new theatre complex – stage@ leeds – at the heart of the University's city centre campus. Students are not only given the opportunity to work on their own writing projects but are also encouraged to collaborate with colleagues on the other MAs in Performance, Choreography and Cultural Entrepreneurship, with a view to seeing their work produced and staged.

University of Sheffield

Institute for Lifelong Learning, 196–198 West Street, Sheffield S1 4ET
☎ 0114 222 7000   📠 0114 222 7001
www.shef.ac.uk/till

Certificate in *Creative Writing* (Degree Level 1) is part-time and open to all. Courses in poetry, journalism, scriptwriting, short story writing, travel writing. Brochures and information available from the address above.

Ireland

Dingle Writing Courses

Ballintlea, Ventry, Co. Kerry
☎ 00 353 66 915 9815
info@dinglewritingcourses.ie
www.dinglewritingcourses.ie
Directors *Abigail Joffe, Nicholas McLachlan*

An autumn programme of weekend courses for beginners and experienced writers alike. Tutored by professional writers the courses include poetry, fiction, starting to write, writing for children and writing for theatre as well as special themed courses. Recent tutors have included Anne Enright, Maurice Riordan, Claire Kilroy, Carlo Gébler, Paula Meehan, Kate Thompson, Andrew O'Hagan, Mary O'Malley, Marina Carr and Niall Williams. Also organizes tailor-made courses for schools, writers' groups or students on a *Creative Writing* programme.

INKwell Writers' Workshops

The Old Post Office, Kilmacanogue, Co. Wicklow
☎ 00 353 1 276 5921/00 353 87 2835382
info@inkwellwriters.ie
www.inkwellwriters.ie
Contact *Vanessa O'Loughlin*

Offers a series of one-day (Saturday) intensive fiction writing workshops with bestselling Irish writers, held at the Fitzpatrick's Castle Hotel, Killiney, Co. Dublin. 'Designed to inspire and guide new and accomplished writers, giving them invaluable access to authors at the top of their field.' February to May 2009 workshops included: *Start Writing*: Julie Parsons and Sarah Webb; *Womens' Fiction*: Sinead Moriarty and Sarah Webb; *Short Stories*: Eilis ni Dhuibhne and Martina Devlin; *Memoir:Tell your own Story*: Ferdia MacAnna and Martina Devlin; *Web Wise*: *writing for the Web*: Beth Morrissey; *Commercial Literary Fiction*: Kate Holmquist and Ivy Banister; *Writing the Breakout Book*: Vanessa O'Loughlin and Sarah Webb; *Getting Published*: Paula Campbell of Poolbeg Ireland, Ciara Considine of Hachette Ireland, Sheila Crowley of Curtis Brown, London and Patricia O'Reilly. September 5th: *Start Writing* (speakers to be confirmed); October 3rd: *Another World: Fantasy Fiction*: Jonathan Stroud.

Queen's University of Belfast

School of Education, Belfast BT7 1LN
☎ 028 9097 3323   📠 028 9097 1084
openlearning.education@qub.ac.uk
www.qub.ac.uk/edu

Courses have included *Creative Writing* and *Writing for Profit and Pleasure*.

University of Dublin (Trinity College)

Graduate Studies Office, Arts Building, Trinity College, Dublin 2
☎ 00 353 1 896 1166
gradinfo@tcd.ie
www.tcd.ie/graduate_studies
Contact *Admissions*

Offers an MPhil *Creative Writing* course. A one-year, full-time course intended for students who are seriously committed to writing or prospective authors.

The Writer's Academy

Carrig-on-Bannow, Co. Wexford
☎ 00 353 51 561789
thewritersacademy@eircom.net
www.thewritersacademy.net

Distance learning freelance article writing and short story writing courses, critique service and marketing advice. Personal tuition plus expert information and support for writers of all abilities.

Scotland

Arvon Foundation (Inverness-shire)

See entry under Greater London.

University of Dundee

Continuing Education, Nethergate, Dundee DD1 4HN
☎ 01382 384128
www.dundee.ac.uk/learning/conted

A number of daytime and evening short courses are available through the University's Continuing Education Department. Telephone for further details or check the website.

University of Edinburgh

Office of Lifelong Learning/IALS, 11 Buccleuch Place, Edinburgh EH8 9LW
☎ 0131 650 4400   📠 0131 667 6097
oll@ed.ac.uk
www.ed.ac.uk/openstudies

Creative writing courses run throughout the year. Full details available on the website.

University of Glasgow

Department of Adult and Continuing Education, 11 Eldon Street, Glasgow G3 6NH
☎ 0141 330 1835   📠 0141 330 1821
dace-query@educ.gla.ac.uk
www.glasgow.ac.uk/dace

Runs writers' workshops and courses at all levels; all friendly and informal. Daytime and evening meetings. Tutors are all experienced published writers. Call 0141 330 1829 for course brochure.

University of St Andrews

School of English, The University, St Andrews KY16 9AR
☎ 01334 462666   📠 01334 462655
english@st-andrews.ac.uk
www.st-andrews.ac.uk

Offers postgraduate study in *Creative Writing*. Modules in either Writing Fiction or Writing Poetry. Students submit a dissertation of original writing – prose fiction of 15,000 words or a collection of around 30 short poems. Taught by John Burnside, Kathleen Jamie, Meaghan Delahunt and Don Paterson. Also offers PhD in *Creative Writing*.

Wales

Aberystwyth University

Department of English and Creative Writing, Hugh Owen Building, Aberystwyth SY23 3DY
☎ 01970 622534/5   📠 01970 622530
www.aber.ac.uk/english
Secretary *Mrs June Baxter*
Administrator *Ms Julie Roberts*

BA in *English and Creative Writing*, a three-year course whose creative modules are taught by practising writers: novelist and poet Professor Jem Poster; poet Dr Tiffany Atkinson; poet and novelist Dr Matthew Francis; poet and life-writer Dr Kelly Grovier; poet Dr Damian Walford Davies; and poet Dr Richard Marggraf Turley. Also offers a taught MA in *Creative Writing* with modules in writing poetry, writing fiction, research for writers, and writing and publication; and a PhD in *Creative Writing*.

Tŷ Newydd Writers' Centre

Llanystumdwy, Cricieth LL52 0LW
☎ 01766 522811   📠 01766 523095
post@tynewydd.org
www.tynewydd.org

Residential writers' centre set up by the Taliesin Trust with the support of the **Arts Council of Wales** to encourage and promote writing in both English and Welsh. Most courses run from Monday evening to Saturday morning. Each course has two tutors and a maximum of 16 participants. A wide range of courses for all levels of experience. Early booking essential. Fees start at £215 for weekends and £420 for week-long courses, all inclusive. People on low incomes may be eligible for a grant or bursary. Course leaflet available. (See also *Organizations of Interest to Poets*.)

University of Glamorgan

Department of English, Treforest, Pontypridd
CF37 1DL
☎ 0800 716925
www.glam.ac.uk

MPhil in *Writing*: a two-year part-time
Masters degree for writers of fiction and poets.
Established 1993. Contact *Professor Tony Curtis*
at the Faculty of Humanities and Social Sciences
(tcurtis@glam.ac.uk).

BA in *Creative and Professional Writing*:
a three-year course (5–6 years part-time) for
undergraduates.

MA in *Scriptwriting (Theatre, Film, TV or
Radio)*: a two-year part-time Masters degree
for scriptwriters (held at the Cardiff School for
Creative and Cultural Industries). Contact: *Wyn
Mason* (wmason@glam.ac.uk).

University of Wales, Bangor

www.bangor.ac.uk

The School of English (College Road, Bangor
LL57 2DG ☎ 01248 382102 elso29@bangor.ac.uk
postgrad-english@bangor.ac.uk) offers MPhil/
PhD *Creative and Critical Writing*; Diploma/
MA *Creative Studies (Creative Writing)*; BA
(Hons) *English with Creative Writing*; BA (Hons)
English with Songwriting; BA (Hons) *English with
Journalism*.

The Centre for Creative and Performing
Arts offers *Creative Writing* courses from
undergraduate level to MA and PhD.

The School of Creative Studies and Media
(Creative Studies and Creative Industries

Building, John Phillips Hall, College Road, Bangor
LL57 2DG ☎ 01248 383215) offers PhD *Creative
and Critical Writing*; MPhil *Creative Writing*;
MA *Creative Writing*; Diploma *Creative Writing*;
BA *Creative Writing and Communications*; BA
Creative Studies; BA *Music and Creative Writing*;
also, specialist Creative Writing short-courses.

The National Institute for Excellence in the
Creative Industries (NIECI) and the International
Centre for Creative Writing Research (ICCWR)
offer additional key support for these Creative
Writing Programmes. Contact: *Dr Nathan
Abrams*, Director of Graduate Studies
(n.abrams@bangor.ac.uk) or *Dr Iwan Davies*,
Administrator (iwan.davies@bangor.ac.uk;
☎ 01248 388560).

Writers' Holiday at Caerleon

School Bungalow, Church Road, Pontnewydd,
Cwmbran NP44 1AT
☎ 01633 489438
writersholiday@lineone.net
www.writersholiday.net
Contact *Anne Hobbs*

Annual six-day comprehensive conference
including 12 courses for writers of all standards
held in the summer at the University of Wales'
Caerleon Campus. Courses, lectures, concert and
excursion all included in the fee. Private, single
and en-suite, full board accommodation. Courses
have included *Writing for Publication*; *Writing
Poetry*; *Writing Romantic Fiction* and *Writing for
the Radio*.

US Writers' Courses

Alaska

University of Alaska, Fairbanks

English Dept., University of Alaska, Fairbanks, PO Box 755720, Fairbanks AK 99775–5720

☎ 001 907 474 7193

www.alaska.edu/english/

Contact *Gerri Brightwell (ffgab1@uaf.edu)*

MFA in *Creative Writing* (fiction, non-fiction, poetry and dramatic writing). Offers a three-year programme with teaching assistantships available that cover tuition and provide a stipend. A combined degree is also available: MFA/MA. The creative writing programme hosts the Midnight Sun Visiting Writer Series that brings in writers from across North America to read and offer masterclasses to students. Deadline for applications: 1st February. For application information, see the department website.

Arizona

University of Arizona

MFA Program in Creative Writing, Dept. of English, Modern Languages 445, University of Arizona, PO Box 210067, Tucson AZ 85721–0067

☎ 001 520 621 3880

mcooksey@email.arizona.edu

cwp.web.arizona.edu

Established in 1974, the MFA in *Creative Writing* is a two-year programme with concentrations in Poetry, Fiction and Creative Non-Fiction. 'It boasts a faculty of 14 distinguished writers who share an uncommon commitment to teaching and supporting their students in the completion of publishable manuscripts.'

Arkansas

University of Arkansas

Programs in Creative Writing & Translation, Dept. of English, University of Arkansas, 333 Kimpel Hall, Fayetteville AR 72701

☎ 001 479 575 4301   🖷 001 479 575 5919

mfa@uark.edu

www.uark.edu/depts/english/PCWT.html

MFA in *Creative Writing*, offering small, intensive workshops and innovative classes in fiction and poetry.

California

American Film Institute

AFI Conservatory, 2021 N. Western Avenue, Los Angeles CA 90027

☎ 001 323 856 7740   🖷 001 323 467 4578

admissions@afi.com

www.afi.com

Contact *Admissions Manager*

Screenwriting at AFI focuses on narrative storytelling. In their First Year, Screenwriting Fellows are immersed in the production process in order to learn how screenplays are visualized. Initially writing short screenplays – one of which will be the basis for a first year production – writers collaborate with Producing and Directing Fellows to see their work move from page to screen. The remainder of the first year is devoted to the completion of a feature-length screenplay. Second Year Fellows may write a second feature-length screenplay, or develop materials for television, including biopics, television movies and spec scripts for sitcoms and one-hour dramas. They also have the opportunity to work closely with other disciplines by writing a Second Year thesis script.

California Institute of the Arts

Office of Admissions, MFA Writing Program, School of Critical Studies, California Institute of the Arts, 24700 McBean Pa, Valencia CA 91355–2340

☎ 001 661 255 1050

writing@calarts.edu

www.calarts.edu/~writing

Director, Writing Program *Brighde Mullins*

The two-year MFA *Writing Program* is a flexible programme giving writers the opportunity to work in the areas of cultural commentary, fiction, criticism, writing for performance, poetry, playwriting and writing for interactive media. Although the MFA Writing Program covers traditional forms of literary and scholarly writing in depth, students are encouraged to work across genres and often work in various multi-media forms in addition to their literary production.

Chapman University

Graduate Admissions Office, Chapman University, One University Drive, Orange CA 92866

☎ 001 714 997 6711

www.chapman.edu

Contact *Graduate Admissions*

The programme is a two-year Master of Fine Arts (MFA) degree in *Creative Writing* intended for graduate students who wish to specialize in writing literary fiction. There is an emphasis on fiction but secondary studies include screenwriting, playwriting and poetry.

Saint Mary's College of California

MFA Program in Creative Writing, Saint Mary's College of California, PO Box 4686, Moraga CA 94575–4686

☎ 001 925 631 4762

writers@stmarys-ca.edu

www.stmarys-ca.edu

Program Coordinator *Thomas Cooney*

Two-year MFA course in *Creative Writing* with Fiction, Non-Fiction or Poetry. The core of the programme is the Writing Workshop which provides the opportunity for students to work with established writers.

San Francisco State University

Creative Writing Department, College of Humanities, San Francisco State University, 1600 Holloway Avenue, San Francisco CA 94132

☎ 001 415 338 1891

cwriting@sfsu.edu

www.sfsu.edu/~cwriting

Offers BA and MA in *English:* with a concentration on *Creative Writing*; and MFA in *Creative Writing*. Undergraduate classes include feature writing, short story writing, the craft of poetry and the craft of playwriting while graduate classes include advanced story writing, advanced poetry writing, advanced playwriting, experimental fiction, playwright's theatre workshop and workshops in fiction, poetry and playwriting.

University of Southern California (USC)

School of Cinematic Arts, Attn: Writing for Screen & Television Program, USC, University Park, 900 W. 34th Street, Los Angeles CA 90089

☎ 001 213 740 3303

writing@cinema.usc.edu

www-cntv.usc.edu

Offers a Bachelor of Fine Arts or Master of Fine Arts degree in *Writing for Screen and Television.* There is a strong international presence, having accepted students from over 44 countries.

Colorado

Colorado State University

Creative Writing Program, 1773 Campus Delivery, Colorado State University, 359 Eddy Hall, Fort Collins CO 80523–1773

☎ 001 970 491 2403

english@lamar.colostate.edu

www.colostate.edu/Depts/English/programs/mfa.htm

Contact *Assistant to the Director of Creative Writing or Director of Creative Writing*

Three-year Master of Fine Arts (MFA) programme in creative writing with concentrations in fiction or poetry. The programme offers a balance of intimate and intensive writing and translation. Course work culminates in a thesis – a collection of poetry or short stories or a novel. Students have the opportunity to teach introductory creative-writing courses, intern with literary journals including the *Colorado Review* and be a part of a thriving community.

District of Columbia

American University

Dept. of Literature, American University, 237 Battelle-Tompkins, 4400 Massachusetts Avenue, N.W., Washington DC 20016–8047

☎ 001 202 885 2971

www.american.edu/cas/lit/mfa-lit.cfm

Contact *Jenny Napolitano (Graduate Programs Assistant)*

Established in 1980, the MFA in *Creative Writing* is a 48-semester-hour programme allowing students to concentrate in Poetry, Fiction or Creative Non-Fiction. The programme is allied with the University's MA in Literature which offers numerous courses and master's-level seminars enabling creative writing students to explore both classic texts and contemporary literature.

Florida

Florida International University

English Dept., Florida International University, Biscayne Bay Campus, 3000 NE 151st Street, North Miami FL 33181

☎ 001 305 919 5857

crwriting@fiu.edu
www.fiu.edu/crwriting
Director *Les Standiford*

The MFA Program in *Creative Writing* includes writing workshop, literature, form and theory and thesis (most students complete the course in about three years; completion of study within eight years is required). There is no language requirement. Graduate workshops include short fiction, the novel, popular fiction, screenwriting, creative non-fiction and poetry. Great emphasis is placed upon preparation and completion of a book-length thesis. Admission is based primarily on the strength of the applicant's submitted writing sample. Application deadline: January 15.

University of Florida

MFA@FLA Program in Creative Writing, University of Florida, 4008 Turlington Hall, PO Box 117310, Gainesville FL 32611–7310
☎ 001 352 392 6650   🖷 001 352 392 0860
crw@english.ufl.edu
www.english.ufl.edu/crw
Co-Directors, Program in Creative Writing
David Leavitt, Padgett Powell

MFA@FLA, the graduate program in *Creative Writing* at the University of Florida, is one of the oldest writing programs in the United States, begun in 1948. 'We require an equal interest in writing and in reading literature. We don't believe in any particular school of writing; we have no wish to foster or found one. All the members of our faculty – four fiction writers and four poets – live in Gainesville and are tenured. We put out *Subtropics*, a new and important literary magazine.' In the autumn of 2008 the programme changed from a two-year to a three-year format; all students will receive full tuition waivers and teaching assistantships.

University of West Florida

MA in Creative Writing, Dept. of English, University of West Florida, 11000 University Parkway, Building 50, Pensacola FL 32514
jfink@uwf.edu
www.uwf.edu
Director of Creative Writing *Jonathan Fink*

The MA degree in English/Creative Writing is a two-year residency offering concentrations in the writing of fiction, poetry, creative non-fiction and dramatic scripts for theatre and film, as well as courses in the pedagogy of writing and hands-on editorial experience on the staff of *Panhandler*, a journal receiving international submissions and with national distribution. All MA candidates are required to make a scholarly or creative presentation to their peers and the faculty, and to submit a master's thesis – which, for Creative Writing students, customarily means a collection of poems, a performable script, a group of short stories or significant progress toward the completion of a novel. Classes are kept small. A Visiting Writers series provides regular additional attention. Contact the department via e-mail.

Illinois

Chicago State University

MFA Program in Creative Writing, Division of Graduate Studies, Chicago State University, 9501 South King Drive, LIB338, Chicago IL 60628
☎ 001 773 995 2189
www.csu.edu/GraduateSchool/

Offers an MFA with courses in Creative Writing; writing workshops in fiction, poetry, creative non-fiction, playwriting and scriptwriting. Students undertake coursework in African American Literature and non-African American Literature. Students will show competency in the genre of their choice, choosing from fiction, creative non-fiction and poetry. Students may take course work in playwriting and scriptwriting for film and television as electives.

Indiana

Purdue University

MFA Program in Creative Writing, Dept. of English, Purdue University, West Lafayette IN 47907
☎ 001 765 494 3740
www.cla.purdue.edu/mfacw
Program Director *Porter Shreve*

MFA in *Creative Writing* offering 'full funding, generous stipends and editorial experience with the award-winning literary journal *Sycamore Review*'. The three-year programme in either fiction or poetry is small and intensive and includes workshops, literature and craft courses, and thesis tutorials toward a book-length manuscript.

Taylor University

Dept. of English, Taylor University, 1025 West Rudisill Blvd, Fort Wayne IN 46807–2170
☎ 001 260 744 8647
DNHensley@TaylorU.edu
Contact *Dr Dennis E. Hensley*

Summer Honours Programme one-week seminars in *Professional Writing*; evening courses in *Freelance Writing* and *Fiction Writing*; one year college certificate in *Professional Writing*. Dr Hensley is a professor of English and director of the professional writing major.

University of Notre Dame

Creative Writing Program, Dept. of English, University of Notre Dame, Notre Dame IN 46556

☏ 001 574 631 7526   🖷 001 574 631 4795

creativewriting@nd.edu

www.nd.edu/~alcwp

The MFA in *Creative Writing* is a two-year degree programme centred around workshops in poetry and fiction, offering literature courses, translation, a literary publishing course and twelve credits of thesis preparation with an individual faculty adviser.

Louisiana

Louisiana State University

English Department, Louisiana State University, 260 Allen Hall, Baton Rouge LA 70803–5001

☏ 001 225 578 5922

www.lsu.edu/creativewriting/

Director of Creative Writing *James Wilcox*

Master of Fine Arts in *Creative Writing* course with a focus in Poetry, Fiction, Drama, Screenwriting. Includes opportunity to edit literary journals.

Maryland

Goucher College

Welch Center for Graduate Studies, Goucher College, 1021 Dulaney Valley Road, Baltimore MD 21204–2794

☏ 001 410 337 6200

www.goucher.edu/mfa

Program Director *Patsy Sims*

The two-year MFA Program in *Creative Nonfiction* is a limited-residency programme that allows students to complete most of the requirements off campus while developing their skill as non-fiction writers under the close supervision of a faculty mentor. Provides instruction in the following areas: narrative non-fiction, literary journalism, memoir, the personal essay, travel/nature/science writing and biography/profiles.

Massachusetts

Boston University

Creative Writing Program, Boston University, 236 Bay State Road, Boston MA 02215

☏ 001 617 353 2510   🖷 001 617 353 3653

crwr@bu.edu

www.bu.edu/writing

Director *Leslie Epstein*

Contact *Matthew Yost*

Intensive Master of Fine Arts in *Creative Writing*: Fiction, Poetry or Drama, can be completed in one or two years, depending on a student's needs. The Boston University Creative Writing Program is one of the oldest in the country. Students participate in workshops or seminars concentrating on their chosen specialization and are expected to balance that work with an equal number of graduate literature courses. Internships at the literary journal, *AGNI*, assistantships at the Boston Arts Academy, as well as teaching fellowships are available on a limited basis.

Emerson College

120 Boylston Street, Boston MA 01226–4624

☏ 001 617 824 8610   🖷 001 617 824 8614

gradapp@emerson.edu

www.emerson.edu/graduate_admission

Graduate Program Director *Douglas Whynott*

Offers MFA in *Creative Writing* with a focus on writing fiction, non-fiction and poetry. Includes writing workshops, internship opportunities, a student reading series and student teaching opportunities. Publishes literary magazines, *Ploughshares* and *Redivider*. Contact the Graduate Admission Office for more information.

Lesley University

MFA Program in Creative Writing, Lesley University, Cambridge MA 02138

☏ 001 617 349 8369

jvanderv@lesley.edu

www.lesley.edu/gsass/creative_writing/index.html

Contact *Jana Van der Veer, Assistant Director*

Director *Steve Cramer*

The MFA Program in *Creative Writing* is a low-residency programme that allows students, with the oversight of a faculty mentor, to design their own concentrations in fiction, poetry, creative non-fiction, writing for stage and screen, or writing for young people. A nine-day residency begins each semester with a programme of seminars, workshops, readings, and the chance to design the semester's programme of study. Students work independently during the semester, under the guidance of their faculty mentor. Students also design individual, 3-credit interdisciplinary projects for the first three of their four distance-learning semesters. The options for these projects are multiple – in some cases, students take courses; others develop independent studies or work as publishing interns, teachers, or teaching assistants. In their final semester, students work on their creative thesis as well as prepare a craft seminar to be presented to continuing students during the following residency. The programme consists of four full residencies preceding four semesters of work, plus partial attendance at a fifth residency, for a total of 48 credits. Most students complete the programme in two years of full-time study.

Michigan

Western Michigan University

Dept. of English, Western Michigan University, 1903 W. Michigan Avenue, Kalamazoo MI 49008-5331
☎ 001 269 387 2584
english-graduate@wmich.edu
www.wmich.edu/english
Director of Graduate Studies *Jana Schulman*

BA English Major/Minor with *Creative Writing Emphasis* (poetry, fiction, playwriting). The English Major aims at giving students intensive practice in writing and criticism in various genres in a workshop format; for general writing careers or for prospective candidates for the MFA in Creative Writing.
MFA in *Creative Writing* (poetry, fiction, playwriting, non-fiction) For students who wish to become professional writers of poetry, fiction, drama or non-fiction. Qualifies them to teach the craft at college or university level.
PhD in *English, with an Emphasis on Creative Writing* (poetry, fiction, playwriting, non-fiction). Qualifies students to teach creative writing and literature at university level.

Minnesota

Minnesota State University, Mankato

Dept. of English, Minnesota State University, Mankato, 230 Armstrong Hall, Mankato MN 56001
☎ 001 507 389 2117　🖷 001 507 389 5362
richard.robbins@mnsu.edu
www.english.mnsu.edu/cw
Program Director *Richard Robbins*

The MFA Program in *Creative Writing* meets the needs of students who want to strike a balance between the development of individual creative talent and the close study of literature and language. The programme gives appropriate training for careers in freelancing, college-level teaching, editing and publishing and arts administration.

Minnesota State University, Moorhead

Graduate Studies Office, MSUM, 1104 7th Avenue South, Moorhead MN 56563
☎ 001 218 477 2344
graduate@mnstate.edu
www.mnstate.edu/finearts
Contact *MFA Coordinator*

MFA in *Creative Writing*. Students specialize in poetry, fiction, scriptwriting or creative non-fiction and are encouraged to experiment in more than one genre. The programme offers the opportunity to take workshops, seminars and tutorials in chosen areas and to work with New Rivers Press (www.newriverspress.com) or with *Red Weather*, the campus literary magazine.

Missouri

University of Missouri

MFA Program, Dept. of English, University of Missouri, One University Drive, St Louis MO 63121
☎ 00 314 516 6845
www.umsl.edu/~mfa
Contact *Mary Troy*

The MFA in *Creative Writing* provides opportunities for growth in the writing of fiction and poetry (with some work in non-fiction) as well as practical training in literary editing. While normally a studio/academic programme mixing the study of literature and criticism with workshops and independent study and editing, the plan of study is flexible and individual. Students ordinarily specialize in one genre, either fiction or poetry, and regular workshops in these forms are at the heart of the programme. Five workshops, at least four in the student's chosen genre, are required for the degree though more may be taken as electives. Students also take from five to nine courses on offer, choosing from graduate courses in literary journal editing; in poetry and fiction form, theory and technique; in literature and literary criticism; in composition theory; and in linguistics. A creative thesis of six hours completes the 390-hour programme.

University of Missouri-Columbia

Creative Writing Program, Dept. of English, University of Missouri-Columbia, 107 Tate Hall, Columbia MO 65211–1500
☎ 011 573 884 7773
creativewriting@missouri.edu
creativewriting.missouri.edu
Contact *Sharon Fisher*

Creative Writing MA and PhD in English programmes in fiction, poetry and non-fiction. Access the website for further information.

New York

Adelphi University

MFA Program in Creative Writing, Dept. of English, Adelphi University, PO Box 701, Garden City NY 11530
☎ 001 516 877 4044

mfa@adelphi.edu
www.adelphi.edu/mfacw
Director of Creative Writing *Judith Baumel*

The MFA in *Creative Writing* offers students the opportunity to specialize in three major genres: fiction, poetry and dramatic writing. Its unique Professional Development Practicum introduces students to the professional and practical life of writers across many disciplines and prepares them for careers in writing, teaching and/or more advanced graduate studies through training in creative writing, language and literary studies, research and teaching. Students are required to complete 37 credits in a plan of study that includes writing workshops and literature classes. A student thesis is a degree requirement.

Brooklyn College of the City University of New York

MFA Program in Creative Writing, Dept. of English, Brooklyn College, 2900 Bedford Avenue, Brooklyn NY 11210–2889
☎ 001 718 951 5914
profmsp@msn.com
academic.brooklyn.cuny.edu/english
Deputy Chairperson for Graduate Studies *Mark Patkowski*

The MFA in *Creative Writing* is a small, highly personal two-year programme which confers degrees in fiction, poetry and playwriting. Admission is highly competitive. Students must complete 12 courses (four workshops, four tutorials and four literature courses), submit a book-length work in their chosen genre near the end of the second year and pass a comprehensive examination that tests their knowledge and expertise in literature. The workshops, seminars and one-on-one tutorials with established writers particularly emphasize relationships between instructors and students.

Hunter College of the City University of New York

MFA Program in Creative Writing, Dept. of English, Hunter College, 695 Park Avenue, New York NY 10065
☎ 001 212 772 5164
mfa@hunter.cuny.edu
www.hunter.cuny.edu/creativewriting
Program Coordinator *Sue Nacey*

Located in the heart of New York City, Hunter offers two-year MFA programs in fiction, memoir and poetry. All students take a workshop and a craft class every semester, plus two literature classes in their first year and one-on-one supervisions with a member of faculty throughout their second year. Hunter's Distinguished Writers Series brings world-class writers to teach and give readings. Their Hertog Fellowship in fiction

and memoir pays students to work as researchers for distinguished writers (Toni Morrison, Jonathan Safran Foer, Jonathan Franzen, E.L. Doctorow) and their Norma Lubetsky Friedman Scholarships pay students to work for an established poet (Philip Levine, Marie Howe, Jean Valentine). Students may also apply to teach the undergraduate course, Introduction to Creative Writing, in their second year.

New York University

NYU Creative Writing Program, Lillian Vernon Creative Writers House, 58 West 10th Street, New York NY 10011
☎ 001 212 998 8816
creative.writing@nyu.edu
www.cwp.fas.nyu.edu
Program Coordinator *Allison Brotherton*

Offers MFA in *English with a Concentration in Creative Writing* and MA in *Creative Writing* in the genres of either poetry or fiction. Includes writing workshops and craft courses, literary outreach programmes, a public reading series, student readings, special literary seminars and student teaching opportunities. Publishes a literary journal, *Washington Square*. Contact the department via e-mail.

North Carolina

University of North Carolina at Wilmington

Dept. of Creative Writing, University of North Carolina at Wilmington, 601 S. College Road, Wilmington NC 28403
☎ 001 910 962 3070
mfa@uncw.edu
www.uncw.edu/writers
Contact *MFA Coordinator*

An intensive studio-academic programme in the writing of fiction, poetry and creative non-fiction leading to either the Master of Fine Arts or Bachelor of Fine Arts degree in *Creative Writing*. Courses include workshops in the three genres, special topics and forms courses as well as a range of courses in literature. Course work in publishing and editing is also offered in the Publishing Laboratory, a university press imprint which supports local, regional and national publishing projects. The MFA programme is home to the literary journal, *Ecotone: reimagining place.*

University of North Carolina at Greensboro

MFA Writing Program, 3302 HHRA Building, UNC Greensboro, Greensboro NC 27402
☎ 001 336 334 5459
TLKenned@uncg.edu
www.uncg.edu/eng/mfa
Contact *Terry Kennedy*

MFA in *Creative Writing (Poetry, Fiction)*. One of the oldest of its kind in the country, the MFA Writing Program at Greensboro is a two-year residency with an emphasis on providing students with studio time in which to study the writing of poetry or fiction. Gives a flexibility that permits students to develop their particular talents through small classes in writing, literature and the arts.

Ohio

Bowling Green State University

Creative Writing Program, English Dept., Bowling Green State University, Bowling Green OH 43403
☎ 001 419 372 8370   🖷 001 419 372 6805
www.bgsu.edu/departments/creative-writing/

Undergraduate BFA programme: a four-year programme which offers comprehensive and rigorous training in the art of writing and develops students' skills in preparation for numerous post-graduate careers. Graduate MFA programme: a two-year studio/academic composite, mostly work in writing itself in either poetry or fiction. Also offers ten new teaching assistantships each autumn. See the website for details on both programmes.

Oregon

University of Oregon

Creative Writing Program, 5243 University of Oregon, Eugene OR 97403–5243
☎ 001 541 346 3944
crwrweb@darkwing.uoregon.edu
www.uoregon.edu

Offers MFA in *Creative Writing* in poetry or fiction. Includes writing workshops, craft courses, individualized tutorials with faculty, public reading series, student readings and student teaching opportunities. Admission is highly competitive.

Philadelphia

Seton Hill University

1 Seton Hill Drive, Greensburg PA 15601
☎ 001 724 838 4283
bartos@setonhill.edu
www.setonhill.edu
Contact *Director of Graduate Studies*

The MA in *Writing Popular Fiction* program at Seton Hill University allows students to earn a graduate degree by writing fiction that people *actually* read. Students attend week-long residencies in January and June and complete

writing projects off-campus, working with a faculty mentor who is a published author in the chosen genre. Students can choose to specialize in science fiction, fantasy, horror, children's literature, romance or mystery.

Wilkes University

Graduate Creative Writing Programs, Wilkes University, 84 West South Street, Wilkes-Barre PA 18766
☎ 001 570 408 4534   🖷 001 570 408 3333
cwriting@wilkes.edu
www.wilkes.edu/creativewriting
Assistant Director, Graduate Creative Writing Programs *James Warner*

The low-residency MA/MFA *Graduate Creative Writing Program* of Wilkes University offers five areas of study: poetry, fiction, creative nonfiction, screenwriting and playwriting. Throughout the programme, students study the personal life of a writer, the craft, technique and analysis of creative writing and the art delivery method for one's work. Low-residency means that students are on the campus of Wilkes University for two eight-day residencies per year, in January and June. The regular coursework will be delivered through a password-protected website. Students can complete their MA in a year-and-a-half and their MFA in two-and-a-half years. The faculty includes writers such as Colum McCann, Toi Derricotte, William Kennedy, Larry Heinemann, H.L. Hix, Lenore Hart, David Poyer and Kaylie Jones. For more information, please visit the website or send an e-mail.

Rhode Island

Brown University

Literary Arts, Box 1923, Brown University, Providence RI 02912
☎ 001 401 863 3260   🖷 001 401 863 1535
writing@brown.edu
www.brown.edu/cw
Contact *Director of Literary Arts*

Two-year MFA in *Literary Arts*. Students attend two courses per semester: a writing workshop and an elective. In their final semester, graduate students work independently with a faculty advisor on thesis preparation. The thesis is a substantial work of electronic arts, drama, fiction, or poetry – or may cross genre boundaries. Admission is highly selective and is based primarily on the quality of the applicant's writing. Application deadline: 15th December.

Tennessee

University of Memphis

MFA Creative Writing Program, University of Memphis, English Dept., 471A Patterson Hall, Memphis TN 38152

☎ 001 901 678 4692

creativewriting@memphis.edu

www.mfainmemphis.com

Contact *MFA Coordinator*

The MFA *Creative Writing Program* offers students a wide choice in the courses they can take. It has four options: Studio Option: only creative writing courses are required, for those who want to focus entirely on their creative work; Literary Studies Option: combines the study of literature with creative writing, for those who want to build a background in literature while they write; Creative/Professional Writing Option: for those who want to gain an excellent preparation for a flexible career in writing; and Cross Disciplinary Option: allows a wide range of options for taking courses in other departments, so that a student can use other interests to inform their writing and study foreign languages, history, art, theatre and other disciplines.

Texas

University of Houston

Creative Writing Program, University of Houston, Dept. of English, Room 229, Roy Cullen Building, Houston TX 77204-3015

☎ 001 713 743 3015   ℻ 001 713 743 3697

cwp@uh.edu

www.uh.edu

Office Coordinator *Shatera Anderson*

Offers MFA and PhD in *Creative Writing*. Admission is highly competitive; students can apply for admission in the genres of poetry, fiction and non-fiction. Normally admits 20 students a year. The MFA is a three-year degree programme; the PhD normally takes five years. Fellowships and teaching assistantships are available. In collaboration with Inprint, it hosts the Inprint/Brown Reading series and the Inprint Studio series. The programme offers opportunities to collaborate with artists in other disciplines. Publishes a literary journal, *Gulf Coast*.

University of Texas at Austin

Michener Center for Writers, University of Texas at Austin, MFA Program in Writing, 702 East Dean Keeton Street, Austin TX 78705

☎ 001 512 471 1601

mcw@www.utexas.edu

www.utexas.edu/academic/mcw

Contact *Graduate Coordinator*

A three-year interdisciplinary MFA programme in fiction, poetry, playwriting and screenwriting in which students work in two fields of writing, with thesis emphasis in a primary field. Courses include workshops and readings/studies courses. All students must be enrolled full-time, in residence, and are supported by annual fellowships of $25,000, plus remission of tuition and fees. Admission is extremely competitive and applicants must submit all materials by an annual 15th December deadline for the following autumn.

Virginia

George Mason University

Graduate Creative Writing Program, George Mason University, 4400 University Drive, 3E4, Fairfax VA 22030

☎ 001 703 993 1180

writing@gmu.edu

creativewriting.gmu.edu

Contact *Graduate Coordinator*

The MFA in *Creative Writing* offers three concentrations – fiction, non-fiction, and poetry – each of which require 48 semester hours and takes at least three years to complete. Course work blends writing workshops with craft seminars and the study of literature. Each concentration also requires completion of a thesis (a book-length manuscript). In addition, the poetry concentration requires completion of a comprehensive exam. The fiction and non-fiction concentrations offer several options, including an oral exam, completion of a reading list coupled with an annotated bibliography, and designing an individualized project to meet a student's long-term goals.

Hollins University

MFA in Creative Writing, Hollins University Graduate Center, PO Box 9603, Roanoke VA 24020–1603

☎ 001 540 362 6575   ℻ 001 540 362 6288

hugrad@hollins.edu

www.hollins.edu/grad

Manager, Graduate Studies *Cathy Koon*

MFA in *Creative Writing*: a small, highly selective two-year full-time programme which allows students to specialize in poetry, fiction or both, while also offering opportunities to study and write screenplays and creative non-fiction. 'The programme is characterized by an individual approach, a lively community of writers and a stimulating combination of challenge and support'. Also three summer-term MA/MFA programmes in *Children's Literature*; *Playwriting*;

and *Screenwriting & Film Studies*. Students typically study for three to five six-week terms.

Old Dominion University

Batten Arts and Letters 5000, Old Dominion University, Norfolk VA 23529
001 757 683 4770
sreynolds@odu.edu
www.al.odu.edu/english/mfacw/
Program Director *Sheri Reynolds*

The MFA Program in *Creative Writing* registers 25–35 students each semester in fiction, poetry and non-fiction. In the three-year, 54-hour programme, students graduate not only as writers but also as individuals competent in literary studies. The Writers-in-Community Program gives MFA students the opportunity to facilitate writing workshops in schools, hospitals, homeless shelters, housing projects and retirement homes in the area. Students can also participate in the reading series at Prince Books in downtown Norfolk and assist with ODU's internationally recognized annual literary festival. ODU's Writer-in-Residency Program (offered in both the autumn and spring semesters) allows students to work one-on-one with respected poets and writers from around the country.

Virginia Commonwealth University

Dept. of English, Virginia Commonwealth University, PO Box 842005, Richmond VA 23284–2005
001 804 828 1329
tndidato@vcu.edu
www.has.vcu.edu/eng
Graduate Programs Coordinator *Thom Didato*

Three-year MFA in *Creative Writing* with tracks in fiction and poetry, and workshops in short fiction, the novel, poetry, drama, screenwriting and creative non-fiction. Opportunities to work with *Blackbird*, an online journal of literature and the arts (www.blackbird.vcu.edu), *Stand Magazine*, the Levis Reading Prize and the First Novelist Award.

Washington

Eastern Washington University

Creative Writing, Inland Northwest Center for Writers, Eastern Washington University, 501 N. Riverpoint Blvd., Suite 425, Spokane WA 99202
001 509 359 4956   001 509 455 7252
prussell@ewu.edu
www.ewumfa.com
Contact *Program Director*

The MFA *Creative Writing* Program is an intensive two-year, pre-professional course of study with an emphasis on the practice of literature as a fine art. Includes course work in the study of literature from the vantage point of its composition and history, but the student's principal work is done in advanced workshops and in the writing of a book-length thesis of publishable quality in fiction, literary non-fiction or poetry. Students have the opportunity to work as interns on several projects: the literary magazine, *Willow Springs*; the community outreach programme, Writers in the Community; and Eastern Washington University Press. The MFA is a terminal degree programme.

Seattle Pacific University

MFA Program in Creative Writing, Seattle Pacific University, 3307 Third Avenue West, Seattle WA 98119
001 206 281 2727   001 206 281 2335
mfa@spu.edu
www.spu.edu/mfa
Contact *Gregory Wolfe, Program Director*

The low-residency MFA at SPU is a two-year programme for apprentice writers who not only want to pursue excellence in the craft of writing but also place their work within the larger context of the Judeo-Christian tradition of faith. Students work closely with a faculty mentor in their chosen genre (poetry, fiction or creative non-fiction) toward the completion of a creative thesis. They also write a critical paper each quarter and read a minimum of 62 books over the course of the programme. Students attend two intensive ten-day residencies per year, one on Whidbey Island in Washington State and one in Santa Fe, New Mexico. Graduating students attend a final residency in which they deliver a public reading and a lecture.

Wyoming

University of Wyoming

MFA in Creative Writing, Dept. of English – 3353, University of Wyoming, 1000 E. University Avenue, Laramie WY 82071–2000
001 307 766 2867
cw@uwyo.edu
www.uwyo.edu/creativewriting
Contact *Rachel Ferrell (Program Assistant)*

The two-year Master of Fine Arts in *Creative Writing* is an intensive 40-hour studio degree in Poetry, Fiction or Creative Non-Fiction. Special features include opportunities for interdisciplinary study, supported by a wide range of university departments, and a required professional internship on campus, in the community or further afield, ensuring the acquisition or polishing of 'real-life' writing skills.

Writers' Circles and Workshops

Directory of Writers' Circles, Courses and Workshops

☎ 07850 709596
diana@writers-circles.com
www.writers-circles.com
Editor *Diana Hayden*

Directory of UK writers' circles, courses and workshops.

Camelford Poetry Workshops

▷ entry under Organizations of Interest to Poets

Carmarthen Writers' Circle

Lower Carfan, Tavernspite, Whitland SA34 0NP
☎ 01994 240441
Contact *Jenny White*

Founded 1989. The Circle meets monthly on the second Monday of the month at the Indoor Bowls Centre, Carmarthen. All levels and genres welcome. Holds occasional workshops.

Chiltern Writers' Group

Wendover Library, High Street, Wendover HP22 6DU
info@chilternwriters.org
www.chilternwriters.org

Guest speakers, workshops, manuscript critiques and monthly meetings, held at Wendover Library every second Thursday of the month at 8.00 pm. Regular newsletter. SUBSCRIPTION £17 p.a.; £12 (concessions); £3 (non-members, meetings).

Coleg Harlech WEA

Harlech, LL46 2PU 01766 781900
www.harlech.ac.uk

Coleg Harlech WEA was formed in 2001 with the merger of Coleg Harlech and the Workers' Educational Association North Wales.

The Cotswold Writers' Circle

Vestry Cottage, Bell Lane, Minchinhampton GL6 9BP
granne@waitrose.com
Patron *Katie Fforde*

Membership Secretary & Competition Secretary *Mrs Anne Brookes*
The Circle meets every Tuesday morning (except second Tuesday of the month) in Cirencester. Activities include workshops with well-known authors and an annual International Open Writing Competition; closing date: 31 January (send s.a.e. for details). Winners published in the Circle's anthology, *Pen Ultimate*.

Creuse Writers' Workshop & Retreat

45 Browning Drive, Lincoln LN2 4HF
☎ 07724 191077 ☒ 01522 880565
info@creusewritersworkshopandretreat
paul@creusewritersworkshopandretreat
www.creusewritersworkshopandretreat
Director of Studies *Sandra Doran*
Bookings & Information *Paul Bett*

One week writers' workshops held twice yearly in Creuse, central France. Run by a team of professional writers to encourage creativity in writers of all grades. The week includes guest speakers, fun practical discussions and social events. Opportunities to write exercises, discuss work as a group and one-to-one. Check the website for special offers and details.

Cumbrian Literary Group

'Calgarth', The Brow, Flimby, Maryport CA15 8TD
☎ 01900 813444
President *Glyn Matthews*
Secretary *Joyce E. Fisher*

Founded 1946 to provide a meeting place for readers and writers in Cumbria. The Group meets once a month (April to November) in Keswick. Invites speakers to meetings and holds annual competitions for poetry and prose. Publishes *Bookshelf* magazine. Further details from the Secretary at the address above.

Sean Dorman' Manuscript Society

3 High Road, Britford, Salisbury SP5 4DS
Director *Jan Smith*

Founded 1957. Provides mutual help among writers and aspiring writers in England, Wales and Scotland. By means of circulating manuscript parcels, members receive constructive comment on their own work and read and comment on the work of others. Full details and application forms available on receipt of s.a.e.

East Anglian Writers

chair@eastanglianwriters.org.uk
www.eastanglianwriters.org.uk
Chair *Victoria Manthorpe*

A group for professional writers living in Norfolk, Suffolk, Essex, Cambridgeshire and Bedfordshire. Affiliated to the **Society of Authors**. Aims to help writers socialize and promote their work. Informal social events, speakers' evenings and contact point for professional writers in the area.

éQuinoxe TBC

43 rue Beaubourg, 75003 Paris, France
☎ 00 33 1 47 55 11 89   📠 00 33 1 48 04 52 58
equinoxetbc@equinoxetbc.fr
www.equinoxetbc.fr
President & Artistic Director *Noëlle Deschamps*

Week-long intensive screenwriting workshops, held twice a year in English and French, open to experienced European screenwriters. Participants are selected by an international jury: candidates must have written a script for a feature film, which is sufficiently developed and ready to go into production. Entry form available on the website.

Euroscript Ltd

64 Hemingford Road, London N1 1DB
☎ 07958 244656
ask@euroscript.co.uk
www.euroscript.co.uk

Established in 1995 by the Screenwriters' Workshop with funding by the EU's Media II programme. Script development organization offering creative and editorial input to writers and production companies. Develops screenplays through an intensive consultancy programme, including residential script workshops; also offers a broad programme of short courses, provides script reports and runs an annual Screen Story Competition. Open to writers of any nationality. E-mail, telephone, or access the website for further information.

Fiction Writing Workshops and Tutorials

5 Queen Elizabeth Close, London N16 0HL
☎ 020 8809 4725
henrietta@wittenwords.net
www.WrittenWords.net
Contact *Henrietta Soames*

Fiction workshops for beginner and advanced writers led by author Henrietta Soames. 'Lively discussion, stimulating exercises, valuable feedback.' Fiction tutorials for students who require concentated attention and ongoing support. Short stories/novels, full-length mss welcomed.

Foyle Street Writers Group

> See Sunderland City Library and Arts Centre under Library Services

Susan Grossman – Writing Workshops

☎ 020 7794 0288
susangrossman@tiscali.co.uk
www.susangrossman.co.uk
Contact *Susan Grossman*

Established travel writer, freelance journalist and editor, Susan Grossman runs journalism workshops for freelancers and authors looking to increase their writing and selling ability. Workshops cover how to focus on an idea, write compelling synopses and approach editors with confidence. Selling Freelance Features and Travel Writing workshops are held regularly in Central London. One-to-one sessions are also available.

Guildford Writers

> The Guildford Institute under UK and Irish Writers' Courses

Historical Novel Folio

187 Baker Street, Alvaston, Derby DE24 8SG
Contact *Mrs Heidi Sullivan*

An independent postal workshop – single folio dealing with any period before World War II. Send s.a.e. for details.

'How to Self-Publish Your Book' Workshops

Faculty of Research and Knowledge Transfer, University of Winchester, Winchester SO22 4NR
☎ 01962 827238
barbara.large@winchester.ac.uk
www.writersconference.co.uk
Contact *Barbara Large, MBE, FRSA, HFUW*

One-day workshops to inform writers who wish to self-publish their books about the digital and lithographic processes. Participants will learn the skills of researching, writing and revising their mss; the use of photographs, maps, diagrams, illustrations, design and formatting; types of paper, fonts, sizes, ISBN numbers and costs, and marketing their books. 2009 workshops: Thursday, 7th May and Thursday, 17th September.

Bernard Kops and Tom Fry Drama Writing Workshops

41B Canfield Gardens, London NW6 3JL
☎ 07976 352837/020 7624 2940/8533 2472
artwrite@dsl.pipex.com
bernardkops@blueyonder.co.uk
Tutors *Bernard Kops, Tom Fry*

Three 11-week terms per year. Small group drama writing workshops for stage, screen and television with actors in attendance. Tutorials also available. Call for details.

Leicester Writers' Club

www.leicesterwriters.org.uk

Founded 1958. With over 50 members, the club meets weekly in central Leicester. Offers manuscript sessions, workshops, speaker evenings and competitions.

New Writing South

9 Jew Street, Brighton BN1 1UT
☎ 01273 735353
admin@newwritingsouth.com
www.newwritingsouth.com
Contact *Chris Taylor*

Works throughout the South East to create an environment in which creative writers and their works can flourish. Offers resources and development and also builds partnerships between writers and those able to produce their work. Membership open to professional, emerging and aspiring creative writers in the region and to companies with a professional interest in new writing.

North West Playwrights (NWP)

18 Express Networks, 1 George Leigh Street, Manchester M4 5DL
☎ 0161 237 1978
newplaysnw@hotmail.com
www.newplaysnw.co.uk

NWP is the regional development agency for new theatre writing in the north west of England. Operates a script-reading service, provides script development and other support to writers at all stages of their careers, and presents a range of classes. Also collaborates with theatres and other bodies in a variety of writing-related projects. Bi-monthly subscription newsletter, *The Lowdown*. Services available to writers in the region only.

The Original Writers Group

info@theoriginalwriter.com
www.originalwriters.ning.com
Contact *Rupert Davies-Cooke*

'We welcome writers from all walks of life as everyone has a point of view, and the more varied these opinions, the richer the experience.' Currently, has a membership of over 100: poets, novelists, playwrights, scriptwriters, historians, philosophers. Usually six to eight writers turn up each evening and there is time to discuss three to four pieces. 'Even if you are just starting out and are just interested in writing, but don't have anything to read, come along and experience the evening; as sometimes the best motivation to writing is to hear how others got there before you.' The group meets at Caffè Nero, 1–3 Lavender Hill, Battersea, London SW11 every first and third Wednesday of the month 6.30 pm to 8.30 pm; no entrance fee. NB Change of venue due; check website for details.

Pitstop Refuelling Writers' Weekend Workshops

Faculty of Research and Knowledge Transfer, University of Winchester, Winchester SO22 4NR
☎ 01962 827238
barbara.large@winchester.ac.uk
www.writersconference.co.uk
Director *Barbara Large, MBE, FRSA, HFUW*

Following on from the **Winchester Writers' Conference** in Winchester, these are small-group writing workshops under the guidance of professional writers. Join *Writing Marketable Children's Fiction or Non-Fiction*; *Editing the Final Draft of Your Novel and Marketing It*; *Lifewriting and Research*. 2009 workshops: in March and 23rd–25th October

Player–Playwrights
▷ entry under Professional Associations

Queer Writing South

Brighton Writers Centre, 49 Grand Parade, Brighton BN2 9QA
☎ 01273 571700
QueerWritingSouth@gmail.com
Contact *Maria Jastrzebska*

Queer Writing South is a network for queer-identified writers in the South East of England. Runs writing workshops, queer-themed literary events and last year produced their first anthology of new writing from LGBT writers in the region. Events and activities take place on a quarterly basis at various venues around Brighton and further afield.

Script Yorkshire

Communications Officer: 409 Gateway West, East Street, Leeds LS9 8DZ
members@scriptyorkshire.co.uk
www.scriptyorkshire.co.uk
Membership *Joanna Loveday*

'Script Yorkshire is the region's only support and advocacy organization for scriptwriters across performance and broadcast media. Run democratically by writers for writers, it puts the needs of members at the very core of its activity, developing both their practice and their industry connections. Script Yorkshire aims to empower Yorkshire's writers for performance and broadcast media by giving them the skills and insights they need to develop their own careers.' Send an e-mail or visit the website for information on becoming a member.

Short Story Writers' Folio/Children's Novel Writers' Folio

5 Park Road, Brading, Sandown, Isle of Wight
PO36 0HU
☎ 01983 407697
nott.dawn-wortley@tiscali.co.uk
Contact *Mrs Dawn Wortley-Nott*

Postal workshops; members receive constructive criticism of their work and read and offer advice on fellow members' contributions. Send an s.a.e. for further details.

Society of Sussex Authors

Dolphin House, 51 St Nicholas Lane, Lewes
BN7 2JZ
☎ 01273 470100   📠 01273 470100
sussexbooks@aol.com
Contact *David Arscott*

Founded 1968. Six meetings per year, held in Lewes, plus social events. Membership restricted to Sussex-based writers only.

Southwest Scriptwriters

info@southwest-scriptwriters.co.uk
www.southwest-scriptwriters.co.uk
Artistic Director *Tim Massey*

Founded 1994 to offer support to regional drama writers for all media. The group meets at the Naval Volunteer pub in King Street, Bristol. Annual subscription: £6.

Speakeasy – Milton Keynes Writers' Group

PO Box 5948, Stoke Hammond, Milton Keynes
MK10 1GX
☎ 01908 663860
speakeasy@writerbrock.co.uk
www.mkweb.co.uk/speakeasy
Chairman *Martin Brocklebank*

Monthly meetings on the first Friday of each month at 7.30 pm. Full and varied programme includes Local Writer Nights where work can be read and peformed and Guest Nights where writers, poets and journalists are invited to speak. Mini-workshops, critique nights and information nights are also in the programme. Invites entries to Open Creative Writing Competitions closing 31 October each year; see website for entry form and rules. Phone, e-mail or send s.a.e. for details to address above.

Spread the Word

77 Lambeth Walk, London SE11 6DX
☎ 020 7735 3111   📠 020 7735 2666
annette@spreadtheword.org.uk
www.spreadtheword.org.uk
Programme Coordinator *Annette Brook*

'Spread the Word is a catalyst for developing writers.' Provides a range of services and opportunities for all levels of writers including advice and information, networking, mentoring, publishers' events and work in the community.

Sussex Playwrights' Club

2 Brunswick Mews, Hove BN3 1HD
☎ 01273 730106
www.sussexplaywrights.com
Secretary *Dennis Evans*

Founded 1935. Monthly readings of members' work by experienced actors. Membership open to all. Meetings held at New Venture Theatre, Bedford Place, Brighton.

Ver Poets

☎ 01727 864898
daphneschiller@ntlworld.com
www.verpoets.org.uk
Secretary *Daphne Schiller*

Founded 1966. Postal and local members. Meets in St Albans, runs evening meetings and daytime workshops and organizes competitions.

Walton Wordsmiths

27 Braycourt Avenue, Walton on Thames
KT12 2AZ
☎ 01932 702874
wendy@stickler.org.uk
www.waltonwordsmiths.co.uk
Contact *Wendy Hughes*

Founded in 1999 by Wendy Hughes. Supports writers of all grades and capabilities by offering constructive criticism and advice.

Workers' Educational Association

National Office: 3rd Floor, 70 Clifton Street, London EC2A 4HB
☎ 020 7426 3450
national@wea.org.uk
www.wea.org.uk

Founded in 1903, the WEA is a charitable provider of adult education with students drawn from all walks of life. It runs writing courses and workshops in many parts of the country which are open to everyone. Contact your local WEA office for details of courses in your region.

Eastern Cintra House, 12 Hills Road, Cambridge, B2 1JP eastern@wea.org.uk

East Midlands 39 Mapperley Road, Mapperley Park, Nottingham NG3 5AQ ☎ 0115 962 8400 eastmidlands@wea.org.uk

London 4 Luke Street, London EC2A 4XW ☎ 020 7426 1950 london@wea.org.uk

North East 21 Portland Terrace, Newcastle upon Tyne NE2 1QQ ☎ 0191 212 6100 northeast@wea.org.uk

North West Suite 405–409, The Cotton Exchange, Old Hall Street (Bixteth Street

Entrance), Liverpool L3 9JR ☎ 0151 243 5340
northwest@wea.org.uk

Southern Unit 57 Riverside 2, Sir Thomas
Longley Road, Rochester ME2 4DP
☎ 01634 298600 southern@wea.org.uk

South West Bradninch Court, Castle Street,
Exeter EX4 3PL ☎ 01392 490970
southwest@wea.org.uk

West Midlands 4th Floor, Lancaster House,
67 Newhall Street, Birmingham, B3 1NQ
westmidlands@wea.org.uk

Yorkshire and Humber 6 Woodhouse Square,
Leeds LS3 1AD ☎ 0113 245 3304
yorkshumber@wea.org.uk

Workers' Educational Association North Wales
▷ Coleg Harlech WEA

Workers' Educational Association Northern Ireland

1–3 Fitzwilliam Street, Belfast BT9 6AW
☎ 028 9032 9718　🖷 028 9023 0306
info@wea-ni.com
www.wea-ni.com

Autonomous from the WEA National
Association, WEA Northern Ireland was founded
in 1910.

Workers' Educational Association Scotland

Riddles Court, 322 Lawnmarket, Edinburgh
EH1 2PG
☎ 0131 226 3456
hq@weascotland.org.uk
www.weascotland.org.uk
Scottish Secretary *Joyce Connon*

WEA Scotland (registered charity no: 2806910)
was founded in 1905 and is part of the national
Workers' Educational Association.

Workers' Educational Association South Wales

7 Coopers Yard, Curran Road, Cardiff CF10 5NB
☎ 029 2023 5277　🖷 029 2023 3986
weasw@swales.wea.org.uk
www.swales.wea.org.uk

Autonomous from the WEA National
Association, WEA South Wales has been active in
community education for over 100 years.

Writers in Oxford

wio_membership@yahoo.co.uk
www.writersinoxford.org
Membership Secretary *Cherry Mosteshar*

Chair *Frank Egerton*
SUBSCRIPTION £25 p.a.

Founded 1992. Open to published authors,
playwrights, poets and journalists. Literary
seminars and social functions. Publishes *The
Oxford Writer* newsletter.

Bursaries, Fellowships and Grants

Arts Council of Wales ACW's Creative Wales Awards
> entry under Arts Councils and Regional Offices

The Authors' Contingency Fund
The Society of Authors, 84 Drayton Gardens, London SW10 9SB
☎ 020 7373 6642   ✉ 020 7373 5768
info@societyofauthors.org
www.societyofauthors.org

This fund makes modest grants to published authors who find themselves in sudden financial difficulties. Contact *Sarah Baxter* at the **Society of Authors** for an information sheet and application form.

The Authors' Foundation
The Society of Authors, 84 Drayton Gardens, London SW10 9SB
☎ 020 7373 6642   ✉ 020 7373 5768
info@societyofauthors.org
www.societyofauthors.org

The Foundation provides grants to writers to assist them while writing books. Application is by letter and there are two rounds a year. The deadlines are 30th April and 30th September. Contact *Paula Johnson* at the **Society of Authors** for guidelines.

The K. Blundell Trust
The Society of Authors, 84 Drayton Gardens, London SW10 9SB
☎ 020 7373 6642   ✉ 020 7373 5768
info@societyofauthors.org
www.societyofauthors.org

Grants to writers to fund travel, important research or other expenditure. Author must be under 40, has to submit a copy of his/her previous book and the work must 'contribute to the greater understanding of existing social and economic organization'. Application by letter. Contact the **Society of Authors** for full entry details. Final entry dates: 30th April and 30th September.

Alfred Bradley Bursary Award
c/o BBC Radio Drama, Room 2130, New Broadcasting House, Oxford Road, Manchester M60 1SJ
☎ 0161 244 4253   ✉ 0161 244 4248
Contact *Coordinator*

Established 1992. Biennial award in commemoration of the life and work of the distinguished radio producer Alfred Bradley. Aims to encourage and develop new radio writing talent in the BBC North region. Entrants must have been born, currently resident or previously lived in the north of England for at least five years. The award is given to help writers to pursue a career in writing for radio. Next award was launched winter 2008. (Further information can be found on bbc.co.uk/writersroom) Previous winners: Mark Shand, Anthony Cropper, Julia Copus, Michael Stewart, Ben Tagoe, Katie Douglas.
💰 Award of up to £6,000 over two years; potential BBC Radio Drama commissions and opportunities for mentoring with producers.

British Academy Small Research Grants
10 Carlton House Terrace, London SW1Y 5AH
☎ 020 7969 5217   ✉ 020 7969 5414
grants@britac.ac.uk
www.britac.ac.uk
Contact *Research Grants Department*

Award to further original academic research at postdoctoral level in the humanities and social sciences. Entrants must be resident in the UK. See website for final entry dates.
💰 Award £7,500 (maximum).

Cholmondeley Awards
The Society of Authors, 84 Drayton Gardens, London SW10 9SB
☎ 020 7373 6642   ✉ 020 7373 5768
info@societyofauthors.org
www.societyofauthors.org

Founded in 1965 by the late Dowager Marchioness of Cholmondeley. Annual honorary awards to recognize the achievement and distinction of individual poets. 2008 winners: John Burnside, John Greening, David Harsent, Sarah Maguire. Awards £6,000 (total).

Olive Cook Prize
▷ Tom-Gallon Trust Award

The Economist/Richard Casement Internship

The Economist, 25 St James's Street, London SW1A 1HG
☏ 020 7830 7000
casement@economist.com
www.economist.com
Contact *Geoffrey Carr, Science Editor (re. Casement Internship)*

For an aspiring journalist to spend three months in the summer writing for *The Economist* about science and technology. Applicants should write a letter of introduction along with an article of approximately 600 words suitable for inclusion in the Science and Technology Section. 'Our aim is more to discover writing talent in a scientist or science student than scientific aptitude in a budding journalist.' Competition details normally announced in the magazine late January or early February and 4–5 weeks allowed for application.

European Jewish Publication Society

PO Box 19948, London N3 3ZJ
☏ 020 8346 1776   ☐ 020 8346 1776
colinshindler@gmail.com
www.ejps.org.uk
Contact *Professor Colin Shindler*

Established in 1995 to help fund the publication of books of European Jewish interest which would otherwise remain unpublished. Helps with the marketing, distribution and promotion of such books. Publishers who may be interested in publishing works of Jewish interest should approach the Society with a proposal and manuscript in the first instance. Books which have been supported recently include: *The History of Zionism* Walter Laqueur; *Photographing the Holocaust* Janina Struk; *The Arab-Israeli Cookbook* Robin Soans; *Daughters of Sarah: An Anthology of Jewish Women Writing in French* eds. Eva Martin Sartori and Madeleine Cottenet-Hage; *Whistleblowers and the Bomb: Vanunu, Israel and Nuclear Secrecy* Yoel Cohen. Also supports the publication of poetry, and translations from and into other European languages.
Grant £3,000 (maximum).

Fulbright Awards

The Fulbright Commission, Fulbright House, 62 Doughty Street, London WC1N 2JZ
☏ 020 7404 6880   ☐ 020 7404 6834
programmes@fulbright.co.uk
www.fulbright.co.uk
Contact *Awards Assistant*

The Fulbright Commission offers a number of scholarships given at postgraduate level and above, open to any field of study/research to be undertaken in the USA. Length of award is typically an academic year. The application deadline for awards is 31 May of the preceding year of study. Full details and application forms are available on the Commission's website.

Tony Godwin Memorial Trust

c/o 38 Lyttelton Court, Lyttelton Road, London N2 0EB
☏ 020 8209 1613
info@tgmt.org.uk
www.tgmt.org.uk
Chairman *Iain Brown*

Biennial award established to commemorate the life of Tony Godwin, a prominent publisher in the 1960s/70s. Open to all young people (under 35 years old) who are UK nationals and working, or intending to work, in publishing. The award provides the recipient with the means to spend at least one month as the guest of an American publishing house in order to learn about international publishing. The recipient is expected to submit a report upon return to the UK. Next award: 2010; final entry date: 31st December 2009. Previous winners: George Lucas (Hodder), Clive Priddle (Fourth Estate), Richard Scrivener (Penguin), Lisa Shakespeare (Weidenfeld & Nicolson), Fiona Stewart (HarperCollins).
Bursary of approx. US$5,000.

Eric Gregory Trust Fund

The Society of Authors, 84 Drayton Gardens, London SW10 9SB
☏ 020 7373 6642   ☐ 020 7373 5768
info@societyofauthors.org
www.societyofauthors.org

Annual awards of varying amounts are made for the encouragement of poets under the age of 30 on the basis of a submitted collection. Open only to British-born subjects resident in the UK. Final entry date: 31st October. Contact the **Society of Authors** for full entry details. 2008 winners: Emma Berry, Rhiannon Hooson, James Midgley, Adam O'Riordan, Heather Phillipson.
Awards £20,000 (total).

The Guardian Research Fellowship

Nuffield College, Oxford OX1 1NF
☏ 01865 278542   ☐ 01865 278666
Contact *The Academic Administrator*

One-year fellowship endowed by the Scott Trust, owner of *The Guardian*, to give someone working in the media the chance to put their experience into a new perspective, publish the outcome and give a *Guardian* lecture. Applications welcomed from journalists and management members, in newspapers, periodicals or broadcasting. Research or study proposals should be directly related to experience of working in the media. Accommodation and meals in college will be provided and a stipend. Advertised biennially in November.

Hawthornden Literary Institute

Hawthornden International Retreat for Writers, Lasswade EH18 1EG

01313 440 2180   0131 440 1989

Contact *The Administrator*

Established 1982 to provide a peaceful setting where published writers can work in silence. The Castle houses up to six writers at a time, who are known as Hawthornden Fellows. Writers from any part of the world may apply for the fellowships. No monetary assistance is given, nor any contribution to travelling expenses, but once arrived at Hawthornden, the writer is the guest of the Retreat. Applications on forms provided must be made by the end of September for the following calendar year. Previous winners include: Les Murray, Alasdair Gray, Helen Vendler, David Profumo, Hilary Spurling.

Francis Head Bequest

The Society of Authors, 84 Drayton Gardens, London SW10 9SB

020 7373 6642   020 7373 5768

info@societyofauthors.org

www.societyofauthors.org

Provides grants to published British authors over the age of 35 who need financial help during a period of illness, disablement or temporary financial crisis. Contact *Sarah Baxter* at the **Society of Authors** for an information sheet and application form.

Hosking Houses Trust

33 The Square, Clifford Chambers, Stratford-upon-Avon CV37 8HT

sarahhosking@btinternet.com

www.hoskinghouses.co.uk

Contact *The Secretary*

Established in 2001 to offer women writers over the age of 40 a period of financially-protected, domestic peace. Residence of Church Cottage two miles from Stratford-upon-Avon; furnished and equipped; offered for between two months and one year, all bills paid (except BT) also a personal bursary. No duties except to complete personal work, report to the Trustees every three months, to acknowledge the Trust in all work produced during tenure or published as a result of it. Advertised as available: appointed on merit and need to an established, published woman writer on any subject, and who has the legal right to be in the UK.

Jerwood Awards

The Royal Society of Literature/Jerwood Awards

Journalists' Charity

Dickens House, 35 Wathen Road, Dorking RH4 1JY

01306 887511   01306 888212

enquiries@journalistscharity.org.uk

www.journalistscharity.org.uk

Director/Secretary *David Ilott*

Aims to relieve distress among journalists and their dependants. Continuous and/or occasional financial grants; also retirement homes for eligible beneficiaries. Further information available from the Director.

Ralph Lewis Award

University of Sussex Library, Brighton BN1 9QL

01273 678158   01273 873413

a.timoney@sussex.ac.uk

www.sussex.ac.uk/library/

Established 1985. Occasional award set up by Ralph Lewis, a Brighton author and art collector who left money to fund awards for promising manuscripts which would not otherwise be published. The award is given in the form of a grant to a UK-based publisher in respect of a publication of literary works by new authors. No direct applications from writers. Previous winners: **Peterloo Poets**; **Serpent's Tail**.

The Elizabeth Longford Grants

The Society of Authors, 84 Drayton Gardens, London SW10 9SB

020 7373 6642   020 7373 5768

info@societyofauthors.org

www.societyofauthors.org

Biannual grant, sponsored by Flora Fraser and Peter Soros, to a historical biographer for essential research, travel or other necessary expenditure. Contact the **Society of Authors** for full details. Final entry dates: 30th April and 30th September.

Grant £2,500.

The John Masefield Memorial Trust

The Society of Authors, 84 Drayton Gardens, London SW10 9SB

020 7373 6642   020 7373 5768

info@societyofauthors.org

www.societyofauthors.org

This trust makes occasional grants to professional poets (or their immediate dependants) who are faced with sudden financial problems. Contact *Sarah Baxter* at the **Society of Authors** for an information sheet and application form.

The Somerset Maugham Awards

The Society of Authors, 84 Drayton Gardens, London SW10 9SB

☎ 020 7373 6642   🖷 020 7373 5768

info@societyofauthors.org

www.societyofauthors.org

Annual awards designed to encourage writers under the age of 35 to travel. Given on the basis of a published work of fiction, non-fiction or poetry. Open only to British-born subjects resident in the UK. Final entry date: 20th December. 2008 winners: Gwendoline Riley *Joshua Spassky* (Jonathan Cape); Steven Hall *The Raw Shark Texts* (Canongate); Nick Laird *On Purpose* (Faber & Faber); Adam Thirwell *Miss Herbert* (Jonathan Cape).

💷 Awards £10,000 (total).

The Airey Neave Trust

PO Box 36800, 40 Bernard Street, London WC1N 1WJ

☎ 020 7833 4440

hanthoc@aol.com

www.theaireyneavetrust.org.uk

Contact *Hannah Scott*

Initiated 1989. Annual research fellowships for up to three years – towards a book or paper – for serious research connected with national and international law, and human freedom. Preferably attached to a particular university in Britain.

North East Literary Fellowship

Arts Council England, North East, Central Square, Forth Street, Newcastle upon Tyne NE1 3PJ

☎ 0191 255 8533   🖷 0191 230 1020

www.artscouncil.org.uk

Contact *Literature Officer*

A competitive fellowship in association with the Universities of Durham and Newcastle upon Tyne. Contact **Arts Council England, North East** for details.

Northern Writers' Awards

▷ New Writing North under Professional Associations and Societies

PAWS (Public Awareness of Science) Drama Fund

The PAWS Office, OMNI Communications, First Floor, 155 Regents Park Road, London NW1 8BB

pawsomni@btconnect.com

www.pawsdrama.com

Contacts *Andrew Millington, Andrée Molyneux*

Established 1994. PAWS encourages and supports television drama drawing on science, engineering and technology. It offers a range of activities to bring science and technology to television writers and producers in an easily accessible way.

Events: cover developments in science and technology and enable writers to meet scientists and engineers; Contacts and Information Service: provides specialist advice for writers to help them develop their television programme ideas; Periodic Funding Support: intended to help writers with some successful television drama experience bring their work closer to production; Annual Awards: for the best European programmes with scientific content in drama, documentary and all AV media.

Pearson Playwrights' Scheme

c/o Pearson Plc, 80 Strand, London WC2R 0RL

playwrightscheme@tiscali.co.uk

www.pearson.com

Administrator *Jack Andrews, MBE*

Awards up to five bursaries to playwrights annually, each worth £7,000. Applicants must be sponsored by a theatre which then submits the play for consideration by a panel. Each award allows the playwright a 12-month attachment. Applications invited via theatres in October each year. A further award of £10,000, for Best Play, is on offer each year to previous bursary winners. E-mail for up-to-date information.

The Charles Pick Fellowship

School of Literature and Creative Writing, University of East Anglia, Norwich NR4 7TJ

☎ 01603 592286   🖷 01603 507728

charlespickfellowship@uea.ac.uk

www1.uea.ac.uk/cm/home/schools/hum/lit/awards/pick

Contact *The Fellowship Administrator*

Founded in 2001 by the Charles Pick Consultancy in memory of the publisher and literary agent who died in 2000, to support a new unpublished writer of fictional or non-fictional prose. This residential Fellowship is awarded annually and lasts six months. There is no interview; candidates will be judged on the quality and promise of their writing, the project they describe and the strength of their referee's report. 2008 winner: David Sornig. Deadline for applications: 31st January each year.

💷 Award £10,000 plus accommodation on campus.

Peggy Ramsay Foundation

Hanover House, 14 Hanover Square, London W1S 1HP

☎ 020 7667 5000   🖷 020 7667 5100

laurence.harbottle@harbottle.com

www.peggyramsayfoundation.org

Contact *G. Laurence Harbottle*

Founded 1992 in accordance with the will of the late Peggy Ramsay, the well-known agent. Grants are made to writers for the stage who have some experience and who need time and resources to make writing possible. Grants are also made for writing projects by organizations connected with the theatre. The Foundation does not support production costs or any project that does not have a direct benefit to playwriting. Writers applying must have had one full-length play professionally produced for adults.

🄴 Grants total £150,000 to £200,000 per year.

The Royal Literary Fund

3 Johnson's Court, off Fleet Street, London EC4A 3EA

☎ 020 7353 7159   🖷 020 7353 1350
egunnrlf@globalnet.co.uk
www.rlf.org.uk
Secretary *Eileen Gunn*

Grants and pensions are awarded to published authors of several works in financial need, or to their dependants. Examples of author's works are needed for assessment by Committee. Contact the Secretary for further details and application form.

The Royal Society of Literature/Jerwood Awards

The Royal Society of Literature, Somerset House, Strand, London WC2R 1LA

☎ 020 7845 4676   🖷 020 7845 4679
paulaj@rslit.org
www.rslit.org
Submissions *Paula Johnson*

The awards offer financial assistance to authors engaged in writing their first major commissioned works of non-fiction. Three awards: one of £10,000 and two of £5,000 will be offered annually to writers working on substantial non-fiction projects. Open to UK and Irish writers and writers who have been resident in the UK for at least three years. Applications for the awards should be submitted by mid-October. Further details available on the website.

Scottish Arts Council Creative Scotland Awards

Scottish Arts Council, 12 Manor Place, Edinburgh EH3 7DD

☎ 0131 226 6051/ Help Desk: 0845 603 6000
🖷 0131 225 9833
help.desk@scottisharts.org.uk
www.scottisharts.org.uk

The Scottish Arts Council is currently reviewing the scope and opportunities offered by the Creative Scotland Awards in their present format in order to build on their existing success and to widen the range of potential applicants, as well as developing stronger links with successful applicants. It is anticipated that the Awards will be re-launched during 2009. Further information

will be available on the website and through SAC's 'Information Bulletin'.

Scottish Arts Council Writers' Bursaries

Scottish Arts Council, 12 Manor Place, Edinburgh EH3 7DD

☎ 0131 226 6051   🖷 0131 225 9833
gavin.wallace@scottisharts.org.uk
www.scottisharts.org.uk
Head of Literature *Dr Gavin Wallace*

Bursaries to enable published writers of literary work, playwrights and storytellers to devote more time to their writing. Around 20 bursaries of up to £15,000 awarded annually; deadline for applications in July and January. Application open to writers based in Scotland.

Scottish Book Trust Mentoring Scheme

Sandeman House, Trunk's Close, 55 High Street, Edinburgh EH1 1SR

☎ 0131 524 0160   🖷 0131 524 0161
www.scottishbooktrust.com
Contact *Caitrin Armstrong (Writer Development Coordinator)*

The mentoring scheme aims to support published writers living and working in Scotland who have a specific project on which they'd like some focused, dedicated support and advice. Successful applicants will be paired with another writer or industry professional who has appropriate experience and will support them as they work together over an intensive period of nine months. Applicants should see the website for full details.

Scottish Book Trust New Writers' Awards

Sandeman House, Trunk's Close, 55 High Street, Edinburgh EH1 1SR

☎ 0131 524 0160   🖷 0131 524 0161
www.scottishbooktrust.com
Contact *Caitrin Armstrong (Writer Development Coordinator)*

(Formerly the Scottish Arts Council New Writers' Bursaries.) Eight awards available for writers living and working in Scotland who have little or no previous publication or production track record. Successful applicants will receive a cash award of £2,000 plus a bespoke professional development package which may include nine months working with a mentor, plus networking opportunities with publishers/agents. The award is open to writers of fiction, literary non-fiction, poets and playwrights who can demonstrate their commitment to writing. A c.v., sample of work and referee will be required. Applicants must see website for full application details.

Laurence Stern Fellowship

Department of Journalism, City University, Northampton Square, London EC1V 0HB

☎ 020 7040 4036

a.r.mckane@city.ac.uk
www.city.ac.uk/journalism
Contact *Anna McKane*

Founded 1980. Awarded each year to a young journalist experienced enough to work on national stories. It gives a chance to work on the national desk of the *Washington Post*. A senior executive from the newspaper selects from a shortlist drawn up in February/March. 2008 winner: Holly Watt of *The Sunday Times*. Full details about how to apply available on the City University website in January.

Tom-Gallon Trust Award and the Olive Cook Prize

The Society of Authors, 84 Drayton Gardens, London SW10 9SB
📞 020 7373 6642　📠 020 7373 5768
info@societyofauthors.org
www.societyofauthors.org

An award of £1,000 is made on the basis of a submitted story to fiction writers of limited means who have had at least one short story accepted for publication. Both awards are biennial and are awarded in alternate years. Contact the **Society of Authors** for an entry form. Final entry date: 31st October.

The Betty Trask Awards

The Society of Authors, 84 Drayton Gardens, London SW10 9SB
📞 020 7373 6642　📠 020 7373 5768
info@societyofauthors.org
www.societyofauthors.org

These annual awards are for authors who are under 35 and Commonwealth citizens, awarded on the strength of a first novel (published or unpublished) of a traditional or romantic nature. The awards must be used for a period or periods of foreign travel. Final entry date: 31st January. Contact the **Society of Authors** for an entry

form. 2008 winners: David Szalay *London and the South-East* (Jonathan Cape); Ross Raisin *God's Own Country* (Viking); Thomas Leveritt *The Exchange Rate Between Love and Money* (Harvill Secker); Anna Ralph *The Floating Island* (Hutchinson).
💷 Awards £20,000 (total).

The Travelling Scholarships

The Society of Authors, 84 Drayton Gardens, London SW10 9SB
📞 020 7373 6642　📠 020 7373 5768
info@societyofauthors.org
www.societyofauthors.org

Annual honorary grants to established British writers. No submissions. 2008 winners: Marina Lewycka, Ruth Padel, Colin Thubron.
💷 Award £1,800 each.

The David T.K. Wong Fellowship

School of Literature and Creative Writing, University of East Anglia, Norwich NR4 7TJ
📞 01603 592286　📠 01603 507728
davidtkwongfellowship@uea.ac.uk
www1.uea.ac.uk/cm/home/schools/hum/lit/awards/wong
Contact *The Fellowship Administrator*

Named for its sponsor, Mr. David Wong (retired Hong Kong businessman, teacher, journalist, senior civil servant and writer of short stories); this unique and generous award was launched in 1997 to enable a fiction writer who wants to write in English about the Far East. This residential Fellowship is awarded annually and lasts one academic year. Unpublished or established published writers are welcome to apply. There is no interview; candidates will be judged on the quality and promise of their writing and the project they describe. 2008 winner: Nam Wi. Deadline for applications: 31st January each year.
💷 Award £26,000.

Prizes

Academi Cardiff International Poetry Competition

Academi, PO Box 438, Cardiff CF10 5YA
☎ 029 2047 2266 📠 029 2049 2930
post@academi.org
www.academi.org
Contact *Peter Finch (Chief Executive)*

Established 1986. An annual competition supported by Cardiff Council for unpublished poems in English of up to 50 lines. Closing date in January each year.
💷 Prizes total £6,000; £5,000 (1st); £500 (2nd); £250 (3rd); five runners-up receive £50 each.

J.R. Ackerley Prize

English PEN, 6–8 Amwell Street, London EC1R 1UQ
☎ 020 7713 0023 📠 020 7837 7838
enquiries@englishpen.org
www.englishpen.org

Commemorating the novelist/autobiographer J.R. Ackerley, this prize is awarded for a literary autobiography, written in English and published in the year preceding the award. Entry restricted to nominations from the Ackerley Trustees only ('please do not submit books'). 2008 winner: Miranda Seymour *In My Father's House* (Simon & Schuster).
💷 Prize of £1,000, plus silver pen.

Aldeburgh First Collection Prize
▷ Jerwood Aldeburgh First Collection Prize

Alexander Prize

Royal Historical Society, University College London, Gower Street, London WC1E 6BT
☎ 020 7387 7532 📠 020 7387 7532
royalhistsoc@ucl.ac.uk
www.royalhistoricalsociety.org/grants.htm
Contact *Administrative Secretary*

Established 1897. Awarded for a published journal article or essay based upon original research. The article/essay must have been published during the period 1st January 2007 to 31st December 2008. Competitors may choose their own subject for the essay. Closing date: 31 December.
💷 Prize of £250.

The And/or Book Awards

sarah.jackson@airdesk.com
www.kraszna-krausz.org.uk
Coordinator *Sarah Jackson*

Formerly known as the Kraszna-Krausz Book Awards, established in 1985. Annual awards to encourage and recognize oustanding achievements in the publishing and writing of books on the art, practice, history and technology of still photography and the moving image (film, television, video and related screen media). Books in English, distributed in the UK, are eligible. Entries must be submitted by publishers only.

Hans Christian Andersen Awards

IBBY, Nonnenweg 12, Postfach, CH-4003 Basel, Switzerland
☎ 00 41 61 272 2917 📠 00 41 61 272 2757
ibby@ibby.org
www.ibby.org
Deputy Director of Administration *Forest Zhang*
Director, Member Services, Communications & New Projects *Liz Page*

The highest international prizes for children's literature: The Hans Christian Andersen Award for Writing, established 1956; The Hans Christian Andersen Award for Illustration, established 1966. Candidates are nominated by National Sections of IBBY (The International Board on Books for Young People). Biennial prizes are awarded, in even-numbered years, to an author and an illustrator whose body of work has made a lasting contribution to children's literature. 2008 winners: Award for Writing: Jürg Schubiger (Switzerland); Award for Illustration: Roberto Innocenti (Italy).
💷 Awards: gold medals.

Angus Book Award

Angus Council, Educational Resources Service, Bruce House, Arbroath DD11 3TL
☎ 01241 435045 📠 01241 435038
cularbers@angus.gov.uk
www.angus.gov.uk/bookaward
Contact *Moyra Hood, Educational Resources Librarian*

Established 1995. Designed to try to help teenagers develop an interest in and enthusiasm for reading. Shortlisted books are read and voted on by third-year schoolchildren in all eight Angus secondary schools. Authors visit schools and take part in the award ceremony in May of each year. 2008 winner: Kate Cann *Leaving Poppy* (Scholastic).

£ Prize of £500 cheque, plus trophy in the form of a replica Pictish stone.

Annual Theatre Book Prize
▷ The Society for Theatre Research Annual Theatre Book Prize

Arts Council Children's Award
▷ Brian Way Award

Arvon Foundation International Poetry Competition

2nd Floor, 42a Buckingham Palace Road, London SW1W ORE

☎ 020 7931 7611   📠 020 7963 0961
poetry@arvonfoundation.org
www.arvonfoundation.org
Contact *The London Office*

Established 1980. Biennial competition (next in 2010) for poems written in English and not previously broadcast or published. There are no restrictions on the number of lines, themes, age of entrants or nationality. No limit to the number of entries. Entry fee: £7 per poem. Previous winners: Paul Farley *Laws of Gravity*; Don Paterson *A Private Bottling*.

£ Prize of £5,000 (1st), and £5,000 worth of other prizes.

Asham Award for Women

Asham Literary Endownment Trust, c/o Town Hall, High Street, Lewes BN7 2QS

☎ 01273 483159
carole.buchan@btinternet.com
www.ashamaward.com
Contact *Carole Buchan*

Founded in 1996, the award (named after the house in Sussex where Virginia Woolf lived) is a biennial short story competition for women and is administered by the Asham Literary Endownment Trust which encourages new writing through competitions, mentoring, training and publication. Candidates must be female, over 18, resident in the UK and must not have had a novel or complete anthology of work published previously. The winners of the current award (sponsored by the John. S. Cohen Foundation and Much Ado Books of Alfriston, East Sussex) will be announced to coincide with the launch of the new Asham anthology in September 2009. Information about the 2010 Award will be available on the website in the autumn of 2009.

£ Prize money up to £3,600; 12 winning writers

are published by Bloomsbury alongside specially commissioned stories by leading professional writers, including Margaret Atwood and Esther Freud.

Authors' Club First Novel Award

Authors' Club, 40 Dover Street, London W1S 4NP
☎ 020 7499 8581
stella@theartsclub.co.uk
Contact *Stella Kane*

Established 1954. This award is made for the most promising work published in Britain by a British author, and is presented at a dinner held at The Arts Club. Entries for the award are accepted from publishers from August of the year in question and must be full-length (short stories are not eligible). For further details/application form, contact *Stella Kane* at the e-mail address above or access the website. 2008 winner: Laura Beatty *Pollard* (Chatto & Windus).

The BA/Nielson BookData Author of the Year Award

Booksellers Association Ltd, 272 Vauxhall Bridge Road, London SW1V 1BA
☎ 020 7802 0801   📠 020 7802 0803
naomi.gane@booksellers.org.uk
www.booksellers.org.uk
Contact *Naomi Gane*

Founded as part of the BA Annual Conference to involve authors more closely in the event. Authors must be British or Irish. Not an award open to entry but voted on by the BA's membership. 2007 winner: Anthony Horowitz.

£ Award of £1,000, plus trophy.

Banipal Prize
▷ The Translators Association Awards

Verity Bargate Award

Soho Theatre, 21 Dean Street, London W1D 3NE
☎ 020 7287 5060   📠 020 7287 5061
www.sohotheatre.com

The award was set up to commemorate the late Verity Bargate, co-founder and director of the **Soho Theatre**. This national award is presented biennially for a new and unperformed play. Access details on the website. Previous winners include: In-Sook Chappell, Matt Charman, Shan Khan, Fraser Grace, Lyndon Morgans, Adrian Pagan, Diane Samuels, Judy Upton and Toby Whithouse.

The BBC National Short Story Prize

Room 316, BBC Henry Wood House, 3–6 Langham Place, London W1B 3DF
www.theshortstory.org.uk
www.bbc.co.uk/radio4

Launched at the 2005 Edinburgh International Book Festival, this annual short story award – the largest for a single story – is funded by the

BBC and is administered in partnership with the **Booktrust** and **Scottish Book Trust**. It was formerly known as the National Short Story Prize. Open to authors with a previous record of publication who are UK nationals or residents, aged 18 years or over. Entries can be stories published during 2008 or previously unpublished and must not be more than 8,000 words. Entrants must have a prior record of publication and submit original work that does not infringe the copyright or any other rights of any third party. The stories must be written in English and only one will be accepted per author. Full entry details from the address above or via the website. 2008 winner: Clare Wigfall *The Numbers*.

Prizes of £15,000 (overall); £3,000 (runner-up).

BBC Wildlife Magazine Nature Writing Award

BBC Wildlife Magazine, 14th Floor, Tower House, Fairfax Street, Bristol BS1 3BN
0117 927 9009   0117 934 9008
wildlifemagazine@bbcmagazinesbristol.com
www.bbcwildlifemagazine.com
Editor *Sophie Stafford*

This competition aims to revive the art of nature writing. Entrants should submit an original and unpublished 800-word essay on any aspect of the natural world that interests them, aiming for originality, story-telling and a sense of connection. The top prize is an expedition with Earthwatch and publication in the magazine. See the March issue and BBC Wildlife website for the call for entries and entry form.

Award of an Earthwatch expedition and publication in the magazine.

BBC Wildlife Magazine Poet of the Year Awards

BBC Wildlife Magazine, 14th Floor, Tower House, Fairfax Street, Bristol BS1 3BN
0117 927 9009   0117 934 9008
wildlifemagazine@bbcmagazinesbristol.com
www.bbcwildlifemagazine.com
Contact *Sophie Stafford*

Annual award for a poem, the subject of which must be the natural world and/or our relationship with it. Entrants may submit one poem only of no more than 50 lines with the entry form which is published in the April issue. Closing date for entries varies from year to year. See the magazine and website for entry details.

Poet of the Year Award: a wildlife break, publication in the magazine, plus the possibility of a reading of your poem on Radio 4's *Poetry Please*; four young poets awards, categories: age 7 and under, 8–11, 12–14, 15–17.

BBC Wildlife Magazine Travel Writing Award

BBC Wildlife Magazine, 14th Floor, Tower House, Fairfax Street, Bristol BS1 3BN

0117 927 9009   0117 934 9008
wildlifemagazine@bbcmagazinesbristol.com
www.bbcwildlifemagazine.com
Editor *Sophie Stafford*

This competition looks for travel writing that transports the reader to another place. Entrants should submit an 800-word essay that includes a true account of intimate encounter with wildlife, either local or exotic, and conveys a sense of travel, discovery and something of the essence of the natural world. The top prize is a wildlife 'holiday of a lifetime' and publication in the magazine. See the January issue and BBC Wildlife website for the call for entries and entry form.

The BBCFour Samuel Johnson Prize for Non-Fiction

Colman Getty, Middlesex House, 28 Windmill Street, London W1T 2JJ
020 7631 2666   020 7631 2699
lois@colmangetty.co.uk
Contact *Lois Tucker*

Established 1998. Annual prize sponsored by BBCFour. Eligible categories include the arts, autobiography, biography, business, commerce, current affairs, history, natural history, popular science, religion, sport and travel. Entries submitted by publishers only. 2008 winner: Kate Summerscale *The Suspicions of Mr Whicher: or the Murder at Road Hill House* (Bloomsbury).

Prize of £30,000; £1,000 to each shortlisted author.

Adrienne Benham Award

c/o Theatre Centre, Shoreditch Town Hall, 380 Old Street, London EC1V 9LT
020 7729 3066
admin@theatre-centre.co.uk
www.theatre-centre.co.uk

The award is made to a writer who satisfies the following qualities: demonstrates interest and enthusiasm for young people's theatre; proves a professional commitment to playwriting; displays imagination, skills, personality and verve in their writing. Applicants must be resident in the UK or Republic of Ireland and should also provide proof of a professionally performed play since August 2007 and submit the first ten pages of a script written in the last twelve months.

Award of £2,000.

David Berry Prize

Royal Historical Society, University College London, Gower Street, London WC1E 6BT
020 7387 7532   020 7387 7532
royalhistsoc@ucl.ac.uk
www.royalhistoricalsociety.org/grants.htm
Contact *Administrative Secretary*

Annual award for an essay of not more than 10,000 words on Scottish history. Closing date: 31st December.
🏵 Prize of £250.

Besterman/McColvin Awards

3 High Street, Eynsham, Witney OX29 4HA
📞 01865 810210
ISGRefAwards@cilip.org.uk
www.cilip.org.uk
Contact *Awards Administrator (at address above)*

ISG (CILIP) Reference Awards. The Besterman McColvin awards are given annually for an outstanding reference work first published in the UK during the preceding year. Consists of two categories: printed and electronic. Works eligible for consideration include: encyclopedias, general and special dictionaries; annuals, yearbooks and directories; handbooks and compendia of data; atlases. Nominations are invited from members of **CILIP**, publishers and others. 2006 winners: *Oxford African American Studies Centre*, published by Oxford University Press (electronic category); *Dictionary of Pastellists Before 1800* Neil Jeffares (printed category).
🏵 Cash prize for each category.

The Biographers' Club Prize

119A Fordwych Road, London NW2 3NJ
📞 020 8452 4993
anna@annaswan.co.uk
www.biographersclub.co.uk
Contact *Anna Swan*

Established 1999 by literary agent, biographer and founder of the Biographers' Club, Andrew Lownie, to finance and encourage first-time writers researching a biography. Sponsored by the *Daily Mail*. Open to previously unpublished writers, applicants should submit a proposal of no more than 20 pages, including outline, sample chapter, a note on the market for the book, sources consulted, competing books and a c.v. Entry fee: £10 (cheques made payable to the Biographers' Club). 2008 winner: Michael Bundock.
🏵 Prize of £2,000.

Birdwatch Bird Book of the Year

c/o Birdwatch Magazine, B403A The Chocolate Factory, 5 Clarendon Road, London N22 6XJ
📞 020 8881 0550
editorial@birdwatch.co.uk
www.birdwatch.co.uk
Contact *Dominic Mitchell*

Established in 1992 to acknowledge excellence in ornithological publishing – an increasingly large market with a high turnover. Annual award. Entries, from publishers, must offer an original and comprehensive treatment of their particular ornithological subject matter and must have a broad appeal to British-based readers. 2008 winner: D.W. Yalden and U. Albarella *The History of British Birds* (Oxford University Press).

Bisto Book of the Year Awards
▷ CBI Bisto Book of the Year Awards

James Tait Black Memorial Prizes

University of Edinburgh, David Hume Tower, George Square, Edinburgh EH8 9JX
📞 0131 650 3619 📠 0131 650 6898
www.ed.ac.uk/about/2.419/tait-black/intro
Contact *Department of English Literature*

Established 1918 in memory of a partner of the publishing firm of **A.&C. Black Ltd**. Two prizes, one for biography and one for fiction. Closing date for submissions: 1 December. Each prize is awarded for a book first published in Britain in the previous calendar year (1st January–31st December). 2007 winners: Rosalind Belben *Our Horses in Egypt* (fiction); Rosemary Hill *God's Architect: Pugin and the Building of Romantic Britain* (biography).
🏵 Prizes of £10,000 each.

Blue Peter Children's Book Awards

Booktrust, Book House, 45 East Hill, London SW18 2QZ
📞 020 8516 2973
www.bbc.co.uk/bluepeter
www.booktrust.org.uk
Contacts *Rosa Anderson, Helen Hayes*

Judged by a panel of adults and children. The three categories are: Book I Couldn't Put Down, Most Fun Story with Pictures and Best Book with Facts. Books must be published in paperback between 1st November and 31st October the following year. Winners are announced in the spring in a special broadcast on Blue Peter. 2009 winners: Book I Couldn't Put Down and winner of Blue Peter Book of the Year 2009: *Shadow Forest* by Matt Haig (Corgi); Most Fun Story with Pictures: *Mr Gum and the Dancing Bear* by Andy Stanton, illustrated by David Tazzyman (Egmont); Best Book with Facts: *Horrible Geography Handbooks – Planet in Peril* by Anita Ganeri, illustrated by Mike Phillips (Scholastic).

Boardman Tasker Award

Pound House, Llangennith, Swansea SA3 1JQ
📞 01792 386215 📠 01792 386215
margaretbody@lineone.net
www.boardmantasker.com
Contact *Maggie Body, Honorary Secretary*

Established 1983, this award is given for a work of fiction, non-fiction or poetry, whose central theme is concerned with the mountain environment and which can be said to have made an outstanding contribution to mountain

literature. Authors of any nationality are eligible, but the book must have been published or distributed in the UK for the first time between 1st November 2007 and 31st October 2008. Entries from publishers only. 2008 winner: Andy Kirkpatrick *Psychovertical* (Hutchinson).
£ Prize of £3,000 (at Trustees' discretion).

Bollinger Everyman Wodehouse Prize

Colman Getty, 28 Windmill Street, London
W1T 2JJ
☎ 020 7631 2666
katym@colmangetty.co.uk
Contact *Katy Macmillan-Scott*

Annual prize, launched in 2000 on the 25th anniversary of the death of P.G. Wodehouse. Presented at the **Guardian Hay Festival**, awarded to the best comic novel published in the preceding twelve months. 2008 winner: Will Self *The Butt* (Bloomsbury).

The Booker Prize for Fiction
▷ The Man Booker Prize for Fiction

Booktrust Early Years Awards

Booktrust, Book House, 45 East Hill, London
SW18 2QZ
☎ 020 8516 2972   ☒ 020 8516 2978
tarryn@booktrust.org.uk
www.booktrust.org.uk
Contact *Tarryn McKay*

Formerly the Sainsbury's Baby Book Award, established in 1999. Annual awards with three categories: the Baby Book Award, the Best Book for Pre-School Children (up to the age of 5), and an award for the Best Emerging Illustrator. Books to be submitted by publishers only; authors and illustrators must be of British nationality or other nationals who have been resident in the UK for at least five years. 2008 winners, Baby Book: Georgie Birkett *Is This My Nose?* (Red Fox); Pre-School: Jeanne Willis, illus. Gwen Millward *The Bog Baby* (Puffin); Best Emerging Illustrator: Tim Hopgood *Here Comes Frankie!* (Macmillan).
£ Prizes of £2,000 for each category; in addition the Best Emerging Illustrator receives a specially commissioned piece of artwork.

Booktrust Teenage Prize

Book House, 45 East Hill, London SW18 2QZ
☎ 020 8516 2973   ☒ 020 8516 2978
megan@booktrust.org.uk
www.bookheads.org.uk
Contact *Megan Farr*

Established 2003. Annual prize that recognizes and celebrates the best in writing for teenagers. Funded and administered by Booktrust. Open to works of fiction for young adults in the UK, the books to be published between 1st July 2008 and 30th June 2009. Final entry date in March. 2008

winner: Patrick Ness *The Knife of Never Letting Go* (Walker).
£ Prize of £2,500.

Harry Bowling Prize

c/o 16 St Briac Way, Exmouth EX8 5RN
☎ 01395 279659
marina@marina-oliver.net
www.harrybowlingprize.net
Contacts *Margaret James, Marina Oliver*

Established 2000 in honour of Harry Bowling, 'the king of Cockney sagas' (died 1999). Biennial award (next in 2010), sponsored by **Headline Book Publishing**, to encourage writers of adult fiction set in London. Open to anyone who has not been published previously. Final entry date: 31st March. Entry fee charged; forms from address above (enclose s.a.e.) or via the website. 2008 winner: Kathie Highley.
£ Prize winner receives £1,000; runners-up, £100.

The Branford Boase Award

8 Bolderwood Close, Bishopstoke, Eastleigh
SO50 8PG
☎ 01962 826658   ☒ 01962 856615
anne.marley@tiscali.co.uk
www.branfordboaseaward.org.uk
Administrator *Anne Marley*

Established in 2000 in memory of children's novelist, Henrietta Branford and editor and publisher, Wendy Boase. To be awarded annually to encourage and celebrate the most promising novel by a new writer of children's books, while at the same time highlighting the importance of the editor in nurturing new talent. 2008 winner: Jenny Downham *Before I Die* (author); David Fickling, **David Fickling Books** (editor).
£ Award of a specially commissioned box, carved and inlaid in silver with the Branford Boase Award logo; winning author receives £1,000.

The Bridport Prize

PO Box 6910, Bridport DT6 9BQ
☎ 01308 428333
frances@bridportprize.org.uk
www.bridportprize.org.uk
Contact *Frances Everitt, Administrator*

Annual competition for poetry and short story writing. Unpublished work only, written in English. Winning stories are read by a literary agent, the winning poems are put forward to the **Forward Prize**, and an anthology of winning entries is published. Final entry date: 30th June. Send s.a.e. for entry forms.
£ Prizes of £5,000, £1,000 & £500 in each category, plus 10 supplementary prizes of £50 each. Winning stories are submitted to the National Short Story Prize.

Katharine Briggs Folklore Award

The Folklore Society, c/o The Warburg Institute,
Woburn Square, London WC1H 0AB
☏ 020 7862 8564
enquiries@folklore-society.com
www.folklore-society.com
Contact *The Convenor*

Established 1982. An annual award in November
for the book, published in Britain and Ireland
between 1 June in the previous calendar year and
30th May, which has made the most distinguished
non-fiction contribution to folklore studies.
Intended to encourage serious research in the
field which Katharine Briggs did so much to
establish. The term folklore studies is interpreted
broadly to include all aspects of traditional and
popular culture, narrative, belief, custom and folk
arts. 2008 winner: Richard Bebb *Welsh Furniture
1250–1950* (Saer Books).
💷 Prize of £200, plus engraved goblet.

Bristol Festival of Ideas Best Book of Ideas Prize

Bristol Cultural Development, Leigh Court,
Abbots Leigh, Bristol BS8 3RA
☏ 07778 932778
ideas@gwebusinesswest.co.uk
www.ideasfestival.co.uk
Contact *Vicky Washington*

This annual award was established to mark the
fifth anniversary of the Bristol Festival of Ideas in
2009 and was founded by Andrew Kelly, Director
of the Festival, in association with Blackwell and
Arts & Business. The award is given to the book
which is considered to present new, important
and challenging ideas, which is rigorously argued,
and which is engaging and accessible. Submitted
books must be first published in the year prior
to each award (2008 for the 2009 award). Final
entry date: end October (for books published in
November and December proofs and manuscripts
can be provided). Submissions by publishers only.
💷 Prize of £10,000.

British Association for Applied Linguistics (BAAL) Book Prize

BAAL Publications Secretary, Department of
Linguistics and English Language, Lancaster
University, Lancaster LA1 4YT
☏ 01524 594642   💷 01524 843085
v.koller@lancaster.ac.uk
www.baal.org.uk
Contact *Veronika Koller*

Annual award made by the British Association
for Applied Linguistics to an outstanding book
in the field of applied linguistics. Final entry in
mid-December . Nominations from publishers

only. 2008 winner: Alastair Pennycook *Global
Englishes and Transcultural Flows* (Routledge).

British Book Awards
▷ Galaxy British Book Awards

British Czech & Slovak Association Prize

The BCSA Prize Administrator, 24 Ferndale,
Tunbridge Wells TN2 3NS
☏ 01892 543206
prize@bcsa.co.uk
www.bcsa.co.uk
Contact *Prize Administrator*

The British Czech & Slovak Association offers an
annual prize for the best piece of original writing,
in English, on the links between Britain and
the Czech and Slovak Republics, or describing
society in transition in those Republics since the
Velvet Revolution in 1989. Entries can be fiction
or factual, should not have been previously
published and should be up to 2,000 words in
length. Submissions are invited from individuals
of any age, nationality or educational background.
Closing date: 30th June. Entry details for 2009
should be checked with the Prize Administrator.
2008 winners: *A Scottish Harvest in Bohemia
– Edwin Muir's Return to Hope* Clarice Cloutier
(1st prize); *The Leander Bud* Zuzana Demcakova
(2nd).
💷 Prize of £300, presented at the BCSA's annual
dinner in London; 2nd prize, £100. The winning
entry is published in the *British Czech & Slovak
Review*.

British Fantasy Awards

23 Mayne Street, Hanford, Stoke on Trent
ST4 4RF
bfsawards@britishfantasysociety.org
www.britishfantasysociety.org
Awards Administrator *David Sutton*

Awarded by the **British Fantasy Society** by
members at its annual conference for Best Novel
and Best Short Story categories, among others.
Not an open competition. Previous winners
include: Ramsey Campbell, Dan Simmons,
Michael Marshall Smith, Thomas Ligotti.

British Press Awards

Press Gazette, WBI Media & Entertainment,
Paulton House, 8 Shepherdess Walk, London
N1 7LB
☏ 020 7549 8666   💷 020 7549 8668
apont@wilmington.co.uk
www.pressgazette.co.uk

'The Oscars of British journalism.' Open to all
British national newspapers and news agencies.
March event. Run by *Press Gazette*.

British Science Fiction Association Awards

61 Ivycroft Road, Warton, Nr Tamworth B79 0JJ
awards@bsfa.co.uk
www.bsfa.co.uk
Awards Administrator *Donna Scott*

Established 1970. Categories for novel, short fiction, non-fiction and artwork. The awards are announced and presented at the British National Science Fiction Convention every Easter. 2007 winners: Ian McDonald *Brasyl* (novel); Ken MacLeod *Lighting Out* (short fiction); Andy Bigwood *Cracked World* (artwork – for cover of *disLocations*); BSFA Fiftieth Anniversary Award – Best Novel of 1958: *Non-Stop* by Brian Aldiss.

British Sports Book Awards

National Sporting Club, Café Royal, 68 Regent Street, London W1B 5EL
☏ 020 7437 0144   🖷 020 7437 5441
david@nationalsportingclub.co.uk
www.nationalsportingclub.co.uk
Chairman *David Willis*

Established 2003. Annual awards presented at a Café Royal lunch in London in March. Seven categories: Best Autobiography, Best Biography, Best New Writer, Best Illustrated Title, Best Cricket Book, Best Football Book, Best Rugby Book. Nominations received in November each year by e-mail (address above). Sponsors include Skysports, *The Times*, Llanllyr Spring Water, Littlehampton Book Services and Ladbrokes.

Buxton Poetry Competition

c/o Buxton Festival, 3 The Square, Buxton SK17 6AZ
☏ 01298 70395   🖷 01298 72289
claire@buxtonfestival.co.uk
www.derby.ac.uk/buxtonpoetrycompetition
Contact *Claire Rhodes*

Founded in 2008 by Buxton Festival and the University of Derby, the competition aims to develop interest in creative writing across the country and particularly in Derbyshire. Also aims to encourage children and young people to get involved with poetry by giving free entry to everyone under 19 years old. Schools in Derbyshire are offered poetry workshops to help the children get their poems ready for the competition (booking information on the website). Final entry date: 1st April. All poems must be submitted by post with a completed entry form and cheque (where applicable); entry details and form available on the website. Open Poetry: for poets over 18 years old; entry fee: £5 per poem; Young People's Poetry: for poets aged between 12 and 18; free entry; Children's Poetry: for poets aged under 12 years; free entry. 2008 Open Poetry winner: Lesley Saunders with her poem entitled *Rill*.

💷 Prizes – Open Poetry: cash prizes of £300, £200 and £100; Young People's Poetry and Children's Poetry: Waterstone's book tokens. All winning poets have the chance to have their poems displayed in an exhibition at the Devonshire Dome in Buxton during the Festival.

The Caine Prize for African Writing

51a Southwark Street, London SE1 1RU
☏ 020 7378 6234   🖷 020 7378 6235
info@caineprize.com
www.caineprize.com
Administrator *Nick Elam*
Secretary *Jan Hart*

Annual award founded in 1999 in memory of Sir Michael Caine, former chairman of Booker plc, to recognize the worth of African writing in English. Awarded for a short story by an African writer, published in English anywhere in the world. An 'African writer' is someone who was born in Africa, or who is a national of an African country, or whose parents are African, and whose work has reflected African sensibilities. Final entry date: 31st January; submissions by publishers only. 2008 winner: Henrietta Rose-Innes *Poison* (Spearhead).
💷 Prize of £10,000.

The Calouste Gulbenkian Prize
▷ The Translators Association Awards

James Cameron Award

City University, Department of Journalism, Northampton Square, London EC1V 0HB
☏ 020 7040 8221   🖷 020 7040 8594
H.Stephenson@city.ac.uk
Contact *Hugh Stephenson*

Annual award for journalism to a reporter of any nationality, working for the British media, whose work is judged to have contributed most during the year to the continuance of the Cameron tradition. Previous winner: Patrick Cockburn.

Canadian Poetry Association (Annual Poetry Contest)

Canadian Poetry Association, 331 Elmwood Dr., Suite 4-212, Moncton, New Brunswick E1A 1X6, Canada
riverbones@live.ca
www.canadianpoetryassoc.com

Annual contest now open to non-members of the Association. See the website for entry details. All winning poems are published on the CPA website and in *Poemata*.
💷 Three cash prizes.

Cardiff International Poetry Competition
▷ Academi Cardiff International Poetry Competition

Carey Award

Society of Indexers, Woodbourn Business Centre, 10 Jessell Street, Sheffield S9 3HY

☎ 0114 244 9561 ℻ 0114 244 9563

admin@indexers.org.uk

www.indexers.org.uk

Secretary *Judith Menes*

A private award made by the Society to a member who has given outstanding services to indexing. The recipient is selected by the Executive Board with no recommendations considered from outside the Society.

Carnegie Medal

▷ CILIP: The Chartered Institute of Library and Information Professionals Carnegie Medal

CBI Bisto Book of the Year Awards

17 North Great Georges Street, Dublin 1, Republic of Ireland

☎ 00 353 1 872 7475 ℻ 00 353 1 872 7476

info@childrensbooksireland.ie

www.childrensbooksireland.ie

Contact *Jenny Murray*

Founded 1990 as Bisto Book of the Decade Awards. This led to the establishment of an annual award made by the Irish Children's Book Trust, later to become Children's Books Ireland. Open to any author or illustrator of children's books born or resident in Ireland; open to English or Irish languages. 2007/08 winners: Bisto Book of the Year: Siobhan Dowd *The London Eye Mystery* (David Fickling Books); Eilís Dillon Award: Tom Kelly *The Thing with Finn* (Macmillan).
💷 Prizes of €10,000 (Bisto Book of the Year); €3,000 (Eilís Dillon Award); three honour awards of €1,000 each.

The Centaur Essay Prize

centaurfound@gmail.com

A new prize which marks the first stage in founding a charitable educational trust, the Centaur Foundation, supporting teachers and parents wanting innovative humane standards encouraged from primary school onwards. The submitted essay of 2,000–3,000 words should be on a theme of non-violent solutions and methods for the 21st century and beyond. The essay may be academic, even partly anecdotal, with a religious or secular slant. Entry deadline: 30th November 2009. Winners will be contacted by 28th February 2010. All essayists *must* in the first instance e-mail the address above to request a copy of the entry guidelines(essay submissions are not accepted by e-mail).
💷 Prize of £500 and two runners-up prizes of £250.

Sid Chaplin Short Story Competition

Shildon Town Council, Civic Hall Square, Shildon DL4 1AH

☎ 01388 772563 ℻ 01388 775227

Contact *Mrs J.M. Stafford*

Established 1986. Annual short story competition. Maximum 3,000 words; £2.50 entrance fee (juniors free). All stories must be unpublished and not broadcast and/or performed.
💷 Prizes of £300 (1st); £150 (2nd); £75 (3rd); £30 (Junior).

Chapter One Promotions Children's Story Competition

19–35 Sylvan Grove, London SE15 1PD

☎ 0845 456 5364

kidskorner2@chapteronepromotions.com

www.chapteronepromotions.com

Contacts *Johanna Bertie, Preetha Leela*

Established in 2007, the competition is open to all young writers between the ages of seven and 18 to submit unpublished short stories. Five titles or themes are given to choose from. Maximum length 1,000 words. Contact details on a separate sheet. Submissions and payment accepted online or by post. Closing date: 31 October.
💷 Prizes of £150, £75, £50 in each of the four age categories, plus publication of the winners and runners-up in the Children's Story Book. Best school entry receives £200 worth of books plus a trophy.

Chapter One Promotions International Short Story Competition

19–35 Sylvan Grove, London SE15 1PD

☎ 0845 456 5364

info@chapteronepromotions.com

www.chapteronepromotions.com

Contacts *Johanna Bertie, Preetha Leela*

Established in 2005, the competition is open to new and established writers to submit unpublished short stories of fewer than 2,500 words. Author details on separate sheet. Submissions and payment accepted online. Closing date: 14th January. 2008 winner: Jendi Reiter.
💷 Prizes of £2,500, £1,000, £500, plus publication of winners, runners-up and highly recommended in anthology.

Chapter One Promotions Open Poetry Competition

19–35 Sylvan Grove, London SE15 1PD

☎ 0845 456 5364

info@chapteronepromotions.com

www.chapteronepromotions.com

Contacts *Johanna Bertie, Preetha Leela*

Established in 2005. Unpublished poems of no more than 30 lines requested. Judge decides on the top 20 poems which are displayed on the website and the public votes for their favourite poem. The three poems with the most votes win. Closing date: 1st June. Online voting period: 1st–15th July. Winning poems appear on the website in August. Submissions and payment accepted online. 2008 winner: Sarah Redford.
£ Prizes £1,000, £500, £250.

Children's Book Circle Eleanor Farjeon Award
▷ Eleanor Farjeon Award

The Children's Laureate

Booktrust, Book House, 45 East Hill, London SW18 2QZ
☎ 020 8516 2976   📠 020 8516 2978
childrenslaureate@booktrust.org.uk
www.childrenslaureate.org

Established 1998. The Laureate is awarded biennially to an eminent British writer or illustrator of children's books both in celebration of a lifetime's achievement and to highlight the role of children's book creators in inspiring, informing and entertaining young readers. 2007–09 winner: Michael Rosen.
£ Award winner receives a medal and a bursary.

CILIP: The Chartered Institute of Library and Information Professionals Carnegie Medal

7 Ridgmount Street, London WC1E 7AE
☎ 020 7255 0650   📠 020 7255 0651
ckg@cilip.org.uk
www.ckg.org.uk

Established 1936. Presented for an outstanding book for children written in English and first published in the UK during the preceding year. Fiction, non-fiction and poetry are all eligible. 2008 winner: Philip Reeve *Here Lies Arthur* (Scholastic).
£ Award winner receives a medal.

CILIP: The Chartered Institute of Library and Information Professionals Kate Greenaway Medal

7 Ridgmount Street, London WC1E 7AE
☎ 020 7255 0650   📠 020 7255 0651
ckg@cilip.org.uk
www.ckg.org.uk

Established 1955. Presented annually for the most distinguished work in the illustration of children's books first published in the UK during the preceding year. 2008 winner: Emily Gravett *Little Mouse's Big Book of Fears* (Macmillan).
£ Award winner receives a medal. The Colin Mears Award (£5,000 cash) is given annually to the winner of the Kate Greenaway Medal.

Arthur C. Clarke Award for Science Fiction

246e Bethnal Green Road, London E2 0AA
clarkeaward@gmail.com
www.clarkeaward.com
Administrator *Tom Hunter*

Established 1986. The Arthur C. Clarke Award is given annually to the best science fiction novel with first UK publication in the previous calendar year. Both hardcover and paperback books qualify. Made possible by a generous donation by the late Arthur C. Clarke, this award is selected by a rotating panel of judges nominated by the **British Science Fiction Association**, the **Science Fiction Foundation** and the website SFCrowsnest.com. 2008 winner: Richard Morgan *Black Man* (Gollancz).
£ Award of £2,008 (award increases by £1 per year), plus trophy.

CLPE Poetry Award

Centre for Literacy in Primary Education, Webber Street, London SE1 8QW
☎ 020 7902 2287   📠 020 7928 4624
ann@clpe.co.uk
www.clpe.co.uk
Contact *Ann Lazim*

Founded in 2003, the CLPE Poetry Award honours excellence in poetry written for children. It is presented annually for a collection or anthology of poetry for children, published for the first time in the UK or the Republic of Ireland in the preceding year. Entry deadline: end of February. 2008 winner Jackie Kay *Red Cherry Red* (Bloomsbury).

David Cohen Prize for Literature

Arts Council England, 14 Great Peter Street, London SW1P 3NQ
☎ 0845 300 6200
www.artscouncil.org.uk/davidcohenprize

Established in 1993 by the Arts Council and awarded biennially, the David Cohen Prize for Literature is one of the most distinguished literary prizes in Britain. It recognizes writers who use the English language and who are citizens of the UK and the Republic of Ireland, encompassing dramatists, novelists, poets and essayists. The prize is for a lifetime's achievement and is donated by the John S. Cohen Foundation. The Clarissa Luard Award (a £12,500 award provided by Arts Council England) is awarded by the winner, to encourage new work, with the dual aim of encouraging young writers and readers. Submissions are not accepted for either award. Previous winners of the David Cohen Prize for Literature: William Trevor, Dame Muriel Spark, Harold Pinter, Doris Lessing, Beryl Bainbridge and Thom Gunn; Derek Mahon. 2009 winner: Seamus Heaney.

£ Award of £40,000, plus £12,500 towards new work.

The Commonwealth Writers Prize

Commonwealth Foundation, Marlborough House, Pall Mall, London SW1Y 5HY

☎ 020 7747 6262

j.sobol@commonwealth.int

www.commonwealthfoundation.com/
culturediversity/writersprize/

Established 1987. An annual award to reward and encourage the upsurge of new Commonwealth fiction. Any work of prose or fiction is eligible, i.e. a novel or collection of short stories. No drama or poetry. The work must be first written in English by a citizen of the Commonwealth and be first published in the year before its entry for the prize. Entries must be submitted by the publisher to the region of the writer's Commonwealth citizenship. The four regions are: Africa, Europe and South Asia, South East Asia and South Pacific, Caribbean and Canada. 2008 winners: Best Book Award: Lawrence Hill (Canada) *The Book of Negroes* (HarperCollins); Best First Book Award: Tahmima Anam (Bangladesh) *A Golden Age* (John Murray).

£ Prizes of £10,000 for Best Book; £5,000 for Best First Book; 8 prizes of £1,000 for each best and best first book in four regions.

The Duff Cooper Prize

54 St Maur Road, London SW6 4DP

artemiscooper@btopenworld.com

Contact *Artemis Cooper*

An annual award for a literary work of biography, history, politics or poetry, published by a recognized publisher (member of the **Publishers Association**) during the previous 12 months. The book must be submitted by the publisher, not the author. Financed by the interest from a trust fund commemorating Duff Cooper, first Viscount Norwich (1890–1954). 2008 winner: Kei Bird and Martin J. Sherwin *American Prometheus: The Triumph and Tragedy of J. Robert Oppenheimer* (Atlantic Books).

£ Prize of £5,000.

Costa Book Awards

The Booksellers Association, Minster House, 272 Vauxhall Bridge Road, London SW1V 1BA

☎ 020 7802 0802 📠 020 7802 0803

naomi.gane@booksellers.org.uk

www.costabookawards.com

Contact *Naomi Gane*

Established 1971. Formerly the Whitbread Book Awards. The awards celebrate and promote the best contemporary British writing. They are judged in two stages and offer a total of £50,000 prize money. The awards are open to novel, first novel, biography, poetry and children's book, each judged by a panel of three judges. The winner of each award receives £5,000. The Costa Book of the Year (£25,000) is chosen from the category winners. Writers must have lived in Britain and Ireland for three or more years. Submissions received from publishers only. Closing date: early July. Sponsored by Costa. 2008 winners: Novel and Overall Winner: Sebastian Barry *The Secret Scripture* (Faber & Faber); First Novel: Sadie Jones *The Outcast* (Vintage); Biography: Diana Athill *Somewhere Towards the End* (Granta Books); Poetry: Adam Foulds *The Broken Word* (Jonathan Cape); Children's: Michelle Magorian *Just Henry* (Egmont).

The Coventry Inspiration Book Awards

Coventry Schools Library & Resource Service, Central Library, Smithford Way, Coventry CV1 1FY

☎ 02476 832338 📠 02476 832358

sls@coventry.gov.uk

www.myvotescoventry.org

Children & Young People's Categories and Management of Awards *Joy Court*

Adult Categories *Mandy Hayward*

Established 2006/07. Annual award inspired by the Family Reading Campaign uniquely has an award for every age-group of children and young people and by genre categories for adults. The awards go live in October during Children's Book Week and every shortlisted title is open for reader's comments and votes. From January each week the two books with the least votes in each category get 'eliminated'. Voting then starts again each week so fans have to keep voting to keep their favourites in. The final winners are announced in February half term. An Awards ceremony is held later in the year to celebrate the winners in each category and to present individual trophies. Books in all C&YP categories should not have won major national awards before. The exception is the 'Our Special Favourites' which by its very nature and the nomination process could not make that distinction.

Rose Mary Crawshay Prize

The British Academy, 10 Carlton House Terrace, London SW1Y 5AH

☎ 020 7969 5200 📠 020 7969 5300

events@britac.ac.uk

www.britac.ac.uk

Contact *British Academy Chief Executive and Secretary*

Established 1888 by Rose Mary Crawshay, this prize is given for an historical or critical work to a woman of any nationality on English literature, with particular preference for a work on Keats,

Byron or Shelley. The work must have been published in the preceding three years.

Crime Writers' Association (Cartier Diamond Dagger)

PO Box 273, Boreham Wood WD6 2XA
secretary@thecwa.co.uk
www.thecwa.co.uk
Contact *The Secretary*

Established 1986. An annual award for a lifetime's outstanding contribution to the genre. 2009 winner: Andrew Taylor.

Crime Writers' Association (The CWA Gold Dagger for Non-Fiction)

PO Box 273, Boreham Wood WD6 2XA
secretary@thecwa.co.uk
www.thecwa.co.uk
Contact *The Secretary*

Biennial award for the best non-fiction crime book published during the year. Nominations from publishers only. 2008 winner: Kester Aspden *Nationality: Wog – The Hounding of David Oluwale* (Jonathan Cape).
£ Award winner receives a Dagger, plus cheque (sum varies).

Crime Writers' Association (The Dagger in the Library)

PO Box 273, Boreham Wood WD6 2XA
secretary@thecwa.co.uk
www.thecwa.co.uk
Contact *The Secretary*

Reinstated 2002. Annual award (sponsored by **Random House**) to the author whose work has given most pleasure to readers. Nominated and judged by librarians. 2008 winner: Craig Russell.
£ Award winner receives a Dagger, plus cheque.

Crime Writers' Association (The Debut Dagger)

The Debut Dagger Competition, PO Box 273, Boreham Wood WD6 2XA
debut.dagger@thecwa.co.uk
www.thecwa.co.uk
Contact *The Secretary*

Annual competition (sponsored by **Orion**) for unpublished writers to submit the first 3,000 words and 500-word outline of a crime novel (entry fee and form required). For full details see website or send s.a.e. to address above. 2008 winner: Amer Anwar *Western Fringes*.
£ Award winner receives a Dagger, plus cheque.

Crime Writers' Association (The Ian Fleming Steel Dagger)

PO Box 273, Boreham Wood WD6 2XA
secretary@thecwa.co.uk
www.thecwa.co.uk

Contact *The Secretary*

Founded 2002. Annual award for the best thriller, adventure or spy novel. Sponsored by Ian Fleming Publications Ltd to celebrate the best of contemporary thriller writing. 2008 winner: Tom Rob Smith *Child 44* (Simon & Schuster).
£ Award winner receives a Dagger, plus cheque.

Crime Writers' Association (The CWA Ellis Peters Award)

PO Box 273, Boreham Wood WD6 2XA
secretary@thecwa.co.uk
www.thecwa.co.uk
Contact *The Secretary*

Sponsored by the Estate of Ellis Peters, Time Warner and **Headline**. Established 1999. Annual award for the best historical crime novel. Nominations from publishers only. 2008 winner: Laura Wilson *Stratton's War* (Orion).
£ Award plus cheque.

Crime Writers' Association (The CWA Gold Dagger)

PO Box 273, Boreham Wood WD6 2XA
secretary@thecwa.co.uk
www.thecwa.co.uk
Contact *The Secretary*

An annual award for the best crime fiction published during the year. Nominations from publishers only. 2008 winner: Frances Fyfield *Blood from Stone* (Little, Brown).

Crime Writers' Association John Creasey (New Blood) Dagger

PO Box 273, Boreham Wood WD6 2XA
secretary@thecwa.co.uk
www.thecwa.co.uk
Contact *The Secretary*

Established in 1973 following the death of crime writer John Creasey, founder of the **Crime Writers' Association**. This award, sponsored by **BBC Audio Books**, is given annually for the best crime novel by an author who has not previously published a full-length work of fiction. Fomerly known as the John Creasey Dagger. Nominations from publishers only. 2008 winner: Matt Rees *The Bethlehem Murders* (Atlantic Books).
£ Award winner receives a Dagger, plus cheque.

The Roald Dahl Funny Prize

Booktrust, Book House, 45 East Hill, London SW18 2QZ
☎ 020 8516 2972   ▣ 020 8516 2978
tarryn@booktrust.org.uk
www.booktrust.org.uk
Contacts *Tarryn McKay, Helen Hayes*

Established in 2008 by Michael Rosen as part of his work as Children's Laureate, and administered by Booktrust. The prize is to honour those books

that simply make us laugh and aims to reward those authors and artists who write and illustrate their books using humour in their stories, poetry and fiction. Books must have been first published in the UK or Ireland between 1st October and 31st September. Authors and illustrators must be British or Irish, or other nationals living in Britain or Ireland at the time of entering. The prize is awarded to the funniest book in two categories: for children aged six and under (picture book) and for children aged 7–14. Inaugural winners: six and under: *The Witch's Children Go to School* Ursula Jones, illustrated by Russell Ayto (Orchard Books); 7–14: *Mr Gum and the Dancing Bear* Andy Stanton, illustrated by David Tazzyman (Egmont Press).
🊱 Prizes of £2,500 for each category.

Daily Mail First Novel Award

Transworld Publishers, 61–63 Uxbridge Road, London W5 5SA
www.dailymail.co.uk/books

Founded 2007. Open competition for writers aged 16 or over, resident in the UK or Republic of Ireland. The winning novel will be published by Transworld in 2009 with an advance of £30,000. Inaugural winner: Roland Vernon *A Dark Enchantment* (Black Swan).

Hunter Davies Prize
▷ Lakeland Book of the Year Awards

Felix Dennis Prize for Best First Collection
▷ The Forward Prizes for Poetry

George Devine Award

9 Lower Mall, Hammersmith, London W6 9DJ
Contact *Christine Smith*

Annual award for a promising new playwright writing for the stage in memory of George Devine, artistic director of the **Royal Court Theatre**, who died in 1965. The play, which can be of any length, does not need to have been produced. Send two copies of the script, plus outline of work, to *Christine Smith*. Closing date: 1st March. Send s.a.e. for the script to be returned if required. Information leaflet available from January on receipt of s.a.e. 2008 winner: Hassan Abdulrazzak.
🊱 Prize of £10,000.

Dingle Prize

PO Box 3401, Norwich NR7 7JF
📞 01603 516236
lucy.tetlow@bshs.org.uk
www.bshs.org.uk

Biennial award made by the British Society for the History of Science to the best book in the history of science (broadly construed) which is accessible to a wide audience of non-specialists. Next award:

2009. 2007 winner: Philip Ball *Elegant Solutions: Ten Beautiful Experiments in Chemistry* (Royal Society of Chemistry).
🊱 Prize of £300.

Dolman Best Travel Book Award

Authors' Club, 40 Dover Street, London W1S 4NP
📞 020 7408 5092
stella@theartsclub.co.uk
Contact *Stella Kane*

William Dolman, a former chairman of the **Authors' Club**, instituted this prize for the most promising book of travel literature. Books published in Great Britain by British writers are eligible and submissions are accepted in January of each year. 2008 winner: John Lucas *92 Acharnon Street* (Eland Publishing).

Drama Association of Wales Playwriting Competition

The Old Library, Singleton Road, Splott, Cardiff CF24 2ET
📞 029 2045 2200   📠 029 2045 2277
aled.daw@virgin.net
Contact *Teresa Hennessy*

Annual competition held to promote the writing of one-act plays in English and Welsh of between 20 and 50 minutes' playing time. Application forms from the address above. Closing date: 31st January.

The John Dryden Competition

School of Literature and Creative Writing, University of East Anglia, Norwich NR4 7TJ
📞 01603 250599
transcomp@uea.ac.uk
www.bcla.org
Competition Organizer *Professor Jean Boase-Beier*

Established 1983. Annual competition open to unpublished literary translations from all languages. Maximum submission: 25 pages.
🊱 Prizes of £350 (1st); £200 (2nd); £100 (3rd); plus publication for all winning entries in the British Comparative Literature Association's journal. Other entries may receive commendations.

The Duke of Westminster's Medal for Military Literature

Royal United Services Institute for Defence and Security Studies, Whitehall, London SW1A 2ET
📞 020 7747 2637   📠 020 7321 0943
www.rusi.org/westminstermedal

Established in 1997, this annual award is sponsored by the Duke of Westminster and aims to mark a notable and original contribution to the study of international or national security, or the military profession. Work must be in English, by a living author, and have been published as a

book, rather than an article, in the preceding or next six months of the closing date for entries. 2008 winner: Christopher Bellamy *Absolute War: The Definitive History of the War on the Eastern Front, 1941–1945* (Macmillan). A special Lifetime Achievement award was also presented to Sir Max Hastings.
Ⓔ Silver medal, plus £1,000. Winning author is invited to give lecture at RUSI where the Duke of Westminster will present the Medal.

Edge Hill Prize for the Short Story

Dept. of English, Edge Hill University, St Helen's Road, Ormskirk L39 4QP
☏ 01695 584121 📠 01695 579997
coxa@edgehill.ac.uk
www.edgehill.ac.uk
Contact *A. Cox*

Annual award founded in 2007 following a one-day conference on the short story. Its aim is to reward high achievement in writing the short story and to promote the genre. Awarded to the author of a published short story collection from the UK or Ireland. Entries are submitted through publishers and must be published in the previous year.
Ⓔ 1st prize of £5,000, plus specially commissioned artwork sponsored by Blackwells; 2nd prize of £1,000; Readers' Choice prize, £1,000.

T.S. Eliot Prize

Poetry Book Society, Fourth Floor, 2 Tavistock Place, London WC1H 9RE
☏ 020 7833 9247 📠 020 7833 9247
info@poetrybooks.co.uk
www.poetrybookshoponline.com
Contact *Chris Holifield*

Established in 1993 to mark the 40th anniversary of the founding of the Poetry Book Society, the T.S. Eliot prize is awarded by the PBS in January of each year for the best new single author collection of poetry published in the previous year. Submissions must be made by the publisher and are for a single-author poetry collection published in the preceding calendar year in the UK or Republic of Ireland. No self-published work is eligible. Final entry date: beginning of August. The prize-winners' cheques are given by Mrs Valerie Eliot, who presents them at an award ceremony in central London. The T.S. Eliot Prize Readings are held on the evening before the ceremony. The Prize is distinctive in having a Shadowing Scheme, run in association with the English and Media Centre's *emagazine*. 2008 winner: Jen Hadfield *Nigh-No-Place* (Bloodaxe Books).
Ⓔ Prizes of £15,000 for the winner and £1,000 for each of the ten shortlisted poets.

T.S. Eliot Prize for Poetry

Truman State University Press, 100 East Normal Street, Kirksville, MO 63501, USA
tsup@truman.edu
tsup.truman.edu
Contact *Nancy Rediger*

An annual award, established in 1997 in honour of native Missourian, T.S. Eliot, to publish and promote contemporary English-language poetry regardless of a poet's nationality, reputation, stage in career or publication history. Entry requirements: 60–100 pages of original poetry with $25 fee. Final entry date: 31st October. See website for full entry details. 2008 winner: Victoria Brockmeier *My Maiden Cowboy Names*.
Ⓔ Prize of $2,000, plus publication.

The Desmond Elliott Prize

The Desmond Elliott Charitable Trust, 84 Godolphin Road, London W12 8JW
☏ 020 8222 6580
ema.manderson@googlemail.com
www.desmondelliottprize.com
Contact *Mrs Emma Manderson*

Established in 2008 in memory of literary agent and publisher, Desmond Elliott, this annual prize is for a first full-length novel, written in English and published in book form in the UK, by an author who is permanently resident in the UK. Final entry date: 30th November 2009. Inaugural winner: Nikita Lalwani *Gifted* (Penguin).
Ⓔ Prize of £10,000.

The Encore Awards

The Society of Authors, 84 Drayton Gardens, London SW10 9SB
☏ 020 7373 6642 📠 020 7373 5768
info@societyofauthors.org
www.societyofauthors.org

Established 1990. Awarded biennially for the best second published novel or novels of the year. Final entry date: 30th November 2008. Details from the **Society of Authors**. 2007 winner: M.J. Hyland *Carry Me Down* (Canongate).
Ⓔ Prize of £10,000.

Envoi Poetry Competition

Ty Meirion, Glan yr afon, Tanygrisiau, Blaenau Ffestiniog LL41 3SU
☏ 01766 832112
Contact *Jan Fortune-Wood*

Run by *Envoi* poetry magazine. Competitions are featured regularly, with prizes of £300, plus three annual subscriptions to *Envoi*. Winning poems along with full adjudication report are published. Send s.a.e. to Competition Secretary at the address above.

The European Prize for Literature

ec.europa.eu/culture/our-programmes-and-
actions/doc627_en.htm

A new literature prize (to run each year between
2009 and 2011) aimed at putting the spotlight
on the creativity and diverse wealth of Europe's
contemporary literature, to promote more
circulation of literature within Europe and
encourage greater interest in non-national literary
works. The prize, which is financed through
the Culture Programme of the European Union
and organized by the European Booksellers
Federation, the European Writers' Council and
the Federation of European Publishers, is for
emerging talents in the field of contemporary
literature (fiction) from each of the participating
countries (Austria, Croatia, France, Hungary,
Ireland, Italy, Lithuania, Norway, Poland,
Portugal, Slovakia and Sweden in 2009). Each
country selects their respective winner.

Geoffrey Faber Memorial Prize

Faber & Faber Ltd, Bloomsbury House, 74–77
Great Russell Street, London WC1B 3DA
📞 020 7927 3800   📠 020 7927 3801
www.faber.co.uk

Established 1963 as a memorial to the founder
and first chairman of **Faber & Faber**, this prize is
awarded in alternate years for the volume of verse
and the volume of prose fiction published in the
UK in the preceding two years, which is judged
to be of greatest literary merit. Authors must be
under 40 at the time of publication and citizens
of the UK, Commonwealth, Republic of Ireland
or South Africa. 2009 winner: Nick Laird *On
Purpose* (Faber) 💷 Prize of £1,000.

The Faculty of Advocates/Saltire Society Scottish Book of the Year
▷ The Saltire Literary Awards

The Alfred Fagon Award

The Royal Court Theatre, Sloane Square, London
SW1W 8AS
www.talawatheatrecompany.co.uk/afa

First presented in 1997. An annual award, in
memory of playwright Alfred Fagon, which is
open to any playwright of Caribbean descent,
resident in the UK, for the best new stage play
in English, which need not have been produced
(television and radio plays and film scripts not
eligible). Two copies of the script (plus s.a.e.
for return), together with a brief history of the
play and c.v. (including author's Caribbean
connection), should be sent to the address above
by 31st August. 2007 winner: Allia V. Oswald.
💷 Award of £5,000.

Eleanor Farjeon Award

childrensbookcircle@hotmail.co.uk
www.childrensbookcircle.org.uk

This award, named in memory of the much-loved
children's writer, is for distinguished services to
children's books by an individual or organization.
It may be given to a librarian, teacher, publisher,
bookseller, author, artist, reviewer, television
producer, or any other person or organization
working with or for children through books.
Nominations from members of the **Children's
Book Circle**. 2008 winner: Chris Brown, educator
and editor, School Library Association.

Fish Fiction Prizes

Fish Publishing, Durrus, Bantry, Co. Cork,
Republic of Ireland
info@fishpublishing.com
www.fishpublishing.com
Contact *Clem Cairns*

Offers several writing contests including the
Short Story Prize, an annual award, founded
in 1994, which aims to discover, encourage and
publish new literary talent. The best 15 stories
are published in an anthology. Also, the Fish
Unpublished Novel Award. Details of these and
other contests can be found on the website.

Sir Banister Fletcher Award

Authors' Club, 40 Dover Street, London W1S 4NP
📞 020 7408 5092
stella@theartsclub.co.uk
Contact *Stella Kane*

Founded in 1954, this award was created by the
late Sir Banister Fletcher, former President of
the **Authors' Club** and the Royal Institute of
British Architects, and is presented for 'the most
deserving book on architecture or the arts'. The
award is open to titles written by British authors
or those resident in the UK and published under
a British imprint. 2008 winner: Andrew Saint
Architect and Engineer: A Study in Sibling Rivalry
(Yale University Press).

The John Florio Prize
▷ The Translators Association Awards

The Paul Foot Award

Private Eye, 6 Carlisle Street, London W1D 3BN
📞 020 7437 4017   📠 020 7437 0705
www.private-eye.co.uk
Contact *Sheila Molnar*

Annual award, established in 2005 by *The
Guardian* and *Private Eye* for campaigning
journalism in memory of former contributor Paul
Foot. Submissions can be made by individual
journalists, teams of journalists or publications
for work appearing between October and August.
Single pieces or entire campaigns are eligible.

Final entry date: 1st September of the year in question. No broadcast material. 'Please do not send original material as no correspondence can be entered into.' 2008 joint winners: Camilla Cavendish (The Times) and Richard Brooks (Private Eye).

💷 Award of £5,000 with five runners-up receiving £1,000.

The Forward Prizes for Poetry

Administrator: Colman Getty, 28 Windmill Street, London W1T 2JJ
📞 020 7631 2666   📠 020 7631 2699
kate@colmangetty.co.uk
www.forwardartsfoundation.co.uk
Contact *Kate Wright-Morris*

Established 1992. Three awards: the Forward Prize for Best Collection, the Felix Dennis Prize for Best First Collection and the Forward Prize for Best Single Poem in memory of Michael Donaghy, which is not already part of an anthology or collection. All entries must be published in the UK or Republic of Ireland and submitted by poetry publishers (collections) or newspaper and magazine editors (single poems). Individual entries of poets' own work are not accepted. 2008 winners: Best Collection: Mick Imlah *The Lost Leader* (Faber & Faber); Best First Collection: Kathryn Simmonds *Sunday at the Skin Laundrette* (Seren); Best Single Poem: Don Paterson *Love Poem for Natalie 'Tusja' Beridze*.
💷 Prizes of £10,000 (best collection); £5,000 (best first collection); £1,000 (best single poem).

The Foyle Young Poets of the Year Award

The Poetry Society, 22 Betterton Street, London WC2H 9BX
📞 020 7420 9880   📠 020 7240 4818
fyp@poetrysociety.org.uk
www.foyleyoungpoets.org
Contact *George Ttoouli*

This annual award was founded by the Poetry Society in 1997 in order to help young poets thrive as readers and writers. Since 2001 the Foyle Foundation has sponsored the award which encourages creativity, imagination and literacy amongst young people at school. Two top poets are selected as judges and they choose 100 of the best young poets, out of almost 5,000 poets who enter each year. Poets must be aged 11–17 on or before the closing date of 31st July each year. Entry is free and can be via the web entry form, or by post. Poets can enter as many poems as they like (unlimited), with no restrictions on length, content, style, form or theme.
💷 Fifteen overall winners aged 11–17, including three international winners, have their poems printed in the winners' anthology; five 11–14-year-olds win a visit to their school by a leading poet;

fifteen 15–17 year-old-winners attend a week-long residential course at one of the prestigious Arvon Centres. The three schools who inspire the most entries receive a special selection of books from **Faber & Faber**, **Bloodaxe Books** and **Salt Publishing**.

The Frances Lincoln Diverse Voices Children's Book Award

30 Lime Street, Ouseburn Valley, Newcastle upon Tyne NE1 2PQ
📞 0845 271 0777   📠 0191 261 1931
diversevoices@sevenstories.org.uk
www.sevenstories.org.uk
Diverse Voices Administrator *Helena McConnell*
Office Administrator *Liz Paton*

Founded in 2009 by **Frances Lincoln Ltd** and Seven Stories, the Centre for Children's Books, an innovative cultural centre for children's literature. The aim of the award, in memory of Frances Lincoln (1945–2001), is to take positive steps to increase the representation of people writing from or about different cultural perspectives; promote new writing for children, especially by or about people whose culture and voice are currently under-represented; and recognize that as children's books shape earliest perceptions of the world and its cultures, promoting writing that represents diversity will contribute to social and cultural tolerance. Awarded to the best work of unpublished fiction for 8-to-12-year-olds by a writer who has not previously published a novel for children (although may have contributed to an anthology of prose or poetry). The work must be written in English and be a minimum of 10,000 words and a maximum of 30,000 words. Entrants must be aged 16 years or over by the closing date (31st January 2010); only one submission per writer will be accepted. Application form is available on the website. Seven Stories is a registered charity number 1056812.
💷 Prize of £1,500, plus the option for Frances Lincoln Children's Books to publish the novel.

The Frogmore Poetry Prize

42 Morehall Avenue, Folkestone CT19 4EF
📞 07751 251689
www.frogmorepress.co.uk
Contact *Jeremy Page*

Established 1987. Awarded annually and sponsored by the Frogmore Foundation. The winning poem, runners-up and short-listed entries are all published in the magazine. Closing date 31st May. Previous winners: David Satherley, Bill Headdon, John Latham, Diane Brown, Tobias Hill, Mario Petrucci, Gina Wilson, Ross Cogan, Joan Benner, Ann Alexander, Gerald Watts, Katy Darby, David Angel, Howard Wright, Caroline

Price, Julie-ann Rowell, Arlene Ang, Peter Marshall, Gill Andrews.
Ⓔ The winner receives 200 guineas and a two-year subscription to the biannual literary magazine, *The Frogmore Papers*.

FT & Goldman Sachs Business Book of the Year Award

Financial Times, One Southwark Bridge, London SE1 9HL
Ⓣ 020 7873 3000   Ⓕ 020 7873 3072
bookaward@ft.com
www.ft.com/bookaward
Contact *Lizzie Allen*

Established in 2005 to identify the book that provides the most compelling and enjoyable insight into modern business issues, including management, finance and economics. Titles must be published for the first time in the English language, or in English translation, between 31st October 2008 and 1st November 2009. Submissions by a publisher or *bona fide* imprint which holds English language rights in the book. 2008 winner: Mohamed El-Erian *When Markets Collide: Investment Strategies for the Age of Global Economic Change* (McGraw-Hill).
Ⓔ Award of £30,000 (winner); shortlisted authors receive £5,000 each.

Galaxy British Book Awards

PN Ltd., 39 Store Street, London WC1E 7DS
Ⓣ 0870 870 2345   Ⓕ 0203 089 4460
mailbox@publishingnews.co.uk
www.galaxybritishbookawards.com

Established 1988. Viewed by the book trade as the one to win, 'The Nibbies' are presented annually. The awards are made in various categories. Each winner receives the prestigious Nibbie and the awards are presented to those who have made the most impact in the book trade during the previous year. 2009 winners included: Michael Palin, Aravind Adiga, Stieg Larsson, Barack Obama, Kate Atkinson For further information contact: Merric Davidson, PO Box 60, Cranbrook TN17 2ZR (Ⓣ/Ⓕ 01580 212041; gbba@mdla.co.uk).

Martha Gellhorn Trust Prize

Crosscombe, Town's Lane, Loddiswell TQ7 4QY
Ⓣ 01548 550344   Ⓕ 01548 550344
sandyandshirlee@tiscali.co.uk

Annual prize for journalism in honour of one of the twentieth century's greatest reporters. Open for journalism published in English, giving 'the view from the ground – a human story that penetrates the established version of events and illuminates an urgent issue buried by prevailing fashions of what makes news'. The subject matter can involve the UK or abroad. Six copies of each entry should be sent to the address above by 31st March. Previous winners include: Hala Jaber, Michael Tierney, Ghaith Abdul-Ahad, Patrick Cockburn, Robert Fisk, Jeremy Harding, Geoffrey Lean.
Ⓔ Prize of £5,000.

The Gladstone History Book Prize

Royal Historical Society, University College London, Gower Street, London WC1E 6BT
Ⓣ 020 7387 7532   Ⓕ 020 7387 7532
royalhistsoc@ucl.ac.uk
www.royalhistoricalsociety.org/grants.htm
Contact *Administrative Secretary*

Established 1998. Annual award for the best new work on any historical subject which is not primarily related to British history, published in the UK in the preceding calendar year. The book must be the author's first (solely written) history book and be an original and scholarly work of historical research. Closing date: 31st December.
Ⓔ Prize of £1,000.

Golden Hearts Awards
▷ Romance Writers of America under Professional Associations and Societies

Golden PEN Award for Lifetime Distinguished Service to Literature

English PEN, 6–8 Amwell Street, London EC1R 1UQ
Ⓣ 020 7713 0023   Ⓕ 020 7837 7838
enquiries@englishpen.org
www.englishpen.org

Awarded to a senior writer, with a distinguished body of work written over many years, who has made a significant and constructive impact on fellow writers, the reading public and the literary world. Nominations by members of English PEN only. Previous winners include: Harold Pinter, Doris Lessing, Michael Frayn, Nina Bawden. 2007 winner: Josephine Pullein-Thompson.

Good Housekeeping Book Awards

The National Magazine Company Ltd, 72 Broadwick Street, London W1F 9EP
Ⓣ 020 7439 5218
natalie.earl@natmags.co.uk
Contact *Natalie Earl*

Annual awards, established in 2007. Categories: Best Fiction, Best Non-Fiction, Best Debut and The Good Housekeeping Reader Award for Best Author. Books are submitted by publishers for the first three categories while readers of the magazine vote for 'Best Author'. 2008 winners: Rose Tremain *The Road Home* (Best Fiction); Sally Brampton *Shoot the Damn Dog* (Best Non-Fiction); Sadie Jones *The Outcast* (Best Debut); Ian McEwan (The Good Housekeeping Reader Award for Best Author).
Ⓔ Prize winners receive a certificate.

Gourmand World Cookbook Awards

Pintor Rosales 50, 28008 Madrid, Spain
☎ 00 34 91 541 6768   ℻ 00 34 91 541 6821
icr@virtualsw.es
www.cookbookfair.com
Contact *Edouard Cointreau*

Founded in 1995 by Edouard Cointreau to reward those who 'cook with words'. The only world competition for food and wine books in all languages. Annual event with the winners of the worldwide local or regional competitions competing for the 'Best Book in the World' award. 107 countries participated in the 2008 awards with the Awards event taking place in London.

The Winston Graham Historical Prize

Royal Cornwall Museum, River Street, Truro
TR1 2SJ
☎ 01872 272205   ℻ 01872 240514
enquiries@royalcornwallmuseum.org.uk
www.royalcornwallmuseum.org.uk
Contact *Prize Administrator*

Established in 2008 the prize has been funded by a generous legacy to the Royal Institution of Cornwall from the author. It will be awarded in its inaugural year for an unpublished work of historical fiction, preferably with a clear connection to Cornwall. Details and conditions of entry can be found on the website or by writing to the Royal Cornwall Museum.
£ Prize 2008/9: £5,000 and potential for publication by **Pan Macmillan**.

Kate Greenaway Medal

▷ CILIP: The Chartered Institute of Library and Information Professionals Kate Greenaway Medal

The Griffin Poetry Prize

6610 Edwards Boulevard, Mississauga, Ontario, Canada L5T 2V6
☎ 001 905 565 5993   ℻ 001 905 564 3645
info@griffinpoetryprize.com
www.griffinpoetryprize.com
Contact *Ruth Smith, Manager*

Annual award established in 2000 by Toronto-based entrepreneur, Scott Griffin, for books of poetry written in or translated into English. Trustees include Margaret Atwood and Michael Ondaatje. Submissions from publishers only.
£ Prizes total C$100,000, divided into two categories: International and Canadian.

The Guardian Children's Fiction Award

The Guardian, Kings Place, 90 York Way, London
N1 9AG
☎ 020 3353 2000
Children's Book Editor *Julia Eccleshare*

Established 1967. Annual award for an outstanding work of fiction for children aged seven and over by a British or Commonwealth author, first published in the UK in the year of the award, excluding picture books. No application form necessary. 2008 winner: Patrick Ness *The Knife of Never Letting Go* (Walker Books).
£ Award of £1,500.

The Guardian First Book Award

The Guardian, Kings Place, 90 York Way, London
N1 9AG
☎ 020 3353 2000
sarah.waldie@guardian.co.uk
www.guardian.co.uk
Contact *Sarah Waldie*

Established 1999. Annual award for first time authors published in English in the UK. All genres of writing eligible, apart from academic, guidebooks, children's, educational, manuals, reprints and TV, radio and film tie-ins. All books must be published between January and December 2008 and have an ISDN number or equivalent. Submissions are only received direct from publishers, not from individual authors. Self-published work is not eligible. 2008 winner: Alex Ross *The Rest Is Noise: Listening to the Twentieth Century* (Fourth Estate).
£ Award of £10,000, plus *Guardian/Observer* advertising package.

The Guild of Food Writers Awards

255 Kent House Road, Beckenham BR3 1JQ
☎ 020 8659 0422
awards@gfw.co.uk
www.gfw.co.uk
Contact *Jonathan Woods*

Established 1985. Annual awards in recognition of outstanding achievement in all areas in which food writers work and have influence. Entry is not restricted to members of the Guild. Entry form available from the address above. 2008 winners: Derek Cooper Award for Campaigning and Investigative Food Writing: Hattie Ellis *Planet Chicken: The Shameful Story of the Bird on Your Plate*; Michael Smith Award for Work on British Food: Hugh Fearnley-Whittingstall and Nick Fisher *The River Cottage Fish Book*; Jeremy Round Award for the Best First Book: Marwood Yeatman *The Last Food of England*; Food Book of the Year: Martin Jones *Feast: Why Humans Share Food*; Cookery Book of the Year: Sarah Raven *Sarah Raven's Garden Cookbook*; Food Journalist of the Year: Bee Wilson, for work in *The Sunday Telegraph*'s *Stella* magazine; Evelyn Rose Award for Cookery Journalist of the Year: Shona Crawford Poole, for work in *Country Living* magazine; Restaurant Reviewer of the Year: Tracey Macleod, for reviews in *The Independent* magazine; Miriam Polunin Award for Work on Healthy Eating: Jill Dupleix *Lighten Up: A New*

Healthier Way to Cook; New Media Award: www.bbc.co.uk/food; Food Broadcast Award: *The Food Programme: China Tea*, BBC Radio 4; Lifetime Achievement Award: Katie Stewart.

Guildford Book Festival First Novel Award

Guildford Book Festival, c/o Tourist Information Centre, 14 Tunsgate, Guildford GU1 3QT
☎ 01483 444334
deputy@guildfordbookfestival.co.uk
www.guildfordbookfestival.co.uk
Contacts *Pamela Thomas, Glenis Pycraft*

This annual award was known originally as The Pendleton May Award. It was initiated to encourage local talent but has now been opened to all writers in the UK. Submissions must be from agents or publishers with accompanying letter (proofs accepted). Books to be published between 1st November and 31st October, i.e. between Festivals. Shortlisted authors must be prepared to attend the presentation dinner held during the Festival. 2008 winner: Ross Raisin *God's Own Country* (Viking).
💰 Prize of £2,500.

Gwobr Llyfr y Flwyddyn
▷ Wales Book of the Year Award

The Anthony Hecht Poetry Prize

The Waywiser Press, 14 Lyncroft Gardens, Ewell KT17 1UR
☎ 020 8393 7055
waywiserpress@aol.com
waywiser-press.com/hechtprize.html
Contact *Philip Hoy, Managing Editor*

Established 2005. Annual prize awarded for an outstanding, unpublished, full collection of poems. Entrants must be at least 18 years of age and may not have published more than one previous collection. Detailed guidelines and entry forms available from August on the website or by sending an A4-sized s.a.e.
💰 Prize of £1,500 and publication by Waywiser in the UK and USA.

Hellenic Foundation Prize
▷ The Translators Association Awards

Felicia Hemans Prize for Lyrical Poetry

University of Liverpool, PO Box 147, Liverpool L69 7WZ
☎ 0151 794 2458   📠 0151 794 2454
wilderc@liv.ac.uk
Contact *The Sub-Dean, Faculty of Arts*

Established 1899. Annual award for published or unpublished verse. Open to past or present members and students of the University of Liverpool. One poem per entrant only. Closing date: 1st May.
💰 Prize of £30.

The Hessell-Tiltman Prize for History

English PEN, 6–8 Amwell Street, London EC1R 1UQ
☎ 020 7713 0023   📠 020 7837 7838
enquiries@englishpen.org
www.englishpen.org

Founded 2002. Awarded for a history book covering any period up to the end of the Second World War, written in English (including translations) and aimed at a wide audience. Submissions cannot be made as books are nominated by an independent judging panel. 2007 winner: Vic Gattrell *City of Laughter* (Atlantic Books).
💰 Prize of £3,000.

William Hill Sports Book of the Year

Greenside House, Station Road, Wood Green, London N22 7TP
☎ 020 8918 3731   📠 020 8918 3728
pressoffice@williamhill.co.uk
www.williamhillmedia.com
Contact *Graham Sharpe*

Established 1989. Annual award introduced by Graham Sharpe of bookmakers William Hill. Sponsored by William Hill and thus dubbed the 'Bookie' prize, it is the first, and only, Sports Book of the Year award. Final entry date: September. 2008 winner: Marcus Trescothick *Coming Back to Me* (HarperCollins).
💰 Prize (reviewed annually) of £22,000 package, including £20,000 cash, hand-bound copy, £2,000 free bet. Runners-up prizes.

HISSAC Annual Short Story Award

20 Lochslin, Balintore, Easter Ross IV20 1UP
☎ 01862 832266   📠 01862 832266
info@hissac.co.uk
www.hissac.co.uk
Judge Chair *Clio Gray*

Established 2003. Annual short story award organized by HISSAC, the Highlands & Islands Short Story Association, an international network of writers aiming to critique and encourage each other's work and encourage all manner of creative writing, including the short story form. Entry fees: £4 per entry, £10 for three. Paypal facilities on website. Maximum wordage: 2,500. No connection to Scotland is needed. Double-spaced, single-sided entries; no children's stories. Final entry date: 31st July 2009. Winning entries are published on the website.
💰 Prizes of £400 (1st); £50 each (2nd & 3rd).

Calvin & Rose G. Hoffman Prize

King's School, 25 The Precincts, Canterbury CT1 2ES
☎ 01227 595544
Contact *The Bursar*

Annual award for distinguished publication on Christopher Marlowe, established by the late Calvin Hoffman, author of *The Man Who was Shakespeare* (1955) as a memorial to himself and his wife. For unpublished works of at least 5,000 words written in English for their scholarly contribution to the study of Christopher Marlowe and his relationship to William Shakespeare. Final entry date: 1st September each year.

David C. Horn Prize
▷ The Yale Drama Series

L. Ron Hubbard's Writers of The Future Contest
PO Box 218, East Grinstead RH19 4GH
Contest Administrator *Andrea Grant-Webb*

Established 1984 by L. Ron Hubbard to encourage new and amateur writers of science fiction and fantasy. Quarterly awards with an annual grand prize. Entrants must submit a short story of up to 10,000 words, or a novelette of fewer than 17,000 words, which must not have been published previously. The contest is open only to those who have not been published professionally. Send s.a.e. for entry form. Previous winners: Roge Gregory, Malcolm Twigg, Tom Brennan, Alan Smale, Janet Barron.
💷 Prizes of £500 (1st); £400 (2nd); £250 (3rd) each quarter; Annual Grand Prize: £2,000. All winners are awarded a trip to the annual L. Ron Hubbard Achievement Awards which include a series of professional writers' workshops, and are published in the *L. Ron Hubbard Presents Writers of the Future* anthology.

The Hugo Award
info@thehugoawards.org
www.thehugoawards.org
Contact *Kevin Standlee*

Founded in 1953, the Hugos are awards for excellence in the field of science fiction and fantasy. Awarded annually since 1953, and voted on by members of the World Science Fiction Convention (Worldcon), they cover written fiction, art, media, editing and writing about the field. The Hugos are awarded for works first published or presented in the previous year. Work may not be submitted for consideration as the nomination process is carried out by about 500 members of the Worldcon.
💷 Award of a trophy in the shape of a classic rocket ship.

The Imison Award
The Society of Authors, 84 Drayton Gardens SW10 9SB
📞 020 7373 6642   📠 020 7373 5768
jmccrum@societyofauthors.org
www.societyofauthors.org

Contact *Jo McCrum*

Annual award established 'to perpetuate the memory of Richard Imison, to acknowledge the encouragement he gave to writers working in the medium of radio, and in memory of the support and friendship he invariably offered writers in general, and radio writers in particular.' Administered by the **Society of Authors** and generally sponsored by the **Peggy Ramsay Foundation**, the purpose is 'to encourage new talent and high standards in writing for radio by selecting the radio drama by a writer new to radio which, in the opinion of the judges, is the best of those submitted.' An adaptation for radio of a piece originally written for the stage, television or film is not eligible. Any radio drama first transmitted in the UK between 1st January and 31st December is eligible, provided the work is an original piece for radio and it is the first dramatic work by the writer(s) that has been broadcast. Submission may be made by any party to the production in the form of three copies of both script and recording (non-returnable), accompanied by a nomination form and 250 word synopsis and c.v. 2008 winner: Adam Beeson *The Magician's Daughter*.
💷 Prize of £1,500.

The Independent Foreign Fiction Prize
c/o Literature Department, Arts Council England, 14 Great Peter Street, London SW1P 3NQ
📞 0845 300 6200
www.artscouncil.org.uk/iffp

The Independent Foreign Fiction Prize celebrates an exceptional work of fiction by a living author which has been translated into English from any other language and published in the United Kingdom in the last year. 2008 winner: *Omega Minor* by Paul Verhaeghen, translated by the author from the Dutch.
💷 Prize of £10,000 shared equally between author and translator.

The International Dagger (in association with Crime Writers' Association)
PO Box 273, Boreham Wood WD6 2XA
secretary@thecwa.co.uk
www.thecwa.co.uk
Contact *The Secretary*

Established 2006. An annual award for the best foreign crime fiction translated into English during the year. Nominations from publishers only. 2008 winner: Dominique Manotti *Lorraine Connection*, translated by Amanda Hopkinson and Ros Schwartz (EuroCrime).

The International Dundee Book Prize
City of Discovery Campaign, 3 City Square, Dundee DD1 3BA
📞 01382 434214   📠 01382 434650

book.prize@dundeecity.gov.uk
www.dundeebookprize.com
Contact *Karin Johnston*

'The Dundee International Book Prize has established itself as the UK's premier prize for emerging novelists. Its £10,000 cash award together with publication by **Birlinn Ltd**, publishers of the Polygon imprint, make the Prize highly valued by tomorrow's great new writers seeking to break into the publishing world.' The award is for an unpublished novel on any theme and in any genre. Telephone or e-mail *Karin Johnston* for further information. 2007 winner: Fiona Dunscombe *The Triple Point of Water*.

The International IMPAC Dublin Literary Award

Dublin City Library & Archive, 138–144 Pearse Street, Dublin 2, Republic of Ireland
📞 00 353 1 674 4802 📠 00 353 1 674 4879
literaryaward@dublincity.ie
www.impacdublinaward.ie

Established 1995. Sponsored by Dublin City Council and US-based productivity improvement firm, IMPAC, this prize is awarded for a work of fiction written and published in the English language or written in a language other than English and published in English translation. Initial nominations are made by municipal public libraries in major and capital cities worldwide, each library putting forward up to three books to the international panel of judges in Dublin. 2008 winner: *De Niro's Game* by Rawi Hage (Old Street Publishing).
💷 Prize of €100,000 (if the winning book is in English translation, the prize is shared €75,000 to the author and €25,000 to the translator).

International Prize for Arabic Fiction

IPAF Office, Sahel Alma, PO Box 280, Jounieh, Lebanon
📞 00 961 9 935333 📠 00 961 9 935333
info@arabicfiction.org
www.arabicfiction.com
Prize Administrator *Joumana Haddad*

Launched in Abu Dhabi, UAE, in April 2007 in association with the Booker Prize Foundation and with the support of the Emirates Foundation. Annual award which aims to reward excellence in contemporary Arabic creative writing and to encourage wider readership of quality Arabic literature internationally. Awarded specifically for prose fiction in Arabic (no short stories or poetry) with each of the six shortlisted authors receiving $10,000 with a further $50,000 going to the winner. Submissions from publishers only. Final entry date: 31st July. 2009 winner: Youssef Ziedan *Beelzebub* (Dar El Shorouk).

International Student Playscript Competition
▶ University of Hull under UK and Irish Writers' Courses

Irish Book Awards

Unique Media Ltd, 600A Cathedral Court, New Street, Dublin 8, Republic of Ireland
📞 00 353 1 522 5200 📠 00 353 1 522 5220
info@uniquemedia.ie
www.irishbooksawards.ie

Established in 2005 to acknowledge the wealth of talent in Irish literature. Annual awards comprising nine categories which are open to all Irish authors who had a book published in the previous twelve months. 2008 winners: Anne Enright *The Gathering* (The Hughes & Hughes Irish Novel of the Year); Diarmaid Ferriter *Judging Dev* (The Argosy Irish Non-Fiction Book of the Year; The Tubridy Show Listeners' Choice Award; Eason Irish Published Book of the Year); Anita Notaro *Take a Look at Me Now* (Galaxy Irish Popular Fiction Book of the Year); *Trevor Brennan: Heart and Soul* Trevor Brennan with Gerry Thornley (Energise Sport Irish Sports Book of the Year); Julia Kelly *With My Lazy Eye* (International Education Services Irish Newcomer of the Year Award); Brendan O'Brien *The Story of Ireland* and Roddy Doyle *Wilderness* (The Dublin Airport Authority Irish Children's Book of the Year – Junior and Senior); William Trevor (Bob Hughes Lifetime Achievement in Irish Literature).

Jerwood Aldeburgh First Collection Prize

The Poetry Trust, The Cut, 9 New Cut, Halesworth IP19 8BY
📞 01986 835950
info@thepoetrytrust.org
www.thepoetrytrust.org
Contact *Naomi Jaffa*

The prize is awarded to the author of what in the opinion of the judges is the best first full collection of poetry published in book form in the UK and the Republic of Ireland in the preceding twelve months. The winner receives £3,000 plus an invitation to read (fee paid) at the **Aldeburgh Poetry Festival** the following year. Please refer to the website for details on how to enter. 2008/9 winner: Ciaran Berry *The Sphere of Birds* (The Gallery Press).

Jewish Quarterly Wingate Prize

PO Box 37645, London NW7 1WB
📞 020 8343 4675
www.jewishquarterly.org
Administrator *Pam Lewis*

Formerly the H.H. Wingate Prize. Annual award (for fiction or non-fiction) for a work that best stimulates an interest in and awareness of themes of Jewish interest. Books must have been

published in the UK in the year of the award (written in English originally or in translation) by an author resident in Britain, the Commonwealth, Israel, Republic of Ireland or South Africa. 2008 winner: Etgar Keret *Missing Kissinger* (Chatto & Windus).
Ⓔ Prize of £4,000.

Samuel Johnson Prize for Non-Fiction
▷ The BBCFour Samuel Johnson Prize for Non-Fiction

Mary Vaughan Jones Award
Cyngor Llyfrau Cymru (Welsh Books Council), Castell Brychan, Aberystwyth SY23 2JB
☎ 01970 624151   Ⓕ 01970 625385
wbc.children@wbc.org.uk
www.wbc.org.uk
Contact *The Administrator*

Triennial award for distinguished services in the field of children's literature in Wales over a considerable period of time. 2006 winner: Mair Wynn Hughes.
Ⓔ Award winner receives a silver trophy.

Keats–Shelley Prize
Keats–Shelley Memorial Association, 117 Cheyne Walk, London SW10 0ES
☎ 020 7352 2180   Ⓕ 020 7352 6705
hello@keats-shelley.co.uk
www.keats-shelley.co.uk
Contact *Harriet Cullen*

Established 1998. Annual award to promote the study and appreciation of Keats and Shelley, especially in the universities, and of creative writing inspired by the younger Romantic poets. Two categories: essay and poem; open to all ages and nationalities. 2008 winners: John Gohorry (poem), Susan Miller (essay).
Ⓔ Prize of £3,000 distributed between the winners of the two categories.

Kelpies Prize
Floris Books, 15 Harrison Gardens, Edinburgh EH11 1SH
☎ 0131 337 2372   Ⓕ 0131 347 9919
floris@florisbooks.co.uk
www.florisbooks.co.uk/kelpiesprize
Contact *Prize Administrator*

Established 2004. Annual prize to encourage and reward new Scottish writing for children. The prize is for an unpublished novel (40–60,000 words) for children aged 9–12, set wholly or mainly in Scotland. The author does not need to be Scottish. Application form, guidelines, terms and conditions are available on the website.
Ⓔ Prize of £2,000 and publication in Floris Books' Kelpies series.

The Petra Kenney Poetry Competition
PO Box 32, Filey YO14 9YG
morgan@kenney.uk.net
www.petrapoetrycompetition.co.uk
Contact *Secretary*

Established 1995. Annual poetry award. Original, unpublished poems up to 80 lines on any theme. Closing date: 1 December. Entry fee: £3 per poem. Send s.a.e. for rules and entry form.
Ⓔ Prizes of £1,000 (1st); £500 (2nd); £250 (3rd) plus inscribed Royal Brierley Crystal Vase; and three highly commended prizes of £125 each. Also prizes for comic verse: £250 (1st) and Young Poets (14–18): £250 (1st); £125 (2nd).

Kent & Sussex Poetry Society Open Competition
13 Ruscombe Close, Southborough, Tunbridge Wells TN4 0SG
☎ 01892 543862
www.kentandsussexpoetrysociety.org
Chairman *Clive R. Eastwood*

Annual competition. Entry fee: £4 per poem, maximum 40 lines. Full details available on the website.
Ⓔ Prizes total £1,350.

Kiriyama Pacific Rim Book Prize
Pacific Rim Voices, 300 Third Street, Suite 822, San Francisco, CA 94107, USA
☎ 001 415 777 1628   Ⓕ 001 415 777 1646
jeannine@kiriyamaprize.org
www.kiriyamaprize.org
Contact *Jeannine Stronach, Prize Manager*

Founded 1996 with the aim of promoting books that contribute to greater understanding and cooperation among the peoples and nations of the Pacific Rim and South Asia, this annual award takes its name from the Reverend Seiyu Kiriyama. Entry details may be obtained from the address above or from the website. 2008 winners: Lloyd Jones *Mister Pip* (fiction); Julia Whitty *The Fragile Edge*.
Ⓔ Prizes of $30,000, divided between both winners.

Kraszna-Krausz Book Awards
▷ The And/or Book Awards

Lakeland Book of the Year Awards
Cumbria Tourism, Windermere Road, Staveley LA8 9PL
☎ 01539 825052   Ⓕ 01539 825076
slindsay@cumbriatourism.org
www.golakes.co.uk
Contact *Sheila Lindsay*

Established in 1984 by local author and broadcaster Hunter Davies in conjunction with the Cumbria Tourist Board, the books entered can be about any aspect of life in the county of Cumbria, from local history books and walking guides to novels and poetry. The contest attracts entries from both new and established authors. Since the establishment of the awards they have grown in importance and are now attracting in the region of 60 entries annually, all competing for the Hunter Davies Prize for the Lakeland Book of the Year.

In addition to the Hunter Davies Prize there are currently five categories including Award for Guides, Walks and Places; Award for People and Social History; Award for Arts and Culture; Award for Heritage and Tradition; and the Best Illustrated Book. Closing date for entries is mid-March. The awards are presented at a charity luncheon in early July.

Ⓔ Awards of £100 for each category together with a framed certificate. Overall winner of the Hunter Davies Prize for the Lakeland Book of the Year also receives a cheque for £100 and a framed certificate.

Lancashire County Library and Information Service Book Award

Lancashire County Library Headquarters, County Hall, PO Box 61, Preston PR1 8RJ
☎ 01772 534751　🖷 01772 534880
jacob.hope@lcl.lancscc.gov.uk
Award Coordinator *Jake Hope*

Established 1986. Annual award, presented in June for a work of original fiction suitable for 12–14-year-olds. The winner is chosen by 13–14-year-old secondary school pupils in Lancashire. Books must have been published between 1st September and 31st August in the previous year of the award and authors must be UK and Republic of Ireland residents. Final entry date: 1st September each year. 2008 winner: Tim Lott *Fearless*.

Ⓔ Prize of £1,000, plus engraved glass decanter.

Lannan Literary Awards and Fellowships

Lannan Foundation, 313 Read Street, Santa Fe, New Mexico 87501, USA
☎ 001 505 986 8160
www.lannan.org

Established 1989. Annual awards given to writers of exceptional poetry, fiction and non-fiction who have made a significant contribution to English-language literature, as well as emerging writers of distinctive literary merit who have demonstrated potential for outstanding future work. On occasion, the Foundation recognizes a writer for lifetime achievement. Candidates for the awards and fellowships are recommended to the Foundation by a network of writers, literary scholars, publishers and editors. Applications

or unsolicited nominations for the awards and fellowships are not accepted.

Le Prince Maurice Prize

Constance Hotels Experience, Poste de Flacq, Mauritius, Indian Ocean
☎ 00 230 402 29 30　🖷 00 230 415 10 82
dircom@constancehotels.com
www.constancehotels.com
Contact *Kelly Desire*

Founded in 2003 and sponsored by one of Mauritius's five-star resorts, Constance Le Prince Maurice. Annual award designed to celebrate the literary love story. Administered in both the UK and France, it is awarded alternately to an English-speaking and French-speaking writer. The aim is to strengthen the cultural links between Mauritius and Europe. Submissions are made by publishers only; closing date for next English award is September 2009. Previous winner: James Meek *We Are Now Beginning Our Descent*.

Ⓔ Prize winner receives a trophy, plus all-expenses-paid two-week 'writer's retreat' at Constance Le Prince Maurice.

Legend Writing Awards

39 Emmanuel Road, Hastings TN34 3LB
www.legendwritingaward.com
Contact *Legend Coordinator*

Established 2001. Annual awards to encourage new fiction writers resident in the UK. The competition is for short stories of 2,000 words maximum, and new from 2009, flash fiction of 100 words exactly (excluding title). There is no set theme. Closing date: 31st August. Entry fees: short story, £7 (first entry); £5 (each subsequent entry); flash fiction, £3 per entry. Rules/entry form available from website or by sending s.a.e. to address above. The competition is organized and judged by Hastings Writers' Group.

Ⓔ Prizes of £500 (1st); £250 (2nd); £100 (3rd); plus three runners-up prizes of £50 for short stories; £50 (1st), £30 (2nd), £20 (3rd) plus three runners-up prizes of £10 for flash fiction.

The Bernard Levin Award

Society of Indexers, Woodbourn Business Centre, 10 Jessell Street, Sheffield S9 3HY
☎ 0114 244 9561　🖷 0114 244 9563
admin@indexers.org.uk
www.indexers.org.uk
Secretary *Judith Menes*

Established 2000 to celebrate the late Bernard Levin, a journalist and author whose writings showed untiring and eloquent support for indexers and indexing. An occasional award for outstanding services to the Society of Indexers. Last awarded in 2007 to Jill Halliday.

The Astrid Lindgren Memorial Award for Literature (ALMA)

Swedish Arts Council, PO Box 27215, SE-102 53 Stockholm, Sweden

☎ 00 46 8 519 264 00 📠 00 46 8 519 264 99
literatureaward@alma.se
www.alma.se
Director *Erik Titusson*

Established 2002 by the Swedish government in memory of the children's author Astrid Lindgren. Administered by the Swedish Arts Council, it is an international award for children's and young people's literature given annually to one or more recipients, irrespective of language or nationality. Writing, illustrating and storytelling, as well as reading promotion activities may be awarded. Selected organizations worldwide are invited to submit nominations once a year; jury members may also contribute nominations. 2008 winner: Sonya Hartnett.
💰 Award of SEK 5 million (approx. £400,000)

John Llewellyn Rhys Prize

Booktrust, Book House, 45 East Hill, London SW18 2QZ

☎ 020 8516 2972 📠 020 8516 2978
tarryn@booktrust.org.uk
www.booktrust.org.uk
Contact *Tarryn McKay*

Established 1942. An annual literary award for a memorable work of fiction, non-fiction, drama or poetry. Entrants must be 35 or under at the time of publication; books must have been published in the UK during the calendar year of the award. The author must be a citizen of the UK or the Commonwealth, writing in English. 2007/8 winner: Henry Hitchings *The Secret Life of Words* (John Murray).
💰 Prize of £5,000 (1st); £500 for shortlisted entries.

Local Poem Competition

United Press, Admail 3735, London EC1B 1JB

☎ 0870 240 6190 📠 0870 240 6191
mail@unitedpress.co.uk
www.unitedpress.co.uk
Contact *Julie Embury*

Free to enter annual poetry competition. Organizers United Press allow up to three poems (20 lines and 160 words max. each) by the annual closing date of 31st December. United Press select one winning poem which must have a local theme and be about a person, place or event.
💰 Prize of £1,000 cash.

The Elizabeth Longford Prize for Historical Biography

The Society of Authors, 84 Drayton Gardens, London SW10 9SB

☎ 020 7373 6642 📠 020 7373 5768
info@societyofauthors.org
www.societyofauthors.org

Established in 2003 in memory of Elizabeth Longford and sponsored by Flora Fraser and Peter Soros. Awarded annually for an historical biography published in the year preceding the prize. No unsolicited submissions. 2008 winner: Rosemary Hill *God's Architect – Pugin and the Building of Romantic Britain* (Allen Lane).
💰 Prize of £5,000.

Longman-History Today Book of the Year Award

c/o History Today, 20 Old Compton Street, London W1D 4TW

☎ 020 7534 8000
www.historytoday.com
Contact *Andy Patterson*

Established 1993. Annual award set up as a joint initiative between the magazine *History Today* and the publisher Longman (**Pearson**) to mark the past links between the two organizations, to encourage new writers, and to promote a wider public understanding of, and enthusiasm for, the study and publication of history. Award for author's first or second book.
💰 Prize of £2,000.

The Clarissa Luard Award
▷ David Cohen Prize for Literature

Sir William Lyons Award

The Guild of Motoring Writers, 40 Baring Road, Bournemouth BH6 4DT

☎ 01202 422424 📠 01202 422424
generalsec@gomw.co.uk
www.guildofmotoringwriters.co.uk
Contact *Patricia Lodge*

An annual competitive award, sponsored by Jaguar, to encourage young people in automotive journalism and to foster interests into motoring and the motor industry. Entrance by two essays and interview with Awards Committee. Applicants must be British, aged 17–23 and resident in UK. Final entry date: 1st October.

McColvin Medal
▷ Besterman/McColvin Medal

W.J.M. Mackenzie Book Prize

Political Studies Association, Dept. of Politics, Newcastle University, Newcastle upon Tyne NE1 7RU

☎ 0191 222 8021 📠 0191 222 3499

psa@ncl.ac.uk
www.psa.ac.uk
PSA Executive Director *Jack Arthurs*

Established 1987. Annual award to best work of political science published in the UK during the previous year. Submissions from publishers only. Final entry date: 31st October. Prizes are judged in the year following publication and awarded the year after that. 2008 winner: Professor Philip Schofield *Utility & Democracy* (Oxford University Press).

McKitterick Prize

Society of Authors, 84 Drayton Gardens, London SW10 9SB
℡ 020 7373 6642   📠 020 7373 5768
info@societyofauthors.org
www.societyofauthors.org
Contact *Awards Secretary*

Annual award for a full-length novel in the English language, first published in the UK or unpublished. Open to writers over 40 who have not had any novel published other than the one submitted (excluding works for children). Closing date: 20th December. 2008 winner: Jennie Walker *24 for 3* (Bloomsbury).
💷 Prize of £4,000.

Enid McLeod Prize

Franco-British Society, 2 Dovedale Studios, 465 Battersea Park Road, London SW11 4LR
℡ 020 7924 3511
www.francobritishsociety.org.uk
Executive Secretary *Kate Brayn*

Established 1982. Annual award to the author of the work of literature published in the UK which, in the opinion of the judges, has contributed most to Franco-British understanding. Any full-length work written in English by a citizen of the UK Commonwealth is eligible. No English translation of a book written originally in any other language will be considered. Nominations from publishers for books published between 1st January and 31st December of the year of the prize. Closing date: 31st December. 2007 winner: Professor Tim Blanning *The Pursuit of Glory*.
💷 Prize winner receives a cheque.

The Macmillan Prize for Children's Picture Book Illustration

Macmillan Children's Books, 20 New Wharf Road, London N1 9RR
℡ 020 7014 6124   📠 020 7014 6142
lindsey.evans@macmillan.co.uk
www.panmacmillan.com
Contact *Lindsey Evans, Macmillan Children's Books*

Established by Macmillan Children's Books in order to stimulate new work from young illustrators in art schools, and to help them start their professional careers. Fiction or non-fiction. Macmillan have the option to publish any of the prize winners.
💷 Prizes of £1,000 (1st); £500 (2nd); £250 (3rd).

The Mail on Sunday Novel Competition

Postal box address may change annually (see below)

Annual award established 1983. Judges look for a story/character that springs to life in the 'tantalizing opening 50–150 words of a novel'. Details of the competition, including the postal box address, are published in *The Mail on Sunday* in July/August. 2008 winner: Isabel Ashdown.
💷 Awards of £400 book tokens and a writing course at the **Arvon Foundation** (1st); £300 tokens (2nd); £200 tokens (3rd); three further prizes of £150 tokens each.

The Man Asian Literary Prize

22nd Floor, Easy Commercial Building, 253–261 Hennessy Road, Hong Kong
℡ 00 852 2877 8444
info@manasianliteraryprize.org
www.manasianliteraryprize.org
Chairman *Peter Gordon*

Initiated by the Man Group plc and the Hong Kong Literary Festival in 2007. An annual award which aims to recognize the best of new Asian literature and to bring it to the attention of the world literary community. Open to a single work of Asian fiction in English (but unpublished in English) of no less than 30,000 words by an Asian author residing in an Asian country or territory. Final entry date: end March. 2008 winner: Miguel Syjuco *Ilustrado*.
💷 Award of US$10,000.

The Man Booker International Prize

Colman Getty, 28 Windmill Street, London W1T 2JJ
℡ 020 7631 2666   📠 020 7631 2699
info@colmangetty.co.uk
www.manbookerinternational.com

Established in 2004 to complement **The Man Booker Prize for Fiction** by recognizing one writer's achievement in literature and their significant influence on writers and readers worldwide. Sponsored by the Man Group, this biennial award is given to a living author who has published fiction either originally in English or whose work is generally available in translation into the English language. Submissions for the prize are not invited. Where the winning author's work has been translated into English an additional prize of £15,000 is awarded to the translator. The winning author chooses who the translator's prize should go to and whether it is to be awarded to one translator or divided between several. 2007 winner: Chinua Achebe.

The Man Booker Prize for Fiction

Colman Getty, 28 Windmill Street, London W1T 4JE

📞 020 7631 2666   📠 020 7631 2699
info@colmangetty.co.uk
www.themanbookerprize.com
Contact *Lois Tucker (Submissions)*

The Booker Prize for Fiction was originally set up by Booker plc in 1968 to reward merit, raise the stature of the author in the eyes of the public and encourage an interest in contemporary fiction. In April 2002 it was announced that the Man Group had been chosen by the Booker Prize Foundation as the new sponsor of the Booker Prize. The sponsorship is due to run until 2011. United Kingdom publishers may enter up to two full-length novels, with scheduled publication dates between 1st October 2008 and 30th September 2009. 2008 winner: Aravind Adiga *The White Tiger* (Atlantic Books).
💷 The winner receives £50,000. The six shortlisted authors each receive £2,500.

Manchester Writing Competition

Writing School, Dept. of English, Manchester Metropolitan University, Geoffrey Manton Building, Rosamond Street West, off Oxford Road, Manchester M16 6LL

📞 0161 247 1787
j.draper@mmu.ac.uk
www.manchesterwritingcompetition.co.uk
Project Manager *James Draper*

An annual competition, launched in 2008 by the Writing School at Manchester Metropolitan University, the Manchester Writing Competition alternates between poetry and fiction. The inaugural prize was for the best portfolio of poems with the winner receiving cash. In addition, a bursary was awarded to an entrant aged 18–25 for study at MMU as part of the Manchester Young Writer of the Year Award. Deadline for the 2009 fiction prize will be 1 August 2009; open to UK and international writers. See website for entry details. 2008 Manchester Poetry Prize joint winners: Lesley Saunders *Christina the Astonishing* and *The Uses of Greek* and Mandy Coe *Pinning the Tail on the Donkey*; Manchester Young Writer of the Year Award: Helen Mort.

Manchester Young Writer of the Year Award
▷ Manchester Writing Competition

Michael Marks Awards for Poetry Pamphlets

Poetry Book Society, Fourth Floor, 2 Tavistock Place, London WC1H 9RA

📞 020 7833 9247   📠 020 7833 5990
pamphlets@poetrybooks.co.uk
www..poetrybookshoponline.com/pamphlets
Contacts *Hilary Davidson, Chris Holifield*

Established in 2009 and run by the British Library and the Poetry Book Society (funded by the Michael Marks Foundation), this competition consists of two awards, one for poetry pamphlets published in the UK, including self-published work and one for UK poetry pamphlet publishers. Final entry date: late March. Eligibility guidelines available from the Poetry Book Society or from the website.
💷 Prize of £5,000 for the most outstanding poetry pamphlet of the year and £5,000 for the most outstanding poetry pamphlet publisher of the year.

Marsh Award for Children's Literature in Translation

The English-Speaking Union, Dartmouth House, 37 Charles Street, London W1J 5ED

📞 020 7529 1550   📠 020 7495 6108
education@esu.org
Contact *Katherine Plummer*

Established 1995 and sponsored by the Marsh Christian Trust, the award aims to encourage translation of foreign children's books into English. It is a biennial award (next award: 2011), open to British translators of books for 4–16-year-olds, published in the UK by a British publisher. Any category will be considered with the exception of encyclopedias and reference books. No electronic books. 2008 winner: Sarah Ardizzone for her translation of *Toby Alone* by Timothee de Fombelle (Walker Books).
💷 Prize of £2,000.

Marsh Biography Award

The English-Speaking Union, Dartmouth House, 37 Charles Street, London W1J 5ED

📞 020 7529 1564   📠 020 7495 6108
gillian_parker@esu.org
www.esu.org
Contact *Gillian Parker*

A biennial award for the most significant biography published over a two-year period by a British publisher. Next award October 2009. 2007 winner: Maggie Fergusson *George Mackay Brown: The Life* (John Murray).
💷 Award: Membership of the ESU and £4,000, plus a silver trophy presented at a gala dinner.

Colin Mears Award
▷ CILIP: The Chartered Institute of Library and Information Professionals Kate Greenaway Medal

Medical Book Awards

The Society of Authors, 84 Drayton Gardens, London SW10 9SB
☎ 020 7373 6642   📠 020 7373 5768
sbaxter@societyofauthors.org
www.societyofauthors.org
Contact *Sarah Baxter, Medical Writers Group*

Annual awards sponsored by the Royal Society of Medicine. Nine categories for medical textbooks published in the twelve months preceding the deadline. Contact *Sarah Baxter* for entry details. Closing date: 20th April.
💷 Prizes total £8,800.

The Mercedes-Benz Award for the Montagu of Beaulieu Trophy

Guild of Motoring Writers, 40 Baring Road, Bournemouth BH6 4DT
☎ 01202 422424   📠 01202 422424
generalsec@gomw.co.uk
www.guildofmotoringwriters.co.uk
Contact *Patricia Lodge*

First presented by Lord Montagu on the occasion of the opening of the National Motor Museum at Beaulieu in 1972. Awarded annually to a member of the **Guild of Motoring Writers** who, in the opinion of the nominated jury, has made the greatest contribution to recording in the English language the history of motoring or motor cycling in a published book or article, film, television or radio script, or research manuscript available to the public.
💷 Cash prize sponsored by Mercedes-Benz UK.

Mere Literary Festival Open Competition

'Lawrences', Old Hollow, Mere BA12 6EG
☎ 01747 860475
www.merelitfest.co.uk
Contact *Mrs Adrienne Howell (Events Organizer)*

Annual open competition which alternates between short stories and poetry. The winners are announced at the **Mere Literary Festival** during the second week of October. The 2009 competition is for poetry. For further details, including entry fees and form, access the website or contact the address above from 1st March with s.a.e.
💷 Cash prizes.

Meyer-Whitworth Award

Playwrights' Studio Scotland, CCA, 350 Sauchiehall Street, Glasgow G2 3JD
☎ 0141 332 4403
info@playwrightsstudio.co.uk
www.playwrightsstudio.co.uk
Contact *Claire Dow*

The Meyer-Whitworth Award is one of the largest annual monetary prizes for playwriting in the UK. Funded by the Royal National Theatre Foundation and managed by the Playwrights' Studio, Scotland in association with the UK Playwrights Network, it is intended to help further the careers of UK playwrights who are not yet established. Nominations by directors of professional theatre companies only. Plays must be original work, be in the English language and have been produced professionally in the UK for the first time between 1st November and 30th November of the previous year. Translations are not eligible. Candidates will have had no more than two of their plays professionally produced, including the play submitted. No writer who has previously won the award may reapply and no play that has previously been submitted for the award is eligible. Writers must be resident in the British Isles or Republic of Ireland. Nomination forms available from the Playwrights' Studio, Scotland.
💷 Award of £10,000.

MIND Book of the Year

Granta House, 15–19 Broadway, London E15 4BQ
☎ 020 8215 2301   📠 020 8215 2269
j.bird@mind.org.uk
www.mind.org.uk

Established 1981. Annual award, in memory of Sir Allen Lane, for the author of a book published in the current year (fiction or non-fiction), which furthers public understanding of mental health problems. 2008 winner: Martin Townsend *The Father I Had*.

The Mitchell Prize for Art History/The Eric Mitchell Prize

c/o The Burlington Magazine, 14–16 Duke's Road, London WC1H 9SZ
☎ 020 7388 1228   📠 020 7388 1229
cronin@burlington.org.uk
(with 'Mitchell Prize' in subject line)
Chairman *Caroline Elam*
Contact *Bébhinn Cronin*

The Mitchell Prize is awarded for a book which has made an outstanding and original contribution to the understanding of the visual arts, and The Eric Mitchell Prize is awarded for the best exhibition catalogue. Both must be on western art, written in English and published in the previous two years. Nominations should be submitted by publishers before the end of April. 2006 winners: Mark P. McDonald *The Print Collection of Ferdinand Columbus* (The Mitchell Prize); Elena Phipps, Johanna Hecht, Christina Esteras Martin *The Colonial Andes: Tapestry & Silverware 1530–1830*, The Metropolitan Museum of Art, New York in association with Yale University Press 2004 (The Eric Mitchell Prize). (NB These prizes are under review.)

Momaya Press Short Story Writing Competition

Momaya Press, Flat 1, 189 Balham High Road, Rear Building, London SW12 9BE

☎ 020 8673 9616
infouk@momayapress.com
www.momayapress.com
Contact *Monisha Saldanha*

Established 2004. Annual short story competition sponsored by Momaya Press. Open to writers in the English language worldwide. Entries, which should not have been published previously, can be in any style and format and on any subject; 2,500 words maximum. Entry fee: £7. Submissions by post or via entry form on the website.
£ Prizes of £110 (1st); £60 (2nd); £30 (3rd); all winners are published in the *Momaya Annual Review*.

Scott Moncrieff Prize
▷ The Translators Association Awards

The Sheridan Morley Prize

c/o Oberon Books, 521 Caledonian Road, London N7 9RH

☎ 020 7607 3637   📠 020 7607 3629
info@oberonbooks.com
Contact *Michelle Marsh*

Established in 2007/08 as a lasting memorial to Sheridan Morley, who during a long and illustrious career as critic, broadcaster and author, wrote more than 30 books. The prize celebrates the art of theatre biography and is awarded every April to the best biography on a theatrical subject, written in the English language and published in the given year. Final entry date: early March. Inaugural winner: *Will & Me. How Shakespeare Took Over My Life* Dominic Dromgoole.
£ Prize of £2,000.

nasen Awards 2009 – Celebrating Inclusive Practice

Nasen House, 4–5 Amber Business Village, Amber Close, Amington, Tamworth B77 4RP

☎ 01827 311 500   📠 01827 313 005
welcome@nasen.org.uk
www.nasen.org.uk

Awards presented in nine categories for 2009: Book to Promote Professional Development; ICT Accessibility (Software) Resource; ICT Accessibility (Hardware) Resource; Book to support teaching and learning – The Pupil Book; Book to support teaching and learning – The Educational Practitioner's Book; Inclusive Resource for Primary Classrooms; Inclusive Resource for Secondary Classrooms; The Inclusive Children's Book; The Special Educational Needs Academic Book. Deadline: 30th April each year. Eligibility for 2009 Awards:

books and resources must have been published in the UK between 22 June 2008 and 30 April 2009. Contact nasen for more information.
£ Prize of £500 for each category.

Melissa Nathan Award for Comedy Romance

Melissa Nathan Foundation and Award, PO Box 56923, London N10 3YU

☎ 020 8671 8424   📠 020 8883 6694
information@melissanathan.co.uk
www.melissanathan.com
Chair *Andrew Saffron*
Committee *Maggie Phillips*

Established in 2007 and named after bestselling novelist, Melissa Nathan, who died of cancer aged 37, to honour her memory and writers of comedy romance novels. Awarded to the novel that best combines laugh-out-loud comedy with believable, heart-warming romance (varying sub-category awards each year). To be eligible, novels must be published in the UK in the year of the award, including paperback version of hardbacks published the previous year. Final entry date: 31st December. 2008 winner: Lisa Jewell *31 Dream Street* (Penguin).
£ Prize of £5,000, plus trophy.

National Association of Writers' Groups (NAWG)

Annual competitions with several categories. Prizes are engraved glass book trophies. See entry under *Professional Associations*.

National Poetry Anthology

United Press, Admail 3735, London EC1B 1JB

☎ 0870 240 6190   📠 0870 240 6191
mail@unitedpress.co.uk
www.unitedpress.co.uk
Contact *Julie Embury*

Free-to-enter annual poetry competition. Organizers United Press allow up to three poems (20 lines and 160 words maximum each) by annual closing date of 30th June. They select around 250 regional winners in the UK. These receive a free copy of the annual book, *The National Poetry Anthology* and vote for the overall winner who receives £1,000 and a trophy to keep for life.

National Poetry Competition

The Poetry Society, 22 Betterton Street, London WC2H 9BX

☎ 020 7420 9895   📠 020 7240 4818
marketing@poetrysociety.org.uk
www.poetrysociety.org.uk
Contact *Competition Organizer (WH)*

One of Britain's major open poetry competitions. Closing date: 31st October. Poems on any theme, up to 40 lines. For rules and entry form send s.a.e.

to the competition organizer at the address above or enter the competition via the website.

Ⓔ Prizes of £5,000 (1st); £1,000 (2nd); £500 (3rd); plus 10 commendations of £50.

Bill Naughton Short Story Competition

Box No. [year of competition], Aghamore, Ballyhaunis, County Mayo, Republic of Ireland

☎ 00 353 9 4936 7019
paulwdrogers@eircom.net
www.aghamoreireland.com/kennynaughton/shortstory.htm
Contact *Paul Rogers*

Established 1994. Annual award organized by the Kenny/Naughton Autumn School. Stories can be on any topic and no more than 2,500 words in length. All work must be unpublished; typed mss only with no name or address appearing on the work. Entry fee: £5, €7 or $8 per story (three stories may be submitted for the price of two). Closing date: first Friday in September. Previous winners: David Oglesby, Toby McGrath, Sarah Evans.

Ⓔ Awards of €200 (1st); €130 (2nd); €65 (3rd).

'The Nibbies'
▷ **Galaxy British Book Awards**

Nielsen Gold & Platinum Book Awards

Nielsen BookScan, 3rd Floor, Midas House, 62 Goldsworth Road, Woking GU21 6LQ

☎ 01483 712222   ⓕ 01483 712220
awards.bookscan@nielsen.com
www.nielsenbookscan.co.uk
Contact *Mo Siewcharran*

Established in 2000 to award sales achieved for actual consumer purchases of a title through UK bookshops. Awarded to any title, in all its editions, that sells more than 500,000 copies (Gold) or one million copies (Platinum) over a period of five consecutive years.

Ⓔ Award winner receives commemorative plaque issued by the publisher to the author.

Nobel Prize

The Nobel Foundation, PO Box 5232, 102 45 Stockholm, Sweden

☎ 00 46 8 663 0920   ⓕ 00 46 8 660 3847
www.nobelprize.org
Contact *Information Section*

Awarded yearly for outstanding achievement in physics, chemistry, physiology or medicine, literature and peace. Founded by Alfred Nobel, a chemist who proved his creative ability by inventing dynamite. In general, individuals cannot nominate someone for a Nobel Prize. The rules vary from prize to prize but the following are eligible to do so for Literature: members of the Swedish Academy and of other academies, institutions and societies similar to it in constitution and purpose; professors of literature and of linguistics at universities or colleges; Nobel Laureates in Literature; presidents of authors' organizations which are representative of the literary production in their respective countries. British winners of the literature prize, first granted in 1901, include Rudyard Kipling, John Galsworthy and Winston Churchill. Winners since 1999: Günter Grass (Germany); Gao Xingjian (France); V.S. Naipaul (UK); Imre Kertész (Hungary); J.M. Coetzee (South Africa); Elfriede Jelinek (Austria); Harold Pinter (UK); Orhan Pamuk (Turkey); Doris Lessing (UK). Nobel Laureate in Literature 2008: Jean-Marie Gustave Le Clézio (France).

The Noma Award for Publishing Africa

PO Box 128, Witney OX8 5XU

☎ 01993 775235   ⓕ 01993 709265
maryljay@aol.com
www.nomaaward.org
Contact *Mary Jay, Secretary to the Managing Committee*

Established 1979. Annual award, founded by the late Shoichi Noma, President of Kodansha Ltd, Tokyo. The award is for an outstanding book, published in Africa by an African writer, in three categories: scholarly and academic; literature and creative writing; children's books. Entries, by publishers only, by 28th February for a title published in the previous year. Maximum number of three entries. Previous winners: Hamdi Sakkut *The Arabic Novel: Bibliography and Critical Introduction 1865–1995*; Elinor Sisulu *Walter and Albertina Sisulu. In Our Lifetime*; Werewere-Liking *La mémoire amputée*; Lebogang Mashile *In a Ribbon of Rhythm*; Shimmer Chinodya *Strife*.

Ⓔ Prize of US$10,000 and presentation plaque.

Frank O'Connor International Short Story Award

The Munster Literature Centre, Frank O'Connor House, 84 Douglas Street, Cork, Republic of Ireland

☎ 00 353 21 431 2955
munsterlit@eircom.net
www.munsterlit.ie
Contact *Patrick Cotter*

Founded in 2005 to reward and encourage writers and publishers of short stories. Annual competition for a collection of original short stories by a single living author published between September of the previous year and August of the award year. Final entry date: 31st August. Vanity press publications not accepted. 2008 winner: Jhumpa Lahiri *Unaccustomed Earth* (Bloomsbury).

Ⓔ Award of €35,000.

Sean O'Faolain Short Story Prize
The Munster Literature Centre, Frank O'Connor House, 84 Douglas Street, Cork, Republic of Ireland
☎ 00 353 21 431 2955
munsterlit@eircom.net
www.munsterlit.ie
Contact *Patrick Cotter*

Founded 2003. Annual competition for an original, unpublished story of under 3,000 words. Final entry date: 31st July. 2008 winner: Julia Van Middlesworth.
£ Awards of €1,500 (1st); €500 (2nd).

Old Possum's Children's Poetry Competition
▷ Poetry Book Society under Organizations of Interest to Poets

C.B. Oldman Prize
Special Libraries & Archives, University of Aberdeen, King's College, Aberdeen AB24 3SW
☎ 01224 274266   ☎ 01224 273891
r.turbet@abdn.ac.uk
Contact *Richard Turbet, Special Collections Cataloguer and Music Subject Specialist*

Established 1989 by the International Association of Music Libraries, UK & Ireland Branch. Annual award for best book of music bibliography, librarianship or reference published the year before last (i.e. books published in 2006 considered for the 2008 prize). Previous winners: Pamela Thompson, Malcolm Lewis, Andrew Ashbee, Michael Talbot, Donald Clarke, John Parkinson, John Wagstaff, Stanley Sadie, William Waterhouse, Richard Turbet, John Gillaspie, David Fallows, Arthur Searle, Graham Johnson, Michael Twyman, John Henderson, David Golby, David Wyn Jones.
£ Prize of £200.

Ondaatje Prize
▷ The Royal Society of Literature Ondaatje Prize

Orange Broadband Prize for Fiction/Orange Broadband Award for New Writers
Booktrust, 45 East Hill, London SW18 2QZ
☎ 020 8516 2972   ☎ 020 8516 2978
tarryn@booktrust.org.uk
www.orangeprize.co.uk
Contact *Tarryn McKay*

Established 1996. Annual award founded by a group of senior women in publishing to 'create the opportunity for more women to be rewarded for their work and to be better known by the reading public'. Awarded for a full-length novel written in English by a woman of any nationality, and published in the UK between 1st April and 31st March of the following year. 2008 winner:

Rose Tremain *The Road Home* (Chatto & Windus).
£ Prize of £30,000 and a work of art (a limited edition bronze figurine known as 'The Bessie' in acknowledgement of anonymous prize endowment).
Established 2005, the **Orange Broadband Award for New Writers** is open to all first works of fiction, written by women of any age or nationality, published in the UK between 1 April and 31 March (short story collections and novellas also eligible). 2008 winner: Joanna Cavenna *Inglorious* (Faber & Faber).

The Orwell Prize
Wiley-Blackwell, Blackwell Publishing, 9600 Garsington Road, Oxford OX4 2DQ
☎ 01865 476255   ☎ 01865 471255
lucie.crowther@wiley.com
www.theorwellprize.co.uk
www.blackwellpublishing.com/poqu
Contact *Lucie Crowther*

Jointly established in 1993 by the George Orwell Memorial Fund and *Political Quarterly* to encourage and reward writing in the spirit of Orwell's 'What I have most wanted to do ... is to make political writing into an art'. Two categories: book or pamphlet; sustained journalism, reporting, comment or blogs (on a theme) in broadcast, online, newspaper, periodicals, features or columns. Submissions by editors, publishers or authors. 2007 winners (for work published in 2007): Johann Hari of *The Independent* (journalism); Raja Shehadeh, *Palestinian Walks* (book); and a Special Award for Writing and Broadcasting was presented to Clive James.
£ Prizes of £1,000 for each category.

The Wilfred Owen Award for Poetry
21 Culverden Avenue, Tunbridge Wells TN4 9RE
☎ 01892 532712   ☎ 01892 532712
mmccrane@ukonline.co.uk
www.wilfredowen.org.uk
Contact *Meg Crane*

Established in 1988 by the **Wilfred Owen Association**. Given to a poet whose poetry reflects the spirit of Owen's work in its thinking, expression and inspiration. Applications are not sought; the decision is made by the Association's committee. Previous winners: Seamus Heaney, Christopher Logue, Michael Longley, Harold Pinter, Tony Harrison.
£ Award winner receives a work of art, suitably decorated and engraved.

OWPG Awards for Excellence
Outdoor Writers & Photographers Guild, PO Box 520, Bamber Bridge, Preston PR5 8LF
☎ 01772 321243   ☎ 0870 137 8888

secretary@owpg.org.uk
www.owpg.org.uk
Secretary *Terry Marsh*

Established 1980. Annual awards by the **Outdoor Writers & Photographers Guild** to raise the standard of outdoor writing, journalism, broadcasting and photography. Categories include guidebook, outdoor book, outdoor feature, travel feature, words and pictures, photography. Open to OWPG members only. Final entry date: April.

The Oxford Weidenfeld Translation Prize

St Anne's College, Oxford OX2 6HS
⌶ 01865 274820   🖷 01865 274895
sandra.madley@st-annes.ox.ac.uk
www.stannes.ox.ac.uk/about/translationprize.html
Contact *The Principals' Secretary*

Established in 1996 by publisher Lord Weidenfeld and now includes St. Anne's, Queen's and New colleges to encourage good translation into English. Annual award to the translator(s) of a work of fiction, poetry or drama written in any living European language. Translations must have been published in the previous calendar year. Submissions from publishers only.
🖷 Prize of £2,000.

PEN Awards

See **J.R. Ackerley Prize; Golden PEN Award for Lifetime Distinguished Service to Literature; The Hessell-Tiltman Prize for History**

Samuel Pepys Award

Haremoor House, Faringdon SN7 8PN
Contact *Paul Gray*

Established 2003. Biennial award (closing date: 31st May 2009) given by the Samuel Pepys Award Trust for a book that makes the greatest contribution to the understanding of Samuel Pepys, his times or his contemporaries. 2007 winner: John Adamson *The Noble Revolt.*
🖷 Award of £2,000, plus commemorative medal.

Poetry Business Competition

Bank Street Arts, 32–40 Bank Street, Sheffield
S1 2DS
office@poetrybusiness.co.uk
www.poetrybusiness.co.uk
Contact *The Competition Administrator*

Established 1986. Annual award which aims to discover and publish new writers. Entrants should submit a short manuscript of poems. Winners will have their work published by the **Poetry Business** under the Smith/Doorstop imprint. Contact for conditions of entry. Previous winners include: Pauline Stainer, Michael Laskey, Mimi Khalvati, David Morley, Moniza Alvi, Selima Hill, Catherine Smith, Daljit Nagra.

🖷 Prizes: Publication of full collection; runners-up have pamphlets; 20 complimentary copies. Also cash prize (£1,000) to be shared equally between all winners.

The Poetry Society's National Poetry Competition
▷ **National Poetry Competition**

The Portico Prize for Literature

The Portico Library, 57 Mosley Street, Manchester M2 3HY
⌶ 0161 236 6785   🖷 0161 236 6803
librarian@theportico.org.uk
www.theportico.org.uk
Contact *Miss Emma Marigliano*

Established 1985. Administered by the Portico Library in Manchester. Biennial awards for a work of fiction and of non-fiction published between the two closing dates. Set wholly or mainly in the North of England. Previous winners include: Anthony Burgess *Any Old Iron*; Jenny Uglow *Elizabeth Gaskell: A Habit of Stories*; Andrew Biswell *The Real Life of Anthony Burgess.*

Practical Art Book of the Year

PO Box 3, Huntingdon PE28 0QX
⌶ 01832 710201   🖷 01832 710488
award@acaward.com
www.acaward.com
www.artists-choice.co.uk
Contact *Henry Malt*

Jointly sponsored by **Artists' Choice** book club and *Leisure Painter* and *The Artist* magazines. A short list is drawn up by the editors and the winner voted by readers. Shortlist will be announced in June and the winner in November.

The Premio Valle Inclán
▷ **The Translators Association Awards**

V.S. Pritchett Memorial Prize

The Royal Society of Literature, Somerset House, Strand, London WC2R 1LA
⌶ 020 7845 4676   🖷 020 7845 4679
info@rslit.org
www.rslit.org

Established 1999. Awarded for a previously unpublished short story of between 2,000 and 5,000 words. For entry forms, please contact the Secretary from April 2009 onwards. Closing date to be confirmed as the prize is being relaunched. Open to UK and Irish writers.
🖷 Prize of £1,000.

The Projection Box Essay Awards 2009–2010

12 High Street, Hastings TN34 3EY
⌶ 01424 204144   🖷 01424 204144
info@pbawards.co.uk
www.pbawards.co.uk

Contact *Mo Heard*

Founded 2007 by small independent publishers, The Projection Box. Annual open competition which in the current year is to encourage new research and thinking into any historical, artistic or technical aspect of popular optical media. The prize is for an essay in English, not previously published, of between 5,000 and 8,000 words (including notes). Further information available on the website.

Ⓔ Prize of £250 plus publication in the journal, *Early Popular Visual Culture*.

Prose & Poetry Prizes

The New Writer, PO Box 60, Cranbrook TN17 2ZR
☎ 01580 212626
admin@thenewwriter.com
www.thenewwriter.com
Contact *Merric Davidson*

Established 1997. Annual award. Open to all poets writing in the English language for an original, previously unpublished poem or collection of six to ten poems. Also open to writers of short stories and novellas/serials, features, articles, essays and interviews. Final entry date: 30th November. Full guidelines on the website. Previous winners: Wes Lee, Graham Clifford, Mario Petrucci, Ros Barber, Celia de Fréine, Andrew McGuinness, Katy Darby, David Grubb.

Ⓔ Prizes total £2,000, plus publication in annual collection published by *The New Writer*.

Pulitzer Prizes

The Pulitzer Prize Board, 709 Journalism Building, Columbia University, 2950 Broadway, New York, NY 10027, USA
☎ 001 212 854 3841   📠 001 212 854 3342
pulitzer@pulitzer.org
www.pulitzer.org

Awards for journalism in US newspapers, and for published literature, drama and music by American nationals. 2008 winners include: Junot Diaz *The Brief Wondrous Life of Oscar Wao* (fiction); John Matteson *Eden's Outcasts* (biography); *The Washington Post* (public service).

Michael Ramsey Prize for Theological Writing

SPCK, 36 Causton Street, London SW1P 4ST
☎ 020 7592 3900
prize@spck.org.uk
www.michaelramseyprize.org.uk
Coordinator *Primavera Quantrill*
Press Enquiries *Mark Casserley*

Founded in 2005 by Archbishop of Canterbury Dr Rowan Williams to encourage the most promising contemporary theological writing and to identify it for a wider Christian readership. This biennial prize commemorates Dr Ramsey (Archbishop of Canterbury, 1961–1974) and his commitment to increasing the breadth of theological understanding of people in general. Awarded to the author of a theological work that is judged to contribute most towards advancing theology and making a serious contribution to the faith and life of the Church. The book must be by a living author; must have been first published between March 2008 and December 2009; must be written in or translated into English and available in the UK (other terms and conditions on website). Nominations are made by invited Anglican Primates worldwide, by leaders of UK Christian church bodies, Anglican Diocesan Bishops and by principals of theological colleges and by the sponsors of the prize, the Lambeth Fund Trustees. Closing date for nominations: 14th March 2010. 2007 winner: *What is the point of being a Christian?* by Fr Timothy Radcliffe OP. The Michael Ramsey Prize is administered by SPCK on behalf of the Archbishop of Canterbury, and is sponsored by the Lambeth Partners.

Ⓔ Prize of £15,000.

The Red House Children's Book Award

The Federation of Children's Book Groups, 2 Bridge Wood View, Horsforth, Leeds LS18 5PE
☎ 0113 258 8910
info@fcbg.org.uk
www.redhousechildrensbookaward.co.uk
National Coordinator *Sinead Kromer*

Established 1980. Three categories. Awarded annually for best book of fiction suitable for children. Unique in that it is judged by the children themselves. 2008 overall winner: Derek Landy *Skulduggery Pleasant* (HarperCollins).

Ⓔ Award winners receive silver bowls and trophy, portfolio of letters, drawings and comments from the children who take part in the judging.

Trevor Reese Memorial Prize

Institute of Commonwealth Studies, School of Advanced Studies, University of London, 28 Russell Square, London WC1B 5DS
☎ 020 7862 8853   📠 020 7862 8820
troy.rutt@sas.ac.uk
commonwealth.sas.ac.uk/reese.htm

Established in 1979 with the proceeds of contributions to a memorial fund to Dr Trevor Reese, Reader in Commonwealth Studies at the Institute and a distinguished scholar of Imperial History (d.1976). Triennial award (next award 2009 for works published 2006, 2007 and 2008) for a scholarly work, usually by a single author, which has made a wide-ranging, innovative and scholarly contribution in the broadly-defined field of Imperial and Commonwealth History. All correspondence relating to the prize should be marked 'Trevor Reese Memorial Prize'.

Ⓔ Prize of £1,000.

Regional Press Awards

Press Gazette, Paulton House, Sheperdess Walk, London N1 7LB

☎ 020 7324 2396 ☏ 020 7549 8668
apont@wilmington.co.uk
www.pressgazette.co.uk

Open to all regional newspapers and regional journalists, whether freelance or staff. June event. Run by the *Press Gazette.*

Renault UK Journalist of the Year Award

Guild of Motoring Writers, 40 Baring Road, Bournemouth BH6 4DT

☎ 01202 422424 ☏ 01202 422424
generalsec@gomw.co.uk
www.guildofmotoringwriters.co.uk
Contact *Patricia Lodge*

Originally the Pierre Dreyfus Award, established in 1977. Awarded annually by Renault UK Ltd in honour of Pierre Dreyfus, president director general of Renault 1955–75, to the member of the **Guild of Motoring Writers** who is judged to have made the most outstanding journalistic effort in any medium during the year. Particular emphasis is placed on initiative and endeavour.

RIBA International Book Awards

c/o 15 Bonhill Street, London EC2P 2EA

☎ 020 7496 8364 ☏ 020 7374 8500
john.morgan@ribabookshops.com
www.ribabookshops.com
Contact *John Morgan*

Founded in 2005 and hosted by RIBA Bookshops and the British Architectural Library (RIBA Trust). Annual open, free competition to recognize and celebrate the best writing on architecture, construction, architectural practice and interior design. Entries must have been published in the year preceding the award year; final entry date: 31st December. Foreign (English language) books, which are encouraged, currently represent around 40 per cent of entries. 2008 winners: RIBA Sir Nikolaus Pevsner Award for Architecture: *Siena: Constructing the Renaissance City* Fabrizio Nevola (Yale University Press); RIBA Sir Robert McAlpine Award for Construction: *The Architecture of Parking* Simon Henley (Thames & Hudson).
£ Money award to author and publishers' trophy (in four categories from 2008).

RITA Awards

⊡ Romance Writers of America under Professional Associations and Societies

Romantic Novel of the Year

www.rna-uk.org
Honorary Administrator *Mrs P. Fenton*

Established 1960. Annual award for the best romantic novel of the year, open to non-members as well as members of the **Romantic Novelists' Association**. Novels must be first published in the UK between specified dates. Full details on the RNA website. 2009 winner: Julia Gregson *East of the Sun* (Orion). Contact the organizer for entry form via the website address.

Rossica Translation Prize

Academia Rossica, 4th Floor, 76 Brewer Street, London W1F 9TX

☎ 020 7287 2614 ☏ 020 7287 5712
rossica-prize@academia-rossica.org
www.academia-rossica.org
Contacts *Svetlana Adjoubei, Ruth Atkinson*

Founded in 2005 by Academia Rossica (AR), a UK registered charity that supports cultural collaboration between Russia and the West. A biennial prize, which is awarded for the best new literary translation from Russian into English, published anywhere in the world. The aim is to promote the best of Russian literature and a better understanding of Russian culture in English-speaking countries. Four copies of the (published) English translation and of the Russian original should be submitted to Academia Rossica. 2007 winner: Joanne Turnbull, translator of *7 Stories* by Sigizmund Krzhizhanovsky.
£ Prizes of £3,000 (translator); £1,000 (publisher).

Royal Economic Society Prize

The Economic Journal, Department of Economics, London Business School, Regent's Park, London NW1 4SA

☎ 020 7000 8413 ☏ 020 7000 8401
econjournal@london.edu
www.res.org.uk/society.resprize.asp
Contact *Heather Daly*

Annual award for the best article published in *The Economic Journal.* Open to members of the Royal Economic Society only. Previous winners: Professors Marcos Rangal, Tilman Börgers, Christian Dustmann, Paul Cheshire, Stephen Sheppard.
£ Prize of £3,000.

Royal Mail Awards for Scottish Children's Books

Scottish Book Trust, Sandeman's House, Trunk's Close, 55 High Street, Edinburgh EH1 1SR

☎ 0131 524 0160 ☏ 0131 524 0161
anna.gibbons@scottishbooktrust.com
www.scottishbooktrust.com
Children's Programme Manager *Anna Gibbons*

Awards are given to new and established authors of published books in recognition of high standards of writing for children in three age group categories: younger children (0–7 years); younger readers (8–11), older readers (12–16). There is also a competition for short stories written in Gaelic by children. A shortlist is drawn up by a panel of children's book experts and the winner in each category is selected by children and young people voting for their favourites in schools and libraries across Scotland. Authors should be Scottish or resident in Scotland, but books of particular Scottish interest by other authors are eligible. Final entry date: 31st January; guidelines available on request. 2008 winners: 0–7 years: Alan Durant and Ross Collins *Billy Monster's Daymare* (OUP); 8–11: D.A. Nelson *DarkIsle* (Strident); 12–16: *Bunker 10* J.A. Henderson (OUP).

🄴 Awards total £15,000.

Royal Mail/Saltire Society Scottish First Book of the Year
▷ The Saltire Literary Awards

Royal Society of Literature Awards

The Royal Society of Literature Ondaatje Prize, V.S. Pritchett Memorial Prize and (under *Bursaries, Fellowships and Grants*) **The Royal Society of Literature/Jerwood Awards**

The Royal Society of Literature Ondaatje Prize

The Royal Society of Literature, Somerset House, Strand, London WC2R 1LA
☎ 020 7845 4676   🖷 020 7845 4679
paulaj@rslit.org
www.rslit.org
Submissions *Paula Johnson*

Administered by the **Royal Society of Literature** and endowed by Sir Christopher Ondaatje, the prize is awarded annually to a book of literary merit, fiction, non-fiction or poetry, best evoking the spirit of a place. All entries must be published within the calendar year 2009 and should be submitted by mid-December. The writer must be a citizen of the UK, Commonwealth or Ireland. Further details available on the website. 2008 winner: Graham Robb *The Discovery of France* (Picador).

🄴 Prize of £10,000.

The Royal Society Prizes for Science Books

The Royal Society, 6–9 Carlton House Terrace, London SW1Y 5AG
☎ 020 7451 2513   🖷 020 7930 2170
sciencebooks@royalsociety.org
http://royalsociety.org
Contact *Katherine Jarrett*

Annual prizes established in 1988 to celebrate the best in popular science writing. Awarded to books

that make science more accessible to readers of all ages and backgrounds. Owned and managed by the Royal Society and supported by the Beecroft Trust. All entries must be written in English and their first publication must have been between 1st January and 31st December; submission by publishers only. Educational textbooks published for professional or specialist audiences are not eligible. 2008 winners: Junior Prize: Rebecca Gilpin and Leonie Pratt *The Big Book of Science Things to Make and Do* (Usborne); General Prize: Mark Lynas *Six Degrees: Our Future on a Hotter Planet* (Fourth Estate).

🄴 Prizes total £30,000; £10,000 to each winner; £1,000 to each of the five shortlisted authors in each prize.

Runciman Award

The Anglo-Hellenic League, c/o The Hellenic Centre, 16–18 Paddington Street, London W1U 5AS
☎ 020 7486 9410
info@anglohellenicleague.org
www.anglohellenicleague.org
Contact *The Administrator*

Established 1985. Annual award, sponsored by the National Bank of Greece. Founded by the Anglo-Hellenic League to promote Anglo-Greek understanding and friendship, for a work wholly or mainly about some aspect of Greece or the world of Hellenism, published in English in any country of the world in its first edition during 2008. It is a condition of the award that shortlisted books should be available for purchase to readers in the UK at the time of the award ceremony. Named after the late Professor Steven Runciman, former chairman of the Anglo-Hellenic League. No category of writing will be excluded from consideration. Works in translation, with the exception of translations from Greek literature, will not be considered. Award presented in May/June. Additional information available on the website.

🄴 Award of £9,000.

Sainsbury's Baby Book Award
▷ Booktrust Early Years Awards

The Saltire Literary Awards

Saltire Society, 9 Fountain Close, 22 High Street, Edinburgh EH1 1TF
☎ 0131 556 1836   🖷 0131 557 1675
saltire@saltiresociety.org.uk
www.saltiresociety.org.uk
Administrator *Kathleen Munro*

Established 1982 and 1988. Annual awards, one for Book of the Year and one for a First Book by an author publishing for the first time. Open to any author of Scottish descent or living in

Scotland, or to anyone who has written a book which deals with either the work and life of a Scot or with a Scottish problem, event or situation. Nominations are invited from editors of leading newspapers, magazines and periodicals. 2008 winners: The Faculty of Advocates/Saltire Society Scottish Book of the Year: James Kelman *Kieron Smith, Boy* (Hamish Hamilton); Royal Mail/Saltire Society Scottish First Book of the Year: Andrew Nicoll *The Good Mayor* (Black & White Publishing).

£ Prizes of £5,000 (Scottish Book); £1,500 (First Book).

Schlegel-Tieck Prize
▷ The Translators Association Awards

Science Writer Awards
www.science-writer.co.uk

Established in 1987, this award is designed to bridge the gap between science and writing. Award winners and runners-up receive cash prizes and have the opportunity to have their pieces published on the science pages of *The Daily Telegraph*. The competition is not taking place in 2009; check the website for further news.

Scottish Arts Council Book of the Year Awards
▷ Sundial Scottish Arts Council Book of the Year Awards

Scottish History Book of the Year
The Saltire Society, 9 Fountain Close, 22 High Street, Edinburgh EH1 1TF
☎ 0131 556 1836   ℻ 0131 557 1675
saltire@saltiresociety.org.uk
www.saltiresociety.org.uk
Administrator *Kathleen Munro*

Established 1965. Annual award in memory of the late Dr Agnes Mure Mackenzie for a published work of distinguished Scottish historical research of scholarly importance (including intellectual history and the history of science). Editions of texts are not eligible. Nominations are invited and should be sent to the Administrator. 2008 winner: Alex Woolf *From Pictland to Alba 789–1070* (Edinburgh University Press).

£ Prize of £1,500.

The Self-Publishing Awards
The David St John Thomas Charitable Trust, PO Box 6055, Nairn IV12 4YB
☎ 01667 453351   ℻ 01667 452365
dsjtcharitynairn@fsmail.net
Contact *Anne Hill*

Established in 1990 and run jointly by the David St John Thomas Charitable Trust and *Writers' News*. Annual awards for books self-published in the previous year. Final entry date: 31 January.

Full details available from the Trust at the address above. Previous winner: Joyce Reed.
£ Awards total £2,250 plus two silver cups.

SES Book Prizes
id5@york.ac.uk
www.soc-for-ed-studies.org.uk
Acting Honorary Secretary *Professor Ian Davies*

Annual awards given by the Society for Educational Studies for the best books on education published in the UK during the preceding year. Nominations by publishers or by individual authors based in the UK.

£ Prizes of £2,000 (1st); £1,000 (2nd); £750 (highly commended).

Bernard Shaw Translation Prize
▷ The Translators Association Awards

André Simon Memorial Fund Book Awards
1 Westbourne Gardens, Glasgow G12 9XE
☎ 0141 342 4929
katie@andresimon.co.uk
Contact *Katie Lander*

Established 1978. Awards given annually for the best book on drink, best on food and special commendation in either. 2008 winners: James Crowden *Ciderland* (Birlinn); Peter Brears *Cooking and Dining in Medieval Britain* (Prospect Books); Commendation: Michael Roux *Pastry* (Quadrille).

£ Awards of £2,000 (best book); £1,000 (special commendation); £200 to shortlisted books.

SJA British Sports Journalism Awards
▷ Sports Journalists' Association of Great Britain under Professional Associations and Societies

The Society for Theatre Research Annual Theatre Book Prize
The Society for Theatre Research, PO Box 53971, London SW15 6UL
theatrebookprize@btinternet.com
www.str.org.uk

Established 1997. Annual award for books, in English, of original research into any aspect of the history and technique of British or British-related theatre. Not restricted to authors of British nationality nor books solely from British publishers. Books must be first published in English (no translations) during the calendar year. Play texts and those treating drama as literature are not eligible. Publishers submit books directly to the independent judges and should contact the Book Prize Administrator for further details. 2008 winner: *Theatre and Globalization* Patrick Lonergan (Palgrave Macmillan).

£ Award of £400.

Sony Radio Academy Awards

Alan Zafer & Associates, 47–48 Chagford Street, London NW1 6EB

☎ 020 7723 0106   ⊠ 020 7724 6163

secretariat@radioawards.org

www.radioawards.org

Contact *The Secretariat*

Established 1982 in association with the **Society of Authors** and Sony UK. Presented in association with the **Radio Academy**. Annual awards to recognize excellence in radio broadcasting. Entries must have been broadcast in the UK between 1st January and 31st December in the year preceding the award. The categories for the awards are reviewed each year and announced in November.

Southport Writers' Circle Open Short Story Competition

16 Ormond Avenue, Westhead L40 6HT

☎ 01695 577938

southportwriterscircle@yahoo.co.uk

Founded 2005. Open to fiction of any genre, up to 2,000 words, the story must be entered anonymously. No entry form required; include a cover sheet supplying title of story, name and contact details (send s.a.e. for results if no e-mail). Fee: £3 per story (payable to Southport Writers' Circle). Final entry date: 31st October.

💷 Prizes of £150 (1st); £75 (2nd); £25 (3rd).

Southport Writers' Circle Poetry Competition

32 Dover Road, Southport PR8 4TB

Contact *Mrs Hilary Tinsley*

For previously unpublished work. Entry fee: £3 per poem. Any subject, any form; maximum 40 lines. Closing date: end April. Poems must be entered anonymously, accompanied by a sealed envelope marked with the title of poem, containing s.a.e. Entries must be typed on A4 paper and be accompanied by the appropriate fee payable to Southport Writers' Circle. No application form is required. Envelopes should be marked 'Poetry Competition'. Postal enquiries only. No calls.

💷 Prizes of £200 (1st); £100 (2nd); £50 (3rd); additional £25 Humour Prize.

Bram Stoker Awards for Superior Achievement

Horror Writers Association, 244 Fifth Ave., Suite 2767, New York, NY 10001, USA

hwa@horror.org

www.horror.org

Founded 1988 and named in honour of Bram Stoker, author of *Dracula*. Presented annually by the **Horror Writers Association** (HWA) for works of horror first published in the English language. Works are eligible during their first year of publication. HWA members recommend works for consideration in eight categories: Novel, First Novel, Short Fiction, Long Fiction, Fiction Collection, Poetry Collection, Anthology and Non-fiction. In addition, Lifetime Achievement Stokers are occasionally presented to individuals whose entire body of work has substantially influenced horror.

Strokestown International Poetry Competition

Strokestown International Poetry Festival Office, Strokestown, Co. Roscommon, Republic of Ireland

☎ 00 353 71 963 3759

www.strokestownpoetry.org

Administrator *Melissa Newman*

Annual poetry festival and competition. The festival takes place over the first weekend in May with readings and competitions. A centrepiece of the festival is the Strokestown International Poetry Competition for unpublished poems not exceeding 70 lines. Entry forms available on the website or call the Festival Office. Entry fee: €5.

💷 Prizes of €4,000, €2,000 and €1,000 for a poem in English; €4,000, €2,000 and €1,000 for a poem in Irish or Scottish Gaelic.

Sunday Times Writer of the Year Award

The Sunday Times, 1 Pennington Street, London E98 1ST

☎ 020 7782 5770   ⊠ 020 7782 5798

www.societyofauthors.org

Established 1987. Annual award to fiction and non-fiction writers. The panel consists of *Sunday Times* journalists and critics. Previous winners: Anthony Burgess, Seamus Heaney, Stephen Hawking, Ruth Rendell, Muriel Spark, William Trevor, Martin Amis, Margaret Atwood, Ted Hughes, Harold Pinter, Tom Wolfe, Robert Hughes. No applications; prize at the discretion of the Literary Editor.

Sunday Times Young Writer of the Year Award

The Society of Authors, 84 Drayton Gardens, London SW10 9SB

☎ 020 7373 6642   ⊠ 020 7373 5768

info@societyofauthors.org

Contact *Awards Secretary*

Established 1991. Annual award given on the strength of the promise shown by a full-length published work of fiction, non-fiction, poetry or drama. Entrants must be British citizens, resident in Britain and under the age of 35. The panel consists of *Sunday Times* journalists and critics. Closing date: 31st October. The work must be by one author, in the English language, and published in Britain. Applications by publishers via the **Society of Authors**. 2008 winner: Adam Foulds *The Truth About These Strange Times* (Weidenfeld).

Sundial Scottish Arts Council Book of the Year Awards

Scottish Arts Council, 12 Manor Place, Edinburgh EH3 7DD

☎ 0131 226 6051   ▣ 0131 225 9833

aly.barr@scottisharts.org.uk

www.scottisharts.org.uk

Literature Officer *Aly Barr*

Sponsored by Sundial Properties Ltd, these annual awards are the largest literary prizes of their kind in Scotland and the third biggest in the UK. Given in recognition of outstanding literary merit in fiction, poetry and literary non-fiction, there are four categories of awards: Fiction (including the short story); Poetry; Literary Non-Fiction; First Book of the Year. Authors should be Scottish or resident in Scotland, but books of Scottish interest by other authors are eligible for consideration. Submissions are made by publishers only, on behalf of their authors. Final closing date: 31st December for books published in that calendar year. Guidelines are available on request. 2008 Book of the Year: *A Book of Lives* Edwin Morgan (Carcanet).

▣ Prize of £5,000 is given to the winning author of each category and the four winning authors are considered for the Sundial Scottish Arts Council Book of the Year, worth an additional £20,000.

Reginald Taylor and Lord Fletcher Essay Prize

Journal of the British Archaeological Association, Institute of Archaelogy, c/o School of Art History, St Andrew's University, 9 The Scorel, St Andrew's KY16 9AR

Contact *Dr Julian Luxford*

A biennial prize, in memory of the late E. Reginald Taylor and of Lord Fletcher, for the best unpublished scholarly essay, not exceeding 7,500 words, on a subject of archaeological, art history or antiquarian interest within the period from the Roman era to AD 1830. The essay should show *original* research on its chosen subject, and the author will be invited to read the essay before the Association. The essay may be published in the journal of the Association if approved by the Editorial Committee. Closing date for entries is 30th April 2010. All enquiries by post, please. No phone calls. Send s.a.e. for details.

▣ Prize of £300 and a medal.

Theakstons Old Peculier Prize for the Crime Novel of the Year

Raglan House, Raglan Street, Harrogate HG1 1LE

☎ 01423 562303   ▣ 01423 521264

crime@harrogatefestival.org.uk

www.harrogate-festival.org.uk/crime

Festival Manager *Sharon Canavar*

Coordinator *Erica Morris*

Established 2005. Sponsored by Theakstons Old Peculier in association with Waterstone's, the award is open to full-length crime or mystery novels by British authors, published in the UK. It is the only crime fiction prize to be voted for by the general public following announcement of a long list in April. 2008 winner: (announced on the opening night of the **Theakstons Old Peculier Harrogate Crime Festival**) Stef Penney *The Tenderness of Wolves*.

▣ Prize of £3,000 cash.

The Dylan Thomas Prize

The Dylan Thomas Centre, Somerset Place, Swansea SA1 1RR

☎ 01792 474051   ▣ 01792 463993

tim@dylanthomasprize.com

www.dylanthomasprize.com

Contact *Tim J. Prosser*

Founded 2004. An award of £60,000 will be given to the winner of this prize, which was established to encourage, promote and reward exciting new writing in the English-speaking world and to celebrate the poetry and prose of Dylan Thomas. Entrants should be the author of a published book (in English), under the age of 30 (when the work was published), writing within one of the following categories: poetry, novel, collection of short stories by one author, play that has been professionally performed, a broadcast radio play, a professionally produced screenplay that has resulted in a feature-length film. Authors need to be nominated by their publishers, or producers in the case of performance art. Final entry date: 30th April. The award will be presented at the **Dylan Thomas Festival** in Swansea in November 2010. Further entry details can be found on the website. 2008 winner: Nam Le *The Boat* (Canongate).

The Times/Chicken House Children's Fiction Competition

The Chicken House, 2 Palmer Street, Frome BA11 1DS

☎ 01373 454488   ▣ 01373 454488

chickenhouse@doublecluck.com

www.doublecluck.com

Established 2007. Annual competition for the story, suitable for a worldwide children's audience aged nine to 16, which, in the opinion of the judges, demonstrates the greatest entertainment value, quality and originality. Manuscripts of no more than 80,000 words should submitted to the address above by the author (*no e-mail submissions*) and must be the original, unpublished work of the entrant. Submission period: between 1st August and 12th October 2009. The winning entry will be announced Easter 2010. Full terms and conditions are available on timesonline.co.uk/childrensauthor or write to The Times/Chicken House Children's Fiction Competition Ts&Cs request, Customer Liaison Department, 1 Pennington Street, London E98 1RL enclosing an s.a.e. 2009 winner: Sophia Bennett *Threads*.

£ The prize is the offer of a worldwide publishing contract with Chicken House, subject to negotiation and completion.

The Tinniswood Award

The Society of Authors, 84 Drayton Gardens, London SW10 9SB
☎ 020 7373 6642   ⊞ 020 7373 5768
jmccrum@societyofauthors.org
www.societyofauthors.org
Contact *Jo McCrum*

Annual award established in 2004 by the **Society of Authors** and the **Writers' Guild of Great Britain** to perpetuate the memory of playwright Peter Tinniswood as well as to celebrate and encourage high standards in radio drama. Producers are invited to send in any radio drama first transmitted within the UK and Northern Ireland during the period 1st January to 31st December 2008. Submissions must come from the producers and are restricted to a maximum of *two entries only* per producer. The work must be an original piece for radio, and may also include the first episode from an original series or serial. If submitting 15-minute episodes from a series or serial, consecutive episodes are required (including the first episode) to make up at least 45 minutes. An adaptation for radio of a piece originally written for any other medium e.g. stage, television, film, novel, poem or a short story will not be eligible. 2008 winner: Stephen Wyatt *Memorials to the Missing*.

£ Prize of £1,500, donated by the **ALCS**.

The Tir Na N-Og Award

Cyngor Llyfrau Cymru (Welsh Books Council), Castell Brychan, Aberystwyth SY23 2JB
☎ 01970 624151   ⊞ 01970 625385
wbc.children@wbc.org.uk
www.wbc.org.uk

An annual award given to the best original book published for children in the year prior to the announcement. There are three categories: Welsh Language Book – Primary Sector; Welsh Language Book – Secondary Sector; Best English Book with an authentic Welsh background.

£ Awards of £1,000 (each category).

The Translators Association Awards

The Translators Association, 84 Drayton Gardens, London SW10 9SB
☎ 020 7373 6642   ⊞ 020 7373 5768
pjohnson@societyofauthors.org
www.societyofauthors.org
Contact *Awards Secretary*

Various awards for published translations into English from Arabic (Banipal Prize), Dutch and Flemish (The Vondel Translation Prize), French (Scott Moncrieff Prize), German (Schlegel-Tieck Prize), Modern Greek (Hellenic Foundation Prize), Italian (The John Florio Prize), Portuguese (The Calouste Gulbenkian Prize), Spanish (The Premio Valle Inclán) and Swedish (Bernard Shaw Translation Prize).

The Betty Trask Awards
▷ **entry under Bursaries, Fellowships and Grants**

Sir Peter Ustinov Television Scriptwriting Award

Foundation of the International Academy of Television Arts & Sciences, 888 Seventh Avenue, 5th Floor, New York, NY 10019, USA
☎ 001 212 489 6969   ⊞ 001 212 489 6557
emmys@iemmys.tv
www.iemmys.tv
Contact *Award Administrator*

Established 1998. The late Sir Peter Ustinov gave his name to the Foundation's Television Scriptwriting Award. This annual competition is designed to motivate novice writers worldwide. The scriptwriter cannot be a United States citizen nor resident and must be below 30 years of age. Further details and entry form available on the website. 2008 winner: Jez Freedman.

£ The award winner will receive US$2,500 and an invitation to the International Emmy Awards Gala.

Ver Poets Open Competition

181 Sandridge Road, St Albans AL1 4AH
☎ 01727 762601
gillknibbs@yahoo.co.uk
www.verpoets.org.uk
Contact *Gillian Knibbs*

Various competitions are organized by **Ver Poets**, the main one being the annual Open for unpublished poems of no more than 30 lines written in English. Entry fee: £3 per poem; £10 for four poems; £2 per poem thereafter. Entry form available from address above and/or website. Two copies of poems typed on A4 white paper. The anthology of winning and selected poems, and the adjudicators' report are normally available from mid-June. Final entry date: 30th April. Back numbers of the anthology are available for £4, post-free; one copy each free to those included.

Includes FREE online access to **www.thewritershandbook.com**

Prizes of £500 (1st); £300 (2nd); £100 (3rd). Young Writers' Prize (16–21), £100.

Vogue Talent Contest

Vogue, Vogue House, Hanover Square, London W1S 1JU

☎ 020 7152 3003   ☏ 020 7408 0559

Contact *Frances Bentley*

Established 1951. Annual award for young writers and journalists (under 25 on 1st January in the year of the contest). Final entry date is in April. Entrants must write three pieces of journalism on given subjects.

Prizes of £1,000, plus a month's paid work experience with *Vogue*; £500 (2nd).

The Vondel Translation Prize
▷ The Translators Association Awards

Wadsworth Prize for Business History

School of Management, University of Southampton, Building 2, Highfield, Southampton SO17 1BJ

R.A.Edwards@soton.ac.uk

Contact *Dr Roy Edwards*

Now in its 33rd year, the Wadsworth Prize is awarded annually by the Business Archives Council to an individual judged to have made an outstanding contribution to the study of British business history. Books are nominated by publishers. 2008 winner: Professor Peter Scott *Triumph of the South. A Regional Economic History of Early Twentieth Century Britain* (Ashgate).

Wales Book of the Year Award

Academi, 3rd Floor, Mount Stuart House, Mount Stuart Square, Cardiff CF10 5FQ

☎ 029 2047 2266   ☏ 029 2047 0691

post@academi.org

www.academi.org

Contacts *Peter Finch (Chief Executive), Lleucu Siencyn (Deputy)*

Annual prize awarded for work of exceptional literary merit during the previous calendar year. There is one major prize in English, the Wales Book of the Year Award, and one major prize in Welsh, Gwobr Llyfr y Flwyddyn.

Prizes of £10,000 for winners in both languages and four runner-up prizes of £1,000 each.

Walford Award

Library Support Services, Holton, Oxford OX33 1QQ

☎ 01865 810210

ISGRefAwards@cilip.org.uk

www.cilip.org.uk

Reference Awards Coordinator *Katharine Spackman*

Part of the ISG (CILIP) Reference Awards. Awarded to an individual who has made a sustained and continual contribution to British bibliography over a period of years. The nominee need not be resident in the UK. The award is named after Dr A.J. Walford, a bibliographer of international repute. 2006 winner: Dr Diana Dixon. Cash prize and certificate.

The Warwick Prize for Writing

The Writer's Room, Millburn House, University of Warwick, Warwick CV4 7AL

☎ 024 761 50708   ☏ 024 765 24578

prizeforwriting@warwick.ac.uk

www2.warwick.ac.uk/fac/cross_fac/prizeforwriting/

Contact *Helen May*

Founded in 2008, the award, which is run by the University of Warwick, is a biennial international award for a substantial piece of writing in the English language, in any genre or form, on a theme that will change with every award. Nominations are made by the current staff of the University, honorary graduates and honorary professors. Self nominations are ineligible. The theme for the inaugural award was Complexity. Inaugural (2009) winner: Naomi Klein *The Shock Doctrine* (Penguin).

Award of £50,000 and the opportunity to take up a short placement at the University of Warwick.

Waterstone's Children's Book Prize

Waterstone's, Capital Court, Capital Interchange Way, Brentford TW8 0EX

☎ 020 8742 3800   ☏ 020 8742 0215

sarah.clarke@waterstones.com

waterstones.com

Contact *Sarah Clarke*

Founded 2005. An annual award which aims to support new children's authors and introduce them to a wide audience while introducing Waterstone's customers to some of the brightest new talent writing today. Books submitted should be fiction for children between the ages of seven and 14 with a focus on text rather than illustration. Open to new authors of any nationality with not more than two previously published fiction titles (adults' or children's). Publishers must declare any titles written under another name, including series fiction. All submitted titles must be available in paperback for a period of shortlist promotion and can be the first part of a series or trilogy but stand alone novel. See website for submission deadline. 2009 winner: Michelle Harrison *The 13 Treasures* (Simon & Schuster).

Prize of £5,000 and major promotion in Waterstone's.

The David Watt Prize

Rio Tinto plc, 2 Eastbourne Terrace, London
W2 6LG
☎ 01985 844613   📠 01985 844002
davidwattprize@riotinto.com
celiabeale@globalnet.co.uk
www.riotinto.com
Contact *The Administrator*

Initiated in 1988 to commemorate the life
and work of David Watt. Annual award, open
to writers currently engaged in writing for
English language newspapers and journals, on
international and national affairs. The winner
is judged as having made 'an outstanding
contribution towards the greater understanding
of national, international or global issues'. Entries
must have been published during the year
preceding the award. Final entry date 31st March.
2008 winner: Jonathan Freedland for his *New
York Review of Books* article 'Bush's Amazing
Achievement'.
💷 Prize of £10,000.

Brian Way Award

c/o Theatre Centre, Shoreditch Town Hall, 380
Old Street, London EC1V 9LT
☎ 020 7729 3066
admin@theatre-centre.co.uk
www.theatre-centre.co.uk
Contact *Theatre Writing Section*

Formerly the Arts Council Children's Award,
founded 2000. An annual award for playwrights
who write for children and young people.
The plays, which must have been produced
professionally, must be at least 45 minutes long.
The playwright must be resident in the UK.
Contact e-mail address/website for full details
and application form. 2008 winner: David Greig
Yellow Moon.
💷 Award of £6,000.

The Harri Webb Prize

10 Heol Don, Whitchurch, Cardiff CF14 2AU
☎ 029 2062 3359
Contact *Professor Meic Stephens*

Established 1995. Annual award to commemorate
the Welsh poet, Harri Webb (1920–94), for a
single poem in any of the categories in which
he wrote: ballad, satire, song, polemic or a first
collection of poems. The poems are chosen by
three adjudicators; no submissions. Previous
winner: Grahame Davies.
💷 Prizes of £100/£200.

The Wellcome Trust Book Prize

Wellcome Trust, 215 Euston Road, London
NW1 2BE
☎ 020 7631 2666
bookprize@wellcome.ac.uk
www.wellcomebookprize.org
Contact *Lois Tucker*

Established in 2008, The Wellcome Trust
Book Prize is an annual award which aims to
stimulate interest, excitement and debate about
medicine and literature and will be awarded to
an outstanding work of fiction or non-fiction
that is centred around the theme of medicine.
The work must have been published in a given
12-month period. A shortlist of six books will be
announced at **The Times Cheltenham Literature
Festival** in 2009 with the winner announced in
early November. Submitted books, in English
or English translation, must be published by
a UK-based publisher during the prize year.
Reissues, anthologies and self-published books
are not eligible for the prize, and authors must
be living at the time of submission. Any book of
fiction or non-fiction will be considered, so long
as medicine or biomedical science is central to
its theme. This can include a collection of short
stories by the same author. Children's fiction may
also be eligible, as long as the book is published
by an adult imprint within the specified dates.
💷 Prize of £25,000.

The Wheatley Medal

Society of Indexers, Woodbourn Business Centre,
10 Jessell Street, Sheffield S9 3HY
☎ 0114 244 9561   📠 0114 244 9563
admin@indexers.org.uk
www.indexers.org.uk
Chair *Ann Kingdom*

The Wheatley Medal is awarded for an
outstanding index and was established by the
Society of Indexers and the Library Association
(now **CILIP**) to highlight the importance of good
indexing and confer recognition on the most
highly regarded practitioners and their publishers.
First awarded in 1963, it has since been presented
for indexes to a wide range of publications, from
encyclopedias to journals.
💷 Cash prize, plus medal and certificate.

Whitbread Book Awards
▷ Costa Book Awards

Whitfield Prize

Royal Historical Society, University College
London, Gower Street, London WC1E 6BT
☎ 020 7387 7532   📠 020 7387 7532
royalhistsoc@ucl.ac.uk
www.royalhistoricalsociety.org/grants.htm
Contact *Administrative Secretary*

Established 1977. An annual award for the best
new work within a field of British or Irish history,
published in the UK in the preceding calendar
year. The book must be the author's first (solely

written) history book and be an original and scholarly work of historical research. Final entry date: 31st December.

£ Prize of £1,000.

Whiting (John) Award
▷ The Peter Wolff Theatre Trust supports the Whiting Award

Wilkins Memorial Poetry Prize
Birmingham & Midland Institute, 9 Margaret Street, Birmingham B3 3BS

☎ 0121 236 3591   📠 0121 212 4577

www.bmi.org.uk

Administrator *P.A. Fisher*

The structure of the Wilkins Memorial Poetry Prize is currently under discussion. For further information, contact the Administrator.

Winchester Writers' Conference Prizes
Faculty of Research and Knowledge Transfer, University of Winchester, Winchester SO22 4NR

☎ 01962 827238

niall.robertson@winchester.ac.uk

www.writersconference.co.uk

Fifteen writing competitions sponsored by major publishers who offer 56 prizes and trophies for entries under categories of: The First Three Pages of the Novel; Poetry; Writing can be Murder; Feature Articles and Scriptwriting; Slim Volume; Small Edition; Short Stories; Shorter Short Story; Haibun; Lifewriting; Retirement and Transition to a Low Carbon Future. Competitions open for all writers in January; closing date: 5th June. Phone or write for booklet giving details of each competition and the entry blanks or access them on the website.

H.H. Wingate Prize
▷ Jewish Quarterly Wingate Prize

Wisden (Cricket) Book of the Year
John Wisden & Co., 13 Old Aylesfield, Froyle Road, Alton GU34 4BY

☎ 01420 83415

almanack@wisdengroup.com

www.wisden.com

Contact *Hugh Chevallier*

Founded 2002. Annual award. The Wisden Book of the Year is chosen exclusively by the Wisden Cricketers' Almanack book reviewer. In each annual *Wisden Almanack*, a different person is commissioned to review all cricket books published in the previous calendar year, and that reviewer selects the Wisden Book of the Year. Any cricket book sent to John Wisden & Co. for review in Wisden Cricketers' Almanack will be considered. Final entry date: 1st December. Previous winner: *Tom Cartwright: The Flame Still Burns* by Stephen Chalke, selected by book reviewer Patrick Collins.

The Peter Wolff Theatre Trust supports the Whiting Award
Hampstead Theatre, Eton Avenue, London NW3 3EU

☎ 020 7749 4200

info@hampsteadtheatre.com

www.hampsteadtheatre.com

Contact *Neil Morris (General Manager)*

The original award was founded in 1965 by the Arts Council to commemorate the life and work of the playwright John Whiting (*The Devils, A Penny for a Song*). In 2008 it was taken over by a consortium of theatres, headed by Hampstead Theatre and funded by the Peter Wolff Theatre Trust. Annual, open award, for a play that is an original piece of writing (no adaptations or pieces of verbatim theatre), produced by a company in the subsidized sector and must have had its first performance in the award time period. The judges look for a play in which the writing is of special quality, of relevance and importance to contemporary life and of potential value to British theatre. Each year a different member of the consortium will host the award. 2008 winner: Bryony Lavery. The 2009 award is hosted by Birmingham Repertory Theatre. For more information, contact *Clare Lovell*, Arts Team Administrator (clare.lovell@birmingham-rep. co.uk).

£ Prize of £6,000.

Wolfson History Prize
Wolfson Foundation, 8 Queen Anne Street, London W1G 9LD

☎ 020 7323 5730   📠 020 7323 3241

www.wolfson.org.uk

Contact *Prize Administrator*

The Wolfson History Prize, established in 1972, is awarded annually to promote and encourage standards of excellence in the writing of history for the general reading public. Submissions are made through publishers. 2007 winners (awarded in 2008): Professor John Darwin *After Tamerlaine*; Rosemary Hill *God's Architect*.

The Writers Bureau Poetry and Short Story Competition
The Writers Bureau, Sevendale House, 7 Dale Street, Manchester M1 1JB

☎ 0161 228 2362   📠 0161 228 3533

studentservices@writersbureau.com

www.writersbureau.com/competition

Competition Secretary *Angela Cox*

Established 1994. Annual award. Poems should be no longer than 40 lines and short stories no more than 2,000 words. £5 entry fee. Closing date: 30th June 2010.

£ Prizes in each category: £1,000 (1st); £400 (2nd); £200 (3rd); £100 (4th); £50 (x 6).

The Writers' Circle Antholoogy Trophy

The David St John Thomas Charitable Trust, PO Box 6055, Nairn IV12 4YB
☏ 01667 453351   ✉ 01667 452365
dsjtcharitynairn@fsmail.net
Contact *Anne Hill*

Established in 1990 and run jointly by the David St John Thomas Charitable Trust and *Writers' News*. Annual prize for an anthology produced by a writers' group. Final entry date: 31 January. Full details available from the Trust at the address above. Previous winner: Ross-shire Writers.
💷 Prize of £100 plus a silver cup.

Writers' Forum Short Story & Poetry Competitions

See *Writers' Forum* under *Magazines*.

The Writer's Handbook 2010 Competition

www.thewritershandbook.com

Launched with the 2010 edition of *The Writer's Handbook*, this new prize offers 50 exclusive creative writing prizes to be won, when you purchase a copy of *The Writer's Handbook 2010*. In order to enter the competition, entrants should submit a 1,500 to 2,000 word article entitled 'What Makes a Bestseller?' All entries will be judged by a panel of experts. Competition deadline: 31st January 2010. For full terms and conditions and further details please visit: www. thewritershandbook.com.
💷 First prize: the winning article will be published in *The Writer's Handbook* 2011 and the author will also receive a free creative writing course of their choice from a wide range of courses from **The Complete Creative Writing Course** at the Groucho Club in Soho, London. Forty-nine randomly selected entrants will receive creative writing prizes.

The Yale Drama Series/David C. Horn Prize

PO Box 209040, New Haven, CT 06520–9040, USA
www.yalebooks.com/drama

Major new annual playwriting competition intended to support emerging playwrights. Submissions must be original, unpublished full-length plays written in English (translations, musicals and children's plays are not accepted). Only one manuscript may be submitted per year; plays that have had professional productions are not eligible. 2010 submissions accepted between 1st June and 15th August 2009. Further information available on the website.
💷 Prize winner receives the David C. Horn Prize of $10,000, publication by Yale University Press and a staged reading at Yale Rep.

Yeovil Literary Prize

The Octagon Theatre, Hendford, Yeovil BA20 1UX
☏ 01935 422884 (messages only)
www.yeovilprize.co.uk
Contact *Penny Deacon*

Annual competition, established in 2004, to stimulate interest in literary pursuits and to provide funds for the Yeovil Community Arts Association. Categories: novel (send synopsis and opening pages; maximum 15,000 words); short story: (maximum 2,000 words); poetry (maximum 40 lines). Entries, which can be sent by post or online, must be in English and not previously published; submissions from overseas welcome. Entrants must be over 18 years of age. Final entry date: end May. 2008 winners: Annette Keen *The Generation Club* (novel); Joel Willans *The Voice, the Fish Woman and the Coconut Girly* (short story); Fionna Inward Allen *Fruit on the Tube* (poetry).
💷 Prizes: Novel: £1,000 (1st); £250 (2nd); £100 (3rd). Short Story/Poetry: £500 (1st); £200 (2nd); £100 (3rd).

Yorkshire Post Book Awards

Yorkshire Post Newspapers Ltd, PO Box 168, Wellington Street, Leeds LS1 1RF
Contact *Peter Charlton, Editor*

Annual awards: **Yorkshire Post Fiction Book of the Year Award**; **Yorkshire Post Non-Fiction Book of the Year Award**; **Yorkshire Post Yorkshire Book Award**. Awarded to the book which, in the opinion of the judges, is the best work published in the preceding year and is the Best Title with Yorkshire Theme.
💷 Prizes of £1,200 (Fiction Book of the Year); £1,200 (Non-Fiction Book of the Year); £500 (Yorkshire Book Award) .

YoungMinds Book Award

Youngminds, 48–50 St John Street, London EC1M 4DG
☏ 020 7336 8445   ✉ 020 7336 8446
hannah.smith@youngminds.org.uk
www.youngminds.org.uk/bookaward
Contact *Hannah Smith*

The award will be given to a work of fiction or biography for young people aged 12+, published between 1st June 2008 and 31st May 2009, which encourages their self esteem and helps them to cope with the stresses and challenges of growing up. 2008 winner: Tabitha Suzuma, *From Where I Stand* (Definitions).
💷 Prize of £2,000.

Professional Associations and Societies

ABSW

Wellcome Wolfson Building, 165 Queen's Gate, London SW7 5HD

☎ 0870 770 3361

absw@absw.org.uk

www.absw.org.uk

Chairman *Ted Nield*

Administrator & EUSJA Board Member *Barbara Drillsma*

MEMBERSHIP £40 (full) p.a.; £36 (associate); £20 (student)

ABSW has played a central role in improving the standards of science journalism in the UK over the last 40 years. The Association seeks to improve standards by means of networking, lectures and organized visits to institutional laboratories and industrial research centres. Puts members in touch with major projects in the field and with experts worldwide. A member of the European Union of Science Journalists' Associations, ABSW is able to offer heavily subsidized places on visits to research centres in most other European countries, and hosts reciprocal visits to Britain by European journalists. Membership open to those who are considered to be *bona fide* science writers/editors, or their film/TV/radio equivalents, who earn a substantial part of their income by promoting public interest in and understanding of science. ABSW is hosting the World Conference of Science Journalists (29th June – 1st July 2009).

Academi (Yr Academi Gymreig)

3rd Floor, Mount Stuart House, Mount Stuart Square, Cardiff CF10 5FQ

☎ 029 2047 2266   ☎ 029 2049 2930

post@academi.org

www.academi.org

Glyn Jones Centre, Wales Millennium Centre, Bute Place, Cardiff Bay, Cardiff CF10 5AL

North Wales office: Tŷ Newydd, Llanystumdwy, Cricieth, Gwynedd LL52 0LW

☎ 01766 522817 Fax 01766 523095

llio@academi.org

West Wales office: Dylan Thomas Centre, Somerset Place, Swansea SA1 1RR

☎ 01792 463980 ☎ 01792 463993

dylanthomas.centre@swansea.gov.uk

Chief Executive *Peter Finch*

Academi is the trading name of Yr Academi Gymreig, the Welsh National Literature Promotion Agency and Society for Writers. Yr Academi Gymreig was founded in 1959 as an association of Welsh language writers. An English language section was established in 1968. Membership, for those who have made a significant contribution to the literature of Wales, is by invitation. The Academi runs courses, competitions (including the **Cardiff International Poetry Competition**), conferences, tours by authors, festivals, and represents the interests of Welsh writers and Welsh writing both inside Wales and beyond. Its publications include *Taliesin*, a quarterly literary journal in the Welsh language, *The Oxford Companion to the Literature of Wales*, *The Welsh Academy English-Welsh Dictionary* and *The Welsh Academy Encyclopaedia of Wales*.

The Academi won the franchise from the Arts Council of Wales to run the Welsh National Literature Promotion Agency. In 2008 Academi won an ACW Beacon Company Award. The new, much enlarged organization now administers a variety of schemes including bursaries, the annual **Wales Book of the Year Award**, critical services, writers' mentoring, Writers on Tour, Writers Residencies and a number of literature development projects. It promotes an annual literary festival alternating between North and South Wales, runs its own programme of literary activity and publishes *A470*, a quarterly literature information magazine. Those with an interest in literature in Wales can become an Associate of the Academi (which carries a range of benefits). Rates are £15 p.a. (waged); £7.50 (unwaged).

ALCS

▷ Authors' Licensing & Collecting Society Limited

Alliance of Literary Societies

59 Bryony Road, Selly Oak, Birmingham B29 4BY

☎ 0121 475 1805

l.j.curry@bham.ac.uk

www.allianceofliterarysocieties.org.uk

Chair *Linda J. Curry*

Founded 1974. Aims to help and support its 100+ member societies and, when necessary, to act as a pressure group. Produces an annual journal (*ALSo*).

Arts & Business (A&B)

Nutmeg House, 60 Gainsford Street, Butlers Wharf, London SE1 2NY
☎ 020 7378 8143   📠 020 7407 7527
head.office@AandB.org.uk
www.AandB.org.uk
Director of Press & Public Affairs *Jonathan Tuchner*
Director of Marketing & Communications *Sebastian Paul*
Media Manager *Sophie Gaskill*

A&B enables business and its people to be more successful by engaging with the arts and to increase resources for the arts from business and its people. It provides a wide range of services, business and arts networks. With the support of its President, HRH The Prince of Wales, it explores and develops new ways for business, the arts and society to interact. Arts & Business has 11 regional offices offering a range of services throughout the UK.

Arvon Foundation
▷ entry under UK and Irish Writers' Courses

ASLS
▷ Association for Scottish Literary Studies

Association for Scottish Literary Studies

c/o Department of Scottish Literature, 7 University Gardens, University of Glasgow, Glasgow G12 8QH
☎ 0141 330 5309   📠 0141 330 5309
office@asls.org.uk
www.asls.org.uk
Contact *Duncan Jones*
SUBSCRIPTION £38 (individual); £67 (institutional)

Founded 1970. ASLS is an educational charity promoting the languages and literature of Scotland. Publishes works of Scottish literature; essays, monographs and journals; and *Scotnotes*, a series of comprehensive study guides to major Scottish writers. Also produces *New Writing Scotland*, an annual anthology of contemporary poetry and prose in English, Gaelic and Scots (see entry under *Magazines*).

Association of American Correspondents in London

c/o People Magazine, Blue Fin Building, 110 Southwark Street, London SE1 0SU
info@theaacl.co.uk
www.theaacl.co.uk
Secretary/Treasurer *Monique Jessen*

SUBSCRIPTION £110 (organizations)

Founded 1919 to serve the professional interests of its member organizations, promote social cooperation among them, and maintain the ethical standards of the profession. (An extra £40 is charged for each department of an organization which requires separate listing on the Association's website and a charge of £10 for each full-time editorial staff listed, up to a maximum of £120 regardless of the number listed.)

Association of American Publishers, Inc

71 Fifth Avenue, 2nd Floor, New York, NY 10003, USA
☎ 001 212 255 0200   📠 001 212 255 7007
www.publishers.org
Also at: 50 F Street, NW, 4th Floor, Washington, DC 20001 ☎ 001 202 347 3375 📠 001 202 347 3690

Founded 1970. For information, visit the Association's website.

Association of Authors' Agents (AAA)

LAW, 14 Vernon Street, London W14 0RJ
www.agentsassoc.co.uk
President *Philippa Milne-Smith*
MEMBERSHIP £150 p.a. (£75 for agencies with fewer than 3 members of staff)

Founded 1974. Membership voluntary. The AAA maintains a code of practice, provides a forum for discussion and represents its members in issues affecting the profession. For a full list of members and a list of frequently asked questions visit the AAA website. The AAA is a voluntary body and unable to operate as an information service to the public.

Association of Authors' Representatives, Inc. (AAR)

676A 9th Avenue #312, New York, NY 10036, USA
aarinc@mindspring.com
www.aar-online.org
Contact *Administrative Secretary*

Founded in 1991 through the merger of the Society of Authors' Representatives and the Independent Literary Agents Association. Membership of this US organization is restricted to agents of at least two years' operation. Provides information, education and support for its members and works to protect their best interests.

Association of British Editors
▷ Society of Editors

Association of British Science Writers
▷ ABSW

Association of Canadian Publishers

174 Spadina Avenue, Suite 306, Toronto, Ontario
M5T 2C2, Canada
☎ 001 416 487 6116   🖷 001 416 487 8815
admin@canbook.org
www.publishers.ca
Executive Director *Carolyn Wood*

Founded 1971. ACP represents 125 Canadian-
owned book publishers country-wide from the
literary, general trade, scholarly and education
sectors. Aims to encourage the writing,
publishing, distribution and promotion of
Canadian books and to support the development
of a 'strong, independent and vibrant Canadian-
owned publishing industry'. The organization's
website has information on getting published
and links to many of their member publishers'
websites. The ACP does not accept manuscripts
and cannot provide assistance to authors who
wish to find a Canadian publisher.

Association of Christian Writers

23 Moorend Lane, Thame OX9 3BQ
☎ 01844 213673
admin@christianwriters.org.uk
www.christianwriters.org.uk
President *Adrian Plass*
Chairman *Lin Ball*
Administrator *Rev. Simon Baynes*
SUBSCRIPTION single: £22 (£18.50 Direct Debit);
joint husband/wife: £25 (£21.50 DD); overseas:
£30 (£25.50 DD on UK a/c)

Founded 1971 'to inspire and equip men
and women to use their talents and skills with
integrity to devise, write and market excellent
material which comes from a Christian
worldview. In this way we seek to be an influence
for good and for God in this generation'. Publishes
a quarterly magazine. Runs three training events
each year, biennial conference, competitions,
postal workshops, area groups, prayer support
and manuscript criticism. Charity No. 1069839.

Association of Freelance Editors, Proofreaders & Indexers (Ireland)

11 Clonard Road, Sandyford, Dublin 16, Republic
of Ireland
☎ 00 353 1 295 2194
brenda@ohanlonmediaservices.com
powerediting@eircom.net
www.afepi.ie
Co-chairs *Brenda O'Hanlon (at address above),
Winifred Power (28 Parklane Avenue, Abbeyside,
Dungarvan, Co Waterford, 00 353 58 48458)*
SUBSCRIPTION €40 p.a. (full); €20 (associate)

The organization was established in Ireland
to protect the interests of its members and to
provide information to publishers on freelancers
working in the relevant fields. Full membership
is restricted to freelancers with experience and/
or references (but the association does not test or
evaluate the skills of members). A new category of
membership, Associate Member, is available for
trainees in proofreading/editing who are taking
the Publishing Training Centre correspondence
courses in Proofreading and Copy-editing.

Association of Freelance Writers

Sevendale House, 7 Dale Street, Manchester
M1 1JB
☎ 0161 228 2362   🖷 0161 228 3533
fmn@writersbureau.com
www.freelancemarketnews.com
Contact *Angela Cox*
SUBSCRIPTION £29 p.a.

Founded in 1995 to help and advise new and
established freelance writers. Members receive
a copy of *Freelance Market News* each month
which gives news, views and the latest advice
and guidelines about publications at home and
abroad. Other benefits include one free appraisal
of prose or poetry each year, reduced entry to **The
Writers Bureau** writing competition, reduced
fees for writing seminars and discounts on books
for writers.

Association of Golf Writers

1 Pilgrims Bungalow, Mulberry Hill, Chilham
CT4 8AH
☎ 01227 732496   🖷 01227 732496
andyfarrell292@btinternet.com
Honorary Secretary *Andy Farrell*

Founded 1938. Aims to cooperate with golfing
bodies to ensure best possible working
conditions.

Association of Illustrators

2nd Floor, Back Building, 150 Curtain Road,
London EC2A 3AT
☎ 020 7613 4328   🖷 020 7613 4417
info@theaoi.com
www.theaoi.com
Contact *Membership Coordinator*

The AOI is a non-profit-making trade association
and members consist primarily of freelance
illustrators as well as agents, clients, students
and lecturers. As the only body to represent
illustrators and campaign for their rights in
the UK, the AOI has successfully increased
the standing of illustration as a profession and
improved the commercial and ethical conditions
of employment. Organizes annual events
programme and provides an advisory service
for members. Publications include: triannual
magazine, *Varoom – the journal of illustration
and made images*; *The Illustrator's Guide to Law
and Business Practice*; *The Illustrator's Guide*

to Success (not guaranteed), client directories (online) and *Images*, the only jury-selected annual of British contemporary illustration.

Association of Independent Libraries

Leeds Library, 18 Commercial Street, Leeds LS1 6AL
☎ 0113 245 3071
enquiries@leedslibrary.co.uk
www.independentlibraries.co.uk
Chairman *Geoffrey Forster*
Contact *Catherine Levy*

Established in 1989 to 'further the advancement, conservation and restoration of a little-known but important living portion of our cultural heritage'. Members include the **London Library**, **Devon & Exeter Institution**, **Linen Hall Library** and **Plymouth Proprietary Library**.

Association of Learned and Professional Society Publishers

9 Stanbridge Road, London SW15 1DX
☎ 020 8789 2394   ☎ 0870 051 1921
ian.russell@alpsp.org
www.alpsp.org
Chief Executive *Ian Russell*
Chief Operating Officer *Nick Evans*
Editor-in-Chief, Learned Publishing *Sally Morris* (*editor@alpsp.org*)

The Association of Learned and Professional Society Publishers (ALPSP) is the international trade association for not-for-profit publishers and those who work with them. It currently has over 360 members in 36 countries. ALPSP provides representation of its sector, cooperative services such as the ALPSP Learned Journals Collection and ALPSP eBooks Collection, a professional development programme of training courses and seminars and a wealth of information and advice. ALPSP runs an annual international conference and awards programme. See the website for further information.

Association of Scottish Motoring Writers

32 Caledonia Crescent, Ardrossan KA22 8LW
☎ 07775 930686
spark@cfpress.co.uk
Secretary *Stephen Park*
SUBSCRIPTION £50

Founded 1961. Aims to co-ordinate the activities of, and provide shared facilities for, motoring writers resident in Scotland. Welcomes applications from active journalists.

Audiobook Publishing Association

c/o 23 Wilcot Road, Pewsey SN9 5EH
☎ 07531 902975
info@theapa.net
www.theapa.net
Chair *Ali Muirden*

Administrator *Laura Briscall (at address above)*

Founded 1994. The Audiobook Publishing Association (formerly the Spoken Word Publishing Association) is the UK trade association for the audiobook industry with membership open to all those involved in the publishing of spoken word audio. Publishes annual *APA Resources Directory*, available from the address above, and now also available online at www.theapa.net (click on 'APA Members').

Australian Copyright Council

PO Box 1986, Strawberry Hills, NSW 2012, Australia
☎ 00 612 8815 9777   ☎ 00 612 8815 9799
www.copyright.org.au
Contact *Customer Service*

Founded 1968. The Council's activities and services include a range of publications, organizing and speaking about copyright at seminars, research, consultancies and free legal advice. Aims include assistance for copyright owners to exercise their rights effectively, raising awareness about the importance of copyright and seeking changes to the law of copyright.

Australian Publishers Association

60/89 Jones Street, Ultimo, NSW 2007, Australia
☎ 00 612 9281 9788   ☎ 00 612 9281 1073
www.publishers.asn.au
Chief Executive *Maree McCaskill*
SUBSCRIPTION Annual membership subscription open to Australian publishers

Founded 1948. The APA initiates programmes that contribute to the development of publishing in Australia, virgorously protects and furthers the interests of copyright owners, agents and licensees and actively represents members' interests to government and other organizations as appropriate. The Association encourages excellence in writing, editing, design, production, marketing and distribution of published works in Australia, protects freedom of expression and manages members' funds to further the interests of the industry.

Australian Society of Authors

PO Box 1566, Strawberry Hills, NSW 2012, Australia
☎ 00 612 9318 0877   ☎ 00 612 9318 0530
asa@asauthors.org
www.asauthors.org
Executive Director *Dr Jeremy Fisher*

Founded in 1963, the Australian Society of Authors is the professional association for Australia's literary creators with 3,000 members in Australia and overseas.

Authors' Club

40 Dover Street, London W1S 4NP
☏ 020 7499 8581
stella@theartsclub.co.uk
www.theartsclub.co.uk
Secretary *Stella Kane*

Founded in 1891 by Sir Walter Besant, the Authors' Club welcomes as members writers, agents, publishers, critics, journalists, academics and anyone involved with literature and the written word. Administers the **Authors' Club Best First Novel Award**, the **Dolman Best Travel Book Award** and **Sir Banister Fletcher Award**, and organizes regular talks and dinners with well-known guest speakers. Membership fee: apply to the Secretary.

Authors' Licensing & Collecting Society Limited (ALCS)

The Writers' House, 13 Haydon Street, London EC3N 1DB
☏ 020 7264 5700   ☏ 020 7264 5755
alcs@alcs.co.uk
www.alcs.co.uk
Chief Executive *Owen Atkinson*
SUBSCRIPTION One-off membership fee: £25 incl. VAT (UK; free to members of Society of Authors, Writers' Guild)

Founded 1977. The UK collective rights management society for writers and their successors, ALCS is a non-profit organization whose principal purpose is to ensure that hard-to-collect revenues due to writers are efficiently collected and speedily distributed. Established to give assistance to writers through the protection and exploitation of collective rights, ALCS has distributed over £170 million in secondary royalties to writers since its creation.

ALCS represents all types of writer, fiction and non-fiction, including educational, research and academic authors, scriptwriters, playwrights, poets, editors and freelance journalists across the print and broadcast media. On joining, members give ALCS a mandate to administer on their behalf those rights which the law determines must be received or which are best handled collectively. Chief among these are: photocopying, cable retransmission, rental and lending rights (but not British Public Lending Right), off-air recording, electronic rights, the performing right and public reception of broadcasts. The society is a prime resource and a leading authority on copyright matters and writers' collective interests. It maintains a watching brief on all matters affecting copyright both in Britain and abroad, making representations to UK government authorities and the EU. Visit the ALCS website

or contact the office for registration forms and further information.

BAPLA (British Association of Picture Libraries and Agencies)

18 Vine Hill, London EC1R 5DZ
☏ 020 7713 1780   ☏ 020 7713 1211
enquiries@bapla.org.uk
www.bapla.org.uk

Everything you need to know about finding, buying and selling pictures. Represents over 400 image supplying members.

BFI

21 Stephen Street, London W1T 1LN
☏ 020 7255 1444   ☏ 020 7436 0439
www.bfi.org.uk
Chair *Greg Dyke*
Director *Amanda Nevill*

'There's more to discover about film and television through the BFI. Our world-renowned archive, cinemas, festivals, films, publications and learning resources are here to inspire you.' To find out more, visit the website.

The Bibliographical Society

c/o Institute of English Studies, University of London, Senate House, Malet Street, London WC1E 7HU
☏ 020 7862 8679   ☏ 020 7862 8720
secretary@bibsoc.org.uk
www.bibsoc.org.uk
President *E. Leedham-Green*
Honorary Secretary *M.L. Ford*
SUBSCRIPTION £33 p.a.

Aims to promote and encourage the study and research of historical, analytical, descriptive and textual bibliography, and the history of printing, publishing, bookselling, bookbinding and collecting; to hold meetings at which papers are read and discussed; to print and publish works concerned with bibliography; to form a bibliographical library. Awards grants and bursaries for bibliographical research. Publishes a quarterly magazine called *The Library*.

Booksellers Association of the UK & Ireland Ltd

Minster House, 272 Vauxhall Bridge Road, London SW1V 1BA
☏ 020 7802 0802   ☏ 020 7802 0803
mail@booksellers.org.uk
www.booksellers.org.uk
Chief Executive *Tim Godfray*

Founded 1895. The BA helps 4,400 independent, chain and multiple retail outlets to sell more books, reduce costs and improve efficiency. It represents members' interests to the UK Government, European Commission, publishers, authors and others in the trade as well as offering marketing assistance, conferences, seminars and exhibitions. Together with **The Publishers Association**, coordinates World Book Day. Publishes directories, catalogues, surveys and various other publications connected with the book trade and administers the **Costa Book Awards**.

Booktrust

Book House, 45 East Hill, London SW18 2QZ
☎ 020 8516 2977 📠 020 8516 2978
query@booktrust.org.uk
www.booktrust.org.uk
Director *Viv Bird*
Head of Promotions *Helen Hayes*

Founded 1925. Booktrust is an independent national charity that encourages readers of all ages and cultures to discover and enjoy reading. Their website includes reviews of thousands of books, as well as information about Children's Book Week and the **Children's Laureate**, both of which are administered by Booktrust. Booktrust also administers literary prizes, including the **Orange Prize for Fiction**, the **John Llewellyn Rhys Prize**, the **Booktrust Teenage Prize** and **The Roald Dahl Funny Prize** (see entries under *Prizes*), and promotes reading through campaigns such as Get London Reading (www.getlondonreading.com), Story (www.theshortstory.org.uk) and a website about translated fiction (www.translatedfiction.org.uk). Booktrust runs Bookstart (www.bookstart.org.uk), the acclaimed national scheme that works through locally based organizations to give a free pack of books to babies and guidance materials for parents and carers. They also run two other gifting schemes for older children (Booktime and Booked Up), as well as a scheme for looked after children (the Letterbox Club), and a programme to encourage writing in schools (Everybody Writes).

British Academy of Composers and Songwriters

2nd Floor, British Music House, 25–27 Berners Street, London W1T 3LR
☎ 020 7636 2929 📠 020 7636 2212
info@britishacademy.com
www.britishacademy.com
Membership Manager *Vicky Hunt*

The Academy represents the interests of music writers of all genres. Campaigns in the UK and Europe on copyright and numerous other issues, provides a wide range of member support services, publishes bi-monthly magazines and administers the annual Ivor Novello Awards and British Composer Awards.

British Association of Communicators in Business (CiB)

Suite GA2, Oak House, Woodlands Business Park, Linford Wood, Milton Keynes MK14 6EY
☎ 01908 313755 📠 01908 313661
enquiries@cib.uk.com
www.cib.uk.com
Chief Executive *Kathie Jones*

The professional body for in-house, freelance and agency staff involved in internal and corporate communications. Services to members include a business support helpline, an online forum, a monthly e-zine, the CiB Freelance Directory, an annual conference and gala dinner, automatic eligibility to FEIEA, a European network of 5,000 communicators.

British Association of Journalists

89 Fleet Street, London EC4Y 1DH
☎ 020 7353 3003 📠 020 7353 2310
office@bajunion.org.uk
www.bajunion.org.uk
General Secretary *Steve Turner*
SUBSCRIPTION National newspaper staff, national broadcasting staff, national news agency staff: £17.50 per month. Other seniors, including magazine journalists and freelance writers and photographers who earn the majority of their income from journalism: £10 p.m. Journalists under 24: £7.50 p.m.

Founded 1992. Aims to protect and promote the industrial and professional interests of journalists.

British Association of Picture Libraries and Agencies
▷ BAPLA

British Centre for Literary Translation

University of East Anglia, Norwich NR4 7TJ
☎ 01603 592785 📠 01603 592737
bclt@uea.ac.uk
www.uea.ac.uk/bclt
Director *Amanda Hopkinson*

Founded 1989, BCLT is funded jointly by Arts Council England and the University of East Anglia. It aims to raise the profile of literary translation in the UK through events, publications, activities and research aimed at professional translators, students and the general reader. Member of the international RECIT literary translation network. Activities include the annual Sebald Lecture, Summer School and www.literarytranslation.com website with the British Council. BCLT has a PhD programme in literary translation as well as an MA in Literary Translation and other units at undergraduate and postgraduate level. Joint sponsor with BCLA of the **John Dryden Translation Prize**. Publishes a journal *In Other Words* and *New Books in German*. Free mailing list.

British Copyright Council

Copyright House, 29–33 Berners Street, London
W1T 3AB
☎ 01986 788122   📠 01986 788847
secretary@britishcopyright.org
www.britishcopyright.org
Contact *Janet Ibbotson*

Works for the national and international acceptance of copyright and acts as a lobby/watchdog organization on behalf of creators, publishers and performers on copyright and associated matters.

British Council

10 Spring Gardens, London SW1A 2BN
☎ 020 7389 3194   📠 020 7389 3199
arts@britishcouncil.org
www.britishcouncil.org/arts

The British Council uses cultural relations to address the need for intercultural dialogue, harnessing the power of education, the arts and creativity, sport, science, English and governance to build relationships of mutual benefit worldwide and to address global issues. 'Our vision says the world needs more cultural relations; our purpose is to respond.'

British Equestrian Writers' Association

Priory House, Station Road, Swavesey, Cambridge
CB24 4QJ
☎ 01954 232084   📠 01954 231362
gnewsumn@aol.com
Contact *Gillian Newsum*
SUBSCRIPTION £15

Founded 1973. Aims to further the interests of equestrian sport and improve, wherever possible, the working conditions of the equestrian press. Membership is by invitation of the committee. Candidates for membership must be nominated and seconded by full members and receive a majority vote of the committee.

British Film Institute
▷ BFI

British Guild of Beer Writers

Woodcote, 2 Jury Road, Dulverton TA22 9DU
☎ 01398 324314
tierneyjones@btinternet.com
www.beerwriters.co.uk
Secretary *Adrian Tierney-Jones*
SUBSCRIPTION £40 p.a.

Founded 1988. Aims to improve standards in beer writing and at the same time extend public knowledge of beers and brewing. Publishes an annual directory of members with details of their publications and their particular areas of expertise; this is then circulated to newspapers, magazines, trade press and broadcasting organizations. Also publishes a monthly newsletter, the *BGBW Newsletter*, edited by Adrian Tierney-Jones. As part of the plan to improve writing standards and to achieve a higher profile for beer, the Guild offers annual awards, including a Travel Bursary, to writers and broadcasters judged to have made the most valuable contribution towards this end in their work.

British Guild of Travel Writers

5 Berwick Courtyard, Berwick St. Leonard, Salisbury SP3 5UA
☎ 01747 820455
secretariat@bgtw.org
www.bgtw.org
Chairman *Melissa Shales*
Secretariat *Judy Robson*

The professional association of travel writers, broadcasters, photographers and editors which aims to serve its members' professional interests by acting as a forum for debate, discussion and 'networking'. The Guild publishes an annual Year Book and has a website giving full details of all its members and useful trade contacts, holds monthly meetings and has a monthly newsletter. Members are required to spend a significant proportion of their working time on travel.

British Science Fiction Association

39 Glyn Avenue, New Barnet EN4 9PJ
www.bsfa.co.uk
SUBSCRIPTION £26 p.a. (reduction for unwaged)

Founded originally in 1958 by a group of authors, readers, publishers and booksellers interested in science fiction. With a worldwide membership, the Association aims to promote the reading, writing and publishing of science fiction and to encourage SF fans to maintain contact with each other. Publishes *Matrix* e-zine, media review newsletter via the web with comment and opinions, news of conventions, etc. Contributions from members welcomed; *Vector* quarterly critical journal – reviews of books and magazines; *Focus* biannual magazine with articles, original fiction and letters column. Also offers postal and online writer's workshop. For further information, contact the Membership Secretary *Peter Wilkinson* at the postal address above or via the website.

British Society of Comedy Writers

61 Parry Road, Ashmore Park, Wolverhampton
WV11 2PS
☎ 01902 722729   📠 01902 722729
comedy@bscw.co.uk
www.bscw.co.uk
Contact *Ken Rock, President*

Founded 1999. The Society aims to develop good practice and professionalism among comedy writers while bringing together the best creative professionals, and working to standards of excellence agreed with the light entertainment industry. Offers a network of industry contacts and a range of products, services and training initiatives including specialized workshops, an annual international conference, script assessment service and opportunities to visit international festivals.

British Society of Magazine Editors (BSME)

137 Hale Lane, Edgware HA8 9QP
☎ 020 8906 4664   📠 020 8959 2137
admin@bsme.com
www.bsme.com
Contact *Gill Branston*

Holds regular industry forums and events as well as an annual awards dinner.

Broadcasting Press Guild

c/o Torin Douglas, Room 7540, BBC Television Centre, Wood Lane, London W12 7RJ
☎ 020 8624 9052
torin.douglas@bbc.co.uk
www.broadcastingpressguild.org
Membership Secretary *Steve Clarke*
Lunch Secretary & Treasurer *Torin Douglas*
SUBSCRIPTION £15 p.a.

Founded 1974 to promote the professional interests of journalists specializing in writing or broadcasting about the media. Organizes monthly lunches addressed by leading industry figures, and annual TV and radio awards. Membership by invitation.

BSME
▷ British Society of Magazine Editors

Bureau of Freelance Photographers

Focus House, 497 Green Lanes, London N13 4BP
☎ 020 8882 3315   📠 020 8886 3933
info@thebfp.com
www.thebfp.com
Membership Secretary *Angela Kidd*
SUBSCRIPTION £54 p.a. (UK); £70 p.a. (overseas)

Founded 1965. Assists members in selling their pictures through monthly *Market Newsletter*, and offers advisory, legal assistance and other services.

Campaign for Press and Broadcasting Freedom

Second Floor, 23 Orford Road, London E17 9NL
☎ 020 8521 5932
freepress@cpbf.org.uk
www.cpbf.org.uk

SUBSCRIPTION £15 p.a. (concessions available); £25 p.a. (institutions/organizations)

Broadly based pressure group working for more accountable and accessible media in Britain. Advises on right of reply and takes up the issue of the portrayal of minorities. Members receive *Free Press* (bi-monthly), discounts on publications and news of campaign progress.

Canadian Authors Association

National Office: Box 419, 320 South Shores Road, Campbellford, Ontario K0L 1L0, Canada
☎ 001 705 653 0323   📠 001 705 653 0593
admin@canauthors.org
www.canauthors.org
Contact *The National Director*

Founded 1921. The CAA was founded to promote recognition of Canadian writers and their works, and to foster and develop a climate favourable to the creative arts. It is Canada's national association for writers of every kind; for those actively seeking to become writers and for those who want to support writers. The Association has branches across the country providing support to local members in the form of advice, local contests, publications and writers' circles. Publishes *National Newsline* (quarterly) and *The Canadian Writer's Guide*.

Canadian Federation of Poets

2431 Cyprus Avenue, Burlington, Ontario L7P 1G5, Canada
info@federationofpoets.com
www.federationofpoets.com
President *Tracy Repchuk*

Founded 2003. Federations and members from around the world can post messages, 'meet' via forums and live chats and learn about themselves and their international colleagues with regard to poetry. Membership includes poetry workshops, anthologies, publishing opportunities, online forums, free book promotion, calendar of events and free subscription to *Poetry Canada* magazine (www.poetrycanada.com). Also offers an online 52-week poetry certification programme to help educate on the various types of poetry and promote the diversity of poetry.

Canadian Publishers' Council

250 Merton Street, Suite 203, Toronto, Ontario M4S 1B1, Canada
☎ 001 416 322 7011   📠 001 416 322 6999
pubadmin@pubcouncil.ca
www.pubcouncil.ca
Executive Director, External Relations *Jacqueline Hushion*

Founded 1910. Trade association of English-language publishers which represents the domestic and international interests of member companies.

Chartered Institute of Journalists

2 Dock Offices, Surrey Quays Road, London SE16 2XU

☎ 020 7252 1187   ℻ 020 7232 2302
memberservices@cioj.co.uk
www.cioj.co.uk
General Secretary *Dominic Cooper*
SUBSCRIPTION £195 p.a.; £16 (monthly)

Founded 1884. The Institute is concerned with professional journalistic standards and with safeguarding the freedom of the media. It is open to writers, broadcasters and journalists (including self-employed) in all media. Affiliate membership (£133) is available to part-time or occasional practitioners and to overseas journalists who can join the Institute's International Division. Non-employing members also belong to the IOJ (TU), an independent trade union which protects, advises and represents them in their employment or freelance work; negotiates on their behalf and provides legal assistance and support. The IOJ (TU) is a certificated independent trade union which represents members' interests in the workplace, and the CIOJ is also a constituent member of the Media Society, the **British Copyright Council** and the Journalists Copyright Fund. Editor of the Institute's quarterly magazine is *Andrew Smith*.

Chartered Institute of Linguists (IoL)

Saxon House, 48 Southwark Street, London SE1 1UN

☎ 020 7940 3100   ℻ 020 7940 3101
info@iol.org.uk
www.iol.org.uk
Chief Executive *John Hammond*
Director of Communications *Cetty Zambrano*
SUBSCRIPTION Rates on application

Founded 1910. Professional association for translators, interpreters and trainers; examining body for languages at degree level and above for vocational purposes; the National Register of Public Service Interpreters is managed by NRPSI Limited, an IoL subsidiary. The Institute's limited company, Language Services Ltd, provides customized assessments of language-oriented requirements, skills, etc.

Children's Book Circle

childrensbookcircle@hotmail.co.uk
www.childrensbookcircle.org.uk
Membership Secretary *Katie Powell*

The Children's Book Circle provides a discussion forum for anybody involved with children's books. Regular meetings are addressed by a panel of invited speakers and topics focus on current and controversial issues. Administers the **Eleanor Farjeon Award** and the Patrick Hardy lecture.

Children's Books Ireland

17 North Great George's Street, Dublin 1, Republic of Ireland

☎ 00 353 1 872 7475   ℻ 00 353 1 872 7476
info@childrensbooksireland.ie
www.childrensbooksireland.ie
Contact *Mags Walsh*
SUBSCRIPTION €30 p.a. (individual); €50 (institution); €20 (student); €45/US$40 (overseas)

Irish national children's books organization. Holds annual conferences as well as other occasional seminars and events. Quarterly magazine, *Inis*. Annual children's book festival in October; **CBI Bisto Book of the Year Award** (see entry under *Prizes*). Partners other organizations to promote and develop ways of bringing young people and books together.

CiB

▷ British Association of Communicators in Business

CILIP: The Chartered Institute of Library and Information Professionals

7 Ridgmount Street, London WC1E 7AE

☎ 020 7255 0500   ℻ 020 7255 0501
info@cilip.org.uk
www.cilip.org.uk
Chief Executive *Bob McKee*

The leading membership body for library and information professionals, formed in 2002 following the unification of The Library Association and the Institute of Information Scientists. **Facet Publishing** (successor to LA Publishing) produces 25–30 new titles each year and has over 150 titles in print from the LA's back catalogue.

CILIP in Scotland (CILIPS)

1st Floor Building C, Brandon Gate, Leechlee Road, Hamilton ML3 6AU

☎ 01698 458888
cilips@slainte.org.uk
www.slainte.org.uk
Director *Elaine Fulton*

CILIP in Scotland (Chartered Institute of Library and Information Professionals in Scotland) aims to bring together everyone engaged in or interested in library work in Scotland. Its main aims are the promotion of library services and the qualifications and status of librarians.

CILIP Wales

Department of Information Studies, Llanbadarn Fawr, Aberystwyth SY23 3AS

☎ 01970 622174   ℻ 01970 622190
cilip-wales@aber.ac.uk
www.dil.aber.ac.uk/cilip_w/index.htm

Development Officer *Mandy Powell (07837 032536)*

The Chartered Institute of Library and Information Professionals Wales was founded in 2002 as a branch of **CILIP**. Publishes a thrice-yearly journal, *Y Ddolen*.

Circle of Wine Writers

5 Ingatestone Hall Cottages, Ingatestone CM4 9NS
julie.arkell@btinternet.com
www.winewriters.org
Administrator *Andrea Warren*
MEMBERSHIP £50 p.a.

Founded 1960. Membership is open to all those professionally engaged in communicating about wines and spirits, with the exception of people primarily doing so for promotional purposes. Aims to improve the standard of writing, broadcasting and lecturing about wines, spirits and beers; to contribute to the growing knowledge and interest in wine; to promote wines and spirits of quality and to comment adversely on faulty products or dubious practices; to establish and maintain good relations with the news media and the wine trade; to provide members with a strong voice with which to promote their views; to provide a programme of workshops, meetings, talks and tastings.

CLE – Irish Book Publishers' Association

Guinness Enterprise Centre, Taylor's Lane, Dublin 8, Republic of Ireland
☎ 00 353 1 639 4868/1 415 1210
info@publishingireland.com
www.publishingireland.com
President *Sean Ó Cearnaigh*
Administrator *Karen Kenny*
Project Manager *Jolly Ronan*

Founded 1970 to promote Irish publishing, protect members' interests and train the industry.

Comhairle nan Leabhraichean/The Gaelic Books Council

22 Mansfield Street, Glasgow G11 5QP
☎ 0141 337 6211   📠 0141 341 0515
brath@gaelicbooks.net
www.gaelicbooks.org
www.ur-sgeul.com
Chairman *Professor Roibeard Ó Maolalaigh*
Director *Ian MacDonald*

Founded 1968 and now a charitable company with its own bookshop. Encourages and promotes Gaelic publishing by giving grants to publishers and writers; providing editorial and word-processing services; retailing Gaelic books; producing a catalogue of all Gaelic books in print and answering enquiries about them; mounting occasional literary evenings and training courses. Stock list on website.

The Copyright Licensing Agency Ltd

Saffron House, 6–10 Kirby Street, London EC1N 8TS
☎ 020 7400 3100/0800 085 6644 (new licences)
📠 020 7400 3101
cla@cla.co.uk
www.cla.co.uk
Chief Executive *Kevin Fitzgerald*

CLA is the UK's reproduction rights organization which looks after the interests of authors, publishers and visual creators in the photocopying and scanning of extracts from books, journals, magazines and digital publications. Founded in 1983 by the **Authors' Licensing and Collecting Society Limited (ALCS)** and the **Publishers Licensing Society Ltd (PLS)** to promote and enforce intellectual property rights of UK rightsholders both at home and abroad. CLA works closely with Reproduction Rights Organizations (RROs) from other countries and has an agency agreement with the **Design and Artists Copyright Society (DACS)** which represents artists and illustrators. A not-for-profit organization, CLA licenses the business, education and government sectors and distributes fees collected to authors, publishers and visual creators. CLA has also developed licences which enable digitization of existing print material. Licences allow photocopying and scanning from original print material as well as copying from born digital publications. Scanning and digital re-use of publications is available for works from the UK, US, Australia, Spain, Ireland, South Africa and Canada. Since its inception in 1983, CLA has distributed over £450 million.

Council for British Archaeology

St Mary's House, 66 Bootham, York YO30 7BZ
☎ 01904 671417   📠 01904 671384
info@britarch.ac.uk
www.britarch.ac.uk
Head of Information & Communications *Dan Hull*
Publications Officer *Catrina Appleby*

Founded in 1944 to represent and promote archaeology at all levels. It is an educational charity working throughout the UK to involve people in archaeology and to promote the appreciation and care of the historic environment for the benefit of present and future generations. The CBA publishes a wide range of academic, educational and general works about archaeology, ranging from monographs and handbooks to online articles and the popular magazine, *British Archaeology* (see **CBA Publishing** under *UK Publishers*).

Crime Writers' Association (CWA)

PO Box 273, Boreham Wood WD6 2XA
secretary@thecwa.co.uk
www.thecwa.co.uk
Hon. Secretary *Liz Evans*
Membership Secretary *Rebecca Tope (Crossways Cottage, Walterstone HR2 oDX)*
MEMBERSHIP £50, £65 (associate); £50/$100 (overseas)

Full membership is limited to professional crime writers, but publishers, literary agents, booksellers, etc. who specialize in crime, are eligible for Associate membership. The Association has regional chapters throughout the country, including Scotland. Meetings are held regularly with informative talks frequently given by police, scenes of crime officers, lawyers, etc., and a weekend conference is held annually in different parts of the country. Produces a monthly newsletter for members called *Red Herrings* and presents various annual awards (see entries under *Prizes*).

The Critics' Circle

c/o Catherine Cooper, 69 Marylebone Lane, London W1U 2PH
☎ 020 7224 1410
www.criticscircle.org.uk
President *Charles Spencer*
Honorary General Secretary *William Russell*
SUBSCRIPTION £25 p.a.

Membership by invitation only. Aims to uphold and promote the art of criticism (and the commercial rates of pay thereof) and preserve the interests of its members: professionals involved in criticism of film, drama, music, dance and art.

Cyngor Llyfrau Cymru
▷ Welsh Books Council

DACS
▷ Design and Artists Copyright Society

Data Publishers Association

Queen's House, 28 Kingsway, London WC2B 6JR
☎ 020 7405 0836
info@dpa.org.uk
www.dpa.org.uk
Contact *Sarah Gooch*
SUBSCRIPTION £450 to £2450 p.a.

Since 1970, the DPA has existed to serve, promote and protect the interests of all companies operating in the directory, data and search publishing sector. Today, the industry contributes well over £1 billion to the UK economy. The DPA membership not only includes mainstream data providers but also a whole host of organizations providing services and products to the data provision and publishing sector, such as printers, software solution providers and contract directory publishers. The DPA's members are predominantly UK companies but it also offers membership to overseas companies with UK interests. It provides its members with opportunities to develop their businesses by providing a complete and diverse range of services, including communications, networking, training and seminars, statistics, representation to government, publicity and promotion, legal support and recognition.

Department for Culture, Media and Sport

2–4 Cockspur Street, London SW1Y 5DH
☎ 020 7211 6000   ⊞ 020 7211 6270
michael.panayi@culture.gsi.gov.uk
Director of Communications *Jane Cooper*

The Department for Culture, Media and Sport has responsibilities for Government policies relating to the arts, museums and galleries, sport, the 2012 Olympic & Paralympic Games, gambling, broadcasting, Press freedom and regulation, the built historic environment, the film and music industries, tourism and the National Lottery. It funds the Arts Council, national museums and galleries, the British Library, the **Public Lending Right** and the Royal Commission on Historical Manuscripts. It is responsible within Government for the public library service in England, and for library and information matters generally, where they are not the responsibility of other departments.

Design and Artists Copyright Society (DACS)

33 Great Sutton Street, London EC1V oDX
☎ 020 7553 9050   ⊞ 020 7336 8822
info@dacs.org.uk
www.dacs.org.uk
Chief Executive *Gilane Tawadros*

DACS is the UK's copyright and collecting society for artists and visual creators. Represents 26,000 fine artists and their heirs as well as photographers, illustrators, craftspeople, cartoonists, architects, animators and designers.

Drama Association of Wales

The Old Library, Singleton Road, Splott, Cardiff CF24 2ET
☎ 029 2045 2200   ⊞ 029 2045 2277
aled.daw@virgin.net
www.amdram.co.uk/daw
Contact *Teresa Hennessy, Member Services Officer*

Runs a large playscript lending library; holds an annual playwriting competition (see entry under *Prizes*); offers a script-reading service (£25 full mss; £17.50 one-act play) which usually takes one month from receipt of play to issue of reports. From plays submitted to the reading service, selected scripts are considered for publication of a short run (70–200 copies). Writers receive a percentage of the cover price on sales and a percentage of the performance fee.

Edinburgh Bibliographical Society

Special Collections, Edinburgh University Library, George Square, Edinburgh EH8 9LJ
warrenmcdougall@aol.com
http://mcs.qmuc.ac.uk/EBS/
Secretary *Warren McDougall*
SUBSCRIPTION £15; £20 (institution); £10 (student)

Founded 1890. Organizes lectures on bibliographical topics and visits to libraries. Publishes an annual journal, which is free to members, and other occasional publications.

Educational Publishers Council
▷ The Publishers Association

Electronic Publishers' Forum
▷ The Publishers Association

The English Association

University of Leicester, University Road, Leicester LE1 7RH
☎ 0116 252 3982   🖷 0116 252 2301
engassoc@leicester.ac.uk
www.le.ac.uk/engassoc
Chief Executive *Helen Lucas*
Administrator *Julia Hughes*

Founded 1906 to promote understanding and appreciation of the English language and its literatures. Activities include sponsoring a number of publications and organizing lectures and conferences for teachers, plus annual sixth-form conferences. Publications include *The Year's Work in Critical and Cultural Theory*; *English, The Use of English*; *English 4–11*; *Essays and Studies* and *The Year's Work in English Studies*.

English PEN

6–8 Amwell Street, London EC1R 1UQ
☎ 020 7713 0023   🖷 020 7837 7838
enquiries@englishpen.org
www.englishpen.org
Director *Jonathan Heawood*
MEMBERSHIP Cheque/standing order: £45/£40 (London/overseas); £40/£35 (country); £15 (students)

English PEN is part of International PEN, a worldwide association of writers and other literary professionals which promotes literature, fights for freedom of expression and speaks out for writers who are imprisoned or harassed for having criticized their governments, or for publishing other unpopular views.

Founded in London in 1921, International PEN now consists of over 130 centres in almost 100 countries. PEN originally stood for poets, essayists and novelists, but membership is now open to all literary professionals. It is also possible to become a 'Friend of English PEN'. A programme of talks and discussions, and other activities such as social gatherings, is supplemented by mailings, website and annual congress at one of the centre countries.

Federation of Entertainment Unions

c/o Equity, Guild House, Upper St Martin's Lane, London WC2H 9EG
☎ 07973 714206
entertainment.unions@gmail.com
Secretary & Coordinator *Paul Evans*

Plenary meetings six times annually. Additionally, there are Training, Equalities & European Committees. Represents the following unions: British Actors' Equity Association; Broadcasting Entertainment Cinematograph and Theatre Union; Musicians' Union; Unite – the union; Professional Footballers' Association; **National Union of Journalists**; The Writers' Guild of Great Britain.

Fellowship of Authors and Artists

PO Box 158, Hertford SG13 8FA
☎ 01992 511697   🖷 0870 116 3398
www.author-fellowship.co.uk
Contact *Graham Irwin*

Founded in 2000 to promote and encourage the use of writing and all art forms as a means of therapy and self healing; to provide a valuable resource and meeting point for all interested parties including, but not limited to, writers, artists, counsellors and healers; to publish as web pages or e-books any suitable works that may help to support or promote the aims of the fellowship.

Film Agency for Wales

33–35 West Bute Street, Cardiff CF10 5LH
☎ 029 2046 7480   🖷 029 2045 7491
adam@filmagencywales.com
www.filmagencywales.com
Contacts *Adam Partridge, Keith Potter*

Founded in July 2006, the Film Agency for Wales is the sole Agency for film in Wales, with a remit to ensure that the economic, cultural and educational aspects of film are effectively represented in Wales, the UK and the world. As a strategic agency, it aims to facilitate the emergence of a viable and sustainable Welsh film industry and to promote a vibrant and dynamic film culture.

Foreign Press Association in London

11 Carlton House Terrace, London SW1Y 5AJ
☎ 020 7930 0445   🖷 020 7925 0469
reception@foreign-press.org.uk
www.foreign-press.org.uk
Director *Christoph Wyld*

ANNUAL SUBSCRIPTION Details available on the website

Founded 1888. Non-profit-making association for foreign correspondents based in London, providing a variety of press-related services. Welcomes enquiries about media-related venue hire.

The Gaelic Books Council
> **Comhairle nan Leabhraichean**

The Garden Media Guild

Katepwa House, Ashfield Park Avenue, Ross-on-Wye HR9 5AX

📞 01989 567393   📠 01989 567676
admin@gardenmediaguild.co.uk
www.gardenmediaguild.co.uk
Contact *Gill Hinton*
SUBSCRIPTION £55 (full members); £75 (associate members); £41.25 (probationary members)

Founded 1990. Aims to raise the quality of gardening communication, to help members operate efficiently and profitably, to improve liaison between garden communicators and the horticultural industry. Administers an annual awards scheme. Operates a mailing service and organizes press training days.

Guidebookwriters.com

27 Camwood, Clayton-le-Woods, Bamber Bridge PR5 8LA

📞 01772 321243   📠 0870 137 8888
tm@wpu.org.uk
www.guidebookwriters.com
Contact *Terry Marsh*
SUBSCRIPTION £50 p.a.

Guidebookwriters.com exists to promote the work of its members, all of whom are professional guidebook writers, to the trade through direct contact with public relations companies, tourist boards, editors, publishers and consultants and, through its website, regular news bulletins and press releases.

Guild of Agricultural Journalists

Isfield Cottage, Church Road, Crowborough TN6 1BN

📞 01892 610628
don.gomery@btinternet.com
www.gaj.org.uk
Honorary General Secretary *Don Gomery*
SUBSCRIPTION £52 p.a.

Founded 1944 to promote a high professional standard among journalists who specialize in agriculture, horticulture and allied subjects. Represents members' interests with representative bodies in the industry; provides a forum through meetings and social activities for members to meet eminent people in the industry; maintains contact with associations of agricultural journalists overseas; promotes schemes for the education of members and for the provision of suitable entrants into agricultural journalism.

Guild of Editors
> **Society of Editors**

The Guild of Food Writers

255 Kent House Road, Beckenham BR3 1JQ
📞 020 8659 0422
gfw@gfw.co.uk
www.gfw.co.uk
Administrator *Jonathan Woods*
SUBSCRIPTION £70

Founded 1985. The objects of the Guild are to bring together professional food writers including journalists, broadcasters and authors, to extend the range of members' knowledge and experience by arranging discussions, tastings and visits, and to encourage new writers through competitions and awards. The Guild aims to contribute to the growth of public interest in, and knowledge of, the subject of food and to campaign for improvements in the quality of food.

Guild of Motoring Writers

40 Baring Road, Bournemouth BH6 4DT
📞 01202 422424   📠 01202 422424
generalsec@gomw.co.uk
www.guildofmotoringwriters.co.uk
General Secretary *Patricia Lodge*

Founded 1944. Represents members' interests and provides a forum for members to exchange information.

Horror Writers Association

244 Fifth Avenue, Suite 2767, New York, NY 10001, USA
hwa@horror.org
www.horror.org
www.horror.org/UK (UK Chapter)

Founded 1987. World-wide organization of writers and publishers dedicated to promoting the interests of writers of horror and dark fantasy. Publishes a monthly newsletter, issues e-mail bulletins, gives access to lists of horror agents, reviewers and bookstores; and keys to the 'Members Only' area of the HWA website. Presents the annual **Bram Stoker Awards** (see entry under *Prizes*).

Independent Alliance

Founded in 2005, the Independent Alliance comprises ten UK publishers and their international partners, linked by a common vision of 'editorial excellence, original, diverse publishing, innovation in marketing and commercial success. Founding publishers were

Faber & Faber, Atlantic Books, Canongate, Icon Books, Profile Books and Short Books, later joined by Quercus Publishing, Serpent's Tail and Granta/Portobello. Faber & Faber provides sales and admin for the organization's publishers and The Book Service Ltd are the sole distributors.

Independent Publishers Guild

PO Box 12, Llain, Whitland SA34 0WU
☎ 01437 563335 📠 01437 562071
info@ipg.uk.com
www.ipg.uk.com
Executive Director *Bridget Shine*

Founded 1962. Actively represents the interests of independent publishers. With over 500 members, and steadily growing with combined revenues of well over £500 million, the IPG provides a networking base whereby members receive regular e-newsletters, training courses and seminars covering important areas for its members, from e-commerce to international trade. Runs a collective stand for members at leading international book fairs including Frankfurt, London and most recently at Beijing. Membership is open to new and established independent publishers. Supplier membership is also available. Check the website for further information.

Independent Theatre Council

12 The Leathermarket, Weston Street, London SE1 3ER
☎ 020 7403 1727 📠 020 7403 1745
admin@itc-arts.org
www.itc-arts.org
Chief Executive *Charlotte Jones*

The Independent Theatre Council (ITC) is the management association and political voice of around 700 performing arts professionals and organizations. ITC provides its members with legal and management advice, training and professional development, networking, regular newsletters and a comprehensive web resource. Additionally ITC initiates and develops projects to enrich, enhance and raise the profile of the performing arts. Publications available from ITC include *The ITC Practical Guide for Writers and Companies*. ITC negotiates contracts and has established standard agreements for theatre professionals with Equity. The rights and fee structure agreement, reached with **The Writers' Guild** in 1991, was updated in December 2002 and rates of pay are reviewed annually. Contact The Writers' Guild or visit ITC's website for further details.

Institute of Copywriting

Overbrook Business Centre, Poolbridge Road, Blackford, Wedmore BS28 4PA
☎ 0800 781 1715 📠 01934 713492

services@inst.org
www.inst.org/copy
Secretary *Lynn Hall*

Founded 1991 to promote copywriters and copywriting (writing publicity material). Maintains a code of practice. Membership is open to students as well as experienced practitioners. Runs training courses (see entry under *UK and Irish Writers' Courses*). Has a list of approved copywriters. Answers queries relating to copywriting. Contact the Institute for a free booklet.

Institute of Translation and Interpreting (ITI)

Fortuna House, South Fifth Street, Milton Keynes MK9 2EU
☎ 01908 325250 📠 01908 325259
info@iti.org.uk
www.iti.org.uk

Founded 1986. The ITI is the only independent professional association of practising translators and interpreters in the UK. Its aim is to promote the highest standards in the profession. It has strong corporate membership and runs professional development courses and conferences. Membership is open to those with a genuine and proven involvement in translation and interpreting (including students). ITI's bi-monthly *Bulletin* and other publications are available from the Secretariat. ITI's *Directory of Members* offers a free referral service for those in need of a professional translator/interpreter and is available online. ITI is a full and active member of FIT (International Federation of Translators).

International Association of Scientific, Technical & Medical Publishers

Prama House, 267 Banbury Road, Oxford OX2 7HT
☎ 01865 339321 📠 01865 339325
www.stm-assoc.org

STM is an international association of scientific, technical, medical and scholarly publishers which aims to create a platform for exchanging ideas and information, and to represent the interest of the STM publishing community in the fields of copyright, technology developments and end user/library relations. The membership consists of large and small companies, not-for-profit organizations, learned societies, traditional, primary, secondary publishers and new entrants to global publishing.

IoL
▷ Chartered Institute of Linguists

Irish Book Publishers' Association
▷ CLÉ

Irish Copyright Licensing Agency Ltd

25 Denzille Lane, Dublin 2, Republic of Ireland
☎ 00 353 1 662 4211   ℻ 00 353 1 662 4213
info@icla.ie
www.icla.ie
Executive Director *Samantha Holman*

Founded 1992 by writers and publishers in Ireland
to provide a scheme through which rights holders
can give permission, and users of copyright
material can obtain permission to copy.

Irish Playwrights and Screenwriters Guild

Art House, Curved Street, Temple Bar, Dublin 2,
Republic of Ireland
☎ 00 353 1 670 9970
info@script.ie
www.script.ie
Contact *David Kavanagh*
SUBSCRIPTION Details available on request

Founded in 1969 to safeguard the rights of
scriptwriters for radio, stage and screen.

Irish Translators' & Interpreters' Association (ITIA)

Irish Writers' Centre, 19 Parnell Square, Dublin 1,
Republic of Ireland
☎ 00 353 87 673 8386   ℻ 00 353 1 87 262 82
itiasecretary@gmail.com
www.translatorsassociation.ie
Chairperson *Annette Schiller*
Honorary Secretary *Mary Phelan*
Treasurer *Mirian Watchorn*
ANNUAL MEMBERSHIP €120 (corporate); €75
(professional); €40 (ordinary); €20 (student)

Founded 1986. The Association is open to all
translators: technical, commercial, literary and
cultural, and to all classes of interpreters. It is also
for those with an interest in translation such as
teachers and third-level students. Information on
the profession of interpreting and translation and
a register of members are available on the website.
Publishes a newsletter, *Translation Ireland* and an
eZine, the *ITIA Bulletin*. Organizes certification
tests (for legal documents) for professional
members and tests for new professional
members.

Irish Writers' Union

Irish Writers' Centre, 19 Parnell Square, Dublin 1,
Republic of Ireland
☎ 00 353 1 872 1302
iwu@ireland-writers.com
www.ireland-writers.com
Chairperson *Helen Dwyer*
SUBSCRIPTION €50 p.a.; €30 (associate)

Founded 1986 to promote the interests and
protect the rights of writers in Ireland.

ITC

▷ Independent Theatre Council

ITI

▷ Institute of Translation and Interpreting

IVCA (International Visual Communications Association)

19 Pepper Street, Glengall Bridge, London E14 9RP
☎ 020 7512 0571   ℻ 020 7512 0591
info@ivca.org
www.ivca.org
Communications Officer *Philip Fey*

The IVCA is a professional association
representing the interests of the users and
suppliers of visual communications. In
particular it pursues the interests of producers,
commissioners and manufacturers involved in
the non-broadcast and independent facilities
industries and also business event companies.
It represents all sizes of company and freelance
individuals, offering information and advice
services, publications, a professional network,
special interest groups, a magazine and a variety
of events including the UK's Film and Video
Communications Festival.

The Library Association

▷ CILIP: The Chartered Institute of Library and
Information Professionals

literaturetraining

PO Box 23595, Leith, Edinburgh EH6 7YX
☎ 0131 553 2210
info@literaturetraining.com
www.literaturetraining.com
Director *Philippa Johnston*

The UK's only dedicated provider of free
information and advice on professional
development for writers and literature
professionals. Draws on the expertise of its
seven partner organizations: **Apples & Snakes**,
Lapidus, the **National Association of Writers
in Education**, the **National Association
for Literature Development**, **Scottish
Book Trust** and **Survivors' Poetry**. Their
online directory contains latest information
on courses, workshops, jobs, residencies,
submissions, competitions, conferences and
events, organizations, books, magazines and
funding for professional development. Other
services include a fortnightly e-bulletin, an
information and help line and a developing
range of info sheets and specially commissioned
features relating to creative and professional
practice. literaturetraining is a partner in the
CreativePeople Network.

The MCPS–PRS Alliance

▷ PRS for Music

Medical Journalists' Association
Fairfield, Cross in Hand, Heathfield TN21 0SH
☎ 01435 868786   📠 01435 865714
secretary@mja-uk.org
www.mja-uk.org
Chairman *David Delvin*
Honorary Secretary *Philippa Pigache*
SUBSCRIPTION £40 p.a. (full); £30 (associate);
£10 (junior)

Founded 1967. Aims to improve the quality and
practice of medical and health journalism and to
improve relationships and understanding between
medical and health journalists and the health
and medical professions. Over 400 members.
Regular meetings with senior figures in medicine
and medico politics; educational workshops
on important subject areas or issues of the day;
debates; awards for medical journalists from
commercial sponsors, plus the MJA's own annual
awards for journalism and books. Publishes a
newsletter five times a year.

Medical Writers' Group
The Society of Authors, 84 Drayton Gardens,
London SW10 9SB
☎ 020 7373 6642   📠 020 7373 5768
s.baxter@societyofauthors.org
www.societyofauthors.org
Contact *Sarah Baxter*

Founded 1980. A specialist group within the
Society of Authors offering advice and help to
authors of medical books.

Mystery Writers of America, Inc.
1140 Broadway, Suite 1507, New York, NY 10001,
USA
☎ 001 212 888 8171   📠 001 212 888 8107
mwa@mysterywriters.org
www.mysterywriters.org
Administrative Manager *Margery Flax*
SUBSCRIPTION $95 (US)

Founded 1945. Aims to promote and protect the
interests of writers of the mystery genre in all
media; to educate and inform its membership
on matters relating to their profession; to uphold
a standard of excellence and raise the profile of
this literary form to the world at large. Holds an
annual banquet at which the 'Edgars' are awarded
(named after Edgar Allan Poe).

National Association for Literature Development
PO Box 49657, London N8 7YZ
☎ 020 7272 8386
director@nald.org
www.nald.org
Contact *Melanie Abrahams*

NALD is the largest membership organization
for literature professionals. The only UK-wide
body for all those involved in developing
writers, readers and literature audiences. Offers
networking advice, training and advocacy
alongside individual and organizational
membership services.

National Association of Writers' Groups
40 Burstall Hill, Bridlington YO16 7GA
☎ 01262 609228
nawg@tesco.net
www.nawg.co.uk
Chairman *Jacqueline Smith*
Secretary *Diane Wilson*
SUBSCRIPTION £35 p.a. + £5 registration (group);
£16 p.a. (associate)

Founded 1995 with the object of furthering
the interests of writers' groups and individuals
throughout the UK. A registered charity (No.
1059047), NAWG is strictly non-sectarian and
non-political. Publishes *Link* magazine and
books for writers (including *Running Good
Writing Groups*); gives free entry to competitions
for group anthologies, poetry, short stories,
articles, novels and sketches; holds an annual
open festival of writing with 21 workshops,
seminars, individual surgeries led by professional,
high-profile writers; annual competitions.
Membership is open to all writers' groups and
individual writers – there are no restrictions or
qualifications required for joining; 160 groups are
affiliated to-date.

National Association of Writers in Education
PO Box 1, Sheriff Hutton, York YO60 7YU
☎ 01653 618429
info@nawe.co.uk
www.nawe.co.uk
Contact *Paul Munden, Director*
SUBSCRIPTION £20 p.a. (individual); £10
(student); £60 (institution); £30 (overseas)

Founded 1991. Aims to promote the contribution
of living writers to education and to encourage
both the practice and the critical appreciation
of creative writing. Has over 1,000 members.
Organizes national conferences and training
courses. A directory of writers who work in
schools, colleges and the community is available
online. Publishes a magazine, *Writing in
Education*, issued free to members three times
per year.

National Campaign for the Arts
1 Kingly Street, London W1B 5PA
☎ 020 7287 3777   📠 020 7287 4777
nca@artscampaign.org.uk
www.artscampaign.org.uk
Director *Louise de Winter*
Membership and Office Manager *Simon Cole*

The UK's only independent campaigning organization representing all the arts. The NCA seeks to safeguard, promote and develop the arts and win public and political recognition for their importance as a key element in the national culture. Its members shape its campaign work, provide its mandate and the core funding, through subscriptions that enable the NCA to act independently on their behalf; the organization receives no public subsidy. Members have access to the NCA's information, advice, seminars, conferences and publications. Membership is open to organizations and individuals working in or with an interest in the arts. Literature subscriptions available.

National Centre for Research in Children's Literature (NCRCL)

Bede House, School of Arts, Digby Stuart College, Roehampton University, Roehampton Lane, London SW15 5PH

ⓣ 020 8392 3000   ⓕ 020 8392 3819

g.lathey@roehampton.ac.uk

l.sainsbury@roehampton.ac.uk

www.ncrcl.ac.uk

The National Centre for Research in Children's Literature facilitates and supports research exchange in the field of children's literature. Runs an internationally acclaimed MA Programme, an annual conference in association with the International Board on Books for Young People (IBBY).

National Literacy Trust

68 South Lambeth Road, London SW8 1RL

ⓣ 020 7587 1842   ⓕ 020 7587 1411

contact@literacytrust.org.uk

www.literacytrust.org.uk

Director *Jonathan Douglas*

The National Literary Trust is an independent charity that changes lives through literacy. 'We have a vision of a society in which everyone has the reading, writing, speaking and listening skills that they need to fulfil their own and, ultimately, the nation's potential. We aim to empower learners, support professionals and influence policy and practice.' Supports those who work with learners through their innovative programmes, information and research, and brings together key organizations to lead literacy promotion in the UK. Programmes include Reading Is Fundamental, UK – providing free books for children to choose and keep; Reading The Game – involving the professional football community; Reading Connects – developing a whole-school approach to reading for pleasure; Reading Champions – putting the boys in charge of their reading; and the Talk To Your Baby campaign which supports parents and professionals.

National Union of Journalists

Headland House, 308 Gray's Inn Road, London WC1X 8DP

ⓣ 020 7278 7916   ⓕ 020 7837 8143

info@nuj.org.uk

www.nuj.org.uk

General Secretary *Jeremy Dear*

SUBSCRIPTION £189 p.a. (freelance) or 1% of annual income if lower; or 0.5% if income less than £13,600 p.a.

Represents journalists in all sectors of publishing, print, broadcast and online media. Responsible for wages and conditions agreements which apply across the industry. Provides advice and representation for its members, as well as administering unemployment and other benefits. Publishes various guides and magazines: *On-Line Freelance Directory*; *On-Line Fees Guide*; *The Journalist* and *The Freelance*.

NCRCL

▷ National Centre for Research in Children's Literature

New Producers Alliance (NPA)

Room 7.03 The Tea Building, 56 Shoreditch High Street, London E1 6JJ

ⓣ 020 7613 0440   ⓕ 020 7729 1852

queries@npa.org.uk

www.npa.org.uk

Founded 1993. National membership and training organization for independent new filmmakers, the NPA provides a forum and focus for over 1,000 members, ranging from students and first timers to highly experienced feature filmmakers and industry affiliates. Services include: producer, writer and director training from 'entrance' to 'advanced' levels; seminars, masterclasses, business breakfasts, themed networking evenings and preview screenings; in addition, the NPA supports its members at film festivals including Cannes, Berlin and Edinburgh with events and industry receptions. Provides free advice services including Legal and Tax & Accounting; the NPA also publishes a monthly newsletter and online directory included in the website for its 1,000+ members.

New Writing North

Culture Lab, Grand Assembly Rooms, Newcastle University, King's Walk, Newcastle upon Tyne NE1 7RU

ⓣ 0191 222 1332   ⓕ 0191 222 1372

mail@newwritingnorth.com

www.newwritingnorth.com

Director *Claire Malcolm*

Administrator *Catherine Robson*

New Writing North is the literature development agency for the Arts Council England, North East region and offers many useful services to writers, organizing events, readings and workshops. NWN has a website and mailing list by which members of the public can receive news of literary events, courses and work opportunities. Administers the Northern Writers' Awards, which include tailored development packages (mentoring and financial help) for local writers at different stages of their careers. NWN works to develop theatre writing and creative radio writing projects within the region, as well as a strong education and community-based programme, working with schools, LEAs and both public and private sector organizations.

The Newspaper Society

St Andrew's House, 18–20 St Andrew Street, London EC4A 3AY
📞 020 7632 7425   📠 020 7632 7401
collette_yates@newspapersoc.org.uk
www.newspapersoc.org.uk
Director *David Newell*

The Newspaper Society is the voice of Britain's regional press, a £4 billion sector read by over 40 million adults a week, It promotes the interests of the regional and local press, provides legal advice and lobbying services, holds a series of conferences and seminars, and runs the annual Local Newspaper Week.

NPA
▷ New Producers Alliance

Outdoor Writers & Photographers Guild

PO Box 520, Bamber Bridge, Preston PR5 8LF
📞 01772 321243   📠 0870 137 8888
secretary@owg.org.uk
www.owpg.org.uk
Secretary *Terry Marsh*
SUBSCRIPTION £70 p.a.

Founded 1980 to promote, encourage and assist the development and maintenance of professional standards among creative professionals working in the outdoors. Members include writers, photographers, broadcasters, filmmakers, editors, publishers and illustrators. Publishes a quarterly journal, *Outdoor Focus*. Presents six Awards for Excellence plus other awards in recognition of achievement. Maintains a website of members' details; informal members' e-mail 'chatline' to exchange news and information; sends out regular media news bulletins; organizes press trips.

Pact (Producers Alliance for Cinema and Television)

3rd Floor, Fitzrovia House, 153–157 Cleveland Street, London W1T 6QW
📞 020 7380 8242   📠 020 7380 8248
clare@pact.co.uk
www.pact.co.uk
Chief Executive *John McVay*
Chief Operations Officer *Andrew Chowns*

Pact is the UK trade association that represents and promotes the commercial interests of independent feature film, television, children's, animation and interactive media companies. Headquartered in London, Pact has regional representation throughout the UK, in order to support its members. A highly effective lobbying organization, it has regular and constructive dialogues with government, regulators, public agencies and opinion formers on all issues affecting its members and contributes to key public policy debates on the media industry, both in the UK and in Europe. Pact negotiates terms of trade with all public service broadcasters in the UK and supports members in their business dealings with cable and satellite channels. It also lobbies for a properly structured and funded UK film industry and maintains close contact with the **UK Film Council** and other relevant film organizations and government departments.

PEN
▷ English PEN

Performing Right Society
▷ PRS for Music

The Personal Managers' Association Ltd

PO Box 6319, London N1P 1HL
📞 0845 602 7191
info@thepma.com
www.thepma.com
Co-chairs *Michelle Kass, Alan Brodie, Nicholas Young*
Secretary *Catherine O'Shea*
SUBSCRIPTION £275 p.a.

An association of artists' and dramatists' agents (membership not open to individuals). Monthly meetings for exchange of information and discussion. Maintains a code of conduct and acts as a lobby when necessary. Applicants screened. A high proportion of play agents are members of the PMA.

The Picture Research Association

c/o 1 Willow Court, off Willow Street, London EC2A 4QB
📞 020 7739 8544   📠 020 7782 0011
chair@picture-research.org.uk
www.picture-research.org.uk
Chair *Veneta Bullen*

Membership Secretary *Frances Topp*
SUBSCRIPTION Members: £50 (introductory); £60 (full); £55 (associate). Magazine only: £16 (inland); £20 (overseas)

Founded in 1977 as the Society of Picture Researchers & Editors. The Picture Research Association is a professional body for picture researchers, managers, picture editors and all those involved in the research, management and supply of visual material to all forms of the media. The Association's main aims are to promote the interests and specific skills of its members internationally; to promote and maintain professional standards; to bring together those involved in the research and publication of visual material; to provide a forum for the exchange of information and to provide guidance to its members. Free advisory service for members, regular meetings, quarterly magazine, monthly newsletter and Freelance Register.

Player-Playwrights

90 Lady Margaret Road, London N19 4ES
carolinelangston@yahoo.co.uk
www.playerplaywrights.co.uk
Presidents *Laurence Marks, Maurice Gran*
Contact *Caroline Langston (at the address above)*
SUBSCRIPTION Members: £10 (joining fee); £6 p.a. thereafter, plus £2 per attendance

Founded 1948. A society giving opportunity for writers new to stage, radio, television and film, as well as others finding difficulty in achieving results, to work with writers established in those media. At weekly meetings (7.30 pm–10.00 pm, Mondays, upstairs at the Horse and Groom, 128 Great Portland Street, London W1), members' scripts are read or performed by actor members and afterwards assessed and dissected in general discussion. New writers and new acting members are always welcome.

Playwrights' Studio, Scotland

350 Sauchiehall Street, CCA, Glasgow G2 3JD
☎ 0141 332 4403
info@playwrightsstudio.co.uk
www.playwrightsstudio.co.uk

The Playwrights' Studio, Scotland is a national initiative designed to engage the people of Scotland directly with new playwriting and raise the standard of plays for presentation to the public. Aims to increase awareness of the wealth of creative talent within Scotland, both at home and internationally, and work to develop the art form of theatre, examining and expanding the playwright's role within it. Delivers a wide range of activities which address the development needs of aspiring, emerging and established playwrights, encourage producers' interest in new plays and

playwrights, and promote Scotland's plays and playwrights to the widest possible markets.

PLR
▷ Public Lending Right

PLS
▷ Publishers Licensing Society Ltd

PMA
▷ The Personal Managers' Association Ltd

Poetry Book Society
▷ under Organizations of Interest to Poets

Poetry Ireland
▷ under Organizations of Interest to Poets

The Poetry Society
▷ under Organizations of Interest to Poets

PPA The Association for Publishers and Providers of Consumer, Customer & Business Media in the UK

Queen's House, 55–56 Lincoln's Inn Fields, London WC2A 3LJ
☎ 020 7404 4166   🖷 020 7404 4167
info1@ppa.co.uk
www.ppa.co.uk

Founded 1913 to promote and protect the interests of its members in particular and the magazine industry as a whole.

Press Complaints Commission

Halton House, 20/23 Holborn, London EC1N 2JD
☎ 020 7831 0022   🖷 020 7831 0025
pcc@pcc.org.uk
www.pcc.org.uk
Director *Tim Toulmin*
Information Officer *Tonia Milton*

Founded in 1991 to deal with complaints from members of the public about the editorial content of newspapers and magazines. Administers the editors' Code of Practice covering such areas as accuracy, privacy, harrassment and intrusion into grief. Publications available: *Code of Practice, How to Complain* and *Annual Review*.

Private Libraries Association

Ravelston, South View Road, Pinner HA5 3YD
dchambrs@aol.com
www.pla.org
Honorary Secretary *Stanley Brett (stanbrett @f2s.com)*
SUBSCRIPTION £25 p.a.

Founded in 1956 for the advancement of the education of the public in the study, production and ownership of books. Publishes work concerned with the objects of the Association, particularly those works which are not commercially viable; holds meetings at which papers on cognate subjects are read and discussed; arranges lectures and exhibitions which are also open to non-members.

Producers Alliance for Cinema and Television
▷ Pact

PRS for Music
Copyright House, 29–33 Berners Street, London W1T 3AB
☎ 020 7580 5544   🖷 020 7306 4455
www.mcps-prs-alliance.co.uk
JOINING FEE Refer to the website for full details

PRS ensures that composers, songwriters and music publishers are paid royalties when their music is used. The music does not need to be published or recorded for eligibility for PRS membership.

Public Lending Right
Richard House, Sorbonne Close, Stockton-on-Tees TS17 6DA
☎ 01642 604699   🖷 01642 615641
authorservices@plr.uk.com
www.plr.uk.com
Registrar *Jim Parker*

Founded 1979. Public Lending Right is funded by the Department for Culture, Media and Sport to recompense authors for books borrowed from public libraries. Authors and books can be registered for PLR only when application is made during the author's lifetime. To qualify, an author must be resident in a European Union member state, or in Iceland, Liechtenstein or Norway, and the book must be printed, bound and put on sale with an ISBN. In 2008, £6.66 million was distributed to over 23,900 authors (£6.81 million in the previous year) with 242 authors reaching the maximum payment threshold of £6,600. The Rate Per Loan was 5.98 pence in 2008. Consult the website for further information.

The Publishers Association
29b Montague Street, London WC1B 5BW
☎ 020 7691 9191   🖷 020 7691 9199
mail@publishers.org.uk
www.publishers.org.uk
Chief Executive *Simon Juden*

The national UK trade association for books, learned journals, and electronic publications, with around 200 member companies in the industry. Very much a trade body representing the industry to Government and the European Commission, and providing services to publishers. Publishes the *Directory of Publishing* in association with **Continuum**. Also home of the Educational Publishers Council (school books), PA's International Division, the Academic and Professional Division, the Trade Publishers Council and the Electronic Publishers' Forum.

Publishers' Association of South Africa
PO Box 106, Green Point, Cape Town, South Africa 8051
☎ 00 27 21 425 2721   🖷 00 27 21 421 3270
pasa@publishsa.co.za
www.publishsa.co.za
Contact *Dudley Schroeder, Chief Executive (dudley@publishsa.co.za)*

Founded in 1992 to represent publishing in South Africa, a key industry sector. With a membership of 162 companies, the Association includes commercial organizations, university presses, one-person privately-owned publishers as well as importers and distributors.

Publishers Licensing Society Ltd
37–41 Gower Street, London WC1E 6HH
☎ 020 7299 7730   🖷 020 7299 7780
pls@pls.org.uk
www.pls.org.uk
Chief Executive *Alicia Wise*
Operations Manager *Tom West*

Founded in 1981, the PLS obtains mandates from publishers which grant PLS the authority to license photocopying of pages from published works. Some licences for digitization of printed works are available. PLS aims to maximize revenue from licences for mandating publishers and to expand the range and repertoire of mandated publishers available to licence holders. It supports the **Copyright Licensing Agency Ltd (CLA)** in its efforts to increase the number of legitimate users through the issuing of licences and vigorously pursues any infringements of copyright works belonging to rights' holders.

Publishers' Publicity Circle
65 Airedale Avenue, London W4 2NN
☎ 020 8994 1881
ppc-@lineone.net
www.publisherspublicitycircle.co.uk
Contact *Heather White*

Enables book publicists from both publishing houses and freelance PR agencies to meet and share information regularly. Meetings, held monthly in central London, provide a forum for press journalists, television and radio researchers and producers to meet publicists collectively. A directory of the PPC membership is published each year and distributed to over 2,500 media contacts.

Publishing Scotland
Scottish Book Centre, 137 Dundee Street, Edinburgh EH11 1BG
☎ 0131 228 6866   🖷 0131 228 3220
enquiries@publishingscotland.org
www.publishingscotland.org
Chief Executive *Marion Sinclair*

Membership Services Manager *Jane Walker*
BooksfromScotland.com Editor *Liam Davison*
Events and Marketing Manager *Vanessa Garden*
Finance and Office Manager *Carol Lothian*
Training and Information Manager *Joan Lyle*

Publishing Scotland, (formerly Scottish
Publishers Association) is the trade, training and
development body for publishing in Scotland. It
represents a wide range of Scottish publishers,
from multinationals to small presses, and has
developed a Network membership from all
sections of the book world, including the **Society
of Authors in Scotland**. Publishing Scotland also
acts as an information centre for the trade and
general public. Publishes annual *Publishing in
Scotland Yearbook*. Attends international book
fairs; runs an extensive training programme;
carries out industry research.

The Radio Academy

5 Market Place, London W1W 8AE
☎ 020 7927 9920   🖷 020 7636 8924
info@radioacademy.org
www.radioacademy.org.uk
Director *Trevor Dann*
SUBSCRIPTIONS £25 p.a. All staff of Patron
members receive free membership

The Radio Academy is a registered charity
dedicated to the encouragement, recognition
and promotion of excellence in UK broadcasting
and audio production. Since 1983 the Academy
has represented the radio industry to outside
bodies including the government and offered
neutral ground where everyone from the national
networks to individual podcasters is encouraged
to discuss the broadcasting, production,
marketing and promotion of radio and audio.

RadioCentre

77 Shaftesbury Avenue, London W1D 5DU
☎ 020 7306 2603   🖷 020 7306 2500
www.radiocentre.org
Chief Executive *Andrew Harrison*

RadioCentre represents commercial radio
stations in the UK. Its role is to maintain and
build a strong and successful commercial radio
industry, both in terms of listening hours and
revenues.

The Reading Agency

2 Tavistock Place, London WC1H 9RA
☎ 020 7833 8846
info@readingagency.org.uk
www.readingagency.org.uk

The Reading Agency was formed in 2002 by
merging three smaller agencies – LaunchPad, The
Reading Partnership and Well Worth Reading.
Runs national partnership schemes, library
reading promotions, projects that target specific

readers, and their own research and publishing
programme. The Agency assists people working
directly with readers by providing training
and advice as well as toolkits, websites and
promotional materials, and uses connections to
join up organizations interested in promoting
reading, for example, helping libraries form
partnerships with major organizations such as
publishers and the BBC. Programmes include
the Summer Reading Challenge for children
and The Vital Link for adults trying to improve
their literacy skills. Active in public policy and
advocacy, working with government departments
and think tanks, the Agency also works with
the corporate sector, through sponsorship or
Corporate Social Responsibility programmes.

Romance Writers of America

14615 Benfer Road, Houston, TX 77069, USA
☎ 001 832 717 5200   🖷 001 832 717 5201
info@rwanational.org
www.rwanational.org
Professional Relations Coordinator *Carol Ritter*
MEMBERSHIP $75 p.a. plus $25 processing fee

Founded 1980. RWA, a non-profit association
with more than 9,000 members worldwide,
provides a service to authors at all stages of their
careers as well as to the romance publishing
industry and its readers. Anyone pursuing a
career in romantic fiction may join RWA. Holds
an annual conference, and provides contests for
both published and unpublished writers through
the RITA Awards and Golden Hearts Awards.

The Romantic Novelists' Association

3 Griffin Road, Thame OX9 3LB
☎ 01844 213947
ianandcatherinejones@btinternet.com
www.rna-uk.org
Contact *Catherine Jones (Chairman)*
SUBSCRIPTION £40 p.a. (full members); £45
(full, non-EU); £50 (associate); full and associate
members who pay by standing order receive a £3
discount

Membership is open to published writers of
romantic fiction (modern or historical), or those
who have had one or more full-length serials
published. Associate membership is open to
publishers, editors, literary agents, booksellers,
librarians and others having a close connection
with novel writing and publishing. Membership
in the New Writers' Scheme is available to a
limited number of writers who have not yet had
a full-length novel published. New Writers must
submit a manuscript each year. The manuscripts
receive a report from experienced published
members and the reading fee is paid with the
annual subscription of £90 (£95 for non-EU
members). Meetings are held in London and the

regions with guest speakers. *Romance Matters* is published quarterly and issued free to members. The Association makes three annual awards: **The Romantic Novel of the Year**; Joan Hessayon New Writers Award for the best published novel by a new writer who has gone through the New Writers' Scheme and remains a member, and the Romance Prize for category romances.

Royal Society of Literature

Somerset House, Strand, London WC2R 1LA
☎ 020 7845 4676   🖷 020 7845 4679
info@rslit.org
www.rslit.org
Chair *Anne Chisholm, FRSL*
Secretary *Maggie Fergusson*
SUBSCRIPTION £40 p.a.

Founded 1820. Membership by application to the Secretary. Fellowships are conferred by the Society on the proposal of two Fellows. Membership benefits include lectures, discussion meetings and poetry readings. Lecturers have included Julian Barnes, Michael Holroyd, Philip Pullman, Zadie Smith and Tom Stoppard. Presents the **Royal Society of Literature Ondaatje Prize**, the **Royal Society of Literature/ Jerwood Awards** and the **V.S. Pritchett Memorial Prize**.

Royal Television Society

Kildare House, 3 Dorset Rise, London EC4Y 8EN
☎ 020 7822 2810   🖷 020 7822 2811
info@rts.org.uk
www.rts.org.uk
Contact *Membership Manager (members@rts. org.uk)*
MEMBERSHIP from £65 p.a.

Founded 1927. Covers all disciplines involved in the television industry. Provides a forum for debate and conferences on technical, social and cultural aspects of the medium. Presents various awards including journalism, programmes, technology and design. Publishes *Television Magazine* ten times a year for members and subscribers.

Science Fiction Foundation

75 Rosslyn Avenue, Harold Wood RM3 0RG
www.sf-foundation.org
Contact *Roger Robinson*

Founded 1970. The SFF is an international academic body for the furtherance of science fiction studies. Publishes a thrice-yearly magazine, *Foundation* (see entry under *Magazines*), which features academic articles and reviews of new fiction. Publishes a series of books on critical studies of science fiction, and arranges conferences, lectures and seminars on related topics. It also has a reference library (see

entry under *Library Services*), housed at Liverpool University.

Scottish Book Trust

Sandeman House, Trunk's Close, 55 High Street, Edinburgh EH1 1SR
☎ 0131 524 0160   🖷 0131 524 0161
info@scottishbooktrust.com
www.scottishbooktrust.com
Contact *Marc Lambert*

Scottish Book Trust is the leading agency for the promotion of literature, reading and writing in Scotland. Its innovative programmes are designed to inspire readers, writers and learning professionals all over the country. Scottish Book Trust funds author events all over Scotland; runs the biggest children's book awards in the UK involving over 15,000 children; supports the professional development of Scottish writers through mentoring, awards and training opportunities; supports teachers and librarians to improve literacy; operates a touring programme for children's authors; develops public participation projects such as Days Like This and promotes other readership development opportunities.

Scottish Daily Newspaper Society

21 Lansdowne Crescent, Edinburgh EH12 5EH
☎ 0131 535 1064   🖷 0131 535 1063
info@sdns.org.uk
Director *J.B. Raeburn*

Founded 1915. Trade association representing publishers of Scottish daily and Sunday newspapers including the major Scottish editions of UK national newspapers.

Scottish Library Association
▷ CILIP in Scotland (CILIPS)

Scottish Newspaper Publishers Association

48 Palmerston Place, Edinburgh EH12 5DE
☎ 0131 220 4353   🖷 0131 220 4344
info@snpa.org.uk
www.snpa.org.uk
Director *S. Fairclough*

The representative body for the publishers of paid-for weekly and associated free newspapers in Scotland. Represents the interests of the industry to government, public and other bodies and provides a range of services including marketing of *The Scottish Weekly Press* and education and training. It is an active supporter of the Press Complaints Commission.

Scottish Print Employers Federation

48 Palmerston Place, Edinburgh EH12 5DE
☎ 0131 220 4353   🖷 0131 220 4344
info@spef.org.uk
www.spef.org.uk

Director *S. Fairclough*

Founded 1910. Employers' organization and trade association for the Scottish printing industry. Represents the interests of the industry to government, public and other bodies and provides a range of services including industry promotion, industrial relations, education, training and commercial activities. Negotiates a national wages and conditions agreement with Unite GPM Sector. The Federation is a member of Intergraf, the international confederation for employers' associations in the printing industry, which represents its interests at European level.

Scottish Publishers Association
▷ Publishing Scotland

Scottish Screen

2nd Floor, 249 West George Street, Glasgow G2 4QE
☎ 0141 302 1700   ▣ 0141 302 1711
info@scottishscreen.com
www.scottishscreen.com
Contact *Linsey Denholm*

Scottish Screen is the national development agency for the screen industries in Scotland. 'We aim to inspire audiences, support new and existing talent and businesses, educate young people, and promote Scotland as a creative place to make great films, award-winning television and world renowned digital entertainment.'

Society for Editors and Proofreaders (SfEP)

Enrico House, 93–99 Upper Richmond Road, London SW15 2TG
☎ 020 8785 5617   ▣ 020 8785 5618
administration@sfep.org.uk
www.sfep.org.uk
Chair *Sarah Price*
Company Secretary *Justina Amenu*
SUBSCRIPTION £88 p.a. (individuals) plus £27.50 joining fee; corporate membership available.

Founded 1988 in response to the growing number of freelance editors and their increasing importance to the publishing industry. Aims to promote high editorial standards by disseminating information through advice and training, and to achieve recognition of the professional status of its members. The Society also supports moves towards recognized standards of training and accreditation for editors and proofreaders. It launched its own Accreditation in Proofreading test in 2002. Following discussions with the SfEP, City & Guilds have developed the Licentiateship in Editorial Skills. This was launched in September 2008. The Licentiateship offers a form of recognition based around Continuing Professional Development.

Society for Technical Communications (STC)

901 N. Stuart Street, Suite 904, Arlington, Virginia 22203, USA
☎ 001 703 522 4114   ▣ 001 703 522 2075
www.stc.org
www.stcuk.org
UK and Ireland Chapter Membership Manager *Karen Lewis (membership@stcuk.org)*

Dedicated to advancing the arts and sciences of technical communication, the STC has more than 15,000 members worldwide, including technical writers, editors, graphic designers, videographers, multimedia artists, web and intranet page information designers, translators and others whose work involves making technical information available to those who need it. The UK and Ireland Chapter, established in 1985, hosts talks and seminars, runs an annual technical publications competition and publishes its own newsletter quarterly.

The Society of Authors

84 Drayton Gardens, London SW10 9SB
☎ 020 7373 6642   ▣ 020 7373 5768
info@societyofauthors.org
www.societyofauthors.org
General Secretary *Mark Le Fanu*
Deputy General Secretary *Kate Pool*
SUBSCRIPTION £90/85 p.a.

The Society of Authors is an independent trade union, chaired by Margaret Drabble, with some 9,000 members. As well as campaigning for the profession, it advises members on negotiations with publishers, agents, broadcasting organizations, theatre managers and film companies and on queries or concerns members may have about the business aspects of their writing. The Society assists with complaints and takes action for breach of contract, copyright infringement, etc. Among the Society's publications are *The Author* (quarterly) and the *Quick Guides* series to various aspects of writing (all free of charge to members). Other services include vetting of contracts, a searchable database of writers, emergency funds for writers, meetings and seminars, and various special discounts, including money off books, hotels and insurance. There are groups within the Society for academic writers, broadcasters, children's writers and illustrators, educational writers, medical writers and translators, as well as some regional groups. Authors under 35 or over 65, not earning a significant income from their writing, may apply for lower subscription rates (after their first year, in the case of over-65s). Contact the Society for an application pack, including a copy of *The Author* or visit the website.

The Society of Authors in Scotland

c/o 99A/2 St. Stephen Street, Edinburgh EH3 5AB

☎ 0131 556 9446   🖷 0131 556 9446

anguskonstam@aol.com

www.societyofauthors.org/

Honorary Secretary *Angus Konstam*

The Scottish branch of the **Society of Authors**, which organizes business meetings, social and bookshop events throughout Scotland.

Society of Children's Book Writers & Illustrators

ra@britishscbwi.org

www.britishscbwi.org

www.scbwi.org

British Isles Regional Adviser *Natascha Biebow*

Founded in 1968 by a group of Los Angeles-based writers, the SCBWI is the only international professional organization for the exchange of information between writers, illustrators, editors, publishers, agents and others involved with literature for young people. With a membership of more than 12,000 worldwide, it holds two annual international conferences plus a number of regional events, publishes a bi-monthly journal and awards grants for works in progress. The British Isles region sponsors workshops and speaker events to help authors and illustrators develop their craft and market their work. It also facilitates local critique groups and publishes a quarterly regional newsletter. For membership enquiries, visit the website or contact membership@britishscbwi.org

Society of Civil and Public Service Writers

17 The Green, Corby Glen, Grantham NG33 4NP

joan@lewis5634.fsnet.co.uk

www.scpsw.co.uk

Membership Secretary *Mrs Joan Lewis*

SUBSCRIPTION £15 p.a.

Founded 1935. Welcomes serving and retired members of the Civil Service, armed forces, Post Office, BT, nursing profession and other public servants who are aspiring or published writers. Offers competitions for short story, article and poetry; postal folios for short story and article; AGM, occasional meetings and luncheon held in London; quarterly magazine, *Civil Service Author*, to which members may submit material; Poetry Workshop (extra £3) offers annual weekend outside London, anthology, newsletter, postal folio, competitions. S.a.e. to the Secretary for details.

Society of Editors

University Centre, Granta Place, Mill Lane, Cambridge CB2 1RU

☎ 01223 304080   🖷 01223 304090

info@societyofeditors.org

www.societyofeditors.org

Executive Director *Bob Satchwell*

Formed by a merger of the Association of British Editors and the Guild of Editors, the Society of Editors has nearly 500 members in national, regional and local newspapers, magazines, broadcasting, new media, journalism education and media law. Campaigns for media freedom and self-regulation. For further information call *Bob Satchwell* (07860 562815).

Society of Indexers

Woodbourn Business Centre, 10 Jessell Street, Sheffield S9 3HY

☎ 0114 244 9561   🖷 0114 244 9563

admin@indexers.org.uk

www.indexers.org.uk

Secretary *Judith Menes*

Administrator *Wendy Burrow*

SUBSCRIPTION £92.50 p.a.; institutions: from £185

Founded 1957. Publishes *The Indexer* (biannual, April and October), a quarterly newsletter, an online directory, *Indexers Available (IA)*, which lists members and their subject expertise. In addition, the Society runs an open-learning course entitled *Training in Indexing* and recommends rates of pay (currently £18.50–£30 per hour or £2–£5 per page). Awards annually **The Wheatley Medal** and **The Bernard Levin Award** (see entries under *Prizes*).

The Society of Medical Writers

Membership Secretary: 30 Dollis Hill Lane, London NW2 6JE

richard_cutler_novelist@yahoo.co.uk

www.somw.org

Contact *Dr R.C. Cutler*

SUBSCRIPTION £45 p.a. (reduced rate for students and retired)

Now in its 21st year, the Society is open to anyone in any branch of medicine or its allied professions who publishes or aspires to publish their written work of whatever nature – fact or fiction, prose or poetry. The Society aims to improve standards of writing; to encourage literacy; to provide meetings for members for the exchange of views, skills and ideas; to provide education on the preparation, presentation and submission of written material for publication; to advise on sources of assistance with regard to technical, legal and financial aspects of writing; to consider questions of ethics relating to writing and publication; to further developments in the art of writing. Biannual conferences held in the spring and autumn at various locations; details published in *The Writer* or available from the secretary (copy for *The Writer* may be submitted to the editor throughout the year).

Society of Women Writers & Journalists

27 Braycourt Avenue, Walton-on-Thames
KT12 2AZ
wendy@stickler.org
www.swwj.co.uk
Membership Secretary *Wendy Hughes*
SUBSCRIPTION £40 p.a. (full); £30 (associate);
£25 (probationary/student); £30 (overseas). £15
joining fee; £15 (Friends of SWWJ)
£35 (town); £30 (country); £25 (overseas). £15
joining fee

Founded 1894. For women writers, the SWWJ
upholds professional standards and literary
achievements through regional group meetings;
residential weekends; postal critique service;
competitions; outings and major London
meetings. For Full members, membership card
doubles as a press card. Publishes *The Woman
Writer*. Male writers are accepted as Associate
Members. A new category has been introduced
for Friends of the Society.

Society of Young Publishers

29b Montague Street, London WC1B 5BW
info@thesyp.org.uk
www.thesyp.org.uk
Contact *Membership Secretary*
SUBSCRIPTION £30 p.a.; £24 (student/unwaged)

Provides facilities whereby members can increase
their knowledge and widen their experience of
all aspects of publishing, and holds regular social
events. Run entirely by volunteers, it is open
to those in related occupations, students, and
associate membership is available for over-35s.
Publishes a bi-monthly newsletter called *In*Print
and holds meetings on the last Wednesday of
each month featuring trade professionals. For
other events see the website. Please enclose an
s.a.e. when writing to the Society.

Spoken Word Publishing Association
▷ Audiobook Publishing Association

Sports Journalists' Association of Great Britain

c/o Start2Finish Event Management, Unit 92,
Capital Business Centre, 22 Carlton Road,
Croydon CR2 0BS
☎ 020 8916 2234   ▪ 020 8916 2235
stevenwdownes@btinternet.com
www.sportsjournalists.co.uk
Secretary *Steven Downes (80 Southbridge Road,
Croydon CR0 1AE; 07710 428562)*
SUBSCRIPTION £30 p.a. incl. VAT (London); £20
(regional)

Founded in 1948 as the Sports Writers'
Association of Great Britain to promote and
maintain a high professional standard among
journalists who specialize in sport in all its

branches – including photography, broadcasting
and online – and to serve members' interests.
Publishes a quarterly *Bulletin* and a Yearbook,
and promotes the annual SJA British Sports
Awards in Nov/Dec, and the SJA British Sports
Journalism Awards and SJA British Sports
Photography Awards every March.

STM
▷ International Association of Scientific, Technical &
Medical Publishers

Tees Valley Arts

Melrose House, Melrose Street, Middlesbrough
TS1 2HZ
☎ 01642 264651   ▪ 01642 264955
info@teesvalleyarts.org.uk
www.teesvalleyarts.org.uk
Acting Executive Director *Rowena Sommerville*
Education Officer *Jane Gray*
Communications Officer *Simon Smith*

TVA is an arts development agency which
operates throughout the Tees Valley,
encompassing all art forms, working with
educational and community organizations and
groups, to add value to existing activities and to
develop new ones. TVA works in partnership
with statutory and voluntary organizations to
embed cultural activity in educational and social
practice, and in neighbourhood renewal and
regeneration initiatives.

Theatre Writers' Union
▷ The Writers' Guild of Great Britain

Trade Publishers Council
▷ The Publishers Association

The Translators Association

84 Drayton Gardens, London SW10 9SB
☎ 020 7373 6642   ▪ 020 7373 5768
sbaxter@societyofauthors.org
www.societyofauthors.org
Contact *Sarah Baxter*

Founded 1958 as a subsidiary group within the
Society of Authors to deal exclusively with the
special problems of literary translators into the
English language. Benefits to members include
free legal and general advice and assistance on
all business matters relating to translators' work,
including the vetting of contracts and advice on
rates of remuneration. Membership is normally
confined to translators who have had their work
published in volume or serial form or produced
in this country for stage, television or radio.
The Association administers translation prizes
(see *Prizes*) and maintains an online database of
members' details for the use of publishers who are
seeking a translator for a particular work.

UK Film Council

10 Little Portland Street, London w1w 7jg
📞 020 7861 7861   📠 020 7861 7863
info@ukfilmcouncil.org.uk
www.ukfilmcouncil.org.uk

The UK's leading film body. Uses lottery money and Government Grant-in-Aid to encourage the development of new talent, skills and creative and technological innovation in UK film; help new and established filmmakers make distinctive British films; support the creation and growth of stable businesses in the film sector; provide access to finance and help the UK film industry to compete in the global marketplace. Also promotes enjoyment and understanding of the cinema and ensures that film's economic and creative and cultural interests are properly represented in public policy making.

Voice of the Listener & Viewer Ltd (VLV)

PO Box 401, Gravesend DA12 9FY
📞 01474 352835
info@vlv.org.uk
www.vlv.org.uk
Chairman *Jocelyn Hay*
Executive Director *Peter Blackman*
Administrative Secretary *Sue Washbrook*

VLV represents the citizen and consumer interests in broadcasting. It is an independent, non-profit-making organization working to ensure independence, quality and diversity in broadcasting. VLV is the only consumer body speaking for listeners and viewers on the full range of broadcasting issues. It is funded by its members and is free from sectarian, commercial and political affiliations. Holds public lectures, seminars and conferences throughout the UK, and has frequent contact with MPs, civil servants, the BBC and independent broadcasters, regulators, academics and other consumer groups. Provides an independent forum where all with an interest in broadcasting can speak on equal terms. Produces a quarterly news bulletin and regular briefings on broadcasting issues. Holds its own archive and those of the former Broadcasting Research Unit (1980–90) and BACTV (British Action for Children's Television). Maintains a panel of speakers, the VLV Forum for Children's Broadcasting, the VLV Forum for Educational Broadcasting, and acts as secretariat for the European Alliance of Listeners' and Viewers' Associations (EURALVA). VLV has responded to all major public enquiries on broadcasting since 1984. The VLV does not handle complaints.

Wales Literature Exchange/Cyfnewidfa Lên Cymru

Mercator Institute for Media, Languages and Culture, Aberystwyth University, Llanbadarn Campus, Aberystwyth SY23 3AS
📞 01970 622544   📠 01970 621524
post@walesliterature.org
post@cyfnewidfalen.org
www.walesliterature.org
www.cyfnewidfalen.org
Director *Sioned Puw Rowlands*

Founded 2000. Wales Literature Exchange was formerly known as Welsh Literature Abroad. Works to facilitate the translation of Wales's literature and to promote exchange between the culture and literatures of Wales and the wider world. Translation grants are available to publishers. Participates in international book fairs, and works with translators, publishers and festivals in Wales and abroad.

WATCH
▷ **Writers, Artists and their Copyright Holders**

Welsh Academy
▷ **Academi**

Welsh Books Council (Cyngor Llyfrau Cymru)

Castell Brychan, Aberystwyth SY23 2JB
📞 01970 624151   📠 01970 625385
castellbrychan@wbc.org.uk
www.wbc.org.uk
www.gwales.com
Director *Gwerfyl Pierce Jones*

Founded 1961 to stimulate interest in books from Wales and to support publishing in Wales. The Council distributes publishing grants and promotes and fosters all aspects of book production in Wales in both Welsh and English. Its Editorial, Design, Marketing and Children's Books departments and wholesale distribution centre offer central services to publishers in Wales. Writers in Welsh and English are welcome to approach the Editorial Department for advice on how to get their manuscripts published or visit the Welsh Book Trade Info website (www.wbti.org.uk).

Welsh Literature Abroad
▷ **Wales Literature Exchange/Cyfnewidfa Lên Cymru**

Welsh National Literature Promotion Agency
▷ **Academi**

West Country Writers' Association

6 The Beals, Greenway, Woodbury, Exeter EX5 1LU
📞 01395 233753
secretarywcwa@aol.com
www.westcountrywriters.com
Secretary *Sue Bury*

Founded 1951 in the interest of published authors living in or writing about the West Country. Annual congress and newsletters. Regional meetings to discuss news and views.

Women in Publishing

info@wipub.org.uk
www.wipub.org.uk
Liaison Officer *Gill Rowley*

Aims to promote the status of women working in publishing and related trades by helping them to develop their careers. Through WiP there are opportunities for members to learn more about every area of the industry, share information and expertise, give and receive support and take part in a practical forum for the exchange of information about different disciplines and industry sectors. Monthly meetings with speakers provide a forum for discussion. Meetings are held on the second Wednesday of each month. See website for further information. Publishes monthly newsletter, *WiPlash*, and *Women in Publishing Directory*.

Writers and Photographers unLimited

27 Camwood, Clayton-le-Woods, Bamber Bridge PR5 8LA
☎ 01772 321243 ⊞ 0870 137 8888
tm@wpu.org.uk
www.wpu.org.uk
Contact *Terry Marsh*
SUBSCRIPTION £52 p.a.

An online marketing service for English-language professional writers and photographers worldwide who specialize in all aspects of travel, tourism, the outdoors, adventure sports, food and drink. The USP of WPu is that all its members are full-time professionals. Promotes the work of its members by direct contact with public relations companies, tourist boards, editors, publishers and consultants and, through its website, regular news bulletins and press releases.

The Writers' Guild of Great Britain

40 Rosebery Avenue, London EC1R 4RX
☎ 020 7833 0777 ⊞ 020 7833 4777
erik@writersguild.org.uk
www.writersguild.org.uk
President *David Edgar*
Chair *Katharine Way*
General Secretary *Bernie Corbett*
Deputy General Secretary *Anne Hogben*

SUBSCRIPTION Full member: 1% of earnings from professional writing (min. £150, max. £1,500); Candidate member: £100; Student member: £20; Affiliate member (agent or writers' group): £275
Founded in 1959 as the Screenwriters' Guild, the Writers' Guild is a trade union, affiliated to the TUC, representing professional writers in television, radio, theatre, film, books and new media. It negotiates Minimum Terms Agreements governing writers' contracts and covering minimum fees; advances; repeat fees; royalties and residuals; rights; credits; number of drafts; script alterations and the resolution of disputes. The most important MTAs cover BBC TV Drama; BBC Radio Drama; ITV Companies; **Pact** (independent TV and film producers); TAC (Welsh language independent TV producers); Theatrical Management Association; **Independent Theatre Council**; and an agreement covering the **Royal National Theatre, Royal Shakespeare Company**, and **Royal Court Theatre**. These agreements are regularly renegotiated and in most cases the minimum fees are reviewed annually.

The Guild advises members on all aspects of their working lives, including contract vetting, legal advice, help with copyright problems and representation in disputes with producers, publishers or other writers.

The Guild organizes seminars, discussions and lectures on all aspects of writing. *UK Writer*, the Guild's quarterly magazine, is free to members and contains features about professional writing, Guild news and details of work opportunities, training courses, literary competitions, etc. A detailed e-mail of news, opportunities and other information is sent to members every Friday.

Full Membership is open to anyone who has received payment for a piece of written work under a contract with terms not less than those negotiated by the Guild. Writers who do not qualify can join as Candidate Members and those on accredited writing courses or theatre attachments can become Student Members.

Writers, Artists and their Copyright Holders (WATCH)

The Library, The University of Reading, PO Box 223, Whiteknights, Reading RG6 6AE
☎ 0118 378 8783 ⊞ 0118 378 6636
www.watch-file.com
Contact *Dr David Sutton*

Founded 1994. Provides an online database of information about the copyright holders of literary authors and artists. The database is available free of charge on the internet and the web. WATCH is the successor project to the Location Register of English Literary Manuscripts and Letters, and continues to deal with location register enquiries.

Yachting Journalists' Association

36 Church Lane, Lymington SO41 3RB
☎ 01590 673894
sec@yja.co.uk
www.yja.co.uk

Contact *Secretary*
SUBSCRIPTION £40 p.a.

To further the interest of yachting, sail and power, and to provide support and assistance to journalists in the field; current membership is just over 260 with 31 from overseas. A handbook, listing details of members and subscribing PR organizations, press facility recommendations, fo\

rthcoming events and other useful information, is published in alternate years, available from the Secretary at the address above. The YJA organizes the Yachtsman of the Year and Young Sailor of the Year Awards, presented annually at the beginning of January.

Yr Academi Gymreig

▷ **Academi**

Literary Societies

Margery Allingham Society

39 Morgan Street, London E3 5AA
www.margeryallingham.org.uk
Honorary Treasurer *Paula Cooze*
SUBSCRIPTION £14 p.a.

Founded in 1988 to promote interest in and study
of the works of Margery Allingham. The Society
publishes two issues of the journal, *The Bottle
Street Gazette*, per year. Contributions welcome.
Two social events a year. Open membership.

Jane Austen Society

9 Nicola Close, South Croydon CR2 6NA
www.janeaustensociety.org.uk
Secretary *Mrs Maureen Stiller*
Membership Secretary *Mrs Rosemary Culley*
SUBSCRIPTION UK: £7.50 (student for 3 years); £15
(standard annual); £12 (annual, members of JAS
branches/groups); £20 (joint); £45 (corporate);
£250 (life); Overseas: £18 (annual) or £50/€75 (for
3 years); £50 (corporate); £300/€450 (life) all by
credit card or UK banker's order only

Founded in 1940 to promote interest in and
enjoyment of Jane Austen's novels and letters.
The Society has the following Branches/Groups:
Bath and Bristol, Cambridge, Midlands, London,
Norfolk, Kent, Hampshire, Isle of Wight,
Southern Circle, Wales and Scotland. There
are independent Societies in North America,
Australia and Japan.

The Baskerville Hounds (The Dartmoor Sherlock Holmes Study Group)

6 Bramham Moor, Hill Head, Fareham PO14 3RU
☎ 01329 667325
baskervillehounds@acd-221b.info
www.acd-221b.info
Chairman *Philip Weller*

Founded 1989. An international Sherlock Holmes
society specializing solely in studies of *The Hound
of the Baskervilles* and its Dartmoor associations.
Organizes many study functions, mostly on
Dartmoor. Currently involved in several large-
scale research projects. Membership is open only
to those who are willing to be actively involved, by
correspondence and/or attendance at functions,
in the group's study activities.

The BB Society

Secretary: 8 Park Road, Solihull B91 3SU
☎ 0121 704 1002
enquiries@roseworldproductions.com
www.roseworldproductions.com
Chairman *Brian Mutlow*
Secretary *Bryan Holden*
SUBSCRIPTION £10 (individual); £17.50 (family);
£50 (corporate); £5 (student); £17.50 (overseas)

Founded in 2000 to bring together devotees
of the writer/illustrator BB (Denys Watkins-
Pitchford). BB's oeuvre included books on
wildfowling, angling, children's fantasy and the
countryside. The Society holds two meetings a
year and publishes newsletters and an annual
journal, *Sky Gypsy*.

The Beckford Society

The Timber Cottage, Crockerton, Warminster
BA12 8AX
☎ 01985 213195
sidney.blackmore@btinternet.com
Secretary *Sidney Blackmore*
SUBSCRIPTION £20 (minimum) p.a.

Founded 1995 to promote an interest in the life
and works of William Beckford (1760–1844)
and his circle. Encourages Beckford studies and
scholarship through exhibitions, lectures and
publications, including an annual journal, *The
Beckford Journal* and occasional newsletters.

Thomas Lovell Beddoes Society

9 Amber Court, Belper DE56 1HG
☎ 01773 828066   📠 01773 828066
john@beddoes.demon.co.uk
www.phantomwooer.org
Chairman *John Lovell Beddoes*
Secretary *Christine Hunkinson*
Journal Editor *Shelley Rees (srees@ossm.edu)*

Formed to research the life, times and work
of poet Thomas Lovell Beddoes (1803–1849),
encourage relevant publications, further the
reading and appreciation of his works by a
wider public and liaise with other groups and
organizations. Publishes an annual journal.

The Adrian Bell Society

Apple Acre, Church Lane, Claxton NR14 7HY
☎ 01508 480665
Chairman *John Ford*
SUBSCRIPTION £5 p.a.

Founded in 1995, the Society aims to promote the study and enjoyment of the writing of Adrian Bell (1901–80); to publish journals twice a year and hold meetings; to help members obtain copies of out of print titles; to assemble a collection of photocopies of his 1500 *Countryman's Notebooks* that appeared in the *Eastern Daily Press* and to index them; to commemorate his career as *The Times* first crossword compiler; to share knowledge of the places and factual background mentioned in his books. The society has 250 members worldwide who keep in touch through the journal. *A Centenary Countryman's Notebook*, published 2001.

Hilaire Belloc Society

1 Hillview, Elsted, Midhurst GU29 0JX
☎ 01730 825575
hilairebelloc1@aol.com
Contact *Dr Grahame Clough*
SUBSCRIPTION £12 p.a. (UK); $20 (US)

Founded in 1996 to promote the life and work of Hilaire Belloc through the publication of newsletters containing rare and previously unpublished material. Organizes walks, social events, a Belloc weekend and a three-day conference. Currently 70 members, of which 30% are from overseas.

Arnold Bennett Society

4 Field End Close, Trentham, Stoke on Trent ST4 8DA
arnoldbennettscty@btinternet.com
www.arnoldbennettsociety.org.uk
Secretary *Mrs Carol Gorton*
SUBSCRIPTION £12 p.a. (single); £14 (family) plus £2 for membership outside the UK

Aims to promote interest in the life and works of 'Five Towns' author Arnold Bennett and other North Staffordshire writers. Annual dinner. Regular functions and talks in and around Burslem, plus annual seminar at Wedgwood College, Barlaston and annual conference at the Potteries Museum & Art Gallery, Stoke on Trent.

E.F. Benson Society

The Old Coach House, High Street, Rye TN31 7JF
☎ 01797 223114
www.efbensonsociety.org
Secretary *Allan Downend*
SUBSCRIPTION £7.50 (UK/Europe); £12.50 (overseas)

Founded in 1985 to promote the life and work of E.F. Benson and the Benson family. Organizes social and literary events, exhibitions, talks and Benson interest walks in Rye. Publishes a quarterly newsletter and annual journal, *The Dodo*, postcards and reprints of E.F. Benson articles and short stories in a series called 'Bensoniana' plus other books of Benson interest. Holds an archive which includes the Seckersen Collection (transcriptions of the Benson collection at the Bodleian Library in Oxford). Also sells second-hand Benson books and organizes walks round Benson's Rye.

The Betjeman Society

6 St Annes Road, Shrewsbury SY3 6AU
☎ 01743 350372
honsecbetjeman@fsmail.net
www.johnbetjeman.com
Honorary Secretary *Colin Wright*
Chairman *David Pattison*
SUBSCRIPTION £10 (individual); £12 (family); £3 (student); £3 additional for overseas members

Aims to promote the study and appreciation of the work and life of Sir John Betjeman. Annual programme includes poetry readings, lectures, discussions, visits to places associated with Betjeman, and various social events. Meetings are held in London and other centres. Regular newsletter and annual journal, *The Betjemanian*.

The Bewick Society

c/o The Hancock Museum, Newcastle upon Tyne NE2 4PT
☎ 0191 232 6386
nhsn@ncl.ac.uk
www.bewicksociety.org.uk
Membership Secretary *June Holmes*
SUBSCRIPTION £10 p.a. (individual); £12 (family)

Founded in 1988 to promote an interest in the life and work of Thomas Bewick, wood-engraver and naturalist (1753–1828). Organizes related events and meetings, and is associated with the Bewick birthplace museum at Cherryburn, Northumberland.

Birmingham Central Literary Association

23 Arden Grove, Ladywood, Birmingham B16 8HG
☎ 0121 454 9352
bruce@bakerbrum.co.uk.
Contact *The Secretary*

Holds fortnightly meetings in central Birmingham to discuss the lives and work of authors and poets. Holds an annual dinner to celebrate Shakespeare's birthday.

The George Borrow Society

60 Upper Marsh Road, Warminster BA12 9PN
www.clough5.fsnet.co.uk/gb.html
Chairman/Bulletin Editor *Dr Ann M. Ridler*
Membership Secretary *Michael Skillman*
Honorary Treasurer *David Pattinson*

Subscription £15 p.a.

Founded in 1991 to promote knowledge of the life and works of George Borrow (1803–81), traveller, linguist and writer. The Society holds biennial conferences (with published proceedings) and informal intermediate gatherings, all at places associated with Borrow. Publishes the *George Borrow Bulletin* twice yearly, containing scholarly articles, reviews of publications relating to Borrow, reports of past events and news of forthcoming events. Member of the **Alliance of Literary Societies** and corporate associate member of the Centre of East Anglian Studies (CEAS) at the University of East Anglia, Norwich (Borrow's home city for many years).

Elinor Brent-Dyer
▷ Friends of the Chalet School and The New Chalet Club

British Fantasy Society

23 Mayne Street, Hanford, Stoke on Trent ST4 4RF
secretary@britishfantasysociety.org
www.britishfantasysociety.org
Honorary President *Ramsey Campbell*
Chairman *Guy Adams*
Secretary *Helen Hopley*
Subscription from £30 p.a.

Founded in 1971 for devotees of fantasy, horror and related fields in literature, art and the cinema. Publishes a regular newsletter with information and reviews of new books and films, plus related fiction and non-fiction magazines. Annual conference at which the **British Fantasy Awards** are presented. These awards are voted on by the membership and are not an open competition.

The Brontë Society

Brontë Parsonage Museum, Haworth, Keighley BD22 8DR
☎ 01535 640195   🖷 01535 647131
bronte@bronte.org.uk
www.bronte.info
Contact *Membership Officer*
Subscription Details from the Membership Officer or the website

Founded 1893. Aims and activities include the preservation of manuscripts and other objects related to or connected with the Brontë family, and the maintenance and development of the museum and library at Haworth. The Society holds regular meetings, lectures and exhibitions; publishes information relating to the family and a triannual *Gazette*. Freelance contributions for either publication should be sent to the Publications Secretary at the address above. Members can receive the journal, *Brontë Studies*, at a reduced Subscription.

The Rupert Brooke Society

40 Belmore Road, Thorpe, Norwich NR7 0PT
☎ 01603 438603
chairman@rupertbrooke.com
www.rupertbrooke.com
Chairman *Lorna Beckett*
Membership Secretary *Wes Smith (membership@rupertbrooke.com)*
Subscription £10 (UK); £15 (overseas)

Founded in 1999 to foster an interest in the work of Rupert Brooke, help preserve places associated with him and to increase the knowledge and appreciation of the village of Grantchester. Members receive an annual journal with in-depth, academic articles relating to Brooke, and a newsletter with news, information about events, book reviews and activities.

The Browning Society

50 Bendemeer Road, London SW15 1JU
☎ 020 8788 0199
v.l.greenaway@btinternet.com
Honorary Secretary *Dr Vicky Greenaway*
Subscription £15 p.a.

Founded in 1969 to promote an interest in the lives and poetry of Robert and Elizabeth Barrett Browning. Meetings are arranged in the London area, one of which occurs in December at Westminster Abbey to commemorate Robert Browning's death.

The John Buchan Society

'The Toft', 37 Waterloo Road, Lanark ML11 7QH
☎ 01555 662103
glennismac2000@yahoo.co.uk
www.johnbuchansociety.co.uk
Secretary *Mrs Glennis McClemont*
Membership Secretary *Diana Durdon*
Subscription £15 (full/overseas); £120 (10 years)

To perpetuate the memory of John Buchan and to promote a wider understanding of his life and works. Holds regular meetings and social gatherings, publishes a journal, and liaises with the John Buchan Centre at Broughton in the Scottish borders.

The Robert Burns World Federation Ltd

Dean Castle Country Park Dower House, Kilmarnock KA3 1XB
☎ 01563 572469   🖷 01563 572469
admin@robertburnsfederation.com
www.worldburnsclub.com
Chief Executive *Shirley Bell*
Office Administrator *Margaret Craig*
Subscription £25 p.a.(individual); £27 (family); £46 (club subscription)

Founded in 1885 to encourage interest in the life and work of Robert Burns and keep alive the old Scottish Tongue. The Society's interests go beyond Burns himself in its commitment to the development of Scottish literature, music and arts in general. Publishes the triannual *Burns Chronicle*.

Randolph Caldecott Society

Blue Grass Cottage, Clatterwick Lane, Little Leigh, Northwich CW8 4RJ
☏ 01606 891303 (day)/781731 (evening)
Honorary Secretary *Kenneth N. Oultram*
SUBSCRIPTION £10–£15 p.a.

Founded in 1983 to promote the life and work of artist/book illustrator Randolph Caldecott. Meetings held in the spring and autumn in Caldecott's birthplace, Chester. Guest speakers, outings, newsletter, exchanges with the Society's American counterpart. (Caldecott died and was buried in St Augustine, Florida.) A medal in his memory is awarded annually in the US for children's book illustration.

Lewis Carroll Society

69 Cromwell Road, Hertford SG13 7DP
☏ 01992 584530
alanwhite@tesco.net
www.lewiscarrollsociety.org.uk
Secretary *Alan White*
SUBSCRIPTION Details on the website

Founded in 1969 to bring together people with an interest in Charles Dodgson and promote research into his life and works. Publishes bi-annual journal, *The Carrollian*, featuring scholarly articles and reviews; a newsletter (*Bandersnatch*) which reports on Carrollian events and the Society's activities; and *The Lewis Carroll Review*, a book-reviewing journal. Regular meetings held in London with lectures, talks, outings, etc.

Lewis Carroll Society (Daresbury)

Blue Grass Cottage, Clatterwick Lane, Little Leigh, Northwich CW8 4RJ
☏ 01606 891303 (day)/781731 (evening)
Honorary Secretary *Kenneth N. Oultram*
SUBSCRIPTION £7–10 p.a.

Founded 1970. To promote the life and work of Charles Dodgson, author of the world-famous *Alice's Adventures*. Holds meetings in the spring and autumn in Carroll's birthplace, Daresbury, Cheshire. Guest speakers, a newsletter and theatre visits.

The New Chalet Club

18 Nuns Moor Crescent, Newcastle upon Tyne NE4 9BE
www.newchaletclub.co.uk
Contact *Membership Secretary*
SUBSCRIPTION £6 p.a. (UK under 18); £10 (UK adult & Europe); £12 (RoW)

Founded in 1995 for all those with an interest in the books of Elinor Brent-Dyer. Publishes a quarterly journal and occasional supplements, including *Children's Series Fiction*, and regularly holds local and national meetings.

Friends of the Chalet School

4 Rock Terrace, Coleford, Bath BA3 5NF
☏ 01373 812705
focs@rockterrace.demon.co.uk
www.chaletschool.org.uk
Contacts *Ann Mackie-Hunter, Clarissa Cridland*
SUBSCRIPTION Details on the website or on application

Founded in 1989 to promote the works of Elinor Brent-Dyer. The society has members worldwide; publishes four magazines a year and runs a lending library.

The Chesterton Society UK

11 Lawrence Leys, Bloxham, Near Banbury OX15 4NU
☏ 01295 720869/07766 711984 ☏ 01295 720869
Honorary Secretary *Rev. Deacon Robert Hughes, KCHS*
SUBSCRIPTION £12.50 p.a.

Founded in 1964 to promote the ideas and writings of G.K. Chesterton.

The Children's Books History Society

26 St Bernards Close, Buckfast, Buckfastleigh TQ11 0EP
☏ 01364 643568
cbhs@abcgarrett.demon.co.uk
Chair/Membership Secretary *Mrs Pat Garrett*
SUBSCRIPTION £10 p.a. (UK/Europe); write for overseas subscription details

Established 1969. Aims to promote an appreciation of children's books and to study their history, bibliography and literary content. Meetings held in London and the provinces, plus a summer meeting to a collection, or to a location with a children's book connection. Three substantial newsletters issued annually, sometimes with an occasional paper. Review copies should be sent to Newsletter co-editor *Mrs Pat Garrett* (address above). The Society constitutes the British branch of the Friends of the Osborne and Lillian H. Smith Collections in Toronto, Canada, and also liaises with **CILIP** (formerly The Library Association). In 1990, the Society established its biennial Harvey Darton Award for a book, published in English, which extends our knowledge of some aspect of British children's literature of the past. 2008 joint winners: Brian Alderson and Felix de Marez Oyens *Be Merry and Wise: Origins of Children's Book Publishing in England 1650–1850*; Tomoko Masaki *A History of Victorian Popular Picture Books*.

The John Clare Society

9 The Chase, Ely CB6 3DR
☎ 01353 668438
www.johnclare.org.uk human.ntu.ac.uk/clare/
clare.html
Honorary Secretary *Miss Sue Holgate*
SUBSCRIPTION £12 (individual); £15 (joint); £10
(retired); £15 (group); £17 (library); £5 (student,
full-time); £17 sterling draft/$34/€28 (eurocheque)
(overseas)

Founded in 1981 to promote a wider appreciation
of the life and works of the poet John Clare
(1793–1864). Organizes an annual festival in
Helpston in July; arranges exhibitions, poetry
readings and conferences; and publishes an
annual society journal and quarterly newsletter.

William Cobbett Society

3 Park Terrace, Tillington, Petworth GU28 9AE
☎ 01798 342008
www.williamcobbett.org.uk
Chairman *Barbara Biddell*
Also: Boynell House, Outlands Lane, Curdridge,
Southampton SO30 2HD ☎ 1489 782453
Contact *David Chun*
SUBSCRIPTION £8 p.a.

Founded in 1976 to bring together those with an
interest in the life and works of William Cobbett
(1763–1835) and to extend the interest to a wider
public. Society activities include an annual
Memorial Lecture; publication of an annual
journal (*Cobbett's New Register*) containing
articles on various aspects of his life and times;
an annual expedition retracing routes taken by
Cobbett on his Rural Rides in the 1820s; visits to
his birthplace and his tomb in Farnham, Surrey.
In association with the Society, the Museum
of Farnham holds bound volumes of *Cobbett's
Political Register*, a large collection of Cobbett's
works, books about Cobbett, and has various
Cobbett artefacts on display.

The Friends of Coleridge

87 Richmond Road, Montpelier, Bristol BS6 5EP
☎ 0117 942 6366
gcdd@blueyonder.co.uk
www.friendsofcoleridge.com
Website Manager *Paul Cheshire (74 Wells Road,
Bath BA2 3AR; friendscol@btconnect.com)*
Editor (Coleridge Bulletin) *Graham Davidson*

Founded in 1987 to advance knowledge about the
life, work and times of Samuel Taylor Coleridge
and his circle, and to support his Nether Stowey
Cottage, with the National Trust, as a centre
of Coleridge interest. Holds literary evenings,
study weekends and a biennial international
academic conference. Publishes *The Coleridge
Bulletin* biannually. Short articles on Coleridge-
related topics may be sent to the editor at the

(Bristol) address above. For membership and
SUBSCRIPTION details, see website.

Wilkie Collins Society

4 Ernest Gardens, London W4 3QU
☎ 020 8747 0115
paul@wilkiecollins.org
www.wilkiecollins.org
Chairman *Andrew Gasson*
Membership Secretary *Paul Lewis (at address
above)*
SUBSCRIPTION £10 (UK/Europe); £18 (RoW;
remittance must be made in UK sterling or by
PayPal to paul@paullewis.co.uk)

Founded in 1980 to provide information on and
promote interest in the life and works of Wilkie
Collins, one of the first English novelists to deal
with the detection of crime. *The Woman in White*
appeared in 1860 and *The Moonstone* in 1868.
Publishes newsletters, reprints of Collins' work
and an annual academic journal.

The Arthur Conan Doyle Society

PO Box 1360, Ashcroft, British Columbia, Canada
V0K 1A0
☎ 001 250 453 2045   📠 001 250 453 2075
sirhenry@telus.net
www.ash-tree.bc.ca/acdsocy.html
Joint Organizers *Christopher Roden,
Barbara Roden*

Founded in 1989 to promote the study and
discussion of the life and works of Sir Arthur
Conan Doyle. Occasional meetings, functions and
visits. Publishes an occasional journal together
with reprints of Conan Doyle's writings.

Joseph Conrad Society (UK)

c/o The Polish Social and Cultural Association
(POSK), 238–246 King Street, London W6 0RF
theconradian@aol.com
www.josephconradsociety.org
Honorary Secretary *Hugh Epstein*
Treasurer/Editor (*The Conradian*) *Allan Simmons*
SUBSCRIPTION £25 p.a. (individual)

Founded in 1973 to promote the study of the
works and life of Joseph Conrad (1857–1924).
A scholarly society, supported by the Polish
Library at the Polish Cultural Association where
a substantial library of Conrad texts and criticism
is held by the **Polish Library**. Publishes a journal
of Conrad studies (*The Conradian*) biannually and
holds an annual International Conference in the
first week of July. Awards an essay prize and travel
and study grants to scholars on application.

The Dartmoor Sherlock Holmes Study Group
▷ **The Baskerville Hounds**

The Rhys Davies Trust

10 Heol Don, Whitchurch, Cardiff CF14 2AU
☏ 029 2062 3359
Contact *Professor Meic Stephens*

Founded in 1990 to perpetuate the literary reputation of the Welsh prose writer, Rhys Davies (1901–78), and to foster Welsh writing in English. Organizes competitions in association with other bodies such as the **Welsh Academy**, puts up plaques on buildings associated with Welsh writers, offers grant-aid for book production, etc.

The Walter de la Mare Society

3 Hazelwood House, New River Crescent, London N13 5RE
☏ 020 8886 1771
fguthrie@talktalk.net
Honorary President *Professor John Bayley, CBE*
Honorary Secretary & Treasurer *Frances Guthrie (3 Hazelwood House, New River Crescent, London N13 5RE; 020 8886 1771)*
SUBSCRIPTION £15 p.a.

Founded in 1997 to honour the memory of Walter de la Mare; to promote the study and deepen the appreciation of his works; to widen the readership of his works; to facilitate research by making available the widest range of contacts and information about de la Mare; and to encourage and facilitate new Walter de la Mare publications. Produces a regular newsletter and organizes events. Membership information from the address above.

The Dickens Fellowship

The Charles Dickens Museum, 48 Doughty Street, London WC1N 2LX
☏ 020 7405 2127
dickens.fellowship@btinternet.com
www.dickens.fellowship.org
Joint Honorary General Secretaries *Mrs Lee Ault, Mrs Joan Dicks*

Founded 1902. The Society's particular aims and objectives are: to bring together lovers of Charles Dickens; to spread the message of Dickens, his love of humanity ('the keynote of all his work'); to remedy social injustice for the poor and oppressed; to assist in the preservation of material and buildings associated with Dickens. Annual conference. Publishes journal called *The Dickensian* (founded 1905 and available at special rate to members) and organizes a full programme of lectures, discussions, visits and conducted walks throughout the year. Branches worldwide.

Dymock Poets
▷ The Friends of the Dymock Poets

Early English Text Society

Lady Margaret Hall, Oxford OX2 6QA

☏ 01865 286581
www.eets.org.uk
Executive Secretary *Professor Vincent Gillespie (at address above)*
Editorial Secretary *Dr H.L. Spencer (Exeter College, Oxford OX1 3DP)*
Membership Secretary *Mrs J.M. Watkinson (12 North End, Durham DH1 4NJ)*
SUBSCRIPTION £20 p.a. (UK); $40 (US)

Founded 1864. Concerned with the publication of early English texts. Members receive annual publications (one or two a year) or may select titles from the backlist in lieu.

The George Eliot Fellowship

12 Fair Isle Drive, Nuneaton CV10 7LJ
blueyorkshirecat@yahoo.co.uk
www.george-eliot-fellowship.com
Secretary *Mrs E. Mellor*

Founded 1930. Exists to honour George Eliot and promote interest in her life and works. Readings, memorial lecture, birthday luncheon and functions. Issues newsletters and an annual journal. Awards an annual prize for a George Eliot essay. Contact the Secretary for SUBSCRIPTION details.

Rev. G. Bramwell Evens
▷ The Romany Society

The John Meade Falkner Society

Greenmantle, Main Street, Kings Newton, Melbourne DE73 8BX
☏ 01332 865315
www.johnmeadefalknersociety.co.uk
Secretary *Kenneth Hillier*
SUBSCRIPTION £5 (UK); $10 (overseas)

Founded in 1999 to promote the appreciation and study of John Meade Falkner's life, times and works. Produces three newsletters a year and an annual journal.

Folly (Fans of Light Literature for the Young)

21 Warwick Road, Pokesdown, Bournemouth BH7 6JW
☏ 01202 432562
folly@sims.abel.co.uk
www.follymagazine.co.uk
Contact *Mrs Sue Sims*
SUBSCRIPTION £9 p.a. (UK); £12 (Europe); £15 (RoW)

Founded in 1990 to promote interest in a wide variety of collectable children's authors, with a bias towards writers of girls' books and girls' school stories. Publishes three magazines a year. Please note that *Folly* does not review any fiction other than girls' school stories, nor any non-fiction save that which is relevant to the history of children's books. 'We do not publish any fiction of any kind.'

The Ford Madox Ford Society

c/o Dr Sara Haslam, English Department, The
Open University, Milton Keynes MK7 6AA
☏ 01908 653453   🖷 01908 653750
s.j.haslam@open.ac.uk
www.open.ac.uk/Arts/fordmadoxford-society/
Chair *Dr Sara Haslam*
Treasurer *Paul Skinner*
SUBSCRIPTION £15 (£12 by standing order) or £8
concession

Founded in 1996 to promote the works of Ford
Madox Ford and to increase knowledge of
his writing and impact on writing in the 20th
century. Meets for academic conferences and
more popular events annually. Distributes
International Ford Madox Ford Studies free to
members. Welcomes all with an interest in Ford
and his works. Has members in the UK, USA,
Italy, France and Germany and holds events in as
many different places as possible.

The Franco-Midland Hardware Company

6 Bramham Moor, Hill Head, Fareham PO14 3RU
☏ 01329 667325
fmhc@acd-221b.info
www.acd-221b.info
Chairman *Philip Weller*

Founded 1989. A federation of societies in four
continents. Two core groups, '221B' (Holmesian/
Sherlockian) and 'The Arthur Conan Doyle Study
Group' (Doylean), form the UK base. Membership
is open only to those interested in being actively
involved in the programme of correspondence
studies and/or functions. Arranges certificated
self-study projects and assists group and
individual research programmes, plus scholarly
gatherings.

The Friars' Club

33 Grenville Court, Chorleywood WD3 5PZ
☏ 01923 283795
petermccall.westlodge@btinternet.com
Secretary *Frances-Mary Blake*
SUBSCRIPTION £10 (first year); £8.50 (renewals)

Founded in 1982 to promote the writings of
Frank Richards (Charles Hamilton), author of
the Greyfriars School stories in *The Magnet* and
creator of Billy Bunter. The Club produces a
quarterly magazine for members. Articles and
comments are welcomed.

The Friends of Shandy Hall (The Laurence Sterne Trust)

Shandy Hall, Coxwold, York YO61 4AD
☏ 01347 868465   🖷 01347 868465
shandyhall@dsl.pipex.com
www.shandean.org
Curator *Patrick Wildgust*

SUBSCRIPTION £7 (annual); £70 (life)

Promotes interest in the works of Laurence
Sterne and aims to preserve the house in which
they were created (open to the public). Publishes
annual journal, *The Shandean*. An Annual
Memorial Lecture is delivered each summer.
Regular exhibitions in gallery. Bookshop.

The Friends of the Dymock Poets

16 Robinson's Meadow, Ledbury HR8 1SE
cateluck2003@yahoo.com
www.dymockpoets.co.uk
Chairman *Roy Palmer*
Hon. Secretary *Catharine Lumby*
Membership Secretary *Jeff Cooper*
SUBSCRIPTION £7 (individual); £12 (couple); £3
(student); £12 (society/family)

Founded 1993. Established to foster an interest in
the work of the Dymock Poets – Edward Thomas,
Robert Frost, Wilfrid Gibson, Rupert Brooke,
John Drinkwater, Lascelles Abercrombie; help
preserve places and things associated with them;
keep members informed of literary and other
matters relating to the poets; increase knowledge
and appreciation of the landscape between May
Hill (Gloucestershire) and the Malvern Hills.
Members are offered lectures, poetry readings,
social meetings, newsletters and annual journal,
guided walks in the countryside around Dymock;
links with other literary societies; and annual
event to commemorate the first meeting between
Edward Thomas and Robert Frost on 6th October
1913.

The Gaskell Society

Far Yew Tree House, Chester Road, Tabley,
Knutsford WA16 0HN
☏ 01565 634668
joanleach@aol.com
www.gaskellsociety.co.uk
Honorary Secretary *Joan Leach*
SUBSCRIPTION £17 p.a.; £22 (joint); £24 (overseas
& institutions)

Founded in 1985 to promote and encourage the
study and appreciation of the life and works of
Elizabeth Cleghorn Gaskell. Meetings held in
Knutsford, Manchester, Bath, London and York;
annual journal and biannual newsletter. On
alternate years holds either a residential weekend
conference or visit.

The Ghost Story Society

PO Box 1360, Ashcroft, British Columbia, Canada
V0K 1A0
☏ 001 250 453 2045   🖷 001 250 453 2075
nebuly@telus.net
www.ash-tree.bc.ca/GSS.html
Joint Organizers *Barbara Roden, Christopher
Roden*

SUBSCRIPTION UK: £25 (airmail); $40 (US$); $45 (Canadian$)

Founded 1988. Devoted mainly to supernatural fiction in the literary tradition of M.R. James, Walter de la Mare, Algernon Blackwood, E.F. Benson, A.N.L. Munby, R.H. Malden, etc. Publishes a biannual journal, *All Hallows*, which includes new fiction in the genre and non-fiction of relevance to the genre.

The Robert Graves Society

3 rue des Grands Augustins, 75006 Paris, France
☎ 00 33 1 43 29 33 48
dunstanward@yahoo.com
www.robertgraves.org/
President *Professor Dunstan Ward*
Vice-President (Europe) *Fran Brearton (f.brearton@qub.ac.uk)*
Secretary/Treasurer *Patrick Villa (pjvilla@aol.com)*
SUBSCRIPTION £17, €30, $40 p.a.

The Society aims to promote interest in, and research on, the life and works of Robert Graves (1895–1985), author of some 140 books of poetry, fiction, biography, criticism, anthropology, social history, mythology, biblical studies, translation and children's books. Organizes an international conference every two years. The 2010 conference will be held in Palma and Deià, Mallorca, 6th–10th July. Publishes annually the leading journal of Graves studies, *Gravesiana*, edited by Professor Dunstan Ward and Dr Fran Brearton. The Society's membership of more than 200 experts and enthusiasts forms the core of an international research community.

Graham Greene Birthplace Trust

17 North Road, Berkhamsted HP4 3DX
☎ 01442 866694
secretary@grahamgreenebt.org
www.grahamgreenebt.org
Secretary *John Partridge*
SUBSCRIPTION £10 (UK, £25 for 3 years); £14 (Europe, £33); £18 (RoW, £45)

Founded on 2nd October 1997, the 93rd anniversary of Graham Greene's birth, to promote the appreciation and study of his works. Publishes a quarterly newsletter, occasional papers, videos and compact discs. Organizes the annual four-day Graham Greene Festival during the weekend nearest to the writer's birthday and administrates the Graham Greene Memorial Awards.

The Ivor Gurney Society

65 Coombe Road, Irby, Wirral CH61 4UW
☎ 0151 648 4957
rolf@rolfjordan.com
www.ivorgurney.org.uk
General Secretary *Rolf Jordan*

SUBSCRIPTION £14 (individual); £18 (joint); £12.50 (group); £10 (retired, individual); £12.50 (retired, joint); £5 (student)

Founded in 1995 to make Ivor Gurney's music and poetry available to a wider audience by way of performances, readings, conferences, recordings and publications; to enhance and promote informed scholarship in all aspects of his life and work through the publication of a regular newsletter and an annual journal, available free to Society members; and to assist the Gloucestershire Archives in maintaining the Ivor Gurney collection. Publishes occasional papers including unpublished letters, poems and compositions. Society events are held each spring and autumn when talks and recitals enable members to explore Gurney's verse and music.

Rider Haggard Society

27 Deneholm, Whitley Bay NE25 9AU
☎ 0191 252 4516   📠 0191 252 4516
rb27allen@blueyonder.co.uk
www.riderhaggardsociety.org.uk
Contact *Roger Allen*
SUBSCRIPTION £9 p.a. (UK); £12 (overseas)

Founded in 1985 to promote appreciation of the life and works of Sir Henry Rider Haggard, English novelist, 1856–1925. News/books exchange and annual meetings. Vast source for Haggard purchases, research and information.

James Hanley Network

Old School House, George Green Road, George Green, Wexham SL3 6BJ
☎ 01753 578632
gostick@london.com
www.jameshanley.inspiron.co.uk/INDEX.HTM
Network Coordinator *Chris Gostick*

An informal international association founded in 1997 for all those interested in exploring and publicizing the works and contribution to literature of the novelist and dramatist James Hanley (1901–1985). Publishes an annual newsletter. Occasional conferences are planned for the future. All enquiries welcome.

The Thomas Hardy Society

PO Box 1438, Dorchester DT1 1YH
☎ 01305 251501   📠 01305 251501
info@hardysociety.org
www.hardysociety.org
Honorary Secretary *Mike Nixon*
SUBSCRIPTION £18 (individual); £25 (corporate); £22.50 (individual, overseas); £30 (corporate, overseas)

Founded in 1967 to promote the reading and study of the works and life of Thomas Hardy. Thrice-yearly journal, events and a biennial conference. An international organization.

The Hazlitt Society

c/o The Guardian, Kings Place, 90 York Way,
London N1 9AG
correspondence@williamhazlitt.org
www.williamhazlitt.org
Founding President *Michael Foot*
Contact *Helen Hodgson (Secretary and Correspondent)*

Established to encourage appreciation of Hazlitt's work and to promote his values. The *Hazlitt Review*, launched in 2008, aims to foster and sustain the restitution of Hazlitt as a canonical author and to stimulate new scholarship in his work.

The Henty Society

205 Icknield Way, Letchworth Garden City
SG6 4TT
davidwalmsley@hentysociety.org
www.hentysociety.org
Honorary Secretary *David Walmsley*
SUBSCRIPTION £15 p.a. (UK); £18 (overseas)

Founded in 1977 to study the life and work of George Alfred Henty, and to publish research, bibliographical data and lesser-known works, namely short stories. Organizes conferences and social gatherings in the UK and North America, and publishes bulletins to members. Published in 1996: *G.A. Henty (1832–1902) a Bibliographical Study* by Peter Newbolt (2nd edition with addenda and corrigenda 2005).

James Hilton Society

49 Beckingthorpe Drive, Bottesford, Nottingham
NG13 0DN
www.jameshiltonsociety.co.uk
Honorary Secretary *J.R. Hammond*
SUBSCRIPTION £10 (UK/EU); £7 concession

Founded in 2000 to promote interest in the life and work of novelist and scriptwriter James Hilton (1900–1954). Publishes *The Hiltonian* (annually) and *The James Hilton Newsletter* (quarterly) and organizes meetings and conferences.

Sherlock Holmes

▷ The Baskerville Hounds and The Franco-Midland Hardware Company

Sherlock Holmes Society (The Musgraves)

Hallas Lodge, Greenside Lane, Cullingworth,
Bradford BD13 5AP
℡ 01535 273468
hallaslodge@btinternet.com
Contact *Anne Jordan*
SUBSCRIPTION £17 p.a.; £18 (joint)

Founded in 1987 to promote enjoyment and study of Sir Arthur Conan Doyle's Sherlock Holmes through publications and meetings. One of the largest Sherlock Holmes societies in Great Britain. Honorary members include Bert Coules, Edward Hardwicke and Clive Merrison. Past honorary members: Dame Jean Conan Doyle, Peter Cushing, Jeremy Brett, Michael Williams and Richard Lancelyn Green. Open membership. Lectures, presentations and consultation on matters relating to Holmes and Conan Doyle available.

Housman Society

80 New Road, Bromsgrove B60 2LA
℡ 01527 874136
info@housman-society.co.uk
www.housman-society.co.uk
Contact *Jim Page*
SUBSCRIPTION £10 (UK); £15 (overseas)

Founded in 1973 to promote knowledge and appreciation of the lives and work of A.E. Housman and other members of his family, and to promote the cause of literature and poetry. Sponsors a lecture at **The Guardian Hay Festival** each year under the title of 'The Name and Nature of Poetry'. Publishes an annual journal and biannual newsletter.

W.W. Jacobs Appreciation Society

3 Roman Road, Southwick BN42 4TP
℡ 01273 596217
Contact *A.R. James*

Founded in 1988 to encourage and promote the enjoyment of the works of W.W. Jacobs, and stimulate research into his life and works. No SUBSCRIPTION charge. Material available for purchase includes *W.W. Jacobs*, a biography published in 1999, price £12, post paid, and *WWJ Book Hunter's Field Guide*, a narrative bibliography published in 2001, price £6, post paid. An expanded biography is to be published in 2009.

Richard Jefferies Society

Pear Tree Cottage, Longcot SN7 7SS
℡ 01793 783040
R.Jefferies_Society@tiscali.co.uk
www.bath.ac.uk/~lissmc/rjeffs.htm
Honorary Secretary *Jean Saunders*
Membership Secretary *Mrs Margaret Evans*
SUBSCRIPTION £7 p.a. (individual); £8 (joint); life membership for those over 60

Founded in 1950 to promote understanding of the work of Richard Jefferies, nature/country writer, novelist and mystic (1848–87). Produces newsletters, reports and an annual journal; organizes talks, discussions and readings. Library and archives. Assists in maintaining the museum in Jefferies' birthplace at Coate near Swindon. Membership applications should be sent to

Margaret Evans, 23 Hardwell Close, Grove, Nr Wantage OX12 0BN.

Jerome K. Jerome Society

c/o Fraser Wood, Mayo and Pinson, 15/16 Lichfield Street, Walsall WS1 1TS
☎ 01922 629000   🖷 01922 721065
tonygray@jkj.demon.co.uk
www.jeromekjerome.com
Honorary Secretary *Tony Gray*
SUBSCRIPTION £12 p.a. (individual); £8 (student); £15 (family); £30 (corporate)

Founded in 1984 to stimulate interest in Jerome K. Jerome's life and works (1859–1927). Meetings, lectures, events and a newsletter, *Idle Thoughts*. Annual dinner in Walsall near Jerome's birth date (2nd May).

The Captain W.E. Johns Appreciation Society

221 Church Road, Northolt UB5 5BE
☎ 07836 680048
brian@zhong-ding.com
Contact (Derby meeting) *Brian Woodruff (at address above)*
Contact (Twyford) *Joy Tilley (01785 240299)*

Society for the appreciation of W.E. Johns, creator of Biggles. Meets twice a year in Derby and Twyford, near Reading (see above for contacts). Biannual magazine, *Biggles Flies Again* (Editor *John Lester*, 01708 446297; johnstuartlester@hotmail.co.uk).

Johnson Society

Johnson Birthplace Museum, Breadmarket Street, Lichfield WS13 6LG
☎ 01543 264972
sjmuseum@lichfield.gov.uk
www.lichfieldrambler.co.uk
Chairman *Mary Baker*
SUBSCRIPTION £10 p.a.; £15 (joint); £100/£150 (life: individual/joint)

Founded in 1910 to encourage the study of the life, works and times of Samuel Johnson (1709–1784) and his contemporaries. The Society is committed to the preservation of the Johnson Birthplace Museum and Johnson memorials.

Johnson Society of London

255 Baring Road, Grove Park, London SE12 0BQ
☎ 020 8851 0173
info@johnson.org.uk
www.johnsonsoc.org.uk
President *Lord Harmsworth*
Honorary Secretary *Mrs Z.E. O'Donnell*
SUBSCRIPTION £20 p.a.; £25 (joint); £15 (student)

Founded in 1928 to promote the knowledge and appreciation of Dr Samuel Johnson and his works. Publishes an annual journal, *New Rambler* and occasional newsletter, *The New Idler*. Regular

meetings from October to April in the meeting room of Wesley's Chapel, City Road, London EC1 on the second Saturday of each month, and a commemoration ceremony around the anniversary of Johnson's death (December) held in Westminster Abbey.

The David Jones Society

22 Gower Road, Sketty, Swansea SA2 9BY
☎ 01792 206144   🖷 01792 470385
anne.price-owen@davidjonessociety.org
www.davidjonessociety.org
Contact *Anne Price-Owen*
SUBSCRIPTION £20 (individual); £35 (corporate)

Founded in 1996, the Society aims to promote and encourage knowledge of the painter-poet David Jones. The Society hosts annual scholarly meetings. The annual SUBSCRIPTION includes a copy of the *David Jones Journal*. Unsolicited material relating to David Jones, art, literature or any of his philosophies may be sent for consideration for publication in the journal.

The Just William Society

Easter Badbea, Dundonnell IV23 2QX
philandpaula@easter-badbea.co.uk
Secretary/Treasurer *Paula Cross*
SUBSCRIPTION £10 p.a. (UK); £12 (overseas); £5 (juvenile/student); £15 (family)

Founded in 1994 to further knowledge of Richmal Crompton's *William* and *Jimmy* books. The *Just William Society Magazines* are published throughout the year and an annual 'William' meeting is held in April.

The Keats–Shelley Memorial Association (Inc)

Registered office: Bedford House, 76A Bedford Street, Leamington Spa CV32 5DT
☎ 01926 427400   🖷 01926 335133
Contact *Honorary Secretary*
SUBSCRIPTION £15 p.a.

Founded in 1903 to promote appreciation of the works of Keats and Shelley, and their circle. One of the Society's main tasks is the preservation of 26 Piazza di Spagna in Rome as a memorial to the British Romantic poets in Italy, particularly Keats and Shelley. Publishes an annual review of Romantic Studies called the *Keats-Shelley Review*, arranges events and lectures for Friends and promotes bursaries and competitive writing on Romantic Studies (see **Keats-Shelley Prize** under *Prizes*). The *Review* is edited by *Professor Nicholas Roe*, c/o Dept. of English, St Andrews University KY16 9AL.

Kenny/Naughton Society

Aghamore, Ballyhaunis,, Co. Mayo, Republic of Ireland
☎ 00 353 94 936 7019

paulwdrogers@eircom.net
www.aghamoreireland.com/kennynaughton/
    society.htm
Chairman *Paul Rogers*

Founded in 1993 for those with a special
interest in the works of two famous writers
with Aghamore links – P.D. Kenny and Bill
Naughton (author of *Alfie*). Society members
go on regular outings and meet monthly with
the view to planning for the Annual Kenny/
Naughton Autumn School, a literary gathering
which honours the memory of Kenny and
Naughton through lectures, talks, readings,
workshops, writing competitions, drama, tours
and entertainment.

Kent & Sussex Poetry Society

39 Rockington Way, Crowborough TN6 2NJ
☎ 01892 662781
john@kentandsussexpoetrysociety.org
www.kentandsussexpoetrysociety.org
Publicity Secretary *John Arnold*
SUBSCRIPTION £15 p.a. (full); £10 (concessionary;
country members living farther afield, senior
citizen, under-16, unemployed)

Founded in 1946 to promote the enjoyment of
poetry. Monthly meetings are held in Tunbridge
Wells, including readings by major poets, a
monthly workshop and an annual writing retreat
week. Publishes an annual folio of members' work
based on Members' Competition, adjudicated
and commented upon by a major poet. Runs an
annual **Open Poetry Competition** (see entry
under *Prizes* and details on website) and Saturday
workshops twice a year with leading poets.

The Kilvert Society

30 Bromley Heath Avenue, Downend, Bristol
BS16 6JP
☎ 0117 957 2030
www.communigate.co.uk/here/kilvertsociety
Honorary Secretary *Alan Brimson*
SUBSCRIPTION £12 p.a.; £15 (two persons at same
address)

Founded in 1948 to foster an interest in the Diary
of the Rev. Francis Kilvert (1870–1879) and the
countryside he loved. The Society publishes three
journals each year and during the spring and
summer holds three weekends of walks, talks and
commemoration services.

The Kipling Society

6 Clifton Road, London W9 1SS
☎ 020 7286 0194   📠 020 7286 0194
jane@keskar.fsworld.co.uk
www.kipling.org.uk
Honorary Secretary *Jane Keskar*

SUBSCRIPTION £24 p.a. (£22 p.a. by standing
order)

Founded 1927. This is a literary society for
all who enjoy the prose and verse of Rudyard
Kipling (1865–1936) and are interested in his
life and times. The Society's main activities are:
maintaining a specialized library at the City
University in Islington, London; answering
enquiries from the public (schools, publishers,
writers and the media); arranging a regular
programme of lectures, especially in London
and in Sussex, and an annual luncheon with
guest speaker; maintaining a small museum and
reference at The Grange, in Rottingdean near
Brighton. The Society also publishes a quarterly
journal. Please contact the Secretary by letter,
telephone, fax or e-mail for further information.

The Kitley Trust

Woodstock, Litton Dale, Litton SK17 8QL
☎ 01298 871564
stottie2@waitrose.com
Contact *Rosie Ford*

Founded in 1990 by a teacher in Sheffield to
promote the art of creative writing, in memory
of her mother, Jessie Kitley. Activities include:
biannual poetry competitions; a 'Get Poetry' day
(distribution of children's poems in shopping
malls); annual sponsorship of a writer for a
school; campaigns; organizing conferences
for writers and teachers of writing. Funds are
provided by donations and profits (if any) from
competitions.

The Charles Lamb Society

BM ELIA, London WC1N 3XX
Chairman *Nicholas Powell*
SUBSCRIPTION UK: £12 (single); £18 (double); £18
(corporate)

Founded 1935. Publishes studies of the life, works
and times of Charles Lamb and his circle. Holds
meetings in London five times a year. *The Charles
Lamb Bulletin* is published four times a year
(Editor: *Professor Richard S. Tomlinson* e-mail:
british_romanticism@comcast.net).

Lancashire Authors' Association

Heatherslade, 5 Quakerfields, Westhoughton,
Bolton BL5 2BJ
☎ 01942 791390
eholt@cwctv.net
General Secretary *Eric Holt*
SUBSCRIPTION £10 p.a.; £13 (joint); £1 (junior)

Founded in 1909 for writers and lovers of
Lancashire literature and history. Aims to foster
and stimulate interest in Lancashire history
and literature as well as in the preservation of
the Lancashire dialect. Meets three times a
year on Saturday at various locations. Publishes

a quarterly journal called *The Record*, which is issued free to members, and holds annual competitions (open to members only) for both verse and prose. Comprehensive library with access for research to members.

The Philip Larkin Society

PO Box 44, Hornsea HU18 1WP
☎ 01964 533324
www.philiplarkin.com
General Secretary *Andrew Eastwood*
SUBSCRIPTION £20 (full rate); £14 (unwaged/ senior citizen); £6 (student); £50 (institutions)

Founded in 1995 to promote awareness of the life and work of Philip Larkin (1922–1985) and his literary contemporaries; to bring together all those who admire Larkin's work as a poet, writer and librarian; to bring about publications on all things Larkinesque. Organizes a programme of events ranging from lectures to rambles exploring the countryside of Larkin's schooldays and publishes a biannual journal, *About Larkin*.

The D.H. Lawrence Society

24 Briarwood Avenue, Nottingham NG3 6JQ
☎ 0115 950 3008
Secretary *Ron Faulks*
SUBSCRIPTION £14; £12 (concession); £16 (European); £19 (RoW); £50 (institutions)

Founded in 1974 to increase knowledge and the appreciation of the life and works of D.H. Lawrence. Monthly meetings, addressed by guest speakers, are held in the library at Eastwood (birthplace of DHL). Organizes visits to places of interest in the surrounding countryside, supports the activities of the D.H. Lawrence Centre at Nottingham University, and has close links with DHL Societies worldwide. Publishes two newsletters and one journal each year, free to members.

The Leamington Literary Society

52 Newbold Terrace East, Leamington Spa CV32 4EZ
☎ 01926 425733
Honorary Secretary *Mrs Margaret Watkins*
SUBSCRIPTION £10 p.a.

Founded in 1912 to promote the study and appreciation of literature and the arts. Holds regular meetings every second Tuesday of the month (from September to June) at the Royal Pump Rooms, Leamington Spa. Also smaller groups which meet regularly to study poetry and the modern novel. The Society has published various books of local interest.

Lewes Monday Literary Club

c/o 1c Prince Edward's Road, Lewes BN7 1BJ
☎ 01273 478512

chris.lutrario@btinternet.com
Contact *Christopher Lutrario*
SUBSCRIPTION £20 p.a.; £5 (guest, per meeting)

Founded in 1948 for the promotion and enjoyment of literature. Seven meetings are held during the winter generally on the last Monday of each month (from October to April) at the Pelham House Hotel in Lewes. The Club attracts speakers of the highest quality and a balance between all forms of literature is aimed for. Guests are welcome to attend meetings.

The George MacDonald Society

38 South Vale, Upper Norwood, London SE19 3BD
☎ 020 8653 8768
r.lines878@btinternet.com
www.george-macdonald.com
Chairman *Richard Lines*
Membership Enquiries *Roger Bardet (r.bardet@ hotmail.co.uk)*
SUBSCRIPTION £10 p.a. (UK); £11 (overseas); US$18 (N. America)

Founded in 1980 to promote interest in the life and works of George MacDonald (1824–1905) and his friends and contemporaries. Organizes conferences, meetings, lectures and outings. Annual newsletter, *Orts*, and an academic journal, *North Wind*, published by St Norbert College, De Pere, WA, USA, are distributed free to members. (Registered Charity No: 1024021.)

The Friends of Arthur Machen

Apt 5, 26 Hervey Road, Blackheath, London SE3 8BS
www.machensoc.demon.co.uk
Contact *Jeremy Cantwell*
SUBSCRIPTION £17 p.a. (UK); £20 (Europe/RoW); $36 (US)

Brings together those who appreciate Machen's writings and generally promotes discussion and research. Publishes two journals and two newsletters annually, plus occasional publications for the membership only.

The Marlowe Society

9 Middlefield Gardens, Hurst Green Road, Halesowen B62 9QH
☎ 0121 421 1482   ☎ 0121 421 1482
www.marlowe-society.org
Membership Secretary *Frieda Barker*
Treasurer *Bruce Young*
SUBSCRIPTION £15 p.a. (individual); £10 p.a. (pensioner/student/unwaged); £20 p.a. (overseas); £200 (group); £20 (joint/couple)

Founded 1955. Holds meetings, lectures and discussions, stimulates research into Marlowe's life and works, encourages production of his plays and publishes a biannual newsletter. At a ceremony on 11th July 2002 the Society celebrated their success in establishing a memorial to the playwright and poet in Poets' Corner in Westminster Abbey.

The John Masefield Society

The Frith, Ledbury HR8 1LW

☎ 01531 633800

ies.sas.ac.uk/cmps/Projects/Masefield/Society/jmsws.htm

Chairman *Peter Carter*

SUBSCRIPTION £5 p.a. (individual); £8 (family, institution, library); £10 (overseas); £2.50 (junior, student)

Founded in 1992 to stimulate the appreciation of and interest in the life and works of John Masefield (Poet Laureate 1930–1967). The Society is based in Ledbury, the Herefordshire market town of his birth and holds various public events in addition to publishing a journal and occasional papers.

William Morris Society

Kelmscott House, 26 Upper Mall, Hammersmith, London W6 9TA

☎ 020 8741 3735   ℻ 020 8748 5207

uk@morrissociety.org

www.morrissociety.org

Curator *Helen Elletson*

SUBSCRIPTION £18 p.a.; £8 (student); £30 (corporate)

Founded in 1955 to promote interest in the life, work and ideas of William Morris (1834–1896), English poet and craftsman. Organizes events and educational programmes. Members receive newsletters and biannual journal. Archive and library open by appointment. Museum open Thursdays and Saturdays, 2.00 pm to 5.00 pm.

The Neil Munro Society

4 Randolph Road, Broomhill, Glasgow G11 7LG

☎ 0141 339 5425

ronnierenton@googlemail.com

www.neilmunro.co.uk

Secretary *Margaret Renton*

SUBSCRIPTION £13 p.a.; £10 (unwaged); £15 (institutional)

Founded in 1996 to encourage interest in the works of Neil Munro (1863–1930), the Scottish novelist, short story writer, poet and journalist. An annual programme of meetings is held in Glasgow and Munro's home-town of Inveraray. Publishes *ParaGraphs*, a twice-yearly magazine, sponsors reprints of Munro's work and is developing a Munro archive.

Violet Needham Society

c/o 19 Ashburnham Place, London SE10 8TZ

☎ 020 8692 4562

richardcheffins@aol.com

violetneedhamsociety.org.uk

Honorary Secretary *R.H.A. Cheffins*

SUBSCRIPTION £7.50 p.a. (UK & Europe); £11 (RoW)

Founded in 1985 to celebrate the work of children's author Violet Needham and stimulate critical awareness of her work. Publishes thrice-yearly *Souvenir*, the Society journal, with an accompanying newsletter; organizes meetings and excursions to places associated with the author and her books. The journal includes articles about other children's writers of the 1940s and 1950s and on ruritanian fiction. Contributions welcome.

The Edith Nesbit Society

26 Strongbow Road, Eltham, London SE9 1DT

www.edithnesbit.co.uk

Contact *Mrs Marion Kennett*

SUBSCRIPTION £7 p.a.; £9 (joint)

Founded in 1996 to celebrate the life and work of Edith Nesbit (1858–1924), best known as the author of *The Railway Children*. The Society's activities include a regular newsletter, booklets, talks and visits to relevant places.

The Wilfred Owen Association

29 Arthur Road, London SW19 7DN

☎ 020 8947 0476

vcedavis@hotmail.com

www.wilfredowen.org.uk

Chairman *Mrs Meg Crane*

SUBSCRIPTION £10 adult (£8 concessions); £15 couple (£13 concessions); £15 overseas (adult); £20 overseas (couple)

Founded in 1989 to commemorate the life and works of Wilfred Owen by promoting readings, visits, talks and performances relating to Owen and his work. The Association offers practical support for students of literature and future poets through links with education, support for literary foundations and information on historical and literary background material. Membership is international with 380 members. Publishes regular newsletters. Speakers are available.

The Elsie Jeanette Oxenham Appreciation Society

32 Tadfield Road, Romsey SO51 5AJ

☎ 01794 517149

abbeybufo@gmail.com

www.bufobooks.demon.co.uk/abbeylnk.htm

Membership Secretary/Treasurer *Ms Ruth Allen*

Editor (The Abbey Chronicle) *Fiona Dyer* (*fionadyer@btopenworld.com*)

SUBSCRIPTION £7.50 p.a.; enquire for overseas rates

Founded in 1989 to promote the works of Elsie J. Oxenham. Publishes a newsletter for members, *The Abbey Chronicle*, three times a year. 500+ members.

Thomas Paine Society

19 Charles Rowan House, Margery Street, London
WC1X 0EH
☎ 020 7833 1305
barb@thecollectedcity.com
www.thomaspainesocietyuk.org.uk
President *The Rt. Hon. Michael Foot*
Honorary Secretary *Barbara Jacobson*
(at address above)
Treasurer *Stuart Wright*
SUBSCRIPTION (minimum) £15 p.a. (UK); $35
(overseas); £5 (unwaged/pensioner/student)

Founded in 1963 to promote the life and work
of Thomas Paine, and continues to expound
his ideals. Meetings, newsletters, lectures and
research assistance. The Society has members
worldwide and keeps in touch with American
and French Thomas Paine associations. Publishes
magazine, *The Journal of Radical History*, twice
yearly (Editor, *R.W. Morrell*, 43 Eugene Gardens,
Nottingham NG2 3LF) and a newsletter. Holds
occasional exhibitions and lectures, including the
biannual Thomas Paine Memorial Lecture and
the annual Eric Paine Memorial Lecture.

The Beatrix Potter Society

The Lodge, Salisbury Avenue, Harpenden AL5 2PS
☎ 01582 769755
beatrixpottersociety@tiscali.co.uk
www.beatrixpottersociety.org.uk
Membership Secretary *Jenny Akester*
SUBSCRIPTION UK: £20 p.a. (individual); £25
(institution); Overseas: £25 (individual); £30
(institution)

Founded in 1980 to promote the study and
appreciation of the life and works of Beatrix
Potter (1866–1943). Potter was not only the
author of *The Tale of Peter Rabbit* and other
classics of children's literature; she was also a
landscape and natural history artist, diarist,
farmer and conservationist, and was responsible
for the preservation of large areas of the Lake
District through her gifts to the National Trust.
The Society upholds and protects the integrity of
the unique work of Potter, her aims and bequests;
holds regular talks and meetings in London and
elsewhere with visits to places connected with
Beatrix Potter. International Study Conferences
are held in the UK and the USA. The Society
has an active publishing programme. Members
receive a quarterly newsletter. (UK Registered
Charity No. 281198.)

Anthony Powell Society

76 Ennismore Avenue, Greenford UB6 0JW
☎ 020 8864 4095   ☎ 020 8864 6109
secretary@anthonypowell.org
www.anthonypowell.org
Patron *John M.A. Powell*

President *Simon Russell Beale*
Honorary Secretary *Dr Keith C. Marshall*
SUBSCRIPTION £22 p.a. (individual); £28
(overseas); £33 (joint, UK); £39 (joint, oveseas);
£13 (student, UK); £19 (student, overseas); £100
(organization); £5 p.a. supplement for overseas
members

Founded in June 2000 to advance education and
interest in the life and works of the English author
Anthony Powell (1905–2000). The Society's major
activity is a biennial Powell conference, the first
four of which were at Eton College (2001), Balliol
College, Oxford (2003), The Wallace Collection,
London (2005) and Bath (2007). The Society
organizes events for members, ranging from
'pub meets' to talks and visits to places of Powell
interest as well as publishing *Secret Harmonies*
(annual journal) and a quarterly *Newsletter*.
Member of the **Alliance of Literary Societies**.

The Powys Society

Flat D, 87 Ledbury Road, London W11 2AG
☎ 020 7243 0168
chris.d.thomas@hotmail.co.uk
www.powys-society.org
Honorary Secretary *Chris Thomas*
SUBSCRIPTION £18.50 (UK); £22 (overseas); £10
(student)

The Society (with a membership of 297) aims to
promote public education and recognition of the
writings, thought and contribution to the arts of
the Powys family; particularly of John Cowper,
Theodore and Llewelyn, but also of the other
members of the family and their close associates.
The Society holds two major collections of
Powys published works, letters, manuscripts
and memorabilia. Publishes the *Powys Society
Newsletter* in April, June and November and *The
Powys Journal* in August. Organizes an annual
conference as well as lectures and meetings in
Powys places.

The J.B. Priestley Society

Eldwick Crag Farm, Otley Road, Bingley BD16 3BB
☎ 01274 563078
reavill@globalnet.co.uk
www.jbpriestley-society.com
President *Tom Priestley*
Chairman *Lee Hanson*
Honorary Secretary *R.E.Y. Slater*
Membership Secretary *Tony Reavill*
SUBSCRIPTION £12 (individual); £7 (concession);
£18 (group/family)

Founded in 1997 to widen the knowledge and
understanding of Priestley's works; promote
the study of his life and his social, cultural and
political influences; provide members of the
Society with lectures, seminars, films, journals
and stimulate education projects; promote public

performances of his works and the distribution of material associated with him. For further information, contact the Membership Secretary at the address above.

The Barbara Pym Society

St Hilda's College, Oxford OX4 1DY
☎ 01865 373753
eileen.roberts@st-hildas.ox.ac.uk
www.barbara-pym.org
Chairman *Clemence Schultz*
Archivist *Yvonne Cocking*
SUBSCRIPTION £18 p.a. (individual); £25 p.a. (household); £10 p.a. (concession)

Aims to promote interest in and scholarly research into the works of Barbara Pym; to bring together like-minded people to enjoy the exploration of all aspects of her novels and to continue to encourage publishers to keep her novels in print. Annual weekend conference at St. Hilda's College in September, focusing on a theme or a particular novel. Annual North American Conference at Harvard Law School. Annual one-day spring meeting in London, with speaker. Publishes a newsletter, *Green Leaves*, biannually. In addition to reports on conferences and society activities, the newsletter includes scholarly papers and various unpublished items from the Pym archives at the Bodleian Library (i.e. short stories).

The Queen's English Society

1 Oban Gardens, Woodley, Reading RG5 3RG
☎ 07766 416900
enquiries@queens-english-society.com
www.queens-english-society.com
Honorary Secretary *Sidney Callis*
SUBSCRIPTION £20 p.a. (ordinary); £25 (joint)

Founded in 1972 to promote and uphold the use of good English and to encourage the enjoyment of the language. Holds regular meetings and conferences to which speakers are invited, an annual luncheon and publishes a quarterly journal, *Quest*, for which original articles are welcome. (Registered Charity No. 272901.)

The Arthur Ransome Society Ltd

Abbot Hall Art Gallery & Museum, Kendal LA9 5AL
tarsinfo@arthur-ransome.org
www.arthur-ransome.org
Trustee Chairman *Florence Galbraith*
Company Secretary *Peter Hyland*
SUBSCRIPTION UK: £5 (junior); £10 (student); £15 (adult); £20 (family); £40 (corporate); Overseas: £5 (junior); £10 (student); £20 (adult); £25

(family) Payable in local currency in US, Canada, Australia, Japan and New Zealand

Founded in 1990 to celebrate the life and to promote the works and ideas of Arthur Ransome, author of *Swallows and Amazons* titles for children, biographer of Oscar Wilde, works on the Russian Revolution and extensive articles on fishing. TARS seeks to encourage children and adults to engage in adventurous pursuits, to educate the public about Ransome and his works, and to sponsor research into his literary works and life.

The Romany Society

10 Haslam Street, Bury BL9 6EQ
☎ 0161 764 7078
phil_shelley@btopenworld.com
www.romanysociety.org.uk
Honorary Secretary *John Thorpe (at address above)*
Publications Officer *Phil Shelley*
SUBSCRIPTION £5 (individual); £9 (family); £10 (institution); £3.50 (student/unwaged)

Promotes the life, work and conservation message of the Rev. G. Bramwell Evens – 'Romany' – the first broadcasting naturalist who influenced generations of listeners with his *Out With Romany* radio programmes on the BBC during the 1930s and 1940s and with his series of natural history books. The Romany Society was established originally following Evens' death at the age of 59 in 1943 and ran until 1965. Revived in 1996, the Society holds an annual members' weekend to areas of significance in the Romany story, erects memorial plaques where appropriate, encourages media coverage of Romany and supports young naturalists with the Romany Memorial Grant.

The Followers of Rupert

31 Whiteley, Windsor SL4 5PJ
☎ 01753 865562
followersofrupert@hotmail.com
www.rupertthebear.org.uk
Membership Secretary *Mrs Shirley Reeves*
SUBSCRIPTION UK: £25 (single); £27 (joint); Europe: £30 (single); £32 (joint); RoW: £35 (single); £37 (joint)

Founded 1983. The Society caters for the growing interest in the Rupert Bear stories, past, present and future. Publishes the *Nutwood Newsletter* quarterly which gives up-to-date news of Rupert and information on Society activities. A national get-together of members – the Followers Annual – is held during the summer.

The Ruskin Society of London

20 Parmoor Court, Summerfield, Oxford OX2 7XB
☎ 01865 310987

Honorary Secretary *Miss O.E. Forbes-Madden*
SUBSCRIPTION £15 p.a.

Founded in 1986 to promote interest in John
Ruskin (1819–1900) and his contemporaries. All
aspects of Ruskinia are introduced. Functions
are held in London. Publishes the annual *Ruskin
Gazette*, a journal concerned with Ruskin's
influence. Affiliated to other literary societies.

The Ruskin Society

49 Hallam Street, London W1W 6JP
☎ 020 7580 1894
cgamble@zen.co.uk
www.lancs.ac.uk/depts/ruskin/links.htm
President *Professor Michael Wheeler*
Chairman *Dr Malcolm Hardman*
Honorary Secretary *Dr Cynthia J. Gamble*
SUBSCRIPTION £10 p.a. (payable on 1st Jan.)

Founded in 1997 to encourage a wider
understanding of John Ruskin and his
contemporaries. Organizes lectures and events
which seek not only to explain to the public at
large the nature of Ruskin's theories but also to
place these in a modern context.

The Malcolm Saville Society

6 Redcliffe Street, London SW10 9DS
mystery@witchend.com
www.witchend.com
Secretary *George Jasieniecki*
SUBSCRIPTION £10 p.a. (UK); £12 (EU); £16 (RoW)

Founded in 1994 to remember and promote
interest in the work of the popular children's
author. Regular social activities, library, contact
directory and four magazines per year.

The Dorothy L. Sayers Society

Rose Cottage, Malthouse Lane, Hurstpierpoint
BN6 9JY
☎ 01273 833444   📠 01273 835988
www.sayers.org.uk
Contact *Christopher Dean*
SUBSCRIPTION £14 p.a. (UK); £16.50 (Europe);
$34 (US)

Founded in 1976 to promote the study of the
life, works and thoughts of Dorothy Sayers;
to encourage the performance of her plays
and publication of her books and books about
her; to preserve original material and provide
assistance to researchers. Acts as a forum and
information centre, providing material for study
purposes which would otherwise be unavailable.
Annual conventions, usually in August and other
meetings. Co-founder of the Dorothy L. Sayers
Centre in Witham. Publishes bi-monthly bulletin,
annual proceedings and other papers. Grants are
currently made to support a young musician and
an actor.

The Shaw Society

Flat F, 10 Compton Road, London N1 2PA
☎ 020 7226 4266
shawsociety@blueyonder.co.uk
Honorary Secretary *Alan Knight*
SUBSCRIPTION £15 p.a. (individual); £22 (joint)

Founded in 1941 to promote interest in the life
and works of G. Bernard Shaw. Meetings are
held on the last Friday of every month (except
July, August and December) at Conway Hall,
Red Lion Square, London WC1 (6.30 pm for
7.00 pm) at which speakers are invited to talk
on some aspect of Shaw's life or works. Monthly
playreadings are held on the first Friday of each
month (except August and January). Over 40
years ago the Society began the annual open-air
performance held at Shaw's Corner, Ayot St
Lawrence in Hertfordshire, on the weekend
nearest to Shaw's birthday (26th July). Publishes a
biannual magazine, *The Shavian*, with a calendar
of Shavian events. (No payment for contributors.)

The Robert Southey Society

1 Lewis Terrace, Abergarwed, Neath SA11 4DL
☎ 01639 711480
Contact *Robert King*
SUBSCRIPTION £10 p.a.

Founded in 1990 to promote the work of Robert
Southey. Publishes an annual newsletter and
arranges talks on his life and work. Open
membership.

The Laurence Sterne Trust
▷ The Friends of Shandy Hall

The Robert Louis Stevenson Club

17 Heriot Row, Edinburgh EH3 6HP
☎ 0131 556 1896   📠 0131 556 1896
mail@stevenson-house.co.uk
Correspondence Secretary *John W.S. Macfie*
SUBSCRIPTION £20 p.a. (individual); £26 p.a.
(overseas); £150 (10-year)

Founded in 1920 to foster interest in Robert Louis
Stevenson's life and works. The Club organizes
an annual lunch and other events. Publishes *RLS
Club News* twice a year.

The Bram Stoker Memorial Association

29 Washington Square West – Penthouse N, New
York City, NY 10011–9180, USA
☎ 001 212 533 5018
bramstoker145@yahoo.com
President/Founder *Dr Jeanne Keyes Youngson*
Executive Director *Jennifer O'Casey*
Vice President (Ireland) *John Moore*
Vice President (USA) *Rev. Edward Connor*

Founded in 1985 by Dr Jeanne Keyes Youngson
who, in 1965, formed the Dracula Fan Club
(now known as The Vampire Empire). News

and announcements of Stoker events and relevant books appear in the club's quarterly newsletter. Society members are invited to join Dr Youngson's 'Bram Stoker Walks' in Dublin, starting at Trinity College and ending at St Michan's Church.

The R.S. Surtees Society

Manor Farm House, Nunney, Near Frome BA11 4NJ
☎ 01373 836937   🖷 01373 836574
rssurtees@fsmail.net
www.r.s.surteessociety.org
Contact *Orders and Membership Secretary (01373 302155)*

Founded in 1979 to republish the works of R.S. Surtees and others.

The Tennyson Society

Central Library, Free School Lane, Lincoln LN2 1EZ
☎ 01522 552862
kathleen.jefferson@lincolnshire.gov.uk
www.tennysonsociety.org.uk
Honorary Secretary *Miss K. Jefferson*
SUBSCRIPTION £10 p.a. (individual); £12 (family); £15 (corporate); £160 (life)

Founded in 1960. An international society with membership worldwide. Exists to promote the study and understanding of the life and work of Alfred, Lord Tennyson. The Society is concerned with the work of the Tennyson Research Centre, 'probably the most significant collection of manuscripts, family papers and books in the world'. Publishes annually the *Tennyson Research Bulletin*, which contains articles and critical reviews; and organizes lectures, visits and seminars. Annual memorial service at Somersby in Lincolnshire.

The Angela Thirkell Society

54 Belmont Park, London SE13 5BN
☎ 020 8244 9339
penny.aldred@ntlworld.com
www.angelathirkellsociety.com
www.angelathirkell.org (N. American branch)
Honorary Secretary *Mrs. P. Aldred*
SUBSCRIPTION £10 p.a.

Founded in 1980 to honour the memory of Angela Thirkell as a writer and to make her works available to new generations. Publishes an annual journal, holds an AGM in the autumn and a spring meeting which usually takes the form of a visit to a location associated with Thirkell. Has a flourishing North American branch which has frequent contact with the UK parent society.

The Dylan Thomas Society of Great Britain

4 Bishops Grove, Swansea SA2 8BE
☎ 01792 206363

Contact *John Rhys Thomas*
SUBSCRIPTION £5 (individual); £8 (two adults from same household)

Founded in 1977 to foster an understanding of the work of Dylan Thomas and to extend members' awareness of other 20th century writers, especially Welsh writers in English. Meetings take place monthly, mainly in Swansea.

The Edward Thomas Fellowship

1 Carfax, Undercliff Drive, St Lawrence, Isle of Wight PO38 1XG
☎ 01983 853366
colingthornton@btopenworld.com
www.edward-thomas-fellowship.org.uk
Hon. Secretary *Colin G. Thornton*
SUBSCRIPTION £10 p.a. (single); £15 p.a. (joint)

Founded in 1980 to perpetuate and promote the memory of Edward Thomas and to encourage an appreciation of his life and work. The Fellowship holds a commemorative birthday walk on the Sunday nearest the poet's birthday (3rd March) and an autumn walk; issues newsletters and holds various events and seminars.

The Tolkien Society

210 Prestbury Road, Cheltenham GL52 3ER
secretary@tolkiensociety.org
www.tolkiensociety.org
Secretary *Sally Kennett*
SUBSCRIPTION £21 p.a. (UK); £24 (EU); £27 (RoW)

An international organization which aims to encourage and further interest in the life and works of the late Professor J.R.R. Tolkien, CBE, author of *The Hobbit* and *Lord of the Rings*. Current membership stands at 1,100. Publishes *Mallorn* annually and *Amon Hen* bi-monthly.

The Trollope Society

Maritime House, Old Town, Clapham, London SW4 0JW
☎ 020 7720 6789   🖷 020 7627 2965
info@trollopesociety.org
www.trollopesociety.org
Contact *Pelham Ravenscroft*

Founded in 1987 to study and promote Anthony Trollope's works. Publishes the complete works of Trollope's novels and travel books.

Wainwright Society

Kendal Museum, Station Road, Kendal LA9 6BT
☎ 01539 721374   🖷 01539 737976
membership@wainwright.org.uk
www.wainwright.org.uk
Membership Secretary *Morag Clement*

SUBSCRIPTION £12 p.a. (per household, from 1st Jan. 2009)

Founded in 2002 'to keep alive the things that Alfred Wainwright (1907–1991) promoted through his guidebooks (*Pictorial Guides to the Lakeland Fells*), started 50 years ago, and the many other publications which were the "labour of love" of a large portion of his life'. Produces a newsletter three times per year, organizes walks, events, annual dinner and lecture.

The Walmsley Society

April Cottage, No 1 Brand Road, Hampden Park, Eastbourne BN22 9PX
☎ 01323 506447
walmsley@haughshw.demon.co.uk
www.walmsleysoc.org
Honorary Secretary *Fred Lane*
SUBSCRIPTION £13 p.a.; £15 (family); £11 (student/senior citizen); £12 (overseas, £20, one year; £35, two years)

Founded in 1985 to promote interest in the writings of Leo Walmsley and to foster an appreciation of the work of his father, the artist Ulric Walmsley. Two annual meetings – one held in Robin Hood's Bay on the East Yorkshire coast, spiritual home of the author Leo Walmsley. Publishes a journal twice-yearly and newsletters, and is involved in other publications which benefit the aims of the Society. A biography of Leo Walmsley is now available.

Sylvia Townsend Warner Society

2 Vicarage Lane, Fordington, Dorchester DT1 1LH
☎ 01305 266028
judithmbond@tiscali.co.uk
www.townsendwarner.com
Contact *Eileen Johnson*
SUBSCRIPTION £10 p.a.; $25 (overseas)

Founded in 2000 to promote a wider readership and better understanding of the writings of Sylvia Townsend Warner.

Mary Webb Society

8 The Knowe, Willaston, Neston CH64 1TA
☎ 07816 424712
suehigginbotham@yahoo.co.uk
www.marywebbsociety.co.uk
Secretary *Sue Higginbotham*
SUBSCRIPTION UK: £11 p.a. (individual); £14 p.a. (joint); Overseas: £14 p.a. (individual); £16 p.a. (joint)

Founded 1972. Attracts members from the UK and overseas who are devotees of the literature of Mary Webb and of the beautiful Shropshire countryside of her novels. Publishes biannual journal, organizes summer schools in various locations related to the authoress's life and works.

Archives; lectures; tours arranged for individuals and groups.

H.G. Wells Society

Flat 3, 27b Church Road, London NW4 4EB
secretaryhgwellsociety@hotmail.com
www.hgwellsusa.50megs.com/
Hon. General Secretary *Mark Egerton*
SUBSCRIPTION £18 (UK/EU); £21 (RoW); £22 (institutional); £12 (concessions)

Founded in 1960 to promote an interest in and appreciation of the life, work and thought of Herbert George Wells. Publishes *The Wellsian* (annual) and *The H.G. Wells Newsletter* (three issues yearly). Organizes meetings and conferences.

The Oscar Wilde Society

22 Edric Road, London SE14 5EL
michael.seeney@btinternet.com
www.oscarwildesociety.co.uk
Chairman *Donald Mead*
Deputy Chairman *Michael Seeney*
SUBSCRIPTION £25 p.a. (UK); £30 p.a. (Europe); £35 p.a. (RoW)

Founded in 1990 to promote knowledge, appreciation and study of the life, personality and works of the writer and wit Oscar Wilde. Activities include meetings, lectures, readings and exhibitions, and visits to locations associated with Wilde. Members receive a journal of Oscar Wilde studies, *The Wildean*, twice-yearly and a newsletter journal *Intentions* (six per year).

The Charles Williams Society

35 Broomfield, Stacey Bushes, Milton Keynes MK12 6HA
☎ 01908 316779
charles_wms_soc@yahoo.co.uk
www.charleswilliamssociety.org.uk
Contact *Honorary Secretary*
SUBSCRIPTION £12.50 p.a.

Founded in 1975 to promote interest in, and provide a means for the exchange of views and information on the life and work of Charles Walter Stansby Williams (1886–1945).

The Henry Williamson Society

7 Monmouth Road, Dorchester DT1 2DE
☎ 01305 264092
zseagull@aol.com
www.henrywilliamson.co.uk
General Secretary *Mrs Sue Cumming*
Membership Secretary *Mrs Margaret Murphy (16 Doran Drive, Redhill RH1 6AX; 01737 763228 mm@misterman.freeserve.co.uk)*
SUBSCRIPTION £12 p.a.; £15 (family); £5 (student)

Founded in 1980 to encourage, by all appropriate means, a wider readership and deeper understanding of the literary heritage left by the 20th-century English writer Henry Williamson (1895–1977). Publishes annual journal.

The P.G. Wodehouse Society (UK)

26 Radcliffe Road, Croydon CR0 5QE
info@pgwodehousesociety.org.uk
www.pgwodehousesociety.org.uk
Membership Secretary *Christine Hewitt*
SUBSCRIPTION £15 p.a. (£20 Dec. to March, for membership extended to May the following year)

Formed in 1997 to promote enjoyment of the works of P.G. Wodehouse, author, lyricist, playwright and journalist, creator of Lord Emsworth, Jeeves & Wooster and many more. Membership of 1,000+. Publications include the quarterly journal, *Wooster Sauce*, and regular *By The Way papers*. Meetings and events are held in London and around the country. The Society is allied to Wodehouse societies around the world.

The Parson Woodforde Society

22 Gaynor Close, Wymondham NR18 0EA
☎ 01953 604124
editor@parsonwoodforde.org
www.parsonwoodforde.org.uk
Membership Secretary *Mrs Ann Elliott*
SUBSCRIPTION £16 (UK); £25 (overseas)

Founded 1968. Aims to extend and develop knowledge of James Woodforde's life and the society in which he lived and to provide the opportunity for fellow enthusiasts to meet together in places associated with the diarist. Publishes a quarterly journal and newsletter. The Society has produced a complete edition of the diary of James Woodforde in 17 volumes covering the period 1759–1802.

The Virginia Woolf Society of Great Britain

Fairhaven, Charnleys Lane, Banks, Southport PR9 8HJ
stuart.n.clarke@btinternet.com
www.virginiawoolfsociety.co.uk
Membership Secretary *Stuart N. Clarke*
SUBSCRIPTION £16 p.a.; £21 (overseas)

Founded in 1998 to promote interest in the life and work of Virginia Woolf, author, essayist and diarist. The Society's activities include trips away, walks, reading groups and talks. Publishes a literary journal, *Virginia Woolf Bulletin*, three times a year.

The Wordsworth Trust

Dove Cottage, Grasmere LA22 9SH
☎ 01539 435544   ☏ 01539 463508
enquiries@wordsworth.org.uk
www.wordsworth.org.uk
Literature Officer *Andrew Forster*

Founded in 1880, the Trust is a living memorial to the life and poetry of William Wordsworth and his contemporaries. As Centre for British Romanticism, the Wordsworth Trust, with its wealth of manuscripts, books, drawings and pictures, provides 'the full context for understanding and celebrating a major cultural moment in history in which Britain played a profound role'. Research facilities are available at the Jerwood Centre; contact: *Jeff Cowton*, Curator. The Wordsworth Trust also runs a Contemporary Poetry Programme. 'Summer Poetry at Grasmere' is a live poetry event that takes place each year (see **The Wordsworth Trust** entry under *Festivals*). (Registered Charity No. 1066184.)

WW2 HMSO PPBKS Society

3 Roman Road, Southwick BN42 4TP
☎ 01273 596217
Contact *A.R. James*

Founded in 1994 to encourage collectors and to promote research into HMSO's World War II series of paperbacks. Most of them were written by well-known authors, though in many cases anonymously. No SUBSCRIPTION charge. Available for purchase: Collectors' Guide (£5); Handbook, *Informing the People* (£10).

The Yeats Society Sligo

Yeats Memorial Building, Hyde Bridge, Sligo, Republic of Ireland
☎ 00 353 71 914 2693   ☏ 00 353 71 914 2780
info@yeats-sligo.com
www.yeats-sligo.com
President *Joe Cox*
SUBSCRIPTION €25 (single); €40 (couple at same address)

Founded in 1958 to promote the heritage of W.B. Yeats and the Yeats family. Attractions include a permanent Yeats Exhibition; films for public viewing throughout the year; the annual Yeats International Summer School in July/August; the Winter School in January/February. The Yeats Festival is also held in the summer, and lectures, poetry readings, workshops, etc. are held in the autumn and spring, together with a variety of outings and other events for members. Visits are made to schools, and contacts are made with Yeats scholars and students worldwide. The 2009 Summer School (25th July to 7th August) is the 50th successive Summer School. 'Early booking is advisable.'

The Charlotte M. Yonge Fellowship

8 Anchorage Terrace, Durham DH1 3DL
www.dur.ac.uk/c.e.schultze/contact.html
Contact *Dr C.E. Schultze*

Founded in 1996 to provide the opportunity for all those who enjoy and admire the work of Charlotte M. Yonge (1823–1901), author of *The Heir of Radclyffe*, to learn more about her life and writings. Holds two meetings annually, one in November in London and the other, the spring AGM, at venues associated with Yonge's life, works or interests, or where buildings or gallery and museum collections relate to Victorian history, art and society. Publishes *Review* twice-yearly and an occasional *Journal*. A loan collection of Yonge's works enables members to borrow publications that are hard to obtain. The Fellowship's archive is held at St Hugh's College, Oxford and material is available to members and *bona fide* researchers.

Yorkshire Dialect Society

51 Stepney Avenue, Scarborough YO12 5BW
www.ydsociety.org.uk
Secretary *Michael Park*

SUBSCRIPTION £10 p.a.

Founded in 1897 to promote interest in and preserve a record of the Yorkshire dialect. Publishes dialect verse and prose writing. Two journals to members annually. Details of publications are available from YDS at address above.

Francis Brett Young Society

92 Gower Road, Halesowen B62 9BT
📞 0121 422 8969
www.fbysociety.co.uk
Honorary Secretary *Mrs Jean Hadley*
SUBSCRIPTION £7 p.a. (individual); £10 (couple sharing a journal); £5 (student); £7 (organization/overseas); £70 (life); £100 (joint, life)

Founded 1979. Aims to provide a forum for those interested in the life and works of English novelist Francis Brett Young and to collate research on him. Promotes lectures, exhibitions and readings; publishes a regular newsletter.

Arts Councils and Regional Offices

Arts Council England

14 Great Peter Street, London SW1P 3NQ
℡ 0845 300 6200 Textphone 020 7973 6564
enquiries@artscouncil.org.uk
www.artscouncil.org.uk
Chairman *Dame Liz Forgan*
Chief Executive *Alan Davey*
Director of Literature Strategy *Antonia Byatt*

Founded 1946. Arts Council England is the national development agency for the arts in England, distributing public money from government and the National Lottery to artists and arts organizations. ACE works independently and at arm's length from government. Information about Arts Council England funding is available on the website, by e-mail or by contacting the enquiry line on 0845 300 6200.

In February 2009, Arts Council England announced plans to reduce staffing by grouping the nine regional offices in four areas: London, East and South East, Midlands and South West, and North. The London office will move to the national offices in Great Peter Street. At the time of writing, the nine regional offices are:

Arts Council England, East
Eden House, 48–49 Bateman Street, Cambridge CB2 1LR
℡ 0845 300 6200 ℻ 0870 242 1271
Textphone 01223 306893

Arts Council England, East Midlands
St Nicholas Court, 25–27 Castle Gate, Nottingham NG1 7AR
℡ 0845 300 6200 ℻ 0115 950 2467

Arts Council England, London
2 Pear Tree Court, London EC1R 0DS
℡ 0845 300 6200 ℻ 020 7608 4100
Textphone 020 7973 6564

Arts Council England, North East
Central Square, Forth Street, Newcastle upon Tyne NE1 3PJ
℡ 0845 300 6200 ℻ 0191 230 1020
Textphone 0191 255 8585

Arts Council England, North West
Manchester House, 22 Bridge Street, Manchester M3 3AB

℡ 0845 300 6200 ℻ 0161 834 6969
Textphone 0161 834 9131

Arts Council England, South East
Sovereign House, Church Street, Brighton BN1 1RA
℡ 0845 300 6200 ℻ 0870 242 1257
Textphone 01273 710659

Arts Council England, South West
Senate Court, Southernhay Gardens, Exeter EX1 1UG
℡ 0845 300 6200 ℻ 01392 498546
Textphone 01392 433503

Arts Council England, West Midlands
82 Granville Street, Birmingham B1 2LH
℡ 0845 300 6200 ℻ 0121 643 7239
Textphone 0121 643 2815

Arts Council England, Yorkshire
21 Bond Street, Dewsbury WF13 1AX
℡ 0845 300 6200 ℻ 01924 466522
Textphone 01924 438585

The Arts Council/An Chomhairle Ealaíon

70 Merrion Square, Dublin 2, Republic of Ireland
℡ 00 353 1 618 0200   ℻ 00 353 1 676 1302
artistsservices@artscouncil.ie
www.artscouncil.ie
Director *Mary Cloake*
Head of Literature *Sarah Bannon*

The Arts Council is the Irish government agency for developing the arts. It provides financial assistance to artists, arts organizations, local authorities and others for artistic purposes. It offers advice and information on the arts to Government and to a wide range of individuals and organizations. As an advocate for the arts and artists, the Arts Council undertakes projects and research, often in new and emerging areas of arts practice, and increasingly in cooperation with partner organizations. Its funding from government for 2009 was €75.7 million.

Of particular interest to individual writers is the Council's free booklet, *Arts Council Awards 2009*, which describes bursaries, awards and schemes on offer and how to apply for them. Applicants to these awards must have been born in, or be resident in, the Republic of Ireland.

The Arts Council of Northern Ireland

MacNeice House, 77 Malone Road, Belfast
BT9 6AQ
☏ 028 9038 5200 🖷 028 9066 1715
dsmyth@artscouncil-ni.org
www.artscouncil-ni.org
Literature and Language Arts Development
Officer *Damian Smyth*

Funds book production by established publishers,
programmes of readings, literary festivals,
mentoring services, writers-in-residence schemes,
network service providers, literary magazines
and periodicals. Annual awards and bursaries
for writers are available. Holds information also
on various groups associated with regional arts,
workshops and courses.

Scottish Arts Council

12 Manor Place, Edinburgh EH3 7DD
☏ 0131 226 6051 🖷 0131 225 9833
help.desk@scottisharts.org.uk
www.scottisharts.org.uk
Acting Chief Executive *Jim Tough*
Head of Literature *Dr Gavin Wallace*
Literature Officers *Aly Barr, Emma Turnbull*
Literature Administrator *Catherine Allan*

Principal channel for government funding of
the arts in Scotland. The Scottish Arts Council
is funded by the Scottish Government and
National Lottery. It aims to develop and improve
the knowledge, understanding and practice
of the arts, and to increase their accessibility
throughout Scotland. It offers grants to artists
and arts organizations concerned with the visual
arts, crafts, dance and mime, drama, literature,
music, festivals and traditional, ethnic and
community arts. Scottish Arts Council's support
for Scottish-based writers with a track record of
publication includes bursaries, writing fellowships
and book awards (see entries under *Bursaries,
Fellowships and Grants* and *Prizes*). Information
offered includes lists of literature awards, literary
magazines, agents and publishers.

The Arts Council of Wales/Cyngor Celfyddydau Cymru

9 Museum Place, Cardiff CF10 3NX
☏ 029 2037 6500 🖷 029 2022 1447
www.artswales.org.uk
www.celfcymru.org.uk

The Arts Council of Wales (ACW) is the
development body for the arts in Wales. ACW's
Creative Wales Awards offer substantial grants
for established writers and playwrights. Funds
the **Welsh Academy** and **Tŷ Newydd Writers'
Centre** to provide services to individual writers,
including bursaries, mentoring, the critical
writers services and writers in residence/on tour.
Responsibility for promoting the publishing and
sales of books from Wales and funding literary
magazines rests with the **Welsh Books Council**
(see entry under *Professional Associations*).

Library Services

Aberdeen Central Library

Rosemount Viaduct, Aberdeen AB25 1GW

☎ 01224 652500   🖷 01224 641985

centrallibrary@aberdeencity.gov.uk

www.aberdeencity.gov.uk

OPEN Central Library Lending Services: 9.00 am
to 7.00 pm Monday to Thursday; 9.00 am to 5.00
pm Friday & Saturday. Business Information,
Community Reference & Local Studies: 9.00 am
to 8.00 pm Monday to Thursday; 9.00 am to 5.00
pm Friday & Saturday. Branch library opening
hours vary.

OPEN ACCESS

General reference and loans. Books, pamphlets,
periodicals and newspapers; videos, CDs, DVDs;
arts equipment lending service; Internet and
Learning Centre for public access; photographs
of the Aberdeen area; census records, maps,
newspapers; online and remote access databases,
patents and standards. The library offers special
services to housebound readers.

Armitt Collection, Museum & Library

Rydal Road, Ambleside LA22 9BL

☎ 015394 41212   🖷 015294 31313

info@armitt.com

www.armitt.com

OPEN Museum: 10.00 am to 5.00 pm every day
(last admission 4.30 pm); Library: times vary but
usually open most weekdays (phone in advance).

A small but unique reference library of rare
books, manuscripts, pictures, antiquarian
prints, maps and museum items, mainly about
the Lake District. It includes early guidebooks
and topographical works, books and papers
relating to Ruskin, H. Martineau, Charlotte
Mason and others; fine art including work by
W. Green, J.B. Pyne, John Harden, K. Schwitters
and Victorian photographs by Herbert Bell; also
a major collection of Beatrix Potter's scientific
watercolour drawings and microscope studies.
Entry charge for museum (£2.50 adults, £2
concessions, £1 children).

The Athenaeum, Liverpool

Church Alley, Liverpool L1 3DD

☎ 0151 709 7770   🖷 0151 709 0418

info@theathenaeum.org.uk

www.theathenaeum.org.uk

OPEN 9.00 am to 4.00 pm Monday and Tuesday;
9.00 am to 9.00 pm Wednesday to Friday

ACCESS To club members; researchers by
application only.

General collection, with books dating from the
15th century, now concentrated mainly on local
history with a long run of Liverpool directories
and guides. SPECIAL COLLECTIONS Liverpool
playbills; William Roscoe; Blanco White; Robert
Gladstone; 18th-century plays; 19th-century
economic pamphlets; Bibles; Yorkshire and other
genealogy. Some original drawings, portraits,
topographical material and local maps.

Bank of England Information Centre

Threadneedle Street, London EC2R 8AH

☎ 020 7601 4715   🖷 020 7601 4356

informationcentre@bankofengland.co.uk

www.bankofengland.co.uk

OPEN 9.00 am to 5.30 pm Monday to
Friday ACCESS For research workers only by
prior arrangement, when material is not readily
available elsewhere.

50,000 volumes of books and periodicals. 2,000
periodicals taken. UK and overseas coverage
of banking, finance and economics. SPECIAL
COLLECTIONS Central bank reports; UK
17th–19th-century economic tracts; government
reports in the field of banking.

Barbican Library

Barbican Centre, London EC2Y 8DS

☎ 020 7638 0569/7638 0568 (24-hr renewals)

🖷 020 7638 2249

barbicanlib@cityoflondon.gov.uk

www.cityoflondon.gov.uk/libraries

Librarian *John Lake, BA, MCLIP*

OPEN 9.30 am to 5.30 pm Monday and
Wednesday; 9.30 am to 7.30 pm Tuesday and
Thursday; 9.30 am to 2.00 pm Friday; 9.30 am to
4.00 pm Saturday

OPEN ACCESS

Situated on Level 2 of the Barbican Centre, this
is the City of London's largest lending library.
Study facilities are available plus free internet

access. In addition to a large general lending department, the library seeks to reflect the Centre's emphasis on the arts and includes strong collections (including DVDs, MP3s, videos, spoken word CDs and CD-ROMs) on painting, sculpture, theatre, cinema and ballet, as well as a large music library with books, scores and CDs (sound recording loans available at a small charge). Also houses the City's main children's library and has special collections on basic skills, materials for young adults, finance, graphic novels and the history of London. Service available for housebound readers. A literature events programme is organized by the Library which supplements and provides cross-arts planning opportunities with the Barbican Centre artistic programme. Reading groups meet in the library on the first Thursday of every month and children's reading group during term time.

Barnsley Public Library

Central Library, Shambles Street, Barnsley S70 2JF
☎ 01226 773930   ℻ 01226 773955
barnsleylibraryenquiries@barnsley.gov.uk
www.barnsley.gov.uk/Libraries
OPEN Lending & Reference: 9.30 am to 7.00 pm Monday and Wednesday; 9.30 am to 5.30 pm Tuesday, Thursday, Friday; 9.30 am to 4.00 pm Saturday. Telephone to check hours of other departments and other branch libraries.
OPEN ACCESS

General library, lending and reference. Archives and Local Studies collection of family history and local material, local authors, Yorkshire and Barnsley collections; large junior library. (Specialist departments are closed on certain weekday evenings and Saturday afternoons.)

BBC Written Archives Centre

Caversham Park, Reading RG4 8TZ
☎ 0118 948 6281   ℻ 0118 946 1145
heritage@bbc.co.uk
www.bbc.co.uk/heritage
Contact *Jacqueline Kavanagh*
OPEN 9.30 am to 5.30 pm Monday to Friday
ACCESS For reference, by appointment only, Wednesday to Friday.

Holds the written records of the BBC, including internal papers from 1922 to the 1990s and published material to date. 20th century biography, social history, popular culture and broadcasting. Charges for certain services.

Beamish, The North of England Open Air Museum
▶ entry under Picture Libraries

Bedford Central Library

Harpur Street, Bedford MK40 1PG
☎ 01234 350931/270102 (Reference Library)

☎ 01234 342163
www.bedfordshire.gov.uk
OPEN 9.00 am to 6.00 pm Monday, Tuesday, Wednesday, Friday; 9.00 am to 1.00 pm Thursday; 9.00 am to 5.00 pm Saturday
OPEN ACCESS

Lending library with a wide range of stock, including books, music, audiobooks, DVDs and videos; reference and information library, children's library, local history library and internet facilities.

Belfast Public Libraries: Central Library

Royal Avenue, Belfast BT1 1EA
☎ 028 9050 9150   ℻ 028 9033 2819
info.belb@ni-libraries.net
www.belb.org.uk
www.ni-libraries.net
OPEN 9.00 am to 8.00 pm Monday to Thursday; 9.00 am to 5.30 pm Friday; 9.00 am to 4.30 pm Saturday
OPEN ACCESS is available to the Central Lending Library and to the stock of the specialist reference departments within the building.

Over two million volumes for lending and reference. SPECIAL COLLECTIONS United Nations depository; complete British Patent Collection; Northern Ireland Newspaper Library; British and Irish government publications. The Central Library provides the following Reference Departments: General Reference; Belfast, Ulster and Irish Studies; Music. A Learning Gateway includes public internet facilities. The Lending Library is one of 19 branches along with a range of outreach services to hospitals, care homes and housebound readers.

BFI National Library

21 Stephen Street, London W1T 1LN
☎ 020 7255 1444   ℻ 020 7436 2338
library@bfi.org.uk
www.bfi.org.uk
OPEN 10.30 am to 5.30 pm Monday and Friday; 10.30 am to 8.00 pm Tuesday and Thursday; 1.00 pm to 8.00 pm Wednesday; Telephone Enquiry Service operates from 10.00 am to 5.00 pm (closed 1.00 pm to 2.00 pm)
ACCESS For reference only; annual, 5-day and limited day membership available.

The world's largest collection of information on film and television including periodicals, cuttings, scripts, related documentation, personal papers. Main library catalogue available via website.

Birmingham and Midland Institute

9 Margaret Street, Birmingham B3 3BS
☎ 0121 236 3591   ℻ 0121 212 4577
admin@bmi.org.uk
www.bmi.org.uk

Administrator & General Secretary *Philip Fisher*
ACCESS Members only.

Established 1854. Later merged with the
Birmingham Library which was founded in 1779.
The Library specializes in the humanities, with
approximately 100,000 volumes in stock. Founder
member of the **Association of Independent
Libraries**. Meeting-place of many affiliated
societies devoted to poetry and literature.

Birmingham Library Services

Central Library, Chamberlain Square,
Birmingham B3 3HQ
☎ 0121 303 4511
central.library@birmingham.gov.uk
www.birmingham.gov.uk/libraries
OPEN 9.00 am to 8.00 pm Monday to Friday; 9.00
am to 5.00 pm Saturday

Over a million volumes. RESEARCH COLLECTIONS
include the Shakespeare Library; War Poetry
Collection; Parker Collection of Children's
Books and Games; Johnson Collection; Milton
Collection; Cervantes Collections; Early and
Fine Printing Collection (including the William
Ridler Collection of Fine Printing); Joseph
Priestley Collection; Loudon Collection;
Railway Collection; Wingate Bett Transport
Ticket Collection; Labour, Trade Union and Co-
operative Collections. PHOTOGRAPHIC ARCHIVES
Sir John Benjamin Stone; Francis Bedford; Francis
Frith; Warwickshire Photographic Survey;
Boulton and Watt Archive. Also, Charles Parker
Archive; Birmingham Repertory Theatre Archive
and Sir Barry Jackson Library; Local Studies
(Birmingham); Patents Collection; Song Sheets
Collection; Oberammergau Festival Collection.

Bournemouth Library

22 The Triangle, Bournemouth BH2 5RQ
☎ 01202 454848   📠 01202 454840
bournemouth@bournemouthlibraries.org.uk
www.bournemouth.gov.uk/libraries
OPEN 10.00 am to 7.00 pm Monday; 9.30 am
to 7.00 pm Tuesday, Thursday, Friday; 9.30 am
to 5.00 pm Wednesday; 10.00 am to 4.00 pm
Saturday
OPEN ACCESS

Main library for Bournemouth with lending,
reference and music departments, plus the
Heritage Zone – local and family history.

Bradford Central Library

Princes Way, Bradford BD1 1NN
☎ 01274 433600   📠 01274 395108
public.libraries@bradford.gov.uk
www.bradford.gov.uk/libraries

OPEN 9.00 am to 7.30 pm Monday to Friday; 9.00
am to 5.00 pm Saturday

Wide range of books and media loan services.
Comprehensive reference and information
services, including major local history collections
and specialized business information service.

Brighton Jubilee Library

Jubilee Street, Brighton BN1 1GE
☎ 01273 290800
libraries@brighton-hove.gov.uk
www.citylibraries.info/libraries/jubilee.asp
OPEN 10.00 am to 7.00 pm Monday and Tuesday;
10.00 am to 5.00 pm Wednesday, Friday and
Saturday; 10.00 am to 8.00 pm Thursday; 11.00
am to 4.00 pm Sunday
ACCESS Stock on open access and in onsite store;
material for reference use and lending

Specializations include art and antiques, history
of Brighton, local illustrations, Hebrew and
Oriental literature, natural history, children's
books, World War Two and large bequests of rare
and historical books.

Bristol Central Library

College Green, Bristol BS1 5TL
☎ 0117 903 7200   📠 0117 922 1081
www.bristol.gov.uk/libraries
OPEN 9.30 am to 7.30 pm Monday, Tuesday and
Thursday; 10.00 am to 5.00 pm Wednesday; 9.30
am to 5.00 pm Friday and Saturday; 1.00 pm to
5.00 pm Sunday
OPEN ACCESS

Lending, reference, art, music, business and
local studies are particularly strong. DVD and
CD collections on site. Facilities available: PCs
(large screen with Jaws and Zoomtext), internet,
printing; videophone on site; black & white and
colour photocopiers.

British Architectural Library

Royal Institute of British Architects, 66 Portland
Place, London W1B 1AD
☎ 020 7580 5533   📠 020 7631 1802
info@inst.riba.org
www.architecture.com
OPEN 10.00 am to 8.00 pm Tuesday; 10.00 am to
5.00 pm Wednesday to Friday; 10.00 am to 1.30
pm Saturday; Closed Sunday, Monday and any
Saturday preceding a Bank Holiday; full details on
the website.
ACCESS Free to RIBA members and non-members
on proof of identity and address; loans available to
RIBA members only.

Collection of books, photographs and periodicals.
All aspects of architecture, current and historical.
Material both technical and aesthetic, covering
related fields including: interior design, landscape
architecture, topography, the construction

industry and applied arts. Brochure available; queries by telephone, letter, e-mail or in person. Charge for research (min. charge £40 + VAT).

RIBA British Architectural Library Drawings & Archives Collections

Drawings and manuscripts can be consulted at the RIBA Study Rooms, Henry Cole Wing, Victoria and Albert Museum: ✆ 020 7307 3708 drawings&archives@inst.riba.org
ACCESS Free to all.

The British Cartoon Archive
▷ entry under Picture Libraries

The British Library

Admission to St Pancras Reading Rooms – British Library Readers' Passes

Everyone is welcome to visit the British Library exhibition galleries or to tour the building. However, to use the reading rooms you will need to apply for a reader's pass, for which identification is required. The British Library issues passes to those who want to use its collections – researchers, innovators and entrepreneurs across all fields of study, and in academic, commerce or personal research – whether or not they are affiliated to a research institution. Two pieces of identification are required (original documents only). One proof of home address, e.g. a utility bill, bank statement or driving licence; and proof of signature, i.e. a bank or credit card, passport, driving licence or national identity card.

It would also be helpful to take with you anything to support your application (such as a student card, business card, professional membership card or details of the items you wish to see).

For further information, visit the website at www.bl.uk or contact Customer Services, The British Library, 96 Euston Road, London NW1 2DB ✆ 01937 546060 ✆ 01937 546333; customer-services@bl.uk

British Library Asia, Pacific and Africa Collections

96 Euston Road, London NW1 2DB
✆ 020 7412 7873 ✆ 020 7412 7641
apac-enquiries@bl.uk
www.bl.uk
OPEN 10.00 am to 5.00 pm Monday; 9.30 am to 5.00 pm Tuesday to Saturday; closed for public holidays.
ACCESS By British Library reader's pass.

An extensive collection of printed volumes and manuscripts in the languages of Africa, the Near and Middle East and all of Asia, plus records of the East India Company and British government in India until 1947. Also prints, drawings and paintings by British artists of India.

For information on British Library collections and services, visit the website.

British Library Business & IP Centre

96 Euston Road, London NW1 2DB
✆ 020 7412 7454 (free enquiry service)
✆ 020 7412 7453 (free enquiry service)
bipc@bl.uk
www.bl.uk/bipc
OPEN 10.00 am to 8.00 pm Monday; 9.30 am to 8.00 pm Tuesday to Thursday; 9.30 am to 5.00 pm Friday and Saturday; closed for public holidays.
ACCESS By British Library reader's pass.

The Business & IP Centre holds the most comprehensive collection of business and intellectual property information in the UK, including market research reports and journals, directories, company annual reports, trade and business journals, up-to-date literature on patents, trade marks, designs and copyright.

British Library Early Printed Collections/ Rare Books and Music Reading Room

96 Euston Road, London NW1 2DB
✆ 020 7412 7564 ✆ 020 7412 7691
rare-books@bl.uk
www.bl.uk/collections/early.html
General enquiries about reader services & advance reservations: ✆ 020 7412 7676 ✆ 01937 546321 reader-services-enquiries@bl.uk
OPEN 10.00 am to 8.00 pm Monday; 9.30 am to 8.00 pm Tuesday to Thursday; 9.30 am to 5.00 pm Friday and Saturday; closed for public holidays.
ACCESS By British Library reader's pass.

Early Printed Collections selects, acquires, researches and provides access to material printed in Britain and Western Europe from the 15th to early 20th centuries. The collections are available in the Rare Books and Music Reading Room at St Pancras which also functions as the focus for the British Library's extensive collection of humanities microforms.

Further information about Early Printed Collections can be found at the British Library website.

British Library Humanities Reading Room

96 Euston Road, London NW1 2DB
✆ 020 7412 7676 ✆ 020 7412 7794
humanities-enquiries@bl.uk
www.bl.uk/resources/humanities
OPEN 10.00 am to 8.00 pm Monday; 9.30 am to 8.00 pm Tuesday, Wednesday, Thursday; 9.30 am to 5.00 pm Friday and Saturday; closed for public holidays.
ACCESS By British Library reader's pass.

This reading room is the focus for the Library's modern collections service in the humanities. It is on two levels, Humanities 1 and Humanities 2 and provides access to the Library's comprehensive collections of books and periodicals in all subjects in the humanities and social sciences and in all languages apart from Oriental. These collections are not available for browsing at the shelf. Material is held in closed access storage and needs to be identified and ordered from store using an online catalogue. A selective open access collection on most humanities subjects can be found in Humanities 1 whilst in Humanities 2 there are open access reference works relating to periodicals and theses, to recorded sound and to librarianship and information science.

To access British Library catalogues, go to the website at http://catalogue.bl.uk

British Library Manuscript Collections

96 Euston Road, London NW1 2DB
☎ 020 7412 7513   📠 020 7412 7745
mss@bl.uk
www.bl.uk
OPEN 10.00 am to 5.00 pm Monday; 9.30 am to 5.00 pm Tuesday to Saturday; closed for public holidays.
ACCESS Reading facilities only, by British Library reader's pass; a written letter of recommendation and advance notice is required for certain categories of material.

Two useful publications, *Index of Manuscripts in the British Library*, Cambridge 1984–6, 10 vols, and *The British Library: Guide to the Catalogues and Indexes of the Department of Manuscripts* by M.A.E. Nickson, help to guide the researcher through this vast collection of manuscripts dating from Ancient Greece to the present day. Approximately 300,000 mss, charters, papyri and seals are housed here.

For information on British Library collections and services and to access British Library catalogues, including the Manuscripts online catalogue, visit the website.

British Library Map Collections

96 Euston Road, London NW1 2DB
☎ 020 7412 7702   📠 020 7412 7780
maps@bl.uk
www.bl.uk/collections/maps
OPEN 10.00 am to 5.00 pm Monday; 9.30 am to 5.00 pm Tuesday to Saturday; closed for public holidays.
ACCESS By British Library reader's pass.

A collection of about 4.5 million maps, charts and globes, manuscript, printed and, increasingly, digital, with particular reference to the history of British cartography. Maps for all parts of the world in a wide range of scales, formats and dates, including the most comprehensive collection of Ordnance Survey maps and plans. SPECIAL COLLECTIONS King George III Topographical

Collection and Maritime Collection, the Crace Collection of maps and plans of London and the cartographic archive of the Ministry of Defence (i.e. GSGS).

For information on British Library collections and services, visit the website.

To access main British Library catalogues, go to the website at http://catalogue.bl.uk

British Library Music Collections

96 Euston Road, London NW1 2DB
☎ 020 7412 7772   📠 020 7412 7751
music-collections@bl.uk
www.bl.uk/collections/music/music.html
OPEN 10.00 am to 8.00 pm Monday; 9.30 am to 8.00 pm Tuesday to Thursday; 9.30 am to 5.00 pm Friday and Saturday; closed for public holidays.
ACCESS By British Library reader's pass.

SPECIAL COLLECTIONS The Royal Music Library (containing almost all Handel's surviving autograph scores), The Zweig Collection of Music & Literary Mss, The Royal Philharmonic Society Archive and the Paul Hirsch Music Library. Also a large collection (about one and a half million items) of printed music (UK via legal deposit) and about 100,000 items of manuscript music, both British and foreign.

The British Library website contains details of collections and services, and provides access to the catalogues.

British Library Newspapers

Colindale Avenue, London NW9 5HE
☎ 020 7412 7353   📠 020 7412 7379
newspaper@bl.uk
www.bl.uk/collections/newspapers.html
OPEN 10.00 am to 5.00 pm Monday to Saturday (last newspaper issued 4.15 pm); closed for public holidays.
ACCESS By British Library reader's pass or Newspaper Library pass (availablefrom and valid only for Colindale). Proof of address and proof of signature required.

Major collections of English provincial, Scottish, Welsh, Irish, Commonwealth and selected overseas foreign newspapers from *c*.1700 are housed here. Some earlier holdings are also available. London newspapers from 1801 and many weekly and fortnightly periodicals are also in stock. (London newspapers pre-dating 1801 are housed at the new library building in St Pancras – 96 Euston Road, NW1 – though many are available at Colindale Avenue on microfilm.) Readers are advised to check availability of material in advance.

For information on British Library Newspapers collections and services, visit the website.

British Library Science, Technology and Business Collections

96 Euston Road, London NW1 2DB
☎ 020 7412 7288 (general enquiries)
🖷 020 7412 7217
scitech@bl.uk
www.bl.uk
Business enquiries: ☎ 020 7412 7454 (Free enquiry service)
OPEN 10.00 am to 8.00 pm Monday; 9.30 am to 8.00 pm Tuesday to Thursday; 9.30 am to 5.00 pm Friday and Saturday; closed for public holidays.
ACCESS By British Library reader's pass.

Engineering, business information on companies, markets and products, physical science and technologies. See also **British Library Business & IP Centre**.

To access British Library catalogues go to the website at http://catalogue.bl.uk

British Library Social Sciences & Official Publications

96 Euston Road, London NW1 2DB
☎ 020 7412 7676   🖷 020 7412 7794
social-sciences@bl.uk
www.bl.uk/collections/social/social.html
OPEN Reading Room: 10.00 am to 8.00 pm Monday; 9.30 am to 8.00 pm Tuesday to Thursday; 9.30 am to 5.00 pm Friday and Saturday; closed for public holidays.
ACCESS By British Library reader's pass.

Provides an information service on the social sciences, law, public administration, and current and international affairs, and access to current and historical official publications from all countries and intergovernmental bodies. Material available in the reading room includes House of Commons sessional papers, UK legislation, UK electoral registers, up-to-date reference books on official publications and on the social sciences, a major collection of statistics and a browsing collection of recent social science books and periodicals; other collections are kept in closed stores and items have to be ordered for delivery to the reading room.

To access British Library catalogues, go to the website at http://catalogue.bl.uk

British Library Sound Archive

96 Euston Road, London NW1 2DB
☎ 020 7412 7676   🖷 020 7412 7441
sound-archive@bl.uk
www.bl.uk/soundarchive
OPEN 10.00 am to 8.00 pm Monday; 9.30 am to 8.00 pm Tuesday to Thursday; 9.30 am to 5.00 pm Friday and Saturday; closed for public holidays.
Listening service (by appointment)
Northern Listening Service
British Library Document Supply Centre, Boston

Spa: 9.15 am to 4.30 pm Monday to Friday
OPEN ACCESS

An archive of over 1,000,000 discs and more than 200,000 hours of tape recordings, including all types of music, oral history, drama, literature, poetry, wildlife, selected BBC broadcasts and BBC Sound Archive material. Produces a twice-yearly newsletter, *Playback*.

For information on British Library Sound Archive collections and services, visit the website.

British Psychological Society Library

c/o Psychology Library, Senate House Library, University of London, Senate House, Malet Street, London WC1E 7HU
☎ 020 7862 8451/8461   🖷 020 7862 8480
enquiries@shl.lon.ac.uk
www.shl.lon.ac.uk
OPEN Term-time: 9.00 am to 9.00 pm Monday to Thursday; 9.00 am to 6.30 pm Friday; 9.30 am to 5.30 pm Saturday (Holidays: 9.00 am to 6.00 pm Monday to Friday; 9.30 am to 5.30 pm Saturday)
ACCESS Members only; non-members £5 day ticket.

Reference library, containing the British Psychological Society collection of periodicals – over 140 current titles housed alongside the University of London's collection of books and journals. Largely for academic research. General queries referred to **Swiss Cottage Library** in London which has a good psychology collection.

Benjamin Britten Collection
▷ **Suffolk County Council – Suffolk Libraries**

Bromley Central Library

High Street, Bromley BR1 1EX
☎ 020 8460 9955   🖷 020 8313 9975
central.library@bromley.gov.uk
www.bromley.gov.uk
OPEN 9.30 am to 6.00 pm Monday, Wednesday, Friday; 9.30 am to 8.00 pm Tuesday and Thursday; 9.30 am to 5.00 pm Saturday
OPEN ACCESS

A large selection of fiction and non-fiction books for loan, both adult and children's. Also DVDs, videos, CDs, cassettes, language courses, open learning packs, CD-ROM and Playstation games for hire. Other facilities include CD-ROMs, computer hire, People's Network Internet, local studies library, 'Upfront' teenage section, large reference library with photocopying, fax, microfiche and film facilities, 'Bromley Knowledge' – online community information, reading groups and Local Links (access to council services). Library and Archives catalogues online, 'Bromley Training Truck' – free computer and basic skills training. SPECIALIST COLLECTIONS include: H.G. Wells, Walter de la Mare, Crystal

Palace, The Harlow Bequest, and the history and geography of Asia, America, Australasia and the Polar regions.

Bromley House Library

Angel Row, Nottingham NG1 6HL
☏ 0115 947 3134
enquiries@bromleyhouse.org
www.bromleyhouse.org
Librarian *Carol Barstow*
OPEN 9.30 am to 5.00 pm Monday to Friday; 9.30 am to 12.30 pm Saturday
ACCESS For members only.

Founded 1816 as the Nottingham Subscription Library. Collection of 35,000 books including local history, topography, biography, travel and fiction.

Robert Burns Collection

▷ The Mitchell Library

CAA Library and Information Centre

Aviation House, Gatwick Airport South, Gatwick RH6 0YR
☏ 01293 573725   ✆ 01293 573181
infoservices@caa.co.uk
www.caa.co.uk
OPEN 9.00 am to 5.00 pm Monday to Friday
OPEN ACCESS

A collection of books, reports, directories, statistics and periodicals on most aspects of civil aviation and related subjects.

Cambridge Central Library (Reference Library & Information Service)

7 Lion Yard, Cambridge CB2 3QD
☏ 0845 045 5225   ✆ 01223 712011
your.library@cambridgeshire.gov.uk
www.cambridgeshire.gov.uk/leisure/libraries

The Central Library is undergoing major redevelopment and is due to re-open in May 2009. During this period a full range of services is available through the network of 31 libraries in the county (details on the website) and by telephone, post, e-mail and online referral.

Large stock of books, periodicals, newspapers, maps, plus comprehensive collection of directories and annuals covering UK, Europe and the world. Microfilm and fiche reading and printing services. Online access to news and business databases. News databases on CD-ROM; internet access. Monochrome and colour photocopiers.

Camomile Street Library

12–20 Camomile Street, London EC3A 7EX
☏ 020 7332 1855
camomile@cityoflondon.gov.uk
www.cityoflondon.gov.uk/camomilestlibrary

OPEN 9.30 am to 5.30 pm Monday, Tuesday, Thursday, Friday; 9.30 am to 6.30 pm Wednesday
OPEN ACCESS

City of London Corporation lending library. Wide range of fiction and non-fiction books and language courses on cassette and CD, foreign fiction, paperbacks, maps and guides for travel at home and abroad, children's books, a selection of large print, and collections of DVDs, videos and music CDs, books in Bengali. Spoken word recordings on cassette and CD. Free internet access.

Cardiff Central Library

John Street, Cardiff CF10 5BA
☏ 029 2038 2116   ✆ 029 2087 1588
rboddy@cardiff.gov.uk
www.cardiff.gov.uk/libraries
OPEN 9.00 am to 6.00 pm Monday, Tuesday, Wednesday, Friday; 9.00 am to 7.00 pm Thursday; 9.00 am to 5.30 pm Saturday

General lending library with the following departments: leisure, music, children's, local studies, information, science and humanities.

Carmarthen Public Library

St Peter's Street, Carmarthen SA31 1LN
☏ 01267 224824   ✆ 01267 221839
wtphillips@carmarthenshire.gov.uk
www.carmarthenshire.gov.uk
Regional Library Manager *William Phillips*
OPEN 9.30 am to 7.00 pm Monday, Tuesday, Wednesday, Friday; 9.30 am to 5.00 pm Thursday and Saturday
OPEN ACCESS

Comprehensive range of fiction, non-fiction, children's books and reference works in English and in Welsh. Large local history library. Free internet access and computer facilities. Large Print books, audiobooks, CDs, CD-ROMs, DVDs and videos available for loan.

Catholic National Library

St Michaels Abbey, Farnborough Road, Farnborough GU14 7NQ
☏ 01252 543818
library@catholic-library.org.uk
www.catholic-library.org.uk
OPEN ACCESS For reference (non-members must sign in; loans restricted to members).

Contains books, many not readily available elsewhere, on theology, religions worldwide, scripture and the history of churches of all denominations.

City Business Library

1 Brewers' Hall Garden, off Aldermanbury Square, London EC2V 5BX
☏ 020 7332 1812/(3803 textphone)

020 7600 1185
cbl@cityoflondon.gov.uk
www.cityoflondon.gov.uk/citybusinesslibrary
OPEN 9.30 am to 5.00 pm Monday to Friday
(except public holidays).
OPEN ACCESS

Local authority free public reference library
run by the City of London Corporation.
Electronic resources catering for the needs
of the business community. Large directory
collection for both the UK and overseas, plus
companies information, market research reports,
management, banking, insurance, investment
and statistics. Free public internet access. No
academic journals or textbooks.

City of London Libraries
▷ Barbican Library; Camomile Street Library; City
Business Library; Guildhall Library

Commonwealth Secretariat Library
Marlborough House, Pall Mall, London SW1Y 5HX
020 7747 6164   020 7747 6168
library@commonwealth.int
www.thecommonwealth.org
Library Services Manager *Catherine Hume*
OPEN 10.00 am to 4.45 pm Monday to Friday
ACCESS For reference only, by appointment.

Extensive reference source concerned with
economy, development, trade, production and
industry of Commonwealth countries; also
human resources including women, youth, health,
management and education. Includes the archives
of the Secretariat.

Coventry Central Library
Smithford Way, Coventry CV1 1FY
024 7683 2314/2395 (Minicom)
024 7683 2440
central.library@coventry.gov.uk
www.coventry.gov.uk
OPEN 9.00 am to 8.00 pm Monday to Friday; 9.00
am to 4.30 pm Saturday; 12.00 am to 4.00 pm
Sunday
OPEN ACCESS

Located in the middle of the city's main shopping
centre. Approximately 120,000 items (books,
cassettes, CDs and DVDs) for loan, and a
reference collection; free computer and internet
access for first hour. SPECIAL COLLECTIONS
Tom Mann Collection (trade union and labour
studies); 'Peoplelink' community information
database available. Local History and archive
resources available in new purpose built History
Centre next to the Museum and Art Gallery.

Crace Collection
▷ British Library Map Collections

Department for Environment, Food and Rural Affairs
Nobel House, 17 Smith Square, London SW1P 3JR
020 7238 3000   020 7238 6591
DEFRA Helpline 08459 335577 (local call
rate): general contact point which can provide
information on the work of DEFRA, either
directly or by referring callers to appropriate
contacts. Available 9.00 am to 5.00 pm Monday
to Friday (excluding Bank Holidays).
OPEN 9.30 am to 5.00 pm Monday to Friday
ACCESS For reference (but at least 24 hours notice
must be given for intended visits).

Derby Central Library
Wardwick, Derby DE1 1HS
01332 641702   01332 369570
www.derby.gov.uk/libraries
OPEN 9.30 am to 7.00 pm Monday, Tuesday,
Thursday, Friday; 9.30 am to 1.00 pm Wednesday;
9.30 am to 4.00 pm Saturday
Local Studies Library
25B Irongate, Derby DE1 3GL
01332 255393 01332 255381
OPEN 9.30 am to 7.00 pm Monday and Tuesday;
9.30 am to 5.00 pm Wednesday, Thursday, Friday;
9.30 am to 4.00 pm Saturday
OPEN ACCESS

General library for lending, information and
Children's Services. The Central Library also
houses specialist private libraries: Derbyshire
Archaeological Society; Derby Philatelic Society.
The Local Studies Library houses the largest
multimedia collection of resources in existence
relating to Derby and Derbyshire. The collection
includes mss deeds, family papers, business
records including the Derby Canal Company,
Derby Board of Guardians and the Derby China
Factory. Both libraries offer free internet access.

Devon & Exeter Institution Library
7 Cathedral Close, Exeter EX1 1EZ
01392 251017
J.P.Gardner@exeter.ac.uk
www.exeter.ac.uk/library
www.devonandexeterinstitution.org
OPEN 9.30 am to 5.00 pm Monday to Friday
ACCESS Members only (temporary membership
available).

Founded 1813. Under the administration of
Exeter University Library. Contains over 36,000
volumes, including long runs of 19th-century
journals, theology, history, topography, early
science, biography and literature. A large
and growing collection of books, journals,
newspapers, prints and maps relating to the
South West.

Doncaster Library and Information Services

Central Library, Waterdale, Doncaster DN1 3JE

☎ 01302 734305   ⊠ 01302 369749

Reference.Library@doncaster.gov.uk

library.doncaster.gov.uk

OPEN 9.00 am to 6.00 pm Monday to Thursday; 10.00 am to 6.00 pm Friday; 9.00 am to 5.00 pm Saturday

OPEN ACCESS

Books, newspapers, periodicals, spoken word cassettes/CDs, DVDs. Reading aids unit for people with visual impairment; activities for children during school holidays and regular storytimes. Events programme, including visits by authors and talks, readers' groups, etc. Also reference library and Local Studies library.

Dorchester Library (part of Dorset County Council)

Colliton Park, Dorchester DT1 1XJ

☎ 01305 224440 (lending)/224448 (reference)

⊠ 01305 266120

dorchesterreferencelibrary@dorsetcc.gov.uk

www.dorsetforyou.com/libraries

OPEN 10.00 am to 7.00 pm Monday; 9.30 am to 7.00 pm Tuesday, Wednesday, Friday; 9.30 am to 5.00 pm Thursday; 9.00 am to 4.00 pm Saturday

OPEN ACCESS

General lending and reference library, including special collections on Thomas Hardy, the Powys Family and William Barnes. Periodicals, children's library, CD-ROMs, free Internet access. Video lending service.

Dundee Central Library

The Wellgate, Dundee DD1 1DB

☎ 01382 431500   ⊠ 01382 431558

central.library@dundeecity.gov.uk

www.dundeecity.gov.uk

OPEN Departments of Leisure Reading, Art & Music, Children's, Reference & Information, and Local History Centre: 9.30 am to 6.00 pm Monday, Tuesday, Friday; 10.00 am to 6.00 pm Wednesday; 9.30 am to 8.00 pm Thursday; 9.30 am to 5.00 pm Saturday. Science & Business: 9.30 am to 9.00 pm Monday, Tuesday, Thursday, Friday; 10.00 am to 9.00 pm Wednesday; 9.30 am to 5.00 pm Saturday

ACCESS Reference services available to all; lending services to those who live, work, study or were educated within Dundee City.

Adult lending, reference and children's services. Art, music, audio, video and DVD lending services. Internet access. Schools service (Agency). Housebound and mobile services.

SPECIAL COLLECTIONS The Wighton Collection of National Music; The Wighton Heritage Centre;
The Wilson Photographic Collection; The Lamb Collection.

Durning-Lawrence Library
⊡ Senate House Library

English Nature
⊡ Natural England

Equality and Human Rights Commission

Arndale House, Arndale Centre, Manchester M4 3EQ

☎ 0161 829 8100   ⊠ 0161 829 8110

englandhelpline@equalityhumanrights.com

www.equalityhumanrights.com

The EHRC is open to the public via the Helpline (0845 604 6610) 9.00 am to 5.00 pm Monday, Tuesday, Thursday, Friday; 9.00 am to 7.45 pm Wednesday.

Essex County Council Libraries

Goldlay Gardens, Chelmsford CM2 0EW

☎ 01245 284981   ⊠ 01245 492780

essexlib@essex.gov.uk

www.essex.gov.uk

Essex County Council Libraries has 73 static libraries throughout Essex as well as 12 mobile libraries. Services to the public include books, newspapers, periodicals, CDs, cassettes, videos, CD-ROMs and internet access. Specialist subjects and collections are listed below at the relevant library.

Chelmsford Library

PO Box 882, Market Road, Chelmsford CM1 1LH

☎ 01245 492758 ⊠ 01245 492536

chelmford.library@essex.gov.uk

OPEN 8.30 am to 7.00 pm Monday to Friday; 8.30 am to 5.30 pm Saturday; 12.30 pm to 4.30 pm Sunday

Colchester Library

Trinity Square, Colchester CO1 1JB

☎ 01206 245900 ⊠ 01206 245901

colchester.library@essex.gov.uk

OPEN 8.30 am to 7.30 pm Monday to Friday; 8.30 am to 5.00 pm Saturday; 12.30 pm to 4.30 pm Sunday

Local studies; Castle collection (18th-century subscription library); Cunnington collection; Margaret Lazell collection; Taylor collection.

Harlow Library

The High, Harlow CM20 1HA

☎ 01279 413772 ⊠ 01279 424612

harlow.library@essex.gov.uk

OPEN 9.00 am to 7.00 pm Monday to Friday; 9.00 am to 5.00 pm Saturday; 1.00 pm to 4.00 pm Sunday

Sir John Newson Memorial collection; Maurice Hughes Memorial collection.

Loughton Library
Traps Hill, Loughton IG10 1HD
☎ 020 8502 0181 📠 020 8508 5041
loughton.library@essex.gov.uk
OPEN 9.00 am to 7.00 pm Monday to Friday; 9.30 am to 5.30 pm Saturday; 11.00 am to 3.00 pm Sunday
National Jazz Foundation Archive.

Saffron Walden Library
2 King Street, Saffron Walden CB10 1ES
☎ 01799 523178 📠 01799 513642
OPEN 9.00 am to 7.00 pm Monday to Friday; 9.00 am to 5.00 pm Saturday; 1.00 pm to 4.00 pm Sunday
Victorian studies collection.

Witham Library
18 Newland Street, Witham CM8 2AQ
☎ 01376 519625 📠 01376 501913
OPEN 9.00 am to 7.00 pm Monday to Friday; 9.00 am to 5.00 pm Saturday; 1.00 pm to 4.00 pm Sunday
Dorothy L. Sayers and Maskell collections.

Family Records Centre
Part of the **National Archives** (see entry)

The Fawcett Library
▷ **The Women's Library**

Edward Fitzgerald Collection
▷ **Suffolk County Council – Suffolk Libraries**

Forestry Commission Library
Forest Research Station, Alice Holt Lodge, Wrecclesham, Farnham GU10 4LH
☎ 01420 22255 📠 01420 23653
library@forestry.gsi.gov.uk
www.forestry.gov.uk
www.forestresearch.gov.uk
OPEN 9.00 am to 5.00 pm Monday to Thursday; 9.00 am to 4.30 pm Friday
ACCESS By appointment for personal visits.

Approximately 20,000 books on forestry and arboriculture, plus 400 current journals. CD-ROMS include Forest Science Database (1939 onwards). Offers a Research Advisory Service for advice and enquiries on forestry (☎ 01420 23000) with a charge for consultations and diagnosis of tree problems exceeding ten minutes.

French Institute Library
Institut français du Royaume Uni, 17 Queensberry Place, London SW7 2DT
☎ 020 7073 1350 📠 020 7073 1363
library@ambafrance.org.uk
www.institut-francais.org.uk
OPEN 12 noon to 7.00 pm Tuesday to Friday; 12 noon to 5.00 pm Saturday; Children's Library: 2.00 pm to 6.00 pm Tuesday to Friday; 12 noon to

6.00 pm Saturday; closed in August and for one week at Christmas.
OPEN ACCESS For reference and consultation (loans restricted to members; leaflet on demand).

A collection of over 55,000 volumes mainly centred on French cultural interests with special emphasis on French as a Foreign Language, literature and history. Books and periodicals, mainly in French, a few in English and some bilingual. Collection of DVDs, videos, CDs (French music), CD-ROMs, audiobooks; special collections: 'France Libre'. Denis Saurat MSSS, Paris. Recordings of lectures, press-cuttings on French current affairs; Campus France; internet access to members; Wifi. Group visits on request. Reading group; documentary screenings; talks. Children's library (8,000 documents). Bistro.

The Froebel Archive for Childhood Studies
▷ **entry under Useful Websites**

John Frost Newspapers
22b Rosemary Avenue, Enfield EN2 0SS
☎ 020 8366 1392/0946 📠 020 8366 1379
andrew@johnfrostnewspapers.com
www.johnfrostnewspapers.co.uk
Contacts *Andrew Frost, John Frost*

A collection of 80,000 original newspapers (1630 to the present day) and 200,000 press cuttings available, on loan, for research and rostrum/stills work (TV documentaries, book and magazine publishers and audiovisual presentations). Historic events, politics, sports, royalty, crime, wars, personalities, etc., plus many in-depth files.

German Historical Institute Library
17 Bloomsbury Square, London WC1A 2NJ
☎ 020 7309 2050 📠 020 7309 2055
library@ghil.ac.uk
www.ghil.ac.uk
OPEN 10.00 am to 5.00 pm Monday, Tuesday, Wednesday, Friday; 10.00 am to 8.00 pm Thursday
OPEN ACCESS Visitors must bring proof of address and recent photograph to be issued with a reader's ticket.

70,000 volumes; around 200 journals; databases. Devoted primarily to German history from the Middle Ages to the present day, with special emphasis on the 19th and 20th centuries, in particular Germany between 1933 and 1945, the development of the two German states after 1945/49 and German unification after 1989.

Gloucestershire County Council Libraries & Information
Quayside House, Shire Hall, Gloucester GL1 2HY
☎ 08452 305420 📠 01452 425042
libraryhelp@gloucestershire.gov.uk
www.libraries.gloucestershire.gov.uk/

Head of Libraries & Information *David Paynter*
OPEN ACCESS

The service includes 39 local libraries and five mobile libraries. The website includes library opening hours, mobile library route schedules; the library catalogue; book renewal/reservations facility; and information about news and events.

Goethe-Institut Library

50 Princes Gate, Exhibition Road, London
SW7 2PH

☎ 020 7596 4040/4044   🖷 020 7594 0230
library@london.goethe.org
www.goethe.de/london
Head of Library Services *Elisabeth Pyroth*
OPEN 1.00 pm to 6.30 pm Monday to Thursday;
1.00 pm to 5.00 pm Saturday

Library specializing in German literature and books/audiovisual material on German culture and history: 20,000 books (4,000 of them in English), 120 periodicals, 13 newspapers, 3,700 audiovisual media (including 1,200 videos/DVDs), selected press clippings on German affairs from the German and UK press, information service, photocopier, video facility. Also German language teaching material for teachers and students of German. See website for membership details.

Goldsmiths' Library
▷ Senate House Library

Greater London Council History Library
▷ London Metropolitan Archives Library Services

Guildford Institute Library

Ward Street, Guildford GU1 4LH
☎ 01483 562142   🖷 01483 451034
library@guildford-institute.org.uk
www.guildford-institute.org.uk
OPEN 10.00 am to 3.00 pm Tuesday to Friday;
10.00 am to 1.00 pm each first Saturday of the month.
OPEN ACCESS to members only but open to enquirers for research purposes.

Founded 1834. Some 14,000 volumes of which 7,500 were printed before the First World War. The remaining stock consists of recently published works of fiction and non-fiction. Newspapers and periodicals also available. SPECIAL COLLECTIONS include an almost complete run of the *Illustrated London News* from 1843–1906, a collection of Victorian and early 20th century local history ephemera albums, and about 400 photos and other pictures relating to the Institute's history and the town of Guildford.

Guildhall Library

Aldermanbury, London EC2V 7HH
☎ See below   🖷 020 7600 3384
www.cityoflondon.gov.uk/guildhalllibrary

ACCESS For reference (but much of the material is kept in storage areas and is supplied to readers on request; proof of identity is required for consultation of certain categories of stock). Free, *limited* enquiry service available. Also a fee-based service for in-depth research: ☎ 020 7332 1854 🖷 020 7600 3384 search.guildhall@cityoflondon. gov.uk www.cityoflondon.gov.uk/search_guildhall

Part of the City of London libraries. Seeks to provide a basic general reference service but its major strength, acknowledged worldwide, is in its historical collections. The library is divided into two sections, each with its own catalogues and enquiry desks. These are: Printed Books; Manuscripts. NB the Print & Maps Collection is now housed at the **London Metropolitan Archives** (see entry).

PRINTED BOOKS
☎ 020 7332 1868/1870
printed books.guildhall@cityoflondon.gov.uk
OPEN 9.30 am to 5.00 pm Monday to Saturday;
NB closes on Saturdays preceding Bank Holidays;
check for details.

Strong on all aspects of London history, with wide holdings of English history, topography and genealogy, including local directories, poll books and parish register transcripts. Also good collections of English statutes, law reports, parliamentary debates and journals, and House of Commons papers. Home of several important collections deposited by London institutions: the Marine collection of the Corporation of Lloyd's, the Stock Exchange's historical files of reports and prospectuses, the Clockmakers' Company library and museum, the Gardeners' Company, Fletchers' Company, the Institute of Masters of Wine, International Wine and Food Society and Gresham College.

MANUSCRIPTS
☎ 020 7332 1862/3
manuscripts.guildhall@cityoflondon.gov.uk
www.history.ac.uk/gh/
OPEN 9.30 am to 5.00 pm Monday to Saturday (no requests for records after 4.30 pm; no manuscripts can be produced between 12 noon and 2.00 pm on Saturdays); NB closes on Saturdays preceding Bank Holidays; check for details.

The official repository for historical records relating to the City of London (except those of the City of London Corporation itself, which are housed at the **London Metropolitan Archives**). Records date from the 11th century to the present day. They include archives of most of the City's parishes, wards and livery companies, and of many individuals, families, estates, schools, societies and other institutions, notably the Diocese of London, St Paul's Cathedral, Trinity

House and the London Stock Exchange, as well as the largest collection of business archives in any public repository in the UK. Although mainly of City interest, holdings include material for the London area as a whole and beyond. Many of the records are of national and international significance.

Guille–Alles Library

Market Street, St Peter Port, Guernsey GY1 1HB
☎ 01481 720392   🖷 01481 712425
ga@library.gg
www.library.gg
OPEN 9.00 am to 5.00 pm Monday, Thursday, Friday, Saturday; 10.00 am to 5.00 pm Tuesday; 9.00 am to 8.00 pm Wednesday
OPEN ACCESS For residents; payment of returnable deposit by visitors. Music CD collection: £10 for two-year subscription.

Lending, reference and information services. Public internet service.

Herefordshire Libraries

Shirehall, Hereford HR1 2HX
☎ 01432 261644   🖷 01432 260744
libraries@herefordshire.gov.uk
www.herefordshire.gov.uk/leisure/
Principal Libraries Officer *James Anthony*
Senior Stock Librarian *Carolyn Huckfield*
OPEN Opening hours vary in the libraries across the county.
ACCESS Information and reference services open to anyone; loans to members only (membership criteria: resident, being educated, working in Herefordshire. Proof of identity and address required.) Temporary membership to visitors is also available.

Information service, reference and lending libraries. Non-fiction and fiction for all age groups, including normal and large print, spoken word cassettes, and CDs, music CDs, DVDs, playstation2. Hereford Library houses the largest reference section and the county local history collection, although some reference material and local history is available at all libraries. Internet access at all libraries. *Special collections* Cidermaking; Beekeeping; Alfred Watkins; John Masefield; Pilley.

Highgate Literary and Scientific Institution Library

11 South Grove, London N6 6BS
☎ 020 8340 3343   🖷 020 8340 5632
librarian@hlsi.net
www.hlsi.net
Librarian *Margaret Mackay*
OPEN 10.00 am to 5.00 pm Tuesday to Friday; 10.00 am to 4.00 pm Saturday (closed Sunday and Monday)

ANNUAL MEMBERSHIP £55 (individual); £90 (household)

25,000 volumes of general fiction and non-fiction, with a children's section and extensive local archives. SPECIAL COLLECTIONS on local history, London, and local poets Samuel Taylor Coleridge and John Betjeman.

Highland Libraries, The Highland Council, Education, Culture and Sport Service

Library Support Unit, 31a Harbour Road, Inverness IV1 1UA
☎ 01463 235713   🖷 01463 236986
libraries@highland.gov.uk
www.highland.gov.uk
OPEN Library opening hours vary to suit local needs. Contact administration and support services for details (8.00 am to 5.00 pm Monday to Friday).
OPEN ACCESS

Comprehensive range of lending and reference stock: books, pamphlets, periodicals, newspapers, compact discs, audio and video cassettes, maps, census records, genealogical records, photographs, educational materials, etc. Free access to the internet in all libraries. Highland Libraries provides the public library service throughout the Highlands with a network of 43 static and 12 mobile libraries.

Paul Hirsch Music Library
▷ British Library Music Collections

Holborn Library

32–38 Theobalds Road, London WC1X 8PA
☎ 020 7974 6345
holbornlibrary@camden.gov.uk
www.camden.gov.uk/libraries
OPEN 10.00 am to 7.00 pm Monday and Thursday; 10.00 am to 6.00 pm Tuesday, Wednesday and Friday; 10.00 am to 5.00 pm Saturday
OPEN ACCESS

London Borough of Camden public library. Includes the London Borough of Camden Local Studies and Archive Centre.

Holcenberg Jewish Collection
▷ City of Plymouth Library and Information Services

Sherlock Holmes Collection (Westminster)

Marylebone Library, Marylebone Road, London NW1 5PS
☎ 020 7641 1206   🖷 020 7641 1019
ccooke@westminster.gov.uk
www.westminster.gov.uk/libraries/special/sherlock.cfm
OPEN 9.30 am to 8.00 pm Monday, Tuesday, Thursday, Friday; 10.00 am to 8.00 pm

Wednesday; closed Saturday and Sunday (unless by prior arrangement).

ACCESS By appointment only.

Located in Westminster's Marylebone Library. An extensive collection of material from all over the world, covering Sherlock Holmes and Sir Arthur Conan Doyle. Books, pamphlets, journals, newspaper cuttings and photos, much of which is otherwise unavailable in this country. Some background material.

House of Commons Research Library
▷ entry under Useful Websites

IHR Library
▷ The Institute of Historical Research

Imperial College Central Library
▷ Science Museum Library

Imperial War Museum
Department of Printed Books, Lambeth Road, London SE1 6HZ
☎ 020 7416 5342   ℻ 020 7416 5246
collections@iwm.org.uk
www.iwm.org.uk
OPEN 10.00 am to 5.00 pm Monday to Saturday (restricted service Saturday; closed on Bank Holiday Saturdays and two weeks during the year for annual stock check).
ACCESS For reference (but at least 24 hours' notice must be given for intended visits).

A large collection of material on British and Commonwealth 20th-century life with detailed coverage of the two World Wars and other conflicts. This collection includes substantial holdings of European language material. Books, pamphlets and periodicals, including many produced for short periods in unlikely wartime settings; also maps, biographies and privately printed memoirs, and foreign language material. Additional research material available in the departments of Art, Documents, Exhibits and Firearms and in the Film, Photographs and Sound Archives. See www.iwmcollections.org.uk for their online catalogues.

The Institute of Historical Research
Senate House, Malet Street, London WC1E 7HU
☎ 020 7862 8760
IHR.Library@sas.ac.uk
www.history.ac.uk/library
OPEN 9.00 am to 8.45 pm Monday to Friday; 9.30 am to 5.15 pm Saturday
ACCESS For reference only. Non-members are advised to telephone to check availability of items before visiting.

Reference collection of printed primary sources, bibliographies, guides to archives, periodicals and works covering the history of Western Europe from the fall of the Roman Empire. Holds a substantial microform collection of mainly British material from repositories outside London.

Instituto Cervantes
102 Eaton Square, London SW1W 9AN
☎ 020 7201 0757   ℻ 020 7235 0329
biblon@cervantes.es
londres.cervantes.es
OPEN 12 noon to 7.30 pm Monday to Thursday; 12 noon to 6.30 pm Friday; 9.30 am to 2.00 pm Saturday
OPEN ACCESS For reference and lending.

Spanish language and literature, history, art, philosophy. The library houses a collection of books, periodicals, videos, DVDs, slides, tapes, CDs, cassettes, CD-ROMs specializing entirely in Spain and Latin America.

Italian Institute Library
39 Belgrave Square, London SW1X 8NX
☎ 020 7396 4425   ℻ 020 7235 4618
www.icilondon.esteri.it
OPEN 10.00 am to 1.30 pm and 2.30 pm to 5.00 pm Monday to Friday. Term-time: 10.00 am to 1.30 pm and 2.30 pm to 5.00 pm Monday, Tuesday, Thursday, Friday; 2.30 pm to 8.00 pm Wednesday; 10.30 am to 4.30 pm Saturday (first and third of each month). The Library is now open to the general public every Wednesday evening and on the first and third Saturday of each month during term time. Call 020 7396 4425 to confirm exact opening times.
OPEN ACCESS For reference.

A collection of over 30,000 volumes relating to all aspects of Italian culture, including DVDs and CDs relating to Italian cinema. Texts are mostly in Italian, with some in English.

Jersey Library
Halkett Place, St Helier, Jersey JE2 4WH
☎ 01534 448700 (Jersey Library)/448701 (Reference)/448702 (Open Learning)/448733 (Branch Library)   ℻ 01534 448730
je.library@gov.je
www.gov.je/library
OPEN 9.30 am to 5.30 pm Monday, Wednesday, Thursday, Friday; 9.30 am to 7.30 pm Tuesday; 9.30 am to 4.00 pm Saturday
OPEN ACCESS

Books, periodicals, newspapers, CDs, DVDs and language packs, cassettes, videos, microfilm, specialized local studies collection, public internet access. Branch Library at Les Quennevais School, St Brelade. Mobile library and homes services. Open Learning Centre.

Kent County Central Library

Kent Libraries and Archives, Springfield,
Maidstone ME14 2LH
☎ 01622 696511   🖷 01622 696494
countycentrallibrary@kent.gov.uk
www.kent.gov.uk/libs
Customer Services Officer *Craig Fox*
OPEN 9.00 am to 6.00 pm Monday to Friday; 9.00
am to 5.00 pm Saturday
OPEN ACCESS

50,000 volumes available on the floor of the
library plus 250,000 volumes of non-fiction,
mostly academic, available on request to staff.
English literature, poetry, classical literature,
drama (including play sets), music (including
music sets). Strong, too, in sociology, art
history, business information and government
publications. Loans to all who live or work in
Kent; those who do not may consult stock for
reference or arrange loans via their own local
library service.

King George III Topographic Collection and Maritime Collection
▷ British Library Map Collections

Leeds Central Library

Calverley Street, Leeds LS1 3AB
☎ 0113 247 8911   🖷 0113 247 8426
enquiry.express@leeds.gov.uk
www.leeds.gov.uk/libraries
OPEN 9.00 am to 8.00 pm Monday, Tuesday,
Wednesday; 9.00 am to 5.00 pm Thursday and
Friday; 10.00 am to 5.00 pm Saturday; 1.00 pm to
5.00 pm Sunday
ACCESS Level access and lift. Open access to
lending libraries and some reference materials.
Other reference materials on request. Free public
internet access.

Lending Library ☎ 0113 247 8270
Adult and children's popular fiction and non-
fiction for loan, including talking books, large
print and books in world languages.

Music Library ☎ 0113 247 8273
Scores, books, DVD and audio.

The Information Centre ☎ 0113 247 8282 or
0113 395 1833
Company and product information, market
research, statistics, directories, journals and
computer-based information. Highlighted
collections covering Law and Rights, Health,
Jobs and Careers, Europe Direct, Education
and Business. Extensive files of newspapers and
periodicals plus all government publications since
1960. Special collections include military history,
Judaic, early gardening books.

Art Library ☎ 0113 247 8247
Major collection of material on fine and applied
arts.

Local and Family History Library
☎ 0113 247 8290
Extensive collection on Leeds and Yorkshire,
including maps, books, pamphlets, local
newspapers, illustrations and playbills. Census
returns for the whole of Yorkshire also available.
International Genealogical Index and parish
registers.

Leeds Library and Information Service has an
extensive network of branch and mobile libraries.

Leeds Library

18 Commercial Street, Leeds LS1 6AL
☎ 0113 245 3071   🖷 0113 245 1191
enquiries@leedslibrary.co.uk
www.theleedslibrary.org.uk
OPEN 9.00 am to 5.00 pm Monday to Friday;
9.30 am to 1.00 pm on the first Saturday of each
month.
ACCESS To members; research use upon
application to the librarian.

Founded 1768. Contains over 125,000 books and
periodicals from the 15th century to the present
day. SPECIAL COLLECTIONS include Reformation
pamphlets, Civil War tracts, Victorian and
Edwardian children's books and fiction, European
language material, spiritualism and psychical
research, plus local material.

Library of the Religious Society of Friends

Friends House, 173 Euston Road, London NW1 2BJ
☎ 020 7663 1135   🖷 020 7663 1001
library@quaker.org.uk
www.quaker.org.uk/library
OPEN 1.00 pm to 5.00 pm Monday, Tuesday,
Thursday, Friday; 10.00 am to 5.00 pm
Wednesday
OPEN ACCESS Reader registration required with
proof of permanent address.

Quaker history, thought and activities from the
17th century onwards. Supporting collections on
peace, anti-slavery and other subjects in which
Quakers have maintained long-standing interest.
Also archives and manuscripts relating to the
Society of Friends.

Lincoln Central Library

Free School Lane, Lincoln LN2 1EZ
☎ 01522 782010   🖷 01522 575011
linclib@lincolnshire.gov.uk
www.lincolnshire.gov.uk
OPEN 9.00 am to 7.00 pm Monday to Friday; 9.00
am to 4.00 pm Saturday
OPEN ACCESS to the library; appointment
required for the Tennyson Research Centre.

Lending and reference library. Special collections include Lincolnshire local history (printed and published material, photographs, maps, directories and census data) and the Tennyson Research Centre (contact *Grace Timmins*, grace.timmins@lincolnshire.gov.uk).

Linen Hall Library

17 Donegall Square North, Belfast BT1 5GB
028 9032 1707   028 9043 8586
info@linenhall.com
www.linenhall.com
OPEN 9.30 am to 5.30 pm Monday to Friday; 9.30 am to 1.00 pm Saturday
OPEN ACCESS For reference (loans restricted to members).

Founded 1788. Contains about 250,000 books in the general lending library. Also houses a number of unique collections including the Irish and Local Studies Collection, which incorporates the world renowned Northern Ireland Political Collection relating to the 'Troubles' and ongoing peace process (c. 300,000 items).

Literary & Philosophical Society of Newcastle upon Tyne

23 Westgate Road, Newcastle upon Tyne NE1 1SE
0191 232 0192   0191 261 4494
library@litandphil.org.uk
www.litandphil.org.uk
Librarian *Kay Easson*
OPEN 9.30 am to 7.00 pm Monday, Wednesday, Thursday; 9.30 am to 8.00 pm Tuesday; 9.30 am to 5.00 pm Friday; 9.30 am to 1.00 pm Saturdays
ACCESS Members; research facilities for *bona fide* scholars on application to the Librarian.

200-year-old library of 140,000 volumes, periodicals (including 130 current titles), classical music on vinyl recordings and CD, plus a collection of scores. Free public lectures, events and recitals. Recent publications include: *The Reverend William Turner: Dissent and Reform in Georgian Newcastle upon Tyne* Stephen Harbottle; *History of the Literary and Philosophical Society of Newcastle upon Tyne, Vol. 2 (1896–1989)* Charles Parish; *Bicentenary Lectures 1993* ed. John Philipson.

Liverpool Libraries and Information Services

William Brown Street, Liverpool L3 8EW
0151 233 5829   0151 233 5886
refbt.central.library@liverpool.gov.uk
www.liverpool.gov.uk/libraries
OPEN 9.00 am to 6.00 pm Monday to Friday; 9.00 am to 5.00 pm Saturday; 12 noon to 4.00 pm Sunday
OPEN ACCESS

Learn Direct Centre – UK Online Centre Free broadband internet access.

Humanities Reference Library A total stock in excess of 120,000 volumes and 24,000 maps, plus book plates, prints and autographed letters. SPECIAL COLLECTIONS Walter Crane and Edward Lear illustrations, Kelmscott Press, Audubon's *Birds of America*.

Business and Technology Reference Library Extensive stock dealing with all aspects of science, commerce and technology, including British and European standards and patents and trade directories.

Audio Visual Library Extensive stock relating to all aspects of music. Includes 128,000 volumes and music scores, over 4,000 CDs, 1,000 videos and 800 DVDs.

Record Office and Local History Department Material relating to Liverpool, Merseyside, Lancashire and Cheshire, together with archive material mainly on Liverpool. Proof of name and address required to obtain reader's ticket.

Lending Library Graphic novels, large print collection, children's collections, audio books, study support collections, Reader friendly displays of bestsellers and out of print titles.

University of the Arts London – London College of Communication

Library and Learning Resources, Elephant and Castle, London SE1 6SB
020 7514 8026   020 7514 6597
www.arts.ac.uk/library
ACCESS Appointment required.

Library and Learning Resources provides books, periodicals, slides, CD-ROMs, DVDs, videos and computer software on all aspects of the art of the book, printing, management, film/photography, graphic arts and retailing. SPECIAL COLLECTIONS History and the art of the western book.

The London Library

14 St James's Square, London SW1Y 4LG
020 7930 7705   020 7766 4766
membership@londonlibrary.co.uk
www.londonlibrary.co.uk
Librarian *Miss Inez Lynn*
OPEN 9.30 am to 7.30 pm Monday, Tuesday, Wednesday; 9.30 am to 5.30 pm Thursday, Friday, Saturday
ACCESS For members only (2009 fees: £395 p.a.; £200 for 16–24-year-olds). Day and weekly reference tickets available for non-members (£10 and £30).

With over a million books and 8,000 members, The London Library 'is the most distinguished private library in the world; probably the largest, certainly the best loved'. Founded in 1841, it is a registered charity and wholly independent of public funding. Its permanent collection embraces most European languages as well as English. Its subject range is predominantly within the humanities, with emphasis on literature, history, fine and applied art, architecture, bibliography, philosophy, religion, and topography and travel. Some 8,000–9,000 titles are added yearly. Over 95 per cent of the stock is on open shelves to which members have free access. Members may take out up to 10 volumes at a time; 15 if they live more than 20 miles from the Library and may hold the books on loan until they are needed by another member. The comfortable Reading Room has an annexe for users of personal computers. There are photocopiers, CD-ROM workstations, free access to the internet, and the Library also offers a postal loans service.

The London Library has embarked on its most ambitious redevelopment project in over a century; it includes the integration of a new contiguous building which will increase the total area of the site by approximately 30%.

Membership is open to all: prospective members are required to submit a refereed application form in advance of admission, but there is at present no waiting list for membership. The London Library Trust may offer reduced cost 'Carlyle Memberships' to those who are unable to afford the full annual fee; details on application.

London Metropolitan Archives Library Services

History Library: 40 Northampton Road, London EC1R OHB
☎ 020 7332 3820   ☏ 020 7833 9136
ask.lma@cityoflondon.gov.uk
www.lma.gov.uk
Contact *The Enquiry Team*
OPEN 9.30 am to 4.45 pm Monday, Wednesday, Friday; 9.30 am to 7.30 pm Tuesday and Thursday; open some Saturdays – call for details.
ACCESS For reference only.

This 100,000 volume library covers all aspects of the life and development of London, with strong holdings on the history and organization of London local government. As the former Greater London Council History Library, the collection covers all subjects of London life, from architecture and biography to theatres and transport. The collection includes London directories from 1677 to the present, Acts of Parliament, statistical returns, several hundred periodical titles and public reports. From April 2009, includes the prints and maps graphic collection formerly held at the **Guildhall Library**.

London's Transport Museum Poster Archives
▷ London's Transport Museum Photographic Library under Picture Libraries

Lord Louis Library
Orchard Street, Newport, Isle of Wight PO30 1LL
☎ 01983 527655/823800 (Reference Library)
☏ 01983 825972
referencelibrary@iow.gov.uk
www.iwight.com/thelibrary
OPEN 9.00 am to 5.30 pm Monday, Tuesday, Wednesday and Friday; 10.00 am to 8.00 pm Thursday; 9.00 am to 5.00 pm Saturday; 10.00 am to 1.00 pm Sunday
OPEN ACCESS

General adult and junior fiction and non-fiction collections; local history collection at Library HQ (☎ 01983 203880 for details). Internet access in all branches of the library on the Isle of Wight. Also the county's main reference library.

Manchester Central Library
St Peters Square, Manchester M2 5PD
☎ 0161 234 1900   ☏ 0161 234 1963
libraries@manchester.gov.uk
www.manchester.gov.uk/libraries
OPEN 9.00 am to 8.00 pm Monday to Thursday; 9.00 am to 5.00 pm Friday and Saturday.
Commercial Library: 9.00 am to 6.00 pm Monday to Thursday; 9.00 am to 5.00 pm Friday and Saturday
OPEN ACCESS

One of the country's leading public libraries with extensive collections covering all subjects. Departments include: Commercial, Science & Humanities, Arts, Music, Local Studies, Chinese, General Readers, Language & Literature. Large lending stock and VIP (visually impaired) service available. Internet access.

Tom Mann Collection
▷ Coventry Central Library

Marylebone Library (Westminster)
▷ Sherlock Holmes Collection

Ministry of Defence Cartographic Archive
▷ British Library Map Collections

The Mitchell Library
North Street, Glasgow G3 7DN
☎ 0141 287 2999   ☏ 0141 287 2815
lil@cls.glasgow.gov.uk
www.glasgowlibraries.org
OPEN 9.00 am to 8.00 pm Monday to Thursday; 9.00 am to 5.00 pm Friday and Saturday
OPEN ACCESS

One of Europe's largest public reference libraries with stock of over 1,200,000 volumes. It subscribes to 48 newspapers and more than 1,200 periodicals. There are collections in microform, records, tapes and videos, as well as CD-ROMs, electronic databases, illustrations, photographs, postcards, etc.

The library contains a number of special collections, e.g. the Robert Burns Collection (5,000 volumes), the Scottish Poetry Collection (12,000 items) and the Scottish Drama Collection (1,650 items).

Morrab Library

Morrab House, Morrab Gardens, Penzance TR18 4DA
☎ 01736 364474
Librarian *Annabelle Read*
OPEN 10.00 am to 4.00 pm Tuesday to Friday; 10.00 am to 1.00 pm Saturday
ACCESS Non-members may use the library for a small daily fee but may not borrow books.

Formerly known as the Penzance Library. An independent subscription lending library of over 40,000 volumes covering virtually all subjects except modern science and technology, with large collections on history, literature and religion. There is a comprehensive Cornish collection of books, newspapers and manuscripts including the Borlase letters; a West Cornwall photographic archive; many runs of 18th and 19th-century periodicals; a collection of over 2,000 books published before 1800.

The National Archives

Kew, Richmond TW9 4DU
☎ 020 8876 3444
www.nationalarchives.gov.uk
www.nationalarchives.gov.uk/contact/form
OPEN 9.00 am to 5.00 pm Monday and Friday; 9.00 am to 7.00 pm Tuesday and Thursday; 10.00 am to 5.00 pm Wednesday; 9.30 am to 5.00 pm Saturday (closed public holidays and for annual stocktaking in early December). From April 2008 the Family Record Centre services have been incorporated at the Kew site.
ACCESS Reader's ticket required for access to original documents, available free of charge on production of two forms of identification, one proof of address and one proof of identity, which must include a valid signature. Further information is available online (www.nationalarchives.gov.uk/registration/#id) or by phone.

The National Archives is the official archive for England, Wales and the central UK government, containing 900 years of history from the Domesday Book to the present, with records ranging from parchment and paper scrolls through to recently created digital files and archived websites. Increasingly, these records are being put online, making them universally accessible.

The National Archives is a government department and an executive agency under the Secretary of State for Justice. Gives detailed guidance to government departments and the public sector on information management, in order to ensure the survival of records in whichever form they are created, be it paper or digital. Also advises custodians throughout the public and private sectors about the care of historical archives.

National Library of Ireland

Kildare Street, Dublin 2, Republic of Ireland
☎ 00 353 1 603 0200   ⊠ 00 353 1 676 6690
info@nli.ie
www.nli.ie
Reading Rooms: OPEN 9.30 am to 9.00 pm Monday, Tuesday, Wednesday; 9.30 am to 5.00 pm Thursday and Friday; 9.30 am to 1.00 pm Saturday
ACCESS Passes are issued for genealogical research or to consult newspapers; Reader's Ticket required for access to other material.

Collections include books, manuscripts, prints and drawings, maps, photographs, newspapers, music, ephemera and genealogical materials. Exhibitions are free and close 30 minutes before reading rooms.

National Library of Scotland

George IV Bridge, Edinburgh EH1 1EW
☎ 0131 623 3700   ⊠ 0131 623 3701
enquiries@nls.uk
www.nls.uk
OPEN Main Reading Room: 9.30 am to 8.30 pm Monday, Tuesday, Thursday, Friday; 10.00 am to 8.30 pm Wednesday; 9.30 am to 1.00 pm Saturday. Map Library: 9.30 am to 5.00 pm Monday, Tuesday, Thursday, Friday; 10.00 am to 5.00 pm Wednesday; 9.30 am to 1.00 pm Saturday
ACCESS to all reading rooms, for research not easily done elsewhere, by reader's ticket.

Collection of over seven million volumes. The library receives all British and Irish publications. Large stock of newspapers and periodicals. Many special collections, including early Scottish books, theology, polar studies, baking, phrenology and liturgies. Also large collections of maps, music and manuscripts including personal archives of notable Scottish persons.

National Library of Wales

Penglais, Aberystwyth SY23 3BU
☎ 01970 632800   ⊠ 01970 615709
holi@llgc.org.uk
www.llgc.org.uk
OPEN 9.30 am to 6.00 pm Monday to Friday; 9.30 am to 5.00 pm Saturday (closed Bank Holidays and the week after Christmas).
ACCESS to reading rooms by reader's ticket, available on application. Open access to a wide-ranging exhibition programme.

Collection of over four million books and including large collections of periodicals, maps, manuscripts and audiovisual material. Particular emphasis on humanities in printed foreign material, and on Wales and other Celtic areas in all collections.

National Meteorological Library and Archive

FitzRoy Road, Exeter EX1 3PB
☎ 01392 884841   ⓕ 0870 900 5050
metlib@metoffice.gov.uk
www.metoffice.gov.uk
OPEN Library: 8.30 am to 5.00 pm Monday to Thursday; 8.30 am to 4.30 pm Friday. Archive: 10.00 am to 6.00 pm
ACCESS By Visitor's Pass available from the reception desk; advance notice of a planned visit is appreciated.

The major repository of most of the important literature on the subjects of meteorology, climatology and related sciences from the 16th century to the present day. The Library houses a collection of books, journals, articles and scientific papers, plus published climatological data from many parts of the world.

The Technical Archive (National Meteorological Archive, Great Moor House, Sowton Industrial Estate, Bittern Road, Exeter EX2 7NL ☎ 01392 360987 ⓕ 0870 900 5050; metarc@metoffice. gov.uk) holds the document collection of meteorological data and charts from England, Wales and British overseas bases, including ships' weather logs. Records from Scotland are stored in Edinburgh and those from Northern Ireland in Belfast.

National Monuments Record
▷ entry under Picture Libraries

Natural England

1 East Parade, Sheffield S1 2ET
☎ 0114 241 8920   ⓕ 0114 241 8921
enquiries@naturalengland.org.uk
www.naturalengland.org.uk
Enquiry Service ☎ 0845 600 3078 ⓕ 01733 568834
OPEN 8.30 am to 5.00 pm Monday to Thursday; 8.30 am to 4.30 pm Friday
ACCESS Bona fide students only (telephone the library for appointment on 01733 455094).

Information on nature conservation, nature reserves, SSSIs, planning, legislation, countryside, recreation, stewardship schemes and advice to farmers.

The Natural History Museum Library

Cromwell Road, London SW7 5BD
☎ 020 7942 5460   ⓕ 020 7942 5559
library@nhm.ac.uk
www.nhm.ac.uk/library/index.html
OPEN 10.00 am to 4.30 pm Monday to Friday

ACCESS To bona fide researchers, by reader's ticket on presentation of identification (telephone first to make an appointment).

The library is in five sections: general; botany; zoology; entomology; earth sciences. The sub-department of ornithology is housed at the Zoological Museum, Akeman Street, Tring HP23 6AP (☎ 020 7942 6156). Resources available include books, journals, maps, manuscripts, drawings and photographs covering all aspects of natural history, including palaeontology and mineralogy, from the 15th century to the present day. Also archives and historical collection on the museum itself.

Norfolk Library & Information Service

Norfolk and Norwich Millennium Library, The Forum, Millennium Plain, Norwich NR2 1AW
☎ 0844 800 8006
libraries@norfolk.gov.uk
www.library.norfolk.gov.uk
OPEN Lending Library, Reference and Information Service and Norfolk Studies: 9.00 am to 8.00 pm Monday to Friday; 9.00 am to 5.00 pm Saturday EXPRESS: 9.00 am to 9.30 pm Monday to Friday; 9.00 am to 8.30 pm Saturday; 10.30 am to 4.30 pm Sunday CHILDREN'S LIBRARY: 9.00 am to 8.00 pm Monday to Friday; 9.00 am to 5.00 pm Saturday; 10.30 am to 4.30 pm Sunday
OPEN ACCESS

Reference lending library (stock merged together) with wide range, including books, recorded music, music scores, plays and videos. Houses the 2nd Air Division Memorial Library (www.2ndair. org.uk) and has a strong Norfolk Heritage Library. Extensive range of reference stock including business information. Online databases. Public fax and colour photocopying, free access to the Internet. EXPRESS (fiction, sound & vision library within a library): selection of popular fiction, videos, CDs and DVDs available, with extended opening hours.

Northamptonshire Libraries & Information Service

PO Box 216, John Dryden House, 8–10 The Lakes, Northampton NN4 7DD
☎ 01604 237959   ⓕ 01604 237937
nlis@northamptonshire.gov.uk
www.northamptonshire.gov.uk/Leisure/Libraries/home.htm

'Northamptonshire Libraries and Information Service provides its users with much more than just book loans. There are clubs to join, events to attend, skills to learn, local history resources to use, CDs and DVDs to rent – to list but a few services.' In addition, writers and readers are supported by 'Literature Northants'. Run by the Literature Development Officer, Literature

Northants (http://litnorthants.wordpress.com) provides access to literature opportunities across the county. The LDO is also the key contact for all the latest literary news and events.

Northumberland County Library

Beechfield, Gas House Lane, Morpeth NE61 1TA
☎ 01670 502218/534503
libraries@northumberland.gov.uk
www.northumberlandlibraries.com
OPEN 9.00 am to 7.30 pm Monday, Tuesday, Wednesday, Friday; 9.00 am to 12.30 pm Saturday (closed Thursday)
OPEN ACCESS

Books, periodicals, newspapers, story cassettes, CD-ROMs, DVDs, CDs, videos, Free internet access, word processing facilities, prints, microforms, vocal scores, playsets. SPECIAL COLLECTIONS Northern Poetry Library: 15,000 volumes of modern poetry (see entry under *Organizations of Interest to Poets*); Cinema: comprehensive collection of about 5,000 volumes covering all aspects of the cinema; Family History.

Nottingham Central Library

Angel Row, Nottingham NG1 6HP
☎ 0115 915 2828   📠 0115 915 2840
central_admin.library@nottinghamcity.gov.uk
enquiryline@nottinghamcity.gov.uk
www.nottinghamcity.gov.uk/libraries
OPEN 9.00 am to 7.00 pm Monday to Friday; 9.00 am to 1.00 pm Saturday
OPEN ACCESS

City Centre Library; includes wide range of stock. Strengths: Business collection; Sound and Vision collection; comprehensive Nottinghamshire Local Studies. Drama and music sets for loan to groups. Free internet access for visitors.

Nottingham Subscription Library
▷ Bromley House Library

Office for National Statistics, National Statistics Information and Library Service

Customer Contact Centre, Room 1.015 Office for National Statistics, Cardiff Road, Newport NP10 8XG
☎ 0845 601 3034   📠 01633 652747
info@statistics.gov.uk
www.statistics.gov.uk
OPEN 9.00 am to 5.00 pm; appointment required; visitors must supply some form of identification. *Also:* National Statistics Information and Library Service, Government Buildings, Cardiff Road, Newport NP10 8XG
OPEN as above.

Wide range of government statistical publications and access to government internet-based data. Census statistical data from 1801; population and health data from 1837; government social survey reports from 1941; recent international statistical data (UN, Eurostat, etc.); monograph and periodical collections of statistical methodology. The library in south Wales holds a wide range of government economic and statistical publications.

Orkney Library and Archive

44 Junction Road, Kirkwall KW15 1AG
☎ 01856 873166   📠 01856 875260
general.enquiries@orkneylibrary.org.uk
archives@orkneylibrary.org.uk
www.orkneylibrary.org.uk
Principal Archivist *Gary Amos*
Principal Librarian *Karen Walker*
OPEN 9.00 am to 7.00 pm Monday to Thursday; 9.00 am to 5.00 pm Friday and Saturday. Archives: 9.00 am to 5.00 pm Monday, Tuesday, Wednesday, Friday; 9.00 am to 7.00 pm Thursday; 9.00 am to 5.00 pm Saturday
OPEN ACCESS

Local studies collection. Archive includes sound and photographic departments.

Oxford Central Library

Westgate, Oxford OX1 1DJ
☎ 01865 815549   📠 01865 721694
oxfordcentral.library@oxfordshire.gov.uk
www.oxfordshire.gov.uk
OPEN 9.00 am to 7.00 pm Monday to Thursday; 9.00 am to 5.30 pm Friday and Saturday

General lending and reference library including Oxfordshire Studies. Also periodicals, audio visual materials, music library and children's library.

Penzance Library
▷ Morrab Library

City of Plymouth Library and Information Services

Central Library, Drake Circus, Plymouth PL4 8AL
☎ 01752 305923
library@plymouth.gov.uk
www.plymouth.gov.uk/libraries
OPEN ACCESS

CENTRAL LIBRARY LENDING DEPARTMENTS:
Lending ☎ 01752 305912
lendlib@plymouth.gov.uk
Children's Department ☎ 01752 305916
childrens.library@plymouth.gov.uk
Music & Drama Department ☎ 01752 305914
music@plymouth.gov.uk

OPEN 9.00 am to 7.00 pm Monday to Friday; 9.00 am to 5.00 pm Saturday

The Lending departments offer books on all subjects; language courses on cassette and foreign language books; the Holcenberg Jewish Collection; books on music and musicians, drama and theatre; music parts and sets of music parts; play sets; DVDs, videos; song index; CDs; public internet access.

CENTRAL LIBRARY REFERENCE DEPARTMENTS:
Reference ☎ 01752 305907/305908
ref@plymouth.gov.uk
Local Studies & Naval History Department
☎ 01752 305909
localstudies@plymouth.gov.uk

OPEN 9.00 am to 7.00 pm Monday to Friday; 9.00 am to 5.00 pm Saturday

The Reference departments include an extensive collection of Ordnance Survey maps and town guides; community and census information; marketing and statistical information; Patents; books on every aspect of Plymouth; naval history; GRO Index on microfiche; Baring Gould manuscript of 'Folk Songs of the West'; public internet access, including some electronic subscriptions.

Plymouth Proprietary Library

Alton Terrace, 111 North Hill, Plymouth PL4 8JY
☎ 01752 660515
Librarian *Shirley Lamerton*
OPEN Monday to Saturday from 9.30 am (closing time varies).
ACCESS To members; visitors by appointment only.

Founded 1810. The library contains approximately 17,000 volumes of mainly 19th and 20th century work. Member of the **Association of Independent Libraries**.

The Poetry Library
▷ entry under Organizations of Interest to Poets

Polish Library POSK

238–246 King Street, London W6 0RF
☎ 020 8741 0474   🖷 020 8741 7724
library@polishlibrary.co.uk
www.posk.org
Librarian *Mrs Jadwiga Szmidt*
OPEN 10.00 am to 8.00 pm Monday and Wednesday; 10.00 am to 5.00 pm Friday; 10.00 am to 4.00 pm Saturday (library closed Tuesday and Thursday)
ACCESS For reference to all interested in Polish affairs; limited loans to members and *bona fide* scholars only through inter-library loans.

Books, pamphlets, periodicals, maps, music, photographs on all aspects of Polish history and culture. SPECIAL COLLECTIONS Emigré publications; Joseph Conrad and related works; Polish underground publications; bookplates.

Poole Central Library

Dolphin Centre, Poole BH15 1QE
☎ 01202 262424   🖷 01202 262442
libraries@poole.gov.uk
www.boroughofpoole.com/libraries
www.poolepartnership.info/displayOrganisation.
    cfm?OrganisationID=9
Information Services Librarian *David Parish*
OPEN 9.00 am to 6.00 pm Monday to Friday; 9.00 am to 5.00 pm Saturday
OPEN ACCESS

General lending and reference library, including Healthpoint health information centre, business information, children's library, periodicals and newspapers, Internet suite and meeting room.

Public Record Office
▷ The National Archives

Racing Collection (Newmarket)
▷ Suffolk County Council – Suffolk Libraries

Reading Central Library

Abbey Square, Reading RG1 3BQ
☎ 0118 901 5950   🖷 0118 901 5954
info@readinglibraries.org.uk
www.readinglibraries.org.uk
OPEN 9.00 am to 5.30 pm Monday and Friday; 9.00 am to 7.00 pm Tuesday and Thursday; 9.00 am to 5.00 pm Wednesday; 9.30 am to 5.00 pm Saturday
OPEN ACCESS

Ground Floor: Fiction, audio-visual material, children's library; First Floor: Newspapers, non-fiction – biography, business, careers, cookery, DIY, engineering, gardening, health, languages, law, mind, body and spirit, pets, science, social studies, sport, transport, travel; Second Floor: LearnDirect, non-fiction – art, music, literature, plays, computing; Third floor: Non-fiction – world history, national history, local history. Magazines and periodicals and internet access on all floors.

RIBA British Architectural Library Drawings & Archives Collection
▷ British Architectural Library

Richmond Central Reference Library

Old Town Hall, Whittaker Avenue, Richmond TW9 1TP
☎ 020 8940 5529   🖷 020 8940 6899
reference.services@richmond.gov.uk
www.richmond.gov.uk
OPEN 9.30 am to 6.00 pm Monday, Tuesday, Thursday, Friday; 9.30 am to 8.00 pm Wednesday; 9.30 am to 5.00 pm Saturday
OPEN ACCESS

General reference library serving the needs of local residents and organizations. Internet access and online databases for public use. Enquiries received by visit, telephone and e-mail.

Royal Geographical Society (with the Institute of British Geographers)

1 Kensington Gore, London SW7 2AR
☎ 020 7591 3044   🖷 020 7591 3001
enquiries@rgs.org
www.rgs.org/collections
Foyle Reading Room: OPEN 10.00 am to 5.00 pm
Monday to Friday (except Bank Holidays and the
period from Christmas to New Year).

Around two million items on geography and
exploration, including monographs, periodicals,
images, maps and manuscripts. Coverage
primarily from 1830 onwards, with particular
strengths in Africa, Central Asia and Antarctica.
For information on commercial Picture Library
services, see entry under *Picture Libraries*.

Royal Institute of Philosophy
▷ Senate House Library, University of London

The Royal Philharmonic Society Archive
▷ British Library Music Collections

Royal Society Library

6–9 Carlton House Terrace, London SW1Y 5AG
☎ 020 7451 2606   🖷 020 7930 2170
library@royalsociety.org
royalsociety.org
OPEN 10.00 am to 5.00 pm Monday to Friday
ACCESS Open to all researchers with an interest
in the history of science, the Fellowship of the
Royal Society and science policy. Researchers are
advised to contact the Library in advance of their
first visit.

History of science, scientists' biographies, science
policy reports, and publications of international
scientific unions and national academies from all
over the world.

RSA (Royal Society for the Encouragement of Arts, Manufactures & Commerce)

8 John Adam Street, London WC2N 6EZ
☎ 020 7930 5115   🖷 020 7839 5805
general@rsa.org.uk
www.theRSA.org
OPEN Library: 8.00 am to 8.00 pm every weekday.
Archive material by appointment.
ACCESS to Fellows of RSA; Archive: by
appointment to all researchers (contact the
Archivist, 020 7451 6847; archive@rsa.org.uk).

Archives of the Society since 1754. A collection
of approximately 15,000 items including minutes
of the Society, correspondence, prints, original
drawings and international exhibition material
and an early library of over 700 volumes.

Royal Society of Medicine Library

1 Wimpole Street, London W1G 0AE
☎ 020 7290 2940   🖷 020 7290 2939

library@rsm.ac.uk
www.rsm.ac.uk
Contact *Director of Library Services*
OPEN 9.00 am to 5.00 pm Monday to Thursday;
9.00 am to 5.30 pm Friday; 10.00 am to 4.30 pm
Saturday
ACCESS For reference only, on introduction by
Fellow of the Society or temporary membership
is available to non-members; £10 per day; £35 per
week; £90 per month. Identification required.

Books, periodicals, databases on postgraduate
biomedical information. Extensive historical
collection dating from the fifteenth century and
medical portrait collection.

Eric Frank Russell Archive
▷ Science Fiction Foundation Research Library

St Bride Library

Bride Lane, Fleet Street, London EC4Y 8EE
☎ 020 7353 4660
nigelroache@stbridefoundation.org
elizabethklaiber@stbridefoundation.org
www.stbridefoundation.org
Librarian *Nigel Roache*
Assistant Librarian *Elizabeth Klaiber*
OPEN 12 noon to 5.30 pm Tuesday and Thursday;
12 noon to 9.00 pm Wednesday
OPEN ACCESS

A public reference library maintained by the St
Bride Foundation. Appointments advisable for
consultation of special collections. Every aspect
of printing and related matters: publishing and
bookselling, newspapers and magazines, graphic
design, calligraphy and typography, papermaking
and bookbinding. One of the world's largest
specialist collections in its field, with over 50,000
volumes, over 3,000 periodicals (200 current
titles), and extensive collection of drawings,
manuscripts, prospectuses, patents and materials
for printing and typefounding. Noted for its
comprehensive holdings of historical and early
technical literature.

Science Fiction Foundation Research Library

Liverpool University Library, PO Box 123,
Liverpool L69 3DA
☎ 0151 794 3142   🖷 0151 794 2681
asawyer@liverpool.ac.uk
www.sfhub.ac.uk
www.sf-foundation.com
Contact *Andy Sawyer*
ACCESS For research, by appointment only
(telephone first).

This is the largest collection outside the US of
English-language science fiction and related
material – including autobiographies and critical
works. SPECIAL COLLECTIONS Runs of 'pulp'
magazines dating back to the 1920s. Foreign-

language material (including a large Russian collection), and the papers of the Flat Earth Society. The collection also features a growing range of archive and manuscript material, including material from Stephen Baxter, Ramsey Campbell and John Brunner. The University of Liverpool also holds the Olaf Stapledon, Eric Frank Russell and John Wyndham archives.

Science Museum Library

Imperial College Road, London SW7 5NH
☏ 020 7942 4242   📠 020 7942 4243
smlinfo@sciencemuseum.org.uk
www.sciencemuseum.org.uk/library
OPEN 10.00 am to 5.00 pm Monday to Friday
OPEN ACCESS Reference only; no loans.

National reference library for the history and public understanding of science and technology, with a large collection of source material. Operates jointly with Imperial College Central Library.

Scottish Drama Collection/Scottish Poetry Collection
▷ The Mitchell Library

Scottish Poetry Library
▷ entry under Organizations of Interest to Poets

Seckford Collection
▷ Suffolk County Council – Suffolk Libraries

2nd Air Division Memorial Library
▷ Norfolk Library & Information Service

Senate House Library, University of London

Senate House, Malet Street, London WC1E 7HU
☏ 020 7862 8461/62 (Information Centre)
📠 020 7862 8480
enquiries@shl.lon.ac.uk (Information Centre)
www.shl.lon.ac.uk
MEMBERSHIP DESK: ☏ 020 7862 843940
userservices@shl.lon.ac.uk
OPEN Term-time: 9.00 am to 9.00 pm Monday to Thursday; 9.00 am to 6.30 pm Friday; 9.45 am to 5.30 pm Saturday. Vacation: 9.00 am to 6.00 pm Monday to Friday; 9.45 am to 5.30 pm Saturday (closed on Sundays and at certain periods during Bank Holidays).

The Senate House Library is a major academic research library predominantly based across the Humanities and Social Sciences. Housed within its 16 floors are some two million titles including 5,500 current periodicals and a wide range of electronic resources. It contains a number of outstanding research collections which, as well as supporting the scholarly activities of the University, attract researchers from throughout the UK and internationally. These include: English (e.g. the Durning-Lawrence Library and Sterling Collection of first editions); Economic and Social

History (the Goldsmiths' Library, containing 70,000 items ranging from 15th to early 19th century); Modern Languages (primarily Romance and Germanic); Palaeography (acclaimed as being the best open access collection in its field in Europe); History (complementary to the **Institute of Historical Research**); Music, Philosophy (acts as the Library of the Royal Institute of Philosophy); Psychology (includes the BPS library); Major area studies collections (Latin-American, including Caribbean; United States and Commonwealth Studies, British Government Publications and maps). The Library has a wide range of Special Collections holdings. Check the website for full details of collections and current access arrangements.

Sheffield Libraries, Archives and Information

Central Library, Surrey Street, Sheffield S1 1XZ
☏ 0114 273 4712   📠 0114 273 5009
libraries@sheffield.gov.uk
www.sheffield.gov.uk (click on 'In your area')

Central Lending Library
☏ 0114 273 4727 (enquiries)/4729 (book renewals)
OPEN 10.00 am to 8.00 pm Monday; 9.30 am to 5.30 pm Tuesday, Thursday and Friday; 9.30 am to 8.00 pm Wednesday; 9.30 am to 5.30 pm Saturday

Books, talking books, large print, language courses, European fiction, books in cultural languages, play sets. Free internet access. Writers' Resource Centre, Wednesday evenings, 5.00 pm to 7.00 pm. Proof of signature and a separate proof of address required to join.

Sheffield Archives
52 Shoreham Street, Sheffield S1 4SP
☏ 0114 203 9395 📠 0114 203 9398
archives@sheffield.gov.uk
OPEN 10.00 am to 5.30 pm Monday; 9.30 am to 5.30 pm Tuesday to Thursday; 9.00 am to 1.00 pm and 2.00 pm to 5.00 pm Saturday (documents should be ordered by 5.00 pm Thursday for Saturday); closed Friday
ACCESS By reader's card.

Holds documents relating to Sheffield and South Yorkshire, dating from the 12th century to the present day, including records of the City Council, churches, businesses, landed estates, families and individuals, institutions and societies. Free internet access.

Arts and Social Sciences and Sports Reference Service
☏ 0114 273 4747/8
artsandsport.library@sheffield.gov.uk
OPEN 10.00 am to 8.00 pm Monday; 9.30 am to 5.30 pm Tuesday, Friday, Saturday; 9.30 am to 8.00 pm Wednesday
ACCESS Mainly reference but some lending material.

A comprehensive collection of books, periodicals and newspapers covering all aspects of the arts (excluding music), sports, social sciences and free internet access.

Music and Video Service
☎ 0114 273 4733
musicandav.library@sheffield.gov.uk
OPEN As for Central Lending Library above.
ACCESS For reference and lending.

An extensive range of books, CDs, cassettes, scores, etc. related to music. Also a video cassette and DVD loan service. Free internet access.

Local Studies Service
☎ 0114 273 4753
localstudies.library@sheffield.gov.uk
OPEN As for Arts & Social Sciences above (except Wednesday 9.30 am to 5.30 pm)
ACCESS For reference.

Extensive material covering all aspects of Sheffield and its population, including maps, photos and videos. Free internet access. Photograph collection available on www.picturesheffield.co.uk

Business, Science and Technology Library
☎ 0114 273 4736/7 or 273 4743 📠 0114 273 5009
businessandtech.library@sheffield.gov.uk
OPEN As for Arts & Social Sciences above.
ACCESS For reference only.

Extensive coverage of science and technology as well as commerce and commercial law. British patents and British and worldwide standards with emphasis on metals. Hosts the World Metal Index. The business section holds a large stock of business and trade directories, and reference works with business emphasis. Free internet access.

Sheffield Information Service
☎ 0114 273 4712 📠 0114 275 7111
sis@sheffield.gov.uk
OPEN 10.00 am to 5.30 pm Monday; 9.30 am to 5.30 pm Tuesday to Saturday

Full local information service covering all aspects of the Sheffield community. Free internet access.

Central Children's Library
☎ 0114 273 4734
kidsandteens.library@sheffield.gov.uk
OPEN 10.30 am to 5.00 pm Monday and Friday; 1.00 pm to 5.00 pm Tuesday, Wednesday, Thursday; 9.30 am to 5.30 pm Saturday

Books, spoken word cassettes, DVDs, videos; under-five play area; teenage reference section; readings and promotions; storytime and Babytime sessions. Free internet access.

Shetland Library
Lower Hillhead, Lerwick ZE1 0EL
☎ 01595 743868 📠 01595 694430
shetlandlibrary@sic.shetland.gov.uk
www.shetland-library.gov.uk
OPEN 9.30 am to 8.00 pm Monday and Thursday; 9.30 am to 5.00 pm Tuesday, Wednesday, Friday, Saturday

General lending and reference library; extensive local interest collection including complete set of *The Shetland Times*, *The Shetland News* and other local newspapers on microfilm and many old and rare books; audio collection including talking books/newspapers. Junior room for children. Disabled access and Housebound Readers Service (delivery to reader's home). Mobile library services to rural areas. Open Learning Service. Same day photocopying service. Publishing programme of books in dialect, history, literature. Learning centre providing access to learning opportunities including the internet.

Shoe Lane Library
Hill House, Little New Street, London EC4A 3JR
☎ 020 7583 7178
shoelane@cityoflondon.gov.uk
www.cityoflondon.gov.uk/shoelanelibrary
OPEN 9.00 am to 5.30 pm Monday, Wednesday, Thursday, Friday; 9.00 am to 6.30 pm Tuesday
OPEN ACCESS

Corporation of London general lending library, with a comprehensive stock of 50,000 volumes, most of which are on display. Free internet access.

Shrewsbury Library and Reference & Information Service
Castlegates, Shrewsbury SY1 2AS
☎ 01743 255300 📠 01743 255309
shrewsbury.library@shropshire.gov.uk
www.shropshire.gov.uk/library.nsf
OPEN 9.30 am to 5.00 pm Monday, Wednesday, Friday; 9.30 am to 8.00 pm Tuesday and Thursday; 9.00 am to 5.00 pm Saturday; 1.00 pm to 4.00 pm Sunday
OPEN ACCESS

The largest public library in Shropshire. Books, cassettes, CDs, talking books, DVDs, videos, language courses. Open Learning and study centre with public use computers for word processing and internet access. Music, literature and art book collection for lending and reference. The West Midlands Literary Heritage Collection is housed at Shrewsbury Library.

Spanish Institute Library
▷ Instituto Cervantes

Olaf Stapledon
▷ Science Fiction Foundation Research Library

Sterling Collection
▷ Senate House Library

Suffolk County Council – Suffolk Libraries

Endeavour House, Russell Road, Ipswich IP1 2BX
☎ 01473 264285 🖷 01473 216847
help@suffolklibraries.co.uk
www.suffolk.gov.uk/LeisureAndCulture/
Libraries/SuffolkLibrariesDirect
OPEN See website for details of individual libraries
or contact Endeavour House. Major libraries open
seven days a week; all libraries open Sunday.
ACCESS A single library card gives access to
the lending service of 43 libraries and resource
centres across the county. Loans can be collected
and/or returned at any service point. Details on
website.

Full range of lending and reference services, free
public access to the internet. Wide range of online
services for registered library users through the
Suffolk Libraries Direct pages, e.g. self-service
reservations, renewals, free access to subscription
services. Suffolk InfoLink Plus database gives
details of local organizations throughout the
county. SPECIAL COLLECTIONS include Suffolk
Archives and Local History Collection; Benjamin
Britten Collection; Edward Fitzgerald Collection;
Seckford Collection and Racing Collection
(Newmarket). The Suffolk Infolink service
gives details of local groups and societies and is
available in libraries throughout the county and
on the website.

Sunderland City Library and Arts Centre

28–30 Fawcett Street, Sunderland SR1 1RE
☎ 0191 514 1235 🖷 0191 514 8444
enquiry.desk@sunderland.gov.uk
www.sunderland.gov.uk/libraries
OPEN 9.30 am to 7.30 pm Monday and
Wednesday; 9.30 am to 5.00 pm Tuesday,
Thursday, Friday; 9.30 am to 4.00 pm Saturday

The city's main library for lending and reference
services. Local studies and children's sections,
plus Sound and Vision department (CDs, DVDs,
talking books). Sunderland Public Libraries also
maintains community libraries of varying size,
offering a range of services, plus mobile libraries.
Free internet access is available in all libraries
across the city. A Books on Wheels service is
available to housebound readers; the Schools
Library Service serves teachers and schools.
The Foyle Street Writers Group meets every
Wednesday, 10.00 am to 12 noon.

Swansea Central Library

The Civic Centre, Oystermouth Road, Swansea
SA1 3SN
☎ 01792 636464 🖷 01792 637103
central.library@swansea.gov.uk
www.swansea.gov.uk/libraries
OPEN 8.30 am to 8.00 pm Tuesday to Friday;
10.00 am to 4.00 pm Saturday and Sunday. The

Library has a lending service open the same
hours, including books, DVDs, videos, music CDs
and talking books.
ONLINE REFERENCE & ENQUIRY SERVICE
OPEN 8.30 am to 6.00 pm Monday to Friday;
10.00 am to 4.00 pm Saturday Enquiries taken
by phone, e-mail, fax or letter (see contact details
above).
ACCESS For reference only (many local studies
items now available on open access; other items
must be requested on forms provided).

General reference material; also Welsh Assembly
information, statutes, company information,
maps, European Community information. Local
studies: comprehensive collections on Wales;
Swansea & Gower; Dylan Thomas. Local maps,
periodicals, illustrations, local newspapers from
1804. B&w and colour photocopying and fax
facilities, free computer and internet access
available (library membership required and
pre-booking recommended) and microfilm/
microfiche copying facility.

Swiss Cottage Central Library

88 Avenue Road, London NW3 3HA
☎ 020 7974 6522 🖷 020 7974 6532
swisscottagelibrary@camden.gov.uk
OPEN 10.00 am to 7.00 pm Monday; 10.00 am to
6.00 pm Tuesday, Wednesday, Friday; 10.00 am to
8.00 pm Thursday; 10.00 am to 5.00 pm Saturday;
11.00 am to 4.00 pm Sunday
OPEN ACCESS

Over 300,000 volumes in the lending and
reference libraries. Home of the London Borough
of Camden's Information and Reference Services.

Tennyson Research Centre
▷ Lincoln Central Library

Thurrock Communities, Libraries & Cultural Services

Grays Library, Orsett Road, Grays RM17 5DX
☎ 01375 413973 🖷 01375 370806
grays.library@thurrock.gov.uk
www.thurrock.gov.uk/libraries
OPEN 10.00 am to 7.00 pm Monday; 9.00 am to
7.00 pm Tuesday and Thursday; 9.00 am to 5.00
pm Wednesday, Friday, Saturday; branch library
opening times vary.
OPEN ACCESS

General library lending and reference through
ten libraries and a mobile library. Services include
books, magazines, newspapers, audiocassettes,
CDs, videos and language courses. Large
collection of Thurrock materials. Internet and
Microsoft Office.

Truro Library

Union Place, Truro TR1 1EP
☎ 01872 322005 lending)/272702

(reference)/0845 607 2119(24hr-renewals)
truro.library@cornwall.gov.uk
www.cornwall.gov.uk/library
OPEN 8.30 am to 6.00 pm Monday, Tuesday,
Thursday, Friday; 9.30 am to 6.00 pm Wednesday;
9.00 am to 4.00 pm Saturday

Books, cassettes, CDs, videos, DVDs and PC
games for loan through branch or mobile
networks. Internet access and IT support
and training. Reference collection. SPECIAL
COLLECTIONS on local studies.

Vin Mag Archive Ltd

84–90 Digby Road, London E9 6HS
☎ 020 8533 7588   ☷ 020 8525 9209
piclib@vinmag.com
www.vinmagarchive.com
Contact *Angela Maguire*

Wholly owned subsidiary of the Vintage
Magazine Company. A collection of original
printed material – magazines, newspapers,
posters, adverts, books, ephemera and movie
material. Also photographic collection; see entry
under *Picture Libraries*.

West Midlands Literary Heritage Collection
▷ Shrewsbury Library and Reference & Information
Service

Western Isles Libraries

Public Library, 19 Cromwell Street, Stornoway
HS1 2DA
☎ 01851 708631   ☷ 01851 708676
www.cne-siar.gov.uk
OPEN 10.00 am to 5.00 pm Monday to
Wednesday; 10.00 am to 6.00 pm Thursday and
Friday; 10.00 am to 5.00 pm Saturday
OPEN ACCESS

General public library stock, plus local history
and Gaelic collections including maps, videos,
printed music, cassettes and CDs; census records
and Council minutes; music collection (CDs and
cassettes). Branch libraries on the isles of Barra,
Benbecula, Harris and Lewis.

City of Westminster Archives Centre

10 St Ann's Street, London SW1P 2DE
☎ 020 7641 5180/4879 (Minicom)
☷ 020 7641 5179
archives@westminster.gov.uk
www.westminster.gov.uk/archives
OPEN 10.00 am to 7.00 pm Tuesday, Wednesday,
Thursday, 10.00 am to 5.00 pm Friday and
Saturday (closed Monday)
ACCESS For reference.

Comprehensive coverage of the history of
Westminster and selective coverage of general
London history. 22,000 books, together with a
large stock of maps, prints, photographs, local
newspapers, theatre programmes and archives.

Westminster Music Library

Victoria Library, 160 Buckingham Palace Road,
London SW1W 9UD
☎ 020 7641 1300   ☷ 020 7641 4281
musiclibrary@westminster.gov.uk
www.westminster.gov.uk/libraries/special/music
OPEN 11.00 am to 7.00 pm Monday to Friday; 10
am to 5.00 pm Saturday
OPEN ACCESS

Located at Victoria Library, this is the largest
public music library in the South of England,
with extensive coverage of all aspects of music,
including books, periodicals and printed scores.
No recorded material, notated only. Lending
library includes a small collection of CDs and
DVDs.

Westminster Reference Library

35 St Martin's Street, London WC2H 7HP
☎ 020 7641 1300   ☷ 020 7641 5247
referencelibrarywc2@westminster.gov.uk
www.westminster.gov.uk/libraries
OPEN: 10.00 am to 8.00 pm Monday to Friday.
10.00 am to 5.00 pm Saturday
ACCESS For reference only, but there are now
within the Art & Design and Performing
Arts Collections several thousand older
books available for loan to library members.
Membership open to anyone with proof of a
permanent address in Britain.

A general reference library with emphasis
on the following: Art & Design – fine and
decorative arts, architecture, graphics and
design; Performing Arts – theatre, cinema, radio,
television and dance; Official Publications – major
collection of HMSO publications from 1947, plus
parliamentary papers dating back to 1906 (TSO
subscription ceased April 2006); Business –online
and hard copy directories, market research,
industry and company data; Periodicals – long
files of many titles. One working day's notice is
required for some government documents and
some older periodicals. Westminster Reference
Library is free and open to everyone 57 hours
a week, with three study floors and strong
collections in business information, performing
arts and art & design.

The Wiener Library

4 Devonshire Street, London W1W 5BH
☎ 020 7636 7247   ☷ 020 7436 6428
info@wienerlibrary.co.uk
www.wienerlibrary.co.uk
Director *Ben Barkow*
Education and Outreach Coordinator *Katherine
Klinger*
OPEN 10.00 am to 5.30 pm Monday to Friday

ACCESS The Library is open to the public. On your first visit please bring proof of address and some form of photo ID.

The Weiner Library is one of the world's leading and most extensive archives on the Holocaust and Nazi era. Founded in 1933 by Dr Alfred Wiener, the Library's unique collection of over one million items includes published and unpublished works, press cuttings, pamphlets, periodicals, photographs, audiovisual material and eyewitness testimony.

The Wighton Collection of National Music
▷ Dundee Central Library

Vaughan Williams Memorial Library
English Folk Dance and Song Society, Cecil Sharp House, 2 Regent's Park Road, London NW1 7AY
☎ 020 7485 2206 exts. 33 or 34   📠 020 7284 0523
library@efdss.org
www.efdss.org
Library Director *Malcolm Taylor, OBE*
OPEN 9.30 am to 5.30 pm Tuesday to Friday; 10.00 am to 4.00 pm 1st and 3rd Saturday (sometimes closed between 1.00 pm and 2.00 pm)
ACCESS For reference to the general public, on payment of a daily fee; EFDSS members may borrow books and use the library free of charge.

A multimedia collection: books, periodicals, manuscripts, tapes, records, CDs, films, videos, photographs. Mostly British traditional culture and how this has developed around the world. Some foreign language material, and some books in English about foreign cultures. Also, the history of the English Folk Dance and Song Society.

Dr Williams's Library
14 Gordon Square, London WC1H 0AR
☎ 020 7387 3727
enquiries@dwlib.co.uk
www.dwlib.co.uk
OPEN 10.00 am to 5.00 pm Monday, Wednesday, Friday; 10.00 am to 6.30 pm Tuesday and Thursday. The Library will be closed for approximately three months from June 2009 for essential building work.
OPEN ACCESS to reading room (loans restricted to subscribers). Visitors required to supply identification and register for a visitor's ticket. Requests to see manuscripts and rare books must be made in advance; separate registration required. Annual subscription: £10; ministers of religion and undergraduate students: £5.

Primarily a library of theology, religion and ecclesiastical history. Also philosophy, history (English and Byzantine). Particularly important for the study of English Nonconformity. Also major manuscript collections. Trustees of Dr Williams's Library manage the Congregational Library on behalf of the Memorial Hall Trustees. The Dr Williams's Centre for Dissenting Studies is a collaboration with Queen Mary, University of London (see www.english.qmul.ac.uk/drwilliams).

Wolverhampton Central Library
Snow Hill, Wolverhampton WV1 3AX
☎ 01902 552025 (lending)/552026 (reference)
📠 01902 552024
libraries@wolverhampton.gov.uk
www.wolverhampton.gov.uk/libraries
OPEN 9.00 am to 7.00 pm Monday to Thursday; 9.00 am to 5.00 pm Friday and Saturday

General lending and reference libraries, plus children's library and audiovisual library holding cassettes, CDs, videos and music scores. Internet access.

Archives & Local Studies Collection
42–50 Snow Hill, Wolverhampton WV2 4AG
☎ 01902 552480
OPEN 10.00 am to 5.00 pm Monday, Tuesday, Friday, 1st and 3rd Saturday of each month; 10.00 am to 7.00 pm Wednesday; closed Thursday

The Women's Library
London Metropolitan University, Old Castle Street, London E1 7NT
☎ 020 7320 2222   📠 020 7320 2333
moreinfo@thewomenslibrary.ac.uk
www.thewomenslibrary.ac.uk
OPEN Reading Room: 9.30 am to 5.00 pm Tuesday to Friday (8.00 pm Thursday); closed Saturday; Exhibition: 9.30 am to 5.30 pm Monday to Friday (8.00 pm Thursday); 10.00 am to 4.00 pm Saturday
OPEN ACCESS

The Women's Library, national research library for women's history, is the UK's oldest and most comprehensive research library on all aspects of women in society, with both historical and contemporary coverage. The Library includes materials on feminism, work, education, health, the family, law, arts, sciences, technology, language, sexuality, fashion and the home. The main emphasis is on Britain but many other countries are represented, especially the Commonwealth and the developing countries. Established in 1926 as the library of the London Society of Women's Service (formerly Suffrage), a non-militant organization led by Millicent Fawcett. In 1953 the Society was renamed after her and the library became the Fawcett Library.

Collections include: women's suffrage, work, education, women and the church, the law, sport, art, music, abortion, prostitution. Mostly British materials but some American, Commonwealth and European works. Books, journals, pamphlets, archives, photographs, posters, postcards, audiovisual materials, artefacts, scrapbooks, albums and press cuttings dating mainly from the 19th century although some materials date from the 17th century.

The Library's new building, which opened in 2002, includes a reading room, exhibition space, café, education areas and a conference room, and is the cultural and research centre for anyone interested in women's lives and achievements.

Worcestershire Libraries and Learning

Cultural Services, Worcestershire County Council, County Hall, Spetchley Road, Worcester WR5 2NP
℡ 01905 822819 🖷 01905 766930
libraries@worcestershire.gov.uk
www.worcestershire.gov.uk/libraries
OPEN Opening hours vary in the 22 libraries, the History Centre and mobile libraries covering the county; all full-time libraries open at least one evening a week until 8.00 pm, and on Saturday until 5.30 pm; part-time libraries vary.
ACCESS Information and reference services open to anyone; loans to members only (membership criteria: resident, being educated, working, or an elector in the county or neighbouring authorities; temporary membership to visitors. Proof of identity and address required). No charge for membership or for borrowing books.

Information service, and reference and lending libraries. Non-fiction and fiction for all age groups, including normal and large print, spoken word cassettes, sound recordings (CD, cassette), videos, maps, local history, CD-ROMs for reference at main libraries, free public internet access in all libraries. Joint Libraries Service/County Record Office History Centre with resources for local and family history. SPECIAL COLLECTIONS Carpets and Textiles; Needles & Needlemaking; Stuart Period; A.E. Housman.

The Wordsworth Trust (Jerwood Centre)
▷ entry under Literary Societies

John Wyndham Archive
▷ Science Fiction Foundation Research Library

York Central Library
Museum Street, York YO1 7DS
℡ 01904 551550 🖷 01904 552835

lending@york.gov.uk
reference.library@york.gov.uk
www.york.gov.uk/libraries

Lending Library
OPEN 9.30 am to 8.00 pm Monday, Tuesday, Friday; 9.30 am to 5.30 pm Wednesday and Thursday; 9.30 am to 4.00 pm Saturday (NB closed until 1.00 pm on the first Thursday of each month).

General lending library including CDs, DVDs, audio books, children's storytapes language courses and printed music. Large print books. Photocopying, internet and fax facilities.

Reference Library
OPEN 9.00 am to 8.00 pm Monday, Tuesday, Wednesday, Friday; 9.00 am to 5.30 pm Thursday; 9.00 am to 4.00 pm Saturday (NB closed until 1.00 pm on the first Thursday of each month).

General reference library; periodicals, newspapers, local newspaper index, EU information, organizations database; local studies library for York and surrounding area; microfilm/fiche readers for national and local newspapers; census returns and family history resource; general reference collection. Maintains strong links with other local history resource centres, namely the Borthwick Institute, York City Archive and York Minster Library. CD-ROM and internet facilities. In addition, 13 branch libraries and one mobile serve the City of York Council area.

Zoological Society Library
Regent's Park, London NW1 4RY
℡ 020 7449 6293 🖷 020 7586 5743
library@zsl.org
www.zsl.org
OPEN 9.30 am to 5.30 pm Monday to Friday
ACCESS To members and staff; non-members require proof of address and photographic ID.

160,000 volumes on zoology including 5,000 journals (1,300 current) and a wide range of books on animals and particular habitats. Slide collection available and many historic zoological prints. Library catalogue available online (library.zsl.org).

The Zweig Collection of Music & Literary Mss
▷ British Library Music Collections

Picture Libraries

actionplus sports images

54–58 Tanner Street, London SE1 3PH
☏ 020 7403 1558   ⊞ 020 7403 1526
info@actionplus.co.uk
www.actionplus.co.uk

Founded 1986. Specialist sports and action library providing complete and creative international coverage of over 300 varieties of professional and amateur sport. Elite, junior, professional, amateur, extreme, disabled, minority, offbeat, major and minor events, adults, children, high and low, fast and slow. Millions of images covering over 30 years of sport.

Lesley & Roy Adkins Picture Library

Ten Acre Wood, Whitestone, Exeter EX4 2HW
☏ 01392 811357
mail@adkinshistory.com
www.adkinshistory.com

Colour and some b&w coverage of archaeology, antiquity, heritage and related subjects in the UK, Europe, Egypt and Turkey. Subjects include towns, villages, housing, landscape and countryside, churches, temples, castles, monasteries, art and architecture, gravestones and tombs, inscriptions, naval, nautical and antiquarian views. Images supplied in digital form. No service charge if pictures are used.

The Advertising Archive Limited

45 Lyndale Avenue, London NW2 2QB
☏ 020 7435 6540   ⊞ 020 7794 6584
suzanne@advertisingarchives.co.uk
www.advertisingarchives.co.uk
Contacts *Suzanne Viner, Larry Viner*

With over one million images, the largest collection of British and American press ads, TV commercial stills and magazine cover illustrations in Europe. Material spans the period from 1850 to the present day. In-house research; rapid service, competitive rates. On-line database and digital delivery available.

akg-images Ltd, The Arts and History Picture Library

5 Melbray Mews, 158 Hurlingham Road, London SW6 3NS
☏ 020 7610 6103   ⊞ 020 7610 6125
enquiries@akg-images.co.uk
www.akg-images.co.uk

250,000 images online with direct access to ten million kept in the Berlin AKG Library. Specializes in art, archaeology, history, topography, music, personalities and film.

Bryan & Cherry Alexander Photography

Higher Cottage, Manston, Sturminster Newton DT10 1EZ
☏ 01258 473006   ⊞ 01258 473333
alexander@arcticphoto.co.uk
www.arcticphoto.com
Contact *Cherry Alexander*

Arctic and Antarctic specialists; indigenous peoples, wildlife and science in polar regions; Canada, Greenland, the Scandinavian Arctic, Siberia and Alaska.

Alpine Garden Society

AGS Centre, Avon Bank, Pershore WR10 3JP
☏ 01386 554790   ⊞ 01386 554801
ags@alpinegardensociety.net
www.alpinegardensociety.net
Contact *Peter Sheasby*

Over 30,000 colour transparencies (35mm) covering plants in the wild from many parts of the world; particularly strong in plants from mountain and sub-alpine regions, and from Mediterranean climates; South Africa, Australia, Patagonia, California and the Mediterranean. Extensive coverage of show alpines in pots and in gardens and of European orchids. Specific slide lists available on request.

Alvey & Towers

A37 The Springboard Centre, Mantle Lane, Coalville LE67 3DW
☏ 01530 450011
office@alveyandtowers.com
www.alveyandtowers.com

Houses one of the country's most comprehensive collections of transport images depicting not only actual transport systems but their surrounding industries as well. Also specialist modern railway image collection.

Andes Press Agency

26 Padbury Court, London E2 7EH
☎ 020 7613 5417 📠 020 7739 3159
apa@andespressagency.com
www.andespressagency.com
Contacts *Val Baker, Carlos Reyes*

80,000 colour transparencies and 300,000 b&w, specializing in social documentary, world religions, Latin America and Britain.

Heather Angel/Natural Visions

Highways, 6 Vicarage Hill, Farnham GU9 8HG
☎ 01252 716700 📠 01252 727464
hangel@naturalvisions.co.uk
www.naturalvisions.co.uk
Contact *Heather Angel*

Constantly expanding worldwide natural history, wildlife and landscapes: polar regions, tropical rainforest flora and fauna, all species of plants and animals in natural habitats from Africa, Asia (huge Green China file), Australasia, South America and USA, urban wildlife, pollution, biodiversity, global warming. Also worldwide gardens and cultivated flowers. Can send a digital lightbox to authors with an e-mail. 'We supply high resolution digital files direct to the publisher.'

Architectural Association Photo Library

36 Bedford Square, London WC1B 3ES
☎ 020 7887 4066 📠 020 7414 0782
www.aaschool.ac.uk/photlib
Contacts *Valerie Bennett (valerie@aaschool. ac.uk), Sarah Franklin (sfranklin@aaschool.ac.uk), Henderson Downing (henderson@aaschool.ac.uk)*

100,000 35mm transparencies on architecture, historical and contemporary. Archive of large-format b&w negatives from the 1920s and 1930s.

ArenaPAL

Thompson House, 42–44 Dolben Street, London SE1 0UQ
☎ 020 7403 8542 📠 020 7403 8561
enquiries@arenapal.com
www.arenapal.com
Contacts *Biddy Hayward, Robert Piwko, Mike Markiewicz*

'Probably the best collection of entertainment images in the world.' Continually updated and covering classical music, opera, theatre, musicals, film, pop, rock, jazz, instruments, festivals, venues, circus, ballet and contemporary dance, props and personalities. Over two million images

from the late 19th century onwards. Phone, fax or e-mail requests.

ArkReligion.com

57 Burdon Lane, Cheam SM2 7BY
☎ 020 8642 3593 📠 020 8395 7230
images@artdirectors.co.uk
www.arkreligion.com
Contacts *Helene Rogers, Bob Turner*

Extensive coverage of all religions.

The Art Archive

2 The Quadrant, 135 Salusbury Road, London NW6 6RJ
☎ 020 7624 3500 📠 020 7624 3355
info@picture-desk.com
www.picture-desk.com

Picture library holding over 80,000 high-resolution images covering fine art, history and ancient civilizations.

Art Directors & Trip Photo Library

57 Burdon Lane, Cheam SM2 7BY
☎ 020 8642 3593 📠 020 8395 7230
images@artdirectors.co.uk
www.artdirectors.co.uk
Contacts *Helene Rogers, Bob Turner*

Extensive coverage, with over 750,000 images, of all countries, lifestyles, peoples, etc. with detailed coverage of all religions. Backgrounds a speciality.

Aspect Picture Library Ltd

40 Rostrevor Road, London SW6 5AD
☎ 020 7736 1998 📠 020 7731 7362
Aspect.Ldn@btinternet.com
www.aspect-picture-library.co.uk

Colour and b&w worldwide coverage of countries, events, industry and travel, with large files on art, namely paintings, space, China, the Middle East, French villages, English villages and Ireland.

Australia Pictures

28 Sheen Common Drive, Richmond TW10 5BN
☎ 020 7602 1989 📠 020 7602 1989
equilibrium.films@virgin.net
Contact *John Miles*

Collection of 4,000 transparencies covering all aspects of Australia: Aboriginal people, paintings, Ayers Rock, Kakadu, Tasmania, underwater, reefs, Arnhem Land, Sydney. Also Africa, Middle East and Asia.

aviation-images.com

42B Queens Road, London SW19 8LR
☎ 020 8944 5225 📠 020 8944 5335
pictures@aviation-images.com
www.aviation-images.com
Contacts *Mark Wagner, Denise Lalonde*

Two million 'stunning' aviation images, civil and military, archive and modern, from the world's best aviation photographers. Member of **BAPLA** and RAeS.

Barnaby's Library
▷ Mary Evans Picture Library

Barnardo's Photographic and Film Archive
Barnardo's House, Tanners Lane, Barkingside, Ilford IG6 1QG
☎ 020 8498 7345   📠 020 8550 0429
stephen.pover@barnardos.org.uk
www.barnardos.org.uk
Image Librarian *Stephen Pover*

The Barnardo's image archive is a unique resource of images of social history dating from 1874. Images can be sent either by post or e-mail.

Colin Baxter Photography Limited
The Old Dairy, Grantown-on-Spey PH26 3NA
☎ 01479 873999   📠 01479 873888
sales@colinbaxter.co.uk
www.colinbaxter.co.uk
Contact *Mike Rensner (Editorial)*

Specializes in Scotland. Publishes guidebooks and maps, plus books, calendars and postcards on landscape, cityscape and natural history, and a range of books and stationery based on reproductons of the designs of Charles Rennie Mackintosh. Also publishers of the World*Life*Library series of natural history books.

The Photographic Library Beamish, The North of England Open Air Museum
Regional Resource Centre, The North of England Open Air Museum, Beamish DH9 ORG
☎ 0191 370 4000   📠 0191 370 4001
museum@beamish.org.uk
www.beamish.org.uk
www.beamishcollections.com

An archive of over 300,000 images representing everyday life from 1860 to the present day. The images cover agricultural, industrial, topographical, advertising and shop scenes, and people at work and play. Major collections include: Durham Advertiser newspaper; Huwood Mining Machinery; ICI Teesside, Consett Iron Company, Farmer's Guardian and Agricultural Gazette. Digital images are available to view by appointment weekdays, and searches can be arranged. A sound archive and reference Library are also available at the Regional Resource Centre.

Francis Bedford
▷ Birmingham Library Services under Library Services

Ivan J. Belcher Colour Picture Library
57 Gibson Close, Abingdon OX14 1XS
☎ 01235 521524   📠 01235 521524

Extensive colour picture library specializing in top-quality medium-format transparencies depicting the British scene. Particular emphasis on tourist, holiday and heritage locations, including famous cities, towns, picturesque harbours, rivers, canals, castles, cottages, rural scenes and traditions photographed throughout the seasons. Mainly of recent origin and constantly updated.

BFI Stills Sales
BFI, 21 Stephen Street, London W1T 1LN
☎ 020 7957 4797   📠 020 7323 9260
stills.films@bfi.org.uk
www.bfi.org.uk/collections/stills/index.html

'BFI Stills Sales is the world's most comprehensive collection of film and television images.' The collection captures on and off screen moments, portraits of the world's most famous stars – and those behind the camera who made them famous – as well as publicity posters, set designs, images of studios, cinemas, special events and early film and TV technologies.

Peter Boardman Collection
▷ Chris Bonington Picture Library

Chris Bonington Picture Library
Badger Hill, Hesket Newmarket, Wigton CA7 8LA
☎ 016974 78286   📠 016974 78238
frances@bonington.com
www.bonington.com
Contact *Frances Daltrey*

Based on the personal collection of climber and author Chris Bonington and his extensive travels and mountaineering achievements; also work by Doug Scott and other climbers, including the Peter Boardman and Joe Tasker Collections. Full coverage of the world's mountains, from British hills to Everest, depicting expedition planning and management stages, the approach march showing inhabitants of the area, flora and fauna, local architecture and climbing action shots on some of the world's highest mountains.

Boulton and Watt Archive
▷ Birmingham Library Services under Library Services

The Bridgeman Art Library
17–19 Garway Road, London W2 4PH
☎ 020 7727 4065   📠 020 7792 8509
research@bridgeman.co.uk
www.bridgemanart.co.uk
Head of Picture Research *Jenny Page*

Fine art photo archive representing more than 1,000 museums, galleries and picture owners around the world. High res digital files and large-format colour transparencies of paintings, sculptures, prints, manuscripts, antiquities, photographs and the decorative arts. From prehistoric sculpture to contemporary paintings. Offices in London, Paris, New York and Berlin. Collections represented include the British Library, National Galleries of Scotland, Boston Museum of Fine Arts, the Wallace Collection, The Courtauld, Detroit Institute of Arts, The Barnes Foundation, the National Army Museum, the National Trust as well as the Giraudon archive.

The British Cartoon Archive

The Templeman Library, University of Kent, Canterbury CT2 7NU
☎ 01227 823127 📠 01227 823127
N.P.Hiley@kent.ac.uk
J.M.Newton@kent.ac.uk
www.cartoons.ac.uk
Contacts *Dr Nicholas Hiley, Jane Newton*

A national research archive of over 120,000 original cartoons and caricatures by 350 British cartoonists, supported by a library of books, papers, journals, catalogues, cuttings and assorted ephemera. A computer database of 120,000 cartoons provides quick and easy catalogued access via the web. A source for exhibitions and displays as well as a picture library service. Specializes in historical, political and social cartoons from British newspapers and magazines.

British Library Images Online

British Library, 96 Euston Road, London NW1 2DB
Images Online ☎ 020 7412 7614 (Research)/7755 (Reproduction rights) 📠 020 7412 7771
imagesonline@bl.uk www.imagesonline.bl.uk

Images Online offers instant access to thousands of images from the British Library's historic worldwide collections which include manuscripts, rare books, music scores and maps spanning almost 3,000 years. A collection of 20,000 images including digital images of drawings, illustrations and photographs. Specialities: historical images.

Brooklands Museum Photo Archive

Brooklands Museum, Brooklands Road, Weybridge KT13 0QN
☎ 01932 857381 📠 01932 855465
info@brooklandsmuseum.com
www.brooklandsmuseum.com
Head of Collections *John Pulford*
General Manager – Operations *Julian Temple*

About 40,000 b&w and colour prints and slides. Subjects include: Brooklands Motor Racing 1907–1939; British aviation and aerospace, 1908 to the present day, particularly BAC, Hawker, Sopwith and Vickers aircraft built at Brooklands.

Hamish Brown MBE, D.Litt, FRSGS Scottish Photographic

26 Kirkcaldy Road, Burntisland KY3 9HQ
☎ 01592 873546 📠 01592 873546
Contact *Hamish Brown*

Coverage of most topics and areas of Scotland (sites, historic, buildings, landscape, mountains), also travel, mountains, general (50,000 items) and Morocco. Commissions undertaken.

Capital Pictures

85 Randolph Avenue, London W9 1DL
☎ 020 7286 2212
sales@capitalpictures.com
www.capitalpictures.com
Contact *Phil Loftus*

Up-to-date celebrity and entertainment photography from around the world and a film stills collection of all the great films and yet to be released blockbusters. The website is fully searchable.

Cephas Picture Library

A1 Kingsway Business Park, Oldfield Road, Hampton TW12 2HD
☎ 020 8979 8647 📠 020 8941 4001
pictures@cephas.com
www.cephas.com
Contact *Mick Rock*

The wine industry and vineyards of the world is the subject on which Cephas has made its reputation. '100,000 images make this the most comprehensive and up-to-date archive in Britain.' Almost all wine-producing countries and all aspects of the industry are covered in depth. Spirits, beer and cider also included. A major food and drink collection now also exists, through preparation and cooking, to eating and drinking.

Giles Chapman Library

25 Homefield Road, Sevenoaks TN13 2DU
☎ 01732 452547
chapman.media@virgin.net
www.gileschapman.com
Contact *Giles Chapman*

Some 250,000 colour and b&w images of cars and motoring, from 1945 to the present day. No research fees.

Christian Aid
▸ **Still Pictures**

Christie's Images Ltd

1 Langley Lane, London SW8 1TJ
☎ 020 7582 1282 📠 020 7582 5632
imageslondon@christies.com
www.christiesimages.com
Contact *Mark Lynch*

The collection spans 80 categories representing every culture, from the distant past to the present day. An unrivalled collection from Old Masters to Impressionists and Twentieth Century masterpieces. Antiquities, Art Nouveau and Decorative Art, Books and Manuscripts, Contemporary Art and Furniture, Music and

Movie Memorabilia, Textiles, Toys. Experienced researchers will search for specific requests.

Classical Collection Ltd

22 Avon, Hockley, Tamworth B77 5QA
☎ 01827 286086/07963 194921   🖷 01827 286086
neil@classicalcollection.co.uk
Managing Director *Neil Arthur Williams, MA (Hum), MA (Mus)*

Archive specializing in classical music ephemera, particularly portraits of composers, musicians, conductors and opera singers, comprising old and very rare photographs, postcards, antique prints, cigarette cards, stamps, First Day Covers, concert programmes, Victorian newspapers, etc. Also modern photos of composer references such as museums, statues, busts, paintings, monuments, memorials and graves. Other subjects covered include ballet, musical instruments, concert halls, opera houses, 'music in art', manuscripts, opera scenes, music-caricatures, bands, orchestras and other music groups. Neil Arthur Williams is a qualified music historian, musicologist and freelance writer of music articles, concert programme notes and CD as well as being a commissionable composer.

John Cleare/Mountain Camera

Hill Cottage, Fonthill Gifford, Salisbury SP3 6QW
☎ 01747 820320   🖷 01747 820320
cleare@btinternet.com
www.mountaincamera.com

A wide portfolio includes specialist coverage in colour and b&w of mountains and wild places, climbing, ski-touring, expeditions and travel, landscapes, geographical features and people from all continents, especially Britain, the Alps, the Himalaya and the Americas. Topics range from reindeer in Lapland and camels in Australia, white-water rafting in Utah to skiing in China. Recent projects have included photographing books on rock climbing and canoeing for youngsters and providing abstract and poetic images for titles on Japanese haiku, Chinese wisdom and Buddhist teaching. Commissions, presentations and consultancy work undertaken. Researchers welcome by appointment. Member of **BAPLA** and the **OWPG**.

Michael Cole Camerawork

The Coach House, 27 The Avenue, Beckenham BR3 5DP
☎ 020 8658 6120   🖷 020 8658 6120
mikecole@dircon.co.uk
Contacts *Michael Cole, Derrick Bentley*

Probably the largest and most comprehensive collection of tennis pictures in the world. Over 50 years' coverage of the Wimbledon Championships. M.C.C. incorporates the tennis

archives of Le Roye Productions, established in 1945. Picture requests and enquiries by e-mail.

Collections

13 Woodberry Crescent, London N10 1PJ
☎ 020 8883 0083   🖷 020 8883 9215
collections@btinternet.com
www.collectionspicturelibrary.co.uk
www.collectionspicturelibrary.com
Contact *Simon Shuel*

Extensive coverage of the British Isles and Ireland, from the Shetlands to the Channel Islands and Connemara to East Anglia, including people, traditional customs, landscapes, buildings old and new – both well known and a bit obscure, 'plus a considerable collection of miscellaneous bits and pieces which defy logical filing'. Images supplied digitally or as original transparencies. Visitors welcome but please call first.

Corbis Images

111 Salusbury Road, London NW6 6RG
☎ 020 7644 7644   🖷 020 7644 7645
sales.uk@corbis.com
www.corbis.com

A unique and comprehensive resource containing more than 65 million images, with over 2.1 million of them available online. The images come from professional photographers, museums, cultural institutions and public and private collections worldwide. Subjects include history, travel, celebrities, events, science, world art and cultures.

Sylvia Cordaiy Photo Library

45 Rotherstone, Devizes SN10 2DD
☎ 01380 728327   🖷 01380 728328
info@sylvia-cordaiy.com
www.sylvia-cordaiy.com

Over 180 countries on file from the obscure to main stock images – Africa, North, Central and South America, Asia, Atlantic, Indian and Pacific Ocean islands, Australasia, Europe, polar regions. Covers travel, architecture, ancient civilizations, world heritage sites, people worldwide, environment, wildlife, natural history, Antarctica, domestic pets, livestock, marine biology, veterinary treatment, equestrian, ornithology, flowers. UK files cover cities, towns villages, coastal and rural scenes, London collection. Transport, railways, shipping and aircraft (military and civilian). Aerial photography. Backgrounds and abstracts. Also the Paul Kaye B/W archive. All images available by digital delivery.

Country Images Picture Library

27 Camwood, Clayton-le-Woods, Bamber Bridge PR5 8LA
☎ 01772 321243   🖷 0870 137 8888

tm@wpu.org.uk
www.countryimages.info
Contact *Terry Marsh*

35mm film and digital colour coverage of landscapes and countryside features generally throughout the UK (Cumbria, North Yorkshire, Lancashire, southern Scotland, Isle of Skye, St Kilda and Scottish islands, Wales, Cornwall), France (Alps, Auvergne, Pyrenees, Provence, Charente-Maritime, Hérault, Aube-en-Champagne, Loire valley, Somme), Madeira, the Azores and Australia. Commissions undertaken.

Country Life Picture Library

Blue Fin Building, 110 Southwark Street, London SE1 0SU
☎ 020 3148 4474
clpicturelibrary@ipcmedia.com
www.clpicturelibrary.co.uk
Contact *Justin Hobson*

Over 150,000 b&w negatives, dating back to 1897, and 100,000 colour transparencies. Country houses, stately homes, churches and town houses in Britain and abroad, interiors of architectural interest (ceilings, fireplaces, furniture, paintings, sculpture), and exteriors showing many landscaped gardens, sporting and social events, crafts, people and animals. Visitors by appointment.

Philip Craven Worldwide Photo-Library

Surrey Studios, 21 Nork Way, Nork, Banstead SM7 1PB
☎ 01737 814646
www.philipcraven.com
Contact *Philip Craven*

Extensive coverage of British scenes, cities, villages, English countryside, gardens, historic buildings and wildlife. Worldwide travel and wildlife subjects on medium and large-format transparencies.

Sue Cunningham Photographic

56 Chatham Road, Kingston upon Thames KT1 3AA
☎ 020 8541 3024
info@scphotographic.com
www.scphotographic.com
Contacts *Patrick Cunningham, Sue Cunningham*

Extensive coverage of many geographical areas: South America (especially Brazil), Eastern Europe from the Baltic to the Balkans, various African countries, Western Europe including the UK. Colour and b&w.

Dale Concannon Collection
▷ Phil Sheldon Golf Picture Library

Dalton–Watson Collection
▷ Ludvigsen Library

The Defence Picture Library

14 Mary Seacole Road, Howeson Court, Plymouth PL1 3JY
☎ 01752 312061 ☎ 01752 312063
picdesk@defencepictures.com
www.defencepictures.com
Contacts *David Reynolds, Jessica Kelly, James Rowlands*

Specialist source of military photography covering all areas of the UK Armed Forces, supported by a research agency of facts and figures. More than one million images. Visitors welcome by appointment.

Patrick Eagar Photography

☎ 020 8940 9269
info@patrickeagar.com
www.patrickeagar.com

The cricket library consists of Patrick Eagar's work over the last 30 years with coverage of over 290 Test matches worldwide, unique coverage of all eight World Cups, countless one-day internationals and player action portraits of over 2,000 cricketers. The wine library consists of vineyards, grapes and festivals from Argentina to New Zealand. France and Australia are specialist areas. Photographs can be supplied by e-mail, on CD or DVD.

Ecoscene

Empire Farm, Throop Road, Templecombe BA8 0HR
☎ 01963 371700
pix@ecoscene.com
www.ecoscene.com
Contact *Sally Morgan*

Expanding colour library of over 80,000 transparencies specializing in all aspects of the environment: pollution, conservation, recycling, restoration, wildlife (especially underwater), habitats, education, industry and agriculture. All parts of the globe are covered with specialist collections covering Antarctica, Australia, North America. Sally Morgan, who runs the library, is a professional ecologist and expert source of information on all environmental topics. Photographic and writing commissions undertaken. Images delivered by e-mail or available for download.

Edifice

Cutterne Mill, Evercreech BA4 6LY
☎ 01749 831400
info@edificephoto.com
www.edificephoto.com
Contacts *Philippa Lewis, Gillian Darley*

Colour material on exteriors of all possible types of building worldwide from grand landmarks to the small and simple. Specialists in domestic architecture, garden features, architectural detail, ornament, historical and period style, building materials and techniques. Website searchable online, but expert help and advice happily given over the phone or by e-mail.

Education Photos

8 Whitemore Road, Guildford GU1 1QT
☎ 01483 511666
johnwalmsley@educationphotos.co.uk
www.educationphotos.co.uk

Formerly the John Walmsley Photo Library. Specialist library of learning/training/working subjects. Comprehensive coverage of learning environments such as playgroups, schools, colleges and universities. Images reflect a multi-racial Britain. Commissions undertaken.

English Heritage Photo Library

Kemble Drive, Swindon SN2 2GZ
☎ 01793 414903
images@english-heritage.org.uk
www.englishheritageimages.com
Contacts *Duncan Brown, Javis Gurr, Jonathan Butler*

Images of English castles, abbeys, houses, gardens, Roman remains, ancient monuments, battlefields, industrial and post-war buildings, interiors, paintings, artifacts, aerials, scenic views, landscapes. Special features for 2010: Aerofilms, Victorian England. See also **National Monuments Record** entry.

Mary Evans Picture Library

59 Tranquil Vale, Blackheath, London SE3 0BS
☎ 020 8318 0034   📠 020 8852 7211
pictures@maryevans.com
www.maryevans.com

Leading historical archive of artwork, photographs, prints and ephemera illustrating and documenting all aspects of the past, from ancient times to the later decades of the 20th century. Emphasis on everyday life plus extensive coverage of events, portraits, transport, and costume. 'Unrivalled material on folklore and paranormal phenomena.' Notable collections include the Weimar Archive documenting the Third Reich, and Barnaby's Library of social documentary photography from the 1930s to the 1970s. Own material complemented by important contributor collections, such as the **Illustrated London News Picture Library** (see entry); the archive of the National Magazine Company; the Harry Price Library of Magical Literature; Sigmund Freud Copyrights; the Women's Library and the Castle Howard Collection of fine art and prints. Arrangements with agencies in France (Rue des Archives); Italy (De Agostini); Germany (Interfoto); Austria (Imagno) and Spain (Aisa)

have added thousands of additional photographs and fine art images to the website. Over 220,000 images searchable online, with 120,000 available in high-resolution for immediate download. Brochure on request. Founder member of **BAPLA**. Compilers of the *Picture Researcher's Handbook* published by Pira International.

Eye Ubiquitous

65 Brighton Road, Shoreham BN43 6RE
☎ 01273 440113   📠 01273 440116
library@eyeubiquitous.com
www.eyeubiquitous.com

General stock specializing in social documentary worldwide, and an extensive travel related collection.

Faces and Places

28 Sheen Common Drive, Richmond TW10 5BN
☎ 020 7602 1989/07930 622964
equilibrium.films@virgin.net
john@facesandplacespix.com
www.facesandplacespix.com

Extensive worldwide library built up over twenty years which covers 40,000 photographs focusing mainly on travel, tourism, unusual locations, indigenous peoples, scenery and underwater photography; Australia, Tibet, Mali, Yemen and Iran.

ffotograff

10 Kyveilog Street, Pontcanna, Cardiff CF11 9JA
☎ 029 2023 6879
ffotograff@easynet.co.uk
www.ffotograff.com
Contact *Patricia Aithie*

Library and agency specializing in travel, exploration, the arts, architecture, traditional culture, archaeology and landscape. Based in Wales but specializing in the Middle and Far East; Africa, Central and South America; Yemen and Wales are strong aspects of the library. Churches and cathedrals of Britain and Crusader castles.

Financial Times Pictures

One Southwark Bridge, London SE1 9HL
☎ 020 7873 3671   📠 020 7873 4606
photosynd@ft.com
Contacts *Nicky Burr, Richard Pigden*

Photographs from around the world ranging from personalities in business, politics and the arts, people at work and other human interests and activities. Delivery via ftp or e-mail.

Fine Art Photographic Library Ltd

2A Milner Street, London SW3 2PU
☎ 020 7589 3127
mail@fineartphotolibrary.com
www.fineartphotolibrary.com

Over 30,000 large-format transparencies, with a specialist collection of 19th and 20th century paintings. CD-ROM available.

Firepix International

68 Arkles Lane, Anfield, Liverpool L4 2SP
☎ 0151 260 0111/0777 5930419   🖷 0151 260 0111
info@firepix.com
www.firepix.com
Contact *Tony Myers*

The UK's only fire photo library. 23,000 images of fire-related subjects, firefighters, fire equipment manufacturers. Website contains 15 categories from industrial fire, domestic, digital images and abstract flame.

Fogden Wildlife Photographs

16 Locheport, North Uist, Western Isles HS6 5EU
☎ 01876 580245   🖷 01876 580777
susan.fogden@fogdenphotos.co.uk
www.fogdenphotos.com
Contact *Susan Fogden*

Natural history collection, with special reference to rain forests and deserts. 'Emphasis on quality rather than quantity'; growing collection of around 25,000 images.

Food Features

Stream House, West Flexford Lane, Wanborough, Guildford GU3 2JW
☎ 01483 810840
frontdesk@foodfeatures.net
www.foodfeatures.net
Contacts *Steve Moss, Alex Barker*

Specialized high-quality food and drink photography, features and tested recipes. Clients' specific requirements can be incorporated into regular shooting schedules. Commissions welcome.

Forest Life Picture Library

231 Corstorphine Road, Edinburgh EH12 7AT
☎ 0131 314 6411
neill.campbell@forestry.gsi.gov.uk
www.forestry.gov.uk/pictures
Contact *Neill Campbell*

The official image bank of the Forestry Commission, the library provides a single source for all aspects of forest and woodland management. The comprehensive subject list includes tree species, scenic landscapes, employment, wildlife, flora and fauna, conservation, sport and leisure.

Werner Forman Archive Ltd

36 Camden Square, London NW1 9XA
☎ 020 7267 1034   🖷 020 7267 6026
wfa@btinternet.com
www.werner-forman-archive.com

Colour and b&w coverage of ancient civilizations, oriental and primitive societies around the world. A number of rare collections. Searchable website.

Formula One Pictures

☎ 00 36 26 322 826/00 36 70 776 9682 (mobile)
jt@f1pictures.com
www.f1pictures.com
Contact *John Townsend*

500,000 35mm colour slides, b&w and colour negatives and digital images of all aspects of Formula One Grand Prix racing from 1980 including driver profiles and portraits.

Robert Forsythe Picture Library

16 Lime Grove, Prudhoe NE42 6PR
☎ 01661 834511   🖷 01661 834511
robert@forsythe.demon.co.uk
www.forsythe.demon.co.uk
Contacts *Robert Forsythe, Fiona Forsythe*

30,000 transparencies of industrial and transport heritage; plus a unique collection of 250,000 items of related publicity ephemera from 1945. Image finding service available. Robert Forsythe is a transport/industrial heritage historian and consultant. Nationwide coverage, particularly strong on northern Britain. A bibliography of published material is available.

Fresh Food Images

20 Blades Court, Deodar Road, Putney, London SW15 2NU
☎ 020 8877 1123   🖷 020 8877 9787
uksales@freshfoodimages.com
www.freshfoodimages.com

'Europe's premier source' of food and wine related images. From the farm and the vineyard to the plate and the bottle. Cooking and kitchens, top chefs and restaurants, country trades and markets, worldwide travel with an extensive Italian section.

The Francis Frith Collection

Frith's Barn, Teffont, Salisbury SP3 5QP
☎ 01722 716376   🖷 01722 716881
julia_skinner@francisfrith.co.uk
www.francisfrith.com
Contact *Julia Skinner, Managing Editor*

Publishers of *Frith's Photographic Memories* series of illustrated local books, all featuring nostalgic photographs from the archive, founded by Frith in 1860. The archive contains over 360,000 images of 7,000 British towns.

John Frost Newspapers

▷ entry under Library Services

Andrew N. Gagg's PHOTO FLORA

Town House Two, Fordbank Court, Henwick
Road, Worcester WR2 5PF
☎ 01905 748515
gaggsphotoflora@googlemail.com
www.gagg.f2s.com/PHOTOFLORA/photoflora.
html
Contact *Andrew Gagg*

Specialist in British and European wild plants,
flowers, ferns, grasses, trees, shrubs, etc. with
colour coverage of most British and many
European species (rare and common) and
habitats; also travel in India, Sri Lanka, Nepal,
Egypt and North Africa, China, Mexico, Thailand,
Tibet, Vietnam and Cambodia.

Galaxy Picture Library

34 Fennels Way, Flackwell Heath, High Wycombe
HP10 9BY
☎ 01628 521338
robin@galaxypix.com
www.galaxypix.com
Contact *Robin Scagell*

Specializes in astronomy, space, telescopes,
observatories, the sky, clouds and sunsets.
Composites of foregrounds, stars, moon and
planets prepared to commission. Editorial service
available.

Garden and Wildlife Matters Photo Library

'Marlham', Henley's Down, Battle TN33 9BN
☎ 01424 830566   ⊠ 01424 830224
carol@gmpix
john@wildlifematters.com
www.gardenandwildlife.com/home.php
Contact *Dr John Feltwell*

Collection of 110,000 6x4" and 35mm images.
General gardening techniques and design;
cottage gardens and USA designer gardens.
10,000 species of garden plants and over 1,000
species of trees. Flowers, wild and house plants,
trees and crops. Environmental, ecological and
conservation pictures, including sea, air, noise
and freshwater pollution, SE Asian and Central
and South American rainforests; Eastern Europe,
Mediterranean. Recycling, agriculture, forestry,
horticulture and oblique aerial habitat shots
from Europe and USA. Digital images supplied
worldwide.

Garden Picture Library

2nd Floor, Waterside House, 9 Woodfield Road,
London W9 2BA
☎ 020 7432 8200   ⊠ 020 7432 8201
sales@gardenpicture.com
creative@gardenpicture.com
www.gardenpicture.com
Creative Director *Lee Wheatley*

'Our inspirational images of gardens, plants
and gardening lifestyle offer plenty of scope for
writers looking for original ideas to write about.'
The collection covers all garden related subjects;
everything from plant portraits to floral graphics
and garden design details to whole landscapes.
Brings together the work of over 100 professional
photographers from across the gardening globe.
Holds approximately 300,000 fully captioned
images of which around 65,000 are available
digitally. Free in-house picture research can
be undertaken on request or searches can be
made online via keyword or subject category
to create lightboxes which can be e-mailed or
downloaded. Visitors to the library are welcome
by appointment and copies of promotional
literature are available on request.

Genesis Space History Photo Library

20 Goodwood Park Road, Bideford EX39 2RR
☎ 01237 477883
tim@spaceport.co.uk
www.spaceport.co.uk
Contact *Tim Furniss*

Historical b&w spaceflight collection including
rockets, spacecraft, spacemen, Earth, moon and
planets. Catalogue available on the website.

Geo Aerial Photography

4 Christian Fields, London SW16 3JZ
☎ 020 8764 6292/0115 981 9418   ⊠ 020 8764
6292/0115 981 9418
geo.aerial@geo-group.co.uk
www.geo-group.co.uk
Contact *Kelly White*

Established 1990 and now a growing collection
of aerial oblique photographs from the UK,
Scandinavia, Asia and Africa – landscapes,
buildings, industrial sites, etc. Commissions
undertaken.

GeoScience Features

6 Orchard Drive, Wye TN25 5AU
☎ 01233 812707   ⊠ 01233 812707
gsf@geoscience.uk.com
www.geoscience.uk.com

Fully computerized and comprehensive library
containing the world's principal source of volcanic
phenomena. Extensive collections, providing
scientific detail with technical quality, of rocks,
minerals, fossils, microsections of botanical and
animal tissues, animals, biology, birds, botany,
chemistry, earth science, ecology, environment,
geology, geography, habitats, landscapes, macro/
microbiology, peoples, sky, weather, wildlife
and zoology. Over 500,000 original colour
transparencies in medium and 35mm-format.
Subject lists and CD-ROM catalogue available.

Incorporates the RIDA photolibrary and Landform Slides.

Geoslides Photography

4 Christian Fields, London SW16 3JZ
☎ 0115 981 9418   🖷 0115 981 9418
geoslides@geo-group.co.uk
www.geo-group.co.uk
Contact *John Douglas*

Established in 1968. Landscape and human interest subjects from the Arctic, Antarctica, Scandinavia, UK, Africa (south of Sahara), Middle East, Asia (south and southeast). Also specialist collections of images from British India (the Raj) and Boer War. Forty years of photography services.

Getty Images

101 Bayham Street, London NW1 0AG
☎ 0800 376 7981/020 7428 6109
www.gettyimages.com
Contact *Sales Department*

'Getty Images is the world's leading imagery company, creating and providing the largest collection of still and moving images to communication professionals around the globe.' From sports and news photography to archival and contemporary imagery. For those who wish to commission photographers to fulfil specific needs, the company maintains a full-service department for custom-shot images.

Martin and Dorothy Grace

Boxwood Cottage, The Row, Lyth, Kendal LA8 8DD
☎ 015395 68569
martin.grace@tiscali.co.uk
Contact *Martin Grace (Partner)*

Colour coverage of natural history, topography and travel in Britain, the Galapagos Islands, France, southern Spain, Peru, Madagascar, the Falklands, South Georgia and the Antarctic. Specializes in trees, shrubs and wild flowers but also cover ferns, birds and butterflies, habitats, landscapes, ecology. High quality digital stock supplied online.

The Ronald Grant Archive

The Master's House, 2 Dugard Way, off Renfrew Road, London SE11 4TH
☎ 020 7840 2200   🖷 020 7840 2299
pixdesk@rgapix.com
www.ronaldgrantarchive.com

Colour and b&w coverage (including stills) of the motion picture industry throughout its history. Smaller collections on theatre, variety, television and popular music.

Sally and Richard Greenhill

357 Liverpool Road, London N1 1NL
☎ 020 7607 8549
sr.greenhill@virgin.net
www.srgreenhill.co.uk
Photo Librarian *Denise Lalonde*

Social documentary photography in colour and b&w of working lives: pregnancy and birth, child development, education, work, old people, medical, urban. Also Modern China, 1971 to the present; most London statues. Some material from Borneo, USA, India, Israel, Philippines and Sri Lanka.

V.K. Guy Ltd

Browhead Cottage, Troutbeck, Windermere LA23 1PG
☎ 015394 33519   🖷 015394 32971
admin@vkguy.co.uk
www.vkguy.co.uk
Contacts *Vic Guy, Mike Guy*

25,000 5x4" UK and Ireland scenic transparencies from Shetland in the north to the Channel Isles in the south, suitable for tourism brochures, calendars, etc. Can be supplied digitally and browsed through via the online database.

Robert Harding World Imagery

58–59 Great Marlborough Street, London W1F 7JY
☎ 020 7478 4000   🖷 020 7478 4161
info@robertharding.com
www.robertharding.com

A leading source of stock photography with over two million colour images covering a wide range of subjects – worldwide travel and culture, geography and landscapes, people and lifestyle, architecture. Rights protected and royalty-free images. Visitors welcome; telephone or visit the website.

Dennis Hardley Photography, Scottish Photo Library

Rosslynn, Ceum Dunrigh, Benderloch, Nr Oban PA37 1ST
☎ 01631 720434   🖷 01631 720434
info@dennishardley.com
www.ineedanimage.com
Contacts *Dennis Hardley, Tony Hardley*

Established 1974. About 30,000 images (6x7", 6x9" format colour transparencies) of Scotland: castles, historic, scenic landscapes, islands, transport, etc. Also English views – Liverpool, Chester, Bath, Weston Super Mare, Sussex, Somerset and Cambridge. All images available to buy and instantly download from the website. Includes Scotland, England and North Wales, covering lifestyle, landscapes, transport, cities, etc.

Jim Henderson Photography

Crooktree, Kincardine O'Neil, Aboyne AB34 4JD

☎ 01339 882149   📠 01339 882149

JHende7868@aol.com

www.jimhendersonphotography.com

Contact *Jim Henderson*

Scenic and general activity coverage of north east Scotland (Aberdeenshire) and Highlands for tourist, holiday and activity illustration. Specialist collection of over 300 Aurora Borealis displays from 1989–2008 on Royal Deeside and co-author of *The Aurora* (published 1997). Author and photographer for *Aberdeen* and *Aberdeen & Royal Deeside* titles for Aberdeen Journals. Recent images of ancient Egyptian sites: Cairo through to Abu-Simbel. **BAPLA** member. Contributor to Alamy & jasonfriendimages Commissions undertaken.

John Heseltine Archive

Mill Studio, Frogmarsh Mill, South Woodchester GL5 5ET

☎ 01453 873792

john@heseltine.co.uk

www.heseltine.co.uk

Contact *John Heseltine*

A new website displays this distinctive collection of mainly black and white photography dealing with a range of subjects from Italy, France, the UK and further afield.

Hobbs Golf Collection

5 Winston Way, New Ridley, Stocksfield NE43 7RF

☎ 01661 842933/07941 445993   📠 01661 842933

info@hobbsgolfcollection.com

www.hobbsgolfcollection.com

Contact *Margaret Hobbs*

Specialist golf collection: players, courses, art, memorabilia and historical topics (1300–present). 40,000+ images, mainly 35mm colour transparencies and b&w prints. All images can be supplied in digital format. Commissions undertaken. Author of 30 golf books.

David Hoffman Photo Library

c/o BAPLA, 18 Vine Hill, London EC1R 5DZ

☎ 020 8981 5041

info@hoffmanphotos.com

www.hoffmanphotos.com

Contact *David Hoffman*

David Hoffman Photo Library is a social issues focused stock archive built up from journalistic and documentary work over more than 30 years. The collection covers drugs and drug use, policing, disorder, race and racists, riots, youth, prostitution, protest, homelessness, housing, environmental demonstrations and events, waste disposal, alternative energy and pollution. Includes a wide range of reportage and general

stock, plus topical issues from flash mobs to SUVs. Primarily UK images plus Europe, USA, Canada, Venezuela, Mexico and Thailand.

Holt Studios International Ltd

▷ Frank Lane Picture Agency Ltd

Houghton's Horses/Kit Houghton Photography

Radlet Cottage, Spaxton, Bridgwater TA5 1DE

☎ 01278 671362

kit@enterprise.net

www.houghtonshorses.com

Contacts *Kit Houghton, Kate Houghton*

Specialist equestrian library of over 300,000 transparencies and digital images on all aspects of the horse world, with images ranging from the romantic to the practical and competition pictures in all equestrian disciplines worldwide. Online picture delivery is now the norm. All transparencies in the process of being scanned. Commissions undertaken.

Chris Howes/Wild Places Photography

PO Box 100, Abergavenny NP7 9WY

☎ 01873 737707

photos@wildplaces.co.uk

Contacts *Chris Howes, Judith Calford*

Expanding collection of over 50,000 colour transparencies, b&w prints and digital photos covering travel, topography and natural history worldwide, plus action sports such as climbing. Supply now fully digital. Specialist areas include caves, caving and mines (with historical coverage using engravings and early photographs), wildlife, landscapes and the environment, including pollution and conservation. Europe (including Britain), Canada, USA, Africa and Australia are all well represented within the collection. Commissions undertaken.

Huntley Film Archive

22 Islington Green, London N1 8DU

☎ 020 7226 9260   📠 020 7359 9337

films@huntleyarchives.com

www.huntleyarchives.com

Contact *Amanda Huntley*

Originally a private collection, the library is now a comprehensive archive of rare and vintage documentary film dating from 1895. 50,000 films on all subjects of a documentary nature. Online catalogue available.

Jacqui Hurst

66 Richford Street, Hammersmith, London W6 7HP

☎ 020 8743 2315/07970 781336

jacquihurst@yahoo.co.uk

www.jacquihurstphotography.co.uk

Contact *Jacqui Hurst*

A specialist library of traditional and contemporary applied arts, regional food producers and markets. The photos form illustrated essays of how something is made and finish with a still life of the completed object. The collection is always being extended and a list is available on request. Commissions undertaken.

Hutchison Picture Library

65 Brighton Road, Shoreham-by-Sea BN43 6RE
📞 01273 440113   📠 01273 440116
library@hutchisonpictures.co.uk
www.hutchisonpictures.co.uk

Worldwide contemporary images from the straight-forward to the esoteric and quirky. Over half a million documentary colour photographs on file, covering people, places, customs and faiths, agriculture, industry and transport. Special collections include the environment and climate, family life (including pregnancy and birth), ethnic minorities worldwide (including Disappearing World archive), conventional and alternative medicine, and music around the world. Search service available.

Illustrated London News Picture Library

c/o Mary Evans Picture Library, 59 Tranquil Vale, Blackheath, London SE3 0BS
📞 020 7805 5585   📠 020 8852 7211
iln@maryevans.com
www.ilnpictures.co.uk

Engravings, photographs and illustrations from 1842 to the present day, taken from magazines published by Illustrated Newspapers: *Illustrated London News*; *Graphic*; *Sphere*; *Tatler*; *Sketch*; *Illustrated Sporting and Dramatic News*; *Illustrated War News 1914–18*; *Bystander*; *Britannia & Eve*. Social history, London, Industrial Revolution, wars, travel. Now housed and managed by **Mary Evans Picture Library** (see entry). Research and scanning services available.

Images of Africa Photobank

11 The Windings, Lichfield WS13 7EX
📞 01543 262898   📠 01543 417154
info@imagesofafrica.co.uk
www.imagesofafrica.co.uk
Contact *David Keith Jones, FRPS*

Images covering Botswana, Chad, Egypt, Ethiopia, Kenya, Lesotho, Madagasca, Malawi, Morocco, Namibia, Rwanda, South Africa, Swaziland, Tanzania, Uganda, Zaire, Zambia, Zanzibar and Zimbabwe. 'Probably the best collection of photographs of Kenya in Europe.' A wide range of topics are covered. Strong on African wildlife with over 80 species of mammals including many sequences showing action and behaviour. Popular animals like lions and elephants are covered in encyclopædic detail. More than 100 species of birds and many reptiles are included. Other strengths include National Parks and reserves, natural beauty, tourism facilities, traditional and modern people. Most work is by David Keith Jones, FRPS; several other photographers are represented. Many thousands of these images can now be accessed online. Go to www.imagesofafrica.co.uk and look for Online Images.

Imperial War Museum Photograph Archive

All Saints Annexe, Austral Street, London SE11 4SJ
📞 020 7416 5333/5338/5309   📠 020 7416 5355
photos@iwm.org.uk
www.iwm.org.uk

A national archive of ten million photographs illustrating all aspects of 20th and 21st century conflict. Emphasis on the two World Wars but includes material from other conflicts involving Britain and the Commonwealth. Majority of material is b&w, although holdings of colour material increase with more recent conflicts. Visitors welcome by appointment, Monday, Wednesday and Friday, 10.00 am to 5.00 pm.

International Photobank

PO Box 6554, Dorchester DT1 9BS
📞 01305 854145   📠 01305 853065
info@internationalphotobank.co.uk
www.internationalphotobank.co.uk
Contacts *Peter Baker, Gary Goodwin*

Over 450,000 images, in digital and transparency format. Colour coverage of travel subjects: places, people, folklore, events. Digital broadband service available for newspapers, magazines and other users. Pictures can also be selected from the comprehensive website.

Robbie Jack Photography

45 Church Road, Hanwell, London W7 3BD
📞 020 8567 9616   📠 020 8567 9616
robbie@robbiejack.com
www.robbiejack.com
Contact *Robbie Jack*

Built up over the last 25 years, the library contains over 500,000 colour images of the performing arts – theatre, dance, opera and music. Includes West End shows, the RSC and Royal National Theatre productions, English National Opera and Royal Opera. The dance section contains images of the Royal Ballet, English National Ballet, the Rambert Dance Company, plus many foreign companies. Also holds the largest selection of colour material from the Edinburgh International Festival. Researchers are welcome to visit by appointment. From 2001 began offering colour transparencies as digital images which are being added to on a daily basis.

Jayawardene Travel Photo Library

7A Napier Road, Wembley HAO 4UA
☏ 020 8795 3581 📠 020 8795 4083
jaytravelphotos@aol.com
www.jaytravelphotos.com
Contact *Rohith Jayawardene (Partner)*

170,000 travel and travel-related images – high end digital and original transparencies. Many places photographed in depth, with more than 800 pictures per country. Over 10,000 images available online.

Paul Kaye B/W Archive
▷ Sylvia Cordaiy Photo Library

David King Collection

90 St Pauls Road, London N1 2QP
☏ 020 7226 0149
davidkingcollection@btopenworld.com
www.davidkingcollection.com
Contact *David King*

250,000 b&w original and copy photographs and colour transparencies of historical and present-day images. Russian history and the Soviet Union from 1900 to the fall of Khrushchev; the lives of Lenin, Trotsky and Stalin; the Tzars, Russo-Japanese War, 1917 Revolution, World War I, Red Army, the Great Purges, Great Patriotic War, etc. Special collections on China, Eastern Europe, the Weimar Republic, John Heartfield, American labour struggles, Spanish Civil War. Open to qualified researchers by appointment, Monday to Friday, 10.00 am to 6.00 pm. Staff will undertake research; negotiable fee for long projects. David King's latest photographic books: *Red Star Over Russia*, a visual history from 1917 to the death of Stalin, *The Commissar Vanishes*, documenting the falsification of photographs and art in Stalin's Russia, and *Ordinary Citizens*, mugshots of victims shot without trial from the archives of Stalin's secret police. A new display of Soviet posters, 1917–1960, from the David King Collection continues at Tate Modern throughout 2009.

The Kobal Collection

2 The Quadrant, 135 Salusbury Road, London NW6 6RJ
☏ 020 7624 3500 📠 020 7624 3355
info@picture-desk.com
www.picture-desk.com

Picture library holding colour and b&w coverage of Hollywood films, portraits, stills, publicity shots, posters and ephemera.

Kos Picture Source Ltd

PO Box 104, Midhurst GU29 1AS
☏ 020 7801 0044 📠 020 7801 0055
images@kospictures.com
www.kospictures.com
Contact *Chris Savage*

Specialists in water-related images including international yacht racing and cruising, classic boats and superyachts, and extensive range of watersports. Also worldwide travel including seascapes, beach scenes, underwater photography and the weather.

Ed Lacey Collection
▷ Phil Sheldon Golf Picture Library

Landform Slides
▷ GeoScience Features

Frank Lane Picture Agency Ltd

Pages Green House, Wetheringsett, Stowmarket IP14 5QA
☏ 01728 860789 📠 01728 860222
pictures@flpa-images.co.uk
www.flpa.co.uk

Colour coverage of natural history, environment, pets and weather. Represents Sunset from France, Foto Natura from Holland, Minden Pictures from the US, Holt Studios and works closely with Eric and David Hosking, plus 270 freelance photographers. Website with 175,000 images.

Last Resort Picture Library

Manvers Studios, 12 Ollerton Road, Tuxford, Newark NG22 0LF
☏ 01777 870166 📠 01777 871739
dick@dmimaging.co.uk
www.dmimaging.co.uk
Contacts *Jo Makin, Dick Makin*

Subject areas include agriculture, architecture, IT, education, food, industry, landscapes, people at work, skiing, trees, flowers, mountains and winter landscapes. Images cover a wide variety of topics, ranging from the everyday to the obscure. 'Bespoke service available through Dick Makin Imaging, our linked photographic studio. Contact us for details.'

LAT Photographic

Teddington Studios, Broom Road, Teddington TW11 9BE
☏ 020 8267 3010
lat.photo@haymarket.com
www.latphoto.co.uk

Motor sport collection of almost 12 million images dating from 1895 to the present day.

Lebrecht Music and Authors

3 Bolton Road, London NW8 0RJ
☏ 020 7625 5341
pictures@lebrecht.co.uk
www.authorpictures.co.uk
Contact *Elbie Lebrecht*

Classical music, opera, jazz, rock, musicians, instruments, performers, writers, manuscripts, book covers, artists, architects, sculptors, historical personalities, politicians, society and glamorous showbusiness personalities.

The Erich Lessing Archive of Fine Art & Culture

c/o akg-images Ltd, The Arts and History Picture Library, 5 Melbray Mews, 158 Hurlingham Road, London sw6 3ns
☎ 020 7610 6103   📠 020 7610 6125
enquiries@akg-images.co.uk
www.akg-images.co.uk

Top quality high resolution scans depicting the contents of many of the world's finest art galleries as well as ancient archaeological and biblical sites available in the UK via the akg-images website.

Library of Art & Design
▷ V&A Images

Lindley Library, Royal Horticultural Society

80 Vincent Square, London sw1p 2pe
☎ 020 7821 3050   📠 020 7821 3022
library.london@rhs.org.uk
www.rhs.org.uk/Learning/Library/
    LindleyLondon/index.htm
Contact *Picture Librarian*

22,000 original drawings and approx. 8,000 books with hand-coloured plates of botanical illustrations. Appointments are necessary to visit the collection of unpublished material; all photography is carried out by their own photographer.

Link Picture Library

41A The Downs, London sw20 8hg
☎ 020 8944 6933
library@linkpicturelibrary.com
www.linkpicturelibrary.com
Contact *Orde Eliason*

100,000 images of South Africa, India, China and Israel. A more general collection of colour transparencies from 100 countries worldwide. Link Picture Library has an international network and can source material not in its file from Japan, Scandinavia, India and South Africa. Original photographic commissions undertaken.

London Aerial Photo Library

Studio D1, Fairoaks Airport, Chobham, Woking gu24 8hu
☎ 01276 855997/855344   📠 01276 855455
info@londonaerial.co.uk
www.londonaerial.co.uk
www.flightimages.com
Contact *Amanda Campbell*

Extensive collections of aerial imagery. 450,000 images including continually updated

photography of London plus excellent coverage of the UK in general. Covers landmarks, industrial/retail properties, sporting venues/football stadiums, conceptual shots. Library searches by expert staff, free of charge. Selection of digital images also available via website. Commissioned photography undertaken.

London Metropolitan Archives

40 Northampton Road, London ec1r 0hb
☎ 020 7332 3820   📠 020 7833 9136
ask.lma@cityoflondon.gov.uk
www.cityoflondon.gov.uk/lma
Contact *The Enquiry Team*

London Metropolitan Archives (LMA) is the largest local authority record office in the UK. Holds over 32 miles of archives – an enormous amount of information about the capital and its people. These include records of London government, hospitals, charities, businesses and parish churches. Types of record range from books and manuscript documents to photographs, maps and drawings. 'Nearly 900 years of London history can be brought to life at LMA.' There is also a 100,000 volume reference library specializing in London history. See entry under *Library Services*.

London's Transport Museum Photographic Library

39 Wellington Street, London wc2e 7bb
☎ 020 7379 6344   📠 020 7565 7252
www.ltmcollection.org/photos/index.html
www.ltmcollection.org/posters/index.html

Photographic collection reflecting London's public transport history from the 1860s to the present day. The core of the collection is made up of the old London Transport photographic archive of over 150,000 b&w photographs. Also includes a smaller archive of colour material, as well as historic albums and prints from all periods. Over 16,000 images are now available to view and purchase on-line, with new photographs added regularly. Each image has a brief caption, stating location and date (where known) to assist retrieval. For more specialist enquiries, please make an appointment to view the photographic collection at Acton Depot (contact: dutycurator@ltmuseum.co.uk).

London Transport Museum Poster Archives: for 100 years, since Frank Pick commissioned the first graphic poster for London Underground, the company and its successors have kept copies of everything they produced. In the early 20th century, under Pick's guidance, London Transport commissioned work from the best artists and designers in the country. By the 1980s, when the collection was transferred to the Museum, it contained more than 5,000 printed posters and nearly 1,000 original artworks. The collection has grown steadily ever since and is now a uniquely comprehensive picture of a century of British graphic design.

Ludvigsen Library

Scoles Gate, Hawkedon, Bury St Edmunds IP29 4AU

📞 01284 789246   📠 01284 789246
library@ludvigsen.com
www.ludvigsen.com
Contact *Karl Ludvigsen*

Extensive information research facilities for writers and publishers. Approximately 400,000 images (both b&w and many colour transparencies) of automobiles and motorsport, from 1890s through 1980s. Glass plate negatives from the early 1900s; Formula One, Le Mans, motor car shows, vintage, antique and classic cars from all countries. Includes the Dalton-Watson Collection and the work of noted photographers such as John Dugdale, Edward Eves, Peter Keen, Max le Grand, Karl Ludvigsen, Rodolfo Mailander, Ove Nielsen, Stanley Rosenthall and others.

Magnum Photos Ltd

Ground Floor, 63 Gee Street, London EC1V 3RS
📞 020 7490 1771   📠 020 7608 0020
magnum@magnumphotos.co.uk
www.magnumphotos.co.uk
Archive Director *Nick Galvin*

Founded 1947 by Henri Cartier Bresson, George Rodger, Robert Capa and David 'Chim' Seymour. Represents over 70 of the world's leading photo-journalists and documentary photographers. Coverage of major world events from the Spanish Civil War to present day, daily life, social issues and reportage. Also a large collection of personalities and travel images.

The Raymond Mander & Joe Mitchenson Theatre Collection

Jerwood Library of the Performing Arts, Trinity College of Music, King Charles Court, Old Royal Naval College, London SE10 9JF
📞 020 8305 4426   📠 020 8305 9426
rmangan@tcm.ac.uk
www.mander-and-mitchenson.co.uk
Contact *Richard Mangan*
Archive Officer/Cataloguer *Kristy Davis*

Enormous collection covering all aspects of the theatre: plays, actors, dramatists, music hall, theatres, singers, composers, etc. Visitors welcome by appointment.

The Roger Mann Collection

Wensley Court, 48 Barton Road, Torquay TQ1 4DW
📞 01803 323868   📠 01803 616448

rogermann48bart@aol.com
www.therogermanncollection.co.uk
Contact *R.F. Mann*

Comprehensive collection of cricket photographs covering the period 1750 to 1945. The photographs feature most of the first-class players, teams, Test match action and overseas tours of the period. This collection includes almost 2,000 original match scorecards, cartoons, images, prints, postcards, cigarette cards, letters and the personal memorabilia of many of the best-known players of the time. Also some coverage of the period 1946 to 1970.

S&O Mathews Photography

Little Pitt Place, Brighstone, Isle of Wight PO30 4DZ
oliver@mathews-photography.com
www.mathews-photography.com

Specialist stock library of fine photographs of botanical and horticultural subjects, plants, plant portraits and plant associations, as well as gardens, including views, details and features.

Institution of Mechanical Engineers

1 Birdcage Walk, London SW1H 9JJ
📞 020 7973 1274   📠 020 7222 8762
library@imeche.org
www.imeche.org

Historical images and archives on mechanical engineering. Open 9.15 am to 5.30 pm, Monday to Friday.

Lee Miller Archives

Farley Farm House, Muddles Green, Chiddingly, Near Lewes BN8 6HW
📞 01825 872691   📠 01825 872733
archives@leemiller.co.uk
www.leemiller.co.uk

The work of Lee Miller (1907–77). As a photo-journalist she covered the war in Europe from early in 1941 to VE Day with further reporting from the Balkans. Collection includes photographic portraits of prominent Surrealist artists: Ernst, Eluard, Miró, Picasso, Penrose, Carrington, Tanning, and others. Surrealist and contemporary art, poets and writers, fashion, the Middle East, Egypt, the Balkans in the 1930s, London during the Blitz, war in Europe and the liberation of Dachau and Buchenwald.

Mirrorpix

22nd Floor, One Canada Square, Canary Wharf, London E14 5AP
📞 020 7293 3700   📠 020 7293 0357
desk@mirrorpix.com
www.mirrorpix.com

Age – 1903 to present. The photographic archive of the Daily Mirror newspaper, rights managed. 250,000 images online, 50 million negatives and prints offline. Ideal for illustrating autobiographical, entertainment, social and political history (domestic and foreign), the arts, culture, fashion, humour, industry, sport and royalty.

Monitor Picture Library

The Forge, Roydon, Harlow CM19 5HH

☎ 01279 792700

sales@monitorpicturelibrary.com
www.monitorpicturelibrary.com

Colour and b&w coverage of 1960s, '70s and '80s personalities and celebrities from: music, entertainment, sport, politics, royals, judicial, commerce etc. Specialist files on Lotus cars. Syndication to international, national and local media.

Motoring Picture Library

National Motor Museum, Beaulieu SO42 7ZN

☎ 01590 614656   🖷 01590 612655

motoring.pictures@beaulieu.co.uk
www.motoringpicturelibrary.com
www.alamy.com/mpl
Contacts *Jonathan Day, Tim Woodcock, Tom Wood*

Three-quarters of a million b&w images, plus over 120,000 colour images covering all forms of motoring history from the 1880s to the present day. Commissions undertaken. Own studio.

Mountain Camera
▷ John Cleare

Moving Image Communications

9 Faversham Reach, Faversham ME14 7LA

☎ 0845 257 2968   🖷 01795 534306

mail@milibrary.com
nathalie@milibrary.com
mike@milibrary.com
www.milibrary.com
Contact *Nathalie Banaigs*

Over 16,000 hours of quality archive and contemporary footage, including: Channel X; TVAM Archive 1983–92; Drummer Films (travel classics, 1950–70); The Freud Archive (1930–39); Film Finders (early cinema); Adrian Brunel Films; Cuban Archives; Stockshots (timelapse, cityscapes, land and seascapes, chroma-key); Space Exploration (NASA); Wild Islands; Flying Pictures.

Museum of Childhood
▷ V&A Images

Museum of London Picture Library

150 London Wall, London EC2Y 5HN

☎ 020 7814 5604/5612   🖷 020 7600 1058

picturelib@museumoflondon.org.uk
www.museumoflondon.org.uk
Library Manager *Sean Waterman*
Picture Researchers *Nikki Braunton, Sarah Williams, Sandra Marquez, Jenna Collins*

The Museum of London picture library holds over 35,000 images illustrating the history of London and the life of its people from prehistoric times to the present day. The images are drawn from the Museum's extensive and unique collections of oil paintings, historic photographs, drawings, prints maps and artefacts including costume, jewellery and ceramics.

The National Archives Image Library

Ruskin Avenue, Kew, Richmond TW9 4DU

☎ 020 8392 5225   🖷 020 8487 1974

image-library@nationalarchives.gov.uk
www.nationalarchives.gov.uk/imagelibrary
Contacts *Paul Johnson, Hugh Alexander*

British and colonial history from the Domesday Book to the 1970s, shown in photography, maps, illuminations, posters, advertisements, textiles and original manuscripts. Approximately 30,000 5x4" and 35mm colour transparencies and b&w negatives. Open 9.00 am to 5.00 pm, Monday to Friday.

National Art Library
▷ V&A Images

National Galleries of Scotland Picture Library

The Scottish National Gallery of Modern Art, 75 Belford Road, Edinburgh EH4 3DR

☎ 0131 624 6258/6260   🖷 0131 623 7135

picture.library@nationalgalleries.org
www.nationalgalleries.org
Contact *Shona Corner*

Over 80,000 b&w and several thousand images in colour of works of art from the Renaissance to present day. Specialist subjects cover fine art (painting, sculpture, drawing), portraits, Scottish, historical, still life, photography and landscape.

National Magazine Company
▷ Mary Evans Picture Library

National Maritime Museum Picture Library

Greenwich, London SE10 9NF

☎ 020 8312 6631/6704   🖷 020 8312 6533

picturelibrary@nmm.ac.uk
www.nmmimages.com
Contact *Doug McCarthy*

Over 400,000 images from the leading collection of maritime art and artefacts, including oil paintings from the 16th century to present day, prints and drawings, historic photographs, ships' plans, models, rare maps and charts, globes, manuscripts and navigation and scientific instruments. Images are available as digital files.

Contact for further information or for picture research assistance.

National Media Museum

▷ Science & Society Picture Library

National Monuments Record

English Heritage, Kemble Drive, Swindon SN2 2GZ

☎ 01793 414600   ℻ 01793 414606
nmrinfo@english-heritage.org.uk
www.english-heritage.org.uk/nmr

The National Monuments Record (NMR), the public archive of English Heritage, holds over ten million photographs, plans, drawings, reports, records and publications covering England's archaeology, architecture, social and local history. The NMR is the largest publicly accessible archive in Britain and is the biggest dedicated to the historic environment.

National Portrait Gallery Picture Library

St Martin's Place, London WC2H 0HE
☎ 020 7312 2474/5/6   ℻ 020 7312 2464
www.npg.org.uk
Contact Head of Rights & Reproductions

Pictures of brilliant, daring and influential characters who have made British history are available for publication. Images can be searched, viewed and ordered on the website. Copyright clearance is arranged for all images supplied.

National Railway Museum Search Engine – Library & Archive

Leeman Road, York YO26 4XJ
☎ 01904 686235   ℻ 01904 611112
www.nrm.org.uk

The National Railway Museum and the **Science and Society Picture Library** are both part of the National Museum of Science and Industry (NMSI) whose other constituents are the Science Museum in London and the National Media Museum in Bradford (formerly the National Museum of Photography, Film & Television). 1.5 million images, mainly b&w, covering every aspect of railways from the 1850s to the present day; over one million engineering drawings; the most comprehensive railway library in the UK; sound and oral history archives; a growing digital video record of today's railway industry; a large collection of British railway posters, graphic art and advertising material from the 1820s onwards. Search engine is open 10.00 am to 5.30 pm, seven days a week. Closed 24th–26th December.

The National Trust Photo Library

Heelis, Kemble Drive, Swindon SN2 2NA
☎ 01793 817700
photo.library@nationaltrust.org.uk
www.ntpl.org.uk

Contact Chris Lacey

The National Trust Photo Library houses a unique collection of contemporary photography which vividly illustrates the rich diversity and historical range of properties in the National Trust's care, throughout England, Wales and Northern Ireland. The Library has an exceptional choice of images, created by commissioned specialist photographers, suitable for a broad range of creative applications. 40,000 images can be searched and ordered online.

Natural History Museum Image Resources

Cromwell Road, London SW7 5BD
☎ 020 7942 5401/5324   ℻ 020 7942 5443
nhmpl@nhm.ac.uk
www.nhm.ac.uk/piclib
Contacts Jennifer Vladimirsky, Sally Jennings, Katie Anderson

Pictures from the Museum's collections, including dinosaurs, man's evolution, extinct species and fossil remains. Also pictures of gems, minerals, birds and animals, plants and insect specimens, plus historical artworks depicting the natural world.

Nature Photographers Ltd

West Wit, New Road, Little London, Tadley RG26 5EU
☎ 01256 850661   ℻ 01256 851157
paul@naturephotographers.co.uk
www.naturephotographers.co.uk
Contact Dr Paul Sterry

Over 150,000 images, digitized to order, on worldwide natural history and environmental subjects. The library is run by a trained biologist and experienced author on his subject.

Nature Picture Library

5a Great George Street, Bristol BS1 5RR
☎ 0117 911 4675   ℻ 0117 911 4699
info@naturepl.com
www.naturepl.com
Contact Helen Gilks

A collection of 300,000 nature photos from around the world, including strong coverage of animal portraits and behaviour. Other subjects covered include plants, pets, landscapes and travel, environmental issues and wildlife film-makers at work. Thousands of images can be viewed online and downloaded direct for reproduction.

Peter Newark's Picture Library

3 Barton Buildings, Queen Square, Bath BA1 2JR
☎ 01225 334213   ℻ 01225 480554

Over one million images covering world history from ancient times to the present day. Includes an extensive military collection of photographs, paintings and illustrations. Also a special collection on American history covering Colonial times, exploration, social, political and the Wild West and Native-Americans in particular. Subject list available. Telephone, fax or write for further information.

NHPA
▷ Photoshot

Offshoot
▷ Skishoot

Only Horses Picture Agency
27 Greenway Gardens, Greenford UB6 9TU
☎ 020 8578 9047   📠 020 8575 7244
onlyhorsespics@aol.com
www.onlyhorsespictures.com

Colour and b&w coverage of all aspects of the horse. Foaling, retirement, racing, show jumping, eventing, veterinary, polo, breeds, personalities.

Oxford Picture Library
15 Curtis Yard, North Hinksey Lane, Oxford OX2 0LX
☎ 01865 723404   📠 01865 725294
opl@cap-ox.com
www.cap-ox.co.uk
Contacts *Chris Andrews, Annabel Matthews*

Specialist collection on Oxford: the city, university and colleges, events, people, spires and shires. Also, the Cotswolds, architecture and landscape from Stratford-upon-Avon to Bath; the Thames and Chilterns, including Henley on Thames and Windsor; Channel Islands, especially Guernsey and Sark. Aerial views of all areas specified above. General collection includes wildlife, trees, plants, clouds, sun, sky, water and teddy bears. Commissions undertaken.

Oxford Scientific (OSF)
2nd Floor, Waterside House, 9 Woodfield Road, London W9 2BA
☎ 020 7432 8200   📠 020 7432 8201
sales@osf.co.uk
www.osf.co.uk

Collection of 350,000 colour transparencies and digital files of wildlife and natural science images supplied by over 300 photographers worldwide, covering all aspects of wildlife plus landscapes, weather, seasons, plants, pets, environment, anthropology, habitats, science and industry, space, creative textures and backgrounds and geology. Macro and micro photography. UK agents for Animals Animals, USA, Okapia, Germany and Dinodia, India. Research by experienced researchers for specialist and creative briefs. Visits welcome, by appointment.

Panos Pictures
1 Honduras Street, London EC1Y 0TH
☎ 020 7253 1424   📠 020 7253 2752
pics@panos.co.uk
www.panos.co.uk

Documentary colour and b&w library specializing in Third World and Eastern Europe, with emphasis on environment and development issues. Leaflet available. Fifty per cent of all profits from this library go to the Panos Institute to further its work in international sustainable development.

Charles Parker Archive
▷ Birmingham Library Services under Library Services

Ann & Bury Peerless Picture Library
St David's, 22 King's Avenue, Minnis Bay, Birchington-on-Sea CT7 9QL
☎ 01843 841428   📠 01843 848321
www.peerlessimages.com
Contacts *Ann Peerless, Bury Peerless*

Specialist collection on world religions: Hinduism, Buddhism, Confucianism, Taoism, Jainism, Christianity, Islam, Sikhism, Zoroastrianism (Parsees of India). Geographical areas covered: India, Afghanistan (Bamiyan Valley of the Buddhas), Pakistan, Bangladesh, Sri Lanka, Cambodia (Angkor), Java (Borobudur), Bali, Malaysia, Thailand, Taiwan, Russia, China, Spain, Poland, Uzbekistan (Samarkand and Bukhara), Vietnam. Basis of collection (35mm colour transparencies), historical, cultural, extensive coverage of art (sculpture and miniature paintings), architecture including Pharaonic Egypt, earlier coverage of significance on Iran, including oil industry of Iran, Kenya, Uganda and Zimbabwe, including the Victoria Falls, also Morocco, Malta, Libya and Tunisia.

Photofusion
17A Electric Lane, London SW9 8LA
☎ 020 7733 3500   📠 020 7738 5509
library@photofusion.org
www.photofusionpictures.org
Contact *Liz Somerville*

Colour and b&w coverage of contemporary social and environmental UK issues including babies and children, disability, education, the elderly, environment, family, health, housing, homelessness, people and work. Brochure available.

The Photolibrary Wales
2 Bro-nant, Church Road, Pentyrch, Cardiff CF15 9QG
☎ 029 2089 0311   📠 029 2089 0311
info@photolibrarywales.com
www.photolibrarywales.com

Contacts *Steve Benbow, Kate Benbow*

Over 100,000 digital images covering all areas and subjects of Wales. Represents the work of 260 photographers, living and working in Wales.

Photoshot

29–31 Saffron Hill, London EC1N 8SW
☎ 020 7421 6000   📠 020 7421 6006
www.photoshot.com
www.nhpa.co.uk
www.staystill.com
www.worldpictures.com
Contacts *Charles Taylor (U.P.P.A.); Emma Hier (Stay Still); Tracey Howells (Starstock); Colin Finlay (World Illustrated); Tim Harris (NHPA & Woodfall Wild Images); David Brenes (World Pictures)*

Photoshot collections include U.P.P.A. (daily national and international business, political and establishment news); Stay Still (exclusive celebrity portraiture and TV publicity images); Starstock (live images of celebrity and entertainment personalities and events from around the world); World Illustrated (the world in pictures, art, culture, environment and heritage); NHPA (wildlife and nature); World Pictures (comprehensive archive of practical and stylish travel images); Woodfall Wild Images (wildlife, landscape and environmental).

PictureBank Photo Library Ltd

Parman House, 30–36 Fife Road, Kingston upon Thames KT1 1SY
☎ 020 8547 2344   📠 020 8974 5652
info@picturebank.co.uk
www.picturebank.co.uk

Over 400,000 images covering people (girls, couples, families, children), travel and scenic (UK and world), moods (sunsets, seascapes, deserts, etc.), industry and technology, environments and general. Commissions undertaken. Visitors welcome. Member of **BAPLA**. New material in digital format welcome.

Pictures Colour Library

10 James Whatman Court, Turkey Mill, Ashford Road, Maidstone ME14 5SS
☎ 01622 609809   📠 01622 609806
enquiries@picturescolourlibrary.co.uk
www.picturescolourlibrary.co.uk

Travel and travel-related images depicting lifestyles and cultures, people and places, attitudes and environments from around the world, including a comprehensive section on Great Britain.

Axel Poignant Archive

115 Bedford Court Mansions, Bedford Avenue, London WC1B 3AG
☎ 020 7636 2555   📠 020 7636 2555
Rpoignant@aol.com
Contact *Roslyn Poignant*

Anthropological and ethnographic subjects, especially Australia and the South Pacific. Also Scandinavia (early history and mythology), Sicily and England.

PPL (Photo Agency) Ltd

Bookers Yard, The Street, Walberton, Arundel BN18 0PF
☎ 01243 555561   📠 01243 555562
ppl@mistral.co.uk
www.pplmedia.com
Contacts *Barry Pickthall, Richard Johnson*

Two million pictures covering watersports, sub-aqua, business and commerce, travel and tourism; pictures of yesteryear and a fast growing archive on Sussex and the home counties. Pictures available in high resolution directly from the website.

Premaphotos Wildlife

Amberstone, 1 Kirland Road, Bodmin PL30 5JQ
☎ 01208 78258
enquiries@premaphotos.com
www.premaphotos.com
Contact *Jean Preston-Mafham, Library Manager*

Natural history worldwide. Subjects include flowering and non-flowering plants, fungi, slime moulds, fruits and seeds, galls, leaf mines, seashore life, mammals, birds, reptiles, amphibians, insects, spiders, habitats, scenery and cultivated cacti. Commissions undertaken. Searchable website. Visitors welcome.

Harry Price Library of Magical Literature
▷ Mary Evans Picture Library

Professional Sport UK Ltd

18–19 Shaftesbury Quay, Hertford SG14 1SF
☎ 01992 505000   📠 01992 505020
pictures@prosport.co.uk
www.professionalsport.com

Photographic coverage of tennis, soccer, athletics, golf, cricket, rugby, winter sports and many minor sports. Major international events including the Olympic Games, World Cup soccer and all Grand Slam tennis events. Also news and feature material supplied worldwide. Online photo archive; photo transmission services available for editorial and advertising.

Public Record Office Image Library
▷ The National Archives

Punch Cartoons

The Mansion, Bletchley Park, Bletchley MK3 6EB

☎ 0845 658 2470

info@createonline.net

www.punchcartoons.com

www.punch.co.uk

Owner *Punch Limited*

Choose online from half a million images published from 1841 onwards. 'The world's largest cartoon archive' with a vast subject range including social history, politics, fashion, famous faces and much more, all drawn by some of the worlds' greatest cartoonists such as Sir John Tenniel, E.H. Shepard and Norman Thelwell.

Redferns Music Picture Library

7 Bramley Road, London W10 6SZ

☎ 020 7792 9914   🖷 020 7792 0921

info@redferns.com

www.redferns.com

Picture library covering every aspect of music, from 18th century classical to present day pop. Over one million archived artists and other subjects including musical instruments, recording studios, crowd scenes, festivals, etc. Brochure available. Over 250,000 images available on the website. 'Contact us for free lightbox searches.'

Report Digital

4 Clarence Road, Stratford-on-Avon CV37 9DL

☎ 01789 262151

info@reportdigital.co.uk

www.reportdigital.co.uk

A collection of 60,000 images. Subjects available online include work issues and occupations, leisure, economy, health, education, politics, social issues, protest, trades union, environmental issues, culture. An increasing coverage of international issues of globalization, migration and climate change. Pictures are available online; search, browse and download high resolution images. Member of **BAPLA**.

Retna Pictures Ltd

Units 1a & 1b, Farm Lane Trading Estate, 101 Farm Lane, London SW6 1QJ

☎ 0845 034 0645   🖷 0845 034 0646

info@retna.co.uk

sales@retna.co.uk

www.retna.co.uk

Established 1978, Retna Pictures Ltd is a leading picture agency with two libraries: celebrity/music and lifestyle. The former specializes in images of international and national celebrities, music from the '60s to the current day, films and personalities. The lifestyle library specializes in people, family life, work, leisure and food. Both libraries are constantly receiving new material from established and up and coming photographers.

Rex Features Ltd

18 Vine Hill, London EC1R 5DZ

☎ 020 7278 7294   🖷 020 7696 0974

enquiries@rexfeatures.com

www.rexfeatures.com

Contact *Sales Manager*

Extensive picture library established in the 1950s. Daily coverage of news, politics, personalities, showbusiness, glamour, humour, art, medicine, science, landscapes, royalty, etc.

RIDA Photolibrary

▷ **GeoScience Features**

Royal Air Force Museum

Grahame Park Way, Hendon, London NW9 5LL

☎ 020 8205 2266   🖷 020 8200 1751

photographic@rafmuseum.org

www.rafmuseum.org

www.rafmuseumphotos.com

About a quarter of a million images, mostly b&w, with around 1,500 colour in all formats, on the history of aviation. Particularly strong on the activities of the Royal Air Force from the 1870s to 1970s. Researchers are requested to enquire in writing only.

The Royal Collection, Photographic Services

St. James's Palace, London SW1A 1JR

☎ 020 7839 1377   🖷 020 7024 5643

picturelibrary@royalcollection.org.uk

www.royalcollection.org.uk

www.royal.gov.uk

Contacts *Shruti Patel, Karen Lawson*

Photographic material of items in the Royal Collection, particularly oil paintings, drawings and watercolours, works of art, and interiors and exteriors of royal residences. Over 35,000 colour transparencies, plus 25,000 b&w negatives.

Royal Geographical Society Picture Library

1 Kensington Gore, London SW7 2AR

☎ 020 7591 3060   🖷 020 7591 3001

www.rgs.org/images

A strong source of geographical images, both historical and modern, showing the world through the eyes of photographers and explorers dating from the 1830s to the present day. The RGS Contempory Collection provides up-to-date transparencies from around the world, highlighting aspects of cultural activity, environmental phenomena, anthropology, architectural design, travel, mountaineering and exploration. The RGS Picture Library offers a professional and comprehensive service for both commercial and academic use.

Royal Photographic Society Collection
▷ Science & Society Picture Library

RSPB Images
The Old Dairy, Broadfield Road, Sheffield s8 0xQ
☎ 0114 258 0001 🖷 0114 258 0101
info@rspb-images.com
www.rspb-images.com

RSPB Images represents some of the UK's leading wildlife photographers, and holds stunning images ranging from birds, butterflies, moths, mammals, reptiles and their habitats to most RSPB reserves, plants and abstract subjects. High resolution downloads are now available for many of the images and a free picture research service is also provided.

RSPCA Photolibrary
RSPCA Trading Limited, Wilberforce Way, Southwater, Horsham RH13 9RS
☎ 01403 793150
pictures@rspcaphotolibrary.com
www.rspcaphotolibrary.com
Picture Requests *Becky Murray*

With over 150,000 colour images, the RSPCA Photolibrary has a comprehensive collection of natural history images whose subjects include mammals, birds, domestic and farm animals, amphibians, insects and the environment, as well as a unique photographic record of the RSPCA's work. Also includes the Wild Images collection. Downloadable website and e-mail lightboxes. No search fees.

Russia and Eastern Images
'Sonning', Cheapside Lane, Denham, Uxbridge UB9 5AE
☎ 01895 833508
easteuropix@btinternet.com
www.easteuropix.com

Architecture, cities, landscapes, people and travel images of Russia and the former Soviet Union. Considerable background knowledge available and Russian language spoken.

Peter Sanders Photography Ltd
24 Meades Lane, Chesham HP5 1ND
☎ 01494 773674/771372 🖷 01494 773674
photos@petersanders.com
www.petersanders.com
Contacts *Peter Sanders, Hafsa Garwatuk*

Specializes in the world of Islam in all its aspects from culture, arts, industry, lifestyles, celebrations, etc. Areas included are north, east and west Africa, the Middle East (including Saudi Arabia), China, Asia, Europe and USA. A continually expanding library.

scenicireland.com
17 Clarence Street, Belfast BT2 8DY
☎ 028 9024 5038 🖷 028 9023 1942
sales@scenicireland.com
www.scenicireland.com
Contact *Christopher Hill*

A comprehensive collection of scenic landscapes of Ireland. Every aspect of Irish life is shown, concentrating on the positive. Also large miscellaneous section. The website contains over 15,000 images available to buy online.

Science & Society Picture Library
Science Museum, Exhibition Road, London SW7 2DD
☎ 020 7942 4400 🖷 020 7942 4401
piclib@nmsi.ac.uk
www.scienceandsociety.co.uk

'The Science & Society Picture Library has one of the widest ranges of photographs, paintings, prints, posters and objects in the world.' The images come from the Science Museum, the **National Railway Museum** and the National Media Museum (formerly the National Museum of Photography, Film & Television) which now includes the Royal Photographic Society Collection. Images are available as high or low resolution files via e-mail, FTP or direct from the website.

Science Museum
▷ Science & Society Picture Library

Science Photo Library
327–329 Harrow Road, London W9 3RB
☎ 020 7432 1100 🖷 020 7286 8668
info@sciencephoto.com
www.sciencephoto.com

Subjects covered include the human body, health and medicine, research, genetics, technology and industry, space exploration and astronomy, earth science, satellite imagery, environment, flowers, plants and gardens, nature and wildlife and the history of science. The whole collection, more than 250,000 images, is available online.

Mick Sharp Photography
Eithinog, Waun, Penisarwaun, Caernarfon LL55 3PW
☎ 01286 872425 🖷 01286 872425
mick.jean@virgin.net
Contacts *Mick Sharp, Jean Williamson*

Archaeology, ancient monuments, buildings, churches, countryside, environment, history, landscape, past cultures and topography. Emphasis on British Isles but material also from other countries. Features the photos of both Mick Sharp and Jean Williamson. Access to other specialist collections on related subjects.

Commissions undertaken. Medium format and 35mm colour transparencies plus b&w prints from 5x4 negatives.

Phil Sheldon Golf Picture Library

40 Manor Road, Barnet EN5 2JQ
☎ 020 8440 1986　📠 020 8440 9348
gill@philsheldongolfpics.co.uk
info@philsheldongolfpics.co.uk
www.philsheldongolfpics.co.uk

An expanding collection of over 600,000 quality images of the 'world of golf'. In-depth worldwide tournament coverage including every Major championship and Ryder Cup since 1976. Instruction, portraits, trophies and over 400 golf courses from around the world. Also the Dale Concannon collection covering the period 1870 to 1940, the classic 1960s collection by photographer Sidney Harris and the Ed Lacey Collection.

Skishoot–Offshoot

Hall Place, Upper Woodcott, Whitchurch RG28 7PY
☎ 01635 255527　📠 01635 255528
pictures@skishoot.co.uk
www.skishoot.co.uk
Contact *Claire Randall*

Skishoot ski and snowboarding picture library has 500,000 images. Offshoot travel library specializes in France. Also produces the ski information website: www.welove2ski.com

The Skyscan Photolibrary

Oak House, Toddington, Cheltenham GL54 5BY
☎ 01242 621357/07941 445993　📠 01242 621343
info@skyscan.co.uk
www.skyscan.co.uk
Contact *Brenda Marks*

As well as the Skyscan Photolibrary collection of unique balloon's-eye views of Britain, the library includes the work of photographers from across the aviation spectrum; air to ground, aviation, aerial sports – 'in fact, anything aerial!' Links have been built with photographers across the world; photographs can be handled on an agency basis and held in house, or as a brokerage where the collection stays with the photographer; terms 50/50 for both. Commissioned photography arranged. Enquiries welcome.

Snookerimages (Eric Whitehead Photography)

7 Brow Close, Bowness on Windermere LA23 2HA
☎ 01539 448894
snooker@snookerimages.co.uk
www.snookerimages.co.uk
www.snookerimages-pictures.com
Contact *Eric Whitehead*

Over 30,000 images of snooker from 1982 to the present day.

Solo Syndication Ltd

17–18 Haywards Place, Clerkenwell, London EC1R 0EQ
☎ 020 7566 0360　📠 020 7566 0388

Three million images from the archives of the *Daily Mail*, *Mail on Sunday* and *Evening Standard*. Hard prints or Mac-to-Mac delivery. 24-hour service.

Sotheby's Picture Library

34–35 New Bond Street, London W1A 2AA
☎ 020 7293 5383　📠 020 7293 5062
piclib@sothebys.com
www.sothebys.com
Contact *Joanna Ling*

The library consists of over one million subjects sold at Sotheby's. Images from the 15th to the 20th century. Oils, drawings, watercolours, prints and decorative items as well as the unique archive of Cecil Beaton photographs. 'Happy to do searches or, alternatively, visitors are welcome by appointment.'

South American Pictures

48 Station Road, Woodbridge IP12 4AT
☎ 01394 383963
info@southamericanpictures.com
www.southamericanpictures.com
www.nonesuchinfo.info (specialist site)
Contact *Marion Morrison*

Colour and b&w images of South/Central America, Cuba, Mexico, New Mexico (USA), Dominican Republic and Haiti, including archaeology and the Amazon. There is an archival section, with pictures and documents from most countries. Now with 40 contributing photographers. All images are supplied digitally.

Starstock
▷ Photoshot

Stay Still
▷ Photoshot

Still Pictures

1 Glen Cottages, Sandy Lane, Abbots Leigh, Bristol BS8 3SE
☎ 01275 375520　📠 0705 061 3938
info@stillpictures.com
www.stillpictures.com
Contact *Julie-anne Wilce*

Founded 1970. High profile photo library specializing in environment, the Third World, social issues and nature. Represents 15 leading European and US agencies and over 400 photographers as well as the United Nations Environment Programme (UNEP) archive, the

Christian Aid collection and a growing selection of images from Woodfall Wild Images. The website has nearly one million images online, ready for instant download.

Stockfile

5 High Street, Sunningdale SL5 0LX
☎ 01344 872249
info@stockfile.co.uk
www.stockfile.co.uk
Contacts *Jill Behr, Steven Behr*

Specialist cycling collection with emphasis on mountain biking. Expanding adventure sports section covering snow, land, air and water activities.

Stockscotland.com

The Grange House, Lochside, Lairg IV27 4EG
☎ 01549 402295
info@stockscotland.com
www.stockscotland.com
Contact *Hugh Webster*

150,000 Scottish images with 10,000 currently available online.

Sir John Benjamin Stone
▷ Birmingham Library Services under Library Services

Joe Tasker Collection
▷ Chris Bonington Picture Library

Tate Images

The Lodge, Tate Britain, Millbank, London SW1P 4RG
☎ 020 7887 8871/8890/4933   ⊞ 020 7887 8805
tate.images@tate.org.uk
www.tate-images.com
Contact *Alison Fern*

Tate is one of the world's leading visual arts organizations and its collection encompasses the national collection of historic British art from 1500, including iconic masterpieces by Gainsborough, Constable, Turner, David Hockney and Henry Moore, and the national collection of international modern art that includes works by Dali, Picasso, Matisse, Rothko, Emin, Hirst, Warhol and Andreas Gursky. This vast collection of art imagery is available from the Tate Images website. Clients can browse the entire Tate collection using a new, keyword-based search engine; select images and save their selections to a personal light box; download low-resolution imagery for layout purposes. E-commerce coming soon.

Theatre Museum
▷ V&A Images

Thoroughbred Photography Ltd

The Hornbeams, 2 The Street, Worlington IP28 8RU
☎ 01638 713944
mail@thoroughbredphoto.com
www.thoroughbredphoto.com
Contacts *Gill Jones, Laura Green*

Photography by Trevor Jones. Extensive library of high-quality colour images depicting all aspects of thoroughbred horseracing dating from 1987. Major group races, English Classics, studs, stallions, mares and foals, early morning scenes, personalities, jockeys, trainers and prominent owners. Also international work: USA Breeders' Cup, Arc de Triomphe, French Classics, Irish Derby, Dubai World Cup Racing and scenes; and more unusual scenes such as racing on the sands at low tide, Ireland, and on the frozen lake at St Moritz. Visitors by appointment.

TopFoto

PO Box 33, Edenbridge TN8 5PF
☎ 01732 863939   ⊞ 01732 860215
admin@topfoto.co.uk
www.topfoto.co.uk
Contact *Alan Smith*

International editorial distributor with over two million pictures available online to download and 12 million in hard copy, representing 40 leading suppliers.

B.M. Totterdell photography

Constable Cottage, Burlings Lane, Knockholt TN14 7PE
☎ 01959 532001   ⊞ 01959 532001
btrial@btinternet.com
whatvolleyball.co.uk
Contact *Barbara Totterdell*

Specialist volleyball library covering all aspects of the sport.

Tessa Traeger Picture Library

7 Rossetti Studios, 72 Flood Street, London SW3 5TF
☎ 020 7352 3641   ⊞ 020 7352 4846
info@tessatraeger.com
www.tessatraeger.com

Food, gardens, travel and artists.

Travel Ink Photo Library

The Old Coach House, 14 High Street, Goring on Thames, Nr Reading RG8 9AR
☎ 01491 873011   ⊞ 01491 875558
info@travel-ink.co.uk
www.travel-ink.co.uk
Library Manager *Frances Honnor*
Sales and Administration *Felicity Bazell*

A collection of over 120,000 travel, tourism and lifestyle images, carefully edited and constantly updated, from countries worldwide. Specialist collections from the UK, Greece, France, Far East and Caribbean. The website offers a fully captioned and searchable selection of over 31,000 images and is ideal for picture researchers.

Tropix Photo Library

44 Woodbines Avenue, Kingston upon Thames KT1 2AY

☏ 020 8546 0823
info@tropix.co.uk
www.tropix.co.uk
Managing Director *Veronica Birley*

Specializes in images of developing nations: travel and editorial pictures emphasizing the attractive and progressive. Fully searchable, keyworded website. Evocative photos concerning the economies, environment, culture and society of 100+ countries across Africa, Central and South America, Caribbean, Eastern Europe, Middle East, Indian sub-continent, South East Asia, CIS and Far East. Worldwide travel collections also include UK, Europe, North America and Antarctica. Assignment photography available. All photos supplied with detailed captions. Established 1982. **BAPLA** member.

True North

Louper Weir, Ghyll Head, Windermere LA23 3LN

☏ 07941 630420
www.northpix.co.uk
Contact *John Morrison*

The collection features the life and landscape of the north of England, photographed by John Morrison. No other photographer's work required.

U.P.P.A.
▷ Photoshot

Ulster Museum Picture Library

Ulster Folk & Transport Museum, Cultra, Holywood BT18 0EU

☏ 028 9039 5109   ☐ 028 9039 5004
michelle.ashmore@nmni.com
www.ulstermuseum.org.uk

Specialist subjects: art (fine and decorative, late 17th–20th century), particularly Irish art, archaeology, ethnography, treasures from the Armada shipwrecks, geology, botany, zoology, local history and industrial archaeology.

United Nations Environment Programme (UNEP)
▷ Still Pictures

Universal Pictorial Press & Agency (U.P.P.A.)
▷ Photoshot

V&A Images

Victoria and Albert Museum, Cromwell Road, London SW7 2RL

☏ 020 7942 2482 (commercial)/2479 (academic)
☐ 020 7942 2482
vaimages@vam.ac.uk
www.vandaimages.com

A vast collection of photographs from the world's largest museum of decorative and applied arts, reflecting culture and lifestyle spanning over 1,000 years of history to the present time. Digital delivery of contemporary and historical textiles, costumes and fashions, ceramics, furniture, metalwork, glass, sculpture, toys, and games, design and photographs from around the world. Unique photographs include 1960s fashion by John French, Harry Hammond's behind the scenes pop idols, Houston Rogers theatrical world of the 1930s to 1970s, images of royalty by Lafayette, Cecil Beaton, and the 19th century pioneer photographers. Images from the National Art Library and Library of Art & Design, and from the Theatre Museum and Museum of Childhood are readily available.

Vin Mag Archive Ltd

84–90 Digby Road, London E9 6HX
☏ 020 8533 7588   ☐ 020 8533 7283
piclib@vinmag.com
www.vinmagarchive.com
Contact *Angela Maguire*

Wholly owned subsidiary of the Vintage Magazine Company. 'A unique source of images covering just about any subject, period, theme and location.' Also a collection of half a million movie, TV, celebrity and sports images available online. Contact *Angela Maguire* for code and password. See also entry under *Library Services*.

John Walmsley Photo Library
▷ Education Photos

Warwickshire Photographic Survey
▷ Birmingham Library Services under Library Services

Waterways Photo Library

39 Manor Court Road, Hanwell, London W7 3EJ
☏ 020 8840 1659   ☐ 020 8567 0605
watphot39@aol.com
www.waterwaysphotolibrary.com
Contact *Derek Pratt*

A specialist photo library on all aspects of Britain's inland waterways. Top-quality 35mm and medium-format colour transparencies, plus a large collection of b&w and an increasing collection of digital photography. Rivers and canals, bridges, locks, aqueducts, tunnels and waterside buildings. Town and countryside scenes, canal art, waterway holidays, boating, fishing, windmills, watermills, watersports and wildlife.

Weimar Archive
▷ Mary Evans Picture Library

Wellcome Images
183 Euston Road, London NW1 2BE
☎ 020 7611 8348 📠 020 7611 8577
images@wellcome.ac.uk
http://images.wellcome.ac.uk
Contact *Venita Paul*

Approximately 180,000 images on the history of medicine and human culture worldwide, including religion, astronomy, botany, genetics, landscape and cell biology.

West Cornwall Photographic Archive
▷ Morrab Library under Library Services

Eric Whitehead Photography
7 Brow Close, Bowness on Windermere LA23 2HA
☎ 01539 448894
snooker@snookerimages.co.uk
www.snookerimages.co.uk
www.snookerimages-pictures.com
Contact *Eric Whitehead*

The agency covers local news events, PR and commercial material, also leading library of snooker images (see **Snookerimages**).

Wild Images
▷ RSPCA Photolibrary

Wild Places Photography
▷ Chris Howes

David Williams Picture Library
Allt-na-Craobh, Old Shore Road, Connel PA37 1PT
☎ 01631 710586 📠 01631 710586
david@scotland-guide.co.uk

Specializes in travel photography with wide coverage of Scotland, Iceland and Spain. Many other European countries also included plus smaller collections of Western USA and Canada. The main subjects in each country are: cities, towns, villages, 'tourist haunts', buildings of architectural or historical interest, landscapes and natural features. The Scotland and Iceland collections include many pictures depicting physical geography and geology. Photographic commissions and illustrated travel articles undertaken. Catalogue available.

Vaughan Williams Memorial Library
English Folk Dance and Song Society, Cecil Sharp House, 2 Regent's Park Road, London NW1 7AY
☎ 020 7485 2206 ext. 29/33 📠 020 7284 0523
library@efdss.org
www.efdss.org

Mainly b&w coverage of traditional/folk music, dance and customs worldwide, focusing on Britain and other English-speaking nations. Photographs date from the late 19th century to the present day.

The Wilson Photographic Collection
▷ Dundee Central Library under Library Services

Woodfall Wild Images
▷ Photoshot

World Illustrated
▷ Photoshot

World Pictures
▷ Photoshot

Yemen Pictures
28 Sheen Common Drive, Richmond TW10 5BN
☎ 020 7602 1989 📠 020 7602 1989
www.facesandplacespix.com

Large collection (4,000 transparencies) covering all aspects of Yemen – culture, people, architecture, dance, qat, music. Also Africa, Australia, Middle East and Asia.

York Archaeological Trust Picture Library
47 Aldwark, York YO1 7BX
☎ 01904 663006 📠 01904 663024
ckyriacou@yorkat.co.uk
www.yorkarchaeology.co.uk

Specialist library of rediscovered artifacts, historic buildings and excavations, presented by the creators of the highly acclaimed Jorvik Viking Centre. The main emphasis is on the Roman, Anglo-Saxon, Medieval and Viking periods.

John Robert Young Collection
Paxvobiscum, 16 Greenacres Drive, Ringmer, Lewes BN8 5LX
☎ 01273 814172
pax@freedom255.com
www.johnrobertyoung.com
Contacts *Jennifer Barrett, John Robert Young*

50,000 colour and b&w images. Military (French Foreign Legion, Spanish Foreign Legion, Royal Marines), religion (Christian), travel, China, personalities of the 1960s and '70s, sixties' fashion, also a growing portfolio of fine art photography.

Press Cuttings Agencies

Cision

Cision House, 16–22 Baltic Street West, London
EC1Y 0UL
☎ 0870 736 0010
info.uk@cision.com
www.cision.com

Formerly Romeike, Cision monitors national
and international dailies and Sundays, provincial
papers, consumer magazines, trade and technical
journals, teletext services as well as national radio
and TV networks. Back research, advertising
checking and internet monitoring, plus analysis
and editorial summary service available.

Durrants

Discovery House, 28–42 Banner Street, London
EC1Y 8QE
☎ 020 7674 0200   ℻ 020 7674 0222
contact@durrants.co.uk
www.durrants.co.uk

Wide coverage of all print media sectors including
foreign press plus internet, newswire and
broadcast monitoring. High speed, early morning
press cuttings from the national press by web or
e-mail. Overnight delivery via courier to most
areas or first-class mail. Well presented, laser
printed, A4 cuttings. Rates on application.

International Press-Cutting Bureau

224–236 Walworth Road, London SE17 1JE
☎ 020 7708 2113   ℻ 020 7701 4489
info@ipcb.co.uk
www.ipcb.co.uk
Contact *Robert Podro*

Covers national, provincial, trade, technical and
magazine press. Cuttings are normally sent twice
weekly by first-class post. Charges start at £75 per
month + £1.10 per cutting.

We Find It (Press Clippings)

40 Galwally Avenue, Belfast BT8 7AJ
☎ 028 9064 6008   ℻ 028 9064 6008
Contact *Avril Forsythe*

Specializes in Northern Ireland press and
magazines, both national and provincial. Rates on
application.

Festivals

Aberdeen Arts Carnival

Aberdeen Arts Centre, 33 King Street, Aberdeen
AB24 5AA
☏ 01224 635208
www.aberdeenartscentre.org.uk
Venue Manager *Paula Gibson*

Performances by local amateurs and professional theatre companies. Arts workshops in drama, music, art, dance and creative writing take place each summer during the school holidays. Drama workshops also take place throughout the year.

University of Aberdeen Writers Festival
▷ Word

The Aldeburgh Literary Festival

44 High Street, Aldeburgh IP15 5AB
☏ 01728 452389   ℻ 01728 452389
johnandmary@aldeburghbookshop.co.uk
www.aldeburghbookshop.co.uk
Festival Organizers *John & Mary James*

Founded 2002. Held on the first weekend in March featuring a programme of talks. Previous speakers have included Beryl Bainbridge, Alan Bennett, Craig Brown, Richard Dawkins, Lady Antonia Fraser, Michael Frayn, Anthony Horowitz, P.D. James, Doris Lessing, David Lodge, Ian McEwan, Harold Pinter, Matt Ridley, Alexander McCall Smith, Claire Tomalin, Salley Vickers, A.N. Wilson.

Aldeburgh Poetry Festival

The Poetry Trust, The Cut, 9 New Cut,
Halesworth IP19 8BY
☏ 01986 835950   ℻ 01986 874524
info@thepoetrytrust.org
www.thepoetrytrust.org
Director *Naomi Jaffa*

Founded 1989. One of the UK's biggest celebrations of international contemporary poetry, held in the small coastal town of Aldeburgh annually in early November (6th–8th in 2009). Features readings, performances, craft talks, public masterclass, workshops, discussions and a family event. Includes a reading by the winner of the annual **Aldeburgh First Collection Prize** (see entry under *Prizes*).

Aspects Literature Festival

North Down Borough Council, Town Hall, The
Castle, Bangor BT20 4BT
☏ 028 9127 8032   ℻ 028 9127 1370
gail.prentice@northdown.gov.uk
Festival Director/Arts Officer *Gail Prentice*

Founded 1992. 'Ireland's Premier Literary Festival' is held at the end of September and celebrates the richness and diversity of living Irish writers with occasional special features on past generations. It draws upon all disciplines – fiction (of all types), poetry, theatre, non-fiction, cinema, song-writing, etc. It also includes a day of writing for young readers and sends writers to visit local schools during the festival. Highlights of recent festivals include appearances by Bernard MacLaverty, Marion Keyes, Alice Taylor, Frank Delaney, Seamus Heaney, Brian Keenan and Fergal Keane.

Aye Write! Bank of Scotland Book Festival

The Mitchell Library, North Street, Charing
Cross, Glasgow G3 7DN
☏ 0141 287 2999/2876   ℻ 0141 287 2815
lil@csglasgow.org
www.ayewrite.com
Festival Director *Karen Cunningham*
Festival Programmer *Andrew Kelly*

Launched 2005. Annual festival, held in March, which aims to increase the use of libraries, encourage a love of books, reading and writing and an awareness of Glasgow's writing heritage. The 2008 Festival guests included Kathleen Turner, Gerry Anderson, Hanif Kureishi, Dame Helena Kennedy, Louis de Bernières, Edwin Morgan, Will Self, Alasdair Gray, Joanne Harris.

Bath Festival of Children's Literature

PO Box 4123, Bath BA1 0FR
☏ 01225 463362 (tickets)   ℻ 01225 445485
info@bathkidslitfest.co.uk
www.bathkidslitfest.co.uk
Festival Directors *John McLay, Gill McLay*

Founded 2007. Annual event celebrating children's books and reading held 18th–27th September in 2009 (17th–26th in 2010). Over 80 events for readers aged up to 16. Guest authors have included Jacqueline Wilson, Eoin Colfer,

Anthony Horowtiz, Michael Morporgo, Michael Rosen, Meg Cabot, Darren Shan, Lauren Child, Louise Rennison, Francesca Simon, Michelle Paver, Charlie Higson, Rick Riordan, Allan Ahlberg and Harry Hill. Sponsored by *The Daily Telegraph* and Waterstone's Booksellers.

Bath Literature Festival

Bath Festivals, Abbey Chambers, Kingston Buildings, Bath BA1 1NT
☎ 01225 462231   ⊞ 01225 445551
info@bathfestivals.org.uk
www.bathlitfest.org.uk
Box Office: Bath Festivals Box Office, 2 Church Street, Abbey Green, Bath BA1 1NL
☎ 01225 463362
boxoffice@bathfestivals.org.uk
Artistic Director *James Runcie*

Founded 1995. This annual festival (held in February and March) programmes over 100 different literary events from debates and lectures to readers' groups and workshops in venues throughout the city. In addition there are a number of events for children and young people. Previous featured writers include Martin Amis, Terry Pratchett, Margaret Drabble, Joanna Trollope, A.C. Grayling, Steven Berkoff, Hermione Lee.

BayLit Festival

Academi, Mount Stuart House, Mount Stuart Square, Cardiff CF10 5FQ
☎ 029 2047 2266   ⊞ 029 2049 2930
post@academi.org
www.academi.org
Chief Executive *Peter Finch*

Annual literature festival held in Cardiff Bay, featuring writers from Wales and beyond. Lectures, readings, performances, workshops and book launches in English and Welsh. Dates and further details available on the Academi website. Writers who appeared in previous festivals include Ian McMillan, Andrew Motion, Gillian Clarke, Fay Weldon, Simon Singh, Howard Marks and Will Self.

Belfast Festival at Queen's
▷ The Ulster Bank Belfast Festival at Queen's

Beverley Literature Festival

Wordquake, Libraries and Information, Council Offices, Skirlaugh HU11 5HN
☎ 01482 392745   ⊞ 01482 392710
john@bevlit.org
www.bevlit.org
Festival Director *John Clarke*

Founded 2002. Annual October festival which has acquired 'a national reputation for commissioning and hosting quality poetry events'. Also covers all other literary genres and mixes readings and performances with author-led readers' groups and creative writing workshops. Held in intimate and historic venues across Beverley.

Birmingham Book Festival

c/o Unit 116, The Custard Factory, Gibb Street, Birmingham B9 4AA
☎ 0121 246 2770   ⊞ 0121 246 2771
jonathan@birminghambookfestival.org
www.birminghambookfestival.org
Contact *Jonathan Davidson*

Founded 1999. Annual literature festival held during October at venues around Birmingham. Also promotes events throughout the year. Includes performances, lectures, discussion events and workshops.

Book Now - Richmond Literature Festival

The Arts Service, Orleans House Gallery, Riverside, Twickenham TW1 3DJ
artsinfo@richmond.gov.uk
www.richmond.gov.uk/literature

Founded 1992. Annual three-week literature festival held in November, delivered by the Arts Service of Richmond Borough. Plays host to leading writers.

Borders Book Festival

Harmony House, St Mary's Road, Melrose TD6 9LJ
☎ 07929 435575
info@bordersbookfestival.org
www.bordersbookfestival.org
Festival Coordinator *Paula Ogilvie*

Founded in 2004. Annual literary festival held over four days during the third weekend in June. The event brings internationally acclaimed and new authors to the gardens of the National Trust for Scotland's splendid Georgian property, Harmony House, and other locations in Melrose, for lively and entertaining discussions and performances to audiences of all ages and tastes. Past speakers include Ian Rankin, Iain Banks, Michael Palin, Rory Bremner, Alexander McCall Smith, Sarah Raven, Roy Hattersley, Claire Messud, Sheila Hancock, William Dalrymple and A.L. Kennedy. A well-stocked Festival bookshop is open throughout the Festival dates. Free mailing list.

Bournemouth Literary Festival

☎ 01202 417535
info@bournemouthliteraryfestival.co.uk
www.bournemouthliteraryfestival.co.uk
Director & Founder *Lillian Avon*

Founded in 2004, the Festival fuses literature with performing arts, celebrating the international world of words in venues all over Bournemouth.

Brighton Festival

Brighton Dome & Festival Ltd, 12a Pavilion Buildings, Castle Square, Brighton BN1 1EE
☎ 01273 700747   📠 01273 707505
info@brightonfestival.org
www.brightonfestival.org
Contact *Festival Office*

Founded 1966. For 24 days every May, Brighton hosts England's largest mixed arts festival. Music, dance, theatre, film, opera, literature, comedy and exhibitions. Literary enquiries will be passed to the literature officer. Deadline October for following May.

Bristol Poetry Festival
▷ Poetry Can under Organizations of Interest to Poets

Broadstairs Dickens Festival

Dickens House, 2 Victoria Parade, Broadstairs CT10 1QD
☎ 01843 861232   📠 01843 871133
cramptontower@hotmail.com
www.broadstairsdickensfestival.co.uk
Festival Coordinator *Peter Shaw*

Founded in 1937 to commemorate the 100th anniversary of Charles Dickens' first visit to Broadstairs in 1837, which he continued to visit until 1859. The Festival lasts for five days in June and events include a parade, a performance of a Dickens play (*Bleak House* in 2008), melodramas, Dickens readings, a Victorian cricket match, Victorian bathing parties, talks, music hall, five-day Victorian country fair. Costumed Dickensian ladies in crinolines with top-hatted escorts promenade during the Festival.

Buxton Festival

3 The Square, Buxton SK17 6AZ
☎ 01298 70395   📠 01298 72289
info@buxtonfestival.co.uk
www.buxtonfestival.co.uk
Chief Executive *Glyn Foley*
Press & Marketing Officer *Johanna Dollerson*

Founded 1979. Held annually in July, this 19-day opera, literature and music festival includes a varied literary programme, attracting those with a broad interest in well-crafted writing, whether in biography, fiction, politics or personal memoir.

Cambridge Wordfest

☎ 01223 264404
admin@cambridgewordfest.co.uk
www.cambridgewordfest.co.uk
Festival Director *Cathy Moore*
Festival Patrons *Dame Gillian Beer, Rowan Pelling, Ali Smith*

An annual spring festival for writers as well as readers which takes place at various venues across Cambridge and the surrounding area.

A weekend event packed with the best of contemporary writing, poetry, political debate, events for children and writing workshops. Also a one-day winter festival during the last weekend in November.

Canterbury Festival

Christ Church Gate, The Precincts, Canterbury CT1 2EE
☎ 01227 452853   📠 01227 781830
info@canterburyfestival.co.uk
www.canterburyfestival.co.uk
Festival Director *Rosie Turner*

Canterbury provides a unique location for the largest festival of arts and culture in the region (17th–31st October in 2009), showcasing local, national and international talent in classical and contemporary music, world theatre, opera, comedy, dance, walks and community events, plus an extended programme of literature and debate.

Carnegie Sporting Words Festival

Harrogate International Festivals, Raglan House, Raglan Street, Harrogate HG1 1LE
☎ 01423 562303   📠 01423 521264
erica@harrogate-festival.org.uk
www.sportingwords.co.uk
Contact *Erica Morris*

Founded in 2008, the Carnegie Sporting Words Festival (CSWF) takes place in Leeds and Harrogate over a weekend in the autumn. It is the UK's first festival celebrating the sporting word as expressed through books, newspapers, broadcast media and the web, and will feature sports celebrities, broadcasters, journalists and authors. Previous events have included a Sports Quiz Dinner hosted by BBC's Mark Pougatch; a seminar exploring the future of digital sports media and a Schools Activity Day at the Headingley Carnegie Stadium in partnership with BBC Sport.

Centre for Creative & Performing Arts Spring Literary Festival at UEA

University of East Anglia, School of Literature and Creative Writing, Norwich NR4 7TJ
☎ 01603 592810
v.striker@uea.ac.uk
www1.uea.ac.uk/cm/home/schools/hum/booksandwriters/litfest
Contact *Val Striker*

Founded 1993. Annual event held in the spring and summer.

Charleston Festival

The Charleston Trust, Charleston, Nr Firle, Nr Lewes BN8 6LL
☎ 01323 811626

info@charleston.org.uk
www.charleston.org.uk
Festival Programmer *Diana Reich*

Annual literary festival held in May over two weeks. Novelists, biographers, travel writers, broadcasters, poets, food writers, actors and artists gather at Charleston, the country home of the Bloomsbury Group. Past speakers include Margaret Atwood, Alan Bennett, Jeanette Winterson, Louis de Bernières, Patti Smith, Paula Rego, Clive James, Ali Smith, Roger McGough, Germaine Greer and Harold Pinter.

Cheltenham Literature Festival
> The Times Cheltenham Literature Festival

Chester Literature Festival
55–57 Watergate Row South, Chester CH1 2LE
☎ 01244 409113   📠 01244 401697
info@chesterlitfest.org
www.chesterfestivals.org.uk/literature
Chairman *Bill Hughes*

Founded 1989. Annual festival held in early October. Events include international and nationally known writers, as well as events by local literary groups. There is a Literary Lunch and a Festival Dinner, events for children, workshops and competitions. Free mailing list.

Children's Books Ireland – Annual Festival
> Children's Books Ireland under Professional Associations

Dartington Literary Festival
> Ways With Words

Derbyshire Literature Festival
c/o Arts Team, Derbyshire County Council, Cultural & Community Services Department, Alfreton Library, Severn Square, Alfreton DE55 7BQ
☎ 01773 832497   📠 01773 831359
ann.wright@derbyshire.gov.uk
www.derbyshire.gov.uk
Festival Organizer *Ann Wright*

Founded 2000. The festival takes place in June every two years and covers the whole county. The festival programming includes all types of live literature events; performance poetry, theatre, talks, readers' groups, workshops, dramatized readings, signings, storytelling and literary trails, as well as a number of cross-art form events. The festival takes place in libraries and many other community venues including heritage centres, industrial buildings, stately homes, parks and moors, churches and schools and is specifically designed to reach as many different communities and geographical areas in the county as possible.

Dorchester Festival
Dorchester Arts Centre, School Lane, The Grove, Dorchester DT1 1XR
☎ 01305 266926
enquiries@dorchesterarts.org.uk
www.dorchesterarts.org.uk
Artistic Director *Sharon Hayden*

Founded 1996. An annual five-day international festival held over late May Bank Holiday which includes performance, live music and visual arts with associated educational and community projects. Also three days of free events in various venues around the town, including literature and poetry. The theme for the 2009 festival is Africa and its wider cultural impact.

The Daphne du Maurier Festival of Arts & Literature
Restormel Borough Council, 39 Penwinnick Road, St Austell PL25 5DR
☎ 01726 223300
rbc@dumaurierfestival.co.uk
www.dumaurierfestival.co.uk

Founded in 1997, this annual arts and literature festival is held in and around Fowey in Cornwall in May and includes a wide range of professional and community events.

Dumfries and Galloway Arts Festival
Gracefield Arts Centre, 28 Edinburgh Road, Dumfries DG1 1JQ
☎ 01387 260447   📠 01387 260447
info@dgartsfestival.org.uk
www.dgartsfestival.org.uk
Festival Organizer *Mrs Barbara Kelly*

Founded 1980. Annual week-long festival held at the end of May with a variety of events including classical and folk music, theatre, dance, literary events, exhibitions and children's events.

Dundee Literary Festival
The Tower Building, Nethergate, Dundee DD1 4HN
☎ 01382 384768
dundeeliteraryfestival@gmail.com
www.literarydundee.co.uk
Festival Director *Anna Day*

Founded 2007. Annual literary festival held in June with a mix of internationally renowned writers, local authors, poetry and new writing. Speakers have included Philip Pullman, Jacqueline Wilson and Kirsty Gunn. E-mail to be added to the mailing list.

Durham Book Festival
Durham City Arts, 2 The Cottages, Fowlers Yard, Back Silver Street, Durham DH1 3RA
☎ 0191 375 0763

enquiries@durhamcityarts.org.uk
www.bookfestival.org.uk
www.durhamcityarts.org
Festival Coordinators *Becca Pelly-Fry, Kate James*

Founded 1989. Annual festival held in September/
October at various locations in the city.
Performances and readings plus workshops,
cabaret, exhibitions and other events.

Edinburgh International Book Festival

5a Charlotte Square, Edinburgh EH2 4DR
☏ 0131 718 5666   ☏ 0131 226 5335
admin@edbookfest.co.uk
www.edbookfest.co.uk
Director *Catherine Lockerbie*
Children and Education Programme Director
Sara Grady
Programme Manager *Roland Gulliver*

Founded 1983. 'The world's largest and most
dynamic annual book event' takes place across
17 days in Edinburgh each August. An extensive
programme of over 750 events featuring 700
or more authors, showcasing the work of the
world's top writers and thinkers to an audience
of over 200,000 adults and children. Featuring
workshops, readings, lectures and a high profile
debate and discussion series, book signings,
bookshops and cafes.

Folkestone Literary Festival

The Block, 65–69 Tontine Street, Folkestone
CT20 1JR
☏ 01303 245799   ☏ 01303 223761
info@folkestonelitfest.co.uk
www.folkestonelitfest.co.uk
Festival Director *Roberta Spicer*

Annual festival. Tours, discussions, talks, live
performances and readings.

Frome Festival

25 Market Place, Frome BA11 1AH
☏ 01373 453889
office@fromefestival.co.uk
www.fromefestival.co.uk

An annual festival held from the first Friday in
July for ten days celebrating all aspects of visual
and performing arts and entertainment and with
a strong literary element. Features music, film,
dance, drama and visual arts plus readings, talks
and workshops. Previous literary guests have
included nationally renowned writers as well as
local authors.

Graham Greene Festival

▷ Graham Greene Birthplace Trust under Literary
Societies

The Guardian Hay Festival

25 Lion Street, Hay-on-Wye HR3 5AD
☏ 0870 787 2848   ☏ 01497 821066

admin@hayfestival.com
www.hayfestival.com
Festival Director *Peter Florence*

Founded 1988. Annual May festival sponsored by
The Guardian. Also has a children's programme
which began in 2003. The event coincides with
half term and caters for young children through
to adult. Guests have included Paul McCartney,
Bill Clinton, Salman Rushdie, Toni Morrison,
Stephen Fry, Joseph Heller, Carlos Fuentes, Maya
Angelou, Amos Oz, Arthur Miller, Jimmy Carter.

Guildford Book Festival

c/o Tourist Information Centre, 14 Tunsgate,
Guildford GU1 3QT
☏ 01483 444334
deputy@guildfordbookfestival.co.uk
www.guildfordbookfestival.co.uk
Book Festival Director *Glenis Pycraft*

Patrons: Elizabeth Buchan, Michael Buerk,
Michael Rosen, Sandi Toksvig, Fay Weldon,
Timothy West and Jacqueline Wilson. Held
annually, during October in venues throughout
the ancient town of Guildford (15th–24th
in 2009). The festival includes events with
established and new writers, workshops,
poetry performances, children's events and the
Guildford Book Festival First Novel Award (see
entry under *Prizes*). This diverse festival aims
to involve, instruct and entertain all who care
about literature. Guest authors in 2008 included
Michael Rosen, Ray Mears, Mark Thomas, Gerald
Scarfe, Lionel Schriver, Katharine Whitehorn.

Harrogate Crime Writing Festival

▷ Theakstons Old Peculier Harrogate Crime Writing
Festival

Harrogate International Festival

Raglan House, Raglan Street, Harrogate HG1 1LE
☏ 01423 562303   ☏ 01423 521264
info@harrogate-festival.org.uk
www.harrogate-festival.org.uk
Ops Director *Sharon Canavar*

Founded 1966. Annual two-week festival at the
end of July and beginning of August. Events
include international symphony orchestras,
chamber concerts, ballet, celebrity recitals,
contemporary dance, opera, drama, world music,
jazz and comedy.

Harwich Festival of the Arts

2A Kings Head Street, Harwich CO12 3EG
☏ 01255 503571
anna@rendell-knights.freeserve.co.uk
Contact *Anna Rendell-Knights*

Founded 1980. Annual ten-day summer festival
held in June/July. Events include concerts,
drama, film, dance, art, exhibitions, historic
town walks. The Festival in 2009 (3rd–12th)

will commemorate the 40th anniversary of the founding of The Harwich Society.

The Hay Festival
▷ The Guardian Hay Festival

Hebden Bridge Arts Festival
New Oxford House, Albert Street, Hebden Bridge HX7 8AH
☎ 01422 842684
hbfestival@gmail.com
www.hebdenbridge.co.uk/festival
Administrator *Rebecca Yorke*
Programming *Stephen May*

Founded 1994. Annual arts festival with increasingly strong adult and children's literature events. Previous guest writers include Roger McGough, Benjamin Zephaniah, Anne Fine, Juliet Barker, Jacqueline Wilson, Quentin Blake, Ian McMillan, Adele Geras, George Alagiah.

Humber Mouth – Hull Literature Festival
City Arts Unit, Central Library, Albion Street, Kingston upon Hull HU1 3TF
☎ 01482 616961   🖷 01482 616827
humbermouth@gmail.com
www.humbermouth.org.uk
Contact *Maggie Hannan*

Founded 1992. Hull's largest festival, held in the summer, and one of the region's liveliest events. Features readings, talks, performances and workshops by writers and artists from around the world and from the city.

Ilkley Literature Festival
The Manor House, 2 Castle Hill, Ilkley LS29 9DT
☎ 01943 601210   🖷 01943 817079
admin@ilkleyliteraturefestival.org.uk
www.ilkleyliteraturefestival.org.uk
Director *Rachel Feldberg*

Founded 1973. Major literature festival in the north held over 17 days every October. Includes children's literature weekend and Festival fringe. Recent guests include Maya Angelou, Simon Armitage, Alan Bennett, Sarah Waters, Claire Tomalin, Benjamin Zephaniah, Richard Ford, Carol Ann Duffy, P.D. James, Donna Tartt, Kate Adie, Germaine Greer, Michael Ondaatje, Chimamanda Adichie, Liz Lochhead, Tony Harrison, Ian Rankin. Telephone or e-mail to join free mailing list.

The International Festival of Mountaineering Literature
130 Nicholson Road, Sheffield S8 9SX
☎ 0114 258 4063
niallgrimes@hotmail.com
www.festivalofmountaineeringliterature.co.uk
Director *Niall Grimes*

Founded in 1987. Annual one-day festival held at Kendal Mountain Festivals, Kendal, Cumbria. Taking place in November, the Festival celebrates recent books, commissions new writing, gives overviews of national literatures, holds debates of issues, book signings. Announces the winner of the festival writing competition run in conjunction with *Climb* magazine. Write to join free mailing list.

International Playwriting Festival
Warehouse Theatre, Dingwall Road, Croydon CR0 2NF
☎ 020 8681 1257   🖷 020 8688 6699
info@warehousetheatre.co.uk
www.warehousetheatre.co.uk/ipf.html
Festival Administrator *Rose Marie Vernon*

The Warehouse Theatre Company's International Playwriting Festival will celebrate 24 successful years in 2009. It is held in two parts: the *first* is the competition with entries accepted between January and June from all over the world and judged by a panel of distinguished theatre practitioners. The *second* is the Festival itself which showcases the selected plays from the competition in the following February. The IPF also showcases the successful plays in Europe in association with its partners Extra Candoni in Italy and Theatro Ena in Cyprus. Rules and entry form on web page above. Previous winners produced at the theatre include: Guy Jenkin *Fighting for the Dunghill*; James Martin Charlton *Fat Souls*; Peter Moffat *Iona Rain*; Kevin Hood *Teach Yourself Italian*; Dino Mahoney *YoYo*; Dominic McHale *The Resurrectionists*; Philip Edwards *51 Peg*; Roumen Shomov *The Dove*; Maggie Nevill *The Shagaround*; Andrew Shakeshaft *Just Sitting*; Mark Norfolk *Knock Down Ginger*; Des Dillon *Six Black Candles*; Alex Evans *Mother Russia*.

Isle of Man Literature Festival
Isle of Man Arts Council, St Andrew's House, Finch Road, Douglas IM1 2PX
☎ 01624 694598   🖷 01624 686860
iomartscouncil@dtl.gov.im
www.iomarts.com
Arts Development Manager *Mrs Emma Callin*

Founded 1997. Regular programme of literature events throughout the year.

King's Lynn, The Fiction Festival
19 Tuesday Market Place, King's Lynn PE30 1JW
☎ 01553 691661 (office hours) or 761919
🖷 01553 691779
tony.ellis@hawkins-solicitors.com
www.lynnlitfests.com
Contact *Anthony Ellis*

Founded 1989. Annual weekend festival held in March. Over the weekend there are readings and discussions, attended by guest writers of which there are usually twelve. Previous guests have included Beryl Bainbridge, Louis de Bernière, Alan Sillitoe, J.P. Donleavy and D.J. Taylor.

King's Lynn, The Poetry Festival

19 Tuesday Market Place, King's Lynn PE30 1JW
☎ 01553 691661 (office hours) or 761919
✉ 01553 691779
tony.ellis@hawkins-solicitors.com
www.lynnlitfests.com
Contact *Anthony Ellis*

Founded 1985. Annual weekend festival held at the end of September, with guest poets (usually eight). Previous guests have included Clive James, Les Murray, Peter Porter, D.M. Thomas, C.K. Williams and Kit Wright.

King's Sutton Literary Festival

Glebe House, St Rumbold's Drive, King's Sutton, Banbury OX17 3PJ
☎ 01295 811473
info@kslitfest.co.uk
Festival Organizer *Mrs Sara Allday*

Founded 2004. Takes place over a weekend in early March. Celebrates reading in a village that has attracted many world-renowned writers to the festival, such as Margaret Drabble, Andrew Motion, Salley Vickers, Justin Cartwright, David Lodge, Rachel Billington. Takes place under one roof in the modern village centre and café with new and second-hand book sales, a programme involving the village primary school. Proceeds to the Parish Church Restoration Fund.

Knutsford Literature Festival

☎ 01565 722738/07050 183417
knutsfordlitfest@yahoo.co.uk
www.knutsfordlitfest.blogspot.com
Contact *Charlotte Peters Rock*

An annual two-week festival, held at the beginning of October to celebrate writing and performance by distinguished national, international and local authors. Events include readings and discussions, a literary lunch and theatrical performances.

Lambeth Readers & Writers Festival

London Borough of Lambeth, Brixton Library, Brixton Oval, London SW2 1JQ
☎ 020 7926 1105
to'dell@lambeth.gov.uk
www.lambethgov.uk/Libraries
Reader Development Officer *Tim O'Dell*

Founded in 2001. Held annually in May, the festival brings together internationally recognized, new and local authors for a month of talks, poetry and writing courses focusing on the reading experience. Previous guests have included Peter Ackroyd, Ben Okri, Armando Iannucci, Kate Mosse, Linton Kwesi Johnson, Doris Lessing, Buchi Emecheta, Jo Brand and Lionel Shriver.

Lancaster LitFest

PO Box 751, Lancaster LA1 9AJ
☎ 01524 62166
all@litfest.org
www.litfest.org

Founded 1978. Regional Literature Development Agency, organizing workshops, readings, residencies, publications. Year-round programme of literature-based events, and annual festival in November featuring a wide range of writers from the UK and overseas. Publishes new Lancashire and Cumbrian prose and poetry writers under the Flax Books imprint.

Ledbury Poetry Festival

Church Lane, Ledbury HR8 1DH
☎ 0845 458 1743
admin@poetry-festival.com
www.poetry-festival.com
Festival Director *Chloe Garner*

Founded 1997. Annual ten-day festival held in June/July. Includes readings, discussions, workshops, exhibitions, music and walks. There is also an extensive year-round community programme and a national poetry competition. Past guests have included James Fenton, Helen Dunmore, Bernard MacLaverty, Andrew Motion, Benjamin Zephaniah, Simon Armitage, Roger McGough, John Hegley. Full programme available in May.

Lewes Live Literature

PO Box 2766, Lewes BN7 2WF
☎ 07972 037612
info@leweslivelit.co.uk
www.leweslivelit.co.uk
Artistic Director *Mark Hewitt*

Lewes Live Lit is a project-based live literature production company working with writers to develop crafted performances and other outcomes as part of curated seasons or festivals of new work. Occasional/year-round.

Lichfield Festival

7 The Close, Lichfield WS13 7LD
☎ 01543 306270   ✉ 01543 306274
info@lichfieldfestival.org
www.lichfieldfestival.org
Festival Director *Richard Hawley*

Annual July festival with events taking place in the 13th century Cathedral, the Lichfield Garrick Theatre and other venues. Mainly music but a growing programme of literary events such as poetry, plays and talks. New dedicated literature weekend each autumn; literary dinner plus two days of writer talks.

Lincoln Book Festival

☎ 01522 873844
sara.bullimore@lincoln.gov.uk
www.lincolnbookfestival.co.uk
Contact *Sara Bullimore (Arts & Cultural Sector Officer)*

Founded 2004. Annual festival held in May, usually for two weeks. Guests in 2008 included Iain Banks, Colin Dexter and Steven Berkoff.

City of London Festival

Fitz-Eylwin House, 25 Holborn Viaduct, London EC1A 2BP
☎ 020 7583 3585   🖷 020 7353 0455
admin@colf.org
www.colf.org
Director *Ian Ritchie*

Founded 1962. Annual three-week festival held in June and July. Features over fifty classical and popular music events alongside poetry and prose readings, street theatre and open-air extravaganzas, in some of the most outstanding performance spaces in the world.

London Literature Festival

Southbank Centre, Belvedere Road, London SE1 8XX
☎ 0871 663 2500 (box office)
www.southbankcentre.co.uk

Established in 2007, events take place at the Southbank Centre over two weeks in early July. 2008 guests included Tony Benn, Mark E. Smith, Wendy Cope, Sebastian Barry, Halima Bashir, Lemn Sissay, James Hopkin, Kate Summerscale, Timothy O'Grady, Rebecca Miller and George Monbiot.

Lowdham Book Festival

The Bookcase, 50 Main Street, Lowdham NG14 7BE
☎ 0115 966 4143
janestreeter@thebookcase.co.uk
www.lowdhambookfestival.co.uk
Contact *Jane Streeter*

Founded 1999. Annual nine-day festival held in June combining a village fête atmosphere with that of a major literature festival. Guests have included Carol Ann Duffy, Polly Toynbee, Ian McMillan, Alan Sillitoe, Jackie Kay, Gary Younge and Carole Blake. Talks on everything from St Kilda to the Jewish roots of rock 'n' roll. Also holds annual 'Winter Weekends' with guest speakers in November or January and, in the first weekend of March, 'Flicks in the Sticks', a mini film festival.

Manchester Literature Festival

Beehive Mill, Jersey Street, Ancoats, Manchester M4 6JG
☎ 0161 236 5555
admin@manchesterliteraturefestival.co.uk
www.manchesterliteraturefestival.co.uk
Contact *Administrator*

Held in October (15th–25th in 2009), Manchester Literature Festival offers 'unique and imaginative literature experiences to its audiences' with a programme that features readings by some of the world's finest authors, freshly commissioned work, and a series of cutting-edge events exploring the crossover between new technology and literature.

Mere Literary Festival

Lawrence's, Old Hollow, Mere BA12 6EG
☎ 01747 860475/861211 (Tourist Information)
www.merelitfest.co.uk
Contact *Adrienne Howell*

Founded 1997. Annual festival held in the second week of October in aid of registered charity, The Mere & District Linkscheme. Events include readings, quiz, workshop, writer's lunch and talks. Finale is adjudication of the festival's writing competition (see entry under *Prizes*) and presentation of awards.

Arthur Miller Centre Literary Festival at UEA

University of East Anglia, School of American Studies, Norwich NR4 8TJ
☎ 01603 592810
v.striker@uea.ac.uk
www1.uea.ac.uk/cm/home/schools/hum/booksandwriters/litfest
Contact *Val Striker*

Founded 1991. Annual festival, held in the autumn.

National Association of Writers' Groups (NAWG) Open Festival of Writing

40 Burstall Hill, Bridlington YO16 7GA
☎ 01262 609228
nawg@tesco.net
www.nawg.co.uk
Festival Administrator *Mike Wilson*

Founded 1995. Annual Festival held at St Aidan's College, University of Durham (4th–6th September in 2009). Three days of creative writing tuition covering poetry, short and long fiction, playwriting, journalism, TV sitcom and many other subjects all led by professional writer-tutors. Workshops, one-to-one tutorials and fringe events. Saturday gala dinner and awards ceremony. Full or part-residential weekend, or single workshops only. Open to all, no qualifications or NAWG membership required.

National Eisteddfod of Wales

40 Parc Ty Glas, Llanishen, Cardiff CF14 5DU
📞 029 2076 3777   📠 029 2076 3737
info@eisteddfod.org.uk
www.eisteddfod.org.uk

The National Eisteddfod, held in August, is the largest arts festival in Wales, attracting over 170,000 visitors during the week-long celebration of more than 800 years of tradition. Competitions, bardic ceremonies and concerts.

National Student Drama Festival
▷ University of Hull under UK Writers' Courses

Northern Children's Book Festival

Schools Library Service, Sandhill Centre, Grindon Lane, Sunderland SR3 4EN
📞 0191 553 8866/7/8   📠 0191 553 8869
schools.library@sunderland.gov.uk
www.ncbf.org.uk
Secretary *Eleanor Dowley*

Founded 1984. Annual two-week festival during November. Events in schools and libraries for children in the North East region. One Saturday during the festival sees the staging of a large book event hosted by one of the local authorities involved.

Off the Page Literature Series

County Library Support Services, Glaisdale Parkway, Nottingham NG8 4GP
📞 0115 982 9027   📠 0115 928 6400
alison.hirst@nottscc.gov.uk
Contact *Alison Hirst*

Author visits aimed at making authors accessible to readers. Programme is available from late summer/early autumn.

Off the Shelf Literature Festival

Central Library, Surrey Street, Sheffield S1 1XZ
📞 0114 273 4716/4400   📠 0114 273 5009
offtheshelf@sheffield.gov.uk
www.offtheshelf.org.uk
Festival Organizers *Maria de Souza, Susan Walker, Lesley Webster*

Founded 1992. Annual festival held over three weeks at the end of October into November. Lively and diverse mix of readings, workshops, children's events, storytelling and competitions. Previous guests have included Michael Frayn, Rabbi Lionel Blue, Germaine Greer, Robert Goddard, Dom Joly, Barry Davies, Michael Wood, Linton Kwesi Johnson, Sir Arnold Wesker, Jeremy Bowen, John Pilger, David Guterson, David Lodge, Andrew Davies, Tariq Ali, Alan Sillitoe, Simon Armitage, Sheila Hancock.

Oundle Festival of Literature

2 New Road, Oundle PE8 4LA
📞 01832 273050
enquiries@oundlelitfest.org.uk
www.oundlelitfest.org.uk
Festival Deputy Chair *Liz Dillarstone*

Founded 2002. Annual festival running in the first two weeks of March for adults and children. Features talks by high profile authors and poets, writing workshops, poetry and prose showcases for local writers, children's writing competitions, school events, drama productions and community readings for young and old.

Oxford Literary Festival
▷ The Sunday Times Oxford Literary Festival

Purple Patch Poetry Convention

25 Griffiths Road, West Bromwich B71 2EH
ppatch66@hotmail.com
www.purplepatchpoetry.co.uk
Contact *Geoff Stevens*

Held at the Barlow Theatre near Birmingham, this event celebrates the strength of small press poets. The Purple Patch Poetry Convention is a three-day event of poetry readings, talks, workshops and discussions. Admission charges are kept very low, payments to participants being low to non-existent. The Convention's aims are quality and an opportunity to participate. Poets at past festivals have included Ray Avery, R.G. Bishop, Tilla Brading, Gerald England, J.F. Haines, Brendan Hawthorne, Martin Holroyd, Mike Hoy, Eamer O'Keeffe, Carolyn King, Paul McDonald, Bob Mee, Les Merton, Michael Newman, Simon Pitt, Andy Robson, Sam Smith and Steve Sneyd. Monthly poetry readings are held at the Barlow Theatre with open mic, guest poet, music and residents Unleaded Petrels.

Redbridge Book and Media Festival

3rd Floor, Central Library & Museum, Clements Road, Ilford IG1 1EA
📞 020 8708 2855   📠 020 8708 2431
anita.verma@redbridge.org.uk
www.redbridge.gov.uk
Contact *Anita Verma (Arts Development Manager)*

Founded in 2003, the festival celebrates literature in all its forms, and takes place annually over two weeks in April/May. Features a programme of author readings, book signings, creative writing classes, school-based workshops, community events, performance poetry evenings, competitions, exhibitions and panel debates.

Richmond Literature Festival
▷ Book Now

Runnymede International Literature Festival

Royal Holloway, University of London, Egham
TW20 0EX
☎ 01273 571700
info@rfest.org.uk
www.rfest.co.uk
Festival Director *Professor Robert Hampson*

Founded 2006. Annual literary festival held
in Runnymede as a partnership beween Royal
Holloway, University of London and Runnymede
Borough Council. Held annually in the spring, the
festival runs for a core weekend with associated
community and education activities. Includes
readings, performances, masterclasses and
community workshops.

Saffron Walden Literary Festivals

Harts, Hart House, Shire Hill, Saffron Walden
CB11 3AQ
☎ 01799 523456 (box office)
events@harts1836.co.uk
www.hartsevents.co.uk
www.hartsbooks.co.uk
Festival Director *Jo Burch*

Founded 2006. Various events in 2009 run
by family-owned Harts Bookshop: 'Words
Around the World' Literature Festival, headed
by Benjamin Zephaniah, held in February and
March; various schools events throughout the
year; and main Book Festival held over a fortnight
in October with events and workshops for adults
and children. Please see events website for full
details of these and previous festivals.

Southwold Festival
▷ Ways with Words

StAnza: Scotland's International Poetry Festival

Registered Office: 57 Lade Braes, St Andrews
KY16 9DA
☎ 01333 360491
info@stanzapoetry.org
www.stanzapoetry.org
Festival Director *Brian Johnstone (admin@
stanzapoetry.org)*
Artistic Director *Eleanor Livingstone (arts@
stanzapoetry.org)*
Press & Media Manager *Annie Kelly (press@
stanzapoetry.org)*

The only regular festival dedicated to poetry in
Scotland, StAnza is international in outlook. Held
annually in March in the ancient university town
of St Andrews, the festival is an opportunity to
hear world class poets reading in exciting and
atmospheric venues. The 2009 festival themes
were *Homecoming* and *Poetry Centre Stage.*
The festival features readings, discussions,
conversations, performance poetry, poetry in

exhibition, workshops and children's poetry. It
has a strong showing of poets from across the
UK and overseas as well as the festival's signature
foreign language readings. Free programmes can
be ordered from Fife Contemporary Art & Craft
on 01334 474610 or by e-mail to mail@fcac.co.uk

Stena Line Wigtown Book Festival

Festival Office, County Buildings, Wigtown
DG8 9JH
☎ 01988 403222
mail@wigtownbookfestival.com
www.wigtownbookfestival.com
Festival Director *Michael McCreath*
Festival Admin *Anne Barclay*

Founded 1999. Annual Festival held over ten days
(last week September/first week October). Author
readings, poetry events, workshops, drama,
music. Children's events and school outreach
programme. Also Spring Festival (founded 2005)
held over first May Bank Holiday weekend.

Stratford-upon-Avon Poetry Festival

The Shakespeare Centre, Henley Street, Stratford-
upon-Avon CV37 6QW
☎ 01789 204016/292176 (box office)
🖷 01789 296083
info@shakespeare.org.uk
www.shakespeare.org.uk
Festival Director *Paul Edmondson*

Sponsored by the Shakespeare Birthplace Trust,
this Festival, now in its 56th year, takes place in
April/May and celebrates poetry past and present
with special reference to the work of William
Shakespeare. Features weekly recitals given by
established poets and by actors who present
themed evenings of verse. The Festival includes
opportunities for young people to write their own
work and have it performed. There is a Poetry
Mass and local poets' events. Full details are
available from January and can be accessed on the
Trust's website.

Strokestown International Poetry Festival
▷ Strokestown International Poetry Competition under Prizes

The Sunday Times Oxford Literary Festival

301 Woodstock Road, Oxford OX2 7NY
☎ 01865 514149
info@sundaytimes-oxfordliteraryfestival.co.uk
www.sundaytimes-oxfordliteraryfestival.co.uk
Directors *Sally Dunsmore, Angela Prysor-Jones*

Founded 1997. Annual week-long festival held
in March/April. Authors speaking about their
books, covering a wide variety of writing: fiction,
poetry, biography, travel, food, art, gardening,
children's books. There is also a range of events
for schools. Previous guests have included
William Boyd, Andrew Motion, Beryl Bainbridge,

Sophie Grigson, Philip Pullman, Doris Lessing, Seamus Heaney, Richard Dawkins, Zandra Rhodes, Kazuro Ishiguro.

Swindon Festival of Literature

Lower Shaw Farm, Shaw, Swindon SN5 5PJ
☎ 01793 771080
swindonlitfest@lowershawfarm.co.uk
www.swindonfestivalofliterature.co.uk
Festival Director *Matt Holland*

Founded 1994. Annual festival held in May, starting with 'Dawn Chorus' at sunrise. Includes a wide range of authors, speakers, discussions, performances and workshops, plus the Swindon Performance Poetry Slam competition. Guests at the 2008 Festival included Andrew Motion, Ben Okri and John Pilger.

Theakstons Old Peculier Harrogate Crime Writing Festival

Raglan House, Raglan Street, Harrogate HG1 1LE
☎ 01423 562303   📠 01423 521264
crime@harrogate-festival.org.uk
www.harrogate-festival.org.uk
Operations Director *Sharon Canavar*
Festival Coordinator *Erica Morris*

Launched 2003. Weekend of events at the end of July each year featuring the best of British and American crime writers. The winner of the **Theakstons Old Peculier Prize for the Crime Novel of the Year** is announced at the Festival (see entry under *Prizes*). Events also include industry 'How to ...' sessions, social events and late night shows. Part of the **Harrogate International Festival**.

Dylan Thomas Festival

Dylan Thomas Centre, Somerset Place, Swansea SA1 1RR
☎ 01792 463980   📠 01792 463993
dylanthomas.lit@swansea.gov.uk
www.dylanthomas.com
Contacts *David Woolley, Jo Furber*

Two weeks of performances, talks, lectures, films, music, poetry, exhibitions and celebrity guests. The 2008 Festival included the award of the second Dylan Thomas Prize for young writers. The Dylan Thomas Centre also runs a year-round programme of literary events; please e-mail for details.

The Times Cheltenham Literature Festival

109 Bath Road, Cheltenham GL53 7LS
☎ 01242 775861
clair.greenaway@cheltenhamfestivals.com
www.cheltenhamfestivals.com
Festival Manager *Clair Greenaway*

Founded 1949. Annual festival held in October (9th–18th in 2009). The first purely literary

festival of its kind, it has over the past decade developed from an essentially local event into the largest and most popular in Europe. A wide range of events including talks and lectures, poetry readings, novelists in conversation, exhibitions, discussions and a large bookshop.

Torbay Weekend Festival of Poetry

6 The Mount, Higher Furzeham, Brixham TQ5 8QY
☎ 01803 851098
pwoxley@aol.com
www.acumen-poetry.co.uk/events
Festival Organizer *Patricia Oxley*

Founded 2001. Held annually at the end of October, Thursday evening until Monday afternoon, with workshops, poetry readings, talks and debates. Programme for adults and children. Encourages active participation through workshops, open mike events, etc. Guest poets include internationally known writers as well as local authors.

Tŷ Newydd Festival

Tŷ Newydd, Llanystumdwy, Cricieth LL52 0LW
☎ 01766 522811   📠 01766 523095
post@tynewydd.org
www.tynewydd.org
Director *Sally Baker*

Biennial, bilingual literature festival held on alternate years. Run by the National Writers' Centre for Wales, it is located at Tŷ Newydd Writers' Centre and other venues near Cricieth. Features writers and poets from Wales (working in both English and Welsh) and worldwide. Tŷ Newydd also holds an annual weekend festival in November devoted solely to Cynghanedd – Welsh strict meter poetry.

The Ulster Bank Belfast Festival at Queen's

8 Fitzwilliam Street, Belfast BT9 6AW
☎ 028 9097 1034   📠 028 9097 1336
g.farrow@qub.ac.uk
www.belfastfestival.com
Director *Graeme Farrow*

Founded 1963. Annual three-week festival held in October (16th–31st in 2009). Organized by Queen's University, the festival covers a wide variety of events, including literature. Programme available in September.

Warwick Words

The Court House, Jury Street, Warwick CV34 4EW
☎ 01926 427056
info@warwickwords.co.uk
www.warwickwords.co.uk

Founded 2002. Annual festival featuring both living writers and those who have had connections with Warwick – Tolkien, Larkin and Landor in particular. Also workshops and an education programme. A festival of literature and spoken word for all the family. Events at the Bridge House Theatre, St Mary's Church, the Lord Leycester Hospital and other historic buildings around Warwick over the first weekend in October.

Ways with Words

Droridge Farm, Dartington, Totnes TQ9 6JG
☎ 01803 867373   🖷 01803 863688
admin@wayswithwords.co.uk
www.wayswithwords.co.uk
Festival Director *Kay Dunbar*

Ways With Words runs three major annual literature festivals: at Dartington Hall, Devon for ten days in July, featuring over 150 writers giving talks, readings, interviews, discussions, performances and workshops; Words by the Water, Keswick, Cumbria in March, a ten-day festival as Dartington; and a five-day festival at Southwold, Suffolk in November. Also organizes writing, reading and painting courses in Italy and France.

Wellington Literary Festival

Civic Offices, Larkin Way, Tan Bank, Wellington, Telford TF1 1LX
☎ 01952 567697   🖷 01952 567690
welltowncl@aol.com
www.wellington-shropshire.gov.uk
Contact *Howard Perkins*

Founded 1997. Annual festival held throughout October. Events include storytelling, writers' forum, 'Pints and Poetry', children's poetry competition, story competition, theatre review and guest speakers.

Wells Festival of Literature

25 Chamberlain Street, Wells BA5 2PQ
☎ 01749 670929
www.somersite.co.uk/wellsfest.htm

Founded 1992. Annual week-long festival held in the middle of October. Main venue is the historic, moated Bishop's Palace. A wide range of speakers caters for different tastes in reading. Recent guests include: John Le Carré, Douglas Hurd, Fay Weldon, Melvyn Bragg, Margaret Drabble, William Dalrymple, Timothy West. Short story and poetry competitions and writing workshops are run in conjunction with the Festival.

Wigtown Book Festival
▷ Stena Line Wigtown Book Festival

Winchester Writers' Conference, Festival, Bookfair
▷ entry under UK and Irish Writers' Courses

Wonderful Words

c/o Redruth Library, Redruth TR15 2QE
☎ 07967 340907
mtwose@cornwall.gov.uk
www.cornwall.gov.uk/library
Festival Organizer *Maureen Twose*

Founded 1994. Biennial literature festival hosted throughout Cornwall. Organized by the Cornwall Library Service, Wonderful Words has grown into a prestigious event for children and adults, offering author talks, writers' workshops, poetry performances, theatre and storytelling. Guest authors have included Doris Lessing, Ruth Rendell, Margaret Drabble, Simon Callow and Dick King-Smith.

Word – University of Aberdeen Writers Festival

Office of External Affairs, University of Aberdeen, King's College, Aberdeen AB24 3FX
☎ 01224 273874
word@abdn.ac.uk
www.abdn.ac.uk/word
Artistic Director *Alan Spence*

Held annually in May, the Festival takes place over six days at the historic King's College Campus and at venues throughout the city. Attracts over 70 authors and over 10,000 visitors to a weekend of readings, lectures, debates, music, art exhibitions and film screenings. Also includes an extended schools' and children's programme and a Festival of Gaelic.

Word Market

PO Box 150, Barrow-in-Furness LA14 3WF
☎ 07812 172193
janice@wordmarket.org.uk
www.wordmarket.org.uk
Project Coordinator *Janice Benson*

Word Market celebrates the written word through a programme of exciting, live events and projects throughout the year. As well as an annual festival, regular workshops, master-classes and performances are held with well-known poets, best-selling authors, publishers, literary agents and theatre directors. Check the website for further details.

Words Around the World Literature Festival
▷ Saffron Walden Literary Festivals

Words by the Water
▷ Ways with Words

The Wordsworth Trust

Dove Cottage, Grasmere LA22 9SH
☎ 015394 35544   🖷 015394 35748
enquiries@wordsworth.org.uk
www.wordsworth.org.uk
Contact *Andrew Forster (Literature Officer)*

Ongoing contemporary poetry programme held over the summer and one-off events held around the country. Past readers have included Simon Armitage, Seamus Heaney, Paul Muldoon, Sharon Olds, Philip Pullman and Don Paterson as well as a 'support' reader at most events.

The Trust also operates an annual poets-in-residence programme, giving time and space for new writers to develop their work in this setting: a cottage and stipend are supplied and the artist is given the space to develop their own work as well as opportunities to become involved in The Wordsworth Trust literary programme. Past artists-in-residence include Henry Shukman, Owen Sheers, Paul Farley, Helen Farish, Jack Mapanje, Rebecca O'Connor, Jacob Polley, Neil Rollinson and John Hartley Williams.

Writenow!

▷ Queen's Theatre, Hornchurch under Theatre Producers

Writing on the Wall

First Floor, Mission Hall, 36 Windsor Street, Liverpool L8 1XE

☎ 0151 703 0020

info@writingonthewall.org.uk

www.writingonthewall.org.uk

Festival Coordinator *Madeline Heneghan*

Annual festival held in Liverpool that works alongside schools, young people, local communities and broader audiences to celebrate writing, diversity, tolerance, story-telling and humour through controversy, inquiry and debate. Performances, readings, workshops, screenings, high profile debates and discussions. Guest speakers have included Noam Chomsky, Irvine Welsh, Howard Marks, Roddy Doyle, Benjamin Zephaniah.

International Book Fairs

Abu Dhabi International Book Fair

P.O. Box 2380, Abu Dhabi, United Arab Emirates
☎ 00 971 2 621 2975　📠 00 971 2 621 5772
info@adbookfair.com
www.adbookfair.com

Week-long event held in March. The 19th fair, held in 2009, attracted over 500 exhibitors from 51 countries.

American Library Association (ALA) Conferences

50 E. Huron, Chicago, IL 60611, USA
☎ 001 800 545 2433
www.ala.org

The American Library Association holds two annual events. The ALA Annual Conference (Chicago, 9th–15th June in 2009) is attended by more than 25,000 librarians, publishers, writers and educators. The ALA Midwinter Meeting (Boston, 15th–20th January in 2010) includes some 450 exhibitors featuring the latest in books, videos, computers and other materials available to today's libraries and their users.

Beijing International Book Fair

International Exhibition Centre, Beijing, China
www.bibf.net

Five-day book fair held in September (3rd–7th in 2009), now in its 16th year.

Bologna Children's Book Fair

Piazza Costituzione, 6, 40128 Bologna, Italy
☎ 00 39 051 282 242　📠 00 39 051 6374 011
bookfair@bolognafiere.it
www.bookfair.bolognafiere.it

Children's book fair held in March, now in its 46th year.

BookExpo America (BEA)

Jacob K. Javits Convention Center of New York, 655 W 34th Street, New York, NY 1001, USA
inquiry@bookexpoamerica.com
www.bookexpoamerica.com

Annual five-day event held in May. From 2009 to 2012, BEA will take place in New York.

Cairo Book Fair

General Egyptian Book Organization, Corniche El-Nil, Cairo, Egypt
www.egyptianbook.org.eg

Run by the General Egyptian Book Organization, the 42nd Cairo Book Fair takes place in January/February 2010.

Frankfurt Book Fair

Ausstellungs- und Messe GmbH, Reineckstr. 3, 60313 Frankfurt am Main, Germany
☎ 00 49 69 2102-0　📠 00 49 69 2102-227/-277
info@book-fair.com
www.frankfurt-book-fair.com

Major annual book fair held in October. For the 2009 event (14th–18th), the 'Guest of Honour' is China.

Kolkata Book Fair

Guild House, 2B Jhama Pukur Lane, Kolkata 700 009, West Bengal, India
☎ 00 91 33 2360 4588　📠 00 91 33 2360 4566
guild@cal2.vsnl.net.in
www.kolkatabookfaironline.com

The 33rd fair, held in 2009, had Scotland as the 'Focal Theme' and the USA as 'Honoured Country'. Tentative dates for 2010 are 27th January–7th February.

London Book Fair

Earls Court Exhibition Centre, Warwick Road, London SW5 9TA
☎ 020 8271 2124　📠 020 8910 7930
www.londonbookfair.co.uk

Annual, three-day event held at Earls Court; 2010 dates: 19th–21st April.

Mumbai International Book Fair

Bandra Kurla Complex, Mumbai, India
bookfair@fairfest.com
www.internationalbookfair.com

Month-long book fair, held in January.

Tokyo International Book Fair

18F Shinjuku-Nomura Bldg., 1-26-2 Nishishinjuku, Shinjuku-ku, Tokyo 163-0570, Japan
☎ 00 81 3 3349 8507　📠 00 81 3 3345 7929
tibf-eng@reedexpo.co.jp
http://web.reedexpo.co.jp/tibf/english/

Asia's leading book fair which attracted 763 exhibitors from 31 countries in 2008. The 16th fair takes place 9th–12th July 2009.

Useful Websites

Many of these and other useful websites for writers can be found in *The Internet for Writers* by Nick Daws (ISBN 1-84025-308-8), one of a series of books published by Internet Handbooks; *The Incredibly Indispensable Web Directory* by Clive and Bettina Zietman, published by Kogan Page (ISBN 0-7494-3617-4); *thegoodwebguide for writers* by Paul Chronnell, published by The Good Web Guide Limited (ISBN 1-903282-38-1).

AbeBooks

www.abebooks.co.uk

Online bookshop for new and second-hand publications. Includes search engine for textbooks and reference books.

Academi (Welsh Academy/Yr Academi Gymreig)

www.academi.org

News of events, publications and funding for Welsh-based literary events. (See entry under *Professional Associations and Societies.*)

Alibris

www.alibris.com

Over 60 million used, new and hard-to-find books online. Bargain books, text and reference listings and rare books. ISBN search engine. Also movies and music.

Alliance of Literary Societies

www.allianceofliterarysocieties.org.uk

Details of societies and events. (See entry under *Professional Associations and Societies.*)

Amazon

www.amazon.co.uk

The online shop.

Ancestry

www.ancestry.co.uk

Family history information – databases, articles and other sources of genealogical data.

Arts Council England

www.artscouncil.org.uk

Includes information on funding applications, publications and the National Lottery. (See entry under *Arts Councils and Regional Offices.*)

Arts Council of Northern Ireland

www.artscouncil-ni.org

Information on funding and awards, events and free E-Newsletter service. (See entry under *Arts Councils and Regional Offices.*)

Arts Council of Wales/Cyngor Celfyddydau Cymru

www.acw-ccc.org.uk

Information on publications, grants, council meetings, the arts in Wales. (See entry under *Arts Councils and Regional Offices..*)

Arvon Foundation

www.arvonfoundation.org

See entry under *UK and Irish Writers' Courses.*

Ask About Writing

www.askaboutwriting.net

Resource site for writers.

Association for Scottish Literary Studies

www.asls.org.uk

The educational charity promoting the languages and literature of Scotland. (See entry under *Professional Associations and Societies.*)

Association of Authors' Agents (AAA)

www.agentsassoc.co.uk

UK agents' organization including list of current members. (See entry under *Professional Associations and Societies.*)

Association of Authors' Representatives (AAR)

www.aar-online.org

US agents' organization including search engine for current members. (See entry under *Professional Associations and Societies.*)

Author-Network
www.author-network.com

Writers' resource site.

Author.co.uk
www.author.co.uk

Links to advice and information for writers.

Authorbank
www.authorbank.com

Registration for a fee enables authors to present book ideas to publishers online and access to a free advice service.

Authors' Licensing and Collecting Society Limited (ALCS)
www.alcs.co.uk

Details of membership, news and publications. (See entry under *Professional Associations and Societies.*)

Bartleby.com
www.bartleby.com

An ever-expanding list of books published online for reference, free of charge.

BBC
www.bbc.co.uk

Access to all BBC departments and services.

bibliofind
www.bibliofind.com

Millions of second-hand and rare books, periodicals and ephemera for sale online via Amazon.com

Bibliomania
www.bibliomania.com

Over 2,000 classic texts, study guides and reference resources available online for free. Also selected books for sale in the Bibliomania shop.

book2book/booktrade.info
www.booktrade.info/index.php

Established by a group of publishers, booksellers, website developers and trade journalists to provide up-to-date news, features and useful information for the book trade.

BookBrunch
www.bookbrunch.co.uk

BookBrunch is a fully searchable website and daily news bulletin for the book industry, in Britain and abroad.

Booktrust
www.booktrust.org.uk

Book information service, guide to prizes and awards, factsheets on getting published. (See entry under *Professional Associations and Societies.*)

British Association of Picture Libraries and Agencies (BAPLA)
www.bapla.org.uk

Website includes the BAPLA Online Database search facility by category or name. (See entry under *Professional Associations and Societies.*)

British Centre for Literary Translation
www.literarytranslation.com

A joint website with the British Council with workshops by leading translators, contacts and networks, and search engine for translation conferences, seminars and events. (See entry under *Professional Associations and Societies.*)

The British Council
www.britishcouncil.org

Information on the Council's English Language services, education programmes, society and science links. (See entry under *Professional Associations and Societies.*)

British Film Institute (bfi)
www.bfi.org.uk

See entry under *Professional Associations and Societies.*

British Library
www.bl.uk

Reader service enquiries, access to main catalogues, information on collections, links to the various reading rooms and exhibitions. (See related entries under *Library Services.*)

Chapter One Promotions
www.chapteronepromotions.com

Provides services to writers at various stages of their writing career: story and poetry critique services, proof reading, 'Kids Korner'.

Characterization Tool for Novelists
www.synergise.com/p4

The 'P4 Personality Mapping tool' provides novelists with the means to quickly create realistic characters based upon accepted psychological types. Works online and offline

Chartered Institute of Linguists (IoL)
www.iol.org.uk

Discussion forum, news on regional societies, job opportunities, 'Find a Linguist' service and *The Linguist* magazine. (See entry under *Professional Associations and Societies.*)

CILIP
www.cilip.org.uk

The professional body for librarians and information professionals. (See entry under *Professional Associations and Societies.*)

CILIP in Scotland (CLIPS)
www.slainte.org.uk

See entry under *Professional Associations and Societies.*

CILIP Wales
www.dil.aber.ac.uk/cilip_w/index.htm

See entry under *Professional Associations and Societies.*

Complete Works of William Shakespeare
www.tech.mit.edu/Shakespeare/works.html

Access to the text of the complete works.

Copyright Licensing Agency Ltd (CLA)
www.cla.co.uk

Copyright information, customer support and information on CLA services. (See entry under *Professional Associations and Societies.*)

Crime Writers' Association (CWA)
www.thecwa.co.uk

Website of the professional crime writers' association. (See entry under *Professional Associations and Societies.*)

Daily Express
www.express.co.uk

Daily Express online.

Daily Mail
www.dailymail.co.uk

Daily Mail online.

Daily Mirror
www.mirror.co.uk

Daily Mirror online.

The Daily Telegraph
www.telegraph.co.uk

The Daily Telegraph online.

Dictionary of Slang
dictionaryofslang.co.uk

A guide to slang 'from a British perspective.' Research information and search facility.

The Eclectic Writer
www.eclectics.com/writing/writing.html

US website offering a selection of articles giving advice for writers and an online discussion board.

The English Association
www.le.ac.uk/engassoc

News, publications, conference and membership information. (See entry under *Professional Associations and Societies.*)

FilmAngels
www.filmangel.co.uk

Established in conjunction with **Hammerwood Films** to create a shop window for writers and would-be film angels alike. Submitted synopses are displayed (for a fee) for a pre-determined period, while would-be angels are invited to finance a production

Filmmaker Store
www.filmmakerstore.com

US site giving scriptwriting resources, listings and advice.

Financial Times
www.ft.com

Financial Times online.

Frankfurt Book Fair
www.book-fair.com/en

Provides latest news and market analysis of the book business plus information on the annual Book Fair. (See entry under *International Book Fairs.*)

Froebel Archive for Childhood Studies
http://studentzone.roehampton.ac.uk/library/ specialcollectionsandarchives/ froebelarchiveforchildhoodstudies/

Collection of children's literature. Supports courses at Roehampton University and is available to *bona fide* researchers. Catalogue available online.

The Good Web Guide Ltd
www.thegoodwebguide.co.uk

Guide to the best websites. Provides thousands of detailed and independent reviews of a wide range of websites.

Google Images
www.images.google.com

'The most comprehensive image search on the web.'

The Guardian/Observer
www.guardian.co.uk

Website of *The Guardian* and *The Observer* newspapers online.

Guide to Grammar and Style
www.andromeda.rutgers.edu/~jlynch/Writing

A guide to grammar and style, organized alphabetically, plus articles and links to other grammatical reference sites.

Hansard

www.parliament.the-stationery-office.co.uk/pa/cm/cmhansrd.htm

The official record of debates and written answers in the House of Commons. The transcript of each day's business appears at noon on the following weekday.

The Herald

www.theherald.co.uk

Scottish daily broadsheet.

House of Commons Research Library

www.parliament.uk/parliamentary_publications_and_archives/research_papers.cfm

Gives access to the text of research reports prepared for MPs on a wide range of current issues.

The Independent

www.independent.co.uk

The Independent newspaper online.

Inpress Books

www.inpressbooks.co.uk

Funded by Arts Council England, Inpress Books provides sales, marketing and distribution for small independent presses.

Institute of Translation and Interpreting (ITI)

www.iti.org.uk/indexMain.html

Website of the professional association of translators and interpreters, with the ITI directory of members, publications, training and membership information. (See entry under *Professional Associations and Societies*.)

Internet Classics Archive

classics.mit.edu

Includes works of classical literature; mostly Greek and Roman with some Chinese and Persian. All are in English translation.

Internet Movie Database (IMDb)

www.imdb.com

Essential resource for film buffs and researchers with search engine for cast lists, screenwriters, directors and producers; film and television news, awards, film preview information, video releases.

The Irish Arts Council/An Chomhairle Ealaíon

www.artscouncil.ie

Monthly e-mail newsletter available giving latest information on grants and awards, news and events, etc. (See entry under *Arts Councils and Regional Offices*.)

Journalism UK

www.journalismuk.co.uk

A website for UK-based print journalists who write for text-based publications. Includes links to newspapers, magazines, e-zines, news sources plus information on training and organizations.

The Library Association
▶ CILIP

Literature North East

www.literaturenortheast.co.uk

Monthly e-newsletter; listings of events and readings in the region; writing courses and local training opportunities.

Literature North West

www.publishingnorthwest.co.uk

Promotional agency for the region's independent presses and literature organizations

literaturetraining

www.literaturetraining.com

Online directory of training and professional development opportunities for UK writers and literature professionals. Includes information on courses, workshops, jobs, residencies, submissions, competitions, organizations and funding for professional development. Also, information sheets and specially commissioned features.

Location Register of 20th-Century English Literary Manuscripts and Letters

www.reading.ac.uk/library/about-us/projects/lib-location-register.asp

Reference source for the study of English literature. Information about the manuscript holdings of repositories of all sizes, from the British Library to small-town museums, of literary authors – from major poets to minor science fiction writers.

The Mail on Sunday

www.mailonsunday.co.uk

The Mail on Sunday online.

Mr William Shakespeare and the Internet

shakespeare.palomar.edu

Guide to scholarly Shakespeare resources on the internet.

National Union of Journalists (NUJ)

www.nuj.org.uk

Represents those journalists who work in all sectors of publishing, print and broadcasting. (See entry under *Professional Associations and Societies*.)

The Never Ending Story

www.TheNeverEndingStory.co.uk

A range of constantly evolving online publications which online users can read and contribute to. Includes stories and poems by well-known writers, personalities and site members (free membership).

New Writers Consultancy

www.new-writers-consultancy.com

Advice for writers and critiques.

New Writing North

www.newwritingnorth.com

Essentially for writers based in the north of England but also a useful source of advice and guidelines. (See entry under *Professional Associations and Societies*.)

New Writing Partnership

www.newwritingpartnership.org.uk

Promotes and supports creative writing in the east of England.

Newnovelist

www.newnovelist.com

Novel-writing software. Aids research, characterization, structure and plot, and breaks down the process of writing a novel into manageable chunks.

PEN

www.englishpen.org

Website of the English Centre of International PEN. News of events, information on prizes, membership details. (See entry under *Professional Associations and Societies*.)

Poets and Writers

www.pw.org

A US site containing information and advice for writers.

Producers Alliance for Cinema and Television (pact)

www.pact.co.uk

Publications, training, production companies, membership details. (See entry under *Professional Associations and Societies*.)

ProfWriting

www.thewritingcentre.com

A resource site for writers in Cornwall, the South West and beyond. Covers all genres from poetry to business writing, from novels to radio comedy. Helps find mentors, run workshops, seminars and short courses. Details of training and listings of magazines and publishers in the region.

The Publishers Association

www.publishers.org.uk

Information about the Association and careers in publishing; also 'Getting Published' pages. (See entry under *Professional Associations and Societies*.)

Publishing North West
▷ Literature North West

Publishing Scotland

www.publishingscotland.org

Links to websites of members of the Association, information on activities and publications. (See entry under *Professional Associations and Societies*.)

RefDesk.com

refdesk.com

Free facts and statistics on every country in the world plus charts and maps, illustrations and related sources.

Royal Society of Literature

www.rslit.org

Information on lectures, discussions and readings; membership details and prizes. (See entry under *Professional Associations and Societies*.)

The Scotsman

www.scotsman.com

The Scotsman newspaper online.

Scottish Arts Council

www.sac.org.uk

Information on funding and events. (See entry under *Arts Councils and Regional Offices*.)

Scottish Book Trust

www.scottishbooktrust.com

See entry under *Professional Associations and Societies*.

Scottish Publishers Association
▷ Publishing Scotland

Screenwriters Online

screenwriter.com/insider/news.html

US website, described as the '*only* professional screenwriter's site run by major screenwriters who get their scripts and screenplays made into movies'. Contains screenplay analysis, expert articles and *The Insider Report*.

The SF Hub

www.sfhub.ac.uk/

Science fiction research website created by the University of Liverpool. Includes links to the Science Fiction Foundation collection and the John Wyndham archives.

Shots Magazine
www.shotsmag.co.uk

Electronic magazine of crime and mystery fiction.

Slainte: Information & Libraries Scotland
www.slainte.org.uk

Links to various services and information on librarianship and information management in Scotland. (See entry under *Professional Associations and Societies*.)

Society for Editors and Proofreaders (SfEP)
www.sfep.org.uk

Information about the Society including online directory of members. (See entry under *Professional Associations and Societies*.)

The Society of Authors
www.societyofauthors.org

Includes FAQs for new writers, diary of events, membership details, prizes and awards. (See entry under *Professional Associations and Societies*.)

Society of Indexers
www.indexers.org.uk/

Indexing information for publishers and authors, 'Find an Indexer' pages. Membership information. (See entry under *Professional Associations and Societies*.)

South Bank Centre, London
www.sbc.org.uk

Links to the Royal Festival Hall, Purcell Room, Queen Elizabeth Hall, the Hayward Gallery and Poetry Library; news of literature events.

Story Wizard
www.storywizard.co.uk

Software to enable children to write creative stories.

The Sun
www.thesun.co.uk

The Sun online.

The Times
www.thetimes.co.uk

The Times online.

trAce Online Writing Centre
tracearchive.ntu.ac.uk

Online archive of work published by the trAce Online Writing Centre between 1995–2005. Also holds articles and transcripts of discussions.

UK Children's Books Directory
www.ukchildrensbooks.co.uk

Website created by Steve and Diana Kimpton to 'increase the profile of UK children's books on the internet'.

The UK Public Libraries Page
dspace.dial.pipex.com/town/square/ac940/
   weblibs.html

Website links to public libraries throughout the UK, compiled by Sheila and Robert Harden.

Webster Dictionary/Thesaurus
www.m-w.com/home.htm

Merriam-Webster Online. Includes a search facility for words in the Webster Dictionary or Webster Thesaurus; word games and daily podcast.

Welsh Academy
▷ Academi

Welsh Books Council (Cyngor Llyfrau Cymru)
www.cllc.org.uk

Information about books from Wales, editorial and design services, publishing grants. (See entry under *Professional Associations and Societies*.)

The Word Pool
www.wordpool.co.uk

Children's book site with information on writing for children and a thriving discussion group for children's writers. Also Word Pool Design (www.wordpooldesign.co.uk): web design for writers, illustrators and publishers.

WordCounter
www.wordcounter.com

Highlights the most frequently used words in a given text. Use as a guide to see what words are overused. Also Political Vocabulary Analysis which measures indications of political leanings in given text.

Write4kids.com
www.write4kids.com

US website for children's writers, whether published or beginners. Includes special reports, articles, advice, news on the latest bestsellers and links to related sites.

Writers Nexus International
www.writersnexus.com

Enables writers to place their work online and for publishers, agents and readers to access it.

Writers' Circles
www.writers-circles.com

Directory of writers' circles, courses and workshops.

The Writers' Guild of Great Britain
www.writersguild.org.uk

Information on contracts, copyright, news, writers' resources and industry regulations. (See entry under *Professional Associations and Societies*.)

Writers, Artists and their Copyright Holders (WATCH)
tyler.hrc.utexas.edu

Database of copyright holders in the UK and North America. (See entry under *Professional Associations and Societies*.)

WritersNet
www.writers.net

Internet directory of writers, editors, publishers and literary agents.

WritersServices
www.WritersServices.com

Established in March 2000 by Chris Holifield, former deputy managing director and publisher at Cassell. Offers factsheets, book reviews, advice, links and other resources for writers including editorial services, contract vetting and self-publishing.

Writing-World.com
writing-world.com

'The online community for readers and writers.'

Miscellany

Apple Coaching (inc. CoachingWriters.co.uk)

8 Feering Road, Billericay CM11 2DR

☎ 01277 632085

eve@applecoaching.com

www.CoachingWriters.co.uk

Contact *Eve Menezes Cunningham*

Helping professional and aspiring writers increase their success and happiness through life coaching, business coaching and NLP. Writing and editorial services also available. Sign up for free e-newsletter on the website. Rates negotiable.

DOI Registration Agency

3rd Floor, Midas House, 62 Goldsworth Road, Woking

☎ 0870 777 8712   ℻ 0870 777 8714

doi.agency@nielsen.com

www.doi.nielsenbookdata.co.uk

Manager, ISBN, SAN & DOI Agencies *Diana Williams*

Senior Manager, Registration Services ISBN, SAN & DOI Agencies *Julian Sowa*

DOIs (Digital Object Identifiers) are used to uniquely identify files or other resources on the internet. The DOI Registration Agency is responsible for issuing DOI prefixes and numbers and can provide help and advice on maintaining DOIs. Nielsen Book also runs the **ISBN** and **SAN** agencies (see entries).

Jacqueline Edwards

104 Earlsdon Avenue South, Coventry CV5 6DQ

twigsbranches@yahoo.co.uk

Contact *Jacqueline Edwards, MA, LLB(Hons)*

Historical research: family, local and 19th and 20th century legal history. Covers Warwickshire, Gloucestershire, Northamptonshire, Worcestershire and the National Archives, Kew, London.

ISBN Agency

3rd Floor, Midas House, 62 Goldsworth Road, Woking

☎ 0870 777 8712   ℻ 0870 777 8714

isbn.agency@nielsen.com

www.isbn.nielsenbookdata.co.uk

Manager, ISBN, SAN & DOI Agencies *Diana Williams*

Senior Manager, Registration Services ISBN, SAN & DOI Agencies *Julian Sowa*

ISBNs are product numbers used by all sections of the book trade for ordering and listing purposes. The ISBN Agency is responsible for issuing ISBN prefixes for publishers based in the UK and Republic of Ireland. Nielsen Book also runs the **SAN** & **DOI** agencies (see entries).

Caroline Landeau

6 Querrin Street, London SW6 2SJ

☎ 07050 600420

Contact *Caroline Landeau*

Experienced research and production – films, multimedia, books, magazines, exhibitions, animation, general interest, art, music, crime, travel, food, film, theatre.

Literature Northants

Arts Development Team, Northamptonshire County Council, Music & Performing Arts Centre, 125–129 Kettering Road, Northampton NN1 4AZ

☎ 01604 824155   ℻ 01604 824158

kwilkinson@northamptonshire.gov.uk

www.northamptonshire.gov.uk/Leisure/Libraries/home.htm

Literature Development Officer *Kate Wilkinson*

Writers and readers are supported by Literature Northants. Run by the Literature Development Officer, it provides access to literature opportunities across the country. The LDO is also the key contact for all the latest literary news and events.

M-Y Books Ltd

187 Ware Road, Hertford SG13 7EQ

☎ 01992 586279

jonathan@m-ybooks.co.uk

www.m-ybooks.co.uk

Contact *Jonathan Miller*

Offers book promotion and marketing services, also provides audio book and e-book production and distribution worldwide.

Julia McCutchen, Writers' Coach & Professional Publishing Consultant

PO Box 3703, Trowbridge
☎ 01380 871331   🖷 01380 871331
Julia@JuliaMcCutchen.com
www.JuliaMcCutchen.com
Contact *Julia McCutchen*

Julia McCutchen is the author of *The Writer's Journey: From Inspiration to Publication* and has 20 years experience of publishing. She specializes in helping writers who want to write a book for publication by offering individual coaching, courses and classes to provide information, guidance, feedback and support for each stage of the writing journey. Primary areas of expertise include how to write a first-class book proposal and how to approach the right people in the right way for the best possible chances of success. See the website for more information and a range of free resources for writers.

Nielsen BookData

3rd Floor, Midas House, 62 Goldsworth Road, Woking GU21 6LQ
☎ 01483 712200   🖷 01483 712201
info.bookdata@nielsen.com
www.nielsenbookdata.co.uk
Commercial Director *Anne Betts*

BookData services disseminate timely, accurate and enriched bibliographic data for English-language books to booksellers and libraries worldwide in a variety of formats including data feed, CD-ROMs and online services.

Nielsen BookNet

3rd Floor, Midas House, 62 Goldsworth Road, Woking GU21 6LQ
☎ 01483 712200   🖷 01483 712201
sales.booknet@nielsen.com
www.nielsenbooknet.co.uk
Head of BookNet Sales *Stephen Long*

BookNet's range of e-commerce services allows electronic trading between booksellers, publishers/distributors, libraries and other suppliers. Services include BookNet Web for booksellers and publishers/distributors, TeleOrdering and EDI messaging.

Nielsen BookScan

3rd Floor, Midas House, 62 Goldsworth Road, Woking GU21 6LQ
☎ 01483 712222   🖷 01483 712220
info.bookscan@nielsen.com
www.nielsenbookscan.co.uk
Publisher Account Manager *Reeta Windsor*

The BookScan service collects transactional data at the point of sale from tills and despatch systems of all the major book retailers in the UK, Ireland, US, South Africa, Spain, Italy, New Zealand and Denmark. Each week data is coded and analyzed, producing complete market information for retailers, publishers, libraries, agents and the media. In the UK these reports are produced within 72 hours of the week ending Saturday.

Ormrod Research Services

Weeping Birch, Burwash TN19 7HG
☎ 01435 882541
richardormrod@aol.com
www.writeonservices.co.uk
Contact *Richard Ormrod*

Established 1982. Comprehensive research service: literary, historical, academic, biographical, commercial. Verbal quotations available. Also editing, indexing, ghost-writing and critical reading.

Patent Research

Dachsteinstr. 12a, D–81825 Munich, Germany
☎ 00 49 89 430 7833
Contact *Gerhard Everwyn*

All world, historical patents for researchers, authors, archives, museums and publishers. Rates on application.

Leda Sammarco

☎ 07930 568516
leda.sammarco@btopenworld.com
www.ledasammarco.com
Contact *Leda Sammarco (Writer's Coach)*

A coaching service for writers in the area of mind body spirit, self-help, popular psychology and health and wellbeing. Gives advice and guidance on the creative process, structure of your book, overcoming writer's block, getting your book published and boosting your profile.

SAN Agency

3rd Floor, Midas House, 62 Goldsworth Road, Woking
☎ 0870 777 8712   🖷 0870 777 8714
san.agency@nielsen.com
www.san.nielsenbookdata.co.uk
Manager, ISBN, SAN & DOI Agencies *Diana Williams*
Senior Manager, Registration Services ISBN, SAN & DOI Agencies *Julian Sowa*

SANs, Standard Address Numbers, are unique identifiers for geographical locations and can be assigned to the addresses of organizations involved in the book selling or publishing industries. The SAN Agency is responsible for managing the scheme on behalf of Book Industry Communication in the UK and Republic of Ireland. Nielsen Book also runs the **ISBN** & **DOI** agencies (see entries).

The United Kingdom Copyright Bureau

110 Trafalgar Road, Portslade BN41 1GS

℡ 01273 277333

info@copyrightbureau.co.uk

www.copyrightbureau.co.uk

Contacts *Ralph de Straet von Kollman,
Petra Ginman*

The UKCB provides a secure copyright service at reasonable cost, enabling multiple copyrights to be registered nominally when required. Prices are advertised on the website including the UKCB's solicitors, etc. Copyrights preferred on floppy disk or CD-ROM; manuscripts are not accepted due to storage space.

Writers Direct

29 Great Southsea Street, Southsea PO5 3BY

℡ 023 9283 1289 ℻ 023 9283 1289

enquiries@writersdirect.net

www.writersdirect.net

'Writers Direct is dedicated to helping authors achieve success selling and promoting their books. We harness the internet to the best advantage of the individual writer's needs, from first-timers to best-sellers.' Designs and manages authors websites, and provides marketing and promotion to keep clients at the top of the search engine rankings.

Index of Entries

Subject Index

Fiction: Publishers (European)

Fiction: Publishers (International)

Includes FREE online access to www.thewritershandbook.com